1989 Recommended Dietary Allowances *(continued)*

Age (yr) or Condition	Water-Soluble Vitamins Vitamin C (mg)	Minerals Iron (mg)	Zinc (mg)	Iodine (μg)	Selenium (μg)
0.0–0.5	30	6	5	40	10
0.5–1.0	35	10	5	50	15
1–3	40	10	10	70	20
4–6	45	10	10	90	20
7–10	45	10	10	120	30
11–14[e]	50	12	15	150	40
15–18[e]	60	12	15	150	50
19–24[e]	60	10	15	150	70
25–50[e]	60	10	15	150	70
51+[e]	60	10	15	150	70
11–14[f]	50	15	12	150	45
15–18[f]	60	15	12	150	50
19–24[f]	60	15	12	150	55
25–50[f]	60	15	12	150	55
51+[f]	60	10	12	150	55
Pregnant	70	30	15	175	65
Lactating[g]	95	15	19	200	75
Lactating[h]	90	15	16	200	75

[e]A male in that age group.
[f]A female in that age group.
[g]First 6 months.
[h]Second 6 months.

1989 Estimated Safe and Adequate Daily Dietary Intakes of Selected Vitamins and Minerals[a]

Category	Age (yr)	Trace Elements[b] Copper (mg)	Manganese (mg)	Chromium (μg)	Molybdenum (μg)
Infants	0–0.5	0.4–0.6	0.3–0.6	10–40	15–30
	0.5–1	0.6–0.7	0.6–1.0	20–60	20–40
Children and adolescents	1–3	0.7–1.0	1.0–1.5	20–80	25–50
	4–6	1.0–1.5	1.5–2.0	30–120	30–75
	7–10	1.0–2.0	2.0–3.0	50–200	50–150
	11+	1.5–2.5	2.0–5.0	50–200	75–250
Adults		1.5–3.0	2.0–5.0	50–200	75–250

[a]Because there is less information on which to base allowances, these figures are not given in the main table of RDA and are provided here in the form of ranges of recommended intakes.
[b]Since the toxic levels for many trace elements may be only several times usual intakes, the upper levels for the trace elements given in this table should not be habitually exceeded.

1997–1998 Dietary Reference Intakes (DRIs)

Age (years)	Recommended Dietary Allowances (RDA)								Adequate Intakes (AI)					
	Thiamin	Riboflavin	Niacin	Vitamin B_6	Folate	Vitamin B_{12}	Phosphorus	Magnesium	Vitamin D	Pantothenic acid	Biotin	Choline	Calcium	Fluoride
	mg	mg	mg[a]	mg	µg DFE[c]	µg	mg	mg	µg	µg	µg	mg	mg	mg
Infants[b]														
0.0–0.5	0.2	0.3	2	0.1	65	0.4	100	30	5	1.7	5	125	210	0.01
0.5–1.0	0.3	0.4	4	0.3	80	0.5	275	75	5	1.8	6	150	270	0.5
Children														
1–3	0.5	0.5	6	0.5	150	0.9	460	80	5	2.0	8	200	500	0.7
4–8	0.6	0.6	8	0.6	200	1.2	500	130	5	3.0	12	250	800	1.0
Males														
9–13	0.9	0.9	12	1.0	300	1.8	1250	240	5	4.0	20	375	1300	2.0
14–18	1.2	1.3	16	1.3	400	2.4	1250	410	5	5.0	25	550	1300	3.0
19–30	1.2	1.3	16	1.3	400	2.4	700	400	5	5.0	30	550	1000	4.0
31–50	1.2	1.3	16	1.3	400	2.4	700	420	5	5.0	30	550	1000	4.0
51–70	1.2	1.3	16	1.7	400	2.4	700	420	10	5.0	30	550	1200	4.0
>70	1.2	1.3	16	1.7	400	2.4	700	420	15	5.0	30	550	1200	4.0
Females														
9–13	0.9	0.9	12	1.0	300	1.8	1250	240	5	4.0	20	375	1300	2.0
14–18	1.0	1.0	14	1.2	400	2.4	1250	360	5	5.0	25	400	1300	3.0
19–30	1.1	1.1	14	1.3	400	2.4	700	310	5	5.0	30	425	1000	3.0
31–50	1.1	1.1	14	1.3	400	2.4	700	320	5	5.0	30	425	1000	3.0
51–70	1.1	1.1	14	1.5	400	2.4	700	320	10	5.0	30	425	1200	3.0
>70	1.1	1.1	14	1.5	400	2.4	700	320	15	5.0	30	425	1200	3.0
Pregnancy	1.4	1.4	18	1.9	600	2.6	700[d]	*	5	6.0	30	450	1000[d]	3.0
Lactation	1.5	1.6	17	2.0	500	2.8	700[d]	**	5	7.0	35	550	1000[d]	3.0

[a]1 mg niacin = 60 mg tryptophan = niacin equivalent (NE); for infants 0–0.5 years RDA as only preformed niacin, not NE.

[b]For all nutrients for infants, an AI was established instead of an RDA.

[c]DFE = dietary folate equivalent = 1 µg food folate = 0.6 µg folic acid (fortified food or supplement) eaten with food = 0.5 µg synthetic folic acid eaten with an empty stomach.

[d]If ≤ 18 years, use the value listed for women of comparable age.

*If ≤ 18 years, 400 mg; 19–30 years, 350 mg, and 31–50 years 360 mg.

**If ≤ 18 years, 360 mg; 19–30 years, 310 mg, and 31–50 years 320 mg.

Source: Adapted from Dietary Reference Intake series, National Academy Press, Washington, D.C. Copyright 1998.

Third Edition

Advanced Nutrition and Human Metabolism

James L. Groff

Georgia State University
University of North Carolina at Asheville

Sareen S. Gropper

Auburn University

Wadsworth
Thomson Learning™

Australia • Canada • Denmark • Japan • Mexico • New Zealand • Philippines • Puerto Rico
Singapore • South Africa • Spain • United Kingdom • United States

Dedication

To my wife Gerda who has been such a great help to me in the preparation of this text, not only through her ongoing encouragement and support, but also for her assistance in locating many of our listed web sites, made possible by her ability to surf the Net "with the best of them."

Jim

To Michelle Lauren and Michael James Gropper who with the second edition of this book were my little readers in the house, but now with this third edition are not so little anymore. I thank the both of you, as well as my husband Daniel, for all your patience, support, and love.

Sareen

Publisher: Peter Marshall
Development Editor: Laura Graham
Editorial Assistant: Keynia Johnson
Marketing Manager: Becky Tollerson
Project Editor: Sandra Craig
Print Buyer: Barbara Britton

Permissions Editor: Susan Walters
Production Service: Tobi Giannone, Michael Bass & Associates
Text Designer: Detta Penna
Copy Editor: Laura E. Larson
Illustrator: Asterisk Group Inc.

Compositor: Pre-Press Company, Inc.
Cover Designer: Norman Baugher
Printer/Binder: R.R. Donnelley/ Crawfordsville

For permission to use this material from this text, contact us:
Web: www.thomsonrights.com
Fax: 1-800-730-2215
Phone: 1-800-730-2214

Printed in the United States of America
1 2 3 4 5 6 7 03 02 01 00 99

Wadsworth/Thomson Learning
10 Davis Drive
Belmont, CA 94002-3098
USA
www.wadsworth.com

International Headquarters
Thomson Learning
290 Harbor Drive, 2nd Floor
Stamford, CT 06902-7477
USA

UK/Europe/Middle East
Thomson Learning
Berkshire House
168-173 High Holborn
London WC1V 7AA
United Kingdom

Asia
Thomson Learning
60 Albert Street #15-01
Albert Complex
Singapore 189969

Canada
Nelson/Thomson Learning
1120 Birchmount Road
Scarborough, Ontario M1K 5G4
Canada

Library of Congress Cataloging-in-Publication Data
Groff, James L..
    Advanced nutrition and human metabolism / James L. Groff, Sareen S. Gropper. — 3rd ed.
        p.   cm.
    Includes bibliographical references and index.
    ISBN 0-534-55521-7
    1. Nutrition   2. Metabolism.   I. Gropper, Sareen Annora Stepnick.   II. Title.
QP141.G76   1999
612.3'9—dc21   99-27331

 This book is printed on acid-free recycled paper.

Brief Contents

Contents

Chapter 8

*Integration and Regulation of Metabolism and
the Impact of Exercise and Sport 220*

SECTION III

The Regulatory Nutrients

Chapter 9

The Water-Soluble Vitamins 245

SECTION IV

Homeostatic Maintenance

SECTION V

Nutrition Knowledge Base

Preface

The science of nutrition continues to move ahead with vigor since we authored the first edition of our *Advanced Nutrition and Human Metabolism* text in 1990. We felt that we captured the dynamics of the field in our second edition in 1995, but during the ensuing four years to the present, as new revelations and concepts continued to emerge, the need for a third edition became quite evident.

We conceived the first edition of *Advanced Nutrition and Human Metabolism* to fill the need that teachers expressed for a text on normal metabolism, designed for upper-division nutrition students. Judging from the positive response we received from our users, we feel that the level of depth and scope of the material presented indeed came very close to satisfying their needs. Continued solicited reviews and comments from faculty and students familiar with our book have encouraged us to retain the scope, level, and organization that has popularized the first two editions. But, motivated in part by our own perceptions for change and in part by the insightful suggestions of our reviewers, we have introduced notable changes in the third edition.

New to This Edition

Retained are the popular Perspectives at the ends of all but a few of the chapters. Each with their own list of references, these Perspectives relate chapter subject matter to an important aspect of human nutrition and health. Many reviewers were of the opinion that the Perspective material is important enough to be included in the chapter text. As a result, much of this material has been incorporated, and we have given a new focus to the Perspective material. While the book concentrates on normal nutrition and physiological function, the Perspectives deal with clinical/pathological aspects germane to the subject of the corresponding chapters. New Perspectives following this theme include the assay of enzymes in the diagnosis of disease (Chap. 1), consequences of drug-induced uncoupling of oxidative phosphorylation (Chap. 3), the question of fact or fiction regarding hypoglycemia (Chap. 4), the role of herbal supplements in disease (Chap. 5), the role of lipids and lipoproteins in atherogenesis (Chap. 6), nutrient controls on gene expression (Chap. 9), and osteoporosis (Chap. 15). The remaining Perspectives, retained from the second edition, carry the pathological theme also.

A number of current topics of interest in nutrition have been added to the third edition. Some of these include sports and exercise nutrition, leptin, Olestra and Orlistat in obesity control, apoptosis, apolipoprotein E phenotypes and *trans* fats in cardiovascular disease risk, and an augmentation of nutrient-gene interaction. All of the text material is annotated with many new and updated references. We have also included a number of appropriate web site resources at the end of each chapter, representing technical databases, health organizations, federal agencies, and some popular general health web sites. All of this will familiarize readers with current research on the many topics and subtopics covered. Additionally, a brief section on nutrition research on the Internet has been added to Chapter 17.

Some changes have been made to the order of chapters and chapter titles in the third edition. Because of the large amount of information covered in the micominerals chapter (previously Chap. 12), we have extended the material into two chapters, divided according to the quantity of mineral required by the body. Chapter 12 will continue to be called "Microminerals"; the new Chapter 13 is titled "Ultratrace Elements." Minerals discussed in Chapter 13 include nickel, silicon, vanadium, arsenic, boron, and cobalt. Additionally, the former Chapters 14 (on body composition) and 15 (on energy balance and weight control) have been consolidated into a new Chapter 15, "Body Composition and Energy Expenditure." All of the other chapters remain the same. These changes will not compromise the quality or quantity of the material in any way, and we feel that a smoother flow of information is the reward.

Presentation

The presentation of the third edition is designed to be even more reader-friendly. When feasible, information is presented in the form of bulleted or numbered lists. This format draws attention to important elements in the text and helps to generate readers' interest.

Because this book focuses on normal human nutrition and physiological function, it is an effective resource for students majoring in either nutrition science or dietetics. Intended for a course in advanced nutrition, this text presumes a sound background in the biological sciences. However, we do provide a review of the basic sciences, particularly biochemistry and physiology, which are important to the understanding of the material. Because this text applies biochemistry to nutrient use from consumption through digestion, absorption, distribution, and cellular metabolism, it is a valuable reference for health care workers. Health practitioners may use this as a resource to refresh their memories of metabolic and physiological interrelationships, and to obtain a concise update on current scientific discoveries and concepts related to human nutrition.

We continue to present nutrition as the science that integrates life processes from the cellular level through the multisystem operation of the whole organism. Our primary goal is to give a comprehensive picture of cell reactions at the tissue, organ, and system levels. Subject matter has been selected for its relevance in meeting this goal. We retain the feature of generous cross-referencing throughout the text to strengthen the reader's access to in-depth discussions of each topic.

Organization

Each of the 17 chapters begins with a topical outline followed by a brief introduction to the chapter's subject matter. This is followed in order by the chapter text, a brief summary that ties together the ideas presented in the chapter, a reference list, and finally a Perspective, with its own list of references. The text has been divided into five sections. Section I (Chaps. 1–3) focuses on the structure, function, and nourishment of the cell, and reviews energy transformation. Section II (Chaps. 4–8) discusses the metabolism of macronutrients. In this section, we review primary metabolic pathways for carbohydrates, lipids, and proteins, emphasizing reactions that have particular relevance for health. We include a chapter on dietary fiber and one on the interrelationships among the macronutrient metabolic pathways. Also included in the section (Chap. 8) is a description of the metabolic dynamics of the feed-fast cycle, along with a newly introduced presentation of exercise and sports nutrition and how physical exertion impacts the body's metabolic pathways.

Section III (Chaps. 9–13) concerns nutrients considered regulatory in nature: the vitamins (water- and fat-soluble) and the minerals, both macro and micro. These chapters cover nutrient features such as digestion, absorption, transport, function, metabolism, excretion, deficiency, and toxicity. We also discuss the new Dietary Reference Intakes (when available) and Recommended Dietary Allowances (RDAs) for each nutrient. Section IV, "Homeostatic Maintenance," includes Chapters 14 through 16. We discuss, in order, body fluid and electrolyte balance, body composition and energy expenditure, and nutrition and the central nervous system.

The last chapter (17) constitutes Section V. It is supplementary to the rest of the book. Titled "Experimental Design and Critical Interpretation of Research," this chapter discusses the types of research and the methodologies by which research can be conducted. It is designed to familiarize the student with research organization and implementation, to point out problems and pitfalls inherent in research, and to help students critically evaluate scientific literature.

Acknowledgments

We cannot overestimate the importance of the helpful comments, encouragement, and patience of those whose lives were touched by our efforts, both at the workplace and at home.

We are particularly appreciative of Dr. Kyle Willian's design of illustrations used throughout Chapters 11 and 12. Others to whom the authors owe special thanks are the reviewers whose thoughtful comments, criticisms, and suggestions were indispensable in shaping the third edition of our text:

Eldon Wayne Askew, Utah State University

Janine T. Baer, University of Dayton

Irene Berman-Levine, University of Pennsylvania

Patricia Brevard, James Madison University

Lou Ann Carden, University of Tennessee at Martin

Richard A. Cook, University of Maine

Jamie Erskine, University of Northern Colorado-Greeley

Coni C. Francis, University of Colorado, School of Medicine

Howard P. Glauert, University of Kentucky

Jessica Hodge, Framingham State College

Rita M. Johnson, Indiana University of Pennsylvania

Mary F. Locniskar, University of Texas at Austin

Barbara Lohse Knous, University of Wisconsin–Stout

Mark Kern, San Diego State University

M. E. Kunkel, Clemson University

Phylis B. Moser-Veillon, University of Maryland

Carol P. Ries, Eastern Illinois University

Neil Shay, University of Illinois

Joanne Slavin, University of Minnesota

May-Choo Wang, San Jose State University

M. K. (Suzy) Weems, Stephen F. Austin State University

Chapter 1

The Cell: A Microcosm of Life

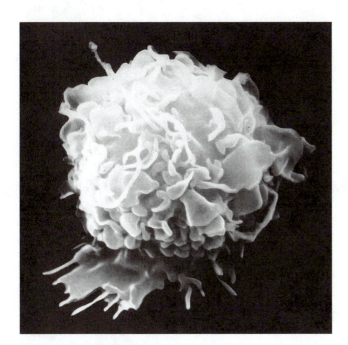

A human scavenger cell

Cells are the very essence of life. They may be defined as the basic, living, structural, and functional units of the human body. Cells vary greatly in size, chemical composition, and function, but each one is a remarkable miniaturization of human life. They move, grow, ingest food and excrete wastes, react to their environment, and even reproduce.

Cells of all multicellular organisms are called *eukaryotic cells* (from the Greek *eu,* meaning "true," and *karyon,* "nucleus"). Eukaryotic cells are known to have evolved from simpler, more primitive cells called *prokaryotic cells.* The major distinguishing feature between the two cell types is that eukaryotic cells possess a defined nucleus, whereas the prokaryotic cells do not. Also, eukaryotic cells are larger and much more complex structurally and functionally than their ancestors. Because this text addresses human metabolism and nutrition, all discussions of cellular structure and function in this and subsequent chapters pertain to eukaryotic cells.

Specialization among cells is a necessity for the living, breathing human, but cells in general have certain basic similarities. All human cells have a plasma membrane and a nucleus (or have had a nucleus), and most of them contain an endoplasmic reticulum, Golgi apparatus, and mitochondrion. For convenience of discussion, a so-called typical cell is considered so that the various organelles and their functions, which characterize cellular life, may be identified. Considering the relationship between the normal functioning of a typical cell and the health of the total organism—the human

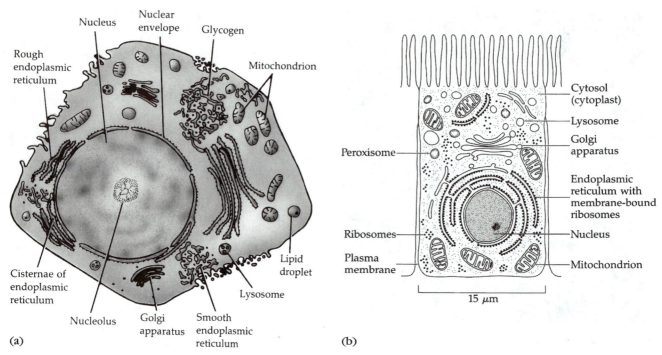

Figure 1.1   The fine structure of a typical animal cell (hepatocyte) is shown in (a), while (b) depicts a typical animal absorptive cell.

being—one is reminded of the old rule: "A chain is only as strong as its weakest link."

Figure 1.1a shows the fine structure of a typical animal cell (hepatocyte), while Figure 1.1b gives a schematic view of a typical animal absorptive cell (such as an intestinal epithelial cell), showing its major components or organelles.

The discussion begins with consideration of the plasma membrane, which forms the outer boundary of the cell, and then moves inward to examine the organelles held within this membrane.

Components of Typical Cells

Plasma Membrane

The *plasma membrane* is the membrane encapsulating the cell. By surrounding the cell, it lets the cell become a unit by itself. The plasma membrane, like other membranes found within the cell, has distinct functions and structural characteristics. Nevertheless, all membranes share some common attributes:

• Membranes are sheetlike structures composed primarily of lipids and proteins held together by noncovalent interactions.

• Membrane lipids consist primarily of phospholipids, which have both a hydrophobic and hydrophilic

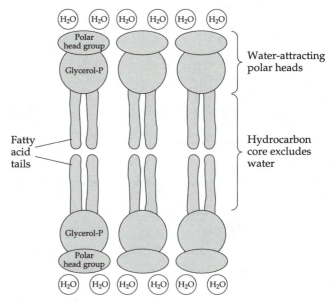

Figure 1.2   Lipid bilayer structure of biological membranes. Structures are schematic representations of phosphoglycerides (glycerophosphatides), which consist of fatty acid tails attached to polar "heads." Contained within the polar heads are glycerol, phosphate (P), and a polar head group, which can be choline, ethanolamine, serine, or inositol.

moiety. Because of this structural property of phospholipids, they spontaneously form bimolecular sheets in water, called lipid bilayers, as shown in Figure 1.2. These bilayer sheets, because of their hydrophobic core, retard the passage of many water-soluble compounds into and out of the cell. Although such an arrangement requires transport systems across the membrane, it allows retention of essential water-soluble substances within the cell.

• Phosphoglycerides and phosphingolipids (phosphate-containing sphingolipids) comprise most of the membrane phospholipids. Of the phosphoglycerides, phosphatidylcholine and phosphatidylethanolamine are particularly abundant in higher animals. Another important membrane lipid is cholesterol, but its amount varies considerably from membrane to membrane. Molecular structures of these lipids are detailed in Chapter 6.

• Membrane proteins confer on biological membranes their functionality: they serve as pumps, gates, receptors, energy transducers, and enzymes. Figure 1.3 schematically illustrates the functions and positioning of some membrane proteins.

• Membranes are asymmetrical. The inside and outside faces of the membrane are different.

• Membranes are fluid structures in which lipid and protein molecules can move laterally with ease and rapidity.

Membranes can no longer be considered entities that are structurally distinct from the aqueous compartments that they surround. For example, the cytoplasm, which is the cell's aqueous ground substance, affords a connection among the various membranes of the cell. Such an interconnection creates a structure that makes it possible for a signal generated at one part of the cell to be transmitted quickly and efficiently to other regions of the unit.

The plasma membrane provides protection to the cellular components while at the same time allowing them sufficient exposure to their environment for stimulation, nourishment, and removal of wastes. Chemical differences that distinguish the plasma membrane from other membranes in the cell are as follows:

• It has a greater carbohydrate content owing to the presence of glycolipids and glycoproteins. Although some carbohydrate is found in all membranes, most of the glycolipids and glycoproteins are associated with the plasma membrane.

• It has a higher content of cholesterol. Cholesterol enhances the mechanical stability of the membrane and regulates its fluidity.

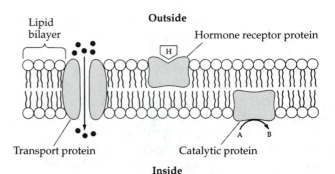

Figure 1.3 Schematic representation of various functions performed by cell membrane proteins.

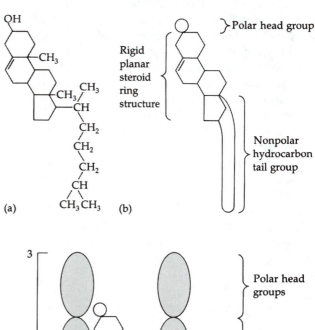

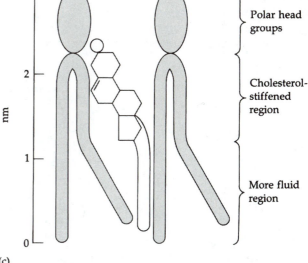

Figure 1.4 The structural role of cholesterol in the cell membrane is represented by a formula (a), a schematic drawing (b), and as it interacts with two phospholipid molecules in a monolayer (c).

Figure 1.4 illustrates the positioning of a cholesterol molecule between two phospholipid molecules. The orientation is such that the hydrocarbon side chain of the cholesterol molecule associates with the hydrocarbon fatty acid tails of the phospholipids, creating a hydrophobic region. The hydroxyl groups of the cholesterol are positioned close to the polar head groups of the phospholipid molecules, resulting in a more hydrophilic region [1]. It is this layering of polar and nonpolar regions that has led to the lipid bilayer notation to describe the plasma membrane structure. The rigid planar steroid rings of cholesterol are positioned so as to interact with and stabilize those regions of the hydrocarbon chains closest to the polar head groups. The rest of the hydrocarbon chain remains flexible and fluid. Cholesterol, by regulating fluidity of the membrane, regulates membrane permeability, thereby exercising some control over what may pass into and out of the cell. Fluidity of the membrane also appears to affect the structure and function of the proteins embedded in the lipid membrane.

The carbohydrate moiety of the glycoproteins and the glycolipids in membranes help maintain the asymmetry of the membrane. This is because the oligosaccharide side chains are located exclusively on the membrane layer facing away from the cytoplasmic matrix. In plasma membranes, therefore, the sugar residues are all exposed to the outside of the cell, forming what is called the *glycocalyx,* the layer of carbohydrate on the cell's outer surface. On the membranes of the organelles, however, the oligosaccharides are directed inwardly into the lumen of the membrane-bound compartment. Figure 1.5 illustrates the glycocalyx and the location of oligosaccharide side chains in the plasma membrane.

Although the exact function of the sugar residues is unknown, it is believed that they act as specificity markers for the cell and as "antennae" to pick up signals for transmission of substances in the cell. The membrane glycoproteins are crucial to the life of the cell, very possibly serving as the receptors for hormones, for certain nutrients, and for various other substances that influence cellular function. Also, glycoproteins may help to regulate the intracellular communication necessary for cell growth and tissue formation. The term *intracellular communication* refers to pathways that convert information from one part of a cell to another in response to external stimuli. Generally, it involves the passage of chemical messengers from organelle to organelle or within the lipid bilayers of membranes. Intracellular communication will be looked at more closely in the section "Receptors and Intracellular Signaling."

Whereas the lipid bilayer determines the structure of the plasma membrane, proteins are primarily responsible for the many membrane functions. The membrane proteins are interspersed within the lipid bilayers, where they mediate information transfer (e.g., receptors), transport ions and molecules (channels, carriers, and pumps), and speed up metabolic activities (enzymes). Figure 1.6 illustrates the functions of these proteins in the transport of molecules into and out of the cell.

Membrane proteins are classified as either integral or peripheral. The *integral proteins* are attached to the

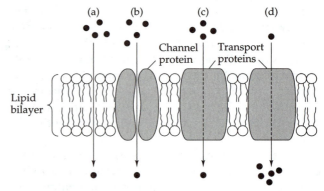

Figure 1.6 Schematic illustration of the role of proteins in transport, showing four ways in which molecules (•) may cross cellular membranes: (a) simple diffusion through the lipid bilayer; (b) diffusion through the membrane pores created by channel proteins; (c) facilitated diffusion or transport, necessitating a carrier protein specific for the molecule; and (d) active transport, a mechanism that, like facilitated transport, requires a specific transporter but that also requires the expenditure of energy in the form of ATP hydrolysis. Molecules crossing membranes by simple or facilitated diffusion are moved only from a region of higher concentration on one side of the membrane to a region of lower concentration on the other. Actively transported molecules, in contrast, can be moved against a concentration gradient.

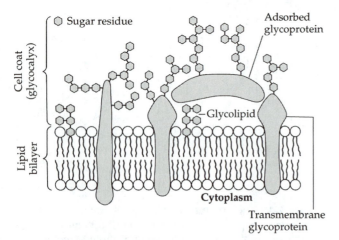

Figure 1.5 Diagram of the cell membrane glycocalyx, illustrating that all carbohydrate is located on the outside of the membrane.

membrane through hydrophobic interactions and are embedded in the membrane. *Peripheral proteins,* in contrast, are associated with membranes via ionic interactions and are located on or near the membrane surface (Fig. 1.7). Peripheral proteins are believed to be attached to integral membrane proteins either directly or through intermediate proteins [1].

Most receptor and carrier proteins are integral proteins, whereas the glycoproteins of the cell recognition complex are peripheral proteins [1]. Functions of membrane proteins, as well as functions of proteins located intracellularly, are discussed later in the chapter.

Cytoplasmic Matrix

A new frontier in the study of cell structure and cell physiology was made possible with the advent of the high-voltage electron microscope. It was able to identify the microtrabecular lattice, a fibrous cord of connective tissue that supports and controls the movement of cell organelles. The microtrabecular lattice is actually an intercommunication system of proteins and other macromolecules held in some state of aggregation. A model of this lattice, as shown in Figure 1.8, shows the microtrabeculae as being continuous with the proteins underlying the plasma membrane, the surface of the endoplasmic reticulum, the microtubules, and the filaments of the stress fibers. The lattice also appears to support certain extracellular extensions emanating from the cell surface. For example, the microvilli, which are extensions of intestinal epithelial cells, are associated with the microtrabeculae. Microvilli are designed to present a large surface area for the absorption of dietary nutrients. Microtubules, together with a network of filaments that interconnect them, form what is called the *cytoskeleton.* The cellular matrix, of which the cytoskeleton is a part, is most commonly called the *cytoplasm.* However, an alternative term that may be more appropriate is the *cytoplast,* because it suggests the structural and functional unity that is such an important feature of the matrix.

Like the microtrabecular lattice, microfilaments, and microtubules are apparently complex polymers of many different proteins, including actin, myosin, and tubulin, the latter being a protein necessary for the formation of microtubules. These structures provide mechanical support for the cell, but they also serve as binding surfaces for soluble macromolecules such as proteins and nucleic acids that are present in the aqueous portion of the cytoplasmic matrix. In fact, there is evidence that the nonfilamentous aqueous phase of the cell contains very few macromolecules, and that many proteins in the cytoplasmic matrix are bound to the filaments for a large portion of their lives. Filling the intertrabecular

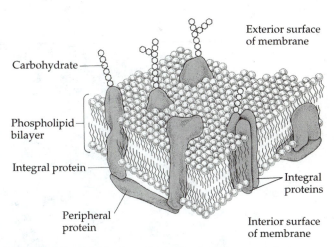

Figure 1.7 Positioning of proteins in plasma membrane. Proteins that penetrate into the interior of the lipid bilayer are called *integral proteins*. Proteins that lie on the surface and do not penetrate the lipid bilayer are called *peripheral* or *extrinsic proteins*. Polysaccharides are frequently attached to membrane proteins and membrane lipids on the exterior side of the membrane.

spaces is the fluid phase of the cytoplasmic matrix, containing small molecules such as glucose, amino acids, oxygen, and carbon dioxide. This arrangement of the polymeric and fluid phases apparently gives the cytoplasm (cytoplast) its gel-like consistency.

The spatial arrangement of the lattice, or polymeric phase, with the aqueous phase improves the efficiency of the many enzyme-catalyzed reactions that take place in the cytoplast. The aqueous phase contacts the polymeric phase over a very broad surface area. For this reason, enzymes that are associated with the polymeric phase are brought into close proximity to their substrate molecules in the aqueous phase, thereby facilitating the reaction (see discussion of enzymes, p. 17). Furthermore, if enzymes catalyzing the reactions of a metabolic pathway were oriented sequentially so that the product of one reaction was released in very close proximity to the next enzyme for which it is a substrate, the velocity of the overall pathway would be greatly enhanced. Evidence indicates that such an arrangement of the enzymes participating in glycolysis does in fact exist.

Possibly all metabolic pathways occurring in the cytoplasmic matrix may be influenced by its structural arrangement. Metabolic pathways of particular significance that occur in the cytoplasmic matrix and that might be affected by the structure include

- glycolysis,
- hexose monophosphate shunt (pentose phosphate pathway),
- glycogenesis and glycogenolysis,

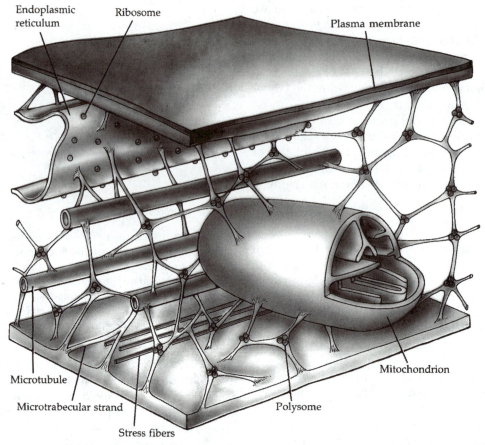

Figure 1.8 A model of the cytoplast (microtrabecular lattice). The cytoplast is shown at about 300,000 times its actual size and was derived from hundreds of images of cultured cells viewed in a high-voltage electron microscope. The model illustrates how the microtrabecular filaments are related to other components of the cell cytoplasm: the substance of the cell outside the cell nucleus. In the model the microtrabeculae suspend the elongated structures of the endoplasmic reticulum, the mitochondria, and the microtubules. At junctions of the microtrabecular lattice are polysomes: organized clusters of ribosomes. Adapted from Porter and Tucker, ("The Ground Substance of the Cell." Copyright © 1981 by Scientific American, Inc. All rights reserved.)

- fatty acid synthesis, including the production of nonessential, unsaturated fatty acids.

Normal intracellular communication among all cellular components is vital for cell activation and survival. The importance of the microtrabecular network is therefore evident, considering that its function is to support and interconnect cellular components.

Mitochondrion

The *mitochondria* are the primary sites of oxygen use in the cell and are responsible for most of the metabolic energy (adenosine triphosphate, or ATP) produced in cells. The size and shape of the mitochondria in different tissues vary according to the function(s) of the tissue. In the muscle tissue, for example, the mitochondria are held tightly among the fibers of the contractile system. In the liver, however, the mitochondria have fewer restraints, appear spherical in shape, and move freely through the cytoplasmic matrix.

The mitochondrion consists of a matrix or interior space surrounded by a double membrane (Fig. 1.9). The mitochondrial outer membrane is relatively porous, whereas the inner membrane is only selectively permeable, thereby serving as a permeability barrier between the cytoplasmic matrix and the mitochondrial matrix. The inner membrane has many invaginations called the *cristae*. These cristae serve to increase the surface area of the inner membrane in which are embedded all the components of the electron transport (respiratory) chain.

The electron transport chain is central to the process of oxidative phosphorylation, the mechanism by which most cellular ATP is produced. Its components act as carriers of electrons in the catalytic oxidation of nutrient molecules by enzymes functioning in

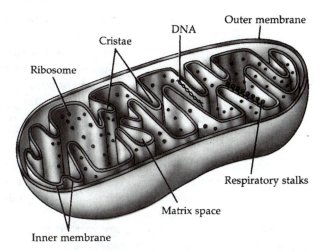

Figure 1.9 The mitochondrion.

the mitochondrial matrix. Flow of electrons through the electron transport chain is strongly exothermic, and the energy released is used in part for the synthesis of ATP, an endothermic process. Molecular oxygen is ultimately, but indirectly, the oxidizing agent in these reactions. The purpose of the electron transport chain is to couple the energy released by nutrient oxidation to the formation of ATP. The precise positioning of the chain components within the inner membrane is fortuitous, because it brings them into close proximity to both the oxidizable products released in the matrix and molecular oxygen. Figure 1.10 shows the flow of major reactants into and out of the mitochondrion.

Among the metabolic enzyme systems functioning in the mitochondrial matrix are those catalyzing reactions of the Krebs cycle and fatty acid oxidation. Other enzymes are involved in the oxidative decarboxylation and carboxylation of pyruvate (pp. 91, 93) and in certain reactions of amino acid metabolism.

While the nucleus contains most of the cell's deoxyribonucleic acid (DNA), the mitochondrial matrix contains a small amount of DNA and a few ribosomes so that limited protein synthesis within the mitochondrion does occur. It is of interest that the genes contained in mitochondrial DNA, unlike those in the nucleus, are inherited only from the mother. The primary function of mitochondrial genes is to code for proteins vital to the production of ATP [2]. Most of the enzymes operating in the mitochondrion, however, are coded by nuclear DNA and synthesized on the rough endoplasmic reticulum (RER) in the cytoplasm (p. 10). Then they are incorporated into existing mitochondria.

All cells in the body, with the exception of the erythrocyte, possess mitochondria. The erythrocyte (p. 296), in the process of maturing, disposes of its mitochondria and must depend solely on the energy produced through anaerobic mechanisms, primarily glycolysis (p. 85).

Nucleus

The nucleus of the cell is the largest of the organelles, and because of its DNA content, it initiates and regulates most cellular activities. Surrounding the nucleus

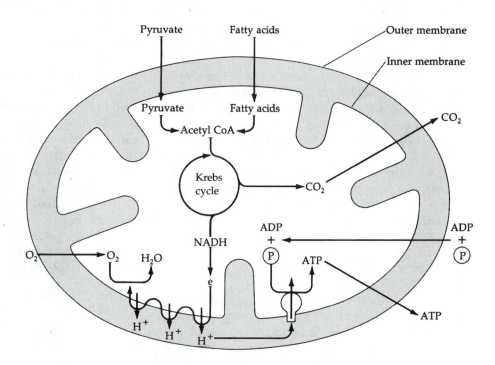

Figure 1.10 The flow of major reactants in and out of the mitochondrion.

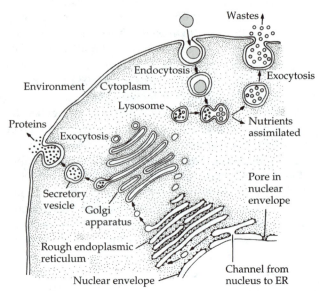

Figure 1.11 Communication between the nucleus and other cell components. Membrane material in the nuclear envelope, the endoplasmic reticulum, the Golgi apparatus, and the plasma membrane are cycled continuously, as illustrated.

is the nuclear envelope, composed of two membranes (an inner and an outer membrane) that appear to be dynamic structures. The dynamic nature of these membranes makes possible communication between the nucleus and the cytoplasmic matrix and allows a continuous channel between the nucleus and the endoplasmic reticulum. At various intervals the two membranes of the nuclear envelope fuse, thereby creating pores in the envelope (Fig. 1.11). It appears that the nucleus and the microtubules of the cytoskeleton are interdependent. The polymerization and the intracellular distribution of the microtubules are controlled by nucleus-based activities. Centers of these activities are clusters of proteins on the outer nuclear membrane. The clusters are called microtubule organization centers (MTOCs), and they begin their polymerization and organization of the microtubules during mitosis. A review of MTOC activity has been published [3].

The matrix held within the nuclear envelope is composed of chromatin plus all the enzymes and minerals necessary for the activity of the nucleus. Condensed regions of the chromatin are called *nucleoli,* in which are found not only DNA and its associated alkaline proteins (histones) but also considerable amounts of RNA (ribonucleic acid). This particular RNA is believed to give rise to the microsomal RNA (i.e., RNA associated with endoplasmic reticulum).

Encoded within the nuclear DNA of the cell are thousands of genes that direct the synthesis of proteins, with each gene specifically encoding a single protein. The cell *genome* refers to the entire genetic DNA

within the cell. Barring mutations that may arise in the DNA, daughter cells, produced from a parent cell by mitosis, possess the identical genomic makeup of the parent. It is the process of DNA *replication* that enables the DNA to be precisely copied at the time of mitosis.

Protein biosynthesis occurs in phases, referred to as *transcription, translation,* and *elongation,* each of which requires DNA and/or RNA activity. These, together with replication, will be reviewed briefly here, but the reader is reminded of the very large scope of this subject and should consult a current, comprehensive textbook of biochemistry for a more thorough treatment of protein biosynthesis.

Nucleic acids (DNA and RNA) are formed from repeating units called *nucleotides,* sometimes referred to as *bases.* Structurally, they consist of a nitrogenous core (either purine or pyrimidine), a pentose sugar (ribose in RNA or deoxyribose in DNA), and phosphate. Five different nucleotides are contained in the structures of the nucleic acids, adenylic acid, guanylic acid, cytidylic acid, uridylic acid, and thymidylic acid. The nucleotides are more commonly referred to by their nitrogenous base core only—namely, adenine, guanine, cytosine, uracil, and thymine, respectively. For convenience, particularly in describing the sequence of the polymeric nucleotides in a nucleic acid, single letter abbreviations (A, G, C, U, and T) are most often used. Adenine (A), guanine (G), and cytosine (C) are common to both DNA and RNA, while uracil (U) is unique to RNA and thymine (T) is found only in DNA. When two strands of nucleic acids interact with each other, as occurs in replication, transcription, and translation, bases in one strand pair specifically with bases in the second strand. A always pairs with T (or U), and G pairs with C, in what is called *complementary base pairing.*

The nucleotides are connected by phosphates esterified to hydroxyl groups on the pentose—that is, deoxyribose (or ribose) component of the nucleotide. The carbon atoms of the pentoses are assigned prime numbers for identification. The phosphate group connects the 3' carbon of one nucleotide with the 5' carbon of the next nucleotide in the sequence. The 3' carbon of the latter nucleotide is in turn connected to the 5' carbon of the next nucleotide in the sequence, and so on. Therefore, nucleotides are attached to each other by 3', 5' diester bonds. The ends of a nucleic acid chain are designated as either the free 3' end or the free 5' end, meaning that the hydroxyl groups at those positions are not attached by phosphate to another nucleotide.

Replication: synthesis of a daughter duplex DNA molecule identical to the parental duplex DNA

At cell division the chromatin is condensed into chromosomes, each chromosome containing a single molecule of DNA. The DNA exists as two large strands of nucleic acid that are intertwined to form a double helix (duplex DNA). During cell division the two unravel at one end, each forming a template for the synthesis of a new strand via complementary base pairing. Incoming nucleotide bases first pair with their complementary bases in the template, then they are connected together through phosphate diester bonds by the enzyme DNA polymerase. The end result of the process is two new DNA chains, which, together with their parent chains, permit the production of two duplex DNA molecules from the one parent molecule. Each new duplex DNA is therefore identical in base sequence to that of the parent, and each new cell of a tissue consequently carries within its nucleus identical information to direct its functioning.

The two strands of double helical DNA are antiparallel, which means that the free 5' end of one strand is paired to the free 3' end of the other. Base-paired DNA strands may be represented as follows:

free 3' A-G-T-C-C-A-T ——— G-G-C-T-A-C-G free 5' end
free 5' T-C-A-G-G-T-A ——— C-C-G-A-T-G-C free 3' end

The broken line in this simplified example represents a very large number of additional base pairs. Although only 14 base pairs at the chain termini are shown, the reader is reminded that, in actuality, millions of base pairs exist in eukaryotic cell DNA molecules.

Transcription: the process by which the genetic information (base sequence) in a single strand of DNA is used to specify a complementary sequence of bases in an mRNA chain

Transcription proceeds continuously throughout the entire life cycle of the cell. In the process, various sections of the duplex DNA molecule unravel, and one strand, designated the *sense strand,* serves as the template for the synthesis of messenger RNA (mRNA). The genetic code of the DNA is transcribed into mRNA via complementary base pairing, as in DNA replication, except that the purine adenine (A) pairs with the pyrimidine uracil (U), instead of with thymine (T). Genes are composed of critically sequenced base pairs along the entire length of the DNA strand that is being transcribed. A gene, on average, is just over 1,000 base pairs in length, compared with the nearly 5×10^6 base pair length of typical chromosomal DNA chains. Although these figures provide a rough estimate of the number of genes per transcribed DNA chain, not all of the base pairs of a gene are transcribed into functional mRNA.

A puzzling feature of eukaryotic genes is that certain regions of nucleotide sequences get transcribed into complementary mRNA sequences but do not code for the ultimate protein product. These segments are called *introns* (intervening sequences), and they have to be removed from the mRNA prior to its translation into protein (the translation process, which follows). The introns are enzymatically excised from the newly formed mRNA, and the ends of the functional, active segments spliced together. This is called *posttranscriptional processing* of the mRNA. The gene segments that get both transcribed and translated into the protein product are called *exons* (expressed sequences). Exons, therefore, require no posttranscriptional processing.

Following processing, the mRNA is exported into the cytoplasmic matrix where it is attached to the ribosomes of the RER or to the freestanding polysomes. On the ribosomes, the transcribed genetic code becomes translated so as to bring amino acids into a specific sequence to produce a protein with a clearly delineated function. This process is called *translation.*

Translation: the process by which genetic information in an mRNA molecule specifies the sequence of amino acids in the protein product

The genetic code for specifying the amino acid sequence of a protein resides in the mRNA in the form of three-base sequences called *codons.* A given codon codes for a single amino acid. For example, any of the codons—CUU, CUC, CUA, or CUG—code for the amino acid leucine. Although a given amino acid may therefore have several codons, a given codon can code for only one amino acid. The transition from nucleic acid base sequence to protein synthesis lies in the translation of the mRNA codons by *anticodons,* which are three-base sequences of nucleotides on molecules of a different RNA called *transfer RNA* (tRNA). The anticodons attach to the codons by, once again, complementary base pairing.

Amino acids are first activated by ATP at their carboxyl end and are then transferred to their specific tRNAs that bear the anticodon complementary to each amino acid's codon. For example, because codons that code for leucine are sequenced CUU, CUC, CUA, or CUG, the only tRNAs to which an activated leucine can be attached would have to have anticodon sequences of GAA, GAG, GAU, or GAC. The tRNAs then bring the code-designated amino acids to the mRNA situated at the protein synthesis site on the ribosomes. After the amino acids are positioned according to codon-anticodon association, peptide bonds are then formed between the aligned amino acids, as the protein chain elongates.

Elongation: extension of the polypeptide chain of the protein product

Each incoming amino acid is connected to the C-terminal end of the growing polypeptide chain by peptide bond formation. New amino acids are incorporated until all the codons (corresponding to one gene)

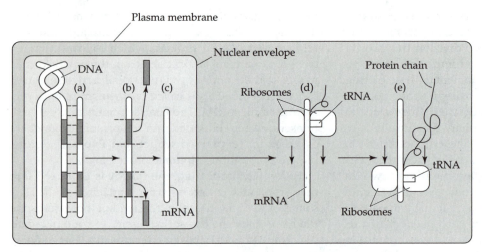

Figure 1.12   A schematic representation of protein biosynthesis in the cell. Information flows in the direction DNA → RNA → protein. (a) Transcription of the sense strand of DNA to produce progenitor mRNA. Introns (see the text) are shown as shaded segments in the DNA and transcribed RNA strands. (b) Processing of progenitor mRNA by the excision of introns and the splicing of translatable (exon) regions. (c) Processed, nuclear mRNA prior to transportation to the cytoplast. (d) mRNA associated with ribosomes of the rough endoplasmic reticulum. Translation of the mRNA as amino acid–bearing tRNA molecules bring code-designated amino acids to mRNA codons via a ribosome. Following each incorporation of a new amino acid, ribosomes translocate to the next codon on the mRNA strand, exposing a new site for the next amino acid to be added. (e) Elongation of the polypeptide chain as peptide bonds are formed between amino acids, and ribosomes continue their process of translocation.

of the mRNA have been translated. At this point, the process stops abruptly, signaled by a "nonsense" codon that does not code for any amino acid. The completed protein dissociates from the mRNA in active form, although some *posttranslational,* chemical modification of the protein is often necessary.

Summarizing, it can be seen that information for protein biosynthesis flows from chromosomal DNA to ribosomal mRNA to protein. Figure 1.12 provides a schematic overview of the process.

Endoplasmic Reticulum and Golgi Apparatus

The *endoplasmic reticulum* (ER) is a network of membranous channels pervading the cytoplast and providing continuity between the nuclear envelope, the Golgi apparatus, and the plasma membrane. This structure, therefore, is a mechanism for communication from the innermost part of the cell to its exterior (Fig. 1.11). The ER cannot be separated from the cell as an entity by laboratory preparation. During mechanical homogenization, its structure is disrupted and re-forms into small spherical particles called *microsomes.*

The ER is classified as either rough (granular) or smooth (agranular). The granularity or lack of granularity is determined by the presence or absence of ribosomes. Rough endoplasmic reticulum (RER), so named because it is studded with ribosomes, abounds in cells

where protein synthesis is a primary function. Smooth endoplasmic reticulum (SER) is found in most cells. However, because it is the site of synthesis for a variety of lipids, it is more abundant in cells that synthesize steroid hormones (e.g., the adrenal cortex and gonads) and in liver cells, which synthesize fat transport molecules (the lipoproteins). In skeletal muscle, SER is called *sarcoplasmic reticulum* and is the site of the calcium ion pump, a necessity for the contractile process.

Ribosomes associated with the RER are composed of ribosomal RNA (rRNA) and structural protein. They are the primary site for protein synthesis. All proteins to be secreted (or excreted) from the cell or destined for incorporation within an organelle membrane in the cell more than likely are synthesized on the RER. The clusters of ribosomes (i.e., polyribosomes or polysomes) that are freestanding in the cytoplast are also the synthesis site for some proteins, but all proteins synthesized here are believed to remain within the cytoplasmic matrix.

Located on the SER of liver cells is a system of enzymes that is very important in the detoxification and metabolism of many different drugs. It consists of a family of cytochromes, designated the P450 system, that functions along with other enzymes. The system is particularly active in the oxidation of drugs, but because its action results in the simultaneous oxidation of other compounds as well, the system is collectively referred to as the *mixed-function oxidase system.* Lipophilic substances—for example, the steroid hormones and numerous drugs—can be made hydrophilic by oxidation, reduction, or hydrolysis. Becoming hydrophilic allows these substances to be excreted easily via the bile or urine.

The Golgi apparatus functions closely with the ER in the trafficking and sorting of proteins synthesized in the cell. It is particularly prominent in neurons and secretory cells. The Golgi apparatus consists of four to eight membrane-enclosed, flattened cisternae that are stacked in parallel (Fig. 1.11). The Golgi cisternae are often referred to as "stacks" because of this arrangement. Tubular networks at either end of the Golgi stacks have been identified:

- The cis-Golgi network acts as an acceptor compartment of newly synthesized proteins coming from the ER.
- The trans-Golgi network is the exit site of the Golgi apparatus. It is responsible for sorting proteins for delivery to their next destination [4].

Cisternae positioned between the cis- and trans-Golgi networks are designated as medial cisternae.

The Golgi apparatus is the site for membrane differentiation and the development of surface specificity. For example, the polysaccharide moieties of mucopolysaccharides and of the membrane glycoproteins are synthesized and attached to the polypeptide during its passage through the Golgi apparatus. Such an arrangement allows for the continual replacement of cellular membranes, including the plasma membrane.

Sorting and trafficking of proteins in the secretory pathway between the ER and the Golgi stacks are not thoroughly understood. We do know, however, that the ER is a quality control organelle, in that it prevents proteins that have not achieved normal tertiary or quaternary structure from reaching the cell surface. The ER can retrieve or retain proteins destined for residency within the ER, or it can target proteins for delivery to the cis-Golgi compartment. Retrieved or exported protein "cargo" is coated with protein complexes called *coatomers,* abbreviated COPs (coat proteins). Some coatomers are structurally similar to the clathrin coat of endocytic vesicles. The choice of what is retrieved/ retained by the ER or what is exported to the Golgi apparatus is probably mediated by signals that are inherent in the terminal amino acid sequences of the proteins in question. Certain amino acid sequences of cargo proteins are thought to interact specifically with certain coatomers [5].

The membrane-bound compartments of the ER and the Golgi apparatus are interconnected by transport vesicles, through which cargo proteins are moved from compartment to compartment. The vesicles leaving a compartment are formed by a budding and pinching off of the compartment membrane, and the vesicles

then fuse with the membrane of the target compartment. The specificity of vesicle-membrane interactions has been the focus of considerable research [5]). The current literature abounds with acronyms for proteins that are reactive in the process, according to hypothesis. Examples include NSF (N-ethylmaleimide-sensitive fusion proteins), SNAP (soluble NSF attachment proteins), and SNAREs (soluble NSF attachment protein receptors). It is beyond the scope of this text to elaborate on the proposed series of reactions and interactions involved in the secretory protein pathway. Instead, the interested reader is referred to informative publications that relate to the subject [5,6]. Briefly, vesicle transport from the ER to the cis-Golgi complex, through the medial Golgi stacks, and from the trans-Golgi network to the plasma membrane is specifically directed by SNAPs and SNAREs (see earlier discussion). Budded vesicles departing a membrane have, among their coatomers, integral membrane proteins called *vesicle SNAREs* (vSNAREs). Target membranes possess target SNAREs (tSNAREs). The targeting of vesicles is mediated by interactions between these types of proteins, as illustrated in Figure 1.13.

Secretion from the cell of products such as proteins can be either constitutive or regulated. If secretion follows a constitutive course, it means the secretion rate remains relatively constant, uninfluenced by external regulation. Regulated secretion, as the word implies, is affected by regulatory factors, and therefore its rate is changeable.

Lysosomes and Peroxisomes

Lysosomes and peroxisomes are cell organelles packed with enzymes. Whereas the lysosomes serve as the cell's digestive system, the peroxisomes perform some specific oxidative catabolic reactions.

Lysosomes are particularly large and abundant in those cells that perform digestive functions—for example, the macrophages and leukocytes. Approximately 36 powerful enzymes capable of splitting complex substances such as proteins, polysaccharides, nucleic acids, and phospholipids are held within the confines of a single, thick membrane. The lysosome, just like a protein synthesized for excretion, is believed to develop through the combined activities of the ER and Golgi apparatus, the result of which is a very carefully packaged group of lytic enzymes (Fig. 1.11). The membrane surrounding these catabolic enzymes has the capacity for very selective fusion with other vesicles so that catabolism (or digestion) may occur as necessary. Wastes thereby produced can be removed from the cells by exocytosis, as indicated in Figure 1.11.

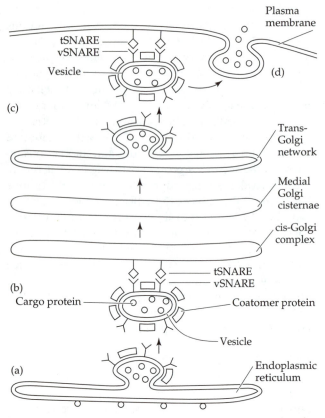

Figure 1.13 Proposed secretory pathway for cellular proteins, illustrating the delivery of ER-synthesized (cargo) proteins to the plasma membrane, and secretion from the cell. Cargo proteins are depicted as small circles. (a) Budding and pinching off of cargo protein-containing vesicle from the ER. (b) Docking of vesicle with cis-Golgi complex. The specificity of the transport is postulated to reside in vSNARE (vesicle snare) and tSNARE (target snare) interaction (see the text). (c) Vesicle formation following budding from the trans-Golgi network and vSNARE-tSNARE interaction at the plasma membrane. Sorting among proteins to determine those to be secreted and those to be retained occurs on the trans-Golgi network. (d) Vesicle fusion with the plasma membrane and release of cargo proteins into the extracellular environment.

Important catabolic activities performed by the lysosomes include participation in phagocytosis, in which foreign substances taken up by the cell may be digested or rendered harmless. An example of digestion by lysosomes is their action in the proximal tubules of the kidney. Lysosomes of the proximal tubule cells are believed to digest the albumin absorbed via endocytosis from the glomerular filtrate. Lysosomal phagocytosis serves as protection against invading bacteria; following a wound or infection, it becomes part of the normal repair process.

A second catabolic activity of lysosomes is autolysis (p. 207), in which intracellular components, including organelles, are digested as a result of degeneration or cellular injury. Autolysis also can serve as a survival mechanism for the cell as a whole. Digestion of dispensable intracellular components can provide the nutrients necessary to fuel functions essential to the life of the cell. The mitochondrion is an example of an organelle whose degeneration requires autolysis. It is estimated that the mitochondria of liver cells must be renewed approximately every 10 days.

Another catabolic activity of the lysosomes is bone resorption, an essential process in the normal modeling of bone. Lysosomes of the osteoclasts promote the dissolution of mineral and the digestion of collagen, both of which actions are necessary in bone resorption and regulation of calcium and phosphorus homeostasis.

Lysosomes, with their special membrane and numerous catabolic enzymes, also function in hormone secretion and regulation. Of particular significance is the role of lysosomes in the secretion of the thyroid hormones (p. 452).

It was in the early 1960s that the *peroxisomes* were first recognized as separate intracellular organelles. These small bodies are believed to originate via "budding" from the SER. They are similar to the lysosomes in that they are bundles of enzymes surrounded by a single membrane. Rather than having digestive action, however, the enzymes within the peroxisomes are catabolic oxidative enzymes. Although the mitochondrial matrix is the major site of fatty acid oxidation to acetyl coenzyme A (acetyl CoA), the peroxisomes can also carry out a similar series of reactions. Acetyl CoA produced in peroxisomes, however, cannot be further oxidized for energy at that site, and must therefore be transported to the mitochondria for oxidation via the Krebs cycle (p. 89).

Peroxisomes are also the site for certain reactions of amino acid catabolism. Some of the oxidative enzymes involved in these pathways catalyze the release of hydrogen peroxide (H_2O_2) as an oxidation product. H_2O_2 is a very reactive chemical that could cause cellular damage if not promptly removed or converted, and for this reason H_2O_2-releasing reactions are segregated within these organelles. Present in large quantities in the peroxisomes is the enzyme catalase, which degrades the potentially harmful H_2O_2 into water and molecular oxygen. Other enzymes in the peroxisomes are important in detoxifying reactions. Particularly important is the oxidation of ethanol to acetaldehyde (p. 99).

Cellular Proteins

Proteins synthesized on the cell's ribosomes may be destined for secretion, or they may remain within the cell to perform their specific structural, digestive, regulatory, or other functions. Among the more interesting

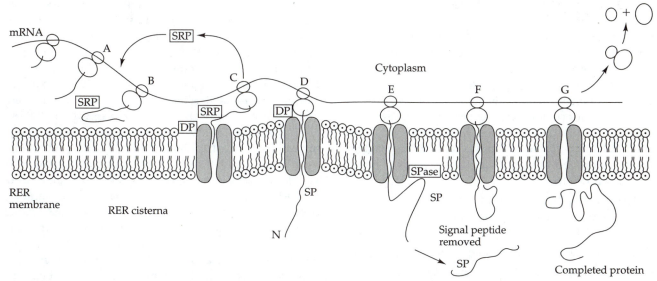

Figure 1.14 The signal (leader) peptide hypothesis. A membrane-free ribosome begins the synthesis of protein on a mRNA specific for the particular protein to be secreted. The signal peptide (consisting of ~80 amino acids) emerges first from the ribosome (A) and is recognized (B, C) by a complex of several proteins called the *signal recognition particle (SRP)*. Translation stops (C) until the ribosome-SRP complex binds to the rough endoplasmic reticulum (RER) membrane via a specific receptor for the SRP called the *docking protein (DP)*. Translation then resumes following the dissociation of the SRP (D). The binding of the SRP to the DP creates a channel in the RER membrane through which the signal peptide (SP) travels, emerging in the RER cisterna, where it is eventually removed from the rest of the protein by a proteolytic enzyme, the membrane-bound signal peptidase, SPase (E). Protein synthesis continues, with the completed protein folding within the RER (G).

areas of biomolecular research has been the attempt to understand how newly synthesized protein that is to remain in the cell finds its way from the ribosomes to its intended destination. The answer lies in leader sequences within the proteins. Leader, or signaling, sequences are amino acid sequences attached to the amino end of the newly synthesized protein to direct it to its ultimate destination. Leader sequences are required for

- proteins destined for secretion from the cell,
- proteins to be retained in the cell's lysosomes and peroxisomes, and
- proteins intended for incorporation into various membranes of the cell.

Interaction between the leader sequences and specific receptors located on the various membranes permits the protein to enter its designated membrane. As the proteins move into their site of localization, the leader sequence is cleaved. Figure 1.14 illustrates the principle of a leader (or signal) sequence for the movement of a secretory protein into the ER from membrane-free ribosomes.

There is a long list of metabolic diseases attributed to a deficiency of, or inactivity of, certain enzymes. Tay-Sachs disease, phenylketonuria, maple syrup urine disease, and the lipid and glycogen storage diseases are a few well-known examples. As a result of research on certain mitochondrial proteins, it is believed that in at least some cases it is not necessarily the enzymes themselves that are inactive or deficient but rather that these enzymes fail to reach their correct destination [7,8].

Cellular proteins of particular interest to the health science student are

- *receptors,* proteins that modify the cell's response to its environment;
- *transport proteins,* those that regulate the flow of nutrients into and out of the cell; and
- *enzymes,* the catalysts for the hundreds of biochemical reactions taking place in the cell.

Receptors and Intracellular Signaling

Receptors are highly specific proteins located in the plasma membrane facing the exterior of the cell. Bound to the outer surface of these specific proteins are oligosaccharide chains, which are believed to act as recognition markers. Membrane receptors act as attachment sites for specific external stimuli such as hormones, growth factors, antibodies, lipoproteins, and certain nutrients. Molecular stimuli such as these, which bind specifically to receptors, are called *ligands.* There are also receptors located on the membrane of cell organelles. Less is known about them, but they appear to be glycoproteins necessary for the correct positioning of newly synthesized cellular proteins.

Although most receptor proteins are probably integral membrane proteins, some may be peripheral. In addition, receptor proteins can vary widely in their composition and mechanism of action. Although composition and mechanism of action of many receptors have not yet been determined, at least three distinct types of receptors are known to exist:

1. those that bind the ligand stimulus and convert it into an internal signal that alters behavior of the affected cell,

2. those that function as ion channels, and

3. those that internalize their stimulus intact.

Following are examples of these three types of receptors.

The internal chemical signal that is most often produced by a stimulus-receptor interaction is 3',5'-cyclic adenosine monophosphate (cyclic AMP, or cAMP). It is formed from adenosine triphosphate (ATP) by the enzyme adenylate cyclase. Cyclic AMP is frequently referred to as the second messenger in the stimulation of target cells by hormones. Figure 1.15 represents a proposed model for the ligand-binding action of receptors, which leads to production of the internal signal cAMP. As shown in the figure, the stimulated receptor reacts with the guanosine triphosphate (GTP)-binding protein (G protein), which is responsible for the activation of adenylate cyclase and the production of cAMP from ATP.

The mechanism of action of cAMP signaling within the cell is complex, but it can be viewed briefly as follows: cAMP is an activator of protein kinases. These are enzymes that phosphorylate (add phosphate groups to) other enzymes, and in doing so, convert the enzymes from inactive forms into active forms. In some cases, the phosphorylated enzyme is the inactive form. Protein kinases that can be activated by cAMP contain two subunits, a catalytic and a regulatory subunit. In the inactive form of the kinase, the two subunits are bound together in such a way that the catalytic portion of the molecule is inhibited sterically by the presence of the regulatory subunit. Phosphorylation of the enzyme by cAMP causes dissociation of the subunits, thereby freeing the catalytic subunit, which regains its full catalytic capacity.

Receptors can also act as ion channels in stimulating a cell. In some cases, the binding of the ligand to its receptor causes a voltage change, which then becomes the signal for an appropriate cellular response. Such is the case when the neurotransmitter acetylcholine is the stimulus. The receptor for acetylcholine appears to function as an ion channel in response to voltage change. Stimulation by acetylcholine becomes a signal for the channels to open, allowing Na ions to pass through an otherwise impermeable membrane [9].

The internalization of a stimulus into a fibroblast via its receptor is illustrated in Figure 1.16. Receptors performing in such a manner exist for a variety of biologically active molecules, including the hormones insulin and triiodothyronine. Low-density lipoproteins (LDLs) are internalized in much the same fashion (see Chapter 6), except that their receptors, rather than being mobile, are already clustered in coated pits. Coated pits, vesicles formed from the plasma membrane, are coated with several proteins, primary among which is clathrin. A coated pit containing the receptor with its ligand soon loses the clathrin coating and forms a smooth-walled vesicle. This vesicle delivers the ligand into the depths of the cell, and then along with the receptor it is recycled into the plasma membrane. If the endocytotic process is for scavenging, the ligand, perhaps a protein, is not used by the cell but instead undergoes lysosomal degradation. This is shown in Figure 1.16 and is exemplified by the endocytosis of LDL.

The reaction of a fibroblast to changes in blood glucose levels is a good example of cellular adjustment to existing environment, made possible through receptor proteins. When blood glucose levels are low, muscular activity leads to release of the hormone epinephrine by the adrenal medulla. Epinephrine becomes attached to its receptor protein on the fibroblast, thereby activating the receptor and causing the receptor to stimulate the G protein and adenylate cyclase, which catalyzes the formation of cAMP from ATP. Then cAMP initiates a series of enzyme phosphorylation modifications, as described earlier in this section, that result in the phosphorolysis of glycogen to glucose 1-phosphate for use by the fibroblast (p. 83).

In contrast, when blood glucose is elevated, the hormone insulin, secreted by the β-cells of the pancreas, reacts with its receptors on the fibroblast and is transported into the cell via receptor-mediated endocytosis (Fig. 1.16). The action of insulin is to allow diffusion of glucose into the cell, but the mechanism of action is not fully understood. It is postulated that insulin increases the number of glucose receptors in the cell membrane and that these in turn promote diffusion of glucose via its transport protein. Glucose transporters will be discussed in Chapter 4. The hormone itself is degraded within the cell [1].

There are many intracellular chemical messengers other than those cited as examples in this section [10]. Listed here, along with cAMP, are several additional examples of intracellular messengers. Page numbers are included for those that will be discussed in more detail in subsequent chapters.

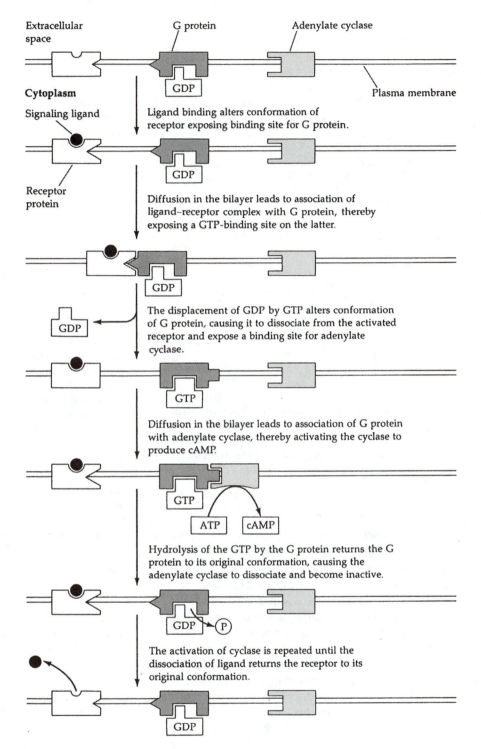

Extracellular space

G protein

Adenylate cyclase

Plasma membrane

Cytoplasm

Signaling ligand

Ligand binding alters conformation of receptor exposing binding site for G protein.

Receptor protein

Diffusion in the bilayer leads to association of ligand–receptor complex with G protein, thereby exposing a GTP-binding site on the latter.

The displacement of GDP by GTP alters conformation of G protein, causing it to dissociate from the activated receptor and expose a binding site for adenylate cyclase.

Diffusion in the bilayer leads to association of G protein with adenylate cyclase, thereby activating the cyclase to produce cAMP.

Hydrolysis of the GTP by the G protein returns the G protein to its original conformation, causing the adenylate cyclase to dissociate and become inactive.

The activation of cyclase is repeated until the dissociation of ligand returns the receptor to its original conformation.

Figure 1.15 A proposed model for production of internal signal, cAMP. The response to an activated receptor protein is much greater than shown in this model because each activated receptor protein activates many molecules of G protein. Also, rather than diffusing independently through the plasma membrane and interacting only after the ligand binds to the receptor protein, the G protein and adenylate cyclase may be permanently associated. (Alberts B, Bray D, Lewis J. *Molecular biology of the cell.* New York: Garland, 1983, p. 385.)

- Cyclic AMP (p. 14)
- Cyclic GMP
- Ca^{++} (p. 378)
- Inositol triphosphate (pp. 132, 379)
- Diacyl glycerol (pp. 132, 379)
- Fructose 2,6-bisphosphate (p. 97)

Transport Proteins

Transport proteins are responsible for regulating the flow of nutrients and other substances into and out of the cell. They may function by acting as carriers (or pumps) for the substances, or they may provide protein-lined passages (pores) through which water-soluble

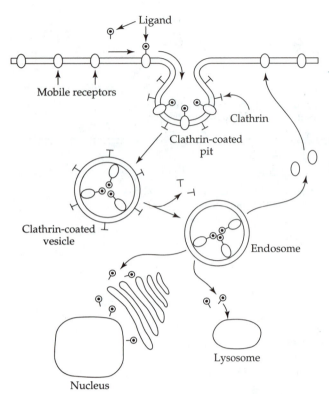

Figure 1.16 Receptor-mediated endocytosis, summarizing the steps involved in the binding, clustering, and entry of a typical ligand into a fibroblast. In this example, the receptor is recycled to the plasma membrane, and the ligand can either be used by the cell or undergo lysosomal degradation.

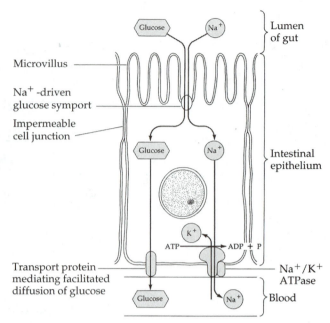

Figure 1.17 The active transport of glucose. Glucose is pumped into the cell's apical membrane by a Na⁺-powered glucose symport, and glucose passes out of the cell by facilitated diffusion. The Na⁺ gradient driving the glucose symport is maintained by the Na⁺/K⁺-ATPase in the basal and lateral plasma membrane, which keeps the internal concentration of Na⁺ low.

materials of small molecular weight may diffuse. Figure 1.6 is an illustration of these various transport proteins.

The active transport protein that has been studied most is the sodium (Na⁺) pump. Not only is the Na⁺ pump essential for the maintenance of ionic and electrical balance, but it also is necessary for the intestinal absorption and renal absorption of certain key nutrients (e.g., glucose and certain amino acids). These nutrients move into the epithelial cell of the small intestine against a concentration gradient, necessitating the need for a carrier and source of energy, both of which are provided by the Na⁺ pump.

The proposed mechanism by which glucose is actively absorbed is termed *symport* because it is a simultaneous transport of two compounds (Na⁺ and glucose) in the same direction. A transport protein with two binding sites binds both Na⁺ and glucose. The attachment of Na⁺ to the carrier increases the transport protein's affinity for the glucose. Sodium, because it is moving down a concentration gradient created by energy released through Na⁺/K⁺ adenosine triphosphatase (ATPase), is able to carry along with it glucose

that is moving up a concentration gradient. When Na⁺ is released inside the cell, the carrier's affinity for glucose is decreased, and glucose can be released into the cell also. Na⁺/K⁺-ATPase then "pumps" the Na ions back out of the cell. The sodium pump is illustrated in Figure 1.17.

Na⁺/K⁺-ATPase works by first combining with ATP in the presence of Na⁺ on the inner surface of the cell membrane. The enzyme then is phosphorylated by breakdown of ATP to adenosine diphosphate (ADP) and is consequently able to move three Na ions out of the cell. On the outer surface of the cell membrane, ATPase becomes dephosphorylated by hydrolysis in the presence of K ions and is then able to return two K ions into the cell. The term *pump* is used because the Na and K ions are both transported across the membrane against their concentration gradients. This pump is responsible for most of the active transport in the body.

Transport of glucose and amino acids into the epithelial cells of the intestinal tract is active in that the carriers needed for their transport are dependent on the concentration gradients achieved by action of Na⁺/K⁺-ATPase at the basolateral membrane. The activity of Na⁺/K⁺-ATPase is the major energy demand of the body at rest.

The process of facilitated (non-energy-dependent) transport is also a very important mechanism for regulating the flow of nutrients into the cell. It is used broadly across a wide range of cell types. Proteins involved in this function are often called *transporters,* probably the most thoroughly studied of which are the glucose transporters, discussed in Chapter 4.

Catalytic Proteins (Enzymes)

Enzymes are distributed throughout all cellular compartments. Those that are components of the cellular membranes usually are found on the inner surface of the membranes. Exceptions are the digestive enzymes: isomaltase, the disaccharidases (lactase, sucrase, and maltase), and certain peptidases that are located on the brush border of the epithelial cells lining the small intestine (p. 28). Membrane-associated enzymes are found distributed throughout the cell organelles, but the greatest concentration is found in the mitochondria. As mentioned earlier, the enzymes of the electron transport chain, where energy transformation occurs, are located within the inner membrane of the mitochondria.

Metabolic processes occurring in the cells are governed by the enzymes that, for the most part, have been synthesized on the cell's RER under the direction of nuclear DNA. The functional activity of most enzymes, however, depends not only on the protein portion of the molecule but also on a nonprotein prosthetic group or coenzyme. If the nonprotein group is an organic compound, it usually contains a chemically modified B-complex vitamin. Very commonly, however, the prosthetic group may be inorganic (i.e., metal ions such as Mg, Zn, Cu, Mn, Fe).

Enzymes have an active center that possesses a high specificity. This means that a substrate must fit perfectly into the specific contours of the active center. The velocity of an enzyme-catalyzed reaction increases as the concentration of the substrate that is available to the enzyme increases. However, this relationship applies only to a concentration of substrate that is less than that which "saturates" the enzyme. At saturation levels of substrate, the enzyme molecule functions at its maximum velocity (V_{max}), and the occurrence of still higher concentration of substrate cannot increase the velocity further.

The K_m, or Michaelis constant, is a useful parameter that aids in establishing how enzymes will react in the living cell. K_m represents the concentration of a substrate that will be found in an occurring reaction when the reaction is at one-half its maximum velocity. If an enzyme has a high K_m value, an abundance of substrate must be present to raise the rate of reaction to half its maximum. In other words, this enzyme has a low affinity for its substrate. An example of an enzyme with a high K_m is *glucokinase,* an enzyme operating in the liver cells. Because glucose can diffuse freely into the liver, the fact that glucokinase has a high K_m is very important in the regulation of blood glucose. This low affinity of glucokinase for glucose prevents too much glucose from being removed from the blood during periods of fasting. Conversely, when the glucose load is high—for example, following a high-carbohydrate meal—the excess glucose will still be able to be converted by glucokinase, which does not function at its maximum velocity when glucose levels are basal. The enzyme can therefore be thought of as a protection against high cellular concentrations of glucose.

The nature of enzyme catalysis can be described by the following reactions:

$$\text{Enzyme (E) + substrate (S)} \longleftrightarrow \text{E-S complex}$$
$$\text{(reversible reaction)}$$
$$\text{E-S} \longleftrightarrow \text{E-P}$$

The substrate activated by combination with the enzyme is converted into an enzyme-product complex through rearrangement of the substrate's ions and atoms.

$$\text{E-P} \longrightarrow \text{E + P}$$

The product is released, and the enzyme is free to react with more of the substrate.

Most biochemical reactions are reversible, meaning that the same enzyme can catalyze a reaction in both directions. The extent to which a reaction can proceed in a reverse direction depends on several factors, the most important of which are relative concentrations of substrate (reactant) and product, and the differences in energy content between reactant and product. In those instances when a very large disparity in energy content or concentration exists between reactant and product, the reaction can proceed in only one direction. Such a reaction is *unidirectional* as opposed to reversible. In unidirectional reactions, the same enzyme cannot catalyze in both directions. Instead, a different enzyme is required to catalyze the reverse direction of the reaction (see Chapter 3). Comparing glycolysis with gluconeogenesis allows us to see how unidirectional reactions may be reversed by the introduction of a different enzyme (p. 94).

Simultaneous reactions, catalyzed by various multienzyme systems or pathways, constitute cellular metabolism. Enzymes are compartmentalized within the cell and function in sequential chains. A good example of such a multienzyme system is the Krebs cycle located in the mitochondrial matrix (p. 89). Each sequential reaction is catalyzed by a different enzyme, and some of the reactions are reversible while others are unidirectional. Although some reactions in almost any pathway are reversible, it is important to understand that the removal of one of the products drives the reaction toward formation of more of that product. Removal (or use) of the product, then, becomes the driving force that causes reactions to proceed primarily in the desired direction.

A very important aspect of nutritional biochemistry is the regulation of metabolic pathways. Anabolic and catabolic reactions must be kept in balance appropriate for life (and perhaps growth) of the organism. Regulation involves primarily the adjustment of the catalytic activity of certain participating enzymes, and there are three major mechanisms for doing this:

1. covalent modification of enzymes via hormone stimulation,
2. modulation of allosteric enzymes, and
3. increasing enzyme concentration by induction.

The first of these mechanisms, *covalent modification* of enzymes, is usually achieved by the addition of, or hydrolytic removal of, phosphate groups to and from the enzyme. This is the mechanism involving cAMP and protein kinase activation previously discussed (p. 14). An example of covalent modification of enzymes is the regulation of glycogenesis and glycogenolysis (p. 97).

The second important regulatory mechanism is that exerted by certain unique enzymes called *allosteric enzymes.* The term *allosteric* refers to the fact that they possess an allosteric or specific "other" site besides the catalytic site. Specific compounds, called *modulators,* can bind to these allosteric sites and influence profoundly the activity of these regulatory enzymes. Modulators may be positive (i.e., causing an increase in enzyme activity), or they may exert a negative effect and inhibit activity. Modulating substances are believed to alter the activity of the allosteric enzymes by changing the conformation of the polypeptide chain or chains of the enzyme, thereby altering the binding of its catalytic site with the intended substrate. Negative modulators are often the end products of a sequence of reactions. As an end product accumulates above a certain critical concentration, it can inhibit, through an allosteric enzyme, its own further production.

An excellent example of an allosteric enzyme is *phosphofructokinase* in the glycolytic pathway. Glycolysis gives rise to pyruvate, which is decarboxylated and oxidized to acetyl CoA and which enters the Krebs cycle by combination with oxaloacetate to form citrate. Citrate is a negative modulator of phosphofructokinase. Therefore, an accumulation of citrate causes an inhibition of glycolysis through its regulation of phosphofructokinase. In contrast, an accumulation of AMP or ADP, which indicates a depletion of ATP, signals the need for additional energy in the cell in the form of ATP. AMP or ADP there-

fore act to modulate phosphofructokinase positively. The result is an active glycolytic pathway that ultimately leads to the formation of more ATP through the Krebs cycle/electron transport chain connection (p. 93).

The third mechanism of enzyme regulation, *enzyme induction,* creates changes in the concentration of certain *inducible* enzymes. Inducible enzymes are *adaptive,* meaning that they are synthesized at rates dictated by cellular circumstances. This is in contrast to *constitutive* enzymes, which are synthesized at a relatively constant rate, uninfluenced by external stimuli. Induction usually occurs through the action of certain hormones such as the steroid hormones and thyroid hormones, and is exerted through changes in the expression of genes encoding the enzymes. Dietary changes can elicit the induction of enzymes necessary to cope with the changing nutrient load. This is a relatively slow regulatory mechanism, however, as opposed to the first two mechanisms discussed, which exert effects very quickly—that is, in terms of seconds or minutes.

Specific examples of enzyme regulation will be cited in subsequent chapters dealing with metabolism of the major nutrients. It should be noted at this point, however, that *enzymes targeted for regulation catalyze essentially unidirectional reactions.* In every metabolic pathway, at least one reaction is essentially irreversible, exergonic, and enzyme limited. That is, the rate of the reaction is limited only by the activity of the enzyme catalyzing it. Such enzymes are frequently the regulatory enzymes, capable of being stimulated or suppressed by one of the mechanisms described. It is logical that an enzyme catalyzing a reaction reversibly at near equilibrium in the cell cannot be regulatory. This is because its up or down regulation would affect its forward and reverse activities equally. This, in turn, would not accomplish the purpose of regulation, which is the stimulation of one direction of a metabolic pathway relative to its reverse direction.

Enzymes participating in cellular reactions are located throughout the cell both in the cytoplasmic matrix (cytoplast) and in the various organelles. Location of specific enzymes depends on the site of the metabolic pathways or metabolic reactions in which these enzymes participate. Enzyme classification therefore is based on the type of reaction catalyzed by the various enzymes. Enzymes fall within six general classifications:

1. *Oxidoreductases* (dehydrogenases, reductases, oxidases, peroxidases, hydroxylases, oxygenases), enzymes catalyzing all reactions in which one compound is oxidized and another is reduced. Good examples of oxidoreductases are the enzymes found in the electron transport chain located on the inner membrane of the mitochondria. Other good examples are the cytochrome P450 enzymes located on the ER of liver cells.

2. *Transferases,* enzymes catalyzing reactions not involving oxidation and reduction in which a functional group is transferred from one substrate to another. Included in this group of enzymes are transketolase, transaldolase, transmethylase, and the transaminases. The transaminases (α-amino transferases), which figure so prominently in protein metabolism, fall under this classification and are located primarily in the mitochondrial matrix.

3. *Hydrolases* (esterases, amidases, peptidases, phosphatases, glycosidases), enzymes catalyzing cleavage of bonds between carbon atoms and some other kind of atom by the addition of water. Digestive enzymes fall within this classification, as do those enzymes contained within the lysosome of the cell.

4. *Lyases* (decarboxylases, aldolases, synthetases, cleavage enzymes, deaminases, nucleotide cyclases, hydrases or hydratases, and dehydratases), enzymes that catalyze cleavage of carbon-carbon, carbon-sulfur, and certain carbon-nitrogen bonds (peptide bonds excluded) without hydrolysis or oxidation reduction. Citrate lyase, which frees acetyl CoA for fatty acid synthesis in the cytoplast, is a good example of an enzyme belonging to this classification.

5. *Isomerases* (isomerases, racemases, epimerases, and mutases), enzymes catalyzing the interconversion of optical or geometric isomers. Phosphohexose isomerase that converts glucose 6-phosphate to fructose 6-phosphate in glycolysis (occurring in the cytoplast) exemplifies this particular class of enzyme.

6. *Ligases,* enzymes that catalyze the formation of bonds between carbon and a variety of other atoms, including oxygen, sulfur, and nitrogen. Formation of bonds catalyzed by ligases requires energy that usually is provided by hydrolysis of ATP. A good example of a ligase is acetyl CoA carboxylase, which is necessary to initiate fatty acid synthesis in the cytoplast. Through the action of acetyl CoA carboxylase, a bicarbonate ion (HCO_3^-) is attached to acetyl CoA to form malonyl CoA, the initial compound in de novo fatty acid synthesis (p. 151).

Apoptosis

It is said that dying is a normal part of living. So it is with the cell. Like every living thing, a cell has a well-defined life span, after which its structural and functional integrity diminishes, and it is phagocytically removed by other cells.

As cells die, they are replaced by new cells that are continuously being formed by cell mitosis. However, both daughter cells formed in the mitotic process do not always enjoy the full life span of the parent. If they did, the number of cells, and consequently tissue mass, could increase inordinately. Therefore, one of the two cells produced by mitosis is generally programmed to die before its sister. In fact, most dying cells are already doomed at the time of their formation. Those targeted for death are usually smaller than their surviving sisters, and their phagocytosis begins even before the mitosis generating them is complete [11]. The processes of cell division and cell death must be carefully regulated to generate the proper number of cells during development. Later, in the mature state, it is important that the desired number of cells be maintained. The mechanism by which naturally occurring cell death arises has, in recent years, been subjected to intense research.

Many terms have been used to describe naturally occurring cell death. It is now most commonly referred to as *programmed cell death,* to distinguish it from pathological cell death, which is not part of a normal physiological process. The emergent term describing programmed cell death is *apoptosis,* a word borrowed from the Greek, meaning to "fall out."

Apoptotic cells are recognizable morphologically. They contract, the chromatin condenses, and the cytoplasmic organelles initially remain intact. Apoptosis also involves activation of a Ca^{++}-dependent endonuclease that cleaves the genome of the cell into fragments of approximately 180 base pairs. Dead cells are removed by phagocytosis. Although the process of apoptosis has been extensively studied [11], it remains unclear as to how many distinct mechanisms are involved in causing cell death.

Cell death appears to be activated by specific genes in dying cells. Genes designated as ced-3 and ced-4 must be expressed within dying cells for death to occur. The ced-3 and ced-4 genes encode products (proteins) that activate cytotoxic activity, and they therefore must be tightly controlled to avoid damage to the wrong cells. A major control factor is a third gene, ced-9, which negatively regulates the ced-3 and ced-4 genes. This has been demonstrated by the fact that mutations to ced-9, which inactivate the gene, kill an animal under study by causing the killing of cells that were otherwise intended to survive [11].

There is also interest in the gene designated p53 for its possible involvement in the apoptotic process. The gene has been garnering a great deal of attention in the cancer field because of the tumor suppressor properties of some of its protein products. The connection of p53 to apoptosis is that it is believed to regulate the reduction/oxidation state of the cell and is able to increase concentrations of reactive oxygen species (ROS). ROS may, in turn, signal the beginning of apoptosis. The mechanisms underlying p53-dependent apoptosis is largely unknown, but a documented property of p53, its ability to activate gene transcription, may be a factor. A proposed sequence of events by which p53 results in apoptosis is as follows:

1. The transcription-induction of redox-related genes

2. Formation of ROS

3. Oxidative degradation of mitochondrial components, resulting in cell death [12]

Recently, the release of cytochrome c from the mitochondrion into the cytoplasmic matrix has been investigated as one of the factors that promote apoptosis. Once cytochrome c has translocated to the cytoplasmic matrix, it activates certain proteolytic enzymes that are products of the ced-3 genes and that are therefore associated with apoptosis [13]. A cytochrome c–apoptosis connection is supported by the discovery of a protein designated Bcl-2 (B-cell lymphoma gene product) that appears to block the release of mitochondrial cytochrome c. By blocking the release of mitochondrial cytochrome c, Bcl-2 therefore interferes with the apoptotic process. Bcl-2 is an integral membrane protein on the outer membrane of the mitochondrion. Two observations are relevant:

• Bcl-2 prevents the efflux of cytochrome c from the mitochondrion to the cytoplasmic matrix.

• Genetic overexpression of Bcl-2 prevents cells from undergoing apoptosis in response to various stimuli.

Therefore, a possible role of Bcl-2 in preventing apoptosis is to block release of cytochrome c from the mitochondrion [14].

Much is to be learned about the mechanisms underlying apoptosis. What is obvious is the intense interest in the mechanism. Once the apoptotic process is understood, can scientific intervention, designed to slow the rate of programmed cell death, be far behind?

Summary

This brief walk through the cell, beginning with its outer surface, the plasma membrane, and moving into its innermost part, where the nucleus is located, has provided a view of how this living entity functions.

Characteristics of the cell that seem particularly notable are as follows:

• The flexibility of the plasma membrane in adjusting or reacting to its environment while protecting the rest of the cell as it monitors what may pass into or out of the cell. Very prominent in the membrane's reaction to its environment are the receptor proteins, which are believed to have been synthesized on the RER and to have moved through the Golgi apparatus to their intended site on the plasma membrane.

• The communication among the various components of the cell made possible through the cytoplast with its microtrabecular network and also through the endoplasmic reticulum and Golgi apparatus. The networking is such that communications flow not only among components within the cell but also between the nucleus and the plasma membrane.

• The efficient division of labor among the cell components (organelles). Each component has its own specific functions to perform, and there is little overlap. Furthermore, much evidence is accumulating to support the concept of an "assembly line" not only in oxidative phosphorylation on the inner membrane of the mitochondrion but also in almost all operations wherever they occur.

• The superb management exercised by the nucleus so that all the proteins needed for a smooth operation are synthesized. Proteins needed as recognition markers, receptors, transport vehicles, and catalysts are available as needed.

• Like all living things, cells must die a natural death. This programmed process is called apoptosis, a particularly attractive focus of current research.

Despite the efficiency of the cell, it is still not a totally self-sufficient unit. Its continued operation is contingent on receiving appropriate and sufficient nutrients. Nutrients needed are not only those that can be used for production of immediate metabolic energy (ATP) or for storage as chemical energy but also those required as building blocks for structural macromolecules. In addition, the cell must have an adequate supply of the so-called regulatory nutrients (i.e., vitamins, minerals, and water).

With a view of the structure of the "typical cell," the division of labor among its component parts, and the location within the cell for many of the key metabolic reactions necessary for the continuation of life, consideration can now be given to how the cell receives its nourishment so that life can continue.

References Cited

1. Berdanier CD. Role of membrane lipids in metabolic regulation. Nutr Rev 1988;46:145–149.

2. Young, P. Mom's mitochondria may hold mutation. Sci News 1988;134:70.

3. Baluska F, Volkmann D, and Barlow P. Nuclear components with microtubule-organizing properties in multicellular eukaryotes: Functional and evolutionary

considerations. In: International Review of Cytology 1997;175:91–135. Kwang W. Jeon, ed.

4. Griffiths G, Simons K. The trans Golgi network: Sorting at the exit site of the Golgi complex. Science, 1986; 234:438–443.

5. Teasdale R and Jackson M. Signal-mediated sorting of membrane proteins between the endoplasmic reticulum and the Golgi apparatus. In: Annu. Rev. Cell Dev. Biol 1996; 12: 27–54.

6. Rothman J. Mechanisms of intracellular protein transport. Nature 1994; 372: 55–63.

7. How do proteins find mitochondria? Science 1985;228: 1517–1518.

8. Wickner WT, Lodish JT. Multiple mechanisms of protein insertion into and across membranes. Science 1985; 230:400–407.

9. A potpourri of membrane receptors. Science 1985;230: 649–651.

10. Barritt GJ. Networks of extracellular and intracellular signals. In: Communication within animal cells. Oxford, England: Oxford University Press, 1992;1–19.

11. Ellis R, Yuan J, and Horvitz H. Mechanisms and functions of cell death. In: Annu Rev Cell Biol 1991; 7: 663–698.

12. Polyak K, Xia Y, Zweler J, Kinzler K, and Vogelstein B. A model for p53-induced apoptosis. Nature 1997; 389: 300–305.

13. Kluck R, Bossy-Wetzel E, Green D, and Newmeyer D. The release of cytochrome c from mitochondria: A primary site for Bcl-2 regulation of apoptosis. Science 1997; 275: 1132–1136.

14. Yang J, Liu X, Bhalla K, et al. Prevention of apoptosis by Bcl-2: Release of cytochrome c from mitochondria blocked. Science 1997; 275: 1129–1132.

Suggested Reading

Alberts B, Bray D, Lewis J, Raff M, Roberts K, and Watson JD. Molecular biology of the cell, 3rd ed. New York, NY: Garland, 1994.

Murray R and Granner D. Membranes: Structure, assembly, and function. In: Harper's biochemistry, 24th ed. Appleton & Lange, Stamford, CT 1996, pp. 483–508.

Barritt GJ. Networks of extracellular and intracellular signals. In: Communication within animal cells. Oxford, England: Oxford University Press, 1992;1–19.

Masters C and Crane D. The peroxisome: A vital organelle. Cambridge, England: Cambridge University Press, 1995. This is a clearly written overview of the multifaceted functions of this important organelle.

Web Sites

www.nlm.nih.gov
  National Library of Medicine: MEDLINE
www.chid.nih.gov/welcome/welcome.html
  National Institutes of Health: Combined Health Information Database

Cellular Enzymes on the Loose: Indicators of Disease

All of the hundreds of enzymes present in the human body are synthesized intracellularly, and most of them function within the cell in which they were formed. These are the enzymes responsible for catalyzing the myriad of metabolic reactions occurring in the cell. As explained in the "Cellular Proteins" section of this chapter, they are directed to specific locations within the cell following their synthesis on the ribosomes. Some enzymes, on the other hand, are secreted from the cell, usually in an inactive form, and are rendered active in the extracellular fluids where they function. Examples of secreted enzymes are the digestive proteases and other hydrolases that are formed in the cells of the pancreas and then secreted into the lumen of the small intestine. Other secreted enzymes function in the bloodstream, and these are called plasma-specific enzymes. Examples include the enzymes involved in the blood clotting mechanism.

Diagnostic enzymology focuses on intracellular enzymes, which, because of a perturbation of the cell structure, egress from the cell and ultimately express their activity in the serum. By measuring the serum activity of the released enzymes, both the site and the extent of the cellular damage can be determined. If the site of the damage is to be determined with reasonable accuracy, the enzyme being measured must exhibit a relatively high degree of organ or tissue specificity. For example, an enzyme having an intrahepatocyte concentration many times greater than its concentration in other tissues, could potentially be a marker for liver damage should its serum activity increase.

The rate at which intracellular enzymes enter the bloodstream is based on two factors:

1. those that affect rates at which enzymes *leak* from cells, and

2. those that affect rates of production either by increased synthesis or accelerated proliferation of enzyme-producing cells.

Leakage Factors

Intracellular enzymes are normally retained within the cell by the plasma membrane. The plasma membrane is metabolically active, and its integrity depends on the cell's energy consumption and therefore its nutritive status. Any process that impairs the cell's use of nutrients can therefore compromise the structural integrity of the plasma membrane. Membrane failure can also arise from its mechanical disruption, such as would occur by a viral attack on the cell. Damage to the plasma membrane manifests as leakiness and eventual death, allowing unimpeded passage of substances, including enzymes, from intracellular to extracellular compartments.

Factors contributing to cellular damage, and resulting in abnormal egress of cellular enzymes, include the following:

- Tissue ischemia. Ischemia refers to an impairment of blood flow to a tissue or part of a tissue. It deprives affected cells of oxidizable nutrients.

- Tissue necrosis

- Viral attack on specific cells

- Damage from organic chemicals such as alcohol and organophosphorus pesticides

- Hypoxia (inadequate intake of oxygen)

Increases in blood serum concentrations of cellular enzymes can be good indicators of even minor cellular damage. This is because the intracellular concentration of enzymes is hundreds or thousands of times greater than in blood, and because enzyme assays are extremely sensitive.

Not all intracellular enzymes are of value in diagnosing damage to the cells in which they are contained. Several conditions must be met for the enzyme to be suitably diagnostic:

- *There must be a sufficiently high degree of organ or tissue specificity of the enzyme.* Suppose that an enzyme is widely distributed among organ or tissue systems. Although an abnormal increase of its activity in serum does indicate a pathological process with cellular damage, it cannot identify precisely the site of the damage. An example is lactate dehydrogenase (LDH). LDH activity is widely distributed among cells of the heart, liver, skeletal muscle, erythrocytes, platelets, and lymph nodes. Therefore, elevated serum activity of LDH can hardly be a specific marker for tissue pathology. In practice, however, LDH does have diagnostic value if it is first separated into its five different isoenzyme forms, and these quantified individually. LDH isoenzyme forms are designated LDH1, LDH2, LDH3, LDH4, and LDH5. Each isoform is more organ specific than total LDH, with LDH1 being associated primarily with heart muscle, and LDH5 with liver cells.

- *There must be a steep concentration gradient of enzyme activity between the interior and exterior of the cells under normal conditions.* If this were not true, small increases in serum activity would not be detectable. Examples of enzymes that are in compliance with this requirement, and that have been useful over the years as disease markers, are prostatic acid phosphatase, having a prostate cell/serum concentration ratio of 10^3 to 1, and alanine aminotransferase, with an hepatocyte/

Table 1 Diagnostically Important Enzymes

Enzymes	Principal Sources	Principal Clinical Significance
Acid phosphatase	Prostate, erythrocytes	Carcinoma of prostate
Alanine amino transferase	Liver, skeletal muscle, heart	Hepatic parenchymal cell disease
Aldolase	Skeletal muscle, heart	Muscle diseases
Amylase	Pancreas, salivary glands	Pancreatitis, carcinoma of pancreas
Cholinesterase	Liver	Organophosphorus insecticide poisoning, hepatic parenchymal cell disease
Creatine kinase (CK-2 isoform)	Heart, skeletal muscle	Myocardial infarction
Gamma glutamyl transferase	Liver, kidney	Alcoholism, hepatobiliary disease
Prostate specific antigen (PSA)	Prostate	Carcinoma of prostate

serum ratio of 10^4 to 1. These enzymes have been useful in the diagnosis of prostatic disease, primarily carcinoma, and viral hepatitis, respectively.

- *The enzyme must function in the cytoplasmic compartment of the cell.* If the enzyme is compartmentalized within an organelle such as the nucleus or mitochondrion, its leakage from the cell is impeded even in the event of significant damage to the plasma membrane. An example of an enzyme not in compliance with this condition is the mitochondrial enzyme, ornithine carbamoyl transferase, functioning within the urea cycle. Although the enzyme adheres rigidly to the two previous conditions (i.e., it is strictly liver specific, and its cell/serum concentration is as high as 10^5 to 1), it provides little value in the diagnosis of hepatic disease.

- *The enzyme must be stable for a reasonable period of time in the vascular compartment.* Isocitrate dehydrogenase has an extremely high activity in heart muscle. Yet, following the resultant damage of a myocardial infarction, the released enzyme is rapidly inactivated upon entering the bloodstream, therefore becoming undeterminable.

Increased Production Factors

The most common cause for increased production of an enzyme, resulting in a spike in its serum concentration, is malignant disease. Substances that occur in body fluids as a result of malignant disease are called *tumor markers.* A tumor marker

- may be produced by the tumor itself or
- may be produced by the host in response to a tumor.

In addition to enzymes and isoenzymes, other forms of tumor markers include hormones, oncofetal protein antigens such as carcinoembryonic antigen (CEA), and products of oncogenes. Oncogenes are mutated genes that encode abnormal, mitosis-signaling proteins that cause unchecked cell division.

Products of malignant cells, such as intracellular enzymes, are predictably increased in their rate of synthesis

because of the nature of the disease process. Should the proliferating cells of the tumor retain their capacity to synthesize the enzyme, the gross output of the enzyme will be markedly elevated. Furthermore, the enzyme can be released into the systemic circulation as a result of tumor necrosis or by the change in permeability of the plasma membrane of the malignant cells.

While tumor markers are present in higher quantities in cancer tissue or blood from cancer patients than in benign tissue or the blood of normal subjects, few markers are specific for the organ in which the tumor is located. This is because most enzymes are not unique to a specific organ. A possible exception is prostate specific antigen (PSA).

PSA is a proteolytic enzyme produced almost exclusively by the prostate gland. Its value as a tumor marker is further enhanced by the fact that metastasized malignant prostate cells produce nearly 10 times as much PSA as normal prostate cells. A significant rise in serum PSA concentration may therefore signal that a tumor has metastasized to other sites in the body, suggesting a different therapeutic approach. With prostate cancer being the leading cause of death among older men, PSA has become a valuable diagnostic tool.

Table 1 offers a list of enzymes that have been used successfully as indicators of organ or tissue pathology. The source of the enzymes and the clinical significance of their serum determination are also included in the table.

Suggested Reading

Moss DW and Henderson AR. Enzymes (Chapter 19), and Chan DW and Sell S. Tumor markers (Chapter 21). In: Tietz, Fundamentals of clinical chemistry, 4th ed. Philadelphia: Saunders, 1996; 297–350. Burtis C and Ashwood E, ed.

Web Sites

http://www.clinchem.org
Clinical Chemistry; Journal of the American Association for Clinical Chemistry

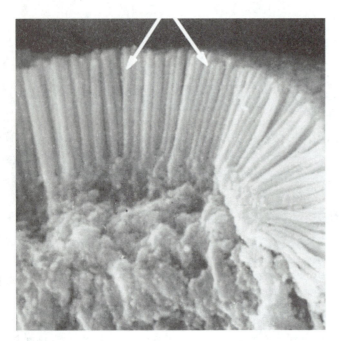

Intestinal microvilli

Chapter 2

The Digestive System: Mechanism for Nourishing the Body

Nutrition is the science of nourishment. Ingestion of foods and beverages provides the body with at least one, if not more, of the nutrients needed to nourish the body. There are six classes of nutrients that the body needs: carbohydrate, lipid, protein, vitamins, minerals, and water. For the body to use carbohydrate, lipid, and protein found in foods, the food must be digested. In other words, the food must be broken down mechanically and chemically. The process of digestion occurs in the digestive tract and on its completion yields nutrients ready for absorption and use by the body.

An Overview of the Structure of the Digestive Tract

The digestive tract, approximately 16 feet in length, includes organs that comprise the alimentary canal, also called the *gastrointestinal tract,* as well as certain accessory organs, primarily the pancreas, liver, and gallbladder. The accessory organs provide secretions that are ultimately delivered to the lumen of the digestive tract and aid in the digestive process. Figure 2.1 illustrates the digestive tract and accessory organs.

The wall of the gastrointestinal tract, shown as a cross section in Figure 2.2, consists primarily of four tunics or layers:

- the mucosa,
- the submucosa,

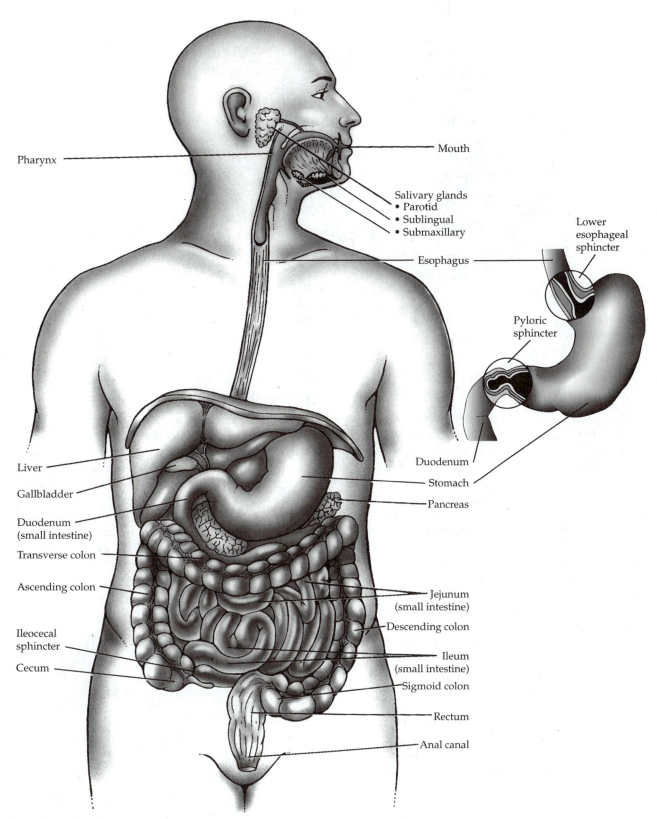

Figure 2.1   The digestive system.

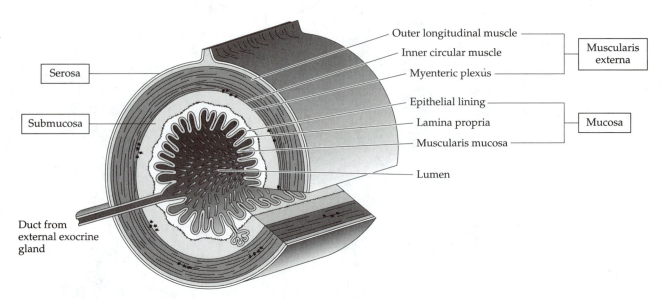

Figure 2.2 Cross section of a digestive tract segment showing the four layers: the mucosa, submucosa, muscularis externa, and serosa. (Adapted from Stalheim-Smith and Fitch, *Understanding Human Anatomy and Physiology,* St. Paul: West, 1993, p. 793.)

- the muscularis externa, and
- the serosa.

The mucosa in turn is made of three sections: the epithelium or epithelial lining, the lamina propria, and the muscularis mucosa. The mucosa epithelium lines the lumen (shown in Fig. 2.2) of the gastrointestinal tract. It is the mucosal epithelium that is in contact with ingested nutrients in the food we eat. The lamina propria lies below the epithelium and consists of connective tissue and small blood and lymphatic vessels. A thin layer of smooth muscle is found in the muscularis mucosa. Protecting this mucosa layer is a group of substances that include mucus (consisting of glycoproteins and glycolipids) and phospholipids, among others. The phospholipids (phosphatidylcholine, phosphatidylethanolamine, and phosphatidylinositol) act as a protective barrier but also serve as precursors for the synthesis of cytoprotective prostaglandins. The tripeptide glutathione may also help protect the mucosa from oxidative damage by scavenging radicals.

Next to the mucosa is the submucosa. The submucosa, the second tunic or layer, is made up of connective tissue and contains a neuronal network called the submucosal plexus. The submucosa serves to bind the first mucosa layer of the gastrointestinal tract to the muscularis externa or third layer of the gastrointestinal tract.

The muscularis externa contains both circular and longitudinal smooth muscle important for peristalsis as well as the myenteric plexus. The myenteric and submucosal plexuses are discussed in more detail on page 31 under "Neural Regulation."

The Structures of the Upper Gastrointestinal Tract

The *mouth* and pharynx constitute the oral cavity and provide the entryway to the digestive tract. Food is passed from the mouth through the pharynx and into the *esophagus.* Secretory glands are found distributed throughout the digestive tract, with the exception of the colon. The secretory glands release secretions into a duct (exocrine) or into the blood (endocrine). Sphincters, which are circular muscles, are also located throughout the digestive tract and allow for the passage of food

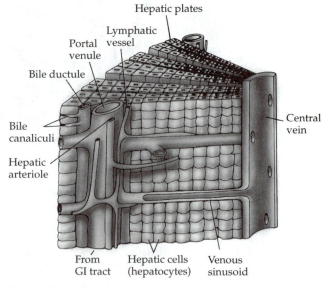

Figure 2.3 Structure of a liver lobule.

from one section of the gastrointestinal tract to another. For example, an area of the esophagus referred to as the lower esophageal sphincter (LES) lies just above the juncture of the esophagus and stomach. Relaxation of this sphincter permits the passage of food from the esophagus into the stomach. The *stomach* is a J-shaped organ located on the left side of the abdomen under the diaphragm, extending from the LES to the duodenum, the upper or proximal section of the *small intestine.* The stomach consists of two major sections, the body and the antrum. The volume of the stomach when empty is about 50 mL (~2 oz), but on being filled, as usually shown in diagrams, it can expand to accommodate 1 to approximately 1.5 L (~37–52 oz). The pyloric sphincter at the distal end of the stomach controls the release of chyme (partially digested food existing as a thick semiliquid mucky mass) from the stomach into the duodenum.

The Structures of the Lower Gastrointestinal Tract and Accessory Organs

The *small intestine* is composed of the duodenum (slightly <1 ft long), and the jejunum and ileum (which together are ~9 ft long). Microscopy is needed to iden-tify where one of these sections of the small intestine ends and the other begins. The small intestine represents the main site for nutrient digestion and absorption. The duodenum receives secretions from the liver, gallbladder, and pancreas, which are accessory organs. These secretions are necessary for the digestion of nutrients.

The *liver,* pictured in Figures 2.1, 2.3, and 2.4, is the largest single internal organ of the body. Liver cells (hepatocytes) synthesize bile, which is necessary for fat digestion. Bile is then released into bile canaliculi, which lie between the hepatocytes in the hepatic plates (Fig. 2.3). Bile drains from the plates and into bile ducts. As shown in Figure 2.4, right and left hepatic bile ducts join to form the common hepatic duct. The common hepatic duct unites with the cystic duct, which leads into and out of the gallbladder, to form the common hepatic bile duct (Fig. 2.4).

The *gallbladder,* a small organ with a capacity of approximately 40 to 50 mL (1.4–1.8 oz), is located on the visceral surface of the liver (Fig. 2.4) and is also considered an accessory organ necessary for digestion. The gallbladder functions to concentrate and store the bile made in the liver until needed in the small intestine for fat digestion. Bile flow into the duodenum is regulated by the intraduodenal segment of the common hepatic bile duct

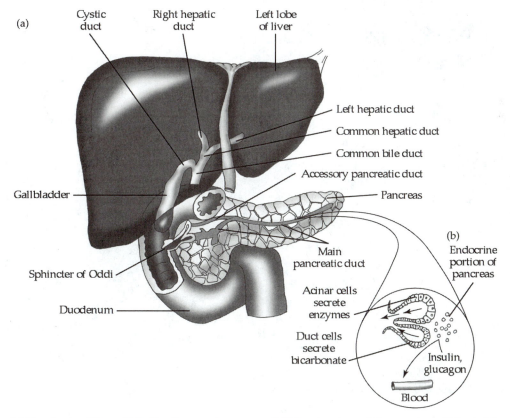

Figure 2.4   (a) The ducts of the gallbladder, liver, and pancreas. (b) The structure of the pancreas. (Adapted from Stalheim-Smith and Fitch, *Understanding Human Anatomy and Physiology,* St. Paul: West, 1993, p. 808.)

and the sphincter of Oddi, located at the junction of the common hepatic bile duct and the duodenum.

The *pancreas,* the third accessory organ, is a slender, elongated organ ranging in length from about 15 to 23 cm (6–9 in). The pancreas is found behind the greater curvature of the stomach, lying between the stomach and the duodenum (Figs. 2.1 and 2.4). Two types of active tissue are found in the pancreas (Fig. 2.4):

- the acini or ducted exocrine tissue that produces the digestive enzymes as well as pancreatic juice, and

- the ductless endocrine tissue that secretes hormones, primarily insulin and glucagon.

The acinar cells of the pancreas are arranged into circular glands that are attached to small ducts. Enzymes are packaged into secretory structures called *granules.* Stimulation of pancreatic acinar cells such as by hormones or the parasympathetic nervous system results in the release of the pancreatic digestive enzymes from the granules by exocytosis. The digestive enzymes along with bicarbonate, cations such as sodium, potassium, and calcium, and the anion chloride in a watery solution consti-

tute pancreatic juice that is secreted into small ducts within the pancreas. These small ducts coalesce to form a large main pancreatic duct, which later joins with the common hepatic bile duct to form the bile pancreatic duct. The bile pancreatic duct empties into the duodenum through the sphincter of Oddi (Fig. 2.4).

Although the structure of the small intestine contains the same layers as identified in Figure 2.2, the epithelial lining or mucosa of the small intestine is structured to maximize surface area. The small intestine has a surface area of approximately 300 m². This area is about equal to a 3-foot-wide sidewalk that is more than three football fields in length. Contributing to this enormous surface area are:

- the large folds of mucosa (folds of Kerckring) that protrude into the lumen of the small intestine;

- the villi, fingerlike projections lined by hundreds of cells (enterocytes, also called absorptive epithelial cells) along with blood capillaries and a central lacteal (lymphatic vessel) for transport of nutrients out of the enterocyte (Fig. 2.5);

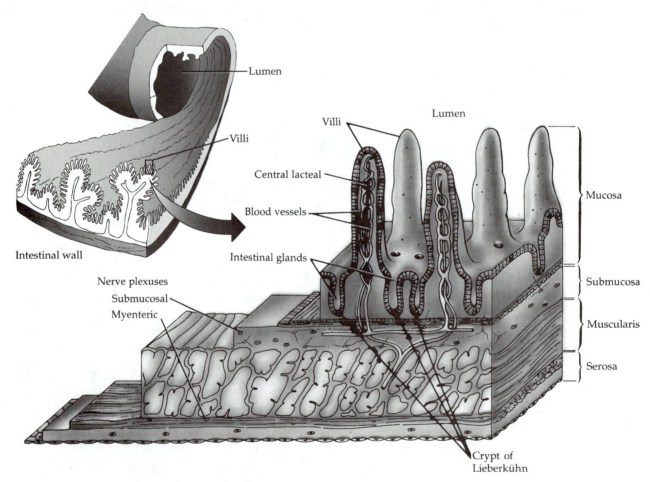

Figure 2.5 Structure of the small intestinal wall.

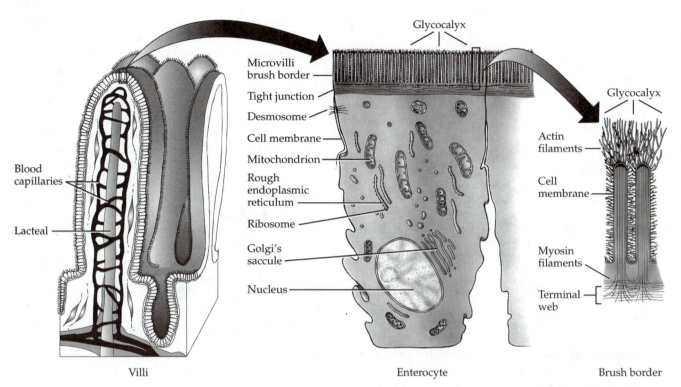

Figure 2.6 Structure of the absorptive cell of the small intestine.

- the microvilli, hairlike extensions of the plasma membrane of the enterocytes that make up the villi.

The microvilli possess a surface coat or glycocalyx, as shown in Figure 2.6, and together these comprise the brush border of the enterocytes. Most of the digestive enzymes produced by the intestinal mucosal cells are found on the brush border, and they function to hydrolyze already partially digested nutrients, mainly carbohydrate and protein. Structurally, the digestive enzymes are glycoproteins. The carbohydrate moiety or glyco- portion of these glycoprotein enzymes may in part make up the glycocalyx. The glycocalyx lines the luminal side of the intestine. The glycocalyx is thought to consist of numerous fine filaments that extend almost perpendicular from the microvillus membrane to which it is attached. Digestion is usually completed on the brush border but may be completed within the cytoplasm of the enterocytes.

The small intestine also contains small pits called crypts of Lieberkühn (Fig. 2.6) that lie between the villi. Epithelial cells in these crypts continuously undergo mitosis; the new cells gradually migrate upward and out of the crypts toward the tips of the villi. Shortly after reaching the tip of the villus, the cells will be sloughed off into the intestinal lumen and excreted in the feces. Intestinal cell turnover is rapid, approxi-

mately every 3 to 5 days. Cells in the crypts include paneth cells that secrete proteins with unclear functions, goblet cells that secrete mucus, and enterochromaffin cells with endocrine function. Cells in the crypts of Lieberkühn also secrete fluid and electrolytes into the lumen of the small intestine. This fluid is typically reabsorbed by the villi. The paneth, goblet, and enterochromaffin cells along with the enterocytes serve additionally as a protective barrier against pathogens.

Also serving as protection throughout the mucosa and submucosa of the small intestine is gut associated lymphoid tissue (GALT). GALT is composed of leukocytes (white blood cells). GALT leukocytes are found between intestinal absorptive epithelial cells, making up approximately 15% of the epithelial mucosa. GALT leukocytes include: T-lymphocytes, Natural Killer (NK) cells, and microfold (M) cells, among others. In addition, Peyer's patches, aggregates of lymphoid tissue located underneath the large M-cells in the mucosa and submucosa, contain more T-lymphocytes, as well as B-lymphocytes and macrophages. The M-cells pass foreign substances to the lymphocytes, which can in turn mount an immune response. Together these leukocytes residing throughout the first two layers of the gastrointestinal tract provide a defense against bacteria or other foreign substances that may have been ingested with consumed food.

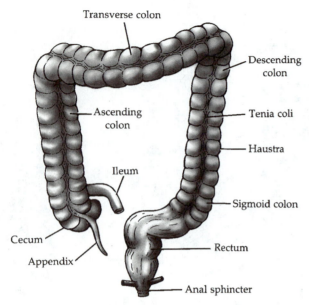

Figure 2.7 The colon.

From the distal or terminal section of the small intestine, the ileum, unabsorbed materials empty via the ileocecal valve into the cecum, the right side of the *colon (large intestine)*. From the cecum, materials move sequentially through the ascending, transverse, descending, and sigmoid sections of the colon (Figs. 2.1 and 2.7), which in its entirety is almost 5 feet long. Rather than being a part of the entire wall of the alimentary canal as it is in the upper digestive tract, the longitudinal muscle in the colon is gathered into three muscular bands or strips called *teniae* (also spelled *taenia* or *teneae*) coli that extend throughout most of the colon. Contraction of a strip of longitudinal muscle along with contraction of circular muscle cause the uncontracted portions of the colon to bulge outward, thereby creating pouches or haustra (Fig. 2.7). Contractions typically occur in one area of the colon then move to a different, nearby area. As described by Guyton [1], the fecal material is slowly dug into and rolled over in the colon as one would spade the earth, so that deeper, moister fecal matter is put in contact with the colon's absorptive surface. This process permits dehydration of fecal matter for defecation while increasing fluid and electrolyte absorption.

Coordination and Regulation of the Digestive Process

Regulatory Peptides

Factors influencing digestion and absorption are thought to be coordinated, in part, by a group of gastrointestinal tract molecules referred to as *regulatory peptides* or more specifically as gastrointestinal hormones and neuropeptides. Together the regulatory peptides affect a variety of digestive functions, including gastrointestinal motility, intestinal absorption, cell growth, and the secretion of digestive enzymes, electrolytes, water, and other hormones.

Some of the regulatory peptides such as gastrin, cholecystokinin (CCK, also called CCK-pancreozymen and abbreviated CCK-PZ), secretin, and gastric inhibitory polypeptide (GIP) are considered hormones. Peptides, such as somatostatin, released by endocrine cells but diffusing through the extracellular space to their target tissues rather than being secreted into the blood for transport are termed *paracrines*. *Neurocrines* are peptides that originate from nerves of the gut and include vasoactive intestinal peptide (VIP), gastrin-releasing peptide (GRP, also called *bombesin*), neurotensin, and substance P.

The functions of regulatory peptides with respect to the gastrointestinal tract and the digestive process are numerous. Some selected functions are as follows:

• Gastrin, synthesized primarily by G-cells in the stomach but also in the proximal small intestine, acts mainly in the stomach. Gastrin release occurs in response to vagal stimulation, ingestion of specific substances or nutrients, gastric distention, hydrochloric acid in contact with gastric mucosa, as well as local and circulating hormones. Principally, gastrin stimulates the release of hydrochloric acid, but it also stimulates gastric and intestinal motility and pepsinogen release. Gastrin stimulates the cellular growth of (has trophic action on) the stomach, and both the small and large intestine.

• CCK, secreted by I-cells of the proximal small intestine into the blood, principally binds to receptors on the pancreas to stimulate the secretion of pancreatic juice and enzymes into the duodenum. It also stimulates the contraction of the gallbladder to facilitate the release of bile into the duodenum. Gastric motility may be slightly inhibited by CCK, whereas intestinal motility is enhanced by CCK.

• Secretin, secreted into the blood by S-cells of the proximal small intestine in response to the release of acid chyme into the duodenum, acts primarily on pancreatic acinar cells to stimulate the release of pancreatic juice into the intestine. Secretin is thought to inhibit motility of most of the gastrointestinal tract, especially the stomach and proximal small intestine.

• GIP, also produced by cells of the small intestine, inhibits gastric secretions and motility. GIP stimulates intestinal secretions and, like the other three hor-

Table 2.1 Actions of Selected GI Hormones

Action	Hormones			
	Gastrin	CCK	Secretin	GIP
Acid secretion	S*	S	I*	I*
Gastric emptying	I	I	I*	I
Pancreatic HCO_3^- secretion	S	S*	S*	O
Pancreatic enzyme secretion	S	S*	S	O
Gallbladder contraction	S	S*	S	?
Gastric motility	S	I	I	I
Intestinal motility	S	S	I	?
Insulin release	S	S	S	S*
Mucosal growth	S*	S	I	?
Pancreatic growth	S	S*	S*	?

Abbreviations: S, stimulates; I, inhibits; O, no effect; ?, information unavailable.

*Particularly important function.

Source: Adapted from Johnson LR. Gastrointestinal physiology 3rd ed. St. Louis: Mosby, 1985:8.

mones, stimulates insulin secretion from the pancreas as well. Table 2.1 lists additional actions of gastrin, CCK, secretin, and GIP.

Other regulatory peptides also affect the digestive process:

• Somatostatin, synthesized by pancreatic and intestinal cells, acts in a paracrine fashion by entering gastric juice. Somatostatin appears to mediate the inhibition of gastrin release as well as the release of GIP, secretin, VIP, and motilin. Gastric acid, gastric motility, pancreatic exocrine secretions and gallbladder contraction are also inhibited by somatostatin.

• Motilin, secreted by the cells of the small intestine, causes contraction of intestinal smooth muscle and may be involved in regulating different phases of the migrating motility complex (MMC). The MMC is important in maintaining gastrointestinal motility and is described in more detail on page 37 of this chapter.

• VIP is present in neurons within the gut; it is not thought to be present in intestinal endocrine cells. VIP stimulates intestinal secretions, relaxes most gastrointestinal sphincters, inhibits gastric acid secretion, and stimulates pancreatic bicarbonate release into the small intestine.

• Gastrin-releasing peptide (GRP), also called bombesin, released from nerves, stimulates the release of hydrochloric acid, gastrin, and CCK.

• Neurotensin, produced by small intestine mucosa, has no physiological role at normal circu-lating con-

centrations; however, it may serve to mediate gastric emptying, intestinal motility, and gastric acid secretion after fat ingestion.

• Substance P, another neuropeptide, increases blood flow to the gastrointestinal tract, inhibits acid secretion and motility of the small intestine, and binds to pancreatic acinar cells associated with enzyme secretion.

Neural Regulation

The nervous system of the gastrointestinal tract is referred to as the *enteric* (relating to the intestine) *nervous system,* and lies in the wall of the gastrointestinal tract beginning in the esophagus and extending to the anus. Neural regulation of the gastrointestinal tract involves a combination of neural plexuses and reflexes. The enteric nervous system can be divided into two neuronal networks or plexuses: the myenteric plexus (or plexus of Auerbach) and the submucosal plexus (or plexus of Meissner). The location and actions of the two plexuses are given here:

Enteric Nervous System

Myenteric Plexus	Submucosal Plexus
• Lies in the muscularis externa between longitudinal and circular muscles • Controls peristaltic activity and/or gastrointestinal motility	• Lies in the submucosa • Controls mainly gastrointestinal secretions and local blood flow

The myenteric plexus is innervated by parasympathetic and sympathetic nervous systems, and it greatly influences gastrointestinal motility. Acetylcholine most often excites gastrointestinal motility. Norepinephrine and epinephrine, in contrast, typically inhibit gastrointestinal activity. The submucosal plexus is also part of the autonomic nervous system and controls gastrointestinal secretions, as well as vascular blood flow. Gastrointestinal reflexes involving the enteric nervous system also control gastrointestinal secretions and peristalsis, as well as other processes involved in digestion. For example, reflexes originating from the intestines inhibit gastric motility and secretions and are called *enterogastric reflexes.* The colonoileal reflex from the colon inhibits the emptying of the contents of the ileum into the colon.

Table 2.2 Digestive Enzymes and Their Actions

Enzyme or Zymogen/Enzyme	Site of Secretion	Preferred Substrate(s)	Primary Site of Action
Salivary α amylase	Mouth	α 1-4 bonds in starch, dextrins	Mouth
Lingual lipase	Mouth	Triacylglycerol	Stomach, small intestine
Pepsinogen/pepsin	Stomach	Carboxyl end of phe, tyr, trp, met, leu, glu, asp	Stomach
Trypsinogen/trypsin	Pancreas	Carboxyl end of lys, arg	Small intestine
Chymotrypsinogen/chymotrypsin	Pancreas	Carboxyl end of phe, tyr, trp, met, asn, his	Small intestine
Procarboxypeptidase/			
carboxypeptidase A	Pancreas	C-terminal neutral amino acids	Small intestine
carboxypeptidase B	Pancreas	C-terminal basic amino acids	Small intestine
Proelastase/elastase	Pancreas	Fibrous proteins	Small intestine
Collagenase	Pancreas	Collagen	Small intestine
Ribonuclease	Pancreas	Ribonucleic acids	Small intestine
Deoxyribonuclease	Pancreas	Deoxyribonucleic acids	Small intestine
Pancreatic α amylase	Pancreas	α 1-4 bonds, in starch, maltotriose	Small intestine
Pancreatic lipase and colipase	Pancreas	Triacylglycerol	Small intestine
Phospholipase	Pancreas	Lecithin and other phospholipids	Small intestine
Cholesterol esterase	Pancreas	Cholesterol esters	Small intestine
Retinyl ester hydrolase	Pancreas	Retinyl esters	Small intestine
Amino peptidases	Small intestine	N-terminal amino acids	Small intestine
Dipeptidases	Small intestine	Dipeptides	Small intestine
Nucleotidase	Small intestine	Nucleotides	Small intestine
Nucleosidase	Small intestine	Nucleosides	Small intestine
Alkaline phosphatase	Small intestine	Organic phosphates	Small intestine
Monoglyceride lipase	Small intestine	Monoglycerides	Small intestine
Alpha dextrinase or isomaltase	Small intestine	α 1-6 bonds in dextrins, oligosaccharides	Small intestine
Glucoamylase, glucosidase, and sucrase	Small intestine	α 1-4 bonds in maltose, maltotriose	Small intestine
Trehalase	Small intestine	Trehalose	Small intestine
Disaccharidases	Small intestine		Small intestine
Sucrase*		Sucrose	
Maltase		Maltose	
Lactase		Lactose	

*Part of an enzyme complex.

The Process of Digestion: Secretions and Enzymes Required for Nutrient Digestion

Table 2.2 provides a partial list of zymogens/enzymes that participate in the digestion of nutrients in foods. Zymogens are often called *proenzymes* or inactive enzymes because they must be chemically altered to function as an enzyme. Each of these enzymes are briefly discussed as nutrient digestion is traced from the oral cavity throughout the rest of the gastrointestinal tract.

The enzymes are discussed again in Chapters 4, 6, and 7, which cover carbohydrates, lipids, and protein, respectively.

The Oral Cavity

On entering the mouth, food is chewed by the action of the jaw muscles and is made ready for swallowing by mixing with secretions (*saliva*) released from the salivary glands. Three pairs of small, bilateral saliva-secreting salivary glands—the parotid, the submandibular, and the sublingual—are located throughout the lining of the oral cavity, along the jaw from the base of the ear to the chin (Fig. 2.8). Secretions from these glands constitute saliva. Specifically, the parotid glands secrete water, electrolytes (sodium, potassium, chloride), and enzymes. The submandibular and sublingual glands secrete water, electrolytes, enzymes, and mucus. Saliva is primarily (99.5%) water, which helps dissolve foods. The principal enzyme of saliva is α amylase (also called *ptyalin*)

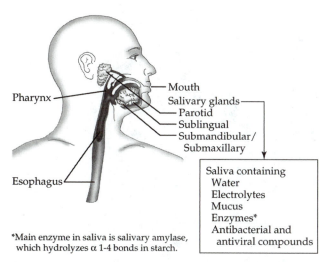

Pharynx

Esophagus

Mouth
Salivary glands
Parotid
Sublingual
Submandibular/
Submaxillary

Saliva containing
Water
Electrolytes
Mucus
Enzymes*
Antibacterial and
antiviral compounds

*Main enzyme in saliva is salivary amylase,
which hydrolyzes α 1-4 bonds in starch.

Figure 2.8 Digestion in the oral cavity.

(Table 2.2). This enzyme hydrolyzes internal α 1-4 bonds within starch. A second digestive enzyme that is produced by lingual serous glands in the mouth is lingual lipase. This enzyme, which hydrolyzes dietary triacylglycerols in the stomach and small intestine, is mostly of importance to infants. Mucus secretions found in saliva contain glycoproteins, known as *mucins,* that lubricate food and protect the oral mucosa. Antibacterial and antiviral compounds, one example being the antibody IgA (immunoglobulin A), along with trace amounts of organic substances (such as urea) and other solutes (i.e., phosphates, bicarbonate) are also found in saliva.

The passage of food from the mouth through the pharynx (the oral cavity) into the esophagus constitutes swallowing. Swallowing, which can be divided into several stages—voluntary, pharyngeal, and esophageal—is a reflex response initiated by a voluntary action and regulated by the swallowing center in the medulla of the brain. To swallow food, the esophageal sphincter relaxes allowing the opening of the esophagus. Food then passes into the esophagus. Simultaneously, the larynx (part of the respiratory tract) moves upward to effect a shift of the epiglottis over the glottis. The closure of the glottis is important to keep food from entering the trachea, which leads to the lungs. Once food is in the esophagus, the larynx shifts downward to allow the reopening of the glottis.

The Esophagus

When food moves into the esophagus, both the striated (voluntary) muscle of the upper portion of the esophagus and the smooth (involuntary) muscle of the distal portion are stimulated by cholinergic (parasympathetic) nerves. The result is peristalsis, a progressive wavelike motion that moves the food (now called a *bo-*

lus) through the esophagus into the stomach, a process that can take less than 10 seconds. At the lower (distal) end of the esophagus, just above the juncture with the stomach, lies the lower esophageal sphincter (LES). The LES, however, may be a misnomer, since there is no consensus about this particular muscle area being sufficiently hypertrophied to constitute a true sphincter. On swallowing, the LES pressure drops. This drop in LES pressure relaxes the sphincter so that food may pass from the esophagus into the stomach.

Multiple mechanisms, including neural and hormonal, regulate LES pressure. The musculature of the LES has a tonic pressure that is normally higher than the intragastric pressure, that is, the pressure within the stomach. This high LES tonic pressure functions to keep the sphincter closed in between meals. Closure of the LES sphincter is important because it prevents gastroesophageal reflux, the movement of chyme from the stomach back into the esophagus. The gastric acid in the chyme when present in the esophagus is an irritant to the esophageal mucosa. The individual experiencing reflux feels a burning sensation in the midchest region, a condition referred to as *heartburn.* Repeated exposure of the esophagus to the acid chyme can cause esophagitis, inflammation of the esophagus. Foods and/or food-related substances can indirectly affect LES pressure and cause reflux. LES incompetence (failure of the LES to maintain an adequate tonic pressure) occurs when LES pressure is decreased or the LES is relaxed, leading to reflux of gastric chyme into the esophagus. Smoking, chocolate, high-fat foods, alcohol, and carminatives such as peppermint and spearmint, for example, promote LES relaxation. LES relaxation typically leads to heartburn and ultimately can cause esophagitis. Reflux esophagitis and its treatment are discussed in the perspective at the end of this chapter.

The Stomach

After passage through the LES, food or the bolus enters the proximal or upper section of the stomach known as the *body,* shown in Figure 2.9.

- The body makes up approximately the first three-quarters of the stomach; it includes a portion of the stomach called the *fundus* and extends from the LES to the angular notch. The body of the stomach serves primarily as the reservoir for swallowed food and is the production site for much of the gastric juice.

The other major section of the stomach is the *antrum,* also shown in Figure 2.9.

- The antrum or distal portion of the stomach extends from the angular notch to the duodenum

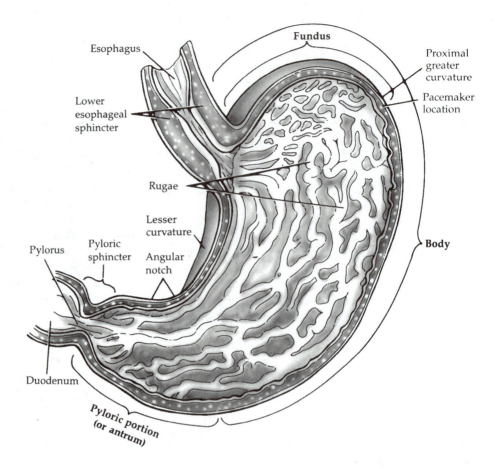

Figure 2.9 Structure of the stomach.

and functions to grind and mix food with the digestive gastric juices to form semiliquid *chyme.* The antrum also provides strong peristalsis for gastric emptying.

Gastric Juices

Mixing of food with gastric juices begins in the body of the stomach. Gastric juice is produced by three functionally different gastric glands that are found deep within the gastric mucosa and penetrate the entire epithelium of the stomach. The three gastric glands include

- the cardiac glands, found in a narrow rim at the juncture of the esophagus and stomach;
- the oxyntic glands, found in the body of the stomach; and
- the pyloric glands, located primarily in the antrum.

Several cell types secreting different substances may be found within a gland. For example, oxyntic glands, depicted in Figure 2.10, contain

- neck (mucus) cells, located close to the surface mucosa, which secrete bicarbonate and mucus;

- parietal (oxyntic) cells, which secrete hydrochloric acid and intrinsic factor;
- chief (peptic or zymogenic) cells, which secrete pepsinogens; and
- endocrine cells, which secrete a variety of hormones.

In contrast to the oxyntic glands, the cardiac glands contain no parietal cells but do possess endocrine and mucus cells. The pyloric glands contain both mucus and parietal cells as well as endocrine G-cells that produce the hormone gastrin. Gastrin, released from the G-cells into the blood, is carried to its target tissues that includes the stomach.

In review, the main constituents of gastric juice produced by the different cells of the gastric glands include water, electrolytes, hydrochloric acid, enzymes, mucus, and intrinsic factor. Some of the main constituents—hydrochloric acid, enzymes, and mucus—of gastric juice are discussed hereafter.

Gastric juice contains an abundance of hydrochloric acid secreted from gastric parietal cells. Parietal cells contain both a potassium chloride transport system as well as a hydrogen (proton) potassium ATPase ex-

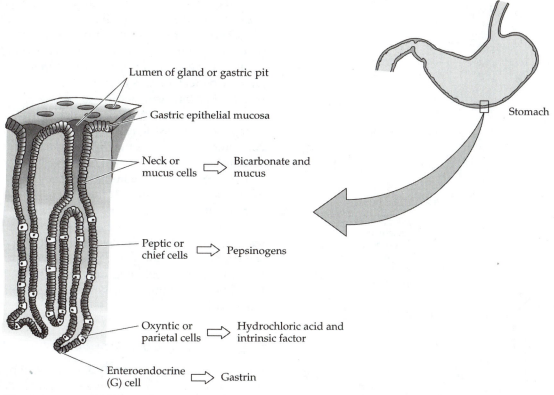

Figure 2.10   An oxyntic gland from the body of the stomach.

change system. The potassium chloride system transports both ions into the gastric lumen. The hydrogen potassium ATPase system (H^+, K^+ ATPase), also referred to as a *proton pump*, allows the exchange of two potassium ions for two hydrogens (protons) for each ATP molecule hydrolyzed. The net effect is the secretion of hydrogen and chloride or hydrochloric acid into the gastric lumen as part of gastric juice. The high concentration of hydrochloric acid in the gastric juice results in its low pH, about 2. pH is the negative logarithm of the hydrogen ion concentration. The lower the pH, the more acidic the solution. Figure 2.11 shows the approximate pHs of selected body fluids and, for comparison, some drinking beverages. Notice that the pH of orange juice and typically all fruit juices is higher than that of gastric juice. Thus, the ingestion of such juices can not lower the gastric pH.

Hydrochloric acid has several functions in gastric juice. Some of these functions include

- conversion or activation of the zymogen pepsinogen to form pepsin;
- denaturation of proteins, which results in the destruction of the tertiary and secondary protein structure and thereby opens interior bonds to the proteolytic effect of pepsin;

- release of various nutrients from organic complexes; and
- bactericide agent that kills many bacteria ingested along with food.

Two enzymes are found in gastric juice. The main enzyme pepsin functions as the principal proteolytic enzyme in the stomach and is derived from either of two pepsinogens, I or II. Pepsinogen I is found primarily in the body of the stomach where most hydrochloric acid is secreted. Pepsinogen II is found both in the body and antrum of the stomach. The distinction between the two groups of pepsinogens has no known implications with respect to digestion; however, higher concentrations of pepsinogen I, which correlate positively with acid secretion, have been associated with an increased incidence of peptic ulcers. Pepsinogens are secreted in granules into the gastric lumen from chief cells upon stimulation by acetylcholine and/or by acid. Pepsinogens can be converted to pepsin, an active enzyme, in an acid environment or in the presence of previously formed pepsin.

Pepsinogen ⟶ Pepsin
acid or pepsin

Pepsin (Table 2.2) functions as a protease—that is, an enzyme that hydrolyzes proteins. More specifically,

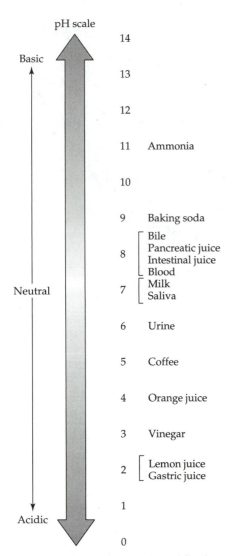

Figure 2.11 Approximate pHs of selected fluids.

pepsin is an endopeptidase, meaning that it hydrolyzes interior peptide bonds within proteins. Specifically, pepsin appears to have an affinity for peptide bonds adjacent to the carboxyl end of the amino acids methionine, leucine, the dicarboxylic amino acids (glutamate and aspartate), and the aromatic amino acids (phenylalanine, tyrosine, and tryptophan). Optimal pepsin activity occurs at about pH 3.5.

The second enzyme present in the gastric juice is α amylase that originates from the salivary glands in the mouth. This enzyme, which hydrolyzes starch, retains some activity in the stomach until it is inactivated by the low pH of gastric juice.

Mucus, secreted by neck or mucus cells, is also found in gastric juice. Mucus, consisting of a network of glycoproteins (mucin), glycolipids, water, and bicarbonate ions (HCO_3^-), serves to lubricate the ingested gastrointestinal contents and to coat and protect the gastric mucosa from mechanical and chemical damage. Tight junctions between gastric cells also help prevent H^+ from penetrating into the gastric mucosa and initiating peptic ulcer formation. Intrinsic factor, secreted into gastric juice by parietal cells, is necessary for the absorption of vitamin B_{12}. Intrinsic factor will be discussed in more detail in Chapter 9, "Water-Soluble Vitamins."

In summary, gastric juice contains several important compounds that aid in the digestive process. However, very little chemical digestion of nutrients occurs in the stomach except for the initiation of protein hydrolysis by the protease pepsin and the limited continuation of starch hydrolysis by salivary α amylase (Fig. 2.12). The only absorption that occurs in the stomach is that of water, a few fat-soluble drugs such as ethyl alcohol and aspirin, and a few minerals. Nourishment and survival, therefore, are possible without the stomach. Nevertheless, a healthy stomach makes adequate nourishment much easier.

Regulation of Gastric Secretions

Gastric secretion and motility are regulated by multiple mechanisms. As described in a general fashion earlier in this chapter in the section "Coordination and Regulation of the Digestive Process," gastrin, CCK, secretin, and GIP all affect to some extent gastric secretions and/or motility. Some additional information regarding regulation of gastric secretions and motility will be presented here.

Inhibitants of gastric secretions include the hormones GIP and secretin as well as the peptide somatostatin. Somatostatin inhibits hydrochloric acid secretion at the parietal cell as well as through the inhibition of gastrin release from the G-cells. In contrast, CCK and gastrin stimulate hydrochloric acid secretion. Gastrin release is stimulated in turn by ingested foods, hormones, and neurotransmitters. Foods, such as coffee and alcohol, as well as nutrients, such as calcium, amino acids, and peptides, present in the gastrointestinal tract lumen stimulate gastrin release. Epinephrine in the blood and gastrin-releasing peptide, released by some nerves, also stimulate gastrin release. Direct mediators of hydrochloric acid release by the parietal cells include

- gastrin, released by G-cells into the blood to in turn stimulate parietal cells;

- acetylcholine, released, for example, from the vagus nerve, to act on parietal cells; and

- histamine, released from gastrointestinal tract mast cells, to bind to H_2 receptors on parietal cells.

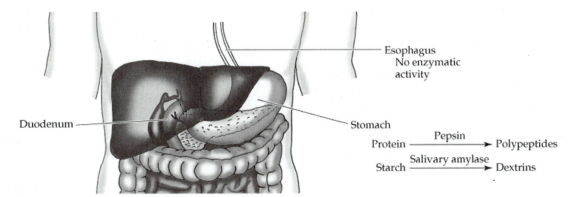

Figure 2.12   Digestion in the stomach.

Gastrin, acetylcholine, and histamine are potent secretogogues (compounds that stimulate secretion) of hydrochloric acid; together, they prompt the release of substantial amounts of hydrochloric acid from parietal cells.

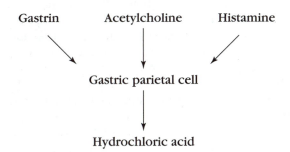

An understanding of how hydrochloric acid production occurs in the body and what stimulates its release is essential to understanding the treatment of peptic ulcers. Several drugs—cimetidine (Tagamet), ranitidine (Zantac), famotidine (Pepcid), and nizatidine (Axid)—used to treat peptic ulcers, act to prevent histamine from binding to H_2 receptors on parietal cells. These drugs, known as H_2 receptor blockers, bind to the H_2 receptors on the parietal cells. Thus, when histamine is released, its binding to the H_2 receptor is prevented (the drug blocks histamine's ability to bind), and acid release from the parietal cell is diminished. Another drug used to treat ulcers, omeprazole (Prilosec) works by binding to the proton pump (discussed on p. 35) at the secretory surface of the parietal cell, and thus directly inhibits hydrogen release into the gastric juice. Drug therapies are quite effective in the treatment of peptic ulcers; however, foods that act as irritants to the gastric mucosa also need to be avoided during acute peptic ulcer episodes. Further medical nutrition therapy may necessitate increased consumption of some nutrients such as protein and iron if the peptic ulcer resulted in bleeding into the gastrointestinal tract.

Gastric Emptying and Motility

When food is swallowed, the proximal portion of the stomach relaxes to accommodate the ingested food. Two processes, receptive relaxation and gastric accommodation, mediated by the vagus nerve, influence this relaxation process, considered to be a reflex. Signals for antral contraction, which occur at regular intervals, begin in the proximal stomach at a point along the greater curvature and migrate distally toward the pylorus, the juncture to the small intestine. Response of the antrum to signals by the pacemaker, located between the fundus and body of the stomach, is thought to be affected by gastrointestinal hormones and neuropeptides. The pacemaker determines the frequency of the contractions that occur. As the food moves into the antrum, the rate of contractions increases so that in the distal portion of the stomach, food is liquefied into chyme.

The migrating motility complex (MMC), a series of contractions with several phases, moves distally like a wave down the gastrointestinal tract, but mainly in the stomach and intestine. The migrating motility complex waves occur approximately every 80 to 120 minutes during interdigestive periods but change during digestive periods. The migrating motility complex functions to sweep gastrointestinal contents and prevent bacterial overgrowth in the intestine.

Although contractions within the stomach promote physical disintegration of solid foods into a liquid form, complete liquefaction is not necessary for the emptying of the stomach contents through the pyloric sphincter into the duodenum. Particles as large as 3 mm in diameter (~1/8 in) can be emptied from the stomach, but solid particles are usually emptied with fluids when they have been degraded to a diameter of about 2 mm or less. Approximately 1 to 5 mL (≤1 tsp) of chyme enter the duodenum about twice per minute. Contraction of the pylorus and proximal duodenum is thought to be coordinated with contraction of the antrum. Receptors in the duodenal bulb (the first few centimeters of the proximal duodenum) are sensitive to the osmolarity and volume of chyme present in the duodenum and the presence of acid and/or irritants in the small intestine. Gastric emptying is also partially affected by the macronutrient composition of the food. Carbohydrate and protein appear to empty at approximately the same rate from the stomach; fat, however, slows gastric emptying into the duodenum. Salts and monosaccharides inhibit gastric emptying, as do many free amino acids like tryptophan and phenylalanine. Complex carbohydrates, especially soluble fiber, decrease (slow down) the rate of gastric emptying. Neural gastrointestinal reflexes, along with the release of regulatory peptides such as secretin by the duodenal bulb, also influence gastric emptying, which following a meal usually takes between 2 and 6 hours.

The Small Intestine

Chyme, once in the small intestine, is mixed and moved through the small intestine by various contractions under nervous system influence. Contractions of longitudinal muscle may be referred to as *sleeve contractions* and function to mix the intestinal contents with the digestive juices. Segmentation, or *standing contractions* of circular muscles, produces bidirectional flow of the intestinal contents, occurs many times per minute, and serves to mix and churn the chyme with digestive secretions in the small intestine. Peristaltic waves, or *progressive contractions,* also accomplished primarily through action of the circular muscles, move the chyme distally along the intestinal mucosa toward the ileocecal valve.

Chyme, moving from the stomach into the duodenum, initially has a pH of about 2 because of its gastric acid content. The duodenum is protected from this gastric acidity by pancreatic secretions with buffering capacity released into the duodenum and by secretions from the Brunner's glands. The Brunner's glands are located in the mucosa and submucosa of the first few centimeters of the duodenum (duodenal bulb). The mucus-containing secretions are viscous and alkaline with a pH approximately 8.2 to 9.3. The mucus itself is rich in glycoproteins (described further in Chapter 7) and helps to protect the epithelial mucosa from damage. Other secretions by intestinal cells include those from glands within the crypts of Lieberkühn. These glands secrete large volumes of intestinal juices, which facilitate digestion of nutrients throughout the small intestine. Distal to the duodenal bulb is a small projection termed the *duodenal papillae* or *papillae of Vater,*

where the bile pancreatic duct carrying secretions from the pancreas and gallbladder empties. These secretions are discussed next.

The Interrelationship Between the Small Intestine and Pancreas in the Digestive Process

The pancreas produces secretions containing fluid, electrolytes, bicarbonate, and enzymes that are released into the duodenum. The enzymes are responsible for the digestion of approximately half (50%) of all ingested carbohydrates, half (50%) of all proteins, and almost all (90%) of ingested fat. Bicarbonate in pancreatic juice released into the duodenum is important for neutralizing the acid chyme passing into the duodenum from the stomach and for maximizing enzyme activity within the duodenum.

Hormones such as secretin and CCK play important roles in the digestive process occurring in the small intestine. Secretin release is triggered by duodenal acidification with gastric hydrochloric acid. Secretin's major action is to increase the pH of the small intestine by stimulating secretion of water and bicarbonate (pancreatic juice) by the pancreas. This hormone further promotes alkalization of intestinal contents by inhibiting gastric acid secretion and gastric emptying. CCK further stimulates the secretion of pancreatic juices and enzymes into the duodenum. CCK is also thought to slightly inhibit gastric motility but to stimulate intestinal motility.

Enzymes produced by the pancreas and necessary for nutrient digestion are listed in Table 2.2, presented earlier. Pancreatic enzymes responsible for the digestion of protein, carbohydrate, and lipid are discussed briefly in the following text, and in detail in Chapters 4, 6, and 7. Proteases—enzymes that digest proteins—found in pancreatic juice and secreted into the duodenum include trypsinogen, chymotrypsinogen, procarboxypeptidases, proelastase, and collagenase.

• Trypsinogen, a zymogen, is converted to its active form, trypsin, by the enzyme enteropeptidase (formerly called *enterokinase*) and by free tryp-sin. Enteropeptidase is stimulated by CCK. Trypsin is an endopeptidase specific for peptide bonds at the carboxyl end of basic amino acids (lysine and arginine).

• Chymotrypsinogen is activated to chymotrypsin by trypsin as well as by enteropeptidase. Chymotrypsin, also an endopeptidase, hydrolyzes interior peptide bonds adjacent to the carboxyl end of aromatic amino acids as well as methi-onine, asparagine, and histidine.

• Procarboxypeptidases are converted to active carboxypeptidases by trypsin. Carboxypeptidase A, an exopeptidase, cleaves carboxyterminal neutral (including aromatic) amino acids, whereas carboxypeptidase B, also an exopeptidase, cleaves carboxyterminal basic amino acids.

- Elastase hydrolyzes fibrous proteins, and collagenase hydrolyzes collagen.

As a group, proteases hydrolyze peptide bonds, either internally or from the ends, and the net result of their collective actions is the production of polypeptides shorter in length than the original polypeptide, oligopeptides (typically 4–10 amino acids in length), tripeptides, dipeptides, and free amino acids. The latter three may be absorbed into the enterocyte. Oligopeptides and some tripeptides may be further hydrolyzed by brush border aminopeptidases prior to absorption.

Pancreatic α amylase secreted by the pancreas into the duodenum is a principal enzyme necessary for starch digestion. Like the α amylase secreted by salivary glands, pancreatic α amylase hydrolyzes α 1-4 bonds within starch to yield oligosaccharides, dextrins, maltotriose, and maltose. The oligosaccharides and dextrins produced from amylase activity may possess α 1-6 branches, which will require further hydrolysis by α dextrinase or isomaltase. The maltotrioses and maltoses also will need further hydrolysis by intestinal brush border enzymes prior to absorption.

Enzymes necessary for lipid digestion are also produced by the pancreas. Pancreatic lipase, the major fat-digesting enzyme, hydrolyzes triacylglycerols, specifically hydrolyzing fatty acids esterified at positions (carbons) 1 and 3 on the glycerol molecule to yield 2-monoacylglycerol and two free fatty acids. With sufficient time, pancreatic lipase may also hydrolyze the fatty acid at the second carbon of the glycerol to yield three free fatty acids and glycerol. In addition to pancreatic lipase, colipase is necessary for some fat digestion. Colipase, secreted in pancreatic juice as a zymogen, is activated by trypsin in the small intestine. Colipase functions by binding to triacylglycerols and pancreatic lipase, thereby displacing bile. Displacement of bile enables pancreatic lipase to better hydrolyze the triacylglycerol.

Phospholipid and cholesterol digestion occur in the small intestine. In the presence of bile, phospholipase A_1 and A_2 hydrolyze fatty acids from phospholipids. Cholesterol esterase hydrolyzes fatty acids from cholesterol esters as well as from triacylglycerols; however, bile is needed for cholesterol esterase activity. The role of bile in fat digestion is discussed in the next section.

The Interrelationships Among the Small Intestine, Liver, and Gallbladder in the Digestive Process: Bile Synthesis, Storage, Function, Recirculation, and Excretion

Bile Synthesis   Bile is composed of mainly bile acids (and/or salts) but also cholesterol, phospholipids, and bile pigments (bilirubin and biliverdin) dissolved in an alkaline solution. The bile acids are synthesized in the hepatocytes from cholesterol, which in a series of reactions is oxidized to chenodeoxycholic acid and cholic acid, the two principal or primary bile acids. Chenodeoxycholate and cholate, once formed, conjugate primarily (~75%) with the amino acid glycine to form the conjugated bile acids glycochenodeoxycholic acid (glycochenodeoxycholate) and glycocholic acid (glycocholate), respectively.

Bile Acid	Amino Acid	Conjugated Bile Acid
cholic acid	+ glycine →	glycocholic acid
chenodeoxycholic acid	+ glycine →	glycochenodeoxycholic acid

Alternately and to a lesser (25%) extent, chenodeoxycholate and cholate conjugate with the amino acid taurine to form two additional primary conjugated bile acids.

Bile Acid	Amino Acid	Conjugated Bile Acid
cholic acid	+ taurine →	taurocholic acid
chenodeoxycholic acid	+ taurine →	taurochenodeoxycholic acid

Conjugation occurs between the carboxyl group of the bile acid and the amino group of the amino acid as shown in Figure 2.13. Conjugation of the bile acids with these amino acids results in better ionization, and thus improved ability to form micelles. The formation and role of micelles are described in the section "The Function of Bile." Chenodeoxycholate and cholate are primary bile acids and make up 80% of the body's total bile acids. The remaining 20% of the bile acids are secondary products that result from bacterial action on chenodeoxycholic acid to form lithocholate and on cholic acid to form deoxycholate.

In addition to being conjugated to amino acids, most conjugated bile acids are present in bile as bile salts owing to bile's pH range. Sodium is the predominant biliary cation, although potassium and calcium bile salts also may be found in the alkaline bile solution.

While bile acids and salts make up a large portion of bile, other substances are also found in bile. These other substances include both cholesterol and phospholipids, especially lecithin, and make up what is referred to as

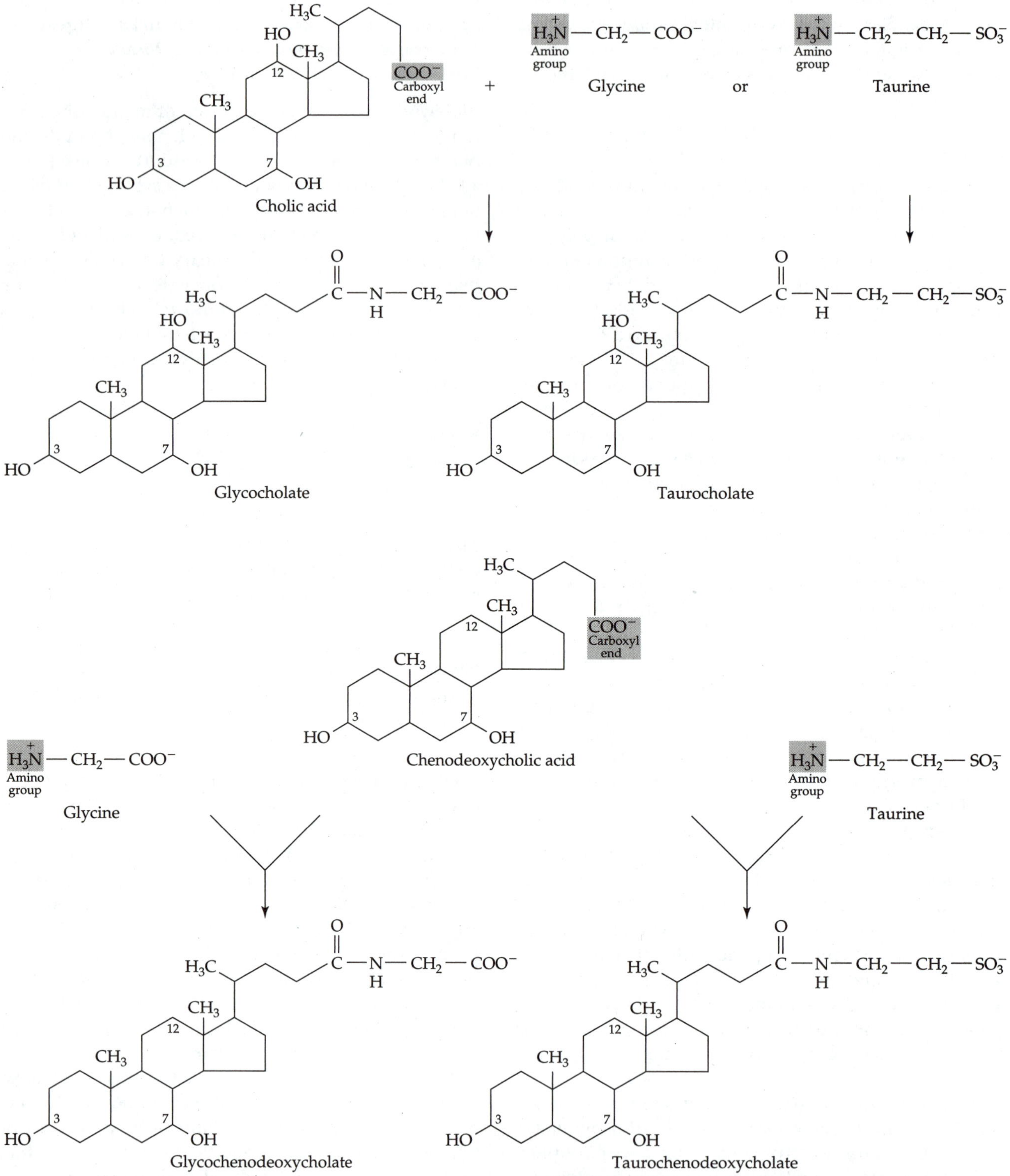

Figure 2.13 The formation of glycoholate, taurocholate, glycochenodeoxycholate, and taurochenodeoxycholate conjugated bile acids.

the *bile acid–dependent* fraction of bile. It is of importance that these bile components remain in proper ratio to prevent gallstone (cholelithiasis) formation. Gallstones are thought to form when bile becomes supersaturated with cholesterol. Cholesterol precipitates out of solution and provides a crystalline-like structure for the deposition of calcium, bilirubin, phospholipids, and other compounds that ultimately form a "stone." Gallstones may reside silently in the gallbladder, or they may irritate the organ causing cholecystitis (inflammation of the gallbladder) or lodge themselves in the common bile duct, thus blocking the flow of bile (choledocholithiasis) into the duodenum. Another constituent of bile is immunoglobulin (Ig) A, synthesized by blood plasma cells adjacent to biliary ducts. IgA, being an antibody, acts as a first line of defense against infectious microorganisms when secreted in the gastrointestinal tract. Water, electrolytes, and bile pigments (mainly bilirubin and/or biliverdin—waste end products of hemoglobin degradation excreted in bile and giving bile its color) conjugated with glucuronic acid are secreted into bile by hepatocytes. This fraction of the bile is referred to as *bile acid–independent*.

Bile Storage   During the interdigestive periods, bile is sent from the liver to the gallbladder, where it is concentrated and stored. The gallbladder concentrates the bile such that as much as 90% of the water, and some of the electrolytes are reabsorbed by the gallbladder mucosa. The fluid reabsorption thus leaves the remaining bile constituents (i.e., bile acids and salts, cholesterol, lecithin, bilirubin, and biliverdin) in a less dilute form. Concentration of the bile permits greater storage of bile produced by the liver between periods of food ingestion. Contraction of the gallbladder is stimulated by CCK released in response to products of protein digestion (primarily tryptophan and phenylalanine) and of fat digestion (especially long-chain fatty acids). Bile is secreted into the duodenum through Oddi's sphincter.

The Function of Bile   Bile acids and bile salts act as detergents. Bile acids and salts are effective as a detergent because they contain both polar (hydrophilic or water "soluble") and nonpolar (hydrophobic or fat "soluble") areas (i.e., they are amphipathic) (see Fig. 2.14). The hydrophobic areas (steroid ring areas) of the bile salts attract the ingested fat molecules, which arrive in the small intestine virtually undigested. This hydrophobic portion of the bile salts surrounds and "melts" or "dissolves" into the surface of the ingested fat molecule, keeping the fat molecule from coalescing with other fat molecules, and helping them to break apart. In other words, the bile functions to decrease the surface ten-

sion of the fat, thus permitting emulsification (i.e., dispersion of the fat or increase in the exposed surface area of the lipids). Surrounding the hydrophobic center are the polar areas of the bile (the hydroxy and carboxyl groups and the amino acid) that direct outward into the watery intestinal juices. This arrangement (hydrophobic interior and hydrophilic exterior) permits the fat to stay in solution. Further, emulsification of fat enables digestion (hydrolysis) of the triacylglycerol molecules to occur by pancreatic and intestinal lipases, which are found in the watery intestinal juice. Once the triacylglycerol molecules have been hydrolyzed, bile acids and salts help in the absorption of these end products of lipid digestion.

Bile acids and salts, along with phospholipids, help in the absorption of lipids by forming small (<10 nm) spherical, cylindrical, or disklike shaped complexes called *micelles*. Micelles can contain as many as 40 bile salt molecules. As with emulsification, to form a micelle, the hydrophobic steroid portion of bile salts and acids, which is mostly fat soluble, position themselves together and surround the monoacylglycerols and fatty acids that formed following the action of lipases. Polar portions of the bile salts, bile acids, and phospholipids project outward from the lipid core of the micelle, thus permitting solubility in the watery digestive fluids and transportation to the intestinal brush border for absorption. Figure 2.14 shows the action of the bile acids in forming micelles and delivering the fatty acids to the enterocytes for absorption. Other lipids in the intestinal lumen, such as fat-soluble vitamins and cholesterol, also may be incorporated into the micelle for delivery to the intestinal brush border. Once at the brush border, the monoacylglycerols and fatty acids diffuse through the unstirred or stagnant water layer lying above the glycocalyx of the microvilli and into the enterocytes; the micelles devoid of the fatty acids and monoacylglycerols diffuse back into the intestinal lumen where they can be reused. Figure 2.15 summarizes the digestion and absorption of triacylglycerol.

The Recirculation and Excretion of Bile   The human body contains a total bile acid pool of about 2.5 to 5.0 g. Greater than 90% of the bile acids and salts secreted into the duodenum are reabsorbed by active transport in the ileum. Small amounts of the bile may be passively reabsorbed in the jejunum and the colon. Of the cholesterol contained within the bile, about half is taken up by the jejunum and is used in the formation of chylomicrons. The remainder of the cholesterol is excreted. Bile that is absorbed in the ileum enters the portal vein and is transported attached to plasma protein albumin in the blood back to the liver. Once in the liver, the reabsorbed bile acids are reconjugated to

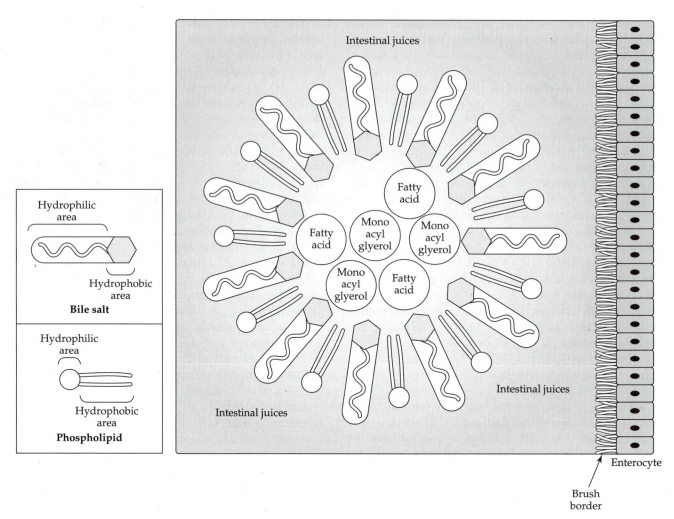

Figure 2.14   A micelle.

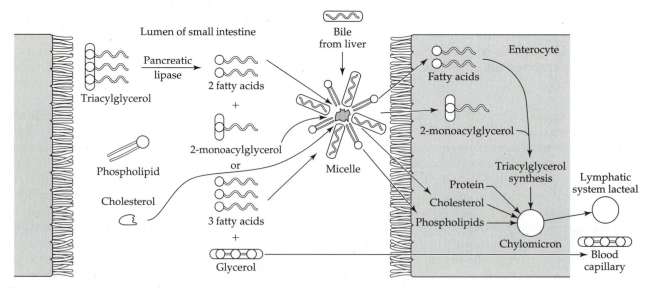

Figure 2.15   Digestion and absorption of dietary fat (triacylglycerols).

amino acids if necessary and secreted into bile along with the newly synthesized bile acids. New bile acids are typically synthesized in amounts about equal to those that are lost in the feces. New bile mixed with re-circulated bile is sent via the cystic duct for storage in the gallbladder. The circulation of bile, termed *entero-hepatic circulation,* is pictured in Figure 2.16. The pool of bile is thought to recycle at least twice per meal.

Of the bile acids that are not reabsorbed in the ileum, some may be deconjugated by bacteria in the colon and possibly terminal ileum to form secondary bile acids (Fig. 2.17). For example, cholic acid, a pri-

mary bile acid, is converted to the secondary bile acid deoxycholic acid, which can be reabsorbed. Chenode-oxycholic acid is converted to the secondary bile acid lithocholic acid, which unlike deoxycholic acid is typi-cally excreted in the feces. The presence of certain di-etary fibers in the gastrointestinal tract, however, may bind to the bile salts and acids and prevent bacterial de-conjugation and conversion to secondary bile acids. About 0.5 g of bile salts are lost daily in the feces.

Knowledge of the recirculation and excretion of bile helps in the understanding of various drug therapies used for some people with high blood cholesterol con-centrations. Drugs—powdered resins such as choles-tyramine (Questran)—have been manufactured with the purpose of binding bile in the gastrointestinal tract to enhance its fecal excretion from the body. The in-creased fecal excretion of the bile necessitates the use of body cholesterol for the synthesis of new bile acids. The goal from the use of such drugs is the lowering of blood cholesterol concentrations and a reduced risk of cardiovascular disease.

The Role of the Intestinal Brush Border in the Digestive Process

Several enzymes necessary for digestion are found on the brush border of the small intestine (Table 2.2). En-zymes are present in the small intestine that are re-sponsible for carbohydrate digestion (Fig. 2.18). Some of these enzymes including isomaltase are active against α 1-6 bonds. Isomaltase is an oligosaccharidase found on the brush border. The enzyme functions as

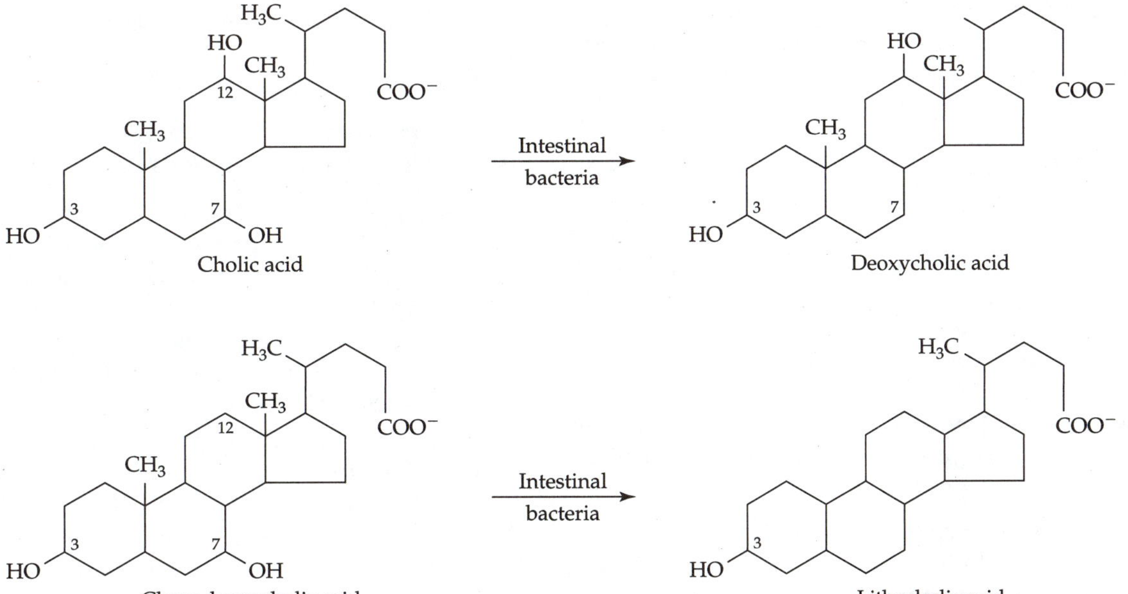

Figure 2.16   Enterohepatic circulation of bile.

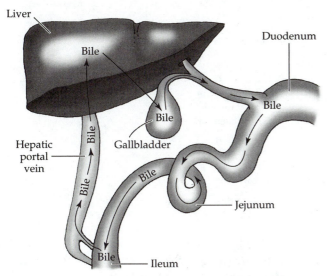

Figure 2.17   The synthesis of secondary bile acids by intestinal bacteria.

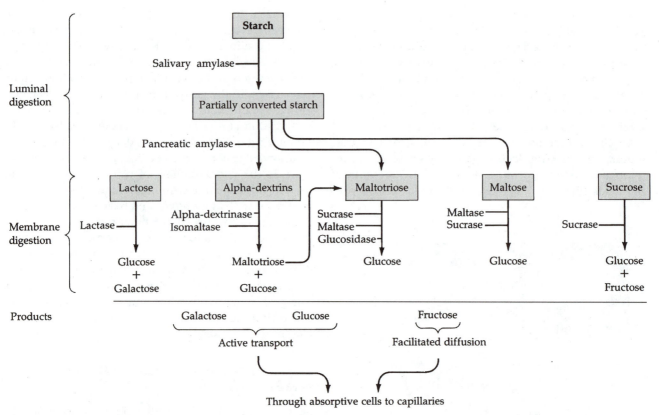

Figure 2.18 Summary of luminal digestion of starch and membrane digestion of disaccharides and oligosacchrides. Also indicated is the mechanism(s) for absorption of digestive products.

part of a complex with sucrase and hydrolyzes α 1-6 bonds in oligosaccharides and dextrins. To hydrolyze oligosaccharides and dextrins, enzymes capable of breaking α 1-4 bonds are necessary.

Glucoamylase, glucosidase, and maltase hydrolyze α 1-4 bonds in oligosaccharides, maltotriose, and maltose. Maltose and maltotriose are generated from the action of pancreatic amylase. Sucrase (invertase), like glucoamylase, glucosidase, and maltase, is also active against maltose and maltotriose as well as hydrolyzing sucrose to yield glucose and fructose. Three disaccharidases hydrolyze the three disaccharides into monosaccharides. Maltase hydrolyzes maltose into two glucose molecules. Lactase hydrolyzes lactose into glucose and galactose. Sucrase hydrolyzes sucrose to yield glucose and fructose. Further and more detailed information on carbohydrate digestion is provided in Chapter 4.

The brush border also contains enzymes for protein digestion (Fig. 2.19). These protein-digesting enzymes include

- several aminopeptidases (with different specificities) that hydrolyze N-terminal amino acids from oligopeptides, tripeptides, and dipeptides;

- tripeptidases that hydrolyze tripeptides into free amino acids and a dipeptide; and

- dipeptidases that hydrolyze dipeptides into two free amino acids.

Further information on protein digestion can be found in Chapter 7.

The Interrelationship Between the Digestive and Absorptive Processes in the Small Intestine

Most nutrients must be digested—that is, broken down into smaller pieces—prior to absorption. Digestion of nutrients occurs both in the lumen of the gastrointestinal tract and on the brush border. Although some absorption of some nutrients may occur in the stomach, absorption of most nutrients begins in the duodenum and continues throughout the jejunum and ileum. Generally, most absorption occurs in the proximal (upper) portion of the small intestine.

For absorption to occur, nutrients typically move through an unstirred or stagnant layer of water lying above the glycocalyx of the microvilli. Digestion and absorption of nutrients within the small intestine are

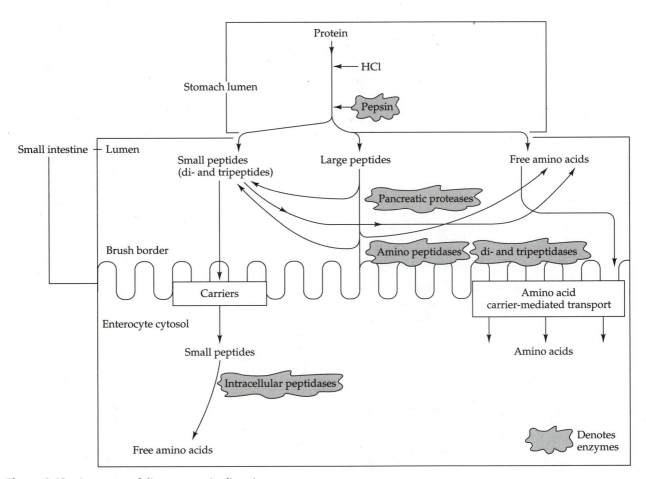

Figure 2.19 Summary of dietary protein digestion.

rapid, with most of the carbohydrate, protein, and fat being absorbed within 30 minutes after chyme has reached the small intestine. The presence of unabsorbed food in the ileum may increase the time period in which material remains in the small intestine, and may therefore increase nutrient absorption.

Absorption of nutrients into enterocytes may be accomplished by diffusion, facilitated diffusion, active transport, or, occasionally, pinocytosis or endocytosis (Fig. 2.20). In addition, a few nutrients may be absorbed by a paracellular (between cells) route. The mechanism of absorption for a nutrient depends on

- the solubility (fat versus water) of the nutrient,
- the concentration or electrical gradient, and
- the size of the molecule to be absorbed.

The absorption and transport of amino acids, peptides, monosaccharides, fatty acids, monoacylglycerols, and glycerol—that is, the end products of macronutrient digestion—are considered in depth in Chapters 4, 6, and 7. The mechanisms of absorption for each of the vitamins and minerals are shown in Figure 2.21, and are discussed in detail in Chapters 9 to 12.

Unabsorbed intestinal contents are passed from the ileum through the ileocecal valve into the colon, although some may serve as substrates for bacteria that inhabit the small intestine. Bacterial counts in the small intestine range up to about 10^3/g intestinal contents, although counts may be larger than this near the ileocecal sphincter. Examples of some of the bacteria that may be found in the small intestine include bacteroides, enterobacteria, lactobacilli, streptococci, and staphylococci. Recommendations for the ingestion of certain foods (known as *probiotics*) such as yogurt with live cultures containing specific strains of bacteria to effect an increase in the counts of specific microflora of the gastrointestinal tract are gaining interest in the health field. The intent of probiotics is for the bacteria to survive the passage through the upper digestive tract and then establish themselves (colonize) in the lower gastrointestinal tract, primarily the colon. Also under study is the use of prebiotics, food ingredients that are not digested or absorbed but can stimulate the growth and/or activity of one or more selected species of bacteria in the colon to improve the health of the host. Short-chain, nondigestible oligosaccharide

Diffusion. Some substances such as water and small lipid molecules cross membranes freely. The concentration of substances that can diffuse across cell membranes tends to equalize on the two sides of the membrane, so that the substance moves from the higher concentration to the lower; that is, it moves down a concentration gradient.

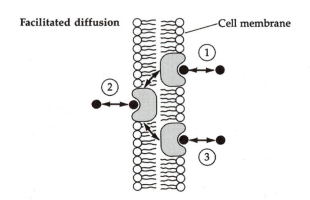

Facilitated diffusion. Other compounds cannot cross cell membranes without a specific carrier. The carrier may affect the permeability of the membrane in such a way that the substance is admitted or it may shuttle the compound from one side of the membrane to the other. Facilitated diffusion, like simple diffusion, allows an equalization of the substance on both sides of the membrane. The figure illustrates the shuttle process:
  1. Carrier loads particle on outside of cell
  2. Carrier releases particle on inside of cell
  3. Reversal of (1) and (2).

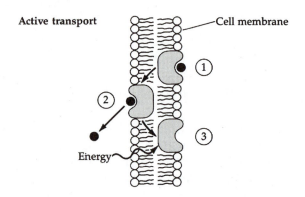

Active transport. Substances that need to be concentrated on one side of the cell membrane or the other require active transport, which involves energy expenditure. The energy is supplied by ATP and Na^+ is usually involved in the active transport mechanism. The figure illustrates the unidirectional movement of a substance requiring active transport:
  1. Carrier loads particle on outside of cell
  2. Carrier releases particle on inside of cell
  3. Carrier returns to outside to pick up another particle.

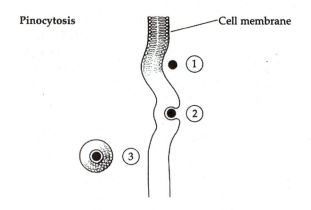

Pinocytosis. Some large molecules are moved into the cell via engulfment by the cell membrane. The figure illustrates the process:
  1. Substance contacts cell membrane
  2. Membrane wraps around or engulfs the substance
  3. The sac formed separates from membrane and moves into the cell.

Figure 2.20 Primary mechanisms for nutrient absorption.

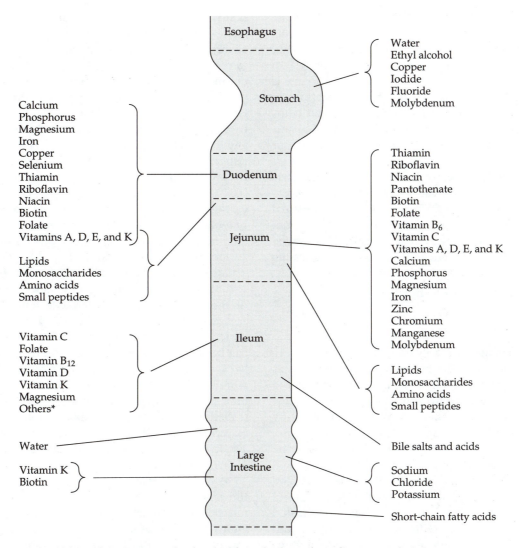

*Many additional nutrients may be absorbed from the ileum depending on transit time.

Figure 2.21 Sites of nutrient absorption in the gastrointestinal tract.

(oligofructose or inulin, for example) ingestion appears to effectively alter selected microbial populations, especially bifido bacteria, in the intestine. The ability to feed prebiotics that selectively promote the growth of certain bacteria, which would simultaneously antagonize the proliferation and growth of other pathogenic strains of bacteria, represents an exciting area of new medical nutrition therapies.

The Colon or Large Intestine

Intestinal material leaving the ileum empties via the ileocecal valve or sphincter into the cecum, the right side of the colon (large intestine; Fig. 2.7 depicts the anatomy of the human large intestine). Initially on en-

tering the colon, the intestinal material is still quite fluid. Contraction of the musculature of the large intestine is coordinated so as to mix the intestinal contents gently and to keep material in the proximal (ascending) colon for sufficient time to allow nutrient absorption to occur. The proximal colonic epithelia absorb sodium, chloride, and water more effectively than does the small intestinal mucosa. For example, about 90% to 95% of the water and sodium entering the colon each day is absorbed. Colonic absorption of sodium is influenced by a number of factors, including hormones. Antidiuretic hormone (ADH), for example, decreases sodium absorption, while glucocorticoids and mineralocorticoids increase sodium absorption in the colon.

Secretions into the lumen of the colon are few. Goblet cells in the crypts of Lieberkühn secrete mucus. Mucus protects the colonic mucosa and acts as a lubricant for fecal matter. Potassium is secreted possibly via an active secretory pathway into the colon. Bicarbonate is also secreted in exchange for chloride absorption. Bicarbonate provides an alkaline environment that helps to neutralize acids produced by colonic anaerobic bacteria. Sodium and hydrogen ion exchanges also occur permitting electrolyte absorption.

Both gram-negative and gram-positive bacteria strains representing over 400 species of at least 40 genera have been isolated from human feces. Intestinal bacterial counts of up to 10^{12}/g gastrointestinal tract contents in the large intestine are reported. Some examples of bacterial flora that inhabit the large intestine include:

- Bacteroides
- Clostridia
- Lactobacteria
- Bifidobacteria
- Coliforms
- Methanogens
- Eubacteria
- Streptococci

These bacteria use primarily dietary carbohydrate, although to a lesser extent amino acids and protein, as substrates needed for their growth. For example, starch that has not undergone hydrolysis by pancreatic amylase may be used by gram-negative bacteroides or by gram-positive bifidobacteria or eubacteria. Glycoproteins (mainly mucins) found in mucus secretions of the gastrointestinal tract may be broken down and used by bacteria such as bacteroides, bifidobacteria, and clostridia. In addition, sugar alcohols such as sorbitol and xylitol, disaccharides such as lactose, undigested oligosaccharides such as raffinose and stacchyose, and fibers such as some hemicelluloses, pectins, and gums may be degraded by selected bacteria found in the large intestine. Even digestive enzymes themselves may serve as substrate for bacteria such as bacteroides and clostridia. The breakdown of carbohydrate and protein by bacteria is an anaerobic process referred to as *fermentation.*

As described earlier, bacteria degrade mostly carbohydrate but also some amino acids and protein, as substrate for the production of substances such as energy and carbon atoms necessary for bacterial growth and maintenance. Acids are one of the principal end products of bacterial carbohydrate fermentation in the large intestine. Specifically, lactate and the short-chain fatty acids—acetate, butyrate, and propionate—are generated from bacterial action. These short-chain fatty acids, formerly referred to as *volatile fatty acids,* serve many purposes. They are thought to stimulate gastrointestinal cell proliferation. The presence of the acids lowers the luminal pH in the colon to affect changes in nutrient absorption and changes in the growth of certain species of bacteria. In addition, they provide substrates for body cell use. Butyrate, for example, may be absorbed via a Na^+/H^+ and/or a K^+/H^+ exchange system in the colon, where it is a preferred energy source for colonic epithelial cells. Butyrate also may regulate gene expression and cell growth. Propionate and lactate are absorbed in the colon and taken up for use by liver cells. Acetate is absorbed and used by muscle and brain cells. Absorption of these acids appears to be concentration-dependent. Other products generated by colonic bacteria include gases such as methane (CH_4), hydrogen (H_2), hydrogen sulfide (H_2S), and/or carbon dioxide (CO_2). Bacterial degradation of branched-chain amino acids generates branched-chain fatty acids such as isobutyrate and isovalerate. Deamination of aromatic amino acids yields phenolic compounds. Amines such as histamine result from the decarboxylation of amino acids such as histidine. Ammonia generated by bacterial deamination of amino acids as well as by bacterial urease action on urea (secreted into the gastrointestinal tract from the blood) can be reabsorbed by the colon and recirculated to the liver, where it can be reused to synthesize urea or amino acids. About 25%, or about 8 g, of the body's urea may be handled in this fashion, and it must be controlled in people with liver disease to prevent hepatic encephalopathy and coma. Uric acid and creatinine also may be released into the digestive tract and metabolized by colonic bacteria.

As intestinal matter passes through the sections of the colon, contractions by longitudinal and circular muscles promote absorption of water and electrolytes. Fluids from the moist fecal matter are absorbed into the mucosa cells of the colon as the fecal matter moves through the ascending, transverse, and descending sections of the colon. The end result, which usually takes 12 to 70 hours, is a progressive dehydration of the unabsorbed materials. Typically, a liter or more of chyme that enters the large intestine each day is reduced to less than 200 g of defecated material containing dead bacteria, inorganic matter, water, small amounts of undigested and unabsorbed nutrients, and constituents of digestive juices.

Summary

An examination of the various mechanisms in the gastrointestinal tract that allow food to be ingested, digested, absorbed, and then its residue to be excreted reveals the complexity of the digestive and absorption processes. Normal digestion and absorption of nutrients depends not only on a healthy alimentary canal but also on integration of the digestive system with the nervous, endocrine, and circulatory systems.

The many factors influencing digestion and absorption, including the dispersion and mixing of ingested food, the quantity and composition of gastrointestinal secretions, enterocyte integrity, the expanse of intestinal absorptive area, and the transit time of intestinal contents, must be coordinated so that nourishment of the body can occur while homeostasis of body fluids is maintained. Much of the coordination required is provided by regulatory peptides, some of which are provided by the nervous system as well as by the endocrine cells of the gastrointestinal tract.

Although the basic structure of the alimentary canal, which consists of the mucosa, submucosa, muscularis externa, and serosa, remains the same throughout, structural modifications allow for more specific functions by various segments of the gastrointestinal tract. Gastric glands that underlie the gastric mucosa secrete fluids and compounds necessary for digestive functions of the stomach. Also, particularly noteworthy is the presence of the folds of Kerckring, the villi, and the microvilli, all of which serve to increase dramatically the surface area exposed to the contents of the intestinal lumen. This enlarged surface area helps to maximize absorption, not only of ingested nutrients but also of endogenous secretions released into the gastrointestinal tract.

Study of the digestive system makes abundantly clear the fact that adequate nourishment of an individual, and therefore his or her health, depends in large measure on a normal functioning gastrointestinal tract. Particularly crucial to nourishment and health is a normally functioning small intestine, where the greatest amount of digestion and absorption occurs. Later chapters expand on the digestion and absorption of the various individual nutrients.

References Cited

1. Guyton AC. Textbook of medical physiology. 8th ed. Philadelphia: Saunders, 1991.

Other References

Gibson GR, Roberfroid MB. Dietary modulation of the human colonic microbiota: Introducing the concept of prebiotics. J Nutr 1995;125:1401-12.

Gibson GR, Willis CL, Loo JV. Nondigestible oligosaccharides and bifidobacteria—implications for health. Int Sugar J 1994;96:381-7.

Web Sites

http://www.nlm.nih.gov/research/visible/
visible_human.html

An Overview of Selected Disorders of the Digestive System with Implications for Nourishing the Body

In Chapter 2, digestion is defined as a process by which food is broken down mechanically and chemically in the gastrointestinal (GI) tract. Digestion ultimately provides nutrients ready for absorption into the body through the cells of the GI tract, principally the cells of the small intestine (enterocytes). Secretions required for digestion of nutrients are produced by multiple organs of the GI tract. These secretions include principally enzymes but also hydrochloric acid for gastric digestion and bicarbonate and bile for digestion in the intestine. Malfunction of one or more of the organs due to pathology (the causes, nature, and effects of disease) can in turn diminish the production and/or release of secretions synthesized by these organs and released into the GI tract. Without secretions or with less than normal amounts of secretions, digestion of nutrients may be impaired and result in nutrient malabsorption.

Many conditions or diseases alter the function of organs of the GI tract and thus affect digestion. For example, some GI tract diseases may cause decreased synthesis and release of secretions needed for nutrient digestion; other conditions or diseases affecting the GI tract can alter motility or clearing of the GI contents through the organs of the GI tract. Malfunction of sphincters, for example, can alter the clearing or passage of the GI contents through the various organs of the GI tract. Clearing problems may cause back fluxes (refluxes) of secretions from, for example, the stomach into the esophagus; normally the contents of the GI tract move from the esophagus to the stomach and not vice versa. Conditions in which the GI mucosa is inflamed or damaged as well as conditions that increase transit time or speed up the movement of GI contents (food and nutrients) through the GI tract typically result in nutrient malabsorption due to diminished time for digestion and absorption of nutrients.

An understanding of the physiology of the GI tract and its accessory organs, and the pathology affecting the GI tract, is essential to the understanding of how to modify an individual's diet from standard dietary recommendations for healthy populations of the United States. Standard dietary recommendations are based on the Recommended Dietary Allowances (RDA) [1], the Food Guide Pyramid, the Dietary and Health Recommendations [2], the Dietary Guidelines for Americans [3], and the Dietary Goals for the United States [4]. This perspective addresses, in a general fashion, three disorders affecting the digestive tract and what implications these conditions have with respect to nourishing the body.

Disorder 1: Reflux Esophagitis (inflammation of the esophagus due to backward flow of gastric contents)

After chewing and swallowing food, food enters the esophagus then passes through the lower esophageal sphincter (LES) into the stomach. Decreases in pressure at the LES, sometimes called *LES incompetence,* can result in reflux esophagitis in which gastric contents including hydrochloric acid flow backward into the esophagus from the stomach and inflame the esophageal mucosa. Recurring reflux of hydrochloric acid into the esophagus can damage the esophageal mucosa. The individual experiencing reflux esophagitis typically complains of heartburn or a burning sensation in the mid-chest region.

To address nutrition implications of this condition, we first need to reexamine some of the foods, nutrients, or substances in foods that influence LES pressure and that may promote increased acid production. Several substances decrease LES pressure [5–9], including the following:

- High-fat meals
- Chocolate
- Smoking (nicotine)
- Alcohol
- Carminatives

Carminatives are loosely and broadly defined as agents that may produce a warm sensation and also relieve symptoms of gas in the GI tract, or more specifically volatile oil extracts of plants, most often oils of spearmint and peppermint. Other offending agents are substances that increase gastric secretions, especially acid production. Alcohol, decaffeinated and caffeinated coffee, and methylxanthine (in tea, for example) stimulate gastric secretions including hydrochloric acid [10–14]. Ingestion of these substances or foods is likely to aggravate irritated esophageal mucosa.

With this knowledge, some of the recommendations for the patient with reflux esophagitis or LES incompetence will make sense to you. Recommendations are aimed at

1. avoiding substances that can further lower LES pressure, which is already low due to the condition, and to a lesser extent

2. avoiding substances that may promote the secretion of acid, which would then be present in higher concentrations than normal if refluxed.

To implement these recommendations individuals with reflux esophagitis must be told which foods or substances need to be avoided. As a dietitian, you would suggest avoiding high-fat foods or meals and also avoiding chocolate, coffee, tea, alcohol, and carminatives such as peppermint and spearmint.

In addition to avoiding substances that lower LES pressure and that promote the secretion of acid, recommendations can include

3. increasing the use of foods or nutrients that increase LES pressure.

Protein acts in this manner; it increases LES pressure [14]. Consequently, a higher than normal protein intake is encouraged; however, excessive protein intakes, especially from foods high in calcium, such as dairy products, are not recommended. The reason for avoiding excessively high intakes of dairy foods relates to the fact that amino acids and peptides (generated from digestion of the protein in the dairy products) and calcium in dairy products are both known to stimulate gastrin release [15]. Although gastrin increases LES pressure, it is also a potent stimulator of hydrochloric acid secretion.

In addition to the previously stated nutrition recommendations for individuals with reflux esophagitis, it is important to remember that reflux is more likely to occur with increased gastric volume (i.e., eating large meals), increased gastric pressure (from, e.g., obesity), and placement of gastric contents near the sphincter (i.e., bending, lying down, or assuming recumbent positions). Thus, recommendations for individuals with reflux esophagitis should include

4. ingestion of smaller meals (versus consumption of larger meals);

5. ingestion of fluids between meals instead of with a meal to help minimize large increases in gastric volume;

6. weight loss, if an individual is overweight or obese;

7. avoidance of tight-fitting clothes; and

8. avoidance of lying down as well as lifting and bending for at least two hours after eating.

Disorder 2: Inflammatory Bowel Diseases

Inflammatory bowel diseases (IBDs) include ulcerative colitis and Crohn's disease or regional enteritis. IBDs cause chronic inflammation of various segments of the GI tract, especially the intestines. Although the causes of IBDs are unknown, nutrient malabsorption is a significant problem. Because of the inflammation of the mucosa, nutrient hydrolysis by brush border disaccharidases and peptidases is diminished. Nutrient transit time is typically decreased or shortened—that is, GI tract contents move through the GI tract quicker than usual. Absorption of many nutrients is impaired as a result of diminished digestion, enterocyte damage, and shortened transit time. People who have IBD generally experience diminished bowel function, both digestive and absorptive processes. Diarrhea and/or steatorrhea (excessive fat in the feces) are common. Blood may also be found in the feces with severe inflammation or ulceration of deeper areas of the GI mucosa. Moreover, food intake is usually decreased, especially during acute attacks.

The nutritional problems of individuals with IBD are multiple. The discussion that follows addresses a few of these problems along with nutrition recommendations. Blood lost in diarrhea results in the loss of iron and protein from the blood into the feces; potassium, other electrolytes, and fluids pulled from the blood are also typically lost with diarrhea. When these losses are coupled with poor nutrient intake, the individual becomes dehydrated, has poor protein and iron status, and usually has electrolyte imbalances. If IBD has affected the ileum, the absorption of vitamin B_{12} may be impaired (absorption of this vitamin occurs in the ileum), reabsorption of bile salts from the ileum may be diminished, and fat malabsorption may occur. Although pancreatic lipase is available to hydrolyze dietary triacylglycerols, the lack of sufficient bile or diminished bile function due to bacterial alteration of bile can decrease micelle formation and thus decrease absorption of fatty acids and fat-soluble vitamins into the enterocyte. Unabsorbed fatty acids bind to calcium and magnesium in the lumen of the intestine; the resulting insoluble complex, sometimes called a soap, is excreted in the feces.

Some dietary recommendations for people with IBD include

1. increased intakes above the RDA for iron due to increased losses with the bloody diarrhea and decreased absorption;

2. a low-fat diet, due to impaired fat absorption;

3. increased intakes of calcium and magnesium due to diminished absorption secondary to soap formation and overall malabsorption with diarrhea;

4. a high-protein diet to improve protein status diminished by protein loss from the blood into the feces with bloody diarrhea and malabsorption of amino acids;

5. fat-soluble vitamin supplements, possibly given in a water-miscible form to improve absorption;

6. increased fluids for rehydration; and

7. increased nutrient intake to meet energy needs.

Easily digestible carbohydrates that are low in fiber, high-protein, low-fat foods with minimal residue, and lactose-free foods will need to provide the bulk of the individual's energy needs if oral intake is deemed appropriate. Medium-chain triacylglycerol (MCT) oil, which is absorbed directly into portal blood and does not need bile for absorption, may be added in small amounts to different foods throughout the day to increase energy intake.

Disorder 3: Pancreatitis (inflammation of the pancreas)

Pancreatitis provides an excellent example of the nutritional ramifications of a condition affecting an accessory organ of the GI tract. Remember, the pancreas produces several enzymes needed for the digestion of all nutrients. People with pancreatitis experience malabsorption, especially of fat and fat-soluble vitamins. Diminished secretion of pancreatic lipase into the duodenum results in malabsorption of fat and fat-soluble vitamins. Fat is malabsorbed because not enough pancreatic lipase is available to hydrolyze the fatty acids from the triacylglycerol. Remember, this hydrolysis is necessary in order for fatty acids and monoacylglycerols to form micelles, the form in which the fatty acids are carried to the enterocyte for absorption. Thus, with pancreatitis, the insufficient enzymes available for fat hydrolysis necessitate the use of a low-fat diet.

In addition to insufficient pancreatic lipase secretion, bicarbonate secretion into the duodenum is also diminished with pancreatitis. Bicarbonate, in part, functions to increase the pH of the small intestine. Intestinal enzymes function best at an alkaline pH, which is provided by the release of bicarbonate into the intestine. Oral supplements of pancreatic enzymes may be needed to replace the diminished output of these enzymes by the malfunctioning, inflamed pancreas. In addition, antacids may need to be taken with the oral pancreatic replacement enzymes. The antacids are taken to replace the role of the bicarbonate, and thus, to help maintain an appropriate pH for enzyme function. Administration of exogenous insulin may also be needed to replace the insulin no longer produced by damaged pancreatic endocrine cells.

These three conditions have been presented to help you understand how pathology affecting the GI tract—malfunction of a sphincter (reflux esophagitis), destruction of enterocyte function (IBD), and malfunction of a GI tract accessory organ (pancreatitis) that provides secretions needed for nutrient digestion—affects the body's ability to digest and absorb nutrients. Furthermore, these three conditions have been presented to help you understand how nutrient intakes need to deviate, in some cases to less than recommended levels and in other cases to greater than recommended levels, depending on the condition. Bidirectional dietary modifications are typical of many conditions affecting not only the gastrointestinal tract, but other organ systems as well.

References Cited

1. Food and Nutrition Board, Commission on Life Sciences, National Research Council. Recommended dietary allowances, 10th ed. Washington DC: National Academy Press, 1989.

2. Committee on Diet and Health, Food and Nutrition Board, Commission on Life Sciences, National Research Council. Diet and health: Implications for reducing chronic disease risk. Washington DC: National Academy Press, 1989.

3. U.S. Department of Agriculture, U.S. Department of Health and Human Services. Dietary Guidelines for Americans, 3rd ed. Washington DC: U.S. Government Printing Office, 1990.

4. U.S. Senate, Select Committee on Nutrition and Human Needs. Dietary goals for the United States, 2nd ed. Washington DC: Government Printing Office, 1977.

5. Babka JC, Castell DO. On the genesis of heartburn: The effects of specific foods on the lower esophageal sphincter. Am J Dig Dis 1973;18: 391–397.

6. Wright LE, Castell DO. The adverse effect of chocolate on lower esophageal sphincter pressure. Digest Dis 1975;20: 703–707.

7. Sigmund CJ, McNally EF. The action of a carminative on the lower esophageal sphincter. Gastroenterology 1969;56:13–18.

8. Dennish GW, Castell DO. Inhibitory effect of smoking on the lower esophageal sphincter. N Engl J Med 1971;284: 1136–1137.

9. Hogan WJ, Andrade SRV, Winship DH. Ethanol-induced acute esophageal motor dysfunction. J Appl Physiol 1972; 32:755–760.

10. Lenz HJ, Rerrari-Taylor J, Isenberg JI. Wine and five percent alcohol are potent stimulants of gastric acid secretion in humans. Gastroenterology 1983;85:1082–1087.

11. Cohen S, Booth GH. Gastric acid secretion and lower esophageal sphincter pressure in response to coffee and caffeine. N Engl J Med 1975;293:897–899.

12. Feldman EJ, Isenberg JI, Grossman MI. Gastric acid and gastrin response to decaffeinated coffee and a peptone meal. JAMA 1981;246:248–250.

13. Thomas FB, Steinbaugh JT, Fromkes JJ, Mekhjian HS, Caldwell JH. Inhibitory effect of coffee on lower esophageal sphincter pressure. Gastroenterology 1980;79:1262–1266.

14. Harris JB, Nigon K, Alonso D. Adenosine-3', 5'-monophosphate: Intracellular mediator for methylxanthine stimulation of gastric secretion. Gastroenterology 1969;57:377–384.

15. Levant JA, Walsh JH, Isenberg JI. Stimulation of gastric secretion and gastrin release by single oral doses of calcium carbonate in man. N Engl J Med 1973;289:555–558.

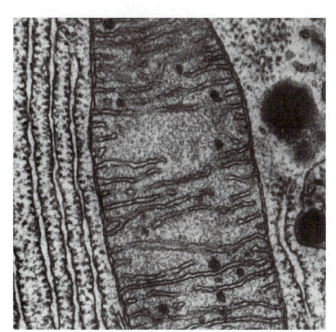

A mitochondrion

Chapter 3

∾ ∾

Energy Transformation

Energy Release or Consumption in Chemical Reactions

The Role of High-Energy Phosphate in Energy Storage

The High-Energy Phosphate Bond

Coupled Reactions in the Transfer of Energy

Formation of ATP
   Substrate-Level Phosphorylation
   Oxidative Phosphorylation

Biological Oxidation and the Electron Transport Chain

∾ **PERSPECTIVE:** If Oxidation and Phosphorylation Disconnect: A Waste of Energy!

Many of the processes that sustain life require energy. To most of us, the term *energy* conjures an image of physical "vim and vigor" and the capacity to exert oneself physically. Indeed, this view of energy does have biochemical foundation because the contraction of muscle fibers associated with mechanical work is an energy-demanding process. The process is accomplished at the expense of adenosine triphosphate (ATP), the major storage form of molecular energy. But there are other equally important, though more subtle, requirements for energy by the living body, including

1. the biosynthetic (anabolic) systems by which substances can be formed from simpler precursors,

2. active transport systems by which compounds or metal ions can be moved across membranes against a concentration gradient (p. 16), and

3. the transfer of genetic information.

Energy used by the body is derived from the energy containing nutrients—carbohydrate, fat, and protein. If the covalent bonds contained within these molecules are cleaved, the bonding forces "relax," and their energy is released. Release of energy may simply be expressed as heat, such as would occur in the combustion of flammable substances. In vitro burning of a mole of glucose therefore liberates heat along with CO_2 and H_2O as products of combustion, as shown:

$$C_6H_{12}O_6 + 6O_2 \longrightarrow 6CO_2 + 6H_2O + \text{heat}$$

53

Cellular metabolism is expressed nearly identically to that of simple combustion. The difference is that in metabolic oxidation, a significant portion of the released energy is salvaged in the form of new, high-energy bonds that represent a usable source of energy for driving energy requiring processes. Such stored energy is generally contained in phosphate anhydride bonds, chiefly those of ATP (Fig. 3.1). The analogy between the combustion and the metabolic oxidation of a typical nutrient (palmitic acid) is illustrated in Figure 3.2.

The unit of energy used throughout this text continues to be the calorie, abbreviated cal. In the expression of the higher caloric values encountered in nutrition, the unit of kilocalories (kcal) is often used: 1 kcal = 1,000 cal. Students of nutrition should be familiar with another unit of energy called the *joule* (J) and its higher-value counterpart, the *kilojoule* (kJ), which has become widely used in biochemistry. Calories can be easily converted to joules by the factor 4.18:

$$1 \text{ cal} = 4.18 \text{ J, or } 1 \text{ kcal} = 4.18 \text{ kJ}$$

Nutrition and the calorie have been so closely linked over the years, however, that we feel that students of nutrition may be more comfortable with the calorie/kilocalorie unit and that digression from this unit may not be appropriate at this time.

Energy Release or Consumption in Chemical Reactions

The potential energy inherent in the chemical bonds of nutrients is released if the molecules undergo oxidation either through combustion or through the controlled oxidation within the cell. This energy is defined as *free energy (G)* if, on its release, it is capable of doing work at constant temperature and pressure. Because the cell does function under conditions of constant temperature and pressure, the energy it uses to drive energy-requiring reactions or processes is therefore free energy.

The products of the complete oxidation of organic molecules containing only carbon, hydrogen, and oxygen are CO_2 and H_2O, and they too have an inherent free energy. However, because energy was released in the course of the oxidation of the organic molecules, the reactants in this case, the free energy of the products would necessarily be lower than that of the reactants. The difference in the free energy between the products and the reactants in a given chemical reaction is a very useful parameter for estimating the tendency for that reaction to occur. This difference is symbolized as follows:

$$G_{\text{products}} - G_{\text{reactants}} = \Delta G \text{ of the reaction}$$

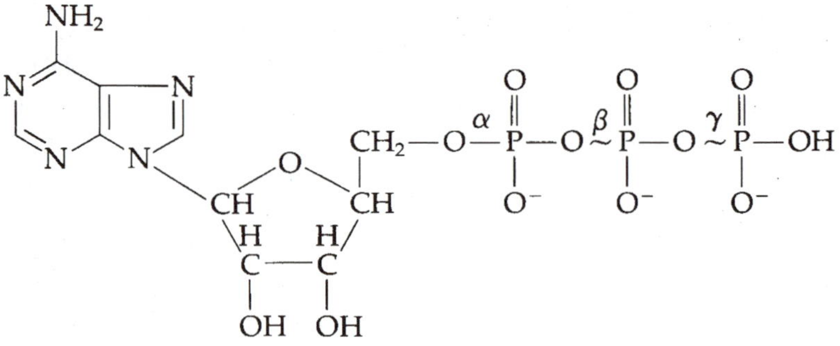

Figure 3.1 Adenosine triphosphate (ATP). The bonds connecting the α- and the β-phosphates and the β- and γ-phosphates are anhydride bonds, which release a large amount of energy when hydrolyzed. The bonds are shown as wavy lines, which are customarily used to denote a high-energy source.

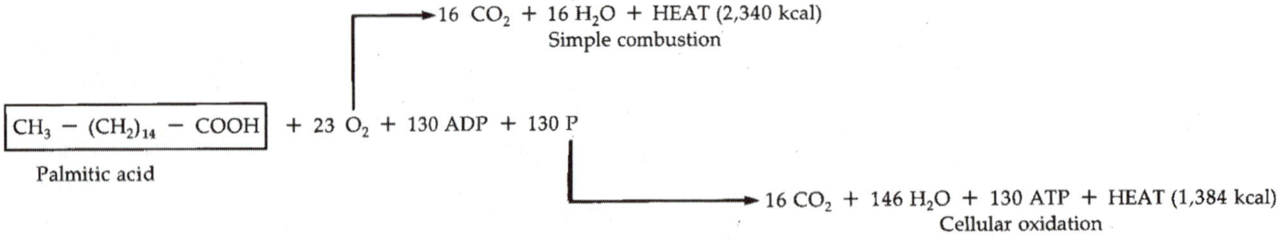

Figure 3.2 A comparison of the simple combustion and the metabolic oxidation of the fatty acid palmitate. The energy liberated from combustion assumes the form of heat only, while approximately 40% of the energy released by metabolic oxidation is salvaged as ATP, with the remainder released in the form of heat.

If the *G* value of the reactants is greater than the *G* value of the products, as in the case of the oxidation reaction, the reaction is said to be *exothermic,* or energy releasing, and the sign of the ΔG is negative. In contrast, a positive ΔG indicates that the *G* of the products is greater than that of the reactants, indicating that energy would have to be supplied to the system to convert the reactants into the higher-energy products. Such a reaction is called *endothermic,* or energy requiring. Exothermic and endothermic reactions are sometimes referred to as *downhill* and *uphill* reactions, respectively, terms that may help to create an image of energy input and release. The free energy levels of reactants and products in a typical exothermic or downhill reaction can be likened to a boulder on a hillside that can occupy the two positions, A and B, as illustrated in Figure 3.3. As the boulder descends to level B from level A, energy capable of doing work is clearly liberated, and the change in free energy is a negative value. The reverse reaction, moving the boulder uphill to level A from level B, necessitates an input of energy, an endothermic process, and a positive ΔG. *It is important to understand that the quantity of energy released in the downhill reaction is precisely the same as the quantity of energy required for the reverse (uphill) reaction.* Only the sign of the ΔG changes.

Although exothermic reactions are favored over endothermic reactions in that they require no external energy input, they do not occur spontaneously. Otherwise, no energy-producing nutrients or fuels would exist throughout the universe, because they would have transformed spontaneously to their lower energy level.

A certain amount of energy must be introduced into reactant molecules to activate them to what is referred to as their *transition state,* a higher energy level or barrier at which the exothermic conversion to products can indeed take place. This energy that must be imposed on the system to promote the reactants to their transition state is called the *activation energy.* Referring again to the boulder-and-hillside analogy of Figure 3.3, the boulder does not spontaneously descend until the required activation energy can dislodge it from its resting place to the brink of the slope.

The cell derives its energy from chemical reactions, each one of which exhibits a free energy change. The reactions occur sequentially as the nutrients are systematically oxidized through a pathway of intermediates, ultimately to CO_2 and H_2O. All the reactions are enzyme catalyzed. Within a given catabolic pathway—for example, the oxidation of glucose to CO_2 and H_2O—some reactions may be energy consuming ($+\Delta G$). However, energy-releasing ($-\Delta G$) reactions will prevail, so that the *net* energy transformation for the entire pathway is exothermic.

Most cellular reactions are reversible, meaning that an enzyme (E) that can catalyze the conversion of hypothetical substance A into substance B can also catalyze the reverse reaction, as shown:

Using the A, B interconversion as an example, the concept of reversibility of a chemical reaction will be

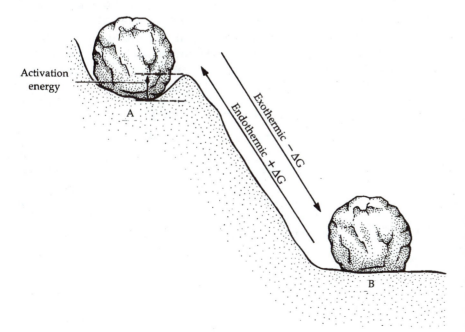

Figure 3.3 The uphill-downhill concept illustrating energy-releasing and energy-demanding processes. Also indicated is the activation energy, which is the amount required to activate the reactant (the boulder, when occupying position A in this case) to its transition state.

reviewed. In the presence of the specific enzyme E, substance A will be converted to substance B. Initially, the reaction is unidirectional, because only A is present. However, because the enzyme is also capable of converting B to A, this reaction becomes significant as B increases in concentration. Therefore, from the moment the reaction is initiated, the amount of A will decrease, while that of B will increase to the point that the rate of the two reactions becomes equal and the concentration of A and B will consequently no longer change. At this point the system is said to be in *equilibrium,* and whether the A ⟶ B reaction or the B ⟶ A reaction is energetically favored is indicated by the relative concentrations of A and B at equilibrium. The equilibrium constant, K_{eq}, of the reaction reflects the difference in concentrations and therefore the favored direction of the reaction. The K_{eq} in this case is simply the ratio of the equilibrium concentration of product B to that of reactant A:

$$K_{eq} = [B]/[A]$$

From this simple relationship, it can be seen that the K_{eq} increases in value to the extent that the A ⟶ B conversion is favored over the B ⟶ A. If the K_{eq} has a value greater than 1, the formation of B from A is favored, whereas a value less than 1 indicates that the equilibrium favors the formation of A from B. It follows that if no bias exists for either reaction, the equilibrium constant becomes equal to 1. The K_{eq} of a reaction can be used to calculate its standard free energy change, which is discussed next.

The standard free energy change (ΔG^0) for a chemical reaction is a constant for that particular reaction. It is defined as the difference between the free energy content of the reactants and the free energy of the products under standard conditions. Standard conditions are defined precisely: a temperature of 25°C (298°K), a pressure of 1.0 atm (atmospheres), and the reactants and products must be present at their standard concentrations—namely, 1.0 mol/L. Under such conditions, the ΔG^0 is mathematically related to the K_{eq} by the equation

$$\Delta G^0 = -2.3 \ RT \log K_{eq}$$

in which R is the gas constant (1.987 cal/mol × degrees K), and T is the absolute temperature, 298°K in this case. In fact, because the factors 2.3, R, and T are constants, their product is equal to −2.3(1.987)(298) or −1,362 cal/mol. The equation therefore simplifies to

$$\Delta G^0 = -1,362 \log K_{eq}$$

We can see from the relationship that the equilibrium constant of a reaction determines the sign and magnitude of the standard free energy change. For example,

referring once again to the A ⟶ B reaction, the log of a K_{eq} value greater than 1.0 will be a positive integer, and the sign of the ΔG will consequently be negative, therefore establishing that the reaction A ⟶ B is energetically favored. Conversely, a K_{eq} of less than 1.0 would have a negative log value and therefore a positive ΔG^0 for the reaction A ⟶ B. The $+\Delta G^0$ in this case indicates that the formation of A from B is favored in the equilibrium. A pH of 7 is, by convention, designated the *standard pH in biochemical reactions.* In this special case, the free energy change of reactions is designated $\Delta G^{0'}$ and will be used henceforth in this discussion.

There is an important difference between the free energy change, ΔG, and the standard free energy change, $\Delta G^{0'}$, of a chemical reaction. The difference in the two values can explain why a reaction having a positive $\Delta G^{0'}$ can proceed exothermically $(-\Delta G)$ in the cell, where standard conditions do not exist. As an example, consider the reaction catalyzed by the enzyme triosephosphate isomerase shown in Figure 3.4. This particular reaction occurs in the glycolytic pathway through which glucose is converted to pyruvate. (The pathway is discussed in detail in Chap. 4.) The reaction catalyzed by the enzyme aldolase produces 1 mol each of dihydroxyacetone phosphate (DHAP) and glyceraldehyde 3-phosphate (G 3-P) from 1 mol of fructose 1,6-bisphosphate. However, of the two products, only the G 3-P is directly degraded in the subsequent reactions of glycolysis, resulting in a significantly lower concentration of this metabolite compared with the DHAP. Therefore, two important conditions within the cell deviate from standard conditions: namely, the temperature (~37°C, or 310°K) and the fact that the reactants and products are not at equal, 1.0 mol/L concentrations. The actual ΔG of the triosephosphate isomerase reaction for the conditions existing in the cell can be calculated from the $\Delta G^{0'}$ of the reaction and the cellular, steady-state concentrations of reactant and product.

$$\Delta G = \Delta G^{0'} + 2.3RT \log [product]/[reactant]$$

The $\Delta G^{0'}$ for the reaction DHAP (reactant) ⟶ G 3-P (product) is +1,830 cal/mol, indicating that under standard conditions the formation of DHAP is preferred over the formation of G 3-P. For the sake of illustration, however, let us assume that the cellular concentration of DHAP is 50 times that of G 3-P for the reason already explained. Substituting these values in the equation:

$$\begin{aligned} \Delta G &= +1,830 + 2.3(1.987)(310°K) \log 1/50 \\ &= +1,830 + 1,416 \log 0.02 \\ &= +1,830 + 1,416 \ (-1.7) \\ &= +1,830 - 2,407 \\ &= -577 \ \text{cal/mol} \end{aligned}$$

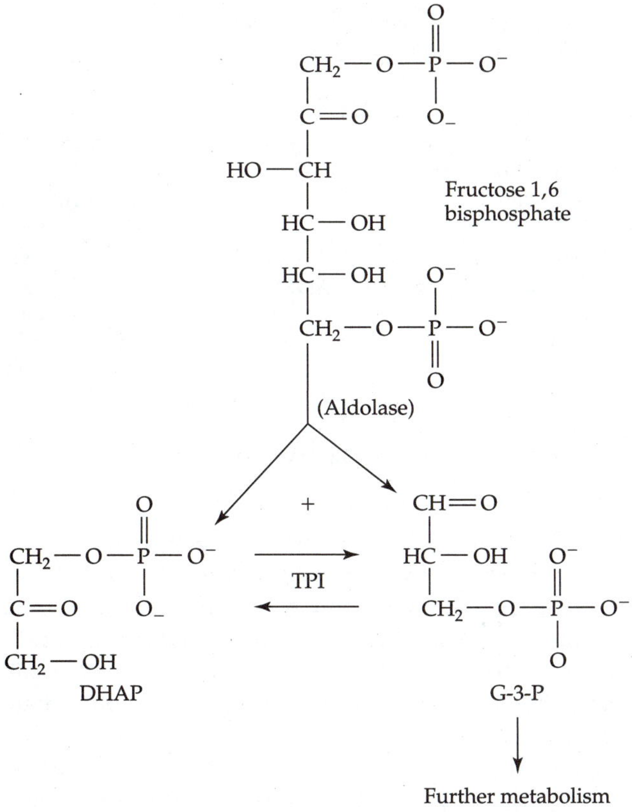

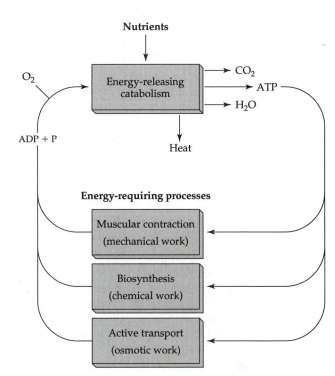

Figure 3.4 A segment of the glycolytic pathway of glucose metabolism, illustrating the interconversion by the enzyme triosephosphate isomerase (TPI) of dihydroxyacetone phosphate (DHAP) and glyceraldehyde 3-phosphate (G 3-P). The two triose phosphates are formed from the splitting of fructose 1,6 bisphosphate by aldolase. Under standard conditions, the TPI equilibrium is shifted toward the formation of DHAP, but in cellular metabolism the further conversion of G 3-P to other metabolites depletes its steady state concentration and causes the rate of the DHAP ⟶ G 3-P reaction to increase.

Figure 3.5 An illustration of how ATP is generated from the coupling of ADP and phosphate through the oxidative catabolism of nutrients and how it is in turn used for energy-requiring processes.

The negative ΔG therefore shows that the reaction is indeed proceeding to the right, as shown, despite the positive $\Delta G^{0'}$ for this reaction.

The Role of High-Energy Phosphate in Energy Storage

The preceding section addressed the fundamental principle of free energy changes in chemical reactions, and the fact that the cell obtains this free energy through the catabolism of nutrient molecules. Furthermore, it was pointed out that this energy must somehow be used to drive the various energy-requiring processes and anabolic reactions so important in normal cell function. Liberated energy can be used to form the phosphate anhy-

dride bonds of ATP, and these bonds can in turn be hydrolyzed to release the energy when needed. Therefore, ATP can be thought of as an energy reservoir, serving as the major linking intermediate between energy-releasing and energy-demanding chemical reactions in the cell. In nearly all cases, the energy stored in ATP is released by the enzymatic hydrolysis of the anhydride bond connecting the β- and γ-phosphates in the molecule (Fig. 3.1). The products of this hydrolysis are adenosine diphosphate (ADP) and a free phosphate group that in certain instances is transferred to various phosphate acceptors, a reaction that activates the acceptors to higher energy levels. The involvement of ATP as a link between the energy-releasing and energy-requiring cellular reactions and processes is summarized in Figure 3.5.

The High-Energy Phosphate Bond

Hydrolysis of the anhydride bond connecting the β- and γ-phosphates of ATP is a highly exothermic reaction. The reaction, which releases ADP and inorganic phosphate (P_i), has a $\Delta G^{0'}$ of $-7,300$ cal/mol. The hydrolysis of the anhydride bond connecting the α- and β-phosphates also has a $\Delta G^{0'}$ of $-7,300$ cal/mol. The bond linking the α-phosphate to the adenosine, on the

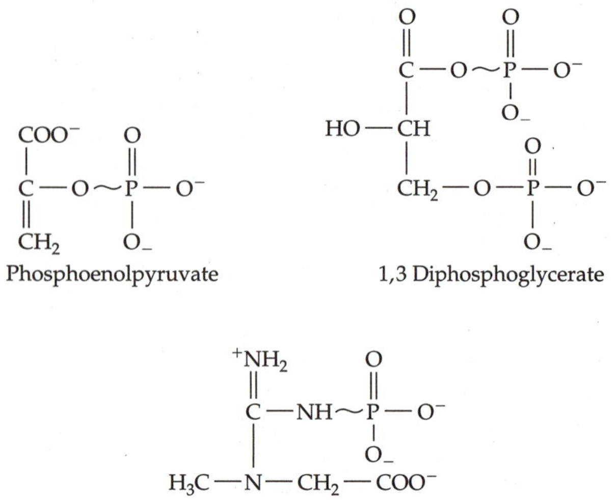

Figure 3.6 Examples of very high-energy phosphate compounds. The hydrolysis of the phosphate bonds represented by the wavy line releases more energy than the terminal phosphate bond of ATP, making it energetically possible to transfer these phosphate groups enzymatically to ADP.

other hand, is a phosphate ester, a bond that characteristically possesses considerably less free energy than a phosphate anhydride bond. The hydrolysis of adenosine monophosphate (AMP) to yield adenosine and phosphate liberates 3,400 cal/mol, typical of phosphate ester hydrolysis, the range for which is generally 2,000 to 5,000 cal/mol. As a family, therefore, phosphate esters rank below ATP and other nucleoside triphosphates on the scale of hydrolytic energy release.

Based on the preceding discussion of phosphate bond energies, it is tempting to think in terms of high-energy and low-energy phosphate bonds, with ATP and other nucleoside triphosphates being high energy and phosphate esters representing low energy. However, the wide range of ester energy and the fact that there are phosphorylated compounds having even higher energy than ATP complicates the high energy–low energy concept. Phosphoenolpyruvate and 1,3-diphosphoglycerate, which occur as intermediates in the metabolic pathway of glycolysis (p. 85), and phosphocreatine, important in the energy of muscle contraction, are examples of compounds having phosphate bond energies significantly higher than ATP. These structures are shown in Figure 3.6. Note that "high-energy" bonds are depicted as a wavy line (~), referred to as a *tilde,* indicating that the free energy of hydrolysis is higher than for the more stable phosphate esters. Table 3.1 lists the standard free energy of hydrolysis of selected phosphate-containing compounds.

Table 3.1 Free Energy of Hydrolysis of Some Phosphorylated Compounds

	$\Delta G^{0'}$ (cal)
Phosphoenolpyruvate	−14,800
1,3-bisphosphoglycerate	−11,800
Phosphocreatine	−10,300
ATP	−7,300
Glucose 1-phosphate	−5,000
Adenosine monophosphate (AMP)	−3,400
Glucose 6-phosphate	−3,300

Coupled Reactions in the Transfer of Energy

It has been pointed out that the $\Delta G^{0'}$ value for the phosphate bond hydrolysis of ATP is intermediate between certain high-energy phosphate compounds and compounds that possess relatively low-energy phosphate esters. Its central position on the energy scale lets ATP serve as an intermediate carrier of phosphate groups. ADP can accept the phosphate groups from high-energy phosphate donor molecules and then, as ATP, transfer them to lower-energy receptor molecules. By receiving the phosphate groups, the acceptor molecules become activated to a higher energy level from which they can undergo subsequent reactions. The end result, therefore, is the transfer of chemical energy from donor molecule through ATP to receptor molecule.

If a given quantity of energy is released in an exothermic reaction, it is important to understand that the same amount of energy must be added to the system if the reaction is to be driven in the reverse direction. For example, the hydrolysis of the phosphate ester bond of glucose 6-phosphate liberates 3,300 cal/mol of energy, whereas the reverse reaction, the addition of the phosphate to glucose to form glucose 6-phosphate, necessitates the input of 3,300 cal/mol. These reactions can be expressed in terms of their standard free energy changes:

1. Glucose 6-phosphate $\longrightarrow$ Glucose + phosphate
$$\Delta G^{0'} = -3{,}300 \text{ cal/mol}$$

2. Phosphate + glucose $\longrightarrow$ Glucose 6-phosphate
$$\Delta G^{0'} = +3{,}300 \text{ cal/mol}$$

The addition of phosphate to a molecule is called a *phosphorylation reaction.* It generally is accomplished by the enzymatic transfer of the terminal phosphate group of ATP to the molecule, rather than by the addition of free phosphate, as suggested by reaction 2. Reaction 2 is hypothetical, designed only to illustrate the energy requirement for phosphorylation of the glucose molecule. In fact, the enzymatic phosphorylation of glucose by ATP is the first reaction that glucose undergoes upon entering the cell. This reaction promotes glucose to a higher energy level, from which it may be indirectly incorporated into glycogen as stored carbohydrate or systematically oxidized for energy (p. 83). Phosphorylation can therefore be viewed as occurring in two reaction steps:

1. the hydrolysis of ATP to ADP and phosphate, and
2. addition of the phosphate to the substrate (glucose) molecule.

A *net energy change for the two reactions coupled together* can be expressed as follows:

$$\text{(Hydrolysis) ATP} \longrightarrow \text{ADP} + P_i$$
$$\Delta G^{0'} = -7{,}300 \text{ cal/mol}$$

$$\text{(P-addition) glucose} + P_i \longrightarrow \text{glucose 6-P}$$
$$\Delta G^{0'} = +3{,}300 \text{ cal/mol}$$

Therefore,

$$\text{ATP} + \text{glucose} \longrightarrow \text{ADP} + \text{glucose 6-P}$$
$$\Delta G^{0'} = -4{,}000 \text{ cal/mol}$$
$$\text{(the coupled reaction)}$$

It is important to understand the significance of these coupled reactions. They show that even though energy is consumed in the endothermic formation of glucose 6-phosphate from glucose and phosphate, the energy released by the ATP hydrolysis is sufficient to force (or drive) the endothermic reaction that "costs" only 3,300 cal/mol. The coupled reactions result in 4,000 cal/mol left over. The reaction is catalyzed by the enzymes hexokinase or glucokinase, both of which hydrolyze the ATP and transfer the phosphate group to glucose. The enzyme brings the ATP and the glucose into close proximity to each other, reducing the activation energy of the reactants and facilitating the phosphate group transfer. The overall reaction, which depicts the activation of glucose at the expense of ATP, is energetically favorable, as evidenced by its high, negative standard free energy change.

Formation of ATP

In the preceding discussion, it was shown that certain molecules can be activated by phosphate group transfer from ATP. The phosphorylation itself is an endothermic reaction but is made possible by the highly exothermic hydrolysis of the terminal phosphate of ATP. From this, it can be seen that the ADP produced by the reaction must be reconverted by phosphorylation back to ATP to maintain the homeostatic concentration of cellular ATP. But how is this accomplished, considering the large amount of energy ($\Delta G^{0'} = +7{,}300$ cal/mol) required for the reaction?

Obviously there must be outside sources of considerable energy that can be linked to the phosphorylation of ADP. Actually, two such mechanisms function in this respect:

- *substrate-level phosphorylation* and
- *oxidative phosphorylation.*

From the standpoint of the amount of ATP produced, oxidative phosphorylation is decidedly the more important of the two mechanisms.

Substrate-Level Phosphorylation

As discussed previously, phosphorylated molecules have a wide range of free energies of hydrolysis of their phosphate groups. Many of them release less energy than ATP, but some release more. The $\Delta G^{0'}$ of hydrolysis of the compounds listed in Table 3.1 is termed the *phosphate group transfer potential,* which is a measure of the compounds' capacities to donate phosphate groups to other substances. The more negative the transfer potential, the more potent the phosphate-donating power. Therefore, a compound that releases more energy on hydrolysis of its phosphate can transfer that phosphate to an acceptor molecule having a relatively more positive transfer potential. For this to occur in actuality, however, there must be a specific enzyme to catalyze the transfer. Therefore, just as a phosphate group can be enzymatically transferred from ATP to glucose, it can be predicted from Table 3.1 that from compounds that have a more negative phosphate group transfer potential than ATP, phosphate can be transferred to ADP, forming ATP. This does, in fact, occur in metabolism. The phosphorylation of ADP by phosphocreatine, for example, represents an important mode for ATP formation in muscle, and the reaction exemplifies a substrate-level phosphorylation.

$$\text{Phosphocreatine} + \text{ADP} \longrightarrow \text{Creatine} + \text{ATP}$$
$$\Delta G^{0'} = -3{,}000 \text{ cal/mol}$$

Oxidative Phosphorylation

This mechanism is the major means by which ATP is formed from ADP. The energy required to do this is tapped from a pool of energy generated by the flow of electrons from a substrate molecule undergoing oxidation. The electrons are then passed through a series of intermediate compounds, ultimately to molecular oxygen, which becomes reduced to H_2O in the process. The compounds participating in this sequential reduction-oxidation constitute the *respiratory chain,* so named because the electron transfer is linked to the uptake of O_2, which is made available to the tissues by respiration. The *electron transport chain* is a more commonly used alternate term and is used throughout this text. The chain functions within the cell mitochondria, and reference to these organelles as the power plants of the cell is founded on the large amount of energy liberated by electron transport (p. 6). Although most of this energy assumes the form of heat to maintain body temperature, much of it is used to form ATP from ADP and inorganic phosphate. Therefore, the term *oxidative phosphorylation* is a descriptive blend of two processes operating simultaneously:

1. the oxidation of a metabolite by O_2 via electron transport and
2. the phosphorylation of ADP.

Cellular oxidation of nutrients, electron transport, and oxidative phosphorylation therefore are unified in function and must be thought of as such. They are considered next in more detail.

Biological Oxidation and the Electron Transport Chain

Oxidation of the energy nutrients is what releases the inherent energy of the energy nutrient molecules. The intimate association of energy release with oxidation is exemplified by the oxidation of the fatty acid palmitate shown in Figure 3.2. Cellular oxidation of a compound can occur by different reactions: the addition of oxygen, the removal of electrons, and the removal of hydrogens. All these reactions are catalyzed by enzymes collectively termed *oxidoreductases* (p. 18). Among these, the *dehydrogenases,* which remove hydrogens and electrons from nutrient metabolites, are particularly important in energy transformation. This is because the hydrogens and electrons removed from metabolites by dehydrogenase reactions pass along the components of the electron transport chain and cause the release of large amounts of energy. In reactions in which oxygen is incorporated into a compound, or hydrogens are removed by other than dehydrogenases, the electron transport chain is not called into play and no energy is released. Such reactions are catalyzed by a subgroup of oxidoreductase enzymes generally referred to as *oxidases* and are not considered further in this section.

The hydrogens removed from a substrate molecule by a dehydrogenase enzyme are transferred to a cosubstrate, such as the vitamin-derived nicotinamide adenine dinucleotide (NAD^+) or flavin adenine dinucleotide (FAD). As a result of this transfer, the metabolite molecule, in its reduced form, which will be designated MH_2,

becomes oxidized to M, while the oxidized forms of the cosubstrates (either NAD^+ or FAD) become reduced to NADH and $FADH_2$, respectively, as a result of their acceptance of the hydrogens and electrons. The hydrogens and electrons are then enzymatically transferred through the electron transport chain components and eventually to molecular oxygen, which becomes reduced to H_2O. Using the example of NAD^+ as hydrogen acceptor, the overall reaction can be summarized as in 3A. The energy given off is derived from the sequence of individual reduction-oxidation (redox) reactions along the electron transport chain, each component having a characteristic ability to donate and accept electrons. The released energy is used in part to synthesize ATP from ADP and phosphate.

(3A)

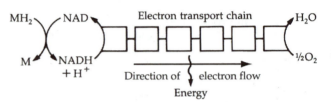

The tendency of a compound to donate and to receive electrons is expressed in terms of its standard reduction potential, $E_{0'}$. More negative values of $E_{0'}$ reflect a greater ability of the compound to donate electrons to another, while increasingly positive values signify an increasing tendency to accept electrons. The reducing capacity of a compound can be expressed by the $E_{0'}$ value of its half reaction, which indicates the compound's electromotive potential of electron donation.

Free energy changes accompany the transfer of electrons between electron donor/acceptor pairs of compounds and are related to the measurable electromotive force of the electron flow. It is important to remember that *in electron transfer, an electron donor reduces the acceptor and in the process becomes oxidized. Consequently, the acceptor, as it is reduced, oxidizes the donor.* The quantity of energy released is directly proportional to the difference in the standard reduction potentials, $\Delta E_{0'}$, between the partners of the redox pair. The free energy of a redox reaction and the $\Delta E_{0'}$ of the interacting compounds are related by the expression

$$\Delta G^{0'} = -n\mathrm{F}\,\Delta E_{0'}$$

where $\Delta G^{0'}$ is the standard free energy change in calories, n is the number of electrons transferred (in cellular oxidation, assumed to be two), and F is a constant called the faraday (23,062 cal absolute volt equivalent). An example of a reduction-oxidation reaction that occurs within the electron transport system is the transfer of hydrogen atoms and electrons from NADH through the flavin mononucleotide (FMN)-linked enzyme called NADH dehydrogenase to oxidized coenzyme Q (CoQ). The half-reactions and $E_{0'}$ values for each of these reactions are

$$NADH + H^+ \longrightarrow NAD^+ + 2H^+ + 2e^-$$
$$E_{0'} = -0.32 \text{ volts}$$

$$CoQH_2 \longrightarrow CoQ + 2H^+ + 2e^-$$
$$E_{0'} = +0.04 \text{ volts}$$

Because the NAD^+ system has a relatively more negative $E_{0'}$ value than the CoQ system, it follows from the preceding discussion that it has a greater reducing potential than the CoQ system, because electrons tend to flow toward the more positive system. The reduction of CoQ by NADH is therefore predictable, and the coupled reaction, linked by the FMN of NADH dehydrogenase, can be written

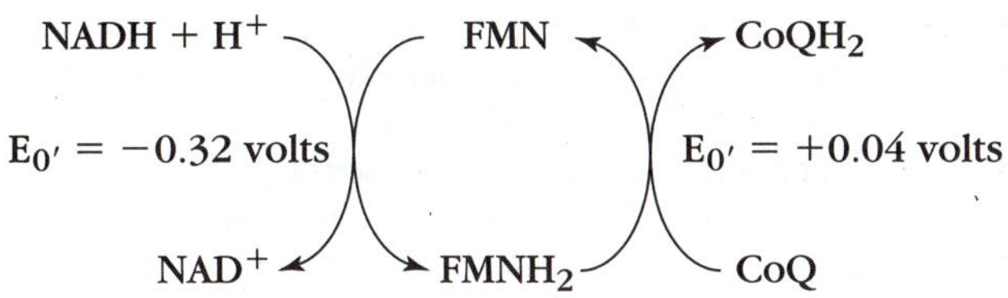

$$\Delta E_{0'} = 0.36 \text{ volts}$$

Inserting this value for $\Delta E_{0'}$ into the energy equation:

$$\Delta G^{0'} = -2(23,062)(0.36) = -16,604 \text{ cal/mol}$$

The amount of energy liberated from this single reduction-oxidation reaction within the electron transport

chain is therefore more than enough to phosphorylate ADP to ATP, which, it will be recalled, requires approximately 7,300 cal/mol. Before considering the electron transport components themselves, it is worthwhile to review the importance of the dehydrogenase reactions that deliver hydrogens to NAD^+ and FAD from various substrates undergoing oxidation. Such reactions are common within the pathways of intermediary metabolism, particularly the energy-rich citric acid (Krebs) cycle, which is detailed in Chapter 4. Two dehydrogenase reactions occurring in the Krebs cycle are offered here as examples of cosubstrate reduction linked to substrate (metabolite) oxidation. The reactions are catalyzed by the enzymes malate dehydrogenase (MDH) and succinate dehydrogenase (SDH) (see 3B).

If dehydrogenation is linked to NAD^+, hydrogens and electrons flow from the substrate undergoing oxidation through the NAD^+ and then the FMN systems. The next component in the sequence is an *iron sulfur center*, which collects only electrons from the FMN system, with the hydrogens being released as protons. The electrons are then passed on to the coenzyme Q (CoQ) system, which also regains two protons, equivalent to the two released at the iron sulfur center. Coenzyme Q is also referred to as *ubiquinone*. Precisely where, within their molecular structure, these substances accept and release hydrogens and electrons is shown by their reduced and oxidized forms in Figures 3.7, 3.8, and 3.9.

The components functioning after CoQ are called *cytochromes,* which are iron-containing, electron-transferring proteins acting sequentially to carry electrons (only) from CoQ to molecular oxygen. The hydro-

(3B)

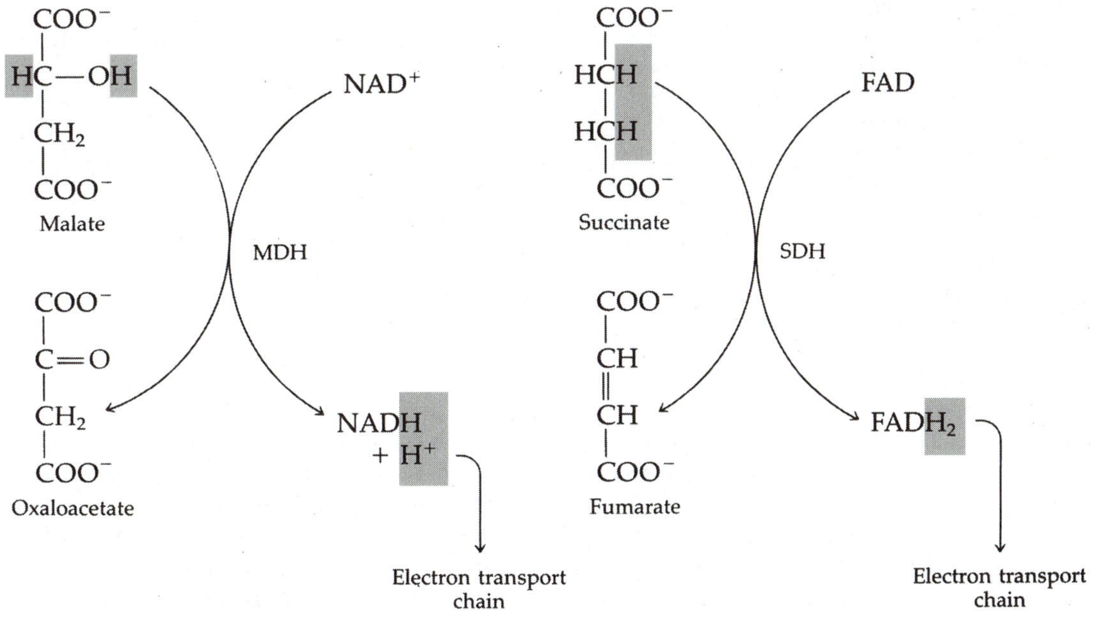

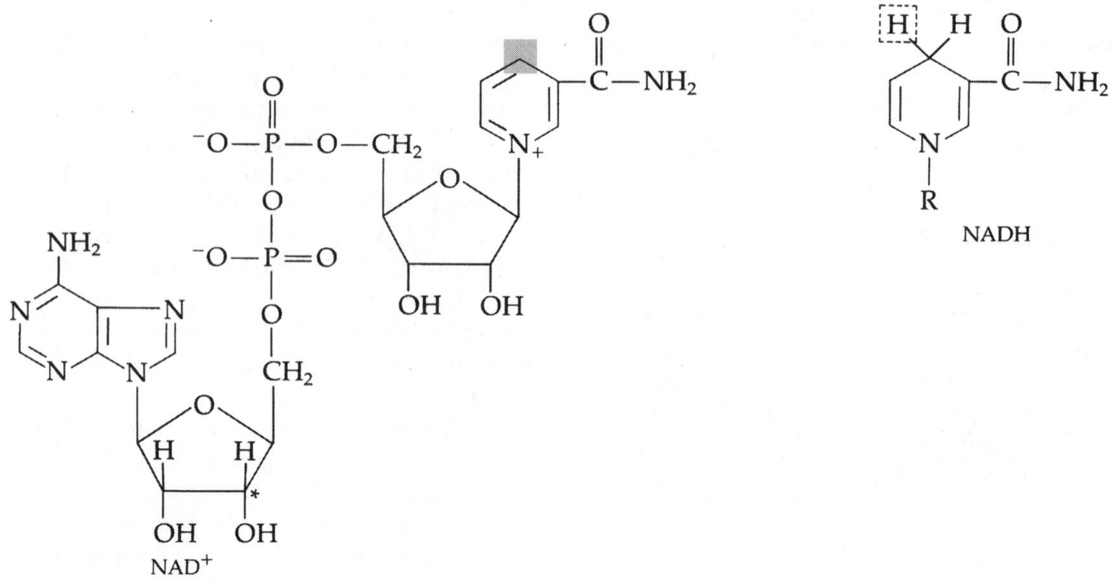

Figure 3.7 Nicotinamide adenine dinucleotide (NAD$^+$) and its reduced form (NADH). The shaded carbon atom is the site of attachment of one of the two hydrogen atoms transferred from a metabolite undergoing dehydrogenation. The hydrogen is actually acquired as a hydride ion (:H$^-$) that attaches to the nicotinamide ring as shown and neutralizes the positive charge on the ring nitrogen. The remainder of the structure, indicated by the symbol R, is unchanged from that of NAD$^+$. The second hydrogen transferred from the metabolite is released as a proton. Another dehydrogenase cosubstrate, NADP, differs from NAD$^+$ only in the attachment of a phosphate group at the position shown by the asterisk.

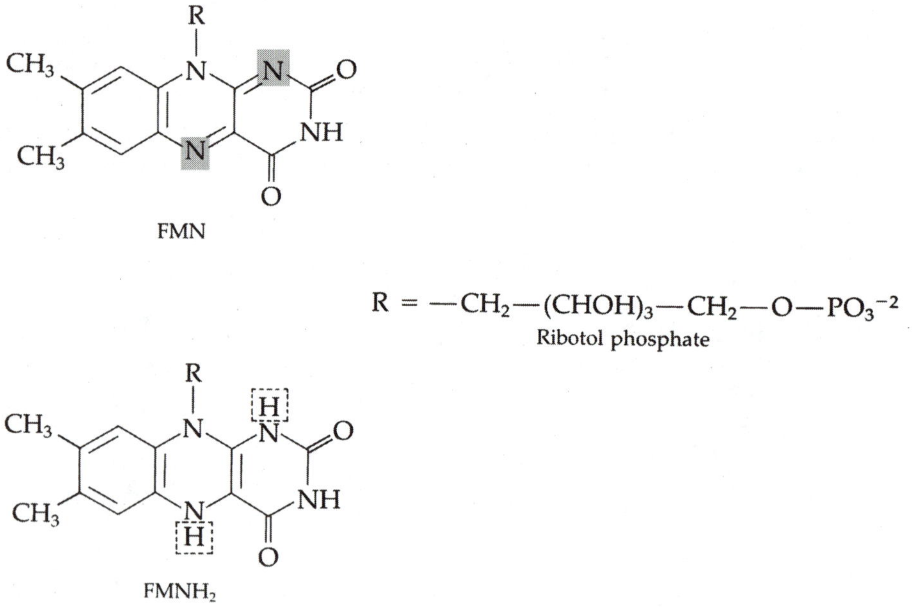

Figure 3.8 Flavin mononucleotide (FMN) and its reduced form (FMNH$_2$). Hydrogens transferred from NADH and H$^+$ attach to the nitrogen atoms shaded in the structures. FMN is the phosphorylated derivative of the vitamin riboflavin (B$_2$). The cosubstrate FAD differs from FMN in that the ribitol phosphate side chain is linked to adenosine monophosphate (AMP) via a pyrophosphate bond. The site of hydrogen atom attachment to FAD is the same as with FMN.

$$R = —CH_2—(CHOH)_3—CH_2—O—PO_3^{-2}$$
Ribotol phosphate

gens of reduced CoQ are released as protons, and therefore no hydrogen transfer takes place within the cytochrome segment of the chain. The protons collected by CoQ are released into the mitochondrial matrix and ultimately combine with O$_2$ to form H$_2$O, as shown later. Specifically, the cytochromes are heme proteins in which the iron is bound to porphyrin, forming the heme moiety such as is found in hemoglobin (p. 167). The heme group is a red-brown pigment, and because of the high concentration of cytochromes in the mitochondrion, this coloration is imparted to that organelle. Cytochromes designated as cyt a-a3, cyt b, cyt c1, and cyt c function as

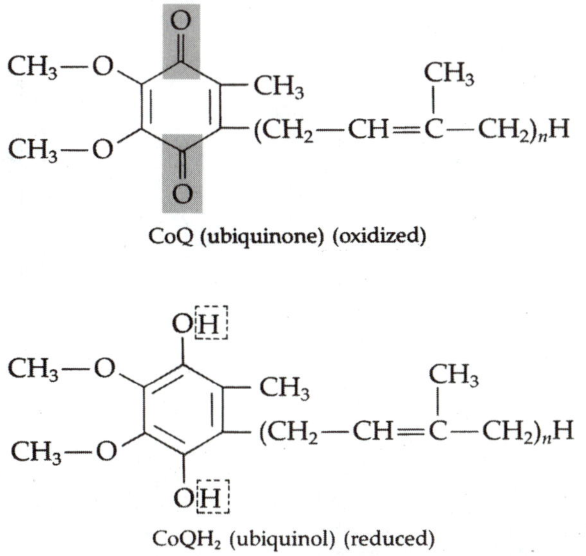

CoQ (ubiquinone) (oxidized)

CoQH$_2$ (ubiquinol) (reduced)

Figure 3.9 Oxidized and reduced forms of coenzyme Q, or ubiquinone. The shaded groups function in the transfer of hydrogen atoms. The subscript n indicates the number of isoprenoid units in the side chain (most commonly 10).

electron carriers, and although subtle structural differences differentiate them, they share the heme moiety. Among them, the structure of cyt c was the first to be defined and is shown in Figure 3.10. The reduction-oxidation of the cytochromes occurs at the iron atom, which alternates between the ferric (Fe^{+3}) and ferrous (Fe^{+2}) forms as it releases and acquires electrons.

$$Fe^{+2} \rightarrow Fe^{+3} + e^-$$

The iron sulfur centers, or iron sulfur proteins, referred to earlier also function as electron carriers in the electron transport chain. The iron in these compounds is not in the form of heme, as it is in the cytochromes, but is associated with inorganic sulfur atoms and/or the sulfur atoms in the cysteine amino acid residues of the protein. Two arrangements in which these iron-sulfur centers exist are shown in Figure 3.11.

All the components of the electron transport chain can be thought of as being compartmentalized into four different complexes, each of which represents a portion of the electron transport chain. Each complex has its own unique composition and catalyzes the transfer of electrons through that portion of the chain. Complexes I and II catalyze electron transfer to CoQ from two different electron donors: NADH (Complex I) and succinate, through FADH$_2$ (Complex II). Complex III carries electrons from CoQ to cytochrome c, and Complex IV completes the sequence by transferring electrons from cytochrome c to O$_2$. Figure 3.12 shows the sequential order of the electron transport intermediates and their division into the four component complexes. The flow of electrons from NADH to O$_2$ is "downhill," meaning that each component has a more negative $E_{0'}$ than its partner on the O$_2$ side and that energy is therefore released at each step in the sequence. However, the $\Delta E_{0'}$ between the redox pairs varies considerably, and, accordingly, so does the quantity of energy released. The $\Delta G^{0'}$ for each reaction site,

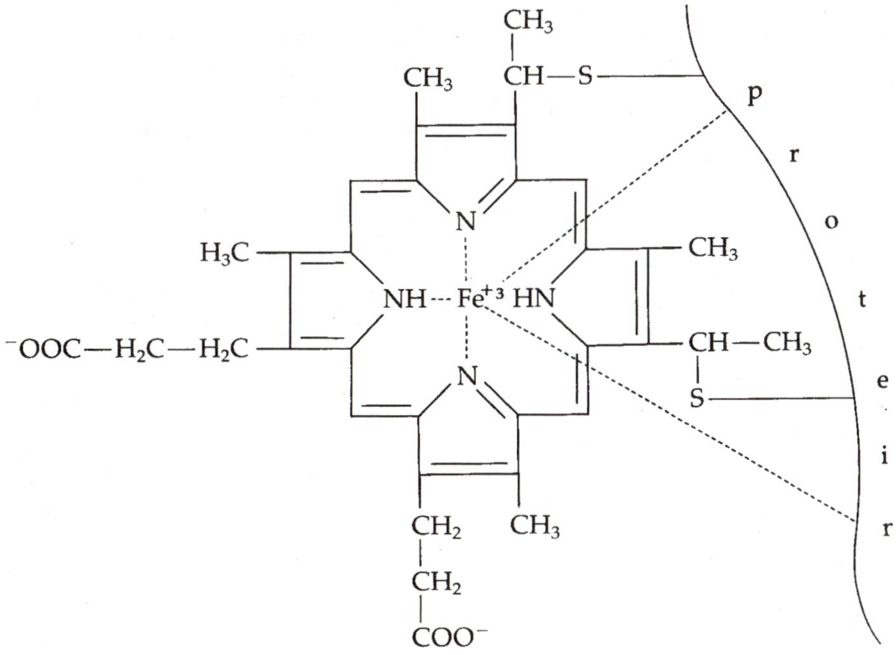

Figure 3.10 Oxidized cytochrome c. In its reduced form, the iron exists as Fe^{+2}.

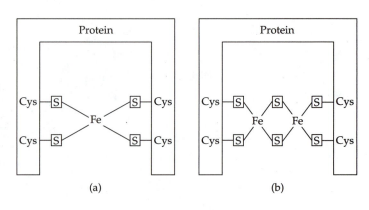

Figure 3.11 A schematic representation of the binding of iron ions in the iron-sulfur centers of the mitochondrial electron transport system. Sulfur atoms can be covalently attached to the protein via cysteine amino acid residues (a), or they can, in addition, be free, as shown in structure (b).

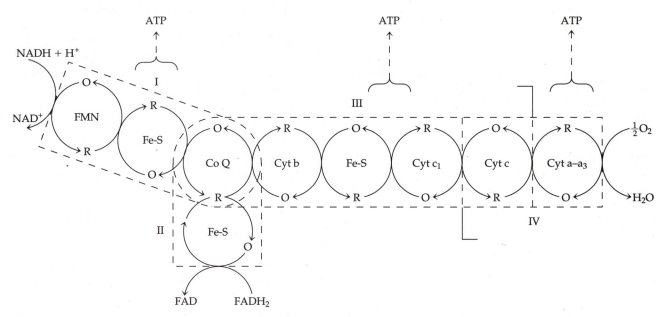

Figure 3.12 The sequential arrangement of the components of the electron transport chain, showing its division into four complexes, I, II, III, and IV. Coenzyme Q (ubiquinone) is shared by Complexes I, II, and III. Cyt c is shared by Complexes III and IV. Reduced cosubstrates, NADH and $FADH_2$, released in the matrix by dehydrogenase enzymes, are ultimately reoxidized by molecular oxygen as the intermediates undergo reversible reduction (R) and oxidation (O). Also indicated are the sites in Complexes I, III, and IV where electron transfers between reduction-oxidation pairs furnish enough energy for the synthesis of ATP from ADP.

calculated from the corresponding $\Delta E_{0'}$ values is shown in Table 3.2.

There are three sites at which the release of energy is sufficient to effect the phosphorylation of ADP, meaning that *for each molecule of NADH reoxidized to NAD^+ by O_2, via the electron transport chain, three ATPs are generated.* These sites of oxidative phosphorylation are indicated in Table 3.2 and also Figure 3.12. It is important to point out that the number of ATPs produced by oxidative phosphorylation is less if the dehydrogenase reaction uses the flavin cosubstrate FAD rather than NAD^+. This is because CoQ collects

Table 3.2 Free Energy Changes at Various Sites within the Electron Transport Chain Showing Phosphorylation Sites

Reaction	$\Delta G^{0'}$ (cal/mol)	ADP Phosphorylation
$NAD^+ \longrightarrow$ FMN	−922	
FMN $\longrightarrow$ CoQ	−15,682	ADP + P $\longrightarrow$ ATP
CoQ $\longrightarrow$ cyt b	−1,380	
cyt b $\longrightarrow$ cyt c1	−7,380	ADP + P $\longrightarrow$ ATP
cyt c1 $\longrightarrow$ cyt c	−922	
cyt c $\longrightarrow$ cyt a	−1,845	
cyt a $\longrightarrow$ ½O_2	−24,450	ADP + P $\longrightarrow$ ATP

electrons directly from FADH$_2$ through an iron-sulfur center, thereby bypassing the NAD$^+$ → FMN → CoQ phosphorylation site. Therefore, the FAD-requiring succinate dehydrogenase reaction generates only two ATPs rather than three.

The components of the electron transport chain are positioned within the inner membrane of the mitochondrion. They are opportunistically situated to facilitate the uptake of electrons from substrates and deliver them to molecular oxygen. Complex I, for example, is oriented so that it can easily interact with the NADH produced by any of the dehydrogenases active in the matrix. The positioning of Complex IV (also called *cytochrome oxidase*) facilitates its picking up of electrons from Complex III and transferring them on to O$_2$ on the matrix side of the membrane. Cytochrome c, in the intermembrane space, shuttles the electrons from Complex III to Complex IV. The spatial arrangement of the electron transport chain is illustrated schematically in Figure 3.13. Notice that electron flow through Complexes I, III, and IV is accompanied by the translocation of protons from the matrix into the intermembrane space.

The difference in the $E_{0'}$ values between the NAD$^+$−NADH and the ½O$_2$−H$_2$O systems makes it possible to calculate the overall change in free energy across the entire electron transport chain:

$$\Delta E_{0'} = E_{0'}(O_2) - E_{0'}(NAD^+)$$
$$= 0.82 - (-0.32) = 1.14 \text{ volts}$$

then

$$\Delta G^{0'} = -2(23,062)(1.14) = -52,581 \text{ cal/mol}$$

Under the standard conditions, 21,900 cal/mol (3 ×7,300) of this total energy is conserved for future use as ATP, while the remaining 30,681 calories, representing 60% of the total, is in the form of heat that is necessary to help to maintain a normal body temperature.

The precise mechanism by which the energy from electron transport is used to synthesize ATP is not entirely understood, but the preferred explanation is the *chemiosmotic theory*, a brief overview of which follows.

Energy released by the downhill (lower to higher $E_{0'}$ values) flow of electrons in Complexes I, III, and IV is used to pump protons against a concentration gradient, from the matrix into the intermembrane space. This produces a marked disparity of both proton concentration and electrical charge on either side of the membrane. Permeating the membrane are proton channels, through which protons passively diffuse back into the matrix from the intermembrane space. The driving force for the diffusion is the large difference in proton concentration in the two compartments. The channels are constructed from protein aggregates and exist as two distinct sectors designated F$_1$ and F$_0$. F$_1$ is the catalytic sector, residing on the matrix side of the membrane, whereas F$_0$ is the membrane sector, involved primarily with proton translocation. Together, these components comprise what is called the *F$_0$F$_1$ ATPase aggregate,* also referred to as *Complex V.*

The return flow of protons furnishes the energy necessary for the synthesis of ATP from ADP and P$_i$. The proton flow is directed by the F$_0$ sector to the F$_1$ head-

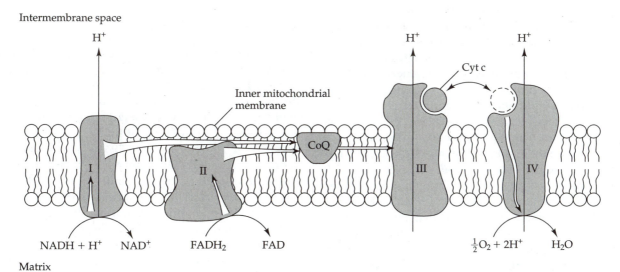

Figure 3.13 The spatial orientation of the complexes of the electron transport chain in the inner membrane of the mitochondrion. The energy of electron flow from reduced cosubstrates (NADH and FADH$_2$) to oxygen, within Complexes I, III, and IV, pumps protons from the matrix to the intermembrane space against a concentration gradient.

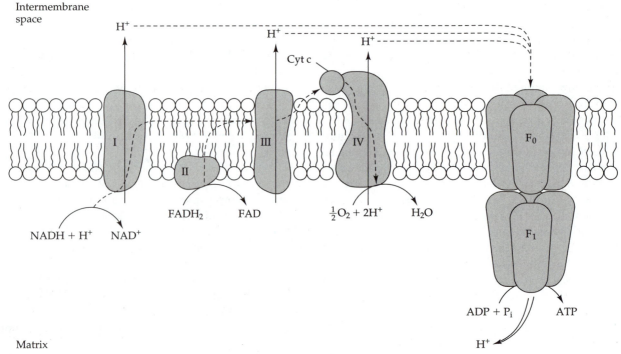

Figure 3.14 An illustration of the chemiosmotic theory of oxidative phosphorylation. Energy from electron transport pumps protons into the intermembrane space from the matrix against a concentration gradient. The passive diffusion of protons back into the matrix through the F_0F_1 ATP-synthase aggregate furnishes the energy to synthesize ATP from ADP and inorganic phosphate. For more details, see the text.

piece, which has binding sites for ADP and P_i. One oxygen atom of the inorganic phosphate is believed to react with two of the energetic protons, eliminating H_2O from the molecule. The precise mechanism of phosphorylation is complicated and not fully understood. It is believed, however, that the energy of oxidative phosphorylation is needed not for the actual coupling of ADP and P_i but for the binding of the substrates (ADP and P_i) to the enzymatic binding sites of the F_1 sector and the subsequent release from the sites of the ATP formed. Earlier, it was stated that the energy available from the oxidation of NADH ($+ H^+$) is sufficient to produce 3 ATPs by the phosphorylation of 3 ADPs. However, the substrates for ATPase—ADP and P_i, must be actively transported into the mitochondrion from the cytoplasm, and the product, ATP, must be actively transported out. The net amount of energy available for ADP phosphorylation is therefore somewhat less, reduced by the quantity of energy that was consumed in the active transport process. It is estimated that the number of ATPs formed from NADH oxidation is therefore more realistically 2.5 rather than 3, and from $FADH_2$, it is 1.5 rather than 2. Throughout this text, however, we will adhere to the convention of ascribing the values of 3 and 2 to the number of moles of ATP produced from the oxidation of molar amounts of NADH and $FADH_2$ respec-

tively. A schematic representation of the chemiosmotic theory of oxidative phosphorylation is shown in Figure 3.14. Comprehensive reviews of electron transport, oxidative phosphorylation, and chemiosmotic coupling are available to the interested reader [1,2,3].

Summary

This chapter has dealt with a subject of vital importance in nutrition: the conversion of the energy contained within nutrient molecules into energy usable by the body. The major portion of this energy is needed to maintain normal body temperature, but much of it is also conserved in the form of high-energy phosphate bonds, principally ATP. The ATP can, in turn, activate various substrates by phosphorylation to higher energy levels from which they can undergo metabolism by specific enzymes. The exothermic hydrolysis of the ATP phosphate is sufficient to drive the endothermic phosphorylation, thereby completing the energy transfer from nutrient to metabolite.

ATP can be generated by two distinct mechanisms:

1. the transfer of a phosphate group from a very high-energy phosphate donor to ADP, a process called substrate-level phosphorylation, and

2. oxidative phosphorylation, by which the energy derived from mitochondrial electron transport is used to unite ADP with an active phosphate.

Oxidative phosphorylation is the major route for ATP production. Electron flow in the electron transport chain is from reduced cosubstrates to molecular oxygen, which therefore becomes the ultimate oxidizing agent and which becomes reduced to H_2O in the process. The downhill flow of electrons generates sufficient energy to effect oxidative phosphorylation at multiple sites along the chain.

Chapter 4 focuses on carbohydrate metabolism, including the energy-releasing, systematic oxidation of glucose to CO_2 and H_2O. Within this pathway, reactions exemplifying both substrate-level and oxidative phosphorylation are encountered. The reader is encouraged to apply the principles of energy transformation learned in this chapter to the metabolic pathways of the energy nutrients in chapters that follow. More specifically, it should be recalled that electron transport is called into play whenever a mitochondrial dehydrogenation reaction occurs. It is important to understand the link between the controlled oxidation of the energy nutrients and how the released energy is ultimately used. Reference to this chapter will be made frequently to strengthen this important union of cellular processes.

References Cited

1. Trumpower B and Gennis R. Energy transduction by cytochrome complexes in mitochondrial and bacterial respiration: the enzymology of coupling electron transfer reactions to transmembrane proton translocation. Annu Rev Biochem 1994;63:675–702.

2. Tyler D. ATP Synthesis in Mitochondria. In: The Mitochondrion in Health and Disease. VCH Publishers, Inc., New York 1992: pp.353–402.

3. Hatefi Y. The mitochondrial electron transport and oxidative phosphorylation system. Annu Rev Biochem 1985; 54:1015–69.

Suggested Reading

Jequier E, Acheson K, Schutz Y. Assessment of energy expenditure and fuel utilization in man. Ann Rev Nutr 1987;7:187–208.
   The assessment of nutrient energy use is emphasized in this review of energy expenditure.
Lehninger AL, Nelson DL, Cox MM. Oxidative Phosphorylation and Photophosphorylation. In: Principles of Biochemistry, 2nd ed. New York: Worth, 1993: Chap. 18.
   This is a clearly written and graphically illustrated survey of energy transduction in the mitochondrion.

Web Sites

www.nlm.nih.gov
   National Library of Medicine: MEDLINE
www.med.jhu.edu
   John Hopkins InfoNet

If Oxidation and Phosphorylation Disconnect: A Waste of Energy!

Revisiting the highlights of Chapter 3, energy transformation in the cell has two major components:

- *Oxidation:* removing electrons (and hydrogens) from a substrate molecule and passing them along the components of the electron transport chain to O_2.

- *Phosphorylation:* harnessing the energy released by the electron flow to synthesize ATP from ADP and P_i and applying "leftover" energy (better than 50% of the total energy released) toward the maintenance of normal body temperature.

According to the chemiosmotic theory of ATP synthesis, a substantial disparity of positive electrical charge exists in the matrix and intermembrane space of the mitochondrion. This is the result of proton pumping from the matrix across the inner membrane into the intermembrane space, the energy for which comes from the electron transport (oxidation) component. But the energy used directly to couple ADP with P_i comes from the return, passive coursing of the protons through the F_0F_1 ATPase channels as they move "down" the concentration gradient back into the matrix compartment.

What if, for any reason, a disruption of the proton gradient in the mitochondrion should arise, so that the difference in proton concentrations between the two compartments is reduced? The consequence would be a reduction in the passive flow of protons through the channels, therefore impeding the synthesis of ATP. In fact, certain chemicals will disrupt the proton gradient. They are called *uncouplers* of oxidative phosphorylation because although they do not interfere with electron flow (oxidation), they do hinder ATP synthesis (phosphorylation). In other words, they uncouple oxidation from phosphorylation.

Certain weak organic acids act as chemical uncouplers. Probably the one studied most thoroughly and the one first identified as an uncoupler is 2,4-dinitrophenol. It is shown here in its acidic (protonated) form and its basic (dissociated) form:

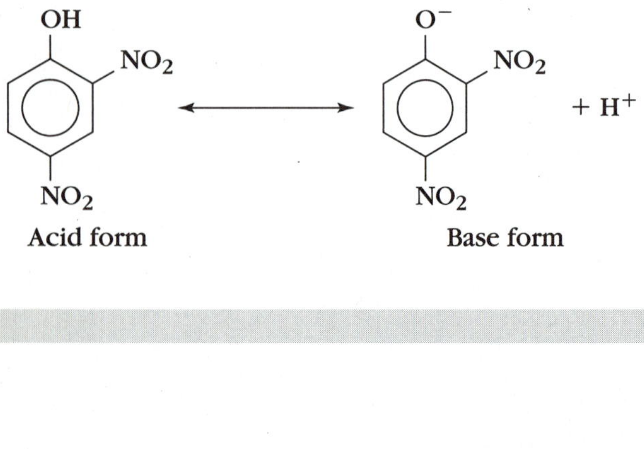

Acid form   Base form

2,4-Dinitrophenol is not of particular interest from a nutritional or medicinal standpoint, however, because it is neither a dietary supplement nor a drug. Therefore, the remainder of this discussion focuses on another, more familiar uncoupler to which any reader of this text can relate, salicylic acid. Salicylic acid is the major metabolite of the wonder drug, aspirin (acetyl salicylic acid). As a weak acid, salicylic acid, like 2,4-dinitrophenol, exists in protonated and dissociated forms.

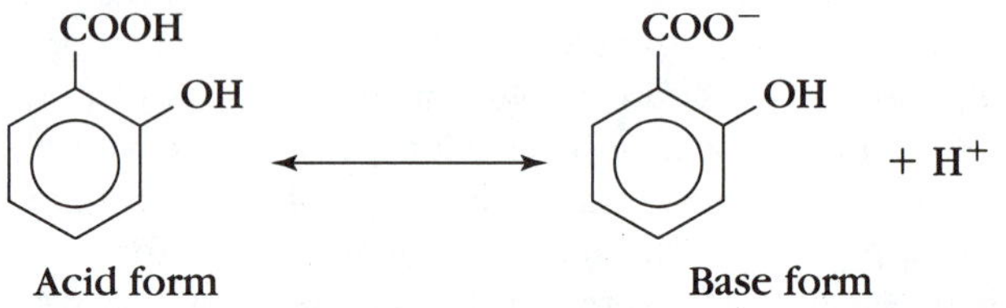

Acid form   Base form

In an environment of low pH (high proton concentration), the basic form of a weak acid acquires a proton and is converted to the acid form. The acid form, in a region of higher pH (low proton concentration), releases its proton and reverts to its basic form. This establishes the premise of how salicylic acid can perturb the delicate proton gradient across the mitochondrial inner membrane. The compound enters the mitochondrion through the outer membrane of the organelle where it encounters the low pH of the intermembrane space. There, it acquires a proton and assumes its acidic, *uncharged* form. As such, it is quite lipid soluble and can readily permeate membranes. The protonated salicylic acid passes freely through the inner membrane, carrying with it, of course, the proton it acquired in the intermembrane space. Upon entering the high pH environment of the matrix, the proton is released as the compound reverts to its basic form. Therefore, the salicylic acid, our example of a chemical uncoupler, acts as a carrier of protons from the intermembrane space across the inner membrane and into the matrix. This is illustrated in Figure 1.

The uncoupler's effect of reducing the disparity of proton concentration on either side of the inner membrane reduces the proton flux through the ATPase channels, thereby inhibiting ATP synthesis. It is important to understand, however, that uncouplers do not interfere with electron transport, the source of energy for pumping protons in the first place. It will be recalled from Chapter 3 that approximately 40% of the total energy released by electron transport is consumed in the synthesis of ATP, the remaining 60% manifesting as heat in the maintenance of body temperature. In the wake of the uncoupler effect, which denies ATP synthesis, essentially all oxidation energy manifests as

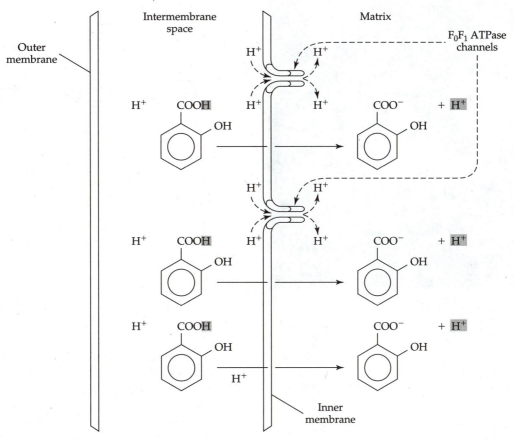

Figure 1 The diffusion of the protonated (acid) form of salicylic acid through the mitochondrial inner membrane and the release of its protons in the matrix. The process reduces the proton concentration gradient across the membrane. This slows the passage of protons through the ATPase channels and interferes with ATP synthesis.

heat. The result is hyperpyrexia (high body temperature)—indeed, a waste of energy and, more important, dangerous physiologically.

It must be emphasized that the hyperpyretic effect of aspirin is incurred only at very high, toxic doses of the drug. However, there is the interesting dichotomy associated with aspirin intake. At normal, therapeutic doses, aspirin acts very effectively as an antipyretic, reducing infection-associated fever. On the other hand, as explained in this Perspective,

toxic levels exhibit the opposite effect. Aspirin overdosing is not common among adults, but it is an all-too-common occurrence with toddlers attracted to its candylike appearance and texture.

Web Sites

www.nlm.nih.gov
National Library of Medicine: MEDLINE

Starch granules in a potato cell

Carbohydrates

The major source of energy fuel in the average human diet is carbohydrate, supplying nearly half of the total caloric intake. Roughly half of dietary carbohydrate is in the form of polysaccharides such as starches and dextrins, derived largely from cereal grains and vegetables. The remaining half is supplied as simple sugars, the most important of which include sucrose, lactose, and, to a lesser extent, maltose, glucose, and fructose.

Carbohydrates are polyhydroxy aldehydes, or ketones, or substances that produce such compounds when hydrolyzed. They are constructed from the atoms of carbon, oxygen, and hydrogen, occurring in proportion that approximates that of a "hydrate of carbon," CH_2O, therefore accounting for the term *carbohydrate.* They exist in three major classes: the monosaccharides, the oligosaccharides, and the polysaccharides.

• *Monosaccharides* are structurally the simplest form of carbohydrate, in that they cannot be reduced in size to smaller units by hydrolysis. For this reason they are sometimes referred to as *simple sugars.* The most abundant monosaccharide in nature and certainly the most important nutritionally is the 6-carbon sugar glucose.

• *Oligosaccharides* consist of short chains of monosaccharide units joined by covalent bonds. The number of units is designated by the prefixes *di-, tri-, tetra-,* and so on, followed by the word *saccharide.* Among the oligosaccharides, the disaccharides, having just two monosaccharide units, are the most abundant. Within this group, sucrose, consisting of a glucose and a fructose residue, is nutritionally most significant, furnishing approximately one-third of total dietary carbohydrate in an average diet.

• *Polysaccharides* are long chains of monosaccharide units that may number from several into the hundreds or even thousands. The major polysaccharides of interest in nutrition are glycogen, found in certain animal tissues, and starch and cellulose, which are both of plant origin. All these polysaccharides consist of repeating units of glucose only.

Structural Features

Monosaccharides and Stereoisomerism

As they occur in nature, or arise as intermediate products in digestion, these simple sugars contain from three to seven carbon atoms and are accordingly termed *trioses, tetroses, pentoses, hexoses,* and *heptoses.* In addition to hydroxyl groups, these compounds possess a functional carbonyl group that can be either an aldehyde or a ketone, therefore necessitating the additional classification of aldoses, those sugars having an aldehyde group, and ketoses, those possessing a ketone. These two classifications are used together with the number of carbon atoms in describing a particular monosaccharide. For example, a five-carbon sugar having a ketone group is called a *ketopentose;* a six-carbon, aldehyde-possessing sugar is an *aldohexose;* and so forth.

Many organic substances, including the carbohydrates, are optically active, which means that if plane-polarized light is passed through a solution of the substances, the plane of light will be rotated to the left (such substances are called *levorotatory*) or to the right (*dextrorotatory*). The direction and extent of the rotation is characteristic of a particular compound and is measurable by an instrument called a *polarimeter.*

The right or left direction of rotation is expressed as [+] (dextrorotatory) or [−] (levorotatory), and the number of angular degrees indicates the extent of rotation. As an example, a solution of glucose is known to have a rotation of +52.7°.

Optical activity is attributed to the presence of one or more asymmetrical or chiral carbon atoms in the molecule. Chiral carbon atoms have four different atoms or groups covalently attached to them. Shown in 4A is a chiral carbon atom with its bonds directed toward the hypothetical atoms or groups W, X, Y, and Z. It can be thought of as being situated in the geometric center of a tetrahedron, with its four bonds directed toward the corners of that tetrahedron. Because the molecule is asymmetrical, it is possible to construct a

similar but not identical figure simply by exchanging the positions of any two of the groups. The two models shown bear the same groups and therefore the same molecular formula, but they are not identical, as evidenced by the reader's inability to superimpose them by mental manipulation. They are, in fact, mirror images of each other and are said to be *enantiomers,* a special class within a broader family of compounds called *stereoisomers.* Stereoisomers are compounds having two or more chiral carbon atoms, which have the same four groups attached to those carbon atoms but are not mirror images of each other.

(4A)

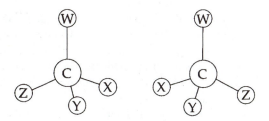

If an asymmetrical substance rotates the plane of polarized light a certain number of degrees to the right, its enantiomer will rotate the light the same number of degrees to the left. Enantiomers exist in D or L orientation, and if a compound is structurally D, its enantiomer will be L. The D and L designation does not predict the direction of rotation of plane-polarized light. Instead, the designation simply refers to structural analogy to the reference compound glyceraldehyde, whose D and L forms are by convention drawn as shown in 4B. The distinction between D and L configurations of enantiomers therefore rests with the direction of the —OH bond on the single, chiral carbon of the molecule. Notice that in the D form it points to the right, and in the L configuration to the left. Remember, these forms are not superimposable.

(4B)

$$HC=O \qquad\qquad HC=O$$
$$HC-OH \qquad\qquad HO-CH$$
$$H_2C-OH \qquad\qquad H_2C-OH$$

D-glyceraldehyde   L-glyceraldehyde

Monosaccharides having more than three carbons will have more than one chiral center, but only the highest numbered chiral carbon is the indicator as to whether the molecule is of the D or L configuration. The configuration is determined by the right or left direction of the hydroxyl group attached to that carbon atom. Carbon

(4C)

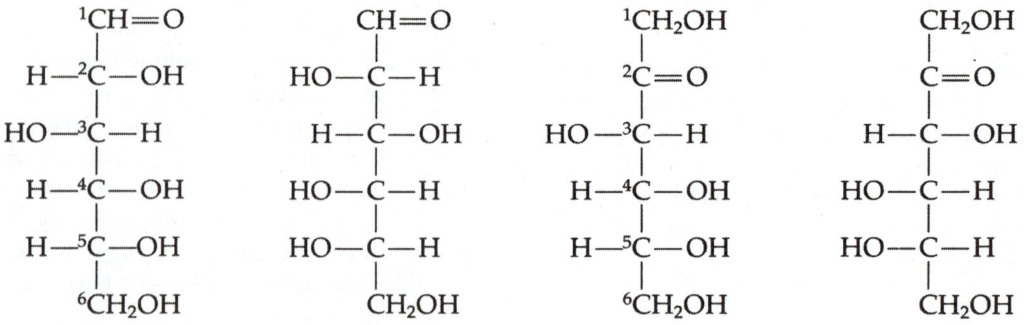

$$^1CH=O \qquad CH=O \qquad ^1CH_2OH \qquad CH_2OH$$

H—^{2}C—OH	HO—C—H	^{2}C=O	C=O
HO—^{3}C—H	H—C—OH	HO—^{3}C—H	H—C—OH
H—^{4}C—OH	HO—C—H	H—^{4}C—OH	HO—C—H
H—^{5}C—OH	HO—C—H	H—^{5}C—OH	HO—C—H
^{6}CH$_2$OH	CH$_2$OH	^{6}CH$_2$OH	CH$_2$OH

D-glucose · L-glucose D-fructose L-fructose

atom numbering and the D versus L designation are illustrated in the structures at the top of this page for the aldohexose glucose and the ketohexose fructose. Because the highest numbered asymmetrical carbon atom in these structures is number 5, this becomes the designator atom for either the D or L stereoisomer. Notice, however, that in going from the D to L isomer, the direction of the hydroxyl groups on all asymmetrical carbons is reversed, not only that on the number 5 carbon (see 4C). Monosaccharides of the D configuration are much more important nutritionally than their L isomers because they exist as such in dietary carbohydrate and are metabolized specifically in that form. This is because the enzymes involved in carbohydrate digestion and metabolism are stereospecific for D sugars, meaning that they convert D sugars only, and are inactive toward the L forms.

The structures in 4C are shown as open chain, or Fischer projection models, in which the carbonyl (aldehyde or ketone) functions are free. The monosaccharides do not generally exist in open-chain form, as explained later, but are shown that way here to clarify the D-L concept and to illustrate the so-called anomeric carbon, which is the carbon atom comprising the carbonyl function. Notice that the anomeric carbon is number 1 in the aldose (glucose) and number 2 in the ketose (fructose).

The fact that the monosaccharides do not exist in open-chain form while in solution is evidenced by the fact that they do not undergo reactions characteristic of true aldehydes and ketones. Actually the molecules cyclize, producing forms called *hemiacetals,* if the sugar is an aldose, and *hemiketals,* if the sugar is a ketose. The hemiacetal or hemiketal bonds are formed by the reaction of an alcohol group with an aldehyde or ketone group, respectively. Therefore, the participating groups within a monosaccharide are the aldehyde or ketone of the anomeric carbon atom and the alcohol group attached to the highest-numbered chiral carbon atom. This is illustrated using the examples of D-glucose and D-fructose (see 4D). The formation of the hemiacetal or hemiketal produces a new chiral center at the anomeric

carbon, designated by an asterisk in the structures, and therefore the bond direction of the newly formed hydroxyl becomes significant. In the cyclized structures shown, the anomeric hydroxyls at position 1 for D-glucose and position 2 for D-fructose are arbitrarily directed to the right, resulting in an alpha (α) configuration. Should the anomeric hydroxyl be directed to the left, the structure would be in a beta (β) configuration. Cyclization to the hemiacetal or hemiketal can produce either the α- or β-isomer; however, in aqueous solution, an equilibrium mixture of the isomers exists, with the concentration of the β-form being roughly twice that of the α-form. If a pure solution of either isomer is prepared, the optical rotation of the solution will change as the equilibrium concentrations of the two forms are approached. This change in optical rotation is called *mutarotation,* and it results from the interconversion of the α- and β-isomers.

(4D)

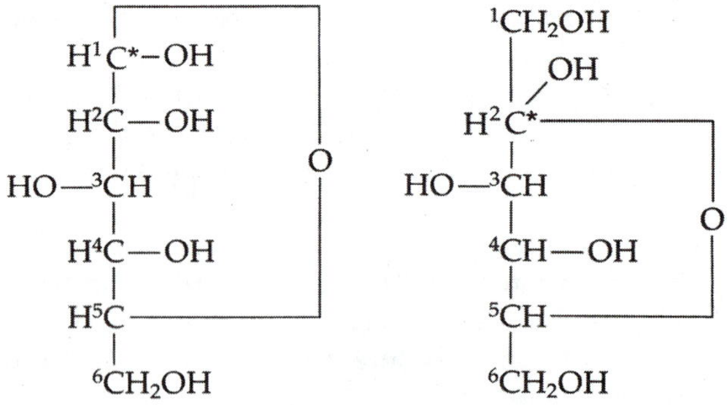

H^{1}C*—OH	^{1}CH$_2$OH
H^{2}C—OH	OH
HO—^{3}CH	H^{2}C*—
H^{4}C—OH	HO—^{3}CH
H^{5}C	^{4}CH—OH
^{6}CH$_2$OH	^{5}CH
	^{6}CH$_2$OH

α-D-glucose α–D-fructose

Stereoisomerism among the monosaccharides and also other nutrients such as amino acids has important metabolic implications due to the stereospecificity of certain metabolic enzymes. An interesting example of stereospecificity is the action of the digestive enzyme α-amylase, which hydrolyzes polyglucose molecules such as the starches, in which the glucose units are connected through an α-linkage (p. 75). Cellulose, also a polymer of glucose but one in which the monomeric glucose residues are connected by β-bonds, is resistant to α-amylase hydrolysis.

The structures of the cyclized monosaccharides are more conveniently and accurately represented by Haworth models. In such models the carbons and oxygen comprising the five- or six-membered ring are depicted as lying in a horizontal plane, with the hydroxyl groups pointing down or up from the plane. Those that are directed to the right in the open-chain structure point down in the Haworth model, and those directed to the left point up. Table 4.1 shows the structural relationship among simple projection and Haworth formulas for the major, naturally occurring hexoses: glucose, galactose, and fructose.

Compared with the hexoses shown in Table 4.1, pentose sugars furnish very little dietary energy because of their low content in the diet. However, they are readily synthesized in the cell from hexose precursors and are incorporated into metabolically important compounds. The aldopentose ribose, for example, is a constituent of key nucleotides such as the adenosine phosphates—adenosine triphosphate (ATP), adenosine diphosphate (ADP), adenosine monophosphate (AMP), cyclic adenosine monophosphate (cAMP), and the nicotinamide adenine dinucleotides (NAD^+, $NADP^+$)

Table 4.1 Various Structural Representations Among the Hexoses: Glucose, Fructose, and Galactose

Hexose	Fischer Projection	Cyclized Fisher Projection	Haworth	Simplified Haworth
α-D-glucose	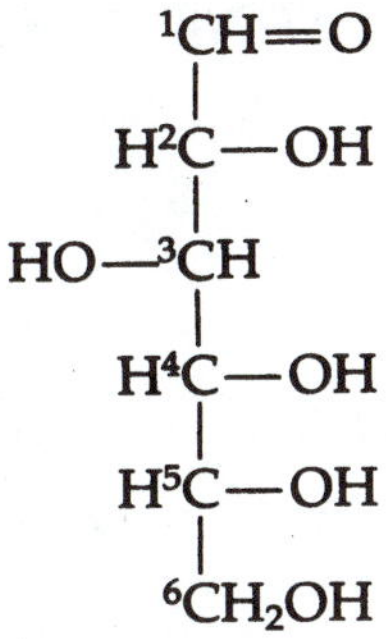	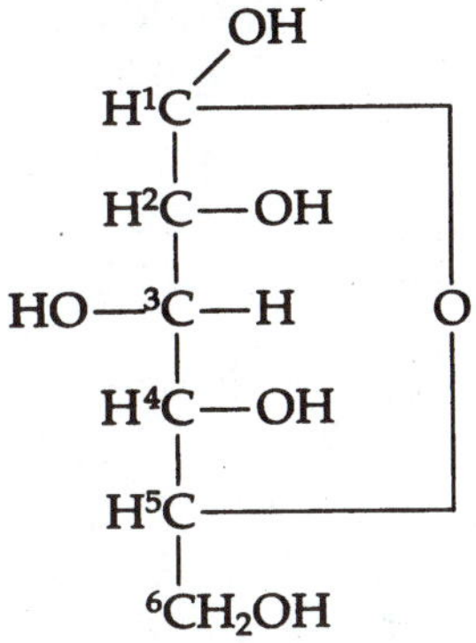	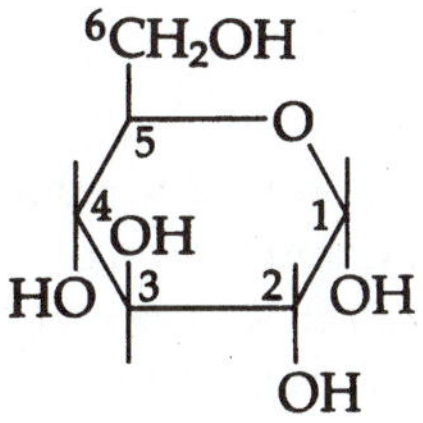	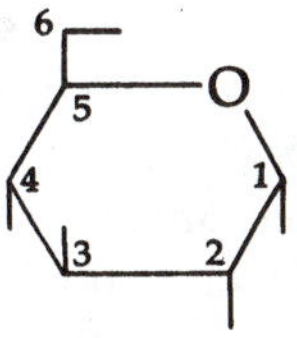
β-D-galactose	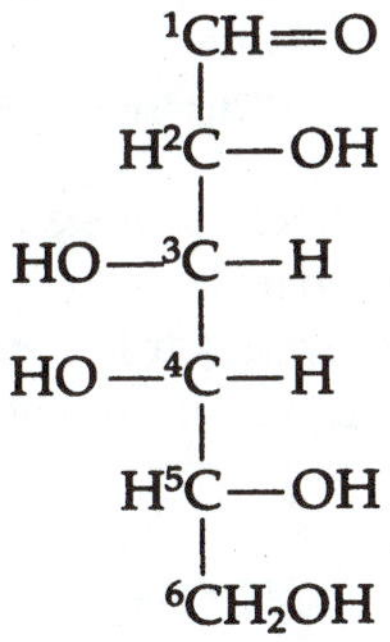	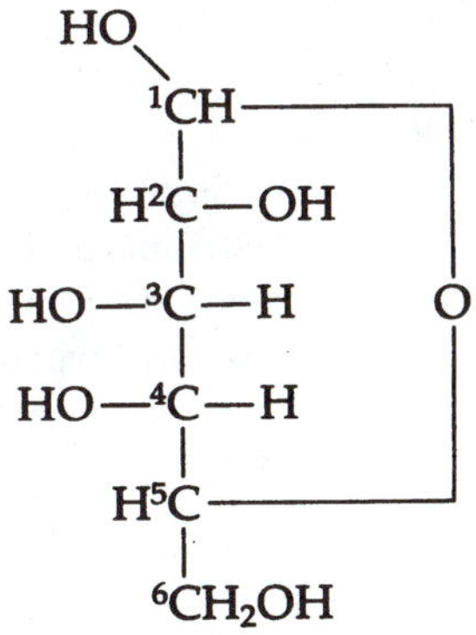	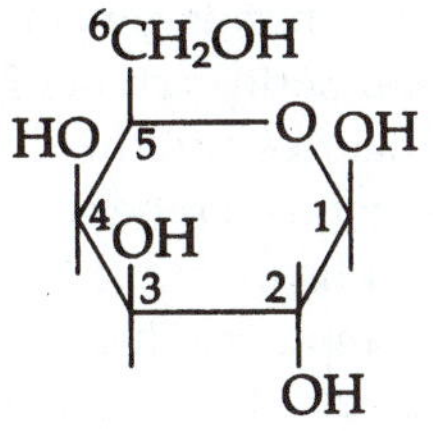	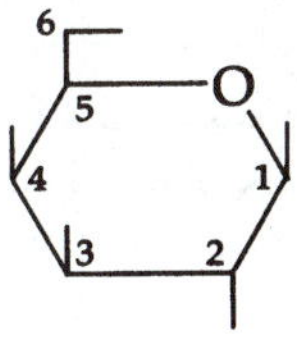
β-D-fructose	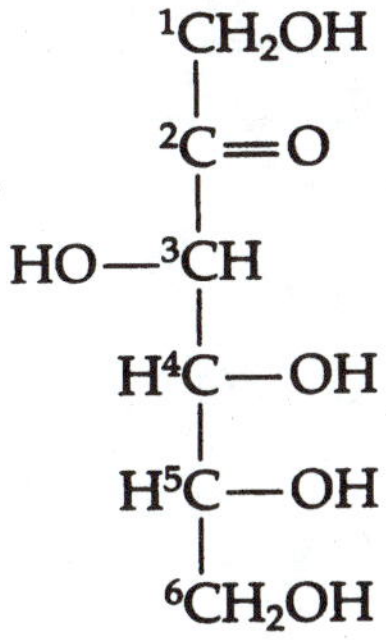	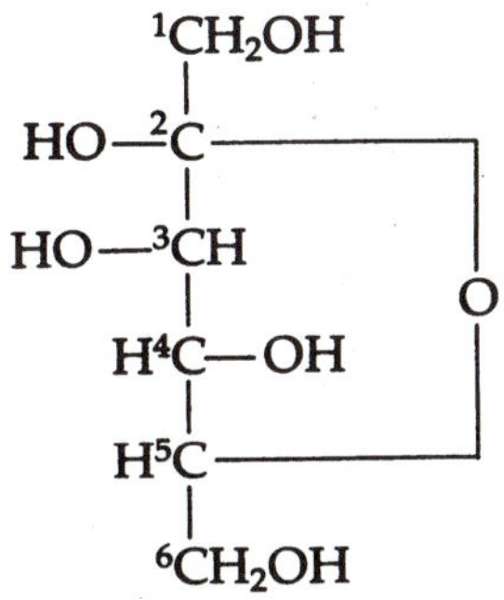	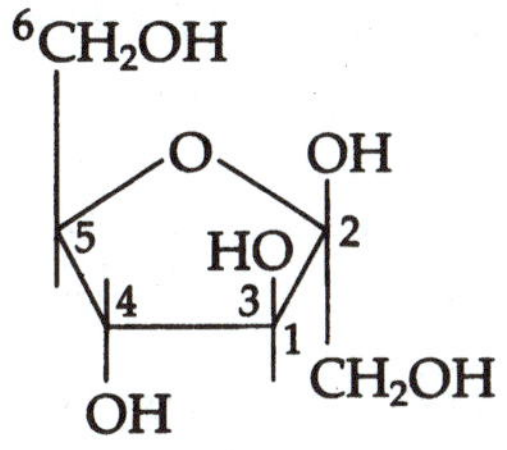	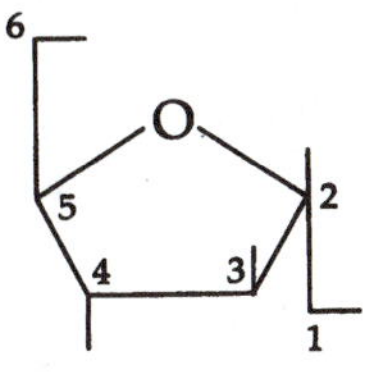

(4E)

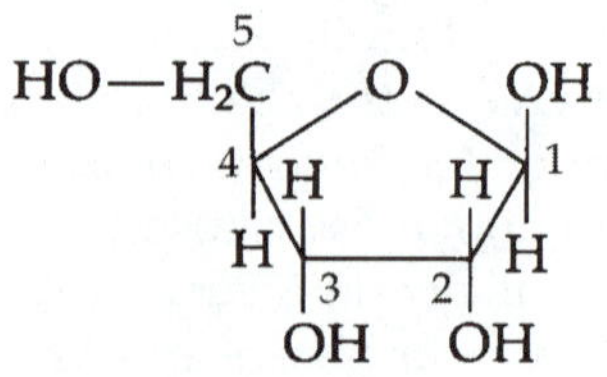

β-D-ribose

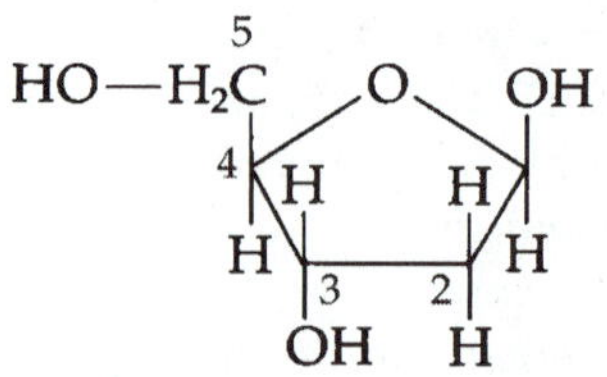

β-D-2-deoxyribose

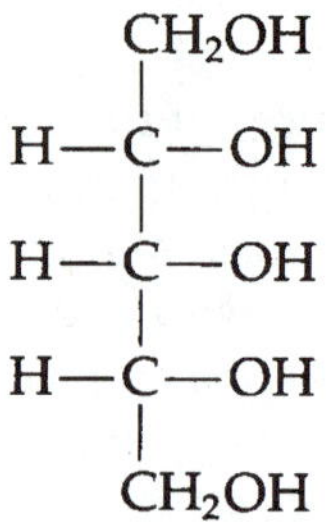

Ribitol

(p. 62). Ribose and its deoxygenated form, deoxyribose, are also part of the structures of ribonucleic acid (RNA) and deoxyribonucleic acid (DNA), respectively. Ribitol, a reduction product of ribose, is a constituent of the vitamin riboflavin and of the flavin coenzymes, flavin adenine dinucleotide (FAD) and flavin mononucleotide (FMN) (see 4E and p. 62).

Oligosaccharides

Oligosaccharides are short-chain polymers of monosaccharide units, generally from 2 to about 10 in number, attached to one another through acetal bonds. Acetal bonds, as they occur in the special case of carbohydrate structures, are also called *glycosidic bonds* and are formed between a hydroxyl group of one monosaccharide unit and a hydroxyl group of the next unit in the polymer, with the elimination of one molecule of water. Disaccharides are the most common members of the oligosaccharide family, and they are major energy-supplying nutrients. The glycosidic bonds generally involve the hydroxyl group on the anomeric carbon of one member of the pair of monosaccharides and the hydroxyl group on carbon 4 or 6 of the second member. Furthermore, the glycosidic bond can be alpha or beta in orientation, depending on whether the anomeric hydroxyl group was alpha or beta before the glycosidic bond was formed. Specific glycosidic bonds may therefore be designated α-1, 4, β-1, 4, α-1, 6, and so on. The most common disaccharides are maltose, lactose, and sucrose (see 4F, 4G, and 4H).

• *Maltose* is formed primarily from the partial hydrolysis of starch and therefore is found in malt beverages

such as beer and malt liquors. It consists of two glucose units linked through an α-1,4 glycosidic bond, as illustrated. The residue on the right is shown in the β-form, although it also may exist in α-form (see 4F).

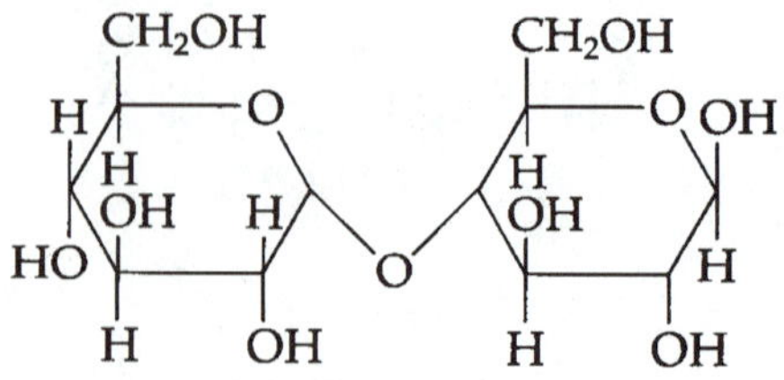

• *Lactose* is found naturally only in milk and milk products. It is composed of galactose, linked by a β-1,4 glycosidic bond to glucose, which can exist in either α- or β-form (see 4G).

(4G)

(Lactose)

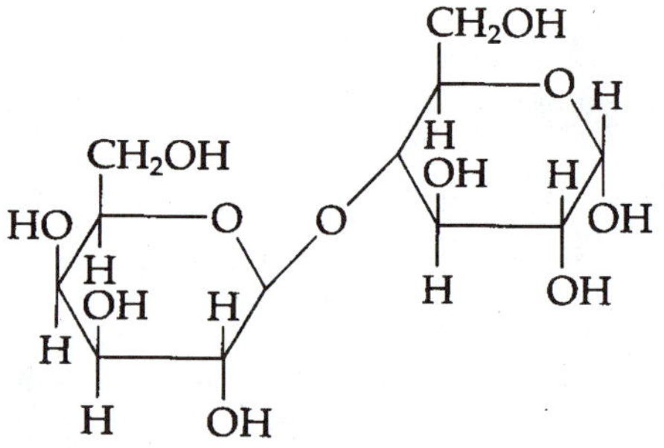

Lactose (α-form)

• *Sucrose* (cane sugar, beet sugar) is the most widely distributed of the disaccharides and is the most commonly used natural sweetener. It is composed of glucose and fructose, and it is structurally unique in that

(4F)

Maltose (β-form)

(4H)

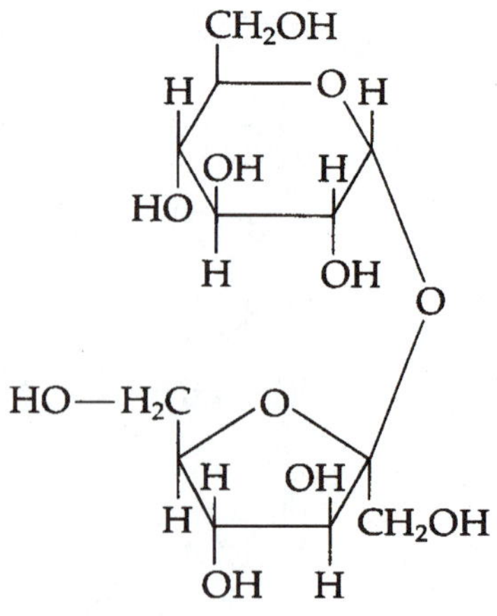

Sucrose

its glycosidic bond involves the anomeric hydroxyl of both the residues. The linkage is α- with respect to the glucose and β- with respect to the fructose residue (see 4H). Having no free hemiacetal or hemiketal function, sucrose is not a reducing sugar.

Polysaccharides

The glycosidic bonding of monosaccharide residues may be repeated many times to form high molecular weight polymers called *polysaccharides.* If the structure is composed of a single type of monomeric unit, it is referred to as a *homopolysaccharide.* If two or more different types of monosaccharides make up its structure, it is called a *heteropolysaccharide.* Both types exist in nature; however, the homopolysaccharide is of far greater importance in nutrition because of the abundance of the substance in many natural foods. The polyglucoses starch and glycogen, for example, are the major storage forms of carbohydrate in plant and animal tissues, respectively. They range in molecular weight from a few thousand to 500,000.

Monosaccharides that are cyclized into hemiacetals or hemiketals are sometimes referred to as *reducing sugars* because they are capable of reducing other substances, such as the copper ion from Cu^{++} to Cu^+. In describing polysaccharide structure, this property is useful in distinguishing one end of a linear polysaccharide from the other end. In a polyglucose chain, for example, the glucose residue at one end of the chain has a hemiacetal group because its anomeric carbon atom is not involved in acetal bonding to another glucose residue. The residue at the other end of the chain is not in hemiacetal form because it *is* attached by acetal bonding to the next residue in the chain. A linear polyglucose molecule therefore has a reducing end (the hemiacetal end) and a nonreducing end, at which no hemiacetal exists. This notation is of use in designating at which end of a polysaccharide certain enzymatic reactions occur.

Starch

The most common digestible polysaccharide in plants is starch, and it can exist in two forms, *amylose* and *amylopectin,* both of which are polymers of D-glucose. The amylose molecule is a linear, unbranched structure in which the glucose residues are attached solely through α-1,4 glycosidic bonds. Amylopectin, however, is a branched-chain polymer, the branch points occurring through α-1,6 bonds, as illustrated in Figure 4.1. Both forms occur in cereals, potatoes, legumes, and other vegetables, with amylose contributing about 15% to 20% and amylopectin 80% to 85% of the total starch content.

Glycogen

The major form of stored carbohydrate in animal tissues is glycogen, which is localized primarily in liver and skeletal muscle. Like amylopectin, it is a highly branched polyglucose molecule and differs from the starch only in the fact that it is more highly branched (see Fig. 4.1b). The glucose residues within glycogen serve as a rich source of energy. When dictated by the body's energy demands, they are sequentially removed enzymatically from the nonreducing ends of the glycogen chains and enter energy-releasing pathways of metabolism. This process (glycogenolysis) is discussed later in this chapter. The high degree of branching in glycogen and amylopectin is a distinct metabolic advantage, because it presents a large number of nonreducing ends from which glucose residues can be cleaved.

Cellulose

Cellulose is the major component of cell walls in plants. Like the starches, it is a homopolysaccharide of glucose but is different in that the glycosidic bonds connecting the residues are β-1,4. This renders the molecule resistant to the digestive enzyme α-amylase, the specificity of which favors α-1,4 linkages. Because cellulose is not digestible by mammalian degradative enzymes, it is defined as a dietary fiber and is considered not to provide energy. However, as a source of fiber, cellulose assumes importance as a bulking agent and potential energy source for intestinal bacteria (p. 108).

Digestion

The dietary carbohydrates that are most important nutritionally are polysaccharides and disaccharides, because free monosaccharides are not commonly present in the diet in significant quantities. However, some free glucose and fructose are in honey, certain fruits, and the carbohydrates that are added to processed foods. The cellular use of carbohydrates depends on their absorption from the gastrointestinal (GI) tract into the bloodstream, a process normally restricted to monosaccharides. Therefore, polysaccharides and disaccharides must be hydrolyzed to their constituent monosaccharide units. The hydrolytic enzymes involved are collectively called *glycosidases* or, alternatively, *carbohydrases.*

Polysaccharides

The key enzyme in the digestion of dietary polysaccharides is α-amylase, a glycosidase that specifically hydrolyzes α-1,4 glycosidic linkages. Resistant to the

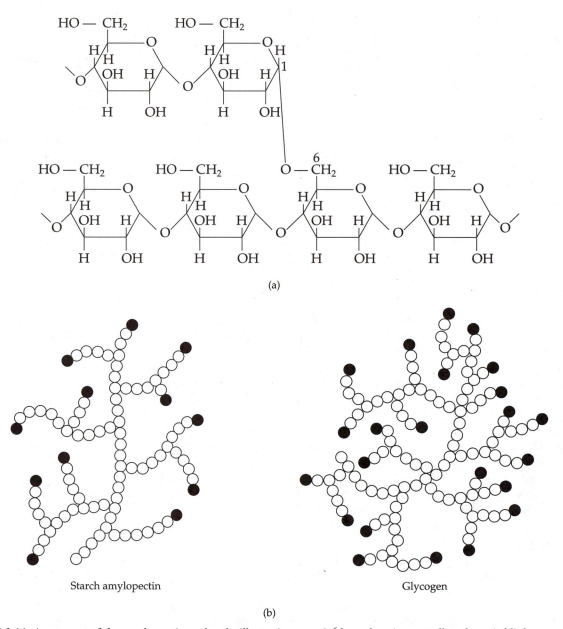

(a)

Starch amylopectin

Glycogen

(b)

Figure 4.1 (a) A segment of the amylopectin molecule, illustrating an α-1,6 branch point as well as the α-1,4 linkages along the linear segments. Branch points may occur as frequently as every 25 to 30 residues. (b) Comparison of the gross structures of starch amylopectin and glycogen, showing the difference in the degree of branching. Circles represent glucose residues. Solid circles are those residues located at nonreducing ends of chains. Only one reducing end is present in each molecule.

action of this enzyme, therefore, are the β-1,4 bonds of cellulose and the α-1,6 linkages that form branch points in starch amylopectin.

Digestion of starches actually begins in the mouth, because α-amylase activity is found in saliva. But considering the short period of time that food is in the mouth prior to being swallowed, this phase of digestion is of little consequence. However, the salivary amy-

lase action continues in the stomach until the gastric acid penetrates the food bolus and lowers the pH sufficiently to inactivate the activity of the enzyme. At this point, partial hydrolysis of the starches has occurred, the major products being dextrins, which are short-chain polysaccharides, and maltose. Further digestion of the dextrins is resumed in the small intestine by the α-amylase of pancreatic origin, which is secreted into

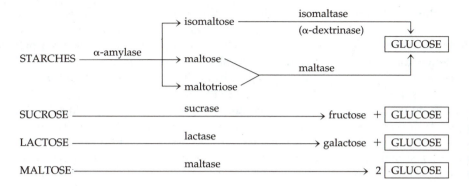

Figure 4.2 The enzymatic hydrolysis of carbohydrates, illustrating the importance of glucose as a component of these major nutrients.

the duodenal contents. Here, the presence of bile and pancreatic bicarbonate elevates the pH to a level favorable for enzymatic function. If the dietary starch form is amylose, which is unbranched, the products of α-amylase hydrolysis are maltose and the trisaccharide maltotriose, which undergoes slower hydrolysis to maltose and glucose. The hydrolytic action of α-amylase on amylopectin, a branched starch, produces glucose and maltose as it did with amylose. However, the α-1,6 bonds linking the glucose residues at the branch points of the molecule cannot be hydrolyzed by α-amylase. Consequently, disaccharide units called *isomaltose,* having α-1,6 glycosidic bonds, are released.

The action of α-amylase on dietary starch, therefore, releases maltose, isomaltose, and glucose as principal hydrolytic products, as illustrated in Figure 4.2. The further breakdown of the disaccharide products to glucose is brought about by specific glycosidases described in the next section.

Disaccharides

Virtually no digestion of disaccharides or small oligosaccharides occurs in the mouth or stomach. In the human, it takes place entirely in the upper small intestine. Unlike α-amylase, disaccharidase activity occurs in the microvilli of the intestinal mucosal cells (the brush border) rather than in the intestinal lumen (p. 29). Among the enzymes located on the mucosal cells are lactase, sucrase, maltase, and isomaltase. Lactase catalyzes the cleavage of lactose to equimolar amounts of galactose and glucose, and sucrase hydrolyzes sucrose to yield one glucose and one fructose residue. Isomaltase (also called an α-dextrinase) hydrolyzes the α-1,6 bond of isomaltose, the branch point disaccharide remaining from the incomplete breakdown of branched starches. The products are two molecules of glucose.

In summary, nearly all dietary starches and disaccharides are ultimately hydrolyzed completely by specific glycosidases to their constituent monosaccharide units. Monosaccharides, together with small amounts of remaining disaccharides, can then be absorbed by the intestinal mucosal cells. The reactions involved in the digestion of starches and disaccharides are summarized in Figure 4.2.

Absorption, Transport, and Distribution

The wall of the small intestine is composed of absorptive mucosal cells and mucus-secreting goblet cells that line projections, called *villi,* that extend into the lumen. The absorptive cells have hairlike projections on the surface on the lumen side called *microvilli* (the *brush border*). A square millimeter of cell surface is believed to have as many as 2×10^5 microvilli projections. The microstructure of the small intestinal wall is illustrated in Figures 2.5 and 2.6. The anatomic advantage of the villi-microvilli structure is that it presents an enormous surface area to the intestinal contents, thereby facilitating absorption. It has been estimated that the absorptive capacity of the human intestine amounts to about 5,400 g/day for glucose and 4,800 g/day for fructose—a capability that would never be challenged in a normal diet.

Glucose and galactose are absorbed into the mucosal cells by active transport, whereas fructose enters by facilitated diffusion. The process of active transport requires energy and the involvement of a specific receptor. The glucose-galactose carrier has been designated SGLT1. It is known to be a protein complex dependent on the Na^+/K^+-ATPase pump (p. 16), which, at the expense of ATP, furnishes energy for the transport of sugar through the mucosal cell. Glucose or galactose cannot attach to the carrier until it has been preloaded with Na^+. Mutation in the SGLT1 gene is associated with glucose-galactose malabsorption.

Glucose appears to leave the mucosal cell at the basolateral surface by three routes. Approximately 15% leaks back across the brush border into the intestinal lumen, about 25% diffuses through the basolateral membrane into the circulation, but the major portion

(~60%) is transported from the cell into the circulation via a carrier in the serosal membrane.

The absorption mechanism for fructose is not completely resolved. There is some evidence for active transport because fructose uptake has been shown to occur against a concentration gradient. This transport is independent of the active, Na^+-dependent transport of glucose, but the rate of uptake is much slower than that of both glucose and galactose. In a large proportion of human subjects studied, there was an inability to absorb completely doses of fructose in the 20- to 50-g range [1]. Fructose, in the absence of glucose, may also be absorbed by facilitated transport involving a specific transporter. In fact, it is likely that GLUT5 (see the next section on glucose transporters) may be the transporter for fructose [2]. The facilitative transport process can only proceed down a concentration gradient, but because fructose is very efficiently trapped and phosphorylated by the liver, there is virtually no circulating fructose in the bloodstream, therefore ensuring its downhill concentration gradient across the intestinal mucosa.

Although fructose absorption takes place more slowly than glucose or galactose, which are actively absorbed, it is absorbed faster than sugars such as sorbitol and xylitol, which are absorbed by purely passive transfer. Although the extent of the contribution of active transport and facilitative transport to fructose absorption is not established, both systems are saturable. This accounts for the observation that its absorption is limited in nearly 60% of normal adults and that intestinal distress, symptomatic of malabsorption, frequently appears following ingestion of 50 g of pure fructose [3]. This level of intake is commonly found in high-fructose syrups used as sweeteners. Interestingly, coconsumption of glucose with fructose accelerates the absorption of fructose and raises the threshold level of fructose ingestion at which malabsorption symptoms appear [3]. This suggests that the pair of monosaccharides might be absorbed by a so-called disaccharidase-related transport system designed to transport the hydrolytic products of sucrose [1].

Following transport across the gut wall, the monosaccharides enter the portal circulation, where they are carried directly to the liver. The liver is the major site of metabolism of galactose and fructose, which are readily taken up by the liver cells via specific hepatocyte receptors. They then enter the cells by facilitated transport and are subsequently metabolized. Both fructose and galactose can be converted to glucose derivatives through pathways that will be discussed later, and then stored as liver glycogen, or they may be catabolized for energy according to the body's energy demand. The blood levels of galactose and fructose are not directly subject to the strict hormonal regulation that is such an important part of glucose homeostasis. However, if their dietary intake is a significantly higher than normal percentage of total carbohydrate intake, they may be regulated indirectly as glucose, because of their metabolic conversion to that sugar.

Glucose is nutritionally the most important monosaccharide, because it is the exclusive constituent of the starches and also occurs in each of three major disaccharides (Fig. 4.2). Like fructose and galactose, glucose is extensively metabolized in the liver, but its uptake by that organ is not as complete as in the case of fructose and galactose. The remainder of the glucose passes on into the systemic blood supply and is then distributed among other tissues such as muscle, kidney, and adipose tissue. It enters these cells by facilitated transport. In skeletal muscle and adipose tissue the process is insulin-dependent, whereas in the liver it is insulin-independent. Because of the nutritional importance of glucose, the facilitated transport process by which it enters the cells of certain organs and tissues warrants a closer look. The following section explores the process in greater detail.

Glucose Transporters

Glucose is effectively used by a wide variety of cell types under normal conditions, and its concentration in the blood must be precisely controlled. The symptoms associated with diabetes mellitus are a graphic example of the consequences of a disturbance in glucose homeostasis. The cellular uptake of glucose requires that it cross the plasma membrane of the cell. This cannot occur by simple diffusion because the highly polar glucose molecule cannot be solubilized in the nonpolar matrix of the lipid bilayer. The cellular use of glucose therefore requires an efficient transport system for moving the molecule into and out of cells. In certain absorptive cells, such as epithelial cells of the small intestine and renal tubule, glucose can cross the plasma membrane (actively) against a concentration gradient, pumped by a Na^+/K^+-ATPase symport system (described in Chap. 1, p. 16). Nearly all cells in the body, however, admit glucose passively by a carrier-mediated transport mechanism that does not require energy. The protein carriers involved in this process are called glucose transporters, abbreviated GLUT.

Considered collectively, all transporter proteins share a unique structure. They are integral proteins, penetrating and spanning the lipid bilayer of the plasma membrane. Most transporters, in fact, span the membrane several times. They are oriented such that hydrophilic regions of the chain protrude into the extracellular and cytoplasmic media while the hydrophobic regions traverse the membrane, juxtaposed with the membrane's lipid matrix. A model for a glucose transporter, reflecting this spatial arrangement of the molecule, is illustrated in Figure 4.3.

In its simplest form, a transporter

- has a specific combining site for the molecule being transported;
- undergoes a conformational change upon binding the molecule, allowing the molecule to be translocated to the other side of the membrane and released;
- has the ability to reverse the conformational changes without the molecule's being bound to the transporter, so that the process can be repeated.

Six isoforms of glucose transporters have been described: GLUT1, GLUT2, GLUT3, GLUT4, GLUT5, and GLUT7 are designated according to the order of their discovery. GLUT6 possesses structural homology to GLUT3 but is not a functioning transporter. All cells express on their plasma membranes at least one of these isoforms. The different isoforms have distinct tissue distribution and biochemical properties, and they contribute to the precise disposal of glucose according to varying physiological conditions. Tissue distribution of the glucose transporters is summarized in Table 4.2.

The GLUT1 and GLUT3 isoforms are believed to be responsible for basal, or constitutive, glucose uptake. Their number on a given area of cell surface is relatively static and does not increase on insulin stimulation. This is also true for the liver isoform (GLUT2), which is responsible for the bidirectional transport of glucose by the hepatocyte. It may also be involved in the movement of glucose out of absorptive epithelial cells into the circulation following the active transport

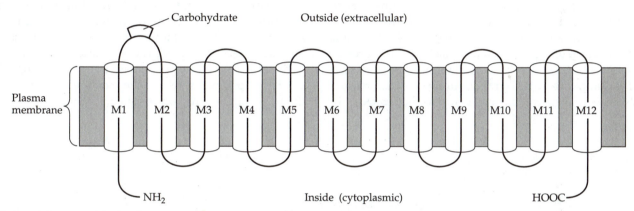

Figure 4.3 A model of a glucose transporter (GLUT1), showing its structural orientation in the plasma membrane. Alpha helical regions of the protein span the membrane 12 times at transmembrane sites designated M1–M12. Transmembrane segments consist largely of hydrophobic amino acids, whereas the loops on the extracellular and cytoplasmic sides of the membrane are primarily hydrophilic. The transporter is a single polypeptide chain composed of approximately 500 amino acid residues.

Table 4.2 Human Glucose Transporters

Designation (Alternative)	M_r^a (residues)	Major Sites of Expression
GLUT1	54,117	Erythrocyte (human), blood-brain barrier,
(erythrocyte/brain, HepG2)	(492)	placenta, fetal tissues in general
GLUT2	57,000	Liver, pancreatic β-cell, kidney, small intestine
(liver)	(524)	
GLUT3	53,933	Brain (neurones)
(brain)	(496)	
GLUT4	54,797	Brown and white adipocytes, heart and
(apidocyte/muscle,	(509)	skeletal muscle
insulin-regulatable)		
GLUT5	54,983	Small intestine
(small intestine)	(501)	
GLUT7	53,000	Endoplasmic reticulum of hepatocytes
(hepatic microsomal)	(528)	

of the glucose into the cells by symporters. GLUT4, in contrast, is quite sensitive to insulin, its concentration on the plasma membrane increasing dramatically in response to the hormone. It follows that the increase in the membrane transporter population is accompanied by an accelerated increase in the uptake of glucose by the stimulated cells. It is therefore the presence of GLUT4 in skeletal muscle and adipose tissue that makes these tissues responsive to insulin. Liver, brain, and erythrocytes lack the GLUT4 isoform and are therefore not sensitive to the hormone. A feature of non-insulin-dependent diabetes mellitus (NIDDM), described in the Chapter 8 Perspective, is a resistance to insulin. The disease is believed to arise from abnormalities in the synthesis or activity of GLUT4. An excellent review of the GLUT isoform family is found in the publication by Baldwin [4].

Synthesis and storage of the insulin-responsive transporter GLUT4, like the other transporter isoforms, occurs as described in Chapter 1 for all proteins. Following its synthesis from mRNA on the ribosomes of the rough endoplasmic reticulum, the transporter enters the compartments of the Golgi apparatus, where it is ultimately packaged in tubulovesicular structures in the trans-Golgi network. In the basal, unstimulated state of the adipocyte, GLUT4 resides in these structures and also, to some extent, in small cytoplasmic vesicles [5].

This subcellular distribution of GLUT4 is also found in skeletal muscle cells [6]. On stimulation by insulin, the tubulovesicle-enclosed transporters are translocated to the plasma membrane, where they become expressed on the cell surface. The proposed mechanism is illustrated in Figure 4.4.

The endothelial tissue of which blood vessel walls are constructed is freely permeable to metabolites such as glucose. Some tissues, however, most notably the brain, possess an additional layer of epithelial tissue between the blood and the cells of that tissue. Epithelial layers are not readily permeable like endothelium, and the passage of metabolites through them requires active transport or facilitative diffusion. For this reason, epithelium is referred to as the blood tissue *barriers* of the body. Among the blood tissue barriers studied, including those of brain, cerebrospinal fluid, retina, testes, and placenta, GLUT1 appears to be the prime isoform for glucose transport [7].

Maintenance of normal blood glucose concentration is a major function of the liver. Regulation is the net effect of the organ's metabolic processes that remove glucose from the blood for either glycogen synthesis or for energy release, and of processes that return glucose to the blood, such as glycogenolysis and gluconeogenesis. These pathways, which are examined in detail in the next section, are hormonally influenced, primarily

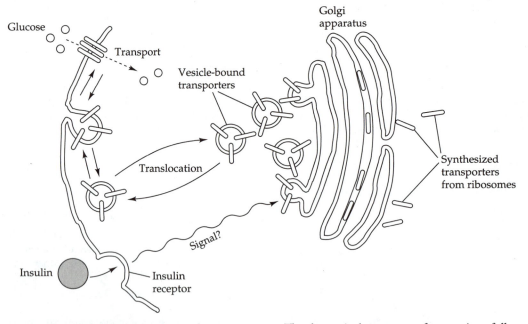

Figure 4.4 Proposed mode of insulin action on glucose transport. The theoretical sequence of events is as follows: (1) The binding of insulin to its specific receptor induces a signal of yet unknown nature that stimulates the translocation of vesicles containing the glucose transporters to the plasma membrane. (2) The vesicles fuse with the plasma membrane, releasing the transporters and allowing them to position themselves in the membrane. (3) On removal of insulin from its receptor, membrane-bound transporters are retranslocated back to the intracellular pool by an endocytosis-like mechanism.

by the antagonistic pancreatic hormones insulin and glucagon and by the glucocorticoid hormones of the adrenal cortex. The rise in blood glucose following the ingestion of carbohydrate, for example, triggers the release of insulin while reducing the secretion of glucagon. This increases uptake of glucose by muscle and adipose tissue, returning blood glucose to homeostatic levels. A fall in blood glucose concentration, conversely, signals the reversal of the pancreatic hormonal secretions—that is, decreased insulin and increased glucagon release. In addition, an increase in the secretion of glucocorticoid hormones, primarily cortisol, occurs in answer to, and to offset, a falling blood glucose level. Glucocorticoids cause an increased activity of hepatic gluconeogenesis, a process described in the following sections.

Integrated Metabolism in Tissues

The metabolic fate of the monosaccharides depends to a great extent on the body's energy demands. According to these demands, the activity of certain metabolic pathways is regulated in such a way that some may be stimulated, whereas others may be suppressed. The major regulatory mechanisms are hormonal (involving the action of hormones such as insulin, glucagon, epinephrine, and the corticosteroid hormones) and allosteric enzyme activation or suppression (p. 96).

Allosteric enzymes are regulatory enzymes, the activities of which can be altered by compounds called *modulators*. A negative modulator of an allosteric enzyme reduces the activity of the enzyme and slows its velocity, whereas a positive modulator increases the activity of an allosteric enzyme, increasing its velocity. The effect of a modulator, negative or positive, is exerted on its allosteric enzyme as a result of increasing concentrations of the modulator. Each of these mechanisms of regulation is discussed in detail in the section dealing with regulation of metabolism.

The metabolic pathways of carbohydrate use and storage consist of glycogenesis, glycogenolysis, glycolysis and hexose monophosphate shunt, the Krebs cycle, and gluconeogenesis. An integrated overview of these pathways is given in Figure 4.5, and a detailed review of their intermediary metabolites, sites of regulation, and, most important, their function in the overall scheme of things are now considered. Reactions within the pathways are numbered to allow elaboration on those that are particularly significant from a nutritional standpoint. Because of the central role of glucose in carbohydrate nutrition, its metabolic fate is featured. The entry of fructose and galactose into the metabolic pathways is introduced later in the discussion.

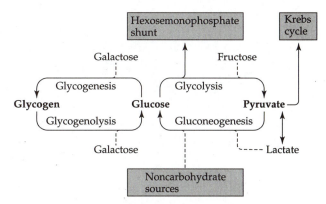

Figure 4.5 An overview of the major pathways of carbohydrate metabolism, emphasizing the fate of glucose but also indicating the sites of entry of galactose and fructose into the pathways.

Glycogenesis

The term *glycogenesis* refers to the pathway by which glucose is ultimately converted into glycogen. This pathway is particularly important in hepatocytes because the liver is the major site of glycogen synthesis and storage. Glycogen accounts for as much as 7% of the wet weight of this organ. The other major site of glycogen storage is skeletal muscle and, to a lesser extent, adipose tissue. In human skeletal muscle, glycogen generally accounts for a little less than 1% of the wet weight of the tissue. The glycogen stores in muscle can be used as an energy source when the body is confronted by an energy demand such as physical exertion or emotional stress, so the glycogenic pathway is of vital importance in ensuring a reserve of instant energy. The pathway is illustrated in Figure 4.6a. The following are comments on selected reactions:

1. Upon entering the cell, glucose is first phosphorylated by hexokinase, producing a phosphate ester at the number 6 carbon of the glucose. In muscle cells, the enzyme catalyzing this phosphate transfer is hexokinase, an allosteric enzyme that is negatively modulated by the reaction product, glucose 6-phosphate. Glucose phosphorylation in the liver is catalyzed primarily by glucokinase (sometimes referred to as hexokinase D). Although the reaction product, glucose 6-phosphate, is the same, interesting differences distinguish it from hexokinase. For example, hexokinase is negatively modulated by glucose 6-phosphate, while glucokinase is not. Also, glucokinase has a much higher K_m than hexokinase, meaning that it can convert glucose to its phosphate form at a higher velocity should the cellular concentration of glucose rise significantly (e.g., after a carbohydrate-rich meal). The much

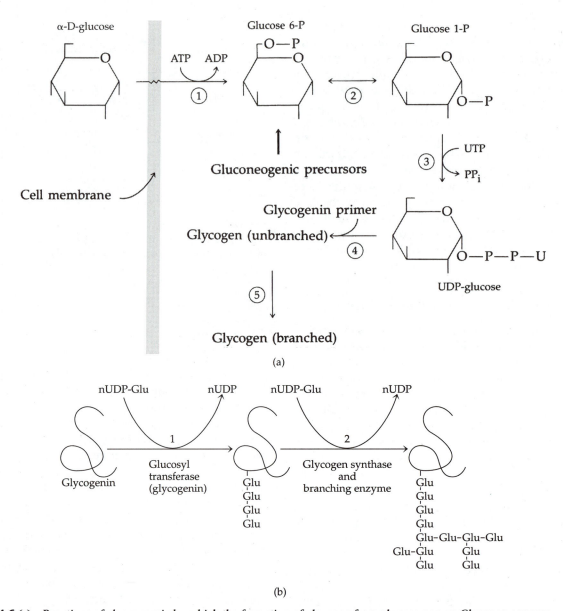

Figure 4.6 (a)   Reactions of glycogenesis, by which the formation of glycogen from glucose occurs. Glycogen appears to be formed principally from gluconeogenic precursor substances rather than from glucose directly [10]. (b) The primer function of glycogenin. The glucosyl transferase activity of glycogenin catalyzes the attachment to itself from two to seven glucose residues transferred from UDP-glucose (reaction 1). Remaining glucose units of the glycogen molecule are incorporated into the molecule by glycogen synthase and linked together by glycogen synthase and the branching enzyme (reaction 2). The letter *n* represents an unspecified number of UDP-glucose molecules.

lower K_m of hexokinase indicates that it is catalyzing at maximum velocity even at average glucose concentrations. Therefore, the liver has the capacity to reduce blood glucose concentration when it becomes high. Furthermore, glucokinase is inducible by the hormone insulin, the activity of which is stimulated by elevated blood glucose levels. In diabetes mellitus, type 1 (see Chap. 8's Perspective), glucokinase activity is therefore below normal values, because such patients are deficient in insulin. This contributes to the cell's inability to take up and metabolize glucose. The concept of K_m and its significance, as it applies to this reaction, has been discussed in Chapter 1. The hexokinase/glucokinase reaction is energy consuming, because the glucose is activated (phosphorylated) at the expense of ATP.

2. The phosphate is transferred from the number 6 carbon of the glucose to the number 1 carbon in a reaction catalyzed by the enzyme phosphoglucomutase.

3. Nucleoside triphosphates sometimes function as activating substances in intermediary metabolism. In this reaction, energy derived from the hydrolysis of the α-β-phosphate anhydride bond of uridine triphosphate allows the coupling of the resulting uridine monophosphate to the glucose 1-phosphate to form uridine diphosphate glucose (UDP-glucose).

4. Glucose is incorporated into glycogen as UDP glucose. The reaction is catalyzed by glycogen synthase, and it requires some preformed glycogen (primer) to which the incoming glucose units can be attached. The nature of the glycogen primer has become understood during only recent years. Before then, the unanswered question was that if glycogen synthase requires a primer, how is a new glycogen molecule initiated? The answer lies in an intriguing protein called *glycogenin*. Glycogenin itself acts as the primer to which the first glucose residue is covalently attached, and it also catalzyes (is the enzyme for) the reaction by which the attachment occurs. It also catalyzes the attachment of additional glucose residues to form chains of up to eight units. At that point, glycogen synthase takes over, extending the glycogen chains further. The role of glycogenin in glycogenesis has been reviewed [8], and a simplified view of the process is shown schematically in Figure 4.6b. Glycogen synthase exists in an active (dephosphorylated) form and a less active (phosphorylated) form. Insulin facilitates glycogen synthesis by stimulating the dephosphorylation of glycogen synthase. The glycogen synthase reaction is the primary target of insulin's stimulatory effect on glycogenesis.

5. Branching within the glycogen molecule is very important because it increases its solubility and compactness. It also makes available many nonreducing ends of chains from which glucose residues can be cleaved and used for energy, the process known as *glycogenolysis,* which is described in the following section. Glycogen synthase cannot form the α-1,6 bonds of the branch points. This is left to the action of the *branching enzyme,* which transfers small oligosaccharide segments from the end of the main glycogen chain to carbon number 6 hydroxyl groups throughout the chain. The overall pathway of glycogenesis, like most synthetic pathways, consumes energy, because an ATP (reaction 1) and a uridine triphosphate (UTP) (reaction 3) are consumed for each molecule of glucose introduced.

Glycogenolysis

The potential energy of glycogen is contained within the glucose residues that comprise its structure. In accordance with the body's energy demands, the residues can be systematically cleaved one at a time from the ends of the glycogen branches and routed through energy-releasing pathways. The breakdown of glycogen into individual glucose units, in the form of glucose 1-phosphate, is called *glycogenolysis.* Like its counterpart, glycogenesis, it is regulated by hormones, most importantly by glucagon, of pancreatic origin, and the catecholamine hormone epinephrine, produced in the adrenal medulla. Both of these hormones stimulate glycogenolysis and are directed at the initial reaction, glycogen phosphorylase. Therefore, they function antagonistically to insulin in regulating the balance between free and stored glucose. The steps involved in glycogenolysis are shown in Figure 4.7. The following are comments on selected reactions:

1. The sequential release of individual glucose units from glycogen is a *phosphorolysis* process by which the glycosidic bonds are cleaved by phosphate addition. The products of one such cleavage reaction are glucose 1-phosphate and the remainder of the intact glycogen chain minus the one glucose residue. The reaction is catalyzed by *glycogen phosphorylase,* an important site of metabolic regulation by both hormonal and allosteric enzyme modulation. Glycogen phosphorylase can exist as phosphorylase a, a phosphorylated active form, or phosphorylase b, a dephosphorylated, inactive form. The two forms are interconverted by other enzymes, which can either attach phosphate groups to the phosphorylase enzyme or remove phosphate groups from it. The enzyme catalyzing the phosphorylation of phosphorylase b to its active "a" form is called *phosphorylase b kinase.* The enzyme that removes phosphate groups from the active "a" form of phosphorylase, producing the inactive "b" form, is called *phosphorylase a phosphatase.* The rate of glycogen breakdown to glucose 1-phosphate therefore depends on the relative activity of these enzymes.

The regulation of phosphorylase activity in the breakdown of liver and muscle glycogen is complex. It can involve covalent regulation, which is the phosphorylation-dephosphorylation regulation just described, and it may also involve allosteric regulation by modulators. These and other mechanisms of regulation are broadly reviewed in the section "Regulation of Metabolism."

• *Covalent regulation* is strongly influenced by the hormones epinephrine, which stimulates glycogenolysis in muscle, and glucagon, which stimulates

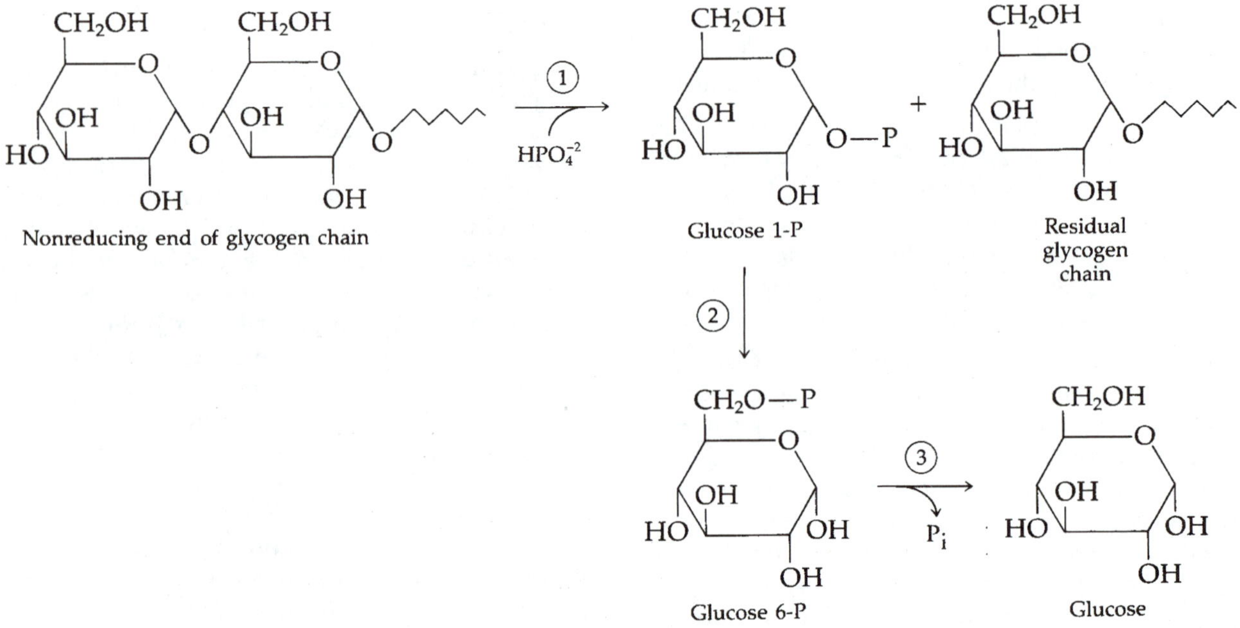

Nonreducing end of glycogen chain

Glucose 1-P

Residual glycogen chain

Glucose 6-P

Glucose

Figure 4.7 The reactions of glycogenolysis, by which glucose residues are sequentially removed from the nonreducing ends of glycogen segments. Reactions 1 and 2 are shared by both liver and muscle cells. In liver cells, glucose 6-phosphate can be converted to free glucose (reaction 3) by glucose 6-phosphatase. Muscle cells lack glucose 6-phosphatase and therefore cannot carry out reaction 3.

glycogenolysis in liver. Both of these hormones exert their effect by stimulating phosphorylase b kinase, thereby promoting the formation of the more active ("a") form of the enzyme. This hormonal activation of phosphorylase b kinase is mediated through cAMP, the cellular concentration of which is increased by the action of the hormones.

• *Allosteric regulation* of phosphorylase generally involves the positive modulator AMP, which induces a conformational change in the inactive "b" form, resulting in a fully active "b" form. ATP competes with AMP for the allosteric site of the enzyme, therefore preventing the shift to its active form and tending to keep it in its inactive form. There is no covalent (phosphorylation) regulation involved in allosteric modulations.

The interconversion of phosphorylases a and b, and the active and inactive forms of phosphorylase b by covalent and allosteric regulation, respectively, are shown in Figure 4.8. For the interested reader, a biochemistry text—for example, Lehninger's *Principles of Biochemistry* [9]—includes a more detailed account of the regulation of the phosphorylase reaction.

Although glycogen phosphorylase cleaves α-1,4 glycosidic bonds, it cannot hydrolyze α-1,6 bonds. Therefore, the enzyme acts repetitively along linear portions of the glycogen molecule until it reaches a point 4 glucose residues from an α-1,6 branch point. Here, the degradation process stops, and it resumes only after an

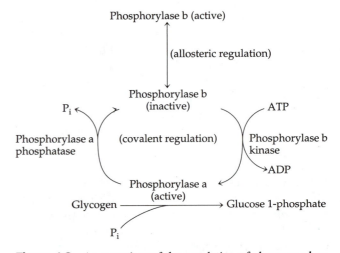

Figure 4.8 An overview of the regulation of glycogen phosphorylase. The enzyme is regulated covalently by phosphorylation to an active "a" form, and dephosphorylation to an inactive "b" form, reactions mediated through cAMP by hormones such as epinephrine and glucagon. It is also regulated allosterically by AMP and ATP, which cause shifts in the equilibrium between inactive and active "b" forms. AMP positively modulates the enzyme by shifting the equilibrium toward its active "b" form. ATP inhibits the effect of AMP, thereby favoring the formation of the inactive "b" form. For more details, see the text earlier under the description for reaction 1.

enzyme called the *debranching enzyme* cleaves the α-1,6 bond at the branch point.

2. At times of heightened glycogenolytic activity, the formation of increased amounts of glucose 1-phosphate shifts the glucose phosphate isomerase reaction toward production of the 6-phosphate isomer.

3. The conversion of glucose 6-phosphate to free glucose requires the action of glucose 6-phosphatase. This enzyme functions in liver and kidney cells, but it is not expressed in muscle cells or adipocytes. Therefore, free glucose can be formed from liver glycogen and transported via the bloodstream to other tissues for oxidation. In this manner, the liver, but not muscle, can control the concentration of glucose in the blood. It follows that although muscle and, to some extent, adipose tissue have stores of glycogen, these stores can be broken down to glucose for use in these sites only.

Glycolysis

Glycolysis is the pathway by which glucose is degraded into two units of pyruvate, a triose. From pyruvate, the metabolic course of the glucose depends largely on the availability of oxygen, and therefore the course is said to be either aerobic or anaerobic. Under anaerobic conditions—that is, in a situation of oxygen debt—pyruvate is converted to lactate. Under otherwise normal conditions, this would occur mainly in times of strenuous exercise when the demand for oxygen by the working muscles, to satisfy their energy needs, exceeds that which is available. In such a case, lactate can accumulate in the muscle cells, contributing in part to the aches and pains associated with overexertion. Under anaerobic conditions glycolysis does, however, release a small amount of usable energy, which can help to sustain the muscles even in a state of oxygen debt. This is the major function of the anaerobic pathway to lactate.

Pyruvate can also follow an aerobic course of reactions, the Krebs cycle, in which it becomes completely oxidized to CO_2 and H_2O. Complete oxidation is accompanied by the release of relatively large amounts of energy, much of which is salvaged as ATP by the mechanism of oxidative phosphorylation (p. 59). Because the glycolytic enzymes function within the cytoplasmic matrix of the cell, while the enzymes catalyzing the Krebs cycle reactions are located within the mitochondrion (p. 6), pyruvate must enter the mitochondrion for oxidation. Glycolysis followed by Krebs cycle activity (aerobic catabolism of glucose) demands an ample supply of oxygen, a condition that is generally met in normal, resting mammalian cells. In a normal, aerobic situation, complete oxidation of pyruvate generally occurs, with only a small amount of lactate being formed. The primary importance of glycolysis in energy metabolism, therefore, is that it provides the initial sequence of reactions (to pyruvate) necessary for the complete oxidation of glucose via the Krebs cycle.

In cells that lack mitochondria, such as the erythrocyte, the pathway of glycolysis is the sole provider of ATP by the mechanism of substrate-level phosphorylation of ADP, as discussed in Chapter 3. Nearly all cell types conduct glycolysis, but most of the energy derived from carbohydrates originates in liver, muscle, and adipose tissue.

The pathway of glycolysis, under both aerobic and anaerobic conditions, is summarized in Figure 4.9. Also indicated in the figure is the mode of entry of dietary fructose and galactose into the pathway for metabolism. The following are comments on selected reactions:

1. The hexokinase/glucokinase reaction consumes 1 mol ATP/mol glucose. Hexokinase (not glucokinase) is negatively regulated by the product of the reaction, glucose 6-phosphate.

2. Glucose phosphate isomerase (also called hexose phosphate isomerase) catalyzes this interconversion of isomers.

3. The phosphofructokinase reaction, an important regulatory site, is modulated negatively by ATP and citrate and positively by AMP and ADP. Another ATP is consumed in the reaction. This important reaction is also regulated hormonally by glucagon, as described later in this chapter.

4. The aldolase reaction results in the splitting of a hexose bisphosphate into two triose phosphates.

5. The isomers glyceraldehyde 3-phosphate and dihydroxyacetone phosphate (DHAP) are interconverted by the enzyme triosephosphate isomerase. In an isolated system, the equilibrium favors DHAP formation. But in the cellular environment it is shifted completely toward the production of glyceraldehyde 3-phosphate, because this metabolite is continuously removed from the equilibrium by the subsequent reaction catalyzed by glyceraldehyde 3-phosphate dehydrogenase.

6. In this reaction, glyceraldehyde 3-phosphate is oxidized to a carboxylic acid, while inorganic phosphate is incorporated as a high-energy anhydride bond. The enzyme is glyceraldehyde 3-phosphate dehydrogenase, which uses NAD^+ as its hydrogen-accepting cosubstrate. Under aerobic conditions, the NADH formed is reoxidized to NAD^+ by O_2 via the electron transport chain in the mitochondria as explained in the following section. The reason that O_2 is not necessary to sustain this reaction under anaerobic conditions is that the NAD^+ consumed is restored by a subsequent reaction (see reaction 11, following).

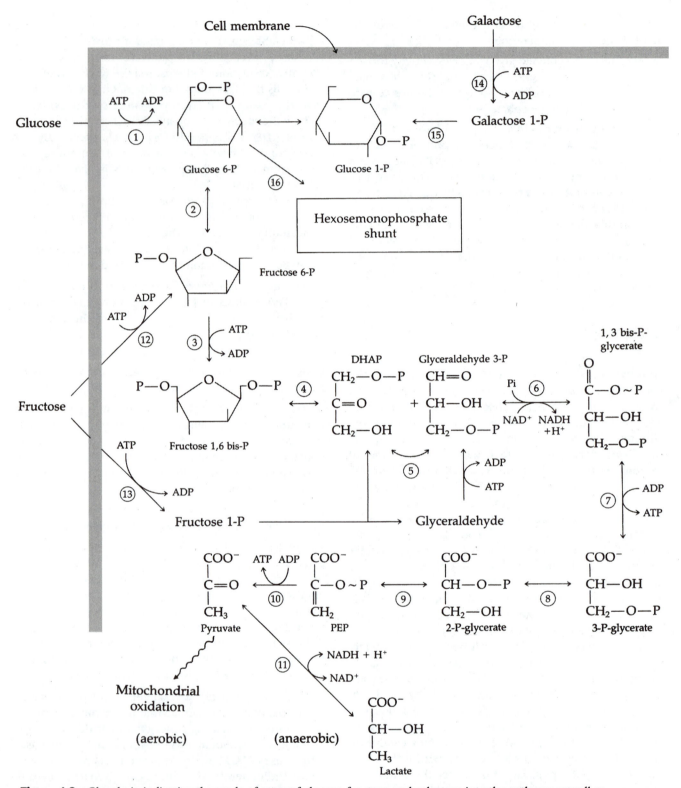

Figure 4.9 Glycolysis, indicating the mode of entry of glucose, fructose, and galactose into the pathway, as well as the alternative entry of glucose 6-phosphate into the hexosemonophosphate shunt. Under conditions of oxygen debt (anaerobic), pyruvate is converted to lactate. If the supply of oxygen is adequate, pyruvate can be oxidized completely in the mitochondrion to carbon dioxide and water.

7. This reaction, catalyzed by phosphoglycerate kinase, exemplifies a substrate-level phosphorylation of ADP. Refer to Chapter 3 for a more detailed review of this mechanism by which ATP can be formed from ADP by the transfer of a phosphate from a high-energy donor molecule (p. 59).

8. Phosphoglyceromutase catalyzes the transfer of the phosphate group from the carbon-3 to the carbon-2 of the glyceric acid.

9. Dehydration of 2-phosphoglycerate by the enzyme enolase introduces a double bond that imparts high energy to the phosphate bond.

10. The product of reaction 9, phosphoenolpyruvate (PEP), donates its phosphate group to ADP in a reaction catalyzed by pyruvate kinase. This is the second site of substrate-level phosphorylation of ADP in the glycolytic pathway.

11. The lactate dehydrogenase reaction transfers two hydrogens from NADH and H^+ to pyruvate, reducing it to lactate. NAD^+ is formed in the reaction and can replace the NAD^+ consumed in reaction 6 under anaerobic conditions. It must be emphasized that this reaction is most active in situations of oxygen debt, as in prolonged muscular activity. Under normal, aerobic conditions, pyruvate enters the mitochondrion for complete oxidation. A third important option available to pyruvate is its conversion to the amino acid alanine by transamination, a reaction by which pyruvate acquires an amino group from the amino acid glutamate (p. 185). This, together with the fact that pyruvate is also the product of the catabolism of various amino acids, makes it an important link between protein and carbohydrate metabolism.

12 and 13. These two reactions provide the means by which dietary fructose enters the glycolytic pathway. Fructose is an important factor in the average American diet, as nearly half of the carbohydrate consumed is sucrose, and high-fructose corn syrup is becoming more popular as a food sweetener. Reaction 12 functions in extrahepatic tissues and involves the direct phosphorylation by hexokinase to form fructose 6-phosphate. This is a relatively unimportant reaction. It is slow and occurs only in the presence of high levels of fructose. Reaction 13 is the major means by which fructose is converted to glycolysis metabolites. The phosphorylation occurs at carbon-1 and is catalyzed by fructokinase, an enzyme found only in liver cells. The fructose 1-phosphate is subsequently split by aldolase (designated aldolase B to distinguish it from the enzyme acting on fructose 1,6-bisphosphate), forming DHAP and glyceraldehyde. The latter can then be phosphorylated by glyceraldehyde kinase (or triokinase) at the expense of

a second ATP to produce glyceraldehyde 3-phosphate. Fructose is therefore converted to glycolytic intermediates, which can follow the pathway to pyruvate formation and Krebs cycle oxidation. Alternatively, they can be used in the liver to produce free glucose by a reversal of the first part of the pathway through the action of gluconeogenic enzymes. Glucose formation from fructose is particularly important if fructose provides the major source of carbohydrate in the diet.

Because the phosphorylation of fructose is essentially the liver's responsibility, eating large amounts of fructose can deplete hepatocyte ATP, reducing the rate of various biosynthetic processes such as protein synthesis.

14. Like glucose and fructose, galactose is first phosphorylated. The transfer of the phosphate from ATP is catalyzed by galactokinase, and the resulting phosphate ester is at carbon 1 of the sugar. The major dietary source of galactose is lactose, from which the galactose is released hydrolytically by lactase.

15. Galactose 1-phosphate can be converted to glucose 1-phosphate through the intermediates uridine diphosphate (UDP)-galactose and uridine diphosphate (UDP)-glucose. The enzyme galactose 1-phosphate uridyl transferase transfers a uridyl phosphate residue from UDP-glucose to the galactose 1-phosphate, yielding glucose 1-phosphate and UDP-galactose. In a reaction catalyzed by epimerase, UDP-galactose can then be converted to UDP-glucose in which form it can be converted to glucose 1-phosphate by the uridyl transferase reaction already referred to, or it can be incorporated into glycogen by glycogen synthase, as described in the "Glycogenesis" section. It can also enter the glycolytic pathway as glucose 6-phosphate, made possible by the reaction series UDP-glucose → glucose 1-phosphate → glucose 6-phosphate. As glucose 6-phosphate, it can also be hydrolyzed to free glucose in liver cells.

16. This indicates the entry of glucose 6-phosphate into another pathway called the *hexose monophosphate shunt* (pentose phosphate pathway), which will be discussed later.

NADH in Anaerobic and Aerobic Glycolysis: The Shuttle Systems

Under *anaerobic* conditions, the NADH produced in the pathway of glycolysis (the glyceraldehyde 3-phosphate dehydrogenase reaction) cannot undergo reoxidation by mitochondrial electron transport because molecular oxygen is the ultimate oxidizing agent in that system. Instead, it is used in the lactate dehydrogenase reduction of pyruvate to lactate, thereby becoming reoxidized to NAD^+ without the involvement of oxygen. In this manner, NAD^+ is restored to sustain the glyceraldehyde 3-phosphate dehydrogenase reaction and allow the pathway to lactate to continue in the absence of oxygen.

When the system is operating *aerobically* and the supply of oxygen is ample to allow total oxidation of incoming glucose, lactic acid is not formed. Instead, pyruvate enters the mitochondrion, as does a carrier molecule of hydrogen atoms that were transferred to it from NADH. NADH cannot enter the mitochondrion directly. Instead, reducing equivalents in the form of carriers of hydrogen atoms removed from the NADH in the cytoplast are shuttled across the mitochondrial membrane. Once in the mitochondrial matrix, the carriers are enzymatically dehydrogenated, and NAD^+ becomes reduced to NADH. The latter can then become oxidized by electron transport and consequently generate three ATPs per mole NADH by oxidative phosphorylation (p. 66). In this manner, 6 mol of ATP are consequently formed per mole of glucose. The result of the shuttle system is therefore equivalent to a transfer of NADH from the cytoplasm into the mitochondrion, although it does not occur directly. Shuttle substances that transport the hydrogens removed from cytoplasmic NADH into the mitochondrion are malate or glycerol 3-phosphate.

The most active shuttle compound, malate, is reoxidized by malate dehydrogenase in the mitochondrion as NAD^+ becomes reduced to NADH, therefore generating the three ATPs per mole. This shuttle system is called the *malate-aspartate shuttle system* because the intramitochondrial malate is eventually converted to aspartate, in which form it returns to the cytoplasm. The *glycerol 3-phosphate shuttle,* in contrast, leads to only two ATPs per mole NADH because the intramitochondrial reoxidation of glycerol 3-phosphate is catalyzed by glycerol phosphate dehydrogenase, which uses FAD instead of NAD^+ as hydrogen acceptor. If the glycerol 3-phosphate shuttle is in effect, therefore, only four ATPs will be formed per mole of glucose by oxidative phosphorylation. Figure 4.10 illustrates how these shuttle systems function in the reoxidation of cytoplastic NADH. The shuttle systems are specific to certain tissues. The more active malate-aspartate shuttle functions in the liver, kidney, and heart, whereas the glycerol 3-phosphate shuttle occurs in the brain and skeletal muscle.

To summarize energy release from glycolysis in terms of ATP produced, a net of two ATPs are formed as a result of substrate-level phosphorylation reactions. If the starting point of glycolysis is glycogen rather than free glucose, the hexokinase reaction is bypassed. The result is that the energy yield is therefore increased by one ATP for glycolysis of glycogen glucose under either aerobic or anaerobic conditions.

If the cell is functioning anaerobically, the two ATPs formed by substrate-level phosphorylation are the total produced. Under aerobic conditions, in contrast, additional ATPs are formed by oxidative phosphorylation. This is because the cytoplasmic NADH produced is not used in the (anaerobic) lactate dehydrogenase reaction but is reoxidized via the shuttle systems to NAD^+ by electron transport and oxygen. The number of additional ATPs formed depends on which shuttle system is used to move the NADH hydrogens into the mitochondrion. If the malate-aspartate shuttle is in effect, six ATPs are produced, bringing the total to eight. In tissues using the glycerol 3-phosphate shuttle, just four ATPs are formed by oxidative phosphorylation, or a glycolytic total of six.

The Hexosemonophosphate Shunt (Pentose Phosphate Pathway)

The purpose of a metabolic shunt is to generate important intermediates not produced in other pathways. Two very important products of the hexosemonophosphate shunt are as follows:

- *Pentose phosphates,* necessary for the synthesis of the nucleic acids, DNA and RNA, and for other nucleotides
- *The reduced cosubstrate NADPH,* used for important metabolic functions, such as the biosynthesis of fatty acids (p. 151), the maintenance of reducing substrates in red blood cells necessary to ensure the functional integrity of the cells, and drug metabolism

The shunt begins with two consecutive dehydrogenase reactions catalyzed by glucose 6-phosphate dehydrogenase (G-6-PD) and 6-phosphogluconate dehydrogenase. Both reactions require $NADP^+$ as cosubstrate, accounting for the formation of NADPH as a reduction product. Pentose phosphate formation is achieved by the decarboxylation of 6-phosphogluconate to form the pentose phosphate, ribulose 5-phosphate, which in turn is isomerized to its aldose isomer, ribose 5-phosphate. Pentose phosphates can subsequently be "recycled" back to hexose phosphates through the transketolase and transaldolase reactions as illustrated in Figure 4.11. This recycling of pentose phosphates, therefore, does not result in pentose utilization, but it does assure generous production of NADPH as the cycle repeats.

The cells of some tissues have a high demand for NADPH, particularly those that are active in the synthesis of fatty acids, such as the mammary gland, adipose tissue, the adrenal cortex, and the liver. These tissues predictably engage the entire pathway, recycling pentose phosphates back to glucose 6-phosphate to repeat

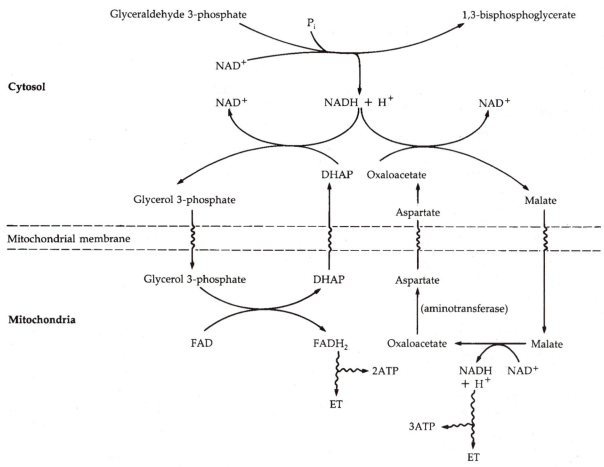

Figure 4.10 The glycerol 3-phosphate and malate-aspartate shuttle systems by which cystosolic NADH can be indirectly reoxidized by mitochondrial electron transport. Reducing equivalents of NADH are reoxidized by either FAD-NAD$^+$-requiring enzymes leading to the production of either two or three ATPs, respectively, per mole of cosubstrate. Abbreviations: DHAP, dihydroxyacetone phosphate. ET, electron transport.

the cycle and to assure an ample supply of NADPH. The pathway reactions that include the dehydrogenase reactions and therefore the formation of NADPH from NADP$^+$ are referred to as the oxidative reactions of the pathway. This segment of the pathway is illustrated on the left in Figure 4.11. The re-formation of glucose 6-phosphate from the pentose phosphates, through reactions catalyzed by transketolase, transaldolase, and hexose phosphate isomerase, are the nonoxidative reactions of the pathway, and are shown as on the right in Figure 4.11. Transketolase and transaldolase enzymes catalyze complex reactions in which three-, four-, five-, six-, and seven- carbon phosphate sugars are interconverted. These reactions are detailed in most comprehensive textbooks of biochemistry.

It should be noted that the reversibility of the transketolase and transaldolase reactions allows the direct conversion of hexose phosphates into pentose phosphates while bypassing the oxidative reactions. There-

fore, cells that undergo a more rapid rate of replication, and consequently have a greater need for pentose phosphates for nucleic acid synthesis, can produce these products in this manner.

The shunt is active in liver, adipose tissue, adrenal cortex, thyroid gland, testis, and lactating mammary gland. Its activity is low in skeletal muscle because of the limited demand for NADPH (fatty acid synthesis) in this tissue and also because of muscle's reliance on glucose for energy metabolism.

The Krebs Cycle

Also referred to as the *tricarboxylic acid cycle* or the *citric acid cycle,* the Krebs cycle is at the forefront of energy metabolism in the body. It can be thought of as the common and final catabolic pathway, because products of carbohydrate, fat, and amino acids that enter the cycle can be completely oxidized to CO$_2$ and H$_2$O, with the

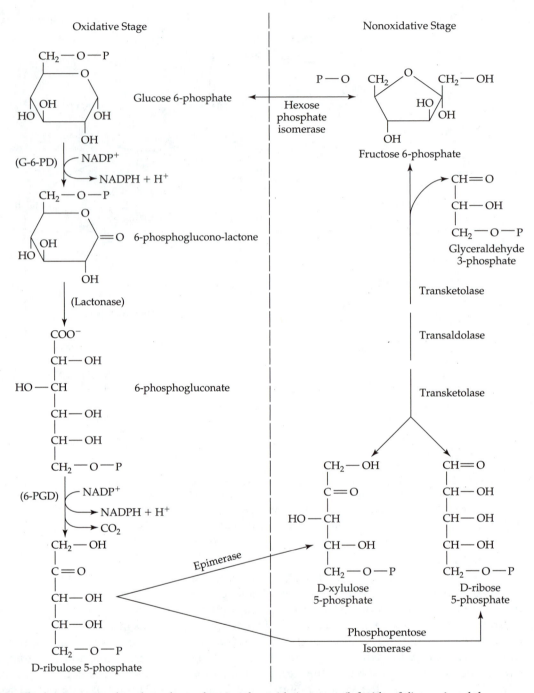

Figure 4.11 The hexosemonophosphate shunt, showing the oxidative stage (left side of diagram) and the nonoxidative stage (right side of diagram). The reversibility of the transketolase and transaldolase reactions makes it possible to form pentose phosphates directly from hexose phosphates (the nonoxidative pathway), while bypassing the dehydrogenation reactions of the oxidative pathway. For an explanation of how the shunt and its stages relate to cell specialization, see the text. Abbreviations: G 6-PD, glucose 6-phosphate dehydrogenase; 6-PGD, 6-phosphogluconate dehydrogenase.

accompanying release of energy. It is estimated that over 90% of the energy released from food occurs as a result of Krebs cycle oxidation. Not all of the substances entering the cycle are totally oxidized, however. Some Krebs cycle intermediates are used to form glucose by the process of gluconeogenesis, which is discussed in the next section, and some can be converted to certain amino acids by transamination (p. 185).

The high energy output of the Krebs cycle is attributed to mitochondrial electron transport, with oxidative phosphorylation being the source of ATP formation, as discussed in Chapter 3. The oxidation reactions occurring in the cycle are actually dehydrogenations in which an enzyme catalyzes the removal of two hydrogens to an acceptor cosubstrate such as NAD^+ or FAD. Because the enzymes of the cycle and the enzymes and electron carriers of electron transport are both compartmentalized within the mitochondria, the reduced cosubstrates, NADH and $FADH_2$ are readily reoxidized by O_2 via the electron transport chain.

In addition to its production of the reduced cosubstrates NADH and $FADH_2$, which furnish the energy through their oxidation via electron transport, the Krebs cycle produces most of the carbon dioxide through decarboxylation reactions. In terms of glucose metabolism, it must be recalled that two pyruvates are produced from one glucose during cytoplasmic glycolysis. These pyruvates are in turn transported into the mitochondria, where decarboxylation leads to the formation of two acetyl CoA units and two molecules of CO_2. The two carbons represented by the acetyl CoA are additionally lost as CO_2 through Krebs cycle decarboxylations. Most of the CO_2 produced is exhaled through the lungs, although some is used in certain synthetic reactions called *carboxylations.*

The Krebs cycle is shown in Figure 4.12. The acetyl CoA, which couples with oxaloacetate to begin the pathway, is formed from numerous sources, including the breakdown of fatty acids, glucose (through pyruvate), and certain amino acids. Its formation from pyruvate is considered now, because pyruvate links cytoplasmic glycolysis to the Krebs cycle.

The reaction shown in 4I is referred to as the *pyruvate dehydrogenase reaction.* Actually, however, the reaction is a complex one requiring a multienzyme system and various cofactors. The enzymes and cofactors are contained within an isolable unit called the *pyruvate dehydrogenase complex.* The cofactors include coenzyme A (CoA), thiamine pyrophosphate (TPP), Mg^{+2}, NAD^+, FAD, and lipoic acid. Four vitamins are therefore necessary for the activity of the complex: pantothenic acid (a component of CoA), thiamine, niacin, and riboflavin. The role of these vitamins and others as precursors of coenzymes is discussed in Chapter 9. The enzymes include pyruvate decarboxylase, dihydrolipoyl dehydrogenase, and dihydrolipoyl transacetylase. The net effect of the complex results in decarboxylation and dehydrogenation of pyruvate, with NAD^+ serving as the terminal hydrogen acceptor. This reaction therefore yields energy, because the reoxidation by electron transport of the NADH produces 3 mol of ATP by oxidative phosphorylation. The reaction is regulated negatively by acetyl CoA and by NADH, and positively by ADP and Ca^{+2}.

The condensation of acetyl CoA with oxaloacetate initiates the Krebs cycle reactions. The following are comments on reactions:

1. The formation of citrate from oxaloacetate and acetyl CoA is catalyzed by citrate synthase. The reaction is regulated negatively by ATP.

2. The isomerization of citrate to isocitrate involves cis aconitate as an intermediate. The isomerization, catalyzed by aconitase, involves dehydration followed by sterically reversed hydration, resulting in the repositioning of the −OH group onto an adjacent carbon.

3. The first of four dehydrogenation reactions within the cycle, the isocitrate dehydrogenase reaction supplies energy through the respiratory chain reoxidation of the NADH. Note that the first loss of CO_2 in the cycle occurs at this site. It arises from the spontaneous decarboxylation of an intermediate compound, oxalosuccinate. The reaction is positively modulated by ADP and negatively modulated by ATP and NADH.

4. The decarboxylation and dehydrogenation of α-ketoglutarate is mechanistically identical to the pyruvate dehydrogenase complex reaction in its multienzyme-cofactor requirement. In the reaction, referred to as the *α-ketoglutarate dehydrogenase* reaction, NAD^+ serves as hydrogen acceptor, and a second carbon is lost as CO_2. The pyruvate dehydrogenase, isocitrate dehydrogenase, and α-ketoglutarate dehydrogenase reactions account for the loss of the three-carbon *equivalent* of pyruvate as CO_2.

(4I)

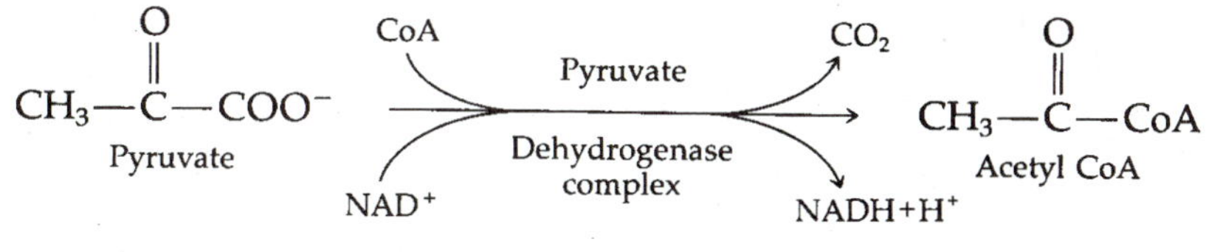

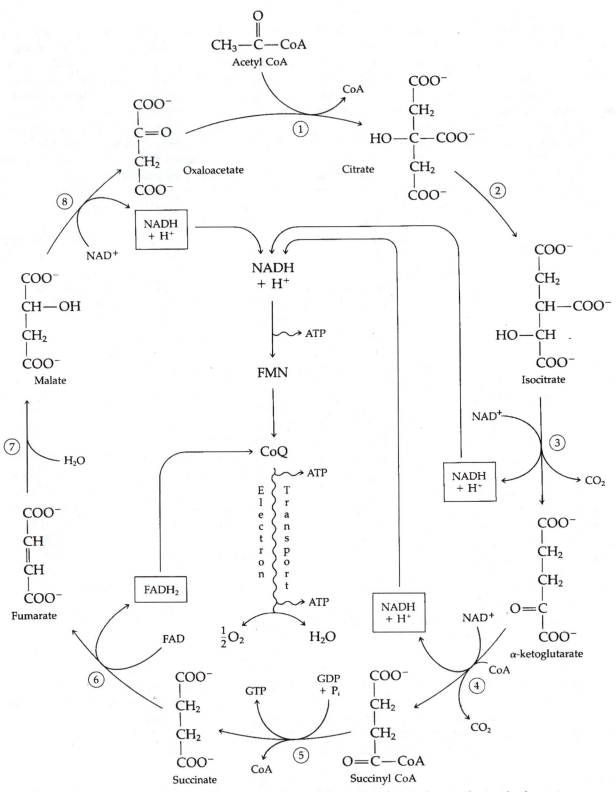

Figure 4.12 The Krebs (citric acid) cycle. This representation of the cycle is designed to emphasize the formation of reduced coenzymes and how their reoxidation by electron transport contributes to the synthesis of ATP.

5. Energy is conserved in the thioester bond of succinyl CoA. The hydrolysis of that bond by succinyl thiokinase releases sufficient energy to drive the phosphorylation of guanosine diphosphate (GDP) by inorganic phosphate. The resulting GTP is a high-energy phosphate anhydride compound like ATP; as such, GTP can serve as phosphate donor in certain phosphorylation reactions. One such reaction occurs in gluconeogenesis (p. 94).

6. The succinate dehydrogenase reaction uses FAD instead of NAD^+ as hydrogen acceptor. The $FADH_2$ is reoxidized by electron transport to O_2, but only two ATPs are formed by oxidative phosphorylation instead of three (pp. 64, 66).

7. Fumarase incorporates the elements of H_2O across the double bond of fumarate to form malate.

8. The conversion of malate to oxaloacetate completes the cycle. NAD^+ acts as a hydrogen acceptor in this dehydrogenation reaction catalyzed by malate dehydrogenase. It is the fourth site of reduced cosubstrate formation and therefore of energy release in the cycle.

In summary, the complete oxidation of glucose to CO_2 and H_2O can be shown by this equation:

$$C_6H_{12}O_6 + 6\,O_2 \longrightarrow 6\,CO_2 + 6\,H_2O + energy$$

This is achieved by the combined reaction sequences of the glycolytic and Krebs cycle pathways. Under aerobic conditions, the amount of released energy conserved as ATP is therefore as follows:

1. The glycolytic sequence, glucose $\longrightarrow$ two pyruvates, produces two ATPs by substrate-level phosphorylation:

(2 ATPs)

2. The two NADHs formed in the glycolytic sequence at the glyceraldehyde 3-phosphate dehydrogenase reaction yield either four or six ATPs depending on the shuttle system for NADH-reducing equivalents. Generally six will be formed, due to the overall greater activity of the malate/aspartate shuttle system.

(6 ATPs)

3. The intramitochondrial pyruvate dehydrogenase reaction yields 2 mol of NADH, one for each pyruvate oxidized and therefore six additional ATPs by oxidative phosphorylation.

(6 ATPs)

The oxidation of 1 mol of acetyl CoA in the Krebs cycle yields a total of 12 ATPs. The sites of formation, indicated by reaction number, follow:

3. 3 ATPs

4. 3 ATPs

5. 1 ATP (as GTP)

6. 2 ATPs

8. 3 ATPs

Total: 12 ATPs × 2 mol of acetyl CoA per mole of glucose = 24 ATPs

The total number of ATPs produced for the complete oxidation of 1 mol of glucose is therefore 38, equivalent to 262.8 kcal. Recall from Chapter 3 that this figure represents only about 40% of the total energy released by mitochondrial electron transport. The remaining 60%, or approximately 394 kcal, is released as heat to maintain body temperature.

As already mentioned, acetyl CoA is produced by fatty acid oxidation and amino acid catabolism as well as from the pyruvate derived from glycolysis, a fact that we address again in Chapters 6 and 7. This clearly leads to an imbalance between the amount of acetyl CoA and oxaloacetate, which condense one to one stoichiometrically in the citrate synthase reaction. It is therefore important that oxaloacetate and/or Krebs cycle intermediates, which can form oxaloacetate, be replenished in the cycle. Such a mechanism does exist. Oxaloacetate, fumarate, succinyl CoA, and α-ketoglutarate can all be formed from certain amino acids, but the single most important mechanism for ensuring an ample supply of oxaloacetate is the reaction by which it is formed directly from pyruvate. This reaction, shown here, is catalyzed by pyruvate carboxylase. The "uphill" incorporation of CO_2 is accomplished at the expense of ATP, and the reaction requires the participation of biotin (see 4J). The conversion of pyruvate into oxaloacetate is called an *anaplerotic* (filling-up) process because of its role in restoring oxaloacetate to the cycle. It is interesting that pyruvate carboxylase is regulated positively by acetyl CoA, thereby accelerating oxaloacetate formation in response to increasing levels of acetyl CoA.

(4J)

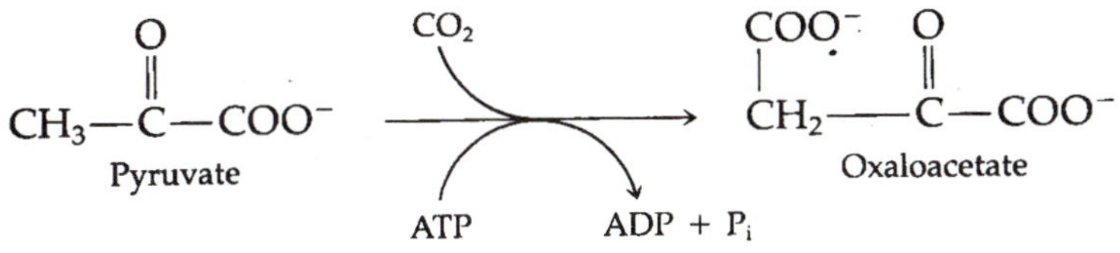

Gluconeogenesis

D-glucose is an essential nutrient for the proper function of most cells. The brain and other tissues of the central nervous system (CNS) and red blood cells are particularly dependent on glucose as a nutrient. When dietary intake of carbohydrate is reduced and blood glucose concentration declines, a hormonal triggering of accelerated glucose synthesis from noncarbohydrate sources occurs. Lactate, pyruvate, glycerol (a catabolic product of triacylglycerols), and certain amino acids represent the important noncarbohydrate sources. The process of producing glucose from such compounds is termed *gluconeogenesis.* The liver is the major site of this activity, although under certain circumstances, such as prolonged starvation, the kidneys become increasingly important in gluconeogenesis.

Gluconeogenesis is essentially a reversal of the glycolytic pathway. Most of the cytoplasmic enzymes involved in the conversion of glucose to pyruvate catalyze their reactions reversibly and therefore provide the means for also converting pyruvate to glucose. There are three reactions in the glycolytic sequence that are *not reversible:* the hexokinase, phosphofructokinase, and pyruvate kinase reactions (sites 1, 3, and 10 in Fig. 4.9). They are unidirectional by virtue of the high, negative-free energy change of the reactions. Therefore, the process of gluconeogenesis requires that these reactions be bypassed or circumvented by other enzyme systems. The presence or absence of these enzymes determines whether a certain organ or tissue is capable or incapable of conducting gluconeogenesis. As shown in 4K, the hexokinase and phosphofructokinase reactions are bypassed by specific phosphatases (glucose 6-phosphatase and fructose 1,6-bisphosphatase, respectively) that remove phosphate groups by hydrolysis.

The bypass of the pyruvate kinase reaction involves the formation of oxaloacetate as an intermediate. Mitochondrial pyruvate can be converted to oxaloacetate by pyruvate carboxylase, a reaction that has been discussed as an anaplerotic process (p. 93). Oxaloacetate can, in turn, be decarboxylated and phosphorylated to phosphoenolpyruvate (PEP) by PEP carboxykinase, thereby completing the bypass of the pyruvate kinase reaction. The PEP carboxykinase reaction is a cytoplasmic reaction, however, and oxaloacetate must therefore leave the mitochondrion to be acted on by the enzyme. The mitochondrial membrane is, however, impermeable to oxaloacetate, which therefore must first be converted to either malate (by malate dehydrogenase) or aspartate (by transamination with glutamate; see Chap. 7), either of which freely traverse the mitochondrial membrane. In the cytoplasm, the malate or aspartate can be converted to oxaloacetate by malate dehydrogenase or aspartate aminotransferase (glutamate oxaloacetate transaminase), respectively.

The reactions of the pyruvate kinase bypass also allow the carbon skeletons of various amino acids to enter the gluconeogenic pathway and lead to a net synthesis of glucose. Such amino acids are accordingly called *glucogenic.* They can be metabolically converted to pyruvate or to various Krebs cycle intermediates. As such, they can ultimately leave the mitochondrion in the form of malate or aspartate, as described. Reactions showing the entry of noncarbohydrate substances into the gluconeogenic system are shown in Figure 4.13, along with the bypass of the pyruvate kinase reaction.

Liver gluconeogenesis accounts for that organ's ability to control the high levels of blood lactate that may accompany strenuous physical exertion. Muscle and adipose tissue, for example, are unable to form free glucose from noncarbohydrate precursors because they lack glucose 6-phosphatase. This means that muscle and adipose lactate cannot serve as a precursor for free glucose within these tissues. Also, muscle cells convert lactate into glycogen only very slowly, especially in the presence of glucose. How, then, is the high level of muscle lactate that can be encountered in situations of oxy-

(4K)

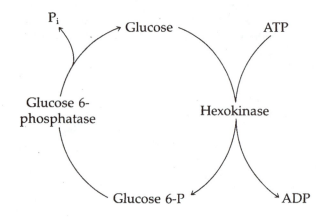

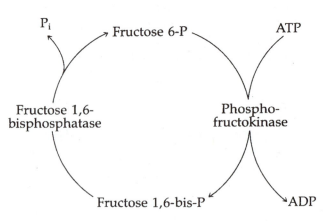

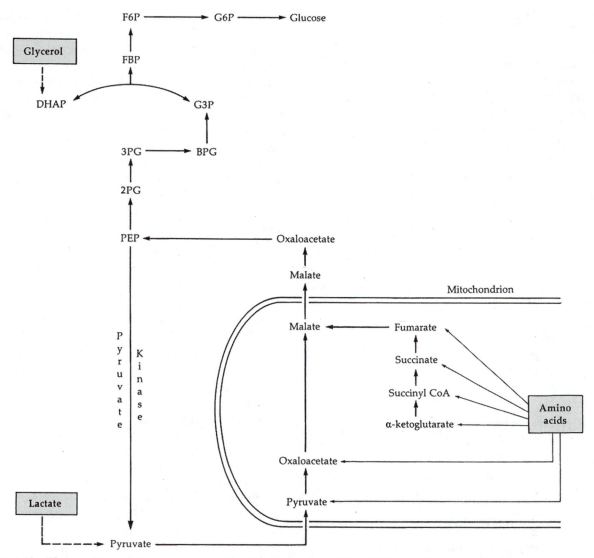

Figure 4.13 The reactions of gluconeogenesis, showing the bypass of the unidirectional pyruvate kinase reaction and the entry into the pathway of noncarbohydrate substances such as glycerol, lactate, and amino acids. Abbreviations: G6P, glucose 6-phosphate; F6P, fructose 6-phosphate; FBP, fructose 1,6-bisphosphate; DHAP, dihydroxyacetone phosphate; G3P, glyceraldehyde 3-phosphate; BPG, 1,3-bisphosphoglycerate; 3PG, 3-phosphoglycerate; 2PG, 2-phosphoglycerate; PEP, phosphoenolpyruvate.

gen debt dealt with? Recovery is accomplished by the gluconeogenic capability of the liver. The lactate leaves the muscle cells and is transported via the general circulation to the liver, where it can be converted to glucose. The glucose can then be returned to the muscle cells to reestablish homeostatic concentrations there. This circulatory transport of muscle-derived lactate to the liver and the return of glucose to the muscle is referred to as the *Cori cycle.*

During the past decade, evidence has emerged from in vitro studies that glucose, as the sole substrate at physiological concentrations, has limited use by the liver and is, in fact, a poor precursor of liver glycogen.

However, use is greatly enhanced if gluconeogenic substances such as fructose, glycerol, or lactate are available along with the glucose. The simple incorporation of glucose into glycogen in vivo, but its limited conversion in vitro, in the absence of other gluconeogenic compounds, has been referred to as the *glucose paradox* [10]. It is now believed that glucose ingested during a meal takes a somewhat roundabout path to glycogen. It is first taken up by red blood cells in the bloodstream and converted to lactate by glycolysis. The lactate is then taken up by the liver and converted to glucose 6-phosphate (and then to glycogen) by gluconeogenesis.

Regulation of Metabolism

The purpose of regulation is to both maintain homeostasis and to alter the reactions of metabolism in such a way as to meet the nutritional and biochemical demands of the body. An excellent example is the reciprocal regulation of the glycolysis and Krebs cycle (catabolic) pathways and the gluconeogenic (anabolic) pathways. Because the glycolytic conversion of glucose to pyruvate liberates energy, the reversal of the process, gluconeogenesis, must therefore be energy consuming. The pyruvate kinase bypass in itself is energetically expensive, considering that 1 mol of ATP and 1 mol of GTP must be expended in converting intramitochondrial pyruvate to extramitochondrial PEP. It follows that among the factors that regulate the glycolysis:gluconeogenesis activity ratio is the body's need for energy. Our discussion focuses on the body's requirements for energy and on how regulation can speed up or slow down the activity of the metabolic pathways contributing to release or consumption of energy.

In a broad sense, regulation is achieved by three mechanisms:

- Negative or positive modulation of allosteric enzymes by effector compounds
- Hormonal activation by covalent modification or induction of specific enzymes
- Directional shifts in reversible reactions by changes in reactant or product concentrations

Allosteric Enzyme Modulation

Allosteric enzymes can be stimulated or suppressed by certain compounds, usually formed within the pathway in which the enzymes function. An allosteric, or regulatory, enzyme is said to be positively or negatively modulated by a substance (modulator) according to whether the effect is stimulation or suppression, respectively. Modulators generally act by altering the conformational structure of their allosteric enzymes, causing a shift in the equilibrium between so-called tight and relaxed conformations of the enzyme. The enzyme is functionally more active in its relaxed form than when it is in its tight form. A positive modulator causes a shift toward the relaxed configuration, whereas a negative modulator shifts the equilibrium toward the tight form.

An important regulatory system in energy metabolism is the cellular concentration ratio of ADP (or AMP) to ATP. The usual breakdown product of ATP is ADP, but as ADP increases in concentration, some becomes enzymatically converted to AMP. Therefore, both ADP and/or AMP accumulation can signify an excessive breakdown of ATP and a depletion of ATP.

AMP, ADP, and ATP all act as modulators of certain allosteric enzymes, but the effect of AMP or ADP will oppose that of ATP. For example, if ATP accumulates, which might occur during a period of muscular relaxation, it negatively modulates certain regulatory enzymes in energy-releasing (ATP-producing) pathways. This acts to reduce the production of additional ATP. An increase in AMP (or ADP) concentration conversely signifies a depletion of ATP and the need to produce more of this energy source. In such a case, AMP or ADP, as their concentration increases, can positively modulate allosteric enzymes functioning in energy-releasing pathways. Two examples of positive modulation by AMP are

1. its ability to bring about a shift from the inactive form of phosphorylase b to an active form of phosphorylase b (see Fig. 4.8) and

2. its stimulation, by a similar mechanism, of the enzyme phosphofructokinase, which catalyzes a reaction in the glycolytic pathway.

It can be reasoned that an enhanced activity of either of these reactions encourages glucose catabolism. The resulting shift in metabolic direction, as signaled by the AMP buildup, therefore releases energy and helps restore depleted stores of ATP.

In addition to being positively modulated by AMP, phosphofructokinase is also positively modulated by ADP and negatively by ATP. So as the store of ATP increases and further energy release is not called for, ATP can signal the slowing of the glycolytic pathway at that reaction. Phosphofructokinase is an extremely important rate controlling allosteric enzyme and is modulated by a variety of substances. Its regulatory function has already been described in Chapter 1.

Other regulatory enzymes in carbohydrate metabolism are modulated by ATP, ADP, or AMP. Pyruvate dehydrogenase complex, citrate synthase, and isocitrate dehydrogenase are negatively modulated by ATP. Pyruvate dehydrogenase complex is positively modulated by AMP, and citrate synthase and isocitrate dehydrogenase are positively regulated by ADP.

The ratio of NADH to NAD^+ also has an important regulatory effect. Certain allosteric enzymes are responsive to an increased level of NADH or NAD^+, which therefore regulate their own formation through negative modulation of those enzymes. Because NADH is a product of the oxidative catabolism of carbohydrate, its accumulation would signal for a decrease in catabolic pathway activity. Conversely, higher proportions of NAD^+ signify that a system is in an elevated state of oxidation readiness and would send a modulating signal to accelerate catabolism. Stated in a different way, the level of NADH in the fasting state is markedly lower than in the fed state because the rate of its reoxidation by electron transport would exceed its formation from substrate oxidation. Fasting, therefore, logically encourages glycolysis and Krebs cycle oxidation of carbohydrates. Dehydrogenase reactions, which involve the interconversion of the reduced and oxidized forms of the cosubstrate, are reversible. If metabolic conditions lead to the accumulation of one form or the other, the equilibrium is shifted so as to consume more of the predominant form. Pyruvate dehydrogenase complex is positively modulated by NAD^+, and pyruvate kinase, citrate synthase, and α-ketoglutarate dehydrogenase are negatively modulated by NADH.

As discussed previously in Chapter 1 (in the section on enzymes), allosteric enzymes catalyze unidirectional, or nonreversible, reactions. This is because modulators of those enzymes must either stimulate or suppress a reaction in one direction only. The stimulation or suppression of an enzyme that catalyzes both the forward and reverse direction of a reaction would have little value.

Hormonal Regulation

Hormones can regulate specific enzymes by either *covalent regulation* or by *enzyme induction.* The term *covalent regulation* refers to the phosphorylation and dephosphorylation of the enzymes being regulated, which converts them into active or inactive forms. In some instances, phosphorylation activates and dephosphorylation inactivates the enzyme. In other cases, the reverse may be true. Examples are found in the covalent regulation of glycogen synthase and glycogen phosphorylase, enzymes that have been discussed under the glycogenesis and glycogenolysis sections, respectively. Phosphorylation inactivates glycogen synthase, whereas dephosphorylation activates it. In contrast, phosphorylation activates glycogen phosphorylase, and dephosphorylation inactivates it.

Another very important example of covalent regulation by a hormone is the control by glucagon of the relative rates of liver glycolysis and gluconeogenesis. The control is directed at the opposing reactions of the phosphofructokinase (PFK) and fructose bisphosphatase (FBPase) site, and it is mediated through a compound called fructose 2,6-bisphosphate. Unlike fructose 1,6-bisphosphate, fructose 2,6-bisphosphate is not a normal glycolysis intermediate but serves solely as a regulator of pathway activity. *Fructose 2,6-bisphosphate stimulates PFK activity and suppresses FBPase activity, therefore stimulating glycolysis and reducing gluconeogenesis.* Cellular concentration of fructose 2,6-bisphosphate is set by the relative rates of its formation and breakdown. It is formed by phosphorylation of fructose 6-phosphate by phosphofructokinase 2

(PFK-2) and is broken down by fructose bisphosphatase 2 (FBPase-2). The designation 2 distinguishes these enzymes from PFK and FBPase, which catalyze the formation and breakdown, respectively, of fructose 1,6-bisphosphate. PFK-2 and FBPase-2 activities are expressed by a single (bifunctional) enzyme, and the relative activity of each is controlled by glucagon. Glucagon stimulates the phosphorylation of the bifunctional enzyme, resulting in a sharply increased activity of FBPase-2 activity and suppression of PFK-2 activity. *Glucagon therefore stimulates hepatic gluconeogenesis and suppresses glycolysis by reducing the concentration of fructose 2,6-bisphosphate, a positive modulator of the glycolytic enzyme PFK.* The end result, therefore, is that in response to falling blood glucose levels, the release of glucagon encourages hepatic gluconeogenesis to restore glucose levels.

Covalent regulation is usually mediated through cAMP, which acts as a second messenger in the hormones' action on the cell. It will be recalled that insulin strongly affects the glycogen synthase reaction positively and that epinephrine and glucagon positively regulate glycogen phosphorylase in muscle and the liver, respectively. Each of these hormonal effects is mediated through covalent regulation.

The control of enzyme activity by hormone induction represents another mechanism of regulation. Enzymes functioning in the glycolytic and gluconeogenic pathways can be divided into the three following groups:

Group 1. *Glycolytic enzymes*
Glucokinase
Phosphofructokinase
Pyruvate kinase

Group 2. *Bifunctional enzymes*
Phosphoglucoisomerase
Aldolase
Triosephosphate isomerase
Glyceraldehyde 3-phosphate dehydrogenase
Phosphoglycerate kinase
Phosphoglyceromutase
Enolase
Lactate dehydrogenase

Group 3. *Gluconeogenic enzymes*
Glucose 6-phosphatase
Fructose biphosphatase
PEP carboxykinase
Pyruvate carboxylase

Groups 1 and 3 are inducible enzymes, meaning that their concentrations can rise and fall in response to molecular signals such as a sustained change in the concentration of a certain metabolite. Such a change might arise through a prolonged shift in the dietary intake of certain nutrients. Induction results in the stimulated transcription of new messenger RNA programmed to produce the hormone. Glucocorticoid hormones are known to stimulate gluconeogenesis by inducing the formation of the key gluconeogenic enzymes, and insulin may stimulate glycolysis by inducing an increased synthesis of key glycolytic enzymes. Group 2 enzymes are not inducible and are produced at a steady rate under the control of constitutive, or basal, gene systems. Noninducible enzymes are required all the time at a relatively constant level of activity, and their genes are expressed at a more or less constant level in virtually all cells. Genes for enzymes that are not inducible are sometimes called *housekeeping genes.*

The interrelationship among pathways of carbohydrate metabolism is exemplified by the regulation of blood glucose concentration. Largely through the opposing effects of insulin and glucagon, fasting serum glucose level is normally maintained within the approximate range of 60 to 90 mg/dL. Whenever hyperglycemia is excessive or sustained owing to an insufficiency of insulin, other insulin-independent pathways of carbohydrate metabolism for lowering blood glucose become increasingly active. Such insulin-independent pathways are indicated in Figure 4.14. The overactivity of these pathways is believed to be partly responsible for the clinical manifestations of diabetes mellitus, type 1 (see Chap. 8's Perspective).

Directional Shifts in Reversible Reactions

Most enzymes catalyze reactions reversibly, and the preferred direction that a reversible reaction is undergoing at a particular moment is largely dependent on the relative concentration of each reactant. A buildup in concentration of one of the reactants will drive or force the reaction toward formation of the other. For example, consider the hypothetical pathway intermediates A and B, which are interconverted reversibly. Reaction 1 (see following diagrams) may represent the reaction in a metabolic steady state in which the formation of A from B is preferred over the formation of B from A. It shows a net formation of A from B as indicated by the size of the directional arrows. Reaction 2 shows that the steady state shifts toward the formation of B from A if some metabolic event or demand causes the concentration of A to rise above its homeostatic levels.

$$\textit{Reaction 1} \qquad \overset{\longleftarrow}{\underset{\longrightarrow}{} } A \longleftarrow B \longleftarrow$$

$$\textit{Reaction 2} \qquad \longrightarrow A \longrightarrow B \longrightarrow \\ \underset{\longleftarrow}{}$$

This concept is exemplified by the phosphoglucomutase reaction, which interconverts glucose 6-phosphate and glucose 1-phosphate, and which functions in the pathways of glycogenesis and glycogenolysis (see Figs. 4.6 and 4.7). At times of heightened glycogenolytic activity (rapid breakdown of glycogen), glucose 1-phosphate concentration rises sharply, driving the reaction toward the formation of glucose 6-phosphate. With the body at rest, glycogenesis and gluconeogenesis are accelerated, increasing the concentration of glucose 6-phosphate. This in turn shifts the phosphoglucomutase reaction toward the formation of glucose 1-phosphate and ultimately glycogen.

Ethyl Alcohol: Metabolism and Biochemical Impact

Ethyl alcohol (ethanol) is a purely exogenous compound in humans, consumed in the form of alcoholic

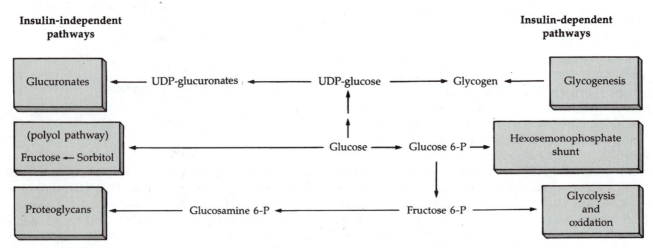

Figure 4.14 Insulin-independent and insulin-dependent pathways of glucose metabolism.

beverages such as beer, wines, and distilled liquors. Although it is not a "natural" nutrient, ethanol does have caloric value, although its calories are "empty," meaning that the substance is devoid of beneficial nutrients. Each gram of ethanol yields 7 kcal, and ethanol may account for up to 10% of the total energy intake of moderate consumers and up to 50% for alcoholics. Because of its widespread consumption and relatively high caloric potency, it commands attention in a textbook of nutrition. Empirically, ethanol's structure (CH_3–CH_2–OH) resembles most closely a carbohydrate; therefore, we have chosen to review it in this chapter.

Ethanol is readily absorbed throughout the entire gastrointestinal tract. It is transported unaltered in the bloodstream and then oxidatively degraded in tissues, primarily the liver, first to acetaldehyde, and then to acetate. In tissues peripheral to the liver, as well as in the liver itself, the acetate is subsequently converted to acetyl CoA and oxidized via the Krebs cycle. At least three enzyme systems are capable of ethanol oxidation:

- alcohol dehydrogenase (ADH),
- the microsomal ethanol-oxidizing system (MEOS; or the cytochrome P-450 system), and
- catalase, in the presence of hydrogen peroxide.

Of these, the catalase-H_2O_2 system is the least active of the three, probably accounting for <2% of in vivo ethanol oxidation. Therefore, it will not be discussed further. Nearly all ingested ethanol is oxidized by hepatic (and to some extent, gastric) alcohol dehydrogenase and hepatic microsomal cytochrome P-450 systems.

The Alcohol Dehydrogenase (ADH) Pathway

ADH is a soluble enzyme functioning in the cytoplasm of liver cells. It is an ordinary NAD^+-requiring dehydrogenase, and it is known to be able to oxidize ethanol to acetaldehyde.

$$\text{CH}_3\text{–CH}_2\text{–OH} + \text{NAD}^+ \xleftrightarrow{\text{ADH}} \text{CH}_3\text{–CHO} + \text{NADH} + \text{H}^+$$
ethanol $\qquad\qquad\qquad$ acetaldehyde

The NADH formed by the reaction can be oxidized by mitochondrial electron transport, by way of the NADH shuttle systems (p. 87), thereby giving rise to ATP formation by oxidative phosphorylation. The K_m of alcohol dehydrogenase for ethanol is approximately 1 mM, or about 5 mg/dL. This means that at that cellular concentration of ethanol, ADH is functioning at one-half of its maximum velocity. At concentrations three or four times the K_m, the enzyme is saturated with the ethanol substrate and is catalyzing at its maximum rate. It follows that any ethanol present in the cell in excess

of the 4 × K_m level cannot be oxidized by ADH. (K_m is reviewed in Chap. 1, in the section dealing with enzymes.) There is, of course, no "normal" concentration of ethanol in the cells or bloodstream. The so-called toxic level of blood ethanol, however, is considered to be in the range of 50 to 100 mg/dL. The high lipid solubility of ethanol allows it to enter cells passively with ease, and if its cellular concentration reaches a level even a third or a fourth of that in blood, ADH would be saturated by the substrate and would be functioning at its maximum velocity. The excess, or "spillover," ethanol must then be metabolized by alternate systems, the most important of which is the microsomal ethanol oxidizing system, described next. Another factor forcing a shift to the microsomal metabolizing system is a depletion of NAD^+, brought about by the high level of activity of ADH. The microsomal system does not require NAD^+ for its oxidative reactions.

Alcohol dehydrogenase activity is also found in gastric mucosal cells, and interestingly, there appears to be a significant gender difference in the level of its activity there. Young (premenopausal) females develop higher blood alcohol levels than male counterparts at equal consumption. They consequently display a lower tolerance for alcohol and run a greater risk for its toxic effects on the liver. It is believed that the lower level of alcohol dehydrogenase activity in the female gastric mucosa accounts for this observation [11,12].

The Microsomal Ethanol Oxidizing System (MEOS)

Despite its name, the microsomal ethanol oxidizing system (MEOS) is able to oxidize a wide variety of compounds in addition to ethanol. These include fatty acids, aromatic hydrocarbons, steroids, and barbiturate drugs. The oxidation occurs through a system of electron transport, similar to the mitochondrial electron transport system described in detail in Chapter 3. Because MEOS is microsomal and is associated with the smooth endoplasmic reticulum, it is sometimes referred to as the "microsomal electron transport system." Another distinction of the system is its requirement for a special cytochrome called *cytochrome P-450,* which acts as an intermediate electron carrier. Cytochrome P-450 is not a singular compound but exists as a family of structurally related cytochromes, the members of which share the property of absorbing light having a wavelength of 450 nm.

Ethanol oxidation by MEOS is linked to the simultaneous oxidation of NADPH by molecular oxygen. Because two substrates are therefore oxidized concurrently, the enzymes that are involved in the oxidations are commonly called mixed-function oxidases. One oxygen atom of the oxygen molecule is used to oxidize NADPH to NADP$^+$, and the second oxidizes the ethanol substrate to acetaldehyde. Both oxygen atoms are reduced to H_2O, and therefore two H_2O molecules are formed in the reactions. Microsomal electron transport of the MEOS is shown in Figure 4.15. Acting as carriers of electrons from NADPH to oxygen are FAD, FMN, and a cytochrome P-450 system.

An important feature of the MEOS is that certain of its enzymes, including the cytochrome P-450 units, are *inducible* by ethanol. This means that ethanol, particularly at higher concentrations, can stimulate (induce) the synthesis of these substances. The result is that the hepatocytes can metabolize the ethanol much more effectively, thereby establishing a state of metabolic *tolerance*. Compared with a normal (nondrinking or light-drinking) subject, an individual in a state of metabolic tolerance to ethanol can ingest larger quantities of the substance before showing the effects of intoxication. In effect, he or she is "tolerant" to it. When enzyme induction occurs, however, it can also accelerate the metabolism of other substances metabolized by the microsomal system. In other words, tolerance to ethanol induced by heavy drinking can also render a person tolerant to other substances as well as ethanol.

Alcoholism: Biochemical and Metabolic Alterations

Excessive consumption of ethanol can lead to alcoholism, defined by the National Council on Alcoholism as that which is capable of producing pathological changes. Alcoholism is a serious socioeconomic and health problem, exemplified by the fact that in the United States, alcohol-related liver disease has been reported as the sixth leading cause of death [13]. The well-known consequences of alcoholism—fatty liver, hepatic disease (cirrhosis), lactic acidosis, and metabolic tolerance—can be explained by the manner in which ethanol is metabolized. Basically, the conse-

quences of excessive alcohol intake are explainable by metabolic effects of (1) acetaldehyde toxicity, (2) elevated NADH:NAD$^+$ ratio, (3) metabolic competition, and (4) induced metabolic tolerance.

A comprehensive review of the association of nutritional and biochemical alterations with alcoholism is available [14].

Acetaldehyde Toxicity

Both the ADH and the MEOS routes of ethanol oxidation produce acetaldehyde, which is believed to exert direct adverse effects on metabolic systems. For example, acetaldehyde is able to attach covalently to proteins forming protein adducts. Should the adduct involve an enzyme, the activity of that enzyme could be impaired. Acetaldehyde has also been shown to impede the formation of microtubules in liver cells and to cause the development of perivenular fibrosis, either of which is believed to initiate the events leading to cirrhosis. These and other possible adverse effects of acetaldehyde are reviewed by Lieber [15].

It was once thought that alcoholic cirrhosis was caused by malnutrition, as the drinker satisfied his or her caloric needs with the "empty" calories of alcohol at the expense of those of a nutritionally balanced diet. In view of the effect of high levels of acetaldehyde on hepatocyte structure and function, however, it is now known that chronic overindulgence can cause cirrhosis in the absence of nutritional deficiency, and even if the alcohol is coingested with an enriched diet.

Alcoholism is commonly associated with vitamin deficiency. This is understandable in light of the fact that the liver is not only a major storage depot for vitamins, but it also converts vitamins into metabolically useful forms. Folic acid deficiency is the most common vitamin deficiency among alcoholics. It is most likely caused by increased demand for nucleic acids needed in the regeneration of injured liver cells, although poor dietary intake of folate also contributes to the problem.

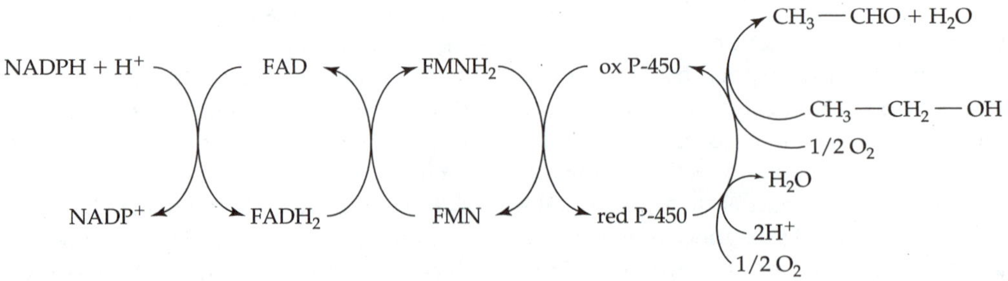

Figure 4.15 The microsomal ethanol oxidizing system (MEOS). Both ethanol and NADPH are oxidized by molecular oxygen in this electron transfer scheme. Because two substrates are oxidized, enzymes involved in the oxidation are referred to as *mixed-function oxidases*.

Other vitamins to which alcohol is antagonistic include vitamin A (see "Substrate Competition"), thiamin, nicotinic acid, and vitamin K.

High NADH:NAD⁺ Ratio

The oxidation of ethanol increases the concentration of NADH at the expense of NAD^+, therefore elevating the $NADH:NAD^+$ ratio. This is because both ADH and acetaldehyde dehydrogenase use NAD^+ as cosubstrate. NADH is an important regulator of certain dehydrogenase reactions. Its rise in concentration represents an overproduction of reducing equivalents, which in turn acts as a signal for a metabolic shift toward reduction—namely, hydrogenation. Such a shift can account for fatty liver and lactic acidemia, which often accompany alcoholism. For example, lactic acidemia can be attributed in part to the direct effect of NADH in shifting the lactate dehydrogenase (LDH) reaction toward the formation of lactate. The reaction, shown as follows, is driven to the right by the high concentration of NADH:

$$\text{Pyruvate} + \text{NADH} \xrightarrow[\underset{\text{LDH}}{\longleftarrow}]{} \text{lactate} + NAD^+$$

Lipids accumulate in most tissues in which ethanol is metabolized, resulting in fatty liver, fatty myocardium, fatty renal tubules, and so on. The mechanism appears to involve both increased lipid synthesis and decreased lipid removal, and is explained in part by the increased $NADH:NAD^+$ ratio. As NADH accumulates, it slows dehydrogenase reactions of the Krebs cycle, such as the isocitrate dehydrogenase and aketoglutarate dehydrogenase reactions, thereby slowing the overall activity of the cycle. This results in an accumulation of citrate, which positively regulates acetyl CoA carboxylase. Acetyl CoA carboxylase converts acetyl CoA into malonyl CoA by the attachment of a carboxyl group. It is the key regulatory enzyme for the synthesis of fatty acids from acetyl CoA (p. 151). The high ratio of $NADH:NAD^+$ therefore directs metabolism away from Krebs cycle oxidation and toward fatty acid synthesis.

Also contributing to the lipogenic effect of alcoholism is the NADH effect on the glycerophosphate dehydrogenase (GPDH) reaction. The reaction, shown as follows, favors the reduction of dihydroxyacetone phosphate (DHAP) to glycerol 3-phosphate if NADH concentration is high.

$$
\begin{array}{ccc}
CH_2-OH & & CH_2-OH \\
| & +H^+ & | \\
C=O \ +NADH & \underset{GPDH}{\rightleftharpoons} & CH_2-OH \ + NAD^+ \\
| & & | \\
CH_2-O-P & & CH_2-O-P \\
\text{DHAP} & & \text{Glycerol 3-P}
\end{array}
$$

Glycerol 3-phosphate provides the glycerol component in the synthesis of triacylglycerols. Therefore, a high $NADH:NAD^+$ ratio stimulates the synthesis of both the fatty acids and glycerol components of triacylglycerols, contributing to the cellular fat accumulation that develops in alcoholism.

The glutamate dehydrogenase (GluDH) reaction also is affected by a rise in NADH concentration, resulting in impaired gluconeogenesis. The GluDH reaction is extremely important in gluconeogenesis because of the role it plays in the conversion of amino acids into their carbon skeletons by transamination, and the release of their amino groups as NH_3. A shift in the reaction toward glutamate by NADH, shown as follows, depletes the availability of α-ketoglutarate, which is the major acceptor of amino groups in the transamination of amino acids.

$$
\begin{array}{ccc}
\textit{Glutamate} & & \textit{α-ketoglutarate} \\
COO^- & & COO^- \\
| & & | \\
CH-NH_3^+ + NAD^+ & \underset{GluDH}{\rightleftharpoons} & C=O + NADH+H^+ \\
| & & | \qquad +NH_3 \\
CH_2 & & CH_2 \\
| & & | \\
CH_2-COO^- & & CH_2-COO^-
\end{array}
$$

Substrate Competition

A well-established nutritional problem associated with excessive alcohol metabolism is a deficiency of vitamin A. Two aspects of ethanol interference on normal metabolism can probably account for this problem. One of these is the effect of ethanol on retinol dehydrogenase, the cytoplasmic enzyme that converts retinol to retinal. Retinol dehydrogenase is thought to be identical to ADH, and therefore ethanol competitively inhibits the hepatic conversion of retinol to retinal. Retinal is required for the synthesis of photopigments used in vision. In addition to this substrate competition effect, ethanol may interfere with retinol metabolism by induced metabolic tolerance.

Induced Metabolic Tolerance

As explained earlier, ethanol can induce enzymes of the MEOS, causing an increased rate of metabolism of substrates oxidized by this system. Retinol, like ethanol, spills over into the MEOS when ADH is saturated and NAD^+ stores are low because of heavy ingestion of ethanol. Ethanol induction of retinol-metabolizing enzymes then can occur. The specific component of the MEOS that is known to be induced by a heavy consumption of ethanol has been designated cytochrome P-450IIE1. Although induction accelerates the hepatic oxidation of retinol, the oxidation product is not retinal but other polar, inert products of oxidation. The hepatic depletion of retinol can therefore be attributed to its accelerated metabolism, secondary to ethanol induction of a metabolizing enzyme. In effect, the alcoholic subject becomes tolerant to vitamin A, necessitating a higher dietary intake of the vitamin to maintain normal hepatocyte concentrations.

Alcohol in Moderation: The Brighter Side

Alcohol is a nutritional "Jekyll and Hyde," and whichever face it flaunts is clearly a function of the extent to which it is consumed. We have focused earlier on the effects of alcohol at high intake levels and the negative impact of alcoholism on metabolism and nutrition. Many studies, however, have suggested that alcohol consumed *in moderation* may have beneficial effects, particularly in its ability to improve plasma lipid profiles and reduce the risk of cardiovascular disease.

Ethanol is known to elevate the level of high-density lipoprotein (HDL) in serum and to lower the amount of serum lipoprotein a [16][17]. Both effects favor a decrease in cardiovascular disease risk. This is because HDL protects against the deposition of arterial fatty plaque (atherogenesis), whereas high levels of lipoprotein a appear to promote it. The effect of these and other lipoproteins on atherogenesis will be discussed in Chapter 6. Other mechanisms for the apparent protective action of alcohol against atherogenesis have been suggested. A well-recognized component of the atherogenic process is the proliferation of smooth muscle cells underlying the endothelium of arterial walls. Studies have shown that alcohol may suppress the proliferation of smooth muscle cells, thereby slowing the atherogenic process [18].

Summary

The major sources of dietary carbohydrates are the starches and the disaccharides. In the course of digestion, these are hydrolyzed by specific glycosidases to their component monosaccharides, which are absorbed into the circulation from the intestine. The monosaccharides are transported to the cells of various tissues, passing through the cells' outer membrane by facilitative transport via transporters. Glucose is transported into the cells of many different tissues by the GLUT family of transporters. In the cells, monosaccharides are first phosphorylated at the expense of ATP and can then follow any of several integrated pathways of metabolism. Glucose is phosphorylated in most cells by hexokinase, but in the liver, by glucokinase. Fructose is phosphory-

lated mainly by fructokinase in the liver. Galactose is phosphorylated by galactokinase, also a liver enzyme.

Cellular glucose can be converted into glycogen, primarily in liver and skeletal muscle, or it can be routed through the energy-releasing pathways of glycolysis and the Krebs cycle in these and other tissues for ATP production. Glycolytic reactions convert glucose (or glucose residues from glycogen) to pyruvate. From pyruvate, either an aerobic course (complete oxidation in the Krebs cycle) or anaerobic course (to lactate) can be followed. Nearly all the energy released by the oxidation of carbohydrates to CO_2 and H_2O occurs in the Krebs cycle, as reduced coenzymes are oxidized via mitochondrial electron transport. Approximately 40% of this energy is retained in the high-energy phosphate bonds of ATP. The remaining energy supplies heat to the body.

Noncarbohydrate substances derived from the other major nutrients, fats and proteins, can be converted to glucose or glycogen by the pathways of gluconeogenesis. Basically, the reactions are the reversible reactions of glycolysis, shifted toward glucose synthesis in accordance with reduced energy demand by the body. Three kinase reactions occurring in glycolysis are not reversible, however, requiring the involvement of different enzymes and pathways to circumvent those reactions in the process of gluconeogenesis. Muscle glycogen provides a source of glucose for energy for that tissue only, because muscle lacks the enzyme glucose 6-phosphatase, which forms free glucose from glucose 6-phosphate. The enzyme is active in liver, however, meaning that the liver can release free glucose from its glycogen stores into the circulation for use by other tissues. The Cori cycle describes the liver's uptake and gluconeogenic conversion of working muscle lactate to glucose.

A metabolic pathway is regulated according to the body's need for energy or for maintaining homeostatic cellular concentrations of certain metabolites. Regulation is mainly exerted through hormones, through substrate concentrations (which can affect the velocity of enzyme reactions), and through allosteric enzymes that can be modulated negatively or positively by certain pathway products.

Ethanol is catabolized ultimately to acetyl CoA, furnishing calories through Krebs cycle oxidation. The nutritional complexities of alcohol abuse are discussed.

In Chapters 6 and 7, we will see that fatty acids and the carbon skeleton of various amino acids also are ultimately oxidized through the Krebs cycle. The amino acids that do become Krebs cycle intermediates may not, however, be completely oxidized to CO_2 and H_2O but instead may leave the cycle to be converted to glucose or glycogen (by gluconeogenesis), should dietary intake of carbohydrate be low. The glycerol portion of triacylglycerols enters the glycolytic pathway at the level of dihydroxyacetone phosphate, from which point it can be oxidized for energy or used to synthesize glucose or glycogen. The fatty acids of triacylglycerols enter the Krebs cycle as acetyl CoA, which is oxidized to CO_2 and H_2O but cannot contribute carbon for the net synthesis of glucose. This topic is considered further in Chapter 6.

These examples of the entrance of noncarbohydrate substances into the pathways discussed in this chapter are cited here to remind the reader that these pathways are not singularly committed to carbohydrate metabolism. Rather, they must be thought of as common ground for the interconversion and oxidation of fats and proteins as well. It is essential to maintain this broadened perspective as we move on into Chapters 6 and 7, in which the metabolism of lipids and proteins, respectively, are examined.

References Cited

1. Riby J, Fujisawa T, Kretchmer N. Fructose absorption. Am J Clin Nutr 1993;58,#5 suppl:748S–53S.

2. Burant C, Takeda J, Brot-Laroche E, Bell G, Davidson N. Fructose transporter in human spermatozoa and small intestine is GLUT5. J Biol Chem 1992;267:14523–6.

3. Truswell AS, Seach JM, Thorburn AW. Incomplete absorption of pure fructose in healthy subjects and the facilitating effect of glucose. Am J Clin Nutr 1988; 48:1424–30.

4. Baldwin SA. Mammalian passive glucose transporters: members of a ubiquitous family of active and passive transport proteins. Biochim et Biophysica Acta 1993;1154:17–49.

5. Bloc J, Gibbs EM, Lienhard GE, Slot JW, Geuze HJ. Insulin-induced translocation of glucose transporters from post-Golgi compartments to the plasma membrane of 3T3-L1 adipocytes. J Cell Biol 1988;106:69–76.

6. Freidman JE, Dudek RW, Whitehead DS, et al. Immunolocalization of glucose transporter GLUT4 within human skeletal muscle. Diabetes 1991;40:150–4.

7. Takata K, Hirano H, and Kasahara M. Transport of glucose across the blood-tissue barriers. Int Rev Cyt 1997;172:1–53.

8. Smythe C, Cohen P. The discovery of glycogenin and the priming mechanism for glycogen biosynthesis. Eur J Biochem 1991;200:625–31.

9. Lehninger AL, Nelson DL, Cox MM. Principles of Biochemistry, 2nd ed. New York: Worth, 1993;428–30.

10. Katz J, McGarry JD. The glucose paradox: is glucose a substrate for liver metabolism? J Clin Invest 1984;74:1901–9.

11. Frezza M, di Padova C, Pozzato G, Terpin M, Baraona E, Lieber C. High blood alcohol levels in women. The role of decreased alcohol dehydrogenase activity and first pass metabolism. N Engl J Med 1990;322:95–9.

12. Thomasson R. Gender differences in alcohol metabolism: physiological responses to ethanol. Recent Dev Alcohol 1995;12:163–79.

13. US Bureau of the Census: Statistical abstract of the United States, 1975. Washington, DC: US Government Printing Office, 1975.

14. Mendenhall C and Weesner R. Alcoholism. In: Clinical Chemistry: Theory, Analysis, Correlation, 3rd ed. St. Louis: Mosby, 1996; Kaplan LA and Pesce AJ eds., pp. 682–95.

15. Lieber CS. Biochemical and molecular basis of alcohol-induced injury to liver and other tissues. N Engl J Med 1988;319:1639–50.

16. V Lim Ki M, Laitinen K, Ylikahri R, et al. The effect of moderate alcohol intake on serum apolipoprotein A-I-containing lipoproteins and lipoprotein(a). Metabolism 1991;40:1168–72.

17. Jackson R, Scragg R, Beaglehole R. Alcohol consumption and risk of coronary heart disease. BMJ 1991;303:211–6.

18. Locher R, Suter P, Vetter W. Ethanol suppresses smooth muscle cell proliferation in the post-prandial state: a new antiatherosclerotic mechanism of ethanol? Am J Clin Nutr 1998;67:338–41.

Suggested Reading

McGarry JD, Kuwajima M, Newgard CB, Foster DW. From dietary glucose to liver glycogen: the full circle round. Ann Rev Nutr 1987;7:51–73.
  The glucose paradox is emphasized, from the standpoint of its emergence, as are the attempts to resolve it.
Pilkis SJ, El-Maghrabi MR, Claus TH. Hormonal regulation of hepatic gluconeogenesis and glycolysis. Ann Rev Biochem 1988;57:755–783.
  This is a brief, clearly presented summary of the effect of certain hormones on the important regulatory enzymes in these major pathways of carbohydrate metabolism.

Web Sites

www.medscape.com
  Medscape
www.cdc.gov
  Centers for Disease Control and Prevention
www.ama-assn.org
  American Medical Association
www.wadsworth.com/nutrition
  Wadsworth Publishing Company

⚭ **PERSPECTIVE**

Hypoglycemia: Fact or Fall Guy?

Maintenance of normal blood glucose concentration (normoglycemia) is essential to good health. The consequences of abnormally high levels (hyperglycemia), such as occurs in diabetes, are well established, and concentrations that are significantly below the normal range (hypoglycemia) introduce a well-recognized syndrome as well. This Perspective deals with the latter situation.

Glucose is our most important carbohydrate nutrient. Its concentration in the blood is established by a balance between processes that infuse glucose into the blood and those that remove it from the blood for use by the cells. The primary sources of blood glucose are

- exogenous (dietary sugars and starches),
- hepatic glycogenolysis, and
- hepatic gluconeogenesis.

The primary glucose-requiring tissues are the brain, erythrocytes, and muscle.

Preprandial serum glucose levels generally range from 70 to 105 mg/dL in patients who have no disorder of glucose metabolism (1 dL = 100 mL). If whole blood is the specimen, the range is somewhat lower, approximately 60 to 90 mg/dL. Should levels drop below this range, glucoreceptors in the hypothalamus stimulate the secretion of counterregulatory hormones to try to return the glucose to homeostatic levels. Such hormones include

- glucagon, which increases hepatic glycogenolysis and gluconeogenesis;
- epinephrine, which inhibits glucose use by muscle and increases muscle glycogenolysis;
- cortisol and growth hormone, although these have a delayed release and do not contribute significantly to acute recovery.

The major hormone acting antagonistically to those listed is insulin, which has the opposing effect of increasing cellular uptake of glucose, thereby reducing its serum concentration.

If, in spite of counterregulatory hormone response, subnormal concentrations of glucose persist, a state of clinical hypoglycemia can possibly result. This can, in turn, give rise to any number of associated symptoms. Symptoms, which may be attributable to the low glucose itself or to the hormonal response to the low glucose, are broadly categorized as adrenergic or neuroglucopenic.

Adrenergic symptoms arise as a result of increased activity of the autonomic nervous system, coincident with accelerated release of epinephrine. Adrenergic symptoms include weakness, sweating/warmth, tachycardia (rapid heart rate), palpitation, and tremor.

Neuroglucopenic symptoms are usually associated with a more severe hypoglycemic state. These include headache, hypothermia, visual disturbances, mental dullness, and seizures. During insulin induced hypoglycemia, adrenergic symptoms may manifest at serum glucose concentrations of about 60 mg/dL, and neuroglucopenic symptoms at approximately 45 to 50 mg/dL.

Many of the symptoms listed here are nonspecific and somewhat vague, and they may arise as a result of any number of unrelated disorders. It is for this reason that hypoglycemia is among the most overdiagnosed ailments. Before it can be diagnosed as bona fide clinical hypoglycemia, a condition must comply with the following contingencies:

1. A low serum glucose level

2. Presence of adrenergic or neuroglucopenic symptoms

3. Relief of symptoms upon ingestion of carbohydrate and a return of glucose levels toward normal

Commonly, patients may have serum glucose concentrations as low as 50 mg/dL and yet be asymptomatic, while others may be normoglycemic but have any number of the symptoms congruent with hypoglycemia. In neither case is true clinical hypoglycemia verifiable.

Two types of hypoglycemia exist:

- *Fasting hypoglycemia.* This is usually caused by drugs, such as exogenous insulin, used in the treatment of type I diabetes, or the sulfonylureas which stimulate the secretion of insulin. It can also be caused by insulinomas (B cell tumors) or excessive intake of alcohol.

- *Fed (reactive) hypoglycemia.* In patients who have not undergone certain gastrointestinal surgical procedures, there are two possible causes for reactive hypoglycemia: *impaired glucose tolerance* (IGT) and the more common *idiopathic postprandial syndrome.*

Some patients with IGT and "mild" diabetes experience postprandial hypoglycemia. This is due to an initially delayed insulin response to food, followed by excessive insulin

release that drives glucose levels down to hypoglycemic concentrations. This condition can be diagnosed by the oral glucose tolerance test (OGTT), in which serum glucose levels are observed following an oral load of glucose.

Much attention has been paid to the putative idiopathic postprandial form of hypoglycemia. It has been difficult to document that adrenergic symptoms occur simultaneously with the analytical finding of hypoglycemia. As discussed earlier, adrenergic symptoms frequently do not correlate with low glucose levels. Furthermore, feeding may not relieve the symptoms or elevate serum glucose. This is the form of hypoglycemia that has been overdiagnosed as such. The actual cause is more complex, and other factors may be players in the game.

Diet therapy is the cornerstone of treatment for all forms of reactive (fed) hypoglycemia. Patients should avoid simple or refined carbohydrates. Patients may also benefit from frequent, small feedings of snacks containing a mixture of carbohydrate, protein, and fat.

Suggested Reading
Andreoli T, Bennett J, Carpenter C, and Plum F. Hypoglycemia. In: Cecil Essentials of Medicine, 4th ed. W. B. Saunders, Philadelphia: 1997: Chap. 72.

Web Sites
http://www.betterhealth.com
Better Health

Chapter 5

Dietary Fiber

Cellulose fibrils of plant cell wall

Dietary fiber was recognized again as an important food component in about the mid-1970s. Yet, the concept of fiber, originally referred to as *crude* fiber or indigestible material, and its extraction from animal feed and forages were introduced in Germany during the 1850s. The crude fiber extraction method has continued to be used, even for human food, into the 1990s, although better methodology is now available. The crude fiber figure, based on food sample residues after treatment with 1.25% sulfuric acid and sodium hydroxide, is believed to underestimate total dietary fiber by at least 40% [1]. Furthermore, because the relationship between the crude fiber and dietary fiber varies among foods, no correction factors can be used. Today, soluble and insoluble dietary fiber fractions may be isolated and analyzed. Such information will perhaps allow expansion of medical nutrition fiber therapy for the treatment and perhaps prevention of many conditions or diseases affecting humans.

Results from extensive research devoted to dietary fiber during the last 25 or so years have implicated dietary fiber as important in various aspects of gastrointestinal tract function and in the prevention and/or management of a variety of disease states. The varying effects of fiber observed by researchers are related to the fact that dietary fiber is made up of different components, each with its own distinctive characteristics. Examining these many components plus their various, distinctive characteristics emphasizes the fact that dietary fiber cannot be considered a single entity. This chapter will address the definition of dietary fiber; the relationship between plants and fiber; and the chemistry, intraplant functions, and properties of fiber. It will also provide recommendations for dietary fiber intake.

Definition of Dietary Fiber

Because dietary fiber is not a single entity, no universally accepted definition for this food component has yet evolved. Probably the most widely accepted definition for dietary fiber is that proposed by Trowell et al. [2]: "plant polysaccharides and lignin which are resistant to hydrolysis by the digestive enzymes of man." Although it is generally agreed that all the nonstarch polysaccharides should be regarded as fiber components, dissatisfaction with this definition exists because

- it fails to include all the indigestible residue from food that may reach the colon, and

- it uses ability to be digested as the basis for the definition when undigested food reaching the colon does not necessarily lack the ability to be digested, nor is it necessarily unavailable to the body [3].

Some starch, for example, may reach the colon in an unaltered state along with nonstarch polysaccharides (NSP). This starch and much of the nonstarch polysaccharides may then undergo fermentation (anaerobic breakdown) by colonic bacteria, thereby producing short-chain fatty acids that may be used for energy by the host. However, despite their presence in colonic residue, starch is not considered by some as dietary fiber because it is digestible by mammalian enzymes, and lignin is not considered by some to be true components of dietary fiber because it is a noncarbohydrate polymer [4].

Plants and Fiber

Dietary fiber, in spite of the controversy about its definition, is derived from plant cells. The plant cell wall is of particular importance because it contains >95% of dietary fiber components including cellulose, hemicellulose, lignin, pectins, as well as some nonstarch polysaccharides. Table 5.1 shows the relationship between the plant cell wall and the plant components generally accepted as comprising dietary fiber.

The plant cell wall consists of both a primary and secondary wall. The primary wall is a thin envelope surrounding the contents of the growing cell; the secondary wall develops as the cell matures. The secondary wall of a mature plant contains many strands of

Table 5.1 Relationship Between Plant Cell Wall and Dietary Fiber

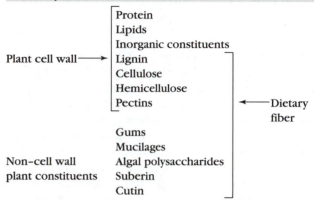

cellulose arranged in an orderly fashion within a matrix of noncellulosic polysaccharides. The primary wall also contains cellulose, but it occurs in smaller amounts and is less well organized. Starch, the energy storage product of the cell, is found within the cell walls.

Lignin deposits form in specialized cells whose function is to provide structural support to the plant. As the plant matures, lignin spreads through the intracellular spaces, penetrating the pectic substances. Pectic substances function as intercellular cement, and are located between and around the cell walls. Lignin continues dispersing through intracellular spaces, but it also permeates the primary wall and then spreads into the developing secondary wall. As plant development continues further, suberin is deposited in the cell wall during the later stages. Cutin, a water-impermeable substance, is secreted onto the plant surface.

The consumption of plant foods provides fiber in the diet. The plant species, the part of the plant (leaf, root, stem) and its maturity all influence the composition (cellulose, hemicellulose, pectin, lignin, etc.) of the fiber that is consumed. For example, consumption of cereal bran such as wheat bran provides primarily hemicellulose as well as lignin. Psyllium provides primarily mucilages but also some nonpolysaccharides. Consumption of fruits and vegetables provides almost equal quantities (~30%) of cellulose and pectin. In contrast, cereals are quite low in cellulose. Each of the components that are collectively known as dietary fiber, their characteristics and intraplant functions, as well as foods rich in the particular fiber component will be reviewed next. Identifying the chemical characteristics and various intraplant functions of these plant cell wall substances (and/or in contact with the wall) promotes conceptualization of how fiber components may affect physiological and metabolic functions in human beings.

Chemistry and Intraplant Functions of Fiber Components

Cellulose

Cellulose, shown in Figure 5.1a, is a long, linear polymer of 1,4 β-linked glucose units that is found in the plant cell wall. Hydrogen bonding between sugar residues in adjacent parallel running cellulose chains imparts a microfibril three-dimensional structure to cellulose. Being a large, linear, neutrally charged molecule, cellulose is water-insoluble, although it can be modified chemically (e.g., sodium carboxymethylcellulose) to be more soluble and used as an additive in foods. Degradation of cellulose by colonic bacteria varies, but generally it is poorly fermented. Some examples of foods high in cellulose relative to other fibers include bran, legumes, peas, root vegetables, vegetables of the cabbage family, outer covering of seeds, and apples.

Hemicellulose

Hemicellulose, like cellulose, is found in the plant cell wall. It consists of a heterogeneous group of polysaccharide substances containing a number of sugars in its backbone and side chains. The sugars, which form a basis for hemicellulose classification, include xylose, mannose, and galactose in the hemicellulose backbone and arabinose, glucuronic acid, and galactose in the hemicellulose side chains. These sugars are shown in Figure 5.1b. The sugars in the side chains also confer important characteristics on the hemicellulose. For example, hemicelluloses that contain acids in their side chains are slightly charged and water-soluble. Other hemicelluloses are insoluble. Fermentability of the hemicelluloses by intestinal microflora is also influenced by the sugars and positions. For example, hexose and uronic acid components of hemicellulose are more accessible to bacterial enzymes than the other hemicellulose sugars. Some foods that are relatively high in hemicellulose are bran and whole grains.

Pectin

Pectic substances, or more commonly called *pectin,* are a complex group of polysaccharides in which galacturonic acid is a primary constituent. The backbone structure of pectin is usually an unbranched chain of α 1,4-linked D-galacturonic acid units, as shown in Figure 5.1c. Other carbohydrate moieties may be linked to the galacturonic acid chain. Additional sugars sometimes found attached as side chains include rhamnose, arabinose, xylose, and fucose. Pectins form part of the primary cell wall of plants and part of the middle lamella.

Pectins are water-soluble and gel forming and have high ion-binding potential. Moreover, they can be almost completely metabolized by colonic bacteria. Apples, strawberries, and citrus fruits are high in pectins. Pectins are also added to commercial products and to some enteral nutrition formulas administered to tube-fed hospitalized patients.

Lignin

Lignin is the main noncarbohydrate component of fiber. It is a three-dimensional polymer composed of phenol units with strong intramolecular bonding. The primary phenols composing lignin, shown in Figure 5.1d, include trans-coniferyl, trans-sinapyl, and trans-p-coumaryl. Lignin forms the structural components of plants and is thought to attach to other noncellulose polysaccharides such as heteroxylans found in plant cell walls. Lignin is insoluble in water, has hydrophobic binding capacity, and is not digested (poorly fermented) by colonic bacterial microflora. Mature root vegetables such as carrots, wheat, and fruits with edible seeds such as strawberries are high in lignin.

Gums

Gums are one of a group of substances that may be referred to as *hydrocolloids.* They are secreted at the site of plant injury by specialized secretory cells and can be exuded from the plants. Gums are composed of a variety of sugars and sugar derivatives. Occurring prominently in the gums are galactose and glucuronic acid as well as uronic acids, arabinose, rhamnose, and mannose, among others. Within the large intestine, gums are highly fermented by colonic bacteria. The structure of the gum arabic is given in Figure 5.1e. It contains a β 1-3 galactose backbone with side chains of galactose, arabinose, rhamnose, glucuronic acid, and/or methyl-glucuronic acid. Gum arabic is the plant hydrocolloid most commonly used as a food additive. Its popularity is due to its physical properties, including high solubility, pH stability, and gelling characteristics. Other water-soluble gums such as guar and locust bean (carob) are referred to as *galactomannans.* Galactomannans contain a mannose backbone in a 2:1 or 4:1 ratio with galactose present in the side chains. Gums are also found naturally in foods such as oatmeal, barley, and legumes. Some gums (xanthan gum and gellan gum) can be synthesized by micro-organisms.

Mucilages and Algal Polysaccharides

Mucilages and algal polysaccharides are hydrocolloids and similar to gums in chemical structure. Because of

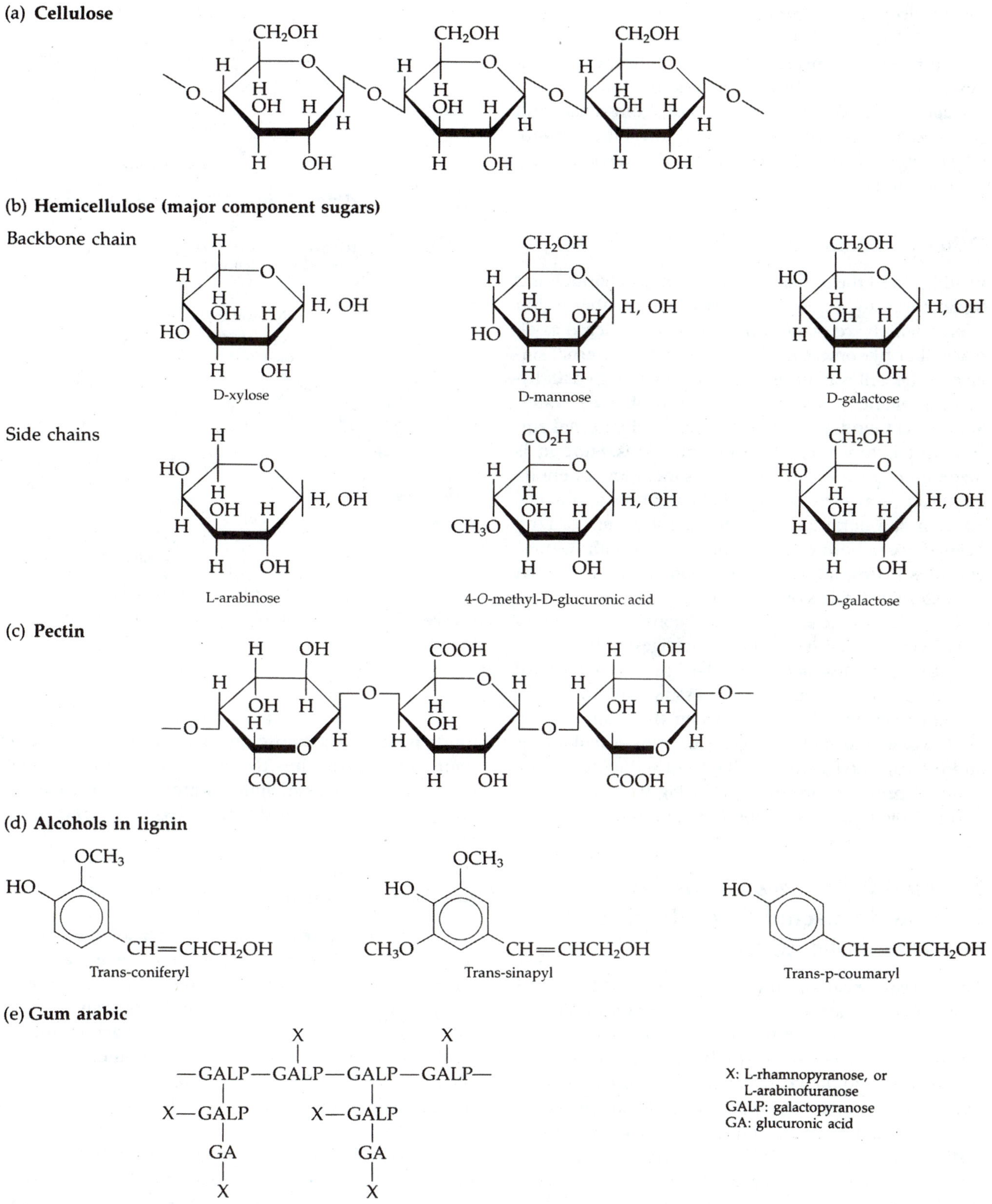

(a) **Cellulose**

(b) **Hemicellulose (major component sugars)**

Backbone chain

D-xylose   D-mannose   D-galactose

Side chains

L-arabinose   4-O-methyl-D-glucuronic acid   D-galactose

(c) **Pectin**

(d) **Alcohols in lignin**

Trans-coniferyl   Trans-sinapyl   Trans-p-coumaryl

(e) **Gum arabic**

```
              X                 X
              |                 |
  —GALP—GALP—GALP—GALP—

  X—GALP          X—GALP
      |               |
     GA              GA
      |               |
      X               X
```

X: L-rhamnopyranose, or
   L-arabinofuranose
GALP: galactopyranose
GA: glucuronic acid

Figure 5.1 Chemical structures of dietary fibers.

their hydrophilic property, these substances are excellent stabilizers. Mucilages are synthesized by plant secretory cells to protect the seed endosperm from desiccation. The algal polysaccharides—for example, carrageenan and agar—are derived from algae and seaweed and are commonly added to food. Both mucilages and algal polysaccharides are degraded by colonic bacteria.

Other Components

In addition to the aforementioned polysaccharides and lignin, three additional substances—cutin, suberin, and waxes, which are plant derived—are considered as dietary fiber by some. Cutin is found on the external surface of the cell wall of plants. It contains long-chain hydroxyaliphatic acids and is impermeable to water. Suberin is found near the plant cell wall external surface, just below the epidermis and skins. Suberin is made up of a variety of substances including phenolic compounds as well as long-chain alcohols and acids. Suberin and cutin are both polymeric esters of fatty acids that are both enzyme- and acid-resistant. Waxes, complex hydrophobic, hydrocarbon compounds, coat the external surfaces of many plants.

Other substances sometimes considered dietary fiber because they are unable to be digested by human digestive enzymes include Maillard products. Maillard products result from food processing and consist of enzyme-resistant linkages between the amino group ($-NH_2$) of amino acids, especially the amino acid lysine, and the carboxyl groups ($-COO^-$) of reducing sugars. Maillard products are formed during heat treatment, particularly in the baking and frying of foods.

Selected Properties of Dietary Fiber and Selected Physiological and Metabolic Effects

The physiological and metabolic effects of fiber vary based on the types of ingested fiber. Significant characteristics of dietary fiber that affect its physiological and metabolic roles include its solubility in water, its hydration or water-holding capacity and viscosity, its adsorptive attraction or ability to bind organic and inorganic molecules, and its degradability or fermentability by intestinal bacteria. Each of these characteristics and their effects on various physiological and metabolic processes will be reviewed; Figure 5.2 diagrams these relationships. However, as you study these characteristics and their effects on the body, it is important to remember that we eat foods with a mixture of dietary fiber, not

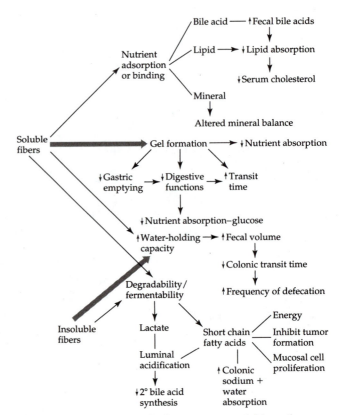

Figure 5.2 Gastrointestinal response to soluble and insoluble fiber.

foods with just cellulose, hemicellulose, pectins, guar gum, and so forth. Thus, the effects on the various body processes are not as straightforward as presented in this chapter and vary considerably based on the foods ingested.

Solubility in Water

Dietary fiber may be classified as water soluble or insoluble (Fig. 5.3); fibers that dissolve in hot water are soluble, and those that will not dissolve in hot water are insoluble. In general, water-soluble fibers include some hemicelluloses, pectin, gums, and mucilages. Cellulose, lignin, and some hemicellulose are classified as insoluble fibers. Generally, vegetables and wheat along with most grain products contain more insoluble fibers than soluble fibers.

Solubility in water also may be used as a basis to broadly divide the characteristics of fibers. For example, generally soluble fibers delay gastric emptying, increase transit time (slower movement) through the intestine, and decrease nutrient (e.g., glucose) absorption. In contrast, insoluble fibers decrease (speed up)

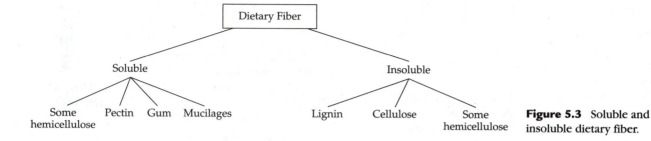

Figure 5.3 Soluble and insoluble dietary fiber.

intestinal transit time and increase fecal bulk. These actions (discussed in the following section) of the soluble and insoluble fibers in turn induce other physiological and metabolic effects.

Water-Holding/Hydration Capacity and Viscosity

Water-holding or hydration capacity of foods refers to the ability of fiber in food to bind water, or think of fiber as a dry sponge moving through the digestive tract hydrating or soaking up water and digestive juices as it moves through the digestive tract. Many of the water-soluble fibers such as pectins, gums, and some hemicelluloses have a high water-holding capacity in comparison with fibers such as cellulose and lignin that have a lower water-holding capacity. In addition, some water-soluble fibers such as pectins, gums, mucilages, and algal polysaccharides form viscous (thick) solutions within the gastrointestinal tract.

Water-holding capacity, however, is not just dependent on the fiber's solubility in water. The pH of the gastrointestinal tract, size of the fiber particles, and/or degree of processing of the foods providing fiber also influence the water-holding capacity and in turn its physiological effects. Coarsely ground bran, for example, has a higher hydration capacity than that which is finely ground. Consequently, coarse bran with large particles holds water, increases fecal volume, and speeds up the rate of fecal passage through the colon. Maintaining the integrity of cells in grains and legumes rather than subjecting them to traditional milling processes also appears to affect the water-holding capacity of fibers.

The gastrointestinal effects of the ingestion of fibers such as pectins, gums, mucilages, algal polysaccharides, and some hemicelluloses that can hold water and create viscous solutions within the gastrointestinal tract include

- delayed (slowed) emptying of food from the stomach,
- reduced mixing of gastrointestinal contents with digestive enzymes,

- reduced enzyme function,
- decreased nutrient diffusion rate and thus delayed nutrient absorption, and
- altered small intestine transit time.

Each of these effects will be discussed.

Delayed (Slowed) Gastric Emptying

When fibers form viscous gels or hydrate within the stomach, the release of the chyme from the stomach into the duodenum (proximal small intestine) is delayed (slowed). Thus, nutrients remain in the stomach longer with these fibers than would occur in the absence of the ingested fiber. This effect creates a feeling of postprandial (after eating) satiety (fullness) as well as slows down the digestion process, because carbohydrates and lipids that remain in the stomach undergo no digestion in the stomach and must move into the small intestine for further digestion to occur.

Reduced Mixing of Gastrointestinal Contents with Digestive Enzymes

The presence of viscous gels or hydrated fiber in the gastrointestinal tract provides a physical barrier that can impair the ability of the nutrients in the food to interact with the digestive enzymes. This interaction is critical for digestion to occur.

Reduced Enzyme Function

Viscous gel-forming fibers have been shown to interfere with the enzymatic hydrolysis of nutrients within the gastrointestinal tract. For example, hydrocolloids may inhibit intestinal peptidases that are necessary for the digestion of peptides to amino acids [5,6]. The activity of pancreatic lipase also has been diminished because of ingestion of viscous gel-forming fibers and has thus resulted in inhibition of lipid digestion [5]. It is unclear whether fiber directly decreases the activity of these digestive enzymes or acts by reducing the rate of enzyme penetration into the food.

Decreased Nutrient Diffusion Rate

Remember for nutrients to be absorbed, they must move from the lumen of the small intestine through a glycoprotein (mucin)-rich water layer lying on top of the enterocytes. The fiber-associated decreased diffusion rate of nutrients through this layer is probably due to an increased thickness of the unstirred water layer. In other words, the unstirred water layer becomes more resistant to nutrient movement, and without this movement nutrients cannot be absorbed into the enterocyte.

Another mechanism may also be responsible for decreased nutrient diffusion. Gums appear to slow glucose absorption by decreasing the convective movement of glucose within the intestinal lumen. Convective currents induced by peristaltic movements are responsible for bringing nutrients from the lumen to the epithelial surface for absorption. Decreasing the convective solute movement also may help to explain the decreased absorption of amino acids and fatty acids caused by viscous fiber [5]. Ingestion of viscous mucilaginous fibers such as guar gum, but also pectin and psyllium, have been shown to slow transit, delay glucose absorption, lower blood glucose concentrations, and affect hormonal response to the absorbed nutrient. Such effects are beneficial to individuals with diabetes mellitus and reduce postprandial blood glucose concentrations and insulin needs/response.

Decreased nutrient diffusion rate may in turn result in nutrients "missing" their normal site of maximal absorption. For example, if a nutrient is normally absorbed in the proximal intestine but because of gel formation is "trapped" as part of the gel, then absorption cannot occur at this site. Should the nutrient be released from the gel, the release is most likely going to occur at a site distal to where the nutrient would normally have been absorbed. The extent to which nutrients are absorbed throughout the digestive tract varies with the individual nutrient.

Altered Small Intestine Transit Time

In general, soluble fibers typically delay (slow down) small intestine transit time versus insoluble fibers, which increase (speed up) transit time within the small intestine. As with decreased diffusion rates, the changes in transit time, especially if it is increased, may result in decreased nutrient absorption due to insufficient time for the nutrients to be in contact with enterocytes.

Adsorption or Binding Ability

Some fiber components, especially lignin, gums, pectins, and some hemicelluloses, but also some Maillard products, have the ability to bind (adsorb) substances such as enzymes and nutrients in the gastrointestinal tract. The ability of these fibers to adsorb substances depends in part on gastrointestinal pH as well as particle size, food processing, and fermentability [7].

The physiological effects of the ingestion of fibers with adsorption properties within the gastrointestinal tract may include

- diminished absorption of lipids,
- increased fecal bile acid excretion,
- lowered serum cholesterol concentrations (hypocholesterolemic properties), and
- altered mineral balance.

The mechanisms by which these effects occur vary considerably and will be reviewed.

Diminished Absorption of Lipids

Soluble fibers (e.g., pectin, guar gum, and oat bran) but also the insoluble fiber lignin may affect lipid absorption by adsorbing fatty acids, cholesterol, and/or bile acids within the digestive tract. Fatty acids and cholesterol that are bound to fiber cannot form micelles and cannot be absorbed in this bound form; only free fatty acids, monoacylglycerols, and cholesterol can be incorporated into micelles. Remember micelles are needed for these end products of fat digestion to be transported through the unstirred water layers and into the enterocyte. Thus, fiber-bound lipids are typically not absorbed in the small intestine and pass into the large intestine where they will be excreted in the feces or degraded by intestinal bacteria.

Increased Fecal Bile Acid Excretion

Adsorption of bile acids to fibers prevents the use of the bile acids for micelle formation. And, like fiber-bound fatty acids, bile acids bound to fiber cannot be reabsorbed and recirculated. Fiber-bound bile acids are typically sent into the large intestine for either fecal excretion or colonic microflora degradation.

Lowered Serum Cholesterol Concentrations (Hypocholesterolemic Properties)

The ability of fibers to lower serum cholesterol concentrations is based on a series of events. First, with the excretion of bile acids in the feces, less bile undergoes enterohepatic recirculation. A decrease in the bile acids returned to the liver necessitates the use of cholesterol for synthesis of new bile acids. The net effect of the process is a lower serum cholesterol. A second proposed mechanism for the hypocholesterolemic (lower blood cholesterol) effect of fiber is the shift of bile acid pools away from cholic acid and toward chenodeoxycholic acid. Chenodeoxycholic acid appears to inhibit 3-hydroxy 3-methylglutaryl (HMG) CoA reductase, a regulatory enzyme necessary for cholesterol biosynthesis [5]. Decreased HMG CoA reductase activity results in reduced hepatic cholesterol synthesis and theoretically lower blood cholesterol concentrations. A third mechanism, shown in animal studies, suggested that production of propionate from bacterial degradation of fiber (discussed in the next section) lowers serum cholesterol concentrations. However, propionate fed to humans has had varying effects on serum cholesterol concentrations.

Studies suggest that psyllium, guar, and oat gum as well as pectin have the greatest potential to lower serum cholesterol. Oat bran and soybean fibers have intermediate effects, while corn, wheat, and rice bran appear to be ineffective [8]. Ingestion of fruits and vegetables that have a mixture of soluble and insoluble fibers has been shown to decrease serum cholesterol concentrations.

Altered Mineral Balance

Some fibers—especially those with uronic acid such as hemicellulose, pectins, and gums—can form cationic bridges with minerals within the gastrointestinal tract. Lignin, which has both carboxyl and hydroxyl groups, is also thought to play a role in mineral adsorption. The overall effect (positive or negative) that fiber has on mineral balance depends to some extent on its degree of fermentability or its accessibility to bacterial enzymes in the colon. Microbial proliferation from slowly fermentable fibers may result in increased binding of minerals within the new microbial cells and result in the loss of minerals from the body, assuming colonic mineral absorption. In contrast, the more rapidly fermentable fibers (such as pectins) appear to have a favorable effect on mineral balance. Calcium, zinc, and iron bound to these fiber components are released as fermentation occurs and may possibly be absorbed in the colon [5].

Degradability or Fermentability

In Chapter 2, the microflora of the gastrointestinal tract were discussed. If you will recall, both the small and large intestines contain numerous microflora, although most are found in the large intestine. Many of these microflora are capable of degrading (fermenting) fiber, especially pectins, gums, mucilages, and algal polysaccharides. In addition to these fibers, some cellulose and some hemicelluloses are also fermentable, but their fer-

mentation is much slower than that of the other fibers. In this section, first fermentable fibers and their effects on the body will be discussed. A discussion of the effects of fiber that are not fermentable will follow.

Fermentable Fibers

The principal metabolites of fermentable fibers (including any starch that has passed into the cecum and been degraded by bacteria) are lactate and short-chain fatty acids (SCFAs), formerly called *volatile fatty acids* (VFAs) because of their volatility in acidic aqueous solutions. Many different fibers including pectin, gums, oat and wheat bran, and psyllium, which is mostly mucilages and nonpolysaccharides, are degraded to short-chain fatty acids. The short-chain fatty acids include primarily acetic, butyric, and propionic acids. In addition to these acids, other products of fiber fermentation are hydrogen, carbon dioxide, and methane gases that are excreted as flatus or are expired by the lungs.

Different fibers are fermented to different short-chain fatty acids in different amounts by different bacteria. For example, ingestion of pectin resulted in higher propionate concentrations in the colon of rats versus wheat bran, which resulted in higher butyrate concentrations [9]. Bacteria that act on pectin include, for example,

- bacteroides that generate acetate, propionate, and succinate;
- eubacteria that yield acetate, butyrate, and lactate; and
- bifidobacteria that produce acetate and lactate.

Some general effects of short-chain fatty acids generated from fiber fermentation by intestinal microbes include

- increased water and sodium absorption in the colon,
- mucosal cell proliferation,
- provision of energy, and
- acidification of luminal environment.

Each of these general effects will be briefly reviewed.

Increased Water and Sodium Absorption in the Colon
Short-chain fatty acids produced by fermentation are rapidly absorbed, and their absorption in turn stimulates water and sodium absorption in the colon.

Mucosal Cell Proliferation   Substrates generated from the degradation of dietary fiber in the colon stimulate the proliferation of mucosal cells in the gastrointestinal tract.

Provision of Energy Short-chain fatty acids provide body cells with substrate for energy production. Butyric acid provides an energy source for colonic epithelial cells. Those fatty acids not used by the colonic cells, primarily the propionic and acetic acids, are carried by the portal vein to the liver, where the propionate and some of the acetate are taken up and metabolized. Most of the acetate, however, passes to the peripheral tissues, where it is metabolized by skeletal and cardiac muscle. Remember from the section on lowered serum cholesterol, it is the propionic acid generated from fiber fermentation that in rats inhibited hepatic cholesterol biosynthesis.

Fermentation of carbohydrates by colonic anaerobic bacteria makes available to the body some of the energy contained in undigested food reaching the cecum. The exact amount of that energy realized depends mostly on the amount and type of dietary fiber that is ingested. It is estimated that in developed countries as much as 10% to 15% of ingested carbohydrate may be fermented in the colon; in the third world (developing) countries, this percentage may be considerably higher [10].

Acidification of Luminal Environment The generation of short-chain fatty acids in the colon from bacterial fermentation of carbohydrate results in a decrease in the pH of the colon's luminal environment. With the more acidic pH, free bile acids become less soluble. Furthermore, the activity of 7 α dehydroxylase diminishes (optimal pH ~6–6.5) and thus decreases the conversion of primary bile acids to secondary bile acids. With the lower pH, calcium also becomes more available (soluble) to bind bile and fatty acids [11]. These latter two changes may be protective against colon cancer.

Nonfermentable Fibers

The fiber components that are nonfermentable, principally cellulose and lignin, or that are more slowly fermentable, such as some hemicelluloses, are particularly valuable in promoting the proliferation of microbes in the colon. Microbial proliferation may be important for both detoxification as well as a means of increasing fecal volume (bulk).

Detoxification The detoxification role is based on the theory that the synthesis of increased microbial cells (i.e., microbial proliferation) could result in the increased microbial scavenging of and sequestering of substances or toxins, which eventually are excreted. Alternately, certain colonic bacteria appear to inhibit proliferation of tumor cells and/or delay tumor formation.

In addition, bacteria such as *L. acidophilus* reduced the activity of enzymes that catalyze the conversion of procarcinogens to carcinogens [12].

Increased Fecal Volume (Bulk) In addition to its detoxifying role, microbial proliferation may promote increased fecal volume or bulk. Fecal bulk consists of unfermented fiber, salts, and water as well as bacterial mass. In general, fecal bulk increases with increased bacterial proliferation. This occurs not only because of the mass of the bacteria but also because bacteria are about 80% water. Thus, with increased fecal bacteria present, there is an increase in mass and in the water-holding capacity of the feces.

In general, fecal bulk increases as fiber fermentability decreases. The rapidly fermentable fiber appears to have little or no effect on fecal bulk. Therefore, choosing the appropriate food source(s) of fiber depends on the specific fiber effect(s) being sought and whether the food contains the fiber components producing this effect(s). Wheat bran is one of the most effective fiber laxatives because it can absorb three times its weight of water, thereby producing a bulky stool. Gastrointestinal responses to wheat bran include

- increased fecal bulk,
- greater frequency of defecation,
- reduced intestinal transit time, and
- decreased intraluminal pressure.

Rice bran has been found to be even more effective than wheat bran in eliciting an increased fecal bulk and a reduced intestinal transit time; both rice and wheat bran are helpful in treating constipation [13].

Other Characteristics

Several of the aforementioned characteristics of fiber are thought to contribute to the proposed mechanisms of action by which fiber has been proposed to prevent colon cancer [5,14–20]. These characteristics and mechanisms are listed and discussed here:

- High bile acid concentrations are associated with a high risk of colon cancer. Thus, fibers that adsorb bile acids to promote fecal excretion serve a protective effect by decreasing free concentration and the availability of bile acids for conversion to secondary bile acids, which are thought to promote colon carcinogenesis.

- Fibers that increase fecal bulk decrease the intraluminal concentrations of carcinogens and thereby reduce the likelihood of interactions with colonic mucosal cells.

- Provision of a fermentable substrate to colonic bacteria alters species and numbers of bacteria and/or their metabolism, which may inhibit proliferation or development of tumor cells or conversion of procarcinogens to carcinogens.

• A shortened fecal transit time decreases the time during which toxins can be synthesized and in which they are in contact with the colon.

• Fiber fermentation to short-chain fatty acids decreases the interluminal pH, thereby decreasing synthesis of secondary bile acids, which have been shown to promote the generation of tumors.

• Degradation of fiber by fermentation may release fiber-bound calcium. The increased calcium in the colon may help eliminate the mitogenic advantage that cancer cells have over normal cells in a low-calcium environment.

• Butyric acid appears to slow the proliferation and differentiation of colon cancer cells.

• Insoluble fibers such as lignin that resist degradation bind carcinogens, thereby minimizing the chances of interactions with colonic mucosal cells.

Not all studies show anticarcinogenic effects with fiber. In some studies, soluble fibers enhance the development of colorectal cancers. Proposed mechanisms for this action include (1) soluble fibers reduce the ability of insoluble fibers to adsorb hydrophobic carcinogens, thus more carcinogens may enter the colon maintained in solution than adsorbed onto insoluble fibers; (2) on degradation of soluble fibers, carcinogens are released and deposited on the colonic mucosal surface; (3) soluble fibers may cross the intestinal epithelium and transport with them carcinogens maintained in solution; (4) soluble fibers may reduce absorption of bile salts and thereby increase the chance for conversion to secondary bile acids [11,20].

There is little agreement among the numerous studies designed to determine the effect of fiber in the development of colon cancer. Most of the evidence for the positive role of fiber in colon cancer prevention has come from epidemiological observations. Unfortunately, in these epidemiological studies, variation in many dietary factors other than fiber intake has been noted. The dietary factors most often identified as being involved in variations in the incidence of colorectal cancer between different population groups are too many total calories, high fat, too much protein, low fiber, low intake of vitamin D and calcium, and a low intake of antioxidants [15]. Meta-analyses, however, of both epidemiological and case-controlled studies that investigated dietary fiber and colon cancer found that fiber-rich diets were associated with a protective effect against colon cancer in the majority of studies [21,22]. Furthermore, risk of colorectal cancer in the United

States is thought to be reducible by up to 31% with a 13-g daily increase in dietary fiber intake [22].

Recommended Intake of Fiber: Implications in Disease Prevention and Management

Recommendations for increasing the amount of dietary fiber in the U.S. diet have come from several governmental and private organizations, each with a concern for improving the health of the U.S. public [23–27]. The importance of an adequate intake of fiber to the improvement of health is demonstrated by some of the physiological effects exerted by its various components. Particularly noteworthy are the hypoglycemic and hypolipidemic effects of soluble fiber. Slowing the absorption rate of carbohydrate can be very helpful to the individual with diabetes mellitus in regulating blood glucose levels. Lowering serum cholesterol levels has significant benefits in the prevention of atherosclerosis. Adequate fiber intake also has been implicated in control of various gastrointestinal disorders, including diverticular disease, gallstones, irritable bowel syndrome, and constipation. The nonfermentable fibers, especially cellulose and lignin, and fibers that are more slowly fermentable, such as some hemicelluloses, have been shown to be helpful in overcoming constipation, particularly constipation associated with symptomatic diverticular disease and/or irritable bowel syndrome. Evidence for the effectiveness of fiber in the control of other diseases appears equivocal; however, populations with high fiber intakes have a lower incidence of these gastrointestinal disorders as well as colon cancer [14–20,28]. Finally, a generous fiber intake appears to be beneficial to some people in their efforts at weight control. The bulk provided by fiber may have some satiety value. High-fiber foods may reduce the hunger associated with caloric restriction while simultaneously delaying gastric emptying and somewhat reducing nutrient utilization [29,30]. The effectiveness of a high-fiber diet as a treatment for obesity remains unclear.

The recommended intake of fiber for the general population ranges from 20 to 40 g/day but may reach as much as 50 g/day for some individuals with specific conditions [23–27]. Guidelines suggest increasing complex carbohydrates to 55% to 65% of total calories and increasing fiber using a combination of high-fiber foods. A minimum fiber intake of about 20 g/day is recommended by the American Dietetic Association [26], the National Cancer Institute [24], and the Federation of American Societies for Experimental Biology (FASEB) [23]. Alternately, 10- to 13-g dietary fiber intake/1,000 kcal also has been suggested [23,26]. FASEB [23], the National Cancer Institute [24], and the American Dietetic Association [26] suggest an upper limit of 35 g/day.

Recommendations for fiber intake for those with specific diseases or conditions are similar to guidelines for healthy individuals. For those with non-insulin-dependent diabetes mellitus, the upper level of recommended fiber consumption should be about 40 g/day, and for obese individuals with non-insulin-dependent diabetes mellitus, about 25 g dietary fiber/1,000 kcal is recommended [27]. For clients with a family history of diet-implicated cancers, daily fiber consumption should regularly be about 35 to 40 g/day [31]. Intakes of up to 50 g/day are recommended for people with hypercholesterolemia [31].

Intakes of fiber beyond the upper range of 40 to 50 g/day may produce problems for some individuals. Intestinal obstruction in susceptible individuals as well as fluid imbalance have occurred with high intakes of fiber, especially those with high water-binding capacity [32].

Excessive insoluble fiber intake may pose the hazard of negative mineral balance, particularly among infants, children, adolescents, and pregnant women whose mineral needs are greater than those for adult men and non-pregnant women [32]. In particular, excessive fiber intake could be particularly detrimental for calcium, zinc, and iron balance if the intake of these minerals was marginal. However, a fiber intake of 30 to 40 g/day when accompanied by an adequate intake of minerals appears to have no detrimental effects, only possible benefits.

Dietary changes encouraged for accomplishing an increased fiber intake are consistent with following the Food Guide Pyramid [33]. Individuals should consume

- fiber-rich legumes, which may serve as the primary source of protein in a meal;
- at least 5 servings of fruits and vegetables per day, and
- at least 2 to 3 servings per day of whole grains as part of the 6 to 11 recommended servings by the Food Guide Pyramid [33].

Notice that recommendations that fiber intake be increased are interpreted in terms of dietary change rather than addition to the diet of fiber supplements, which, more than likely, are devoid of other nutrients. As one incorporates high-fiber foods, the percentage of complex carbohydrates increases in relation to the amount of fat and protein in the diet, and an increase in fiber becomes almost inevitable. It remains important, however, to eat a variety of cereals, legumes, fruits, and vegetables so that variety in dietary fibers is maximized.

Although many tables now provide the soluble and insoluble fiber composition of foods allowing one to calculate total dietary fiber intake, a quick method for calculating typical dietary fiber intakes is available and allows for assessment in a clinical setting from a food history, 24-hour diet recall, or food record [34]. Because fiber-rich foods consist primarily of fruits, vegetables, grains, legumes, nuts, and seeds, the numbers of servings from each of these groups can be multiplied by the mean total fiber content of each food group [34]. For example, numbers of servings (size determined from the U.S. Department of Agriculture data or food label) of fruits (not including juices) and vegetables are each multiplied by 1.5 g. The 1.5 g represents the average amount of dietary fiber per serving of fruit and per serving of vegetable. Numbers of servings of refined grains are multiplied by 1.0 g, and the numbers of servings of whole grains are multiplied by 2.5 g. The totals from each of the four categories are summed and added to food specific fiber values for legumes, nuts, seeds, and concentrated fiber sources; food-specific fiber values are obtained from databases [34]. The values calculated from this quick method were within 10% of the results obtained by looking up each individual food's fiber content [34].

Summary

No definition for dietary fiber has evolved that is entirely satisfactory to the scientific community. Some investigators believe the term should include all food components that reach the cecum unaltered. Others maintain that fiber should refer only to the nonstarch polysaccharides. Nevertheless, there is agreement that dietary fiber is not an entity but a complex made up of several distinctive components. The physiological roles of fiber in the gastrointestinal tract are as varied as the number of dietary fiber components and are determined by the types and amounts present.

The fact that some fiber components ingested by humans can be fermented by colonic microbes into usable short-chain fatty acids is of tremendous interest. No longer can the potential energy in fiber be considered totally unavailable to the human body.

The proportion of a fiber component to the other components varies according to the food source. To obtain all fiber components via the diet, food sources of fiber need to be varied and complementary. Assurance of a good intake of all dietary fiber components requires consumption of a variety of high-fiber foods: whole-grain cereals and breads, legumes, fruits, and vegetables.

References Cited

1. Asp NG, Johansson CG. Dietary fiber analysis. Nutr Abstr Rev 1984;54:736-52.

2. Trowell HC, Southgate DAT, Wolever TMS, et al. Dietary fiber redefined. Lancet 1976;1:967.

3. Heaton KW. Dietary fibre in perspective. Hum Nutr Clin Nutr 1983:37C:151-70.

4. Bingham S. Definitions and intakes of dietary fiber. Am J Clin Nutr 1987;45:1226-31.

5. Ink SL, Hurt HD. Nutritional implications of gums. Food Tech 1987;41:77-82.

6. Jenkins DJ, Jenkins AL, Wolever TMS, Rao AV, Thompson LU. Fiber and starchy foods: gut function and implications in disease. Am J Gastroenterol 1986;81:920-30.

7. Eastwood MA, Passmore R. A new look at dietary fiber. Nutr Today 1984;19:6-11.

8. Anderson JW, Jones AE, Riddell-Mason S. Ten different dietary fibers have significantly different effects on serum and liver lipids of cholesterol-fed rats. J Nutr 1994;124:78-83.

9. Lupton JR, Kurtz PP. Relationship of colonic luminal short chain fatty acids and pH to in vivo cell proliferation in rats. J Nutr 1993;123:1522-30.

10. McNeil NI. The contribution of the large intestine to energy supplies in man. Am J Clin Nutr 1984;39:338-42.

11. Harris PJ, Ferguson LR. Dietary fibre: its composition and role in protection against colorectal cancer. Mut Res 1993;290:97-110.

12. Gorbach SL. Lactic acid bacteria and human health. Ann Med 1990;22:37-41.

13. Tomlin T, Read NW. Comparison of effects on colonic function caused by feeding rice bran and wheat bran. Eur J Clin Nutr 1988;42:857-61.

14. Jenkins DJA, Jenkins AL, Rao AV, et al. Cancer risk: possible role of high carbohydrate, high fiber diets. Am J Gastroenterol 1986;81:931-5.

15. Ausman LM. Fiber and colon cancer: does the current evidence justify a preventive policy? Nutr Rev 1993;51:57-63.

16. Klurfeld DM. Dietary fiber-mediated mechanisms in carcinogenesis. Cancer Res (suppl) 1992;52:2055s-9s.

17. Hill MJ. Bile acids and colorectal cancer. Eur J Canc Prevent 1991;1:69-72.

18. Van Munster IP, Nagengast FM. The influence of dietary fiber on bile acid metabolism. Eur J Cancer Prevent 1991;1:35-44.

19. Potter JD. Colon cancer—do the nutritional epidemiology, the gut physiology and the molecular biology tell the same story? J Nutr 1993;123:418-23.

20. Harris PJ, Roberton AM, Watson ME, Triggs CM, Ferguson LR. The effects of soluble fiber polysaccharides on the adsorption of a hydrophobic carcinogen to an insoluble dietary fiber. Nutr Cancer 1993;19:43-54.

21. Trock B, Lanza E, Greenwald P. Dietary fiber, vegetables, and colon cancer: critical review and meta-analyses of the epidemiologic evidence. J Natl Cancer Inst 1990;82:650-61.

22. Howe GR, Benito E, Castelleto R, Cornee J, Esteve J, Gallagher RP, Iscovich JM, Deng-ao J, Kaaks R, Kune GA et al. Dietary intake of fiber and decreased risk of cancers of the colon and rectum: evidence from the combined analysis of 13 case-controlled studies. J Natl Cancer Inst 1992;84:1887-96.

23. Ad Hoc Expert Panel on Dietary Fiber, Federation of American Societies for Experimental Biology. Physiologic and health consequences of dietary fiber. Rockville, MD: FASEB, 1987.

24. Butrum RR, Clifford CK, Lanza E. NCI dietary guidelines: rationale. Am J Clin Nutr 1988; 48:888-95.

25. Diabetes Care and Education Practice Group, the American Diabetes Association. Nutritional recommendations and principles for individuals with diabetes mellitus. J Am Diet Assoc 1994; 4:504-6.

26. American Dietetic Association. Position of the American Dietetic Association: health implications of dietary fiber. J Am Diet Assoc. 1997;97:1157-9.

27. Beeebe CA, Pastors JG, Powers MA, Wylie-Rosett J. Nutrition management for individuals with noninsulin-dependent diabetes mellitus in the 1990s: a review by the Diabetes Care and Education dietetic practice group. J Am Diet Assoc 1991;91:196-202,205-7.

28. Aldoori W, Giovannucci E, Rockett H, Sampson L, Rimm E, Willett W. A prospective study of dietary fiber types and symptomatic diverticular disease in men. J Nutr 1998;128:714-9.

29. Kritchevsky D. Dietary fiber. Ann Rev Nutr 1988;8:301-28.

30. Anderson JW, Bryant CA. Dietary fiber: diabetes and obesity. Am J Gastroenterol 1986;81:898-906.

31. Floch MH, Maryniuk MD, Bryant C, Franz MJ, Tietyen-Clark J, Marrota RB, Wolever T, Maillet JO, Jenkins AL. Practical aspects of implementing increased dietary fiber intake. Nutr Today 1986;21:27-30.

32. Southgate DAT. Minerals, trace elements, and potential hazards. Am J Clin Nutr 1987;45:1256-66.

33. U.S. Dept. of Agriculture. The Food Guide Pyramid. Washington, DC: Home and Garden Bulletin Number 252, 1992.

34. Marlett JA, Cheung TF. Database and quick methods of assessing typical dietary fiber intakes using 228 commonly consumed foods. J Am Diet Assoc 1997;97:1139-48, 1151.

Suggested Reading

Anderson JW, Deakins DA, Floore TL, Smith BM, Whitis SE. Dietary fiber and coronary heart disease. Critical Review in Food Science and Nutrition 1990; 29:95-147.

Anderson JW, Smith BM, Gustafson NJ. Health benefits and practical aspects of high-fiber diets. Am J Clin Nutr 1994; 59(suppl):1242S-7S.

Friedman M. Dietary impact of food processing. Ann Rev Nutr 1992;12:119-37.

Gibson GR, Willems A, Reading S, Collins MD. Fermentation of non-digestible oligosaccharides by human colonic bacteria. Proc Nutr Soc 1996;55:899-912.

Spiller GA (ed). Handbook of Dietary Fiber in Human Nutrition. Boca Raton: CRC Press, 1993.

Wynder EL, Weisburger JH, Ng SK. Nutrition: The need to define "optimal" intake as a basis for public policy decisions. Am J Pub Hlth 1992;82:346-9.

The Role of Herbal Supplements in Treating or Reducing the Risk of Heart Disease, Cancer, and Gastrointestinal Tract Disorders

If you take a trip to natural food stores or listen to daytime television commercials, it will not take long until you are exposed to a barrage of ads and information on herbal supplements. Herbal supplements have been popular for the last several years. Data dating back to 1993 show that herbal supplements represented the top pharmacy growth category [1]. Today, more than 500 herbs are marketed in the United States [2]. The top five selling herbs (1996 data) in the United States include echinacea, garlic, ginseng, ginkgo biloba, and goldenseal [3]. Other popular herbal preparations or compounds are St. John's wort, glucosamine, and chondritin sulfate.

This perspective will provide an overview of some of the more commonly used herbal supplements in the United States. Although we tried to use original research articles, many articles were published in languages other than English or were published in international journals that could not be obtained. Consequently, reviews and secondary sources are cited when the original source could not be consulted.

Echinacea

Echinacea is derived from a native North American plant species characterized by spiny cone flowering heads, similar in appearance to the daisy. Liquid alcohol based extracts of the rhizome, roots, leaves, and flowers represent the most commonly sold form of the herb and should be made from fresh echinacea (purpurea or angustifolia) roots, leaves, or flowers (tops) either alone or in combination [4]. Active ingredients in echinacea are thought to include high molecular weight polysaccharides, alkamides such as isoburylamide, and cichoric acid [4–6]. These ingredients are thought to stimulate components of the immune system [4–6]. The herb is ingested orally typically for the prevention or treatment of the cold, flu, or other infections and is applied topically to aid in the healing of superficial wounds. It is suggested only as adjuvant therapy for more serious conditions [4–6].

Many studies have investigated echinacea; however, few have been well controlled. A list of some of these clinical reports and references is published elsewhere [4]. In one of the few controlled studies, Braunig [4] reported that 180 drops of an ethanol extract of E. purpurea roots significantly reduced the severity and duration of flulike symptoms in contrast to 90 drops of the same preparation or to an ethanol-sugar placebo. A dose of the herb represents two to three tablets or capsules providing about 500 to 1,000 mg of ground herb, 30 to 40 drops of extract, or 4 to 8 oz of a strong tea brewed using the ground root. Dosages of 300 to 400 mg of ground herb taken three times daily or smaller doses taken every few hours throughout the day have been suggested [4–7]. A consecutive maximum limit of echinacea use of up to 6 to 8 weeks has been suggested, with use for 10 to 14 days thought to be sufficient [5]. Major side effects from use of echinacea have not been reported; however, allergic reactions may occur [6]. Echinacea use by individuals with systemic or immune system dysfunction disorders is contraindicated.

Ginseng

Ginseng consists of dried roots of several different species of the genus *Panax*. The herb is difficult to grow and thus relatively expensive. The roots of a similar plant, *P. pseudoginseng* Wallich (sangui, tienchi, or sanchi ginseng), and the roots of *Eleutherococcus senticosus* Maxim or *Acanthopanax senticosus* also are sold under the general name ginseng. The latter also is sold as Siberian ginseng. Like Siberian ginseng, Brazilian and Indian ginseng are not derivatives of the *Panax* species.

Ginseng is one of the most commonly used herbs in the United States with purported effects in the prevention of cancer and the reduction of fatigue/improvement in stamina (an energy enhancer). Ginseng also is suggested to increase resistance to stress and to generally build vitality. The active ingredients in ginseng root appear to be a group of triterpenoid saponin glycosides and ginsenosides, which are thought to modulate hormones and central nervous system functions [5,6]. Results of analysis of a 100-g ginseng root yield 338 kcal, 12.2 g protein, 70 g carbohydrate, and several B vitamins, vitamins C and A, along with calcium, phosphorus, and iron [8]. In Chinese medicine, American ginseng (*P. quinquefolium* L.) is thought to have cooling (yin) action, whereas Asian ginseng (*P. ginseng* C. A. Meyer) is thought to have heating (yang) action [2]. Results of the few studies that have been conducted are contradictory with only some suggesting improved vitality in the elderly, tendencies toward faster, simple reactions and better abstract thinking, and decreased cancer risk [2,9–11]. Others have reviewed general and specific effects of ginseng on organ systems and performance and have concluded there is little evidence for the efficacy of ginseng [11]. However, such conclusions result in part from inadequate research studies [11]. Efficacy of ginseng clearly requires more well-controlled studies with large numbers of human subjects. In addition, problems with product quality need to be addressed, as products often contain negligible to no ginseng in contrast to label reports [6,12].

Ginseng doses from extracts are 100 to 300 mg three times daily; ginseng extract should contain 7% ginsenoisides [6]. Up to 3 g crude root are suggested [6]. Negative interactions between Siberian ginseng and digoxin, and ginseng and warfarin, have been reported [13,14]. In addition, germanium, an ingredient in some ginseng preparations, has been reported to induce resistance to diuretics [15]. Other side effects of ginseng include diarrhea, insomnia, and nervousness [15].

Ginkgo Biloba

An extract of the leaves of *Ginkgo biloba* L. is another popular herbal preparation sold in the United States. Extracts are concentrated and standardized to contain 24% ginkgo flavone glycosides and 6% terpenes (ginkgolides and bilobalide). Dosages of about 40 to 60 mg two to three times daily are purported to improve arterial and venous blood flow, especially cerebral and peripheral vascular circulation. Conditions in which poor circulation is a factor also are improved such as intermittent claudication, memory and hearing loss, vertigo, and tinnitus [6,16–22]. Ginkgo exhibits antioxidant properties as a scavenger of free radical and inhibitor of cell membrane lipid peroxidation [16]. Controlled clinical studies suggested that ginkgo extracts were effective versus placebo in the treatment of some cognitive disorders, some cerebral disorders, intermittent claudication, and vertigo; however, further clinical trials are needed [6,16–22]. Side effects include headache, dizziness, palpitations, and mild gastrointestinal distress [6,16]. Contact with the whole plant is associated with an allergic skin reaction.

Garlic

Garlic, along with onions, chives, and leeks, are members of a family of vegetables that all contain derivatives of cysteine, a sulfur-containing amino acid. S-allyl-L-cysteine sulfoxide, also known as *alliin,* is the main derivative in garlic. Alliin may be converted to allicin (diallyldisulfide-S-oxide) in the presence of alliinase, exposed when garlic cells are destroyed by cutting or chewing, for example. Allicin degrades to diallyl disulfide, the main component in the odor of garlic. Ajoene also may be formed.

Several studies have found that consumption of garlic promotes positive changes in blood lipid profiles to decrease risk of heart disease. A meta-analysis of well-controlled studies of individuals with total serum cholesterol concentrations >200 mg/dL concluded that garlic (about half to one clove daily) decreased serum total cholesterol concentrations by about 9% [23]. Ajoene in garlic also appears to inhibit platelet aggregation through interrupting

thromboxane synthesis and thus prevent clot formation [24,25]. Recommended daily dosages of garlic include 2 to 5 g fresh garlic or 400 to 1,200 mg dried garlic powder, which should be enteric coated. Excessive garlic consumption may cause heartburn, flatulence, and other gastrointestinal tract problems [6]. Individuals on aspirin therapy or anticoagulants should avoid ingestion of large amounts of garlic.

St. John's Wort

St. John's wort (*Hypericum perforatum* L.) is a perennial herb that produces yellow flowers. Its leaves and flowering tops have been used in the treatment of depression and anxiety [6]. The herb also is used topically to relieve inflammation and promote healing of, for example, burns, hemorrhoids, or minor wounds. The active ingredients in St. John's wort include naphthodianthrons such as hypericin, flavonoids such as quercetin, xanthones, and bioflavonoids that inhibit monoamine oxidase [6,26].

Studies have demonstrated bactericidal actions of hypericin extracts with positive effects on the healing of burns [27]. Preparations of hypericum extracts also have been more effective than placebo and similarly effective as antidepressant therapy in the treatment of mild to moderately severe depressive disorders, as reviewed in a meta-analysis of 23 randomized trials with over 1,750 individuals with depression [26]. Dosages of 2 to 4 g hypericum or 0.2 to 1.0 m hypericin are recommended with therapeutic regimens of 2 to 3 weeks needed before effects are expected [6,27]. Adverse effects may include fatigue, allergic reaction, gastrointestinal distress, and photosensitivity [27]. The photosensitivity occurs with use of high dosages or prolonged use and is manifested as dermatitis and mucous membrane inflammation with sunlight exposure [6]. Further well-designed studies investigating the effects of St. John's wort are necessary.

Goldenseal

Goldenseal consists of rhizome and roots of the *Hydrastis canadensis* L. The herb's activity is thought to arise from alkaloids—namely, hydrastine and berbeine. Teas made from the herb have been used to treat problems affecting the gastrointestinal tract, especially the mouth (canker sores, sore mouth) and stomach but also the urogenital tract [6].

Glucosamine and Chondroitin Sulfate

Glucosamine and or chondroitin sulfate supplementation have been generally shown to significantly decrease the severity of osteoarthritis. As discussed in Chapter 7, glucosamine, derived from glucose, is used as a building block for the synthesis of glycoproteins, glycosaminoglycans, mucopolysaccharides, and proteoglycans. Chondroitin sulfate, a mucopolysaccharide, also is used to make proteoglycans. Proteoglycans function in cartilage to maintain elasticity, strength, and mass. Individuals with osteoarthritis receiving nonsteroidal anti-inflammatory drugs (diclofenac sodium) (50 mg three times daily) reported prompt pain reduction; however, pain resumed when treatment ended. Those individuals receiving chondroitin sulfate (400 mg three times daily) showed slower pain relief, but the relief lasted longer [28]. Response (diminished pain and increased movement) in individuals with osteoarthritis of the knee receiving intramuscular glucosamine sulfate (400 mg twice a week) for 6 weeks was significantly greater than that of those receiving placebo injections [29]. As with other products, further well-controlled testing with large subject numbers are needed.

Regulation of Herbal Supplements

The Dietary Supplement Health and Education Act of 1994 allows herbs and phytomedicinals to be sold as dietary supplements as long as health or therapeutic claims do not appear on the product label. The act defines dietary supplements to include vitamins, minerals, herbal or botanical products, amino acids, metabolites, extracts, and other substances alone or in combination that are added to the diet. Because herbal supplements need not comply with other laws, the consumer and retailer have no assurance that the herb in the supplement corresponds with the label description. Thus, the reputation of the producer becomes extremely important. Herbs may be contaminated or adulterated in the manufacturing process. Incorrect parts of the herb such as the stem versus the root or use of the herb in an incorrect stage of ripeness may be used in production of the supplement [30]. In other words, quality assurance of herb and phytomedicinals is generally lacking in the United States [2]. Several reviews of the Dietary Supplement Health and Education Act of 1994 are available [30,31].

References Cited

1. The right stuff. Drug Store News pricks the categories taking off in '94. Drug Store News 1994;16(2):15.
2. Tyler V. Herbal remedies. J Pharm Tech 1995;11:214–20.
3. Richman A, Witkowski J. A wonderful year for herbs. Foods 1996(Oct);52–60.
4. Hobbs C. Echinacea: a literature review. Herbal Gram 1994;30:33–47.
5. Tyler V. What pharmacists should know about herbal remedies. J Am Pharm Assoc 1996;NS36:29–37.
6. Tyler V. The Honest Herbal. New York: Pharmaceutical Products Press, 1993.
7. Wagner H, Jurcic K. Immunological studies of plant extract combinations: in vitro and in vivo on the stimulation of phagocytosis. Arzneimittel Forschung 1991;41:1072–6.
8. Siegel R. Ginseng abuse syndrome: problems with the panacea. JAMA 1979;241:1614–5.
9. Sorensen H, Sonne J. A double-masked study of the effects of ginseng on cognitive functions. Curr Ther Res 1996;57:959–68.
10. Yun TK. Experimental and epidemiological evidence of the cancer-preventive effects of Panax ginseng C.A. Meyer. Nutr Rev 1996;54:S71–81.
11. Bahrke M, Morgan W. Evaluation of the ergogenic properties of ginseng. Sports Med 1994;18:229–48.
12. Cui J, Garle M, Eneroth P, Bjorkhem I. What do commercial ginseng preparations contain? Lancet 1994;344:134.
13. McRae S. Elevated serum digoxin levels in a patient taking digoxin and siberian ginseng. Can Med Assoc 1996;155:293–5.
14. Janetzky K, Morreale A. Probable interaction between warfarin and ginseng. Am J Health Syst Pharm 1997;54:692–3.
15. Becker B, Greene J, Evanson J, Chidsey G, Stone W. Ginseng-induced diuretic resistance. JAMA 1996;276:606–7.
16. Salvador R. Ginkgo. Can Pharm J 1995;128:39–41,52.
17. Kleijnen J, Knipschild P. Ginkgo biloba. Lancet 1992;340:1136–9.
18. Kleijnen J, Knipschild P. Ginkgo biloba for cerebral insufficiency. Br J Clin Pharmacol 1992;34:352–8.
19. Haguenauer J, Cantenot F, Koshas H, Pierart H. Treatment of equilibrium disorders with Ginkgo biloba. Presse Medicale 1986;15:1569–72.
20. Taillandier J, Ammar A, Rabourdin J, Ribeyre J, Pichon J, Niddam S, Pierart H. Treatment of cerebral aging disorders with ginkgo biloba. Presse Medicale 1986;15:1583–7.
21. Mouren X, Caillard P, Schwartz F. Study of the antiischemic action of EGb 761 in the treatment of peripheral arterial occlusive disease by TcPo2 determination. Angiology 1994;45:413–7.
22. Rai G, Shovlin C, Wesnes K. A double blind, placebo controlled study of Ginkgo biloba extract (tanakan) in elderly outpatients with mild to moderate memory loss. Curr Med Res & Opin 1991;12:350–5.
23. Warshafsky S, Kamer R, Sivak S. Effect of garlic on total serum cholesterol: a meta-analysis. Ann Intern Med 1993;119:599–605.
24. Block E, Ahmad S, Jain M, Crecely R, Apitz-Castro R, Cruz M. (E,Z)-Ajoene: a potent antithrombotic agent from garlic. J Am Chem Soc 1984;106:8295–6.
25. Block E. The chemistry of garlic and onions. Sci Am 1985;252:114–9.
26. Linde K, Ramirez G, Mulrow C, Pauls A, Weidenhammer W, Melchart D. St John's wort for depression—an overview and meta-analysis of randomised clinical trials. BMJ 1996;313:253–8.

∾ **PERSPECTIVE** *(continued)*

27. Upton R, Graff A, Williamson E, et al. American Herbal Pharmacopoeia and therapeutic compendium: St John's wort, Hypericum perforatum. HerbalGram 1997;40:S1–32.

28. Morreal P, Manopulo R, Galati M, Boccanera L, Saponati G, Bocchi L. Comparison of the antiinflammatory efficacy of chondroitin sulfate and diclofenac sodium in patients with knee osteoarthritis. J Rheumatol 1996;23:1385–91.

29. Reichelt A, Forster K, Fischer M, Rovati L, Setnikar I. Efficacy and safety of intramuscular glucosamine sulfate in osteoarthritis of the knee. Arzneimittel Forschung 1994; 44:75–80.

30. Drew A, Myers S. Safety issues in herbal medicine: implications for the health professions. Med J Australia 1997;166:538–41.

31. Dietary supplements: recent chronology and legislation. Nutr Rev 1995;53:31–36.

Additional References

Bartels C, Miller S. Herbal and related remedies. Nutr Clin Pract 1998;13:5–19.

Bisset N. Herbal Drugs and Phytopharmaceuticals. Boca Raton, FL: CRC Press, 1994.

The American Botanical Council, P.O. Box 201660, Austin, TX 78720

Newall C, Anderson L, Phillipson J. Herbal Medicines: A Guide for Health-Care Professionals. London, UK: The Pharmaceutical Press, 1996.

Upton R. The American Herbal Pharmacopoeia. Santa Cruz: AHP.

Web Sites

www.herbs.org www.herbalgram.org/abcmission.html
www.herb.com/herbal.htm www.healthy.com/herbalists
www.mdx.ac.uk/www/pharm www.ars-grin.gov/~ngrslb
http://altmed.od.nih.gov/oam/ www.rheumatology.org/
    hotline/970127.html

Chapter 6

Lipids

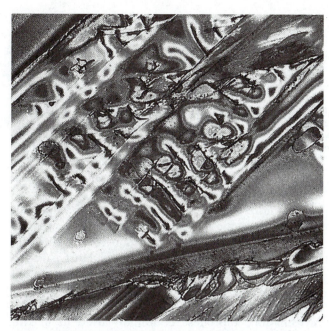

Photomicrograph of crystallized docosahexaenoic acid

The property that sets lipids apart from other major nutrients is their solubility in organic solvents such as ether, chloroform, and acetone. If lipids are defined according to this property, which is generally the case, the scope of their function becomes quite broad. It encompasses not only dietary sources of energy and the lipid constituents of cell and organelle membranes but also the fat-soluble vitamins, corticosteroid hormones, and certain mediators of electron transport such as coenzyme Q.

Among the many compounds classified as lipids, only a small number are important as dietary energy sources or as functional or structural constituents within the cell. The following classification is limited to those lipids germane to this section of the text dealing with energy-releasing nutrients. Fat-soluble vitamins are discussed in Chapter 10.

1. Simple lipids
   a. Fatty acids
   b. Triacylglycerols, diacylglycerols, and monoacylglycerols
   c. Waxes (esters of fatty acids with higher alcohols)
      (1) Sterol esters (cholesterol–fatty acid esters)
      (2) Nonsterol esters (vitamin A esters, and so on)
2. Compound lipids
   a. Phospholipids
      (1) Phosphatidic acids (i.e., lecithin, cephalins)
      (2) Plasmalogens
      (3) Sphingomyelins

123

b. Glycolipids (carbohydrate containing)
c. Lipoproteins (lipids in association with proteins)

3. Derived lipids: derivatives such as sterols and straight-chain alcohols obtained by hydrolysis of those lipids in groups 1 and 2 that still possess general properties of lipids

In the discussion of the structure and physiological function of lipids that follows, they have been arbitrarily grouped according to fatty acids, triacylglycerols (triglycerides), sterols and steroids, phospholipids, and glycolipids.

Structure and Function

Fatty Acids

As a class, the fatty acids are the simplest of the lipids. They are composed of a straight hydrocarbon chain terminating with a carboxylic acid group, therefore creating within the molecules a polar, hydrophilic end, and a nonpolar, hydrophobic end that is insoluble in water. Fatty acids are components of the more complex lipids, which are discussed in this section. They are of vital importance as an energy nutrient, furnishing most of the calories from dietary fat.

The length of the chains of fatty acids found in foods and body tissues ranges from 4 to about 24 carbon atoms. They may be saturated (SFA), monounsaturated (MUFA, possessing one carbon-carbon double bond), or polyunsaturated (PUFA, having two or more carbon-carbon double bonds). PUFAs of nutritional interest

may have as many as six double bonds. Where a carbon-carbon double bond exists, there is an opportunity for either a *cis* or *trans* geometric isomerism that significantly affects the molecular configuration of the molecule. The *cis* isomerism form results in a folding back and kinking of the molecule into a U-like orientation, whereas the *trans* form has the effect of extending the molecule into a linear form similar to that of saturated fatty acids. The following structures illustrate saturation and unsaturation in an 18-carbon fatty acid and show how *cis* or *trans* isomerization affects the molecular configuration (see 6A).

The more carbon-carbon double bonds occurring within a chain, the more pronounced is the bending effect that, in turn, plays an important role in the structure and function of cell membranes. Most naturally occurring unsaturated fatty acids are of the *cis* configuration, although the *trans* form does exist in some natural and partially hydrogenated fats and oils. Partial hydrogenation, a process commonly used in making margarine, is designed to solidify vegetable oils. Double bonds of *cis* orientation, not reduced by hydrogen in the process, undergo an electronic rearrangement to the *trans* form. The availability of *trans* fatty acids in the typical U.S. diet has been estimated to be approximately 8.1 g/person/day, the major source of which is margarines and spreads [1].

There has been concern about the possible adverse nutritional effects of dietary *trans* fatty acids, particularly their reputed role in the etiology of cardiovascular disease. This topic is discussed in this chapter's section dealing with lipoproteins and cardiovascular disease risk.

(6A)

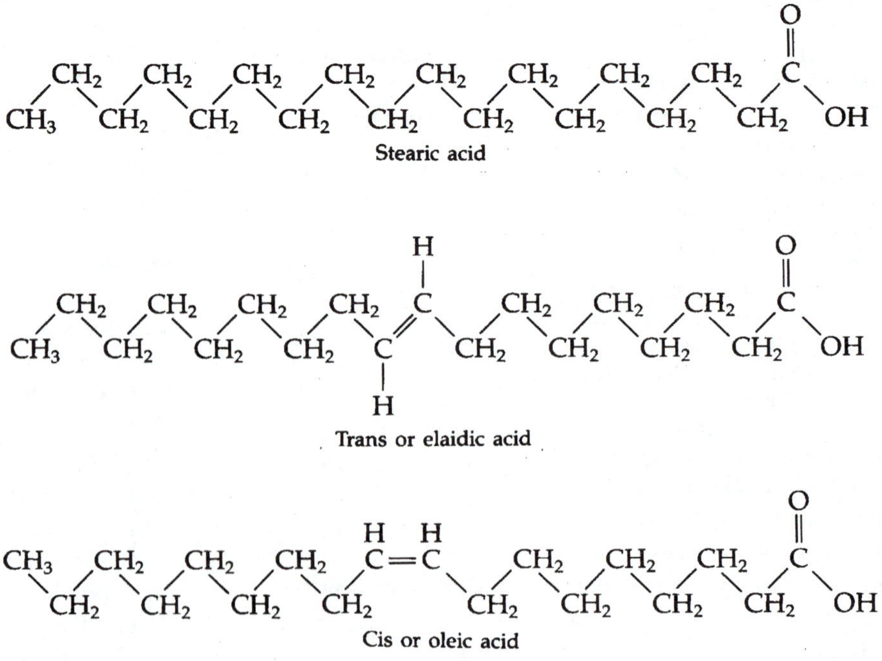

Stearic acid

Trans or elaidic acid

Cis or oleic acid

A notation has been established to denote the chain length of the fatty acids and the number and position of any double bonds that may be present. For example, the notation 18:2 $\Delta^{9,12}$ describes linoleic acid. The first number, 18 in this case, represents the number of carbon atoms; the number following the colon refers to the number of double bonds; and the superscripted numbers following the delta symbol designate the carbon atoms, *numbered from the carboxyl end,* at which the double bond(s) begins. Another commonly used system of notation locates the position of double bonds on carbon atoms *counted from the methyl, or omega (ω), end of the chain.* The system identifies the total number of carbon atoms in the chain, the number of double bonds, and the location (carbon atom number) of the first double bond. Implied in this system of notation is that multiple double bonds are always separated by three carbon atoms—that is,

$$-CH=CH-CH_2-CH=CH-.$$

Therefore, the location of multiple double bonds is unambiguous, given their total number and the location of the one closest to the methyl end of the chain. The omega symbol replaces the delta symbol in this manner of notation. For example, the designation of linoleic acid by each of the two systems is 18:2 $\Delta^{9,12}$ or 18:2 ω-6. The fatty acid α-linolenic acid, which contains three double bonds, is identified as 18:3 $\Delta^{9,12,15}$ or 18:3 ω-3. Substitution of the omega symbol with the letter *n* has been popularized. Using this designation, α-linolenic acid would be expressed as 18:3 n-3.

Table 6.1 lists some naturally occurring fatty acids and their dietary sources. The list includes only those fatty acids having chain lengths of 14 or more carbon atoms, because these are most important nutritionally and functionally. For example, palmitic acid (16:0), stearic acid (18:0), oleic acid (18:1), and linoleic acid (18:2) together account for >90% of the fatty acids in the average U.S. diet. However, it should be understood that shorter-chain fatty acids do occur in nature. Butyric acid (4:0) and lauric acid (12:0) occur in large amounts in milk fat and coconut oil, respectively.

If fat is entirely excluded from the diet of vertebrates, a condition develops characterized by retarded growth, dermatitis, kidney lesions, and early death. Studies have shown that the feeding of certain unsaturated fatty acids such as linoleic, linolenic, and arachidonic acids is effective in curing the condition. It is therefore evident that certain unsaturated fatty acids cannot be synthesized in animal cells and must be acquired in the diet from plant foods. These are the essential fatty acids, of which there are two, linoleic acid (18:2 n-6) and α-linolenic acid (18:3 n-3). From linoleic acid, γ-linolenic and arachidonic acids can be formed in

the body. An intermediate fatty acid in the pathway is eicosatrienoic acid. The pathway is

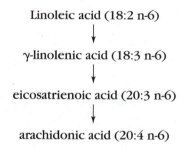

Linoleic acid (18:2 n-6)

↓

γ-linolenic acid (18:3 n-6)

↓

eicosatrienoic acid (20:3 n-6)

↓

arachidonic acid (20:4 n-6)

The essentiality of linoleic and α-linolenic acids is due to the fact that vertebrates lack enzymes called Δ^{12} and Δ^{15} desaturases, which incorporate double bonds at these positions. These enzymes are found only in plants. Vertebrates are therefore incapable of forming double bonds beyond the Δ^9 carbon in the chain. But given a $\Delta^{9,12}$ fatty acid acquired from the diet, additional double bond incorporation (desaturation) at Δ^6 can be accomplished. Fatty acid chains can also be elongated by the enzymatic addition of two carbon atoms at the carboxylic acid end of the chain. Desaturation and elongation reactions are discussed further in the section on fatty acid synthesis (p. 151) and are illustrated later in Figure 6.12.

Most fatty acids have an even number of carbon atoms, the reason for which will be evident from the discussion of fatty acid synthesis. Odd-numbered carbon fatty acids do occur naturally to some extent in some sources. Certain fish, such as menhaden, mullet, and tuna, as well as the bacterium *Euglena gracilis,* contain fairly high concentrations of odd-numbered carbon fatty acids.

Nutritional interest in the n-3 fatty acids has escalated enormously in recent years because of their reported hypolipidemic and antithrombotic effects (p. 145). An n-3 fatty acid of particular interest is eicosapentaenoic acid (20:5 n-3) because it is a precursor of the physiologically important eicosanoids, which are discussed in the following section. Fish oils are particularly rich in these unique fatty acids and are therefore the dietary supplement of choice in research designed to study their effects. Food sources and tissue distribution of a few of the commonly occurring n-3 polyunsaturated fatty acids are shown in Table 6.2.

Eicosanoids: Fatty Acid Derivatives of Physiological Significance

The essentiality of linoleic acid and α-linolenic acid is due to the fact that they act as precursors for some longer, more highly unsaturated fatty acids, which in turn are necessary

Table 6.1 Some Naturally Occurring Fatty Acids

Notation	Common Name	Formula	Source
Saturated Fatty Acids			
14:0	Myristic acid	$CH_3-(CH_2)_{12}-COOH$	Coconut and palm nut oils, most animal and plant fats
16:0	Palmitic acid	$CH_3-(CH_2)_{14}-COOH$	Animal and plant fats
18:0	Stearic acid	$CH_3-(CH_2)_{16}-COOH$	Animal fats, some plant fats
20:0	Arachidic acid	$CH_3-(CH_2)_{18}-COOH$	Peanut oil
24:0	Lignoceric acid	$CH_3-(CH_2)_{22}-COOH$	Most natural fats, peanut oil in small amounts
Unsaturated Fatty acids			
16:1 Δ^9 (n-7)	Palmitoleic acid	$CH_3-(CH_2)_5-CH=CH-(CH_2)_7-COOH$	Marine animal oils, small amount in plant and animal fats
18:1 Δ^9 (n-9)	Oleic acid	$CH_3-(CH_2)_7-CH=CH-(CH_2)_7-COOH$	Plant and animal fats
18:2 $\Delta^{9,12}$ (n-6)	Linoleic acid	$CH_3-(CH_2)_4-CH=CH-CH_2-CH=CH-(CH_2)_7-COOH$	Corn, safflower, soybean, cottonseed, sunflower seed, and peanut oil
18:3 $\Delta^{9,12,15}$ (n-3)	α-Linolenic acid	$CH_3-(CH_2-CH=CH)_3-(CH_2)_7-COOH$	Linseed, soybean, and other seed oils
20:4 $\Delta^{5,8,11,14}$ (n-6)	Arachidonic acid	$CH_3-(CH_2)_3-(CH_2-CH=CH)_4-(CH_2)_3-COOH$	Small amounts animal fats
20:5 $\Delta^{5,8,11,14,17}$ (n-3)	Eicosapentaenoic acid	$CH_3-(CH_2-CH=CH)_5-(CH_2)_3-COOH$	Marine algae, fish oils
22:6 $\Delta^{4,7,10,13,16,19}$ (n-3)	Docosahexaenoic acid	$CH_3-(CH_2-CH=CH)_6-(CH_2)_2-COOH$	Animal fats as phospholipid component, fish oils

Table 6.2 Dietary Sources and Tissue Distribution of the Major n-3 Polyunsaturated Fatty Acids

Fatty Acid Series	Major Members of Series	Tissue Distribution in Mammals	Dietary Sources
n-3	α-linolenic acid 18:3 n-3	Minor component of tissues	Some vegetable oils (soy, canola, linseed, rapeseed) and leafy vegetables
	Eicosapentaenoic acid 20:5 n-3	Minor component of tissues	Fish and shellfish
	Docosahexaenoic acid 22:6 n-3	Major component of membrane phospholipids in retinal photoreceptors, cerebral gray matter, testes, and sperm	Fish and shellfish

- for the formation of cell membranes and
- as precursors of compounds called *eicosanoids.*

Eicosanoids are fatty acids composed of 20 carbon atoms. They include the physiologically potent families of substances called *prostaglandins, thromboxanes,* and *leukotrienes,* all of which are formed from precursor fatty acids by the incorporation of oxygen atoms into the fatty acid chains. Reactions of this sort are often referred to as *oxygenation reactions,* and the enzymes catalyzing the reactions are named *oxygenases.*

The most important fatty acid serving as precursor for eicosanoid synthesis is arachidonate. Its oxygenation follows either of two major pathways:

- the "cyclic" pathway, which results in the formation of prostaglandins and thromboxanes; and
- the "linear" pathway, which produces leukotrienes.

The featured enzyme in the cyclic pathway is *prostaglandin endoperoxide synthase,* sometimes called *cyclo-oxygenase.* It catalyzes the oxygenation of arachidonate together with the cyclization of an internal segment of the arachidonate chain, the hallmark structural feature of the prostaglandins and thromboxanes. The enzyme that converts arachidonate to the leukotrienes in the linear pathway is named *lipoxygenase,* and the pathway is often referred to as the *lipoxygenase pathway.* Figure 6.1 is an overview of the reactions of the cyclic and linear pathways of arachidonate.

Prostaglandins (PG) are 20-carbon fatty acids having a 5-carbon ring in common but displaying modest structural differences among themselves. As shown in Figure 6.1, they are designated PGD, PGE, PGF, PGI, PGG, and PGH. The subscript numbers indicate the number of double bonds, the "2" series being the most important. These compounds, along with the thromboxanes, exhibit a wide range of physiological actions, including the lowering of blood pressure, diuresis, blood platelet aggregation, effects on the immune and nervous systems as well as gastric secretions, and the stimulation of smooth muscle contraction, to name several. They are described as being "hormone-like" in function. However, unlike hormones, which originate from a specific gland and whose actions are the same for all their target cells, prostaglandins are widely distributed in animal tissues but affect only the cells in which they are synthesized. They do appear to alter the actions of hormones, often through their modulation of cAMP levels and the intracellular flow of calcium ions.

Certain combinations of prostaglandins and thromboxanes may exhibit antagonistic effects. For example, prostacyclin (PGI_2) is a potent stimulator of adenylate cyclase and thereby acts as a platelet "antiaggregating"

factor, because platelet aggregation is inhibited by cAMP. Opposing this action is thromboxane A_2, which inhibits adenylate cyclase and consequently serves as a "proaggregating" force. Another example of opposing actions of the prostaglandins is the vasodilation of blood vessels by PGE_2 and their vasoconstriction by PGF_2.

Certain prostaglandins produce a rise in body temperature (fever) and can cause inflammation and therefore pain. The anti-inflammatory and antipyretic (fever-reducing) activity of aspirin, acetaminophen, and indomethacin is due to their inhibitory effect on prostaglandin endoperoxide synthase (cyclo-oxygenase), resulting in reduced prostaglandin and thromboxane synthesis. The 5-lipoxygenase pathway of a leukotriene (LTC_4) formation from arachidonate is shown in Figure 6.1. Although it is not shown in the pathway, LTC_4 is further metabolized to other leukotrienes in the following order:

$$LTC_4 \rightarrow LTD_4 \rightarrow LTE_4 \rightarrow LTF_4$$

The structures of LTA_4 and LTC_4 are shown because these exemplify a leukotriene and a peptidoleukotriene, respectively. Notice that LTC_4 is formed from LTA_4 by incorporation of the tripeptide glutathione (γ-glutamyl-cycteinyl-glycine). LTD_4, LTE_4, and LTF_4 are peptidoleukotrienes produced from LTC_4 by peptidase hydrolysis of bonds within the glutathione moiety. Like the prostaglandins, these substances share structural characteristics but are classified within the A, B, C, D, and E series according to their structural differences. The subscript number represents the number of double bonds in the compound.

Leukotrienes have potent biological actions. Briefly, they contract respiratory, vascular, and intestinal smooth muscles. The effects on the respiratory system include constriction of bronchi and increased mucus secretion. These actions, which are known to be expressed through binding to specific receptors, have implicated the leukotrienes as mediators in asthma, immediate hypersensitivities, inflammatory reactions, and myocardial infarction. In fact, one of the major chemical mediators of anaphylactic shock, the so-called slow-reacting substance of anaphylactic shock, or SRS-A, has been found to be a mixture of the peptidoleukotrienes, LTC_4, LTD_4, and LTE_4. Anaphylactic shock is a life-threatening response to chemical substances, primarily histamine, that are released as a result of a severe allergic reaction.

A necessity for eicosanoid formation is an availability of an appropriate amount of free (unesterified) arachidonate. Cellular concentration of the free fatty acid is not adequate, and it must therefore be released from membrane glycerophosphatides by a specific hydrolytic enzyme called *phospholipase A_2.* Structural features of glycerophosphatides are reviewed in the

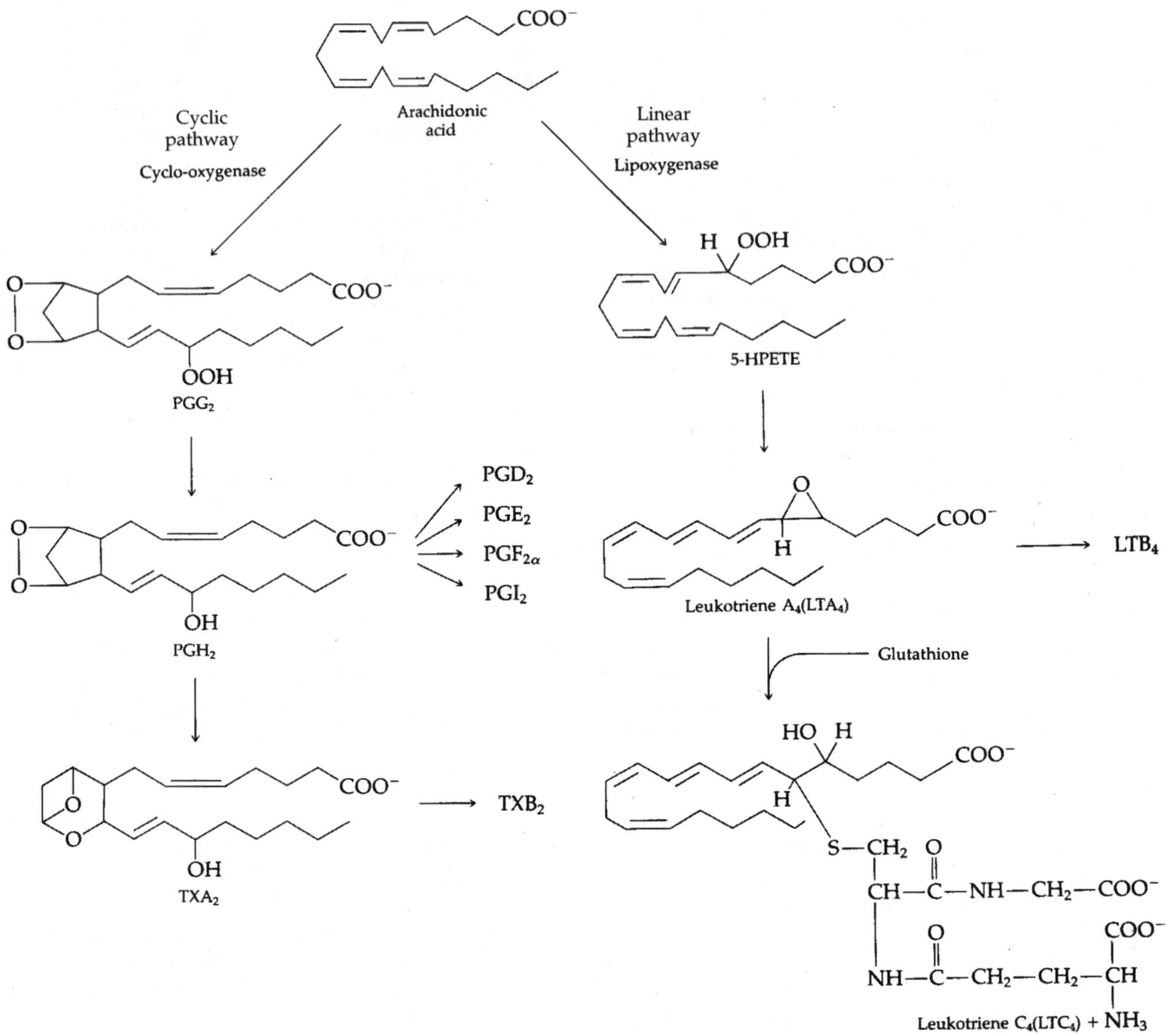

Figure 6.1 The formation of prostaglandins, thromboxanes, and leukotrienes from arachidonic acid via cyclo-oxygenase and lipoxygenase pathways. Abbreviations: PG, prostaglandin; TX, thromboxane; 5-HPETE, 5-hydroperoxy-6, 8, 11, 14-eicosatetraenoic acid.

section on phospholipids (p. 130). The most important glycerophosphatides acting as sources of arachidonate in cells are phosphatidylcholine and phosphatidylinositol. When present in these structures, arachidonate normally occupies the sn-2 position. Stereospecific numbering (sn-) of glycerol-based compounds is discussed in the following section.

The release of arachidonate from membrane glycerophosphatides, for eicosanoid synthesis, is influenced by stimuli. These stimuli are of two main types, physiological (specific) and pathological (nonspecific). Physiological stimulation, a natural occurrence, is brought about by stimulatory compounds such as epinephrine,

angiotensin II, and antigen-antibody complexes. Pathological stimuli, which result in a more generalized release of all fatty acids from the sn-2 position, include mechanical damage, ischemia, and membrane-active venoms. Table 6.3 lists the precursors, site of synthesis, and physiological effects of a few of the major eicosanoid groups.

Triacylglycerols (Triglycerides)

Most stored body fat is in the form of triacylglycerols, which represent a highly concentrated form of energy. They account for nearly 95% of dietary fat. Structurally they are composed of the trihydroxy alcohol, glycerol, to

Table 6.3 Physiological Characteristics of Eicosanoids

Eicosanoid Family	Precursor		Site of Synthesis	Mode of Action
	Arachidonate $(20:4\omega6)$	Eicosapentaenoate $(20:5\omega3)$		
Prostacyclins	PGI_2	PGI_3	Vascular endothelium	Vasodilator
				Platelet antiaggregator
Thromboxanes	TXA_2	TXA_3	Platelet	Vasoconstrictor
				Platelet aggregator
Leukotrienes	LTB_4	LTB_4	Leukocytes	Chemotaxis

Source: Anderson PA, Sprecher HW. Omega-3 fatty acids in nutrition and health. Dietetic currents. Vol 14, No. 2, 1987. Reprinted with permission of Ross Laboratories, Columbus, OH 43216.

(6B)

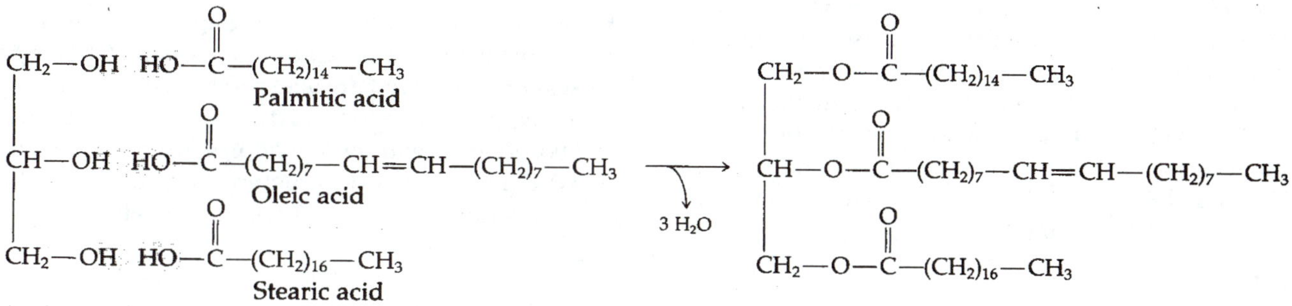

which are attached three fatty acids by ester bonds. The fatty acids may all be the same (a simple triacylglycerol) or different (a mixed triacylglycerol). The linking of the fatty acids palmitate, oleate, and stearate to glycerol with the liberation of three water molecules is shown in 6B.

Acylglycerols composed of glycerol esterified to a single fatty acid (a monoacylglycerol) or to two fatty acids (a diacylglycerol) occur in negligible amounts in tissues; however, they are important intermediates in some metabolic reactions and may be components of other lipid classes. They may also occur in processed foods, to which they can be added as emulsifying agents.

The specific glycerol hydroxyl group to which a certain fatty acid is attached is indicated by a numbering system for the three glycerol carbons. This is complicated somewhat by the fact that the central carbon of the glycerol is asymmetrical when different fatty acids are esterified at the two end carbon atoms, and may therefore exist in either the D or L form (p. 71). Unfortunately, the same monoacylglycerol may therefore be written in two ways (see 6C). To resolve this ambiguity, a system of nomenclature called *stereospecific numbering* (*sn*) has been adopted whereby the glycerol is always written as in 6D, with the C-2 hydroxyl group oriented to the left (L) and the carbons, numbered 1 through 3, beginning at the top. Accordingly, 1-monopalmitoyl glycerol (6E) can be drawn in either of the ways shown in 6C. Using this system, therefore, the naming of the triacylglycerol shown as structure 6B is 1-stearoyl-2-oleoyl-3-palmitoyl-L-glycerol.

(6C)

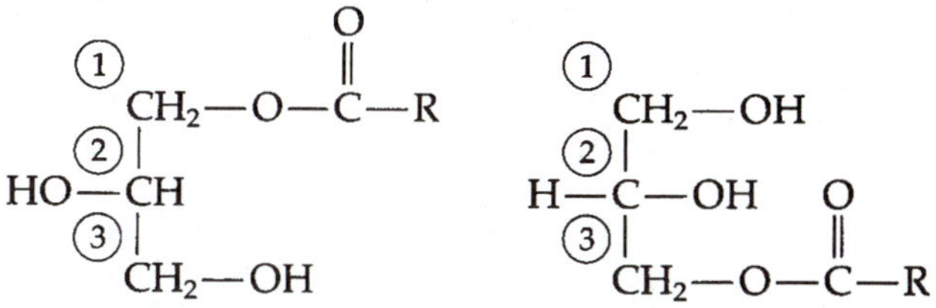

1-acyl-L-glycerol   3-acyl-D-glycerol

Triacylglycerols exist as fats or oils at room temperature according to their physical state, which varies according to the structures of the component fatty acids. Those containing a high proportion of relatively short-chain fatty acids or unsaturated fatty acids tend to be liquid (oils) at room temperature. Saturated fatty acids of longer chain length have a higher melting point and exist as solids at ambient temperatures. When used for energy, fatty acids are released in free form (free fatty acids, FFA) from the triacylglycerols in adipose tissue cells by the activity of lipases and are transported by albumin to various tissues for oxidation.

(6D)

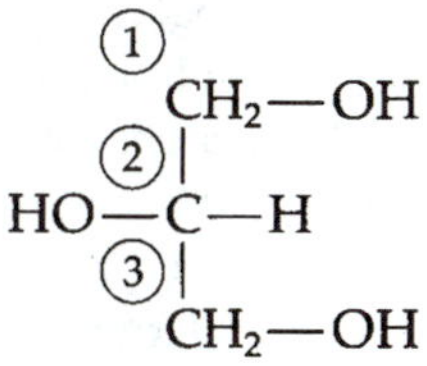

(6E)

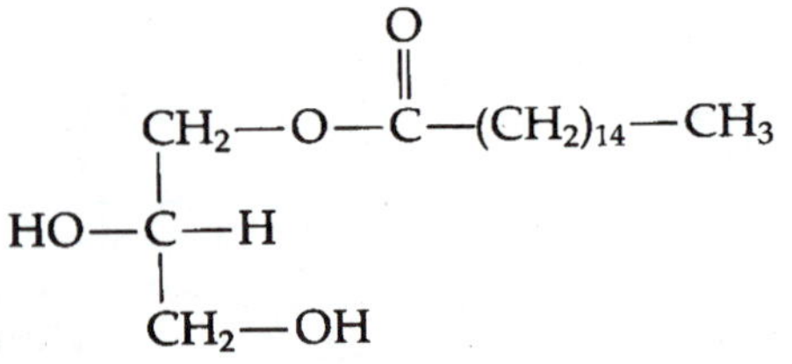

$$CH_2-O-\overset{\overset{\displaystyle O}{\|}}{C}-(CH_2)_{14}-CH_3$$
$$HO-\overset{|}{\underset{|}{C}}-H$$
$$CH_2-OH$$

1-monopalmitoyl
glycerol

Sterols and Steroids

This class of lipid is characterized by a four-ring core structure called the cyclopentanoperhydrophenanthrene, or *steroid,* nucleus. Sterols are monohydroxy alcohols of steroidal structure, cholesterol being the most common example. Cholesterol is present only in animal tissues, and it can exist in free form or can be esterified with a fatty acid. Many other sterols are also found in plant tissues. The structure of cholesterol is shown in 6F, along with the numbering system for the carbons in the steroid nucleus.

(6F)

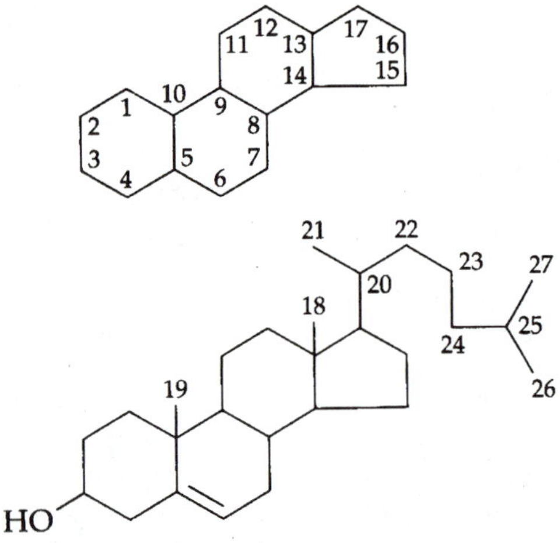

Cholesterol

Meats, egg yolk, and dairy products contain fairly large amounts of cholesterol, and the sterol is an essential component of cell membranes (p. 3), particularly those comprising nerve tissue. Despite the bad press that cholesterol has garnered over the years because of its implication in cardiovascular disease, it serves as the precursor for many other important steroids in the body. Included among these are the bile acids; steroid sex hormones such as estrogens, androgens, and progesterone; the adrenocortical hormones; and the vitamin D of animal tissues (cholecalciferol). These steroids differ from one another in the arrangement of double bonds in the ring system, the presence of carbonyl or hydroxyl groups, and the nature of the side chain at C-17. Such structural modifications are all mediated by enzymes that function as dehydrogenases, isomerases, hydroxylases, or desmolases. Desmolases remove or shorten the length of side chains on the steroid nucleus. The derivation of the various types of steroids from cholesterol is diagrammed in Figure 6.2. Although many physiologically active corticosteroid hormones, sex hormones, and bile acids exist, only representative compounds are shown.

Sterols, together with phospholipids, which will be considered next, comprise only about 5% of dietary lipid.

Phospholipids

As the name implies, lipids belonging to this class contain phosphate as a common component. They also possess one or more fatty acid residues. Phospholipids are categorized into one of two groups called *glycerophosphatides* and *sphingophosphatides,* depending on whether their core structure is glycerol or the amino alcohol sphingosine (6G), respectively.

(6G)

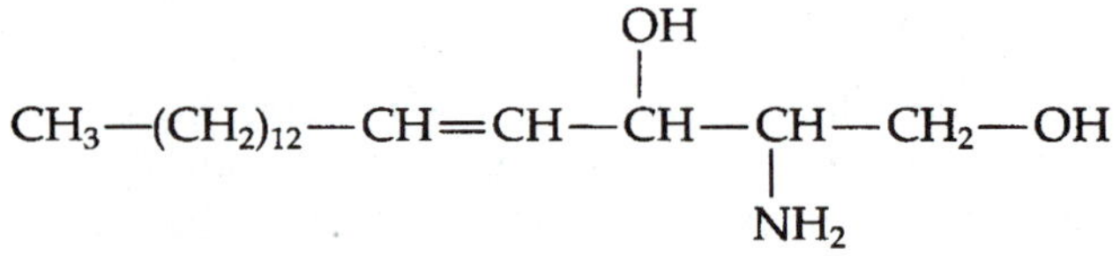

$$CH_3-(CH_2)_{12}-CH=CH-\overset{\overset{\displaystyle OH}{|}}{CH}-\overset{\overset{\displaystyle}{|}}{\underset{\underset{\displaystyle NH_2}{|}}{CH}}-CH_2-OH$$

Sphingosine

Glycerophosphatides

The building block of a glycerophosphatide is *phosphatidic acid,* formed by the esterification of two fatty acids at C-1 and C-2 of glycerol and the esterification of the C-3 hydroxyl with phosphoric acid. The structure in 6H typifies a phosphatidate, a term that does not define a specific structure, because different fatty acids may be involved. The convention of the numbering of the glycerol carbon atoms is the same as that for triacylglycerols. The numbering from top to bottom is *sn*-1, *sn*-2, *sn*-3, provided the glycerol is written in L-configuration so that the C-2 fatty acid constituent is directed to the left as shown in 6H.

(6H)

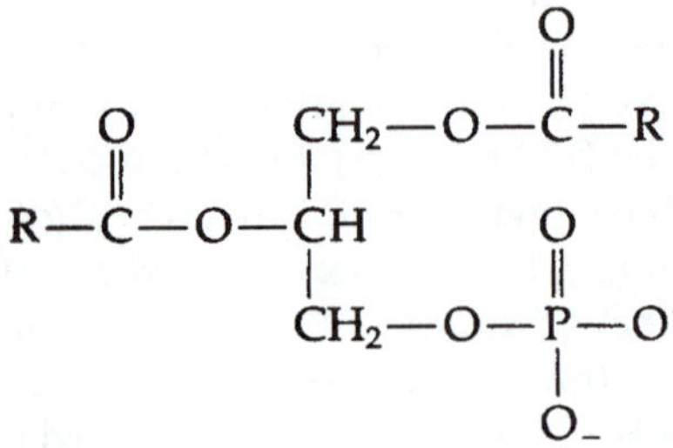

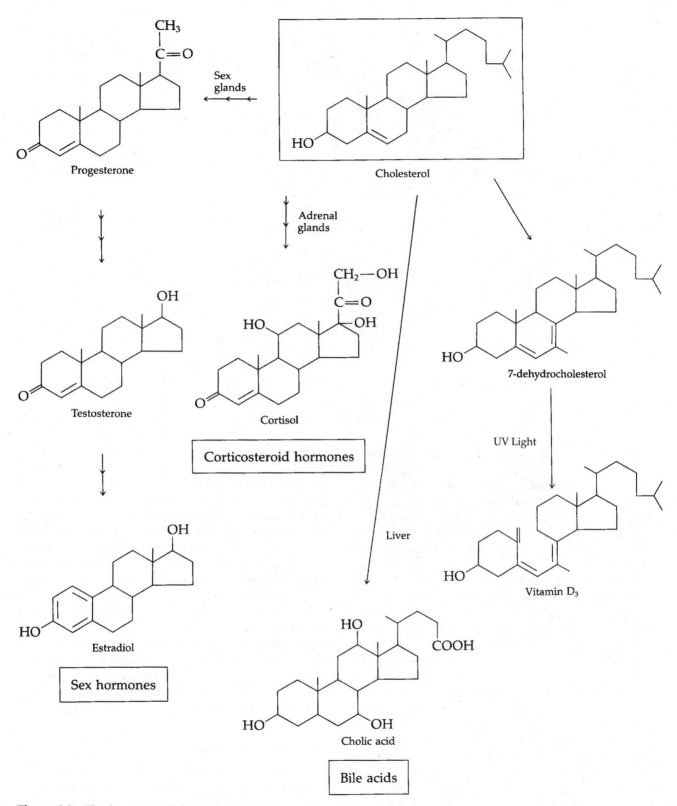

Figure 6.2 The formation of physiologically important steroids from cholesterol. Only representative compounds from each category of steroid are shown.

(6I)

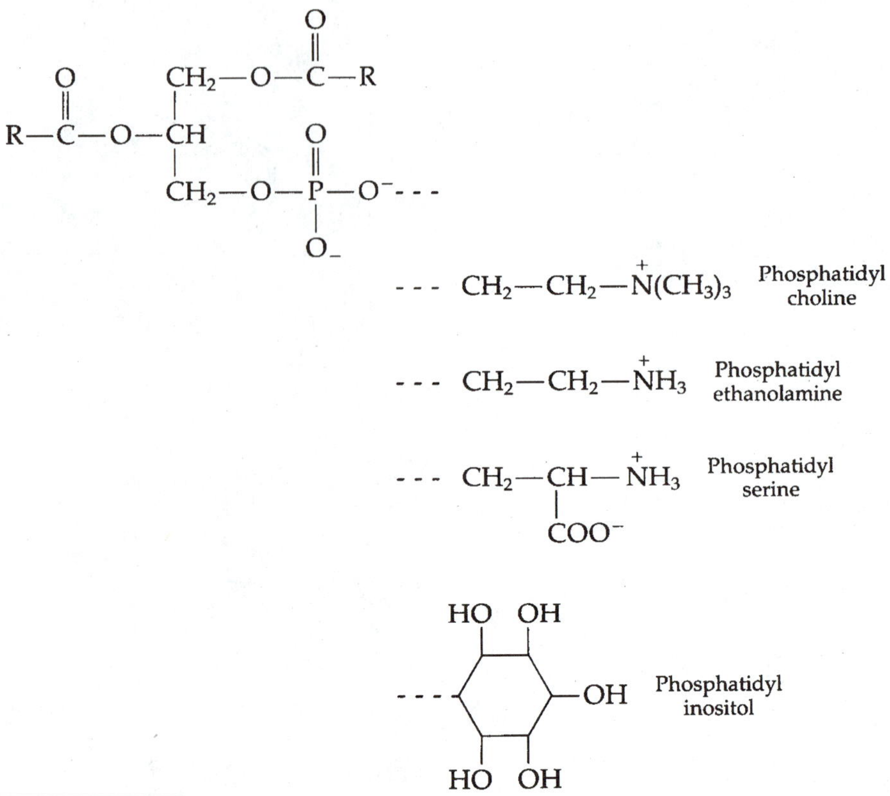

Phosphatidic acids form a number of derivatives with compounds such as choline, ethanolamine, serine, and inositol, each of which possesses an alcohol group through which a second esterification to the phosphate takes place. The compounds are named as the phosphatidyl derivatives of the alcohols, as indicated in 6I. Phosphatidylcholine is probably better known by its common name, *lecithin.*

Glycerophosphatides are very important components of cell membranes. In addition to lending structural support to the membrane, they serve as a source of physiologically active compounds. We have already seen (p. 127) how arachidonate, released on demand from membrane-bound phosphatidylcholine and phosphatidylinositol, is needed for eicosanoid synthesis. Phosphatidylinositol participates in other cell functions as well. For example, it plays a specific role in the anchoring of membrane proteins, when the proteins are covalently attached to lipids. This has been demonstrated by the release of certain membrane proteins when cells are treated with a phosphatidylinositol-specific phospholipase C, which hydrolyzes the ester bond connecting the glycerol to the phosphate. Second, certain hydrolytic products of phosphatidylinositol are active in intracellular signaling and as second messengers in hormone stimulation. A brief discussion of this latter function follows.

Phosphatidylinositol in the plasma membrane can be doubly phosphorylated by ATP, forming phosphatidylinositol-4,5-bisphosphate. Stimulation of the cell by certain hormones activates a specific phospholipase C, which produces inositol-1,4,5-trisphosphate and diacylglycerol from phosphatidylinositol-4,5-bisphosphate. Both of these products function as second messengers in cell signaling. Inositol-1,4,5-trisphosphate causes the release of Ca^{+2} held within membrane-bounded compartments of the cell, triggering the activation of a variety of Ca^{+2}-dependent enzymes and hormonal responses [2]. Diacylglycerol binds to and activates an enzyme, protein kinase C, which transfers phosphate groups to several cytoplasmic proteins, thereby altering their enzymatic activities [3]. This dual signal hypothesis of phosphatidylinositol hydrolysis is represented in Figure 6.3.

Sphingophosphatides

Lipids formed from sphingosine (6G) are categorized into three subclasses: sphingomyelins, cerebrosides, and gangliosides. Of these, only the sphingomyelins are sphingophosphatides. The other two subclasses of sphingolipids contain no phosphate but instead possess a carbohydrate moiety. They are referred to as *glycolipids* and are discussed in the following section.

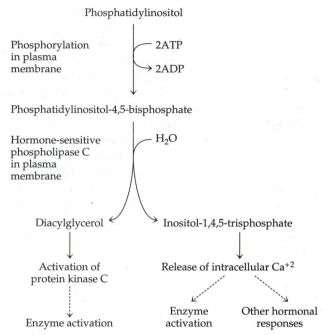

Phosphatidylinositol

Phosphorylation in plasma membrane — 2ATP → 2ADP

Phosphatidylinositol-4,5-bisphosphate

Hormone-sensitive phospholipase C in plasma membrane — H₂O

Diacylglycerol ← → Inositol-1,4,5-trisphosphate

Activation of protein kinase C

Release of intracellular Ca⁺²

Enzyme activation

Enzyme activation   Other hormonal responses

Figure 6.3 Phosphatidylinositol-4,5-bisphosphate, formed in the plasma membrane by phosphorylation of phosphatidylinositol, is hydrolyzed by a specific phospholipase C in response to hormonal signals. Both products of hydrolysis act as intracellular messengers.

Sphingomyelins occur in plasma membranes of animal cells and are found in particularly large amounts in the myelin sheath of nerve tissues. The sphingomyelins contain a fatty acid residue attached in amide linkage to the amino group of the sphingosine. The product of this union is called *ceramide,* which in turn is esterified to phosphorylcholine (see 6J).

Phospholipids are more polar than the triacylglycerols and sterols and therefore tend to attract water molecules, which are also polar. Because of this hydrophilic property, they are commonly expressed on the surface of blood-borne lipid particles, such as chylomicrons (p. 135), thereby stabilizing the particles in the aqueous medium. Furthermore, as a constituent of cell and organelle membranes (p. 2), they serve as a regulator of the passage of water-soluble and fat-soluble materials across the membrane.

Glycolipids

Glycolipids can be subclassified into *cerebrosides* and *gangliosides.* They are so named because they have a carbohydrate component within their structure. Like the phospholipids, their physiological role is principally structural, contributing little as an energy source. They occur in the medullary sheaths of nerves and in brain tissue, particularly the white matter. As in the case of sphingomyelin, the sphingosine moiety provides the backbone for glycolipid structure. It is attached to a fatty acid by an amide bond, forming ceramide, as discussed previously. The glycolipids do not contain phosphate.

A cerebroside is characterized by the linking of ceramide to a monosaccharide unit such as glucose or galactose, producing either a glucocerebroside or galactocerebroside.

Gangliosides resemble cerebrosides except that the single monosaccharide unit of the cerebroside is replaced by an oligosaccharide containing various monosaccharide derivatives such as N-acetyl neuraminic acid and N-acetyl galactosamine. Gangliosides are known to be involved in certain recognition events occurring at the cell surface. For example, they provide the carbohydrate determinants of the human blood groups A, B, and O.

Digestion

Because fats are hydrophobic, their digestion poses a special problem because their digestive enzymes, like all proteins, are hydrophobic, and normally function in an aqueous environment. However, the surface area of dietary lipid, targeted for digestion, is greatly enhanced by a very efficient emulsification process mediated mainly by the bile salts. Consequently, accessibility of the fat to digestive enzymes is greatly increased by bile salt action.

Triacylglycerols, phospholipids (primarily phosphatidylcholine), and sterols (mainly cholesterol) provide the lipid component of the typical Western diet. Of these, triacylglycerols, customarily called *fats,* are by far the major contributor, with a consumption rate of about

(6J)

$$CH_3-(CH_2)_{12}-CH=CH-CH-CH-CH_2-O-\overset{\overset{O}{\|}}{P}-O-CH_2-CH_2-\overset{+}{N}-(CH_3)_3$$

with OH on the third carbon, NH—C—R (with O) branch, and O⁻ on phosphorus

Sphingomyelin   (R = Fatty acid)

150 g daily on the average. Digestive enzymes involved in the breakdown of dietary lipids in the gastrointestinal tract are esterases that cleave the ester bonds within triacylglycerols (lipase), phospholipids (phospholipases), and cholesteryl esters (cholesterol esterase).

Most dietary triacylglycerol digestion is completed in the lumen of the small intestine, although the process actually begins in the stomach. Digestive activity at these two sites is attributed to two different forms of lipases: (1) lingual lipase, secreted by serous glands lying beneath the tongue, and (2) pancreatic lipase. Basal secretion of lingual lipase apparently occurs continuously but can be stimulated by neural (sympathetic agonists), dietary (high fat), and mechanical (sucking and swallowing) factors. Lingual lipase accounts for the limited digestion of fat in the stomach, made possible by the enzyme's particularly high stability at the low pH of the gastric juices. It can readily penetrate milk fat globules without substrate stabilization by bile salts, a feature that makes it particularly important for fat digestion in the suckling infant, whose pancreatic function may not be fully developed. Lingual lipases act preferentially on triacylglycerols containing medium- and short-chain length fatty acids. It preferentially hydrolyzes fatty acids at the *sn*-3 position, releasing fatty acids and 1,2-diacylglycerol as products. This specificity is again advantageous for the suckling infant because in milk triacylglycerols, short- and medium-length fatty acids are usually esterified at the *sn*-3 position. Short- and medium-length fatty acids are metabolized more directly than long chain fatty acids (p. 135).

For dietary fat in the stomach to be hydrolyzed by lingual lipase, some degree of emulsification must occur to expose sufficient surface area of the substrate. Muscle contractions of the stomach, and the squirting of the fat through a partially opened pyloric sphincter, produce shear forces sufficient for emulsification. Also, potential emulsifiers in the acid milieu of the stomach include complex polysaccharides, phospholipids, and peptic digests of dietary proteins. However, quantitative hydrolysis and absorption, especially of the long-chain fatty acids, require less acidity, appropriate lipases, more effective emulsifying agents (bile salts), and specialized absorptive cells. These conditions are provided in the lumen of the upper small intestine. The presence of undigested lipid in the stomach delays the rate of emptying of the stomach contents, presumably by way of the hormone enterogastrone (GIP and secretin; p. 30), which inhibits gastric motility. Fats therefore have a "high satiety value."

The partially hydrolyzed lipid emulsion leaves the stomach and enters the duodenum as fine lipid droplets. Effective emulsification takes place because as mechanical shearing continues, it is complemented by bile that is released from the gallbladder as a result of stimulation by the hormone cholecystokinin (CCK). Bicarbonate is simultaneously released from the pancreas, elevating the pH to a level suitable for pancreatic lipase activity. In combination with triacylglycerol breakdown products, bile salts are excellent emulsifying agents. Their emulsifying effectiveness is due to their amphipathic properties—that is, their possessing both hydrophilic and hydrophobic "ends." Such molecules tend to arrange themselves on the surface of small fat particles with their hydrophobic ends turned inward and hydrophilic regions outward toward the water phase. This chemical action, together with the help of peristaltic agitation, converts the fat into small droplets with a greatly increased surface area. These particles can then be readily acted on by pancreatic lipase.

Pancreatic lipase activation is complex, requiring the participation of the protein colipase, calcium ions, and bile salts. Colipase is formed by the hydrolytic activation by trypsin of procolipase, also of pancreatic origin. It contains approximately 100 amino acid residues and possesses distinctly hydrophobic regions that are believed to act as lipid-binding sites. Colipase has been shown to associate strongly with pancreatic lipase and therefore may act as an anchor, or linking point, for attachment of the enzyme to the bile salt-stabilized micelles described later.

The action of pancreatic lipase on ingested triacylglycerols results in a complex mixture of diacylglycerols, monoacylglycerols, and free fatty acids. Its specificity is primarily toward *sn*-1-linked fatty acids and secondarily to *sn*-3 bonds. Therefore, the main path of this digestion progresses from triacylglycerols to 2,3-diacylglycerols to 2-monoacylglycerols. Only a small percentage of the triacylglycerols is hydrolyzed totally to free glycerol. That which does occur probably follows the isomerization of the 2-monoacylglycerol to the 1-monoacylglycerol, which is then hydrolyzed. Esterified cholesterol, meanwhile, undergoes hydrolysis to free cholesterol and a fatty acid, catalyzed by the enzyme cholesterol esterase. The C-2 fatty acid of lecithin is hydrolytically removed by a specific esterase called *phospholipase A₂*, producing lysolecithin and still another free fatty acid.

The products of the partial digestion of lipids, primarily 2-monoacylglycerols, lysolecithin, cholesterol, and fatty acids, combine with bile salts, forming negatively charged polymolecular aggregates called *micelles*. These have a much smaller diameter (~5 nm) than the unhydrolyzed precursor particles, allowing them access to the intramicrovillus spaces (50–100 nm) of the intestinal membrane.

Absorption

Stabilized by the polar bile salts, the micellar particles are sufficiently water-soluble to penetrate what is called the unstirred water layer bathing the absorptive cells of the small intestine. The absorptive cells are called intestinal mucosal cells, or *enterocytes* (p. 28). Micelles interact at the brush border of these cells, whereupon the lipid contents of the micelles diffuse out of the micelles and into the enterocytes, moving down a concentration gradient. Although this process occurs in the distal duodenum and the jejunum, the bile salts are not absorbed at this point but instead are absorbed in the ileal segment of the small intestine. There they are returned to the liver via the portal vein to be resecreted in the bile. This circuit is referred to as the "enterohepatic circulation of the bile salts" (p. 43).

After the absorption of free fatty acids, 2-monoacylglycerols, cholesterol, and lysophosphatidylcholine into the enterocytes, intracellular re-formation of triacylglycerols, phosphatidylcholine, and cholesteryl esters takes place. The process is, however, a function of the chain length of the fatty acids involved. Fatty acids having more than 10 to 12 carbon atoms are first activated by being coupled to coenzyme A by the enzyme acyl CoA synthetase. They are then reesterified into triacylglycerols, phosphatidylcholine, and cholesteryl esters as mentioned earlier. Short-chain fatty acids, those containing fewer than 10 to 12 carbon atoms, in contrast, pass from the cell directly into the portal blood. In the blood, short-chain fatty acids attach to albumin for transport to other tissues for processing. The different fate of the long- and short-chain fatty acids is due to the specificity of the acyl CoA synthetase enzyme for long-chain fatty acids only.

Lipids resynthesized in the enterocytes, together with fat-soluble vitamins, are collected in the cell's endoplasmic reticulum as large fat particles. While still in the endoplasmic reticulum, the particles receive a layer of protein on their surface, which tends to stabilize the particles in the aqueous environment of the circulation, which they eventually enter. The particles are pinched off as lipid vesicles that then fuse with the Golgi apparatus (p. 11). There, carbohydrate is attached to the protein coat, and the completed particles, called *chylomicrons,* are transported to the cell membrane and exocytosed into the lymphatic circulation. Chylomicrons therefore belong to a family of compounds called *lipoproteins,* acquiring their name from the fact that lipid and protein comprise their composition. A review of the lipoproteins is offered in the next section.

The protein portion only, of any lipoprotein, is called the *apolipoprotein.* Apolipoproteins play a very important role in the structural and functional relationship among the lipoproteins (discussed in the next section). Key features of intestinal absorption of lipid digestion products are depicted in Figure 6.4.

It should be pointed out that triacylglycerols can also be synthesized from α-glycerophosphate in the enterocytes. This metabolite can be formed either from the phosphorylation of free glycerol or from reduction of dihydroxyacetone phosphate, an intermediate in the pathway of glycolysis (see Fig. 4.9). Triacylglycerol synthesis by this route is also shown in Figure 6.4.

Transport

Lipoproteins

Chylomicrons are the primary form of lipoprotein formed from exogenous (dietary) lipids. Lipoproteins other than chylomicrons transport endogenous lipids, which are circulating lipids that do not arise directly from intestinal absorption but are instead processed through other tissues such as the liver. Several types of lipoproteins therefore exist, differing in their chemical composition, physical properties, and metabolic function. The role shared by all the lipoproteins, however, is the transport of lipids from tissue to tissue to supply the lipid needs of different cells. The arrangement of the lipid and protein components of a typical lipoprotein particle is represented in Figure 6.5. It can be seen from the figure that the more hydrophobic lipids are located in the core of the particle, while the relatively more polar proteins and phospholipids are situated on the surface to enhance aqueous stability.

Lipoproteins differ according to the ratio of lipid to protein within the particle as well as having different proportions of lipid types: triacylglycerols, cholesterol and cholesteryl esters, and phospholipids. Such compositional differences influence the *density* of the particle, and this has become the physical characteristic that is used to differentiate and classify the various lipoproteins. In the order of lowest to highest density, the lipoprotein fractions are chylomicrons, very low-density lipoproteins (VLDLs), low-density lipoproteins (LDLs), and high-density lipoproteins (HDLs). There is also an intermediate density particle (IDL), having a density between that of VLDL and LDL. The IDL particles are very short-lived in the bloodstream, however, and have little nutritional or physiological importance. Table 6.4 summarizes several physical and chemical characteristics of the lipoproteins.

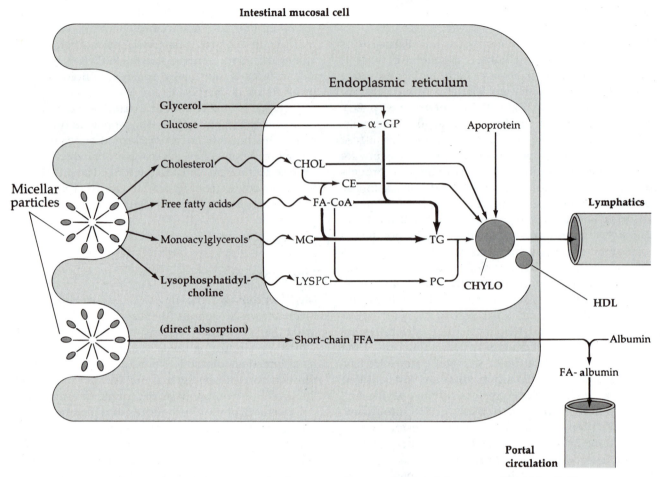

Figure 6.4 The uptake of micellar lipid particles by the intestinal mucosal cell, and the reassembly of the lipid components within the endoplasmic reticulum (shown as white area) of the cell. The reassembly of the lipids produces principally chylomicrons, although small amounts of HDL can be formed in this manner. Following their combination with apoproteins, the lipoprotein particles are released into the lymphatic circulation and ultimately the portal blood. Short-chain free fatty acids enter the portal circulation directly, as shown. Abbreviations: CHOL, cholesterol; CE, cholesteryl ester; FA-CoA, CoA activated fatty acids; MG, monoacylglycerol; LYSPC, lysophosphatidylcholine; α-GP, α-glycerophosphate; CHYLO, chylomicrons; TG, triacylglycerol; PC, phosphatidylcholine.

Apolipoproteins

Apolipoproteins, the protein components of lipoproteins, tend to stabilize the lipoproteins as they circulate in the aqueous environment of the blood. But they have other important functions as well. They confer specificity on the lipoprotein complexes, allowing them to be recognized by specific receptors on cell surfaces. They also stimulate certain enzymatic reactions, which in turn regulate the lipoproteins' metabolic functions.

A series of letters (A to E), with subclasses of each, are now used to identify the various apolipoproteins. For convenience, they are usually abbreviated "apo" followed by the identifying letter—that is, apoA-I, apoB-100, apoC-II, and so on. A partial listing of the apolipoproteins,

together with their molecular weight, the lipoprotein with which they are associated, and their postulated physiological function, are found in Table 6.5.

Distribution of Lipids

The re-formed lipid derived from exogenous sources leaves the enterocytes (intestinal mucosal cells) largely in the form of chylomicrons, which then undergo intravascular conversion to chylomicron remnants (structurally similar to VLDL). To some extent, HDL can be synthesized within the enterocytes and released directly into the mesenteric lymph. The lipoproteins first appear in the lymphatic vessels of the abdominal region and then enter the bloodstream at a slow rate so as to prevent large-scale changes in the lipid content of

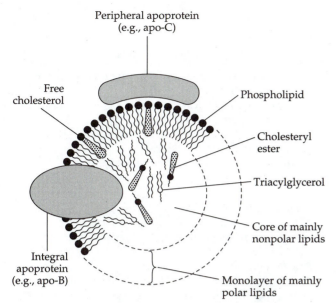

Peripheral apoprotein (e.g., apo-C)

Free cholesterol

Phospholipid

Cholesteryl ester

Triacylglycerol

Core of mainly nonpolar lipids

Integral apoprotein (e.g., apo-B)

Monolayer of mainly polar lipids

Figure 6.5 Generalized structure of a plasma lipoprotein. Note the similarities with the structure of the plasma membrane. A small amount of cholesteryl ester and triacylglycerol are found in the surface layer and a little free cholesterol in the core.

peripheral blood. Entry of chylomicrons into the blood from the lymph continues for up to 14 hours after consumption of a meal rich in fat. The peak level of lipid in blood plasma usually occurs after 30 minutes to 3 hours after a meal and returns to normal within 5 to 6 hours.

Chylomicrons and VLDL, which is formed endogenously in the liver, are transported by the blood throughout all tissues in the body while undergoing intravascular hydrolysis at certain tissue sites. This hydrolysis occurs through the action of the enzyme *lipoprotein lipase,* associated with the endothelial cell surface of the small blood vessels and capillaries within adipose and muscle tissue. Its extracellular action on the circulating particles releases free fatty acids and diacylglycerols, which are quickly absorbed by the tissue cells.

Within the muscle cells, the free fatty acids and those derived from the hydrolysis of the absorbed diacylglycerols are primarily oxidized for energy, with only limited use for the resynthesis and storage of triacylglycerols. In adipose tissue, in contrast, the absorbed fatty acids are largely used for the synthesis of triacylglycerols, in keeping with that tissue's storage role. In this manner, chylomicrons and VLDL are cleared rapidly from the plasma in a matter of minutes and a few hours, respectively, from the time they enter the bloodstream. It is the large, triacylglycerol-laden chylomicrons that account for the turbidity of postprandial plasma. Because lipoprotein lipase is the enzyme that solubilizes these particles by its lipolytic action, it is sometimes referred to as "clearing factor." That which is left of the chylomicron following this lipolytic action is called a *chylomicron remnant*—a smaller particle relatively less rich in triacylglycerol but richer in cholesterol. These are removed from the bloodstream by liver cell endocytosis following interaction of the remnant particles with specific receptors for apolipoprotein E or B/E on the cells [4]. Nascent VLDL of liver origin also undergoes triacylglycerol stripping by lipoprotein lipase at extracellular sites, resulting in the formation of a transient IDL particle and, finally, a cholesterol-rich LDL. ApoC-II is an activator of lipoprotein lipase and a component of both chylomicrons and VLDL, as indicated in Table 6.5. This accounts for the susceptibility of these particles to lipoprotein lipase action, and is an example of the regulatory function of an apolipoprotein. Figure 6.6 summarizes the formation of the lipoproteins and their interconversions, along with their lipid-protein composition.

Table 6.4 Chemical and Physical Properties of Plasma Lipoproteins in Humans

Property	Chylomicrons	VLDL	IDL	LDL	HDL
Density (g/mL)	<1.006	<1.006	1.006–1.019	1.019–1.063	1.063–1.21
Diameter (nm)	80–500	40–80	24.5	20	7.5–12
Lipids (% by wt.)	98	92	85	79	50
Cholesterol	9	22	35	47	19
Triglyceride	82	52	20	9	3
Phospholipid	7	18	20	23	28
Apoproteins (%)	2	8	15	21	50
Major	A-I, A-II	B-100	B-100	B-100	A-I, A-II
	B-48	C-I, II, III	C-I, II, III		C-I, II, III
	C-I, II, III	E	E		
	E				

Table 6.5 Apolipoproteins of Human Plasma Lipoproteins

Apolipoprotein	Lipoprotein	Molecular Mass (Da)	Additional Remarks
A-I	HDL, chylomicrons	28,000	Activator of lecithin: cholesterol acyltransferase (LCAT). Ligand for HDL receptor.
A-II	HDL, chylomicrons	17,000	Structure is two identical monomers joined by a disulfide bridge. Inhibitor of LCAT?
A-IV	Secreted with chylomicrons but transfers to HDL	46,000	Associated with the formation of triacylglycerol-rich lipoproteins. Function unknown.
B-100	LDL, VLDL, IDL	550,000	Synthesized in liver. Ligand for LDL receptor.
B-48	Chylomicrons chylomicron remnants	260,000	Synthesized in intestine.
C-I	VLDL, HDL, chylomicrons	7600	Possible activator of LCAT.
C-II	VLDL, HDL, chylomicrons	8916	Activator of extrahepatic lipoprotein lipase.
C-III	VLDL, HDL, chylomicrons	8750	Several polymorphic forms depending on content of sialic acids.
D	Subfraction of HDL	20,000	Function unknown.
E	VLDL, HDL, chylomicrons, chylomicron remnants	34,000	Present in excess in the β-VLDL of patients with type III hyperlipoproteinemia. The sole apoprotein found in HDL$_c$ of diet-induced hypercholesterolemic animals. Ligand for chylomicron remnant receptor in liver and LDL receptor.

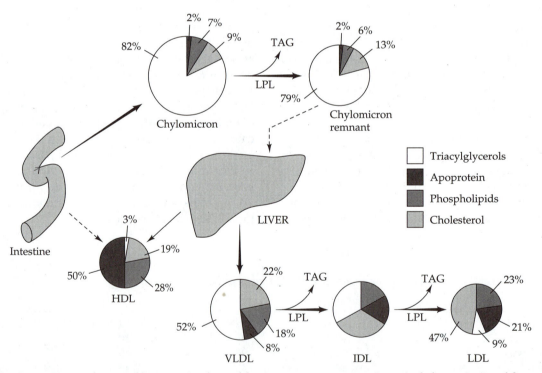

Figure 6.6 Conversion pathways of lipoproteins formed from exogenous sources (intestinal absorption) and from endogenous formation (synthesis in liver). Although small amounts of HDL are synthesized within the enterocytes of the small intestine, its major site of synthesis is the liver. Relative particle size of the lipoproteins is shown along with their approximate composition. Abbreviations: TAG, triacylglycerols; LPL, lipoprotein lipase.

Role of the Liver and Adipose Tissue in Lipid Metabolism

Liver

The liver plays a very important role in the body's use of lipids and lipoproteins. Hepatic synthesis of the bile salts, indispensable for digesting and absorbing dietary lipids, is one of its functions. In addition, the liver is the key player in lipid transport, because it is the site of synthesis of lipoproteins formed from endogenous lipids. It is capable of de novo synthesis of lipids from nonlipid precursors such as glucose and amino acids. It can also take up and catabolize exogenous lipids delivered to it in the form of chylomicron remnants, repackaging their lipids into HDL and VLDL forms. Nutrient metabolism in the liver is summarized in Figure 6.7. Also shown in Figure 6.7 is a glimpse of the reactions by which glucose and amino acids can be converted into lipid. They are included as a reminder to the reader

that pathways of lipid, carbohydrate, and protein metabolism are integrated and cannot stand alone.

In the postprandial state, glucose, amino acids, and short-chain fatty acid concentrations rise in portal blood. In the hepatocyte, glucose is phosphorylated for use, and glycogen is subsequently synthesized until the hepatic stores are repleted. If portal hyperglycemia persists, glucose is converted to fatty acids via acetyl CoA and also to triose phosphates from which the glycerol portion of triglycerides are derived. Amino acids can also serve as precursors for lipid synthesis because they can be metabolically converted to acetyl CoA and pyruvate. The synthesis of fatty acids, triacylglycerols, and glycerophosphatides is described in detail later in this chapter.

In addition to the newly synthesized lipid derived from nonfatty precursors, there is also the exogenous lipid delivered to the liver in the form of chylomicron remnants as well as short-chain fatty acids. The mechanism for the hepatic uptake and hydrolysis of

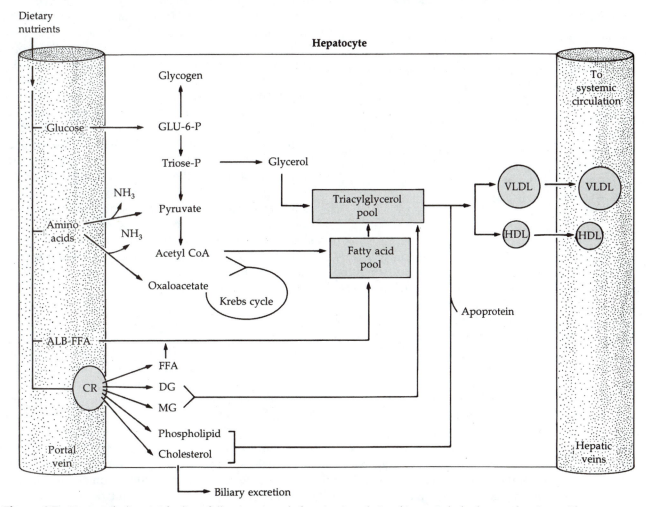

Figure 6.7 Fat metabolism in the liver following a meal, showing its relationship to carbohydrate and amino acid metabolism. Abbreviations: ALB-FFA, albumin-bound free fatty acids; CR, chylomicron remnants; DG, diacylglycerols; MG, monoacylglycerols.

chylomicron remnants is not entirely clear, although these particles are believed to first interact, via the apolipoprotein E on their surface, with specific receptors for apoE in the vascular endothelial cells of the liver. The lipid portion of the chylomicron remnant is hydrolyzed in the hepatocyte to free fatty acids, monoacylglycerols and diacylglycerols, glycerol, and cholesterol; but resynthesis of these compounds promptly occurs once again in a manner analogous to the events in the intestinal mucosal cell. Alternatively, the free fatty acids can undergo oxidation for energy (p. 147).

Exogenous free fatty acids of short-chain length delivered directly to the hepatic tissue can be used for energy or, following chain elongation, for resynthesis of other lipid fractions. Chylomicron remnant cholesterol and cholesteryl esters may be

- converted to bile salts and secreted in the bile,
- secreted into the bile as neutral sterol, or
- incorporated into VLDL or HDL and released into the plasma.

Newly synthesized triacylglycerol is combined with phospholipid, cholesterol, and proteins to form VLDL and HDL, which are released into the circulation. The HDL, small and triacylglycerol-poor, relative to the VLDL, possesses phospholipids and cholesterol as its major lipid constituents. Because triacylglycerols can be formed from glucose, hepatic triacylglycerol production is accelerated when the diet is rich in carbohydrate. This results in VLDL overproduction and may account for the occasional transient hypertriacylglycerolemia in normal people when they consume diets rich in simple sugars.

Adipose Tissue

Adipose tissue shares with the liver an extremely important role in fat metabolism. Unlike the liver, it is not involved in the uptake of chylomicron remnants or the synthesis of endogenous lipoproteins. Adipocytes are instead the major storage site for triacylglycerol, which is in a continuous state of turnover in the cells: lipolysis (hydrolysis), countered by reesterification. These two processes are not simply forward and reverse directions of the same reactions but are different pathways, involving different enzymes and substrates. Each of the processes are regulated separately by nutritional, metabolic, and hormonal factors, the net effect of which determines the level of circulating fatty acids and the extent of adiposity. A single large globule of fat constitutes over 85% by volume of the adipose cell.

In the fed state, metabolic pathways in adipocytes favor triacylglycerol synthesis. As in the liver, adipocyte triacylglycerol can be synthesized from glucose, a process strongly influenced by insulin. Insulin accelerates the entry of glucose into the adipose cells (the liver does not respond to this action of the hormone), and it also increases the availability and uptake of fatty acids by stimulating lipoprotein lipase. Cellular glucose, via its glycolytic breakdown, provides a source of glycerophosphate for reesterification with the fatty acids to form triacylglycerols. Absorbed monoacylglycerols and diacylglycerols also furnish the glycerol building block for this resynthesis. Insulin exerts its lipogenic action further by inhibiting intracellular lipase, which hydrolyzes stored triacylglycerols. Intracellular lipase is hormone sensitive, distinguishing it from the intravascular lipoprotein lipase that functions extracellularly. Postprandial fat metabolism is summarized in Figure 6.8.

To this point, the discussion has dealt with the role of the liver and adipose tissue in the fed state. In the *fasting state,* shifts in the metabolic scheme occur in these tissues. For example, as blood glucose levels diminish, insulin concentration falls, thereby accelerating lipolytic activity in adipose tissue. Free fatty acids derived from adipose tissue circulate in the plasma in association with albumin and are taken up by the liver and oxidized for energy by way of acetyl CoA formation. In the liver, some of the acetyl CoA is diverted to the production of the ketone bodies (p. 150), which can serve as important energy sources for muscle tissue and the brain during fasting and starvation. The liver continues the synthesis of VLDL and HDL and releases them into circulation, although these processes are diminished in a fasting situation. Glucose derived from liver glycogen, and free fatty acids transported to the liver from adipose tissue become the major precursors for the synthesis of endogenous VLDL triacylglycerol. As described previously, this lipoprotein then undergoes catabolism to IDL, transiently, and to LDL by lipoprotein lipase. Most of the plasma HDL is endogenous and is composed mostly of phospholipid and cholesterol along with apoproteins, chiefly of the A series.

Metabolism of Lipoproteins

Chylomicrons and chylomicron remnants are normally not present in the blood serum during the fasting state. The fasting serum concentration of VLDL is relatively quite low compared with its concentration in postprandial serum because of VLDL's rapid conversion to IDL and LDL. Therefore, the major lipoproteins in fasting serum are LDL (derived from VLDL), HDL (synthesized mainly in the liver), and a very small amount of VLDL. As discussed earlier and summarized in Table 6.5, the apolipoproteins may regulate metabolic reactions within the lipoprotein particles and determine to a great extent how the particles interact with each other and with receptors on specific cells.

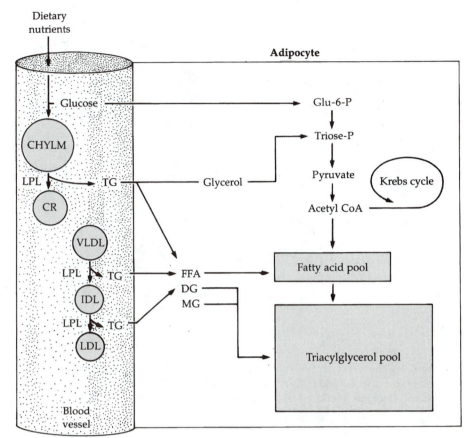

Figure 6.8 Fat metabolism in the adipose cell following a meal. The pathways favor the storage of energy as triacylglycerol. Insulin stimulates lipogenesis by promoting the entry of glucose into the cell and by inhibiting the lipase, which hydrolyzes the stored triacylglycerol to free fatty acids and glycerol. Abbreviations: CHYLM, chylomicrons; TG, triacylglycerol; CR, chylomicron remnant; LPL, lipoprotein lipase; FFA, free fatty acids; DG, diacylglycerol; MG, monoacylglycerol.

Low-Density Lipoprotein (LDL)

The LDL fraction is the major carrier of cholesterol, binding about 60% of the total serum cholesterol. Its function is to transport the sterol to tissues, where it may be used for membrane construction or for conversion into other metabolites such as the steroid hormones. LDL interacts with LDL receptors on cells via its apoB, specifically apoB-100, an event that culminates in the removal of the lipoprotein from the circulation. LDL receptors are located on liver cells as well as on cells of tissues peripheral to the liver, but the liver does not effectively remove the LDL from circulation. The distribution of LDL among tissues may depend on its rate of transcapillary transport as well as on the activity of the LDL receptors on cell surfaces. Once bound to the receptor, the receptor and the LDL particle, complete with its lipid cargo, are internalized together by the cell. The particle's component parts are then degraded by lysosomal enzymes in the cell. The LDL receptor will be examined in greater detail in the following section. Its discovery in the late 1970s and early 1980s was a significant biochemical event.

The LDL Receptor: Structure and Genetic Aberrations

The discovery of the LDL receptor is credited to Michael S. Brown, M.D., and Joseph L. Goldstein, M.D., who received the 1985 Nobel Prize for physiology or medicine. The discovery stemmed from their seeking the molecular basis for the clinical manifestation of hypercholesterolemia, and their research revealed the following facts about LDL and its connection to cholesterol metabolism.

LDL binds to normal fibroblasts (and other cells, particularly the hepatocytes and cells of the adrenal gland and ovarian corpus luteum) with high affinity and specificity. In mutant cells, however, the binding is very inefficient. Although deficient binding of LDL is characteristic of all mutant cells, much variation exists in the binding ability among different patients with familial homozygous hypercholesterolemia.

Membrane-bound LDL is internalized by endocytosis made possible by receptors that cluster in coated pits. Figure 6.9 depicts the fate of the LDL particle following its binding to the membrane receptor. The receptor, having released its LDL, returns to the surface of the

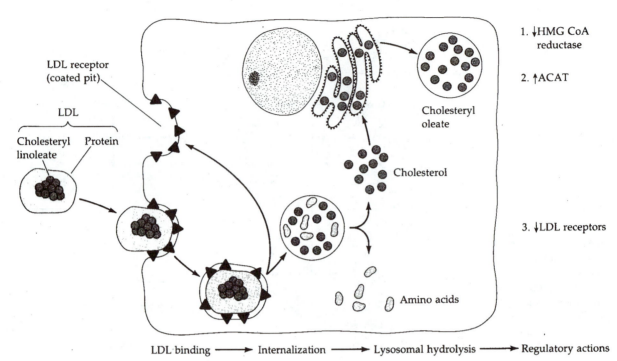

1. ↓HMG CoA reductase

2. ↑ACAT

Cholesteryl oleate

Cholesterol

3. ↓LDL receptors

Amino acids

LDL receptor (coated pit)

LDL

Cholesteryl linoleate   Protein

LDL binding ⟶ Internalization ⟶ Lysosomal hydrolysis ⟶ Regulatory actions

Figure 6.9 The effect of cholesterol on HMG CoA (hydroxymethyl glutaryl CoA) reductase, ACAT (acyl CoA; cholesterol acyltransferase and LDL receptor concentration in the regulation of cholesterol homeostasis. (Brown M, Goldstein J. Receptor mediated endocytosis: insights from the lipoprotein receptor system. (c) The Nobel Foundation 1986.)

cell, making a round trip into and out of the cell every 10 minutes during its 20-hour life span [5]. The dissociated LDL moves into the lysosome, where its protein and cholesteryl ester components are hydrolyzed by lysosomal enzymes into amino acids and free cholesterol, respectively. The resulting free cholesterol exerts the following regulatory functions:

- By lowering the concentration of receptor mRNA, it suppresses synthesis of LDL receptors so as to prevent further entry of LDL into the cell.

- It modulates the activity of two microsomal enzymes, 3-hydroxy 3-methylglutaryl CoA reductase (HMG CoA reductase) and acyl CoA: cholesteryl acyl transferase (ACAT).

Activity of the HMG CoA reductase, the rate-limiting enzyme in cholesterol synthesis, is suppressed through decreased transcription of the reductase gene and the concomitant increased degradation of the enzyme. In contrast, ACAT is activated, thereby promoting formation of cholesteryl esters that can be stored as droplets in the cytoplasm of the cell.

Mutant cells unable to bind and/or internalize LDL efficiently and thereby deprived of the cholesterol needed for membrane synthesis must obtain the needed sterol via de novo synthesis. In these cells HMG CoA reductase is activated while ACAT is depressed.

LDL receptors interact with apoprotein B-100, the protein carried on the surface of the LDL. The interaction between the receptors and the apoprotein B-100 is the key to the cell's internalization of the LDL. The number of receptors synthesized by cells varies according to cholesterol requirements.

The LDL receptor has been found to be a transmembrane glycoprotein that, in the course of its synthetic process, undergoes several carbohydrate-processing reactions. The carbohydrate moiety is important for proper functioning of the receptor, and its location on the molecule has been mapped. Five domains of the LDL receptor have been identified:

- Domain 1, which is furthest from the membrane and contains the NH$_2$ terminal of the receptor protein, is rich in cysteine residues. These residues allow the formation of many disulfide bonds that give stability to the molecule. Many of the other amino acid residues in this cysteine-rich domain have negatively charged side chains. This first domain, then, could be the binding site for apoprotein B-100, with its positively charged lysine and arginine residues. These positively charged residues of this apoprotein are known to be crucial for receptor binding.

- Domain 2 is made up of 350 amino acids and is the possible location for the N-linked glycosylation that occurs during the maturation process of the receptor protein.

• Domain 3 is located immediately outside the plasma membrane and is the site of the O-linked glycosylation. This glycosylation, too, occurs during the maturation process of the receptor.

• Domain 4 is made up of 22 hydrophobic amino acids that, because of their affinity for lipids, are able to span the plasma membrane.

• Domain 5, the final domain, is the COOH terminal end of the protein and projects into the cytoplast. This tail enables the receptors to move laterally, thereby mediating the clustering of the receptors in the coated pits.

Along with the delineation of the structure of the normal LDL receptor, knowledge of the structural defects existing in mutants has developed. Although a gene on chromosome 19 encodes the protein of the LDL receptor, the mutations of the gene are not always the same. How the normal functioning of the receptor is affected depends on what particular domain(s) of the receptor has undergone mutation. Of the 110 familial hypercholesterolemia homozygotes studied, 10 different abnormal forms of the LDL receptors have been identified. These identified mutations can be divided into four classes:

• Class 1, in which no receptors are synthesized

• Class 2, in which precursors of the receptors are synthesized but then are not processed properly and fail to move into the Golgi apparatus

• Class 3, in which the precursors for the LDL receptors are synthesized and processed but the processing is faulty, thereby preventing the receptors from binding LDL normally

• Class 4, mutations that allow production of receptors that reach the surface of the cell and bind LDL but are unable to cluster in the coated pits

Maturation of the LDL receptor precursor proteins, like other proteins synthesized on the endoplasmic reticulum of the cell, occurs in the cell's Golgi apparatus. In the Golgi apparatus the LDL receptors are targeted for their final destination (p. 10). Incomplete or improper processing can prevent the receptor from reaching its proper destination on the plasma membrane.

Relatively few people (1 in 1 million) are homozygous for familial hypercholesterolemia, but many people carry one mutant gene for the disease (1 in 500). Knowledge of the mechanisms of the disease can be of tremendous benefit in treating the latter individuals. There is little doubt as to a causal relationship between hypercholesterolemia and the development of atherosclerosis.

Reducing serum cholesterol through drug therapy can cause an increased transcription for LDL receptors by the one normal gene in the heterozygotes, and serum cholesterol can be normalized. Drug therapy includes both bile acid–binding resins, which increase fecal removal of cholesterol, and HMG-CoA reductase inhibitors, which reduce cholesterol synthesis in the liver.

Many people who exhibit no clear-cut genetic defect also are found to possess an inadequate number of LDL receptors. In this population group, nutrition could be the environmental factor, leading to decreased production of LDL receptors. A diet high in saturated fats and cholesterol appears to be one of the culprits.

To summarize the role of the LDL fraction in normal lipid metabolism, it can be thought of as a depositor of cholesterol and other lipids into peripheral cells possessing the LDL receptor. Included among cells targeted by LDL are the cells of the vascular endothelium, and therefore it follows that a high concentration and activity of LDL have implications in the etiology of cardiovascular disease.

High-Density Lipoprotein (HDL)

Opposing the LDL's cholesterol-depositing role is the HDL fraction. An important function of HDL is to remove unesterified cholesterol from cells and other lipoproteins, where it may have accumulated, and return it to the liver for excretion in the bile. Two key properties of HDL are necessary for this process to occur.

The first key property is its ability to bind to receptors on both hepatic and extrahepatic cells. Receptors may be specific for HDL, but they also include the LDL receptor to which HDL can bind via its apoE component. In other words, the LDL receptor recognizes both apoE and apoB-100 and is consequently referred to as the *apoB,E receptor*. The implication is that HDL can compete with LDL at its receptor site. The second key property of HDL is mediated through its apo A-1 component, which stimulates the activity of the enzyme lecithin: cholesterol acyltransferase (LCAT). This enzyme forms cholesteryl esters from free cholesterol by catalyzing the transfer of fatty acids from the C-2 position of phosphatidylcholine to free cholesterol. The free cholesterol (recipient) substrate is derived from the plasma membrane of cells or surfaces of other lipoproteins. Cholesteryl esters resulting from this reaction can then exchange readily among plasma lipoproteins, mediated by a transfer protein called *cholesteryl ester transfer protein,* or CETP. LCAT, therefore, by taking up free cholesterol and producing its ester form, promotes the net transfer of cholesterol out of nonhepatic cells and other lipoproteins. Cholesteryl esters can then be transported directly to the liver in association with HDL or indirectly by LDL, following CETP transfer from HDL to LDL. It will be recalled that either lipoprotein can bind to LDL (apoB,E) receptors.

Following their deposition in the liver cells, the esters are hydrolyzed by cholesteryl esterase, and the free cholesterol is excreted in the bile as bile salt (Fig. 6.2). This is the major route of cholesterol excretion from the body.

The net effect of these properties of HDL is the retrieving of cholesterol from peripheral cells and other lipoproteins, and returning it, as its ester, to the liver. The process is referred to as *reverse cholesterol transport.* Its benefit to the cardiovascular system is that by reducing the amount of deposited cholesterol in the vascular endothelium, the risk of fatty plaque formation and atherosclerosis is similarly reduced. This topic is reviewed in the next section.

Lipids, Lipoproteins, and Cardiovascular Disease Risk

Atherosclerosis is a degenerative disease of vascular endothelium. The principal players in the atherogenic process are cells of the immune system and lipid material, primarily cholesterol and cholesteryl esters. An early response to arterial endothelial cell injury is an increased adherence of monocytes and T lymphocytes to the affected area. Protein products of the monocytes and lymphocytes, called *cytokines,* mediate the atherogenic process by their chemotactic attraction of phagocytic cells to the area.

Concurrent with the cellular involvement, exposure to a high level of LDL with its subsequent deposition and oxidative modification further promotes the inflammatory process. The process is marked by the uptake of LDL by phagocytic cells that become engorged with lipid, and are termed *foam cells.* Phagocytic uptake is accelerated if the apoB component of the LDL is modified by oxidation. Lipid material in the form of foam cells may then infiltrate the endothelium, and as lipid accumulates, the lumen of the blood vessel involved is progressively occluded. The deposited lipid, known to derive from blood-borne lipids, is called *fatty plaque.* The pathophysiology of atherosclerosis has been reviewed [6], and it is the subject of this chapter's Perspective.

Ever since it was discovered that plaque composition is chiefly lipid, an enormous research effort has been initiated to investigate the possible link between dietary lipid and the development of atherosclerosis. The existence of such a link has come to be called the *lipid hypothesis,* which maintains that dietary lipid intake can alter blood lipid levels, which in turn initiate or exacerbate atherogenesis. Following is a brief account of the alleged involvement of certain dietary lipids and fatty acids and of genetically acquired apolipoproteins in atherogenesis.

Cholesterol

At center stage in the lipid hypothesis controversy is cholesterol. The effect of dietary interventions designed to improve serum lipid profiles are generally measured by the extent to which the interventions raise or lower serum cholesterol. There is some justification for this reasoning in so far that cholesterol is a major component of atherogenic fatty plaque and that many studies have linked cardiovascular disease risk with chronically elevated levels of serum cholesterol. Receiving the greatest attention, however, is not so much the changes in the total cholesterol concentration but how the cholesterol is distributed between its two major transport lipoproteins, LDL and HDL. Because it is commonly and conveniently quantified in clinical laboratories, cholesterol assays can be used to establish LDL:HDL ratios by measuring the amount of cholesterol in each of the two fractions. Assayed cholesterol associated with the LDL fraction is designated by laboratorians as LDLC, and cholesterol transported in the HDL fraction is HDLC.

Because, in the interest of wellness, it is desirable to maintain relatively low serum levels of LDL and relatively high levels of HDL (a low LDL:HDL ratio), the concept of "good and bad cholesterol" emerged. The "good" form is that associated with HDL, and the "bad" is transported as LDL. It is important to understand, however, that the cholesterol per se is not "good" or "bad"; rather, it serves as an indicator of the relative concentrations of LDL and HDL, ratios of which can indeed be good or bad. LDL:HDL ratios are, in fact, determined more reliably by measurements other than cholesterol content. For example, immunological methods for quantifying apoB (the major LDL apoprotein) and apoA (the primary HDL apoprotein) are now widely used. Ratios of apoA to apoB then serve as indicators of cardiovascular disease risk, with risk increasing as ratios decrease.

Among the reasons for cholesterol's "bad press" as it is considered in connection with cardiovascular disease is that cholesterol and especially cholesteryl esters are major components of fatty plaque. Contrary to widespread belief, changing the amount of cholesterol in the diet has only a minor influence on blood cholesterol concentration in most people. This is because compensatory mechanisms are engaged, such as HDL activity in scavenging excess cholesterol, and the down-regulation of cholesterol synthesis by dietary cholesterol (discussed in the "Synthesis of Cholesterol" section). It is well known, however, that certain individuals respond strongly, and others weakly, to dietary cholesterol (hyper- and hyporesponders). This phenomenon may have a genetic basis, but it is further complicated by the observation that a significant within-person variability exists that is independent of diet. This fact clearly confounds the results of intersubject studies.

Several mechanisms may be considered in trying to account for differences in individual responses to dietary cholesterol:

- Differences in absorption or biosynthesis
- Differences in the formation of LDL and its receptor-mediated clearance
- Differences in its rates of removal and excretion

These considerations have been extensively reviewed [7].

Saturated and Unsaturated Fatty Acids

Research dealing with the influence of various kinds of fatty acids on cardiovascular disease risk has focused on the effect that each has on serum levels of cholesterol. The literature dealing with the effect of dietary fats containing primarily saturated fatty acids (SFAs), monounsaturated fatty acids (MUFAs), or polyunsaturated fatty acids (PUFAs) is as extensive as that related to the feeding of cholesterol itself. Early research results generally led to the conclusion that SFAs are hypercholesterolemic and that PUFAs are hypocholesterolemic. Furthermore, it had been assumed that the MUFAs were neutral in this effect, neither increasing nor lowering serum cholesterol.

Current research focuses not so much on total cholesterol effects but rather on how LDLC:HDLC ratios are shifted by the test lipids. For example, studies have shown that diets rich in MUFA were as effective as PUFA-rich diets in lowering LDL cholesterol and triacylglycerols without significant change in HDL [8,9]. There has also been interest in how the position of the double bonds in PUFA (n-3 vs. n-6 species) as well as their *cis/trans* isomerism relates to lipoprotein ratio effects. The effect of *trans* fats is discussed in the next section.

Currently, the consensus among students of the subject is as follows [10]: The consumption of the following shows a *positive* correlation with the risk of cardiovascular disease (CVD), primarily due to a hypercholesterolemic effect or unfavorable shifts in LDLC:HDLC ratios:

- Total fat
- Saturated fatty acids
- Cholesterol

The consumption of the following shows a *negative* correlation with CVD risk, primarily because of a hypocholesterolemic effect or favorable shifts in LDL:HDL:

- Monounsaturated fatty acids (provided adjustments are made for coconsumed cholesterol and saturated fatty acids)
- Polyunsaturated fatty acids (if adjusted for cholesterol and saturated fatty acids). Both n-3 and n-6 types are effective. It was found that the linoleic acid (18:2n-6) content in adipose tissue was inversely associated with CVD risk.
- n-3 Fatty acids. The n-3 PUFA exert antiatherogenic properties by various mechanisms, including the following:

  – Interference with platelet aggregation, due in part to inhibition of thromboxane (TXA_2) production. This is thought to be caused by displacement of the TXA_2 precursor, arachidonic acid, from platelet phospholipid stores by the fatty acids. Eicosapentaenoic acid (EPA)(20:5n-3), docosahexaenoic acid (DHA)(20:6n-3), and α-linolenic acid (18:3n-3) exerted similar antiaggregatory effects [11].
  – Reduction in the release of pro-inflammatory cytokines from cells involved in fatty plaque formation (see this chapter's Perspective).
  – Sharp reduction (25%–30%) in serum triacylglycerol concentration. α-Linoleate was less effective than EPA or DHA, and plant n-3 fatty acids were generally less effective than marine n-3 fatty acids in their capacity to reduce triacylglycerols [12].

There is heterogeneity in the cholesterolemic response to individual fatty acids even within a fatty acid class. This is particularly true among the long-chain saturated fatty acids. Strong evidence indicates that lauric (12:0), myristic (14:0), and palmitic (16:0) acids are all hypercholesterolemic, specifically raising LDLC, with (14:0) being the most potent in this respect. On the other hand, stearic acid (18:0), is neutral in its effect, in fact, is reputed to actually lower total cholesterol and LDLC relative to carbohydrate. Therefore, stearic acid should not be grouped with shorter-chain SFAs with regard to LDLC effects. Oleic acid (18:1) and linoleic acid (18:2n-6) are hypocholesterolemic compared with 12:0 and 16:0 fatty acids. Linoleate 18:2n-6 is the more potent of the two, independently lowering total and LDL cholesterol [13].

Despite years of investigation, the mechanism by which hypercholesterolemic fatty acids exert their effects has not been conclusively defined. However, they have been alleged to operate in one or more of the following ways:

- by suppressing the excretion of bile acids;

- by enhancing the synthesis of cholesterol and LDL, either by reducing the degree of control exerted on the regulatory enzyme HMG-CoA reductase or by affecting apoB synthesis; or

- by retarding LCAT activity or receptor-mediated uptake of LDL.

Trans Fatty Acids

Double-bonded carbon atoms can exist in either a *cis* or *trans* orientation as described on page 124. Most natural fats and oils contain only *cis* double bonds. The much smaller amount of naturally occurring *trans* fats are found mostly in the fats of ruminants, for example in milk fat, which contains 4% to 8% *trans* fatty acids. Much larger amounts are found in certain margarines and margarine-based products, shortenings, and frying fats as a product of the partial hydrogenation of PUFA. The process of hydrogenation imparts to the product a higher degree of hardness and plasticity, which is more desirable to the consumer and food manufacturer. In the process, however, as hydrogen atoms are catalytically added across double bonds, electronic shifts take place that cause remaining, unhydrogenated *cis* double bonds to revert to a *trans* configuration that is energetically more stable.

The most abundant *trans* fatty acids in the diet are elaidic acid and its isomers, which are of an 18:1 structure, and it has been reported that diets rich in these fatty acids are as hypercholesterolemic as saturated fatty acids [14]. In fact, serum lipid profiles following feeding of a diet high in *trans* fatty acids may be even more unfavorable than saturated fatty acids because not only are total cholesterol and LDL cholesterol levels elevated, but HDL cholesterol is decreased. The study cited was criticized for using *trans* elaidic acids obtained by a process not typical of hydrogenation of margarines and shortenings and also for including uncharacteristically large dietary amounts of the *trans* fatty acids in the study diet. However, reports from subsequent studies confirm that *trans* fat consumption elevates serum LDLC while decreasing HDLC and that it raises total cholesterol:HDLC ratios [15]. Another confirmatory investigation, conducted on a large group of normal women over an 8-year period, matched *trans* fat intake with the incidence of nonfatal myocardial infarction or death from coronary heart disease [16]. The study did indicate a positive correlation between *trans* fat intake and coronary heart disease. However, it was not free of criticism, on the basis that the data were obtained by consumer questionnaires rather than by a randomized control study in which intake can be precisely monitored by the researchers [17]. Furthermore, there was a high degree of variability of *trans* fatty acids in the foods consumed by the subjects, and a clear-cut dose-response relationship could not be demonstrated [18].

Some reports exonerate *trans* fat in regard to its alleged hypercholesterolemic properties. In a randomized crossover study involving hypercholesterolemic subjects, the replacement of butter with margarine, in a low-fat diet, actually lowered LDLC and apolipoprotein B by 10%, with HDLC and apolipoprotein A levels being unaffected [19].

It is clear that "the jury is still out" on the putative adverse effects of *trans* fats and that more well-designed research is needed in this important area. The adverse findings on *trans* fats represent an about-face to the long-standing nutrition dogma that unsaturated fats are invariably preferred over saturated fats. It will be difficult to educate the public to such a reversal, which is why recommendations regarding the consumption of *trans* fat–containing foods should not be issued until all the facts are in. Meanwhile, it would seem prudent for those at a high risk of atherosclerosis to avoid a high intake of *trans* fats.

Lipoprotein a

In the 1960s, a genetic variant of LDL was discovered in human serum. The particle differs from normal LDL in that it is attached to a unique marker protein of high molecular weight (513,000 D). The marker protein is currently referred to as apolipoprotein a, or *apo(a),* and the complete lipoprotein particle is called lipoprotein a, or *Lp(a).*

At the time of the discovery of Lp(a), it was evident that not all people had the lipoprotein in their serum, and in many of those that did, its concentration was very low compared with other lipoproteins. Consequently, it was dismissed as having little importance. However, interest in Lp(a) was renewed during the following two decades, when numerous studies suggested a positive correlation between Lp(a) concentration and atherosclerotic disease.

Structurally, Lp(a) is assembled from LDL and the apo(a) protein. The LDL component of the complex possesses apoB-100 as its only protein component. Linkage of the LDL portion of the particle to apo(a) is through a disulfide bond connecting the two proteins apo(a) and apoB-100. A strong structural homology (similar amino acid sequence) has recently been discovered between apo(a) and plasminogen. Plasminogen is the inactive precursor of the enzyme plasmin, which dissolves blood clots by its hydrolytic action on fibrin. This discovery has stimulated extensive research in the genetics, metabolism, function, and clinical significance of Lp(a).

The physiological function of Lp(a) is not yet defined with certainty, although it is tempting to speculate that its role may be linked to the two functional systems from which the particle was derived, a lipid transport system and the blood-clotting system. It has been proposed that Lp(a) may bind to fibrin clots via its plasminogen-like apo(a) and therefore may deliver cholesterol to regions of recent injury and wound healing.

Reviews of Lp(a) cite numerous epidemiological studies [20] that show a positive correlation between Lp(a) concentration and premature myocardial infarction, a blocking of blood vessels in the heart due to clot formation. This has led to the conclusion that Lp(a) may be an independent genetic risk for atherosclerotic disease. But unfortunately, blood levels of the lipoprotein do not respond to dietary intervention and only very weakly to lipid-lowering drugs. One of the many open questions is whether Lp(a) is linked to atherogenesis over an extended period of time because of its lipoprotein properties or whether it plays a role in the sudden development of a clot due to the binding of its plasminogen-like apo(a) component to fibrin. Perhaps both mechanisms apply.

Apolipoprotein E

Recall that the term *apolipoprotein* refers to the protein moiety of a lipoprotein. Studies have shown that one of the apolipoproteins, apolipoprotein E (apoE), may have a role in the etiology of atherogenesis. As shown in Table 6.5, apolipoprotein E is a structural component of VLDL, HDL, chylomicrons, and chylomicron remnant particles. The most important physiological function of apoE is that it is the component recognized by the LDL receptor-related protein (also referred to as the postulated apoE receptor) and the LDL receptor, discussed in detail on page 141. By way of interaction with these receptors, apoE mediates the uptake of apoE-possessing lipoproteins into the liver.

ApoE exists in three isoforms: E2, E3, and E4. They are genetically encoded by three alleles, *E2, E3,* and *E4,* respectively. A single individual inherits one allele from each parent and therefore will become one of the following six possible phenotypes: E2,2; E3,2; E3,3; E3,4; E4,2; and E4,4. The effects of apoE polymorphism on plasma lipids and the predisposition to cardiovascular disease have been well studied [21]. There is evidence that E4 phenotypes have an increased risk of developing cardiovascular disease, and this appears to be due to elevation of the LDL serum level. ApoE phenotype alone is associated with a stepwise increase in LDLC (cholesterol that is contained within the LDL fraction). LDLC increased in the order of phenotypes E3,2 < E3,3 < E4,3. This relationship is independent of the ratio of polyunsaturated to saturated fat that is consumed [22].

Based on the link between the *E4* allele and a predisposition to cardiovascular disease, the allele is considered to be a predictor of latent atherogenesis. A longitudinal study of elderly men conducted over a 5-year period revealed that among those who died of coronary heart disease during the study period there was a doubling of the frequency of the *E4* allele [23].

Integrated Metabolism in Tissues

Catabolism of Triacylglycerols and Fatty Acids

The complete hydrolysis of triacylglycerols yields glycerol and three fatty acids. In the body, this occurs largely through the activity of lipoprotein lipase of vascular endothelium and through an intracellular lipase that is active in the liver and particularly active in adipose tissue. The glycerol portion can be used for energy by the liver and by other tissues having activity of the enzyme glycerokinase, through which glycerol is converted to glycerol phosphate. Glycerol phosphate can enter the glycolytic pathway at the level of dihydroxyacetone phosphate, from which point either energy oxidation or gluconeogenesis can occur (review Fig. 4.9).

Fatty acids are a very rich source of energy, and on an equal-weight basis they surpass carbohydrates in this property. This is because fatty acids exist in a more reduced state than that of carbohydrate and therefore undergo a greater extent of oxidation en route to CO_2 and H_2O. Many tissues are capable of oxidizing fatty acids by way of a mechanism called β-oxidation, which is described next. On entry into the cell of the metabolizing tissue, the fatty acid is first activated by coenzyme A, an energy requiring reaction catalyzed by cytoplasmic fatty acyl-CoA synthetase (see 6K). The reaction actually consumes the equivalent of two ATPs, because the pyrophosphate must also be hydrolyzed to ensure irreversibility of the reaction.

The oxidation of fatty acids is compartmentalized within the mitochondrion. Fatty acids and their CoA derivatives, however, are incapable of crossing the inner mitochondrial membrane, necessitating a membrane transport system. The carrier molecule for this system is *carnitine* (p. 252), which can be synthesized in humans from lysine and methionine, and which is

(6K)

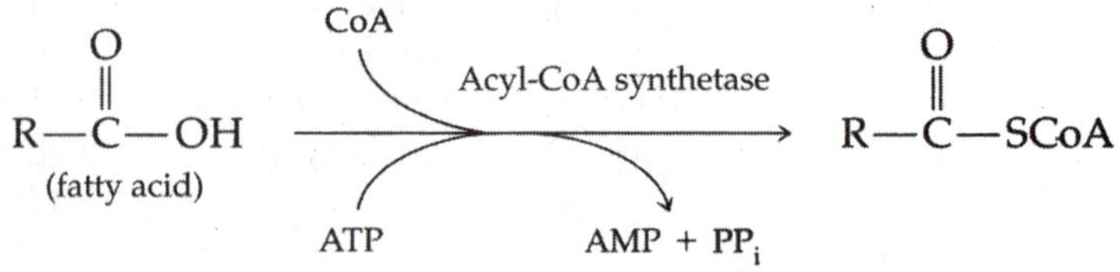

$$R-\underset{\underset{(fatty\ acid)}{\overset{\overset{O}{\parallel}}{C}}}{}-OH \xrightarrow[\text{ATP} \qquad \text{AMP} + \text{PP}_i]{\text{CoA} \quad \text{Acyl-CoA synthetase}} R-\underset{\overset{O}{\parallel}}{C}-SCoA$$

(6L)

Intermembrane space	Inner membrane	Matrix

Fatty acyl CoA ⇀ Carnitine ← ⇀ Fatty acyl CoA

CoA ↽ Carnitine fatty acid ⇁ CoA

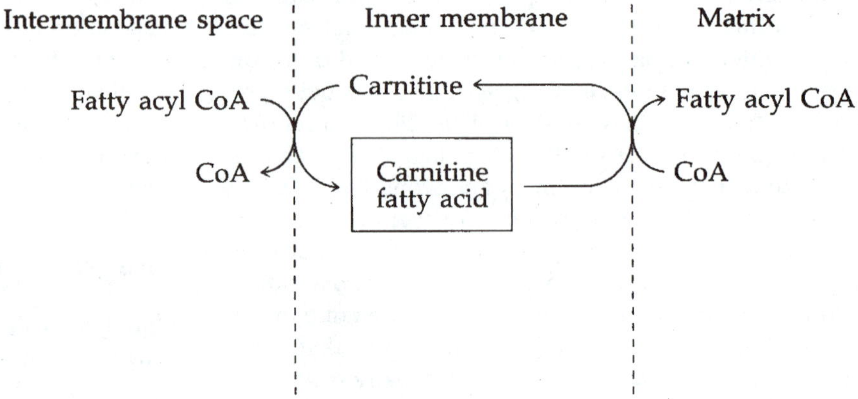

found in high concentration in muscle. The activated fatty acid is joined covalently to carnitine at the cytoplasmic side of the mitochondrial membrane by the transferase enzyme carnitine acyltransferase I (CAT I). A second transferase, acyltransferase II (CAT II), located on the inner face of the inner membrane, releases the fatty acyl CoA and carnitine into the matrix (see 6L).

The oxidation of the activated fatty acid in the mitochondrion occurs via a cyclic degradative pathway, by which two-carbon units in the form of acetyl CoA are cleaved one by one from the carboxyl end. The reactions of β-oxidation are summarized in Figure 6.10. The following comments relate to the reactions:

1. The formation of a double bond between the α- and β-carbons, catalyzed by *fatty acyl CoA dehydrogenase.* There are four such dehydrogenases, each specific for a range of chain lengths. The $FADH_2$ yields its two hydrogens to CoQ and therefore feeds two electrons into the electron transport system (ETS), which, it will be recalled, is also compartmentalized in the mitochondrion. The ETS oxidation of $FADH_2$ yields two ATPs.

2. The unsaturated acyl CoA accepts a molecule of water, a reaction catalyzed by *enoyl CoA hydrase,* sometimes called *crotonase.*

3. The β-hydroxy group is oxidized to the ketone by the NAD^+-requiring enzyme, β-hydroxyacyl CoA dehydrogenase. The ETS oxidation of the NADH leads to the formation of three ATPs.

4. The β-ketoacyl CoA is cleaved by β-*ketothiolase* (also called *acyl CoA thiolase*), resulting in the insertion of CoA and cleavage at the β-carbon. The products

of this reaction are acetyl CoA and a saturated, CoA-activated fatty acid having two fewer carbons than the original fatty acid. This reaction completes one cycle of the degradative pathway. The entire sequence of reactions is repeated, with two carbons being removed with each cycle.

Energy Considerations in Fatty Acid Oxidation

The activation of a fatty acid actually requires two high-energy bonds per mole of fatty acid oxidized. Each cleavage of a carbon-carbon bond yields five ATPs, two by oxidation of $FADH_2$ and three by NADH oxidation via oxidative phosphorylation. The acetyl CoAs produced are oxidized to CO_2 and water in the Krebs cycle, and for each of these oxidized, 12 ATPs (or their equivalent) are produced (p. 93). Using the example of palmitate (16 carbons), the yield of ATP would be summarized as follows:

7 carbon-carbon cleavages	$7 \times 5 = 35$
8 acetyl CoA oxidized	$8 \times 12 = 96$
Total ATPs produced	131
2 ATPs for activation	-2
Net ATPs	129

Nearly one-half of dietary and body fatty acids are unsaturated and therefore must provide a considerable portion of lipid-derived energy. They are catabolized by β-oxidation in the mitochondrion in nearly identical fashion as their saturated counterparts, except for the

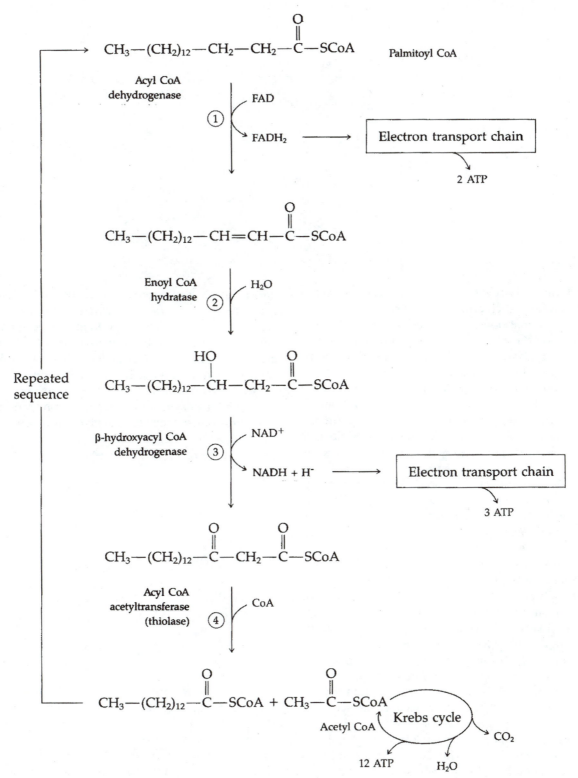

Figure 6.10 The mitochondrial β-oxidation of an activated fatty acid using palmitate as an example. The yield of ATPs from the complete oxidation of the fatty acid is indicated. The name of the enzyme catalyzing each reaction is written to the left of each reaction arrow.

(6M)

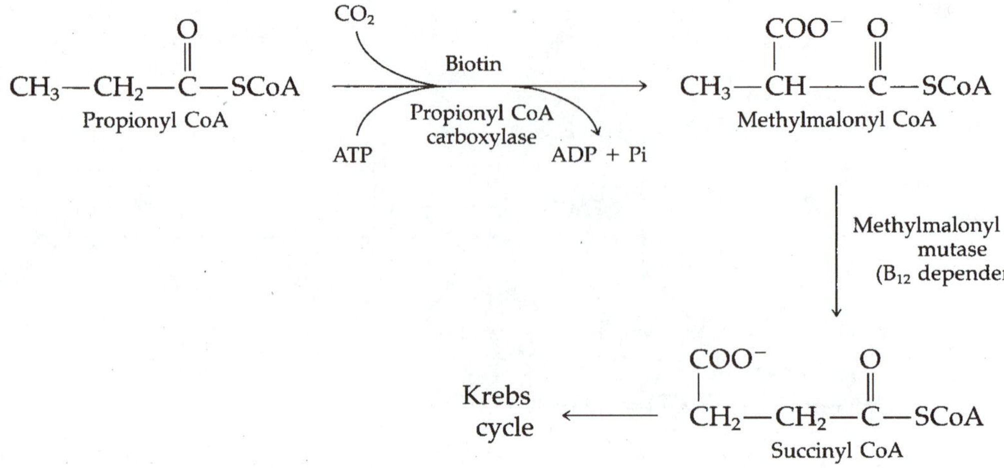

$$CH_3-CH_2-\overset{\displaystyle O}{\overset{\|}{C}}-SCoA$$

Propionyl CoA

CO_2

Biotin

Propionyl CoA
carboxylase

ATP → ADP + Pi

$$CH_3-\overset{\displaystyle COO^-}{\overset{|}{CH}}-\overset{\displaystyle O}{\overset{\|}{C}}-SCoA$$

Methylmalonyl CoA

Methylmalonyl CoA
mutase
(B_{12} dependent)

Krebs
cycle ←

$$\overset{\displaystyle COO^-}{\overset{|}{CH_2}}-CH_2-\overset{\displaystyle O}{\overset{\|}{C}}-SCoA$$

Succinyl CoA

fact that one fatty acyl CoA dehydrogenase reaction is not required for each double bond present. This is because the double bond introduced into the saturated fatty acid by the reaction occurs naturally in unsaturated fatty acids. However, the specificity of the enoyl CoA hydrase reaction requires that the double bond be a Δ^2 in order for the hydration to take place, and the "natural" double bond may not occupy this position. For example, after three cycles of β-carbon oxidation, the position of the double bond in what was originally a Δ^9 monounsaturated fatty acid will occupy a Δ^3 position. The presence of a specific enoyl CoA transisomerase then shifts the double bond from Δ^3 to Δ^2, allowing the hydrase and subsequent reactions to proceed. The oxidation of an unsaturated fatty acid results in somewhat less energy production than a saturated fatty acid of the same chain length, because for each double bond present, one $FADH_2$-producing fatty acyl CoA dehydrogenase reaction is bypassed, resulting in two fewer ATPs.

Although most fatty acids metabolized are composed of an even number of carbon atoms, small amounts of fatty acids having an odd number of carbon atoms are also used for energy. β-oxidation occurs as described, with the liberation of acetyl CoA until a residual propionyl CoA remains. The subsequent oxidation of propionyl CoA requires reactions that use the vitamins biotin and B_{12} in a coenzymatic role; see 6M.

Because the succinyl CoA formed in the course of these reactions can be converted into glucose, the odd-numbered carbon fatty acids are therefore uniquely glucogenic among all the fatty acids.

Formation of the Ketone Bodies

In addition to its direct oxidation via the Krebs cycle, acetyl CoA may follow other catabolic routes in the liver, one of which is the pathway by which the so-called ketone bodies (acetoacetate, β-hydroxybutyrate,

and acetone) are formed. Acetoacetate and β-hydroxybutyrate are not oxidized further in the liver but instead are transported by the blood to peripheral tissues, where they can be converted back to acetyl CoA and oxidized via the Krebs cycle. The steps in ketone body formation occur as shown in 6N. The reversibility of the β-hydroxybutyrate dehydrogenase reaction together with enzymes present in extrahepatic tissues that convert acetoacetate to acetyl CoA (shown by broken arrows in 6N) reveal how the ketone bodies can serve as a source of fuel in these tissues.

(6N)

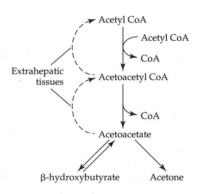

Ketone body formation is actually an "overflow" pathway for acetyl CoA use, providing another way for the liver to distribute fuel to peripheral cells. Normally, the concentration of the ketone bodies is very low in the blood, but it may reach very high levels in situations of accelerated fatty acid oxidation combined with low carbohydrate intake or impaired carbohydrate use. Such a situation would occur in diabetes mellitus, starvation, or simply a very low-carbohydrate diet. The inadequate supply of carbohydrate reduces the pool of oxaloacetate, formed mainly from pyruvate, with which the acetyl CoA normally combines for Krebs cycle oxidation. As carbohydrate use is diminished, oxidation of fatty acids accelerates to provide energy through the production of Krebs cycle substrates (acetyl CoA). This shift to fat catabolism, coupled with reduced oxaloacetate availability, results in an accumulation of acetyl CoA. A sharp increase in ketone body formation follows as would be expected, resulting in the condition known as ketosis. Ketosis can be dangerous in that it can disturb the body's acid-base balance (two of the ketone bodies are, in fact, organic acids). However, the liver's ability to deliver ketone bodies to peripheral tissues such as the brain and muscle is an important mechanism for providing fuel in periods of starvation. It is the lesser of two evils.

Catabolism of Cholesterol

Unlike the triacylglycerols and fatty acids, cholesterol is not an energy-producing nutrient. Its four-ring core structure remains intact in the course of its catabolism and is eliminated as such through the biliary system.

Cholesterol, primarily in the form of its ester, is delivered to the liver chiefly in the form of chylomicron remnants, as well as in the form of LDL and HDL. That which is destined for excretion is hydrolyzed by esterases to the free form, which is secreted into the bile canaliculi directly, or it is first converted into bile acids prior to entering the bile. It is estimated that neutral sterol, most of which is cholesterol, represents about 55% and bile acids (salts) 45% of total sterol excreted. The key metabolic changes in the cholesterol-to-bile acid transformation are

- reduction in the length of the hydrocarbon side chain at C-17,
- addition of a carboxylic acid group on the shortened chain, and
- addition of hydroxyl groups to the ring system of the molecule.

The effect of these reactions is to enhance the water solubility of the sterol, facilitating its excretion in the bile. Cholic acid, the structure of which is shown in Figure 6.2, has hydroxyl groups at C-7 and C-12 in addition to the C-3 hydroxyl of the native cholesterol. The other major bile acids differ from cholic acid only in the number of hydroxyls attached to the ring system. For example, chenodeoxycholic acid has hydroxyls at C-3 and

C-7, deoxycholic acid at C-3 and C-12, and lithocholic acid at C-3 only. Other bile acids are formed from the conjugation of these compounds with glycine or taurine, which attach through the carboxyl group of the steroid.

It has already been pointed out that the enterohepatic circulation can return absorbed bile salts to the liver. Bile salts returning to the liver from the intestine repress the formation of an enzyme catalyzing the rate-limiting step in the conversion of cholesterol into bile acids. If the bile salts are prevented from returning to the liver, the activity of this enzyme increases, thus stimulating the conversion of cholesterol and therefore its excretion. This effect is exploited therapeutically in the treatment of hypercholesterolemia by the use of unabsorbable, cationic resins that bind bile salts in the intestinal lumen and prevent their return to the liver.

Synthesis of Fatty Acids

Aside from linoleic acid and α-linolenic acid, which are essential and must be acquired from the diet, the body is capable of synthesizing fatty acids from simple precursors. An overview of the process is the sequential assembly of a "starter" molecule of acetyl CoA with units of malonyl CoA, the CoA derivative of malonic acid. Ultimately, however, all the carbons of a fatty acid are contributed by acetyl CoA, because malonyl CoA is formed from acetyl CoA and CO_2. This reaction occurs in the cytoplasm. It is catalyzed by acetyl CoA carboxylase, a complex enzyme containing *biotin* as its prosthetic group. The role of biotin in *carboxylation reactions,* such as this one, which involves the incorporation of a carboxyl group into a compound, will be discussed in Chapter 9. ATP furnishes the driving force to attach the new carboxyl group to acetyl CoA (see 6O).

Nearly all the acetyl CoA formed in metabolism occurs in the mitochondria, formed there from pyruvate oxidation (p. 91), from the oxidation of fatty acids (p. 147), and from the degradation of the carbon skeletons of some amino acids (p. 188). The synthesis of fatty acids is localized in the cytoplast, but acetyl CoA as such is not able to pass through the mitochondrial membrane. The major mechanism for the transfer of acetyl CoA to the cytoplast is by way of its passage across the mitochondrial membrane in the form of

(6O)

Acetyl CoA $\xrightarrow[\text{ATP} \quad \text{ADP + Pi}]{\text{Acetyl CoA carboxylase}}$ Malonyl CoA

(6P)

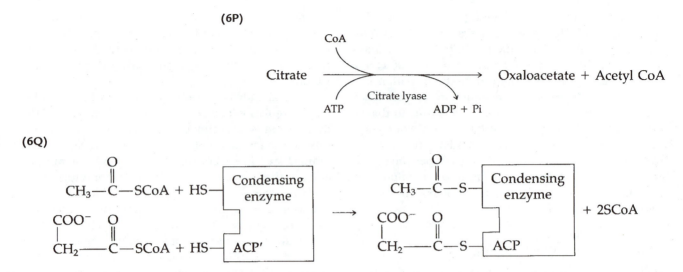

(6Q)

citrate. Once in the cytoplast, *citrate lyase* converts the citrate to oxaloacetate and acetyl CoA, a reaction that is essentially the reversal of the citrate synthetase reaction of the Krebs cycle, except that it requires expenditure of ATP (see 6P).

The enzymes involved in fatty acid synthesis are arranged in a complex called the *fatty acid synthase system.* Key components of this complex are the *acyl carrier protein* (ACP) (p. 281) and the *condensing enzyme* (CE), both of which possess free --SH groups to which the acetyl CoA and malonyl CoA building blocks attach. ACP is structurally similar to CoA (Fig. 9.21) in that they both possess a 4'-phosphopantetheine component that is composed of *pantothenic acid* coupled through β-alanine to thioethanolamine (which contributes the free --SH group) and phosphate. The free --SH of the condensing enzyme is contributed by the amino acid cysteine.

Before the actual steps in the elongation of the fatty acid chain can begin, the two sulfhydryl groups must be "loaded" correctly with malonyl and acetyl groups. Acetyl CoA is transferred to ACP, with the loss of CoA, to form acetyl ACP. The acetyl group is then transferred again to the --SH of the condensing enzyme, leaving available the ACP--SH, to which malonyl CoA then attaches, again, with the loss of CoA. This loading of the complex can be represented as shown in 6Q.

The extension of the fatty acid chain then proceeds through the following sequential steps, which are also shown schematically in Figure 6.11. The enzymes catalyzing these reactions are also a part of the fatty acid synthase complex along with ACP and CE. The following are comments on these reactions:

1. The condensation reaction in which the carbonyl carbon of the acetyl group is coupled to C-2 of malonyl ACP with the elimination of the malonyl carboxyl group as CO_2. The reaction is catalyzed by β-*ketoacyl-ACP synthase.*

2. The β-ketone is reduced, with NADPH serving as hydrogen donor. The reduction is catalyzed by β-*ketoacyl-ACP reductase.*

3. The alcohol is dehydrated, yielding a double bond. The reaction is catalyzed by β-*hydroxyacyl-ACP dehydratase.*

4. The double bond is reduced to butyryl-ACP, again with NADPH acting as reducing agent. *Enoyl-ACP reductase* catalyzes the reaction.

5. The butyryl group is transferred to the CE, exposing the ACP sulfhydryl site to a second molecule of malonyl CoA.

6. Malonyl ACP forms for a second time.

7. A second condensation reaction takes place, with coupling of the butyryl group on the CE to C-2 of the malonyl ACP. The six-carbon chain is then reduced and transferred to CE in a repetition of steps 2 through 5. A third molecule of malonyl CoA attaches at ACP-SH, and so forth. The completed fatty acid chain is hydrolyzed from the ACP without transfer to the CE.

The normal product of the fatty acid synthase system is palmitate. It can in turn be lengthened by *fatty acid elongation* systems to stearic acid and even longer saturated fatty acids. Elongation occurs by the insertion of two-carbon units at the carboxylic acid end of the chain. Furthermore, by *desaturation* reactions, palmitate and stearate can be converted to their corresponding Δ^9 monounsaturated fatty acids-palmitoleic and oleic acids, respectively. Fatty acid desaturation reactions are catalyzed by enzymes referred to as *mixed-function oxidases,* so called because two different substrates are oxidized, the fatty acid (by removal

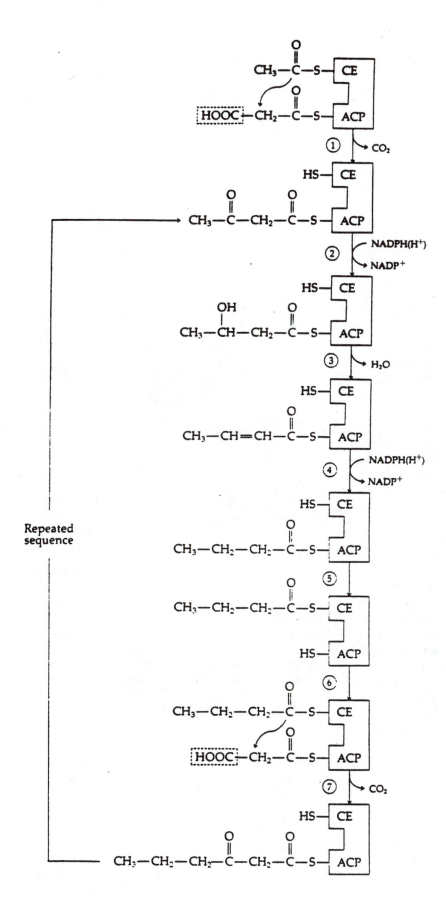

Figure 6.11 The steps in the synthesis of fatty acid. CE (condensing enzyme) and ACP (acyl carrier protein) are members of a complex of enzymes referred to as the *fatty acyl synthase system.*

of hydrogen atoms to form the new double bond) and NADPH. Oxygen is the terminal hydrogen and electron acceptor.

As pointed out previously, human cells cannot introduce additional double bonds beyond the Δ^9 site because they lack enzymes called Δ^{12} and Δ^{15} desaturases. That is why linoleic acid (18:2 $\Delta^{9,12}$) and α-linolenic acid (18:3 $\Delta^{9.12.15}$) are essential fatty acids. They can be acquired from plant sources because plant cells do have the desaturase enzymes. Once linoleic acid is acquired, longer, more highly unsaturated fatty acids can be formed from it by the combination of elongation and desaturation reactions. Figure 6.12 illustrates the elongation and desaturation of palmitate and linoleate.

The rate of fatty acid synthesis can be influenced by diet. Diets that are high in simple carbohydrates and low in fats induce in the liver a set of lipogenic enzymes. The induction is exerted through the process of transcription, leading to elevated levels of the mRNA for the enzymes. The transcriptional response is triggered by an increase in glucose metabolism, and the triggering substance, although not positively identified, is thought to be glucose 6-phosphate [24]. Other studies have confirmed that a very low-fat/high-sugar diet causes an increase in fatty acid synthesis and palmitate-rich, linoleate-poor VLDL triacylglycerols. Furthermore, the effect may be reduced if starch is substituted for the sugar, possibly owing to the slower absorption of starch glucose and a lower postprandial insulin response [25].

Synthesis of Triacylglycerols (Triglycerides)

The biosynthesis of triacylglycerols and glycerophosphatides share common precursors and can be considered together in this section. The precursors are CoA-activated fatty acids and glycerol-3-phosphate, the latter being produced either from the reduction of dihydroxyacetone phosphate or the phosphorylation of glycerol. These and subsequent reactions of the pathways are shown in Figure 6.13. The de novo pathway of lecithin synthesis is the major route; however, the importance of the salvage pathway increases when there is a deficiency of the essential amino acid methionine.

Synthesis of Cholesterol

Nearly all the tissues in the body are capable of synthesizing cholesterol from acetyl CoA. The liver accounts for about 20% of endogenous cholesterol, and among the extrahepatic tissues, responsible for the remaining 80% of synthesized cholesterol, the intestine is probably the most active. The cholesterol production rate,

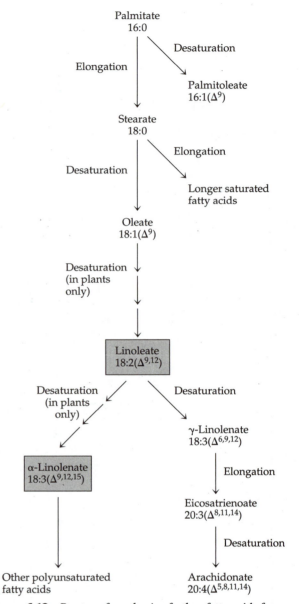

Figure 6.12 Routes of synthesis of other fatty acids from palmitate as precursor. Elongation and desaturation reactions allow palmitate to be converted into longer and more highly unsaturated fatty acids. Mammals cannot convert oleate into linoleate or α-linolenate (shaded in figure). These fatty acids are therefore essential and must be acquired in the diet.

which includes both absorbed cholesterol and endogenously synthesized cholesterol, approximates 1 g/day. The average cholesterol intake is considered to be about 600 mg/day, only about one-half of which is absorbed. Endogenous synthesis therefore accounts for greater than two-thirds of the total cholesterol store.

At least 26 steps are known to be involved in the formation of cholesterol from acetyl CoA. The pathway can be thought of as occurring in three stages:

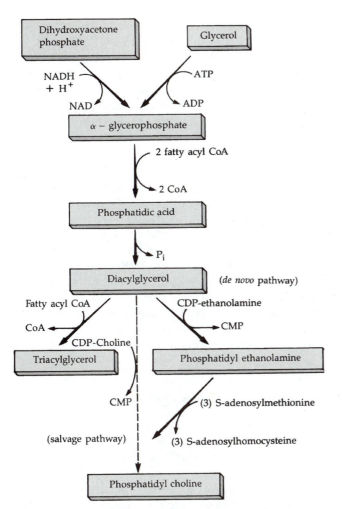

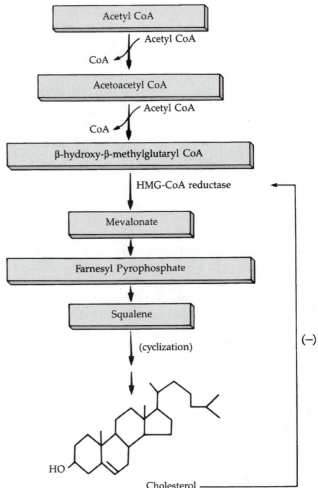

Figure 6.13 A schematic summary of the synthesis of triacylglycerols and lecithin showing that precursors are shared. In lecithin formation, three moles of activated methionine (S-adenosylmethionine) introduce three methyl groups in the de novo pathway, and choline is introduced as CDP (cytidine diphosphate) choline in the so-called salvage pathway.

1. a cytoplasmic sequence by which 3-hydroxy-3-methyl glutaryl CoA (HMG-CoA) is formed from 3 mol of acetyl CoA,

2. the conversion of HMG-CoA to squalene, including the important rate-limiting step of cholesterol synthesis in which HMG-CoA is reduced to mevalonic acid by HMG-CoA reductase, and

3. the formation of cholesterol from squalene.

As total body cholesterol increases, the rate of synthesis tends to decrease, and this is known to be due to a negative feedback regulation of the HMG-CoA reductase reaction. This suppression of cholesterol synthesis by dietary cholesterol seems to be unique to the liver and is not evident in other tissues to a great extent. The

Figure 6.14 An overview of the pathway of cholesterol biosynthesis in the hepatocyte indicating the negative regulatory effect of cholesterol on the HMG-CoA reductase reaction.

effect of feedback control of biosynthesis depends to a great extent on the amount of cholesterol absorbed. The suppression is not sufficient to prevent an increase in the total body pool of cholesterol when dietary intake is high. A brief scheme of hepatic cholesterolgenesis and its regulation is shown in Figure 6.14.

Regulation of Lipid Metabolism

The regulation of fatty acid oxidation is closely linked to carbohydrate status. Fatty acids formed in the cytoplast of liver cells can either be converted into triacylglycerols and phospholipids or transported via carnitine into the mitochondrion for oxidation. Carnitine acyl transferase I (see 6L), which catalyzes the transfer of fatty acyl groups to carnitine, is specifically inhibited by malonyl CoA. Recall that malonyl CoA is the first intermediate in the synthesis of fatty acids, and therefore it is logical that an increase in its concentration would promote fatty acid synthesis while inhibiting fatty acid oxidation. Malonyl CoA increases in concentration whenever the person is well supplied with carbohydrate because excess glucose that cannot be oxidized or stored as glycogen is converted into triacylglycerols for storage, thereby increasing the demands for malonyl CoA. Therefore, glucose-rich cells do not actively oxidize fatty acids for energy. Instead, a switch to lipogenesis is stimulated, accomplished in part by inhibition of the entry of fatty acids into the mitochondrion.

Blood glucose levels can affect lipolysis and fatty acid oxidation by other mechanisms as well. Hyperglycemia triggers the release of insulin, which promotes glucose transport into the adipose cell and therefore lipogenesis. The hormone also exerts a pronounced antilipolytic effect. Hypoglycemia results in a reduced intracellular supply of glucose, therefore suppressing lipogenesis. Furthermore, the low level of insulin accompanying the hypoglycemic state would favor lipolysis, with a flow of free fatty acids into the bloodstream. Low glucose levels also stimulate the rate of fatty acid oxidation in the manner described in the section dealing with the ketone bodies. In this case, accelerated oxidation of fatty acids follows the reduction in Krebs cycle activity, which in turn results from inadequate oxaloacetate availability.

The key enzyme for the mobilization of fat is *hormone-sensitive triacylglycerol lipase,* found in adipose tissue cells. Lipolysis is stimulated by such hormones as epinephrine and norepinephrine, adrenocorticotropic hormone (ACTH), thyroid-stimulating hormone (TSH), glucagon, growth hormone, and thyroxine. Insulin, as mentioned earlier, antagonizes the effects of these hormones by inhibiting the enzymatic activity.

A very important allosteric enzyme involved in the regulation of fatty acid biosynthesis is acetyl CoA carboxylase, which forms malonyl CoA from acetyl CoA (p. 151). The enzyme, which functions in the cytoplast, is positively stimulated by citrate, and in the absence of this modulator the enzyme is barely active. Citrate is continuously produced in the mitochondrion as a Krebs cycle intermediate, but its concentration in the cytoplast is normally low. However, should mitochondrial citrate concentration increase, the compound can escape to the cytoplast, where it acts as a positive allosteric signal to acetyl CoA carboxylase, thereby increasing the rate of formation of malonyl CoA. Also it may be recalled that citrate is the precursor for cytoplasmic acetyl CoA. Therefore the result of citrate accumulation is that excess acetyl CoA is diverted to fatty acid synthesis and away from Krebs cycle activity.

Acetyl CoA carboxylase can be modulated negatively by palmitoyl CoA, which is the product of fatty acid syn-

thesis. This situation would most likely arise when fatty acid concentrations increase as a result of insufficient glycerophosphate, with which they must combine to form triacylglycerols. Deficient glycerophosphate levels would likely be due to inadequate carbohydrate availability, and in such a situation, regulation would logically favor fatty acid oxidation rather than synthesis.

A great deal of interest in serum cholesterol levels has been generated because of the alleged correlation of the compound to the predisposition to cardiovascular disease. The regulation of cholesterol homeostasis is associated with its effect on LDL receptor concentration and on the activity of regulatory enzymes such as acyl CoA:cholesterol acyltransferase (ACAT) and hydroxymethyl glutaryl CoA (HMG-CoA) reductase. Cholesterol's feedback suppression of HMG-CoA reductase has been discussed and is shown in Figure 6.14. The effect of increasing ACAT activity (the conversion of free cholesterol to cholesteryl esters) and decreasing the amount of LDL receptors reduces the accumulation of cholesterol in vascular endothelial and smooth muscle cells. Figure 6.9 illustrates these mechanisms of control.

Brown Fat Thermogenesis

Brown adipose tissue obtains its name from its high degree of vascularity and the abundant mitochondria present in the adipocytes. The mitochondria, it will be recalled, are pigmented, owing to the cytochromes and perhaps other oxidative pigments associated with electron transport. Not only are there larger numbers of mitochondria in brown fat compared with white fat, but they are also structurally different so as to promote *thermogenesis* at the expense of phosphorylation. Thermogenesis is the dissipation of energy derived from ingested food as heat.

Brown fat mitochondria have special H^+ pores in their inner membrane, formed by an integral protein called *thermogenin* or the uncoupling protein (UCP). UCP is a translocator of protons, allowing the external H^+ pumped out by electron transport to flow back into the mitochondria, rather than through the F_0F_1 ATP-synthase site of phosphorylation. Figure 6.15 illustrates how the proposed mechanism of brown fat thermogenesis relates to the chemiosmotic theory of oxidative phosphorylation. Protons within the mitochondrial matrix are pumped outside the inner membrane by the electron transport energy. Then the downhill flow of protons through the F_0F_1 aggregate channels provides the energy for ADP phosphorylation. Membrane pores of brown fat, which allow the futile cycling of protons, appear to be regulated by the 32,000-dalton UCP.

Two types of external stimuli trigger thermogenesis: (1) the ingestion of food and (2) prolonged exposure to

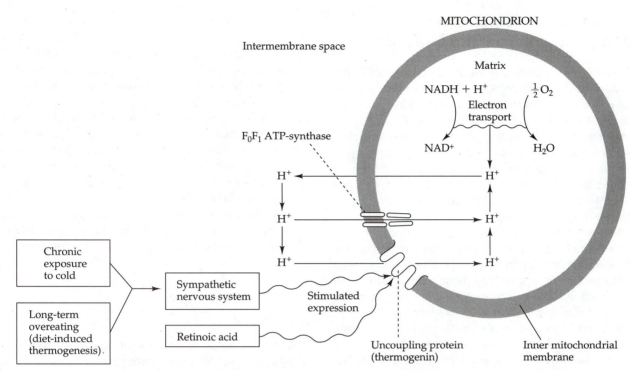

Figure 6.15 The stimulatory effect of some mediators on brown adipocyte thermogenesis. Retinoic acid has recently been shown to act independently of the sympathetic nervous system. Enhanced activity of the uncoupling protein, which translocates protons from the intermembrane space of the mitochondrion to the matrix, shifts the energy released from the proton flow toward thermogenesis and away from ATP synthesis.

cold temperature, both of which stimulate the tissue via sympathetic innervation via the hormone norepinephrine. The sympathetic signal has a stimulatory and hypertrophic effect on brown adipose tissue. This results in enhanced expression of the UCP in the inner membrane of the mitochondrion, as well as accelerated synthesis of lipoprotein lipase and glucose transporters [26]. An additional pathway for enhanced UCP activity, independent of the sympathetic nervous system, involves retinoic acid stimulation. A 27–base pair sequence in the UCP gene has been identified as the retinoic acid receptor. Upon binding to its receptor in brown fat adipocytes, retinoic acid stimulates the transcriptional activity of the gene and therefore the synthesis of UCP [27]. Figure 6.15 includes the pathways for the stimulation of UCP expression. The higher UCP concentration allows a greater proton flux into the matrix, which in turn encourages greater electron transport activity in answer to the reduced proton pressure in the intermembrane space. Enhancement of lipoprotein lipase and glucose transporters provide fuel (fatty acids and glucose, respectively) to meet the higher metabolic demand. The end result of this stimulation is that the phosphorylation of ADP by electron transport in the mitochondria of the brown fat cells becomes

"uncoupled," resulting in less ATP formation but considerably more heat production. As ATP production diminishes, the dynamics of the catabolic breakdown and biosynthesis of stored nutrients would shift to catabolism in an effort to replenish the ATP.

Theoretically, therefore, weight reduction should accompany a higher activity of brown fat, and, indeed, a possible link between obesity and deficient brown fat function has been researched. For example, evidence indicates that thermogenesis is defective in instances of obesity. However, direct evidence that the defect resides in brown adipose tissue is tenuous. Studies using thermographic skin measurements have shown defective thermogenesis in the brown fat depots of obese subjects who had received catecholamine hormone. But such measurements are imprecise because of the insulation provided by the thick, subcutaneous adipose tissue of obese people. In fact, some researchers are of the opinion that a simple, single test to assess appropriate brown fat function in obese subjects is lacking. Despite this shortcoming, the demonstration of the catecholamine stimulation of brown fat thermogenesis deserves further investigation. It raises the tempting possibility that this tissue may relate to human obesity both in a causal way and as a focus for therapy.

Therapeutic Inhibition of Fat Absorption: Olestra and Orlistat

Because of the affiliation between dietary fat and obesity, and even cardiovascular disease, reduction in fat intake is a logical nutritional recourse. This has been difficult to achieve, however, because fat enhances palatability, which is a major driving force in food selection. The synthetic compound *olestra* was developed as a fat replacement to impart palatability to a diet without the consequences of high-calorie fat. *Orlistat* has a different mode of action, interfering with the absorption of natural dietary fat.

Olestra is a mixture of hexa-, hepta-, and octaesters of sucrose with long-chain fatty acids. It imparts taste essentially indistinguishable from fat, yet it is not hydrolyzable by the pancreatic lipases and therefore has no caloric value. Technically, it can replace fat in a wide variety of foods and can be used to reduce fat-derived calories in cooked, baked, or fried foods. Undigested food entering the bowel can, of course, cause intestinal irritability, and olestra was viewed unfavorably as a probable contributor to such discomfort. In fact, some early reports did link olestra with colonic irritability, with accompanying gas and diarrhea. While this remains a debatable issue, most well-designed studies have exonerated olestra in this connection. In one such investigation, substitution of up to 30 g of olestra in a meal of 45 g fat had no effect on gastric, small intestinal, or colonic transit time [28]. This has been corroborated in numerous other studies. Furthermore, there have been positive findings on the overall safety of olestra, not only from the standpoint of its harmlessness to the gastrointestinal tract but also regarding such areas as metabolism and absorption, mutagenicity, carcinogenicity, and nutrition [29].

Still under study, however, is if reduction in fat energy intake may positively influence appetite, so that an energy-compensatory mechanism stimulates an enhanced intake of carbohydrate. Accompanying severe reductions in energy from fat, pronounced biobehavioral responses as compensation may result, making adherence to the low-fat diet difficult to maintain [30].

Orlistat is a semisynthetic derivative of lipstatin, which is a naturally occurring, potent inhibitor of pancreatic lipase. Recall that although hydrolytic products of triacylglycerols—monoacylglycerols and free fatty acids—are absorbed across the intestinal epithelium, intact triacylglycerols are not. The rationale for Orlistat use, therefore, is that by restricting the hydrolysis of triacylglycerols, the compound will sharply reduce absorption of fat.

Orlistat acts by binding covalently to the serine residue of the active site of pancreatic lipase, and it shows little or no inhibitory activity against α-amylase, trypsin, chymotrypsin, or the phospholipases. Dose-response curves show that a plateau of inhibition of dietary fat absorption occurs at high levels of the drug (>400 mg/day), which corresponds to approximately 35% inhibition. At therapeutic doses (300–400 mg/day), taken in conjunction with a well-balanced, mildly hypocaloric diet, Orlistat inhibits fat absorption to the extent of approximately 30%, contributing to a caloric deficit of about 200 calories. Preliminary tests have shown that it may indeed be promising as an alternative approach in the management of obesity.

Orlistat does not cause significant gastrointestinal disturbances and does not appear to affect gastric emptying and acidity, gallbladder motility, bile composition, biliary stone formation, or systemic electrolyte balance [31].

Summary

The hydrophobic character of lipids makes them unique among the major nutrients, requiring special handling in the body's aqueous milieu. Ingested fat must be finely dispersed in the intestinal lumen to present a sufficiently large surface area for enzymatic digestion to occur. In the bloodstream, reassembled lipid must be associated with proteins to ensure its solubility in that environment while undergoing transport.

The major sites for the formation of lipoproteins are the intestine, which produces them from exogenously derived lipids, and the liver, which forms lipoproteins from endogenous lipid. Central to the processes of fat transport and storage is adipose tissue, which accumulates fat as triacylglycerol when the intake of energy-producing nutrients is greater than the body's caloric requirement. When the energy demand so dictates, fatty acids are released from storage and transported to other tissues for oxidation. The mobilization follows the adipocyte's response to specific hormonal signals that stimulate the activity of the intracellular lipase.

Fatty acids are a rich source of energy. Their mitochondrial oxidation furnishes large amounts of acetyl CoA for Krebs cycle catabolism, and in situations of low carbohydrate intake or use, as in starvation or diabetes, the rate of fatty acid oxidation increases significantly with concomitant acetyl CoA accumulation. This causes a rise in the level of the ketone bodies—organic acids that can be deleterious through their disturbance of acid-base balance but that are also beneficial as sources of fuel to tissues such as muscle and brain in periods of starvation.

Although the lipids are thought of first and foremost as energy sources, some can be identified with intriguing hormone-like functions ranging from blood pressure alteration and platelet aggregation to an enhancement of immunological surveillance. These potent, bioactive substances are the prostaglandins, thromboxanes, and leukotrienes, all of which are derived from the fatty acids, arachidonate, and certain other long-chain PUFAs.

Dietary lipid has been implicated in atherogenesis, the development of the degenerative cardiovascular disease called *atherosclerosis*. A major consideration in the prevention and control of this disease has been the concentration of cholesterol in the blood serum and the relative hypocholesterolemic or hypercholesterolemic effect of certain diets. Saturated fatty acids having medium-length chains, and *trans* unsaturated fatty acids are alleged to be hypercholesterolemic, while mono- and polyunsaturated *cis* fats tend to lower serum cholesterol.

Fats can be synthesized by cytoplasmic enzyme systems when energy production by carbohydrate is adequate. The synthesis begins with simple precursors such as acetyl CoA and can be triggered by hormonal signals or by elevated levels of citrate, which acts as a regulatory substance. Blood glucose concentration also acts as a sensitive regulator of lipogenesis, which is stimulated when a hyperglycemic state exists.

Protein metabolism will be surveyed in Chapter 7. There, it will be seen that amino acids, like carbohydrates and lipids, can furnish energy through their oxidation, or they can be metabolically converted into other substances of biochemical importance. Once again it should be emphasized that the metabolic pathways of the energy nutrients are linked through common metabolites. Gluconeogenesis, discussed in Chapter 4, illustrates this integration through the formation of carbohydrate from the glycerol moiety of lipids and certain amino acids. Chapter 7 will demonstrate how protein can be converted into fat through common intermediates. "A rose is a rose is a rose" applies quite appropriately to intermediary metabolites. For example, if acetyl CoA is the substrate for an enzyme, the enzyme will convert it with no regard as to whether it originated through carbohydrate, fat, or protein metabolism.

References Cited

1. Hunter JE, Applewhite TH. Reassessment of *trans* fatty acid availability in the US diet. Am J Clin Nutr 1991;54:363–9.

2. Berridge M. Inositol triphosphate and calcium signalling. Nature 1993; 361:315–25.

3. Berridge MJ. Inositol triphosphate and diacylglycerol: two intersecting second messengers. Ann Rev Biochem 1987;56:159–93.

4. Borensztajn J, Getz GS, Kotlar TJ. Uptake of chylomicron remnants by the liver: further evidence for the modulating role of phospholipids. J Lipid Res 1988; 29:1087–96.

5. Brown MS, Goldstein JL. A receptor-mediated pathway for cholesterol homeostasis. Science 1986;232: 34–47.

6. Ross R. The pathogenesis of atherosclerosis: a prospective for the 90s. Nature 1993;362:801–9.

7. Beynen AC, Katan MB, Van Zutphen LFM. Hypo- and hyperresponders: individual differences in the response of serum cholesterol concentration to changes in diet. Adv in Lipid Res 1987;22:115–71.

8. Mensink RP, Katan MB. Effect of a diet enriched with monounsaturated or polyunsaturated fatty acids on levels of low-density and high-density lipoprotein cholesterol in healthy women and men. N Engl J Med 1989;321: 436–41.

9. Berry EM, Eisenberg S, Haratz D, et al. Effects of diets rich in monounsaturated fatty acids on plasma lipoproteins—the Jerusalem Nutrition Study: high MU-FAs vs. high PUFAs. Am J Clin Nutr 1991;53:899–907.

10. Caggiula A, Mustad V. Effects of dietary fat and fatty acids on coronary artery disease risk and total and lipoprotein cholesterol concentrations: epidemiologic studies. Am J Clin Nutr 1997;65(suppl): 1597S–610S.

11. Freese R, Mutanen M. α-Linolenic acid and marine long chain n-3 fatty acids differ only slightly in their effects on hemostatic factors in healthy subjects. Am J Clin Nutr 1997;66:591–8.

12. Harris W. n-3 Fatty acids and serum lipoproteins: human studies. Am J Clin Nutr 1997;65(suppl): 1645S–54S.

13. Kris-Etherton P, Yu Shaomei. Individual fatty acid effects on plasma lipids and lipoproteins: human studies. Am J Clin Nutr 1997;65(suppl):1628S–44S.

14. Mensink RP, Katan MB. Effect of dietary *trans* fatty acids on high density and low density lipoprotein cholesterol levels in healthy subjects. N Engl J Med 1990; 323:439–45.

15. Ascherio A, Willett W. Health effects of *trans* fatty acids. Am J Clin Nutr 1997;66(suppl):1006S–10S.

16. Willett W, Stampfer M, Manson J, Colditz G, et al. Intake of *trans* fatty acids and risk of coronary heart disease among women. Lancet 1993;341:581–5.

17. Shapiro S. Do *trans* fatty acids increase the risk of coronary artery disease? A critique of the epidemiologic evidence. Am J Clin Nutr 1997;66(suppl):1011–7.

18. Kris-Etherton P, Dietschy J. Design criteria for studies examining individual fatty acid effects on cardiovas-

cular disease risk factors: human and animal studies. Am J Clin Nutr 1997;65(suppl):1590S-6S.

19. Chisholm A, Mann J, Sutherland W, Duncan A, et al. Effect on lipoprotein profile of replacing butter with margarine in a low fat diet: randomized crossover study with hypercholesterolaemic subjects. BMJ 1996;312:931-4.

20. Harris E. Lipoprotein (a): a predictor of atherosclerotic disease. Nutr Rev 1997;55:61-4.

21. Lehtinen S, Lehtimki T, Sisto T, Salenius J, et al. Apolipoprotein E polymorphism, serum lipids, myocardial infarction and severity of angiographically verified coronary artery disease in men and women. Atherosclerosis 1995;114:83-91.

22. Cobb M, Teitlebaum H, Risch N, Jekel J, Ostfeld A. Influence of dietary fat, apolipoprotein E phenotype, and sex on plasma lipoprotein levels. Circulation 1992;86:849-57.

23. Stengard J, Zerba K, Pekkanen J, Ehnholm C, et al. Apolipoprotein E polymorphism predicts death from coronary heart disease in a longitudinal study of elderly Finnish men. Circulation 1995;91:265-9.

24. Towle H, Kaytor E, Shih H. Regulation of the expression of lipogenic enzyme genes by carbohydrate. Annu Rev Nutr 1997;17:405-33.

25. Hudgins L, Seidman C, Diakun J, Hirsch J. Human fatty acid synthesis is reduced after the substitution of dietary starch for sugar. Am J Clin Nutr 1998;67:631-9.

26. Himms-Hagan J. Brown adipose thermogenesis: interdisciplinary studies. FASEB J 1990;4:2890-8.

27. Alvarez R, DeAndres J, Yubero P, et al. A novel regulatory pathway of brown fat thermogenesis: retinoic acid is a transcriptional activator of the mitochondrial uncoupling protein. J Biol Chem 1995;270:5666-73.

28. Aggarwal AM, Camilleri M, Phillips SF, Schlagheck TG, et al. Olestra, a nondigestible, nonadsorbable fat. Effects on gastrointestinal and colonic transit. Dig Dis Sci 1993;38:1009-14.

29. Bergholz CM. Safety evaluation of olestra, a nonabsorbed, fatlike fat replacement. Crit Rev Food Sci Nutr 1992;32:141-6.

30. Cotton J, Weststrate J, Blundell J. Replacement of dietary fat with sucrose polyester: effects on energy intake and appetite control in non-obese males. Am J Clin Nutr 1996;63:891-6.

31. Guerciolini R. Mode of action of Orlistat. Int J Obes Relat Metab Disord 1997;21,suppl 3:S12-S23.

Suggested Reading

Brown MS, Goldstein JL. A receptor-mediated pathway for cholesterol homeostasis. Science 1986; 232:34-47.
This is an excellent step-by-step review of the research resulting in the delineation of the LDL receptor, its mechanisms of cholesterol homeostasis, and the therapeutic implications of these mechanisms.
Budowski P. Omega-3 fatty acids in health and disease. In: Bourne GH, ed. World Review of Nutrition and Dietetics. Basel, Switzerland: Karger, 1988;57:214-74.
The structural aspects, sources, and antithrombotic actions of the omega-3 fatty acids are thoroughly reviewed, along with the eicosanoids and their multiplicity of physiological functions.

Web Sites

http://www.nal.usda.gov/fnic
U.S. Government Food and Nutrition Information Center
http://www.amhrt.org
American Heart Association
www.eatright.org
Amreican Diabetes Association

The Role of Lipids and Lipoproteins in Atherogenesis

Atherogenesis is a degenerative systemic process involving arteries. If unchecked, the process becomes symptomatic, as vascular damage and impeded blood flow develop. The resulting disease state is called *atherosclerosis*. Although the disease is generally associated with old age, lesions begin to develop many years before symptoms become evident.

The innermost layer of the arterial wall, that which is in direct contact with the flowing blood, is called the *intima*. The intima consists of a layer of endothelial cells that act as a barrier to blood-borne cells and other substances. Underlying the intima is the *media,* consisting of layers of smooth muscle cells that make up the muscular component of the arterial wall. The atherogenic process is believed to begin in the endothelium, with subsequent or concurrent involvement of the medial smooth muscle cells.

The mechanism of atherogenesis is complex and not yet fully understood. However, it appears that two major components are implicated:

- cells of the immune system, primarily monocytes and macrophages that are phagocytic cells, and T lymphocytes, and

- lipids and lipoproteins, most importantly LDL, oxidized or otherwise modified.

These components function together in the disease process.

It has been known for a considerable time that a high level of circulating cholesterol is a major risk factor for cardiovascular disease. In more recent years, however, convincing experimental evidence has implicated LDL more specifically as the prime contributor to the process. LDL is the major transporter of cholesterol in the serum. The concentration of circulating LDL is controlled by two factors:

- Its rate of formation from VLDL (see this chapter). Normally, about one-third of VLDL is converted to LDL, and two-thirds removed by the liver via apoE receptors.

- Its fractional clearance rate. LDL is removed by LDL receptors primarily in the liver (75%) but also in other tissues. A small amount is removed by cellular endocytosis, the so-called scavenger process.

Initiation of the atherogenic process may begin in response to some form of endothelial cell injury. This could result from mechanical stress such as hypertension or to a high level of oxidized LDL, known to be toxic to endothelial cells. There follows an increased adherence of monocytes and T lymphocytes to the affected area along with infiltra-tion of platelets. Activation of these cells is believed to occur as a result of concurrent penetration of the endothelium by LDL. The following events, also illustrated in Figure 1, then occur, although not necessarily sequentially in the order listed:

- Growth factors released from platelets stimulate the proliferation of smooth muscle cells in the arterial media. The smooth muscle cells accumulate lipid presented to them as LDL, transforming them into lipid-laden *foam cells.*

- Monocytes (or macrophages) also take up LDL particles, becoming foam cells. This uptake of LDL at the arterial wall is the scavenger pathway of LDL removal. It works independently of the LDL receptor.

- Macrophages, stimulated by the uptake of LDL, release additional growth factors and chemotactic factors that attract more macrophages to the site, therefore creating more foam cells.

- Most of the lipid in foam cells is cholesterol and cholesteryl esters. As the foam cells proliferate, their lipid contents accumulate in the form of fatty streaks. These ultimately enlarge, occluding, to a matter of degree, the arterial lumen, which can become narrowed to the extent that blood flow is compromised.

LDL modified by oxidation is a stronger contributor to atherogenesis than native LDL. Uptake of oxidized LDL by macrophages is much more rapid than native LDL. In fact, cultured macrophages do not take up native LDL in vitro. Also, oxidized LDL has a considerably stronger proliferative effect on smooth muscle cells, and it is more toxic to endothelial cells [1]. It has been proposed that oxidized LDL may play a central role in atherogenesis in at least three additional ways [2]:

- It acts as a chemoattractant for the blood-borne monocytes to enter the subendothelial space.

- It causes the transformation of monocytes into macrophages.

- It causes the trapping of macrophages in the endothelial spaces by inhibiting their motility.

The oxidizing power in the cells involved has not been identified, but metal ions such as Cu^{+2} and Fe^{+3}, superoxide radicals, and heme-containing compounds have been suggested. Oxidation results in peroxidation of double bonds in the lipid moiety of the LDL particle. Cholesterol and cholesteryl esters can be converted into 7-keto derivatives, and unsaturated fatty acids are oxidatively fragmented into shorter chain aldehydes. The toxicity of 7-ketocholesterol is believed

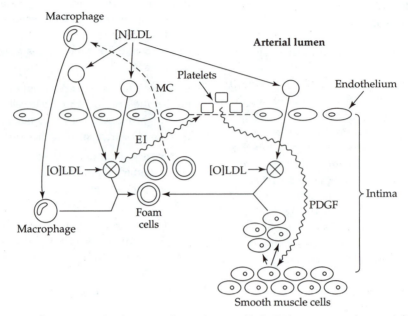

Figure 1 Proposed events in the process of artherogenesis. Native LDL ([N] LDL) penetrates the arterial intima, becoming oxidized ([O] LDL) or otherwise modified to a form readily phagocytosed by macrophages that become lipid-engorged foam cells. Oxidized LDL is also cytotoxic to endothelial cells, creating an endothelial injury (EI) that attracts platelets. Platelets release platelet-derived growth factor (PDGF) that stimulates the proliferation of smooth muscle cells, which can also take up oxidized LDL by endocytosis and become foam cells. Macrophages, activated by the phagocytosis of oxidized LDL, release macrophage (or monocyte) chemotaxins (MC), which attract additional macrophages to the site, thereby perpetuating the process. Eventually, the proliferation of foam cells and smooth muscle cells impacts on the arterial lumen, narrowing it to the point that blood flow is compromised.

to induce apoptosis (p. 19) in smooth muscle cells, a process associated with the release of oxidizing species that can contribute further to lipid peroxidation [3]. It has been suggested that the shorter-chain aldehydes resulting from peroxidation of unsaturated fatty acids may attach covalently to the apoprotein B-100 component of the particle through lysine side chains. In its chemically modified form, but not in its unaltered form, the protein is recognized by receptors on scavenger cells such as macrophages. Therefore, uptake of LDL, and consequently foam cell production, is accelerated [2].

Evidence for the participation of oxidized LDL in the atherogenic process is certainly convincing. Although the benefit of supplementation with antioxidants such as vitamin E to slow the process, and therefore reduce the risk of cardiovascular disease, has yet to be determined, the rationale for doing so is indeed sound. The role of dietary intervention in reducing oxidized LDL and its atherogenic effects has been reviewed [4].

References Cited

1. Aug N, Pieraggi MT, Thiers JC, Ngre-Salvayre A, et al. Proliferative and cytotoxic effects of mildly oxidized low-density lipoproteins on vascular smooth-muscle cells. Biochem J 1995;309:1015–20.

2. Steinberg D, Parthasarathy S, Carew T, Kjoo J, Witztum J. Beyond cholesterol: modifications of low density lipoprotein that increase its atherogenicity. N Engl J Med 1989;320:915–24.

3. Nishio E, Arimura S, Watanabe Y. Oxidized LDL induces apoptosis in cultured smooth muscle cells: a possible role for 7-ketocholesterol. Biochem Biophys Res Commun 1996;223:413–8.

4. Reaven P, Witztum J. Oxidized low density lipoproteins in atherogenesis: role of dietary intervention. Annu Rev Nutr 1996;16:51–71.

Web Sites

http://www.cspinet.org
    Center for Science in the Public Interest
http://www.amhrt.org
    American Heart Association

Photomicrograph of crystallized glutamine

Protein

The importance of protein in nutrition and health cannot be overemphasized. It is quite appropriate that the Greek word chosen as a name for this nutrient is *proteos,* meaning "primary" or "taking first place." Proteins are essential nutritionally because of their constituent amino acids, which the body must have to synthesize its own variety of proteins and nitrogen containing molecules that make life possible. Each body protein is unique in the characteristics and pattern of sequencing of the amino acids that comprise its structure.

In a review of protein, an appropriate focus is the functional roles and structures of various body proteins. Also needed is an examination of the digestion of protein and the subsequent absorption and metabolism of amino acids. Protein needs as they are affected by tissue protein synthesis and catabolism are addressed. Lastly, recommended intakes of protein and protein quality are reviewed.

Functional Categories

The molecular architecture and activity of living cells are largely dependent on proteins, which make up over half of the solid content of cells and which show great variability in size, shape, and physical properties. Their physiological roles also are quite variable, and because of this variability, a categorization of proteins according to their functions can be helpful in the study of human metabolism. This type of categorization demonstrates the body's dependence on properly functioning proteins and provides a basis for understanding the significance of protein structure.

Enzymes

Enzymes are protein molecules (generally designated by the suffix -ase) that act as catalysts to change the rate of reactions occurring in the body. Enzymes are frequently classified according to the type of reactions that they catalyze. For example:

- hydrolases cleave compounds;
- isomerases transfer atoms within a molecule;
- ligases (synthases) join compounds;
- oxidoreductases transfer electrons;
- transferases move functional groups.

Enzymes are necessary for sustaining life. They are constructed so that they combine selectively with other molecules (called *substrates*) in the cell. It is the *active site* on the enzyme (a small region usually in a crevice of the enzyme) where the enzyme and substrate bind and the product is generated. Some enzymes, however, require a cofactor or coenzyme for carrying out the reaction. Minerals such as zinc, iron, and copper function as cofactors for some enzymes. B vitamins serve as coenzymes for many enzymes. Most human physiological processes require enzymes for promotion of chemical changes that could not otherwise occur. Some examples of physiological processes dependent on enzyme function include digestion, tissue energy production, blood coagulation, and excitability and contractibility of neuromuscular tissue.

Hormones

Hormones are chemical messengers that are synthesized and secreted by endocrine tissue (glands) and trans-ported in the blood to target tissue(s) or organ(s) where they bind to protein receptors. Hormones generally regulate metabolic processes, such as through promotion of the synthesis of enzymes or effects on enzyme activity.

Whereas some hormones are derived from cholesterol and classified as a steroid hormone, others are derived from one or more amino acids. The amino acid tyrosine, for example, is used along with the mineral iodine for the synthesis of the thyroid hormones. Tyrosine is also used for the synthesis of the catecholamines, including dopamine, norepinephrine, and epinephrine. The hormone melatonin is derived in the brain from the amino acid tryptophan. Other hormones are made up of one or more polypeptide chains. Insulin, for example, consists of two polypeptide chains linked by a disulfide bridge. Glucagon, parathyroid hormone, and calcitonin each consist of a single polypeptide chain. There are many other peptide hormones, such as adrenocorticotropic hormone (ACTH), somatotropin (growth hormone), and vasopressin (also known as antidiuretic hormone, ADH), that have important roles in human metabolism and nutrition. These hormones are discussed throughout this chapter and the book.

Structural Proteins

Proteins with structural roles include

- the contractile proteins and
- the fibrous proteins.

The two main contractile proteins include actin and myosin, both found in muscles. Fibrous proteins include collagen, elastin, and keratin, and they are found in bone, teeth, skin, tendons, cartilage, blood vessels, hair, and nails. Collagen is a group of well-studied proteins. Each type of collagen is made of three polypeptide chains that are cross-linked for strength. The amino acid composition of the chains is high in the amino acids glycine and proline. In addition, collagen contains two amino acids, hydroxylysine and hydroxyproline, that are not found in other proteins. Collagen polypeptides are also attached to carbohydrate chains and are thus considered to be glycoproteins. Other structural proteins such as elastin are associated with proteoglycans. Both glycoproteins and proteoglycans are conjugated proteins and are discussed further in the section "Other Roles."

Immunoproteins

Immunoproteins may also be referred to as *immunoglobulins* (Ig) or *antibodies* (Ab). Immunoglobulins, of which there are five major classes—IgG, IgA, IgM, IgE, and IgD—are Y-shaped proteins made of four polypeptide chains. The immunoglobulins are produced by plasma cells derived from B-lymphocytes, a type of white blood cell. Immunoglobulins function by binding to antigens and inactivating them. Antigens typically consist of foreign substances such as bacteria or viruses that have entered the body. By complexing with antigens, immunoglobulin-antigen complexes can be recognized and destroyed through reactions with either complement proteins or cytokines. The complement proteins (approximately 20) are produced primarily in the liver and circulate in the blood and extracellular fluid. Cytokines are produced by white blood cells such as T-helper (CD_4) cells and macrophages. In addition, white blood cells such as macrophages and neutrophils also destroy foreign antigens through the process of phagocytosis.

Transport Proteins

Transport proteins are a diverse group of proteins that combine with other substances in the blood to provide a mode of transport for the substances. Some transport proteins of particular importance are

- albumin, which transports a variety of nutrients such as calcium, zinc, and vitamin B_6;
- transthyretin (formerly called *prealbumin*), which complexes with another protein, retinol-binding protein, for the transport of retinol (vitamin A);
- hemeproteins, iron-containing proteins, which bind and/or transport oxygen;
- transferrin, an iron transport protein; and
- ceruloplasmin, a copper transport protein.

Albumin and transthyretin are discussed further in this chapter under the heading "Amino Acid Metabolism." Hemeproteins such as hemoglobin and myoglobin, as well as the transport proteins transferrin and ceruloplasmin are discussed in more detail in Chapter 12, "Microminerals," under the topics of iron and copper.

Other Roles

Proteins, because of their constituent amino acids, can serve as a buffer in the body. A buffer is a compound that ameliorates a change in pH that would otherwise occur in response to the addition of alkali or acid to a solution. The pH of the blood and other body tissues must be maintained within an appropriate range. Blood pH ranges from about 7.35 to 7.45, whereas cellular pH levels are often more acidic. For example, the pH of red blood cells is about 7.2 and that of muscle cells is about 6.9. The H^+ concentration within cells is buffered by both the phosphate system and proteins. For example, the protein hemoglobin functions as a buffer in red blood cells. In the plasma and extracellu-

lar fluid, proteins and the bicarbonate system serve as buffers. Amino acids act as acids or bases in aqueous solutions such as in the body by releasing and accepting hydrogen ions, and they thereby contribute to the buffering capacity of proteins in the body. The buffering ability of proteins can be illustrated by the reaction: $H^+ + protein \longleftrightarrow Hprotein$.

Conjugated proteins also play important and diverse roles in the body. Conjugated proteins are proteins that are conjugated (joined) to nonprotein components. Examples of some conjugated proteins include glycoproteins, proteoglycans, lipoproteins, flavoproteins, and metalloproteins. *Glycoproteins* consist of a protein covalently bound to a carbohydrate component. The carbohydrate in glycoproteins generally includes short chains of glucose, galactose, mannose, fucose, N-acetylglucosamine, N-acetylgalactosamine, and acetylneuraminic (sialic) acid, present at the terminal end of the oligosaccharide chain. The carbohydrate portion of the glycoprotein can make up as much as 85% of the weight of the glycoprotein. The carbohydrate component is bound typically through an N-glycosidic linkage with asparagine's amide group in its side chain or through an O-glycosidic linkage with the hydroxy group in serine or threonine's side chain. Glycoproteins are found in body secretions such as mucus, connective tissue such as collagen and elastin, and bone matrix. In fact, the continuous secretion of mucus glycoproteins from the lungs and small intestine represents a constant drain on an individual's amino acid supplies, especially threonine. Many plasma proteins such as transthyretin, and hormones such as thyrotropin (TSH) are glycoproteins.

Proteoglycans are macromolecules with proteins covalently conjugated via O-glycosidic or N-glycosylamine linkages to glycosaminoglycans (formerly called *mucopolysaccharides*). Glycosaminoglycans consist of long chains of repeating disaccharides and comprise up to 95% of the weight of the proteoglycan. Proteoglycans make up, in part, the extracellular matrix (ground substance) that surrounds many mammalian tissues or cells such as skin, bone, and cartilage. Examples of proteoglycans include hyaluronic acid, chondroitin sulfate, keratan sulfate, dermatan sulfate, and heparan sulfate.

Lipoproteins consist of a hydrophobic core of lipids including cholesterol and triglycerides surrounded by phospholipids and a protein "coat." The proteins in the lipoproteins are actually a group of about 10 different apoproteins that enable transport of the lipid in the blood and help direct the lipoproteins to target cells for use by body tissues. There are four main types of lipoproteins: chylomicrons, very low-density, low-density, and high-density lipoproteins (discussed in detail in Chap. 6).

Flavoproteins generally refer to protein enzymes bound to flavin mononucleotide (FMN) or flavin adenine nucleotide (FAD); flavoproteins are discussed in more detail in Chapter 9, under the vitamin riboflavin, page 271. *Metalloproteins,* as their name implies, are proteins to which minerals are complexed. Some metalloproteins have enzymatic activity and some serve as a transport or storage protein for the metal.

Protein Structure and Organization

The functional role of protein is determined by its basic structure and organization. The primary, secondary, and tertiary structures of proteins illustrate their three key levels of organization. Some proteins have an additional fourth level of organization, the quaternary structure.

Primary Structure

The primary structure of a protein is the sequencing and strong covalent bonding of amino acids occurring as the polypeptide chain is synthesized on the ribosomes. The primary structure of a protein is shown in Figure 7.1. The various amino acids making up the polypeptide are labeled in sequence and represent the primary structure. The side chain of one amino acid differs from that of another amino acid, thus making each amino acid different. Polypeptide backbones do not differ between polypeptide chains. The side chains of the amino acids of the polypeptide chain or chains making up the total protein molecule account for the differences among proteins. Moreover, the side chains affect the coiling and folding of a protein on itself to (in effect) help determine the final form (structure) of the protein molecule.

Secondary Structure

The secondary structure of the protein is achieved through weaker bonding, such as hydrogen bonding, than that which characterizes the primary structure. Weak repeating linkages between nearby amino acids account for this second level of protein organization.

One type of secondary structure of proteins is the α-helix, a cylindrical-like shape formed by a coiling of the polypeptide chain on itself with interactions (H bonds between the hydrogen atom attached to the amide nitrogen and carbonyl oxygen atom) occurring at every fourth peptide linkage (Fig. 7.2a). The side chains of the amino acids in the α-helix structure extend outward. Varying degrees of the α-helix appear in widely divergent proteins, depending on their function. In those regions where it occurs, the α-helix provides some rigidity to that portion of the molecule.

Another type of secondary protein structure is the β-conformation, or β-pleated sheet. In this structure, the polypeptide chain is fully stretched out with the side chains positioned either up or down. The stretched polypeptide can fold back on itself with its segments packed together, as shown in Figure 7.2b. Both this structure and the α-helix are quite stable and provide strength and rigidity to proteins. These two secondary structures, α-helix and β-pleated sheet, are particularly abundant in proteins with structural roles such as collagen, elastin, and keratin. Collagen, for example, is a triple helix composed of three long polypeptide α-chains. Each of the long polypeptide chains is "twisted" and covalently cross-linked within and between the triple helix units. The structure resembles a tough, rodlike structure.

The random coil is the third type of secondary structure (Fig. 7.2c). No real stability exists in this structure because of the presence of certain amino acids whose side chains interfere with one another.

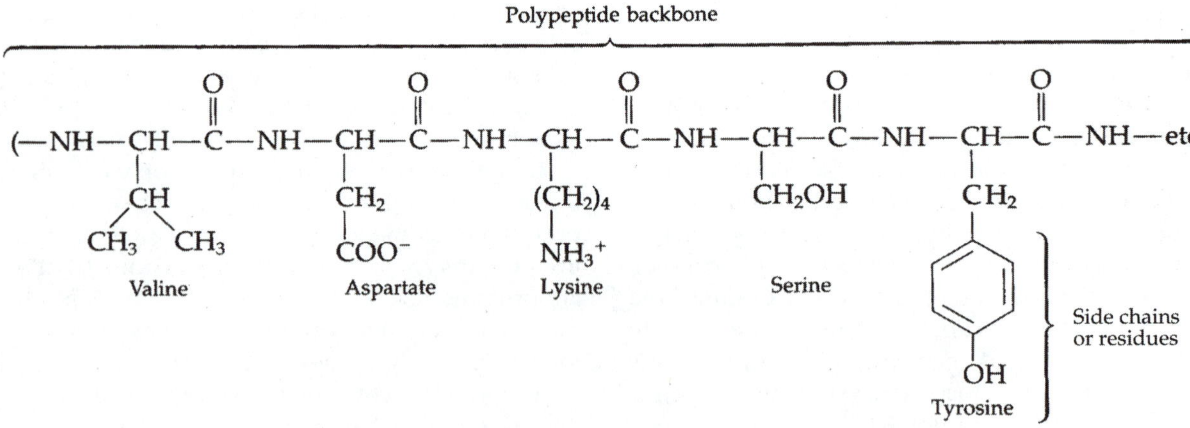

Figure 7.1 The primary structure of proteins.

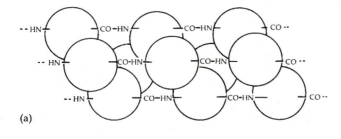

(a)

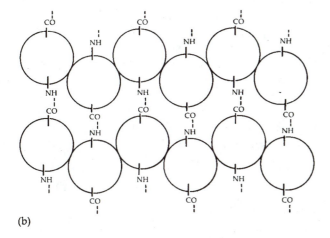

(b)

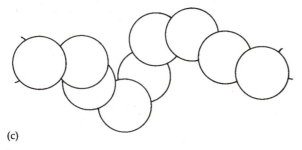

(c)

Figure 7.2 The secondary structure of proteins: (a) the α-helix, (b) the β-pleated sheet, and (c) the random coil.

The interference can be due to the large size of the side chains or the fact that the side chains are carrying similar charges.

Tertiary Structure

The third level of organization in proteins is the tertiary structure. Many proteins such as enzymes, transport proteins, and immunoproteins exist in a tertiary structure in the body. This structure results from interactions occurring among amino acid residues or side chains that are located at considerable linear distances from each other along the peptide chain. These interactions produce a globular-like structure due to the binding and looping of the protein molecule, the result of

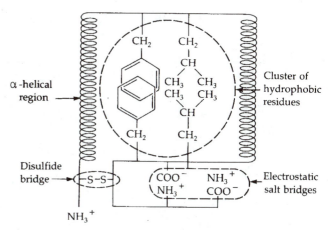

Figure 7.3 Tertiary structure of proteins.

which can be seen in Figure 7.3. Interactions producing this third level of organization include the

- clustering together of hydrophobic amino acids toward the center of the protein;
- electrostatic attraction of oppositely charged amino acid residues, such as lysine ($^+1$) and glutamate ($^-1$); and
- strong covalent bonding between cysteine residues where the -SH groups are oxidized to form disulfide bridges (-S-S-).

Other weaker attractions among amino acid residues may also occur along the chain. Taken together, these interactions among the amino acid residues determine the protein's overall shape and, therefore, the particular function of the protein in the cell.

Quaternary Structure

The final level of protein organization, quaternary structure, involves anywhere from two to several polypeptide chains. Proteins with a quaternary structure most commonly are composed of either two or four polypeptide chains, and the aggregate formed is called an *oligomer*. The polypeptides making up the oligomer are commonly termed *subunits* and are held together by hydrogen bonds and electrostatic salt bridges. Oligomeric proteins are particularly important in intracellular regulation because the subunits can assume different spatial orientations relative to each other and in so doing change the properties of the oligomer. Hemoglobin (Fig. 7.4), an oligomer with four subunits, illustrates this point. Each subunit of hemoglobin can bind 1 mol of oxygen. Rather than acting independently, however, the subunits cooperate by conformational changes so as to enhance the affinity of hemoglobin for oxygen in the lungs and to increase its ability to unload oxygen in the peripheral tissues.

Table 7.1 Structural Classification of Amino Acids

Amino Acids

1. *With aliphatic side chains*

   Glycine (Gly)

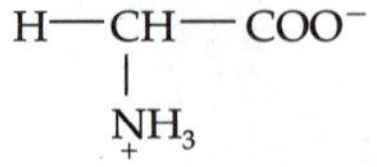

   $$H-\underset{\underset{+}{\overset{|}{NH_3}}}{CH}-COO^-$$

   Alanine (Ala)

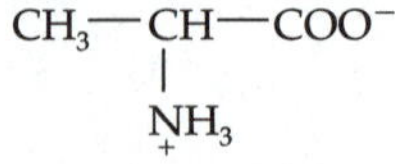

   $$CH_3-\underset{\underset{+}{\overset{|}{NH_3}}}{CH}-COO^-$$

   Valine (Val)

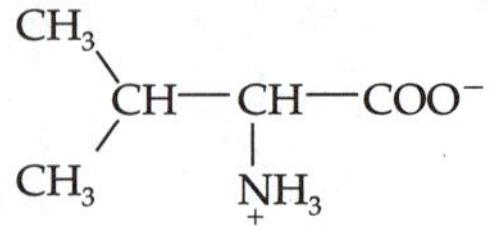

   $$\underset{CH_3}{\overset{CH_3}{\diagdown}}CH-\underset{\underset{+}{\overset{|}{NH_3}}}{CH}-COO^-$$

   Leucine (Leu)

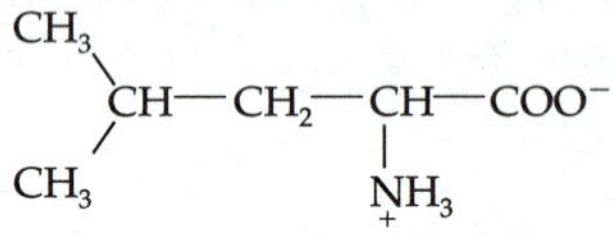

   $$\underset{CH_3}{\overset{CH_3}{\diagdown}}CH-CH_2-\underset{\underset{+}{\overset{|}{NH_3}}}{CH}-COO^-$$

   Isoleucine (Ile)

   $$\underset{CH_3}{\overset{CH_3}{\overset{|}{CH_2}}}CH-\underset{\underset{+}{\overset{|}{NH_3}}}{CH}-COO^-$$

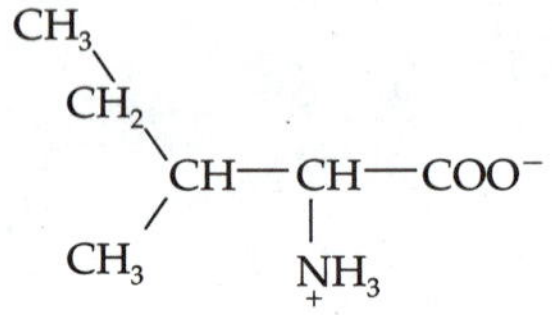

2. *With side chains containing hydroxylic (OH) groups*[a]

   Serine (Ser)

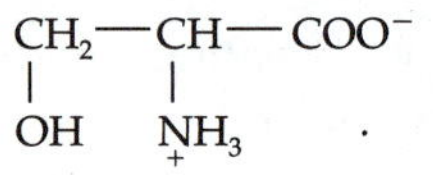

   $$\underset{OH}{\overset{|}{CH_2}}-\underset{\underset{+}{\overset{|}{NH_3}}}{CH}-COO^-$$

   Threonine (Thr)

   $$CH_3-\underset{OH}{\overset{|}{CH}}-\underset{\underset{+}{\overset{|}{NH_3}}}{CH}-COO^-$$

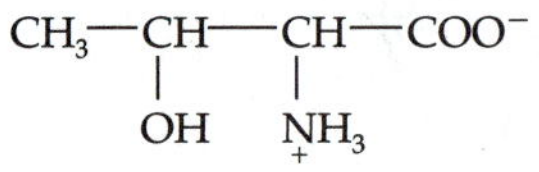

3. *With side chains containing sulfur atoms*

   Cysteine (Cys)

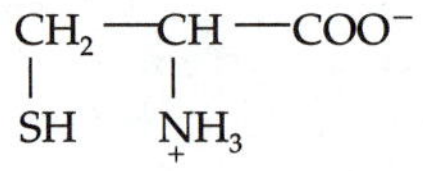

   $$\underset{SH}{\overset{|}{CH_2}}-\underset{\underset{+}{\overset{|}{NH_3}}}{CH}-COO^-$$

   Methionine (Met)

   $$\underset{S-CH_3}{\overset{|}{CH_2}}-CH_2-\underset{\underset{+}{\overset{|}{NH_3}}}{CH}-COO^-$$

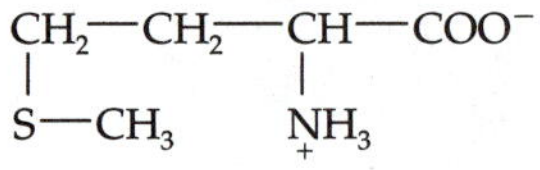

4. *With side chains containing acidic groups or their amides*

   Aspartic acid (Asp)

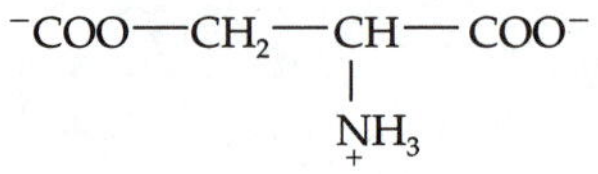

   $$^-COO-CH_2-\underset{\underset{+}{\overset{|}{NH_3}}}{CH}-COO^-$$

   Glutamic acid (Glu)

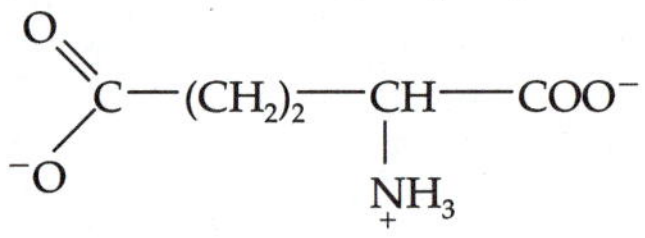

   $$\underset{^-O}{\overset{O}{\diagup}}C-(CH_2)_2-\underset{\underset{+}{\overset{|}{NH_3}}}{CH}-COO^-$$

   Asparagine (Asn)

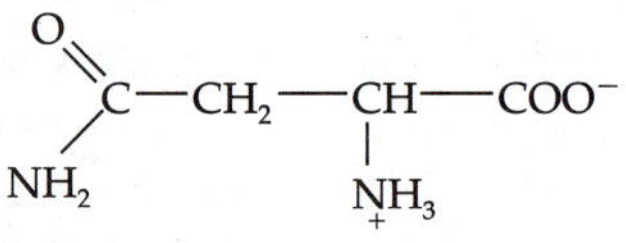

   $$\underset{NH_2}{\overset{O}{\diagup}}C-CH_2-\underset{\underset{+}{\overset{|}{NH_3}}}{CH}-COO^-$$

   Glutamine (Gln)

   $$\underset{NH_2}{\overset{O}{\diagup}}C-(CH_2)_2-\underset{\underset{+}{\overset{|}{NH_3}}}{CH}-COO^-$$

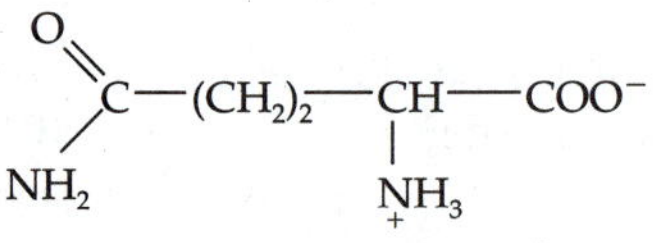

5. *With side chains containing basic groups*

   Arginine (Arg)

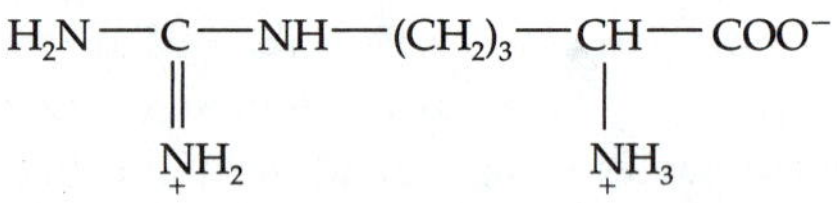

   $$H_2N-\underset{\underset{+}{\overset{||}{NH_2}}}{C}-NH-(CH_2)_3-\underset{\underset{+}{\overset{|}{NH_3}}}{CH}-COO^-$$

   Lysine (Lys)

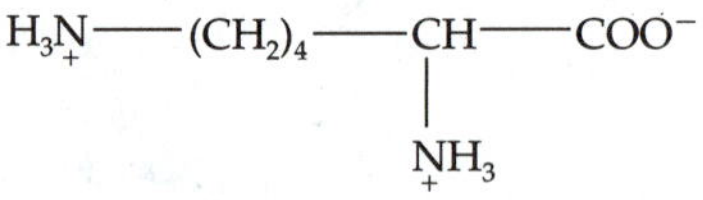

   $$\underset{+}{H_3N}-(CH_2)_4-\underset{\underset{+}{\overset{|}{NH_3}}}{CH}-COO^-$$

   Histidine (His)

   $$\underset{\underset{H}{N\diagdown_C\diagup NH}}{HC=C}-CH_2-\underset{\underset{+}{\overset{|}{NH_3}}}{CH}-COO^-$$

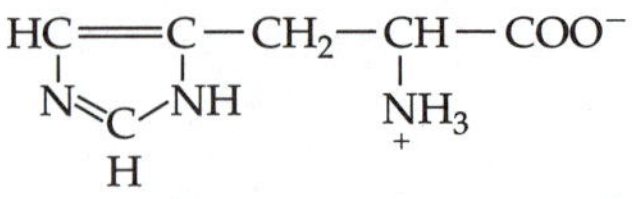

6. *With side chains containing aromatic ring*

   Phenylalanine (Phe)

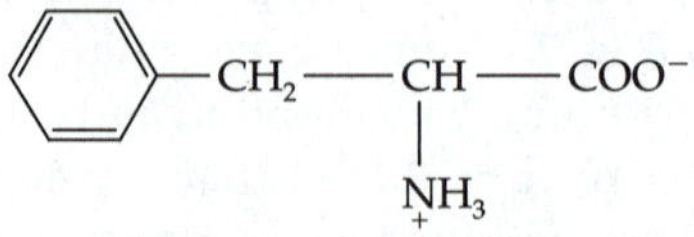

   $$\langle\!\!\!\!\bigcirc\!\!\!\!\rangle-CH_2-\underset{\underset{+}{\overset{|}{NH_3}}}{CH}-COO^-$$

   Tyrosine (Tyr)

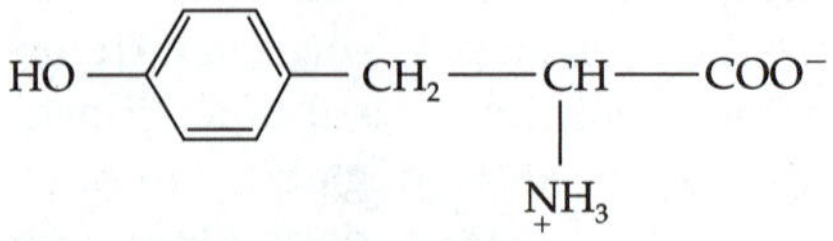

   $$HO-\langle\!\!\!\!\bigcirc\!\!\!\!\rangle-CH_2-\underset{\underset{+}{\overset{|}{NH_3}}}{CH}-COO^-$$

   Tryptophan (Trp)

   $$\text{(indole)}-CH_2-\underset{\underset{+}{\overset{|}{NH_3}}}{CH}-COO^-$$

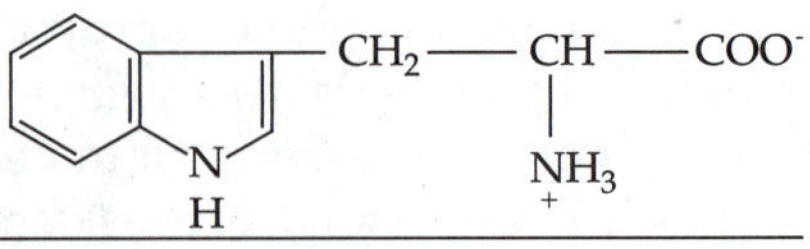

[a]Although tyrosine contains a hydroxyl group, it is classified as an amino acid containing an aromatic ring (see Group 6).

Table 7.1 *(continued)*

Amino Acids

7. *Imino acids*
    Proline (Pro)

$$CH_2—CH_2$$
$$H_2C \underset{\underset{H_2}{N^+}}{\diagdown} CH \diagdown COO^-$$

8. *Amino acids formed posttranslationally*
    Cystine (Cys-S-S-Cys)

$$^-OOC—CH—CH_2—S—S—CH_2—CH—COO^-$$
$$\underset{\underset{+}{NH_3}}{|} \qquad\qquad\qquad \underset{\underset{+}{NH_3}}{|}$$

Hydroxylysine (Hyl)

$$CH_2—CH—CH_2—CH_2—CH—COO^-$$
$$\underset{\underset{+}{NH_3}}{|}\ \underset{}{OH} \qquad\qquad \underset{\underset{+}{NH_3}}{|}$$

Hydroxyproline (Hyp)

$$HO—$$

3-methylhistidine (3-meHis)

$$—CH_2—CH—COO^-$$
$$N \diagdown\diagup N—CH_3 \qquad \underset{\underset{+}{NH_3}}{|}$$

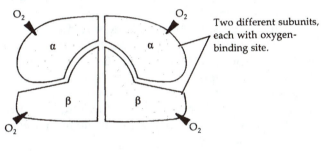

Figure 7.4 Quaternary structure of proteins.

Two different subunits, each with oxygen-binding site.

Hemoglobin tetramer

Other very important oligomers, such as regulatory enzymes, similarly undergo conformational changes on interaction with substrate molecules. In so doing, they enhance the formation of enzyme-substrate complexes whenever the concentration of substrate presented to the cell begins to increase or inhibit the formation of complexes when the substrate concentration falls to a low level.

Amino Acid Classification

Amino acids may be classified in a variety of manners such as structure, net charge, polarity, and essentiality. Each of these classifications will be addressed.

Structure

Structurally all amino acids have a central carbon (C), at least one amino group (-NH$_2$), at least one carboxy (acid) group (-COOH), and a side chain or residue (R) that makes each amino acid unique. The generic amino acid may be represented as follows:

$$H_2N—CH—COOH$$
$$|$$
$$R$$

The distinctive characteristics of the side chains of the amino acids making up a polypeptide bestow on a protein its structure and consequently its functional role in the body. These same distinctive characteristics determine whether certain amino acids can be synthesized in the body or must be ingested. Furthermore, these characteristics program the various amino acids for their specific metabolic pathways in the body. The differences among the side chains of the amino acids that are commonly found in body proteins are shown in the structural classification of amino acids in Table 7.1. The division of amino acids based on structural similarities is one approach used for classification of amino acids.

Net Electrical Charge

The amino acids in Table 7.1 and discussed here are shown based on structure as they would exist in an aqueous solution at the physiological pH, approximately 6 to 8, of the human body. Amino acids in aqueous solutions are ionized. The term *zwitterion* or *dipolar ion* is applied to amino acids with no carboxy or amino groups in their side chain to generate an additional charge to the molecule; zwitterions have no net electrical charge because their side chains are not charged and the one positive and one negative charge from the amino and carboxy group, respectively, in their base structure cancel each other out. Amino acids with no net charge do not migrate substantially if placed in an electric field.

$$H_3N^+ - CH - COO^-$$
$$|$$
$$R$$

Amino acids with neutral side chains have no net electrical charge; these include the following:

Neutral Amino Acids			
Alanine	Glycine	Methionine	Threonine
Asparagine	Hydroxylproline	Phenylalanine	Tryptophan
Cysteine	Isoleucine	Proline	Tyrosine
Glutamine	Leucine	Serine	Valine

The division of amino acids based on the presence or absence of net charge is another classification system used for amino acids.

Two groups of amino acids exhibit a net charge. Because of the presence of additional carboxy(lic) groups in the side chains, the dicarboxylic amino acids aspartic acid and glutamic acid exhibit a net negative charge. Amino acids or proteins with a high content of dicarboxylic amino acids exhibit a negative charge and will migrate toward the anode if placed in an electric field. In contrast, because of the presence of additional amino groups in the side chains, the basic amino acids (lysine, arginine, histidine, hydroxlysine) exhibit a net positive charge.

Negatively Charged Amino Acids		Positively Charged Amino Acids	
Aspartic acid	Glutamic acid	Arginine	Hydroxylysine
		Histidine	Lysine

Polarity

The tendency of an amino acid to interact with water at physiological pH—that is, its polarity—represents another means of classifying amino acids. Polarity is dependent on the side chain/R group of the amino acid. Amino acids are classified as polar or nonpolar, although there are varying levels of polarity. Both the di-

carboxylic (aspartic acid and glutamic acid) and basic (lysine, arginine, histidine, including hydroxlysine) amino acids are polar; that is, they interact with water. The neutral amino acids interact act with water to different degrees and thus can be divided into polar, nonpolar, and relatively nonpolar categories.

Polar Neutral Amino Acids	Nonpolar Neutral Amino Acids	Relatively Nonpolar Amino Acids
Asparagine	Alanine	Phenylalanine
Cysteine	Glycine	Tryptophan
Glutamine	Isoleucine	Tyrosine
Methionine	Leucine	
Serine	Proline	
Threonine	Valine	

The polar neutral amino acids contain functional groups in their side chains such as the hydroxyl group for serine and threonine, the sulfur atom for cysteine and methionine, and the amide group for asparagine and glutamine that can interact through hydrogen bonds with water; thus, we categorize them as polar. In contrast, the nonpolar amino acids listed in the second column contain side chains that do not interact with water. Thus, these amino acids typically do not interact with water and are called *hydrophobic* (water fearing). The aromatic amino acids are relatively nonpolar. Tyrosine because of its hydroxyl group on the phenyl ring, for example, can form, to a limited extent, hydrogen bonds with water—hence the use of the terminology "relatively nonpolar."

Essentiality

While amino acids can be classified based on structure or properties such as net charge or polarity, back in 1957, Rose [1] categorized the amino acids found in proteins as nutritionally *essential* (*indispensable*) or nutritionally *nonessential* (*dispensable*). At that time only eight amino acids—leucine, isoleucine, valine, lysine, tryptophan, threonine, methionine, and phenylalanine—were found to be essential for adult humans. Histidine was later added as an essential amino acid. We now know that if we give an α-keto or hydroxy acid of leucine, isoleucine, valine, tryptophan, methionine, phenylalanine, or histidine (as may be done with individuals with renal failure), the α-keto or hydroxy acid form of these amino acids can be transaminated to form the respective amino acid. Only two amino acids—lysine and threonine—cannot undergo transamination to an appreciable extent. Thus, lysine and threonine are totally indispensable.

Identifying amino acids strictly as dispensable or indispensable, however, is an inflexible classification that allows no gradations, even with decidedly different and/or changing physiological circumstances. Newer categories added to the essential/indispensable and nonessential/dispensable categories include *conditionally* or *acquired indispensable* amino acids. A dispensable amino acid may become indispensable should an organ fail to function properly as in the case of infants born prematurely or in the case of disease associated organ malfunction. For example, neonates born prematurely often have immature organ function and are unable to synthesize many nonessential amino acids such as cysteine and proline [2]. Immature liver function or liver malfunction due to cirrhosis, for example, impairs phenylalanine and methionine metabolism, which occurs primarily in the liver. Consequently, the amino acids tyrosine and cysteine normally synthesized from phenylalanine and methionine catabolism, respectively, become indispensable until normal organ function is established. In some kidney diseases, serine becomes indispensable because it cannot be synthesized in sufficient quantity by the diseased kidneys. Inborn errors of amino acid metabolism resulting from genetic disorders in which key enzymes in amino acid metabolism lack sufficient enzymatic activity also illustrates a situation in which dispensable amino acids become indispensable. Individuals with classical phenylketonuria (PKU) exhibit little to no phenylalanine hydroxylase activity. This enzyme converts phenylalanine to tyrosine. Without hydroxylase activity, tyrosine is not synthesized in the body and must be provided completely by diet; it is indispensable. In other inborn errors of metabolism, amino acids such as cysteine become indispensable. Thus, amino acids that are normally dispensable may become indispensable under certain physiological conditions.

Sources of Protein

Ingested or exogenous proteins serve as sources of the essential amino acids and are the primary source of the additional nitrogen needed for the synthesis of the nonessential amino acids and nitrogen-containing compounds. Dietary or exogenous sources of protein include

- animal products such as meat, poultry, fish, and dairy products, with the exception of butter, sour cream, and cream cheese; and
- plant products such as grains, grain products, legumes, and vegetables.

A discussion of protein quality is found at the end of this chapter, page 208.

Endogenous proteins presented to the digestive tract represent another source of amino acids and nitrogen to the body, and they mix with exogenous nitrogen sources [3]. Endogenous proteins include

- desquamated mucosal cells, which generate about 50 g of protein per day, and
- digestive enzymes and glycoproteins, which generate about 17 g of protein per day.

The digestive enzymes and glycoproteins such as mucus are derived from digestive secretions of the salivary glands, stomach, intestine, biliary tract, and pancreas. Most of these endogenous proteins, which may total 70 g or more per day, are digested and provide amino acids available for absorption. Digestion of protein and absorption of amino acids are crucial for protein nurniture and metabolism.

Digestion and Absorption

This section of the chapter will focus first on the digestion of protein in the digestive tract. Next, the mechanisms by which the end products of protein digestion are transported across the brush border membrane of the intestinal cell will be reviewed. From within the intestinal cell, amino acids must cross the basolateral membrane to gain access to the blood for circulation to tissues. Thus, the transport systems for the ferrying of amino acids across the basolateral membrane and other extraintestinal tissues are presented. Lastly, because not all amino acids entering the intestinal cells get into the blood, the use of amino acids by the intestinal cells themselves is discussed.

Protein Digestion

Although macronutrient digestion has been covered in Chapter 2, this chapter outlines digestion solely with respect to protein within the gastrointestinal tract organs.

Mouth and Esophagus

No appreciable digestion of protein occurs in the mouth or esophagus.

Stomach

The digestion of exogenous protein begins in the stomach, with the action of hydrochloric acid in gastric juice. Hydrochloric acid release is stimulated by a variety of compounds, including the hormone gastrin, the neuropeptide gastrin-releasing peptide (GRP), the neurotransmitter acetylcholine, and the amine histamine. Hydrochloric acid denatures the quaternary, tertiary, and secondary structures of protein and begins the activation of pepsinogen to pepsin. Denaturants such as hydrochloric acid break apart hydrogen and electro-static bonds to "unfold or uncoil" the protein; peptide bonds, however, are not affected by the hydrochloric acid. Pepsin, once formed, is catalytic against pepsinogen as well as other proteins.

$$\text{Pepsinogen} \xrightarrow{\text{HCl}} \text{Pepsin}$$

Pepsin functions as an endopeptidase at a pH <3.5 to hydrolyze peptide bonds in proteins or polypeptides. Pepsin attacks peptide bonds adjacent to the carboxyl end of a relatively wide variety (i.e., pepsin has low specificity) of amino acids including leucine, methionine, aromatic amino acids consisting of phenylalanine, tyrosine, and tryptophan, and the dicarboxylic amino acids glutamate and aspartate. The end products of gastric protein digestion with pepsin include primarily large polypeptides along with some oligopeptides and free amino acids. These end products in an acid chyme are emptied through the pyloric sphincter into the duodenum for further digestion.

Small Intestine

The end products in the acid chyme that are delivered into the duodenum further serve to stimulate the release of regulatory peptides such as secretin and cholecystokinin (CCK) from the mucosal endocrine cells. Secretin and CCK are carried by the blood to the pancreas where the acinar cells are stimulated to secrete alkaline pancreatic juice along with digestive proenzymes.

Digestive proenzymes or zymogens secreted by the pancreas, and further responsible for protein and polypeptide digestion, include

- trypsinogen,
- chymotrypsinogen,
- procarboxypeptidases A and B,
- proelastase,
- collagenase.

Within the small intestine, these inactive zymogens must be chemically altered to be converted into their respective active enzymes capable of substrate hydrolysis. The following reactions occur in the small intestine to activate the zymogens:

$$\text{Trypsinogen} \xrightarrow{\text{enteropeptidase}} \text{Trypsin}$$

Enteropeptidase (an endopeptidase formerly known as *enterokinase*) is secreted from the intestinal brush border in response to CCK and secretin. Once trypsin is formed, it can act on more trypsinogen as well as on chymotrypsinogen to yield active proteolytic enzymes.

$$\text{Trypsinogen} \xrightarrow{\text{trypsin}} \text{Trypsin}$$

$$\text{Chymotrypsinogen} \xrightarrow{\text{trypsin}} \text{Chymotrypsin}$$

Trypsin and chymotrypsin are both endopeptidases. Trypsin is specific for peptide bonds adjacent to dibasic amino acids (lysine and arginine). Excess free trypsin generated from trypsinogen also acts by negative feedback to inhibit pancreatic cell trypsinogen synthesis, thereby regulating pancreatic zymogen secretion [5]. Chymotrypsin is specific for peptide bonds adjacent to aromatic amino acids (phenylalanine, tyrosine, and tryptophan), as well as for methionine, asparagine, and histidine.

Both elastase, an endopeptidase derived from proelastase, and collagenase hydrolyze polypeptides into smaller fragments such as oligopeptides and tripeptides. Procarboxypeptidases are converted to carboxypeptidases by trypsin and serve as exopeptidases.

$$\text{Procarboxypeptidases} \xrightarrow{\text{trypsin}} \text{Carboxypeptidases}$$

These exopeptidases attack peptide bonds at the C-terminal end of polypeptides to release free amino acids. Carboxypeptidases are zinc-dependent enzymes, specifically requiring zinc at its active site. Carboxypeptidase A hydrolyzes peptides with C-terminal aromatic neutral or aliphatic neutral amino acids. Carboxypeptidase B cleaves dibasic amino acids from the C-terminal, generating free dibasic amino acids as end products.

Several peptidases are produced by the brush border of the small intestine, including the ileum, enabling peptide digestion and amino acid absorption to occur in the distal small intestine.

Some of these peptidases include

- aminopeptidases, which vary in specificity and cleave amino acids from the N-terminal end of oligopeptides;
- dipeptidlyaminopeptidases, some of which are magnesium-dependent and hydrolyze N-terminal amino acids from dipeptides; and
- tripeptidases, which are specific for selected amino acids and hydrolyze tripeptides to yield a dipeptide and a free amino acid.

Some tripeptides, such as trileucine, undergo brush border hydrolysis, whereas other tripeptides such as triglycine or proline-containing peptides are absorbed intact and hydrolyzed within the intestinal cell. Amino acids (an end product of protein digestion) have also been shown to inhibit the activity of brush border peptidases (end product inhibition). Table 7.2 outlines the process of protein digestion, which yields two main end

Table 7.2 Some Enzymes Responsible for the Digestion of Protein

Zymogen	Enzyme or Activator	Enzyme	Site of Activity	Substrate (Peptide Bonds Adjacent to)
Pepsinogen	HCl or pepsin ⟶	Pepsin	Stomach	Most amino acids, including aromatic, dicarboxylic, leu, met
Trypsinogen	Enteropeptidase ⟶ or Trypsin	Trypsin	Intestine	Dibasic amino acids
Chymotrypsinogen	Trypsin ⟶	Chymotrypsin	Intestine	Aromatic amino acids, met, asn, his
Procarboxypeptidases	Trypsin ⟶	Carboxypeptidase A B	Intestine	C-terminal neutral amino acids C-terminal dibasic amino acids
		Aminopeptidases	Intestine	N-terminal amino acids

products: *peptides,* principally dipeptides and tripeptides, and free *amino acids.* To be utilized by the body, these end products must now be absorbed across the brush border of the intestinal epithelial mucosal cells.

Brush Border Absorption

Amino acid absorption, the passage from the lumen into the intestinal cell shown in generic fashion without the roles of sodium and pump systems in Figure 7.5, occurs along the entire small intestine, but sites of maximal absorption of peptides and their component amino acids differ. In general, most amino acids are absorbed in the proximal (upper) small intestine. Less than 1% (10 g) of ingested protein is excreted daily in the feces. Moreover, colonic bacterial proteases have little effect on protein digestion and amino acid absorption.

Amino Acids

Multiple energy-dependent (requiring ATP hydrolysis) transport systems with overlapping specificity for amino acids have been demonstrated in the intestinal brush border [4]. Both sodium-dependent and sodium-independent transport systems exist. Figure 7.6 represents a schematic of sodium-dependent amino acid transport, which can be broken down into five stages [5]:

1. First, sodium binds to the carrier.

2. Second, the binding of sodium appears to increase the affinity of the carrier for the amino acid.

3. Next, a sodium-amino acid-cotransporter complex forms.

4. Fourth, a conformational change in the complex that results in the delivery of the sodium and amino acid into the cytoplasm of the enterocyte.

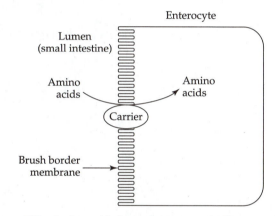

Figure 7.5 Amino acid absorption: a generic diagram without showing the roles of sodium and pump systems.

5. Lastly, sodium is pumped out of the cell by Na^+/K^+-ATPase.

Carrier systems are thought to be present in enterocytes for the dibasic amino acids lysine and arginine plus cystine and ornithine, the dicarboxylic amino acids aspartate and glutamate, as well as the neutral amino acids. In fact, several neutral transport systems are thought to exist in the brush border of the intestine.

Transport systems for amino acids are designated using a lettering system with a further distinction that uppercase letters be used for sodium dependence and lowercase letters be used for sodium independence [6]; however, not all systems, such as the L (which is sodium-independent), follow this rule. Some of the different brush border amino acid transport systems include the L, B (formerly called *NBB*), IMINO, y^+, and PHE systems. Not all systems, however, have been identified in humans. Table 7.3 lists some of the systems transporting amino acids in the intestinal cell,

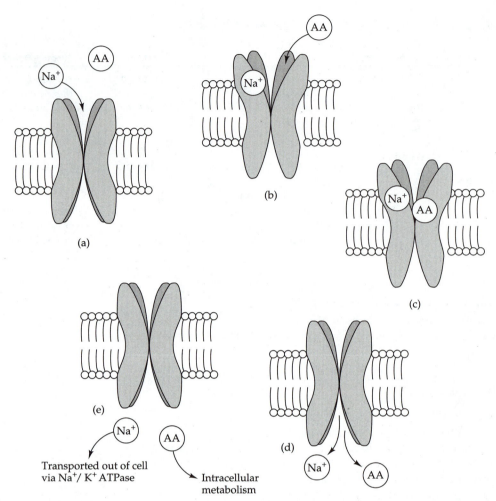

Figure 7.6 Conceptual schematic of a membrane-bound sodium-dependent amino acid carrier. (a) In the "preferred random" binding model, amino acid AA transport is favored when the activator sodium ion (Na$^+$) binds preferentially to the transporter. (b) Initial binding of the sodium ion increases the affinity of the extracellularly oriented cotransporter binding site for the subsequent amino acid attachment. (c) A sodium/amino acid/cotransporter complex is formed. (d) A conformational change of the complex results in delivery of the sodium ion and the amino acid ion into the cytoplasm of the cell. (e) The translocated sodium ion is transported out of the cell by the sodium/potassium adenosine triphosphatase (Na$^+$/K$^+$-ATPase) transporter to maintain the electrochemical gradient, whereas the amino acid is metabolized within the cell. *Source:* Souba WW, Pacitt AJ. How amino acids get into cells: mechanisms, models, menus, and mediators. J Parent Enter Nutr 1992;16:569–78. (c) by the American Society of Parenteral and Enteral Nutrition.

Table 7.3 Systems Transporting Amino Acids across the Intestinal Cell Brush Border Membrane

Amino Acid Transport Systems	Sodium Required	Examples of Substrates Carried
L	No	Leucine, other neutral amino acids
B	Yes	Threonine, neutral amino acids
IMINO	Yes	Proline
y+[a]	No	Dibasic amino acids
PHE	Yes	Phenylalanine, methionine

[a]In the presence of sodium, neutral amino acids can competitively inhibit the y$^+$ system and participate in exchange reactions with the dibasic substrates [4].

the system's requirement for sodium, and examples of amino acids carried by the transport system.

Competition between amino acids for transport by a common carrier has been documented. In addition, regulation (both induced de novo synthesis of specific amino acid carriers and decreased carrier synthesis) of transport carriers has been shown and helps to ensure adequate absorptive capacity [4].

The affinity (K_m) of a carrier for an amino acid is influenced by both the hydrocarbon mass of the amino acid's side chain and by the net electrical charge of the amino acid. As the hydrocarbon mass of the side chain increases, affinity increases [7]. Thus, the branched-chain amino acids typically are absorbed faster than smaller amino acids. Neutral amino acids also tend to be absorbed at higher rates than dibasic or dicarboxylic amino acids. Essential amino acids are absorbed faster than nonessential amino acids, with methionine, leucine, isoleucine, and valine being the most rapidly absorbed [7]. The most slowly absorbed amino acids are the two dicarboxylic amino acids, glutamate and aspartate, both of which are nonessential [7].

Ingesting free, crystalline L-amino acids is thought by many athletes to be superior to ingesting natural foods containing protein for muscle protein synthesis. However, amino acids using the same carrier system compete with each other for absorption. Thus, ingesting one or a particular group of amino acids that use the same carrier system may create, depending on the amount ingested, a competition between the amino acids for absorption. The result may be such that the amino acid present in highest concentration is absorbed but also may impair the absorption of the other less concentrated amino acids carried by that same system. Thus, amino acid supplements may result in impaired or imbalanced amino acid absorption. Furthermore, absorption of peptides (which are obtained from digestion of natural protein-containing foods) is more rapid than absorption of an equivalent mixture of free amino acids. And nitrogen assimilation following ingestion of protein-containing foods is superior to that following ingestion of free amino acids. In other words, free amino acids have no absorptive advantage. Moreover, the supplements are usually expensive, typically taste terrible, and may cause gastrointestinal distress.

Peptides

Peptide (primarily dipeptide and tripeptide) transport into the enterocyte occurs by transport systems different from those that transport amino acids. The number of transport systems for peptides has not been determined, but peptides appear to be absorbed more rapidly than free amino acids. Sodium has been shown to be required for one of the carrier systems that transports both dipeptides and tripeptides, but not for other systems. In addition, peptide transport across the brush border membrane appears to be associated with the movement of protons and thus depolarization of the brush border membrane. An area of low pH lying adjacent to the brush border surface of the enterocyte provides the driving force for the H^+ gradient. Thus, as shown in Figure 7.7, as the dipeptide or tripeptide is transported into the enterocyte, a H^+ ion also enters the enterocyte in exchange for Na^+. The transport of the H^+ into the enterocyte

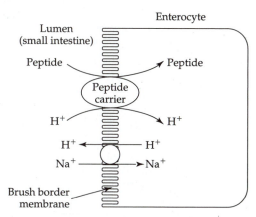

Figure 7.7 Peptide transport across the brush border membrane of the intestinal epithelial mucosal cell.

results in an intracellular acidification. A Na^+/K^+-ATPase allows for Na^+ extrusion at the basolateral membrane to maintain the gradient [8,9].

Affinity of the carrier for the peptide appears to be influenced by stereoisomerism, length of side chain of the N-terminal amino acid, substitutions on the N- and C-terminals, as well as by the number of amino acid residues in the peptide [10,11]. For example, as length of the peptide increases above three amino acids, affinity for transport decreases. In addition, like amino acids, peptides compete with one another for transporters.

Peptide transport is thought to represent the primary system for amino acid absorption. Sixty-seven percent of amino acids are absorbed in the form of small peptides, with the remaining 33% absorbed as free amino acids [12]. Peptides, once within the enterocytes, are hydrolyzed by cytoplasmic peptidases to generate free intracellular amino acids.

Basolateral Absorption

The transport of the amino acids through the basolateral membrane of the enterocyte into the interstitial fluid appears to be the same as the transport of amino acids across the membrane of nonepithelial cells. Amino acid transport across the basolateral membrane is shown in Figure 7.8. Some of the basolateral transport systems include those shown in Table 7.4.

Diffusion and sodium-independent transport are thought to be the primary modes of basolateral membrane transport in the enterocyte. Sodium-dependent pathways are quantitatively important when the amino acid concentrations are low in the gut lumen. Active transport of amino acids into the enterocytes is necessary to provide the enterocyte with amino acids to meet its own needs.

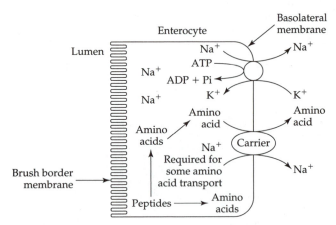

Figure 7.8 Amino acid transport across the basolateral membrane of the intestinal mucosal cell.

Intestinal Cell Amino Acid Use

Although the preceding sections have covered amino acid absorption across the brush border and basolateral membranes, it must be remembered that not all amino acids are transported out of the intestinal cell and into circulation. Many of the amino acids absorbed following protein digestion are used along the villus for protein synthesis. Within the intestinal cell, amino acids may be used for energy or to synthesize compounds such as

- apoproteins necessary for lipoprotein formation,
- new digestive enzymes,
- hormones, or
- nitrogen-containing compounds.

Glutamine is used extensively by intestinal cells as a primary source of energy; it also appears to have trophic (growth) effects, such as stimulating cell proliferation, on the gastrointestinal mucosa cells [13,14]. These roles of glutamine in the gastrointestinal tract have prompted several companies to add or enrich enteral nutrition products with glutamine. Glutamine is being added also to parenteral (intravenous) nutritional mixtures and used for hospitalized patients. When glutamine is fed through tube feedings, over 50% of glutamine is extracted by the splanchnic bed. In addition to dietary glutamine, skeletal muscles produce much of the body's glutamine, which is then released and taken up by the intestinal cells (among other cells). Glutamine is only partially catabolized within the intestinal cells [15] and generates both ammonia and glutamate. Ammonia enters the portal blood for uptake by the liver. Glutamate may be used for glutathione production [16] or may undergo transamination in which its amino group is removed and α-ketoglutarate, an intermediate in the Krebs cycle, remains. The amino group

is transferred to the compound pyruvate (which is present in the intestinal cell from glucose metabolism) to form the amino acid alanine.

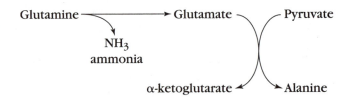

Alanine, once formed, leaves the intestinal cell, enters portal blood, and is subsequently taken up along with ammonia by periportal hepatocytes capable of urea synthesis. Glutamate not used for alanine synthesis may be used in the intestinal cell to synthesize proline, as shown in the reactions

$$\text{Glutamate} \longrightarrow \text{Glutamate semialdehyde} \longrightarrow$$
$$\text{Pyrroline 5-carboxylate} \longrightarrow \text{Proline}$$

or glutamate may be used to synthesize glutathione.

Glutathione (a tripeptide) is made in the enterocyte from three amino acids: glutamate, glycine, and cysteine (Fig. 7.9). Glutathione functions with the selenium-dependent enzyme glutathione peroxidase to destroy hydrogen peroxides (H_2O_2), reactive oxygen intermediates or species (e.g., $O_2^{-}\cdot$ and $OH\cdot$), and other toxic compounds that find their way into the intestinal cell and react with polyunsaturated fatty acids in intestinal cell membranes to cause membrane peroxidation and cell necrosis (death). Glutathione is discussed in more detail later in the chapter on page 179 and in Chapter 12 under selenium, page 443. In addition, the Perspective at the end of Chapter 10 also provides a discussion of glutathione and other cell antioxidant systems.

In addition to amino acid catabolism and synthesis, two of the five enzymes (carbamoyl phosphate synthetase I and ornithine transcarbamoylase) of the urea cycle are present in the intestinal mucosa. Carbamoyl

Table 7.4 Systems Transporting Amino Acids across the Intestinal Cell Basolateral Membrane

Amino Acid Transport Systems	Sodium Required	Examples of Substrates Carried
L	No	Leucine, other neutral amino acids
y⁺	No	Dibasic amino acids
A	Yes	Alanine, other short-chain, polar, neutral amino acids
ASC	Yes	Alanine, cysteine, serine, other three- and four-carbon amino acids
asc	No	Same substrates as ASC

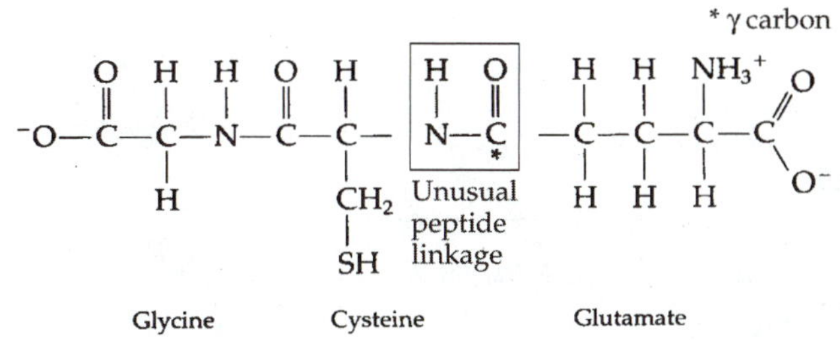

Glycine Cysteine Glutamate

Figure 7.9 The structure of reduced glutathione (GSH).

phosphate synthetase I catalyzes the synthesis of carbamoyl phosphate from ammonia, HCO_3^- and ATP in mucosal epithelial cells.

$$NH_3 + HCO_3^- + ATP \longrightarrow \text{Carbamoyl phosphate}$$

Ornithine transcarbamoylase synthesizes citrulline from ornithine and carbamoyl phosphate.

$$\text{Carbamoyl phosphate} + \text{Ornithine} \longrightarrow \text{Citrulline}$$

Citrulline, once made, is released into blood and then typically taken up by the kidney, which has a high capacity for arginine synthesis.

Amino Acid Absorption into Extraintestinal Tissues

Amino acids not used by the intestinal cell are transported across the basolateral membrane of the enterocyte into interstitial fluid, where they enter the capillaries of the villi and eventually the portal vein for transport to the liver. Most peptides that have been absorbed intact into the intestinal cell undergo hydrolysis by proteases present within the cytoplasm of the enterocyte. Thus, primarily free amino acids are found in portal circulation. Occasionally, however, small oligopeptides can be found in splanchnic circulation and are thought to have entered circulation by paracellular or intercellular routes—that is, passing through tight junctions of mucosal cells or by transcellular endocytosis [17]. The ability to administer peptides directly into the blood (parenteral nutrition) that can be used by body tissues is of nutritional significance in many clinical conditions in which amino acids (e.g., tyrosine, cysteine, and glutamine) need to be provided but cannot be given because they are insoluble or unstable in their free form. The ability to provide these insoluble or unstable amino acids in peptide form that can be utilized by tissue would allow nutrient provision in situations in which traditional free amino acid parenteral mixtures are ineffective.

Hydrolysis of peptides may occur by peptidases or proteases in the plasma, at the cell membrane—especially the liver, kidney, and muscle—or intracellularly in the cytosol or various organelles following transport as an intact peptide [10,17,18]. Peptide transport in renal tubular cells, for example, has been demonstrated and is influenced by molecular structure and the lipophilicity (hydrophobicity) of the amino acids at both the N- and C-terminal of the peptide [19,20]. Peptide with either basic or acidic amino acids at either the N- or C-terminal have lower affinity for transport than peptides with neutrally charged side chains at these positions.

Amino acid transport into liver cells (hepatocytes) occurs by some carrier systems similar to those within the intestinal basolateral membrane. Additional systems also transport amino acids into the liver. The sodium-dependent N system transports glutamine and histidine into hepatocytes [4]. Hormones and cytokines, such as interleukin-1, and tumor necrosis factor alpha, have been shown to influence amino acid transport. System A in hepatocytes, for example, is induced by glucagon [4] and provides amino acid substrates for gluconeogenesis. System Gly is sodium-dependent and specific for glycine; two sodium ions are transported for each glycine. Extrahepatic tissues such as the kidneys are also thought to transport amino acids by systems similar to those described for the intestinal basolateral membrane. However, an additional system, the γ-glutamyl cycle, is thought to be important in transporting amino acids through membranes of renal tubular cells, erythrocytes, and perhaps neurons.

In the γ-glutamyl cycle, glutathione acts as a carrier of selected neutral amino acids into cells. Glutathione, found in most cells of the body, is a tripeptide consisting of glycine, cysteine, and glutamate. As shown in Figure 7.9, an unusual peptide linkage occurs in glutathione between the γ-carboxyl group of glutamate and the α-amino group of cysteine. In the γ-glutamyl cycle (Fig. 7.10), glutathione in its reduced form (GSH) reacts with γ-glutamyl transpeptidase located in cell membranes to form a γ-glutamyl enzyme complex. The glutamate part of the glutathione molecule remains with the enzyme complex; cysteinylglycine is released into the cell cytoplasm and is eventually cleaved into its constituent amino acids by a cytosolic peptidase. The γ-glutamyl enzyme complex functions by binding to a neutral amino acid at the cell surface and carries it via a γ-carboxyl peptide linkage into the cell. Within the cell, γ-glutamyl cyclotransferase can cleave the peptide bond between the neutral amino acid and the γ-carbon of glutamate. Glutathione is resynthesized within the cell from cysteine, glutamate, and glycine in a series of energy-dependent reactions; the neutral amino acid that has just been released within the cell may function in the synthesis of new proteins or nitrogen-containing molecules or may be catabolized.

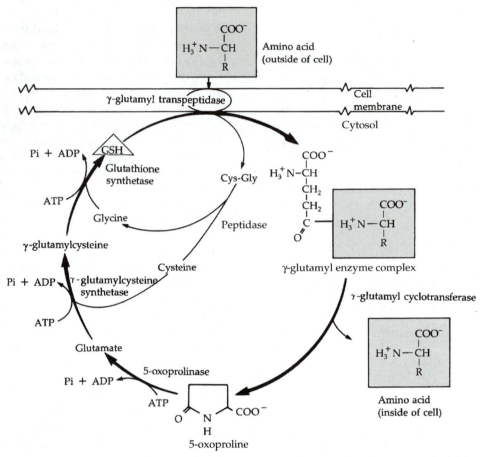

Figure 7.10 The γ-glutamyl cycle for transport of amino acids. Abbreviation: GSH, glutathione in reduced form.

Amino Acid Metabolism

The liver is the primary site for the uptake of most of the amino acids following ingestion of a meal. The liver is thought to monitor the absorbed amino acids and to adjust the rate of their metabolism according to the needs of the body. Typically, of the amino acids entering the liver, about 20% are used for the synthesis of proteins and nitrogen-containing compounds. Of this 20% of amino acids used for synthesis, 14% of what is synthesized will remain in the liver, and the rest (6%) will be released into the plasma. The majority, approximately 57%, of amino acids taken up by the liver are typically catabolized in the liver. The remaining (~23%) of the amino acids (primarily the branched-chain amino acids [BCAAs]: isoleucine, leucine, and valine) are released into systemic circulation. Each of these areas—the hepatic synthesis of proteins, nitrogen-containing compounds, and purine and pyrimidine bases, the catabolism of amino acids, and the metabolism of amino acids not taken up by the liver—will be addressed in this section.

Hepatic Synthesis of Plasma Proteins, Nitrogen-Containing Compounds, and Purine and Pyrimidine Bases

The liver cells, like other body cells, use the amino acids that they receive from portal blood and general circulation to synthesize protein [21]. Many of the proteins, such as enzymes, that are made in the liver will remain in the liver; other proteins, however, are released into the plasma. The concentration of total protein in human plasma typically ranges up to 7.5 g/dL. The proteins found in plasma consist of primarily glycoproteins but also include simple proteins and lipoproteins. These plasma proteins perform a variety of functions.

Plasma Proteins

Several of the proteins in the plasma are used to assess an individual's protein status, specifically visceral (internal organ) protein status. Albumin, the most abundant of the plasma proteins, is synthesized by the liver and released into the blood. Albumin is used quite extensively as an indicator of visceral protein status. Albumin functions in the plasma to maintain oncotic pressure as well as to transport nutrients such as vitamin B_6; minerals including zinc, calcium, and small amounts of copper; nutrients such as fatty acids; and the amino acid tryptophan. Some drugs and hormones such as the thyroid hormones are also transported by albumin. Because of albumin's relatively long half-life (~14–18 days), it is not as good or as sensitive an indicator of visceral protein status as some of the other plasma proteins. The half-life is the time that it takes for 50% of the amount of a protein such as albumin (or nonprotein compound) to be degraded.

Other proteins synthesized by the liver and released into plasma include transthyretin (formerly called *prealbumin*), retinol-binding protein (complexed together and involved with retinol and thyroid hormone transport), blood-clotting proteins, and globulins. There are several classes of globulins:

• α1-globulins include various glycoproteins and high-density lipoproteins (HDLs).

• α2-globulins include various glycoproteins, haptoglobin for free hemoglobin transport, ceruloplasmin for copper transport and oxidase activity, prothrombin for blood coagulation, and very low-density lipoproteins (VLDLs).

• β-globulins include transferrin for iron and other mineral transport and low-density lipoproteins (LDLs).

• γ-globulins include immunoglobulins or antibodies.

Transthyretin and retinol-binding protein, like albumin, are used as indicators of visceral protein status. However, because these two proteins have relatively shorter half-lives (~2 days and 12 hours, respectively) than albumin, they are more sensitive indicators of changes in visceral protein status than albumin.

Nitrogen-Containing Compounds

Nitrogen-containing compounds or molecules (Table 7.5), of which there are several, are also synthesized in the liver (often other sites, too) from amino acids.

• Glutathione is a tripeptide synthesized from three amino acids: glycine, cysteine, and glutamate. Glutathione is referred to as a thiol because it contains a sulfhydryl (-SH) group. It is found in its reduced form (GSH), as shown in Figure 7.9, in most cells of the body and has several functions. Glutathione transports amino acids as part of the γ-glutamyl cycle (Fig. 7.10). It participates in the synthesis of

Table 7.5 Sources of Nitrogen for Some Nitrogen-Containing Compounds

Nitrogen-Containing Compound	Constituent Amino Acids
Glutathione	Cys, gly, glu
Carnitine	Lys, met
Creatine	Arg, gly, met
Carnosine	His, β-ala
Choline	Ser

leukotriene (LT) LTC4, which mediates the body's response to inflammation. Glutathione, with the enzyme glutathione peroxidase, protects cells from toxic oxygen radicals. Specifically, glutathione reduces peroxidative damage by reacting with hydrogen peroxides (H_2O_2) and lipid hydroperoxides (LOOHs) as well as reactive oxygen species ($O_2\cdot$ and OH·). Glutathione thus acts as a scavenger of free radicals to protect cells and cell membranes. However, with inadequate protein intake, inflammation, or other pathological conditions, hepatic GSH concentrations, as well as mucosal and systemic GSH concentrations, reportedly decline [22]. This function of glutathione is discussed in further detail in a discussion of selenium and glutathione peroxidase in Chapter 12, page 443, as well as in the Perspective at the end of Chapter 10.

• Carnitine, another nitrogen-containing compound (Fig. 7.11), is made in the liver from lysine, which has been methylated using methyl groups derived from the amino acid methionine. Iron and vitamins B_6, C, and niacin participate in the synthesis of carnitine (shown in Chap. 9). Carnitine is needed for the transport of long-chain fatty acids across the inner mitochondrial membrane for oxidation. This role of carnitine is discussed in more detail on page 147. In muscle, carnitine also may serve as a buffer for free coenzyme (Co) A and may be involved in branched-chain amino acid metabolism. Carnitine is also thought to be involved with immune system function [23]. Carnitine is found, in addition to synthesis in the liver and kidney, in foods such as beef and pork. In these foods, carnitine may be free or bound to long- or short-chain fatty acid esters such as acetylcarnitine. Carnitine from food is absorbed in the proximal small intestine by sodium-dependent active transport and possibly also by passive diffusion. Approximately 54% to 87% of carnitine intake is absorbed. Muscle represents the primary carnitine pool, although no carnitine is made in muscle.

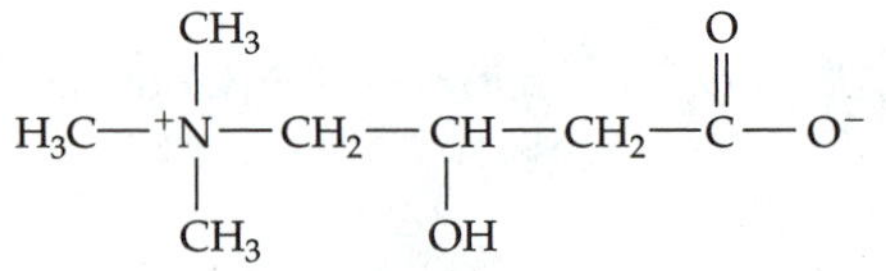

Figure 7.11 Carnitine.

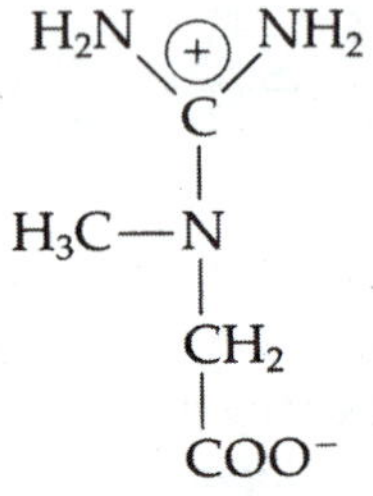

Figure 7.12 Creatine.

Carnitine homeostasis is maintained principally by the kidney, with >84% of filtered carnitine being reabsorbed. Carnitine deficiency results in impaired energy metabolism [23]; however, carnitine deficiency is quite rare. Advertisements marketing carnitine supplements to help burn fat or give one energy are making false claims. Furthermore, although use of carnitine supplements has been shown to increase plasma and muscle carnitine, studies have not uniformly shown improved physical performances [24–27].

• Creatine (Fig. 7.12), a key component of the energy compound creatine phosphate, also called *phosphocreatine,* can be obtained from foods (primarily meat and fish) or can be synthesized in the body. The first step in the synthesis of creatine is in the kidney where arginine and glycine react to form guanidoacetate. In this reaction, the guanidinium group of arginine is transferred to glycine; the remainder of the arginine molecule is released as ornithine. The next step in the synthesis of creatine is the methylation of guanidoacetate. This step occurs in the liver using SAM (S-adenosyl methionine), which is made from the amino acid methionine as a methyl donor (see also Fig. 7.26). Once synthesized, creatine is released into the blood for transport to tissues. About 95% of body creatine is in muscle, with the remaining 5% in organs such as the kidneys and brain. In tissues, creatine is found both in free form as creatine and in its phosphorylated form. The phosphorylation of creatine to form phosphocreatine is shown here:

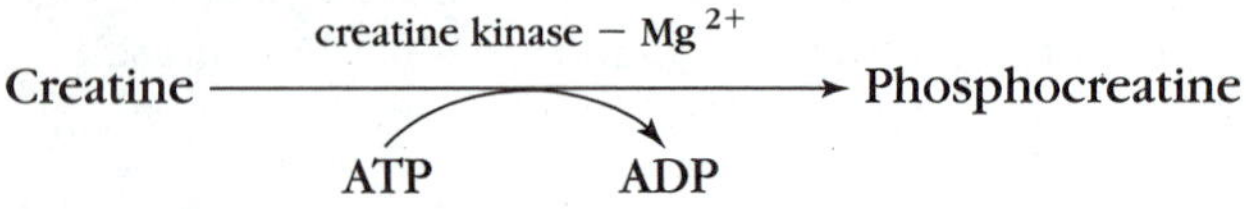

$$\text{Creatine} \xrightarrow[\text{ATP} \quad \text{ADP}]{\text{creatine kinase} - \text{Mg}^{2+}} \text{Phosphocreatine}$$

The function of phosphocreatine is as a "storehouse for high-energy phosphate." In other words, phosphocreatine replenishes ATP in a muscle that is rapidly contracting. Remember, muscle contraction requires energy. This energy is obtained with the hydrolysis of ATP. However, ATP in muscle can suffice for only a fraction of a second. Phosphocreatine, stored in the muscle and possessing a higher phosphate group transfer potential than ATP, can transfer a phosphoryl group to ADP, thereby forming ATP and providing energy for muscular activity. In fact, over one half of creatine in muscle at rest is as phosphocreatine. Phosphocreatine assists in ATP regeneration by the action of creatine kinase, which catalyzes the following reaction in an active muscle:

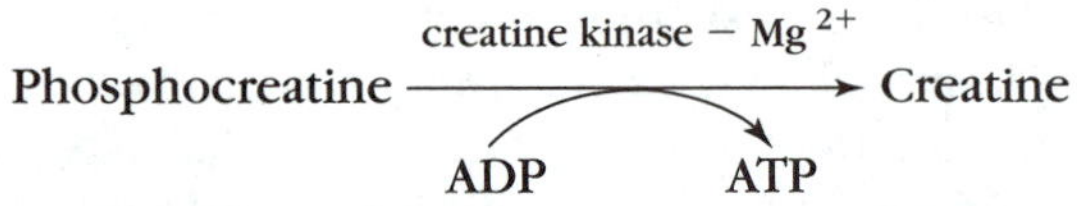

$$\text{Phosphocreatine} \xrightarrow[\text{ADP} \quad \text{ATP}]{\text{creatine kinase} - \text{Mg}^{2+}} \text{Creatine}$$

The availability and use of phosphocreatine by muscle are also thought to delay the breakdown of muscle glycogen stores, which upon further catabolism, also can be used for energy by muscle. Phosphocreatine may also act as an energy transporter.

Creatine supplements (as creatine monohydrate) have been shown in some studies to increase (~20%–50%) muscle total creatine concentrations and to increase the amount of short-duration maximal exercise that can be performed [28–31]. Typical dosages were 5 g creatine monohydrate taken four times per day for a total of 20 g/day; supplements were consumed generally for 5 or 6 days. Ingestion of a carbohydrate solution with creatine supplements increased muscle creatine accumulation greater than ingestion of creatine alone [32]. Some short-term positive effects of creatine included reduced decline in peak muscle torque production during repeat bouts of high-intensity isokinetic contractions, higher peak isokinetic torque production sustained during repeat bouts of maximal voluntary contraction, and increased whole-body exercise performance during two initial bouts of maximal isokinetic cycling lasting 30 seconds [29,31,33]. Other studies, for example in endurance athletes and highly trained swimmers, however, have reported no effects on performance [30,34,35]. Furthermore, side effects associated with long-term use of creatine are unknown.

• Carnosine (Fig. 7.13), derived from histidine and β-alanine, functions in nerve transmission and possibly as a neurotransmitter and activator of myosin ATPase.

• Choline (Fig. 7.14) is made in the body from methylation of the amino acid serine. Choline is found as a component of the neurotransmitter acetylcholine, and the phospholipid phosphatidyl choline, commonly called *lecithin.* Like betaine, choline also functions as a

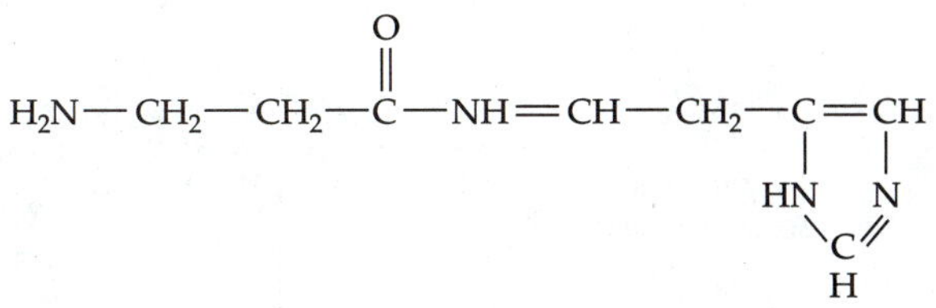

Carnosine

Figure 7.13 Carnosine.

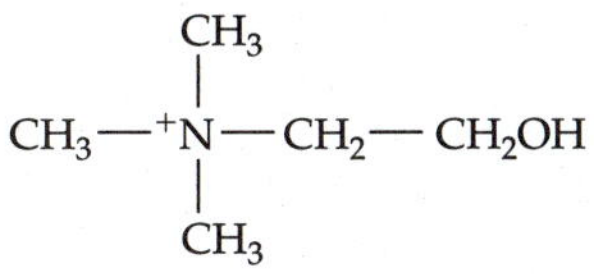

Choline

Figure 7.14 Choline.

methyl donor in the body. The Food and Nutrition Board, Institute of Medicine, National Academy of Sciences, has suggested that an adequate intake of 425 mg and 550 mg choline for adult females and males, respectively; such dietary intake levels are based on observed or experimentally determined approximations of nutrient intake by a group of healthy people and may be used as goals for individual intake [36]. Such intakes are easily obtained through dietary consumption of animal products and foods containing fats. A tolerable upper intake level of 3.5 g daily for choline also has been set [36]. The tolerable upper intake level represents the highest level of daily intake that is likely to pose no risks of adverse health effects to most individuals in the general population [36]. Choline is discussed in more detail in Chapter 16 on "Nutrition and the Central Nervous System."

Purine and Pyrimidine Bases

Another group of compounds derived in part from amino acids are purine and pyrimidine bases. A brief review of purine and pyrimidine synthesis and degradation follows.

The synthesis of the nitrogen-containing bases of nucleotides occurs for the most part de novo in the liver. Synthesis of the pyrimidines uracil, cytosine, and thymine is initiated by the formation of carbamoyl phosphate from glutamine, CO_2, and ATP. The enzyme carbamoyl phosphate synthetase II catalyzes this reaction in the cytoplasm and is distinct from carbamoyl phosphate synthetase I, which is needed in the initial step of urea synthesis and is found in the mitochondria. Carbamoyl phosphate reacts with the amino acid aspartate to form N-carbamoylaspartate. Aspartate trans-

carbamoylase catalyzes the reaction, which is the committed step in pyrimidine biosynthesis. Synthesis of the pyrimidine bases is illustrated in Figure 7.15. Once uridine monophosphate (UMP) is formed, UMP may react with other nucleoside di- and triphosphates. The formation of d-thymidine monophosphate (dTMP) from deoxyuridine monophosphate (dUMP) should be noted particularly. ATP can phosphorylate dTMP to form dTTP. In the synthesis of DNA, dTTP (deoxythymidine triphosphate) replaces UTP, uridine triphosphate.

Purines, adenine and guanine, are synthesized de novo as nucleoside monophosphates by sequential addition of carbons and nitrogens to ribose-P that has originated from the hexose monophosphate shunt. As shown in Figure 7.16a, ribose 5-P reacts with ATP to form phosphoribosyl pyrophosphate (PRPP). Glutamine then donates a nitrogen to form 5-phosphoribosylamine. This step represents the committed step in purine nucleotide synthesis. Next in a series of reactions, nitrogen and carbon atoms from glycine are added, formylation occurs by methenyl tetrahydrofolate (THF, a form of the B vitamin folate), another nitrogen atom is donated by the amide group of glutamine, and ring closure occurs. Another set of reactions occurs involving the addition of carbons from CO_2 and from N^{10} formyl THF (from folate) and a nitrogen from aspartate. Thus, the purine ring as shown in Figure 7.16b is derived from amino acids, including glutamine, glycine, and aspartate, as well as from folate and CO_2.

The formation of individual purine bases and nucleotides is shown in Figure 7.16c. Inosine monophosphate (IMP) is used to synthesize adenosine monophosphate (AMP) and guanosine monophosphate (GMP). AMP and GMP are phosphorylated to ADP and GDP, respectively, by ATP. The deoxyribotides are formed at the diphosphate level by converting ribose to deoxyribose, thereby producing dADP and dGDP. ADP can be phosphorylated to ATP via oxidative phosphorylation; the remaining nucleotides are phosphorylated to their triphosphate form by ATP. Purine nucleotides can also be synthesized by the salvage pathways, whereby purine bases react with PRPP to form the mononucleotides.

Degradation of pyrimidines involves the sequential hydrolysis of the nucleoside triphosphates to mononucleotides, nucleosides, and, finally, free bases. This process can be accomplished in most cells by lysosomal enzymes. During catabolism of pyrimidines, the ring is opened with the production of CO_2 and ammonia from the carbamoyl portion of the molecule. The ammonia can be converted to urea and excreted. Malonyl CoA and methylmalonyl CoA, produced from the remainder of the ring, follow their normal metabolic pathways, therefore requiring no special excretion route.

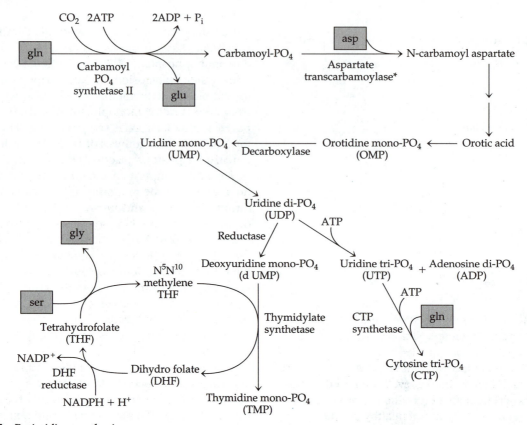

Figure 7.15 Pyrimidine synthesis.
*Aspartate transcarbamoylase catalyzes the committed step in pyrimidine biosynthesis.

Purines are progressively oxidized primarily in the liver, yielding hypoxanthine, xanthine, and uric acid. Xanthine dehydrogenase or oxidase, molybdenum- and iron-dependent enzymes, catalyze these two reactions shown in Figure 7.17. Uric acid normally is excreted in the urine; however, up to 200 mg of uric acid may be excreted daily into the digestive tract. Under pathological conditions such as gout or renal failure, the uric acid may accumulate in the blood and deposit in and around joints. Furthermore, under conditions of oxygen deprivation, as with a myocardial infarction or intestinal ischemia, xanthine dehydrogenase may be converted to xanthine oxidase. Reperfusion of the tissue with oxygen can result in increased hydrogen peroxide and free radical concentrations. Hydrogen peroxide and free radicals may further damage the injured tissues. Research involving introduction of enzymes and antioxidant nutrients to help minimize tissue damage with reoxygenation is ongoing.

An overview of amino acid use for anabolism is shown in Figure 7.18. A summary of the use of selected amino acids for the synthesis of nitrogen containing compounds, and selected hormones and neuromodulators is depicted in Figure 7.19.

Amino Acid Catabolism

Liver cells have a high capacity for the uptake and catabolism of all amino acids. Over half (~57%) of amino acids taken up by the liver are catabolized. Moreover, the liver is the main site for the catabolism of indispensable amino acids with the exception of the branched-chain amino acids. The rate of hepatic catabolism for the amino acids, however, differs. Branched-chain amino acids, for example, are catabolized much slower in the liver than in muscle. Furthermore, not all amino acids are catabolized in the same regions of the liver. Periportal hepatocytes, for example, catabolize all amino acids with the exception of glutamate and aspartate that are metabolized by perivenous hepatocytes. This section on amino acid catabolism will first focus on the reactions that occur as amino acids are broken down in liver cells including first the transamination and deamination of amino acids and the urea cycle; next the uses of the carbon skeleton of amino acids.

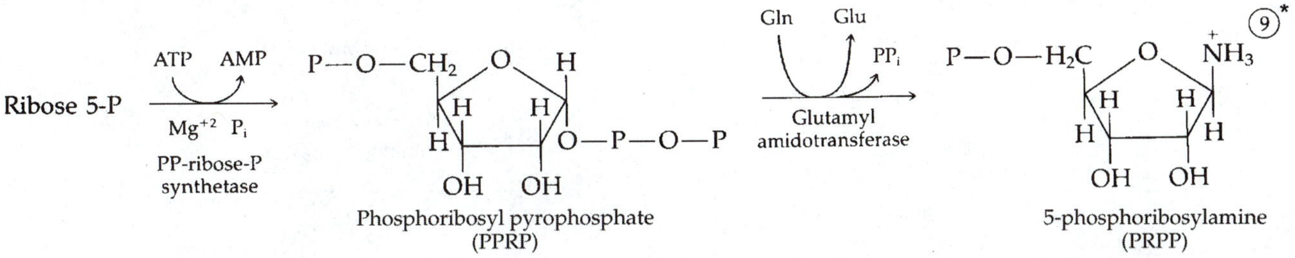

a. Biosynthesis is initiated by reaction between PPRP and glutamine (Gln) as shown above.

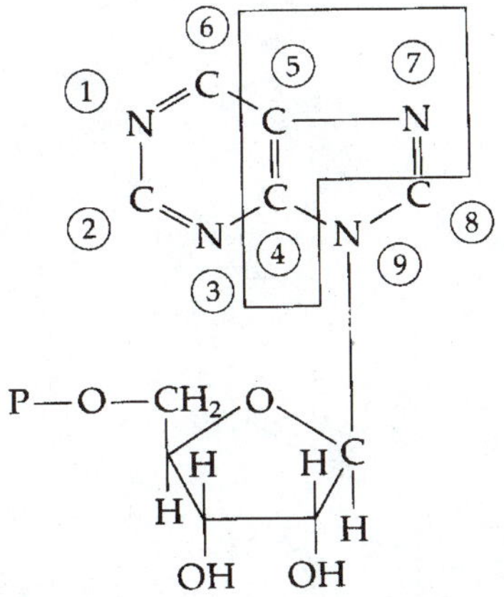

*Nitrogen ⑨ is donated by amide group from *glutamine*. Donors of other atoms in order of introduction:

a. *Glycine*: N ⑦ and C ④ and ⑤ introduced as single unit.

b. *Formate*: C ⑧ donated via N^{10} formyl THF (tetrahydrofolic acid)

c. *Glutamine* amide group: N ③

d. *Respiratory CO_2* as "active CO_2": C ⑥

e. *Aspartic acid*: N ①

f. *Formate*: C ② via N^{10} formyl THF

b. Sources of carbon and nitrogen atoms in purine ring.

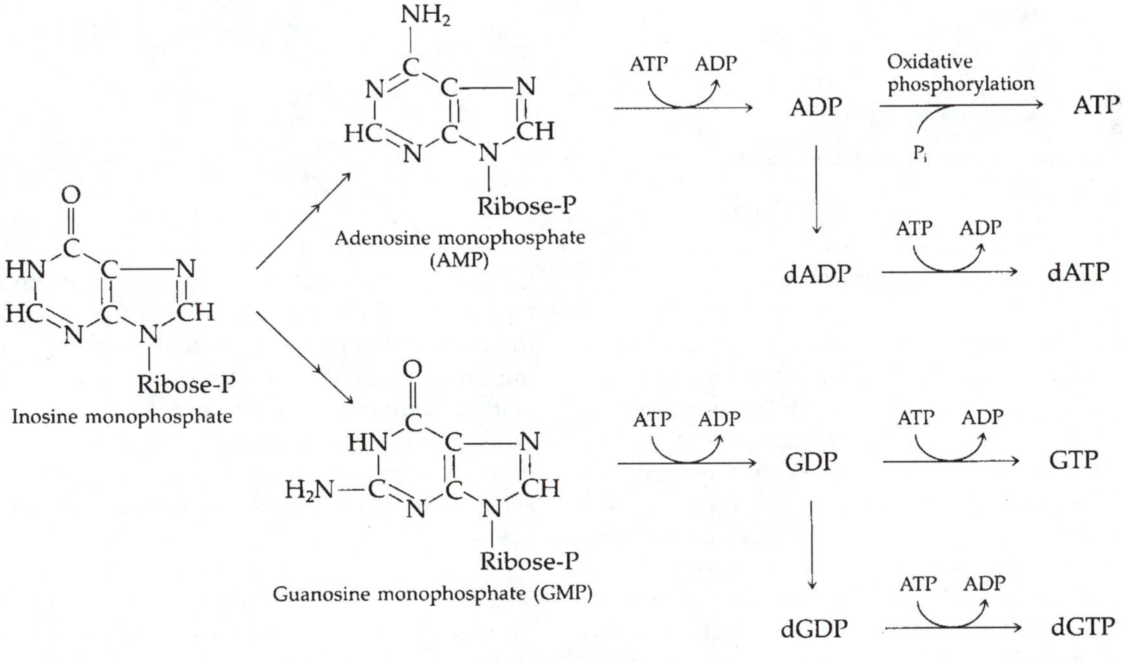

c. Formation of individual purine bases and nucleotides.

Figure 7.16 Synthesis of purines and sources of carbon and nitrogen atoms.

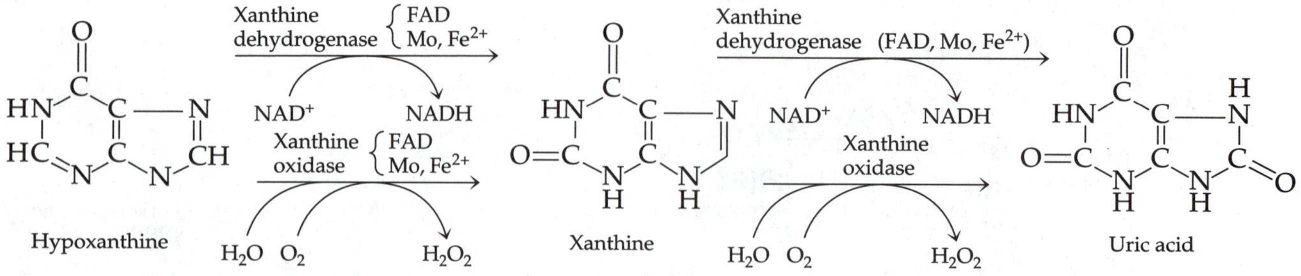

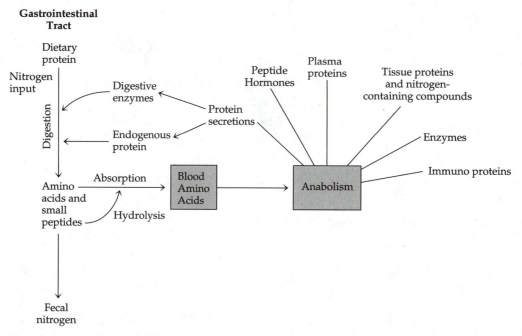

Figure 7.17 Purine catabolism yields uric acid.

Figure 7.18 Use of amino acids for anabolism.

Transamination and Deamination

Usually, the first step in the metabolism of amino acids not used for the synthesis of proteins or nitrogen-containing compounds is the removal of the amino group from the amino acid. Amino acids can undergo either transamination or deamination to remove amino groups. Transamination reactions involve the transfer of an amino group from one amino acid to an amino acid carbon skeleton or α-keto acid (an amino acid without an amino group). The carbon skeleton/α-keto acid that gains the amino group becomes an amino acid, and the amino acid that loses its amino group becomes an α-keto acid. In the transamination reaction shown in the bottom of Figure 7.20, the amino group from the amino acid aspartate is being transferred to α-ketoglutarate, an α-keto acid, to form the amino acid

glutamate. The removal of the amino group from aspartate generates oxaloacetate, another α-keto acid. In the top portion of Figure 7.20, the amino group is being taken from alanine and given to another α-keto acid to generate a new amino acid and pyruvate. Transamination reactions typically require vitamin B_6 in its coenzyme form, pyridoxal phosphate (PLP), and are important for the synthesis of dispensable amino acids.

Deamination reactions involve only the removal of an amino group with no direct transfer to another compound. Figure 7.21 shows the deamination of the amino acid threonine by threonine dehydratase to form α-ketobutyrate (another α-keto acid) and ammonium. Ammonia is readily used by periportal hepatocytes for urea synthesis. The synthesis of urea in the liver is addressed in the next subsection, "The Urea Cycle."

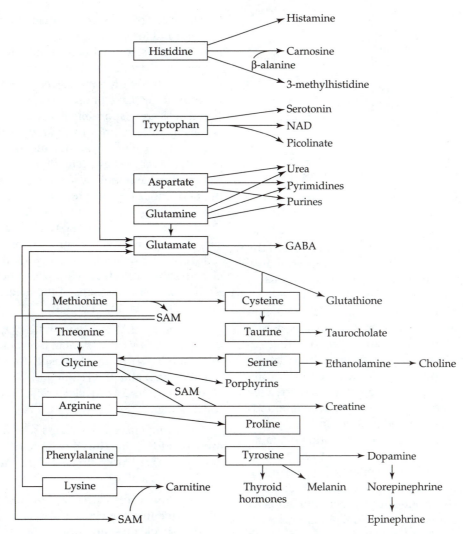

Figure 7.19 A summary of the use of selected amino acids for the synthesis of nitrogen-containing compounds and selected hormones and neuromodulators.

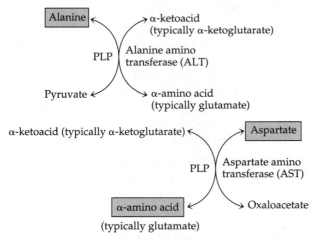

Figure 7.20 Transamination reactions. Transfer of an amino group from alanine (top) and aspartate (bottom) to form other amino acids from their respective α-keto acids.

Many transamination reactions are catalyzed by enzymes called aminotransferases. The most active aminotransferases in cells are alanine aminotransferase (ALT) (formerly called *glutamate pyruvate transaminase* and abbreviated GPT) and aspartate aminotransferase (AST) (formerly called *glutamate oxaloacetate transaminase* and abbreviated GOT). These aminotransferases, shown in Figure 7.20, involve three key amino acids: alanine, glutamate, and aspartate; require PLP for activity; and are found in varying concentrations in different tissues. For example, AST is found in higher concentrations in the heart than the liver, muscle, and other tissues. In contrast, ALT is found in higher concentrations in the liver than the heart but is also found in moderate amounts in the kidney and small amounts in other tissues. Normal serum concentrations of these enzymes are low; however, with trauma or disease to an organ, serum enzyme concentrations rise and serve as

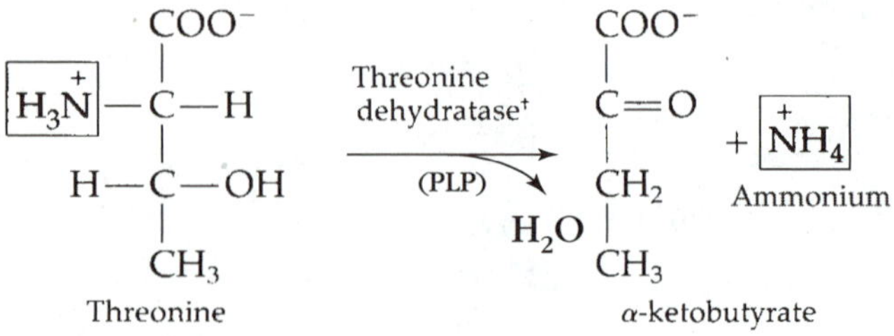

Deamination reaction. †The enzyme is called *dehydratase* rather than *deaminase* because the reaction proceeds by loss of elements of water.

Figure 7.21 The deamination of the amino acid threonine.

an indicator of which organ has been damaged and the severity of the organ damage. Reactions catalyzed by ALT and AST are shown in Figure 7.20. ALT transfers amino groups from alanine to an α-keto acid (e.g., α-ketoglutarate) forming pyruvate and another amino acid (e.g., glutamate), respectively. AST transfers amino groups from aspartate also to an α-keto acid (e.g., α-ketoglutarate), yielding oxaloacetate and another amino acid (glutamate), respectively. These reactions are reversible. Because glutamate and α-ketoglutarate readily transfer and/or accept amino groups, these compounds play central roles in amino acid metabolism.

In summary, the first step in the utilization of amino acids, for functions other than the direct synthesis of proteins, requires either transamination or deamination. Transamination reactions can generate dispensable amino acids from indispensable amino acids or create one dispensable amino acid from another dispensable amino acid. Remember the only exception to this is lysine and threonine, which do not participate in such reactions. Ammonia generated from oxidative deamination reactions must be safely removed from the system; this is accomplished by the actions of the urea cycle, which is discussed next.

The Urea Cycle

The urea cycle, which is found in the liver, is important for the removal of ammonia from the body. Too much ammonia in the body is toxic and can lead to brain malfunction and coma. Some of the sources of ammonia in the body include

- ammonia formed in the body from chemical reactions such as deamination,

- ammonia ingested and absorbed from the foods we eat, and

- ammonia generated in the gastrointestinal tract from bacterial lysis of urea and amino acids and subsequently absorbed through the enterocyte.

The liver has two systems in place to deal with ammonia. First and foremost, periportal hepatocytes are active in ureagenesis. Ammonia from the diet or from intestinal bacterial synthesis enters portal blood and comes first in contact with hepatocytes, specifically periportal hepatocytes capable of urea synthesis. These same periportal cells are responsible for almost all amino acid catabolism, so ammonia generated during amino acid degradative reactions can be immediately taken up for urea synthesis. However, should the periportal cells fail to use all the ammonia, a second group of hepatocytes, the perivenous hepatocytes, are capable of utilizing the ammonia for glutamine synthesis. The perivenous cells thus provide a "back-up" system for ammonia that escaped involvement in urea production [37].

Figure 7.22 reviews key compounds of the urea cycle and shows its relationship with amino acids and the Krebs cycle. The reactions of the urea cycle are broken down in the text that follows:

- Ammonia (NH_3) combines with CO_2 or HCO_3^- to form *carbamoyl phosphate* in a reaction catalyzed by mitochondrial carbamoyl phosphate synthetase I (CPS-I) and using 2 mol of ATP and Mg^{2+}. N-acetylglutamate (NAG) is required as an allosteric activator to allow ATP binding.

- *Carbamoyl phosphate* reacts with *ornithine* in the mitochondria, using the enzyme ornithine transcarbamoylase (OTC) to form *citrulline*. Citrulline in turn inhibits OTC activity.

- *Aspartate* reacts with *citrulline* once it has been transported into the cytosol. This step, catalyzed by argininosuccinate synthetase, is the rate-limiting step of the cycle. ATP (two high-energy bonds) and Mg^{2+} are required for the reaction, and *argininosuccinate* is formed. Argininosuccinate, arginine, and AMP + PP$_i$ inhibit the enzyme.

- *Argininosuccinate* is cleaved by argininosuccinase in the cytosol to form *fumarate* and *arginine*. Both fumarate and arginine can inhibit argininosuccinase activity. Argininosuccinase is found in a variety of tissues throughout the body, especially the liver and kidney. High concentrations of arginine increase the synthesis of N-acetylglutamate (NAG), which is needed for the synthesis of carbamoyl phosphate in the mitochondria.

- *Urea* is formed and *ornithine* is re-formed from the cleavage of *arginine* by arginase, a manganese-requiring enzyme. Arginase activity is inhibited by both ornithine and lysine and may become rate limiting under conditions that limit manganese availability or that alter its affinity for manganese [38].

Overall, the urea cycle uses four high-energy bonds. Oxidations in the Krebs cycle coupled with phos-

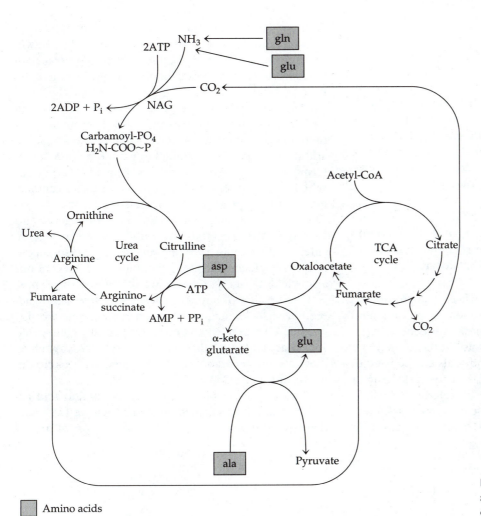

Figure 7.22 Interrelationships of amino acids and the urea and TCA cycles in the liver.

Amino acids

phorylation through the electron transport chain can provide the ATP required for urea synthesis. The urea molecule derives one nitrogen from ammonia, a second nitrogen from aspartate, and its carbon from CO_2/HCO_3^-. Once formed, urea typically travels via the blood to the kidneys for excretion in the urine; however, up to about 25% of urea may be secreted from the blood into the intestinal lumen, where it may be degraded by bacteria to yield ammonia.

Activities of urea cycle enzymes fluctuate with diet and hormone concentrations. For example, with low-protein diets or acidosis, urea synthesis (the amount of mRNA for each of the enzymes) diminishes and urinary urea nitrogen excretion decreases significantly. In the healthy individual with a normal protein intake, blood urea nitrogen (BUN) concentrations range from 8 to 20 mg/dL, and urinary urea nitrogen represents about 80% of total urinary nitrogen. Glucocorticoids and glucagon typically increase mRNA for the urea cycle enzymes [38]. Several urea cycle enzyme-deficient disorders have been characterized. Defects in any one of the en-

zymes of the urea cycle are possible. Urea cycle enzyme defects typically result in high levels of blood ammonia (hyperammonemia) and necessitate a nitrogen-restricted diet, which may be coupled with supplements of carnitine or single amino acids, among other compounds.

The Carbon Skeleton or α-Keto Acid

Once an amino group has been removed from an amino acid, the remaining molecule is referred to as a *carbon skeleton* or *α-keto acid*.

Amino acid $\longrightarrow NH_3$ + Carbon skeleton/α-keto acid

Carbon skeletons of amino acids can be further metabolized with the potential for multiple uses in the cell. An amino acid's carbon skeleton, for example, can be used for the production of

- energy,
- glucose,

- ketone bodies,
- cholesterol, and
- fatty acids.

The potential use of the carbon skeleton depends in part on the original amino acid from which it was derived. Whereas all amino acids can be completely oxidized to generate energy, not all amino acids can be used for synthesis of glucose. Furthermore, the fate of the amino acid's carbon skeleton depends on the physiological nutritional state of the body.

Energy Generation   The complete oxidation of amino acids generates energy, CO_2/HCO_3^- and NH_4^+. Amino acids are used for energy in the body when diets are inadequate in energy (measured in kilocalories).

Glucose and Ketone Body Production   The production of glucose from a noncarbohydrate source such as amino acids is known as gluconeogenesis. Gluconeogenesis occurs primarily in the liver but also in the kidney. The carbon skeletons of several amino acids can be used to synthesize glucose. Oxaloacetate, the carbon skeleton of aspartate, and pyruvate, the carbon skeleton of alanine, may be utilized to produce glucose in body cells through the process of gluconeogenesis, also discussed in

Chapter 3. In addition, the carbon skeleton of asparagine can be converted into oxaloacetate, and the carbon skeletons of glycine, serine, cysteine, and tryptophan can be converted into pyruvate for glucose production in the liver.

Figure 7.23 shows the general fates of amino acid carbon skeletons with respect to key intermediates of metabolism. Some amino acids, such as phenylalanine and tyrosine, can be degraded to form fumarate (an intermediate of the Krebs/tricarboxylic acid [TCA] cycle) which can be used to form glucose but also acetoacetate, which can be used to synthesize ketone bodies. Thus, these two amino acids are both glucogenic and ketogenic. Valine, methionine, and threonine are all considered glucogenic yielding succinyl CoA; isoleucine is partially glucogenic also generating succinyl CoA but also yielding acetyl CoA upon its catabolism. Thus, isoleucine, like phenylalanine and tyrosine, is considered partially ketogenic. Tryptophan and lysine are also considered partially ketogenic and partially glucogenic amino acids. Tryptophan yields acetyl CoA as well as pyruvate upon catabolism; lysine generates α-ketoglutarate and acetyl CoA.

Thus, to be considered a glucogenic amino acid, catabolism of the amino acid must yield selected intermediates of the Krebs cycle. The conversion of amino

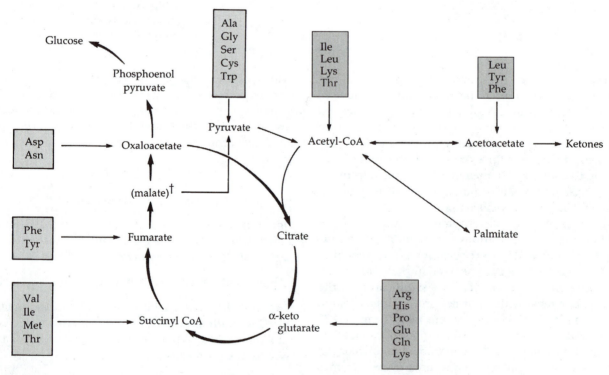

Figure 7.23   The fate of amino acid carbon skeletons. Ketogenic: Leu; partially ketogenic and glucogenic: Phe, Ile, Trp, Tyr, Lys; glucogenic: Ala, Gly, Cys, Ser, Thr, Asp, Asn, Glu, Gln, Arg, Met, Val, His, Pro.

acids to glucose is accelerated by high glucagon:insulin ratios and by glucocorticoids such as cortisol. Such hormones are elevated when people are not receiving sufficient energy or carbohydrate in the diet, in times of illness such as with infection or trauma, or in certain disease states such as untreated diabetes mellitus and liver disease, to name a few.

For an amino acid to be considered ketogenic, the catabolism of the amino acid must generate the non–Krebs cycle intermediates acetyl CoA or acetoacetate, which are used for the formation of ketone bodies. Amino acids are catabolized to generate ketone bodies generally during times when an individual is not consuming an adequate carbohydrate intake. Leucine is the only totally ketogenic amino acid, for its catabolism forms acetyl CoA and acetoacetate.

Cholesterol Production Leucine is also the only amino acid whose catabolism generates HMG CoA, an important intermediate in the synthesis of cholesterol. Although other amino acids generate acetyl CoA, which can be metabolized in the liver for cholesterol production [39].

Fatty Acid Production In times of excess energy and protein intakes coupled with adequate carbohydrate intake, the carbon skeleton of amino acids may be used to synthesize fatty acids. The details of the metabolism of selected amino acids and the formation of Krebs cycle and non–Krebs cycle intermediates will be shown in the sections to follow. The metabolism of the amino acids is categorized according to the structural classification of amino acids.

Hepatic Metabolism of Aromatic Amino Acids

The catabolism of aromatic, along with sulfur (S)-containing amino acids, occurs primarily in the liver. In fact, in end-stage liver disease, the inability of the liver to take up and catabolize these amino acids is evidenced by the increased plasma concentrations of both the aromatic amino acids—phenylalanine, tyrosine, and tryptophan—and the S-containing amino acids methionine and cysteine.

As shown in Figure 7.24, phenylalanine and tyrosine are partially glucogenic because they are degraded to fumarate. In addition, phenylalanine and tyrosine are catabolized to acetoactetate and thus are partially ketogenic. The first step in the degradation of phenylalanine is specific to the liver and, to a smaller extent, the kidney.

• Phenylalanine is converted to tyrosine by the hepatic enzyme phenylalanine hydroxylase or monooxygenase. This enzyme is iron-dependent, and vitamin C and tetrahydrobiopterin are required for the reaction. A genetic absence or deficient activity of phenylalanine hydroxylase results in the genetic disorder phenylketonuria (PKU) and necessitates a phenylalanine-restricted diet.

Further catabolism of tyrosine is not specific to the liver; however, the reactions occur primarily in the liver and yield many compounds significant to metabolism (Fig. 7.24).

• Tyrosine hydroxylase, an iron-dependent enzyme, catalyzes the first step in tyrosine metabolism to generate 3,4 dihydroxyphenylalanine (L-dopa).

• Subsequent reactions with L-dopa yield the catecholamines (dopamine, norepinephrine, and epinephrine), as well as in other cells, such as the skin, eye and hair cells, and melanin (a pigment that gives color to skin, eyes, and hair). In the thyroid gland, tyrosine is taken up and used with iodine to synthesize thyroid hormones.

Another aromatic amino acid principally metabolized by the liver is tryptophan. Its metabolism is shown in Figure 7.25. Tryptophan is partially glucogenic as it is metabolized to form pyruvate; it is also partially ketogenic and forms acetyl CoA.

• The first step in tryptophan metabolism yields N-formylkynurenine. The enzyme tryptophan oxygenase is iron-dependent.

• Further metabolism of N-formylkynurenine yields kynurenine and 3-hydroxykynurenine, which upon catabolism generates the amino acid alanine and 3-hydroxyanthranilate. Alanine formed from tryptophan metabolism can be transaminated to form pyruvate, hence the glucogenic nature of tryptophan. The 3-hydroxyanthranilate is oxidized to form 2-amino 3-carboxymuconic 6-semialdehyde. This compound is further metabolized to produce many more compounds, including picolinic acid (a possible binding ligand for minerals) and niacin as nicotinamide as well as its coenzyme form nicotinamide adenine dinucleotide (NAD) phosphate (NADP). When not catabolized or used for protein synthesis, tryptophan may be used for the synthesis of serotonin and melatonin, described later in the chapter.

Hepatic Metabolism of Sulfur (S)-Containing Amino Acids

The metabolism of methionine, an S-containing essential amino acid, occurs to a large extent in the liver and generates other S-containing nonessential amino acids. While methionine metabolism is important for the generation of cysteine and taurine, methionine also functions in the body as a donor of methyl groups. The

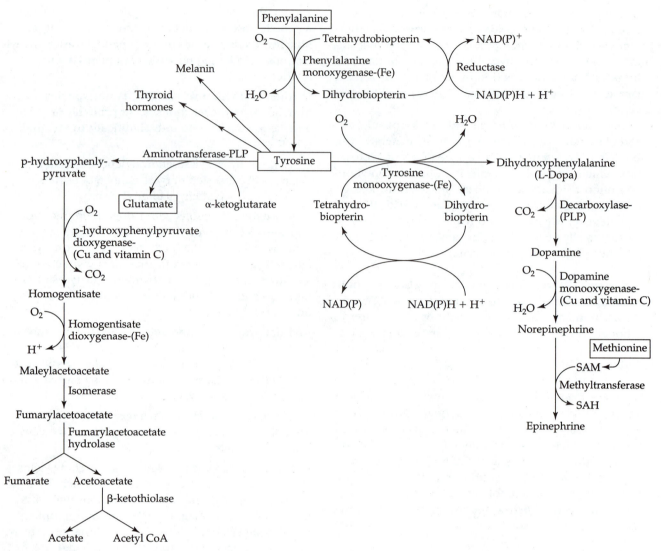

Figure 7.24 Phenylalanine and tyrosine metabolism.

enzyme required for this methyl donor role, methionine adenosyl transferase, is present in abundance in the liver. Methionine metabolism, shown in Figure 7.26, is briefly described.

The first step in methionine metabolism required for the use of methionine's methyl groups is the conversion of methionine to S-adenosyl methionine (SAM) by methionine adenosyl transferase in an ATP-requiring reaction. SAM is the principal methyl donor in the body and is required for the synthesis of carnitine, creatine, epinephrine, purines, and nicotinamide. The removal of or donation of the methyl group from SAM yields the compound S-adenosyl homocysteine (SAH). SAH can then be converted to homocysteine. Homocysteine can be converted back to methionine in a betaine-dependent reaction or a vitamin B_{12} (as methyl-cobalamin)

and folate (as 5-methyl THF)-dependent reaction. In this reaction (Fig. 7.26) methylcobalamin directly provides the methyl group to remethylate homocysteine to form methionine; however, methylcobalamin receives the methyl group from N^5 methyl-tetrahydrofolate (a coenzyme form of the B vitamin folate). Elevated levels of homocysteine in the blood have been found as a risk factor for heart disease. A discussion of the importance of adequate folate, vitamin B_{12}, and vitamin B_6 nutriture and heart disease is found in Chapter 9 in the sections on folic acid and vitamin B_{12}.

To be further metabolized in the body, homocysteine must react with the amino acid serine, and in a series of vitamin B_6 (PLP)–dependent reactions, homocysteine can be converted into α-ketobutyrate. The presence of the vitamin B_6 in its coenzyme is necessary

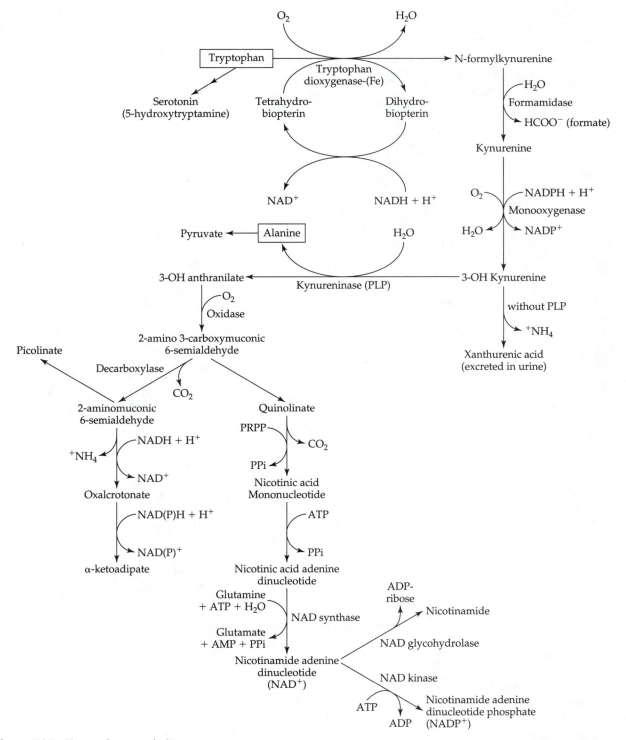

Figure 7.25 Tryptophan metabolism.

for these reactions to occur, hence the need for adequate vitamin B_6 status to prevent elevated blood homocysteine concentrations. α-ketobutyrate is further metabolized to propionyl CoA and subsequently through two more reactions to succinyl CoA. The reaction (shown in Fig. 7.26) generating α-ketobutyrate also yields the nonessential amino acid cysteine. Cysteine is required for the synthesis of both protein and glutathione. Cysteine may be further metabolized to form cystine and another amino acid, taurine. Taurine, a

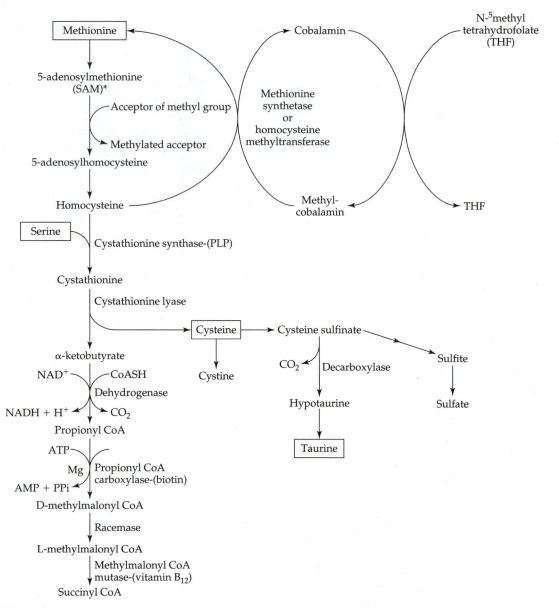

Figure 7.26 Methionine and cysteine metabolism.
*SAM is needed for the synthesis of carnitine, creatine, epinephrine, nicotinamide, and purines.

β-amino sulfonic acid, is concentrated in muscle and the central nervous system. While taurine is not involved in protein synthesis, it is important

- in the retina for vision,

- in membrane stability where it is a scavenger of peroxidative (e.g., oxychloride) products,

- as a bile salt taurocholate, and

- as an inhibitory neurotransmitter.

Genetic defects in methionine metabolism have been documented and include homocystinuria, due to de-fects in cystathionine synthase activity, and cystathio-ninuria due to cystathionase inactivity. Both defects necessitate a methionine-restricted diet.

Metabolism of Other Amino Acids

Several other reactions of amino acid metabolism are confined primarily to the liver. Lysine metabolism generates both acetyl CoA and glutamate, the latter transaminating to α-ketoglutarate (Fig. 7.27). Remember lysine is also used for the synthesis of carnitine. Threo-

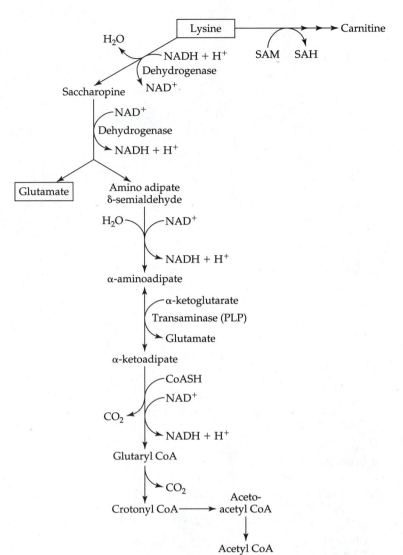

Figure 7.27 Lysine metabolism.

nine is metabolized to succinyl CoA but also generates glycine (Fig. 7.28). Glycine and serine are produced from one another in a reversible reaction requiring folate. Glycine, which is converted to serine mainly in the kidney (Fig. 7.28), is also needed for the synthesis of other important body compounds including creatine, heme (see Chap. 12), and the bile salt glycocholate. Serine is used for the synthesis of ethanolamine and choline for phospholipids. Threonine is found in fairly high quantities relative to other amino acids in mucus glycoproteins. Arginine and glutamate may be metabolized to form proline (Fig. 7.29); both arginine and proline degrade to form glutamate. Arginine is also used for creatine synthesis, as well as for the nitric oxide production in endothelial cells, cerebrellar neurons, and neutrophils. Nitric oxide is involved in the regulation of a variety of physiological processes including regulation

of blood pressure (relaxation of vascular smooth muscle) and intestinal motility, inhibition of platelet aggregation, and macrophage function, among others. Splanchnic arginine catabolism generates some NO_3 for urinary excretion [40]. Histidine degradation is also shown in Figure 7.29. Histidine may be catabolized to form glutamate or may combine with β-alanine to generate carnosine (a nitrogen-containing compound). Through a vitamin B_6–dependent decarboxylation reaction, the amine histamine also can be formed from histidine (Fig. 7.29). Histamine is found in neurons, in cells of the gastric mucosa, and in mast cells. Histamine release causes dilation of capillaries (flushing of the skin), constriction of bronchial smooth muscle, and increased gastric secretions. Figure 7.30 provides an overview of the fates of amino acids not used for the synthesis of body proteins.

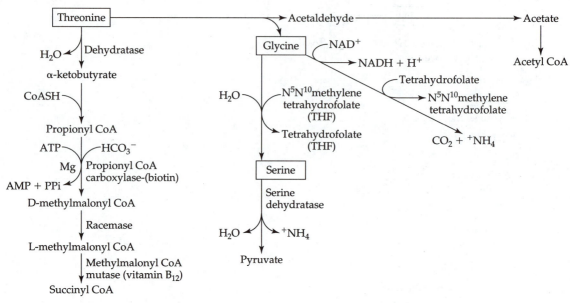

Figure 7.28 Threonine, glycine, and serine metabolism.

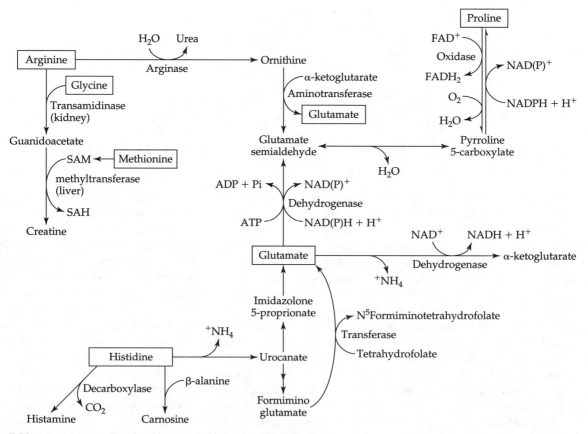

Figure 7.29 Arginine, proline, histidine, and glutamate metabolism.

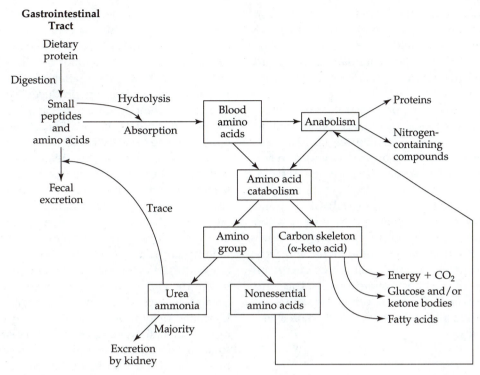

Figure 7.30 Possible fates of amino acids upon catabolism.

Amino Acids Not Taken Up by the Liver: Plasma Amino Acids and Amino Acid Pool(s)

The liver releases 23% of ingested amino acids without metabolism, and of this amount, about 70% of the free amino acids leaving the liver via the plasma are the branched-chain amino acids (BCAAs) [7]. The BCAAs are primarily taken up and transaminated by the skeletal muscles and, to a lesser extent, by tissues such as the heart, brain, and kidneys. Transferases (transaminases) needed in the first step of BCAA catabolism are limited in the liver. Hepatic transferases, however, increase in response to glucocorticoid (cortisol) release, which occurs in situations such as stress, trauma, burns, and sepsis. Insulin release, triggered primarily by a rise in blood glucose concentration that occurs following meals, may dramatically decrease plasma BCAA concentrations by promoting tissue, especially muscle, BCAA uptake. The metabolism of the branched-chain amino acids is detailed later in this chapter in the section "Skeletal Muscle."

Following protein-containing meals, plasma amino acid concentrations rise steadily for several hours, then return to basal concentrations. In basal situations or between meals, plasma amino acid concentrations are relatively stable and are species-specific; however, there is variability (large coefficients of variation) among people with respect to absolute concentrations of specific amino acids in the plasma.

Amino acids circulating in the plasma and found within cells arise from digestion and absorption of dietary protein as well as from the breakdown of existing body tissues. These amino acids of endogenous source intermingle with exogenous amino acids to form a "pool" totaling about 100 g in the plasma along with some smaller pools in various tissues of the body. Reuse of endogenous amino acids is thought to represent the primary source of amino acids needed for protein synthesis. Despite differences in protein intake and rate of degradation of tissue protein, the pattern of the amino acids in the free amino acid pool appears to remain relatively constant, although the pattern is quite different from that found in body proteins. The total amount of the essential amino acids found in the pool is less than that of the nonessential amino acids. The essential amino acids found in greatest concentration are lysine and threonine, both of which are totally indispensable. Of the nonessential amino acids, those found in greatest concentration are alanine, glutamate, aspartate, and glutamine. The nonessential amino acids may function to conserve the essential amino acids, with the exception of lysine and threonine, through the reamination of α-keto acids of the essential amino acid as discussed earlier in this chapter on page 184.

Amino acids within the pool, regardless of source, are metabolized in response to various stimuli such as hormones and physiological state. Amino acids, in excess of need for the synthesis of protein and/or nitrogen-containing compounds, are likely to be oxidized. Remember, if energy intake is inadequate, amino acids can be degraded to generate energy (ATP). Glucose and ketone body production from amino acids occur when carbohydrate intake is insufficient. Cholesterol synthesis can also occur from some amino acids, and fatty acids made be made from amino acids when energy intake is in excess of the body's energy needs. The ratio of glucagon to insulin as well as other hormones are important determinants of the fate of the carbon skeleton of amino acids.

Interorgan "Flow" of Amino Acids and Organ-Specific Metabolism

From the plasma, tissues extract amino acids for synthesis of nonessential amino acids, protein, nitrogen-containing compounds, glucose, fatty acids, ketones, or for energy production, depending on the person's nutritional status and hormonal environment. A brief review of the flow of amino acids between selected organs and organ-specific amino acid metabolism follows.

Glutamine and Alanine

Whereas ammonia arising in the liver from amino acid reactions can be readily shuttled into the urea cycle, this is not true in other tissues. In extrahepatic tissues, ammonia (NH_3) or ammonium ions (NH_4^+) generated in the cell from amino acid reactions generally com-

bine with the amino acid glutamate to form glutamine (Fig. 7.31). This reaction is catalyzed by glutamine synthetase and requires ATP and magnesium (Mg^{2+}) or manganese (Mn^{2+}). Glutamine synthesis occurs within all tissues including the brain and adipose [41], but especially large amounts are produced by the muscle and lungs. Glutamine functions to carry the generated ammonia safely out of the cell. Remember, too much free ammonia is toxic to cells!

Glutamine freely leaves tissues, and travels principally to the liver, kidney, and intestine but also to organs such as the pancreas. Whereas the cells of the gastrointestinal tract, as well as immune system (lymphocytes and macrophages) rely on glutamine catabolism for energy production, glutamine in the liver and kidney is catabolized by glutaminase, which removes the amide nitrogen to yield glutamate and ammonia (Fig. 7.32). The fate of the ammonia varies. *In the absorptive state (or during periods of alkalosis)*, liver glutaminase activity increases, yielding ammonia for the urea cycle. Remember, in the urea cycle, ammonia reacts with HCO_3^-/CO_2 to form carbamoyl phosphate, which is made approximately in proportion to the ammonia concentration. Should the periportal hepatocytes fail to capture the ammonia for ureagenesis, the ammonia is quickly taken up by perivenous hepatocytes for glutamine synthesis. *In an acidotic state,* the use of glutamine for the urea cycle diminishes, and the liver releases glutamine into the blood for transport to and uptake by the kidney. Glutamine is catabolized by renal tubular glutaminase to yield ammonium and glutamate. Glutamate may be further catabolized by glutamate dehydrogenase to yield α-ketoglutarate plus another ammonium (Fig. 7.32).

Ammonium concentrations are in equilibrium with cell ammonia and H^+. Ammonia, which is lipid soluble, may diffuse into the urine and react with H^+ to form ammonium for excretion. Renal glutaminase activity and ammonium excretion increase with acidosis and decrease with alkalosis. Thus, glutamine by virtue of its ubiquitous synthesis in cells and its ability to diffuse out of cells for transport to tissues is a major carrier of nitrogen between cells. Glutamine utilization increases dramatically *during hypercatabolic conditions* such as

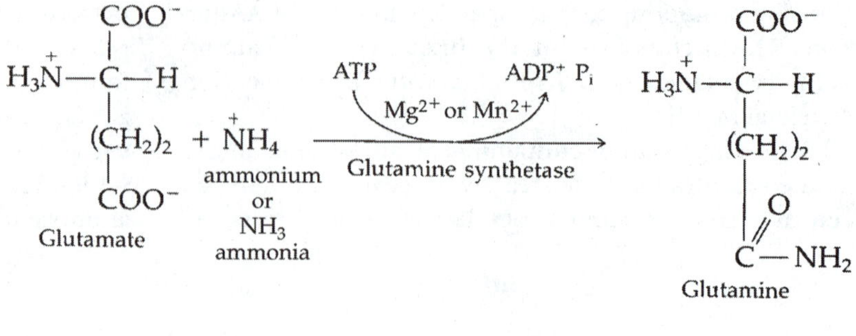

Figure 7.31 The synthesis of glutamine from glutamate and ammonium ion.

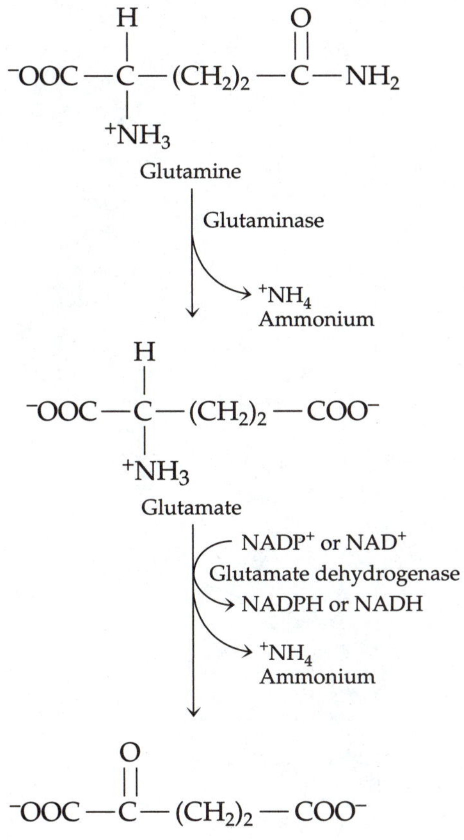

$$^-OOC - \overset{\overset{\displaystyle H}{|}}{\underset{\underset{\displaystyle {}^+NH_3}{|}}{C}} - (CH_2)_2 - \overset{\overset{\displaystyle O}{\|}}{C} - NH_2$$

Glutamine

Glutaminase

$\searrow$ $^+NH_4$
Ammonium

$$^-OOC - \overset{\overset{\displaystyle H}{|}}{\underset{\underset{\displaystyle {}^+NH_3}{|}}{C}} - (CH_2)_2 - COO^-$$

Glutamate

$\big($ NADP$^+$ or NAD$^+$
Glutamate dehydrogenase
$\rightarrow$ NADPH or NADH

$\searrow$ $^+NH_4$
Ammonium

$$^-OOC - \overset{\overset{\displaystyle O}{\|}}{C} - (CH_2)_2 - COO^-$$

α-ketoglutarate

Figure 7.32 Glutamine oxidation.

sepsis and trauma. In these conditions muscle gluta-mine release increases but cannot meet cellular de-mands. Thus, glutamine stores become depleted and cell functions become compromised [16].

In addition to glutamine, another amino acid, ala-nine, also is important in the intertissue (between tis-sue) transfer of amino groups generated from amino acid catabolism. For example, amino groups generated from branched-chain amino acid transamination in tis-sues such as the skeletal muscle can combine with α-ketoglutarate to form glutamate. In Figure 7.33, leucine is transaminated with α-ketoglutarate to form the α-keto acid α-ketoisocaproate and the amino acid gluta-mate. Glutamate may accept another amino group to form glutamine or transfer its amino group to pyruvate, generated from glucose metabolism, to form alanine (Fig. 7.33). From extrahepatic tissues, such as muscle, alanine may travel to the liver. Within the liver, alanine may undergo transamination. Alanine transamination with α-ketoglutarate produces glutamate. Glutamate may be deaminated to yield ammonia for the urea cy-cle, or it may be transaminated with oxaloacetate to form aspartate (Fig. 7.34). Aspartate is used in the syn-

thesis of pyrimidines and purines and is one of the amino acids directly involved in urea generation by the urea cycle. Alternately, within the liver, alanine can be converted to glucose. The glucose that is generated from the alanine is subsequently released into the blood and is available to be taken up again and used by muscle. Muscle cells use the glucose through glycolysis and generate pyruvate. The formed pyruvate is again available for transamination with glutamate. This cycle (Fig. 7.35), known by several names (including the ala-nine, glucose alanine, or alanine glucose cycle), serves to transport nitrogen to the liver for conversion to urea while also allowing the regeneration of needed sub-strates.

Skeletal Muscle

Uptake of amino acids by the skeletal muscles readily occurs following ingestion of a protein-containing meal, and during this time, skeletal muscles typically ex-perience a net protein synthesis. With respect to amino acid degradation, six amino acids (aspartate, as-paragine, glutamate, leucine, isoleucine, and valine) ap-pear to be catabolized in the skeletal muscle. The ca-tabolism of aspartate, asparagine, and glutamate have been presented earlier in the chapter. A brief review of branched-chain amino acid catabolism will follow and is outlined in Figure 7.36.

Muscle, as well as the heart, kidney, diaphragm, and other organs, possess BCAA transferases, located in both the cytosol and mitochondria. The transferases (transaminases) are needed for the transamination of the three BCAAs. Following transamination, the α-keto acids of the BCAA may remain within muscle for further oxidation or may be transported bound to albumin in the blood to other tissues for reamination or for further catabolism. The next step in BCAA catabolism is decar-boxylation (irreversible reaction) by the branched-chain α-keto acid dehydrogenase (BCKAD) complex, which is made up of three subunits: E1α, E1β, and E2. This enzyme complex is found in the mitochondria of many tissues, including liver, muscle, heart, kidney, intes-tine, and the brain. The enzyme complex is highly regu-lated through phosphorylation (inactivation) and de-phosphorylation (activation) mechanisms involving kinase and phosphatase proteins that act on the E1α subunit. This enzyme operates in a fashion similar to pyruvate dehydrogenase in that it requires thiamin in its coenzyme form TDP/TPP, niacin as NADH, and Mg^{2+} and CoA(SH) from pantothenic acid, and it is affected in some tissues by changes in dietary protein intake [42]. A genetic defect diminishing branched-chain α-keto acid dehydrogenase complex activity results in maple syrup urine disease (MSUD). MSUD necessitates a diet

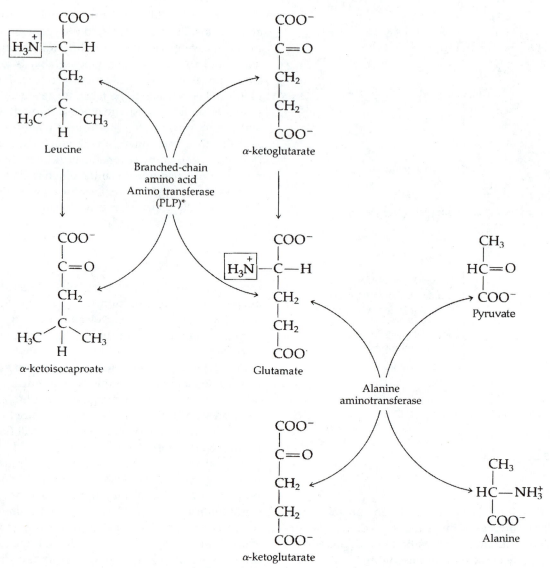

Figure 7.33 Alanine generation from leucine in body cells.

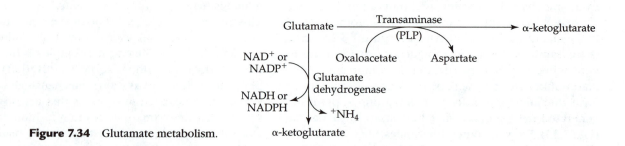

Figure 7.34 Glutamate metabolism.

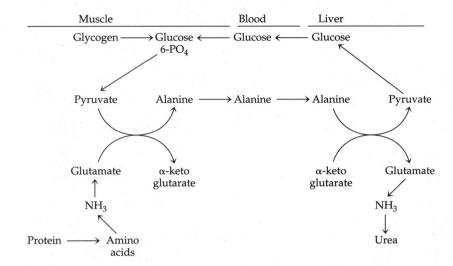

Figure 7.35 The alanine-glucose cycle: alanine generation in muscle, glucose generation in the liver.

restricted in leucine, isoleucine, and valine intakes. Complete oxidation of valine yields succinyl CoA while the end products of isoleucine catabolism are succinyl CoA and acetyl CoA. Complete oxidation of leucine results in acetyl CoA and acetoacetate formation.

Leucine is the only amino acid that is completely oxidized in the muscle for energy. Leucine, a totally ketogenic amino acid, is oxidized in a manner similar to fatty acids, and its oxidation results in the production of 1 mol of acetyl CoA and 1 mol of acetoacetate. During fasting, leucine rises to high levels in the blood and muscle, while the capacity of the muscle to degrade leucine increases concurrently. By supplying the muscle with the equivalent of 3 mol of acetyl CoA per molecule of leucine oxidized, the acetyl CoA produces energy for the muscle while simultaneously inhibiting pyruvate oxidation. Pyruvate is then used for the synthesis of lactate, which is released from the muscles. Thus, the oxidation of leucine spares essential gluconeogenic precursors. Pyruvate, along with lactate, can be returned to the liver, the former either being transported as pyruvate per se or (more likely) being converted to alanine for transport.

A great deal of interest in protein metabolism within the skeletal muscle was generated through the discovery that during starvation the amounts of the various amino acids released from the muscle could not reflect proteolysis alone. In particular, much more alanine and glutamine were appearing in the blood than could be attributed to muscle protein content. We now know that the alanine results from a series of transamination reactions and that glutamine results following amination of glutamate generated during a transamination reaction. Transamination of the BCAAs occurs primarily with α-ketoglutarate to form glutamate, which may either donate its amino group to pyruvate to form ala-

nine (previously shown in Fig. 7.33) or may incorporate free ammonia to form glutamine. The relative amounts of glutamine and alanine produced depend largely on the concentration of ammonia within the tissue; high ammonia concentrations promote glutamine synthesis and decrease alanine production.

Other amino acids released in lesser quantities from muscle (forearm and/or leg) in a postabsorptive state include phenylalanine, methionine, lysine, arginine, histidine, tyrosine, proline, tryptophan, threonine, and glycine [43,44]. Further studies investigating the effects of meals containing all three energy nutrients on amino acid uptake and output by muscle are needed.

Creatine, a nitrogen-containing compound that is made in the kidney and liver from the amino acids arginine and glycine with methyl groups donated from the amino acid methionine, functions in skeletal muscle as an energy source. Remember from earlier in this chapter, creatine is phosphorylated in muscle by ATP to form phosphocreatine (also called *creatine phosphate*), which can replenish ATP in muscles that are actively contracting (as with exercise). Phosphocreatine works in muscle by reacting with ADP generated in the muscle from the hydrolysis of ATP. When phosphocreatine reacts with ADP, creatine and ATP are formed. Creatine kinase catalyzes this important reaction that enables the generation of ATP for muscle contraction. Creatine and creatine phosphate, however, do not remain indefinitely in muscle; both slowly but spontaneously cyclize (Fig. 7.37) because of nonreversible, nonenzymatic dehydration. This cyclization of creatine and phosphocreatine forms creatinine. Once formed, creatinine leaves the muscle, passes across the glomerulus of the kidney, and is excreted in the urine. Small amounts of creatinine may be excreted into the gut and, like urea, metabolized by bacterial flora in the

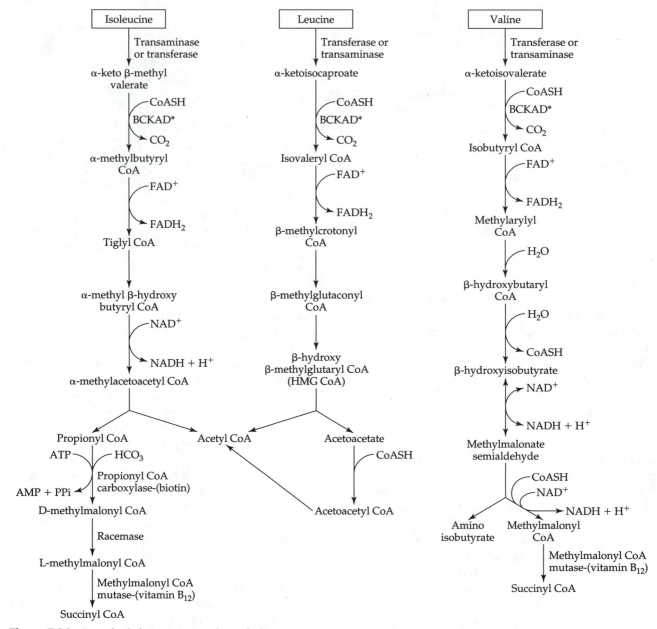

Figure 7.36 Branched-chain amino acid metabolism.
*BCKAD: branched-chain α-keto acid dehydrogenase, requiring thiamin as TPP, niacin as NADH, Mg^{2+} and CoA from pantothenate.

intestine. Creatinine clearance is frequently used as a means of estimating kidney function.

Nitrogen-Containing Compounds as Indicators of Muscle Mass and Muscle Catabolism

Urinary excretion of creatinine and 3-methylhistidine are used as indicators of the amount of existing muscle mass and the rate of muscle degradation, respectively. Urinary creatinine excretion is considered to be a reflection of muscle mass because it is the degradation product of creatine, which makes up approximately 0.3% to 0.5% of muscle mass by weight. The creatinine excreted in the urine reflects about 1.7% of the total creatine pool per day. However, urinary creatinine excretion is not considered to be a completely accurate

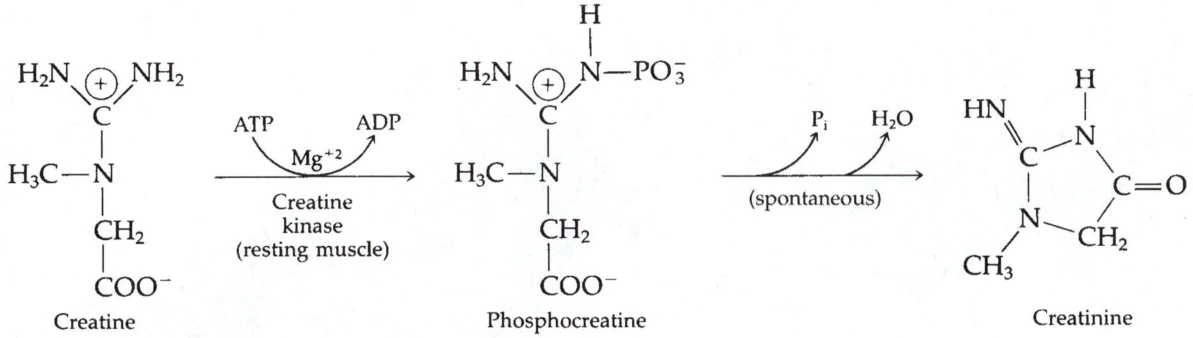

Figure 7.37 Conversion of creatine to phosphocreatine in the muscle and its spontaneous cyclization to creatinine.

indicator of muscle mass because of the variation that occurs in muscle creatine content.

Excretion of 3-methylhistidine provides an indicator of muscle catabolism. On proteolysis, 3-methylhistidine (shown in Table 7.1) is released and is a nonreusable amino acid because the methylation of histidine occurs posttranslationally. Because 3-methylhistidine is found primarily in actin, this compound has been used to estimate muscle protein degradation. Excretion of the compound is used as an index of muscle degradation. However, actin is not just found in muscle. It appears to be widely distributed in the body, including tissues such as the intestine and platelets, which have high turnover rates. Thus, the urinary 3-methylhistidine excretion represents not only an index of muscle breakdown but more inclusively an index of protein breakdown for many body tissues in the body.

Kidneys

Studies in humans as well as animals suggest that the kidneys preferentially take up a number of amino acids including, for example, glycine, alanine, glutamine, glutamate, and aspartate. Metabolism of amino acids within the kidney (Fig. 7.38) includes

- serine synthesis from glycine,
- glycine catabolism,
- histidine synthesis from carnosine degradation,
- arginine synthesis from aspartate and citrulline, and
- guanidoacetate formation from arginine and glycine for creatine synthesis.

The kidney is considered to be the major site for arginine, histidine, and serine production. In addition, the kidney is the only organ besides the liver that has the enzymes necessary for gluconeogenesis. α-keto acids

formed from the transamination or deamination of amino acids brought to the kidney can be used for glucose production or can be used for energy. Renal metabolism of amino acids becomes particularly significant during fasting. Gluconeogenesis can raise the amount of glucose available to the body for energy, while the ammonia (formed from deamination of amino acids, especially glutamine) can help to normalize the pH of the blood, which typically decreases with fasting. Acidosis occurs with fasting because of the resulting rise in ketone concentration in the blood. There is also a loss of sodium and potassium as these minerals are excreted in the urine along with the ketones. Renal glutamine uptake for ammonia production during periods of acidosis (or low bicarbonate concentrations) is increased, while uptake by the intestine, liver, and other organs is diminished. Ammonia generated from amino acids, especially glutamine deamination, enters the filtrate and combines with H^+ ions to form ammonium ions. The ammonium ions cannot be reabsorbed, and thus are excreted in the urine. The loss of the H^+ from the body serves to increase pH from an acidotic state toward a normal value of about 7.35 to 7.45.

The role of the kidney in nitrogen metabolism cannot be overemphasized because of the organ's responsibility in ridding the body of nitrogenous wastes that have accumulated in the plasma. Moreover, enzymes particularly active in the kidney and involved in removal of nitrogenous compounds from the body include the amino transferases, glutamate dehydrogenase, and glutaminase, all of which catalyze the removal of ammonia from glutamate and glutamine. Kidney glomeruli act as filters of blood plasma, and all constituents in plasma move into the filtrate with the exception of plasma proteins. (Some albumin normally moves into the filtrate but is reabsorbed by the proximal tubular cells.) Essential nutrients such as sodium (Na^+), amino acids, and glucose are actively reabsorbed, as the filtrate moves through the

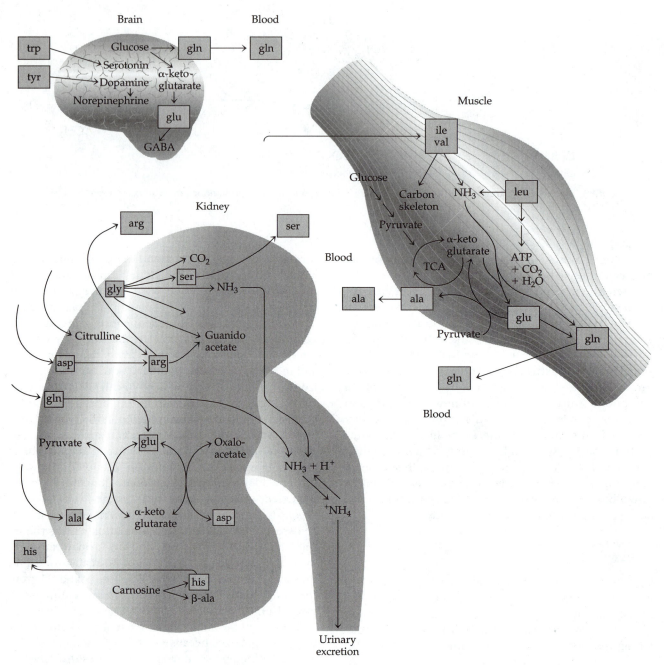

Figure 7.38 Amino acid metabolism in selected organs.

tubules. Many other substances are not actively reabsorbed, and if they move into the tubular cells, they must either move along an electrical gradient or move osmotically with water. The amount of these substances that enters the tubular cells, then, depends on how much water moves into the cells and how permeable the cells are to the specific substances. The cell membranes are relatively impermeable to urea and uric acid, while membranes are particularly impermeable to creatinine, none of which is reabsorbed.

Figure 7.39 shows the normal range of nitrogenous wastes found in the urine. About 80% of nitrogen is lost in the urine as urea under normal conditions. Additional nitrogen may be lost through the skin as urea, and the loss of hair and skin cells also results in insensible nitrogen losses. In acidotic conditions, as occur with fasting, urinary urea nitrogen losses decrease and the percentage of nitrogen lost as urea also decreases. Urinary ammonia excretion rises in absolute terms as well as in percentage terms.

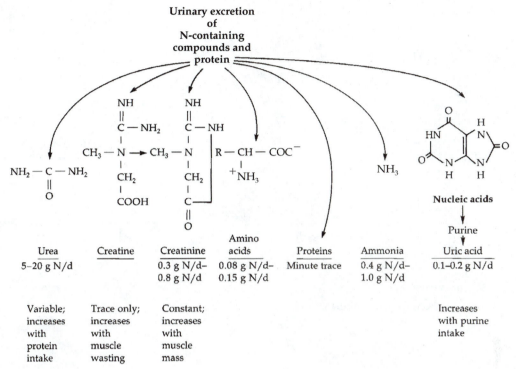

Figure 7.39 Nitrogenous wastes in normal urine.
Source: Reproduced with permission from WC McMurray, *Essentials of Human Metabolism,* 2nd ed., 1983, Harper & Row, p. 261.

Brain and Accessory Tissues

Because the brain has the capacity for active transport of the indispensable amino acids, neurons have access to a higher concentration of these amino acids than is found in the plasma. The brain has transport systems for neutral, dibasic, and dicarboxylic amino acids, and, as in other cells, competition occurs between amino acids for uptake by a common carrier system.

Amino Acids Used in Neurotransmitter/Hormone Synthesis

The uptake of two aromatic amino acids, tryptophan and tyrosine, by the brain is particularly important because these amino acids act as precursors for a variety of neurotransmitters or modulators of nerve function (Fig. 7.38) and hormones.

- Tryptophan is used to synthesize the hormone melatonin or N-acetyl 5-methoxytryptamine, and the excitatory neurotransmitter serotonin, alternately called *5-OH tryptamine* (Fig. 7.40).

- Tyrosine is used to synthesize dopamine, norepinephrine, and epinephrine (Fig. 7.24), referred to as *catecholamines,* because they are derivatives of catechol (Fig. 7.41).

Neurotransmitters are stored in the nerve axon terminal as vesicles or granules until stimuli arrive to cause their release. After their action on the cell membranes, the neurotransmitters are inactivated. The fastest mechanism for inactivation is uptake of the neurotransmitter by adjacent cells, where mitochondrial monoamine oxidase (MAO) remove the amine group. Each of the catecholamines (dopamine, norepinephrine, epinephrine) as well as serotonin can be inactivated by MAO. A slower mechanism of inactivation requires that the catecholamines be carried by the blood to the liver where they are methylated by catechol-O-methyltransferase (COMT).

Foods also contain amines, especially tyramine and dopamine, that must be inactivated in the body by MAO. Examples of foods very high in tyramine include aged cheeses (cheddar, Camembert, Stilton, boursalt), yeast extracts, and Brewer's yeast. Smoked, salted, or pickled fish such as herring or cod, as well as sausage, salami, pepperoni, corned beef, and bologna, also are high in tyramine. Foods moderately high to high in tyramine include meat extracts; tenderizers; red wines, including chianti, vermouth, sherry, and burgundy; and cheeses such as blue, natural brick, Brie, Gruyère, mozzarella, Parmesan, Romano, and Roquefort. Broad beans (fava, Chinese pea pods), chocolate, large amounts of caffeine, liver (chicken or beef), and selected fruits may

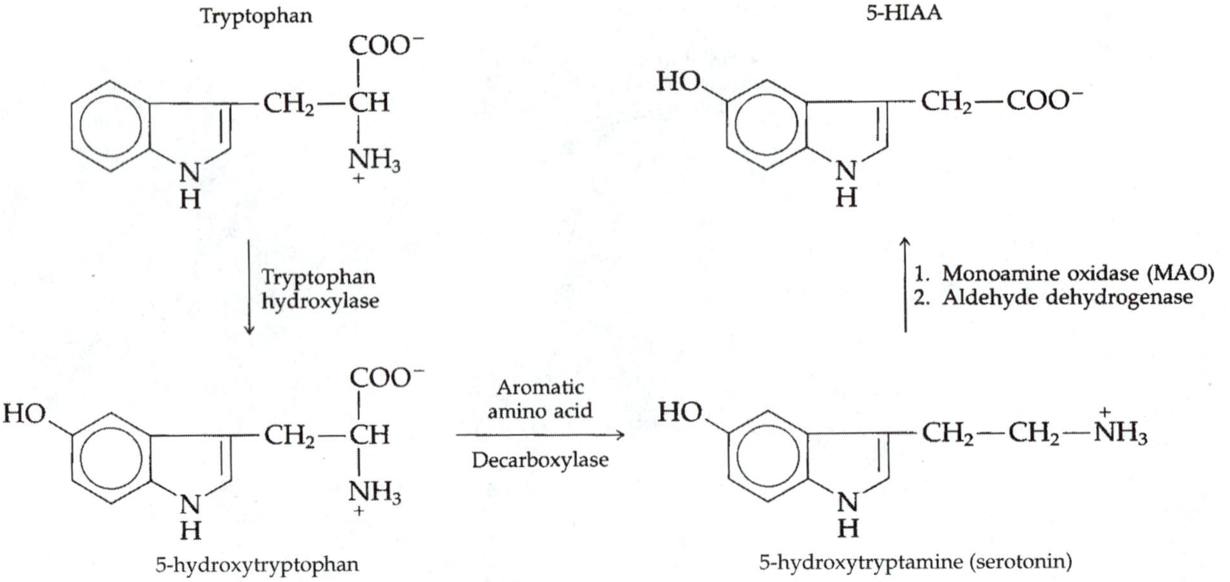

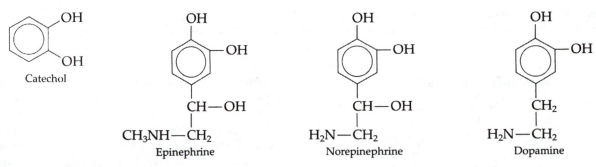

Figure 7.40 Serotonin synthesis from tryptophan.

Figure 7.41 The catecholamines.

also contain high amounts of tyramine [45]. Consumption of these foods ordinarily presents no problem to individuals because the tyramine or dopamine may be quickly inactivated by MAO. However, there are a group of drugs known as MAO inhibitors used in the treatment of depression, hypertension (high blood pressure), or certain cancers. MAO inhibitors prevent MAO from catabolizing both amines in diet and catecholamines made endogenously. The problem arises when people on MAO inhibitors eat foods high in tyramine. High dietary tyramine coupled with endogenous norepinephrine may result in excessive vasoconstriction, manifested as acute hypertension or a hypertensive crisis. Individuals on MAO-inhibitor drugs are counseled against ingesting foods high in tyramine.

Other amino acids also function as neurotransmitters in the brain:

- Glycine acts as an inhibitory neurotransmitter.
- Taurine is also thought to function as inhibitory neurotransmitters.
- Aspartate is thought to act as an excitatory neurotransmitter in the central nervous system (CNS).
- Glutamate acts as an excitatory neurotransmitter or may be converted into γ-amino butyric acid (GABA), an inhibitory neurotransmitter.

Aspartate is derived chiefly from glutamate through AST activity common in neural tissue.

GABA is believed to be the neurotransmitter for cells that exert inhibitory effects on other cells in the central nervous system (CNS). The conversion of glutamate to GABA involves the removal of the α-carboxyl group of glutamate by the enzyme glutamate decarboxylase in a vitamin B_6-(PLP)–dependent reaction

(Fig. 7.42). Glutamate uptake by the brain is typically low; thus, synthesis of glutamate from glucose represents the primary source of brain glutamate.

Glutamate is of significance to the brain not just as a neuromodulator or for the synthesis of GABA but also as a means of ridding the brain of ammonia. The starting point for glutamate metabolism is the synthesis in neurons of α-ketoglutarate from glucose that has been transported from the blood across the blood-brain barrier. α-ketoglutarate can be converted to glutamate via reductive amination. Whenever excessive ammonia is present in the brain, glutamine is formed through the action of glutamine synthetase, which is highly active in neural tissues (Fig. 7.31). Glutamine is freely diffusible and can move easily into the blood or cerebrospinal fluid, thereby allowing the removal of 2 mol of toxic ammonia from the brain. Any condition that causes an unusual elevation of blood ammonia can interfere with the normal handling of amino acids by the brain. Treatment of hepatic encephalopathy (dysfunction of the brain associated with liver disease, which characteristically results in elevated blood ammonia concentrations) is aimed at normalizing the effects of altered amino acid metabolism on the CNS.

Use of amino acid supplements has been promoted with claims that upon ingestion the amino acids will be utilized in the body for the synthesis of compounds necessary to evoke the desired response. For example, use of tryptophan supplements to promote sleep has been promoted, as has supplements of melatonin, which is also made from tryptophan in the pineal gland, which lies about in the center of the brain. Melatonin plays a role in the regulation of sleep. Yet, melatonin supplements of 2 to 500 mg for use as a sleep aid have yielded variable results. Somnolence has been reported in some, especially those with an abnormal sleep wake cycle, and diminished jet lag has been demonstrated in airline personnel on eastward transcontinental flights; however, the effects of long-term use of melatonin and amino acids, as well as effective dosing and administration, remain unknown [46,47].

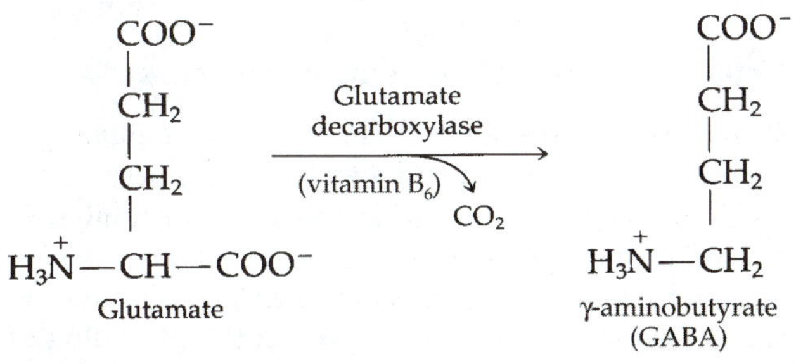

Neuropeptides

The CNS abounds in peptides, termed *neuropeptides.* Many of the same peptides that were mentioned in Chapter 2 with respect to digestion and that were found associated with the intestinal tract are also found associated with the CNS. These neuropeptides are of varying lengths and possess a variety of functions. Some peptides act as hormone-releasing factors, such as ACTH involved with cortisol release; some have endocrine effects, such as somatotropin or growth hormone; some have modulatory actions on transmitter functions, mood, or behavior, such as the enkephalins. The enkephalins and endorphins, although similar to natural opiates, possess a wide range of functions including affecting pain sensation, blood pressure and body temperature regulation, governance of body movement, secretion of hormones, control of feeding, and modulation of learning ability.

The neurosecretory cells of the hypothalamus are foremost in the secretion of the neuropeptides. Those that have hormone action move out of the axons of the nerve cells into the pituitary from which they are secreted. This linkage between the nervous system and pituitary is of great significance in the overall control of metabolism because the pituitary gland is primary in coordinating the various endocrine glands scattered throughout the body.

The neuropeptides are believed to be synthesized from their constituent amino acids via the DNA-coding, messenger RNA (mRNA), ribosomes, and transfer RNA (tRNA) system. Because the nucleus and ribosomes are found in the cell body and dendrites, the peptides must travel to the end of the axon to be stored in vesicles for future release. The neuropeptides are stored as inactive precursor polypeptides, which must be cleaved to generate an active neuropeptide, as shown here:

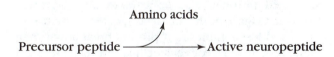

Following synthesis of the active neuropeptide, it is released by exocytosis and performs its function at the membrane. After performing its function, the neuropeptide is hydrolyzed to its constituent amino acids. A further discussion of nutrients that serve as precursors of neurotransmitters is provided in Chapter 16, "Nutrition and the Central Nervous System."

Protein Turnover: Synthesis and Catabolism of Tissue Proteins

Food intake and the nutritional status of the organism affect protein turnover that is mediated through

Figure 7.42 GABA synthesis.

changes in hormone balance [48]. Hormone balance is particularly sensitive to changes in plasma amino acid concentrations, and amino acid catabolism is closely related to plasma and cellular amino acid concentrations. The secretion of insulin, glucagon, growth hormone, and glucocorticoids increases in response to elevated concentrations of selected amino acids. Therefore, the protein component of food intake, up to a certain threshold level, appears to be of particular importance. When this threshold level of protein (or plasma amino acids) has been reached, however, any further increase in intake has little or no effect on the rate of protein synthesis, probably because of the counterregulatory effects of the various hormones being secreted. It is believed that this threshold level for protein intake corresponds roughly to dietary protein requirements [49].

In general, protein synthesis and positive nitrogen balance are promoted by insulin, whereas the counterregulatory hormones, glucagon, catecholamines, and glucocorticoids have an opposite effect, promoting protein degradation and a negative nitrogen balance. Growth hormone, although counterregulatory, is an anabolic agent, like insulin. Prostaglandins and thyroid hormones also are affected by dietary nutrient intakes, and can promote changes in protein turnover. Nitric oxide has been shown to inhibit hepatic protein synthesis [50]. The effects of a hormone on protein turnover, however, may differ depending on the tissue. Generally, following ingestion of a protein-containing meal, net protein anabolism results owing to both an increase in protein synthesis and a decrease in protein degradation.

Protein turnover is also influenced by amino acid oxidation. The catabolism of amino acids is influenced to a large extent by whether conditions are present for protein synthesis to occur. The molecular form of the consumed nitrogen also appears to affect protein turnover. Oligopeptides induced greater protein synthesis and less inhibition of protein breakdown than whole protein [51]. For each cellular protein, there is a specific and characteristic rate of synthesis. Protein synthesis is dictated in part by specific mRNA produced in the nucleus. Transcription and translation of DNA and RNA are under multiple influences such as hormonal (steroid or peptide) and nutrient, which are discussed in the perspective at the end of Chapter 9. Protein synthesis is also affected by

- the amount of mRNA;

- the ribosome number (amount of ribosomal RNA, rRNA);

- the activity of the ribosomes (rapidity of translation, or peptide formation);

- the presence of the amino acids, both essential and nonessential, in the appropriate concentrations to charge the tRNA; and

- the hormonal environment.

Should amino acids not be present or be present, but in insufficient quantity, amino acid oxidation increases. In other words, the supply of amino acids must meet the demand. The rate of amino acid oxidation is sensitive to a surplus or deficit of specific amino acids as well as to hormonal factors, and thus regulates the amino acid pool(s). Amino acid pool(s) remain fairly constant in size and pattern in the body. Further, the pool(s) serves as the connecting link between two cycles of nitrogen metabolism:

protein turnover (protein synthesis vs. degradation)	and	nitrogen balance (nitrogen intake vs. nitrogen output).

These two cycles operate somewhat independently, but if either cycle gets out of balance, the other is affected to some degree. For example, during growth, protein synthesis exceeds degradation, and nitrogen intake exceeds excretion, resulting in a positive nitrogen balance. The increase in protein turnover, however, is much greater than is reflected by the change in nitrogen balance. The amino acid pool(s), regulated in some manner, is acting as ballast between protein turnover and nitrogen balance.

When the body is considered as a whole, it is estimated that protein synthesis and degradation (turnover) account for 10% to 25% of resting energy expenditure [52]. Protein degradation, for example, requires energy for proteolysis, metabolism of amino acids not reincorporated into protein, protein transport across membranes, and RNA metabolism [52]. Rates of protein turnover vary among tissues, as is evidenced in the more rapid turnover of visceral protein as compared with skeletal muscle. Protein synthesis and protein degradation are under independent controls. Rates of synthesis can be quite high as with protein accretion during growth. Alternately, protein degradation can be quite high, as during fever.

Cellular Protein Degradation Systems

Degradation of proteins, either made intracellularly or brought into the cell by endocytosis, occurs primarily by the action of proteases, either lysosomal or nonlysosomal, present in the cytosol. The contributions of lysosomal and nonlysosomal cytosolic pathways in proteolysis vary depending on the cell type and the physiological status [53]. Nonetheless, the constant degradation of intracellular proteins is of prime importance to the life of the cells because it ensures a flux of proteins (amino acids) through the cytosol that can be used for cellular growth and/or maintenance.

Lysosomes, cell organelles with an internal pH of about 4 to 5 and containing a variety of digestive enzymes, represent the primary means of protein digestion. Lysosomes are found in all mammalian cell types with the exception of the erythrocyte. While skeletal muscles do not contain many lysosomes, they are particularly abundant in liver cells where phagic processes occur in vacuoles that have fused with lysosomes.

The lysosomal proteases include endopeptidases and exopeptidases known as *cathepsins.* Numerous cathepsins have been isolated; some examples include protease cathepsins B, H, L, which are referred to as *cysteine proteases,* and protease cathepsin D, which is an *aspartate protease.* Each of the cathepsin proteases vary in specificity [54]. Lysosomal proteases appear to be responsible for degradation of extracellular proteins brought into the cell by endocytosis, and for intracellular protein degradation under conditions of nutrient deprivation and pathological conditions [53,54]. The extent to which lysosomal proteases influence muscle proteolysis is unknown but believed to be fairly small [55].

The process by which lysosomal proteases digest proteins requires that the cell engulf the designated protein. Following engulfment, the protein designated for degradation becomes contained in a vacuole in the cytoplasm. The autophagic vacuoles containing the intracellular protein fuse with lysosomes. The protein in turn is completely digested by the lysosomal proteases; the process is often referred to as *macroautophagy* and is enhanced by glucagon and suppressed by insulin as well as amino acids [56]. This high glucagon:insulin ratio is consistent with conditions in which cells would be nutritionally (amino acid) deprived.

In well-nourished cells, slow, continuous degradation of intracellular proteins by the lysosomal autophagic enzyme system typically occurs, a process referred to as *basal autophagy or microautophagy* [54,56]. Microautophagy needs some mechanism by which selected intracellular proteins could enter the lysosomes more rapidly than other proteins. The conformation of intracellular proteins as a determinant of intracellular degradation rates has been proposed [54]. In the liver, microautophagy may function in the degradation of proteins with long half-lives and that are not good substrates for other proteolytic pathways [53].

In addition to lysosome-mediated cellular protein degradation, nonlysosomal protease systems also de-

grade proteins. The ubiquitin system is nonlysosomal and mediates the degradation of many cellular proteins. In the ubiquitin system, proteins that are to be degraded are ligated to the 76 amino acid polypeptide ubiquitin in an ATP-requiring reaction. The linkage of ubiquitin to protein first necessitates activation of ubiquitin. Ubiquitin is activated by the enzyme (E) 1. E1 is a subunit of the ubiquitin enzyme system, which hydrolyzes ATP to form a thiol ester with the carboxyl end of ubiquitin. This activated ubiquitin is transferred to another enzyme protein E2. Next the carboxyl end of ubiquitin is ligated by E3 to the protein substrate that is ultimately to be degraded. One or more (typically five) ubiquitin proteins may bind to a protein substrate. Once ubiquitins are ligated to the protein to be degraded, proteases present as a proteasome or other multienzyme complex degrade the ubiquitinated proteins in a series of reactions. Following proteolysis, ubiquitin is released for reuse. Signals in proteins for ubiquitin-mediated degradation include N-terminal recognition. Specificity of binding by E3α, a ubiquitin protein ligase of the ubiquitin protein degradation system, to selected N-terminal amino acid residues has been demonstrated [57]. E3 has two distinct sites that interact with specific N-terminal residues. Proteins with valine, methionine, glycine, alanine, serine, threonine, and cysteine residues at their N-terminal position are relatively stable [58]. In contrast, proteins with N-terminal leucine, glutamate, histidine, tyrosine, glutamine, aspartate, asparagine, phenylalanine, leucine, tryptophan, lysine, and arginine are typically unstable, with short half-lives [57,58]. Proteins with acetylated N-termini are not degraded by the ubiquitin system [57].

Most proteins with short half-lives, those that are mislocated, or proteins that are abnormal due to misfolding or denaturation are degraded by cytosolic ATP-dependent nonlysosomal pathways that may or may not require ubiquitin conjugation [53,59]. Activity of nonlysosomal ATP-ubiquitin-dependent pathways also appears to be increased during pathological conditions such as sepsis, cancer, and trauma, among others [55]. The energy- and ubiquitin-dependent pathway also may be involved in muscle protein degradation associated with starvation and other pathological conditions [60]. How specific proteins are selected by the proteolytic system is not fully understood, but the major component of selectivity appears to reside within the structural properties of the protein [54]. The structure can either increase or decrease the vulnerability of the protein to the degradative system. Newly synthesized proteins that have not yet been integrated into their appropriate cell location are particularly vulnerable to degradation. Whether these proteins can escape degradation seems to depend largely on their primary structure and consequently on their final conformation and function. Abnormal or aberrant proteins appear to be recognized and degraded rapidly, but certain normal proteins also undergo rapid degradation. Some normal proteins with extremely short half-lives are the regulatory enzymes, which need to be adjusted quickly in response to appropriate stimuli such as excess substrate. Characteristics of short-lived proteins include large size, acidic net charge, hydrophobicity, and rapid inactivation by a low pH or high temperature [54]. Rapidly degraded proteins may also possess a common amino acid sequence; the PEST hypothesis suggests that proteins with particularly short half-lives contain regions rich in proline (P), glutamic acid (E), serine (S), and threonine (T) [61]. However, neither the mechanism by which the PEST signal targets proteins for degradation, nor the system that recognizes the signal, has been identified [55,57].

In addition to the nonlysosomal ubiquitin pathway, other nonlysosomal, nonubiquitin protease systems have been identified in muscle. The calpain proteolytic pathway is nonlysosomal and non-energy-dependent but instead requires calcium for its function. Two proteases, μ-calpain and m-calpain, comprise the calpain proteolytic pathway and differ in their need for calcium. The overall role of the calpains in protein turnover is unclear; however, the proteases may be responsible for the degradation of specific target proteins. Proteasomes, multicatalytic, multisubunit, nonlysosomal protease complexes, are found in the cytosol, as well as the nucleus. The complex exhibits optimal activity at a neutral to alkaline cellular pH and is capable of hydrolyzing peptide bonds on the carboxyl terminal end of hydrophobic, acidic, and basic amino acids. Substrates for proteasomes include oxidized proteins and myofibrillar proteins [60]. Proteasomes can operate with or without ubiquitin.

The requirement for energy for protein catabolism is a paradox that has been recognized for decades. ATP-dependent proteases have been demonstrated in the cytosolic degradation system. Based on their work with an ATP-dependent protease found in *Escherichia coli*, Waxman and Goldberg [62] have shown that this enzyme is activated by both ATP and by substrate proteins and that the two have an additive effect. When the protein is activated by ATP, it will degrade various small hydrophobic peptides that are bound to its active site, but concomitant hydrolysis of the ATP molecule does not occur. When substrate protein activates the protease, it evidently does so by binding to two sites on the enzyme: the active site, which has a preference for specific amino acid sequences and a regulatory site, which somehow recognizes that the protein is unfolded and vulnerable. When activated by both ATP and a substrate protein,

the enzyme is believed to undergo a distinct structural change that enhances its capacity for peptide cleavage and also causes hydrolysis of ATP. A protein that binds itself to both sites of the enzyme can induce its own destruction, but other proteins that are unable to bind the enzyme at both sites will remain unaffected. Furthermore, the adenosine diphosphate (ADP) produced as a result of the proteolysis acts as an inhibitor of continued hydrolysis, thereby helping to prevent excessive proteolysis once the enzyme has been activated. Should an analogous mechanism exist in mammalian cells, this scheme could go far in explaining not only the observed heterogeneous selectivity of protein catabolism but also the function of ATP in this catabolic process [63].

Protein Quality and Protein Intake

Dietary protein is required by humans because it contributes to the body's supply of indispensable amino acids and to its supply of nitrogen for the synthesis of the dispensable amino acids. The quality of a protein depends to some extent on its digestibility but primarily on its indispensable amino acid composition. Both the specific amounts and proportions of these amino acids are important to the quality of a protein.

Protein containing foods can be divided into two categories:

- high-quality or complete proteins and
- low-quality or incomplete proteins.

A complete protein contains all the indispensable amino acids in the approximate amounts needed by humans. Sources of complete proteins are foods of animal origin such as milk, yogurt, cheese, eggs, meat, fish, and poultry. The exception is gelatin, which is derived from animal origin but does not have the indispensable amino acid tryptophan. Incomplete proteins or low-quality proteins are derived from plant foods such as legumes, vegetables, cereals, and grain products. These foods tend to have too little of one or more particular indispensable amino acids. The term *limiting amino acid* is used to describe the indispensable amino acid that is present in the lowest quantity in the food. Listed here are examples of incomplete protein-containing foods and their limiting amino acid(s):

Food Source of Incomplete Protein	Limiting Amino Acid(s)
Wheat, rice, corn, other grains and grain products	Lysine, threonine (sometimes), and tryptophan (sometimes)
Legumes	Methionine

Unless carefully planned, a diet containing only low-quality proteins will result in inadequate availability of selected amino acids and inhibit the body's ability to synthesize its own body proteins. The body cannot make a protein with a missing amino acid.

To ensure that the body receives all the indispensable amino acids, certain proteins can be ingested together or combined so that their amino acid patterns become complementary. Most, but not all, agree that the complementary proteins should be consumed at the same meal. For the vegan, this knowledge is particularly valuable. For example, legumes, with their high content of lysine but low content of sulfur-containing amino acids, complement the grains, which are more than adequate in methionine and cysteine but limited in lysine. The lacto-ovo vegetarian should have no problems with protein adequacy because when milk and eggs are combined, even in small amounts, with plant foods, the indispensable amino acids are supplied in adequate amounts. One exception is the combination of milk with legumes. Although milk contains more methionine and cysteine per gram of protein than do the legumes, it still fails to meet the standard of the ideal pattern for the sulfur-containing amino acids.

The formula for vegan protein balance is as follows: 60% of protein from grains, 35% from legumes, 5% from leafy greens [64]. The 70-kg adult man whose RDA for protein is 56 g could obtain his needed indispensable amino acids by consuming

- four servings (four slices) whole-grain bread;
- five servings (2 ½ cups) grain from oatmeal, brown rice, cracked wheat;
- one serving (¼ cup) nuts or seeds;
- 2 ½ servings (1 ¼ cups) beans or (2 cups soy milk and ⅓ cup navy beans); and
- four servings (2 cups) vegetables, two of which are leafy greens.

Plant protein foods contribute about 65% of the per capita supply of protein worldwide and about 32% in North America [65].

Evaluation of Protein Quality

Several methodologies are available to determine the protein quality of foods containing protein. A few of these methodologies will be discussed in this section.

Chemical Score

Chemical score (also called *amino acid score*) involves determination of the amino acid composition of a test protein. This procedure is done in a chemical laboratory using either an amino acid analyzer or high-perfor-

mance liquid chromatography techniques. Only the indispensable amino acid content of the test protein is determined, and then compared with that of egg protein (considered to have a score of 100) or with an ideal reference pattern of amino acids. The chemical score of a food protein can be calculated as follows:

$$\text{Score of test protein} = \frac{\text{Content of each indispensable amino acid in food protein (mg/g protein)}}{\text{Content of same amino acid in reference protein or reference pattern (mg/g protein)}}$$

The amino acid with the lowest score on a percentage basis in relation to the reference protein (egg) or pattern becomes the first limiting amino acid; the one with the next lowest score is the second limiting amino acid; and so on. For example, if a test protein only has 85% of the lysine that is present in the reference protein, the test protein's chemical score would be 85.

Table 7.6 gives the amino acid pattern in whole egg, and the amino acid requirement reference pattern for adults for comparison. The amino acid requirement patterns represent the requirements of individual

Table 7.6 Amino Acid Reference Patterns [66]

Indispensable Amino Acids	Whole-Egg Pattern	Requirement Pattern for Adults
Histidine	22[a]	(11)+
Isoleucine	54	13
Leucine	86	19
Total sulfur amino acids (met and cys[b])	57	17
Total phenylalanine and tyrosine[b]	93	19
Lysine	70	16
Threonine	47	9
Tryptophan	17	5
Valine	66	13
Total indispensable amino acids[c] (without histidine)	490	111

[a]Values are milligrams amino acid/gram protein.
[b]Cysteine and tyrosine are considered "conditionally indispensable" because they spare methionine and phenylalanine.
[c]Value is imputed. Requirement for histidine has not been quantified beyond infancy but probable requirement is estimated to be between 8 and 12 mg/kg body weight/day for adults.

Source: Reprinted with permission from Recommended Dietary Allowances: 10th Edition. Copyright 1989 by the National Academy of Science. Courtesy of the National Academy Press, Washington, D.C.

indispensable amino acids (in milligrams) for adults divided by the corresponding recommended intake of protein (in grams). Comparison of the quality of different food proteins against the standard of whole-egg protein can be valuable but probably is not nearly so important to adequate protein nutriture as their comparison with reference patterns for the various population groups.

Protein Efficiency Ratio

The protein efficiency ratio (PER) represents body weight gained on a test protein divided by the grams of protein consumed. To calculate the PER of proteins, young growing animals are typically placed on a standard diet with about 10% (by weight) of the diet as test protein. Weight gain is measured for a specific time period and compared to the amount of the protein consumed. PER for the protein is then calculated using the following formula:

$$PER = \frac{\text{Gain in body weight (in grams)}}{\text{Grams of protein consumed}}$$

To illustrate, the PER for casein (a protein found in milk) is 2.5; thus, rats gain 2.5 g of weight for every 1 g of casein consumed. However, a food with a PER of 5 does not mean that it has doubled the protein quality of casein with a PER of 2.5. Furthermore, although PER allows determination of which proteins promote weight gain (per gram of protein ingested) in growing animals, no distinction is made regarding the composition (fat versus muscle/organ) of the weight gain.

Nonetheless, PER has been used in nutrition labeling. The Food and Drug Administration uses the milk protein casein as a standard for comparison of protein quality. If a test protein has a protein quality equal or better than casein—that is, if the PER is ≥2.5—then 45 g of protein is considered as meeting 100% U.S. RDA. If a test protein is lower in quality than casein—that is, the PER is <2.5—then 65 g of protein is needed to meet 100% U.S. RDA.

In addition to chemical score and protein efficiency ratio, two additional methods, biological value and net protein utilization (use), used to determine protein quality. These methods, however, involve nitrogen balance studies. Thus, the topic of nitrogen balance will be briefly reviewed in the following section, with a discussion of biological value and net protein utilization to follow.

Nitrogen Balance

Nitrogen balance studies involve the measurement and summation of nitrogen losses from the body. Protein contains approximately 16% nitrogen. Thus, to calculate grams of protein from grams of nitrogen the following calculation can be done.

$$0.16 \times \text{protein (measured in grams)} = \text{nitrogen (measured in grams)}$$

Expressed alternately,

protein (grams) = nitrogen (grams) $\times$ 100/16 or
protein (grams) = nitrogen (grams) $\times$ 6.25

Nitrogen balance studies are conducted when subjects consume a diet with a protein (nitrogen) intake that is

- at or near a predicted adequate amount,
- less than (including protein-free nitrogen) a predicted adequate amount, and
- greater than a predicted adequate amount.

Nitrogen losses are measured in the urine (U), feces (F), and skin (S). For example, in the urine, nitrogen is found mainly as urea but also as creatinine, amino acids, ammonia, and uric acid. In the feces, nitrogen may be found as amino acids and ammonia. Gastrointestinal tract losses of amino acids represent 14% to 61% of essential amino acid requirements [67].

The calculation of nitrogen balance is shown in this formula. Stately simply, nitrogen losses are summed and subtracted from nitrogen intake (I_n):

$$\text{Nitrogen balance} = I_n - [(U - U_e) + (F - F_e) + S]$$

The subscript *e* (as U_e and F_e) shown in the equation stands for *endogenous* (also called *obligatory*) and refers to losses of nitrogen that occur when the subject is on a nitrogen-free diet.

Unfortunately, nitrogen balance studies have been criticized for overestimating true nitrogen retention rates in the body due to incomplete collection and/or measurement of losses. In addition, nitrogen balance does not necessarily mean amino acid balance. In other words, a person may be in nitrogen balance but in amino acid imbalance.

Biological Value

The biological value (BV) of proteins is another method used to assess protein quality. BV is a measure of how much nitrogen is *retained* in the body for maintenance and/or growth versus the amount of nitrogen absorbed. BV is most often determined in experimental animals, but it can be determined in humans. Subjects are fed a nitrogen-free diet for a period of about 7 to 10 days, then fed a diet containing the test protein in an amount equal to their protein requirement for a similar time period. Nitrogen that is excreted in the feces and in the urine is analyzed and compared during the time period when subjects consumed the nitrogen-free diet versus when the subjects consumed the test protein. In other words, the change in urinary and fecal nitrogen excretion between the two diets is calculated. BV of the test protein is determined through the use of the following equation:

$$\text{BV of test protein } = \frac{I - (F - F_0) - (U - U_0)}{I - (F - F_0)}$$

$$\times 100 = \frac{\text{Nitrogen retained}}{\text{Nitrogen absorbed}} \times 100$$

where I is intake of nitrogen; F is fecal nitrogen while subjects are consuming a test protein; F_0 is endogenous fecal nitrogen when subjects are maintained on a nitrogen-free diet; U is urinary nitrogen while subjects are consuming a test protein; and U_0 is endogenous urinary nitrogen when subjects are maintained on a nitrogen-free diet. Foods with a high BV are those that provide the amino acids in amounts that are consistent with body amino acid needs. The body will retain much of the absorbed nitrogen, if the protein is of high BV. Although BV provides useful information, the equation fails to account for losses of nitrogen through insensible routes such as the hair and nails. This criticism is true of any method involving nitrogen balance studies. A further consideration is that proteins will exhibit a higher BV when fed at levels below the amount necessary for nitrogen equilibrium, and as protein intake approaches or exceeds adequacy, retention decreases.

Net Protein Utilization

Another measure of protein quality involving nitrogen balance studies is net protein utilization or use (NPU). NPU measures retention of food nitrogen consumed rather than retention of food nitrogen absorbed. NPU is calculated from the following equation:

$$\text{NPU of test protein } = \frac{I - (F - F_0) - (U - U_0)}{I}$$

$$\times 100 = \frac{\text{Nitrogen retained}}{\text{Nitrogen consumed}} \times 100$$

where I is intake of nitrogen; F is fecal nitrogen while subjects are consuming a test protein; F_0 is endogenous fecal nitrogen when subjects are maintained on a nitrogen-free diet; U is urinary nitrogen while subjects are consuming a test protein; and U_0 is endogenous urinary nitrogen when subjects are maintained on a nitrogen-free diet.

Although NPU can be measured in humans through nitrogen balance studies in which two groups of well-matched experimental subjects are used, a more nearly accurate measurement is made on experimental animals through direct analysis of the animal carcasses. In either case, one experimental group is fed the test protein, while the other group receives an isocaloric, protein-free diet. When experimental animals are used as subjects, carcasses can be analyzed for nitrogen directly (total carcass nitrogen, or TCN) or indirectly at the end of the feeding period. The indirect measurement of nitrogen is made by water analysis. Given the amount of water removed from the carcasses, an approximate nitrogen content can be calculated. NPU involving animal studies is calculated from the following equation:

$$\text{NPU} = \frac{\begin{array}{c}\text{TCN on} \\ \text{test protein}\end{array} - \begin{array}{c}\text{TCN on} \\ \text{protein-free diet}\end{array}}{\text{Nitrogen consumed}}$$

Proteins of higher quality typically cause a greater retention of nitrogen in the carcass than poor-quality proteins and would have a higher NPU.

Net Dietary Protein Calories

The net dietary protein calories percentage (NDpCal%) is particularly helpful in evaluation of human diets in which the protein:calorie ratio may vary greatly. The formula is as follows:

$$\text{NDpCal\%} = \text{Protein kcal/Total kcal intake} \times 100 \times \text{NPU}_{op}$$

NPU_{op} is NPU when protein is fed above the minimum requirement for nitrogen equilibrium.

Recommended Protein Intake

Protein and amino acid requirements of humans are influenced by age, body size, physiological state, as well as the level of energy intake. Nitrogen balance studies as well as the factorial method have been used to determine protein and amino acid needs. As with nitrogen balance studies, the factorial method involves the measurement and summation of all losses of nitrogen from the body when the diet is devoid of protein. These losses represent endogenous (obligatory) nitrogen losses. Obligatory nitrogen losses from adults typically range from 41 to 59 mg of nitrogen per kilogram of body weight, with an average of about 55 mg of nitrogen per kilogram of body weight. Obligatory protein losses range from 0.26 to 0.43 g of protein per kilogram of body weight, with an average of about 0.34 g of protein per kilogram of body weight. A summary of the estimates of indispensable amino acid requirements at different ages is given in Table 7.7, while Figure 7.41 illustrates the relationship between these amino acid requirements and total protein need.

Table 7.7 A Summary of the Estimates of Amino Acid Requirements at Different Ages, as Proposed in 1985 by the Food and Agriculture Organization, World Health Organization, United Nations [68]

Amino Acid	Infants (3–4 months)	Children (2 years)	Schoolboys (10–12 years)	Adults
Histidine	28[a]			(8–12)
Isoleucine	70	31	28	10
Leucine	161	73	44	14
Lysine	103	64	44	12
Methionine and cystine	58	27	22	13
Phenylalanine and tyrosine	125	69	22	14
Threonine	87	37	28	7
Tryptophan	17	12.5	3.3	3.5
Valine	93	38	25	10
Total	714	352	216	84

[a]Values are milligrams/kilograms/day.

Source: Reprinted with permission from Recommended Dietary Allowances: 10th Edition. Copyright 1989 by the National Academy of Sciences. Courtesy of the National Academy Press, Washington, D.C.

Figure 7.41 makes clear the currently accepted concept that maintenance of existing body tissue is much less dependent on protein quality than is growth. There is an inconsistency, however. Although the growth component in the older school-aged child appears to be only a small consideration in total protein need, protein quality still assumes much more importance in this group than in the adult [68]. The 1989 Recommended Dietary Allowances (RDA) [66] for protein for all age groups are given on the inside front cover. The RDA for protein encompasses the average requirement (for adults, 0.6 g/kg of body weight) for high-quality protein, and, based on a coefficient of variation of 12.5%, 25% (2 standard deviations) was added to the requirement. For example, for adults the RDA for protein is calculated as 0.6 g/kg of body weight × 1.25, which totals 0.75 g/kg of body weight, and is rounded to 0.8 for RDA purposes. The RDA for protein is estimated to meet the needs of 97.5% of the U.S. population.

Munro et al. [69] introduced a simple method by which the adequacy of a food protein can be roughly estimated. By calculating the percentage of calories provided by the recommended protein intake when adequate energy is supplied, one can estimate the percentage of protein calories that should be found in a food used as a primary protein source. For instance, the protein and energy RDA for a 25-year-old, 79-kg male are 63 g of protein (79 kg × 0.8 g/kg of body weight) and 2,900 kcal. To calculate kilocalories from protein, multiply grams of protein by 4:

63 g of protein × 4 kcal/g = 252 kcal from protein.

Then to determine the percentage total kilocalories from protein, kilocalories from protein are divided by total kilocalories:

252 kilocalories from protein/2,900 total kilocalories × 100 = 8.7% kilocalories provided by dietary protein.

Any food, therefore, used as a primary source of protein and energy but providing <8.7% of its calories as protein, is probably unable to meet maintenance needs for indispensable amino acids. Foods particularly suspect are cassava flour (1.8 kcal from protein/100 kcal of flour) and plantain (3.1 kcal from protein/100 kcal of plantain). Both these foods are known to be staples in some developing countries.

The influence of energy intake on protein requirement is particularly significant. Excess energy intake fosters nitrogen retention, whereas insufficient calories from carbohydrate and/or fat mandate the oxidation of some protein to supply energy needs. The promotion of nitrogen retention by a high caloric intake appears to have been a factor in estimating human requirements for the indispensable amino acids [1,70,71]. The high energy (caloric) intake, causing weight gain of experimental subjects, may have permitted nitrogen equilibrium at an amino acid intake lower than the actual requirement [70]. The many difficulties associated with the nitrogen balance method, both in logistics and interpretation, have led Young et al. [70,71] and numerous other researchers [72–75] to use a continuous, stable isotope infusion technique for estimating amino acid requirements in human adults. Their results from studies of adult requirements for leucine, valine, lysine, and threonine suggest that adult needs for indispensable amino acids are two to three times greater than the currently accepted levels. The debate over amino acid requirements continues in the literature. Further research is being conducted, and ultimately consensus on amino acid requirements may be achieved.

Summary

Proteins in foods become available for use by the body after they have been broken down into their component amino acids. Nine of these amino acids are considered essential; therefore, the quality of dietary proteins correlates with its content of these indispensable amino acids.

An important concept in protein metabolism is the amino acid pool(s), which contains amino acids of dietary origin plus those contributed by the breakdown of body tissue. The amino acids comprising the pool(s) are used in a variety of ways: (1) for synthesis of new proteins for growth and/or replacement of existing body proteins; (2) for production of important nonprotein nitrogen-containing molecules; (3) for oxidation as a source of energy; and (4) for glucose, ketones, or fatty acid synthesis.

The liver is the primary site of amino acid metabolism, but no clear picture of the body's overall handling of nitrogen can emerge without consideration of amino acid metabolism in a variety of tissues and organs. Of particular significance is the metabolism of the branched-chain amino acids in the skeletal muscle and the production of the ammonium ion in the kidney. In addition, current research on neuropeptides spotlights the importance of amino acid metabolism in neural tissue.

Of the nonessential amino acids, glutamate assumes particular importance because of its versatility in overall metabolism of the amino acids. Glutamate and its α-keto acid make possible many crucial reactions in various metabolic pathways for amino acids. An appreciation for the functions performed by glutamate makes one realize that the connotation of "dispensable" as applied to this amino acid may be misleading.

Protein metabolism is particularly responsive to hormonal action, and this action can vary according to the tissue effect. Protein metabolism as regulated by hormonal action is particularly significant during periods of stress (see the Perspective).

Protein plays many roles in the body: it provides structure, it can be used as a source of energy, and many protein molecules, such as enzymes and neuropeptides, serve in a regulatory capacity. Because of its contribution to both energy production and synthesis of regulatory molecules, protein provides an excellent transition from the energy-producing nutrients to those with regulatory functions.

References Cited

1. Rose WC. The amino acid requirements of adult man. Nutr Abstr Rev 1957;27:631–43.

2. Jaksic T, Jahoor F, Reeds PJ, Heird WC. The determinants of amino acid synthesis in human neonates using a glucose-stable isotope tracer. Surg Forum 1993;44:642–4.

3. Mahe S, Roos N, Benamouzig R, Sick H, Baglieri A, Huneau JF, Tome D. True exogenous and endogenous nitrogen fractions in the human jejunum after ingestion of small amounts of ^{15}N-labeled casein. J Nutr 1994;124:548–55.

4. Kilberg MS, Stevens, BR, Novak DA. Recent advances in mammalian amino acid transport. Ann Rev Nutr 1993; 13:137–65.

5. Souba WW, Pacitti AJ. How amino acids get into cells: mechanisms, models, menus, and mediators. JPEN 1992;16:569–78.

6. Christensen HN. Naming plan for membrane transport systems for amino acids. Neurochem Res 1984; 9:1757–8.

7. Adibi SA, Gray S, Menden E. The kinetics of amino acid absorption and alteration of plasma composition of free amino acids after intestinal perfusion of amino acid mixtures. Am J Clin Nutr 1967;20:24–33.

8. Thwaites DT, Hirst BH, Simmons NL. Substrate specificity of the di/tripeptide transporter in human intestinal epithelia (Caco-2): identification of substrates that undergo H$^+$-coupled absorption. Br J Pharmacol 1994;113:1050–6.

9. Ganapathy V, Leibach FH. Is intestinal peptide transport energized by a proton gradient? Am J Physiol 1995;249:G153–60.

10. Vazquez JA, Daniel H, Adibi SA. Dipeptides in parenteral nutrition: from basic science to clinical applications. Nutr Clin Prac 1993;8:95–105.

11. Grimble GK, Rees RG, Keohane PP, Cartwright T, Desreumaux M, Silk DBA. Effect of peptide chain length on absorption of egg protein hydrolysates in the normal human jejunum. Gastroenterology 1987;92:136–42.

12. Zaloga GP. Physiological effects of peptide-based enteral formulas. Nutr Clin Prac 1990; 5:231–7.

13. Scheppach W, Loges C, Bartram P, Christl SU, Richter F, Dusel G, Stehle P, Fuerst P, Kasper H. Effect of free glutamine and alanyl-glutamine dipeptide on mucosal proliferation of the human ileum and colon. Gastroenterology 1994;107:429–34.

14. Matthews DE, Marano MA, Campbell RG. Splanchnic bed utilization of glutamine and glutamic acid in humans. Am J Physiol 1993;264:E848–54.

15. Watford M. Glutamine metabolism in rat small intestine: synthesis of three-carbon products in isolated enterocytes. Biochim Biophys Acta 1994;1200:73–78.

16. Welbourne TC, King AB, Horton K. Enteral glutamine supports hepatic glutathione efflux during inflammation. J Nutr Biochem 1993;4:236–42.

17. Gardner MLG. Gastrointestinal absorption of intact proteins. Ann Rev Nutr 1988;8:329–50.

18. Backwell FRC. Peptide utilization by tissues: current status and applications of stable isotope procedures. Proc Nutr Soc 1994;53:457–64.

19. Daniel H, Morse EL, Adibi SA. Determinants of substrate affinity for the oligopeptide/H$^+$ symporter in the renal brush border membrane. J Biol Chem 1992;267:9565–73.

20. Minami H, Daniel H, Morse EL, Adibi SA. Oligopeptides: mechanism of renal clearance depends on molecular structure. Am J Physiol 1992;263:F109–15.

21. Berthold HK, Jahoor F, Klein PD, Reeds PJ. Estimates of the effect of feeding on whole-body protein degradation in women vary with the amino acid used as tracer. J Nutr 1995;125:2516–27.

22. Jahoor F, Wykes LJ, Reeds PJ, Henry JF, Del Rosario MP, Frazer ME. Protein-deficient pigs cannot maintain reduced glutathione homeostasis when subjected to the stress of inflammation. J Nutr 1995;125:1462–72.

23. Tanphaichitr V, Leelahagul P. Carnitine metabolism and carnitine deficiency. Nutr 1993;9: 246–54.

24. Colombani P, Wenk C, Kunz I, Krahenbuhl S, Kuhnt M, Arnold M, Frey-Rindova P, Frey W, Langhans W. Effects of L-carnitine supplementation on physical performance and energy metabolism of endurance-trained athletes: a double blind crossover field study. Eur J Physiol 1996;73:434–9.

25. Marconi C, Sassi G, Carpinelli A, Cerretelli P. Effects of L-carnitine loading on the aerobic and anaerobic performance of endurance athletes. Eur J Appl Physiol 1985;54:131–5.

26. Cerretelli P, Marconi C. L-carnitine supplementation in humans: the effects on physical performance. Int J Sports Med 1990;11:1–14.

27. Vukovich MD, Costill DL, Fink WJ. Carnitine supplementation: effect on muscle carnitine and glycogen content during exercise. Med Sci Sports Exerc 1994;26:1122–9.

28. Harris RC, Soderlund K, Hultman E. Elevation of creatine in resting and exercised muscle of normal subjects by creatine supplementation. Clin Sci 1992;83:367–74.

29. Birch R, Noble D, Greenhaff PL. The influence of dietary creatine supplementation on performance during repeated bouts of maximal isokinetic cycling in man. Eur J Appl Physiol 1994;69:268–70.

30. Balsom P, Harridge S. Creatine supplementation per se does not enhance endurance exercise performance. Acta Physiol Scan 1993;149:521–3.

31. Hultman E, Soderlund K, Timmons JA, Cederblad G, Greenhaff PL. Muscle creatine loading in men. J Appl Physiol 1996;81:232–7.

32. Green AL, Hultman E, Macdonald IA, Sewell DA, Greenhaff PL. Carbohydrate ingestion augments skeletal muscle creatine accumulation during creatine supplementation in humans. Am J Physiol 996;271: E821–6.

33. Greenhaff, P, Casey A, Short A, Harris R, Soderlund K, Hultman E. Influence of oral creatine supplementation on muscle torque during repeated bouts of maximal voluntary exercise in man. Clin Sci Lond 1993; 84:565–71.

34. Cooke WH, Grandjean PW, Barnes WS. Effect of oral creatine supplementation on power output and fatigue during bicycle ergometry. J Appl Physiol 1995; 78:670–3.

35. Mujika I, Chatard JC, Lacoste L, Barale F, Geyssant A. Creatine supplementation does not improve sprint performance in competitive swimmers. Med Sci Sports Exerc 1996;28: 1435–41.

36. Yates A, Schlicker S, Suitor C. Dietary reference intakes: the new basis for recommendations for calcium and related nutrients, B vitamins and choline. J Am Diet Assoc 1998;98:699–706.

37. Watford M. The urea cycle: a two-compartment system. Essays in Biochemistry 1991; 26:49–58.

38. Morris SM. Regulation of enzymes of urea and arginine synthesis. Ann Rev Nutr 1992;12:81–101.

39. Kurowska EM, Carroll KK. Hypercholesterolemic responses in rabbits to selected groups of dietary essential amino acids. J Nutr 1994;124:364–70.

40. Castillo L, deRojas TC, Chapman TE, Vogt J, Burke JF, Tannenbaum SR. Splanchnic metabolism of dietary arginine in relation to nitric oxide synthesis in normal adult man. Proc Natl Acad Sci USA 1993;90: 193–7.

41. Kowalski TJ, Watford M. Production of glutamine and utilization of glutamate of rat subcutaneous adipose tissue in vivo. Am J Physiol 1994;266:E151–4.

42. Chinsky JM, Bohlen LM, Costeas PA. Noncoordinated responses of branched-chain α-ketoacid dehydrogenase subunit genes to dietary protein. FASEB J 1994;8:114–20.

43. Abumrad NN, Rabin D, Wise KL, Lacy WW. The disposal of an intravenously administered amino acid load across the human forearm. Metabolism 1982; 31:463–70.

44. Wahren J, Felig P, Hagenfeldt L. Effect of protein ingestion on splanchnic and leg metabolism in normal man and in patients with diabetes mellitus. J Clin Invest 1976;57: 987–99.

45. McCabe BJ. Dietary tyramine and other pressor amines in MAOI regimens: a review. J Am Diet Assoc 1986; 86:1059–64.

46. Brzezinski A. Melatonin in humans. N Engl J Med 1997;336:186–95.

47. Cavallo A. The pineal gland in human beings: relevance to pediatrics. J Pediatr 1993; 123:843–51.

48. Pacy PJ, Price GM, Halliday D, Quevedo MR, Millward DJ. Nitrogen homeostasis in man: the diurnal re-

sponses of protein synthesis and degradation and amino acid oxidation to diets with increasing protein intakes. Clin Sci 1994;86:103–18.

49. Fuller MF, Garlick PJ. Human amino acid requirements: can the controversy be resolved? Ann Rev Nutr 1994;14:217–41.

50. Curran RD, Ferrari FK, Kispert PH, Stadler J, Stuehr DJ, Simmons RL, Billiar TR. Nitric oxide and nitric oxide-generating compounds inhibit hepatocyte protein synthesis. FASEB J 1991;5:2085–92.

51. Collin-Vidal C, Cayol M, Obled C, Ziegler F, Bommelaer G, Beaufrere B. Leucine kinetics are different during feeding with whole protein or oligopeptides. Am J Physiol 1994;267:E907–14.

52. Welle S, Nair KS. Relationship of resting metabolic rate to body composition and protein turnover. Am J Physiol 1990;258:E990–8.

53. Dice JF. Peptide sequences that target cytosolic proteins for lysosomal proteolysis. Trends Biol Sci 1990; 15:305–9.

54. Beynon RJ, Bond JS. Catabolism of intracellular protein: molecular aspects. Am J Physiol 1986; 251:C141–52.

55. Attaix D, Taillandier D, Temparis S, Larbaid D, Aurousseau E, Combaret L, Voisin L. Regulation of ATP-ubiquitin-dependent proteolysis in muscle wasting. Reprod Nutr Dev 1994;34:583–97.

56. Mortimore GE, Poso AR. Intracellular protein catabolism and its control during nutrient deprivation and supply. Ann Rev Nutr 1987;7:539–64.

57. Hershko A, Ciechanover A. The ubiquitin system for protein degradation. Ann Rev Biochem 1992;61:761–807.

58. Bachmair A, Finley D, Varshavsky A. In vivo half-life of a protein is a function of its amino-terminal residue. Science 1986;234:179–86.

59. Rechsteiner M. Natural substrates of the ubiquitin proteolytic pathway. Cell 1991;66:615–8.

60. Tiao G, Fagan JM, Samuels N, James JH, Hudson K, Lieberman M, Fischer JE, Hasselgren P. Sepsis stimulates nonlysosomal, energy dependent proteolysis and increases ubiquitin mRNA levels in rat skeletal muscle. J Clin Invest 1994;94:2255–64.

61. Rogers S, Wells R, Rechsteiner M. Amino acid sequences common to rapidly degraded proteins: the PEST hypothesis. Science 1986;234:364–8.

62. Waxman L, Goldberg AL. Selectivity of proteolysis: protein substrates activate the ATP- dependent protease (La). Science 1986;232:500–3.

63. Goldberg AL. The mechanism and functions of ATP-dependent proteases in bacterial and animal cells. Eur J Biochem 1992;203:9–23.

64. Robertson L, Flinder C, Godfrey B. Laurel's Kitchen—A Handbook for Vegetarian Cookery and Nutrition. Petaluma, CA: Nilgiri Press, 1984.

65. Young VR, Pellett PL. Plant proteins in relation to human protein and amino acid nutrition. Am J Clin Nutr 1994;59(suppl):1203S–12S.

66. Food and Nutrition Board, National Research Council. Recommended Dietary Allowances, 10th ed. Washington, DC: National Academy Press, 1989.

67. Fuller MF, Milne A, Harris CI, Reid TMS, Keenan R. Amino acid losses in ileostomy fluid on a protein-free diet. Am J Clin Nutr 1994;59:70–73.

68. Energy and protein requirements. Report of a joint FAO/WHO/UN expert consultation. WHO technical report series 724. Geneva: World Health Organization, 1985.

69. Munro HN, Crim MC. The proteins and amino acids. In: Shils ME, Young VR, eds. Modern nutrition in health and disease. 7th ed. Philadelphia: Lea and Febiger, 1988;1–37.

70. Young VR. 1987 McCollum award lecture: kinetics of human amino acid metabolism: nutritional implications and some lessons. Am J Clin Nutr 1987; 46:709–25.

71. Young VR, Bier DM. A kinetic approach to determination of human amino acid requirements. Nutr Rev 1987;45:289–98.

72. Mequid MM, Matthews DE, Bier DM, Meredith CN, Soeldner JS, Young VR. Leucine kinetics at graded leucine intakes in young men. Am J Clin Nutr 1986; 43:770–80.

73. Mequid MM, Matthews DE, Bier DM, Meredith CN, Young VR. Valine kinetics at graded valine intakes in young men. Am J Clin Nutr 1986;43:781–6.

74. Ahao K, Wen A, Meredith CN, Matthews DE, Bier DM, Young VR. Threonine kinetics at graded threonine intakes in young men. Am J Clin Nutr 1986;43:795–802.

75. Meredith CN, Wen A, Bier DM, Matthews DE, Young VR. Lysine kinetics at graded lysine intakes in young men. Am J Clin Nutr 1986;43:787–94.

Protein Turnover: Starvation versus "Stress"

Starve a cold, feed a fever? Or feed a cold, starve a fever? In the healthy adult, protein synthesis approximately balances protein degradation. Together, protein synthesis and degradation make up what is referred to as *protein turnover*. Protein turnover in humans is correlated to one's metabolic mass ($W^{0.75}$), where W = body weight in kilograms. Daily protein turnover in humans is calculated to be approximately 4.6 g/kg of body weight. Thus, for a 70-kg male, protein turnover would approximate 320 g daily [1]. However, such calculations only approximate truth in a healthy adult. With illness, such as infection or sepsis (the presence of pathogenic micro-organisms or their toxins in the blood and/or body tissues), or during starvation and malnutrition, protein synthesis and degradation are not in balance. An imbalance between protein synthesis and degradation is also found with injury, including surgery, trauma, and burns; however, this imbalance exceeds that found in fasting (starvation) conditions. Conditions of illness or injury comprise what is referred to as "stress" in the title of this perspective. This perspective will compare and contrast what happens to one's protein status during starvation (malnutrition) and illness or injury (i.e., stress).

In malnutrition, protein synthesis decreases. This decrease occurs because of a reduction in mRNA needed for the translation of proteins, and because of a decreased rate of peptide bond formation (or RNA "activity"). Malnutrition to the point of starvation is also characterized by decreased protein synthesis. Even those proteins with very rapid turnover, such as plasma proteins, are synthesized at a rate 30% to 40% below normal. In muscle, protein synthesis rates drop even lower. However, protein degradation rates decrease concurrently so that in chronic starvation daily losses of nitrogen become quite small. For example, a person fully adapted to starvation can survive at a cost of 3 to 4 g of his or her body protein per day [2].

The principal mechanism of adjustment to starvation is a change in hormone balance. In particular, there is a sharp decrease in insulin production. In addition, the muscle and adipocytes become somewhat resistant to the action of insulin so that whatever insulin is circulating is ineffective in promoting cellular nutrient uptake for protein synthesis and lipogenesis. Decreased insulin activity, coupled with increased synthesis of counterregulatory hormones such as glucagon, promotes fatty acid mobilization from adipose tissue, production of ketones, and the availability of amino acids for gluconeogenesis. The glucocorticoids are important in gluconeogenesis because they promote catabolism

of muscle protein to provide substrates for gluconeogenesis. However, an increased adjustment to starvation is characterized by a decrease in the secretion of glucocorticoids. An additional hormonal change facilitating adjustment to starvation includes decreased synthesis of tri-iodothyronine (T_3, a thyroid hormone), which thus results in a lowered metabolic rate.

In the initial stages or first few days of fasting or starvation, glycogen in the liver is depleted. Muscles undergo proteolysis. Urinary 3-methylhistidine excretion increases to reflect myofibrillar protein catabolism. Muscles undergoing proteolysis also release a mixture of amino acids containing relatively high alanine and glutamine concentrations. Alanine is a preferred substrate for gluconeogenesis and serves to stimulate the secretion of the gluconeogenic hormone glucagon. Alanine released from muscle is taken up by the liver, where the nitrogen is removed and converted to urea for excretion by the kidney, and the pyruvate formed can enter the gluconeogenic pathway. Glucose formed in the liver may be released into the blood for cellular uptake and metabolism. Glutamine released from muscles circulates in the blood for uptake and metabolism primarily by the gastrointestinal tract and kidney, and not the liver. Within the liver, glutamine synthesis is increased and urea synthesis diminishes. These hepatic changes help to direct glutamine to the kidneys for maintenance of acid-base balance.

As fasting or starvation continues, tissues continue to use fatty acids and glucose for energy but also begin to use ketones formed in the liver from fatty acid oxidation. A decrease in protein catabolism and gluconeogenesis occurs concurrently with the brain's and other tissues' adaptation to ketones as a source of energy. Glutamine metabolism in the kidney increases as starvation continues and acidosis occurs. Within the kidney, glutamine catabolism generates ammonia, which helps to correct the acidosis.

Figure 1 (left) illustrates how adaptation to starvation allows the conservation of body protein. Fatty acids are shown generating ketones that are then used for energy by muscle. The utilization of the ketones means that less glucose is needed and allows the sparing of lean body mass for glucose production. In other words, because less carbohydrate is required by the body, less protein must be broken down to supply amino acids for gluconeogenesis. Amino acids resulting from proteolysis of muscle tissue can be used for the synthesis of crucial visceral proteins such as plasma proteins, which have more rapid turnover rates than muscle. It is estimated that under normal conditions visceral protein has a turnover rate three times greater than that of muscle protein [3].

The right side of Figure 1 depicts the substrate utilization in sepsis. With sepsis (an example of "stress"), as with star-

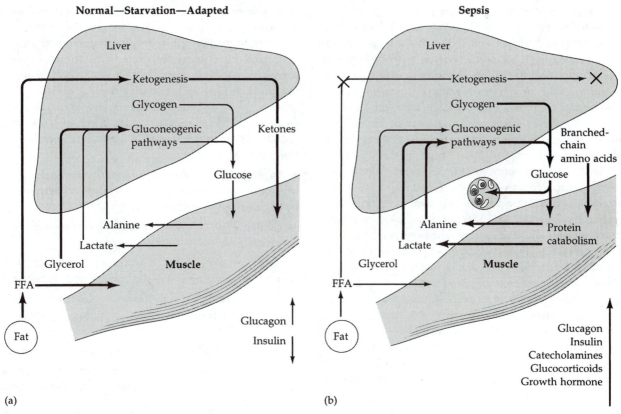

Figure 1 Gluconeogenesis during starvation and sepsis. During starvation (a), fat mobilization and ketogenesis provide energy, thereby sparing body protein and decreasing gluconeogenesis. In contrast, sepsis (b) causes stimulation of gluconeogenesis and inhibition of ketogenesis. Body protein is catabolized with no sparing effects from ketones. Fat mobilization occurs but to a lesser extent than with starvation. Increased responses during starvation and sepsis are shown by the heavy arrows. (Beisel WR, Wannemacher RW. Gluconeogenesis, ureagenesis and ketogenesis during sepsis. J Parent Enter Nutr 1980;4:278. © by the American Society of Parenteral and Enteral Nutrition.

vation, adipose tissue undergoes lipolysis. Although the body can partially defend itself against starvation through conservation of energy and adaptation to ketosis, it has no such defense against injury, trauma, surgery, and/or infection. Unlike during starvation, ketogenesis is inhibited by insulin during sepsis, burns, injury or trauma, and surgery. Without the utilization of ketones, body protein must continue to be degraded to supply amino acids for glucose synthesis (gluconeogenesis). Thus, protein degradation in sepsis, as well as in injury, trauma, and burns, exceeds that in starvation. Each gram of nitrogen lost can be translated into the breakdown of approximately 30 g of hydrated lean tissue [4].

The differences in substrate utilization between starvation and stress result in part from differences in hormone concentrations. Figure 2 demonstrates the metabolic stress response. As shown in this diagram, with stress, including sepsis, trauma, surgery, and burns, glucocorticoids (primarily cortisol), catecholamines (e.g., epinephrine), insulin, and glucagon release increase. However, the glucagon:insulin ratio favors glucagon. Consequently, tissues become resistant to insulin action, and hyperglycemia (high blood glucose concentrations) persists. In addition, cortisol concentrations may remain elevated in the blood for prolonged periods following severe trauma or stress events. High blood cortisol promotes proteolysis and hyperglycemia. In addition to hormone release, cytokines contribute to differences in substrate utilization during stress. Cytokines are low-molecular-weight peptides that evoke a number of varied reactions in the body, and are used by primarily immune cells to communicate with each other. Cytokines such as interleukin-1 (IL-1) and tumor necrosis factor (TNF)-α

∽ PERSPECTIVE *(continued)*

produced from macrophages in part mediate proteolysis and the hormonal response [5]. Inflammation involves similar cytokines such as IL-1 and TNF-α but also interleukin (IL)-6, interleukin (IL)-8, and interferon γ (IFN-γ). During sepsis, interleukin-1 has also been shown to induce proteolysis, although the mechanism of action is unknown [6].

Additional hormonal changes associated with "stress," illustrated in Figure 2, include the release of aldosterone and antidiuretic hormone. Aldosterone promotes renal sodium and fluid reabsorption, thus increasing blood volume. Antidiuretic hormone (ADH) inhibits diuresis (urination) to also effect an increase in blood volume. Both aldosterone and ADH help restore circulation if it has been depressed by shock or loss of blood fluids by hemorrhage associated with injury or surgery. The hormones thus help diminish total fluid losses, which may be high with skin loss from burns or with increased dermal losses from fever.

While surgery, sepsis, burns, and trauma are associated with continued protein degradation to supply amino acids for glucose synthesis, protein turnover also occurs because of the immune response and the acute phase response. The acute phase response is characterized by fever, hormonal changes, blood cell count changes, as well as changes in protein turnover. During the acute phase response, certain

body proteins such as muscle protein are preferentially degraded (by nonlysosomal energy-dependent systems [7]); however, in the liver, protein synthesis predominates. Glucocorticoids, in part, are thought to stimulate the observed increase in hepatic protein synthesis. Proteins that are synthesized in the liver during these stress situations include primarily a group of proteins referred to as *acute phase reactant* (APR) *proteins* or *acute phase response proteins* (APRPs). Some examples of APR proteins and their functions include

- haptoglobin, a protein that binds free hemoglobin (hemoglobin not in the red blood cell) that has been released by hemolysis of the red blood cell;

- ceruloplasmin, a copper containing protein with oxidase activity and the ability to scavenge radicals;

- alpha 2 macroglobulin, a protease inhibitor that functions to effect changes in tissue damage and restructuring;

- alpha 1 antitrypsin, a protease inhibitor that minimizes further tissue damage associated with phagocytosis of micro-organisms;

- fibrinogen, a protein required for blood coagulation;

- C-reactive protein, a protein that stimulates phagocytosis by white blood cells and activates complement proteins,

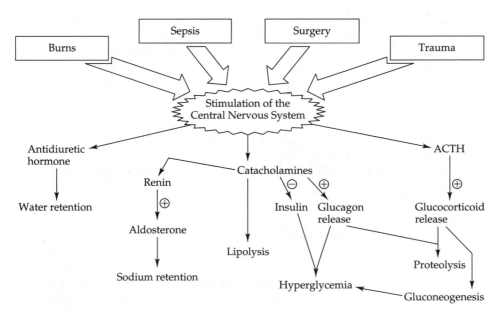

Figure 2 Metabolic stress response.

which are needed for antibody-induced destruction of micro-organisms;

• alpha 1 acid glycoprotein, a protein necessary for wound healing; and

• serum amyloid A, a protein involved in the oxidative burst.

In addition to the synthesis of these proteins, more metallothionein (a zinc-containing protein, see Chap. 12) is also made in the liver with sepsis and inflammation. Consequently, hepatic zinc concentrations increase while plasma zinc concentrations decrease. Similarly, in sepsis, the concentrations of iron and iron-containing proteins in the blood diminish and hepatic iron stores as ferritin rise.

However, when whole-body protein turnover is considered, the increased rate of hepatic synthesis of protein with sepsis and other stress situations is insignificant when compared with the rate of protein degradation. Whole body protein catabolism predominates over synthesis. Moreover, protein deficiency can diminish the magnitude and pattern of the acute phase response as well as the ability of the body to synthesize antioxidant defense compounds such as glutathione [8–10]. Even a delay in or the lack of enteral (meaning by way of or into the gastrointestinal tract) nutrition, which may accompany severe illness, can result in the atrophy of intestinal mucosa. When enterocytes atrophy, bacteria or toxins can more easily translocate from the lumen of the intestinal tract through the deformed enterocyte and into the blood [6]. Such bacterial translocation further increases the risk of sepsis, especially gram-negative. Glutamine has been shown to be a vital fuel for enterocytes. Yet, during illness, the rate of glutamine production and release from body sites (primarily muscle, lungs, adipose tissue) does not meet the intestinal cells' need for glutamine [6,11,12]. Further, lack of sufficient glutamine may contribute to inadequate glutathione production observed in stress [13]. Glutamine-enriched enteral nutrition products are advocated for patients hospitalized with such catabolic illnesses to help provide enterocytes with their needed metabolic fuels.

It has been suggested [4] that the body places a high priority on wound repair and host defense, gambling that convalescence will occur before depletion of tissues becomes a threat to survival. To improve recovery, feed a cold and feed a fever!

References Cited

1. Waterlow JC. Protein turnover with special reference to man. Q J Exp Physiol 1984; 69:409–38.

2. Tepperman J, Tepperman HM. Metabolic and Endocrine Physiology, 5th ed. Chicago: Year Book Medical, 1987.

3. Anon. Measuring human muscle protein synthesis. Nutr Rev 1989; 47:77–9.

4. Kinney JM, Elwyn DH. Protein metabolism and injury. Ann Rev Nutr 1983;3:433–66.

5. Hasselgren P, Fischer JE. Cytokines and protein metabolism. In: Gussler JD, ed. Cytokines in Critical Illness. Columbus, OH: Ross Laboratories, 1992:39–46.

6. Souba WW. Glutamine: a key substrate for the splanchnic bed. Ann Rev Nutr 1991;11: 285–308.

7. Tiao G, Fagan JM, Samuels N, James JH, Hudson K, Lieberman M, Fischer JE, Hasselgren P. Sepsis stimulates nonlysosomal, energy dependent proteolysis and increases ubiquitin mRNA levels in rat skeletal muscle. J Clin Invest 1994;94; 2255–64.

8. Doherty JF, Golden MNH, Raynes JG, Griffin GE. Acute phase protein response is impaired in severely malnourished children. Clin Sci 1993;84:169–75.

9. Jennings G, Bourgeois C, Elia M. The magnitude of the acute phase protein response is attenuated by protein deficiency in rats. J Nutr 1992;122:1325–31.

10. Grimble RF, Jackson AA, Persaud C, Wriede MJ, Delers F, Engler R. Cysteine and glycine supplementation modulate the metabolic response to tumor necrosis factor in rats fed a low protein diet. J Nutr 1992;122:2066–73.

11. Askanazi J, Carpentier YA, Michelsen CB, et al. Muscle and plasma amino acids following injury. Ann Surg 1980; 192:78–85.

12. Lacey JM, Wilmore DW. Is glutamine a conditionally essential amino acid? Nutr Rev 1990;48:297–309.

13. Welbourne TC, King AB, Horton K. Enteral glutamine supports hepatic glutathione efflux during inflammation. J Nutr Biochem 1993;4:236–42.

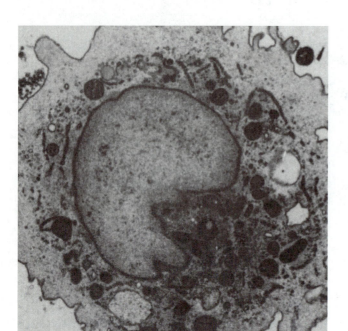

A phagocytic cell (macrophage)

Chapter 8

Integration and Regulation of Metabolism and the Impact of Exercise and Sport

Chapters 4, 6, and 7 featured carbohydrate, lipid, and protein metabolism at the level of the individual cell, with emphasis on metabolic pathways common to nearly all eukaryotic cells. We also discussed in those chapters how the pathways are regulated at the level of certain regulatory enzymes by substrate availability, allosteric mechanisms, and covalent modification such as phosphorylation.

For their significance to be fully appreciated, metabolic pathways, and the specific metabolic roles of different organs and tissues, must be viewed in the context of the whole organism. Therefore, in this chapter we will examine (1) how the major organs and tissues interact through integration of their metabolic pathways and (2) hormonal regulation of these metabolic processes in maintaining homeostasis. The pathways themselves are not reproduced again in this chapter, but when appropriate, the reader will be referred to pertinent sections in previous chapters where the pathways are described. A section on the currently attractive field of sports nutrition is included at the end of this chapter. It is presented at this point in the text because the dynamics of substrate utilization in supplying energy for physical exercise provide a practical example of how the various pathways of metabolism interrelate.

Interrelationship of Carbohydrate, Lipid, and Protein Metabolism

If ingested in sufficient amounts, any of the three energy-producing nutrients—carbohydrate, fat, and protein (amino acids)—can provide the body with its needed energy on a short-term basis. Within certain limitations, anabolic interconversion among the nutrients also occurs. For example, certain amino acids can be synthesized in the body from carbohydrate or fat, and, conversely, most amino acids can serve as precursors for carbohydrate or fat synthesis. The considerable metabolic interconversion among the nutrients is illustrated as an overview in Figure 8.1. Not evident from Figure 8.1, but very important to recall, is that the Krebs cycle is an *amphibolic pathway*, meaning that it functions not only in the oxidative catabolism of carbohydrates, fatty acids, and amino acids, but also provides precursors for many biosynthetic pathways, particularly gluconeogenesis (p. 94). Along with pyruvate, several Krebs cycle intermediates, including α-ketoglutarate, succinate, fumarate, and oxaloacetate, can be formed from the carbon skeletons of certain amino acids and can function as gluconeogenic precursors.

The fact that animals can be fattened on a predominantly carbohydrate diet is evidence of the apparent ease by which carbohydrate can be converted into fat. However, it is believed that in the human, lipogenesis

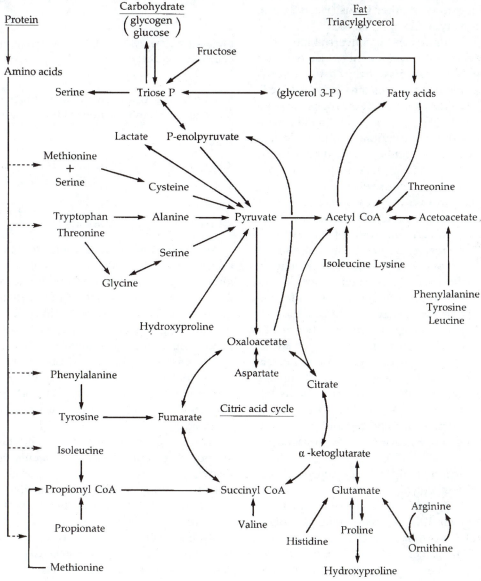

Figure 8.1   Interconversion of the macronutrients.

from glucose may be much less efficient than previously proposed [1] and that weight gain from carbohydrate is probably due to a sparing of lipolysis rather than direct carbohydrate lipogenesis [2]. Glucose is the precursor for both the glycerol and the fatty acid components of triacylglycerols. The glycerol portion can be formed from dihydroxyacetone phosphate (DHAP), a three-carbon intermediate in glycolysis (see Fig. 4.9). Reduction of DHAP by glycerol 3-phosphate dehydrogenase and NADH produces glycerol 3-phosphate, to which CoA-activated fatty acids attach in the course of triacylglycerol synthesis (see Fig. 6.13). A most significant reaction linking glucose metabolism to fatty acid synthesis is the reaction of the pyruvate dehydrogenase complex, which converts pyruvate to acetyl CoA by dehydrogenation and decarboxylation. Acetyl CoA is the starting material for the synthesis of long-chain fatty acids as well as a variety of other lipids (Fig. 8.2) (see also "Synthesis of Fatty Acids," p. 151).

Although carbohydrate can be converted into both the glycerol and fatty acid components of fat, only the glycerol portion of fat can be converted to carbohydrate. The conversion of fatty acids into carbohydrate is not possible because *the pyruvate dehydrogenase reaction is not reversible*. This prevents the direct conversion of acetyl CoA, the sole catabolic product of even-numbered carbon fatty acids, into pyruvate for gluconeogenesis. In addition, gluconeogenesis from acetyl CoA as a Krebs cycle intermediate also cannot occur. This is because for every two carbons in the form of acetyl CoA entering the cycle, two carbons are lost by decarboxylation in early reactions of the pathway (see Fig. 4.12). Therefore, there can be no net conversion of acetyl CoA to pyruvate or to the gluconeogenic intermediates of the cycle. Consequently, acetyl CoA produced from whatever source must be used for energy, lipogenesis, cholesterogenesis, or ketogenesis.

Although fatty acids having an even number of carbons are degraded exclusively to acetyl CoA and therefore are not glucogenic (gluconeogenic) for the reasons mentioned, fatty acids possessing an odd number of carbon atoms are partially glucogenic. This is because propionyl CoA ($CH_3-CH_2-COSCoA$), ultimately formed by β-oxidation, is carboxylated and rearranged to succinyl CoA, a glucogenic Krebs cycle intermediate (see 6M, p.150).

The glycerol portion of all triacylglycerols is glucogenic, entering the glycolytic pathway at the level of DHAP (see Fig. 4.9). Following its release from triacylglycerol by lipase hydrolysis, glycerol can be phosphorylated to glycerol 3-phosphate by glycerokinase, then oxidized to DHAP by glycerol 3-phosphate dehydrogenase (8A). During the fasting state, when fat catabolism

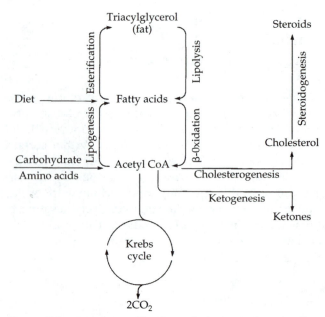

Figure 8.2   Overview of lipid metabolism, emphasizing the central role of acetyl CoA.

is accelerated, this conversion assumes greater importance in maintaining a normal level of blood glucose.

(8A)

$$CH_2-OH \quad \xrightarrow[]{ATP \quad ADP} \quad CH_2-OH \quad \xrightarrow[]{NAD^+ \quad NADH} \quad CH_2-OH$$
$$CH-OH \qquad\qquad CH-OH \qquad\qquad C=O$$
$$CH_2-OH \qquad\qquad CH_2-O-P \qquad\qquad CH_2-O-P$$
$$\text{Glycerol} \qquad\qquad \text{Glycerol 3-P} \qquad\qquad \text{DHAP}$$

Metabolism of the amino acids gives rise to a variety of amphibolic intermediates, some of which produce glucose while others produce the ketone bodies via their conversion to acetyl CoA or acetoacetyl CoA (see 6N, p. 150). Those amino acids that can be used for production of glucose are termed *glucogenic*, and those producing ketones are called *ketogenic*. Because they can be catabolized to acetyl CoA or acetoacetyl CoA, ketogenic amino acids are therefore potentially lipogenic. Several amino acids are both glucogenic and ketogenic, meaning that their carbon skeletons are convertible into both glucose and fats. Only the amino acids leucine and lysine are purely ketogenic. These amino acids cannot be gluconeogenic precursors. The dispensable glucogenic amino acids are usually interconverted with carbohydrate, but like the ketogenic amino acids, they also can be converted (indirectly, however) into fatty acids via acetyl CoA. The fatty acids, however, cannot be converted into the glucogenic amino acids for the same reason that fatty acids cannot form glucose—namely, the irreversibility of the pyruvate dehydrogenase reaction. Although entirely possible, the conversion of the glucogenic amino acids into fat is rather uncommon. Only when protein is supplying a high percentage of the calories would one expect glucogenic amino acids to be used in fat synthesis. Leucine and lysine, along with those amino acids that are partially ketogenic, will give rise directly to acetyl CoA, but the reversal of the process—that is, the synthesis of these amino acids from acetyl CoA—is impossible. All the amino acids producing acetyl CoA directly (isoleucine, threonine, phenylalanine, tyrosine,* lysine, and leucine) are indispensable.

The interconversion of the energy-producing nutrients appears to be skewed toward providing the organism with an energy source that can be easily stored (fat), thereby providing for times when food is not readily available.

Energy released by the catabolic processes of the major nutrients must be shared by the energy-requiring synthetic pathways discussed earlier. On reaching the cells, the energy-producing nutrients can be catabolized to produce phosphorylative energy (ATP) and/or reductive energy (NADH, NADPH, $FADH_2$). Alternatively, they may be synthesized into more complex organic compounds and/or macromolecules that become cellular components. For synthesis of a cellular component to occur, however, chemical energy must be provided. Therefore, when the cell places priority on the synthesis of a particular component, this synthesis is accomplished at the expense of another substance being catabolized. The common energy pool within a cell is finite, and all anabolic and endergonic processes must compete for this energy. For example, when the liver needs to produce more glucose by reversing glycolysis (i.e., gluconeogenesis), it cannot simultaneously synthesize lipids and proteins. Instead, some of the existing cellular proteins or lipids are hydrolyzed, and the resulting amino acids or fatty acids are oxidized to generate the NADH and ATP needed for gluconeogenesis. Likewise, when hepatic lipogenesis occurs, glucose must be used so as to produce the NADPH and ATP necessary for the conversion of acetyl CoA to fatty acids. The final common catabolic pathway for carbohydrate, fat, and protein is the Krebs cycle and oxidative phosphorylation via the electron transport chain (Figs. 4.12 and 3.12). In addition to releasing energy, these mitochondrial processes are crucial for many other metabolic sequences:

- CO_2 produced by oxidation of acetyl CoA is a source of cellular carbon dioxide for carboxylation reactions that initiate fatty acid synthesis and gluconeogenesis (pp. 94, 151). This CO_2 also supplies the carbon of urea and certain portions of the purine and pyrimidine rings (Figs. 7.15, 7.16, 7.22).

- Intermediates in the cycle allow cross-linkages between lipids, carbohydrate, and protein metabolism, as illustrated in Figure 8.1. Particularly notable intermediates are α-ketoglutarate and oxaloacetate. Another interrelationship not shown in Figure 8.1 is that between heme and an intermediate of the Krebs cycle, succinyl CoA. The initial step in heme biosynthesis is the formation of σ-aminolevulinic acid from "active" succinate and glycine.

- Cycle intermediates—citrate and malate—intermesh with lipogenesis. Citrate can move from the mitochondria into the cytoplast where citrate lyase cleaves it into oxaloacetate and acetyl CoA, the initiator of fatty acid synthesis. Malate, in the presence of $NADP^+$-linked malic enzyme, may provide a portion of the $NADPH^†$ required for reductive stages of fatty acid synthesis.

The Central Role of the Liver in Metabolism

Each tissue and organ of the human body has a specific function that is reflected in its anatomy and its metabolic activity. For example, skeletal muscle uses metabolic energy to perform mechanical activity, the brain uses energy to pump ions against concentration gradients in the transfer of electrical impulses, and adipose tissue serves as a depot for stored fat, which on release provides fuel for the rest of the body. Central

---

*Tyrosine is formed by hydroxylation of phenylalanine; therefore, its carbon skeleton cannot be synthesized in the body but must be obtained from food.

†A large portion of the NADPH is formed in the hexose-monophosphate shunt, an alternate pathway in the anaerobic oxidation of glucose. Acetyl CoA resulting from catabolism of carbohydrate and fat enters the Krebs cycle in combination with oxaloacetate. Some of the protein breakdown products also enter the cycle as acetyl CoA, but many enter as α-keto acids. In any event, the oxidative degradation of the product(s) transfers hydrogen from the product(s) to flavin adenine dinucleotide (FAD) fixed in the membranes of the mitochondria or to the mobile NAD^+ molecules. Any $FADH_2$ produced is reoxidized by the coenzyme Q of the electron transport chain. The NADH formed, however, may transfer its hydrogen to the electron transport chain, or its reducing equivalents can be shuttled out of the mitochondria, to be used in reductive syntheses (e.g., in gluconeogenesis). A key factor in determining the direction taken by NADH will be the ADP/ATP ratio. High ADP levels, which signify low energy reserves, will stimulate oxidative phosphorylation, thereby causing NADH to be drawn into the electron transport chain. High ATP levels will inhibit oxidative phosphorylation and cause NADH to be used for reductive syntheses.

to these processes is the liver. It plays the key role of processor and distributor in metabolism, furnishing by way of the bloodstream a proper combination of nutrients to all the other organs and tissues. It warrants special attention in a discussion of tissue-specific metabolism.

Figures 8.3, 8.4, and 8.5 illustrate the fate of glucose 6-phosphate, amino acids, and fatty acids, respectively, in the liver. In these figures, anabolic pathways are shown pointing upward; catabolic pathways, down; and distribution to other tissues, horizontally. The pathways indicated are described in detail in Chapters 4, 6, and 7, which deal with carbohydrate, lipid, and protein metabolism, respectively.

Glucose entering the hepatocytes is phosphorylated by glucokinase to glucose 6-phosphate. Other dietary monosaccharides (fructose, galactose, and mannose) are also phosphorylated and rearranged to glucose 6-phosphate. Figure 8.3 shows the possible metabolic routes available to glucose 6-phosphate. It is likely that liver glycogenesis occurs primarily from gluconeogenic precursors delivered to the hepatocytes from peripheral tissues rather than through glucose directly [3] (see also Fig. 4.6). This finding is referred to again in the following section.

Figure 8.4 reviews the particularly active role of the liver in amino acid metabolism. The liver is the site of synthesis of many different proteins, both structural and plasma-borne, from amino acids. Amino acids can also be converted into nonprotein products such as nucleotides, hormones, and porphyrins. Amino acids can be transaminated and degraded to acetyl CoA and other Krebs cycle intermediates, and these can in turn be oxidized for energy or converted to glucose or fat. Glucose formed from gluconeogenesis can be transported to muscle for use by that tissue, and synthesized fatty acids can be mobilized to adipose tissue for storage or used as fuel by muscle. Hepatocytes are the exclusive site for the formation of urea, the major excretory form of amino acid nitrogen. The fate of fatty acids entering the liver is represented in Figure 8.5. Fatty acids can be assembled into liver lipids or released into the circulation as plasma lipoproteins. In humans, most fatty acid synthesis takes place in the liver rather than in adipocytes. Adipocytes store triacylglycerols arriving from the liver primarily in the form of plasma VLDLs and from the lipoprotein lipase action on chylomicrons (p. 137). Under most circumstances, fatty acids are the major oxidative fuel in the liver. Acetyl CoA not used for energy can be used for the formation of the ketone bodies, which are very important fuels for certain peripheral tissues such as the brain and heart muscle, particularly during prolonged fasting.

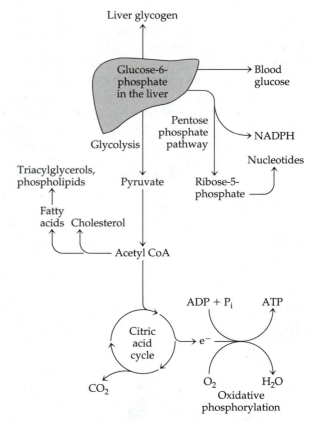

Figure 8.3 Metabolic pathways for glucose 6-phosphate in the liver. Here and in Figures 8.4 and 8.5, which follow, anabolic pathways are shown pointing upward; catabolic pathways, downward; and distribution to other tissues, horizontally.

Tissue-Specific Metabolism during the Feed-Fast Cycle

Carbohydrate and Fat Metabolism

The best way to learn to appreciate the interrelationship of metabolic pathways and the involvement of different organs and tissues in metabolism is to gain an understanding of the feed-fast cycle. Because glucose is the major fuel for most tissues, it is very important that its homeostasis is maintained whether food has just been consumed or a state of fasting exists. If the fasting state is prolonged, other fuels gain in importance. The extent to which different organs are involved in carbohydrate and fat metabolism varies within the feed-fast cycles that underlie the eating habits of the human being. Food consumption often occurs at a level 100 times greater than the basal caloric requirement, allowing us to survive from meal to meal without nibbling continuously. Excess calories are stored as glycogen and fat, and these can be used as needed. A feed-fast cycle can be divided into states, or phases:

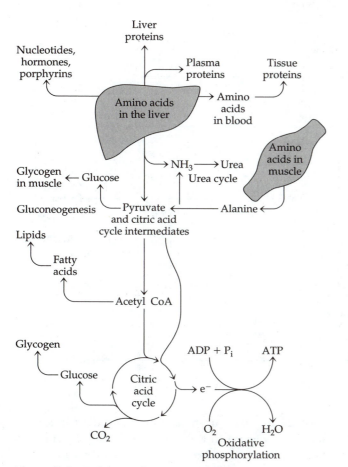

Figure 8.4 Pathways of amino acid metabolism in the liver.

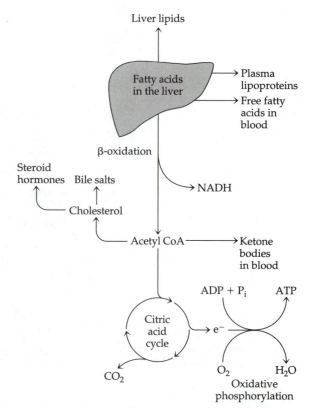

Figure 8.5 Pathways of fatty acid metabolism in the liver.

1. The fed state, lasting about 3 hours after the ingestion of a meal

2. The postabsorptive or early fasting state, occurring during a time span of from 3 to about 12 to 16 hours following the meal

3. The fasting state, lasting up to 2 days without additional intake of food

4. The starvation state, marked by prolonged food deprivation of several weeks' duration

Clearly, in a normal eating routine only the fed and early fasting states apply. Time frames of the phases cited are only approximate and are strongly influenced by factors such as activity level, the caloric value and nutrient composition of the meal, and the subject's metabolic rate.

The Fed State

Figure 8.6 illustrates the disposition of glucose, fat, and amino acids among the various tissues during the fed state. The red blood cells (RBCs) and the central nervous system (CNS) have no metabolic mechanisms by which glucose can be converted into energy stores. Glucose available to these tissues is oxidized immediately to produce energy. In the liver, in contrast, some of the glucose can be converted directly to glycogen. Contrary to the conventional view of liver glycogenesis, however, research indicates that most of the liver glycogen is synthesized indirectly from gluconeogenic precursors (pyruvate and lactate) returning to the liver from the periphery rather than directly from glucose entering the liver via the portal vein [3]. A likely source of lactate for the liver is the red blood cell, as indicated in Figure 8.6. The reason that glucose is not used well as a direct precursor of glycogen has been attributed to the low phosphorylating activity of the liver at physiological concentrations of glucose [4].

The liver is the first tissue to have the opportunity to use dietary glucose. In the liver, glucose can be converted into glycogen, and when available glucose or its gluconeogenic precursors exceed the glycogen storage capacity of the liver, excess glucose can be metabolized in a variety of ways. This is shown in Figure 8.3 and in somewhat more detail in Figure 8.6. The conversion of glucose to fatty acids and glycogen is important because both represent the storage of glucose carbon. The potential conversion of excess glucose to fatty acids is particularly crucial because these fatty acids,

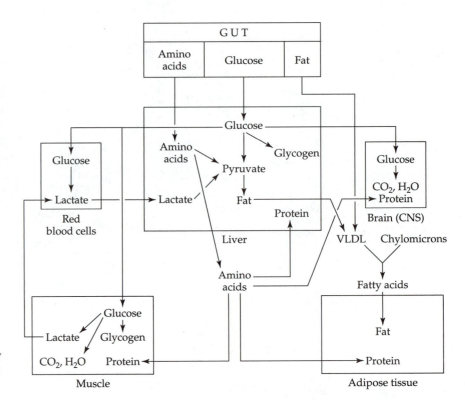

Figure 8.6 Disposition of dietary glucose, amino acids, and fat in the fed state.

along with those removed from the chylomicrons and VLDL by lipoprotein lipase, can be stored in the adipose tissue, thereby providing a ready source of fuel for most body tissues during the postabsorptive and fasting states.

Some exogenous glucose, that coming from the intestine, escapes the liver and circulates to other tissues. The brain and other tissues of the central nervous system are almost solely dependent on glucose as an energy source. Other major users of glucose include (1) the RBCs, which, lacking mitochondria, convert it glycolytically to lactate for the small amount of energy the cell requires, and also use it as a source of NADPH via the hexosemonophosphate shunt; (2) adipose tissue, which can use it to some extent as a precursor for both the glycerol and fatty acid components of triacylglycerols, although, in the human, most triacylglycerol is synthesized by the liver and transported to the adipose tissue; and (3) muscle, which uses glucose for the synthesis of glycogen. With the exception of the RBCs, all the tissues included in Figure 8.6 actively catabolize glucose for energy via the Krebs cycle.

In considering fat delivery to the tissues, it is necessary to differentiate between exogenous and endogenous fat. Dietary fat, except for short-chain fatty acids, enters the bloodstream as chylomicrons, which are promptly acted on by lipoprotein lipase from the vascular endothelium, releasing free fatty acids and glycerol. Chylomicron remnants remaining from this hy-

drolysis are taken up by the liver, and their lipid contents transferred to the very low-density lipoprotein fraction. The fatty acids are taken up by the adipocytes, reesterified with glycerol to form triacylglycerols, and stored as such as large fat droplets within the cells.

The Postabsorptive or Early Fasting State

With the onset of the postabsorptive state, tissues can no longer derive their energy directly from ingested glucose and other nutrients but must begin to depend on other sources of fuel. During the short period of time marking this phase (a few hours after eating), hepatic glycogenolysis is the major provider of glucose as fuel to other tissues. The synthesis of glycogen and triacylglycerols is diminished, and the de novo synthesis of glucose (gluconeogenesis) begins to help to maintain blood glucose levels. Lactate, formed in and released by RBCs and muscle tissue, becomes an important carbon source for hepatic gluconeogenesis. The alanine-glucose cycle, in which carbon in the form of alanine returns to the liver from muscle cells, also becomes important. The alanine is then converted to pyruvate as the first step in the gluconeogenic conversion of alanine in the liver.

Glucose provided to the muscle by the liver comes primarily from the recycling of lactate and alanine and, to a lesser extent, from hepatic glycogenolysis. Muscle glycogenolysis provides glucose as fuel only for those

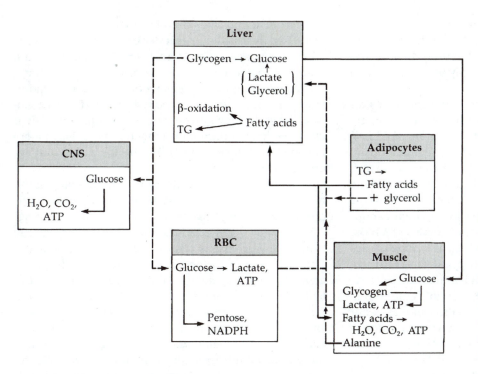

Figure 8.7 The primary postabsorption flow of substrates among the liver, CNS, adipose tissue, muscle, and erythrocytes. *Source:* Modified from Zakim D, Boyer T. Integration of energy metabolism by the liver. In: Hepatology: A Textbook of Liver Disease, edited by M. Greene. Philadelphia: WB Saunders, 1982:78. With permission.

muscle cells in which the glycogen is stored, because muscle lacks the enzyme glucose 6-phosphatase. Glucose, once phosphorylated in the muscle, is trapped there.

The brain and other tissues of the CNS are extravagant consumers of glucose, oxidizing it for energy and releasing no gluconeogenic precursors in return. Therefore, the rate of glucose use is greater than that of gluconeogenesis, and the stores of liver glycogen begin to diminish rapidly. In the course of an overnight fast, nearly all reserves of liver glycogen and most of the muscle glycogen have been depleted. Figure 8.7 shows the shifts of metabolic pathways occurring in the tissues during the postabsorptive state.

The Fasting State

The postabsorptive state evolves into the fasting state after 48 hours of no food intake. Particularly notable in the liver is the de novo glucose synthesis (gluconeogenesis) occurring in the wake of glycogen depletion. Amino acids from muscle protein breakdown provide the chief substrate for gluconeogenesis, although the glycerol from lipolysis and the lactate from anaerobic metabolism of glucose also are used to some extent.

The shift to gluconeogenesis during prolonged fasting is signaled by the secretion of the hormone glucagon and the glucocorticosteroid hormones in response to low levels of blood glucose. Proteins are hydrolyzed in muscle cells at an accelerated rate to provide the glucogenic amino acids. Of all the amino acids, only leucine

and lysine cannot contribute at all to gluconeogenesis because, as noted previously, these amino acids are totally ketogenic. However, ketogenic amino acids released by muscle protein hydrolysis serve a purpose as well. Because they are converted into ketones—that is, acetyl CoA, acetoacetyl CoA, or acetoacetate—they allow the brain, heart, and skeletal muscle to adapt to the use of these substrates should the nutritive state continue to deteriorate into a state of frank starvation.

The fasting state is accompanied by large daily losses of urinary nitrogen, in keeping with the high rate of gluconeogenic conversion of muscle protein to provide substrates for hepatic gluconeogenesis.

The Starvation State

If the fasting state persists and progresses into a starvation state, a metabolic fuel shift occurs again in an effort to spare body protein. This new priority is justified by the vital physiological importance of body proteins. Proteins that must obviously be conserved for life to continue include antibodies, needed to fight infection; enzymes, which catalyze life-sustaining reactions; and hemoglobin, for the transport of oxygen to tissues. The protein-sparing shift at this point is from gluconeogenesis to lipolysis, as the fat stores become the major supplier of energy. The blood level of fatty acids increases sharply, and these replace glucose as the preferred fuel of heart, liver, and skeletal muscle tissue that oxidize them for energy. The brain cannot use fatty acids for energy because fatty acids cannot cross the blood-brain barrier. However, the shift to fat breakdown also releases a large amount of glycerol, which becomes the major gluconeogenic precursor, rather than amino acids. This assures a continued supply of glucose as fuel for the brain.

Eventually, the use of Krebs cycle intermediates for gluconeogenesis depletes the supply of oxaloacetate. Low levels of oxaloacetate, coupled with the rapid production of acetyl CoA from fatty acid catabolism, causes acetyl CoA to accumulate, favoring the formation of acetoacetyl CoA and the ketone bodies. Ketone body concentration in the blood then rises (ketosis) as these fuels are exported from the liver, which cannot use them, to skeletal muscle, heart, and brain, which oxidize them instead of glucose. As long as ketone bodies are maintained at a high level by hepatic fatty acid oxidation, the need for glucose and gluconeogenesis is reduced, thereby sparing valuable protein. Figure 8.8 illustrates the changes in energy metabolism that occur in various tissues during fasting and starvation states.

Survival time in starvation, therefore, depends on the quantity of fat stored before starvation. Stored triacylglycerols in the adipose tissue of an individual of normal weight and adiposity can provide enough fuel to sustain basal metabolism for about 3 months. A very obese adult could probably endure a fast of more than a year, but physiological damage and even death could result from the accompanying extreme ketosis. When fat reserves are gone, the degradation of essential protein begins, leading to loss of liver and muscle function and, ultimately, death [5].

Amino Acid Metabolism

Organ interactions in amino acid metabolism, illustrated in Figure 8.9, are largely coordinated by the liver. The pathways shown undergo regulatory adjustments after consumption of a meal containing protein.

In the fed state, absorbed amino acids pass into the liver, where the fate of most of them is determined in relation to needs of the body; amounts in excess of need are degraded. Only the branched-chain amino acids (BCAAs) are not regulated by the liver according to the body's need. Instead, the BCAAs pass to the periphery, primarily to the muscles and adipose tissue, where they may be metabolized. Of particular interest is the fate of the BCAAs reaching the muscle. These amino acids are usually much in excess of need for muscle protein synthesis. It is believed that the excess is used to synthesize the dispensable amino acids that are needed for the increase in protein synthesis that occurs after a protein meal.

The liver is the site of urea synthesis, the primary mechanism for disposing of the excess nitrogen derived from amino acids used for energy or gluconeogenesis. The liver is the primary site for gluconeogenesis, where α-ketoacids (amino acids from which the amine group has been removed) serve as the chief substrate. During fasting, gluconeogenesis becomes a very important metabolic pathway in the regulation of plasma glucose levels. Kidney gluconeogenesis supplements liver gluconeogenesis during prolonged fasts. Kidney gluconeogenesis is accompanied by the formation and excretion of ammonia.

The importance of the liver to the functioning of the muscle during the fasting state or very vigorous exercise is exemplified in the alanine-glucose cycle (Fig. 7.35). Alanine, formed in the muscle, results primarily from pyruvate that is transaminated with glutamate. The pyruvate is formed mainly from muscle glycogenolysis but may also have been recycled from the liver. The alanine thus formed provides a disposal route for the nitrogen produced from catabolism of muscle amino acids. Alanine returning to the liver is transaminated with α-ketoglutarate, re-forming pyruvate. The transaminated nitrogen enters the urea cycle while the pyruvate is converted once again to glucose through the gluconeogenic pathway. The glucose can then be returned to muscle for energy, thereby completing the cycle. Alanine can also be formed in intestinal mucosal cells by transamination involving pyruvate, thereby serving as a carrier of amino acid nitrogen from that site to the liver.

Glutamine also plays a central role in the transport and excretion of amino acid nitrogen (p. 196). Many tissues, including the brain, combine ammonia, released primarily by the glutamate dehydrogenase reaction, with glutamate to form glutamine. The reaction is catalyzed by glutamine synthetase. In the form of glutamine, ammonia can then be carried to the liver or kidneys for excretion as urea or ammonium ion, respectively. In those tissues, glutamine is acted on by the enzyme glutaminase, releasing the ammonia for excretion, and re-forming glutamate. An overview of organ cooperativity in these and other aspects of amino acid metabolism is illustrated in Figure 8.9. See Chapter 7 for a more detailed discussion of amino acid metabolism in general.

System Integration and Homeostasis

Integration of the metabolic processes, as outlined in the preceding sections, allows the "constancy of the internal milieu" of humans and other multicellular organisms that was described by the French physiologist Claude Bernard about a century ago. This integration of metabolism at the cellular and the organ and tissue levels, which is essential for the survival of the entire

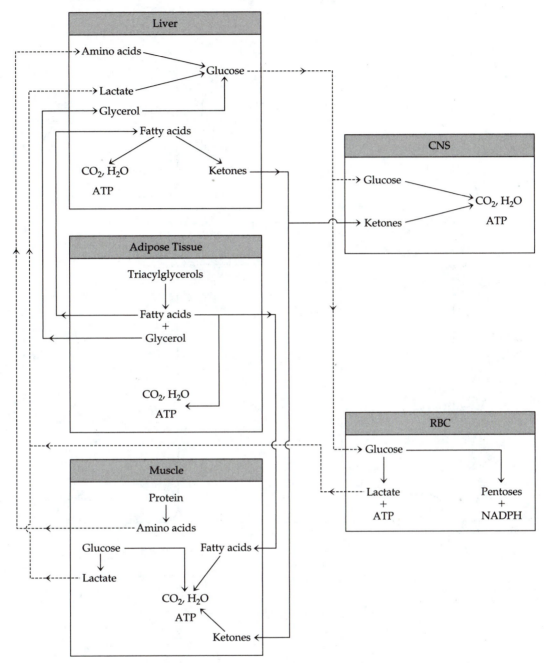

Figure 8.8 Flow of substrates among the liver, brain and CNS, adipose tissue, muscle, and red blood cells during fasting and starvation states. Broken lines indicate major substrate flow during fasting of about 2 days' duration. Solid lines reflect predominant substrate flow during the starvation state. Notice the shift to fatty acid and glycerol export from adipose tissue and the increased use of fatty acids and ketones as fuel, during starvation, by brain and muscle.

organism, receives its direction via body systems. Through the integration of body systems, communication is made possible among all parts of the body.

The three major systems that direct activities of the cells, tissues, and organs to ensure their harmony with the whole organism are the nervous, endocrine, and vascular systems.

The *nervous system* is considered the primary communication system because it not only has receiving mechanisms to assess the body's status in relation to its environment but also has transmitting processes to relay appropriate commands to various tissues and organs. The nervous system can inform the body of such conditions as hunger, thirst, pain, and lack of oxygen.

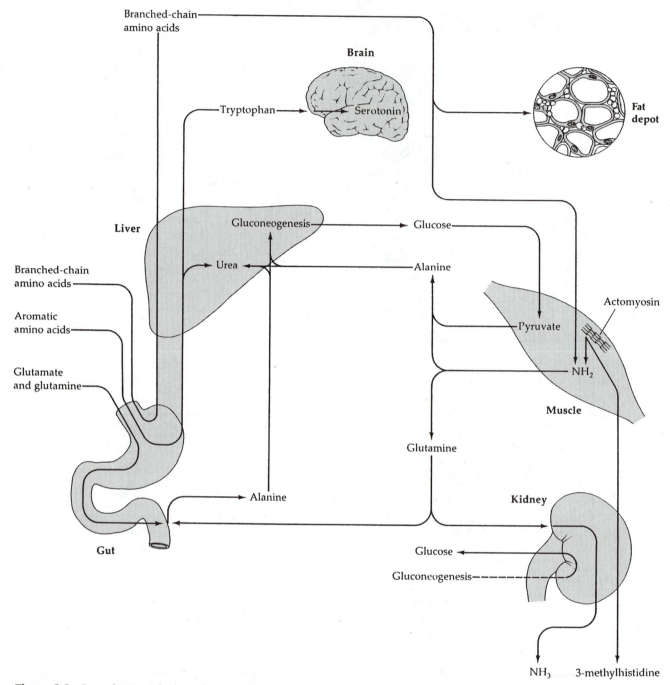

Figure 8.9 Interchanges of selected amino acids and their metabolites among body organs and tissues. *Source:* Modified from Munro HN. Metabolic integration of organs in health and diseases. J Parent Enter Nutr 1982;6:271–9. © by the American Society of Parenteral and Enteral Nutrition. Used with permission.

This allows organs to adjust to external changes, and appropriate behavior by the whole organism may be initiated. Tepperman and Tepperman [6] compare the nervous system to an elaborate system of telegraphy that has a "wire" connection from the source of message initiation to the place where message reception has its needed effect.

The *endocrine system* [6] is compared to a wireless system that transmits messages via highly specialized substances called *hormones*. The endocrine system depends on the vascular system to carry messages to target tissues.

The *vascular system* is the primary transport mechanism for the body, not only delivering specialized chemical substances but also carrying oxygen, organic nutrients, and minerals from the external environment to the cells throughout the body. It also transports the waste products of metabolism from the cells, carrying them to the lungs and/or kidneys for elimination.

The concentration of solutes in the blood must be regulated within a narrow range. Among the most prominent sentinel cells that monitor and regulate solute concentration are those that synthesize and secrete hormones. Although hormone synthesis and secretion occur primarily in the endocrine system, considerable overlap exists between the endocrine system and the CNS. With the recent discovery of a variety of neuropeptides and recognition of the hormonal action of many of these peptides, it has become apparent that the CNS and endocrine system are functionally interdependent [6,7].

Tissues and cells that respond to hormones are called the "target" tissues and cells of the hormones. These hormone-responsive cells have been preprogrammed by the process of differentiation to respond to the presence of hormones by acting in a predictable way. Not only do hormone-responsive cells respond to hormones via specific receptors, but their metabolic pathways also can be affected by the concentration of available substrates. Hormone-responsive cells live in a complex and continually changing environment of fuels and ions. Their ultimate response to these changes is the net result of both hormonal and nonhormonal information brought to them by the extracellular fluids in which they are bathed [6].

Endocrine Function in Fed State

Endocrine organs are distributed throughout the body, and most of these organs are involved primarily with nutrient ingestion—that is, the gastrointestinal (GI) tract. Interspersed among the absorptive and exocrine secretory cells of the upper GI tract are the highly specialized endocrine cells. These cells present a sensor face to the lumen and secrete granule-stored hormones into the bloodstream. Each of these cells is stimulated to secrete by a different combination of chemical messages. Chemical messages include, for example, glucose, amino acids, fatty acids, and alkaline or acid pH. Hormones secreted by these stimulated GI cells (GIP, CCK, gastrin, secretin) (Table 2.1, p. 31) then enter the bloodstream and sensitize appropriate cells of the endocrine pancreas for response to the approaching nutrients. The primary action of the GI hormones, secreted in response to a mixed diet, is to amplify the response of the pancreatic islet β-cells to glucose [6].

Insulin, secreted by the β-cells, is the hormone primarily responsible for the direction of energy metabolism during the fed state (Fig. 8.6). Its effects can be categorized as (1) very fast, occurring in a matter of seconds; (2) fast, occurring in minutes; (3) slower, occurring in minutes to several hours; (4) slowest, only occurring after several hours or even days. An example of a *very fast action* of insulin is membrane changes stimulated by the hormone. These changes occur in specific cells where glucose entry depends on membrane transport (see "Glucose Transporters," in Chap. 4). The *fast action* of insulin involves the activation or inhibition of many enzymes, with anabolic actions being accentuated. For example, the hormone stimulates glycogenesis, lipogenesis, and protein synthesis, while it inhibits opposing catabolic actions. Several metabolic effects of insulin and the corresponding target enzymes involved are listed in Table 8.1. Insulin favors glycogenesis through the activation of a phosphatase that dephosphorylates phosphorylase and glycogen synthase. This dephosphorylation activates glycogen synthetase while inhibiting the phosphorylase that initiates glycogenolysis (p. 97). The fast effect of insulin on protein synthesis is not as clear-cut as its influence on lipogenesis and glycogenesis. Nevertheless, protein synthesis is promoted and appears related to stimulation of the translation process [6].

One *slower action* of insulin involves a further regulation of enzyme activity. This regulation is accomplished through the selective induction or repression of enzyme synthesis. The induced enzymes are the key rate-limiting enzymes for anabolic reaction sequences, while the repressed enzymes are those crucial to control of opposing catabolic reactions. An example of selective induction is the effect of insulin on glucokinase activity. Insulin increases the synthesis of glucokinase

Table 8.1 Metabolic Effects of Insulin and Its Action on Specific Enzymes

Metabolic Effect	Target Enzyme
↑ Glucose uptake (muscle)	↑ Glucose transporter
↑ Glucose uptake (liver)	↑ Glucokinase
↑ Glycogen synthesis (liver, muscle)	↑ Glycogen synthase
↓ Glycogen breakdown (liver, muscle)	↓ Glycogen phosphorylase
↑ Glycolysis, acetyl CoA production (liver, muscle)	↑ Phosphofructokinase-1 ↑ Pyruvate dehydrogenase complex
↑ Fatty acid synthesis (liver)	↑ Acetyl CoA carboxylase
↑ Triacylglycerol synthesis (adipose tissue)	↑ Lipoprotein lipase

by promoting transcription of the glucokinase gene. Another slower action of insulin is its stimulation of cellular amino acid influx. The *slowest effect* of insulin is its promotion of growth through mitogenesis and cell replication. The passage of a cell through its various phases before it can replicate is a relatively slow process that requires 18 to 24 hours for completion. Its complexity precludes a significant rate increase in the overall process.

Endocrine Function in Postabsorptive or Fasting State

Metabolic adjustments that occur in response to food deprivation operate on two time scales: acutely, measured in minutes (such adjustments can be seen in a postabsorptive state), and chronically, measured in hours and days (adjustments occurring during fasting or starvation). In contrast to the fed state, in which insulin is the hormone primarily responsible for the direction of energy metabolism, the body deprived of food requires a variety of hormones to regulate its fuel supply.

Figure 8.7 depicts the postabsorptive state in which hepatic glycogenolysis is providing some glucose to the body, while increased use of fatty acids for energy is decreasing the glucose requirement of cells. Also, gluconeogenesis is being initiated, with lactate and glycerol serving as substrates.

Hepatic glycogenolysis is initiated through the action of glucagon, secreted by the α-cells of the endocrine pancreas, and of epinephrine (adrenaline) and norepinephrine, synthesized primarily in the adrenal medulla and sympathetic nerve endings, respectively. Epinephrine is considerably more potent in this metabolic action than norepinephrine, which mainly functions as a neurotransmitter. Epinephrine and norepinephrine are called the *catecholamine hormones* because they are derivatives of the aromatic alcohol catechol. Although influencing hepatic glycogenolysis somewhat, the catecholamines exert their effect primarily on the muscles. The action of glucagon and the catecholamines is mediated through cAMP and protein kinase phosphorylation. (This mechanism is described in the section on glycogenolysis in Chap. 4; see also Fig. 4.8.) Through the action of glucagon on the liver, phosphorylase and glycogen synthetase are phosphorylated, in direct opposition to the action of insulin; consequently, phosphorylase is activated and glycogen synthetase is inhibited. As a result, glycogen is broken down, giving rise to glucose 6-phosphate, which then can be hydrolyzed by the specific liver phosphatase (glucose 6-phosphatase) to produce free glucose.

In contrast, muscle glycogenolysis, stimulated by the catecholamines, provides glucose only for the muscles in which the glycogen has been stored. This is because muscle tissue lacks glucose 6-phosphatase and cannot release free glucose into the circulation. The catecholamines, however, do raise the blood glucose indirectly because they stimulate the secretion of glucagon and also inhibit the uptake of blood glucose by the muscles.

Glycogenolysis can occur within minutes and meets an acute need for raising the blood glucose level. However, because so little glycogen is stored in the liver (~60–65 g), blood glucose cannot be maintained over a prolonged period. In prolonged glucose deprivation such as occurs during fasting, the liver employs another mechanism for supplying the body with glucose: gluconeogenesis. Although lactate and glycerol serve as substrates for gluconeogenesis, the primary substrates are the amino acids derived from protein tissues, principally from muscle mass. Gluconeogenesis is fostered by the same hormones that initiate glycogenolysis (glucagon and epinephrine), but the amino acids needed as substrate are made available through the action of the glucocorticoids secreted by the adrenal cortex. Glucocorticoid hormones stimulate gluconeogenesis. Alanine, generated in the muscle from other amino acids and from pyruvate by transamination, and serving as the principal gluconeogenic substrate, also acts as a stimulant of gluconeogenesis via its effect on the secretion of glucagon. In fact, alanine is the prime stimulator of glucagon secretion by α-cells that have been sensitized to the action of alanine by the glucocorticoids.

Low levels of circulating insulin not only decrease the use of glucose but also promote lipolysis and a rise in free fatty acids. Contributing to this effect is the increase in glucagon during the fasting period. Glucagon raises the level of cyclic AMP in adipose cells, and the cyclic AMP then activates a lipase that hydrolyzes stored triacylglycerols. The muscles, inhibited by the catecholamines from taking up glucose, now begin to use fatty acids as the primary source of energy. This increased use of fatty acids by the muscles represents an important adaptation to fasting. Growth hormone and the glucocorticoids foster this adaptation because they, like the catecholamines, inhibit in some manner the use of glucose by the muscles.

As starvation is prolonged, less and less glucose is used, thereby reducing the amount of protein that must be catabolized to provide substrate for gluconeogenesis. As glucose use decreases, hepatic ketogenesis increases and the brain adapts to the use of ketones (primarily β-hydroxybutyrate) as a partial source of energy. After 3 days of starvation, about one-third of the energy needs of the brain are met by ketones; with prolonged starvation, ketones become the major fuel source for

Table 8.2 Fuel Metabolism in Starvation

Fuel Exchanges and Consumption	Amount Formed or Consumed in 24 Hours (grams)	
	Day 3	Day 40
Fuel use by the brain		
Glucose	100	40
Ketones	50	100
Fuel mobilization		
Adipose tissue lipolysis	180	180
Muscle protein degradation	75	20
Fuel output of the liver		
Glucose	150	80
Ketones	150	150

Source: Adapted from Stryer L. Biochemistry. 3rd ed. New York: Freeman 1988:640.

the brain. Under conditions of continued carbohydrate shortage, ketones are oxidized by the muscles in preference not only to glucose but also to fatty acids. During starvation, the use of ketones by the muscles as the preferred source of energy spares protein, thereby prolonging life. Although Figure 8.8 depicts fuel metabolism during starvation, it does not show some of the |adjustments in energy substrates that occur when starvation is prolonged. These adjustments are shown in Table 8.2. As mentioned previously, the duration of starvation compatible with life depends to large degree on depot fat status.

Sports Nutrition

Humans have courted the challenge of athletic performance and competition since the days of the early Greeks. Much later, the science of nutrition emerged, spurred by the expanding knowledge of metabolism and the biochemistry on which it is based. Because the energy for physical performance must derive from nutrient intake, it was only a matter of time that these areas of interest would link. The heavy emphasis on the enhancement of health and physical performance in today's society has led to the emergence of sports nutrition as an important science. Nutrition, as a means of positive impact on physical performance, has become a topic of great interest to all those involved in human performance, the scientist as well as the athlete and athletic trainer.

The human body converts the potential energy of nutrients into usable energy, part of which drives the contraction of muscle, a process that is fundamental to athletic prowess. As the body's demand for energy fluctuates—for example, as exertion levels change among resting, mild exercise, and strenuous exercise—shifts in

the rate of catabolism of the different stored forms of nutrients occur also. It follows that an understanding of sports nutrition requires an understanding of the integration of the metabolic pathways that furnish the needed energy. In this respect, therefore, the energy demands of sport resemble the feed-fast cycle described earlier in this chapter, so a discussion of sports nutrition at this point in the text seems appropriate.

Biochemical Assessment of Physical Exertion

To understand how nutrition relates to physical performance at the cellular level, determining various biochemical parameters is necessary. Two commonly applied parameters are the respiratory quotient (RQ) and the maximal oxygen consumption (VO_2 max). A newer generation of procedures has been developed to measure the relative contribution of substrates to energy supply during exercise. An example of such a procedure is the isotope infusion method. These will be described briefly.

The *respiratory quotient* is the ratio of the volume of CO_2 expired to the volume of O_2 consumed. It has served for nearly a century as the basis for determining the relative participation of carbohydrates and fats in exercise [8].

$$RQ = CO_2/O_2$$

For carbohydrate catabolism, the RQ is 1:

$$C_6H_{12}O_6 \text{ (glucose)} + 6O_2 \longrightarrow 6CO_2 + 6H_2O$$

For fat catabolism, the RQ is approximately 0.7:

$$C_{16}H_{32}O_2 \text{ (palmitic acid)} + 23O_2 \longrightarrow 16CO_2 + 16H_2O$$

The RQ for protein is about 0.8. The amount of protein being oxidized can be estimated from the amount of urinary nitrogen produced, and the remainder of the metabolic energy must be a combination of carbohydrate and fat. Should the principal fuel source shift from mainly fat to carbohydrate, the RQ correspondingly increases, while a shift from carbohydrate to fat lowers the RQ. During the past 20 years, however, such knowledge has been advanced by invasive techniques such as arteriovenous measurements and the use of needle biopsies to quantify tissue stores of the energy nutrients.

The *VO_2 max* is the workload that places the highest possible demand on the working muscle of that subject, and it is generally used to monitor intensity of exercise. Exertional output of a particular workload is most commonly expressed in terms of the percentage of the VO_2 max that it induces. Studies are generally carried out within the exercise intensity range of 70% to 85% of the VO_2 max.

Isotope infusion can quantify the contribution of the major energy substrates, plasma glucose and fatty acids, and muscle triacylglycerols and glycogen to energy expenditure during exercise. It involves the intravenous infusion of stable isotope (e.g., ^{2}H [deuterium])-labeled glucose, palmitate, and glycerol during periods of rest and exercise. By monitoring the uptake of infused labeled glucose and palmitate, and knowing whole body substrate oxidation, the contribution of muscle triacylglycerol and glycogen to overall energy supply can be estimated [9].

Energy Sources during Exercise

The hydrolysis of the terminal phosphate group of ATP ultimately provides the energy for conducting biological work. In terms of physical performance, the form of work that is of the greatest interest is the mechanical contraction of skeletal muscles. It is therefore obvious that physical exertion depends on a reservoir of ATP, which is in an ever-changing state of metabolic turnover. Whereas ATP is consumed by physical exertion, its stores are repleted during periods of rest.

The key to optimizing physical performance lies in nutritional strategies that maximize cellular levels of stored nutrients as fuels for ATP production. There are three energy systems that supply ATP during different forms of exercise [10]:

- the ATP-CP (creatine phosphate) system,
- the lactic acid system, and
- the aerobic system.

The ATP-CP (Phosphagen) System

The ATP-CP system is a cooperative system in muscle cells utilizing the high-energy phosphate bond of creatine phosphate (CP) together with ATP. When the body is at rest, energy needs are fulfilled by aerobic catabolism (see "Aerobic System," later) because the low demand for oxygen can be met by matched delivery by the cardiovascular system. The ATP-CP system also operates continuously during this time, although at a slow pace because it is not stressed. If physical activity is initiated, however, energy source shifts from the slower aerobic system to the ATP-CP system with its readily available store of ATP. Despite the availability of this source of ATP as the immediate answer to the energy demands of contracting muscle, its stores in muscle are limited, providing enough energy for only a few seconds of maximal exercise. As ATP is depleted, high-energy phosphate group transfer from CP to ADP can occur rapidly to bolster the supply of available ATP. However, muscle cell concentration of CP is only four to five times greater

than ATP, and therefore all energy furnished by this system is expendable in approximately 10 seconds of strenuous exercise. Performance demands of high intensity and short duration such as weightlifting, 100-m sprinting, football, and various field events benefit most from this system. Lower-intensity activity may allow one to use this system for up to 3 minutes.

The Lactic Acid System

This system involves the glycolytic pathway by which ATP is produced in skeletal muscle by the incomplete breakdown of glucose anaerobically into 2 mol of lactate. The source of glucose is primarily muscle glycogen and to a lesser extent, circulating glucose, and it can generate ATP quickly for high-intensity exercise. The lactate system is not efficient from the standpoint of the quantity of ATP produced. However, because the process is so rapid, the small amount of ATP is produced quickly and absolutely by substrate-level phosphorylation of ADP (p. 59). The lactate produced by this system quickly crosses the muscle cell membrane into the bloodstream from which it can be cleared by other tissues for aerobic production of ATP. In the event that the rate of production of lactate exceeds its clearance rate, blood lactate accumulates. Under these circumstances, exercise cannot be continued for long periods. The lactic acid system is engaged when an inadequate supply of oxygen prevents the aerobic system from furnishing sufficient ATP to meet the demands of exercise. Although the system is operative upon the onset of strenuous exercise, it becomes the primary supplier of energy only after the depletion of CP stores in the muscle. As a back-up to the ATP-CP system, it becomes very important in high-intensity anaerobic power events lasting 20 seconds to a few minutes. Examples are longer sprints of up to 800 m and swimming events of 100 or 200 m.

The Aerobic System

This system involves the Krebs cycle, through which carbohydrates, fats, and proteins are completely oxidized to CO_2 and H_2O. The system, which requires oxygen, is highly efficient from the standpoint of the quantity of ATP produced. Since oxygen is necessary for the system to function, an individual's VO_2 max becomes an important factor in his or her performance capacity. Contributing to the VO_2 max is the cardiovascular system's ability to deliver blood to exercising muscle, pulmonary ventilation, oxygenation of blood, and the utilization of the oxygen by skeletal muscle mitochondria. Matching these contributors to the cellular need for oxygen in exercising muscle is complex because a low efficiency of any of them becomes rate limiting for the entire process. In terms of cellular metabolism, the aerobic pathway is slow to become activated and dominate during the course of activity, and it is probably not used to a significant degree until after at least five minutes of continuous activity. The aerobic system is the predominant supplier of energy for forms of exercise lasting longer than 3 or 4 minutes. These are the so-called endurance feats, of which there are many examples.

Current thinking is that the three systems do not simply take turns serially or that a system can be skipped in meeting the demands of exercise. Rather, it is a fact that all systems function at all times and that as one predominates, the others participate to varying degrees. The interaction of the three systems over the course of the first 2 minutes of exercise is complex but appears to involve energy contributions as follows:

* Decreasing ATP-CP from most contributory to least contributory during the period
* Lactic acid steady with high contribution
* Aerobic progressing from least to highest contribution during the period

Energy sources from aerobic and anaerobic systems for long-term activity are shown in Figure 8.10.

Fuel Sources during Exercise

Carbohydrate, fat, and protein are the dietary sources that provide the fuel for energy transformation. At rest, and during normal daily activities, fats are the primary source of energy, providing 80% to 90% of the energy. Carbohydrates provide 5% to 18%, and protein, 2% to 5% of energy during the resting state [11].

During exercise, the oxidation of amino acids contributes only minimally to the total amount of ATP used by working muscles. Therefore, the four major endogenous sources of energy during exercise are

* muscle glycogen,
* blood glucose,
* plasma fatty acids, and
* intramuscular triacylglycerols.

The extent to which each of these substrates contribute energy for exercise depends on several factors, including

* the intensity and duration of exercise,
* level of exercise training,
* initial muscle glycogen levels, and
* supplementation with carbohydrates during exercise.

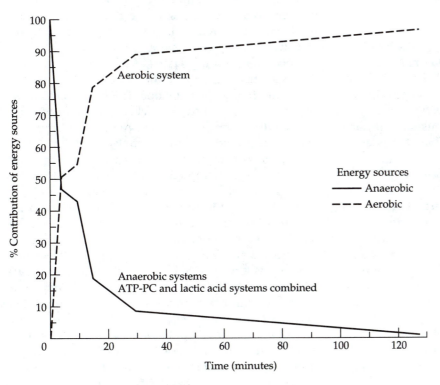

Figure 8.10 Primary energy sources for long-duration activity. *Source:* Adapted from Fox EL, Bowers RW, Foss ML, *The Physiological Basis for Exercise and Sports, 5th ed.,* Dubuque, IA: Brown and Benchmark, 1989:37. Reproduced with permission of The McGraw-Hill Companies.

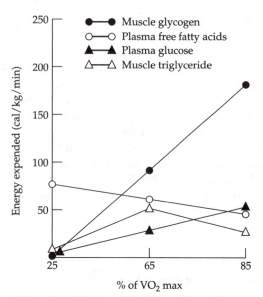

Figure 8.11 Contribution of the four major substrates to energy expenditure after 30 minutes of exercise at 25%, 65%, and 85% VO₂ max.

The relationship among these factors and the "substrate of choice" for energy supply will be discussed. A graphical representation of the contribution of these substrates at 25%, 65%, and 85% VO₂ max is shown in Figure 8.11.

Exercise Intensity and Duration

In the fasted state, nearly all of the energy required for low-intensity levels of exercise (25%–30% VO₂ max) is derived from plasma fatty acid oxidation with a small contribution from plasma glucose. The pattern does not change significantly over a period of up to 2 hours at this exercise level, which is equivalent to walking. During this time, the consumed plasma fatty acids are replaced by mobilization of fatty acids from the large triacylglycerol stores in adipocytes throughout the body. As exercise intensity increases, however, to 65% and on up to 85% VO₂ max, the return of adipocyte fatty acids into the plasma is reduced, resulting in a decrease in the concentration of plasma fatty acids. This occurs despite a continuing high rate of lipolysis in adipocytes. The decreased replacement of plasma fatty acids from fat stores at higher levels of exercise has been attributed to insufficient blood flow and albumin delivery of fatty acids from adipose tissue into the systemic circulation [12]. Therefore, it would be predicted that fatty acids become trapped in adipose tissue and accumulate there during high levels of exercise, a theory supported by research [9].

At moderate intensities of exercise (~65% VO₂ max) equivalent to running for 1 to 3 hours, total fat oxidation increases despite the reduced rate of return of adipose fatty acids into the circulation. This is attributed to an increase in the oxidation of muscle triacylglycerols. In fact, as shown in Figure 8.11, plasma fatty acids and muscle triacylglycerols contribute equally to energy expenditure at this level of exertion in endurance-trained athletes. Within the exertion range of 60% to 75% VO₂ max, however, fat cannot be oxidized at a rate sufficiently high to provide needed energy, and therefore nearly half of the required energy must be furnished by carbohydrate oxidation.

As exercise intensity increases to 85% VO₂ max, the relative contribution of carbohydrate oxidation to total metabolism increases sharply (see Fig. 8.11). At VO₂ max, carbohydrate in the form of blood glucose (derived from glycogenolysis of hepatic glycogen stores) and muscle glycogen become essentially the sole suppliers of energy. Like muscle glycogen, the concentration of blood glucose also falls progressively during prolonged, strenuous exercise. This is due to the fact that glucose uptake by working muscle may increase to as much as 20-fold or more above resting levels, while hepatic glucose output decreases with exercise duration. Interestingly, however, hypoglycemia is not always observed at exhaustion, particularly at exercise intensities >70% VO₂ max. Hypoglycemia following liver glycogen depletion can apparently be postponed by an inhibition of glucose uptake and by accelerated gluconeogenesis in the liver, using the glycerol produced in lipolysis, and by lactate and pyruvate produced by the glycolytic activity of the working muscles.

Accompanying high rates of carbohydrate catabolism is a rise in the production of lactic acid, which accumulates in muscle and blood. This is particularly evident in situations of oxygen debt, in which insufficient oxygen disallows the complete oxidation of pyruvate and favors instead its reduction to lactate.

The essentiality of carbohydrate as an energy substrate at moderate to high levels of exercise is due to the limited ability of muscle to oxidize fat. Muscle fatigue occurs with muscle glycogen depletion and hypoglycemia. To prevent this from happening, the individual must reduce workload to the lowest intensity that matches his or her ability to oxidize fat predominantly, possibly as low as 30% VO₂ max. Even if plasma fatty acid concentrations are high, carbohydrate depletion fatigue remains a factor, indicating that muscle has limited ability to oxidize fatty acids. The reason for this limitation, and therefore the dependence of muscle on carbohydrate as an energy source, remains unclear. However, traditional thinking is that the limitation may be based on the facts that (1) oxidation of fatty acids is limited by the enzyme carnitine acyltransferase (CAT), which catalyzes the transport of fatty acids across the mitochondrial membrane (p. 148), and (2) CAT is known to be inhibited by malonyl CoA. When availability of carbohydrate to the muscle is high, fatty acid oxidation may be reduced by the inhibition of CAT by glucose-derived malonyl CoA [13].

Level of Exercise Training

Endurance training increases an athlete's ability to perform more aerobically at the same absolute exercise intensity as the lesser trained. This is presumed to be due to an increased mitochondrial volume density in trained muscle together with an increase in cardiovascular capacity. It has been shown that the activity of oxidative enzymes in endurance trained subjects is 100% greater than in untrained subjects at 65% VO_2 max.

Endurance training results in an increased utilization of fat as an energy source during submaximal exercise. This is thought to be due to an adaptive increase in mitochondrial enzymes required for fatty acid oxidation. In skeletal muscle, fatty acid oxidation has an inhibitory effect on glucose uptake and glycolysis. For this reason, the trained athlete benefits from the carbohydrate-sparing effect of enhanced fatty acid oxidation during competition because of slower depletion of muscle glycogen and plasma glucose. This largely accounts for the training-induced increase in endurance for exercise over a prolonged period.

It has been reported that trained individuals have a lower plasma fatty acid concentration combined with reduced adipose tissue lipolysis compared to untrained counterparts at similar exercise intensity. This suggests that the primary source of fatty acids used by the trained individual is intramuscular triacylglycerol stores rather than adipocyte triacylglycerols.

There appears to be an increased capacity for muscle glycogen storage as a result of endurance training. Therefore, the trained athlete benefits not only from a slower utilization of muscle glycogen as explained earlier but also from the capacity to have higher glycogen stores at the onset of competition.

Initial Muscle Glycogen Levels

The ability to sustain prolonged moderate to heavy exercise is largely dependent on the initial content of skeletal muscle glycogen, and it is the depletion of muscle glycogen that is the single most consistently observed factor contributing to fatigue. High muscle glycogen levels allow exercise to continue longer at a submaximal workload. Even in the absence of carbohydrate loading (see the following section), a strong positive correlation exists between initial glycogen level and time to exhaustion and/or performance during exercise periods lasting at least 1 hour. The correlation does not apply at low levels of exertion (25%–35% VO_2 max) or high levels of exertion for short time periods because glycogen depletion is not a limiting factor under these conditions. It has been suggested that the importance of initial muscle glycogen stores is related to the inability of glucose and fatty acids to cross the cell membrane rapidly enough to provide adequate substrate for mitochondrial respiration [14].

Carbohydrate Supplementation (Supercompensation)

Because muscle glycogen was identified as the limiting factor for the capacity to exercise at intensities requiring 70% to 85% VO_2 max, dietary manipulation to maximize glycogen stores followed naturally. The most popular subject for research of this nature has been the marathon runner, because of the prolonged physical taxation of the event and the fact that the athlete's performance is readily measurable by the time required to complete the course. There emerged the major dietary concern in the endurance training of marathon runners as to how to elevate muscle glycogen to above-normal (*supercompensated*) levels. In sporting vernacular, maximizing glycogen content by dietary manipulation is referred to as "carbohydrate loading."

The so-called classical regimen for carbohydrate loading resulted from investigations in the late 1960s by Scandinavian scientists [15]. This involved two sessions of intense exercise to exhaustion separated by 2 days of low-carbohydrate diet (<10%), to "starve" the muscle of carbohydrate. This was followed by 3 days of a high-carbohydrate diet (>90%) and rest. The event would be performed on day 7 of the regimen. After this regimen, muscle glycogen levels approached 220 mmol/kg wet weight (expressed as glucose residues), more than double the athlete's resting level. However, because of various undesirable side effects of the classical regimen, such as irritability, dizziness, and a diminished exercise capacity, a less stringent regimen of diet and exercise has evolved that produces comparably high muscle glycogen levels. In this modified regimen, runners perform "tapered-down" exercise periods over the course of 5 days, followed by 1 day of rest. During this time, 3 days of a 50% carbohydrate diet are followed by 3 days of a 70% carbohydrate diet, generally achieved by the consumption of large quantities of pasta and rice or bread. The modified regimen, which can increase muscle glycogen stores 20% to 40% above normal, has been shown to be as effective as the classical approach with less adverse side effects. Figure 8.12 illustrates graphically the

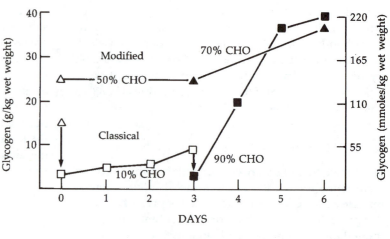

Figure 8.12 Schematic representation of the "classical" regime of muscle glycogen supercompensation described by Scandinavian investigators and the "modified" regime of muscle glycogen supercompensation, which has been shown to elevate muscle glycogen stores to comparably high levels with "normal" diets and a "tapering-down" sequence of exercise. *Source:* From "Carbohydrate, Muscle Glycogen, and Muscle Glycogen Supercompensation" by W. M. Sherman. In *Ergogenic Aids in Sport* (p. 14) by M. H. Williams (ed.), 1983, Champaign, IL: Human Kinetics Publishers. Reprinted by permission.

amount of muscle glycogen formed as a result of each regimen. Predictably, the supercompensation of muscle glycogen by whatever approach has been shown to improve performance in trained runners during races of 30 km and longer. It did not improve performance in shorter races (<21 km) owing to the fact that glycogen depletion is not the limiting factor in such an event.

The timing of the final meal before intense exercise is crucial because fasting results in a reduction of the labile glycogen stores of liver, whereas carbohydrate meals consumed too close in time to the event may cause hyperinsulinemia. In the latter situation, a rapid reduction in plasma glucose ensues, and work capacity is significantly impaired. In addition, elevated plasma insulin inhibits liver glucose output and the normal rise of plasma free fatty acids. Under such conditions, excessive muscle glycogen degradation occurs, resulting in early fatigue.

The form of carbohydrate ingested is also an important consideration in optimizing endurance performance, the principal factor being the glycemic index (GI) of the food. The GI refers to the rate at which the dietary carbohydrate is digested and metabolized so as to stimulate a concomitant increase in blood glucose. High GI carbohydrate includes the sugars glucose, fructose, and sucrose, while complex carbohydrates such as pasta and rice exhibit a low GI. Potato starch is considered to have a relatively high GI, although not as high as the simple sugars. Generally,

low to moderate GI carbohydrate loading prior to the performance is preferred to high GI carbohydrate intake. This is because of the hyperinsulinemic effect of the high GI food, which, as mentioned earlier, results in a rapid reduction in blood glucose, suppressed release of fatty acids from store, and inhibition of hepatic glycogenolysis.

Ergogenic Supplementation

The word *ergogenic* is derived from the Greek word for "work," *ergon,* and is defined as increasing work or the potential to do work. Nutritional ergogenic supplements, or ergogenic aids, are usually substances that are part of a normal diet, or they may be cellular metabolites that are ingested in an effort to enhance the capacity for sport, exercise, and physical performance. These are to be distinguished from the ergogenic drugs such as the anabolic steroids, which have prompted the enactment and enforcement of laws prohibiting their use in strength or endurance competition. The compulsion for improved performance among athletes has prompted an enormous increase in the testing and use of nutritional ergogenic aids. As expected, the literature dealing with the subject has expanded with equal zeal. Therefore, the information presented here must be restricted to a brief overview of what is known about the effectiveness of ergogenic supplementation.

There appears to be a dichotomy between the widespread public use of supplements and the corresponding widespread lack of scientific support for such use. A problem for researchers is the common perception of subjects under study of simply "feeling better" as a result of supplementation, even though actual physiological changes may not be documented by research. In other words, psychological effects are adding a new dimension to the testing of ergogenic aids. These must be considered along with true physiological effects, because as an athlete's mood and mental outlook improve, so does his or her physical performance, and this is the bottom line for supplementing in the first place.

Following is a listing of micronutrient ergogenic supplements that have been consumed on a broad basis. The supplements chosen for citing were selected on the basis of their reputed efficacy from a much longer list of hit-or-miss trial substances. In most instances, research results do not totally support or totally refute supplement efficacy, but instead they are divided in their findings. Reference will occasionally be made to the number of "pro and con" study conclusions to help the reader to evaluate efficacy in his or her own mind. Although specific references will not be included, they, and much more pertinent information, are available to the interested reader [16].

Amino Acids

Arginine/Ornithine These have been shown to elevate serum somatotropin (growth hormone) and insulin, although the practical effects of such hormone increases are unknown. They may be beneficial in resistance training programs by accelerating weight and body fat loss.

Aspartate Salts These are believed to increase the time to exhaustion in untrained subjects.

Branched-Chain Amino Acids BCAAs are reported to increase time to exhaustion, increase anabolic hormone release, prevent lean body mass loss, and protect against muscle damage.

Herbs

The Ginsengs The most widely used and studied herbs are the ginsengs. Some purported ergogenic benefits of *Panax* (Chinese/Korean) ginseng include the following:

- Increased run time to exhaustion (three out of seven studies)
- Increased muscle strength (one out of two studies)
- Improved recovery from exercise (three out of four studies)
- Improved oxygen metabolism during exercise (seven out of nine studies)
- Reduced exercise-induced lactate (five out of nine studies)
- Improved auditory and visual reaction times (six out of seven studies)
- Improved vitality and feelings of well-being (six out of nine studies)

Caffeine

Literature reviews reveal 39 studies showing caffeine having an ergogenic effect and 35 studies showing no ergogenic effect. Apparently a large individual variation exists. Among studies showing definite performance enhancement, CNS stimulation may account for enhancement rather than metabolic alteration.

Intermediary Metabolites

Bicarbonate Since 1994, 46 studies report exercise enhancement, and 37 report no enhancement. Loading may be of benefit for intense, short-term, anaerobic events. Risk of nausea, bloating, and alkalosis comes with excessive bicarbonate use.

Carnitine Supplemented to enhance fatty acid oxidation, seven recent studies show enhancement, and five show no enhancement. Although plasma carnitine levels increase with supplementation, muscle cell concentrations do not. Benefits are not convincing.

Coenzyme Q_{10} Consumed because of its obligatory role in ATP production by oxidative phosphorylation, coenzyme Q_{10} has showed positive effects on many parameters of physical performance in studies outnumbering 12 to 6 those showing no effect. It seems most beneficial for long-term, aerobic exertion (running, cycling).

Creatine Creatine is taken in an attempt to increase muscle cell concentration of creatine phosphate, important as an energy reserve in exercising muscle. It has become a well-supported dietary supplement for enhancement of muscle strength and size and for short-term, explosive sporting events. Twelve studies report significant improvements in performance, with six finding none.

Other Certain minerals have also been tested for their ergogenic properties, including calcium, magnesium, zinc, iron, phosphates, chromium, boron, and vanadium. Reviews of mineral supplements [16] suggest that performance enhancement is not well established and that the major benefit of mineral supplementation lies in the correction of deficiencies should they exist.

General problems of research design remain as the popularity of nutritional ergogenic supplements surges on. As indicated previously, many ergogenic effects may be attributed to mental and psychological changes, and it behooves future researchers to rule these out in an effort to establish strictly physiological effects. The fact that the number of studies finding "for" performance enhancement is nearly equaled by those finding "against" enhancement testifies to the difficulty in researching this important field.

Summary

Animal survival depends on a constant internal environment maintained through specific control mechanisms. Controls, operative at all levels (cellular, organ, and system), integrate energy metabolism and allow the body to adapt to a wide variety of environmental conditions. Primary among the mechanisms of adaptation is the regulation of metabolism, through the cooperative input of the nervous, endocrine, and vascular systems. In normal operation of these systems, metabolic pathways may be stimulated, maintained, or inhibited, depending on the conditions imposed on the body. A poignant example of metabolic adaptation is the shifts that occur in substrate utilization and metabolic pathways in answer to changes in the body's nourishment status (i.e., fed, fasting, and starvation states).

The physical stress of exercise and sport presents an interesting challenge to the regulatory capacity of the body to provide the needed additional energy to the exercising muscles. Substrates fueling the energy include plasma free fatty acids, plasma glucose, muscle glycogen, and muscle triacylglycerols, and their utilization varies according to the intensity and duration of the exercise. Many substances have been tested for their ergogenic properties in attempts to improve performance in high-intensity and endurance sports. In most cases, test results are still controversial, and more research is needed to establish which of the reputed ergogenic aids produce true physiological improvement.

References Cited

1. Role of fat and fatty acids in modulation of energy exchange. Nutr Rev 1988;46:382–4.

2. Hellerstein M, Schwarz J-M, Neese R. Regulation of hepatic de novo lipogenesis in humans. Annu Rev Nutr 1996;16:523–57.

3. McGarry JD, Kuwajima M, Newgard CB, et al. From dietary glucose to liver glycogen: the full circle round. Ann Rev Nutr 1987;7:51–73.

4. Foster DW. From glycogen to ketones and back. Banting Lecture 1984. Diabetes 1984;33:1188–99.

5. Lehninger AL, Nelson DL, Cox MM. Principles of Biochemistry, 2d ed. New York: Worth, 1993:757–8.

6. Tepperman J, Tepperman HM. Metabolic and Endocrine Physiology. 5th ed. Chicago: Year Book, 1987.

7. Turner, AJ, ed. Neuropeptides and Their Peptidases. New York: VCH, 1987.

8. Hermansen L, Hultman E, Saltin B. Muscle glycogen during prolonged severe exercise. Acta Physiol Scand 1967;71:129–39.

9. Romijn J, Coyle E, Sidossis L, et al. Regulation of endogenous fat and carbohydrate metabolism in relation to exercise intensity. Am J Physiol 1993; 265: E380–91.

10. McArdle W, Katch F, and Katch V. Exercise Physiology: Energy, Nutrition, and Human Performance, 3rd ed., Lea and Febiger, Philadelphia, 1991, Chaps. 1, 6, 11.

11. Wolinsky I. Nutrition in Exercise and Sport, 3rd ed., CRC Press, Boca Raton, 1998, Chaps. 6, 13.

12. Hodgetts A, Coppack SW, Frayn KN, and Hockaday TDR. Factors controlling fat mobilization from human subcutaneous adipose tissue during exercise. J Appl Physiol 1991; 71: 445–51.

13. Elayan I , Winder W. Effect of glucose infusion on muscle malonyl CoA during exercise. J Appl Physiol 1991;70:1495–9.

14. Saltin B and Gollnick P. Fuel for muscular exercise: role of carbohydrate, in Exercise, Nutrition, and Energy Metabolism. Horton E and Terjung R, Eds., MacMillan Publishing Co., New York, 1988, chap. 4.

15. Bergstrom J, Hultman E. A study of the glycogen metabolism during exercise in man. Scand J Clin Lab Invest 1967;19:218–28.

16. Wolinsky I. Nutrition in Exercise and Sport, 3rd ed., CRC Press, Boca Raton, 1998.

Suggested Reading

Coyle EF. Substrate utilization during exercise in active people. Am J Clin Nutr 1995; 61(suppl): 968S–79S.
A very useful review of the hierarchy of substrates as they are used for energy release in exercise.

Harris RA, Crabb DW. Metabolic interrelationships. In: Textbook of Biochemistry with Clinical Correlations, 3rd ed. (Devlin TM, ed.). New York: Wiley, 1992:576–606.

This integration of human metabolic pathways is written primarily for the medical student. Information is presented so as to be relevant for the health practitioner.

Tepperman J, Tepperman HM. Metabolic and Endocrine Physiology, 5th ed. Chicago: Year Book, 1987.

This is a well-illustrated, easy-to-read explanation of the regulatory role of the endocrine system in human metabolism.

Wolinsky I. Nutrition in Exercise and Sport, 3rd ed., CRC Press, Boca Raton, 1998.

A thorough treatment of what is known and what is not in sports nutrition.

Web Sites

http://www.nal.usda.gov/fnic/pubs/bibs/gen/97-sp-hp.htm
National Agricultural Library: Sports Nutrition Resource List for Health Professionals

http://www.umass.edu/cnshp/index.html
Center for Nutrition in Sport and Human Performance at the University of Massachusetts

www.faseb.org/ajcn
American Journal of Clinical Nutrition

Diabetes: Metabolism Out of Control

Diabetes mellitus, the disease characterized by the body's inability to metabolize glucose, manifests as one of two types: type 1, or insulin-dependent diabetes mellitus, ketosis prone (IDDM); and type 2, non-insulin-dependent diabetes mellitus (NIDDM). The two types are mechanistically very different and will be discussed separately. Current theories on the etiology and characteristics of these two classifications of diabetes are shown in Figure 1.

Non-Insulin-Dependent Diabetes Mellitus (Type 2)

Type 2 diabetes accounts for 80% to 90% of all reported cases of the disease. The cause of type 2 diabetes has not been completely resolved, but it appears to be associated with insulin resistance in peripheral target tissue. This condition is caused not by a failure of target cells to bind insulin but by a postbinding abnormality, arising somewhere in the sequence of events that follows the binding of insulin to its receptor and leading to the cell's normal response to that signal. There is experimental evidence that a primary cause for the interrupted insulin signal is compromised synthesis or mobilization of the cell's glucose transporters (refer to "Glucose Transporters," p. 78).

In skeletal muscle cells, insulin resistance associated with NIDDM has been shown to be caused by a reduction in glucose transporter activity, specifically the failure of the vesicles to translocate in response to insulin (see Fig. 4.4). The error can be thought of as a block or short-circuit in the insulin signal that normally initiates the trans-location process. The result is a reduced concentration of transporters at the cell surface and a consequent reduction in the rate of glucose uptake. Although a similar defect was found in adipocytes of NIDDM patients, it is not the major cause of the insulin resistance in these cells. Rather, the consequence of NIDDM in adipocytes is a marked depletion of mRNA encoding the GLUT4 transporter, resulting in depleted intracellular stores of the protein [1]. This describes a pretranslational defect, meaning that it interferes with protein synthesis at a level before the translation process, the step that requires mRNA as template. Therefore, even if the vesicle translocation process were not compromised, there would still be an inadequate number of surface receptors expressed upon insulin stimulation.

Insulin resistance has also been described in obesity as well as in NIDDM. Insulin resistance in obesity is mechanistically similar to the NIDDM effect on adipocytes. Reduction in GLUT4 mRNA in obese subjects results in a decrease in de novo synthesis of the transporter. Furthermore, the ex-

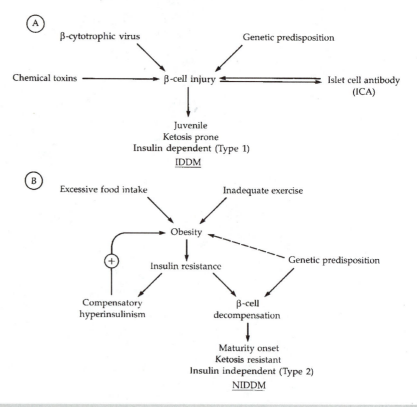

Figure 1 Overview of present theories of diabetes mellitus etiology. (A) depicts the factors impinging on the development of diabetes mellitus that requires exogenous insulin. This type of diabetes presently is most commonly designated as insulin-dependent diabetes mellitus (IDDM). (B) illustrates the interaction of factors that may result in non-insulin-dependent diabetes mellitus (NIDDM).

tent to which mRNA expression is suppressed appears to relate directly to increasing adiposity.

In summary, NIDDM is characterized by insulin resistance in peripheral target tissues, because of a diminished population of functional glucose transporters. In muscle cells, the defect appears to arise from a failure, on insulin stimulation, of vesicle-bound transporters to translocate to the plasma membrane. In adipocytes, translocation is also compromised, but the major mechanism for insulin resistance in these cells, in both NIDDM and obesity, is a pre-translational depletion of GLUT4 mRNA.

Insulin-Dependent Diabetes Mellitus (Type 1)

The hyperglycemia of type 1 diabetes can be attributed to a primary failure of the β-cells of the pancreas to produce and secrete insulin. It is regarded as an autoimmune disease in which the pancreatic islet cells, which are composed largely of β-cells, become targets of an immune response. This ultimately causes cellular dysfunction of the β-cells with an inability of the cells to produce insulin. Factors that trigger the immune attack remain unknown.

Figure 2 emphasizes the crucial role of insulin in the regulation of metabolism, and the metabolic events set into motion by insulin lack. An absence of insulin not only inhibits the use of glucose by muscles and adipose tissue but also sets into motion a sequence of events that, without effective intervention, will result in coma and death of the affected animal or human. Insulin has a variety of actions on metabolism, most of which have the effect of lowering blood glucose. Such actions include decreasing hepatic glucose output, while increasing glucose oxidation, glycogen deposition, lipogenesis, protein synthesis, and cell replication. In the absence of insulin, all the hormones favoring catabolism and the raising of blood glucose operate without opposition. The direction of metabolism in response to these catabolic hormones is that seen in Figure 8.8, which depicts the body's adaptation to fasting. In diabetes, however, the responses are much more violent than those occurring in the body's adaptation to fasting or starvation, during which the purpose is maintenance of a blood glucose level sufficient to meet the crucial demands of the CNS and RBCs. The unrestrained action of the catabolic hormones in

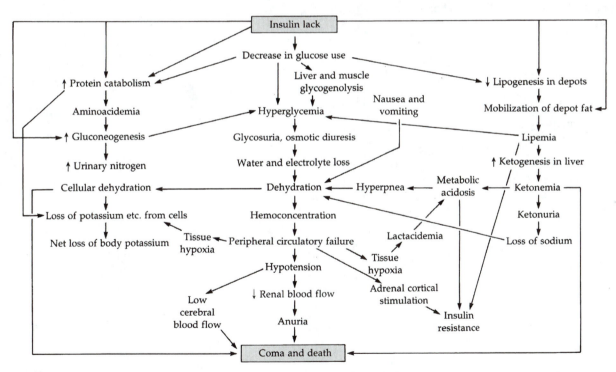

Figure 2 Composite summary of pathophysiology of diabetic acidosis. Particularly striking are the aberrations in the metabolism of carbohydrates, lipids, and protein caused by an insulin lack and the interconnections among these altered metabolic pathways. *Source:* Tepperman D, Tepperman H. Metabolic and Endocrine Physiology. 5th ed. Chicago: Year Book, 1987:284.

∽ **PERSPECTIVE** (continued)

the absence of insulin, along with the dramatically decreased use of glucose caused by an insulin lack, results in aberrations in metabolism. Not only is carbohydrate, fat, and protein metabolism affected, but water and electrolyte imbalance occurs also.

Hyperglycemia, the hallmark of diabetes, which is due to decreased glucose use and increased hepatic glucose output, results in an osmotic diuresis that proves fatal if uninterrupted (Fig. 2). The water and electrolytes lost through this diuresis lead to a dehydration compounded by increased insensible water loss due to the hyperpnea of metabolic acidosis. Metabolic acidosis results from the excessive ketogenesis occurring in the liver.

Peripheral circulatory failure, a consequence of severe hemoconcentration, leads to tissue hypoxia with a consequent shift of the tissues to anaerobic metabolism. Anaerobic metabolism raises the concentration of lactic acid in the blood, thereby worsening the metabolic acidosis.

The ketonuria along with glucosuria associated with acidosis causes an excessive loss of sodium from the body; loss of this extracellular cation further compromises body water balance. A net loss of potassium, the chief intracellular cation, accompanies increased protein catabolism and cellular dehydration, both of which characterize uncontrolled diabetes.

The normal flow of substrates following food intake, as depicted in Figure 8.6, is largely dependent on the secretion of insulin. Insulin exerts a potent, positive effect on anabolism, emphasized in the figure, while inhibiting catabolic pathways. Figure 2, in contrast, shows metabolism out of control when the inhibiting effect of insulin is lacking and conservation of energy is impossible. Diabetes is a vivid negative example that emphasizes the integration of metabolism and the importance of metabolic regulation (homeostasis) to continuance of life.

References Cited

1. Garvey WT, Maianu L, Huecksteadt TP, et al. Pretranslational suppression of a glucose transporter protein causes insulin resistance in adipocytes from patients with non-insulin-dependent diabetes mellitus. J Clin Invest 1991;87:1072–81.

Web Sites

http://www.diabetes.org
    American Diabetes Association
http://www.jdfcure.org
    Juvenile Diabetes Foundation

Photomicrograph of crystallized thiamin

The Water-Soluble Vitamins

The early part of the twentieth century marks the most exciting era in the history of nutrition science. It was during this time that the discovery of vitamins, or "accessory growth factors," began. Researchers found that for life and growth, animals required something more than a chemically defined diet consisting of purified carbohydrate, protein, fat, minerals, and water. The first of these dietary essentials discovered was an antiberiberi substance isolated from rice polishings by Funk, a Polish biochemist. Funk gave it the name *vitamine* because the substance was an amine and necessary for life. Very shortly thereafter McCollum and Davis extracted a factor from butter fat that they called *fat-soluble A* to distinguish it from the water-soluble antiberiberi substance. These two essential factors became known as vitamine A and vitamine B. As each additional vitamin was discovered, it was assigned a letter; the *e* on *vitamine* was dropped to give the general name *vitamin* because only a few of the essential substances were found to be amines.

As the chemical structure of a vitamin became known through its isolation and synthesis, it was given a chemical name. When the chemical name was assigned, it was assumed that the name applied to one substance with one specific activity. Now it is evident that a vitamin may have a variety of functions and that vitamin activity may be found in several closely related compounds known as *vitamers*. An excellent example of this is vitamin A, which has several seemingly unrelated functions and encompasses not only retinol but also retinal and retinoic acid.

Vitamins can be defined as essential organic compounds required in very small amounts (micronutrients) that are involved in fundamental functions of the body, such as growth, maintenance of health, and metabolism. Because these substances must be supplied wholly or partially by the diet, their discovery came about because of their absence in the diet. Although in the case of a deficiency the clinician should be able to recognize the syndrome caused by a lack of the vitamin, it is more relevant in this country of abundant and varied food supply for the nutrition professional to think in terms of what a specific vitamin does rather than what disease it prevents. Unfortunately, it is often impossible to relate the function of the vitamin directly to its deficiency syndrome.

Vitamins, for the most part, are not related chemically and differ in their physiological roles. The broad classifications of water-soluble vitamins and fat-soluble vitamins are made because of certain properties common to each group. The fat-soluble vitamins are discussed in Chapter 10. The body handles the water-soluble vitamins differently from the way it handles the fat-soluble vitamins. They are absorbed into portal blood, in contrast to fat-soluble vitamins and, with the exception of cyanocobalamin (vitamin B_{12}), they cannot be retained for long periods by the body. Any storage occurring results from their binding to enzymes and transport proteins. Water-soluble vitamins are excreted in the urine whenever plasma levels exceed renal thresholds.

Water-soluble vitamins, with the exception of vitamin C (ascorbic acid), are members of the B complex. Most of the B-complex group can be further divided according to general function: energy releasing or hematopoietic. Other vitamins cannot be classified this narrowly because of their wide range of functions. Figure 9.1 shows the classification of vitamins.

In this chapter, discussions of the vitamins are grouped similarly. For each vitamin, consideration is given, when precise information is available, to structure, sources, absorption (also digestion, where applicable), transport, functions and mechanisms of action, metabolism and excretion, recommended dietary allowance, deficiency, toxicity, and assessment of nutriture. Specific interrelationships with other nutrients are also noted for selected vitamins. Table 9.1 contains a summary of the discovery, functions, deficiency syndrome, sources, and the Recommended Dietary Allowance (RDA) or adequate intake of each of the water-soluble vitamins. The inside book covers provide the 1989 RDA or 1997–1998 Dietary Reference Intakes (DRIs), when available, for all nutrients and for all age

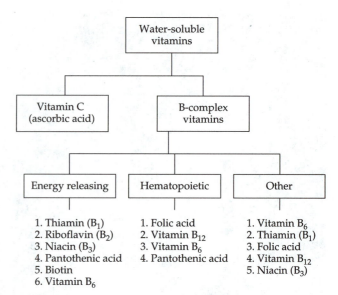

Figure 9.1 The water-soluble vitamins.

groups. DRIs represent quantitative approximations of nutrient intakes for the purpose of planning and assessing diets of healthy people. DRIs include RDA as well as *Adequate Intakes* (AIs), *Tolerable Upper Intake Levels* (ULs), and *Estimated Average Requirements* (EARs). RDAs represent the average daily dietary intake level that is sufficient to meet the nutrient requirements of about 97% of healthy individuals. They are based on EARs. EARS are the amounts of nutrients thought to meet the nutrient requirements of 50% of the healthy individuals in a specified age and gender group. RDAs are set higher than EARs by either two standard deviations or a coefficient of variation for the EAR. AIs are provided for nutrients instead of RDAs when scientific data are insufficient to calculate EAR for the given nutrients. ULs provide the highest intake level for a nutrient that is unlikely to cause any risks of adverse health to almost all individuals in the age- or gender-specified groups.

Suggested Reading

McCollum EV. A History of Nutrition. Boston: Houghton Mifflin, 1957.

Vitamin C (Ascorbic Acid)

The human being is one of the few mammals unable to synthesize vitamin C, also known as ascorbic acid or ascorbate. Other animals that are unable to synthesize vitamin C include primates, fruit bats, guinea pigs, and some birds. The inability to synthesize vitamin C results

Table 9.1 The Vitamins: Discovery, Functions, Human Deficiency Syndromes, Food Sources, and Recommended Intake

			Water-Soluble Vitamins			
Vitamin	Discovery	Coenzymes	Biochemical or Physiological Function	Deficiency Syndrome or Symptoms	Good Food Sources in Rank Order	RDA[a]
Thiamin (vitamin B_1)	Casimir Funk (1912)	Thiamin diphosphate (TDP) or thiamin pyrophosphate (TPP)	Oxidative decarboxylation of α-keto acids and 2-keto sugars	*Beriberi,* muscle weakness, anorexia, tachycardia, enlarged heart, edema	Yeast, pork, sunflower seeds, legumes	1.1 mg 1.2 mg
Riboflavin (vitamin B_2)	Kuhn, Szent-György, and Wagner-Jaunergy (1933)	Flavin adenine dinucleotide (FAD); flavin mononucleotide (FMN)	Electron (hydrogen) transfer reactions	*Cheilosis,* glossitis, hyperemia and edema of pharyngeal and oral mucous membranes, angular stomatitis, photophobia	Beef liver, braun-schweiger sausage, lean sirloin steak, mushrooms, ricotta cheese, nonfat milk, oysters	1.1 mg 1.3 mg
Niacin (vitamin B_3) (nicotinic acid, nicotinamide)	Elvehjem et al. (1937)	Nicotinamide adenine dinucleotide (NAD); nicotinamide adenine dinucleotide phosphate (NADP)	Electron (hydrogen) transfer reactions	*Pellagra,* diarrhea, dermatitis, mental confusion, or dementia	Tuna, beef liver, chicken breast, beef, halibut, mushrooms	14 mg 16 mg
Pantothenic acid	R. J. Williams (1933)	Coenzyme A	Acyl transfer reactions	Deficiency very rare: numbness and tingling of hands and feet, vomiting, fatigue	Widespread in foods; exceptionally high amounts in egg yolk, liver, kidney, yeast	5 mg[b]
Biotin	Szent-György (1940)	N-carboxybiotinyl lysine	CO_2 transfer reactions; carboxylation reactions	Deficiency very rare, usually induced by ingestion of large amounts of raw egg whites containing avidin; anorexia, nausea, glossitis, depression, dry, scaly dermatitis	Synthesized by microflora of digestive tract; yeast, liver, kidney	30μg[b]
Vitamin B_6 (pyridoxine, pyridoxal, pyridoxamine)	Szent-György, Kuhn (1938)	Pyridoxal phosphate (PLP)	Transamination and decarboxylation reactions	Dermatitis, glossitis, convulsions	Sirloin steak, navy beans, potato, salmon, banana	13 mg

(continued)

Table 9.1 *(continued)*

			Water-Soluble Vitamins			
Vitamin	**Discovery**	**Coenzymes**	**Biochemical or Physiological Function**	**Deficiency Syndrome or Symptoms**	**Good Food Sources in Rank Order**	**RDA**[a]
Folic acid (folacin)	Mitchell et al. (1941)	Derivatives of tetrahydrofolic acid $N^{5,10}$ methylidyne THF N^{10} formyl THF N^5 formimino THF $N^{5,10}$ methylene THF N^5 methyl THF	One-carbon transfer reactions	*Megaloblastic anemia,* diarrhea, fatigue, depression, confusion	Brewer's yeast, spinach, asparagus, turnip greens, lima beans, beef liver	400 μg
Vitamin B$_{12}$ (cobalamin)	Riches, Folkers, et al. (1948)	Methylcobalamin, adenosyl cobalamin (cobalamides)	Methylation of homocysteine to methionine; conversion of methylmalonyl CoA to succinyl CoA	*Megaloblastic anemia,* degeneration of peripheral nerves, skin hypersensitivity, glossitis	Meat, fish, shellfish, poultry, milk	2.4 μg
Ascorbic acid (vitamin C)	Szent-György[c] (1928) King[a] (1932)	None	Antioxidant, cofactor of hydroxylating enzymes involved in synthesis of collagen, carnitine, norepinephrine	*Scurvy,* loss of appetite, fatigue, retarded wound healing, bleeding gums, spontaneous rupture of capillaries	Papaya, orange juice, cantaloupe, broccoli, brussels sprouts, green peppers, grapefruit juice, strawberries	60 mg[c]

[a]Adults aged 19–50 years, females and males respectively, 1997–98 DRI RDA.
[b]Adequate Intake.
[c]Szent-György and King are considered to be codiscoverers of vitamin C.

from the lack of gulonolactone oxidase, the last enzyme in the vitamin C synthetic pathway. The synthetic pathway and structure of the vitamin are shown in Figure 9.2, which indicates that vitamin C is a six-carbon compound. It is the L isomer of the vitamin that is biologically active in man.

Sources

The best food sources of vitamin C include asparagus, papaya, oranges, orange juice, cantaloupe, cauliflower, broccoli, brussels sprouts, green peppers, grapefruit, grapefruit juice, kale, lemons, and strawberries. Of these foods, citrus products are most commonly cited as significant sources of the vitamin. Supplements supply vitamin C typically as free ascorbic acid, calcium ascorbate, sodium ascorbate, and ascorbyl palmitate. Rose hip (*Rosa*), a seed capsule found in roses, also contains vita-min C and is used commercially in vitamin C supplements; vitamin C from rose hips does not appear to be superior to other vitamin C sources such as orange juice.

Absorption, Transport, and Storage

Absorption of ascorbate occurs primarily by active transport. Transport systems for ascorbate are saturable and dose-dependent [1–5]. Simple diffusion or carrier-mediated transport may also contribute to a small extent to uptake of the vitamin from the mouth and stomach [2,3,5]. Most absorption, however, occurs in the ileum by sodium-dependent, active transport.

Prior to absorption, ascorbate may be oxidized (two electrons and two protons are removed) to form dehydroascorbate (Fig. 9.3), which may be absorbed by passive diffusion or by use of glucose transporters [1]. Absorption of dehydroascorbate is thought to occur to a

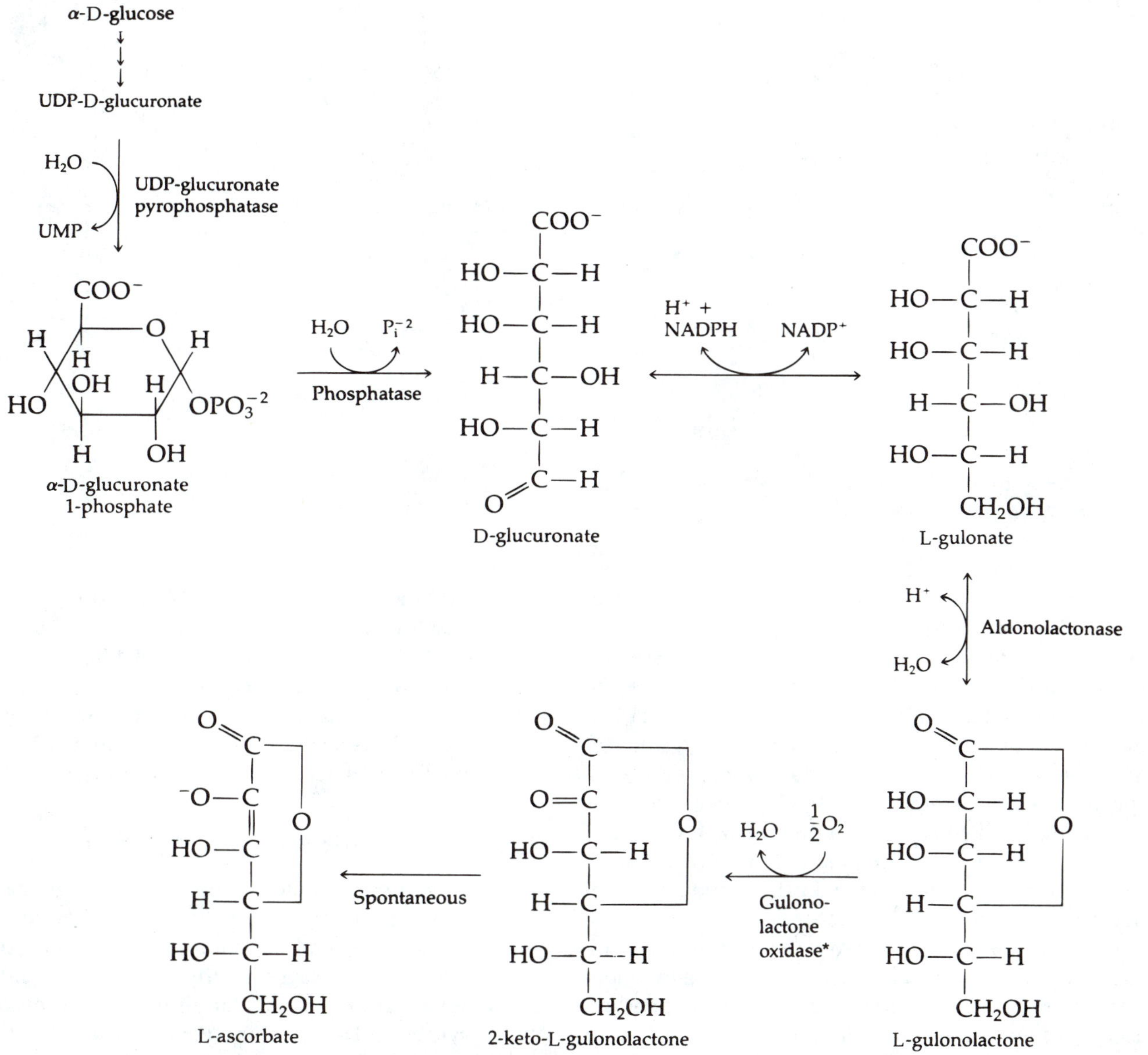

α-D-glucose

↓
↓

UDP-D-glucuronate

H_2O ⟍ UDP-glucuronate pyrophosphatase

UMP ↙

COO⁻ ... O ... H

H ⟍ H ⟋ H

HO ⟍ OH H ⟋ OPO_3^{-2}

H   OH

α-D-glucuronate
1-phosphate

H_2O P_i^{-2} → Phosphatase

COO⁻
HO—C—H
HO—C—H
H—C—OH
HO—C—H
O=C—H

D-glucuronate

$H^+ + NADPH$ $NADP^+$

COO⁻
HO—C—H
HO—C—H
H—C—OH
HO—C—H
CH_2OH

L-gulonate

H^+ ↙ Aldonolactonase
H_2O ↙

O=C
HO—C—H
HO—C—H   O
H—C
HO—C—H
CH_2OH

L-gulonolactone

H_2O $\frac{1}{2}O_2$ → Gulono-lactone oxidase*

O=C
O=C
HO—C—H   O
H—C
HO—C—H
CH_2OH

2-keto-L-gulonolactone

← Spontaneous

O=C
⁻O—C
‖   O
HO—C
H—C
HO—C—H
CH_2OH

L-ascorbate

Figure 9.2 Synthesis of ascorbic acid. Primates lack the gulonolactone oxidase* that catalyzes the final enzymatic reaction.

greater extent than absorption of ascorbate [1,4,5]. Yet, within the intestinal cells (but also other cells), dehydroascorbic acid generally is rapidly reduced back to ascorbic acid (Fig. 9.3) by the enzyme dehydroascorbate reductase, which requires reduced glutathione (GSH). During the reduction of dehydroascorbate, glutathione is oxidized (GSSG). Glutathione is deemed essential for vitamin C metabolism—that is, the reduction of dehydroascorbate; gluthathione has been shown to spare vitamin C and also improve the antioxidant protection capacity of blood [6,7]. Notice in Figure 9.3 that during the oxidation of ascorbate, a free radical called *semidehydroascorbate* (also called *ascorbate free radical, ascorbyl,* or *monodehydroascorbate radical*) is formed. The ascorbate free radical is thought to have a short half-life either being oxidized to dehydroascorbate or reacting with another semidehydroascorbate radical to form ascorbate and dehydroascorbate. Furthermore, the semidehydroascorbate radical reacts poorly with oxygen and thus does not typically generate superoxides.

The degree of vitamin C absorption decreases with increased vitamin intake. Absorption can vary from 16% at high intakes (~12 g) to 98% at low intakes (<20 mg) [4]. Over a range of usual intakes from food (20–120 mg/day), the average, overall absorption is about 80% to 95% [2,8,9]. Unabsorbed vitamin C may be metabolized by intestinal flora.

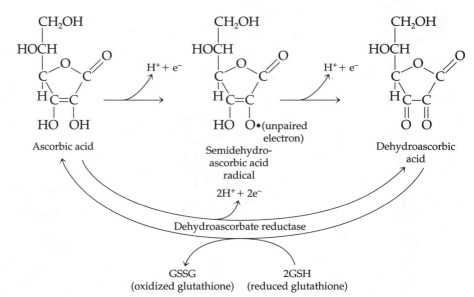

Figure 9.3 The interconversion of ascorbate and dehydroascorbate.

Substances that may impair the absorption of vitamin C include pectin (14.2 g/day) and zinc (9.3 mg/day) [2]. The degree of absorption as suggested by the urinary excretion of vitamin C appears to be adversely affected by both pectin and zinc; however, the mechanism of action has not been elucidated, and the use of urinary vitamin C excretion as a measure of absorption is not standard [2]. In the gastrointestinal tract, a high iron concentration present with vitamin C may result in the oxidative destruction of the vitamin yielding diketogulonic acid and other products without vitamin C activity [2].

Transport of vitamin C across the basolateral membrane of the intestinal cell occurs by sodium-independent carrier-mediated transport systems. Absorbed ascorbic acid is transported in the plasma primarily in free form, although small amounts (~5%) of vitamin C may be present in the plasma as dehydroascorbate [9]. Ascorbate readily equilibrates with the body pool of the vitamin [8]. The size of the pool varies with intake.

Ascorbate and dehydroascorbate concentrations are much greater in some tissues than in others. The highest concentrations of vitamin C are found in the adrenal and pituitary glands (each possessing ~30–50 mg/100 g of wet tissue) [8]. Intermediate levels of vitamin C are found in the liver, spleen, heart, kidneys, lungs, pancreas, and leukocytes or white blood cells, while smaller amounts occur in the muscles and red blood cells [8].

Tissue concentrations of vitamin C usually exceed the plasma level by 3 to 10 times, the degree of concentration depending on the specific tissue. Tissue concentration and plasma level are related to each other, and both are related to intake. Blood cell and tissue ascorbate concentrations appear to be saturated at vitamin C intakes of about 100 mg and 140 mg/day, respec-

tively [10]. Intakes of ascorbic acid at 200 mg/day or higher result in steady state plasma vitamin C concentrations [11]. Increasing vitamin C intake from 200 mg to 2,500 mg resulted in a mean steady state plasma vitamin C increase from about 12 mg/L to only 15 mg/L [12]. Such findings suggest that a daily vitamin C intake of 200 mg is a practical upper limit [12,13].

Functions and Mechanisms of Action

Despite its uncomplicated structure, vitamin C has a very complex functional role in the body. Ascorbic acid is required in several reactions involved in body processes such as collagen synthesis, carnitine synthesis, tyrosine synthesis and catabolism, and neurotransmitter synthesis. Each of these processes and the role of vitamin C will be reviewed.

Collagen Synthesis

Ascorbate functions in a number of hydroxylation reactions. Two hydroxylation reactions requiring vitamin C are necessary for collagen formation. For the collagen molecule to aggregate into its triple-helix configuration, selected proline residues on newly synthesized collagen α chains must be hydroxylated by *proline hydroxylase,* also called *dioxygenase.* Formation of the triple helix is important because it is in this triple helix configuration that fibroblasts and osteoblasts secrete procollagen. The importance of the lysine hydroxylation by *lysine hydroxylase,* also called *dioxygenase,* is not as clear as proline hydroxylase, but hydroxylysyl residues permit cross-linking of collagen and other posttranslational modifications, such as glycosylation and phosphorylation.

Figure 9.4 Ascorbate functions in the hydroxylation of peptide-bound proline and lysine in procollagen. The reaction is driven by α-ketoglutarate decarboxylation. One atom of oxygen* appears in the hydroxyl group of the product and the other in succinate.

The role of vitamin C in the hydroxylation reactions relates to the iron cofactor in the two dioxygenase enzymes. Prolyl hydroxylase and lysyl hydroxylase both require iron bound as a cofactor; although, silicon also may be needed for maximal prolyl hydroxylase activity (see Fig. 13.1). During the reactions, both dioxygenase enzymes catalyze reactions in which one of two atoms of O_2 becomes incorporated into the product, and the second of the two atoms of O_2 becomes incorporated into the cosubstrate α-ketoglutarate. In these reactions (Fig. 9.4), one oxygen atom is incorporated into α-ketoglutarate to form the new carboxyl group of succinate, and the second oxygen atom is found in the other substrate. During the hydroxylation reactions, the iron cofactor in the enzymes is oxidized; that is, it is converted from a ferrous (2+) state to a ferric (3+) state. Ascorbate is needed to function as the reductant, thereby reducing iron back to its ferrous state (2+) in the prolyl and lysyl hydroxylases.

Although these reactions may be seemingly simple, normal development of cartilage, bone, and dentine depend on an adequate supply of vitamin C. Also, the basement membrane lining the capillaries, the "intracellular cement" holding together the endothelial cells, and the scar tissue responsible for wound healing all require the presence of vitamin C for their formation and maintenance.

Carnitine Synthesis

Ascorbate is involved in two reactions required for the synthesis of carnitine [14]. Carnitine synthesis begins with the substrate trimethyllysine.

• The first reaction involving vitamin C requires the iron-containing enzyme *trimethyllysine* dioxygenase (also called *hydroxylase*), which catalyzes the conversion of trimethyllysine to 3-hydroxy trimethyllysine.

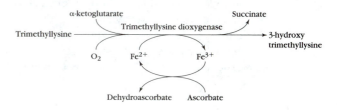

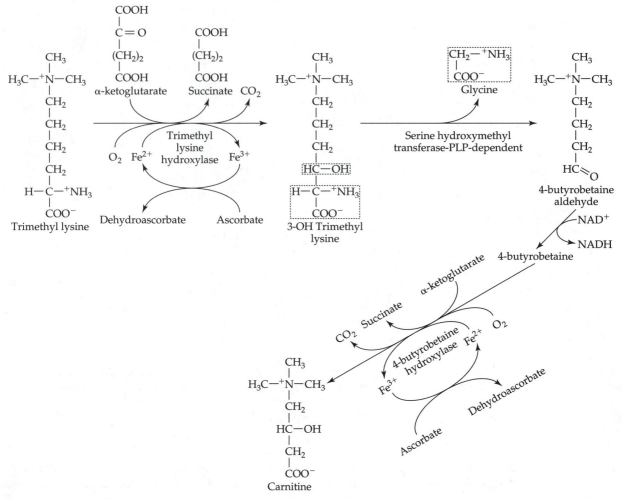

Figure 9.5 The function of vitamin C in carnitine synthesis.

- Next, in the last step of carnitine synthesis, the iron-containing enzyme *4-butyrobetaine dioxygenase* (also called *hydroxylase*) catalyzes the conversion of 4-butyrobetaine to carnitine.

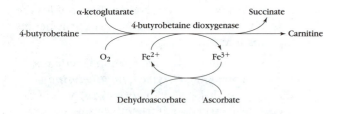

These reactions involving ascorbate are hydroxylations that require α-ketoglutarate and are almost identical to those for proline and lysine hydroxylation depicted in Figure 9.5. Vitamin C is the preferred reducing agent in carnitine synthesis; however, other substances may be able to replace the ascorbate [5,14]. Sufficient produc-

tion of carnitine is of significance in fat metabolism because carnitine is essential for the transport of long-chain fatty acids from the cell cytoplasm into the mitochondrial matrix where β-oxidation occurs. Additional information regarding carnitine can be found on page 179.

Tyrosine Synthesis and Catabolism

Tyrosine is synthesized in the body from the essential amino acid phenylalanine. Tyrosine synthesis requires hydroxylation of phenylalanine via the iron dependent enzyme *phenylalanine mono-oxygenase* (also called *hydroxylase*). The reaction (Fig. 9.6) occurs mostly in the liver and possibly kidney and requires O_2 and the cosubstrate tetrahydrobiopterin. Vitamin C is thought to function in the regeneration of tetrahydrobiopterin from dihydrobiopterin [15].

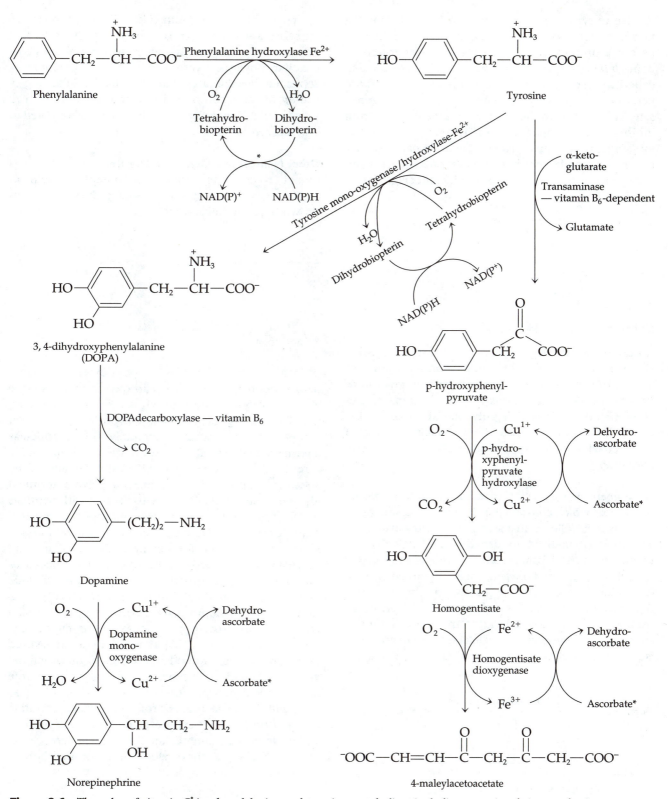

Figure 9.6 The roles of vitamin C* in phenylalanine and tyrosine metabolism, including norepinephrine synthesis.

Also in tyrosine catabolism is another hydroxylation in which ascorbate participates (Fig. 9.6). Ascorbate is a preferred reductant for the copper-dependent *p-hydroxyphenylpyruvate hydroxylase,* also called *dioxygenase,* the enzyme necessary for conversion of p-hydroxyphenylpyruvate to homogentisate. Vitamin C is thought to protect the enzyme from inhibition by its substrate [5]. The conversion of homogentisate to 4-maleylacetoacetate is catalyzed by the iron-dependent *homogentisate dioxygenase* and is dependent on vitamin C as a reductant (Fig. 9.6).

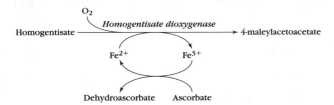

Neurotransmitter Synthesis

Ascorbate is also involved in neurotransmitter synthesis through two Cu^{1+}-dependent mono-oxygenases, *dopamine mono-oxygenase* and *peptidylglycine α-amidating mono-oxygenase.* Mono-oxygenases catalyze reactions in which only one atom of O_2 becomes incorporated into the product. Neither of these enzymes requires a cosubstrate such as α-ketoglutarate.

Norepinephrine In the case of *dopamine mono-oxygenase,* which converts dopamine to norepinephrine (Fig. 9.6), it is believed that ascorbate donates hydrogens and thus is oxidized to dehydroascorbate [5]. The hydrogen donated from ascorbate is thought to be used to reduce the atom of oxygen, not incorporated into the dopamine, to water. The copper atom in the enzyme is thought to act as an intermediate, accepting electrons from ascorbate as it is reduced to the cuprous ion and subsequently transferring these electrons to oxygen as it is reoxidized back to the cupric ion [5].

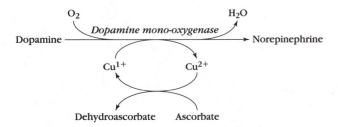

Serotonin Vitamin C, tetrahydrobiopterin, and oxygen are also involved in the hydroxylation of tryptophan for the synthesis of the neurotransmitter serotonin (5-hydro-

xytryptamine) in the brain (Fig. 9.7). Tryptophan hydroxylase, also called mono-oxygenase, catalyzes the first step in serotonin synthesis, whereby tryptophan is converted to 5-hydroxytryptophan in a tetrahydrobiopterin-dependent reaction; ascorbate may function in the regeneration of tetrahydrobiopterin from dihydrobiopterin. Subsequently, 5-hydroxytryptophan is decarboxylated to generate serotonin.

Other Neurotransmitters and Hormones Ascorbate is thought to keep the copper atom in peptidylglycine α-amidating mono-oxygenase in its reduced state, as shown here and in Figure 9.8:

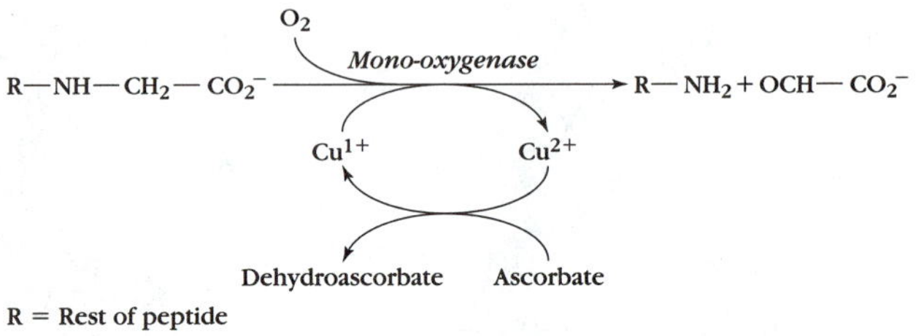

$$R\text{—}NH\text{—}CH_2\text{—}CO_2^- \xrightarrow[\text{Mono-oxygenase}]{O_2} R\text{—}NH_2 + OCH\text{—}CO_2^-$$

$$Cu^{1+} \quad Cu^{2+}$$

$$\text{Dehydroascorbate} \quad \text{Ascorbate}$$

R = Rest of peptide

In the reaction catalyzed by peptidylglycine α-amidating mono-oxygenase, the carboxyl-terminal residue is oxidatively cleaved through use of molecular oxygen. The cleavage is not a simple hydrolytic cleavage of a peptide bond because the amino group is retained as a terminal amide while the rest of the oxidized residue is released as glyoxylate (Fig. 9.8). Although most of the substrate peptides for this enzyme have a terminal glycine residue, the enzyme is also active with peptides terminating in other amino acids. Many of the amidated peptides resulting from this reaction, which has been shown to occur in, for example, the pituitary and adrenal glands, are active as hormones, hormone-releasing factors, or neurotransmitters. Examples include bombesin or gastrin-releasing peptide (GRP), calcitonin, cholecystokinin (CCK), corticotropin-releasing factor, gastrin, growth hormone-releasing factor, oxytocin, and vasopressin [15]. If ascorbate is the favored reductant for the required amidating enzyme, then the vitamin assumes an important, although indirect, role in many regulatory processes.

In addition to ascorbate's roles in collagen, carnitine and neurotransmitter synthesis, and tyrosine synthesis and catabolism, vitamin C functions in a general capacity as a reducing agent or electron donor and thereby has antioxidant activity.

Antioxidant Activity

Ascorbic acid acts as a reducing agent in aqueous solutions such as the blood and within cells. Stated slightly

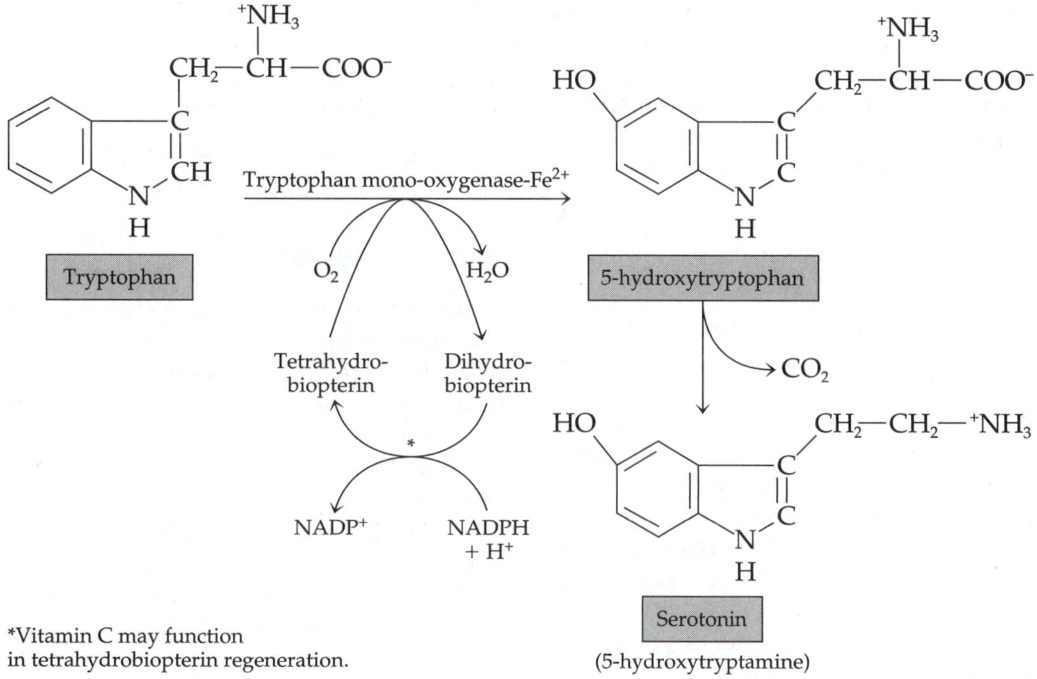

Figure 9.7 Serotonin synthesis.

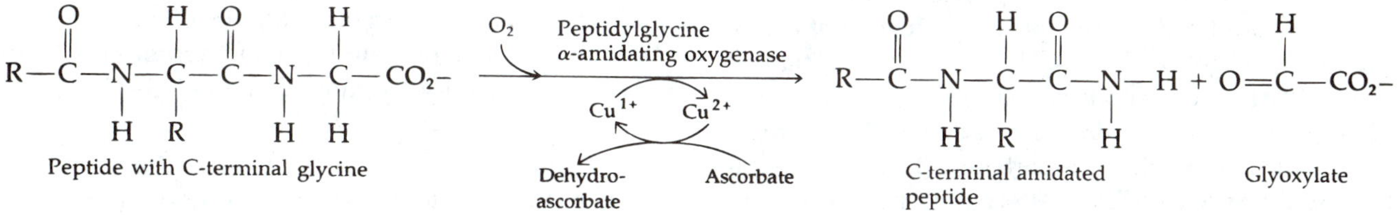

*Vitamin C may function in tetrahydrobiopterin regeneration.

Figure 9.8 Amidation of peptides with C-terminal glycine requires vitamin C.

differently, ascorbate is an antioxidant in that it reverses oxidation. Reducing agents or antioxidants such as ascorbate may reverse oxidation by donating electrons and hydrogen ions. The reduction potential of ascorbate is such that it readily donates electrons/hydrogen ions to regenerate vitamin E as well as to numerous free radicals and reactive oxygen species.

As an antioxidant, ascorbate (expressed as AH_2) may react in an aqueous solution (blood or intracellular) with a variety of reactive oxygen species and give to the radicals an electron in the form of a hydrogen ion. Free radicals exist independently and contain one or more unpaired electrons in an outer orbital surrounding the nucleus of the atom. Remember from chemistry courses that electrons are usually found in pairs in an orbital. Free radicals and other reactive oxygen species are formed during normal cellular metabolism; this process is discussed in more detail in the perspective at the end of Chapter 10.

Examples of some reactive oxygen species that vitamin C may reduce include the following:

- Hydroxyl radical (OH·), a very reactive oxygen-centered radical
- Hydroperoxyl radical (HO_2·), an oxygen-centered radical
- Superoxide radical (O_2^-), an oxygen-centered radical
- Alkoxyl radical (RO·), an oxygen-centered radical
- Peroxyl radical (RO_2·), an oxygen-centered radical

Hydrogen peroxide, H_2O_2, a nonradical because it has no unpaired electrons in its orbital, is an example of a reactive oxygen species that, like hypochlorous acid (HOCl) and singlet oxygen ($^1O_2$), are scavenged by vitamin C.

Once formed, free radicals and reactive oxygen species attack nucleic acids in DNA, polyunsaturated fatty acids in phospholipids, and proteins in cells. Ascorbic acid has been shown to interact with oxidants in the aqueous phase before they initiate damage especially to cell lipids [16,17]. Furthermore, ascorbic acid appears to be superior to other water-soluble antioxidants such as bilirubin, uric acid, and protein thiols [17,18].

Some examples of reactions involving ascorbate as an antioxidant include [18]:

$$\text{Ascorbate} + \text{OH} \cdot \longrightarrow \text{semidehydroascorbate radical} + \text{H}_2\text{O}$$

$$\text{Ascorbate} + \text{O}_2 \cdot \longrightarrow \text{dehydroascorbate} + \text{H}_2\text{O}_2$$

$$\text{Ascorbate} + \text{H}_2\text{O}_2 \longrightarrow \text{dehydroascorbate} + 2\text{H}_2\text{O}$$

The role of vitamin C and other antioxidants as a defense against oxidative damage to the cell is discussed in the Perspective at the end of Chapter 10.

Although ascorbate is a powerful reducing agent and may be the preferred reductant in certain oxidation reduction reactions, its action may be nonspecific. The function of ascorbate in the cells may be to balance or to set the redox potential (i.e., relative states of oxidation or reduction) of other cellular water-soluble substances such as glutathione [7]. In addition, vitamin C may transfer electrons to tocopherol radicals in membranes [19].

As an antioxidant, ascorbate provides electrons and becomes oxidized in the process. Regeneration of ascorbic acid from semidehydroascorbate radical and from dehydroascorbate is crucial. To regenerate ascorbic acid, two semidehydroascorbate radicals may react as follows: 2 semidehydroascorbate radicals ⟶ ascorbate + dehydroascorbate. Alternately, reductases are found in most tissues to reduce the semidehydroascorbate radical to ascorbate. Glutathione in its reduced state (GSH) and niacin as nicotinamide adenine dinucleotide (NADH) or nicotinamide adenine dinucleotide phosphate (NADPH) function in this capacity as shown here [5,18]:

$$\text{2 semidehydroascorbate radicals} + 2\text{GSH} \longrightarrow \text{2 ascorbate} + \text{GSSG}$$

$$\text{Dehydroascorbate} + 2\text{GSH} \longrightarrow \text{ascorbate} + \text{GSSG}$$

$$\text{2 semidehydroascorbate radical} + \text{NADH} + \text{H}^+ \longrightarrow \text{2 ascorbate} + \text{NAD}^+$$

Pro-oxidant

Paradoxically, vitamin C also may act as a pro-oxidant. Vitamin C can reduce transition metals, such as cupric ions (Cu^{2+}) to cuprous (Cu^{1+}), and ferric ions (Fe^{3+}) to ferrous (Fe^{2+}) while itself becomes oxidized to semidehydroascorbate, as shown here:

$$\text{Ascorbate (AH}_2) + \text{Fe}^{3+} \text{ or } \text{Cu}^{2+} \longrightarrow \text{semidehydroascorbate radical (AH}^-) + \text{Fe}^{2+} \text{ or } \text{Cu}^{1+}$$

The products—Fe^{2+} and Cu^{1+}—generated from these reactions can proceed to cause cell damage through the generation of reactive oxygen species and free radicals. Examples of some of these reactions include:

$$\text{Fe}^{2+} \text{ or } \text{Cu}^{1+} + \text{H}_2\text{O}_2 \longrightarrow \text{Fe}^3 \text{ or } \text{Cu}^{2+} + \text{OH}_2 + \text{OH}^{\cdot}$$

$$\text{Fe}^{2+} \text{ or } \text{Cu}^{1+} + \text{O}_2 \longrightarrow \text{Fe}^{3+} \text{ or } \text{Cu}^{2+} + \text{O}_2^{-}$$

Other Functions

Many other diverse biochemical functions for vitamin C have been proposed. Experimental evidence supporting these functions varies considerably. Experimental results often conflict, and the mechanism by which ascorbate may be involved is generally unclear. Possible functions for vitamin C include roles in microsomal hydroxylation reactions of noncholesterol steroids and drugs; regulation of cellular nucleotide concentrations; roles in lipid metabolism; immune function, including complement synthesis; prevention of oxidative destruction of other vitamins, such as vitamin A; sulphation for proteoglycan synthesis; and endocrine systems [4,5,20–22].

Vitamin C and Disease

Much attention has been directed toward vitamin C intakes and disease ranging from the common cold to cancer and heart disease, among others [23–43].

Colds The possible pharmacological effects of vitamin C on the incidence, severity, and duration of the common cold have been almost totally refuted some investigators [23]. High doses of ascorbate appear to be only weakly prophylactic, if at all, and of little or no use in the treatment of human infections [23]. Vitamin C (1 g/day) had no more effect in protection against or in combating the common cold than 50 mg/day. Further, there were no significant differences between the groups for the number of colds, their severity, or their duration [23]. Other reports, however, conflict and suggest a considerable decrease in the duration of cold episodes and the severity of the symptoms [24]. Vitamin C is hypothesized to react with oxidized products released from white blood cell phagocytosis. The reaction of vitamin C with the oxidized products minimizes the inflammatory effects of the oxidized products and thus the severity of the cold [24].

Cancer Epidemiological studies provide evidence that increased intakes of fruits and vegetables are associated with a decreased risk of some cancers [25–29]. The association between high vitamin C intake as a protective effect against cancer is especially strong with cancers of the oral cavity, pharynx, esophagus, and stomach [4,25,29]. Less strong evidence exists between vitamin C and cancer of the lung, colon, and cervix [4,25,29]. Other studies [27–30] both support some and negate some of these associations. In clinical trials, whereas some researchers have shown that the survival time in cancer patients could be prolonged through massive doses of vitamin C, no such success has been demonstrated by others [4,25,26,32–34].

Possible mechanisms of ascorbate action against cancer development include a role in immunocompetence, an ability to act as a free radical scavenger or antioxidant, and an ability to detoxify carcinogens or to block carcinogenic processes [4,25,27,28,34]. The fact that vitamin C in amounts of about 1 g when ingested with nitrates and/or nitrites can prevent formation of carcinogenic nitrosamines supports this latter theory and has lent credence that the vitamin is somewhat protective against stomach and/or esophageal cancers [33,35,36]. Vitamin C is not unique in this regard; other reducing agents and some food components are also effective in preventing nitrosocarcinogens [34].

Cardiovascular Disease   Analysis of National Health and Nutrition Examination Survey (NHANES) data suggests that increased vitamin C intake appears to have a protective effect on pulmonary function as well as cardiovascular disease mortality [37,38]. Studies in selected animals show that vitamin C regulates cholesterol metabolism to bile acid. For example, cholesterol 7α-hydroxylase, found in the microsomes of the liver, is required for the initial step in the synthesis of bile acids from cholesterol. The hydroxylation of cholesterol's steroid nucleus by cholesterol 7-α-hydroxylase has been shown to be diminished in vitamin C–deficient guinea pigs [4]. These findings, which imply impaired vitamin C status may result in increased plasma cholesterol concentrations, however, have not been demonstrated in humans.

Many human studies report a relationship between low vitamin C status and increased blood total cholesterol concentrations, and they suggest that those individuals with total serum cholesterol concentrations in excess of 200 mg/dL and below full tissue vitamin C saturation increase vitamin C intake [31]. High plasma vitamin C concentrations also have been associated with lower blood pressure and with higher plasma high-density lipoprotein cholesterol concentrations, both protective against heart disease.

The mechanisms by which vitamin C may protect against heart disease in humans are not clearly identified. Animal studies suggest impaired vitamin C alters cholesterol metabolism. Impaired vitamin C status also may result in increased oxidation of low-density lipoproteins (LDLs), which have been shown to be more atherogenic than nonoxidized LDL [39,40]. Vitamin C is a potent scavenger of superoxide radicals (O_2^-) and singlet oxygen, and deficiency of the vitamin in guinea pigs has been shown to be associated with myocardial lipid peroxidative damage [39]. The ability of vitamin C to work synergistically with vitamin E to prevent oxidation of lipids in solution is another mechanism by which vitamin C may help to protect against heart disease [19,41]. The addition of vitamin C to low-density lipoproteins undergoing oxidation in vitro mediated by aqueous peroxyl radicals ($ROO^·$) results in diminished utilization of vitamin E and diminished LDL oxidation [41]. Further studies should help to clarify the role of vitamin C and heart disease.

Other Diseases or Conditions   Another condition in which high vitamin C intakes are thought to be beneficial is *cataracts.* Cataracts is one of the major causes of blindness, especially in older individuals. Cataracts results in part from oxidative damage to proteins in the lens of the eye. The damaged proteins aggregate and precipitate causing the lenses to become cloudy. While some epidemiological studies suggest a protective effect of vitamin C against cataracts, the effect cannot be attributed solely to vitamin C because subjects in many of the studies were consuming a multivitamin preparation [32,42,43]. Vitamin E is also thought to be protective against cataracts development [42,43].

Interactions with Other Nutrients

Vitamin C interacts with several minerals including iron, lead, and copper. A possible relationship with folate has also been demonstrated.

The interaction between the mineral *iron* and vitamin C is related not only to the vitamin's effect on intestinal absorption of nonheme iron but also on the distribution of iron in the body. Specifically, ascorbate enhances the intestinal absorption of nonheme iron by either reducing iron to a ferrous (Fe^{2+}) form from a ferric (Fe^{3+}) form or by forming a soluble complex with the iron in the alkaline pH of the small intestine to thereby enhance iron's absorption. Excessive iron in the presence of vitamin C, however, can accelerate the oxidative catabolism of vitamin C, thus negating the enhancing effects of vitamin C on iron absorption. Incorporation of iron into ferritin, the storage form of iron, and stabilization of ferritin by ascorbate have also been demonstrated [44]. The effect of the vitamin in the distribution and mobilization of storage iron is uncertain. Ascorbic acid supplements can cause a change in the distribution of iron in patients suffering with iron overload, but not necessarily in other individuals [8,45]. Vitamin C initiated free radical generation from mobilization of storage iron has been suggested but also refuted [17,32,46,47].

In addition to its effects on iron, ascorbic acid may also increase the absorption and excretion of heavy metals, such as *lead,* from the body by forming chelates with the metals [21].

With respect to *copper,* vitamin C intakes of 1.5 g daily for about 2 months resulted in decreased serum copper and ceruloplasmin, a copper-containing protein with oxidase activity; however, despite the decrease, serum copper levels remained within normal range [48]. Dietary vitamin C intakes in excess of 600 mg daily have also been shown to decrease the oxidase activity of ceruloplasmin. Ascorbate may cause copper dissociation from ceruloplasmin or may influence the binding of copper to enzymes [33,49]. Human cells treated with vitamin C have exhibited enhanced copper uptake from ceruloplasmin [33]. Decreased intestinal absorption of copper by ascorbic acid has been observed in several animal species. A proposed mechanism of interaction for this effect suggested that vitamin C stimulated iron mobilization and the mobilized iron in turn inhibited copper absorption [49]. In addition, vitamin C may inhibit the binding of copper to metallothionein, a protein found in the intestinal cells and other body cells. The delayed binding has been proposed to inhibit copper transport across the intestinal cell [49].

Ascorbic acid also appears necessary for *folate* metabolism. Specifically, vitamin C is thought to be needed to maintain folate in a reduced state, as either tetrahydrofolate, the active form of the vitamin, or dihydrofolate.

Doses of 500 mg or more of vitamin C when ingested with a meal or up to an hour after a meal may destroy *vitamin B$_{12}$* or diminish its absorption [2]. Furthermore, incubation of 200 mg of ascorbate with 25 μg of vitamin B$_{12}$, 400 μg of folate, and 15 mg of iron for 30 minutes in gastric juice (pH 5) resulted in a significant destruction of vitamin B$_{12}$ [50].

Metabolism and Excretion

As previously discussed and shown in Figure 9.3, vitamin C is readily oxidized to dehydroascorbate. Although this oxidized form of the vitamin may be reduced by glutathione or by niacin as nicotinamide adenine dinucleotide (NADH) or nicotinamide adenine dinucleotide phosphate (NADPH) to regenerate ascorbate, it also may be further catabolized for excretion.

Additional postulated biochemical steps in the metabolism of ascorbic acid are given in Figure 9.9; these steps occur primarily in the liver but to some extent in the kidney. Following conversion of ascorbate to dehydroascorbate, further oxidation occurs to hydrolyze (open) the ring structure and yields 2,3-diketogulonic acid. Diketogulonic acid possesses no vitamin C activity and may be further hydrolyzed in the body. Diketogulonate is cleaved by separate pathways (Fig. 9.9) into either oxalic acid and a four-carbon sugar threonic acid or into a variety of five-carbon sugars (xylose, xylonate, and lyxonate). The four- and five-carbon sugars can be converted into cellular compounds or can be oxidized and excreted as CO_2 and water.

Vitamin C metabolites including dehydroascorbate, diketogulonate, oxalic acid, 2-O-methyl ascorbate, ascorbate 2-sulfate, and 2-ketoascorbitol plus excess ascorbate are excreted in the urine. About 25% of vitamin C intake is excreted in the form of oxalic acid [9]. The kidneys can reduce dehydroascorbate to ascorbate and can conserve ascorbic acid and dehydroascorbate through reabsorption by the kidney tubules so long as the body pool of the vitamin is less than, or approximates, 1,500 mg [8,15]. The specific amount of vitamin C filtered and then reabsorbed by the kidneys depends on plasma vitamin C concentrations. Plasma ascorbate levels of about 0.8 to 1.4 mg/dL constitute the renal threshold whereby vitamin C in amounts in excess of this level will not be reabsorbed and will thus be excreted in the urine [4]. When the body pool of ascorbic acid is <1,500 mg, little or no ascorbic acid appears in the urine; only its metabolites are excreted. As the pool increases above 1,500 mg, the efficiency of kidney reabsorption of the vitamin decreases, and ascorbate along with its metabolites are excreted.

Dietary Reference Intakes and Recommended Dietary Allowances

Multiple approaches to establish vitamin C needs by humans exist; which approach is best remains controversial, as does what is the optimal dietary vitamin C intake [11–13,15]. In the United States, the RDA for ascorbate for men has ranged from 75 mg/day in 1943 to 45 mg/day in 1974. The 1989 RDA [51] for vitamin C is 60 mg for adults and is based on maintenance of a body pool of 1,500 mg. This level represents the amount of the vitamin that could be held within the body tissues and fluids without loss via the kidneys and is the same level recommended in the 1980 edition of the RDA [51]. In the 1989 RDA, cigarette smokers for the first time were singled out for an increased vitamin C requirement; recommendations are 100 mg daily for ciga-

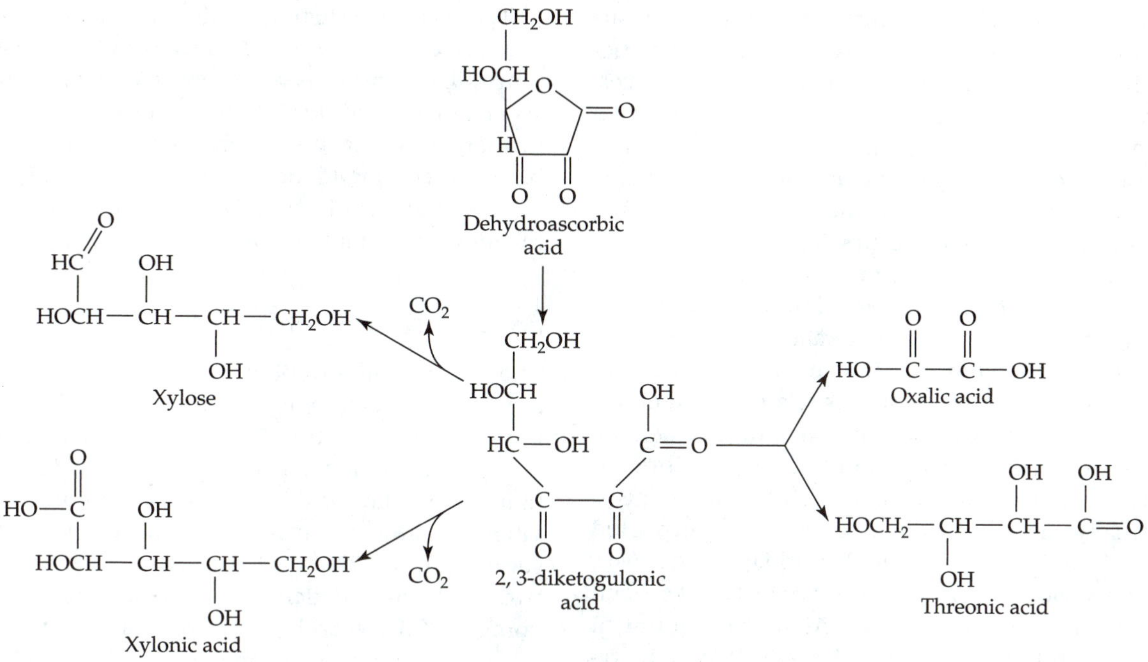

Figure 9.9 Vitamin C metabolism.

rette smokers. Smoking appears to deplete the body's ascorbate pool [52]. Recommendations from a 1980 to 1985 Committee on RDA, referred to as Recommended Dietary Intakes (RDIs), suggest an RDI for adult males and females for vitamin C of 40 mg and 30 mg, respectively [8]. The RDI for vitamin C is based on a body pool of 900 mg, a level believed by some investigators to exceed by 300 mg the amount needed by normal people under ordinary circumstances [8]. With the availability of new information including vitamin C pharmacokinetics, among other data, new recommendations for ascorbic acid intake are being promoted. A maximum recommended daily intake of vitamin C of 200 mg has been suggested by some [11].

Deficiency: Scurvy

Vitamin C intakes of <10 mg daily may result in scurvy. Scurvy is typically manifested when the total body pool of vitamin C falls below about 300 mg [15]. Scurvy may be characterized by a multitude of signs and symptoms. Most notable signs and symptoms include bleeding gums, small skin discolorations due to ruptured small blood vessels (petechiae), sublingual hemorrhages, easy bruising (ecchymoses and purpurae), impaired wound and fracture healing, joint pain (arthralgia), loose and decaying teeth, and hyperkeratosis of hair follicles especially on the arms, legs, and but-

tocks [53]. The four Hs—*hemorrhagic* signs, *hyperkeratosis* of hair follicles, *hypochondriasis* (psychological manifestation), and *hematologic* abnormalities (associated with impaired iron absorption)—are often used as a mnemonic device for remembering scurvy signs [15].

Although scurvy is rare in the United States, low plasma vitamin C levels have been observed in the elderly, especially if institutionalized. People who have poor diets, especially if coupled with alcoholism or drug abuse, are likely to be deficient, as are people with diseases, such as diabetes mellitus and some cancers that increase the turnover rate of the vitamin.

Toxicity

Daily intakes of up to 1 g vitamin C are routinely consumed without adverse effects; 1 g represents the no observed adverse effect level (NOAEL) based on an absence of convincing patterns of effects shown in humans [47]. Thus, vitamin C is considered safe in doses of up to 1 g daily [11]. More vitamin C is absorbed, and thus theoretically toxicity more likely, with ingestion of several large (> 1 g) doses of the vitamin throughout the day than with one single dose. The greater absorption occurs because intestinal vitamin C absorption is saturable and dose-dependent. Therefore, maximal absorption of vitamin C occurs more frequently with ingestion of several doses of vitamin C throughout the day versus one more massive dose (>1 g). Yet, although many potentially harmful effects have been attributed to excessive intakes of ascorbic acid, the frequency of recorded toxicity is quite low [8,54]. One side effect from ingestion of large doses (~2 g) of vitamin C is diarrhea. It is the unabsorbed vitamin C in the intestinal tract that may produce an osmotic diarrhea [2,4,33,54].

A second side effect frequently mentioned is increased possibility in the development of kidney stones (nephrolithiasis), either oxalic acid or uric acid based. The purported mechanism for the development of oxalic acid–based stones relates to vitamin C metabolism in the body. Because vitamin C is metabolized in the body to oxalate and because calcium oxalate is a common constituent of kidney stones, ingestion of large doses of vitamin C has been purported as an etiologic factor in nephrolithiasis. However, although doses up to 10 g of vitamin C have been shown to increase oxalate excretion, the amount of oxalate excreted (generally <50 mg) typically remained within a normal and safe range [55–57]. Nevertheless, some suggest that those people predisposed to calcium oxalate kidney stones avoid high doses (≥500 mg) of vitamin C [4,54–57]. Furthermore, because of interactions in the kidney between vitamin C and uric acid, which also is a constituent of kidney stones, ingestion of large doses (up to ~4 g) of ascorbic acid should be avoided by individuals with uric acid kidney stones [8,58]. Specifically, vitamin C competitively inhibits renal reabsorption of uric acid to increase uric acid excretion. The resulting urine acidification along with the excessive amount of uric acid being excreted could cause precipitation of urate crystals and urate kidney stones [54,58]. The actual clinical importance of uricosuria (high uric acid in the urine) with regard to stone formation is unknown [54].

Chronic high doses of vitamin C are also purported to be unsafe for those people with disorders involving iron metabolism, including individuals with hemochromatosis, thalassemia, and sideroblastic anemia [26,46,54]. However, others contend that pro-oxidant effects of vitamin C on mobilization of iron stores do not occur in vivo [17,32,47].

Excessive ascorbate excretion in the urine and feces can interfere with a variety of clinical laboratory tests. Vitamin C in the urine, for example, may act as a reductive agent and thus interfere with the diagnostic tests using redox chemistry [4]. False-negative tests for fecal occult blood may be generated and occult blood in the urine may not be detected. Tests for glucose in the urine can be rendered invalid [4,33,54].

The issue of systemic conditioning to high intakes of vitamin C is currently deemed doubtful. Although scurvy-like symptoms were reported in a few individuals on abrupt withdrawal of large intakes of vitamin C, further substantiation of conditioned scurvy is needed before recommendations can be made [47]. Others suggest that people are advised to gradually diminish the intake of vitamin C over a 2- to 4-week period [4].

Assessment of Nutriture

The measurement of blood, serum, or plasma levels of ascorbate is a commonly used procedure for determining vitamin C nutriture. Blood, plasma, and serum vitamin C concentrations respond to changes in dietary vitamin C intakes and are thus used to assess recent vitamin C intake. Plasma vitamin C levels are deemed superior to blood levels because the latter is less sensitive to vitamin C deficiency [4]. White blood cell (wbc) content of the vitamin better reflects body stores, but this measurement is technically more difficult to perform. Plasma levels of vitamin C considered to be deficient are those below 0.2 mg/dL. Marginal plasma vitamin C concentrations are 0.2 to 0.39 mg/dL, whereas adequate concentrations are 0.4 to 0.99 mg/dL. Leukocyte vitamin C concentrations of 10 μg/10^8 wbc or less and between 11 and 19 μg/10^8 wbc are considered deficient and marginal, respectively. Vitamin C concentrations of 20 to 30 μg/10^8 wbc are considered in the normal range [59].

References Cited for Vitamin C

1. Goldenberg H, Schweinzer E. Transport of vitamin C in animal and human cells. J Bioenergetics Biomembranes 1994;26:359–67.

2. Sauberlich HE. Bioavailability of vitamins. Prog Food Nutr Sci 1985;9:1–33.

3. Rose R. Intestinal transport of vitamins. J Inherited Metab Dis 1985;8(suppl):13–16.

4. Jacob R. Vitamin C. In: Shils ME, Olson JA, Shike M., eds. Modern Nutrition in Health and Disease, 8th ed. Philadelphia: Lea and Febiger, 1994:432–48.

5. Basu T, Schorah C. Vitamin C in Health and Disease. Westport, CT: AVI, 1982.

6. Johnston C, Meyer C, Srilakshmi J. Vitamin C elevates red blood cell glutathione in healthy adults. Am J Clin Nutr 1993;58:103–5.

7. Martensson J, Han J, Griffith O, Meister A. Glutathione ester delays the onset of scurvy in ascorbate-deficient guinea pigs. Proc Natl Acad Sci 1993;90: 317–21.

8. Olson A, Hodges R. Recommended dietary intakes (RDI) of vitamin C in humans. Am J Clin Nutr 1987;45:693–703.

9. Bender D. Nutritional Biochemistry of the Vitamins. New York: Cambridge University Press, 1992:360–93.

10. Jacob R, Skala J, Omaye S. Biochemical indices of human vitamin C status. Am J Clin Nutr 1987;46: 818–26.

11. Levine M, Cantilena-Conry C, Wang Y, Welch R, Washko P, Dhariwal K, Park J, Lazarev A, Graumlich J, King J, Cantilena L. Vitamin C pharmocokinetics in healthy volunteers: evidence for a recommended requirement. Proc Natl Acad Sci USA 1996;93:3704–9.

12. Blanchard J, Tozer T, Rowland M. Pharmacokinetic perspective on megadoses of ascorbic acid. Am J Clin Nutr 1997;66:1165–71.

13. Shane B. Vitamin C pharmacokinetics: it's déjà vu all over again. Am J Clin Nutr 1997;66:1061–2.

14. Rebouche C. Ascorbic acid and carnitine biosynthesis. Am J Clin Nutr 1991;54:1147S–52S.

15. Levine M. New concepts in the biology and biochemistry of ascorbic acid. N Engl J Med 1986; 314:892–902.

16. Niki E. Action of ascorbic acid as a scavenger of active and stable oxygen radicals. Am J Clin Nutr 1991;54:1119S–24S.

17. Frei B. Ascorbic acid protects lipids in human plasma and low density lipoprotein against oxidative damage. Am J Clin Nutr 1991;54:1113S–8S.

18. Stadtman E. Ascorbic acid and oxidative inactivation of proteins. Am J Clin Nutr 1991;54:1125S–8S.

19. Niki E, Noguchi N, Tsuchihashi H, Gotoh N. Interaction among vitamin C, vitamin E and β-carotene. Am J Clin Nutr 1995;62:1322S–6S.

20. Gey K, Moser U, Jordan P, Stahelin H, Eichholzer M, Ludin E. Increased risk of cardiovascular disease at suboptimal plasma concentrations of essential antioxidants: an epidemiological update with special attention to carotene and vitamin C. Am J Clin Nutr 1993;57(suppl.):787S–97S.

21. Moser U, Bendich A. Vitamin C. In: Machlin LJ, ed. Handbook of Vitamins, 2nd ed. New York: Dekker, 1991:195–232.

22. Levine M, Morita K. Ascorbic acid in endocrine systems. Vitamins & Hormones 1985;42:2–64.

23. Briggs M. Vitamin C and infectious disease: a review of the literature and the results of a randomized, double-blind, prospective study over 8 years. In: Briggs MH, ed. Recent Vitamin Research. Boca Raton, FL: CRC Press, 1984:39–81.

24. Hemila H. Vitamin C and the common cold. Br J Nutr 1992;67:3–16.

25. Block G. Vitamin C status and cancer: epidemiologic evidence of reduced risk. Ann NY Acad Sci 1992;669:280–90.

26. Block G. Vitamin C and cancer prevention: the epidemiologic evidence. Am J Clin Nutr 1991;53:270S–82S.

27. Block G, Patterson B, Subar A. Fruit, vegetables, and cancer prevention: a review of the epidemiological evidence. Nutr Cancer 1992;18:1–29.

28. Block G. The data support a role for antioxidants in reducing cancer risk. Nutr Rev 1992;50:207–13.

29. Byers T, Guerrero N. Epidemiologic evidence for vitamin C and vitamin E in cancer prevention. Am J Clin Nutr 1995;62(suppl):1385S–92S.

30. Gershoff S. Vitamin C (ascorbic acid): new roles, new requirements? Nutr Rev 1993;51:313–26.

31. Simon J. Vitamin C and cardiovascular disease: a review. J Am Coll Nutr 1992;11:107–25.

32. Bendich A, Langseth L. The health effects of vitamin C supplementation: a review. J Am Coll Nutr 1995;14:124–36.

33. Davies M, Austin J, Partridge D. Vitamin C. Its Chemistry and Biochemistry. Cambridge, England: Royal Society of Chemistry, 1991.

34. Carpenter M. Roles of vitamins E and C in cancer. In: Laidlaw SA, Swendseid ME. Contemporary Issues in Clinical Nutrition. New York: Wiley-Liss, 1991: 61–90.

35. Tannenbaum S, Wishnok J, Leaf C. Inhibition of nitrosamine formation by ascorbic acid. Am J Clin Nutr 1991;53:247S–250S.

36. Schorah C, Sobala G, Sanderson M, Collis N, Primrose J. Gastric juice ascorbic acid: effects of disease and implications for gastric carcinogenesis. Am J Clin Nutr 91;53:287S–93S.

37. Schwartz J, Weiss S. Relationship between dietary vitamin C intake and pulmonary function in the First National Health and Nutrition Examination Survey (NHANES I). Am J Clin Nutr 1994;59:110–14.

38. Enstrom J, Kanim L, Klein M. Vitamin C intake and mortality among a sample of the United States population. Epidemiology 1992;3:194–202.

39. Chakrabarty S, Nandi A, Mukhopadhyay C, Chatterjee I. Protective role of ascorbic acid against lipid peroxidation and myocardial injury. Molec Cell Biochem 1992;111:41–47.

40. Jialal I, Grundy S. Effect of combined supplementation with alpha-tocopherol, ascorbate, and beta caro-

tene on low-density lipoprotein oxidation. Circulation 1993;88:2780–6.

41. Thomas S, Neuzil J, Mohr D, Stocker R. Coantioxidants make α-tocopherol an efficient antioxidant for low-density lipoprotein. Am J Clin Nutr 1995;62(suppl): 1357S–64S.

42. Taylor A, Jacques P, Epstein E. Relations among aging, antioxidant status, and cataract. Am J Clin Nutr 1995;62(suppl):1439S–47S.

43. Jacques P, Taylor A, Hankinson S. Long term vitamin C supplement use and prevalence of early age-related lens opacities. Am J Clin Nutr 1997;66:911–6.

44. Hoffman K, Yanelli K, Bridges K. Ascorbic acid and iron metabolism: alterations in lysosomal function. Am J Clin Nutr 1991;54:1188S–92S.

45. Cook J, Watson S, Simpson K, Lipschitz D, Skikne B. The effect of high ascorbic acid supplementation on body iron stores. Blood 1984;64:721–6.

46. Herbert V, Shaw S, Jayatilleke E. Vitamin C–driven free radical generation from iron. J Nutr 1996;126: 1213S–20S.

47. Hathcock J. Vitamins and minerals: efficacy and safety. Am J Clin Nutr 1997;66:427–37.

48. Finley E, Cerklewski F. Influence of ascorbic acid supplementation on copper status in young adult men. Am J Clin Nutr 1983;37:553–6.

49. Harris E, Percival S. A role of ascorbic acid in copper transport. Am J Clin Nutr 1991;54: 1193S–7S.

50. Herbert V. Anti-hyperhomocysteinemic supplemental folic acid and vitamin B$_{12}$ are significantly destroyed in gastric juice if co-ingested with supplemental vitamin C and iron. Blood 1996;88(suppl 1): 492a.

51. National Research Council. Recommended Dietary Allowances, 10th ed. Washington, D.C.: National Academy Press, 1989:115–24.

52. Lykkesfeldt J, Loft S, Nielsen J, Poulsen H. Ascorbic acid and dehydroascorbic acid as biomarkers of oxidative stress caused by smoking. Am J Clin Nutr 1997; 65:959–63.

53. Hodges R, Baker E, Hood J, Sauberlich H, March S. Experimental scurvy in man. Am J Clin Nutr 1969; 22:535–48.

54. Alhadeff L, Gualtieri C, Lipton M. Toxic effects of water-soluble vitamins. Nutr Rev 1984;42:33–40.

55. Tsao C, Salimi S. Effect of large intake of ascorbic acid on urinary and plasma oxalic acid levels. Internatl J Vit Nutr Res 1984;54:245–9.

56. Hughes C, Dutton S, Truswell A. High intakes of ascorbic acid and urinary oxalate. J Hum Nutr 1981;35:274–80.

57. Urivetzky M, Kessaris D, Smith A. Ascorbic acid overdosing: a risk factor for calcium oxalate nephrolithiasis. J Urol 1992;147:1215–18.

58. Sutton J, Basu T, Dickerson J. Effect of large doses of ascorbic acid in man on some nitrogenous components of urine. Human Nutr Appl Nutr 1983;37A: 136–40.

59. Jacob R. Assessment of human vitamin C status. J Nutr 1990;120:1480–5.

Thiamin (Vitamin B$_1$)

Thiamin (vitamin B$_1$), the structural formula of which is shown in Figure 9.10, consists of a pyrimidine ring and a thiazole moiety (meaning one of two parts), which are linked by a methylene (CH$_2$) bridge.

Sources

Thiamin is widely distributed in foods, including meat (especially pork), legumes, and whole or enriched grain products. Yeast and wheat germ also contain significant amounts of the vitamin. In supplements, thiamin is found mainly as thiamin hydrochloride or as thiamin mononitrate salt.

Digestion, Absorption, Transport, and Storage

In plants, thiamin exists in its free form. However, in animal products, >95% of thiamin occurs in a phosphorylated form, primarily *thiamin diphosphate* (TDP), also called *thiamin pyrophosphate* (TPP). Intestinal phosphatases hydrolyze the phosphates from the thiamin diphosphate prior to absorption.

Absorption of thiamin from foods is thought to be high. Occasionally, however, antithiamin factors may be present in the diet. For example, thiaminases present in raw fish catalyze the cleavage of thiamin, thereby destroying its activity. These thiaminases are thermolabile; thus, cooking fish renders the enzymes inactive. Other antithiamin factors include polyhydroxyphenols such as tannic and caffeic acids. Polyhydroxyphenols are thermostable and may be found in coffee, tea, betel nuts, and certain fruits and vegetables such as blueberries, black currants, brussels sprouts, and red cabbage [1]. These polyhydroxyphenols inactivate thiamin by an oxyreductive process. Calcium, magnesium, and other divalent cations assist in the precipitation of thiamin by tannic acid. Thiamin destruction may be prevented by the presence of reducing compounds such as vitamin C and citric acid.

Free thiamin, not phosphorylated thiamin, is absorbed into the intestinal cells. Absorption of thiamin

2,5-dimethyl 6-aminopyridine 4-methyl 5-hydroxyethyl-thiazole

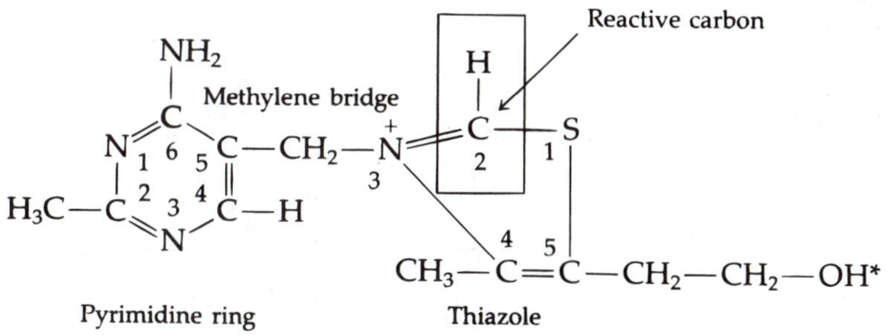

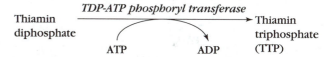

*Diphosphate addition occurs here to form the active coenzyme thiamin diphosphate (TDP).

Figure 9.10 Structure of thiamin.

can be both active and passive, depending on the amount of the vitamin presented in the intestine for absorption. At low physiological concentrations, thiamin absorption is active and sodium-dependent. Absorption occurs primarily in the upper jejunum but can occur in the duodenum and ileum [1–4]. When intakes of thiamin are high, absorption is predominantly passive.

Within the mucosal cells, thiamin may be phosphorylated (i.e., converted into a phosphate ester). Thiamin transport across the basolateral membrane is sodium- and energy-dependent [5]. Ethanol ingestion, however, interferes with active transport of thiamin from the mucosal cell across the basolateral membrane, but not the brush border membrane [3]. Thiamin appearing on the serosal side of the enterocyte is not initially bound to phosphates.

Thiamin in the blood is typically in its free form, bound to albumin, or found as thiamin monophosphate (TMP). Most (~90%) thiamin in the blood, however, is not in the plasma but is present within the blood cells. Transport of thiamin into red blood cells is thought to occur by facilitated diffusion, whereas transport into other tissues requires energy. Only free thiamin or TMP is thought to be able to cross cell membranes. In red blood cells, most thiamin exists as TDP with smaller amounts of free thiamin and TMP.

The human body contains approximately 30 mg of thiamin, with relatively high but still small concentrations found in the skeletal muscles, heart, liver, kidney, and brain. In fact, skeletal muscles are thought to contain about half of the body's thiamin.

Following absorption, most free thiamin is taken up by the liver and phosphorylated, such that thiamin is converted to its coenzyme phosphorylated form, thiamin diphosphate (TDP). Conversion of thiamin to TDP requires adenosine triphosphate (ATP) and *thiamin pyrophosphokinase,* an enzyme found in the liver, brain, as well as other tissues. About 80% of the total thiamin in the body exists as TDP [2].

$$\text{Thiamin} \xrightarrow[\text{ATP} \;\rightarrow\; \text{AMP}]{\textit{Thiamin pyrophosphokinase}} \text{Thiamin diphosphate (TDP)}$$

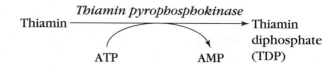

Another form of thiamin, thiamin triphosphate (TTP) represents about 10% of total body thiamin and is synthesized by action of a TDP-ATP phosphoryl transferase that phosphorylates TDP [2,3,5].

$$\text{Thiamin diphosphate} \xrightarrow[\text{ATP} \;\rightarrow\; \text{ADP}]{\textit{TDP-ATP phosphoryl transferase}} \text{Thiamin triphosphate (TTP)}$$

The terminal phosphate on TTP may be hydrolyzed by thiamin triphosphatase to yield TDP.

$$\text{Thiamin triphosphate} \xrightarrow{\textit{thiamin triphosphatase}} \text{thiamin diphosphate}$$

TTP, as well as TDP and TMP, can be found in small amounts in several tissues, including the brain, heart, liver, and kidney. TMP is thought to be derived from the catabolism of the terminal phosphate on TDP and is believed to be inactive [5].

Functions and Mechanisms of Action

At the cellular level, thiamin plays essential roles in the body including roles in

1. energy transformation,
2. synthesis of pentoses and NADPH (nicotinamide adenine dinucleotide phosphate), and
3. membrane and nerve conduction.

Each of these three roles will be discussed.

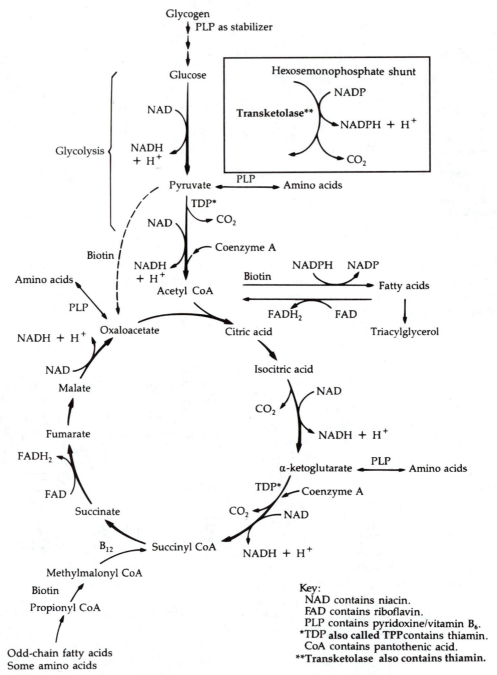

Figure 9.11 Various vitamin cofactors and their action sites in energy metabolism. The role of thiamin as TDP is shown by an asterisk.

Energy Transformation

Thiamin as TDP functions as a coenzyme necessary for the oxidative decarboxylation of both pyruvate and α-ketoglutarate. These reactions (Fig. 9.11) are instrumental in generating energy, ATP. Inhibition of these decarboxylation reactions prevents synthesis of ATP, and of acetyl CoA needed for the synthesis of, for example, fatty acids, cholesterol, and other important compounds, and results in the accumulation of pyruvate, lactate, and α-ketoglutarate in the blood.

The steps that occur in the oxidative decarboxylation of pyruvate are shown in Figure 9.12 and require a multienzyme complex, known as the *pyruvate dehy-*

drogenase complex, which is bound to the mitochondrial membrane. Three enzymes make up the pyruvate dehydrogenase complex: Enzyme 1, a TDP-dependent pyruvate decarboxylase; Enzyme 2, a lipoic acid–dependent dihydrolipoyl transacetylase; and Enzyme 3, an FAD-dependent dihydrolipoyl dehydrogenase. The roles of four vitamins—thiamin (TDP), riboflavin (FAD), niacin (NAD$^+$), and pantothenic acid (CoA)—in this decarboxylation process can be identified here and in Figure 9.12.

Five basic steps, outlined here (see also Fig. 9.12), are involved in the oxidative decarboxylation of α-keto acids such as pyruvate and α-ketoglutarate.

Step 1: The pyruvate decarboxylase enzyme 1-TDP complex reacts with the keto acid (R-CO-COOH).

$$\text{Enzyme 1}-\text{TDP}-\text{H} + \text{R}-\overset{\overset{\displaystyle O}{\|}}{\text{C}}-\text{COOH} \longrightarrow$$

$$\text{Enzyme 1}-\text{TDP}-\overset{\overset{\displaystyle OH}{|}}{\text{CH}}-\text{R} + CO_2$$

During the reaction, the enzyme 1-TDP complex covalently binds to the decarboxylated keto acid forming hydroxyethyl TDP; carbon dioxide (CO_2) is released in the process.

Step 2: The enzyme 1-TDP-decarboxylated complex is acted on by Enzyme 1-lipoic acid (LA)-dependent dihydrolipoyl transacetylase, which transfers an acetyl group from Enzyme 1-TDP to a sulfur atom in lipoic acid.

$$\text{Enzyme 1}-\text{TDP}-\overset{\overset{\displaystyle OH}{|}}{\text{CH}}-\text{R} + \text{Enzyme 2}-\text{LA}\overset{\displaystyle S}{\underset{\displaystyle S}{|}} \longrightarrow$$

$$\text{Enzyme 1}-\text{TDP}-\text{H} + \text{Enzyme 2}-\text{LA}\overset{\displaystyle S-\overset{\overset{\displaystyle O}{\|}}{C}-R}{\underset{\displaystyle SH}{}}$$

Although lipoic acid functions as a prosthetic group, it is not considered a vitamin. Lipoic acid is classified as a sulfur-containing fatty acid; uncertainty exists as to whether it is synthesized in the body or is obtained by diet. Lipoate functions similarly to biotin at the end of the long flexible side chain of the enzyme; specifically, it rotates from one active site to another site on the multisubunit enzyme. Lipoamide picks up the acetyl group from Enzyme 1-TDP to form acetyl lipoamide.

Step 3: The acetyl group is transferred from acetyl lipoamide to CoA with the formation of acetyl CoA and dihydrolipoamide.

$$\text{Enzyme 2}-\text{LA}\overset{\displaystyle S-\overset{\overset{\displaystyle O}{\|}}{C}-R}{\underset{\displaystyle SH}{}} + \text{CoA}-\text{SH} \longrightarrow$$

$$\text{Enzyme 2}-\text{LA}\overset{\displaystyle SH}{\underset{\displaystyle SH}{}} + \text{CoA}-\text{S}-\overset{\overset{\displaystyle O}{\|}}{C}-R$$

Step 4: The dehydrolipoamide is then oxidized back to the original lipoamide by an Enzyme 3-FAD-dependent dihydrolipoyl dehydrogenase.

$$\text{Enzyme 2}-\text{LA}\overset{\displaystyle SH}{\underset{\displaystyle SH}{}} + \text{Enzyme 3}-\text{FAD} \longrightarrow$$

$$\text{Enzyme 2}-\text{LA}\overset{\displaystyle S}{\underset{\displaystyle S}{|}} + \text{Enzyme 3}-\text{FADH}_2$$

Step 5: FADH$_2$ is oxidized by NAD$^+$

$$\text{Enzyme 3-FADH}_2 + \text{NAD}^+ \longrightarrow \text{Enzyme 3-FAD} + \text{NADH} + \text{H}^+$$

The decarboxylation of α-ketoglutarate by the α-ketoglutarate dehydrogenase complex is similar to that for pyruvate. The α-ketoglutarate dehydrogenase complex serves to decarboxylate α-ketoglutarate and forms succinyl CoA.

A key feature of TDP (Fig. 9.13) is that the carbon atom between the nitrogen and sulfur atoms in the thiazole ring is more acidic than most CH groups [2,6]. It ionizes (deprotonizes) to form a carbanion at carbon 2 of the thiazole ring. The carbanion is stabilized by the positively charged nitrogen in the thiazole ring [2,6]. The carbanion combines with the 2-carbonyl group of pyruvate, α-ketoglutarate, and other α-keto acids to form a covalent bond [6].

A group of decarboxylation reactions is also thiamin-dependent. Decarboxylation of the branched-chain α-keto acids, which arise from the transamination of valine, isoleucine, and leucine, is an oxidative process that also requires thiamin as TDP. Failure to oxidize the α-keto acids, α-ketoisocaproic, α-keto β-methyl valeric, and α-ketoisovaleric acids from leucine, isoleucine, and valine, respectively, results in the accumulation of both the branched-chain amino acids and their α-keto acids in blood and other body fluids. Such findings are characteristic of maple syrup urine disease (MSUD). MSUD results from a genetic (inborn error of metabolism) absence or insufficient activity of the branched-chain α-keto acid dehydrogenase enzyme complex. People with MSUD must avoid meat, poultry, fish, and dairy products to limit intakes

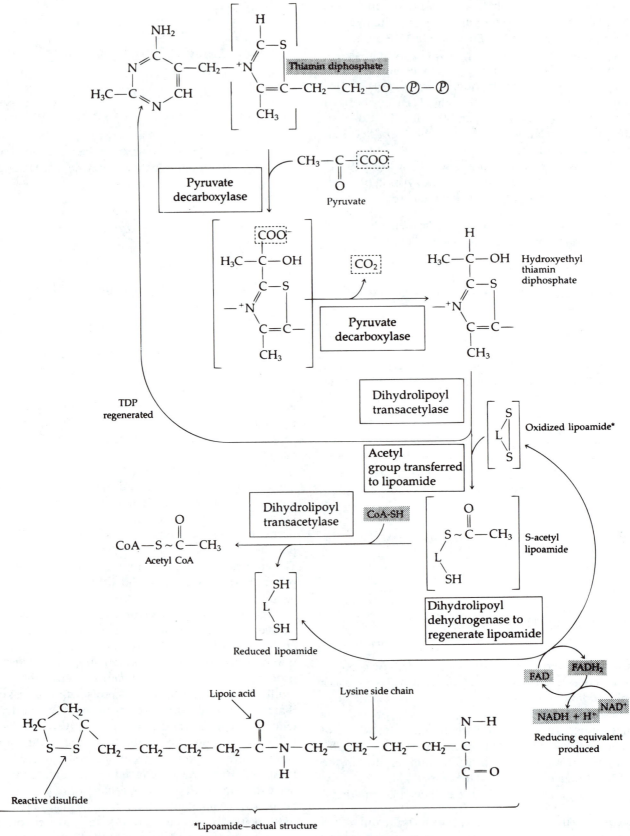

Figure 9.12 Oxidative decarboxylation of pyruvate by pyruvate dehydrogenase complex.

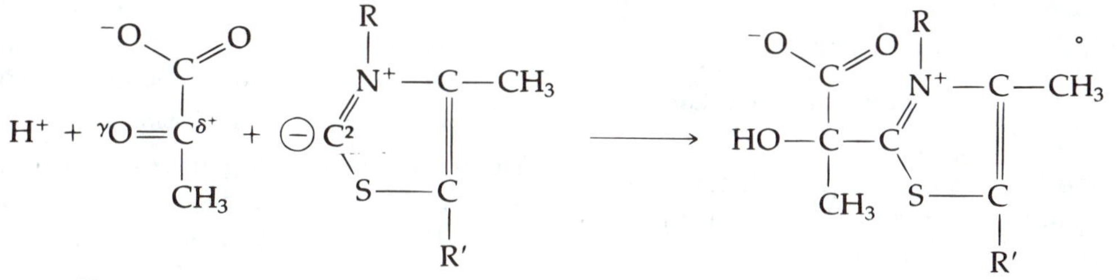

Figure 9.13 Combination of TDP carbanion with substrate pyruvate.

of leucine, isoleucine, and valine. Medical foods devoid of these three amino acids provide the majority of nutrient intake.

Synthesis of Pentoses and NADPH

Thiamin as TDP also functions as a loosely bound prosthetic group of transketolase, a key cytosolic enzyme in the hexose monophosphate shunt. The hexose monophosphate shunt is the pathway in which sugars of varying chain lengths are interconverted. The shunt is essential for the generation of pentoses for nucleic acid synthesis and of NADPH, which is needed, for example, for fatty acid synthesis. TDP forms a carbanion (as in Fig. 9.13) that acts to transfer an activated aldehyde from a donor ketose substrate to an acceptor. The acceptor in the hexose monophosphate shunt is xylulose. *Transketolase* hydrolyzes the carbon to carbon bond in xylose 5-P, sedoheptulose 7-P, and fructose 6-P (i.e., ketoses) and transfers the two carbon fragment (carbons 1 and 2 of the ketoses) to an aldose receptor [6]. The transketolase-activated reaction that requires Mg^{2+} can be written as follows:

$$\textit{transketolase}$$
xylulose 5-P + ribose 5-P ⟷ sedoheptulose 7-P + glyceraldehyde 3-P

$$\textit{transketolase}$$
xylulose 5-P + erythrose 4-P ⟷ glyceraldehyde 3-P + fructose 6-P

Membrane and Nerve Conduction

Exactly how thiamin functions or in what form it functions in nerve membranes and in nerve conduction is unclear. However, it is believed that in the nervous system thiamin as either TDP or TTP exerts its action in some manner other than as a coenzyme. TDP and TTP are rapidly interconvertible; therefore, uncertainty exists about which is the neurophysiologically active

form [7]. But evidence is emerging that TTP may be the form involved in nerve membrane function and nerve transmission possibly through action on ion channels [5,7–9].

Aberrations in nerve function may be due to a lack of energy; a decreased amount of acetylcholine, the synthesis of which requires TDP; or a reduced nerve impulse transmission. In the last instance it is postulated that thiamin, as TDP or TTP, may occupy a site on the nerve membrane. The site is either a sodium channel or is proximal to the channel. In this location thiamin could regulate nerve impulse transmission. Initiation of a nerve impulse could result in dephosphorylation of the thiamin ester, a reaction that somehow would allow sodium to cross the membrane freely, thereby permitting conduction of the nerve impulse [7]. Alternately, thiamin may potentiate acetylcholine release and has been shown to activate maxichloride channels through control of numbers of functional channels [8].

Metabolism and Excretion

Thiamin in excess of tissue needs and storage capacity is metabolized for urinary excretion. When intake of the vitamin is adequate, the excretion products are almost equally divided between thiamin and its metabolites especially TMP, but also small amounts of TPP [3]. Free plasma thiamin is typically filtered by the kidney and excreted in the urine. Degradation of thiamin begins with the cleavage of the molecule into its pyrimidine and thiazole moieties; the two rings are then further metabolized, generating 20 or more metabolites. Urinary thiamin is thought to reflect recent thiamin intake, not body thiamin status.

Some thiamin appears to be catabolized at a fairly constant rate regardless of the level of thiamin intake [10]. The pathway for this degradation may not necessarily be related to physiological function [10]. These losses represent obligatory losses that must be met by dietary thiamin.

Dietary Reference Intakes and Recommended Dietary Allowances

Determination of the 1989 RDA [11] for thiamin has been based on

- the relationship between varying thiamin intake levels and occurrence of clinical signs of deficiency,
- level of excretion of thiamin and/or its metabolites, and
- degree of erythrocyte transketolase activity.

Because of the importance of thiamin in energy metabolism, needed intake varies according to energy (caloric) intake. The RDA for adults is 0.5 mg/1,000 kcal; however, an intake of no less than 1 mg/day is advised [11]. The 1998 Dietary Reference Intakes RDA for individual intake for thiamin for adult men aged 19 years and older is 1.2 mg/day and for adult women aged 19 years and older 1.1 mg daily [12]. Thiamin intakes with pregnancy and lactation are increased to 1.4 and 1.5 mg/day, respectively [12].

Deficiency: Beriberi

Despite the known functional roles of thiamin at the cellular level, it has as yet been impossible to explain all the pathophysiological manifestations in animals or humans that are associated with thiamin deficiency, beriberi. One of the first symptoms of thiamin deficiency is a loss of appetite (anorexia) and thus weight. As deficiency worsens, cardiovascular system involvement (such as hypertrophy and altered heart rate) and neurological symptoms appear.

Three types of beriberi have been identified. *Dry beriberi* is found predominantly in older adults; it is thought to result from a chronic low thiamin intake especially if coupled with a high carbohydrate intake. Dry beriberi is characterized by muscle weakness and wasting especially in the lower extremities [5,6]. *Wet beriberi* results in more extensive cardiovascular system involvement than dry beriberi; right-side heart failure leads to respiratory involvement with edema [5,6]. Lastly, *acute beriberi,* seen mostly in infants, has been documented in countries such as Japan.

In the United States and in Western countries, thiamin deficiency associated with alcoholism is common and is referred to as Wernicke's encephalopathy or Wernicke-Korsakoff syndrome. Individuals with dependency on alcohol are particularly prone to thiamin deficiency because of

- decreased intake of the vitamin due to decreased food consumption;
- an increased requirement in the case of liver damage (decreased liver function impairs TDP formation and, consequently, vitamin use); and
- decreased thiamin absorption [11,13].

Wernicke's encephalopathy, often manifested in those with a history of alcohol abuse, is characterized by ophthalmoplegia, nystagmus, ataxia, loss of recent memory, and confusion [5, 6, 13]. Treatment consists of therapeutic doses (~100 mg) thiamin. Typically some aspects of confusion and ophthalmoplegia begin to improve with the massive thiamin doses [13].

Elderly populations are also at risk for thiamin deficiency. People with diseases that impair absorption of the vitamin (e.g., some cancers, biliary disease, inflammatory bowel diseases) are also at greater risk of developing deficiency.

Folate and protein deficiencies that impair enterocyte turnover also diminish thiamin absorption. Excess glucose infusion intravenously and ingestion of diets that are made up primarily of refined, unenriched grain products necessitate increased thiamin intake.

Toxicity

There appears to be little danger of thiamin toxicity associated with oral intake of large amounts (500 mg daily for 1 month) of thiamin [11,12,14,15]. Excessive (100 times recommendations) thiamin, administered parenterally (intravenous or intramuscular), however, has been associated with headache, convulsions, cardiac arrythmia, and anaphylactic shock, among other signs [6].

Pharmacological levels of thiamin are used in the treatment of certain inborn errors of metabolism. One variant form of MSUD has been shown to respond to oral thiamin supplements (up to 500 mg daily). Other metabolic diseases that may respond to large doses of the vitamin are thiamin-responsive megaloblastic anemia and thiamin-responsive lactic acidosis. In the latter condition, large doses of thiamin can increase the activity of pyruvate dehydrogenase in the liver, thereby decreasing the level of lactic acid as more pyruvate is decarboxylated for entry into the Krebs cycle. How administration of the vitamin helps to correct thiamin-responsive megaloblastic anemia has not been elucidated [16].

Assessment of Nutriture

Adequacy of thiamin nutriture may be assessed by measurement of erythrocyte transketolase activity in hemolyzed whole blood or by measurement of thiamin in the blood [17]. Blood thiamin concentrations are thought to be reduced before enzyme activity is impaired; however, further investigations are necessary [18]. Remember, transketolase is the thiamin-dependent enzyme of the hexose monophosphate shunt. With deficiency of thiamin, the enzyme increases activity with the addition of thiamin to the incubation medium. An increase in activity of >25% is indicative of thiamin deficiency.

	Thiamin Status		
	Deficient	Marginal	Adequate
Transketolase (effect of TDP)	>25%	25%-15%	<15%

References Cited for Thiamin

1. Sauberlich HE. Bioavailability of vitamins. Prog Food Nutr Sci 1985;9:1–33.

2. Tanphaichitr V. Thiamin. In: Shils ME, Olson JA, Shike M., eds. Modern Nutrition in Health and Disease. 8th ed. Philadelphia: Lea and Febiger, 1994:359–65.

3. Bender D. Nutritional Biochemistry of the Vitamins. New York: Cambridge University Press, 1992:128–55.

4. Sauberlich H. Vitamins—how much is for keeps? Nutr Today 1987;22:20–28.

5. Gubler C. Thiamin. In: Machlin LJ., ed. Handbook of Vitamins. 2nd ed. New York: Dekker, 1991:233–81.

6. Combs G. The Vitamins. San Diego, CA: Academic Press, 1992:251–69.

7. Haas R. Thiamin and the brain. Ann Rev Nutr 1988;8:483–515.

8. Bettendorff L. Thiamine in excitable tissues: reflections on a non-cofactor role. Metab Brain Dis 1994;9:183–209.

9. Bettendorff L, Kolb H, Schoffeniels E. Thiamine triphosphate activates an anion channel of large unit conductance in neuroblastoma cells. J Membr Biol 1993;136:281–8.

10. Ariaey-Nejad MR, Balaghi M, Baker EM, Sauberlich HE. Thiamin metabolism in man. Am J Clin Nutr 1970;23:764–78.

11. National Research Council. Recommended Dietary Allowances, 10th ed. Washington, D.C.: National Academy Press, 1989:125–32.

12. Yates A, Schlicker S, Suitor C. Dietary reference intakes: the new basis for recommendations for calcium and related nutrients, B vitamins and choline. J Am Diet Assoc 1998;98:699–706.

13. Wood B, Currie J. Presentation of acute Wernicke's encephalopathy and treatment with thiamine. Metab Brain Dis 1995;10:57–71.

14. Council on Scientific Affairs, American Medical Association. Vitamin preparations as dietary supplements and as therapeutic agents. JAMA 1987; 257: 1929–36.

15. Alhadeff L, Gualtieri C, Lipton M. Toxic effects of water-soluble vitamins. Nutr Rev 1984;42:33–40.

16. Tanphaichitr V, Wood B. Thiamin. In: Olson RE, Broquist HP, Chichester CO, Darby WJ, Kolbye A, Stalvey R. Present Knowledge in Nutrition, 5th ed. Washington, D.C.: The Nutrition Foundation, 1984:273–84.

17. Finglass P. Thiamin. Int J Vitam Nutr Res 1994;63: 270–4.

18. Warnock L, Prudhonme C, Wagner C. The determination of thiamin pyrophosphate in blood and other tissues, and its correlation with erythrocyte transketolase activity. J Nutr 1979;108:421–7.

Riboflavin (Vitamin B$_2$)

Riboflavin consists of flavin (isoalloxazine ring), to which is attached a ribitol (sugar alcohol) side chain. The structure of riboflavin along with its two coenzyme derivatives are given in Figure 9.14.

Sources

Riboflavin is found in a wide variety of foods but especially those of animal origin. Milk and milk products such as cheeses are thought to contribute the majority of dietary riboflavin. Eggs, meat, and legumes also provide riboflavin in significant quantities. Fruits, vegetables, and cereal grains are minor contributors of dietary riboflavin.

The form of riboflavin in food varies. Free or protein-bound riboflavin is found in milk, eggs, and enriched breads and cereals. In most other foods the vitamin occurs as one or the other of its coenzyme derivatives, FMN (flavin mononucleotide) or FAD (flavin adenine dinucleotide), although phosphorus-bound riboflavin is also found in some foods.

Digestion, Absorption, Transport, and Storage

Riboflavin attached noncovalently to proteins may be freed by the action of hydrochloric acid secreted within the stomach and by gastric and intestinal enzymatic hydrolysis of the protein. Riboflavin in foods as FAD, FMN, and riboflavin phosphate must also be freed prior to absorption. Within the intestinal lumen, FAD pyrophosphatase converts FAD to FMN; FMN is converted to free riboflavin by FMN phosphatase.

$$\text{FAD} \xrightarrow[\text{pyrophosphatase}]{\textit{FAD}} \text{FMN} \xrightarrow[\text{phosphatase}]{\textit{FMN}} \text{riboflavin}$$

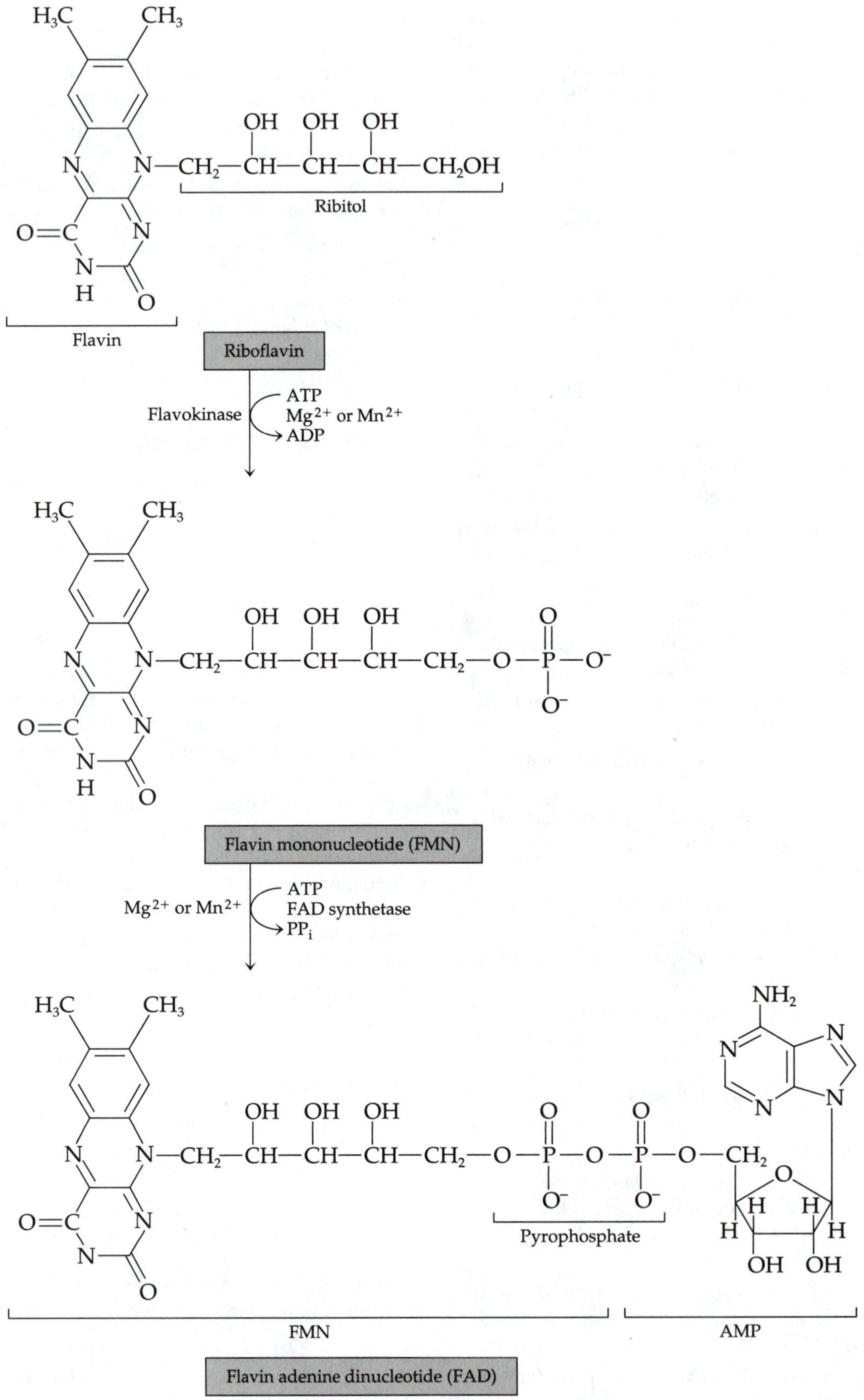

Figure 9.14 Structure of riboflavin and its coenzyme forms.

Other intestinal phosphatases such as nucleotide diphosphatase and alkaline phosphatase may also hydrolyze riboflavin from riboflavin phosphate.

Not all bound riboflavin is hydrolyzed and available for absorption. A small amount (~7%) of FAD is covalently bound to either of two amino acids, histidine or cysteine. For example, following consumption of foods containing succinate dehydrogenase or monoamine oxidase, these proteins are degraded; however, the riboflavin remains bound, typically to histidine or cysteine residues [1,2]. Absorption of the histidine- and cysteine-bound riboflavin does not usually occur; however, should absorption occur via amino acid transport systems, the complex is excreted unchanged in the urine [2].

Generally, animal sources of riboflavin are thought to be better absorbed than plant sources. Divalent metals such as copper, zinc, iron, and manganese have been shown to chelate riboflavin and FMN and inhibit riboflavin absorption [3]. Ingestion of alcohol also impairs riboflavin digestion and absorption.

Free riboflavin is absorbed via a saturable, sodium-dependent carrier mechanism primarily in the proximal small intestine. Riboflavin absorption has been shown to require ATP [4,5]. Absorption rate is proportional to dose, but the rate levels off at approximately 25 mg [1,6]. Peak concentrations of the vitamin in the plasma correlate with intakes of 15 to 20 mg [2].

On absorption into the mucosal cells, riboflavin is phosphorylated into FMN, a reaction catalyzed by flavokinase and requiring ATP, as shown here and in Figure 9.14.

At the serosal surface most of the FMN is probably dephosphorylated to riboflavin, which enters the portal system. The vitamin is carried to the liver, where it is converted again to FMN and to its other coenzyme derivative FAD (Fig. 9.14).

$$\text{Flavin mononucleotide} \xrightarrow[\text{ATP} \quad \text{PP}_i]{\textit{FAD synthetase}} \begin{array}{l}\text{Flavin adenine}\\ \text{dinucleotide}\\ \text{(FAD)}\end{array}$$

Most flavins in systemic plasma are found as riboflavin rather than as one of its coenzyme forms although all three may be present. Riboflavin, FMN, and FAD are transported in the plasma by a variety of proteins including albumin, fibrinogen, and globulins (principally immunoglobulins) [2,5,7]. Albumin appears to be the primary transport protein. The extent to which the flavins are bound to plasma proteins is not believed to be crucial in regulating tissue availability of the vitamin except that the proteins may decrease losses of the vitamin during glomerular filtration [7].

Regardless of the form in which the vitamin reaches the tissues, it is free riboflavin that traverses most cell membranes by a carrier-mediated process. An exception to this is riboflavin uptake by the brain, which is thought to occur by a high-affinity transport system for riboflavin as its coenzyme FAD. The tight binding of FAD to these transport proteins could help explain why the concentration of FAD in the brain does not decline appreciably even in a severe riboflavin deficiency [9].

Riboflavin is found in small quantities in a variety of tissues. The greatest concentrations of riboflavin are found in the liver, kidney, and heart. Intracellular phosphorylation of free riboflavin is necessary to prevent diffusion out of the tissue [2]. Thus, within cells, riboflavin is typically converted to its coenzyme forms by flavokinase and FAD synthetase, both of which are widely distributed in tissues [8]. Synthesis of FMN and FAD appear to be under hormonal regulation. Hormones shown to be particularly important in this regulation are ACTH, aldosterone, and the thyroid hormones, all of which accelerate the conversion of riboflavin into its coenzyme forms apparently by increasing the activity of flavokinase [1]. The coenzyme form of the vitamin is then bound to the apoenzyme. FMN and FAD function as prosthetic groups for enzymes involved in oxidation reduction reactions. These enzymes are called flavoproteins.

Functions and Mechanisms of Action

FMN and FAD function as cofactors for a wide variety of oxidative enzyme systems and remain bound to the enzymes during the oxidation-reduction reactions. Flavins can act as oxidizing agents because of their ability to accept a pair of hydrogen atoms. The isoalloxazine ring is reduced by two successive one-electron transfers with the intermediate formation of a semiquinone free radical, as shown in Figure 9.15. Reduction of the isoalloxazine ring yields the reduced forms of the flavoprotein, which can be found in $FMNH_2$ and $FADH_2$.

Flavoproteins

Flavoproteins exhibit a wide range of redox potentials and therefore can play a wide variety of roles in intermediary metabolism. Some examples of these roles are discussed here:

• The role of flavoproteins in the *electron transport chain* is provided on pages 60–66.

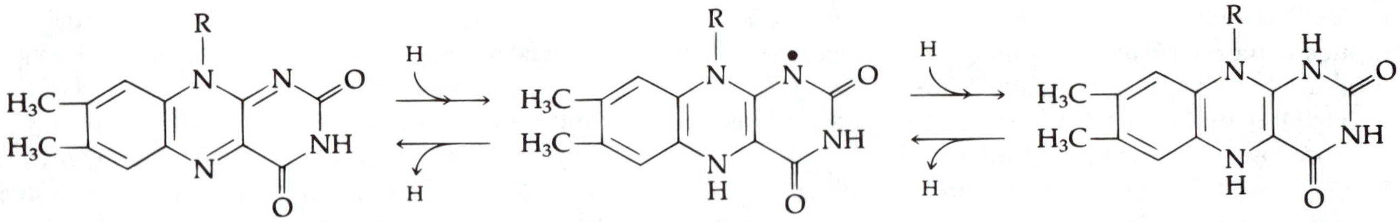

Figure 9.15 Oxidation and reduction of isoalloxazine ring.

• In the *oxidative decarboxylation* of *pyruvate* (Fig. 9.12) and *α-ketoglutarate,* FAD serves as intermediate electron carrier, with NADH being the final reduced product.

• *Succinate dehydrogenase* is an FAD flavoprotein that removes electrons from succinate to form fumarate, and forms $FADH_2$ from FAD (Fig. 9.11). The electrons are then passed into the electron transport chain via coenzyme Q (Fig. 3.12, p. 64).

• In fatty acid oxidation, *fatty acyl CoA dehydrogenase* requires FAD.

• *Sphinganine oxidase*, in sphingosine synthesis, requires FAD.

• As a coenzyme for an oxidase such as *xanthine oxidase*, FAD transfers electrons directly to oxygen with the formation of hydrogen peroxide. Xanthine oxidase, which contains both iron and molybdenum, is necessary for purine catabolism in the liver. The enzyme converts hypoxanthine to xanthine and then xanthine to uric acid (see Fig. 7.17; also see Chap. 12, p. 463).

• Similarly, *aldehyde oxidase* using FAD converts aldehydes, such as pyridoxal (vitamin B_6), to pyridoxic acid, an excretory product, and retinal (vitamin A) to retinoic acid (Fig. 10.4) while also passing electrons to oxygen and generating hydrogen peroxide.

• Also in vitamin B_6 metabolism (Fig. 9.40), *pyridoxine phosphate oxidase*—which converts pyridoxamine phosphate (PMP) and pyridoxine phosphate (PNP) to pyridoxal phosphate (PLP), the primary coenzyme form of vitamin B_6—is dependent on FMN.

• Synthesis of an *active form of folate*, N^5 methyl THF, requires $FADH_2$ (Fig. 9.31).

• In *choline catabolism*, several enzymes (such as choline dehydrogenase, dimethylglycine dehydrogenase, and monomethylglycine dehydrogenase) require FAD.

• Some neurotransmitters (such as dopamine) and other amines (tyramine and histamine) require FAD-dependent *monoamine oxidase* for metabolism.

• Reduction of the oxidized form of glutathione (GSSG) to its reduced form (GSH) is also dependent on FAD-dependent *glutathione reductase*. This reaction forms the basis of one of the assays used to assess riboflavin status (see the section "Assessment of Nutriture").

Metabolism and Excretion

Riboflavin undergoes limited metabolism prior to excretion in the urine; primarily free riboflavin is found in the urine. Some of the riboflavin metabolites are thought to arise from tissue degradation of covalently bound flavins or are thought to be a reflection of bacterial action in the intestinal tract. It is believed that the metabolites formed in the intestinal tract can be absorbed and then excreted in the urine [10].

Riboflavin and riboflavin phosphate that are not bound to proteins in the plasma are filtered by the glomerulus. The phosphate is removed from the riboflavin prior to excretion of the riboflavin by the kidney [2]. Riboflavin bound to cysteine and histidine is also found in the urine if absorbed in such form from the gastrointestinal tract or if generated in body cells from the degradation of flavoenzymes such as succinate dehydrogenase and monoamine oxidase [1].

Urinary excretion of riboflavin may be noticeable a couple of hours following oral ingestion of the vitamin. Riboflavin is a fluorescent yellow compound. Thus, following riboflavin intake in a quantity such as 1.7 mg, similar to that found in a vitamin pill, a color change of the urine occurs whereby the urine will deepen in color from a typical light yellow to a brighter orangish yellow.

In addition to urine, small amounts of free riboflavin may be secreted in the bile and thus found in the feces

[2,5]. Fecal riboflavin metabolites may also arise from metabolism of riboflavin by intestinal flora [5].

Dietary Reference Intakes and Recommended Dietary Allowances

The level of riboflavin intake commensurate with adequate nutriture has been estimated through various studies involving urinary excretion of riboflavin, relationship of dietary intake to clinical signs of deficiency, and the activity of erythrocyte glutathione reductase. The 1989 RDA for riboflavin is given in milligrams per 1,000 kcal. The recommended allowance for people of all ages is 0.6 mg/1,000 kcal with a minimum intake of 1.2 mg for persons whose caloric intake may be <2,000 kcal. Through the years the recommended allowances for riboflavin have been calculated in relation to (1) protein requirement, (2) energy intake, and (3) metabolic body size. Because of the interdependence of these three variables, allowances calculated by the various methods have not differed significantly [11].

The 1998 DRI RDA for individual intake for riboflavin is similar to the 1989 RDA. The DRI RDA for adult men and women aged 19 years and older for individual intake for riboflavin is 1.3 and 1.1 mg/day, respectively [12]. With pregnancy and lactation the DRI for daily riboflavin intake increases to 1.4 and 1.6 mg, respectively [12].

Deficiency

A deficiency of riboflavin rarely occurs in isolation; most often it is accompanied by other nutrient deficits [13]. No clear riboflavin deficiency disease has been characterized; however, clinical symptoms of deficiency after almost four months of inadequate intake include lesions on the outside of the lips (cheilosis) and corners of the mouth (angular stomatitis), inflammation of the tongue (glossitis), redness or bloody (hyperemia) and swollen (edema) mouth cavity, dermatitis, and peripheral nerve dysfunction (neuropathy), among other signs. Severe deficiency of riboflavin may diminish the synthesis of the coenzyme form of vitamin B_6, and niacin (NAD) synthesis from tryptophan.

Conditions and populations associated with increased need for riboflavin intake are many. Because of limited dietary intake, people with congenital heart disease, some cancers, and excess alcohol intake may develop deficiency. Riboflavin metabolism is altered with thyroid disease. Excretion of riboflavin is enhanced with diabetes mellitus, trauma, and stress. Women on oral contraceptives are also more likely to develop deficiency than women not taking these drugs.

Toxicity

Toxicity associated with large oral doses of riboflavin have not been reported, but neither can any benefit be ascribed to megadosing by the well-nourished individual [12,14].

Assessment of Nutriture

The most sensitive method for determining riboflavin nutriture is the measurement of the activity of erythrocyte *glutathione reductase,* an enzyme requiring FAD as a coenzyme. The method is based on the following reaction:

$$NADPH + H^+ + GSSG \xrightarrow{\text{glutathione reductase-FAD}} NADP^+ + 2\ GSH$$

Glutathione in its oxidized form is designated as GSSG and in its reduced form as GSH. In cases of a riboflavin deficiency or marginal riboflavin status, the activity of glutathione reductase is limited and less NADPH is used to reduce the oxidized glutathione. In vitro enzyme activity in terms of "activity coefficients" (AC) is determined both with and without the addition of FAD to the medium. Activity coefficients represent a ratio of the enzyme's activity with FAD to the enzyme's activity without FAD. When addition of FAD stimulates enzyme activity to generate an AC of greater than about 1.2 or 1.3, then riboflavin status is considered inadequate.

References Cited for Riboflavin

1. McCormick D. Riboflavin. In: Shils ME, Olson JA, Shike M., eds. Modern Nutrition in Health and Disease, 8th ed. Philadelphia: Lea and Febiger, 1994:366–75.

2. Bender D. Nutritional Biochemistry of the Vitamins. New York: Cambridge University Press, 1992:156–83.

3. Cooperman J, Lopez R. Riboflavin. In: Machlin LJ., ed. Handbook of Vitamins, 2nd ed. New York: Dekker, 1991:283–310.

4. McCormick D. Riboflavin. In: Brown ML, ed. Present Knowledge in Nutrition, 6th ed. Washington, D.C.: The Nutrition Foundation, 1990:146–54.

5. Combs G. The Vitamins. San Diego, CA: Academic Press, 1992:271–87.

6. Sauberlich H. Vitamins—how much is for keeps? Nutr Today 1987;22:20–28.

7. White H, Merrill A. Riboflavin-binding proteins. Ann Rev Nutr 1988;8:279–99.

8. Merrill A, Lambeth J, Edmondson D, et al. Formation and mode of flavoproteins. Ann Rev Nutr 1981;1: 281–317.

9. Rivlin R. Riboflavin. Nutrition Reviews' Present Knowledge in Nutrition, 5th ed. Washington, D.C.: The Nutrition Foundation, 1984:285–302.

10. Pike RL, Brown ML. Nutrition: An Integrated Approach, 3rd ed. New York: Wiley, 1984: 92–97.

11. National Research Council. Recommended Dietary Allowances, 10th ed. Washington, D.C.: National Academy Press, 1989:132–7.

12. Yates A, Schlicker S, Suitor C. Dietary reference intakes: The new basis for recommendations for calcium and related nutrients, B vitamins and choline. J Am Diet Assoc 1998;98:699–706.

13. McCormick D. Riboflavin. In: Shils ME, Young VR, eds. Modern Nutrition in Diet and Disease, 7th ed. Philadelphia: Lea and Febiger, 1988:362–9.

14. Council on Scientific Affairs, American Medical Association. Vitamin preparations as dietary supplements and as therapeutic agents. JAMA 1987;257: 1929–36.

Niacin (Vitamin B₃)

The term *niacin* is considered a generic term for *nicotinic acid* and *nicotinamide,* also called *niacinamide* (Fig. 9.16). The vitamin activity of niacin is provided by both nicotinic acid and nicotinamide.

Sources

The best sources of niacin include tuna, halibut, beef, chicken, turkey, pork, and other meats. Cereal grains, seeds, and legumes also contain appreciable amounts of niacin. Niacin is also found in coffee and tea. In supplements, niacin is generally found as nicotinamide (niacinamide).

In animals, niacin occurs mainly as the nicotinamide nucleotides nicotinamide adenine dinucleotide (NAD) and nicotinamide adenine dinucleotide phosphate (NADP). However, following the slaughter of animals, NAD and NADP are thought to undergo hydrolysis; thus, meats provide niacin as free nicotinamide [1]. Figure 9.17 shows the structures of NAD and NADP.

In some foods, niacin may be bound covalently to complex carbohydrates and called *niacytin,* or it may

be bound to small peptides and called *niacinogens.* Biologically unavailable niacin has been found primarily in corn but also in wheat and a variety of cereal products [2]. Chemical treatment with bases such as lime water can improve availability of some bound niacin. Some niacin is also thought to be released from niacytin on exposure to gastric acid. At most, however, only about 10% of the niacin in maize is thought to be available for absorption [1].

In addition to dietary sources of niacin, NAD may be synthesized in the liver from the amino acid tryptophan. This biosynthetic pathway is depicted in Figure 9.18. Only about 3% of the tryptophan that is metabolized follows the pathway to NAD synthesis. It is estimated that 1 mg niacin (called *niacin equivalent* and abbreviated NE) can be expected from the ingestion of 60 mg dietary tryptophan. FAD and vitamin B_6 are required as coenzymes in several of the reactions involved in the conversion of tryptophan to NAD. In fact, deficiency of vitamin B_6 can impair NAD synthesis. Moreover, synthesis will vary with different physiological states.

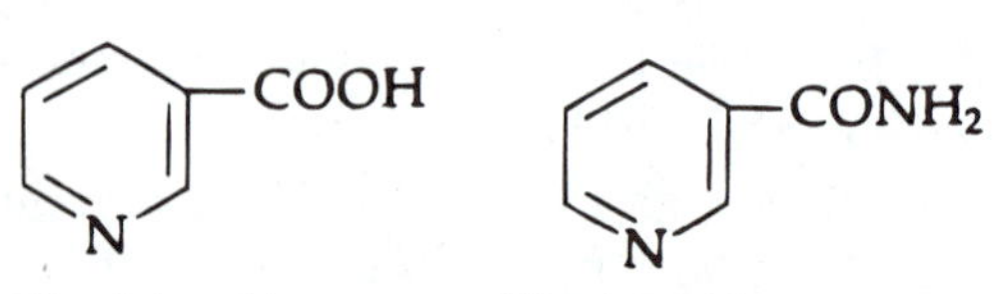

Digestion, Absorption, Transport, and Storage

NAD and NADP may be hydrolyzed within the intestinal tract by glycohydrolase to release free nicotinamide.

$$\text{NAD and NADP} \xrightarrow{\textit{glycohydrolase}} \text{nicotinamide}$$

Nicotinamide and nicotinic acid can be absorbed in the stomach, but they are more readily absorbed in the small intestine [1,2].

The absorption of niacin in the small intestine occurs primarily by a sodium-dependent, saturable system. Concentration of the vitamin in the lumen of the intestine, however, appears to determine its mode of absorption. At low concentrations niacin is absorbed via sodium-dependent, carrier-mediated, facilitated diffusion, while at high concentrations it is absorbed by passive diffusion [1,2].

Once in the intestinal cell, nicotinic acid is believed to be converted into nicotinamide. For this conversion to occur, nicotinic acid probably must first be incorporated into NAD and then released as the amide through NAD hydrolysis [3].

In the plasma, niacin is found primarily as nicotinamide, but nicotinic acid may also be found. Approximately 15% to 30% of nicotinic acid in the plasma is bound to plasma proteins [4]. From the blood, nicotinamide and nicotinic acid move across cell membranes by simple diffusion; however, nicotinic acid transport

Figure 9.16 Nicotinic acid and nicotinamide.

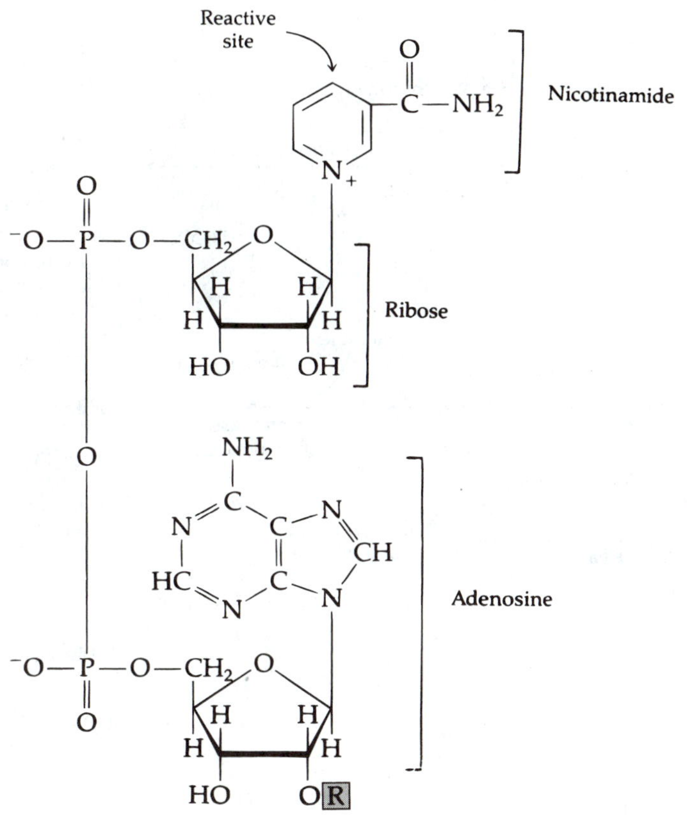

R = H for NAD$^+$ nicotinamide adenine dinucleotide
R = PO$_3^{-2}$ for NADP$^+$ nicotinamide adenine dinucleotide phosphate

Figure 9.17 The structures of NAD and NADP.

into the kidney tubules and red blood cells requires a sodium-dependent carrier system [5].

Nicotinamide serves as the primary precursor of NAD, which is synthesized in all tissues. Nicotinic acid in the liver, however, also may be used to synthesize NAD. As NAD or NADP, the vitamin is trapped within the cell. Intracellular concentrations of NAD typically predominate over those of NADP. NAD may be degraded to yield nicotinamide. It is primarily as NAD and NADP that niacin functions in the body.

Functions and Mechanisms of Action

Approximately 200 enzymes, primarily dehydrogenases, require NAD and NADP, which act as a hydrogen donor or electron acceptor. Figure 9.19 demonstrates the oxidation reduction that may occur in the nicotinamide moiety of the coenzymes. In their oxidized forms, NAD and NADP possess a positive charge and therefore may alternatively be written NAD$^+$ and NADP$^+$.

Although NAD and NADP are very similar and undergo reversible reduction in the same way, their func-

tions are quite different in the cell. The major role of NADH, formed from NAD, is to transfer its electrons from metabolic intermediates through the electron transport chain (pp. 60–66), thereby producing adenosine triphosphate (ATP). NADPH, in contrast, acts as a reducing agent in many biosynthetic pathways such as fatty acid, cholesterol, and steroid hormone synthesis but also in other pathways. Both NAD and NADP coenzymes are not tightly bound to their apoenzymes and can easily transport hydrogen atoms from one part of the cell to another. Reactions in which they participate occur both in the mitochondria and the cytoplasm.

Oxidative reactions in which NAD participates and is reduced include glycolysis (p. 85), oxidative decarboxylation of pyruvate (p. 91), oxidation of acetyl CoA via the Krebs cycle (p. 92), β-oxidation of fatty acids (p. 149), and oxidation of ethanol (p. 99). NAD is also required by aldehyde dehydrogenase for catabolism of vitamin B$_6$ as pyridoxal to its excretory product, pyridoxic acid.

The hexose monophosphate shunt (p. 90) produces NADPH, which can be used in a variety of reductive biosyntheses, including

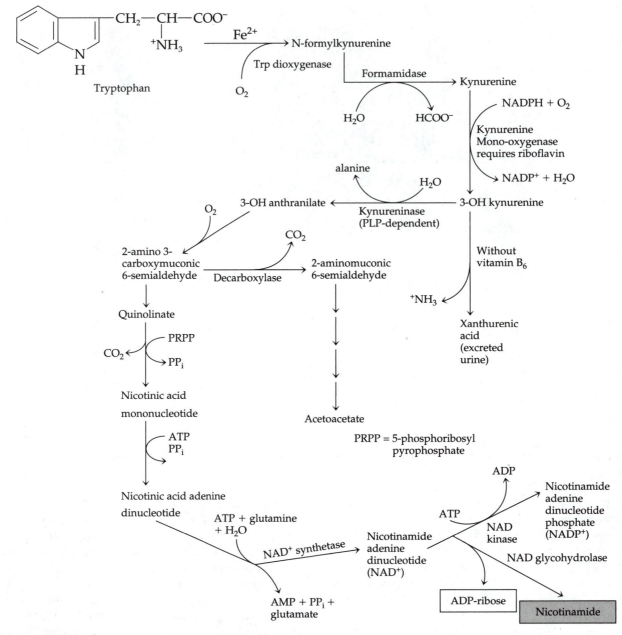

Figure 9.18 NAD$^+$ synthesis from tryptophan.

- fatty acid synthesis (p. 153),
- cholesterol and steroid hormone synthesis,
- oxidation of glutamate (p. 197), and
- synthesis of deoxyribonucleotides (precursors of DNA).

NADPH also may be used to reduce dehydroascorbate, the oxidized form of vitamin C. Enzymes such as glutathione reductase (p. 273), which functions to reduce glutathione from its oxidized state, also require NADPH. Conversion of folate to dihydrofolate (DHF) and tetrahydrofolate (THF) as well as synthesis of N^5 methyl THF and N^5, N^{10} methylene THF, active forms of folate, require NADPH (Fig. 9.33).

Some nonredox functions of NAD have been discovered in recent years. First, in relation to protein synthesis, NAD acts as a donor of adenosine diphosphate ribose (ADP-ribose) (Fig. 9.18) for the posttranslational modification of proteins. The enzyme catalyzing the reaction is a polymerase that attaches ADP-ribose onto various chromosomal proteins. Proteins associated with chromosomes include

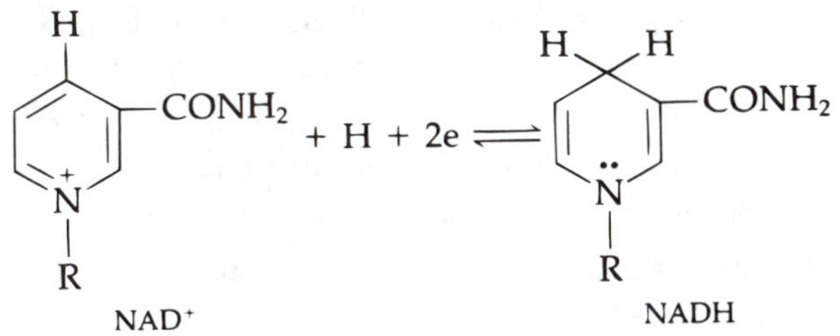

Figure 9.19 (a) The oxidation and reduction in the nicotinamide moiety; (b) the role of NAD in dehydrogenation reactions. One H of the substrate goes to NAD.

both histone and nonhistone proteins. It is the ADP-ribosyl moiety of the NAD that is posttranslationally attached by the polymerase to chromosomal proteins. In some cases, multiple (poly-) ADP-ribosyl moieties may be attached to some of the proteins. A second nonredox function of niacin is as a suspected component of glucose tolerance factor (GTF). The exact function of niacin in GTF is not clear; however, GTP is thought to potentiate the action of insulin (see p. 448 and Fig. 12.17).

Metabolism and Excretion

NAD, generated from nicotinamide or produced in the liver from tryptophan, and NADP can be degraded by glycohydrolase into nicotinamide and ADP-ribose (Fig. 9.18). The released nicotinamide is methylated and is then oxidized into a variety of products that are excreted in the urine. There is typically little excretion of nicotinic acid or nicotinamide because both compounds may be actively reabsorbed from glomerular filtrate.

The primary metabolites of nicotinamide are N' methyl nicotinamide (representing ~20%–30% of niacin) and N' methyl-2-pyridone-5-carboxamide (representing ~40%–60%). Nicotinic acid is metabolized to N' methylnicotinic acid. These metabolites may be used as a basis for assessing niacin status.

Dietary Reference Intakes and Recommended Dietary Allowances

Estimation of niacin requirements is complicated by the uncertain factor of the tryptophan- derived NAD.

The efficiency of the conversion of tryptophan to the vitamin is affected by a variety of influences, including the amount of tryptophan and niacin ingested, protein and energy intake, and vitamin B_6 and riboflavin nutriture [6]. As previously mentioned, 60 mg of tryptophan is thought to be equivalent to 1 mg of niacin and is regarded as a niacin equivalent (NE) [6]. Thus, total niacin represents the sum of milligrams of both nicotinic acid and nicotinamide and 1/60 mg of tryptophan. The average U.S. diet usually contains at least 800 mg of tryptophan a day [4].

Although recommendations are given in niacin equivalents, food composition tables report only preformed niacin. A rough estimate of niacin equivalents from a protein can be made by assuming that for every 1 g high-quality (complete) protein in the diet, 10 mg of tryptophan are provided; 1 g of complete, high-quality protein = 10 mg of tryptophan. Using this information, an intake of 60 g of complete protein, for example, would provide 600 mg of tryptophan; 10 mg of tryptophan/1 g of protein × 60 g of protein = 600 mg of tryptophan. Then, because it takes 60 mg of tryptophan to generate 1 mg of NE, 60 g of protein would generate about 10 NEs; 600 mg of tryptophan × 1 mg of NE/60 mg of tryptophan = 10 NEs.

Information used in estimating niacin requirements has come from human depletion and repletion studies conducted in the 1950s. Requirements for adult subjects appear to range from 9.2 to 13.3 niacin equivalents per day. Allowances for niacin are related to energy intake because of the involvement of NAD and NADP in the oxidation of energy-producing nutrients. The 1989 RDA for adults is 6.6 niacin equivalents per 1,000 kcal with an intake of not less than 13 niacin equivalents daily should caloric intake fall <2,000 kcal/day [6]. The DRI RDA for individual intake of niacin (as niacin equivalents) for adult men and women aged 19 years and older are slightly higher at 16 and 14 mg/day, respectively [7]. With pregnancy and lactation, the DRI RDA for niacin increases to 18 and 17 mg, respectively [7]. A tolerable upper intake level for adults aged 19 years and older of 35 mg of niacin per day also has been suggested [7].

Deficiency: Pellagra

Classical deficiency of niacin results in a condition known as pellagra. The four Ds—*dermatitis, dementia, diarrhea,* and *death*—are often used as a mnemonic device for remembering signs of pellagra [5]. The dermatitis is similar to sunburn at first and appears on areas exposed to sun such as the face and neck and on the extremities such as the back of the hands, wrists, elbows, knees, and feet. Neurological manifestations include peripheral neuritis, paralysis of extremities, and dementia or delirium. Gastrointestinal manifestations include glossitis, cheilosis, stomatitis, nausea, vomiting, and diarrhea. If untreated, death occurs.

A niacin deficiency also can result from the use of the antituberculosis drug isoniazid. Isoniazid binds with PLP and thereby reduces PLP-dependent kynureninase activity required for niacin synthesis. Hartnup disease results in impaired tryptophan absorption and thus decreases concentrations of the precursor tryptophan needed for niacin synthesis. Malabsorptive disorders (chronic diarrhea, inflammatory bowel diseases, some cancers) may impair niacin and tryptophan absorption and result in the increased likelihood of niacin deficiency. Individuals with poor nutrient intakes such as those who consume excessive amounts of alcohol are at risk for deficiency. In addition, people with stress, trauma, or prolonged fever may have increased needs for niacin.

Toxicity

Large doses of nicotinic acid (up to 3 g/day, but given in divided doses such as 1 g three times a day) are used in the treatment of hypercholesterolemia (high blood cholesterol). These pharmacological doses have been shown to significantly lower total serum cholesterol and low-density lipoproteins (LDLs) while causing an increase in high-density lipoproteins (HDLs) [8]. Although the mechanism of action is unclear, it is proposed that nicotinic acid decreases the levels of cAMP in the adipocytes, thereby decreasing lipase activity. Decreased lipase activity results in a decreased mobilization of fatty acids from the adipocytes and, therefore, a decreased substrate for synthesis of very low-density lipoproteins (VLDLs) in the liver. Decreased production of VLDLs lowers triacylglycerol levels, because VLDLs contain relatively high amounts of triacylglycerols. Furthermore, with decreased VLDLs there is less synthesis of LDL and thus lower serum cholesterol levels. An increase in the HDL appears to be due to a decrease in their breakdown within the liver.

Despite the therapeutic benefits of nicotinic acid, many undesirable side effects are associated with its use as a drug especially in doses of 1 g or more per day. Some of these side effects include

- release of histamine, which causes an uncomfortable flushing and which may be injurious to people with asthma and/or peptic ulcer disease (note that doses as low as 10 mg may cause this reaction);
- possible injury to the liver, as indicated by elevated serum levels of enzymes of hepatic origin

(e.g., transaminases and alkaline phosphatases) and by obstruction of normal bile flow from the liver to the small intestine;
- competition of niacin with uric acid for excretion, thereby raising serum uric acid levels;
- development of dermatological problems such as itching; and
- elevation of plasma glucose levels [8,9].

Gastrointestinal problems typically associated with intakes of large doses of nicotinic acid include heartburn and nausea with possible vomiting. Whether the beneficial effects of niacin in reducing blood lipids compensate for its possible toxic effects is a debatable question [8–10]. Intakes up to 500 mg/day appear to be tolerated with little adverse reactions [8]. Thus, a no observed adverse effect level (NOAEL) for niacin is 500 mg daily [11]. A lowest observed adverse effect level (LOAEL) for niacin has been set at 1,000 mg/day [11,12].

Nicotinamide in large doses does not exhibit toxic effects, but neither does it reduce blood lipids. Because of their stimulatory effects on the central nervous system (CNS), both nicotinic acid and nicotinamide have been tried during the past 30 years as therapeutic agents for some mental disorders. The current consensus among experts is that no improvement in brain function accrues from large doses of the vitamin [4,5].

Assessment of Nutriture

A commonly used method for assessing niacin nutriture is measurement of urinary N' methyl nicotinamide (in milligrams) per gram creatinine during a period of 4–5 hours after a 50-mg test dose of nicotinamide. Although differences in urinary creatinine excretion confound accuracy of the interpretation, the method is still used. Guidelines for interpretation are presented here [13]:

	Urinary N' methyl Nicotinamide Excretion (mg/g creatinine)		
	Deficient	Marginal	Adequate
All ages	<0.5	0.5–1.59	>1.6

Erythrocyte NAD concentrations or the ratio of NAD to NADP may be representative of niacin status [14,15]. Alternately, measurement of urinary nicotinamide metabolites may be useful. The ratio of urinary N' methyl-2-pyridone-5-carboxamide (pyridone) to N' methyl nicotinamide (NMN) has been used; however, the ratio may not be sensitive enough to detect marginal intakes of niacin and may reflect dietary protein adequacy and not niacin status [16]. Nonetheless, a ratio <1 is thought to be indicative of niacin deficiency.

References Cited for Niacin

1. Bender D. Nutritional Biochemistry of the Vitamins. New York: Cambridge University Press, 1992:184–222.

2. Sauberlich H. Bioavailability of vitamins. Prog Food Nutr Sci 1985;9:1–33.

3. Jacob R, Swendseid M. Niacin. In: Brown ML, ed. Present Knowledge in Nutrition, 6th ed. Washington, D.C.: The Nutrition Foundation, 1990:163–9.

4. van Eys J. Niacin. In: Machlin LJ, ed., Handbook of Vitamins, 2nd ed. New York: Dekker, 1991:311–40.

5. Combs G. The Vitamins. San Diego, CA: Academic Press, 1992:289–309.

6. National Research Council. Recommended Dietary Allowances, 10th ed. Washington, D.C.: National Academy Press, 1989:137–42.

7. Yates A, Schlicker S, Suitor C. Dietary reference intakes: the new basis for recommendations for calcium and related nutrients, B vitamins and choline. J Am Diet Assoc 1998;98:699–706.

8. McKenney J, Proctor J, Harris S, Chinchili V. A comparison of the efficacy and toxic effects of sustained—vs immediate—release niacin in hypercholesterolemic patients. JAMA 1994;271:672–7.

9. Alhadeff L, Gualtieri C, Lipton M. Toxic effects of water-soluble vitamins. Nutr Rev 1984;42:33–40.

10. Council on Scientific Affairs, American Medical Association. Vitamin preparations as dietary supplements and as therapeutic agents. JAMA 1987;257: 1929–36.

11. Hathcock J. Vitamins and minerals: efficacy and safety. Am J Clin Nutr 1997;66:427–37.

12. Hathcock J. Safety limits for nutrients. J Nutr 1996;126:2386S–9S.

13. Gibson RS. Principles of Nutrition Assessment. New York: Oxford University Press, 1990:437–44.

14. Jacob R, Swendseid M, McKee R. Biochemical markers for assessment of niacin status in young men: urinary and blood levels of niacin metabolites. J Nutr 1989;119:591–8.

15. Fu C, Swendseid M, Jacob R, McKee R. Biochemical markers for assessment of niacin status in young men: levels of erythrocyte niacin coenzymes and plasma tryptophan. J Nutr 1989;119:1949–55.

16. Shibata K, Matsuo H. Effect of supplementing low protein diets with the limiting amino acids on the excretion of N' methylnicotinamide and its pyridones in the rat. J Nutr 1989;119:896–901.

Pantothenic Acid

Pantothenic acid or pantothenate, an amide, consists of β-alanine and pantoic acid joined together by a peptide bond. The structure of pantothenate is shown at the top of Figure 9.20 and as part of coenzyme A in Figure 9.21.

Sources

The Greek word *pantos* means everywhere, and the vitamin pantothenic acid, as its name implies, is found widely distributed in nature. Because this vitamin is present in virtually all plant and animal foods, a deficiency is quite unlikely. Meats (particularly liver), egg yolk, legumes, whole-grain cereals, mushrooms, broccoli, and avocados, among others, are good sources of the vitamin. Royal jelly from bees also provides large amounts of pantothenate. In supplements, pantothenate is usually found as calcium pantothenate.

Digestion, Absorption, Transport, and Storage

Most, about 85%, of the pantothenic acid in food occurs as a component of coenzyme A, abbreviated CoA (Fig. 9.21) [1]. During the digestive process, CoA is hydrolyzed to pantetheine and then to pantothenic acid.

Pantothenic acid is thought to be absorbed principally in the jejunum by passive diffusion [1,2], although animal studies suggest that when present in low concentrations, pantothenate may be absorbed by a sodium-dependent active process [3]. Approximately 40% to 61%, mean 50%, of the ingested pantothenic acid appears to be available for absorption [4,5]. Panthenol, an alcohol form of the vitamin used in multivitamins, may also be absorbed and converted to pantothenate. However, pantothenate absorption has been shown to decrease to about 10% when pantothenate ingestion approaches 10 times recommended intakes in pill form.

From the intestinal cell, pantothenate enters portal blood for transport to body cells. Pantothenic acid is found in whole blood, plasma, serum, and red blood cells. It passively diffuses into and out of red blood cells where it is found not only as pantothenate, but also as 4-phosphopantothenate and pantetheine [6]. Free pantothenic acid is found in serum and plasma, but higher concentrations are found in red blood cells [3].

Uptake of pantothenate by tissues differs. Heart, muscle, and liver cells take up pantothenate by sodium-dependent active transport [1,2,7]. Central nervous system, adipose, and renal uptake of pantothenate is by

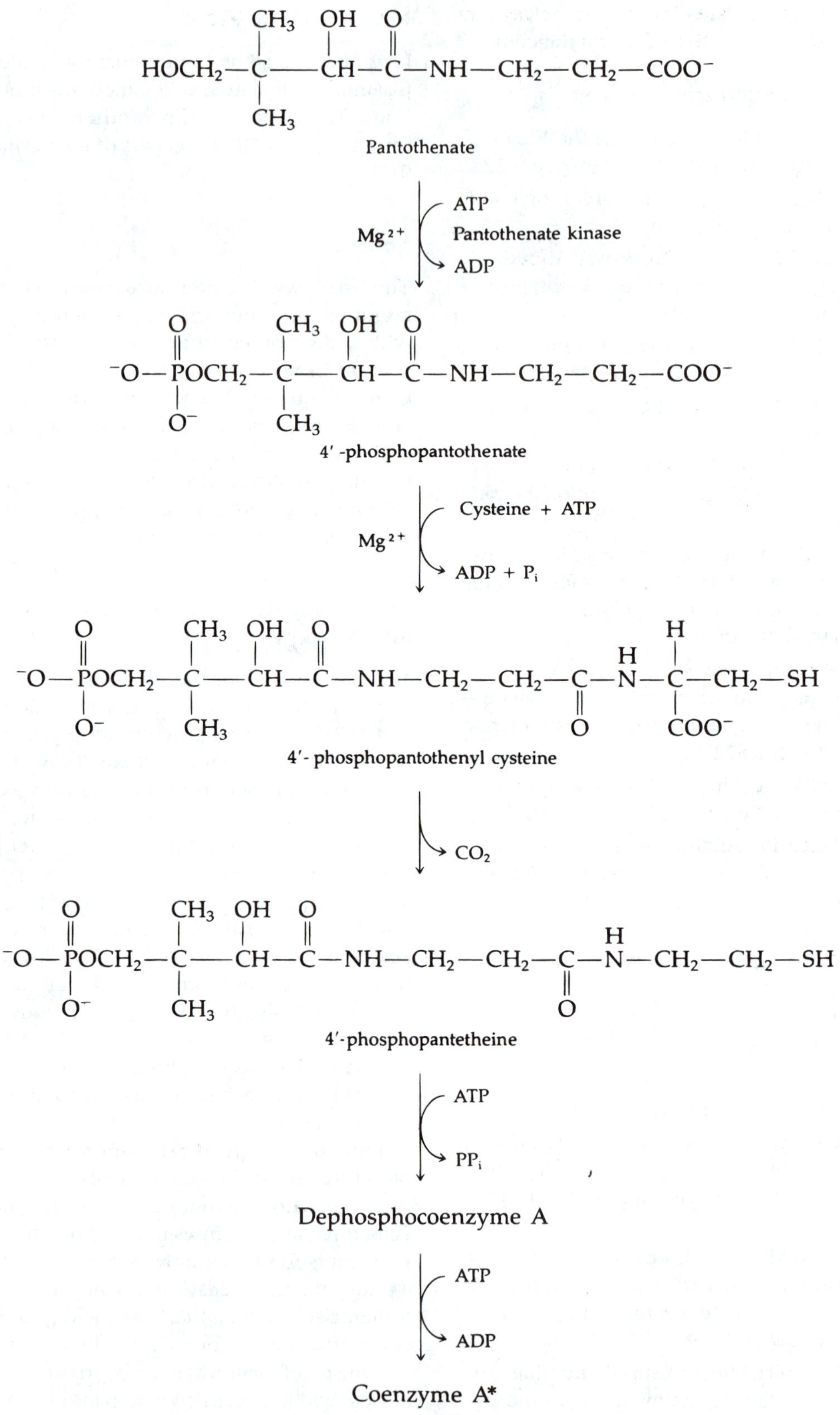

$$\text{HOCH}_2-\underset{\underset{\text{CH}_3}{|}}{\overset{\overset{\text{CH}_3}{|}}{\text{C}}}-\underset{\overset{|}{\text{OH}}}{\text{CH}}-\overset{\overset{\text{O}}{\|}}{\text{C}}-\text{NH}-\text{CH}_2-\text{CH}_2-\text{COO}^-$$

Pantothenate

Mg^{2+} — ATP, Pantothenate kinase → ADP

$$^-\text{O}-\overset{\overset{\text{O}}{\|}}{\underset{\underset{\text{O}^-}{|}}{\text{P}}}\text{OCH}_2-\underset{\underset{\text{CH}_3}{|}}{\overset{\overset{\text{CH}_3}{|}}{\text{C}}}-\underset{\overset{|}{\text{OH}}}{\text{CH}}-\overset{\overset{\text{O}}{\|}}{\text{C}}-\text{NH}-\text{CH}_2-\text{CH}_2-\text{COO}^-$$

4'-phosphopantothenate

Mg^{2+} — Cysteine + ATP → ADP + P$_i$

$$^-\text{O}-\overset{\overset{\text{O}}{\|}}{\underset{\underset{\text{O}^-}{|}}{\text{P}}}\text{OCH}_2-\underset{\underset{\text{CH}_3}{|}}{\overset{\overset{\text{CH}_3}{|}}{\text{C}}}-\underset{\overset{|}{\text{OH}}}{\text{CH}}-\overset{\overset{\text{O}}{\|}}{\text{C}}-\text{NH}-\text{CH}_2-\text{CH}_2-\overset{\overset{\text{H}}{|}}{\underset{\underset{\text{O}}{\|}}{\text{C}}}-\text{N}-\underset{\underset{\text{COO}^-}{|}}{\overset{\overset{\text{H}}{|}}{\text{C}}}-\text{CH}_2-\text{SH}$$

4'-phosphopantothenyl cysteine

→ CO$_2$

$$^-\text{O}-\overset{\overset{\text{O}}{\|}}{\underset{\underset{\text{O}^-}{|}}{\text{P}}}\text{OCH}_2-\underset{\underset{\text{CH}_3}{|}}{\overset{\overset{\text{CH}_3}{|}}{\text{C}}}-\underset{\overset{|}{\text{OH}}}{\text{CH}}-\overset{\overset{\text{O}}{\|}}{\text{C}}-\text{NH}-\text{CH}_2-\text{CH}_2-\overset{\overset{\text{H}}{|}}{\underset{\underset{\text{O}}{\|}}{\text{C}}}-\text{N}-\text{CH}_2-\text{CH}_2-\text{SH}$$

4'-phosphopantetheine

→ ATP → PP$_i$

Dephosphocoenzyme A

→ ATP → ADP

Coenzyme A*

Figure 9.20 Synthesis of coenzyme A (*structure shown in Fig. 9.21) from pantothenate.

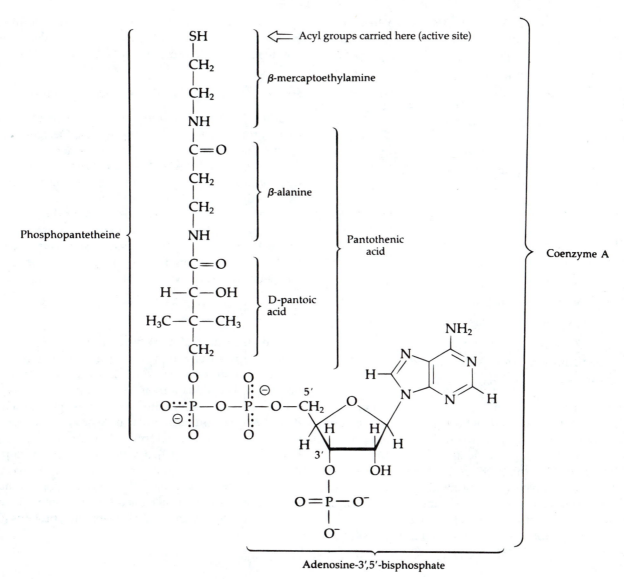

Figure 9.21 Structure of coenzyme A and identification of components.

facilitative diffusion [1,2]. Within cells, pantothenate may accumulate and is typically used to synthesize or resynthesize CoA. CoA is found in the largest concentrations in the liver, adrenal gland, kidney, brain, and heart [7].

Functions and Mechanisms of Action

One of the primary functions of pantothenic acid relates to its role as a component of CoA, although 4'-phosphopantotheine, derived from CoA, may also function bound to a protein. The synthesis of CoA from pantothenate is depicted in Figure 9.20. The synthesis requires pantothenic acid, cysteine, and ATP.

• The synthesis of coenzyme A starts with the rate-limiting phosphorylation of pantothenic acid by *pantothenate kinase* and the formation of 4'-phosphopantothenate. ATP and Mg^{2+} are required for this reaction.

• Next in another ATP- and Mg^{2+}-requiring reaction, cysteine reacts with the 4'-phosphopantothenate. A peptide bond is formed between the carboxyl group of the 4'-phosphopantothenate and the amino group of cysteine.

• Third, a carboxyl group from the cysteine moiety is removed to generate 4'-phosphopantotheine. An adenylation occurs whereby adenosine monophosphate (AMP) is added to the 4'-phosphopantotheine to form dephosphocoenzyme A.

• Lastly, phosphorylation with ATP of the 3'-hydroxyl group of the dephosphocoenzyme A produces CoA.

Figure 9.21 gives the structure of coenzyme A, identifying its active site and its constituents. Synthesis of CoA is inhibited by acetyl CoA, malonyl CoA, and propionyl CoA as well as by other longer-chain acyl CoAs. CoA metabolism has been reviewed in depth by Robishaw and Neely [8].

As a component of CoA, pantothenic acid becomes essential for production of energy from carbohydrate,

$$\overset{O}{\underset{||}{}}$$

fat, and protein. CoA can form thio esters (-S-C-R) with carboxylic acids and can transfer the acyl groups, typically 2 to 13 carbons, as needed for condensation and additional reactions. Examples of acids activated by CoA include

- acetic (two carbons),
- malonic (three carbons),
- propionic (three carbons),
- methylmalonic (four carbons), and
- succinic (four carbons).

Propionic acid is found naturally in some fish and is derived from the catabolism of both amino acids such as methionine, threonine, and isoleucine and from the catabolism of odd-chain fatty acids. Succinate is found as an intermediate in the Krebs cycle as well as generated from the catabolism of amino acids such as methionine, threonine, isoleucine, and valine. Succinyl CoA is also necessary along with the amino acid glycine for the initial vitamin B_6–dependent step in heme synthesis (Figs. 9.42 and 12.5).

Energy Metabolism

A crucial reaction in energy metabolism is the formation of acetyl CoA, which condenses with oxaloacetate to thereby introduce acetate for oxidation via the Krebs cycle (Fig. 9.12). Acetyl CoA, the common compound formed from the three energy-producing nutrients, holds the central position in the transformation of energy. Pantothenic acid, then, joins the other B vitamins thiamin, riboflavin, and niacin, in the oxidative decarboxylation of pyruvate (Fig. 9.12) and α-ketoglutarate. On the synthetic side of metabolism, condensation of acetyl CoA with activated CO_2 to form malonyl CoA is the first step in fatty acid synthesis (p. 151). Therefore pantothenic acid plays an important role in energy storage as well as energy release. Other synthetic reactions include the reaction of acetyl CoA and acetoacetyl CoA to form HMG CoA (Fig. 6.14), important in cholesterol

synthesis and ketogenesis. Phospholipid and sphingomyelin production from phosphatidic acid and sphingosine respectively also use acyl CoA.

Prosthetic Group for Acyl Carrier Protein

Another function of pantothenic acid is as the prosthetic group for acyl carrier protein (ACP). Figure 9.20 shows that 4'-phosphopantotheine is necessary as a prosthetic group for ACP. ACP acts as the acyl carrier in the synthesis of fatty acids and is a necessary component of the fatty acid synthase complex (p. 153). The sulfhydryl group in the 4'-phosphopantotheine and a sulfhydryl group in the protein are the active sites in the ACP. These groups are located close to one another so that the acyl chain being synthesized can be transferred between the two of them.

Protein Acetylation

Pantothenic acid is also involved in the modification of proteins. Specifically, the vitamin is involved in the protein acetylation process, which in turn affects protein functions. The donation of long-chain fatty acids or acetate by CoA to proteins occurs posttranslationally [9,10]. Acetylation of peptides may protect them from degradation and may determine activity, location, and function in the cell [9,10]. Acetylation of the N-terminal amino acids has been shown to affect resistance to ubiquitin-mediated proteolysis [9]. Other proteins that may undergo acetylation include microtubules of the cell's cytoskeleton, histones, and other DNA-binding proteins [9].

Other

Other roles of pantothenic acid include metabolism of certain drugs and xenobiotics by acetylation and synthesis of acetylcholine. Pantothenic acid, based on animal studies, also may accelerate the normal healing process following surgery [11]. The exact mechanism of the effect of pantothenate is unclear; however, an increase in cellular multiplication during the first postoperative period has been proposed [11].

Metabolism and Excretion

During metabolism, CoA is dephosphorylated and through a series of subsequent reactions generates pantothenate. Pantothenate is excreted as such primarily in the urine; no metabolites of the vitamin have been identified in humans. Fecal pantothenate excretion also occurs.

Urinary excretion of pantothenate is thought to reflect dietary intake. Whenever urinary excretion of pantothenate is <1 mg/day, deficiency might be suspected. An excretion of <1 mg/day is thought to correspond to an intake of <4 mg daily [1].

Dietary Reference Intakes and Recommended Dietary Allowances

In 1989, an Estimated Safe and Adequate Daily Dietary Intake range for pantothenic acid of about 4 to 6 mg daily was thought to be sufficient [1,12]. The 1998 DRI *Adequate Intake* (AI) recommendation for adults aged 19 years and older for pantothenate has been set at 5 mg [13]. Adequate Intake is used instead of RDA when there are insufficient data available to establish an estimated average requirement (EAR) and subsequent RDA [13]. An AI for pantothenic acid of 6 and 7 mg/day are suggested for women during pregnancy and lactation, respectively [13].

Deficiency

Pantothenate deficiency in humans has been reported in people with severe malnutrition [3]. "Burning feet syndrome" characterized by abnormal skin sensations, exacerbated by warmth and diminished with cold, of the feet and lower legs has been reported and is thought to result from a pantothenic acid deficiency. The syndrome can be corrected with calcium pantothenate administration. Other symptoms of deficiency include vomiting, fatigue, and weakness. A metabolic inhibitor of pantothenate, omega methylpantothenate, has been used in studies to induce low pantothenate status in humans. Conditions and populations associated with increased need for intake include people with alcoholism, diabetes mellitus, and inflammatory bowel diseases. Increased excretion of the vitamin has been shown in people with diabetes mellitus. Absorption is likely to be impaired with inflammatory bowel diseases. Intake of the vitamin is typically low in people with excessive alcohol intake.

Toxicity

Pantothenate toxicity has not been reported to date in humans. Intakes of 100 mg panthothenate may increase niacin excretion [14]. Intakes of about 10 g pantothenate as calcium pantothenate daily for up to 6 weeks have resulted in no problems [1,7]. Intakes up to 20 g may cause mild intestinal distress and diarrhea [12].

Assessment of Nutriture

Plasma pantothenic acid concentrations <100 mg/dL are thought to reflect low dietary pantothenate intakes [3]. Urinary pantothenate excretion of <1 mg/day is considered to be low [1].

References Cited for Pantothenic Acid

1. Bender D. Nutritional Biochemistry of the Vitamins. New York: Cambridge University Press, 1992:341–59.

2. Olson R. Pantothenic acid. In: Brown ML, ed. Present Knowledge in Nutrition, 6th ed. Washington, D.C.: The Nutrition Foundation, 1990:208–11.

3. Fox H. Pantothenic acid. In: Machlin LJ., ed., Handbook of Vitamins, 2nd ed. New York: Dekker, 1991:429–51.

4. Tarr J, Tamura T, Stokstad E. Availability of vitamin B_6 and pantothenate in an average American diet in man. Am J Clin Nutr 1981;34:1328–37.

5. Sauberlich H. Bioavailability of vitamins. Prog Food Nutr Sci 1985;9:1–33.

6. Annous K, Song W. Pantothenic acid uptake and metabolism by red blood cells of rats. J Nutr 1995; 125:2586–93.

7. Combs G. The Vitamins. San Diego, CA: Academic Press, 1992:345–56.

8. Robishaw J, Neely J. Coenzyme A metabolism. Am J Physiol 1985;248:E1–E9.

9. Plesofsky-Vig, N. Pantothenic acid and coenzyme A. In: Shils ME, Olson JA, Shike M., eds. Modern Nutrition in Health and Disease, 8th ed. Philadelphia: Lea and Febiger, 1994:395–401.

10. Plesofsky-Vig N, Brambl R. Pantothenic acid and coenzyme A in cellular modification of proteins. Ann Rev Nutr 1988;8:461–82.

11. Aprahamian M, Dentinger A, Stock-Damge C, Kouassi J, Grenier J. Effects of supplemental pantothenic acid on wound healing: Experimental study in rabbit. Am J Clin Nutr 1985;41:578–89.

12. National Research Council. Recommended Dietary Allowances, 10th ed. Washington, D.C.: National Academy Press, 1989:169–73.

13. Yates A, Schlicker S, Suitor C. Dietary reference intakes: the new basis for recommendations for calcium and related nutrients, B vitamins and choline. J Am Diet Assoc 1998;98:699–706.

14. Clarke J, Kies C. Niacin nutritional status of adolescent humans fed high dosage pantothenic acid supplements. Nutr Rep Internl 1985;31:1271–9.

Biotin

Biotin was once referred to as vitamin H. Its structure consists of two rings, a ureido ring joined to a thiophene ring, with an additional valeric acid side chain (Fig. 9.22).

Sources

Biotin is found widely distributed in foods. Good food sources of the vitamin are liver, soybeans, and egg yolk. Cereals, legumes, and nuts also contain relatively high amounts of biotin. Within many foods, biotin is found bound to protein or as biocytin, which is also called *biotinyllysine* (Fig. 9.23). Biotin is also produced by bacteria within the colon.

Avidin, a glycoprotein in raw egg whites, may irreversibly bind biotin in a noncovalent bond and prevent biotin absorption. Because avidin is heat-labile (unstable with heat), ingestion of cooked egg whites does not compromise biotin absorption.

Digestion, Absorption, Transport, and Storage

Protein-bound biotin requires digestion by proteolytic enzymes prior to absorption. Proteolysis by proteases yields free biotin, biocytin, or biotinyl peptides. Biocytin and biotinyl peptides can be further hydrolyzed by biotinidase. Biotinidase, on the intestinal brush border or in pancreatic or intestinal juice, hydrolyzes the biocytin or biotinyl peptides to release free biotin, lysine, and possibly other amino acids from the peptides.

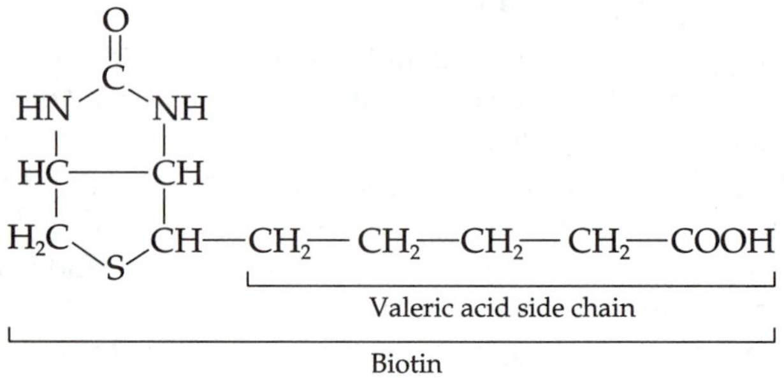

Undigested biocytin may also be absorbed as such into the body and may be acted on by the biotinidase present in plasma or other body tissues such as the liver, kidney, and adrenal glands. Any biocytin not metabolized by biotinidase is excreted in the urine [1]. Biotinidase is active over a wide pH range and is specific for the biotinyl moiety, hydrolyzing at amide or ester linkages [2]. Biotinidase deficiency due to an inborn error of metabolism has been documented in infants and children. Clinical features associated with the genetic disorder, as well as deficiency, include seizures, ataxia, skin rash, alopecia (hair loss), acidosis, among others [2]. The bioavailability of biotin contained in foods is highly variable, ranging from 100% in corn to near 0% in wheat.

Little is known about the mechanism of absorption of biotin in humans. Based on experimental studies in animals, it is proposed that most absorption of the vitamin occurs in the upper one-third to one-half of the small intestine. Absorption in the duodenum is thought to be greater than in the jejunum [3], which is thought to be greater than that in the ileum [4]. Absorption in the proximal and midtransverse colon occurs [4] but is small relative to biotin absorption in the small intestine [5]. Absorption in the small intestine is carrier mediated and thought to require sodium. Absorption may or may not require energy [4–6]. A small amount of biotin absorption may also occur by passive diffusion [5,6].

Both free biotin and protein-bound biotin are found in the plasma. Albumin and α- and β-globulins bind biotin, as does a plasma biotinidase, which has two binding sites for biotin and is thought to arise from the liver [4,6]. Biotin is stored in small quantities in the muscle, liver, and brain. The rate and uptake of biotin by tissues is related to the need of the cells for the vitamin. When biotin enters the cells, its distribution corresponds to localization of the carboxylases, the enzymes requiring biotin as a coenzyme [7].

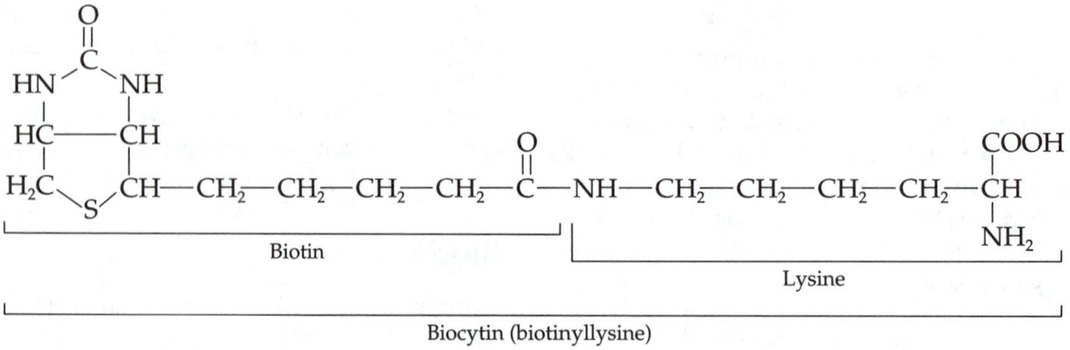

Figure 9.22 The structure of biotin.

Figure 9.23 The structure of biocytin, also called biotinyllysine.

Functions and Mechanisms of Action

Biotin functions in cells covalently bound to enzymes. Biotin uptake into cells is by active transport. Within the cell, biotin reacts in a Mg^{2+}-requiring reaction with ATP to form biotinyl 5'-adenylate, also referred to as *activated biotin*. In subsequent reactions, holoenzyme (carboxylase) synthetase joins the biotinyl moiety with one of several apoenzymes to form a holoenzyme carboxylase with the release of AMP. Multiple carboxylase deficiencies due to inborn errors of metabolism of holoenzyme synthetase have been documented. Some of the genetic disorders benefit from biotin supplements (10 mg or more daily) and dietary amino acid restrictions to limit substrates for the missing or deficient enzymes.

Examples of biotin-dependent enzymes include *acetyl CoA carboxylase, pyruvate carboxylase, propionyl CoA carboxylase,* and *β-methylcrotonyl CoA carboxylase.* Table 9.2 lists the enzymes and their roles in metabolism. Knowles [8] has reviewed the mechanism of action of biotin-dependent enzymes.

The carboxylases are multisubunit enzymes to which biotin is attached by an amide linkage. The carboxyl terminus of biotin is linked to the epsilon amino group of a specified lysine residue in the apoenzyme. The chain connecting biotin and the apoenzyme is long and flexible, thereby allowing the biotin to move from one active site of the carboxylase to another. Figure 9.24 depicts the attachment of biotin to the enzyme, emphasizing the long flexible chain and the amide linkage between the vitamin and the lysine residue of the enzyme. One active site on the apoenzyme generates the carboxybiotin enzyme, while the other transfers the activated carbon dioxide to a reactive carbon on the substrate. Figure 9.25 illustrates the formation of the CO_2 biotin enzyme complex.

Pyruvate Carboxylase

Pyruvate carboxylase is a particularly interesting and important enzyme because of its regulatory function. For its activation pyruvate carboxylase requires the presence of acetyl CoA as well as ATP and Mg^{2+}. Acetyl CoA serves as an allosteric activator, and its presence indicates the need for increased amounts of oxaloacetate. Specifically, pyruvate carboxylase catalyzes the carboxylation of pyruvate to form oxaloacetate (Fig. 9.26). If there is a surplus of ATP in the cell, the oxaloacetate is then used for gluconeogenesis. If, however, there is a deficiency of ATP, the oxaloacetate will enter the Krebs cycle on condensation with acetyl CoA (p. 92). Deficiency of biotin has been shown to impair the activity of pyruvate carboxylase in mice [9].

Table 9.2 Biotin-Dependent Enzymes

Enzyme	Role	Significance
1. Pyruvate carboxylase	Converts pyruvate to oxaloacetate	Replenishes oxaloacetate for Krebs cycle Necessary for gluconeogenesis
2. Acetyl CoA carboxylase	Forms malonyl CoA from acetate	Commits acetate units to fatty acid synthesis
3. Propionyl CoA carboxylase	Converts propionate to succinate	Provides mechanism for metabolism of some amino acids and odd-numbered chain fatty acids. Succinate formed enters Krebs cycle
4. β-methylcrotonyl CoA carboxylase	Converts β-methylcrotonyl CoA to β-methylglutaconyl CoA	Allows catabolism of leucine and certain isoprenoid compounds

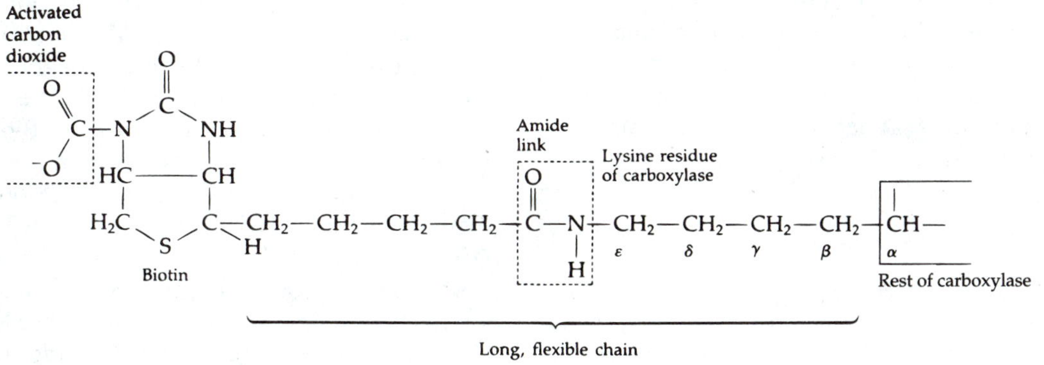

Figure 9.24 Biotin bound to the lysine residue of carboxylase and functioning as a carrier of activated CO_2.

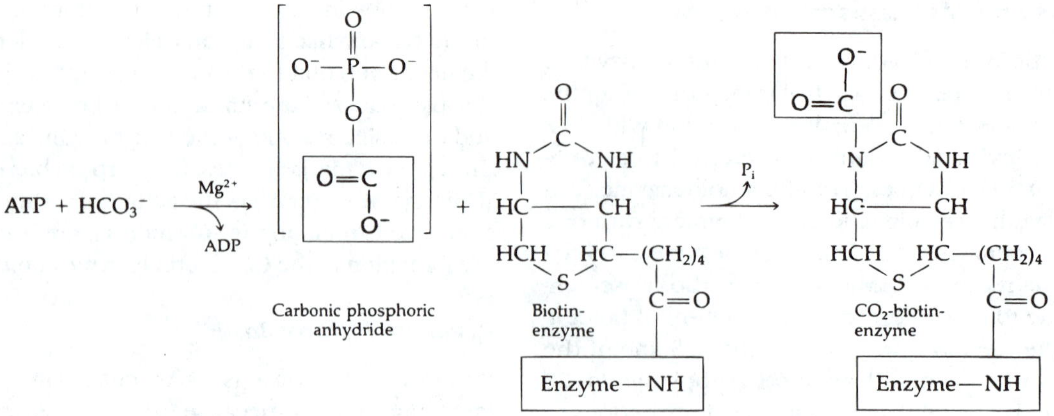

$$ATP + HCO_3^- \xrightarrow[ADP]{Mg^{2+}}$$

Carbonic phosphoric
anhydride

Biotin-
enzyme

Enzyme—NH

P_i

CO$_2$-biotin-
enzyme

Enzyme—NH

Figure 9.25 The formations of the CO$_2$-biotin-enzyme complex.

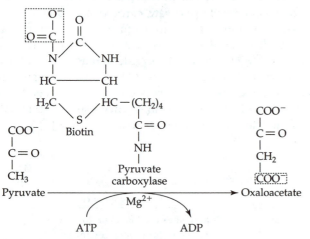

Biotin

Pyruvate

Pyruvate
carboxylase

Mg^{2+}

ATP   ADP

Oxaloacetate

Figure 9.26 The role of biotin in the synthesis of oxaloacetate from pyruvate.

Acetyl CoA Carboxylase

The importance of biotin in energy metabolism is further exemplified by its role in the initiation of fatty acid synthesis (p. 151); that is, the formation of malonyl CoA from acetyl CoA by the regulatory and rate-limiting enzyme acetyl CoA carboxylase. This enzyme is allosterically activated by citrate and isocitrate and inhibited by long-chain fatty acyl CoA derivatives. ATP and Mg^{2+} are required for the reaction.

Propionyl CoA Carboxylase

Propionyl CoA carboxylase is important for the catabolism of isoleucine, threonine, and methionine, which each generate propionyl CoA. It also arises from the catabolism of odd-number chain fatty acids found, for example, in some fish. Propionyl CoA carboxylase catalyzes

the carboxylation of propionyl CoA to methylmalonyl CoA (Fig. 9.27). The reaction requires ATP and Mg^{2+}. Like pyruvate carboxylase, the activity of propionyl CoA carboxylase decreases with biotin deficiency in mice [9]. Methylmalonyl CoA, after being acted on by racemase, generates succinyl CoA in a vitamin B$_{12}$–dependent reaction (Fig. 9.27).

β-methylcrotonyl CoA Carboxylase

β-methylcrotonyl CoA carboxylase is important in the catabolism of the amino acid leucine. During leucine catabolism (Fig. 9.28), β-methylcrotonyl CoA is formed. This compound is carboxylated in an ATP-, Mg^{2+}-, and biotin-dependent reaction by β-methylcrotonyl CoA carboxylase to form β-methylglutaconyl CoA, which is further catabolized to generate acetoacetate and acetyl CoA.

Metabolism and Excretion

Catabolism of the biotin holoenzymes or holocarboxylases by proteases yields biocytin. The biocytin is then degraded by biotinidase to yield free biotin. Some of this biotin is reused, and some is degraded. With respect to biotin catabolism (Fig. 9.29), the valerate side chain of biotin may be degraded by β-oxidation to form bisnorbiotin. Little catabolism of the ring system of the vitamin occurs in humans [5,7]. Biotin is primarily excreted in the urine, as such, or as bisnorbiotin [10]. Small amounts of other metabolites, including biotin sulfoxide, biotin sulfone (Fig. 9.27), bisnorbiotin methyl ketone, and tetranorbiotin-1-sulfoxide, may also be formed and excreted [4,5,11]. In addition, biocytin that has not been hydrolyzed by biotinidase may also be excreted in the urine [1].

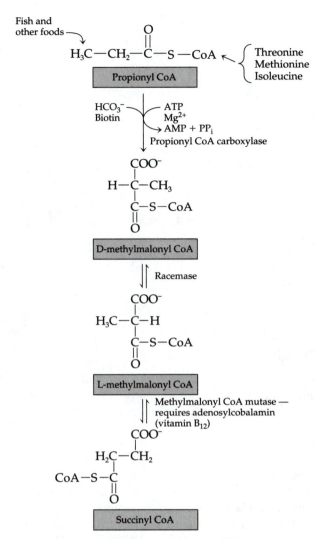

Figure 9.27 The oxidation of propionyl CoA and the role of biotin.

Free biotin may be reabsorbed by the kidney, although saturation of the renal transport system results in urinary excretion [4]. Urinary biotin excretion ranges from <6 to 111 mg/day. United States diets supplying 28 to 42 mg of biotin per day resulted in a urinary biotin excretion of 20 to 24 mg/day with no indication of inadequate status associated with this level of intake [12]. Adults ingesting 100 to 200 mg of biotin daily excreted 18 to 46 mg of biotin in the urine [7].

Unabsorbed biotin and biotin synthesized by intestinal bacteria are excreted in the feces. Excretion in the feces appears to be independent of dietary intake [5]. Urinary and fecal excretion of biotin are often greater than dietary intake presumably because of intestinal bacteria's contribution of biotin to the body. Urinary biotin excretion is typically less than dietary intake [3].

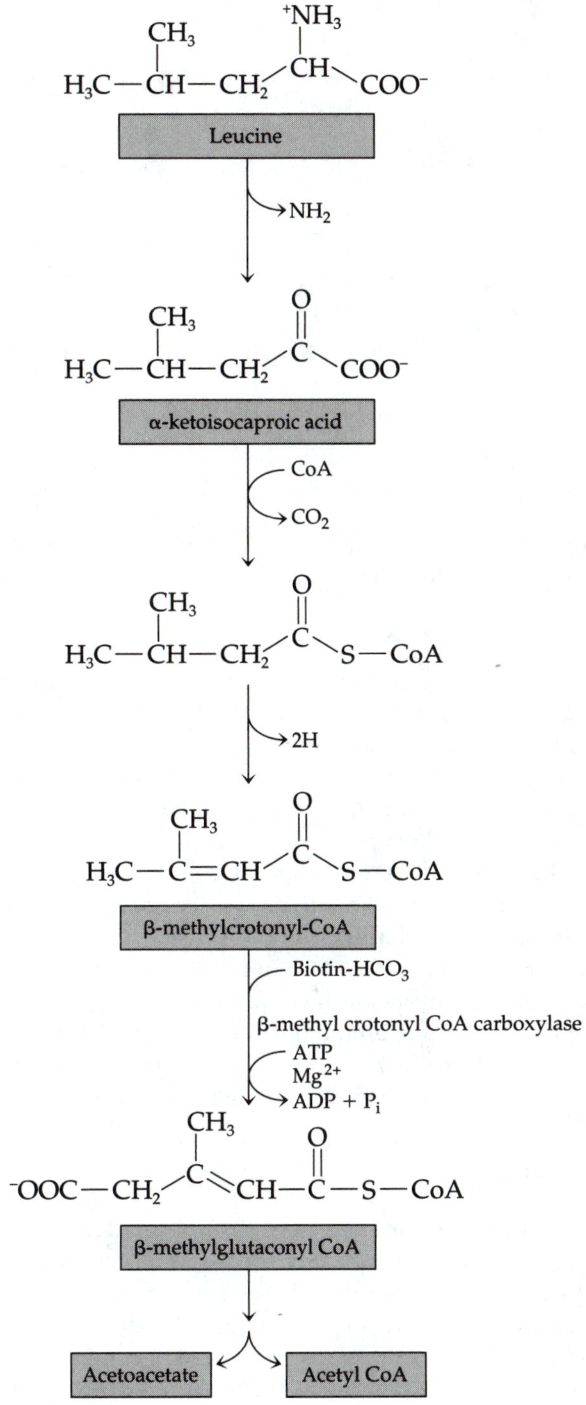

Figure 9.28 The role of biotin in leucine catabolism.

Dietary Reference Intakes and Recommended Dietary Allowances

Because of the uncertain contribution of intestinal synthesis to total biotin intake and the incomplete knowledge about the bioavailability of food biotin, a range of 30 to 100 mg/day is provisionally recommended in the

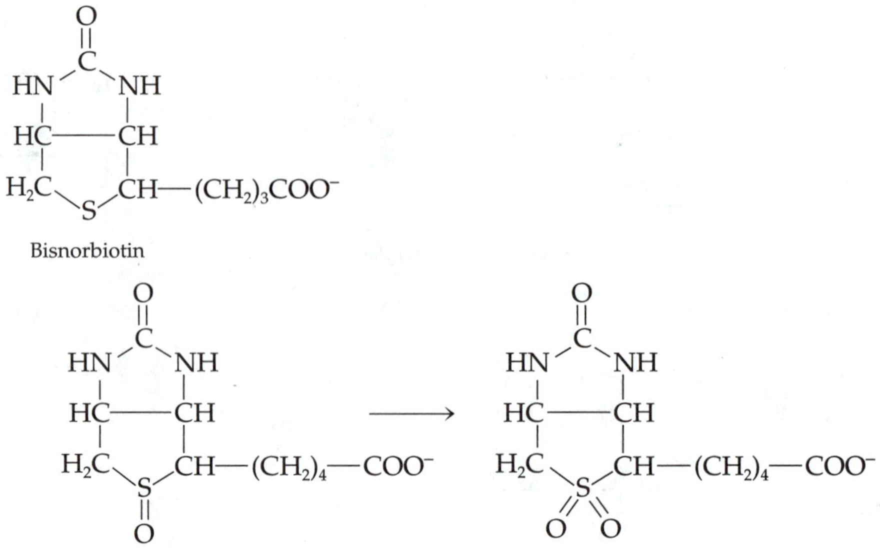

Figure 9.29 Selected metabolites from biotin degradation.

Bisnorbiotin

Biotin sulfoxide

Biotin sulfone

1989 RDA for children 11 years or older and adults [12]. Based on the appearance of a biotin deficiency in studies of children with biotinidase deficiency not receiving supplements of free biotin, it has been concluded that intestinally synthesized biotin is not sufficient to maintain normal biotin status [3]. An Adequate Intake (AI) recommendation for adults aged 19 years and older of 30 µg of biotin per day has been suggested [13]. As discussed initially under panthothenic acid, an AI is used instead of RDA when there are insufficient data available to establish an estimated average requirement (EAR) and subsequent RDA [13]. An AI for biotin of 30 and 35 µg per day are suggested for women during pregnancy and lactation, respectively [13].

Deficiency

Biotin deficiency in humans is characterized by depression, hallucinations, muscle pain, localized paresthesia, anorexia, nausea, alopecia (hair loss), and scaly dermatitis. Deficiency in rats is associated with decreased hepatic ornithine transcarbamylase mRNA and activity; this enzyme is important in the urea cycle [14]. People ingesting raw eggs in excess amounts are likely to develop biotin deficiency due to impaired biotin absorption. Impaired biotin absorption may occur also with gastrointestinal disorders, such as inflammatory bowel disease and achlorhydria (lack of hydrochloric acid in gastric juices), or with people on anticonvulsant drug therapy. Excessive alcohol ingestion and sulfonamide therapy also increase the risk for deficiency.

Toxicity

Toxicity of biotin has not been reported. Biotin given in oral doses of 60 mg daily for over 6 months has not been shown to produce side effects.

Assessment of Nutriture

Evaluation of biotin in blood, plasma, or serum as well as analysis in urine are most often used to assess biotin status. Plasma biotin concentrations <1.02 nmol/L are thought to indicate deficiency [5]. Normal ranges for plasma biotin are 0.82 to 2.87 nmol/L [5]. Urinary biotin excretion is thought to reflect and has been used to assess biotin status [7,10]. Urinary biotin levels of <20 mg/day are thought to be inadequate. In animals, the hepatic and splenic activities of pyruvate carboxylase and propionyl CoA carboxylase significantly decrease with biotin deficiency [9].

References Cited for Biotin

1. Sauberlich H. Vitamins—how much is for keeps? Nutr Today 1987;22:20–28.

2. Wolf B, Heard G, McVoy J, Grier R. Biotinidase deficiency. Ann NY Acad Sci 1985;447:252–62.

3. Bonjour J. Biotin. In: Machlin LJ., ed., Handbook of Vitamins, 2nd ed. New York: Dekker, 1991:393–427.

4. Bender D. Nutritional Biochemistry of the Vitamins. New York: Cambridge University Press, 1992:318–40.

5. Dakshinamurti K. Biotin. In: Shils ME, Olson JA, Shike M., eds., Modern Nutrition in Health and Disease, 8th ed. Philadelphia: Lea and Febiger, 1994:426–31.

6. Combs G. The Vitamins. San Diego, CA: Academic Press, 1992:329–43.

7. McCormick D. Biotin. In: Shils ME, Young VR, eds., Modern Nutrition in Health and Disease, 7th ed. Philadelphia: Lea and Febiger, 1988:436–39.

8. Knowles J. The mechanism of biotin-dependent enzymes. Ann Rev Biochem 1989;58:195–221.

9. Baez-Saldana A, Diaz G, Espinoza B, Ortega E. Biotin deficiency induces changes in subpopulations of spleen lymphocytes in mice. Am J Clin Nutr 1998; 67:431-7.

10. Mock D, Lankford G, Cazin J. Biotin and biotin analogs in human urine: biotin accounts for only half of the total. J Nutr 1993;123:1844–51.

11. Zempleni J, McCormick D, Mock D. Identification of biotin sulfone, bisnorbiotin methyl ketone, and tetra-norbiotin-1-sulfoxide in human urine. Am J Clin Nutr 1997;65:508–11.

12. National Research Council. Recommended Dietary Allowances, 10th ed. Washington, D.C.: National Academy Press, 1989:165–9.

13. Yates A, Schlicker S, Suitor C. Dietary reference intakes: the new basis for recommendations for calcium and related nutrients, B vitamins and choline. J Am Diet Assoc 1998;98:699–706.

14. Maeda Y, Kawata S, Inui Y, Fukuda K, Igura T, Matsuzawa Y. Biotin deficiency decreases ornithine transcarbamylase activity and mRNA in rat liver. J Nutr 1996;126:61–66.

Folic Acid

Folate and folacin are generic terms for compounds that have similar chemical structures and nutritional properties similar to those of folic acid, which is also called pteroylglutamate or pteroylmonoglutamate [1]. Folic acid is made up of three distinct parts, all of which must be present for vitamin activity. Figure 9.30 shows the structure of folic acid. As shown in Figure 9.30, pterin, also called *pteridine,* is conjugated to para-aminobenzoic acid (PABA) to form pteroic acid. The carboxy group of PABA is peptide-bound to the α amino group of glutamate to form folic acid. Although mammals can synthesize all the component parts of the vitamin, they do not have the enzyme necessary for the coupling of the pterin molecule to PABA to form pteroic acid [2]. In the body, metabolically active folate has multiple glutamic acid residues attached.

Sources

Good food sources of folate include mushrooms, green vegetables such as spinach, brussels sprouts, broccoli, asparagus, turnip greens, among others, as well as

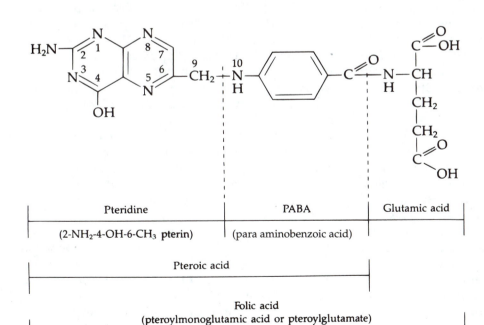

Figure 9.30 Structural formula of folic acid.

legumes (especially lima beans) and liver. Raw foods are typically higher in folate than cooked foods because of folate losses incurred with cooking. Fruits and nonorgan meats are typically poor sources of folate. Fortification of flours, grains, and cereals with folate (140 μg of folate per 100 g of product) was initiated in 1998. Folic acid from fortified grain products appears to be absorbed as well as supplemental synthetic folate [3].

Folate in foods exists primarily in the form of pteroylpolyglutamates containing up to nine glutamate residues instead of the one glutamate as shown in Figure 9.30. The principal pteroylpolyglutamates in foods are N^5 methyl tetrahydrofolate (THF), N^{10} formyl THF [4], although over 150 different forms of folate have been reported [5,6].

Folate bioavailability in individual foods varies from as high as 96% in cooked lima beans to as low as 25% in romaine lettuce [5]. Moreover, reduced forms of pteroylpolyglutamates in foods are labile and easily oxidized. It is estimated that in food preparation or processing, 50% to 95% of the folates originally present can be lost [5,6]. The effect of thermal processing on folate bioavailability depends to a great extent on the form of folate present in food. Availability of folate from food is also affected by inhibitory compounds known as conjugase inhibitors. The conjugase inhibitors prevent the digestion of folate, which is necessary for folate absorption [7].

In contrast, synthetic folate in fortified foods or in supplements, especially if consumed with an empty stomach, is more bioavailable than food folate. The difference in absorbability has led to the use of folate equivalents whereby 1 dietary folate equivalent (DFE) = 1 μg of food folate = 0.6 μg of folic acid from fortified food or supplement = 0.5 μg of synthetic folate supplement taken on an empty stomach [8].

Digestion, Absorption, Transport, and Storage

Before the polyglutamate forms of folate in foods can be absorbed, they must be hydrolyzed to the monoglutamate form. This hydrolysis is performed by the γ-glutamylcarboxypeptidases, also called *conjugases*. The group of conjugases exhibits separate activities in the human jejunal mucosa: one soluble and the other membrane bound and concentrated in the intestinal brush border [5,9]. The conjugases are also found in the pancreatic juice and bile. The brush border conjugase has been characterized; it is a zinc-dependent exopeptidase that stepwise-cleaves the polyglutamate into monoglutamate. Zinc deficiency can impair conjugase activity and diminish digestion and thus absorption of folate [2,3]. In addition, chronic alcohol ingestion can diminish conjugase activity to impair folate absorption. Conjugase inhibitors in foods such as legumes, lentils, cabbage, and oranges also prevent the digestion of polyglutamate forms of folate to the monoglutamate form. This digestion is necessary for folate to be absorbed [7].

The carrier system responsible for transporting folate across the cell membrane is believed to be saturable and pH, energy, and sodium-dependent [4,5,9]. Folate-binding proteins (FBPs) are found associated with the intestinal brush border and are believed to form a part of the transport system; the FBPs may be derived from the cellular membranes in which they serve transport functions [4,6,10]. Absorption is possible throughout the small intestine but is most efficient in the jejunum [2,7], where optimal transport occurs between a pH of 5 and 6 [7]. The folate in milk, which is bound to a high-affinity folate-binding protein, however, appears to be absorbed more avidly in the ileum. Diffusion may also account for 20% to 30% of folate absorption regardless of concentration [4]. Efficiency of folate absorption from foods is estimated at up to 50%; folate absorption from supplements is higher.

Within the intestinal cell, folate is typically reduced to THF and methylated to N^5 methyl THF or formylated. The reduction of various folate forms to THF occurs stepwise through action of NADPH-dependent dihydrofolate reductase (Fig. 9.31). Four additional hydrogens are added at positions 5, 6, 7, and 8. Most of the folate found in portal circulation is N^5 methyl THF, although dihydrofolate and formylated forms are also present.

Within the liver, about 33% of folate is present as THF, 37% of N^5 methyl THF, 23% as N^{10} formyl THF and 7% as N^5 formyl THF [2]. Bound to the N^5 methyl THF and N^{10} formyl THF are glutamates varying in length from typically 4 to 7. Dihydrofolate taken up by the liver is typically reduced and either conjugated for storage or converted to N^5 methyl THF [2]. Most of the N^5 methyl THF, along with N^{10} formyl THF, is secreted into the bile and then reabsorbed by way of enterohepatic circulation. This recirculation process may account for as much as 50% of the total folate that reaches peripheral tissues [9]. Intracellular folate-binding proteins in the liver have been identified and may serve a storage role [10].

In the blood folate is found as a monoglutamate. Almost two-thirds of the folate in blood plasma is bound to protein [9,10]; free folate accounts for the other one-third. Folate-binding proteins have been identified in plasma and within blood cells. Folate-binding protein binds folate with high affinity. Albumin and α-2 macroglobulin also bind folate, but with relatively low affinity. Monoglutamate forms of folate in the blood include THF, N^5 methyl THF, and N^{10} formyl THF, among others [5].

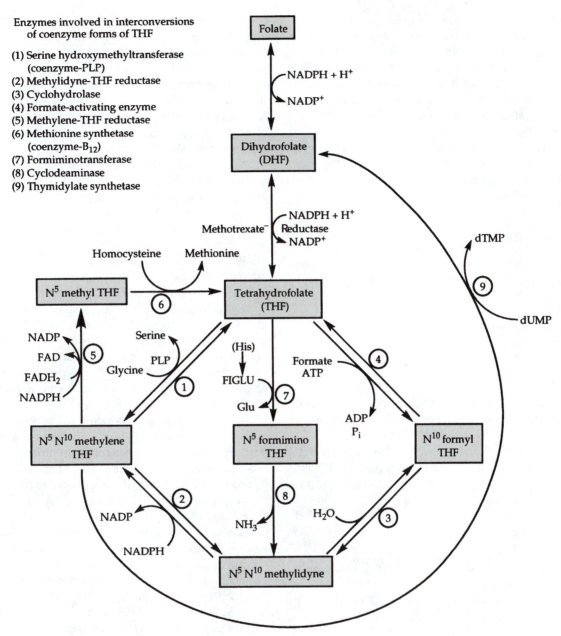

Figure 9.31 Interconversions of coenzyme forms of THF.

Folate transport into tissue cells occurs by a carrier-mediated process that may or may not require ATP. Folate-binding proteins associated with renal and hepatic cell membranes have been identified [10]. Within cells, THF is converted into a polyglutamate form to become a functional coenzyme, accepting and transferring one-carbon fragments. Intracellular demethylation is required, along with formation of polyglutamates by pteroylpolyglutamate synthetase (PPS). Five to seven glutamate residues are usually added to the monoglutamate in ATP requiring reactions. The glutamates are added to the vitamin by peptide bonds. Adding these glutamate residues not only allows the production of the various folate coenzyme forms but also traps the folate in the cell. For example, red blood cells accumulate considerable amounts of folate as polyglutamates. Total body folate levels range from 5 to 10 mg, with the liver storing most of the vitamin [11].

Functions and Mechanisms of Action

Folic acid and subsequently dihydrofolate are both reduced by dihydrofolate reductase, a cytosolic enzyme, to generate THF (Fig. 9.31). THF accepts one-carbon groups from various degradative reactions in amino acid metabolism. These THF derivatives then serve as donors of one-carbon units in a variety of synthetic reactions. The one-carbon group accepted by THF is bonded to its N^5 or N^{10} atoms or to both (Fig. 9.31). The coenzyme forms are interconvertible, except that N^5 methyl THF cannot be converted back to N^5, N^{10} methylene (Fig. 9.31).

The five THF derivatives, which participate in a variety of reactions, are illustrated as follows. The first three derivatives represent the most oxidized forms of folate, while N^5 methyl THF is the most reduced form.

N^5 formyl THF

N^{10} formyl THF

N^5 formimino THF

N^5,N^{10} methylidyne/methenyl THF

N^5,N^{10} methylene THF

N^5 methyl THF

Amino Acid Metabolism

Folate is involved in the metabolism of several amino acids including serine, glycine, methionine, and histidine.

Serine and Glycine   Folate as N^5, N^{10} methylene THF is required for serine synthesis from glycine. N^5, N^{10} methylene THF contributes a hydroxy methyl group to glycine to produce serine. Vitamin B_6 as pyridoxal phosphate (PLP) is required for serine hydroxymethyltransferase activity.

$$\text{Glycine} \xleftarrow{\quad\textit{Serine hydroxymethyltransferase}\quad} \text{Serine}$$
$$N^5, N^{10} \text{ methylene THF} \qquad \text{THF}$$

This reaction is reversible such that glycine may be synthesized from serine in a THF-dependent reaction. Serine represents a major source of one-carbon units used in folate reactions. Glycine degradation also requires THF.

$$\text{Glycine} \longrightarrow CO_2 + NH_4^+$$
$$\text{THF} \qquad N^5, N^{10} \text{ methylene THF}$$
$$NAD^+ \qquad NADH + H^+$$

Methionine   Methionine regeneration from homocysteine also involves folate as N^5 methyl THF. As shown in Figure 9.32, methionine adenosyl transferase catalyzed the conversion of methionine to S-adenosyl methionine (SAM) in an ATP-requiring reaction. SAM is involved in many methylation reactions in the body. Myelin maintenance and neural function are dependent on methylation reactions using SAM. S-adenosyl homocysteine (SAH) is generated from removal of a methyl group from SAM. Removal of the adenosyl group from SAH yields homocysteine. Remethylation of homocysteine to form methionine requires N^5 methyl THF as a methyl donor and vitamin B_{12} in the form of methylcobalamin as a prosthetic group for *homocysteine methyltransferase,* also called *methionine synthetase.* Another homocysteine methyltransferase requiring betaine, which is formed from choline metabolism, and not requiring methylcobalamin, has also been demonstrated [2].

For homocysteine methyltransferase to transfer a methyl group from N^5 methyl THF to homocysteine, cobalamin must be tightly bound to the enzyme. Cobalamin, while bound to the apoenzyme, picks up the methyl group from N^5 methyl THF to generate methylcobalamin and THF. Methylcobalamin then serves as the methyl donor for converting homocysteine to methionine.

The roles of folate and vitamin B_{12} as well as vitamin B_6 in the conversion of homocysteine to methionine have been receiving considerable attention because intakes of these three vitamins have been inversely associated with plasma homocysteine concentrations and because elevated blood homocysteine concentrations have been associated with premature coronary artery disease as well as premature occlusive vascular disease and cerebral or peripheral vascular disease [12,13]. In fact, research suggests that a 5 μmol/L increase in plasma homocysteine concentrations increases the risk for heart disease as much as a 0.5 mmol/L (20 mg/dL) increase in plasma cholesterol concentrations [13]. Supplementation (folic acid, vitamin B_{12}, and vitamin B_6) of healthy people with hyperhomocysteinemia (high blood homocysteine concentrations) for 6 weeks has been shown to normalize blood homocysteine concentrations and thus decrease a risk factor for heart disease [14]. Another study providing a placebo or supplements of folate (650 μg), vitamin B_{12} (400 μg), vitamin B_6 (10 mg), or a combination of the folate, vitamin B_{12}, and vitamin B_6 in 100 men with hyperhomocysteinemia found that folate supplementation resulted in the largest decrease in plasma homocysteine concentrations [15]. Folate supplementation reduced plasma homocysteine concentrations by 42%, and vitamin B_{12} reduced concentrations by 15%. Vitamin B_6 supplements had no significant effects on plasma homocysteine concentrations. The three vitamin combination supple-

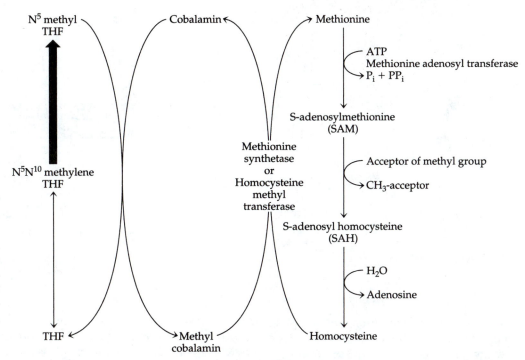

Figure 9.32 The resynthesis of methionine from homocysteine, showing the roles of folate and vitamin B_{12}.

ment resulted in a 50% reduction in plasma homocysteine concentrations, which did not significantly differ from the reduction achieved by only folate supplements [15]. These findings clearly suggest a need for folate supplementation in individuals with hyperhomocysteinemia.

Histidine Histidine metabolism also requires THF (Fig. 9.33). Deamination of histidine generates urocanic acid, which can undergo further metabolism to yield formiminoglutamate (FIGLU). The formimino is removed from FIGLU with the help of formiminotransferase to generate glutamic acid; THF receives the formimino to yield N^5 formimino THF (Fig. 9.33). This reaction, FIGLU to glutamate, can be used as a basis for determination of folate deficiency whereby subjects are given an oral histidine load and FIGLU excretion is measured in the urine. FIGLU accumulates with folate deficiency, because if THF were available FIGLU would be converted into glutamate.

Purine and Pyrimidine Synthesis

The involvement of THF derivatives in purine and pyrimidine synthesis (Figs. 7.15, 7.16) makes folate essential for cell division. Synthesis of cells with short life spans, such as enterocytes, are particularly dependent on adequate levels of folate. In pyrimidine synthesis (Fig. 7.15, p. 182), thymidylate synthase uses N^5, N^{10} methylene THF to convert dUMP to dTMP and dihydrofolate (DHF) (Fig. 9.31). dTMP is required for DNA synthesis. To regenerate N^5, N^{10} methylene THF, DHF is converted by dihydroreductase to THF in a reaction requiring NADPH. The THF is converted to N^5, N^{10} methenyl THF as serine is converted to glycine by serine hydroxymethyl transferase, a vitamin B_6–dependent reaction. Both thymidylate synthetase and dihydrofolate reductase are active enzymes in cells undergoing cell division. Inhibitors of dihydrofolate reductase, such as the chemotherapeutic drug Methotrexate, which binds to the enzyme's active site, have been employed in the treatment of cancer to prevent synthesis of THF needed for actively dividing cancer cells.

Folate as N^{10} formyl THF is needed for purine ring formation (Fig. 7.16, p. 183). C8 of the purine atom involves the formylation of glycinamide ribotide (GAR) to form formylglycinamidine ribotide (FGAR). N^{10} formyl THF donates the formyl group in this reaction. Purine ring atom C2 is acquired by formylation of 5-aminoimidazole 4-carboxamide ribonucleotide (AICAR). N^{10} formyl THF formylates AICAR to generate 5-formaminoimidazole 4-carboxamide ribotide (FAICAR).

Interactions with Other Nutrients

Ascorbic acid, with its reducing capability, has been shown to protect folate from oxidative destruction. The relationship between folate and *zinc* remains unclear.

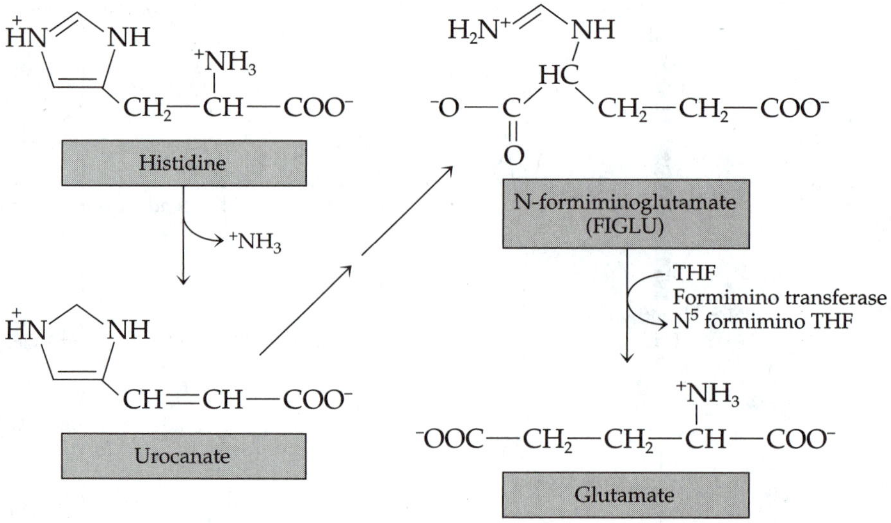

Figure 9.33 The role of folic acid in histidine catabolism.

Folate intakes of about 150 μg/day along with supplements containing 400 μg folate taken every other day have been shown to diminish zinc absorption, although dietary zinc intake averaged only 7.5 mg daily [16]. Rat studies suggest that in the presence of high concentrations of folate, folate and zinc form a complex in the intestinal lumen. This complex then inhibits zinc absorption. Zinc absorption after ingestion of 200 mg of zinc sulfate (50 mg of zinc) was diminished in adults who had been receiving 350 μg of oral folate daily [17]. Yet oral zinc (25 mg as zinc sulfate) when ingested with and without 10 mg of folate did not decrease serum zinc concentrations [18]. And, diets providing 3.5 and 14.5 mg of zinc along with 800 μg of folate for two 25-day periods did not affect folate utilization in 12 men [19]. Thus, further research concerning interactions between folate and zinc are needed.

A synergistic relationship exists between folate and *vitamin B$_{12}$*, also called cobalamin; this relationship is sometimes called the *"methyl-folate trap"* whereby without vitamin B$_{12}$ the methyl group from N^5 methyl THF can't be removed and is thus trapped. The following sequence of events leads to the methyl folate trap. Tracing the reactions shown in Figures 9.31 and 9.32 will be helpful. Serine donates single carbons through conversion to glycine, and in the process THF is converted to N^5, N^{10} methylene THF. N^5, N^{10} methylene THF is readily reduced to N^5 methyl THF by a reductase (whose activity is inhibited by its end product, N^5 methyl THF, and by SAM). N^5 methyl THF is required for methionine synthesis from homocysteine. Methyl groups are transferred by the enzyme methionine synthetase from N^5 methyl THF to vitamin B$_{12}$. Adequate vitamin B$_{12}$ must be present for the activity of methionine synthetase. The addition of the methyl group to cobalamin generates methylcobalamin, which serves as the methyl donor for converting homocysteine to methionine. Without cobalamin to accept the methyl group from N^5 methyl THF, the N^5 methyl THF accumulates, is trapped, and THF is not regenerated.

The THF, resulting from the synthesis of methionine, is important as a substrate for pteroylpolyglutamate synthetase, which adds the glutamate residues to the THF. The polyglutamate form of THF can now be used or converted into its various coenzyme forms. N^{10} formyl THF is needed for purine synthesis; N^5, N^{10} methylene THF is needed for thymidylate synthesis, which in turn must be present for DNA synthesis. Thus, the synergism between folate and vitamin B$_{12}$ is very important for support of rapidly proliferating cells.

Another relationship between folate and vitamin B$_{12}$ relates to the ability of folate when ingested in large amounts (~5 or more milligrams per day) to mask the signs of pernicious anemia, causing the vitamin B$_{12}$ neurological deficiency disorder to progress undetected [20].

Metabolism and Excretion

Under normal conditions, the body appears to hold on tenaciously to absorbed folate. Folate-binding proteins are present in the renal brush border and coupled with tubular reabsorption of folate in the kidney, very little folate is excreted in the urine.

Catabolism of folate occurs through the action of carboxypeptidase G. Two metabolites of folate are para-acetamidobenzoate and para-acetamidobenzoylglutamate, with the latter found in the greater amounts. The relative proportions of the various metabolites suggests that the principal route of catabolism occurs through oxidative cleavage of the folate molecule between positions 9 and 10 [7]. Acetylation of these compounds in the liver occurs prior to urinary excretion [7,11], although the excretion of folate metabolites is minimal.

Although much of the absorbed dietary folate is secreted by the liver into the bile, most of this folate is reabsorbed via enterohepatic recirculation, and losses in the stool are minimal [9]. Folate from microbial origin may, however, appear in the feces in relatively high amounts.

Availability of folate to crucial tissues where rapid cell division is occurring appears to be carefully regulated when the supply of dietary folate is limited. The mechanisms of regulation are unclear, but regulation seems to occur through rate of synthesis of polyglutamates. The less metabolically active tissues return monoglutamates to the liver; the liver then redistributes the folate to the actively proliferating cells. How circulating folates are directed to specific tissues is uncertain, but one possibility is that folate-binding proteins and membrane-associated binding proteins could provide tissue-specific uptake [9].

Dietary Reference Intakes and Recommended Dietary Allowances

The minimal daily folate requirement is estimated at 50 μg based on intravenous administration, whereas 100 μg of dietary folate is thought to prevent folate deficiency [1,7]. The 1989 RDA of 3 μg/kg body weight or 200 μg and 180 μg for adult males and females, respectively, has been suggested as being inadequte [21–23]. Sauberlich et al [22] suggest nonpregnant women need 200 to 250 μg of folate/day to restore or maintain normal plasma folate concentrations.

The Centers for Disease Control (CDC) and Prevention [24] recommend 400 μg of folate/day for women capable of becoming pregnant because of the accumulating evidence that folate supplementation during periconceptional period of pregnancy may reduce the incidence of neural tube defects.

The 1998 DRI RDA for folate consider its bioavailability and is consistent with the CDC recommendations. The DRI RDA for adults for folate is 400 μg/day as dietary folate equivalents (DFE) [8]. An additional 200 and 100 μg of folate per day are added for pregnancy and lactation, respectively [8]. One DFE is equal to 1 μg of food folate, which is equal to 0.6 μg of folate from a supplement, or fortified food consumed with a meal, which is equal to 0.5 μg of folate from a supplement taken without food (empty stomach) [8]. A tolerable upper intake level for adults aged 19 years and older of 1,000 μg for synthetic folate has been suggested based on the ability of folate to mask the neurological manifestations of vitamin B_{12} deficiency [8].

Deficiency: Megaloblastic, Macrocytic Anemia

Marginal folate deficiency is characterized initially by low plasma folate and hypersegmentation of polymorphonuclear leukocytes. Red blood cell folate concentrations diminish after about 4 months of low folate intake [25]. After approximately 4 to 5 months, bone marrow cells become megaloblastic and anemia occurs [25].

Megaloblastic anemia—the release into circulation of large immature erythrocytes—due to folate deficiency is relatively common in the United States. However, megaloblastic anemia also occurs because of a deficiency of vitamin B_{12}. The anemia results from decreased DNA synthesis and failure of the cells to divide properly, coupled with the continued formation of RNA. The quantity of RNA becomes greater than normal, leading to excess production of other cytoplasmic constituents, including hemoglobin. The result is immature, enlarged cells often containing excessive hemoglobin.

A review of the formation and maturation of erythrocytes is given in Figure 9.34 and may help to better illustrate the effects of a folate and vitamin B_{12} deficiency. Briefly, the proerythroblast develops from stem cells in bone marrow under the stimulation of hypoxia (low blood oxygen) via erythropoietin (a hormone produced in the kidney). In the proerythroblast, active DNA and RNA synthesis occur and cell division begins. Cells resulting from first division are termed basophilic erythroblasts because they stain with basic dyes due to the many organelles present within the cell. During this stage, hemoglobin synthesis begins. The next generation of cells consists of the polychromatophil erythroblasts in which hemoglobin synthesis intensifies. The concentration of hemoglobin influences DNA synthesis and cell division. Cell division usually continues into the orthochromatic stage. The orthochromatic erythrocytes are characterized by continued hemoglobin synthesis, discontinuation of DNA synthesis, a slowing of RNA synthesis, and migration of the nucleus to the cell wall in preparation for extrusion. The cell now becomes the reticulocyte in which hemoglobin synthesis continues up to a concentration of approximately 34%. Once this concentration is reached, the ribosomes disappear, and the cells pass into blood capillaries by squeezing through pores of the membrane. In about 2 to 3 days, when the rest of the cell organelles have disappeared, reticulocytes become erythrocytes. The erythrocyte, or mature red blood cell, is all cytoplasm packed with he-

Genesis of RBC

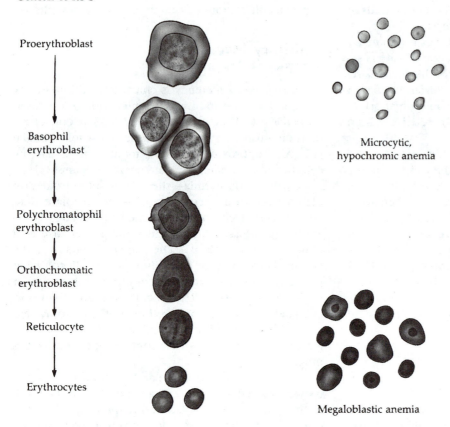

Proerythroblast

Basophil
erythroblast

Polychromatophil
erythroblast

Orthochromatic
erythroblast

Reticulocyte

Erythrocytes

Microcytic,
hypochromic anemia

Megaloblastic anemia

Figure 9.34 Genesis and maturation of the red blood cell; red blood cells characteristic of microcytic and megaloblastic anemias.

moglobin. Glycolysis and hexose monophosphate shunt are the only metabolic pathways occurring in the erythrocyte. A deficiency of folate and/or vitamin B_{12} interferes with normal cell division. Large, malformed, and sometimes nucleated, red blood cells result.

Other symptoms of folate deficiency are neuropsychiatric in nature. Depression, dementia, and peripheral neuropathy as well as megaloblastic anemia have been reported with folate deficiency [26]. Poor response to antidepressant agents also seems to be associated with impaired folate status [27].

Increased risk of heart disease is associated with hyperhomocysteinemia, which appears to be related to poor folate status, as well as vitamins B_{12} and B_6. The relationship between folate and homocysteine concentrations has been previously addressed on page 292.

Folate deficiency is also suspected in the development (initiation) of cancer. Folate deficiency in cells and tissues is thought to increase the potential for neoplastic changes in normal cells during the early stages of cancer [25,28].

Some conditions and populations associated with increased need for folate intake include the elderly and people with excessive alcohol ingestion, achlorhydria, inflammatory bowel diseases, malignancies, oral con-

traceptive users, as well as pregnant and lactating women. Folate deficiency has been observed in people treated with diphenylhydantoin or phenytoin, anticonvulsants used in the treatment of epilepsy. Folate and phenytoin inhibit the cellular uptake of one another in the gastrointestinal tract and possibly in the brain [1,29]. However, although folate supplements of 5 to 30 mg/day corrected the hematologic signs associated with the folate deficiency, seizure activity increased in some patients [7]. Other drugs, such as cholestyramine, used to treat high cholesterol concentrations, and sulfasalazine, used to treat inflammatory bowel diseases, have also been shown to interact with folate to create potential folate deficiency. Malabsorption of folate occurs with inflammatory bowel diseases and excessive alcohol ingestion.

Toxicity

Folate supplements in amounts >0.4 mg daily are considered to be pharmacological doses [7]. Studies suggest no adverse effects of folate supplements, 400 mg/day for 5 months, 10 mg for 4 months, or 10 mg/day for 5 years, in adults [1,7]. Toxicity of oral folic acid in moderate doses is reportedly virtually nonexistent [20,30]. Other studies, however, indicate that folate intakes up to 15 mg daily are problematic. Problems include insomnia, malaise, irritability, diminished zinc status, and gastrointestinal distress [29,31]. Folate supplementation at levels of ≥5,000 μg can mask a vitamin B_{12} deficiency [32]. Folate supplements alleviate the megaloblastic anemia due to a vitamin B_{12} deficiency, while the neurological damage due to a vitamin B_{12} deficiency progresses undetected. A no observed adverse effect level (NOAEL) and lowest observed adverse effect level (LOAEL) for folate are 1,000 μg and 5,000 μg, respectively [32].

Assessment of Nutriture

Folate status is most often assessed through measurement of folate levels in the plasma, serum, or red blood cells [7]. Serum or plasma folate levels reflect recent dietary intake; thus, true deficiency must be interpreted through repeated measures of serum or plasma folate. Red blood cell folate levels, more reflective of folate tissue status than serum folate, represent vitamin status at the time the red blood cell was synthesized [1,21,33]. Red blood cell folate may indicate liver folate stores. A low red blood cell folate may, however, occur with a vitamin B_{12} deficiency [33].

	Folate Status		
	Deficient	Marginal	Adequate
Serum folate (ng/mL)	<3	3–6	>6
Red blood cell folate (ng/mL)	<140	140–160	>160

The deoxyuridine suppression test, another method to assess folate status, measures the availability of folate for de novo thymidine synthesis. In this test, the activity of thymidylate synthetase is measured in cultured lymphocytes or bone marrow cells. The reaction catalyzed by thymidylate synthetase is dependent on folate and indirectly on vitamin B_{12}; therefore, the change in activity elicited by the addition of one or the other vitamin allows identification of the deficiency. In case of a deficiency of both vitamin B_{12} and folate, normalization of enzyme activity would be possible only after the addition of both vitamins.

N-formiminoglutamate (FIGLU) excretion may also be used to measure folate nutriture, because folate as THF must be available for the formimino group to be removed from FIGLU and glutamate to be formed (see Fig. 9.33). FIGLU excretion is measured in a 6-hour urine collection after ingestion of 2 to 5 g of oral L-histidine. Normal FIGLU excretion is about 5 to 20 mg, whereas with folate deficiency FIGLU excretion is 5 to 10 times above normal. A deficiency of vitamin B_{12}, however, will also cause an elevated FIGLU excretion.

Deoxyuridine Suppression Test (Thymidylate Synthetase Activity)[34]

	Deficiency		
	B_{12}	Folate	Folate and B_{12}
Add B_{12}	Normalized	Abnormal	Abnormal
Add folate	Abnormal	Normalized	Abnormal
Add B_{12} and folate	Normalized	Normalized	Normalized

Source: Simko MD, Cowell C, Gilbride JA. Nutrition Assessment, 2nd ed. Rockville, MD: Aspen Publishers, 1995. Reprinted with permission of Aspen Publishers, Inc.

References Cited for Folic Acid

1. National Research Council. Recommended Dietary Allowances, 10th ed. Washington, D.C.: National Academy Press, 1989;150–8.

2. Bender D. Nutritional Biochemistry of the Vitamins. New York: Cambridge University Press, 1992:269–317.

3. Pfeffer C, Rogers L, Bailey L, Gregory J. Absorption of folate from fortified cereal grain products and of supplemental folate consumed with or without food determined using a dual label stable isotope protocol. Am J Clin Nutr 1997;66:1388–97.

4. Coombs G. The Vitamins. San Diego, CA: Academic Press, 1992:357–76.

5. Sauberlich H. Bioavailability of vitamins. Prog Food Nutr Sci 1985;9:1–33.

6. Sauberlich H. Vitamins—how much is for keeps? Nutr Today 1987;22:20–28.

7. Brody T. Folic acid. In: Machlin LJ., ed. Handbook of Vitamins, 2nd ed. New York: Dekker, 1991:453–89.

8. Yates A, Schlicker S, Suitor C. Dietary reference intakes: the new basis for recommendations for calcium and related nutrients, B vitamins and choline. J Am Diet Assoc 1998;98:699–706.

9. Steinberg S. Mechanisms of folate homeostasis. Am J Physiol 1984;246:G319–24.

10. Wagner C. Cellular folate binding proteins: function and significance. Ann Rev Nutr 1982;2:229–48.

11. Herbert V, Das K. Folic acid and vitamin B_{12}. In: Shils ME, Olson JA, Shike M., eds. Modern Nutrition in Health and Disease, 8th ed. Philadelphia: Lea and Febiger, 1994:402–25.

12. Shimakawa T, Nieto F, Malinow M, Chambless L, Schreiner P, Szklo M. Vitamin intake: a possible determinant of plasma homocyst(e)ine among middle-aged adults. Ann Epidemiol 1997;7:285–93.

13. Boushey C, Beresford S, Omenn G, Motulsky A. A quantitative assessment of plasma homocysteine as a risk factor for vascular disease: probable benefits of increasing folic acid intakes. JAMA 1995;274:1049–57.

14. Ubbink J. Vitamin B_{12}, vitamin B_6, and folate nutritional status in men with hyperhomocysteinemia. Am J Clin Nutr 1993;57:47–53.

15. Ubbink J, Vermaak W, Merwe A, Becker P, Delport AR, Potgieter H. Vitamin requirements for the treatment of hyperhomocysteinemia in humans. J Nutr 1994;124:1927–33.

16. Milne D, Canfield W, Mahalko J, Sandstead H. Effect of oral folic acid supplements on zinc, copper, and iron absorption and excretion. Am J Clin Nutr 1984;39:535–9.

17. Simmer K, Iles C, James C, Thompson R. Are iron-folate supplements harmful? Am J Clin Nutr 1987;45:122–5.

18. Keating J, Wada L, Stokstad E, King J. Folic acid: effect on zinc absorption in humans and in the rat. Am J Clin Nutr 1987;46:835–9.

19. Kauwell G, Bailey L, Gregory J, Bowling D, Cousins R. Zinc status is not adversely affected by folic acid supplementation and zinc intake does not impair folate utilization in human subjects. J Nutr 1995; 125:66–72.

20. Butterworth C, Tamura T. Folic acid safety and toxicity: a brief review. Am J Clin Nutr 1989;50:353–8.

21. Herbert V. Recommended dietary intakes (RDI) of folate in humans. Am J Clin Nutr 1987;45:661–70.

22. Sauberlich H, Kretsch M, Skala J, Johnson H, Taylor P. Folate requirement and metabolism in nonpregnant women. Am J Clin Nutr 1987;46:1016–28.

23. Bailey L. Evaluation of a new recommended dietary allowance for folate. J Am Diet Assoc 1992; 92:463–8,471.

24. Centers for Disease Control and Prevention. Recommendations for the use of folic acid to reduce the number of cases of spina bifida and other neural tube defects. MMWSR 1992;41:1–7.

25. Hine R. Folic acid: Contemporary clinical perspective. Persp Appl Nutr 1993;1:3–14.

26. Shorvon S, Carney M, Chanarin I, Reynolds E. The neuropsychiatry of megaloblastic anemia. Brit Med J 1980;281:1036–8.

27. Alpert J, Fava M. Nutrition and depression: the role of folate. Nutr Rev 1997;55:145–9.

28. Folate, alcohol, methionine, and colon cancer risk: is there a unifying theme? Nutr Rev 1994;2:18–20.

29. Alhadeff L, Gualtieri C, Lipton M. Toxic effects of water-soluble vitamins. Nutr Rev 1984;42:33–40.

30. Krumdieck C. Folic Acid. In: Brown ML, ed., Present Knowledge in Nutrition, 6th ed. Washington, D.C.: The Nutrition Foundation, 1990:179–88.

31. Zimmerman M, Shane B. Supplemental folic acid. Am J Clin Nutr 1993;58:127–8.

32. Hathcock J. Vitamins and minerals: efficacy and safety. Am J Clin Nutr 1997;66:427–37.

33. Bailey L. Folate status assessment. J Nutr 1990; 120:1508–11.

34. Simko M, Cowell C, Gilbride J. Nutrition assessment. Rockville, MD: Aspen, 1995.

Vitamin B_{12} (Cobalamins)

Vitamin B_{12} is considered a generic term for a group of compounds called *corrinoids* because of their corrin nucleus. The corrin is a macrocyclic ring made of four reduced pyrrole rings linked together. The corrin of vitamin B_{12} has an atom of cobalt in the center of it to which is attached, at almost right angles, a nucleotide, 5,6-dimethylbenzimidazole. Also attached to the cobalt atom in vitamin B_{12} is one of the following:

Group Attached	Resulting Compound
— CN	Cyanocobalamin
— OH	Hydroxocobalamin
— H_2O	Aquocobalamin
— NO_2	Nitritocobalamin
5'-deoxyadenosyl	5'-deoxyadenosylcobalamin
— CH_3	Methylcobalamin

Cyanocobalamin is shown in Figure 9.35. Only two cobalamins, 5'-deoxyadenosylcobalamin (subsequently called *adenosylcobalamin*) and methylcobalamin, are active as coenzymes. The human body has the biochemical ability to convert most of the other cobalamins into an active coenzyme form of the vitamin.

Sources

The only dietary sources of vitamin B_{12} for humans are from animal products, which have derived their cobalamins from micro-organisms. All naturally occurring vitamin B_{12} is produced by micro-organisms. Any vitamin B_{12} found in plant foods could probably be traced to contamination with micro-organisms contained in manure or, in the case of legumes, to the presence of nitrogen-fixing bacteria in the plant root nodules [1]. Contaminated hands taking foods to the mouth may also provide vitamin B_{12}.

The best sources of the cobalamins are meat and meat products, poultry, fish, shellfish (especially clams and oysters), and eggs (especially the yolk); the cobalamins in these products are predominantly adenosyl- and hydroxocobalamin. Milk and milk products such as cheese and yogurt contain less of the vitamin, mainly as methyl- and hydroxocobalamins [2,3]. Cyanocobalamin

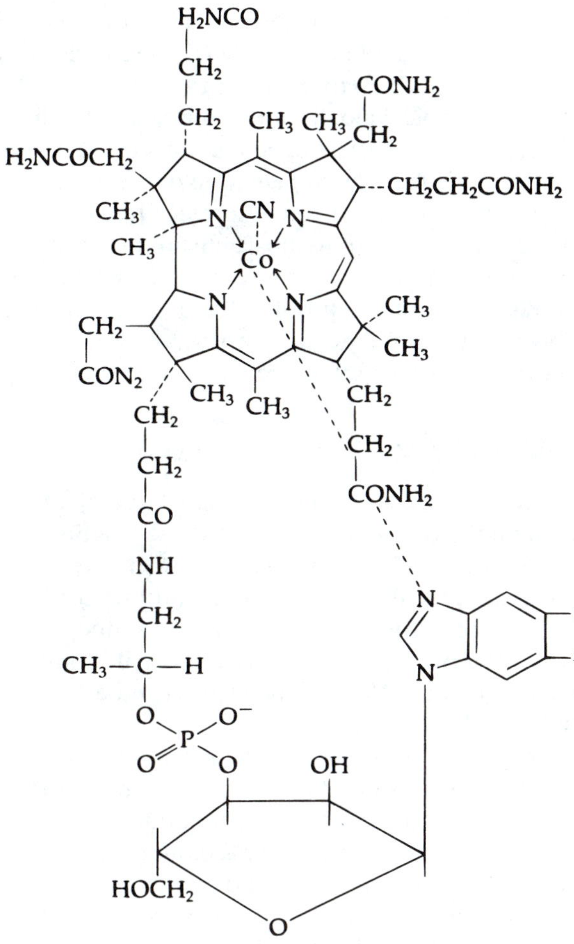

Figure 9.35 Structural formula of vitamin B_{12} (cyanocobalamin).

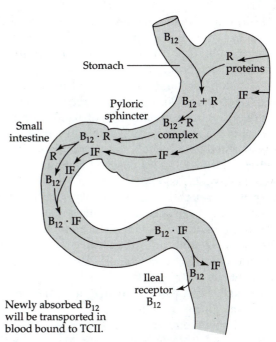

Newly absorbed B_{12} will be transported in blood bound to TCII.

Figure 9.36 Vitamin B_{12} absorption.

may be found in a few foods as well as tobacco; it is also the form, along with hydroxocobalamin [4], that is commercially available in, for example, vitamin preparations. Within the body, cyanocobalamin is converted to aquo- or hydroxocobalamin.

Bioavailability of vitamin B_{12} may be impaired by vitamin C. Vitamin C in doses of 500 mg or more, taken with meals or up to 1 hour after a meal, may diminish vitamin B_{12} availability from food or destroy the vitamin [5–7]. Furthermore, incubation of 200 mg of ascorbate with 25 μg of vitamin B_{12}, 400 μg of folate, and 15 mg of iron for 30 minutes in gastric juice (pH 5) resulted in a significant destruction of vitamin B_{12} [8].

Digestion, Absorption, Transport, and Storage

The digestion and absorption of vitamin B_{12} is believed to proceed according to the scheme depicted in Figure 9.36. Ingested cobalamins must first be released from the polypeptides to which they are linked in foods. This release usually occurs through the action of the gastric proteolytic enzyme pepsin in the stomach.

Once released from foods, vitamin B_{12} absorption involves contact with two proteins, intrinsic factor (IF) and R proteins. IF is a glycoprotein synthesized by the gastric parietal cells. Although it is made and released in the stomach, IF functions in the small intestine. R proteins, known collectively as *cobalophilins* or *haptocorrins* (HCs), are found in most body fluids, including saliva and gastric juice [1,6,9]. R proteins have a high affinity for cobalamins. Free cobalamin released from food combines with R protein; the complex moves from the stomach into the small intestine. Within the duodenum, the R protein is hydrolyzed by pancreatic proteases, and free cobalamin is released. A pancreatic insufficiency could interfere with the release of cobalamin from the R protein and reduce the amount of the vitamin available for absorption. R proteins may also serve to protect vitamin B_{12} from bacterial use [7,10].

In the proximal small intestine, IF, which escapes the catabolic action of the proteases, binds the cobalamin, any of the forms, once released from the R proteins. The cobalamin-IF complex travels to the ileum, where receptor sites for vitamin B_{12} are present. Absorption of the vitamin occurs throughout the entire ileum, especially the distal third [1].

The cobalamin absorption process is complex and poorly understood. It is known that calcium is needed for absorption to occur, and presently the belief is that calcium has some sort of specific action on the receptor site [6]. Whether cellular uptake of IF occurs with the vitamin is unclear.

Absorption of vitamin B_{12} is slow; after attachment of the IF-cobalamin complex to the receptor, there is a delay of 3 to 4 hours before the cobalamin appears in circulation. Peak levels of the vitamin in the blood may not be reached for 8 to 12 hours after ingestion [4].

When pharmacological doses of vitamin B_{12} are ingested, passive diffusion can account for much of the absorption throughout the intestinal tract. However, passive diffusion can account for only about 1% and up to 3% of the total absorption of the vitamin when it is being obtained from ordinary dietary sources [7,10]. Absorption rate of the vitamin decreases with increased intake. At low levels of intake (0.1 mg), absorption averages 80%, whereas at higher intakes, the absorption rate drops to 3% [4].

Enterohepatic circulation is very important in vitamin B_{12} nutriture, accounting in part for the long biological half-life of cobalamin. When enterohepatic circulation is effective, much of the cobalamin in the bile and in other intestinal secretions can be reabsorbed. Malabsorption syndromes not only cause a decrease in absorption of ingested cobalamin but also interfere with enterohepatic circulation, thereby increasing the amount of vitamin B_{12} required to meet body needs.

Following intestinal absorption, cobalamins bind to one of three transcobalamins (TC), designated as TCI, TCII, or TCIII in the blood. It is not known whether attachment to TC occurs within the enterocyte or at the serosal surface. The transcobalamins also are considered R proteins. The exact functions of TCI and TCIII are unknown. TCIII may function in the delivery of cobalamin from peripheral tissues back to the liver [11]. About 90% of vitamin B_{12} is bound to TCI, which may function as a circulating storage form of the vitamin [11].

TCII is the main protein that carries, in a one-to-one ratio, newly absorbed cobalamin to the tissues. TCII is important for normal cobalamin metabolism and is thought to be synthesized in the liver [12]. In the blood, methylcobalamin comprises about 60% to 80%, and adenosylcobalamin may account for up to 20% of total plasma cobalamin. Other forms of cobalamin in the blood include cyanocobalamin and hydroxocobalamin.

TCII also assists uptake of the vitamin by tissues; all tissues appear to have receptors for TCII. Cobalamin, along with TC, are taken up by endocytosis. Within the lysosome, TCII is degraded and hydroxocobalamin is released. This form of the vitamin may undergo cytosolic methylation to generate methylcobalamin or may undergo reduction and subsequent reaction with ATP in the mitochondria to yield adenosylcobalamin [1].

Vitamin B_{12}, unlike other water-soluble vitamins, can be stored and retained in the body for long periods of time, even years. The vitamin is stored mainly in the liver; however, small amounts are also found in the muscle, bone, kidneys, heart, brain, and spleen. Adenosylcobalamin is the primary storage form of the vitamin in the liver, and possibly in other organs; however, hydroxocobalamin and methylcobalamin are also stored to a lesser extent [4]. Haptocorrin represents the circulating storage form of vitamin B_{12} that is in equilibrium with body stores of the vitamin. Hepatocytes have receptors for the uptake of both haptocorrin and for TCII.

Functions and Mechanisms of Action

Three enzymatic reactions requiring vitamin B_{12} have been recognized in humans; one of these reactions requires methylcobalamin, while the other two must have adenosylcobalamin. Adenosyl- and methylcobalamin are formed by a complex reaction sequence resulting in the production of a carbon-cobalt bond between the cobalt nucleus of the vitamin and either the methyl or 5'-deoxyadenosyl ligand.

The reaction requiring methylcobalamin as a coenzyme is the conversion of homocysteine into methionine (Fig. 9.32). This reaction occurs in the cytoplasm of the cell. To form the methylcobalamin needed in methionine synthesis, cobalamin, bound to the methionine synthetase (homocysteine methyl transferase) apoenzyme, picks up the methyl group from N^5 methyl THF and transfers it to homocysteine, thereby producing methionine and free THF. THF can then be converted into any of its coenzyme forms. This reaction explains in large part the synergism between folate and vitamin B_{12}. Because the formation of N^5 methyl THF is irreversible, a vitamin B_{12} deficiency traps body folate in the methyl form. This is known as the folate methyl trap hypothesis (see p. 295).

Nitrous oxide has been shown to inhibit the activity of methionine synthetase; it reacts with cobalamin, converting the cobalt from a +1 to a +3 oxidation state. Vitamin B_{12}, to function as methylcobalamin coenzyme, must contain cobalt present in its reduced state, +1 [1]. Individuals who are vitamin B_{12}–deficient may exhibit deterioration of nervous system function following nitrous oxide anesthesia [13,14].

Two reactions require adenosylcobalamin (Figs. 9.37 and 9.38). These reactions are catalyzed by mutases and occur in the mitochondria. First, adenosylcobalamin is needed for methylmalonyl CoA mutase, which converts L-methylmalonyl CoA to succinyl CoA (Fig. 9.37). L-methylmalonyl CoA is generated from propionyl CoA. Propionyl CoA, which arises from the oxidation of methionine, isoleucine, threonine, and odd-chain fatty acids, is converted into D-methylmalonyl CoA in an ATP-, Mg^{2+}-, and biotin-dependent reaction (Fig. 9.37). L-methylmalonyl CoA is made from the D form through the action of racemase. Methylmalonyl CoA mutase requires adenosylcobalamin to convert L-methylmalonyl CoA to succinyl CoA, the Krebs cycle intermediate (Fig. 9.37). With a deficiency of vitamin B_{12}, mutase activity is impaired, and methylmalonyl CoA and methylmalonic acid, formed from hydrolysis of methylmalonyl CoA, accumulate in body fluids. Genetic defects in methylmalonyl CoA mutase and adenosylcobalamin synthesis have also been demonstrated and result in the accumulation of methylmalonyl CoA and methylmalonic acid.

Leucine aminomutase also requires adenosylcobalamin (Fig. 9.38). This enzyme isomerizes L-leucine and β-leucine. β-leucine generated from intestinal bacteria may be converted to L-leucine within the body [11]. Alternately, β-leucine generated from L-leucine may undergo subsequent transamination in a vitamin B_6 (PLP)–dependent reaction and provide an alternate pathway for leucine catabolism.

Metabolism and Excretion

Very little evidence exists to support any extensive degradation of cobalamin. Whole-body turnover of vitamin B_{12} is approximately 0.1%/day and loss of the vitamin is due primarily to fecal excretion, not catabolism [12]. Most of cobalamin excretion occurs via the bile. Little urinary excretion of vitamin B_{12} occurs.

Dietary Reference Intakes and Recommended Dietary Allowances

Measuring with accuracy the vitamin B_{12} body pool size and determining what amount of the vitamin constitutes an ideal pool have not been accomplished. Nevertheless, studies of vitamin B_{12} nutriture and turnover rate in healthy subjects have led to the belief that 1 µg/day can be expected to sustain normal people [2,12]. The 1980 RDA for vitamin B_{12} was set at 3 µg/day and allowed maintenance of an upper-limit body pool size [15]. The 1980–1985 RDA Committee formulated a RDI of 2 µg/day for vitamin B_{12}. This lower level was chosen because no proven advantage has been associated with maintenance of a higher-than-normal body pool size [2,12]. Two micrograms daily was also chosen for the 1989 RDA for vitamin B_{12} [2].

The DRI RDA recommended intake level for adults aged 19 years and older is 2.4 µg of vitamin B_{12} per day [16]. Increases of 0.2 and 0.4 µg/day are suggested during pregnancy and lactation, respectively [16]. It is further suggested that individuals aged 51 years and older consume foods fortified with the vitamin or consume supplements containing vitamin B_{12}; this recommendation is based on the fact that about 10% to 30% of older individuals do not absorb food-bound forms of the vitamin [16].

Deficiency: Megaloblastic, Macrocytic Anemia

Inadequate absorption of the vitamin rather than inadequate dietary intake is responsible for the majority of

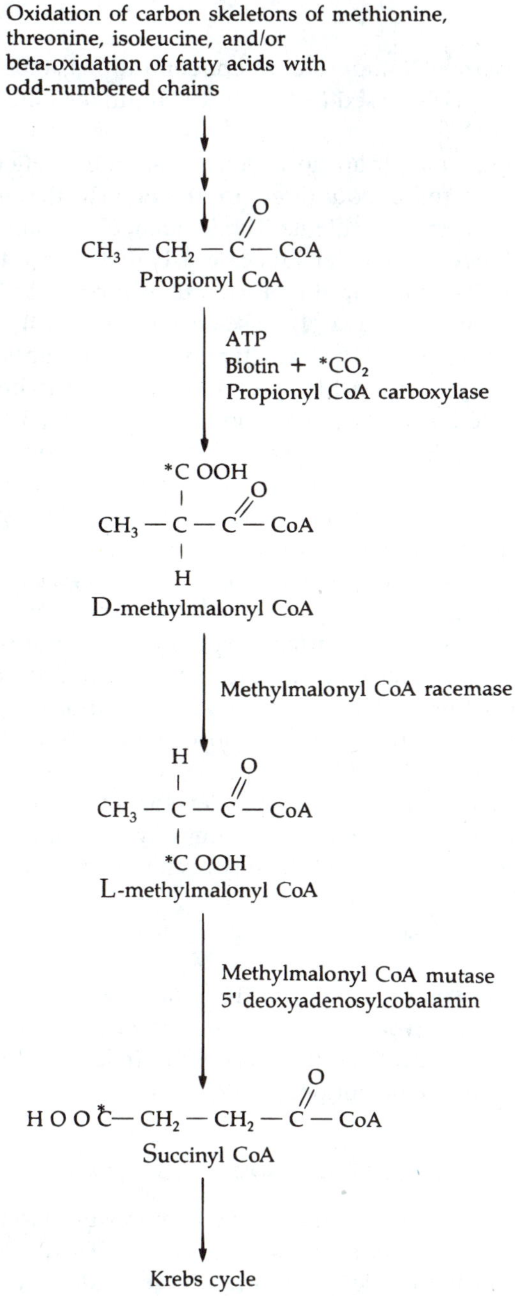

Figure 9.37 Role of vitamin B_{12} in oxidation of odd-numbered chain fatty acids and selected amino acids.

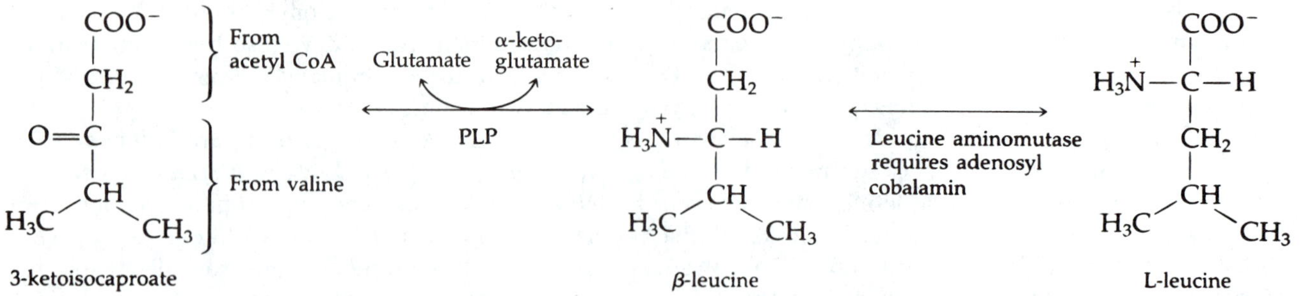

Figure 9.38 Isomerization of leucine by a vitamin B$_{12}$-dependent enzyme.

the vitamin B$_{12}$ deficiency seen in the United States. Although a strict vegetarian diet can produce a deficiency of the vitamin, clinical symptoms may not appear for up to 20 to 30 years on such a diet [6]. An exception would be the infant and/or very young child maintained on unfortified foods of plant origin; vitamin B$_{12}$-fortified tofu provides an excellent source of the vitamin.

Vitamin B$_{12}$ deficiency occurs in stages. Initially, serum concentrations diminish as indicated by low holotranscobalamin II. Second, cell concentrations of the vitamin diminish. Third, biochemical deficiency occurs, as evidenced by decreased DNA synthesis and by elevated homocysteine and methylmalonic acid concentrations in the serum. Finally, anemia occurs [17].

The megaloblastic anemia associated with a vitamin B$_{12}$ deficiency is detailed under folic acid deficiency (p. 295) because deficiencies of both vitamins result in megaloblastic anemia. In fact, the megaloblastic anemia due to a vitamin deficiency can be corrected with large doses of folate [18]. The neuropathy, characterized by demyelination of nerves, caused by a lack of vitamin B$_{12}$ is not responsive to folate therapy, however. The cause of the neuropathy may be related to the availability of methionine [6,19]. The neuropathy can be ameliorated through increased exogenous methionine or an accelerated production of methionine from homocysteine, a reaction that requires vitamin B$_{12}$. An inadequate amount of methionine caused by a deficiency of vitamin B$_{12}$ decreases the availability of S-adenosylmethionine (SAM). SAM is required for methylation reactions, essential to the myelin maintenance and thus neural function. SAM deficiency in the nervous system (i.e., cerebrospinal fluid) has been suggested in the pathogenesis of cobalamin neuropathy [18]. In addition, plasma vitamin B$_{12}$ concentrations have been reported to be inversely associated with plasma homocysteine concentrations [20–25]. Elevated plasma homocysteine concentrations are thought to play a role in early-onset ($\leq$50 years) coronary heart disease, and may be lowered with vitamin B$_{12}$ and/or folate supplements [20,24,25]. This relationship between vitamin B$_{12}$, fo-late, plasma homocysteine concentrations, and heart disease is discussed in further detail under folic acid, page 292.

Population groups in which a vitamin B$_{12}$ deficiency is most often encountered are the elderly, alcoholics, and gastrectomy patients. The incidence of vitamin B$_{12}$ deficiency in the elderly may be as high as 15%, and the vitamin B$_{12}$ content of multivitamin preparations is not sufficient to adequately raise serum vitamin B$_{12}$ concentrations [26,27]. Oral vitamin B$_{12}$ in amounts of at least 6 µg and possibly up to 300 µg appear to be necessary to correct deficiency in the elderly [27]. In most groups with vitamin B$_{12}$ deficiency, absorption is usually impaired owing to pernicious anemia, atrophic gastritis, or hypochlorhydria. Other conditions and populations associated with increased need for intake include those with a lack of IF secretion (gastrectomy and destruction of gastric mucosa), those with decreased absorptive surface (blind loop syndrome associated with intestinal surgery, ileal resection, celiac and tropical sprue, ileitis, Zollinger-Ellison syndrome), and those who may ingest a diet low in vitamin B$_{12}$ (strict vegetarians). In addition, people with parasitic infections such as tapeworms may develop a vitamin B$_{12}$ deficiency due to use of the vitamin by the parasite and consequent limited availability to the infected person.

Toxicity

Although no clear toxicity from massive doses of vitamin B$_{12}$ has ever been recorded, neither has there been noted any benefit from an excessive intake of the vitamin by nondeficient people [2].

Assessment of Nutriture

Serum B$_{12}$ concentrations are commonly used to assess nutriture. Concentrations in the serum of <100 pg/mL are considered deficient and are generally accompanied by metabolic evidence of deficiency despite the absence of clinical manifestations [26]. Red blood cell vitamin B_{12} concentrations can also be used but are less specific, decreasing with both cobalamin and folate deficiencies. Another test, the deoxyuridine suppression test, used to assess vitamin B_{12} nutriture has been discussed previously under "Assessment of Nutriture" in the folic acid section (p. 297).

The Schilling test may be used to determine problems of vitamin B_{12} absorption (i.e., IF insufficiency). The test involves oral administration of radioactive vitamin B_{12}. Urinary excretion of the vitamin is measured. Below-normal urinary excretion of the vitamin suggests impaired absorption. Elevated concentrations of methylmalonic acid in the urine are also used to detect cobalamin deficiency.

The use of serum holotranscobalamin (TCI and TCII) as an indicator of cobalamin deficiency also has been proposed [28,29]. However, further investigation is required to valid its use [30,31].

References Cited for Vitamin B_{12}

1. Seatharam B, Alpers D. Absorption and transport of cobalamin (vitamin B_{12}). Ann Rev Nutr 1982;2:343–69.

2. National Research Council. Recommended Dietary Allowances, 10th ed. Washington, D.C.: National Academy Press, 1989;158–65.

3. Sandberg D, Begley J, Hall C. The content, binding and forms of vitamin B_{12} in milk. Am J Clin Nutr 1981;34:1717–24.

4. Ellenbogen L, Cooper BA. Vitamin B_{12}. In: Machlin LJ., ed. Handbook of Vitamins, 2nd ed. New York: Dekker, 1991:491–536.

5. Herbert V. Vitamin B_{12}. In: Brown ML, ed. Present Knowledge in Nutrition, 6th ed. Washington, D.C.: The Nutrition Foundation, 1990:170–8.

6. Davis R. Clinical chemistry of vitamin B_{12}. Adv Clin Chem 1984;24:163–216.

7. Sauberlich H. Bioavailability of vitamins. Prog Food Nutr Sci 1985;9:1–33.

8. Herbert V. Anti-hyperhomocysteinemic supplemental folic acid and vitamin B_{12} are significantly destroyed in gastric juice if co-ingested with supplemental vitamin C and iron. Blood 1996;88 (suppl 1):492a.

9. Herbert V. Recommended dietary intakes (RDI) of vitamin B_{12} in humans. Am J Clin Nutr 1987;45:671–8.

10. Sauberlich H. Vitamins—how much is for keeps? Nutr Today 1987;22:20–28.

11. Bender D. Nutritional Biochemistry of the Vitamins. New York: Cambridge University Press, 1992:269–317.

12. Herbert V, Das K. Folic acid and vitamin B_{12}. In: Shils ME, Olson JA, Shike M., eds. Modern Nutrition in Health and Disease, 8th ed. Philadelphia: Lea and Febiger, 1994:402–25.

13. Metz J. Cobalamin deficiency and the pathogenesis of nervous system disease. Ann Rev Nutr 1992;12: 59–79.

14. Flippo T, Holder W. Neurologic degeneration associated with nitrous oxide anesthesia in patients with vitamin B_{12} deficiency. Archives Surg 1993;128: 1391–5.

15. National Research Council. Recommended Dietary Allowances, 9th ed. Washington, D.C.: National Academy of Sciences, 1980;113–20.

16. Yates A, Schlicker S, Suitor C. Dietary reference intakes: the new basis for recommendations for calcium and related nutrients, B vitamins and choline. J Am Diet Assoc 1998;98:699–706.

17. Herbert V. Staging of vitamin B_{12} (cobalamin) status in vegetarians. Am J Clin Nutr 1994;59(suppl): 1213S–22S.

18. Council on Scientific Affairs, American Medical Association. Vitamin preparations as dietary supplements and as therapeutic agents. JAMA 1987;257: 1929–36.

19. Metz J. Pathogenesis of cobalamin neuropathy: deficiency of nervous system S-adenosylmethionine. Nutr Rev 1993;51:12–15.

20. Pancharuniti N, Lewis C, Sauberlich H, Perkins L, Go R, Alvarez J, Masaluso M, Acton R, Copeland R, Cousins A, Gore T, Cornwell P, Roseman J. Plasma homocyst(e)ine, folate, and vitamin B_{12} concentrations and risk for early-onset coronary artery disease. Am J Clin Nutr 1994;59:940–8.

21. Mansoor M, Ueland P, Svardal A. Redox status and protein binding of plasma homocysteine and other aminothiols in patients with hyperhomocysteinemia due to cobalamin deficiency. Am J Clin Nutr 1994; 59:631–5.

22. Ubbink J. Vitamin B_{12}, vitamin B_6, and folate nutritional status in men with hyperhomocysteinemia. Am J Clin Nutr 1993;57:47–53.

23. Shimakawa T, Nieto F, Malinow M, Chambess L, Schreiner P, Szklo M. Vitamin intake: possible determinant of plasma homocyst(e)ine among middle-aged adults. Ann Epidemiol 1997;7:285–93.

24. Boushey C, Beresford S, Omenn G, Motulsky A. A quantitative assessment of plasma homocysteine as a risk factor for vascular disease: probably benefits of increasing folic acid intakes. JAMA 1995;274:1049–57.

25. Ubbink J, Vermaak W, Merwe A, Becker P, Delport AR, Potgieter H. Vitamin requirements for the treatment of hyperhomocysteinemia in humans. J Nutr 1994;124:1927–33.

26. Carmel R. Cobalamin, the stomach, and aging. Am J Clin Nutr 1997;66:750-9.

27. Stabler S, Lindenbaum J, Allen R. Vitamin B_{12} deficiency in the elderly: current dilemmas. Am J Clin Nutr 1997;66:741-9.

28. Flynn M, Herbert V, Nolph G, Krause G. Atherogenesis and the homocysteine-folate-cobalamin triad: do we need standardized analyses. J Am Coll Nutr 1997; 16:258-67.

29. Herbert V. The elderly need oral vitamin B_{12}. Am J Clin Nutr 1998;67:739.

30. Carmel R. Reply to V. Herbert. Am J Clin Nutr 1998;67:739-40.

31. Stabler S. Reply to V. Herbert. Am J Clin Nutr 1998;67:740.

Vitamin B_6

Vitamin B_6 exists as several vitamers, the structural formulas of which are given in Figure 9.39. These vitamers are interchangeable and comparably active (Fig. 9.40). Pyridoxine represents the alcohol form, pyridoxal the aldehyde form, and pyridoxamine the amine form. Each has a 5'-phosphate derivative.

Sources

All vitamers are found in food. Pyridoxine, the most stable of the compounds, is found almost exclusively in plant foods. Very small amounts, if any, of pyridoxal and pyridoxamine or its phosphorylated form are present in plant foods. Excellent sources of vitamin B_6 in commonly consumed foods are bananas, navy beans, and walnuts. The other vitamers, primarily pyridoxal phosphate and pyridoxamine phosphate, are found in animal products, with sirloin steak, salmon, and the light meat of chicken being rich sources [1,2]. Vitamin B_6 in supplements is found generally as pyridoxine hydrochloride.

In some plants, vitamin B_6 is found in a conjugated form, pyridoxine β-glucoside. Mammalian glycosidase is not thought to be able to free the pyridoxine from the glucoside. Thus, unless hydrolyzed by glucosidase from intestinal flora, this form of the vitamin is not well utilized [2].

The bioavailability of vitamin B_6 from different food sources is also influenced by the extent and type of processing to which the foods are subjected. Much of the vitamin originally present in foods can be lost through processing, including, for example, heating, canning, milling of wheat, sterilization, and freezing [1-3].

Digestion, Absorption, Transport, and Storage

For absorption of vitamin B_6 to occur, the phosphorylated vitamers must be dephosphorylated. Alkaline phosphatase, found at the intestinal brush border, or other intestinal phosphatases hydrolyze the phosphate to yield either pyridoxine (PN), pyridoxinal (PL), and pyridoxamine (PM).

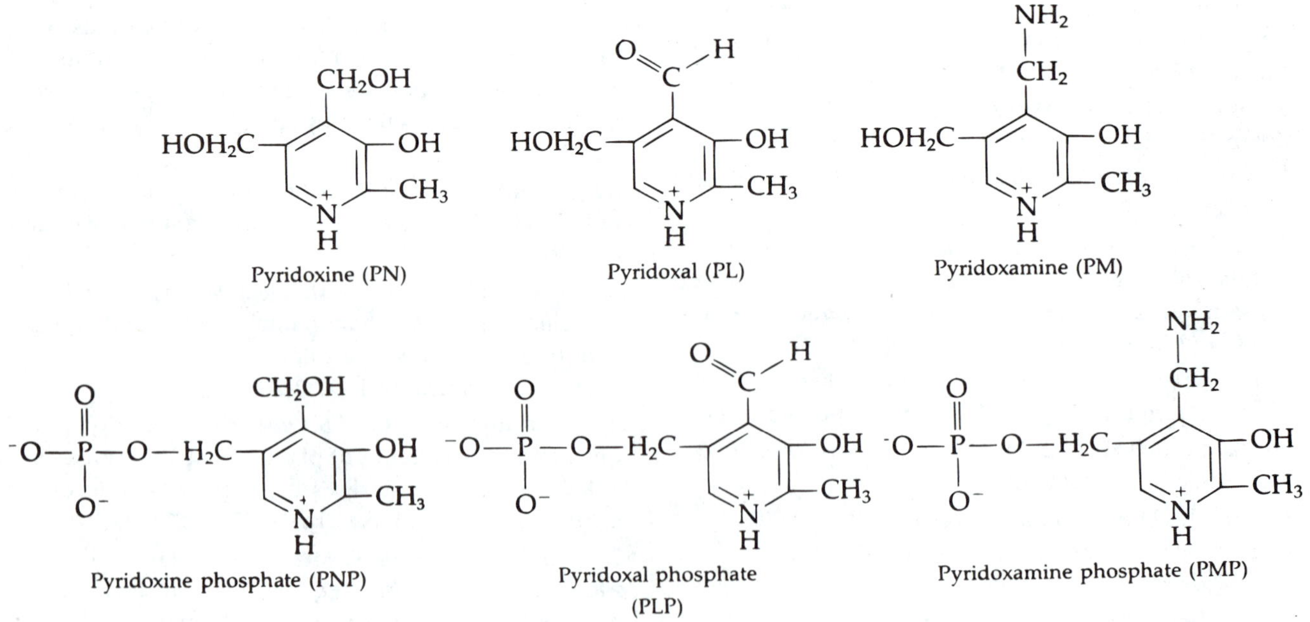

Figure 9.39 Vitamin B_6 structures.

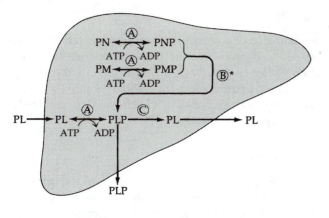

Enzyme
Ⓐ Kinase—ATP dependent
Ⓑ PMP and PNP oxidase*—FMN-dependent
Ⓒ Phosphatase

Figure 9.40 Vitamin B_6 metabolism in the liver. *Oxidase is found mainly in the liver and enterocytes.

Absorption of PL, PN, and PM occurs primarily in the jejunum by passive diffusion. At physiological intakes, the vitamin is absorbed rapidly in its free form; however, when the phosphorylated vitamers are ingested in high concentrations, some of these compounds may be absorbed per se [3]. Absorption of pyridoxine glucosides may also occur by passive diffusion; however, the complex will be excreted unchanged in the urine [4]. Overall absorption of vitamin B_6 furnished by the average U.S. diet ranges from 71% to 82%.

Within the intestinal cell, PN may be converted into pyridoxine phosphate (PNP) by the action of pyridoxine kinase using ATP.

$$\text{Pyridoxine (PN)} \xrightleftharpoons[\text{ATP}]{\textit{Kinase}} \text{Pyridoxine phosphate (PNP)}$$

PL is typically converted to pyridoxal phosphate (PLP) also through the action of kinase and ATP.

$$\text{Pyridoxal (PL)} \xrightleftharpoons[\text{ATP}]{\textit{Kinase}} \text{Pyridoxal phosphate (PLP)}$$

PNP may be converted to PLP through the action of pyridoxine phosphate oxidase, which requires riboflavin as FMN.

$$\text{Pyridoxine phosphate} \xrightarrow{\substack{\textit{oxidase} - \\ \text{FMN-dependent}}} \text{pyridoxal phosphate}$$

PLP is the main form (~60% of the total) of the vitamin found in the blood; it does not cross cell mem-

branes without hydrolysis to PL by extracellular alkaline phosphatase [4]. Other forms of the vitamin, especially PL (comprising ~14% of blood vitamin B_6 content) also may be present in the blood. PL may be found in both red blood cells and in plasma. In the erythrocytes, PL binds to the α chain of hemoglobin and PLP binds to the β chain [5]. In the plasma, both PLP and PL are bound to albumin during transport. Plasma PLP, however, is very tightly bound to albumin and probably is unavailable for use by the tissues. Another 15% of the vitamin circulates as PM, which also can be used after tissue uptake.

The liver is the main organ that takes up by passive diffusion the newly absorbed vitamin B_6. The liver stores about 5% to 10% of the vitamin [6]. Unphosphorylated forms of the vitamin are typically phosphorylated within the cytoplasm of the hepatocyte (liver cell), as shown in Figure 9.40. PNP and PMP are then generally converted to PLP. Figure 9.40 depicts the interconversion of the B_6 vitamers, a process that occurs mainly in the liver. From the liver, PLP and PL are released for transport to extrahepatic tissues.

Extrahepatic tissues, especially muscles, possess the majority (75%–80%) of PLP [6]. Only PL is taken up by these tissues and, thus, PLP in blood must be hydrolyzed before cellular uptake. Within the cell, PL is phosphorylated by pyridoxine kinase. Pyridoxine kinase is found in almost all tissues, and phosphorylation traps the vitamin in the cells. Most tissues, however, lack sufficient PNP/PMP oxidase, which converts PNP and PMP into the coenzyme form of the vitamin, PLP. PNP/PMP oxidase is found mainly in the liver and intestine but also in the kidney, brain, and red blood cell, although the activity in the latter cells is low [4]. PNP/PMP oxidase, crucial to activity of vitamin B_6, is a flavin mononucleotide (FMN)–dependent enzyme. Thus, normal vitamin B_6 metabolism is closely interrelated with riboflavin.

Functions and Mechanisms of Action

The coenzyme form of vitamin B_6 is associated with a vast number of enzymes, the majority of which are involved in amino acid metabolism. PLP, through the formation of a Schiff base (the product formed by an amino group and an aldehyde), labilizes all of the bonds around the α-carbon of the amino acid. The specific bond that is broken is determined by the catalytic groups of the particular enzyme to which PLP is attached. The covalent bonds of an α-amino acid that can be made labile by its binding to specific PLP-containing enzymes are given in Figure 9.41.

Reactions catalyzed by PLP include transamination (which can also be catalyzed by PMP), decarboxylation, transulfhydration and desulfhydration, cleavage,

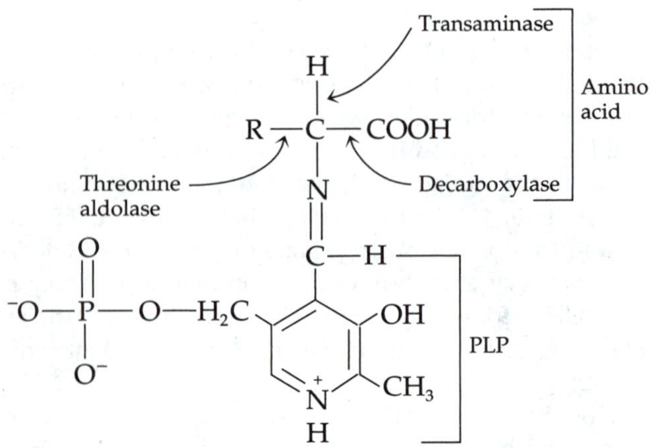

Figure 9.41 The covalent bonds of an α amino acid that can be made labile by its binding to PLP-containing enzymes.

synthesis, and racemization. Glycogen metabolism and the action of steroid hormones also appear to involve vitamin B_6.

Transamination

Of particular importance are the transamination reactions in which PMP as well as PLP can serve as a coenzyme. The most common aminotransferases for which PLP (or PMP) is a coenzyme are glutamate oxaloacetate transaminase (GOT) (also called aspartic amino transferase (AST)) and glutamate pyruvate transaminase (GPT) (also called alanine aminotransferase [ALT]) (Fig. 7.20). Figure 9.42a and b show the two phases of transamination and demonstrate how PLP forms a Schiff base. In the first phase, the corresponding α-keto acid of the amino acid is produced along with PMP. In the second phase, the transamination cycle is completed as a new α-keto acid substrate receives the amino group from the PMP. The corresponding amino acid is generated, along with regeneration of PLP.

Decarboxylation

Common decarboxylation reactions include the formation of γ-aminobutyric acid (GABA) from glutamate (Fig. 7.42) and the production of serotonin from 5-hydroxytryptophan (Fig. 7.40).

Transulfhydration and Desulfhydration

PLP is required for transulfhydration reactions in which cysteine is synthesized from methionine (Fig. 9.43). Both cystathionine synthase and cystathionine lyase require PLP. Cysteine undergoes desulfhydration followed by transamination to generate pyruvate.

Cleavage

An example of a cleavage reaction in which PLP is required is the removal of the hydroxy-methyl group from serine. In this reaction PLP is the coenzyme for a transferase that transfers the hydroxy-methyl group of serine to tetrahydrofolate (THF) so that glycine is formed (p. 291, Fig. 9.31).

Racemization

PLP is required by racemases that catalyze the interconversion of D- and L- amino acids. Although such reactions are more prevalent in bacterial metabolism, some occur in humans.

Synthesis of Various Compounds

Vitamin B_6 is also necessary in the synthesis of heme. PLP is required for delta-aminolevulinic acid synthetase, which catalyzes the condensation of glycine with succinyl CoA to form δ-aminolevulinic acid (ALA) in the mitochondria of the cell (Fig. 9.44). ALA moves into the cytosol of the cell, where it is used to synthesize porphobilinogen (PBG), the parent pyrrole compound in porphyrin synthesis. Through a series of reactions, PBG is converted into protoporphyrin IX. Protoporphyrin IX with the addition of Fe^{2+} by ferrocheletase forms heme. (Heme synthesis is shown in Fig. 12.5.)

Niacin (NAD) synthesis from tryptophan also requires an important PLP-dependent reaction. Specifically, kynureninase required for the conversion of 3-hydroxykynurenine to 3-hydroxyanthranilate requires vitamin B_6 as a coenzyme (Fig. 9.18).

Other compounds synthesized in the body in vitamin B_6–dependent reactions include histamine from the amino acid histidine, carnitine, a nitrogen-containing compound and compounds, such as taurine and dopamine, with neuromodulatory functions.

Glycogen Catabolism

The function of PLP in glycogen degradation is poorly understood. Glycogen is catabolized by glycogen phosphorylase to form glucose 1-PO_4 (p. 84); vitamin B_6 is required for glycogen phosphorylase activity. The mechanism of action of the coenzyme appears to be different from that exerted with other enzymes. The phosphate of the coenzyme is believed to be involved as a proton shuttle or buffer to stabilize the compound and permit covalent bonding of the phosphate to form glucose 1-PO_4 [4]. Most of vitamin B_6 found in muscle is present as PLP, which is in turn bound to glycogen phosphorylase [7].

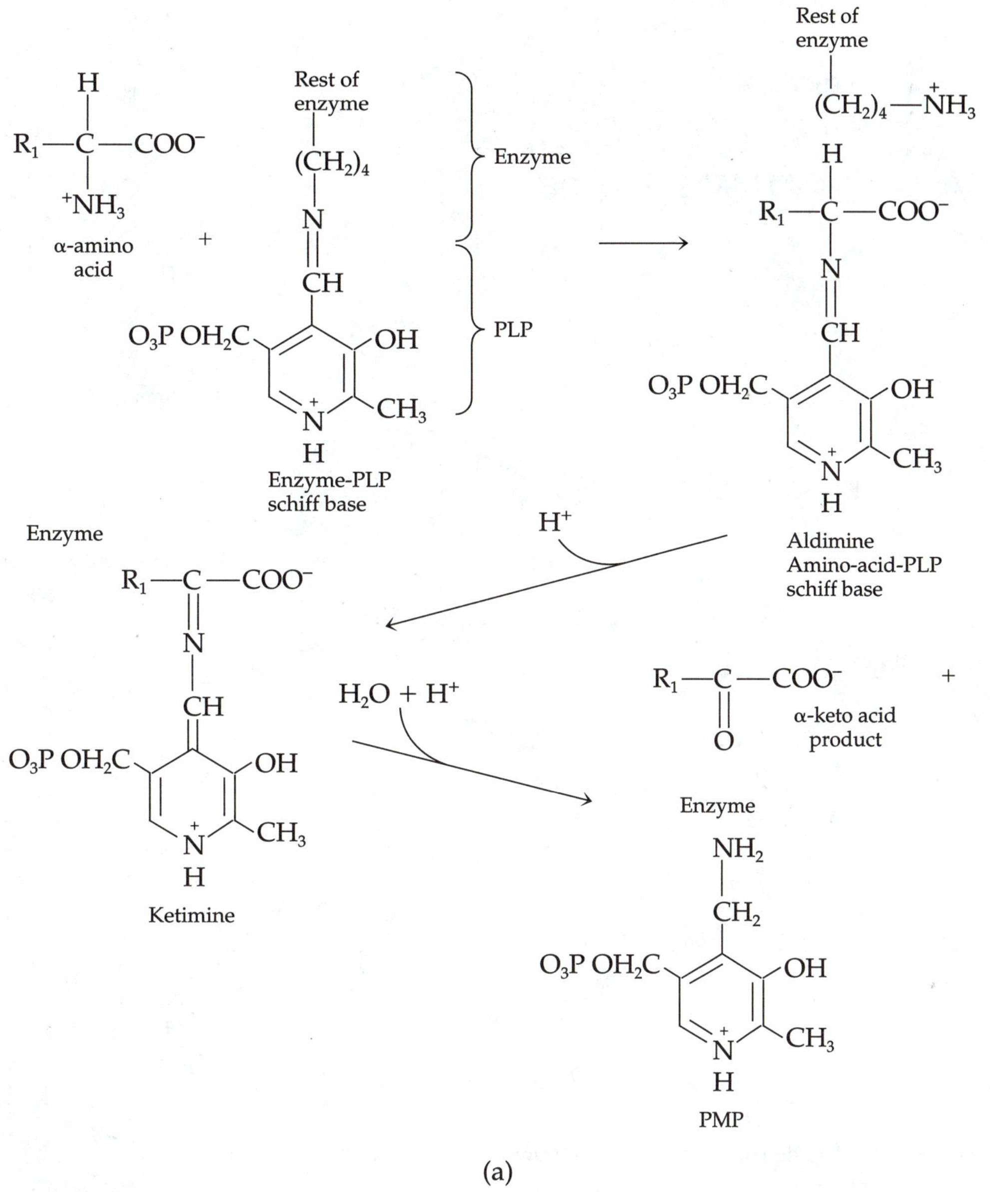

(a)

Figure 9.42 (a) The role of vitamin B$_6$ in transamination, Phase 1.

Steroid Hormone Action

Vitamin B$_6$ as PLP has been shown to react with lysine residues in steroid hormone receptor proteins to prevent or interfere with hormone binding. These receptor proteins mediate nuclear uptake of the steroid hormone and the interaction of the nucleoproteins with the DNA [6]. PLP has also been shown to bind to receptors on steroids [5]. Thus, vitamin B$_6$ appears to be able to diminish the actions of steroids. Diminishing the action of, for example, glucocorticoid hormones can in turn influence metabolism of protein, carbohydrate, and lipid.

Metabolism and Excretion

The intracellular level of PLP is believed to be controlled by enzymatic hydrolysis. The suggested mechanism for control lies primarily with the concentration of the PLP-binding proteins in the cells. When these proteins are saturated, then the newly synthesized PLP will be hydrolyzed by intracellular phosphatase. Another possibility for regulating PLP formed in the cell is product inhibition of PNP/PMP oxidase, operative primarily in the liver and intestinal cells (Fig. 9.42) [3].

Pyridoxic acid (PIC) is the major excretory product resulting from the oxidation of PL by either NAD-

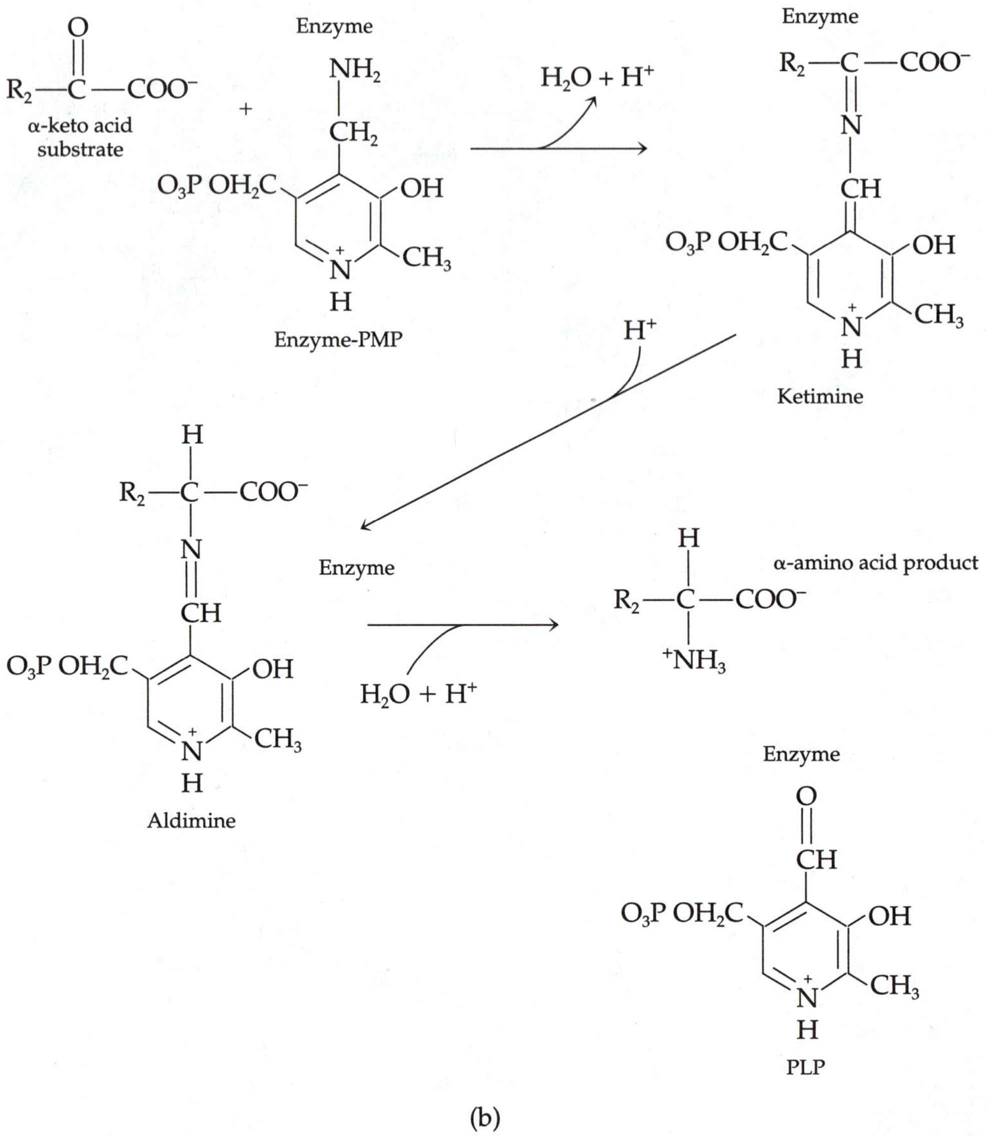

(b)

Figure 9.42 (continued) (b) The role of vitamin B_6 in transamination, Phase 2.

dependent aldehyde dehydrogenase, found in all tissues, or FAD-dependent aldehyde oxidases found in the liver and kidneys. The amount of PIC excreted is thought to be more indicative of recent vitamin intake than of vitamin stores because newly formed PLP is not freely exchangeable with endogenous PLP [3]. Newly formed PLP is instead quickly converted to PL and PIC and is released into the plasma.

The form in which the vitamin is ingested appears to influence the percentage of intake that is excreted as PIC. When large doses (100 mg) of the vitamin were given as PL, PM, or PN, 90% of the PL and PM appeared in the urine as PIC within 36 hours. In contrast, only 70% of the PN was excreted as PIC; much of the PN was excreted as such in the urine within 2 hours. It appears that when PN is administered at high levels, the kidney tubules reduce plasma content of the vitamer by secreting some of it into the urine [8].

Dietary Reference Intakes and Recommended Dietary Allowances

Adequate intake of vitamin B_6 has been estimated through depletion and repletion studies. The requirement for vitamin B_6 has been found to be related to the level of protein intake [9]. Therefore, in formulating the 1989 RDA, consideration was given to the average protein intake among the United States population. Given

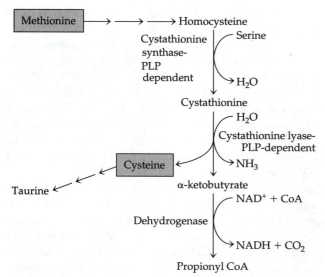

Figure 9.43 Cysteine synthesis from methionine requires vitamin B_6.

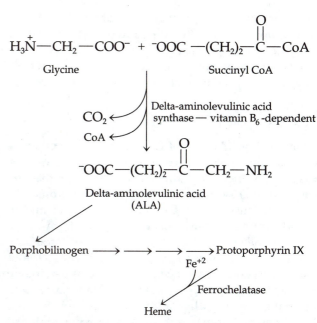

Figure 9.44 Synthesis of delta-aminolevulinic acid by vitamin B_6–dependent delta-aminolevulinic acid synthase.

the assumption that 126 g of protein daily represents the upper limit of a customary intake by adult males, the 1989 RDA for vitamin B_6 for this population group is set at 2.0 mg daily. Because women are expected to consume a little less protein (upper limit 100 g daily), the RDA for vitamin B_6 for this group is reduced to 1.6 mg daily [10]. An intake of 0.016 mg of vitamin B_6/1 g protein is considered sufficient to meet the needs of adults under normal conditions [10].

The 1998 DRI RDA for individual intake of vitamin B_6 are lower for both men and women than the 1989 RDA [11]. DRI RDA for vitamin B_6 for adult men aged 19 to 50 years is 1.3 mg and for men aged 51 years and older is 1.7 mg. DRI RDA for vitamin B_6 for adult women aged 19 to 50 years is also 1.3 mg and for women aged 51 years and older is 1.5 mg. With pregnancy and lactation, the DRI for the vitamin increases to 1.9 and 2.0 mg, respectively. Tolerable upper intake levels for vitamin B_6 are 100 mg/day for adults, including pregnant and lactating women, aged 19 years and older. With pregnancy in women 18 years of age or younger, the tolerable upper intake level for vitamin B_6 drops slightly to 80 mg/day [11].

Deficiency

Vitamin B_6 deficiency is relative rare in the United States. In the 1950s, deficiency occurred in infants because of severe heat treatment of infant milk. The heat processing resulted in a reaction between the PLP and the epsilon amino group of lysine in the milk proteins to form pyridoxyl- lysine, which possesses little vitamin activity. Signs of vitamin B_6 deficiency [12] include sleepiness, fatigue, cheilosis, glossitis, stomatitis in adults, and neurological problems such as abnormal EEGs, seizures, and convulsions in infants. A hypochromic, microcytic anemia may also result from a vitamin B_6 deficiency due to impaired heme synthesis. Deficiency also alters calcium and magnesium metabolism [13] and impairs niacin synthesis from tryptophan [7].

Groups particularly at risk for vitamin B_6 deficiency are

- breastfed infants born with low plasma vitamin B_6 levels;
- the elderly, who have a poor intake of the vitamin and may also have accelerated hydrolysis of PLP and oxidation of PL to PIC;
- people who consume excessive amounts of alcohol (alcohol can impair conversion of PN and PM to PLP, and the presence of acetaldehyde formed from ethanol metabolism may enhance hydrolysis of PLP to PL with subsequent formation of PIC in blood) [1,2];
- renal patients on maintenance dialysis, which causes an abnormal loss of vitamin B_6; and
- people on a variety of drug therapies that inhibit activity of the vitamin, primary among which are those employing isoniazid, penicillamine, corticosteroids, and/or anticonvulsants.

Pregnant women exhibit increased xanthurenic acid excretion with a tryptophan load, but it is uncertain whether this is an indication of a vitamin insufficiency or is a normal physiological condition [8–10]. Other conditions and populations with possible increased needs for the vitamin include those with hyperthyroidism, high protein intake, liver disease, or stress.

Toxicity

Pharmacological doses of vitamin B_6 have been advocated for the prevention or treatment of a variety of disease states including atherosclerotic heart disease, carpal tunnel syndrome, premenstrual syndrome, depression, muscular fatigue, paresthesia, and autism [7,14]. Although some beneficial results from megadoses of the vitamin have been noted in selected individuals, indiscriminate use of the vitamin is not without risk. Adults chronically ingesting 1 to 6 g of pyridoxine per day have been reported to suffer from sensory and peripheral neuropathy. A dose response relationship has been demonstrated for pyridoxine induced neuropathy [15]. Some symptoms include unsteady gait, numbness of the feet and hands, impaired tendon reflexes, and paresthesia [7,16]. Excessive amounts of pyridoxine appear to cause degeneration of dorsal root ganglia in the spinal cord, loss of myelination, and degeneration of sensory fibers in peripheral nerves [4,14,16]. The minimal dosage at which toxicity occurs is not clear; however, daily intakes of 500 mg are thought to be associated with neurotoxicity [7,17]. The no observed adverse effect level (NOAEL) is 200 mg, whereas the lowest observed adverse effect level (LOAEL) is 500 mg [17].

Assessment of Nutriture

A commonly used index of vitamin B_6 nutriture is a functional test measuring xanthurenic acid excretion following tryptophan loading (100 mg of tryptophan/kg body weight). Abnormally high xanthurenic acid (>25 mg in 6 hours) excretion is found in vitamin B_6 deficiency because 3-hydroxykynurenine, an intermediate in tryptophan metabolism, cannot lose its alanine moiety and be converted to 3-hydroxyanthranilate, as should occur (Fig. 9.18). Instead, 3-hydroxykynurenine becomes xanthurenic acid. Interpretation of this test is sometimes difficult owing to factors other than vitamin B_6 in tryptophan metabolism. Acceptable xanthurenic acid excretion following the tryptophan load is <25 mg/6 hours.

Plasma PLP assays are being used more frequently to assess vitamin B_6 nutriture, although intakes of both the vitamin and protein among other factors will affect plasma concentrations of PLP. Plasma PLP concentrations in excess of 30 nmol/L are thought to suggest adequate vitamin status [18].

Urinary vitamin B_6 and pyridoxic acid have also been used to assess status of vitamin B_6. Urinary vitamin B_6 excretion measured over several 24-hour urine collections for a period of 1 to 3 weeks is recommended to more accurately assess vitamin B_6. Urinary vitamin B_6 excretion in comparison to creatinine excretion is also used such that urinary levels of <20 mg/g creatinine suggests B_6 deficiency, while excretion >20 mg/g creatinine suggests acceptable vitamin B_6 status. Urinary pyridoxic acid excretion is considered to be a short-term indicator of vitamin B_6 status [18].

Measurement of erythrocyte transaminase activity before and after vitamin B_6 addition are also useful in determining vitamin B_6 nutriture; however, because of a variety of limitations with the assays, these tests are better used as an adjunct to other tests. Erythrocyte transaminase index examines activity of erythrocyte glutamic oxaloacetic transaminase (EGOT) (also called *aspartic amino transferase,* or EAST) after the addition of vitamin B_6. This assay and the assay discussed next are thought to represent long-term vitamin status. Deficient vitamin B_6 status is suggested by activity of >1.8 following the addition of the vitamin [18]. Similarly, if activity of erythrocyte glutamic pyruvic transaminase (EGPT) (also called *alanine aminotransferase,* or EALT) increases >1.25, then B_6 deficiency is suggested, whereas activity of <1.25 indicates adequate status [18,19].

References Cited for Vitamin B_6

1. Sauberlich H. Bioavailability of vitamins. Prog Food Nutr Sci 1985;9:1–33.

2. Sauberlich H. Vitamins—how much is for keeps? Nutr Today 1987;22:20–28.

3. Ink S, Henderson L. Vitamin B_6 metabolism. Ann Rev Nutr 1984;4:455–70.

4. Bender D. Nutritional Biochemistry of the Vitamins. New York: Cambridge University Press, 1992:223–68.

5. Leklem J. Vitamin B_6. In: Shils ME, Olson JA, Shike M., eds. Modern Nutrition in Health and Disease, 8th ed. Philadelphia: Lea and Febiger, 1994:383–94.

6. Allgood V, Cidlowski J. Novel role for vitamin B_6 in steroid hormone action: a link between nutrition and the endocrine system. J Nutr Biochem 1991;2:523–34.

7. Leklem J. Vitamin B_6. In: Machlin LJ, ed. Handbook of Vitamins, 2nd ed. New York: Dekker, 1991:341–92.

8. Henderson L. Vitamin B_6. In: Olson R, Chairman, Broquist H, Chichester CO Darby, Kolbye A, Jr, Stalvey R, eds. Present Knowledge in Nutrition, 5th ed. Washington, D.C.: The Nutrition Foundation, 1984:303–17.

9. Committee on Dietary Allowances, Food and Nutrition Board. Human Vitamin B_6 requirements. Washington, D.C.: National Academy of Sciences, 1978.

10. National Research Council. Recommended Dietary Allowances, 10th ed. Washington, D.C.: National Academy Press, 1989;142–50.

11. Yates A, Schlicker S, Suitor C. Dietary reference intakes: the new basis for recommendations for calcium and related nutrients, B vitamins and choline. J Am Diet Assoc 1998;98:699–706.

12. Coombs G. The Vitamins. San Diego, CA: Academic Press, 1992:311–28.

13. Turlund J, Betschart A, Liebman M, Kretsch M, Sauberlich H. Vitamin B_6 depletion followed by repletion with animal or plant source diets and calcium and magnesium metabolism in young women. Am J Clin Nutr 1992;56:905–10.

14. Alhadeff L, Gualtieri C, Lipton M. Toxic effects of water-soluble vitamins. Nutr Rev 1984;42:33–40.

15. Council on Scientific Affairs, American Medical Association. Vitamin preparations as dietary supplements and as therapeutic agents. JAMA 1987; 257: 1929–36.

16. Berger A, Schaumburg H, Schroeder C, Apfel S, Reynolds R. Dose response, coasting and differential fiber vulnerability in human toxic neuropathy: a prospective study of pyridoxine neurotoxicity. Neurology 1992;42:1367–70.

17. Hathcock J. Vitamins and minerals: efficacy and safety. Am J Clin Nutr 1997;66:427–37.

18. Leklem J. Vitamin B_6: a status report. J Nutr 1990;120:1503–7.

19. Simko M, Cowell C, Gilbride J. Nutrition Assessment. Rockville, MD: Aspen, 1995.

Additional Reference

Driskell J. Vitamin B_6 requirements of humans. Nutr Res 1994;14:293–324.

Nutrient Controls of Gene Expression

Research continues to reveal that the old adage "You are what you eat" is really true. While we know that the nutrients we eat are utilized by the body for energy, for structural roles, for membranes, for enzyme and cofactor functions, and the like, certain nutrients also appear to be able to control gene expression. This perspective will review briefly gene expression—that is, the synthesis of proteins from genes—and then discuss the effects of iron, zinc, copper, and retinoic acid and vitamin D on the expression of selected genes.

Chromosomes, Genes, DNA, and Proteins

The nucleus of our cells contains chromosomes, which in turn are made up of genes. Genes are made up of deoxyribonucleic acid (DNA). DNA consists of two strands of nucleic acids that intertwine to form a double helix. Each DNA strand consists of a backbone of deoxyribose sugar molecules linked to phosphates; nitrogenous bases then attach to each of the sugar molecules. The nitrogenous bases link the two DNA strands together through specific base pairing. For example, DNA pyrimidine base pairs link with purine base pairs as follows: pyrimidine thymidine—purine adenine, and pyrimidine cytosine—purine guanine. In terms of DNA, genes represent DNA containing hundreds to thousands of base pairs. The sequence of bases in the DNA of a gene corresponds with the sequence of amino acids that make up a specific protein. Each amino acid is specified by a sequence of three bases, known as a triplet. Genes differ from each other in their different sequences of bases in their DNA and thus result in the formation of different proteins made up of differing amino acid sequences.

Review of Transcription and Translation

The expression of a gene, that is the formation of a protein, involves many cellular processes. Two principal events are *transcription* and *translation*. Transcription is the process by which sequences of bases in the DNA of a gene are copied to form another molecule called ribonucleic acid (RNA). Transcription requires the unwinding of segments of the DNA molecule such that one of the two DNA strands is exposed and serves as a template for RNA synthesis. The enzyme RNA polymerase II catalyzes gene transcription until a specific termination sequence in the DNA is reached. The sequence of bases in the DNA is transcribed into RNA using the base (purine-pyrimidine) pair-

ing principle with the exception of adenine, which in DNA base pairs with thymidine but in RNA pairs with uracil. The resulting RNA thus carries a complementary base sequence code as that of the DNA. In other words, base triplets in the DNA strand are transcribed into complementary base triplets in the RNA. Following processing in which, for example, sequences of the RNA may be spliced or sequences of nucleotides may be added (capped) at the ends, the RNA known as messenger (m) RNA leaves the nucleus and enters the cytoplasm for translation. Translation converts the sequence of information provided for in the base triplets of the mRNA into a sequence of amino acids specific for a protein.

Levels of Nutrient Control of Gene Expression

Genes can be controlled at several levels including transcription, posttranscription, and posttranslation. Transcription may be affected by the presence or absence of transcription factors. Remember, although genes contain DNA base triplets that are transcribed into complementary base RNA triplets, not all segments of the DNA are transcribed. A promoter region located in the 5' region upstream from the gene to be transcribed along with an initiator sequence complex serve as the site where the enzyme RNA polymerase II and various transcription factors bind. In the absence of activators, transcription occurs in relatively low (basal) levels [1]. Activators or binding proteins can inhibit or stimulate the transcription rate several hundredfold [1]. In addition to control of transcription, posttranscription and posttranslation also can be controlled. Posttranscription and posttranslation, for example, may be regulated through alterations in processing and in mRNA turnover as well as in protein half life, respectively.

Nutrient Effects on Gene Expression

Iron and the Expression of the Transferrin Receptor Gene

The role of iron in the transcription and posttranscription regulation of the transferrin receptor protein will be reviewed as one example of nutrient gene interactions (Fig. 1). The gene for transferrin receptor proteins may be found on chromosome 3. The promoter for the gene lies in the 5' region upstream from the gene and contains regions or elements that allow for transcription regulation by iron. Messenger RNA for the transferrin receptor contains five iron response or regulatory elements (IRE) in the 3' untranslated region. Iron regulatory or response elements are

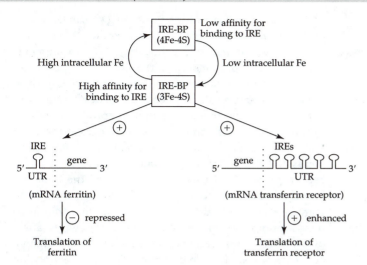

Figure 1 Influence of intracellular iron on translation of ferritin mRNA and transferrin receptor mRNA.

thought of as stem loop structures of about 30 nucleotides found in the mRNA. *Trans*-acting iron response (or regulatory) element binding proteins (IRE-BP), consisting of 889 amino acids and multiple iron-sulfur clusters and found in the cell cytoplasm, respond to the cell's iron status and bind or not bind IREs on the transferrin receptor mRNA as appropriate [2–5]. For example, in a low cellular iron situation, the *trans*-acting IRE-BP contains a 3Fe-4S cluster and readily binds to *cis*-acting 3' IRE, thereby stabilizing the transferrin receptor mRNA. The stabilized transferrin receptor mRNA exhibits a longer half life and consequently, more transferrin receptor mRNA is translated into transferrin receptor proteins [2,4]. Once made, these transferrin receptor proteins become embedded in the cell's plasma membrane to promote cellular iron uptake. Thus, in times of low cellular iron, transferrin receptor synthesis is increased. Now, what happens in the opposite situation, that of relatively high intracellular iron? With adequate to high cellular iron, iron binds to the *trans*-acting IRE-BP. The additional iron is thought to convert the iron clusters in the IRE-BP from a 3Fe-4S structure to a 4Fe-4S structure. This newly shaped 4Fe-4S structure is thought to have lower affinity for the 3' IRE section of the transferrin receptor mRNA [2,4,6]. Without the IRE-BP bound to the IRE of transferrin receptor mRNA, the mRNA is not as stable and is more quickly degraded. This decreased stability and increased degradation in turn diminishes the translation of the mRNA and results in the production of fewer transferrin receptor proteins. The synthesis of fewer transferrin re-

ceptor proteins means that fewer receptors will be available on the cell surface and less iron will be brought into the cell. Thus, the level of transferrin receptor expression is indicative of the cell's need for iron uptake [7].

Iron and the Expression of the Ferritin Gene
A second example of the regulatory role of nutrients at the molecular level involves iron and ferritin synthesis (Fig. 1). Iron controls ferritin synthesis at the level of translation. As with regulation of the transferrin receptor, a metalloregulatory protein IRE-BP responds to the cell's iron status and either binds or doesn't bind to IREs located in the 5' untranslated region of ferritin mRNA [3,4,6]. In the presence of iron below some specific threshold value, the IRE-BP binds to the IRE in the 5' untranslated region of ferritin mRNA. In such "low-iron situations" the IRE-BP is thought to contain a 3Fe-4S cluster and readily binds to the IRE in ferritin mRNA. However, unlike the influence of IRE-BP on the transferrin receptor mRNA, the binding of IRE-BP to the ferritin mRNA acts as a repressor protein to inhibit the translation of the ferritin protein [3,6,7]. Thus, less ferritin protein is made in cells when cellular iron content is low. This makes sense from a physiological standpoint since ferritin functions to store iron; not much ferritin would be needed if the cell's iron content was low. Under the opposite conditions in which the cell has a relatively high iron content, IRE-BP exhibit diminished mRNA binding and appear to be found as free cytoplasmic proteins (possibly aconitase) with a 4Fe-4S cluster [4,6,8]. Without the binding of the

repressor (the IRE-BP), the ferritin mRNA undergoes translation. Thus, more ferritin protein is made in cells when cellular iron concentrations are high. From these descriptions, it should be apparent that the expression of the transferrin receptor and ferritin genes is coordinated and regulated by cellular iron, which influences the iron-sulfur clusters in the iron response element–binding proteins and by iron response elements in both the transferrin receptor and ferritin mRNAs. Iron, however, is not the only metal that influences gene expression. Several other minerals also influence gene expression.

Zinc and Metallothionein Gene Expression

Metallothionein, although having several functions in the cell, serves as a cellular storage site for metals, primarily zinc but also copper and cadmium (see Chap. 12, p. 423). Metallothionein genes lie mostly on chromosome 16, and contain *cis*-acting regulatory or metal response elements (MRE) in the gene's 5' untranslated region. During periods of high cellular zinc concentrations, transcription of the metallothionein gene is promoted when zinc (or selected other metals) interacts with cytosolic *trans*-acting binding proteins (also called *metal binding transcription factors*); the complex then enters the nucleus and interacts with the metal response element located upstream from the metallothionein gene [9]. Metallothionein gene expression in the intestine directly correlates with dietary zinc intake [9].

Copper and Vitamins A and D and Gene Expression

Copper is another mineral that affects gene expression. Specifically, copper appears to stimulate transcription of the superoxide dismutase gene, although it also may increase stability of superoxide dismutase mRNA to control translation. A transcription factor protein, ACE1, in the presence of copper appears to bind upstream of the superoxide dismutase gene and stimulate its transcription [10].

Vitamin A in the form of retinoic acid also influences gene transcription. Receptor proteins that bind retinoic acid and serve as transcription factors mediate the process. The binding of retinoic acid to retinoic acid receptors (RAR), of which there are three, allows for interaction with retinoic acid response elements (RARE) present on particular genes [11,12]. The binding of the retinoic acid to the RAR and then to the retinoic acid response elements on the DNA controls the transcription of the DNA into RNA. Other receptors referred to as retinoic X receptors (RXR) bind 9-*cis* retinoic acid have been identified and control gene expression. RXRs may interact with RARs or function independently [11].

Control of gene expression by *vitamin D* is similar to that of the other nutrients. Vitamin D, also known as *calcitriol* or *1,25 dihydroxycholecalciferol,* is thought to bind to vitamin D receptor (VDR) proteins. Next, the vitamin-receptor complex binds to *cis*-acting vitamin D elements (VDEs) which are found near the calcitriol-regulated genes. Interactions between the VDE and the vitamin-receptor complex are proposed to induce conformational changes in the chromatin structure to alter transcription rates by RNA polymerase II [13]. For example, binding of calcitriol to a receptor protein and then to the VDE of the gene for intestinal calcium binding protein calbindin is thought to stimulate gene transcription severalfold. In addition to transcriptional upregulation, calcitriol is thought to affect posttranscriptional events to increase intestinal calbindin concentrations. The role of vitamin D in gene expression not only includes a role in mineral homeostasis but extends to the areas of differentiation, proliferation, and development [13].

While the molecular details of nutrient control of gene expression are rapidly evolving and expanding, at present it is clearly evident that vitamins and minerals do not just regulate metabolism as coenzymes or cofactors for enzymes. Instead, metals such as iron, zinc, and copper and vitamins such as D and retinoic acid control the transcription of genes as well as posttranscription and posttranslation events. Future research will clarify and expand these relationships between nutrients and gene expression, as well as further expand the relationship to address the role of mutations and human disease [14].

References Cited

1. Johnson P, Sterneck E, Williams S. Activation domains of transcriptional regulatory proteins. J Nutr Biochem 1993; 4:386–98.
2. Lash A, Saleem A. Iron metabolism and its regulation. Ann Clin Lab Sci 1995;25:20–30.
3. Beard J, Dawson B, Pinero D. Iron metabolism. Nutr Rev 1996;54:295–317.
4. Leibold E, Guo B. Iron-dependent regulation of ferritin and transferrin receptor expression by the iron-responsive element binding protein. Ann Rev Nutr 1992;12:345–68.
5. Winzerling J, Law J. Comparative nutrition of iron and copper. Ann Rev Nutr 1997;17:501–26.
6. O'Halloran T. Transition metals in control of gene expression. Science 1993;261:715–25.
7. Klausner R, Rouault T, Harford J. Regulating the fate of mRNA: the control of cellular iron metabolism. Cell 1993; 72:19–28.

∾ **PERSPECTIVE** (continued)

8. Clarke S, Abraham S. Gene expression: nutrient control of pre- and posttranscriptional events. FASEB J 1992;6:3146–52.

9. Bremner I, Beattie J. Metallothionein and the trace minerals. Ann Rev Nutr 1990;10:63–83.

10. Gralla E, Thiele D, Silar P, Valentine J. ACE1, a copper-dependent transcription factor, activates expression of the yeast copper, zinc superoxide dismutase gene. Proc Natl Acad Sci 1991;88:8558–62.

11. Mangelsdorf D. Vitamin A receptors. Nutr Rev 1994; 52:S32–S44.

12. Ross A, Ternus M. Vitamin A as a hormone: recent advances in understanding the actions of retinol, retinoic acid and beta carotene. J Am Diet Assoc 1993;93:1285–90.

13. Lowe K, Maiyar A, Norman A. Vitamin D–mediated gene expression. Crit Rev Eukaryotic Gene Expression 1992; 2:65–109.

14. Semenza G. Transcriptional regulation of gene expression: mechanisms and pathophysiology. Human Mutation 1994; 3:180–99.

Web Site

www.ncbi.nlm.gov/science96

Photomicrograph of crystallized β-carotene

Chapter 10

The Fat-Soluble Vitamins

Vitamin A and Carotenoids

Vitamin D

Vitamin E

Vitamin K

For each vitamin, the following subtopics (when applicable) are discussed:

    Sources
    Digestion, Absorption, Transport, and Storage
    Functions and Mechanisms of Action
    Interactions with Other Nutrients
    Metabolism and Excretion
    Dietary Reference Intakes and Recommended Dietary
    Allowances
    Deficiency
    Toxicity
    Assessment of Nutriture

  PERSPECTIVE: The Antioxidant Nutrients, Reactive Species, and Disease

This chapter addresses each of the four fat-soluble vitamins, A, D, E, and K. The reader is referred to Chapter 9 for an overview of vitamins and information pertaining to the water-soluble vitamins. The absorption and transport of the fat-soluble vitamins, in contrast to that of the water-soluble vitamins, are closely associated with the absorption and transport of lipids. As with dietary lipids, optimal fat-soluble vitamin absorption requires the presence of bile salts. Similarly, the transport of the fat-soluble vitamins in the body occurs initially by chylomicrons. Moreover, the fat-soluble vitamins are stored in body lipids, although the amount stored varies widely among the four fat-soluble vitamins. Table 10.1 provides an overview of the discovery, functions, deficiency syndrome, food sources, and the Recommended Dietary Allowance (RDA) or Adequate Intake (AI) of each of the fat-soluble vitamins. The RDA and AI for all nutrients and for all age groups are provided on the inside cover of the book.

Vitamin A and Carotenoids

The term *vitamin A* is used to refer to retinol (an alcohol) and retinal (the aldehyde form) (Fig. 10.1a and 10.1b, respectively). Retinoic acid (Fig. 10.1c) is a metabolite of retinal. The retinoids consist of isoprenoid units joined in a head to tail manner. The term *provitamin A* refers to β-carotene (Fig. 10.1d) and other carotenoids that exhibit the biologic activity of

316

Table 10.1 The Fat-Soluble Vitamins: Discovery, Function, Deficiency Syndrome, Food Sources, and Recommended Dietary Allowance (RDA) or Adequate Intake (AI)

Vitamin	Discovery	Biochemical or Physiological Function	Deficiency Syndrome or Symptoms	Good Sources in Rank Order	RDA or AI [a]
Vitamin A (retinol, retinal, retinoic acid) Provitamins Carotenoids, particularly β-carotene	McCollum (1916)	Synthesis of rhodopsin and other light receptor pigments; metabolites involved in growth failure, growth and differentiation of epithelia, nervous, bone tissue and immune function	Children: poor dark adaptation, xerosis, keratomalacia Adults: night blindness, xeroderma	Beef liver, sweet potato, carrots, spinach, butternut squash, dandelion greens	1,000 μg RE 800 μg RE
Vitamin D Provitamins Ergosterol 7-dehydrocholesterol Vitamin D_2 (ergocalciferol) Vitamin D_3 (cholecalciferol)	McCollum (1922)	Regulator of bone mineral metabolism, primarily calcium	Children: rickets Adults: osteomalacia	Synthesized in skin exposed to ultraviolet light; fortified milk is a reliable good source	5 μg
Vitamin E Tocopherols Tocotrienols	Evans and Bishop (1922)	Antioxidant	Infants: anemia Children and adults: neuropathy and myopathy	Vegetable seed oil are major source widely distributed in foods	10 mg[a] TE 8 mg[a] TE
Vitamin K Phylloquinones Menaquinones Menadione	Dam (1935)	Activates blood-clotting factors II, VII, IX, X by γ-carboxylating glutamic acid residues; carboxylates bone and kidney proteins	Children: hemorrhagic disease of newborns Adults: defective blood clotting	Synthesized by intestinal bacteria; green leafy vegetables, soy beans, beef liver	70 μg[b] 60 μg[b]

[a]Adults, aged 19 to 50 years, males and females.
[b]For males and females, respectively, aged 19 to 24 years, then increases to 80 μg and 65 μg for males and females, respectively, aged 25 to 50 years.

β-carotene. α-carotene (Fig. 10.1e), γ-carotene (Fig. 10.1f), lycopene (an open-chain analog of β-carotene) (Fig. 10.1g), and oxycarotenoids such as canthaxanthin (Fig. 10.1h), lutein (Fig. 10.1i), and zeaxanthin, represent some of the more than 600 carotenoids in nature. Carotenoids typically possess at least one unsubstituted β-ionone ring. Many, but not all, carotenoids can be converted into retinol.

Sources

Vitamin A (as retinyl esters such as retinyl palmitate; Fig.10.1j) is found primarily in selected foods of animal origin, especially liver and dairy products including whole milk, cheese, and butter, as well as fish such as tuna, sardines, and herring. Liver oils of fish (such as cod liver oil) are also high in vitamin A. In pharmaceutical preparations, retinyl acetate is commonly used. Aquasol A, a water-miscible form of the vitamin, is also available for those individuals with a fat malabsorptive disorder.

Carotenoids, the red, orange, and yellow pigments synthesized by a wide variety of plants, are many in number (over 600), but <10% of these pigments have vitamin A activity. The pigment with the greatest vitamin A activity is β-carotene. Two other commonly occurring carotenoids with vitamin A activity are α- and γ-carotene. α- and γ-carotene activity as compared with that of β-carotene ranges from 50% to 54% and 42% to 50%, respectively [1]. β-carotene and canthaxanthin are approved by the Food and Drug Association (FDA) and used as a food color additive. Carotenoids are found naturally in a variety of fruits and vegetables. Lycopene, for example, is a carotenoid that is red in color and found in fairly substantial quantities in tomatoes. Canthaxanthin is a red-orange carotenoid found in plants as well as fish and seafood such as sea trout and crustaceans. In general, yellow, orange and red (brightly colored) fruits and vegetables such as carrots, papaya, tomato, squash, and pumpkin provide significant amounts of carotenoids. Green vegetables also contain some carotenoids, although the pigment cannot be seen because it is masked by chlorophyll.

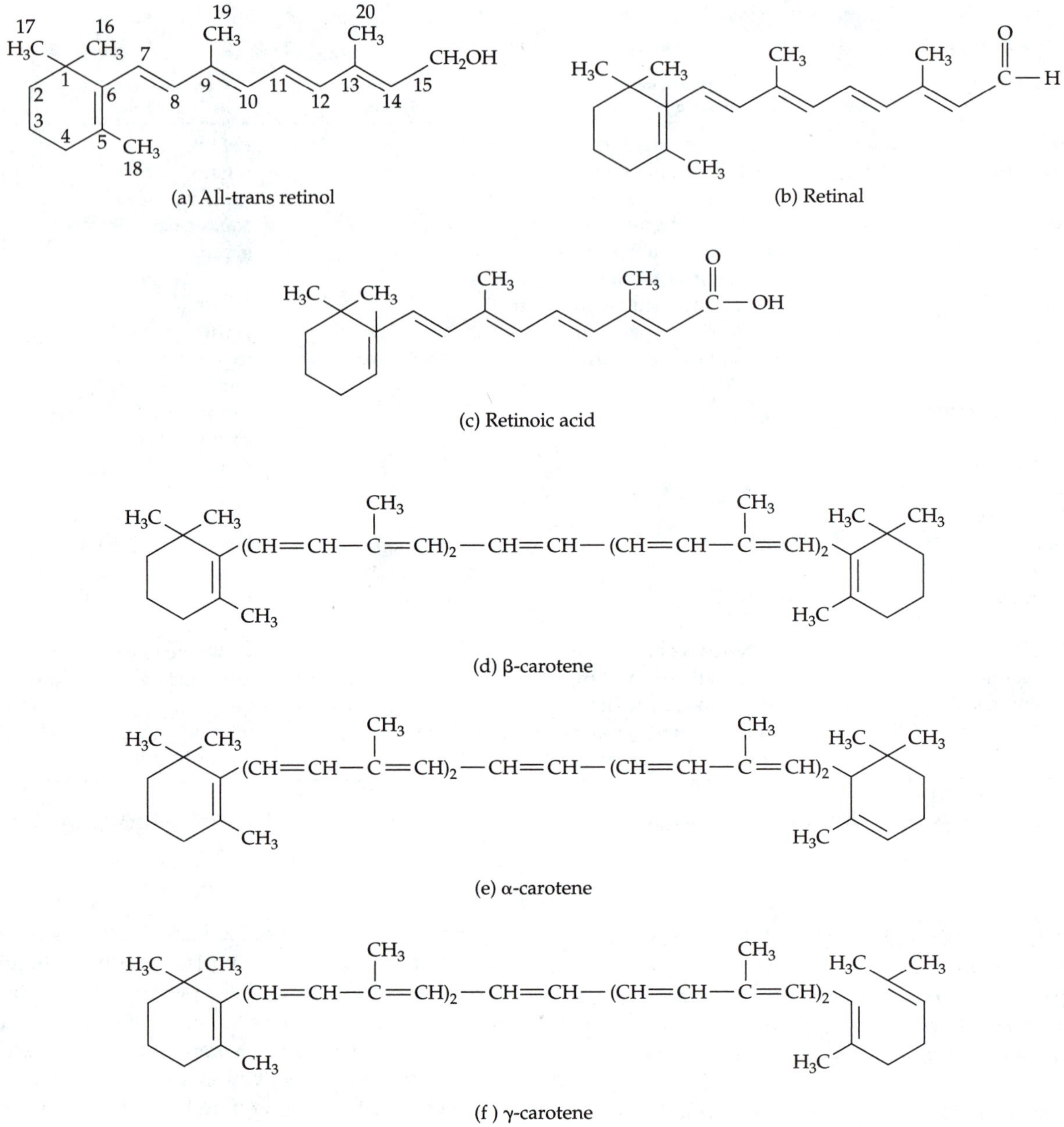

Figure 10.1 Vitamin A and carotenoid structures.

Digestion, Absorption, Intestinal Cell Metabolism, Transport, and Storage

Retinol is not generally found free in foods but is typically present bound to fatty acid esters, among which the most commonly occurring is retinyl palmitate (Fig. 10.1j). Furthermore, retinyl esters and carotenes in foods are often complexed with protein from which they must be released. Hydrolysis from protein occurs by the action of pepsin in the stomach and other proteolytic enzymes in the proximal small intestine. Heating of plant foods before consumption is thought to fa-

cilitate bioavailability by weakening protein-carotenoid complexes [2]. Hydrolysis of retinyl and carotenoid esters by various esterases occurs at the same time triacylglycerols, phospholipids, and cholesteryl esters are being hydrolyzed by pancreatic enzymes. Pancreatic lipase, cholesterol ester hydrolase, as well as esterases from the intestinal brush border are thought to be responsible for deesterification [3–6].

The released carotenoids and retinols in the small intestine are solubilized into micellar solutions along with the other fat-soluble food components. The micellar solutions containing the carotenoids and vitamin A

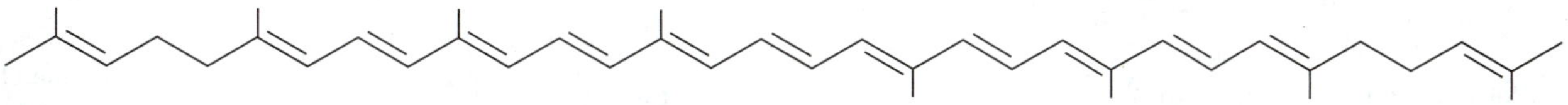

(g) Lycopene

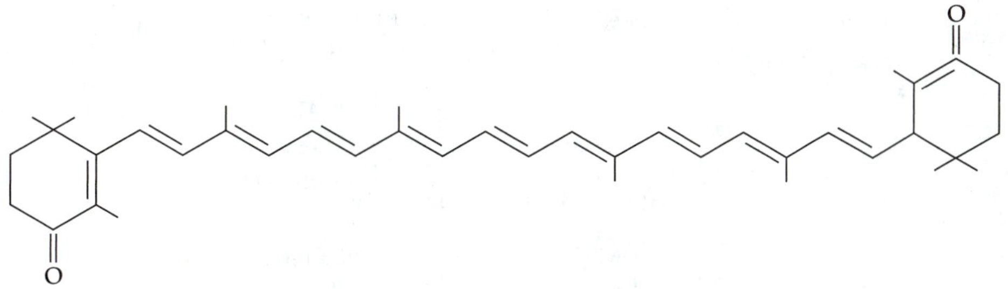

(h) Canthaxanthin

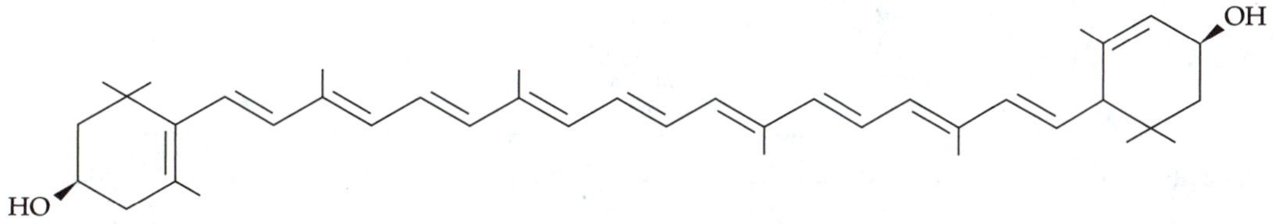

(i) Lutein

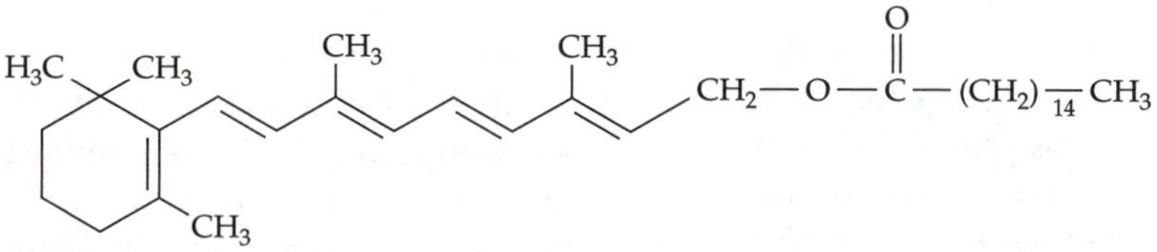

(j) Retinyl palmitate

Figure 10.1 (continued) Vitamin A and carotenoid structures.

diffuse through the glycoprotein layer surrounding the microvilli of the duodenum and jejunum and into the enterocyte. Approximately 70% to 90% of retinol from the diet is absorbed as long as the meal is adequate (~10 g or more) in fat [7,8]. Carotenoid absorption from the diet ranges from about 20% to 50%, but carotenoid absorption may be as low as 5% [4,5,8,9]. Carotenoid absorption decreases as carotenoid intake increases [9]. Figure 10.2 depicts the absorption of vitamin A and carotenoids.

Within the intestinal mucosal cell (and to a very small extent the liver), β-carotene 15,15'-dioxygenase can convert β-carotene into retinal (Fig. 10.3). Retinal then binds to cellular retinoid binding protein (CRBP) Type II. Retinal, formed from β-carotene, can subsequently be converted to retinol by retinal reductase, a NADH/NADPH-dependent enzyme. β-carotene theoretically can produce 2 mol of retinol, but in vivo this does not occur because activity of β-carotene 15,15'-dioxygenase is relatively low. Thus, some (up to 30%) β-carotene may leave the intestine without oxidation. It is estimated that 6 mg β-carotene are required to produce

the vitamin A activity of 1 mg retinol. Further, 12 mg of other provitamin A carotenoids such as α-carotene and γ-carotene are required to produce vitamin A activity of 1 mg retinol. Carotenoids not converted to retinol may be absorbed and transported in the blood to tissues.

Although the retinal is interconvertible with retinol, some of the retinal may be irreversibly oxidized into retinoic acid (Fig. 10.4) within the intestinal cell. Retinoic acid, in contrast to retinol, is picked up by the portal vein and transported in the plasma bound tightly to albumin.

Retinol, formed from the oxidation of carotenoids, follows the same metabolic pathways of reesterification in the intestinal cell as retinol originating from dietary retinyl esters. Retinol in the intestinal cell must be esterified for further use. One of two metabolic pathways may be followed for retinol reesterification in the enterocyte:

1. The primary pathway involves cellular retinol-binding protein (CRBP) II, whose synthesis may depend on retinoic acid [4,5]. CRBPs are part of a group

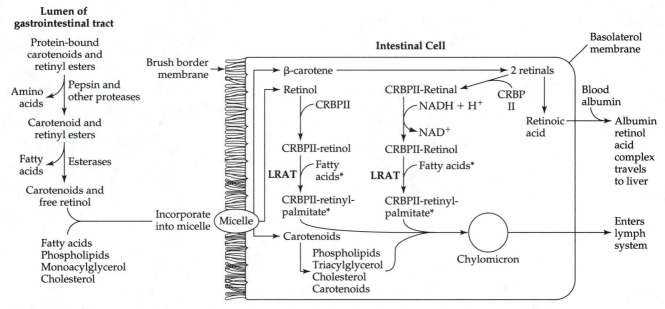

Figure 10.2 Absorption of carotenoids and vitamin A, and reesterification of retinol in the intestinal cell using CRBPII and LRAT.

of low-molecular-weight lipid-binding proteins. CRBPs are thought to help regulate retinol use in cells. CRBP II binds both retinol and retinal and is present in the cytoplasm of epithelial cells of the small intestine [3]. CRBP II directs the reduction of retinal and subsequent esterification [10]. CRBP II-bound retinol is esterified by lecithin retinol acyl transferase (LRAT) to form mainly retinyl palmitate, but also retinyl stearate and retinyl oleate, among others.

2. The minor second pathway for reesterification involves binding of retinol to a cellular protein that is nonspecific, with subsequent reesterification by acyl CoA retinol acyl transferase (ARAT) [3]. ARAT may serve to esterify retinol when large doses of the vitamin are ingested [11].

The newly formed retinyl esters, along with a small amount of unesterified retinol, and any carotenoids that have been absorbed unchanged, are incorporated into chylomicrons containing cholesterol esters, phospholipid, triacylglycerols, and apoproteins; these chylomicrons are then carried first into the lymphatic system and then into general circulation.

Chylomicrons deliver retinyl esters, some unesterified retinol, and carotenoids to many extrahepatic tissues such as bone marrow, blood cells, spleen, adipose tissue, muscle, lungs, and kidneys [12]. Chylomicron remnants deliver retinyl esters and a portion of the carotenoids not taken up by peripheral tissue to the liver.

Carotenoids reaching the liver can follow three routes:

1. a small portion may be cleaved to form retinol;
2. some may be incorporated into the very low-density lipoproteins (VLDLs) synthesized in the liver, and then be released as part of VLDLs for circulation to various tissues of the body; and
3. some can be stored in the liver.

Excess carotenoids not stored in the liver will be deposited in body lipids. Serum carotene levels are reflective of recent intake, and not body stores. The most common serum carotenoids include β-carotene, α-carotene, lycopene, lutein, zeaxanthin, and cryptoxanthin. Carotenoids such as β-carotene and lycopene are thought to concentrate in the hydrophobic core of lipoproteins for serum transport while carotenoids with polar groups would be found partly on the lipoprotein surface [2]. Further, β-carotene, α-carotene, and lycopene distribution among lipoproteins is similar whereby low-density lipoproteins (LDLs) carry 58% to 73%, high-density lipoproteins (HDL) carry 17% to 26%, and VLDLs carry 10% to 16% [2]. In contrast, lutein and zeaxanthin (diOH carotenoids) are carried predominantly (53%) by HDL but also by LDL (31%) and VLDL (16%) [2].

The handling of retinyl esters reaching the liver is shown in Figure 10.5; however, most cells of the body are able to metabolize retinol generated from the retinyl esters through a number of metabolic pathways. Hydrolysis of the retinyl esters occurs following their uptake by the hepatic parenchymal cells. Within the cell, retinol binds with a cellular retinol-binding

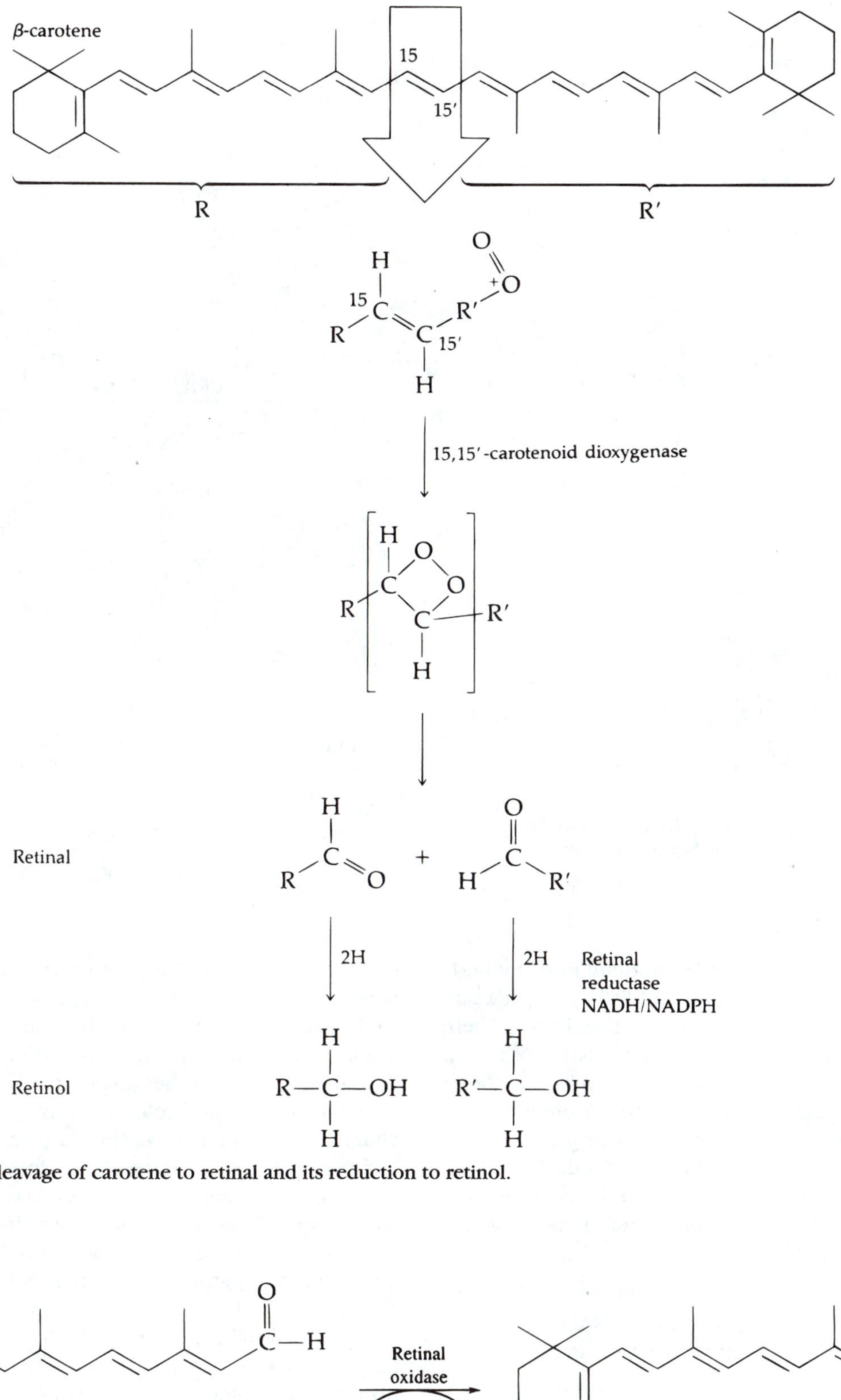

Figure 10.3 Cleavage of carotene to retinal and its reduction to retinol.

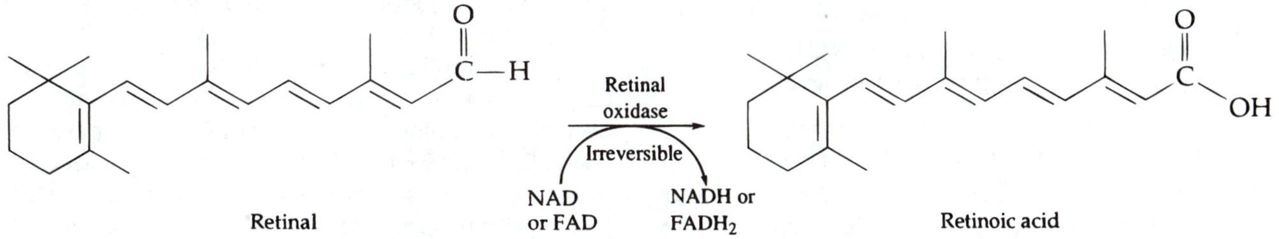

Figure 10.4 The irreversible oxidation of retinal to retinoic acid.

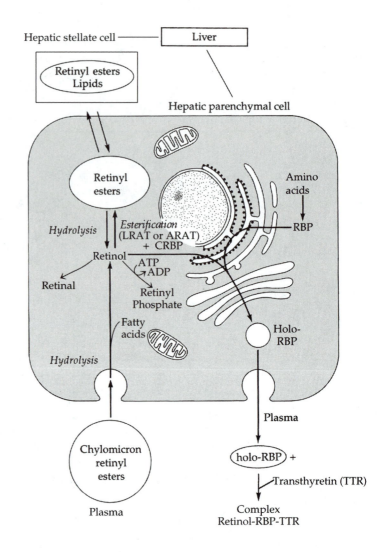

Figure 10.5 Diagram summarizing the metabolism of vitamin A and RBP in the liver.

protein (CRBP). CRBPs have been found in many body cells, especially the intestine (CBRP II), the liver, and the kidney. CRBP is thought to function both to help control concentrations of free retinol within the cell cytoplasm and thus prevent its oxidation, and to direct the vitamin through a series of protein-protein interactions to specific enzymes of metabolism [10,13,14]. CRBP may also assist in the transfer of retinal from the microsomal retinol dehydrogenase to the cytosolic retinal dehydrogenase [13]. Enzymatic metabolism of retinol includes possible

- esterification by enzymes such as LRAT or ARAT,
- oxidation of retinol to retinal by NAD(P)H-dependent retinol dehydrogenase, and
- phosphorylation of retinol to retinyl phosphate by ATP for glycoprotein functions [14].

Retinol not metabolized or transported from the liver may be stored following reesterification. Some storage of retinol occurs in the parenchymal cells, but about 80% to 95% of the retinol is stored in small perisinusoidal cells called *stellate cells* (also known as *Ito cells*). Vitamin A is stored in these stellate cells along with lipid droplets [15]. When liver stores of vitamin A are adequate, the stellate cells will store recently ingested vitamin A as retinyl esters (primarily retinyl palmitate, but also as retinyl stearate, oleate, and linoleate). For a given person, plasma vitamin A levels remain quite constant over a wide range of dietary intakes and liver stores. Only after the hepatic stellate cells can accept no more retinol for storage does hypervitaminosis A (see "Toxicity," p. 330) occur.

Retinol mobilization from the liver and delivery to target tissues are dependent on the synthesis and secretion of retinol-binding protein (RBP) by the parenchymal cells (Fig. 10.6). Each mole of retinol released by a hydrolase from its ester storage form combines with 1 mol of RBP to form holo-RBP. In the plasma, the holo-

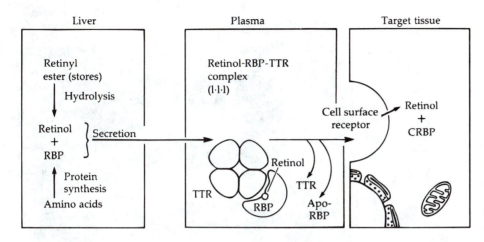

Figure 10.6 Schematic summary of processes involved in vitamin A mobilization from the liver, transport in plasma, and delivery and uptake into target cell.

RBP also interacts with a molecule of transthyretin (TTR), a protein formerly known as *prealbumin* that also binds to one thyroxine (T4) per tetramer. The retinol-RBP-TTR complex circulates in the plasma with a half-life of about 11 hours (Fig. 10.6); it is not filtered by the glomerulus. Some tissues that take up retinol from the RBP-TTR complex include adipose, skeletal muscle, kidney, white blood cells, and bone marrow [4,5]. There is extensive recycling of retinol (in rats 7 to 13 times prior to degradation) among plasma, extrahepatic tissues, and the liver, with both the parenchymal and stellate cells of the liver taking up retinol directly from its complexed form in the plasma [4,5,11].

Entry of the retinol into target cells (Fig. 10.6) is believed to involve slow release from RBP with subsequent association with the target cell [16]. It may also be mediated by specific cell surface receptors that recognize the RBP and internalize the retinol, but not the RBP [16]. The apo-RBP that remains after retinol release can no longer bind to TTR, and this apo-RBP is typically catabolized by the kidney.

In contrast to retinol, which is mobilized from the liver for transport to other tissues, retinoic acid is thought to be produced in small amounts by individual cells. Whether central production of retinoic acid occurs by the intestine or liver for transport to other tissues is unclear. Plasma retinoic acid concentrations are, however, typically low. Within the cell cytoplasm, retinoic acid binds to cellular retinoic acid-binding proteins (CRABPs). CRABPs are thought to function in a capacity similar to that described for CRBPs. Both CRBPs and CRABPs are often found in the same tissues; however, their relative distribution in the tissues differs [17]. CRABP, like CRBP, functions to help control concentrations of free retinoic acid within the cell and thus prevent its catabolism and to direct the usage of retinoic acid intracellularly.

Functions and Mechanisms of Action

Vitamin A

Vitamin A is recognized as being essential for vision, and for systemic functions including cellular differentiation, growth, reproduction, bone development, and the immune system. These functions will each be reviewed, and then those of the carotenoids will be addressed.

Visual Cycle Retinol transported to the retina via the RBP-TTR complex appears to move into the pigment epithelium of photoreceptor rod cells. The photoreceptor cell is shown in Figures 10.7 and 10.8.

The movement of retinol into the outer rod segments, where the visual cycle occurs, involves at least two proteins specific to the retina, cellular retinal-binding protein (CRALBP), and interstitial or interphotoreceptor retinol-binding protein (IRBP) [15]. In addition, CRBP and CRABP are also found in the retina [15].

Within the retina, retinol may be converted into a retinyl ester and stored. The retinyl ester is hydrolyzed as needed to release retinol, which is oxidized in the rod cells by an NAD-activated dehydrogenase to generate all *trans* retinal. This reaction occurs in the rod outer segments. The all-*trans* retinal produced is equilibrated with its 11-*cis* isomer either spontaneously or by an isomerase. The 11-*cis* retinal binds as a protonated Schiff base to a lysine amino acid residue in the protein opsin (Fig. 10.9) to produce the compound rhodopsin. Rhodopsin is embedded in disks, located in the rod's outer segment, which is enclosed within a restricted compartment of the retina created by tight junctions between cells (Figs. 10.7 and 10.8) [18]. The cells on the blood side are one layer thick and form the pigment epithelium. The "outer limiting membrane" is

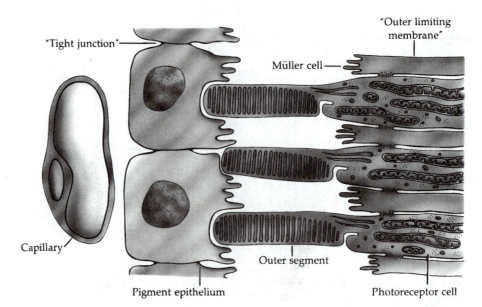

Figure 10.7 The photoreceptor cell and its restrictive boundaries.

formed on the vitreal side of the photoreceptor cells by specific junctions between the photoreceptor cells and the retinal glial cells of Müller (Fig. 10.7) [18]. Despite the barriers around the photoreceptor cells, vitamin A moves into and within the cells so that rhodopsin can be generated. CRBP and CRALBP are believed to assist in vitamin A transport through the pigment epithelium, whereas IRBP is required for transport of the various forms of vitamin A between the cell types [18,19]. Specifically, IRBP, a glycolipoprotein, resides within the retinal interphotoreceptor space that lies between the pigment epithelium and the photoreceptor cells. IRBP transports two molecules of retinol between the tissues [19].

Rod cells with rhodopsin detect small amounts of light; thus, they are important for night vision. When a quantum of light (hv) hits the rhodopsin (Fig. 10.10), rhodopsin breaks down in a series of reactions. The term *bleaching* is often used because a loss of color occurs as the rhodopsin splits. During the degradation of rhodopsin, all-*trans* retinal is generated. These processes also lead to an electrical signal to the optic nerve. A transmitter is required for carrying through the cell to the plasma membrane the message that light has hit the rhodopsin, and the result is that the sodium channels in the plasma membrane are blocked, and the rod cell hyperpolarizes.

Recovery of the rod's dark current and thus vision in dim light is believed to be made possible by the phosphorylation of opsin. With this phosphorylation, the cascade of light-activated enzymes is terminated. The all-*trans* retinal formed as a result of light must be converted back to 11-*cis* retinal; the exact steps involved and the location of the reaction (outer segment versus pigment epithelium) are unclear. The visual cycle is, however, completed when all-*trans* retinal is converted back to 11-cis retinal and bound once again to rhodopsin.

Cellular Differentiation Retinoic acid acts as a hormone to affect gene expression and thus control cell development. Retinoic acid may also act to diminish degradation of retinol by affecting CRBP II synthesis. Retinoic acid or 9-*cis* retinoic acid (generated from 9-*cis* retinol) are transported to the nucleus bound to CRABP. Within the nucleus, retinoic acid and 9-*cis* retinoic acid bind to one or more of three retinoic acid receptors (RAR) or to one or more of three retinoid X receptors (RXR), respectively. Binding of retinoic acid or 9-*cis* retinoic acid to RAR or the binding of 9-*cis* retinoic acid to RXR followed by RAR-RXR dimerization permits interaction with other transcription factors or directly with specific nucleotide sequences of nuclear DNA to regulate (stimulate or inhibit) gene transcription, with the potential to affect a wide variety of body proteins and thus body processes (Fig. 10.11).

Only a few of the vast number of processes affected by the binding of retinoic acid (or 9-cis retinoic acid) to nuclear receptors are unknown. Retinoic acid is thought to act as a morphogen in embryonic developments [19]. Nuclear retinoic acid receptors appear in different cells during different times of development. Thus, retinoic acid may serve to signal morphogenesis [17].

Retinoic acid is needed by epithelial cells found in such places as the lungs, trachea, skin, and gastrointestinal tract, among others. Retinoic acid helps to maintain both the normal structure and the functions of the epithelial cells [14]. For example, retinoic acid directs the

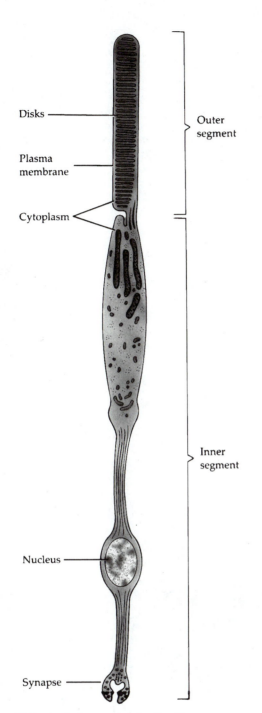

Disks

Plasma
membrane

Cytoplasm

Outer
segment

Inner
segment

Nucleus

Synapse

Figure 10.8 A photoreceptor (rod) cell.

differentiation of keratinocytes (immature skin cells) into mature epidermal cells. Retinoic acid appears to have more specific effects on cellular differentiation than 9-*cis* retinoic acid [17]. Retinoic acid is thought to act as a signal to "switch on" the genes for keratin proteins [14]. Vitamin A also appears to direct the synthesis of keratins, with genes for smaller (versus larger)

keratin molecules transcribed and translated in the presence of vitamin A [16,20]. Vitamin A, in vitro, directs differentiation of squamous epithelial keratinizing cells into mucus-secreting cells [20].

Growth Vitamin A deficiency has long been characterized in animals by impaired growth that can be stimulated with replacement by either retinol or retinoic acid. Specifically, vitamin A has been shown to stimulate the growth of epithelial cells. Cell growth is stimulated, in part, by growth factors that bind to specific receptors on cell surfaces. Retinoic acid appears to increase the number of specific receptors for growth factors [20]. The mechanism by which vitamin A affects growth is unclear; however, the vitamin may act by increasing the synthesis of cell surface components such as glycoproteins (see the next section). Retinyl β-glucuronide, formed in a variety of tissues from retinol and UDP-GA, has been shown to actively support growth and differentiation [21]. Retinoyl β-glucuronide has also been shown to improve acne lesions in patients [22].

Cell Surface Functions: Glycoproteins One cell surface function of vitamin A is thought to be mediated through glycoproteins, principal cell surface constituents involved in, for example, cell communication, cell recognition, cell adhesion, and cell aggregation. Vitamin A is thought to play a role in the synthesis of glycoproteins. The possible mode of action involves the formation of retinyl phosphate, formed from the conjugation of retinol and ATP (Fig. 10.12). Retinyl phosphate can be converted into retinylphosphomannose (also called *mannosyl retinyl-phosphate*) in the presence of GDP-mannose. Retinyl phosphomannose can in turn transfer the mannose to a glycoprotein acceptor. The glycoprotein receptor, on receipt of the mannose, becomes a mannosylated glycoprotein [21]. Such changes in the glycan portion of the glycoprotein in turn can greatly affect dif-ferentiation of cells or tissues through their effects on cell communication, recognition, adhesion, and cell aggregation.

Retinoic acid has also been shown to affect cell membranes by increasing the number of junctions between cells (called *gap junctions*). These junctions are important for cell-to-cell communication and cell adhesion. Vitamin A and retinoic acid can modify cell surfaces, possibly again through increasing glycoprotein synthesis at the gene level or by improving attachment of glycoproteins to cell surfaces to induce cell adhesion [20].

Other Functions Vitamin A, as retinol but not retinoic acid, is essential for *reproductive processes* in both males

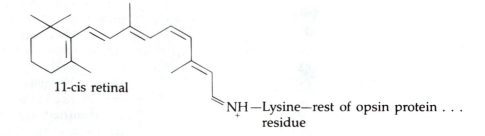

Figure 10.9 11-*cis* retinal bound to the protein opsin to form rhodopsin.

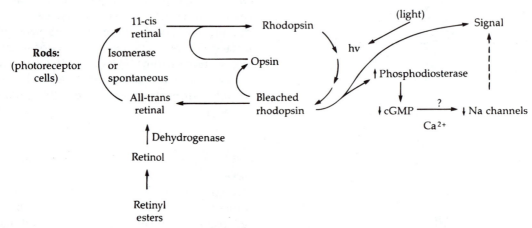

Figure 10.10 The visual cycle.

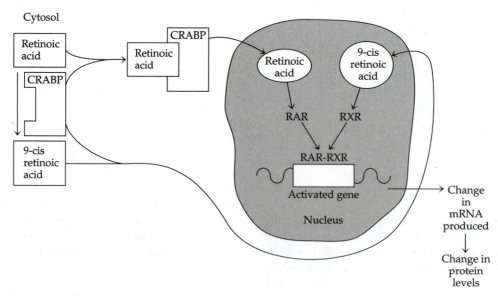

Figure 10.11 Hypothesized mode of action for retinoic acid.

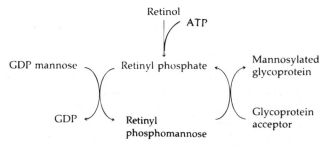

Figure 10.12 The role of vitamin A in the mannosylation of glycoproteins.

and females, although the mechanism(s) of its action(s) are unclear at present. *Bone development and maintenance* also requires vitamin A. Vitamin A is necessary for bone metabolism through involvement with osteoblasts and osteoclasts. Although the mechanism of action is unclear, vitamin A deficiency results in excessive deposition of bone by osteoblasts and reduced osteoclasts. Several aspects of *immune system function* also appear to be influenced by vitamin A. Depletion studies suggest that vitamin A appears to be needed for T-lymphocyte function and for antibody response to viral, parasitic, and bacterial infections [23]. Natural killer cell activity and phagocytosis are also impaired with vitamin A deficiency [17,23].

Carotenoids

Carotenoids are thought to be present in the interior membranes of cells as well as in lipoproteins. Carotenoids structurally possess an extended system (often nine or more) of conjugated double bonds that enable solubility in lipids and the quenching of singlet molecular oxygen ($^1O_2$) and of free radicals such as peroxyl radicals. In other words, carotenoids are thought to function as antioxidants, because they possess the ability to react with and quench free-radical reactions in lipid membranes or compartments and possibly in solution.

Singlet Oxygen and Carotenoids Quenching is a process by which electronically excited molecules, such as singlet molecular oxygen, are inactivated [24]. Singlet molecular oxygen possesses higher energy and is more reactive than ground state molecular oxygen, which exists typically in triplet ($^3O_2$) versus singlet ($^1O_2$) form. Singlet oxygen is generated from lipid peroxidation of membranes, transfer of energy from light (photochemical reactions), or the respiratory burst occurring in neutrophils, for example (enzymatic reactions). Singlet molecular oxygen readily reacts with organic molecules such as protein, lipids, and DNA and thus can damage cellular components unless removed.

Carotenoids, such as β-carotene or lycopene, can react with (quench) singlet oxygen, and it is the conjugated double-bond systems within the carotenoids, which permits the quenching. β-carotene, for example, bound to lymphocytes taken from human blood has been shown to directly quench singlet molecular oxygen in vitro [25]. Lycopene appears to have the highest rate constant and is a more effective quencher of singlet oxygen than other carotenoids [26]. The singlet oxygen ($^1O_2$) transfers its excitation energy and returns to the ground state ($^3O_2$), while the carotenoid receiving the energy enters an excited state. Resonance states in the excited carotenoid allow some stabilization. Carotenoids then release the energy in the form of heat and thus do not need to be regenerated.

$$^1O_2 + \beta\text{-carotene} \longrightarrow {}^3O_2 + \text{excited } \beta\text{-carotene} \longrightarrow$$
$$\beta\text{-carotene} + \text{heat}$$

Free Radicals and Carotenoids In addition to quenching singlet oxygen, β-carotene and other carotenoids have the ability to react directly with peroxyl radicals involved in lipid peroxidation [27]. This ability has been demonstrated in the range of partial pressures of oxygen that exist under physiological conditions. Studies suggest that β-carotene works synergistically with vitamin E in scavenging radicals and inhibiting lipid peroxidation, although vitamin E has higher reactivity toward peroxyl radicals than β-carotene [28,29]. β-carotene is thought to function in the interior of the membrane, whereas α-tocopherol functions on or at the surface of the membrane [28]. Lipid peroxidation has been significantly reduced in smokers (who have chronic oxidative damage due to the effects of smoking) receiving only 20 mg β-carotene for 4 weeks [30]. β-carotene supplementation (50–100 mg daily for 3 weeks) has been shown to shorten the lag phase of metal (copper)-dependent lipid peroxidation of LDLs. In other words, β-carotene supplementation helps to make the LDL more resistant to metal-induced lipid oxidation [31]. Lipid peroxidation is an indicator of free radical activity and reflects damage to membranes and possibly other organelles or DNA. Supplements providing 120 mg β-carotene also significantly reduced lipid peroxidation in humans [32]. Carotene depletion in humans increased the presence of substances associated with a significant decrease by day 29 of serum carotene concentrations and with oxidative damage, measured by the increased presence of thiobarbituric acid reactive substances, hexanal, pentanal, and pentane [33]. Repletion studies providing β-carotene (15 or 120 mg) have been shown to decrease the concentrations of circulating peroxides [33,34]. In addition, a significant inverse relationship between serum β-carotene and lipid peroxide concentrations with 16% of the variability in lipid peroxides attributed to serum β-carotene concentration was noted [34].

Because of the ability of carotenoids to react with free radicals and quench singlet oxygen, carotenoids are thought to be protective against several diseases such as cardiovascular disease and cancer, among others. Epidemiological studies have shown that those individuals with high intakes of fruits and vegetables, which are also rich in carotenoids, or those individuals with higher serum carotenoid concentrations have a lower incidence of diseases such as cardiovascular disease and cancer [35–39]. In the mechanism(s) thought to lead to the development of atherosclerosis, oxidation of cholesterol in LDLs increases the likelihood of monocyte or macrophage uptake in comparison with native (unoxidized cholesterol). Oxidized cholesterol also impairs arterial nitric oxide release, which may contribute to vasospasms and platelet adhesion associated with heart disease [40]. Oxidized LDL cholesterol stimulates the binding of monocytes to blood vessel endothelium. These monocytes become embedded in the endothelium as do native LDLs, which have potential to become oxidized within the subendothelial space. Macrophages in the blood vessel endothelia continue to take up oxidized LDL cholesterol via a specific cell surface scavenger receptor, and, when filled with cholesterol, become foam cells. Fatty streaks (an accumulation of foam cells) lead to atherosclerotic plaque in blood vessels [41]. Carotenoids, such as β-carotene, lycopene, lutein, α-carotene, among others, are thought to prevent the oxidation of LDL cholesterol and other cell membrane lipids to thus prevent or slow the development of atherosclerosis [42,43]. α-tocopherol (vitamin E), however, may be more effective than β-carotene in inhibiting LDL oxidation [44]. Supplementation with 800 IU of vitamin E, 1,000 mg of vitamin C, and 24 mg of β-carotene in individuals with coronary artery disease has been shown to effectively reduce LDL susceptibility to oxidation [45].

As a scavenger of free radicals and singlet oxygen, carotenoids also are thought to protect against cancer. Remember, oxidative stress is thought to contribute in part to carcinogenesis. Carotenoids such as canthaxanthin and β-carotene have been shown in vitro to inhibit carcinogen-induced neoplastic transformation and inhibit lipid oxidation in plasma membranes [46]. Carotenoids appear to vary in their abilities to inhibit oxidation and cell transformation. Canthaxanthin appears to be more effective than β-carotene followed by α-carotene and then lycopene in inhibiting chemically induced neoplastic transformation [47]. In addition, some carotenoids up-regulate the expression of connexin 43, a gene that codes for a structural unit of a gap junction [46]. Some carcinogens appear to inhibit gap junction

communications, thus the ability of carotenoids to protect this activity may be of significance in prevention of carcinogenesis [48].

Results of intervention trials using β-carotene have failed to show protective effects. Intervention trials with β-carotene (BC) (20 mg) along with α-tocopherol (AT) in Finland (referred to as the *ATBC trial*) as well as intervention with 30 mg of β-carotene and 25,000 IU of vitamin A (*CARET trial*) were not shown to be beneficial versus placebo in disease prevention in asbestos exposed workers and in individuals with a long history of smoking [49,50]. β-carotene taken with the vitamin A even appeared to increase the risk of lung cancer and increase the risk of mortality from cancer and heart disease in the high-risk (smokers and asbestos exposed) populations [50]. These two intervention trials appear to suggest that β-carotene supplementation increases risk of cancer; however, others suggest that the populations were at high risk for disease with a long history of smoking (one pack or more cigarettes daily) or asbestos exposure along with alcohol consumption, and it is only with these cofactors that β-carotene supplementation increases risk [51,52].

Interactions with Other Nutrients

Vitamin A interacts with both *vitamins E* and *K*. Cleavage of β-carotene into retinal requires vitamin E. Vitamin E is probably necessary to protect the substrate and the product from oxidation; however, large doses (10 times the RDA) of vitamin E may inhibit β-carotene absorption or conversion to retinol in the intestine [1,7]. Excess vitamin A also appears to interfere with vitamin K absorption.

Protein status also influences vitamin A status and transport. The activity of the enzyme carotenoid dioxygenase that cleaves β-carotene is depressed by inadequate protein intake. Overall vitamin A metabolism is closely related to protein status because transport and use of the vitamin depend on several vitamin A–binding proteins synthesized in the body.

A *zinc* deficiency interferes with vitamin A metabolism. Its effect appears to operate at two levels. First, a general reduction in growth accompanied by decreased food intake and a reduction in the synthesis of plasma proteins, particularly RBP, which is made in the liver. Thus, with zinc deficiency, plasma retinol concentrations decrease, and liver retinol concentrations increase. Second, with zinc deficiency there is a decreased hepatic mobilization of retinol from its storage as retinyl esters. The activity of the enzyme retinyl ester hydrolase, which releases the vitamin from its storage form, may be inhibited by the lack of zinc or possibly by vitamin E [53]. In peripheral tissue, alcohol dehydrogenase, which converts retinol into retinal, is also dependent on zinc.

Iron status is also interrelated with vitamin A. Vitamin A deficiency may result in microcytic anemia. Vitamin A supplementation in turn corrects the anemia with observed increases in indices of iron status [53]. Vitamin A may be directly acting on iron metabolism or storage or may be affecting differentiation of the red blood cell [17,53].

Metabolism and Excretion

Retinol conversion to retinal is reversible, but the oxidation of retinal to retinoic acid is irreversible. Retinoic acid does not accumulate in the liver or any other tissues in appreciable amounts [54].

The major pathway of retinoic acid catabolism is oxidation to 4-hydroxy (OH) retinoic acid in a NADPH-dependent reaction (Fig. 10.13). This compound is subsequently converted to 4-oxoretinoic acid in an NAD-requiring reaction. The latter compound is further oxidized to a variety of metabolites for excretion.

The oxidized products of vitamin A that contain intact chains are conjugated to glucuronide (Fig. 10.13) and excreted primarily via the bile into the feces. About 70% of the vitamin A metabolites is excreted in the feces. Some of the polar products, however, can be absorbed and returned to the liver via enterohepatic

circulation. This recycling mechanism helps to conserve the body's supply of vitamin A. Urinary excretion of vitamin A metabolites accounts for about 30% of vitamin A excretion [8], whereas a small amount is expired by the lungs as CO_2 [17,53].

Carotenoids, newly absorbed and not stored or converted to retinal or retinol, may be metabolized into a variety of compounds depending on the individual carotenoid. Carotenoid metabolites are excreted into the bile [22].

Recommended Dietary Allowances

The 1989 recommended dietary allowances (RDA) [8] for vitamin A are 800 mg RE for adult women and 1,000 mg RE for adult men; these recommendations do not differ from those published in 1980 [55]. Some tables and sources report vitamin A in terms of international units (IU). For conversion purposes, 1 IU = 0.3 µg retinol or 3.33 = 1 µg retinol = 1 RE [8].

Recommendations for adequate intake of vitamin A have been based on the amounts needed to correct night blindness among vitamin A–deficient subjects and to raise plasma vitamin A levels to a normal level in depleted subjects. In addition, estimates of vitamin need have been based on amounts of the vitamin needed to maintain a given body pool size in well-nourished subjects. A concentration of 20 mg retinol/g liver was used

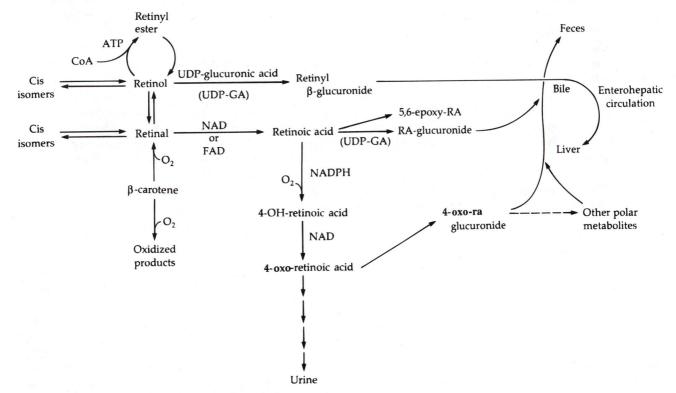

Figure 10.13 Vitamin A metabolism and metabolite excretion.

to represent a satisfactory total body reserve [7]. With this latter approach, the recommended level for vitamin A intake was reduced to 700 RE for adult men and 600 RE for adult women; however, the recommendations (referred to as *Recommended Dietary Intakes,* or RDIs) of the committee were never officially accepted [7]. Studies of vitamin A utilization rates in adult men, although few in number, generated wide variations in vitamin A utilization. The minimum requirement was determined as 600 µg retinol; however, neither liver stores nor plasma retinol concentrations are optimal at this level. A comparison of the 1987 RDI and 1989 RDA for vitamin A in different population groups is given in Table 10.2. Dietary Reference Intakes have not yet been published for vitamin A [56].

Deficiency

Vitamin A deficiency is less common in the United States than in developing countries, where inadequate intake is fairly common in children under 5 years of age. Increased mortality is associated with both clinically evident vitamin A deficiency in children as well as with children with inadequate vitamin A stores but no clinical signs of deficiency [57]. Selected signs and symptoms of deficiency include anorexia, retarded growth, increased susceptibility to infections, obstruction and enlargement of hair follicles, and keratinization of epithelial (mucous) cells of the skin with (thus) failure of normal differentiation [17,21]. Night blindness results from impaired production of rhodopsin in the outer segments of the rods. Xerophthalmia occurs with vitamin A deficiency and is characterized by abnormalities of the conjunctiva and cornea of the eye. Conjunctival changes include the disappearance of goblet cells in the conjunctiva, epithelial cells become enlarged and keratinized, and Bitot's spots appear overlying the keratinized epithelia of the conjunctiva. The Bitot's spots are white accumulations of sloughed cells thought to result from decreased retinol and glycoproteins in tear fluid as well as slower diffusion of retinol from plasma to the epithelial layer [17,53]. Keratomalacia may occur if the changes in the cornea become severe and irreversible (such as with corneal perforation and loss of aqueous humor).

Conditions and populations associated with increased need for vitamin A include those with malabsorptive disorders such as those with steatorrhea, pancreatic, liver, or gallbladder diseases. People with chronic nephritis, acute protein deficiency, intestinal parasites, or acute infections may also become vitamin A–deficient. Measles infections in developing countries are associated with high mortality. Measles is thought to depress vitamin A status, which may be already low in children from developing countries. Vitamin A supplements are recommended by WHO and UNICEF for children with measles and living in a country with measles fatality rates of 1% or greater [23]. Serious infections increase urinary vitamin A excretion [58].

Toxicity: Hypervitaminosis A

Being fat-soluble, vitamin A has a relatively long biological half-life, accumulating mostly in the liver but also other tissues. Thus, high doses ingested over a short time frame as well as chronic intakes of lower doses (but still in excess of recommendations) may lead to hypervitaminosis A. That toxicity can result from an excessive intake of vitamin A has been brought to the public's attention in recent years by the teratogenic effects of 13-*cis* retinoic acid (Acutane), which is used extensively in the treatment of acne. Use of this compound by women in the early months of their pregnancy has resulted in a number of birth defects among the infants born to these women [17,53]. Further, one infant in 57 born to women who ingested daily more than 10,000 IU of vitamin A was deemed to have a malformation attributed to the vitamin A supplement [59]. Consequently, many dermatologists advise against Acutane usage for women who are or may become pregnant, thus prescribing contraceptives for patients in their childbearing years taking the drug.

In adults, a chronic intake of vitamin A in amounts 10 times greater than the RDA (10 mg RE) can result in hypervitaminosis, manifested by a variety of maladies (anorexia, dry, itchy and desquamating skin, alopecia and coarsening of the hair, ataxia, headache, bone and muscle pain, conjunctivitis, among others). Most manifestations of toxicity appear to subside gradually once excessive intake of the vitamin is discontinued [60].

When vitamin A intake is in excess, serum retinol levels may rise above 200 mg/dL (normal is 45–65 mg/dL). Retinol is no longer transported exclusively by

Table 10.2 Comparison of 1987 RDIs and 1989 RDAs for Vitamin A in Different Population Groups

Population Groups	RDI (µg)	RDA (µg)
Infants, 0–24 mo	375	375–400
Children, 2–6 yr	400	400–500
Children, 6–9 yr	500	500–700
Males, 10–11 yr	600	700–1,000
Males, 12–70+ yr	700	1,000
Females, 10–70+ yr	600	700–800
Pregnancy, 6–9 mo	+200	800
Lactation, 0–5 mo	+400	1,300
Lactation, 6+ mo	+320	1,200

Note: Population groups are divided differently in various recommendations.

RBP but can be carried to the tissues by plasma lipoproteins. It has been suggested that when retinol is presented to the cell membranes in a form other than in a RBP complex, the released retinol produces potentially toxic effects [16]. Effects on the liver, the primary storage site for vitamin A are multiple. Some effects on the liver include fat-storing cell hyperplasia and hypertrophy, fibrogenesis, sclerosis of veins, portal hypertension, and congestion in perisinusoid cells, which lead to hepatocellular damage and cirrhosis or a cirrhosis-like hepatic disorder [60–62]. Based on reports of toxicity, a NOAEL (no observed adverse effect level) of 3 mg and a LOAEL (lowest observed adverse effect level) of 6.5 mg have been suggested for vitamin A [63].

β-carotene is listed on the Generally Recognized as Safe (GRAS) list with the FDA for use as a dietary and nutrient supplement as well as for use as a colorant in foods, drugs, and cosmetics [64]. Ingestion by adults in amounts as high as 180 mg of β-carotene daily for several months appears to pose no serious side effects [65]. Hypercarotenosis in individuals ingesting about 30 mg or more of β-carotene daily can, however, cause a yellow discoloration of the skin occurring especially in the fat pads or fatty areas of the palms of the hands and soles of the feet [66]. The condition usually disappears following removal of the carotenoids from the diet. An intake of 25 mg of β-carotene has been deemed safe for adults except those who smoke heavily [63].

Assessment of Nutriture

Vitamin A status may be assessed in a variety of ways including clinical assessment for Bitot's spots, measurements of dark adaptation threshold, and electrophysiological measurements including electroretinograms to measure the level of rhodopsin and its rate of regeneration [67].

The conjunctival impression cytology (CIC) method measures the reduction in goblet cells and the derangement of epithelia on the conjunctiva of the eye, which occurs with impaired vitamin A status [68].

Plasma retinol concentrations are frequently measured as a biochemical indicator of vitamin A status. Plasma retinol levels reflect status best if the individual has exhausted their stores of the vitamin, as with deficiency, or if their stores are filled to capacity, as with toxicity.

Measurement of changes in plasma retinol concentrations before and 5 hours after oral administration of retinyl esters (450–1,000 mg) in oils, a process referred to as *relative dose response* (RDR), is also done to assess vitamin A status.

$$RDR\ (\%) = \frac{5\ hour\ plasma\ retinol - initial\ plasma\ retinol}{5\ hour\ plasma\ retinol\ concentration} \times 100$$

Criteria for RDR and plasma vitamin A status are as follows [8,17,67]:

	Vitamin A Status			
	Deficient	Marginal	Acceptable	Better
Plasma vitamin A (mg/dL)	<10	10–20	>20	>30
RDR (%)	>50	50–20	<20	

References Cited for Vitamin A

1. Erdman J, Poor C, Dietz J. Factors affecting the bioavailability of vitamin A, carotenoids, and vitamin E. Food Tech 1988;42:214–21.

2. Parker R. Absorption, metabolism, and transport of carotenoids. FASEB J 1996;10:542–51.

3. Ong D. Retinoid metabolism during intestinal absorption. J Nutr 1993;123:351–5.

4. Blomhoff R, Green M, Norum K. Vitamin A: Physiological and biochemical processing. Ann Rev Nutr 1992;12:37–57.

5. Norum K, Blomhoff R. McCollum Award Lecture, 1992: Vitamin A absorption, transport, cellular uptake, and storage. Am J Clin Nutr 1992;56:735–44.

6. Rigtrup K, McEwen L, Said H, Ong D. Retinyl ester hydrolytic activity associated with human intestinal brush border membranes. Am J Clin Nutr 1994; 60:111–6.

7. Olson J. Recommended dietary intakes (RDI) of vitamin A in humans. Am J Clin Nutr 1987;45:704–16.

8. National Research Council. Recommended Dietary Allowances, 10th ed. Washington, D.C.: National Academy Press, 1989:78–92.

9. Brubacher G, Weiser H. The vitamin A activity of beta-carotene. Internatl J Vitam Nutr Res 1985; 55: 5–15.

10. Ong D. Cellular transport and metabolism of vitamin A: roles of the cellular retinoid-binding proteins. Nutr Rev 1994;52:S24–S31.

11. Blomhoff R. Transport and metabolism of vitamin A. Nutr Rev 1994;52:S13–S23.

12. Schmitz H, Poor C, Wellman R, Erdman J. Concentrations of selected carotenoids and vitamin A in human liver, kidney and lung tissue. J Nutr 1991;121:1613–21.

13. Napoli J. Biosynthesis and metabolism of retinoic acid: Roles of CRBP and CRABP in retinoic acid: roles of CRBP and CRABP in retinoic acid homeostasis. J Nutr 1993;123:362–6.

14. Ross A, Ternus M. Vitamin A as a hormone: Recent advances in understanding the actions of retinol,

retinoic acid, and beta carotene. J Am Diet Assoc 1993;93:1285-90.

15. Ross A. Overview of retinoid metabolism. J Nutr 1993;123:346-50.

16. Creek K, St. Hilaire P, Hodam J. A comparison of the uptake, metabolism and biologic effects of retinol delivered to human keratinocytes either free or bound to serum retinol-binding protein. J Nutr 1993;123:356-61.

17. Olson J. Vitamin A. In: Shils ME, Olson JA, Shike M, eds. Modern Nutrition in Health andDisease, 8th ed. Philadelphia: Lea and Febiger, 1994:287-307.

18. Ong D. Vitamin A-binding proteins. Nutr Rev 1985;43:225-32.

19. Wolf G. The intracellular vitamin A-binding proteins: an overview of their functions. Nutr Rev 1991;49:1-12.

20. Wolf G. Multiple functions of vitamin A. Physiol Rev 1984;64:873-937.

21. Combs G. The Vitamins. New York: Academic Press, 1992:119-50.

22. Olson J. 1992 Atwater Lecture: the irresistible fascination of carotenoids and vitamin A. Am J Clin Nutr 1993;57:833-39.

23. Ross A. Vitamin A and protective immunity. Nutr Today 1992;27 (July-Aug):18-26.

24. Mascio P, Murphy M, Sies H. Antioxidant defense systems: the role of carotenoids, tocopherols, and thiols. Am J Clin Nutr 1991;53:194S-200S.

25. Bohm F, Haley J, Truscott T, Schalch W. Cellular bound β-carotene quenches singlet oxygen in man. J Photochem Photobiol B Biol 1993;21:219-21.

26. Conn P, Schalch W, Truscott T. The singlet oxygen and carotenoid interaction. J Photochem Photobiol B 1991;11:41-47.

27. Burton G, Ingold K. β-carotene: an unusual type of lipid antioxidant. Science 1984;224:569-73.

28. Niki E, Noguchi N, Tsuchihashi H, Gotoh N. Interaction among vitamin C, vitamin E, and β-carotene. Am J Clin Nutr 1995;62(suppl):1322S-26S.

29. Palozza P, Krinsky N. β-carotene and α-tocopherol are synergistic antioxidants. Arch Biochem Biophys 1992;297:184-7.

30. Allard J, Royall D, Kurian R, Muggli R, Jeejeebhoy K. Effects of β-carotene supplementation on lipid peroxidation in humans. Am J Clin Nutr 1994;59:884-90.

31. Gaziano J, Hatta A, Flynn M, Johnson E, Krinsky N, Ridker P, Hennekens C, Frei B. Supplementation with β-carotene in vivo and in vitro does not inhibit low density lipoprotein oxidation. Atherosclerosis 1995;112:187-95.

32. Gottlieb K, Zarling E, Mobarhan S, Bowen P, Sugerman S. β-carotene decreases markers of lipid peroxidation in healthy volunteers. Nutr Cancer 1993; 19:207-12.

33. Dixon Z, Burri B, Clifford A, Frankel E, Schneeman B, Parks E, Keim N, Barbieri T, Wu M, Fong A, Kretsch M, Sowell A, Erdman J. Effects of a carotene-deficient diet on measures of oxidative susceptibility and superoxide dismutase activity in adult women. Free Radic Biol Med 1994;17:537-44.

34. Mobarhan S, Bowen P, Andersen B, Evans M, Sapuntzakis M, Sugerman S, Simms P, Lucchesi D, Friedman H. Effects of β-carotene repletion on β-carotene absorption, lipid peroxidation, and neutrophil superoxide formation in young men. Nutr Cancer 1990; 14:195-206.

35. Block G, Patterson B, Subar A. Fruit, vegetables, and cancer prevention: a review of the epidemiological evidence. Nutr Cancer 1992;18:1-29.

36. Ziegler R. Vegetable, fruits and carotenoids and risk of cancer. Am J Clin Nutr 1991;53 (suppl):251S-9S.

37. Gaziano J, Hennekens C. The role of beta carotene in the prevention of cardiovascular disease. Ann N Y Acad Sci 1993;69:148-54.

38. Morris D, Kritchevsky S, Davis C. Serum carotenoids and coronary heart disease. JAMA 1994;272:1439-41.

39. Poppel G, Goldbolm R. Epidemiologic evidence for β-carotene and cancer prevention. Am J Clin Invest 1995;62(suppl):1393S-1402S.

40. Levine G, Keaney J, Vita J. Cholesterol reduction in cardiovascular disease. N Eng J Med 1995;332:512-21.

41. Luc G, Fruchart J-C. Oxidation of lipoproteins and atherosclerosis. Am J Clin Nutr 1991;53:206S-9S.

42. Prince M, Frisoli J. Beta-carotene accumulation in serum and skin. Am J Clin Nutr 1993;57:175-81.

43. Krinsky N. Actions of carotenoids in biological systems. Ann Rev Nutr 1993;13:561-87.

44. Abbey M, Nestel P, Baghurst P. Antioxidant vitamins and low-density-lipoprotein oxidation. Am J Clin Nutr 1993;58:525-32.

45. Mosca L, Rubenfire M, Mandel C, Rock C, Tarshis T, Tsai A, Pearson T. Antioxidant nutrient supplementation reduces the susceptibility of low density lipoprotein to oxidation in patients with coronary artery disease. J Am Coll Cardiol 1997;30:392-9.

46. Bertram J, Bortkiewicz H. Dietary carotenoids inhibit neoplastic transformation and modulate gene expression in mouse and human cells. Am J Clin Nutr 1995;62(suppl):1327S-36S.

47. Bertram J, Pung A, Churley M, Kappock T, Wilkins L, Cooney R. Diverse carotenoids protect against chemically induced neoplastic transformation. Carcinogenesis 1991;12:671–8.

48. Acevedo P, Bertram J. Liarozole potentiates the cancer chemopreventive activity of and the up-regulation of gap junctional communication and connexin 43 expression by retinoic acid and beta-carotene in 10T1/2 cells. Carcinogenesis 1995;16:2215–22.

49. Alpha-tocopherol, beta-carotene (ATBC) cancer prevention study group. The effect of vitamin E and beta carotene on the incidence of lung cancer and other cancers in male smokers. N Engl J Med 1994;330:1029–35.

50. Omenn G, Goodman G, Thomquist M, Balmes J, Cullen M, Glass A, Keogh J, Meyskens F, Valanis B, Williams J, Barnhart S, Hammar S. Effects of a combination of beta carotene and vitamin A on lung cancer and cardiovascular disease. N Engl J Med 1996; 334:1150–5.

51. Omenn G, Goodman G, Thornquist M, Balmes J, Cullen M, Glass A, Keogh J, Meyskens F, Valanis B, Williams J, Barnhart S, Cherniack M, Brodkin C, Hammar S. Risk factors for lung cancer and for intervention effects in CARET, the beta-carotene and retinol efficacy trial. J Natl Cancer Inst 1996;88:1550–9.

52. Mayne S, Handelman G, Beecher G. β-carotene and lung cancer promotion in heavy smokers—a plausible relationship? J Natl Cancer Inst 1996;88:1513–5.

53. Olson J. Vitamin A. In: Machlin LJ. Handbook of Vitamins, 2nd ed. New York: Dekker, 1991:1–57.

54. Goodman D. Vitamin A and retinoids in health and disease. N Engl J Med 1984;310:1023–31.

55. National Research Council. Recommended Dietary Allowances, 9th ed. Washington, D.C.: National Academy of Sciences, 1980:55–60.

56. Yates A, Schlicker S, Suitor C. Dietary reference intakes: the basis for recommendations for calcium and related nutrients, B vitamins, and choline. J Am Diet Assoc 1998;98:699–706.

57. Olson J. Hypovitaminosis A: contemporary scientific issues. J Nutr 1994;124:1461S–6S.

58. Stephensen C, Alvarez J, Kohatsu J, Hardmeier R, Kennedy J, Gammon RB. Vitamin A is excreted in the urine during acute infection. Am J Clin Nutr 1994; 60:388–92.

59. Rothman K, Moore L, Singer M, Nguyen U, Mannino S, Milunsky A. Teratogenicity of high vitamin A intake. N Engl J Med 1995;333:1369–73.

60. Stimson W. Vitamin A intoxication in adults. Report of a case with a summary of the literature. N Engl J Med 1961;265:369–73.

61. Russell R, Boyer J, Bagheri S, Hruban Z. Hepatic injury from chronic hypervitaminosis A resulting in portal hypertension and ascites. N Engl J Med 1974;291: 435–40.

62. Geubel A, Galocsy C, Alves N, Rahier J, Dive C. Liver damage caused by therapeutic vitamin A administration: estimate of dose-related toxicity in 41 cases. Gastroenterology 1991;100:1701–9.

63. Hathcock J. Vitamins and minerals: efficacy and safety. Am J Clin Nutr 1997;66:427–37.

64. Life Sciences Research Office FDA Contract No. 223-75-2004. Evaluation of the health aspects of carotene (β-carotene) as a food ingredient. Bethesda, MD: Federation of American Societies for Experimental Biology, 1979.

65. Matthews-Roth M. Beta-carotene therapy for erythropoietic protoporphyria and other photosensitivity diseases. Biochimie 1986;68:875–84.

66. Diplock A. Safety of antioxidant vitamins and β-carotene. Am J Clin Nutr 1995;62(suppl):1510S–16S.

67. Underwood B. Methods for assessment of vitamin A status. J Nutr 1990;120:1459–63.

68. Olson J. Needs and sources of carotenoids and vitamin A. Nutr Rev 1994;52:S67–S73.

Vitamin D

Through the years vitamin D has been associated with skeletal growth and strong bones. This association arose because early in the twentieth century it was shown that rickets, a childhood disease characterized by improper development of bones, could be prevented by a fat-soluble factor D in the diet or by body exposure to ultraviolet light. The emphasis was placed on the dietary factor; therefore, any compound with curative action on rickets was designated as vitamin D.

Sources

Dietary vitamin D is provided primarily by foods of animal origin especially liver, beef, veal, and eggs, dairy products such as milk, cheese and butter, and some saltwater fish including herring, salmon, tuna, and sardines. In the United States, selected foods, such as milk and margarine, are fortified with vitamin D. Table 10.3 provides information on major food sources of the vitamin. Dietary vitamin D is a stable compound not prone to cooking, storage, or processing losses [1].

In plants a commonly occurring steroid, ergosterol, can be activated by irradiation to form ergocalciferol (also called *vitamin D₂* or *ercalciol*) (Fig. 10.14), the antirachitic compound most commonly

Table 10.3 Major Food Sources of Vitamin D

Food	Approximate Vitamin D Content (μg/100 g)
Fortified	
Milk	0.8–1.3
Margarine	8.0–10.0
Nonfortified	
Butter	0.3–2.0
Milk	<1.0
Cheese	<1.0
Liver	0.5–4.0
Fish*	5.0–40.0

*Fatty fish such as herring, salmon, sardines, and tuna.

sold commercially. No ergosterol occurs in animals, but another steroid, 5,7-cholestradienol, commonly called *7-dehydrocholesterol,* is found in animals and humans. 7-Dehydrocholesterol is synthesized in the sebaceous glands of the skin, secreted onto the skin's surface, and may be reabsorbed into the various layers of the skin. This steroid appears to be uniformly distributed throughout the epidermis and dermis. The conjugated set of double bonds (five to seven) in ring B of 7-dehydrocholesterol allows the absorption of specific wavelengths of light found in the ultraviolet range. Thus, during exposure to sunlight, some of the epidermal cutaneous reservoir of 7-dehydrocholesterol is converted to previtamin D_3 (also called *precalciferol*) (Fig. 10.14). Lumisterol is also produced from 7-dehydrocholesterol in the presence of ultraviolet light; tachysterol is generated by further irradiation of previtamin D_3 [2,3]. Much of the previtamin D_3 is thermally isomerized within 2 to 3 days into vitamin D_3, also called *cholecalciferol* or *calciol.* Cholecalciferol diffuses from the skin into the blood with transport in the blood occurring by a transport α-2 globulin vitamin D-binding protein (DBP) (also called *transcalciferin*) that is synthesized in the liver [2,3]. Neither lumisterol, tachysterol, nor previtamin D_3 has much affinity for the DBP. Therefore, rather than entering the blood, they are sloughed off during turnover of the skin [4].

Absorption, Transport, and Storage

Vitamin D from the diet is absorbed from a micelle, in association with fat and with the aid of bile salts, by passive diffusion into the intestinal cell. About 50% of dietary vitamin D is absorbed. Although the rate of absorption is most rapid in the duodenum, the largest amount of vitamin D is absorbed in the distal small intestine.

Within the intestinal cell, vitamin D is incorporated primarily into chylomicrons. These chylomicrons enter the lymphatic system with subsequent entry into the blood. Chylomicrons transport about 40% of the cholecalciferol in the blood, although some vitamin D may be transferred from the chylomicron to DBP for delivery to extrahepatic tissues. Chylomicron remnants deliver the vitamin to the liver.

Cholecalciferol, which slowly diffuses from the skin into the blood, is picked up for transport by DBP. About 60% of plasma cholecalciferol is bound to DBP for transport. The vitamin D bound to DBP travels to the liver; however, much of the vitamin is deposited in muscle and adipose tissue prior to hepatic uptake [5]. Thus, the difference in the transport mechanism for cholecalciferol formed in the skin and that absorbed from the digestive tract impacts on the distribution of the vitamin in the body.

Cholecalciferol reaching the liver either by way of chylomicron remnants or by DBP is hydroxylated at carbon 25 to form 25-OH D_3 (also called *calcidiol*) (Fig. 10.15). The efficiency of the liver 25-hydroxylase (also called *monooxygenase*) in converting cholecalciferol into 25-OH D_3 appears related to vitamin D status. The NADPH-dependent 25-hydroxylase enzyme is more efficient during periods of vitamin D deprivation than when normal amounts of the cholecalciferol are available. The relative activity of the enzyme in various organs including lung, intestine, and kidney [2], and the distribution of activity may be species-dependent [2]. 25-Hydroxylase is poorly regulated; thus, blood levels of 25-OH D_3 are thought to represent vitamin D status, and 25-OH D_3 is the main form of the vitamin in the blood [6]. Lower calcidiol concentrations are reported among healthy individuals in the winter months because of diminished exposure to the sun by many individuals and are inversely associated with parathyroid hormone (PTH) concentrations [7,8]. Serum calcidiol concentrations have been shown to significantly correlate with vitamin D intake [8].

Most of the 25-OH D_3 synthesized in the liver is secreted into the blood and transported by DBP. Because little 25-OH D_3 remains in the liver and very little of this metabolite is taken up by the extrahepatic tissues, the blood is the largest single pool (storage site) of 25-OH D_3, which has a half-life of about 3 weeks [9]. When the 25-OH D_3 pool has been depleted during vitamin D deprivation, maintenance of vitamin D activity is made possible for variable time periods through the release of cholecalciferol from its skin reservoir and from other sites in muscle and adipose tissues [10].

Following hydroxylation in the liver, 25-OH D_3 bound to DBP is released into the blood and taken up

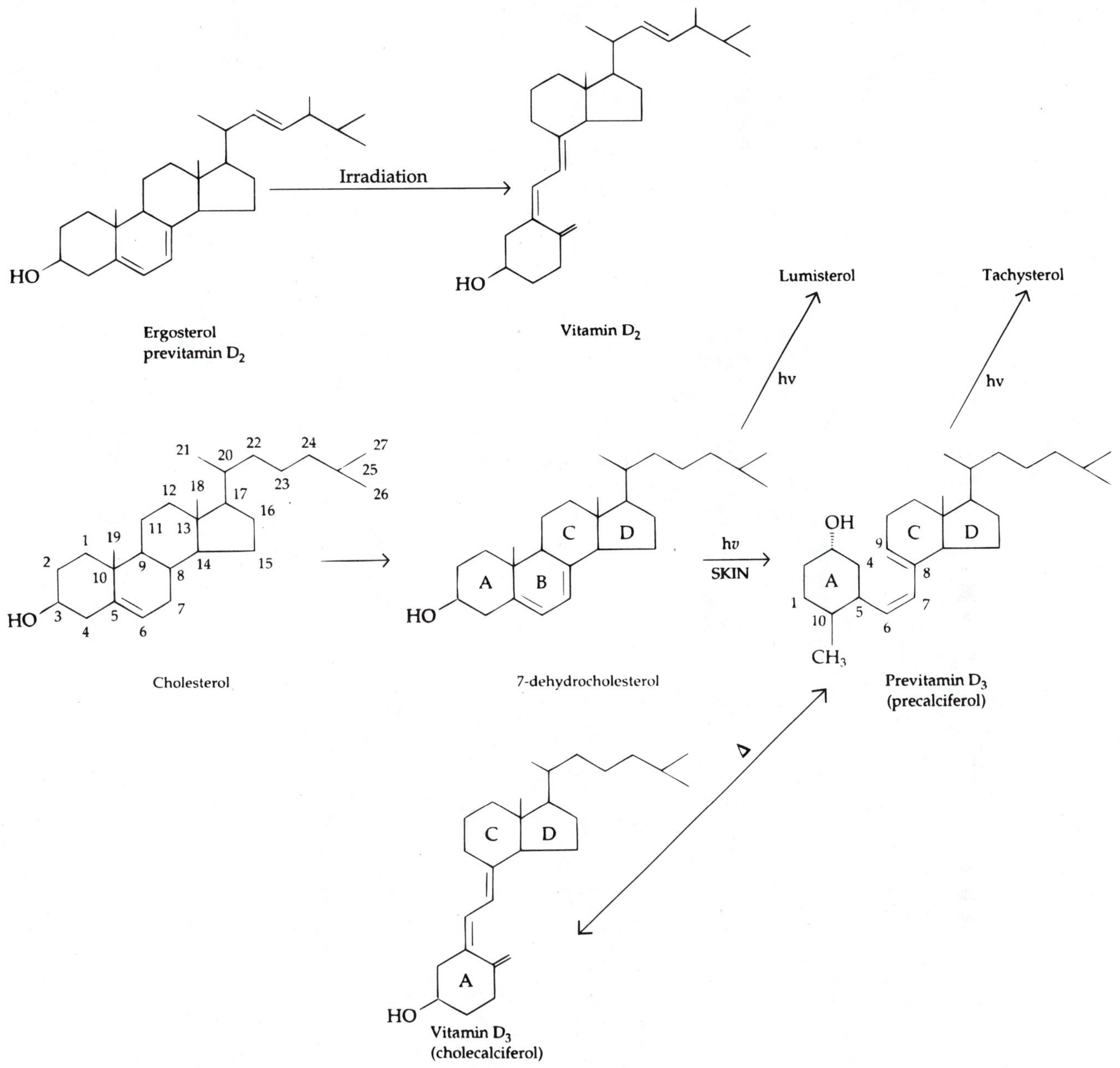

Figure 10.14 Production of vitamin D_3 in skin via previtamin D_3.

by the kidney. In the kidney, a second hydroxylation of 25-OH D_3 occurs at the 1 position, resulting in 1,25-$(OH)_2$ D_3 (also called *calcitriol*), which is considered the active vitamin. Calcitriol (Fig. 10.15) formation in the kidney tubules occurs through the action of another NADPH-dependent enzyme, 25-OH D_3 1 α-hydroxylase (also called *1-hydroxylase* or *1-monooxygenase*), a mitochondrial mixed-function oxidase. This enzyme is also present in macrophages and some cancer cells. Calcitriol in the blood has a half-life of about 4 to 6 hours.

The activity of 1-hydroxylase is influenced by a variety of factors. Parathyroid hormone (PTH) and low plasma calcium concentrations stimulate 1-hydroxylase activity. The concentration of the enzyme's end product, 1,25-$(OH)_2$ D_3-calcitriol, also influences the enzyme's activity whereby high concentrations inhibit enzyme activity and low concentrations stimulate 1-hydroxylase activity. Dietary phosphorus intake impairs calcitriol production by 1-hydroxylase. A high intake of phosphorus causes a decrease in serum 1,25-$(OH)_2$ D_3, whereas a low phosphorus intake stimulates produc-

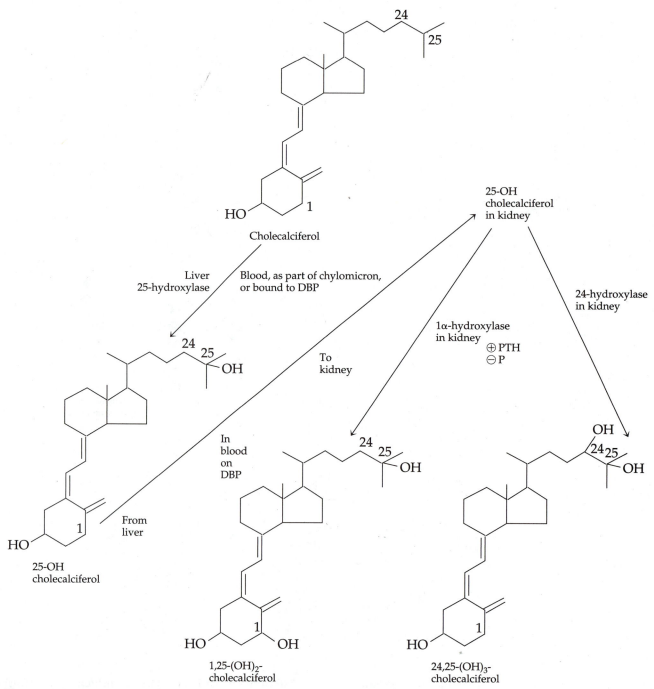

Figure 10.15 Hydroxylation of vitamin D.

tion of calcitriol [11]. When sufficient amounts of calcitriol are present, the activity of 1-hydroxylase in the kidney is decreased significantly, and the activity of another enzyme, 24-hydroxylase, is increased in the kidney and possibly other tissues such as cartilage and the intestine. 24,25-(OH)₂ D₃ is formed from 24-hydroxylation of 25-OH D₃ by 24-hydroxylase (Fig. 10.15), and it may be involved in bone mineralization or may repre-

sent a step in the degradation of the vitamin [12]. Production of 24,25-(OH)₂ D₃ appears to increase during periods of adequate vitamin D status and calcium homeostasis.

Once synthesized, calcitriol is released from the kidney and bound loosely to DBP. Consequently, on reaching its target tissues, the hormone is easily released from the DBP and is quickly bound by its receptor.

Functions and Mechanisms of Action

Calcitriol, 1,25-(OH)$_2$ D$_3$ synthesized in the kidney, is considered the active form of vitamin D and functions like a steroid hormone. Remember hormones are synthesized in one organ and act on a target organ(s). Initially the target tissues of the vitamin were believed to be limited to the intestine, bone, and kidney. The presence of specific cell membrane receptors for the hormone in many other tissues, however, supports the present belief that calcitriol acts in a wide variety of tissues, not just intestine, kidney, and bone but also cardiac, muscle, brain, skin, hematopoietic, and immune system tissues, among others [9,12,13]. Expression of cell receptors in tissues may be related to the cell's stage of differentiation.

Vitamin D not only interacts with cell membrane receptors, it also interacts with nuclear vitamin D receptor proteins called *VDR* to influence gene transcription. Nuclear receptors for the vitamin have been found in over 30 organs such as bone, intestine, kidney, lung, muscle, and skin, among others. In these organs, calcitriol appears to bind to VDR, which initiates a conformational change that in turn increases the receptor's affinity to vitamin D response elements, termed *VDRE* (Fig. 10.16). The vitamin D response elements are found in the promoter regions of specific target genes [13]. Once the VDR-calcitriol complex is bound to the VDRE, transcription of genes for specific mRNA coding for proteins may be either enhanced or inhibited [13]. For example, calcitriol-VDR complex binds to a *cis*-acting DNA sequence within the promoter region of the osteocalcin gene to stimulate transcription of mRNA, which codes for the protein osteocalcin [14]. Osteocalcin is secreted by osteoblasts. The osteocalcin protein, once its three glutamic acid residues are carboxylated to γ-carboxyglutamate (Gla), is able to bind calcium and may be involved in bone mineralization (see vitamin K, p. 354).

As a hormone, calcitriol functions in the body with parathyroid hormone (PTH) in the homeostasis of blood calcium concentrations. In performing this function, calcitriol and PTH impact several tissues including the intestine, bone, and kidney (Fig. 10.17). Hypocalcemia stimulates secretion of PTH from the parathyroid gland. The PTH, in turn, stimulates 1-hydroxylase in the kidney such that 25-OH D$_3$ is converted to calcitriol. Calcitriol then acts alone or with PTH on its target tissues, causing serum calcium and phosphorus concentrations to rise. The effects of calcitriol and PTH on intestine, kidney, and bone will be discussed along with its role in cell differentiation, proliferation, and growth.

Calcitriol and the Intestine

A more thoroughly investigated target tissue of calcitriol is the intestine (Fig. 10.18). The primary function of calcitriol in the intestine is increased absorption of calcium and phosphorus. In this function the vitamin is believed to function as a steroid hormone as well as to function in signal transduction. Calcitriol's involvement in calcium and phosphorus absorption in the intestine will be presented.

Calcium Calcitriol, as a hormone, interacts with high-affinity receptors in the enterocyte and is carried to the nucleus, where it interacts with specific genes encoding for proteins involved in calcium transport [12]. As the result of this interaction, a selective DNA transcription occurs that results in biosynthesis of new messenger RNA (mRNA) molecules. These mRNA molecules are then translated on the endoplasmic reticulum into selected proteins. These proteins may act at the brush border, in the cytoplasm, and/or at the basolateral membrane of the intestinal cells to promote calcium absorption. Calbindin, a calcium-binding protein in the intestinal mucosa, has been shown to be synthesized in response to the action of calcitriol. Although the amount of this calcium-binding protein is positively correlated

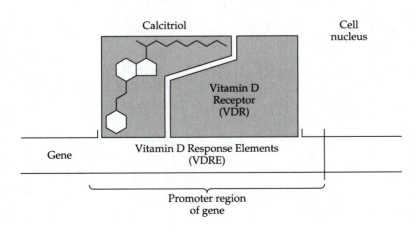

Figure 10.16 Binding of calcitriol to vitamin D receptor proteins allows interaction with the promoter region of specific genes to either stimulate or repress transcription of the gene.

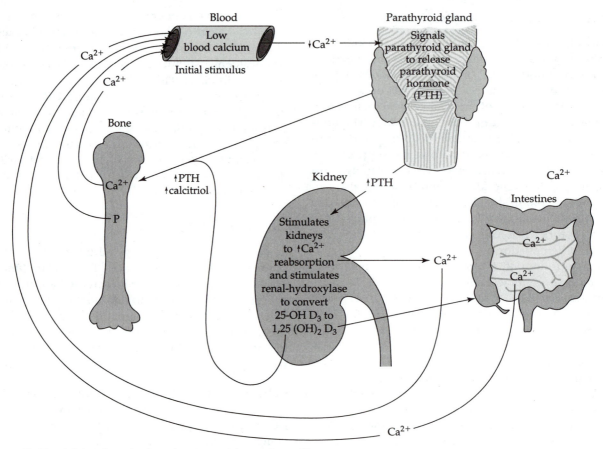

Figure 10.17 Calcitriol synthesis and actions with parathyroid hormone.

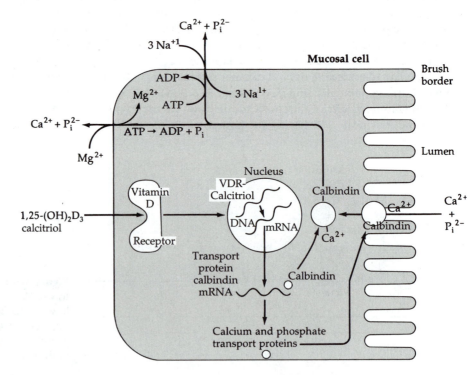

Figure 10.18 The actions of vitamin D on the intestinal absorption of calcium.

with calcium absorption and transport, its exact role in the process is not yet defined.

Also with respect to calcium homeostasis, calcitriol is thought to induce changes in brush border composition and topology to increase calcium absorption. Vitamin D rapidly initiates intestinal calcium absorption, a process referred to as *transcaltachia* and not thought to be mediated through effects on gene expression [15]. The transcaltachic response is thought to result from the opening of voltage-gated calcium channels [15].

Final extrusion of calcium from the intestinal cell and into the blood requires a different transport system from that used in the brush border membrane. Two systems have been identified:

- One system is a Ca^{2+}-Mg^{2+}ATPase with magnesium entering the cell as calcium exits and ATP supplying the energy. This system is thought to be present and active mainly in the duodenum [4].

- In the jejunum and ileum, a sodium-calcium exchange system is thought to exist whereby three Na^+ are exchanged for one Ca^{2+} [4,6].

Calcium release also may involve protein kinase C and cyclic AMP-dependent protein kinases [15,16]. Calcitriol may initiate activation of protein kinase [15].

Phosphorus With respect to phosphorus, calcitriol is thought to increase the activity of brush border alkaline phosphatase, which hydrolyzes phosphate ester bonds allowing phosphorus absorption. Calcitriol is also thought to modulate the number of carriers available for sodium-dependent phosphorus absorption at the brush border membrane [4].

Calcitriol and the Kidney

Calcitriol appears to be involved in the parathyroid hormone stimulation of calcium and phosphorus reabsorption in the distal renal tubule (Fig. 10.17) [6]. Although more research is needed to clarify the actions of calcitriol in the kidney, vitamin D is thought to have more effect on increasing phosphorus reabsorption than on calcium reabsorption in the kidney [4].

Calcitriol and the Bone

With respect to bone, PTH, alone or with calcitriol, directs the mobilization of calcium and phosphorus from bone to help to achieve a normal blood calcium concentration (Fig. 10.17). This process may be mediated by calcitriol-induced cell differentiation of hemopoietic cells to osteoclasts. Osteoclasts in turn mediate bone resorption. Alternately, the process may be mediated by calcitriol-induced increases in osteoclast activity [11].

Should blood calcium levels begin to rise above normal concentrations, calcitonin (a hormone produced by endocrine cells located in the connective tissue of the thyroid gland) is released and promotes the deposition (mineralization) of calcium and phosphorus in bones. Calcitriol or the metabolite 24,25-$(OH)_2$ D_3 may also be involved in bone mineralization and suppression of PTH [17,18]. Elevated serum calcitriol and elevated ionized serum calcium in turn cause a decrease in PTH production through feedback loops. The long feedback loop is an indirect one due to the inhibitory effect of elevated ionized serum calcium on PTH secretion. The short feedback loop is direct; the calcitriol decreases the transcription of the gene for preparathyroid hormone, presumably by interacting with the vitamin D receptor in the parathyroid tissue and influencing the regulatory region of the PTH gene [11].

Modeling and remodeling of bone may also involve calcitriol. Calcitriol appears important in the synthesis of a prominent noncollagenous protein, osteocalcin, found in bone. Vitamin D stimulates osteoblasts to synthesize osteocalcin, a γ-carboxyglutamate vitamin K-dependent protein found in bone matrix and dentine. Osteocalcin (also discussed on p. 354) is associated with new bone formation and when found in circulation is thought to be a sensitive marker of vitamin D action and bone disease.

Calcitriol and Cell Differentiation, Proliferation, and Growth

Calcitriol affects cell differentiation, proliferation, and growth in a variety of different tissues. For example, calcitriol stimulates differentiation of hematopoietic and intestinal epithelial cells as well as osteoblasts, among others. The effects will vary considerably depending on the tissue and may be mediated through control of proto-oncogenes [14,19]. Calcitriol-induced cell differentiation of stem cells to osteoclasts leads to increased blood calcium concentrations because osteoclasts mediate bone resorption and release of calcium into blood [20]. This stimulation is believed to be due to an increase in the number of osteoclasts, derived from hematopoietic cells with the increase in differentiation attributed to vitamin D [11]. In addition, stimulation is due to increased activity of the osteoclasts (short-term effect) whereby calcitriol induces the release of osteoblast-derived resorption factors that stimulate osteoclast activity [11].

Whereas differentiation appears to be stimulated by calcitriol, proliferation of cells such as fibroblasts, keratinocytes, and lymphocytes is reduced by calcitriol.

Proliferation of abnormal intestinal, lymphatic, mammary, and skeletal cells (to name a few) also is diminished by vitamin D. Calcitriol appears to be able to inhibit cancer cell proliferation and growth. Vitamin D's ability to stimulate skin epidermal cell differentiation, while preventing proliferation, provides potential for the treatment of skin disease(s). These functions of the vitamin have been applied in vitamin D treatment of psoriasis (a disorder in which there is proliferation of the keratinocytes and a failure to differentiate rapidly) [2,11,12,18,20]. Vitamin D helps to decrease proliferation associated with psoriasis and enhance differentiation of the epidermis [21]. Potential uses of vitamin D and or vitamin D analogs in the treatment of bone disease, hyperparathyroidism, cancer, as well as other conditions are under investigation.

Interactions with Other Nutrients

Discussion of vitamin D metabolism is impossible without noting the interrelationships existing among this vitamin or hormone and *calcium, phosphorus,* and *vitamin K.* The relationships with calcium and phosphorus are shown in Figures 10.17 and 10.18 and are discussed in the text corresponding to each figure. Interaction of calcitriol and vitamin K is provided on page 356. Also speculated is a decrease in vitamin D absorption as a result of *iron* deficiency [22].

Metabolism and Excretion

Calcitriol hydroxylation at carbon 24 generates the metabolite 1,24,25-(OH)$_3$ D$_3$ (Fig. 10.19), which may be further oxidized to 1,25-(OH)$_2$ 24-oxo D$_3$. Subsequent reactions, including side-chain cleavage, yield calcitroic

acid (Fig. 10.19), a major excretory product that is excreted into the bile [12]. Other vitamin D metabolites are also formed after hydroxylation and oxidation. These other vitamin D metabolites may be conjugated and then excreted primarily in the bile. Less than 5% of the metabolites are excreted in the urine.

Dietary Reference Intakes and Recommended Dietary Allowances

Although the exact requirement for vitamin D has not been elucidated, the 1997 DRI suggests that an AI for infants over 6 months of age, children, and adolescents is 5 μg or 200 IU daily. One IU is defined as the activity contained in 0.025 μg of cholecalciferol [10]. The AI for adults over 19 years of age is 5 μg (200 IU) daily; this amount of the vitamin is thought to be obtainable by exposure to sunlight. In the continental United States about 1.5 IU of vitamin D/cm^2/hour during the winter, and about 6 IU/cm^2/hour during the summer can be synthesized in the skin [17]. Alternately, it takes about 10 minutes of summer sun on the face and hands to produce 10 μg (400 IU) of cholecalciferol [18]. A vitamin D supplement providing 400 to 800 IU may be necessary for most elderly, particularly those who drink little milk and are partially or totally housebound. AI for those aged 51 to >70 years are 10 μg and 15 μg, respectively.

Deficiency: Rickets and Osteomalacia

In infants and children, vitamin D deficiency results in rickets. Rickets is characterized by failure of bone to mineralize. In vitamin D–deficient infants, epiphyseal cartilage continues to grow and enlarge without re-

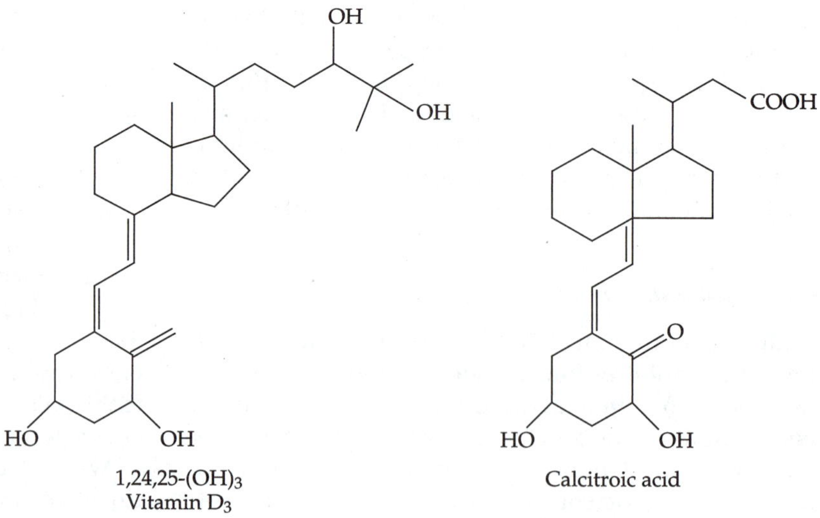

Figure 10.19 Some metabolites of vitamin D catabolism.

1,24,25-(OH)$_3$
Vitamin D$_3$

Calcitroic acid

placement by bone matrix and minerals. Long bones of the legs bow, and knees knock as weight-bearing activity such as walking begins. The spine becomes curved, and pelvic and thoracic deformities occur.

In adults, deprivation of vitamin D leads to impaired calcium—and possibly phosphorus—absorption. Serum calcium homeostasis occurs in part through the action of PTH, which may remain elevated for prolonged time periods. With a marginal phosphorus intake and an insufficiency of vitamin D, an adequate serum phosphorus level may also be impossible to maintain, thereby further jeopardizing bone mineralization. Without sufficient serum calcium and phosphorus, the mineralization of bones under the direction of calcitonin cannot occur. Thus, in vitamin D–deficient adults, as bone turnover occurs, the bone matrix is preserved, but remineralization is impaired. The bone matrix becomes progressively demineralized, resulting in bone pain and osteomalacia (soft bone).

Natural exposure to sunlight maintains adequate vitamin D nutrition for most of the world's population [3]. However, certain diseases, conditions, and populations may be at risk for vitamin D deficiency. The elderly represent one population group that typically has insufficient vitamin D intake [23–25]. The elderly also typically have low sunlight exposure. In addition, aging reduces synthesis of cholecalciferol in the skin and reduces the activity of renal 1-hydroxylase in response to PTH [24,25]. Impaired vitamin D absorption may occur in disorders in which there is fat malabsorption, such as tropical sprue and Crohn's disease. Disorders affecting the parathyroid, liver, and/or kidney will impair the synthesis of active form of the vitamin. People with insufficient sun exposure may be at risk for vitamin D deficiency. People on anticonvulsant drug therapy may develop an impaired response to vitamin D and exhibit problems with calcium metabolism. Infants may be at risk for deficiency because human milk is low in vitamin D and infant's exposure to sunlight typically is minimal.

Vitamin D supplements are widely prescribed for individuals with renal disease because the kidneys of these individuals are typically unable to synthesize calcitriol. Rocaltrol (Hoffman-LaRoche) is a commonly used oral supplement, whereas Calcijex (Abbott Laboratories) is given intravenously to individuals with renal disease in the United States. Other preparations (Calderol—Organon USA) providing calcidiol also are available for use.

Toxicity

Although excessive exposure to sunlight may be the primary risk factor in development of skin cancer, it poses no risk of toxicity through overproduction of endogenous cholecalciferol. Extensive whole-body irradiation with ultraviolet light will generally raise the level of circulating 25-OH D_3 to 40 to 80 ng/mL; levels >150 ng/mL are associated with possible toxicity [9]. Photochemistry regulates the cutaneous production of vitamin D_3, thus protecting people excessively exposed to sunlight from vitamin D intoxication [3].

Exogenous dietary ingestion of vitamin D is the most likely of all vitamins to cause overt toxic reactions when the RDA is chronically exceeded [26]. Even small multiples of the RDA ingested on a continuous basis can be toxic [26]. With excessive dietary ingestion of vitamin D, the vitamin is absorbed and incorporated into chylomicrons, the remnants of which deliver the vitamin to the liver. Here the vitamin is hydroxylated in position 25 and released to the blood. Although the efficiency of 25-hydroxylase is decreased when the vitamin is in abundance, the enzyme 25-hydroxylase is not well regulated, thus an excessive amount of the metabolite can be produced with over supplementation. Calcidiol in high concentrations may stimulate some of the same actions as calcitriol.

In the 1950s, an epidemic of "idiopathic hypercalcemia" among English infants was traced to an intake of vitamin D between 2,000 and 3,000 IU/day. Symptoms of toxicity in the infants included anorexia, nausea, vomiting, hypertension, renal insufficiency, and failure to thrive [26]. Eight individuals experienced hypervitaminosis from consuming milk (1/2–3 cups daily) from a local dairy that contained up to 232,565 IU of vitamin D_3 per quart. The individuals displayed hypercalcemia and hypercalcidiolemia [27].

For adults the lowest safe intake of vitamin D is unclear. Intakes of as little as 2,000 IU/day of vitamin D (or five times the RDA) may pose a risk for adults if this level of intake is prolonged [28]. Some authorities suggest that intakes should not exceed 1,200 IU [29]. Dosages of 10,000 IU/day for several months have resulted in hypercalcemia—with possible calcification of soft tissues (calcinosis) such as the kidney, heart, lungs, and blood vessels, hyperphosphatemia, hypertension, anorexia, nausea, weakness, polyuria, polydypsia, azotemia, nephrolithiasis, renal failure, and, in some cases, death [26].

Assessment of Nutriture

No ready method of measuring vitamin D status exists. Elevations in plasma alkaline phosphatase released from osteoclasts have been used to detect preclinical rickets [20]. The plasma concentration of 25-OH D_3 (calcidiol) is thought to provide an index of vitamin D status. Normal plasma concentration of 25-OH D_3 ranges from 8 to

60 ng/mL [9] (with an average between 25 and 30 ng/mL) [9]. A level <10 or 12 ng/mL is regarded as an indicator for vitamin D therapy and levels >150 ng/mL are associated with possible toxicity [8,9]. However, because cholecalciferol can be stored in extrahepatic tissues such as fat and muscles, measurement of serum 25-OH D$_3$ will not fully reflect pools of the vitamin [5]. Osteocalcin released into circulation is thought to represent a measure of vitamin D action but is not always present in the blood in detectable quantities even if the individual has adequate vitamin D status.

References Cited for Vitamin D

1. Sauberlich H. Vitamins—how much is for keeps? Nutr Today 1987;22:20–28.

2. Henry H, Norman A. Vitamin D: Metabolism and biological actions. Ann Rev Nutr 1984;4:493–520.

3. Webb A, Holick M. The role of sunlight in cutaneous production of vitamin D$_3$. Ann Rev Nutr 1988;8:375–99.

4. Lawson E. Vitamin D. In: Diplock AT, ed. Fat Soluble Vitamins. Lancaster, PA: Technomic, 1985:76–153.

5. Fraser D. Vitamin D. In: Olson RE, Broquist HP, Chichester CO, Darby WJ, Kolbye AC Jr, Stalvey RM, eds. Nutrition Reviews' Present Knowledge in Nutrition, 5th ed. Washington, D.C.: The Nutrition Foundation, 1984:209–25.

6. Combs G. The Vitamins. New York: Academic Press, 1992:151–78.

7. Dawson-Huges B, Harris S, Dallal G. Plasma calcidiol, season and serum parathyroid hormone concentrations in healthy elderly men and women. Am J Clin Nutr 1997;65:67–71.

8. Kinyamu H, Gallagher J, Balhorn K, Petranick K, Rafferty K. Serum vitamin D metabolites and calcium absorption in normal young and elderly free-living women and in women living in nursing homes. Am J Clin Nutr 1997;65:790–7.

9. Holick M. The use and interpretation of assays for vitamin D and its metabolites. J Nutr 1990;120:1464–9.

10. National Research Council. Recommended Dietary Allowances, 10th ed. Washington, D.C.: National Academy Press, 1989:92–98.

11. Reichel H, Koeffler H, Norman A. The role of the vitamin D endocrine system in health and disease. N Engl J Med 1989;320:980–91.

12. Haussler M. Vitamin D receptors: Nature and function. Ann Rev Nutr 1986;6:527–62.

13. Hannah S, Norman A. 1α25(OH)$_2$ Vitamin D$_3$–regulated expression of the eukaryotic genome. Nutr Rev 1994;52:376–82.

14. Pike J. Vitamin D$_3$ receptors: structure and function in transcription. Ann Rev Nutr 1991;11:189–216.

15. Lowe K, Maiyar A, Norman A. Vitamin D–mediated gene expression. Crit Rev Eukaryotic Gene Expression 1992;2:65–109.

16. DeBoland A, Norman A. Evidence for involvement of protein kinase C and cyclic adenosine 3'5' monophosphate–dependent protein kinase in the 1,25-dihydroxyvitamin D$_3$-mediated rapid stimulation of intestinal calcium transport (transcaltachia). Endocrinology 1990;127:39–45.

17. Collins E, Norman A. Vitamin D. In: Machlin LJ. Handbook of Vitamins, 2nd ed. New York: Dekker, 1991:59–98.

18. DeLuca H. Vitamin D: 1993. Nutrition Today 1993;28(Nov–Dec):6–11.

19. Minghetti P, Norman A. 1,25(OH)$_2$ -vitamin D$_3$ receptors: gene regulation and circuitry. FASEB J 1988;2:3043–53.

20. Holick M. Vitamin D. In: Shils ME, Olson JA, Shike M, eds. Modern Nutrition in Health and Disease, 8th ed. Philadelphia: Lea and Febiger, 1994:308–25.

21. Bikle D. Vitamin D: new actions, new analogs, new therapeutic potential. Endocrin Rev 1992; 13: 765–84.

22. Heldenberg D, Tenenbaum G, Weisman Y. Effect of iron on serum 5–hydroxyvitamin D and 24,25-dihydroxy-vitamin D concentrations. Am J Clin Nutr 1992;56:533–6.

23. Omadahl J, Garry P, Hunsaker L, Hunt W, Goodwin J. Nutritional status in a healthy elderly population: vitamin D. Am J Clin Nutr 1982;36:1225–33.

24. MacLaughlin J, Holick M. Aging decreases the capacity of human skin to produce vitamin D$_3$. J Clin Invest 1985;76:1536–8.

25. Zeghoud F, Vervel C, Guillozo H, Walrant-Debray O, Boutignon H, Garabedian M. Subclinical vitamin D deficiency in neonates: definition and response to vitamin D supplements. Am J Clin Nutr 1997;65:771–8.

26. Council on Scientific Affairs, American Medical Association. Vitamin preparations as dietary supplements and as therapeutic agents. JAMA 1987;257:1929–36.

27. Jacobus C, Holick M, Shao Q, Chen T, Holm I, Kolodny J, Fuleihan G, Seely E. Hypervitaminosis D associated with drinking milk. N Engl J Med 1992;326:1173–7.

28. Marshall C. Vitamins and Minerals: Help or Harm? Philadelphia: Stickley, 1983.

29. National Institutes of Health Consensus Development Panel. Osteoporosis. JAMA 1984;252:799–802.

Vitamin E

Vitamin E includes eight compounds (vitamers) synthesized by plants. These compounds fall into two classes:

- the tocols, which have saturated side chains, and
- the tocotrienols (also called trienols), which have unsaturated side chains.

Each class is composed of four vitamers that differ in the number and location of methyl groups on the chromanol (chroman) ring. Vitamers in both classes are designated as α, β, γ, or δ and possess characteristic biological activity. Another compound, all-rac α-tocopheryl acetate, with vitamin E activity is used in fortification of foods.

Figure 10.20 gives the basic structure of the compounds with vitamin E activity and defines the biological activity of each. Vitamin E activity is greatest in α-tocopherol, followed by β, which is greater than γ and δ. Of the tocotrienols, only the α vitamer has significant activity; α-tocotrienol has an activity slightly less than β-tocopherol (0.4 vs. 0.3, respectively) [1]. The biological activities of δ-tocotrienol and γ-tocotrienol are unknown.

Sources

Vitamin E is found in both plant and animal foods. Plant foods, especially oils from plants, are considered the richest sources of vitamin E. Foods such as vegetable oils as used in cooking, salad dressings, mayonnaise, and margarine along with fruits and vegetables provide the majority of vitamin E in the diet [2,3]. In fact, the level of vitamin E best correlates with the level of polyunsaturated fat in the food. Tocopherols are found in leafy plant foods. The leaves and other green (chloroplast) portions of plants contain mostly α-tocopherol with small amounts of γ-tocopherol. γ-, δ-, and β-tocopherols are found mainly in nonchloroplast regions of plants.

In contrast to the tocopherols, the tocotrienols are found in legumes and cereal grains such as wheat,

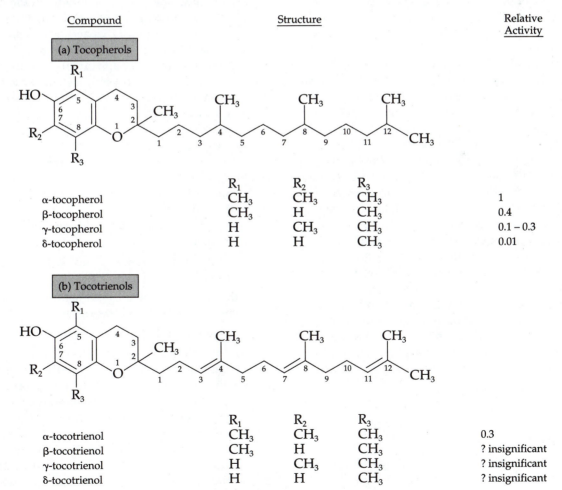

Figure 10.20 Structure and activity of natural tocopherols and tocotrienols. (Diplock AT. Vitamin E. In: Diplock AT., ed. Fat soluble vitamins. Lancaster, PA: Technomic, 1984:156.)

Table 10.4 Approximate Vitamin E Content of Foods as α-Tocopherol Equivalents

Food	mg/100 g
Oils:	
Wheat germ	192
Corn	21
Cottonseed	38
Peanut	13
Safflower	43
Soybean	18
Sunflower	51
Margarine	15
Nuts	0.69–9
Breads	0.4
Vegetables	0.1–2.0
Fruits	0.1–1.1
Meat, fish	1
Eggs	1

Source: www.nal.usda.gov/fnic/foodcomp

barley, rice and oats. The bran and germ sections of cereals are especially rich in tocotrienols. Thus, wheat germ oil and wheat bran represent significant sources of tocotrienols.

In foods of animal origin, vitamin E, primarily α-tocopherol, is found concentrated in fatty tissues of the animal [4]; however, in comparison with plants, animal products represent an inferior source of vitamin E. Fish oils, for example, are low in tocopherols [1]. Processing and storage of foods may affect the food's vitamin E content.

Table 10.4 lists the approximate α-tocopherol equivalents found in commonly consumed foods; 1 α-tocopherol equivalent (TE) has the vitamin E activity of 1 mg α-tocopherol [5]. Use of α-tocopherol equivalent is thought to be more accurate than milligrams because it describes the biological activity of the combination of E vitamers [2]. Most vitamin E in foods is found as α-tocopherol. However, with the increased consumption of oils from, for example, soybeans, the γ-tocopherol content, and to a lesser extent the δ-tocopherol content, of the U.S. diet is increasing relative to α-tocopherol [2,3]. Unfortunately, individuals limiting fat intake also limit foods that are high in vitamin E and may compromise their ability to meet dietary intake recommendations for vitamin E.

Digestion, Absorption, Transport, and Storage

Whereas the tocopherols are found free in foods, the tocotrienols are found esterified and must be hydrolyzed prior to absorption. Pancreatic esterase and/or duodenal mucosal esterase are thought to function in the lumen or at the brush border of the intestine to hydrolyze tocotrienol for absorption.

Absorption of vitamin E as free alcohols occurs primarily in the jejunum by nonsaturable, passive (requiring no carrier) diffusion. Both bile salts and pancreatic juice are needed for micelle formation allowing the vitamin to diffuse through the unstirred water layer and the enterocyte membrane. Simultaneous digestion and absorption of dietary lipids, including medium-chain triacylglycerols, with vitamin E improves absorption of vitamin E [6]. A specific factor not associated with general fat absorption may also be required for absorption of vitamin E as suggested by studies in vitamin E–deficient patients who have no problems absorbing fat [7].

The absorption of vitamin E varies from about 20% to 50% [8] and possibly as high as 80% [9,10]. As vitamin intake increases, vitamin E absorption decreases whereby absorption of pharmacological doses, 200 mg, of vitamin E is <10% [1].

Absorbed tocopherol is incorporated into chylomicrons in the enterocyte and is transported through the lymph into circulation. The tocopherol in the chylomicrons equilibrates with the other plasma lipoproteins, including the HDLs and LDLs and interchanges with erythrocytes [5,11,12]. A specific tocopherol transfer protein made in the liver appears to be necessary for the transfer of tocopherol into VLDLs, which enable distribution of the vitamin to tissues. Tocopherol found in erythrocytes appears to be largely localized in the cell membrane [5,11]. Membrane-binding proteins for tocopherol have been identified. Plasma tocopherol has been shown to be highly correlated with total plasma lipids.

Tocopherol is distributed to the tissues primarily by the LDLs and may play a role in protecting the LDLs from oxidation in the process. Uptake of vitamin E into cells can occur

- as LDL receptor-mediated uptake occurs [9,13], or
- through lipoprotein lipase-mediated hydrolysis of chylomicrons and VLDLs [8], and
- possibly by other mechanisms [9].

Within the cell cytoplasm as well as other parts of the cells including the nucleus, vitamin E appears to bind to specific proteins, tocopherol-binding proteins, for transport [8,14]. Vitamin E is found within the cell primarily located in cell membranes such as the plasma, mitochondrial, and microsomal membranes. As described by Machlin [14], the vitamin is likely to be oriented with the chromanol "head" group toward the surface of the membrane near the phosphate region of the phospholipid, and with the hydrophobic phytyl "tail" buried within the hydrocarbon region.

There is no single storage organ for vitamin E. The largest amount of the vitamin is concentrated in an un-esterified form in the adipose tissue, with smaller amounts in liver, lung, heart, muscle, adrenal glands, and brain. The concentration of vitamin E in adipose tissues increases linearly with dosage of vitamin E, whereas the other tissues maintain a constant concentration or increase only at a very slow rate [1,12]. In times of low intake, withdrawal of tocopherol occurs slowly from adipose (thus it is not readily available), whereas withdrawal from the liver and plasma is rapid [11]. Depletion of vitamin E stored in the heart and muscle occurs at an intermediate rate [14].

Functions and Mechanisms of Actions

The principal function of vitamin E is the maintenance of membrane integrity in body cells. Some investigators believe that vitamin E may also provide physical stability to membranes [11]. The mechanism by which vitamin E functions to protect the membranes from destruction is through its ability to prevent the oxidation (peroxidation) of unsaturated fatty acids contained in the phospholipids of the cellular membranes. The phospholipids of the mitochondrial membrane and endoplasmic reticulum contain more unsaturated fatty acids than the cell's plasma membrane. These membranes, therefore, are at greater risk for oxidation with a vitamin E deficiency than is the membrane surrounding the cell [11]. Cell membranes are, however, still vulnerable to oxidation. Tissues with cell membranes especially susceptible to oxidation include the lungs, brain, and erythrocytes. Erythrocyte membranes, for example, are high in polyunsaturated fatty acids and are exposed to high concentrations of oxygen. Because vitamin E prevents oxidation, it is referred to as an antioxidant. A discussion of vitamin E's role as an antioxidant follows with a brief description of the generation of carbon-centered and peroxyl radicals. More information on how free radicals are generated and how they can damage cell membranes may be found at the end of this chapter in the Perspective, "The Antioxidant Nutrients, Reactive Oxygen Species, and Disease."

Free Radicals and the Antioxidant Vitamin E

Vitamin E effectively reacts and terminates carbon-centered radicals. Carbon-centered radicals may be generated in the body from reactions between organic compounds and other radicals, especially hydroxyl radicals. Hydroxy radicals (OH·) are very highly reactive, rapidly taking electrons from the surroundings. Often the electron taken by the reactive free hydroxy radical is from nearby organic molecules. If the organic molecule is a polyunsaturated fatty acid (PUFA) present in phospholipid portion of the cell membrane, damage to the membrane occurs. Membrane lipid oxidation is thought to represent a primary event in oxidative cellular damage [16]. A series of reactions follows that exemplifies the formation of a carbon centered radical and a peroxyl radical from the attack of a hydroxyl radical on a polyunsaturated fatty acid lipid designated as LH.

- The reaction between organic lipid compounds (LH) and free hydroxy radicals leads to the formation of a lipid carbon-centered radical (L·) and water, as shown here and in Figure 10.21:

$$LH + OH· \rightarrow L· + H_2O$$

This step is sometimes referred to as *initiation*.

- Alternately, organic lipid compounds (LH) can react with molecular oxygen to generate lipid carbon-centered radicals and hydroperoxyl radical, $HO_2·$:

$$LH + O_2 \rightarrow L· + HO_2·$$

Once lipid carbon-centered radicals are formed, they may react to form additional radicals in *propagation* reactions.

- Lipid carbon-centered radicals can react with molecular oxygen in a propagation reaction to form lipid peroxyl radicals, LOO·, shown as follows and in Figure 10.21:

$$L· + O_2 \rightarrow LOO· \text{ (also written } LO_2·)$$

- Lipid peroxyl radicals can also be generated from reactions between Fe^{3+} and LOOH (lipid hydroperoxide).

$$Fe^{3+} + LOOH \rightarrow Fe^{2+} + LOO·$$

Once formed, lipid peroxy radicals (LOO·) can abstract a hydrogen atom from other organic compounds including more polyunsaturated fatty acids in membranes or in lipoproteins to generate a chain reaction with the L·. This reaction is shown as follows and in Figure 10.21:

$$LOO· + LH \rightarrow L· + LOOH$$

Chain reactions involving L· must be terminated to minimize cellular damage. Prevention of damage from oxygen radicals depends on a complex protective system of which vitamin E is a part.

Vitamin E located in or near membrane surfaces can react with nonlipid peroxyl radicals (ROO·) before they interact with cell membranes. The vitamin may also break the chain of radical attack by reacting with the peroxyl radical (LOO·) and preventing further abstraction of hydrogen (H) from the fatty acids (LH) or other organic (R-) compounds. Thus, vitamin E terminates chain-propagation reactions. Vitamin E is less effective, however, in terminating peroxidation that

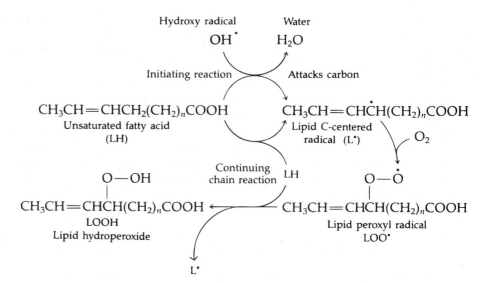

Figure 10.21 The initiating and chain reactions caused by hydroxy free radical attack on an unsaturated fatty acid.

generates free hydroxy radicals, OH·, or alkoxyl radicals (LO· or RO·).

Vitamin E (EH, reduced state), because of the reactivity of the phenolic hydrogen on its carbon 6 hydroxyl group and the ability of the chromanol ring system to stabilize an unpaired electron [8], can provide a hydrogen for the reduction of nonlipid peroxyl radicals as shown here:

$$ROO· + EH \longrightarrow ROOH + E·$$

or react with the lipid peroxy radical as follows:

$$LOO· + EH \longrightarrow LOOH + E·$$

E· represents oxidized vitamin E. The process is sometimes referred to as *"free-radical scavenging."* Termination is achieved when two free radicals combine to form a molecule that is not a free radical and cannot continue the reaction.

The oxidized vitamin E (E·) must be regenerated. The regeneration requires vitamin C [17], reduced glutathione (GSH), and NADPH (Fig. 10.22).

Vitamin E is only one line of defense against oxidative tissue damage. Other parts of the protective system include vitamin C, carotenoids, and enzymes that require a variety of trace or microminerals (iron, selenium, zinc, copper, and manganese) for their activation; therefore, an interrelationship exists among vitamins E and C, carotenoids, and these minerals involved in antioxidant activities. Vitamin C and E appear to work synergistically in inhibiting oxidation [18]. The relationship between vitamin E and these other nutrients with antioxidant functions is reviewed in the Perspective at the end of this chapter.

Clinical trials with vitamin E alone as well as with other antioxidants suggest that the vitamin may decrease susceptibility of LDL to oxidation by free radi-

cals [18–22]. Evidence of high intakes of vitamin E associated with a lower risk of coronary heart disease has been demonstrated in large cohort studies involving men [22] and women [21]. Supplementation with 800 IU of vitamin E, 1 g of vitamin C, and 24 mg of β-carotene significantly reduced susceptibility of LDL to oxidation in patients with cardiovascular disease [23]. α-tocopherol (800 IU) alone was found to be as effective as a combination of ascorbate (1 g), β-carotene (30 mg), and α-tocopherol (800 IU) in decreasing LDL oxidation [24]. Supplementation with 800 IU α-tocopherol in another group of patients with symptomatic heart disease reduced the rate of nonfatal heart attacks after 1 year of treatment [25]. These findings have implications in preventing *atherosclerosis.* Briefly, atherosclerosis is thought to begin with the accumulation of lipid-laden foam cells in the arterial intima. Radical induced oxidation of apoprotein B100, for example, in LDL is thought to be involved in promoting scavenger receptor-mediated uptake of the LDL by macrophages. Macrophages, which develop into foam cells, are thought to more readily take up oxidized LDL versus nonoxidized LDL. With continued accumulation, fatty streaks develop and represent the initial steps in atherosclerosis. Oxidized LDL may also reduce macrophage motility in the arterial intima, increase monocyte accumulation in endothelial cells, and increase cytotoxicity of endothelial cells to contribute to atherogenicity [21,22]. Thus, vitamin E's ability to prevent or decrease LDL oxidation thwarts the development of atherosclerotic lesions.

Vitamin E has been suggested for the treatment or prevention of other disorders. *Cataracts* results in part from oxidative damage to proteins that then aggregate and precipitate in the lens to cause lens opacities or

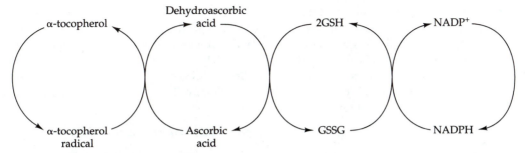

Figure 10.22 The regeneration of vitamin E.

cloudiness. Oxygen and oxyradicals are thought to contribute to the development of cataracts. Because of its antioxidant function, vitamin E is being tested for prevention and treatment of cataracts [26–30]. Multivitamin use, especially with vitamins C and E, appears to be associated with a lower incidence of cataracts [26–28]. Moreover, plasma vitamin E concentrations are significantly inversely associated with cataracts [26].

Whereas high intakes of fruits and vegetables are associated with reduced risks of *cancer,* diets high in vegetable oils, and thus vitamin E, have less consistently been linked with reduction in the risk of cancer [31]. Vitamin E, administered as a supplement with other vitamins, did not influence lung cancer, as discussed in the section on vitamin A. Studies suggest vitamin E can inhibit cell proliferation of smooth muscle through activation of transcription factors [32,33]. Such studies are promising and may yield significant findings for the use of vitamin E for prevention or treatment of cancer in the future.

Vitamin E may also be useful in the treatment of *iron toxicity,* which leads typically to lipid peroxidation through production of free radicals and excessive damage to organs, especially the liver. Specifically, vitamin E may help to protect against the iron-induced cellular damage [34].

Individuals with *diabetes mellitus* may benefit from vitamin E supplementation, which appears to improve insulin action [35,36]. Vitamin E supplementation (900 mg α-tocopherol) by individuals with non-insulin-dependent diabetes mellitus improved metabolic control and increased the plasma GSH-to-GSSG ratio to provide more reduced glutathione (GSH) [36]. Vitamin E was thought to perhaps improve plasma membrane structure and its related activities required for glucose transport and/or metabolism (and thus metabolic control) [36]. Vitamin E by diminishing lipid peroxidation and increasing GSH availability could help to maintain cell membrane fluidity, which in turn may improve glucose transporter function and thus insulin-dependent cellular glucose uptake [37].

Singlet Molecular Oxygen and Vitamin E

Singlet molecular oxygen, $^1O_2$, generated from lipid peroxidation of membranes, transfer of energy from light (photochemical reactions), or the respiratory burst occurring in neutrophils (enzymatic reactions), for example, is another very reactive and destructive compound that may be formed in the body. Singlet molecular oxygen readily reacts with organic molecules such as protein, lipids, and DNA and thus can damage cellular components unless removed. Quenching, a process by which electronically excited molecules, such as singlet molecular oxygen, are inactivated [38], has been already discussed in this chapter in the section on functions of carotenoids (p. 327). However, carotenoids are not alone in their ability to quench singlet oxygen. Vitamin E also has oxygen-quenching abilities. α-tocopherol was found to be as or more effective in the physical quenching of singlet molecular oxygen than β-tocopherol, followed in descending order by γ-tocopherol, then δ-tocopherol [39]. Physical quenching occurs when the singlet excited oxygen is deactivated without light emission and generally involves electron energy transfer [39]. The ability to physically quench singlet oxygen is related to the free hydroxyl group in position 6 of vitamin E's chromane ring. The chemical quenching of singlet oxygen generates a variety of oxidative products and was shown to be higher for α-tocopherol followed in descending order by γ-tocopherol, δ-tocopherol, and β-tocopherol [39]. The $^1O_2$-quenching ability of the carotenoids lycopene and β-carotene is about two orders of magnitude greater than that of vitamin E; however, given the lower plasma concentrations of carotenoids, vitamin E's role in quenching singlet oxygen is of physiological significance [39].

Nonantioxidant Roles of Vitamin E

Other nonantioxidant roles of vitamin E have been demonstrated for tocotrienols. Tocotrienols appear to affect cholesterol metabolism. Suppression of the activity of the rate-limiting enzyme in cholesterol synthesis 3-hydroxy 3-methyl glutaryl (HMG) CoA reductase by tocotrienol has been shown in vitro [40]. These findings are consistent with the observations in animals and in humans that tocotrienols reduce plasma cholesterol concentrations [41]. Suppression of tumor growth and cell proliferation also has been attributed to tocotrienols, although in general diets high in vitamin E have not been associated with lower risk of cancers [32,33,41,42]. α-tocopherol can inhibit protein kinase C, important for signal transduction and cell growth and differentiation [33].

Interactions with Other Nutrients

Because the function of vitamin E in the body is closely tied to that of glutathione peroxidase as discussed in the Perspective at the end of this chapter, an interrelationship exists between vitamin E and *selenium*. Selenium functions as an integral part of glutathione peroxidase that converts a lipid peroxide into a lipid alcohol. To a lesser extent, interrelationships also exist between vitamin E and *sulfur-containing amino acids* (S-aa). Cysteine, an S-aa generated from another S-aa, methionine, is necessary for synthesis of glutathione, which serves as the reducing agent in the glutathione peroxidase reaction. Superoxide dismutase (SOD) activity to remove superoxide radicals requires *zinc, manganese,* and *copper.*

High intakes of vitamin E can interfere with the functions of the other fat-soluble vitamins [8,12]. At dosages exceeding 1 g/day, vitamin E has been shown to be antagonistic to the action of *vitamin K* and to enhance the effect of oral coumarin anticoagulant drugs [14,23]. Vitamin E or its quinone (α-tocopheryl quinone) may block the oxidation of vitamin K and may affect prothrombin formation [12,34]. Vitamin E may also impact vitamin K absorption [14]. Problems with bone mineralization involving *vitamin D* have been reported in animals given high doses of vitamin E [8].

Another relationship is that between vitamin E and *vitamin A*. In a vitamin A deficiency, vitamin E is able to lower the rate at which vitamin A is depleted from the liver. Although the mechanism of this interaction is controversial, it appears to be unrelated to the prevention of lipid peroxidation [11]. Cleavage of β-carotene into retinal also requires vitamin E. Vitamin E is probably necessary to protect the substrate and product from oxidation; however, large doses (10 times the RDA) of vitamin E may inhibit β-carotene absorption or conversion to retinol in the intestine [4,43].

The relationship between vitamin E and dietary PUFA is particularly strong because requirement for the vitamin increases or decreases as the dietary intake of PUFA rises or falls. Some investigators believe that the dietary level of PUFA needs to be specified for a minimal vitamin E requirement to be determined [5,11,44].

Metabolism and Excretion

The metabolic fate of vitamin E in humans is largely unknown. α-tocopherol, in a nonpolar solvent, is oxidized to a tocopheroxy radical, which can be reduced back to active vitamin E [11]. However, in polar solvents such as water this tocopheroxy radical is not formed [11]. In polar solvents the chromanol ring of α-tocopherol appears to be irreversibly oxidized into tocopheryl quinone [11]. Metastable tocopheroxide is formed as an intermediary product in this oxidation reaction, and it is speculated that tocopheroxide may be the primary tissue oxidation product of vitamin E [11].

The major route of α-tocopherol excretion is via the feces. Fecal tocopherol arises from vitamin E that was not absorbed, from secretion of the vitamin from enterocytes back into the intestinal lumen, from desquamation of intestinal epithelial cells, and from secretion into the bile [11]. Vitamin E excreted in bile exists as a presently unidentified metabolite conjugated with glucuronic acid [11].

Two water-soluble metabolites (α-tocopheronic acid and α-tocopheronolactone) resulting from the oxidation of the side chain of α-tocopherol can be conjugated to glucuronic acid and excreted in the urine [8]. Normally these metabolites represent no more than 1% of the α-tocopherol intake [6,8,14]. However, with high α-tocopherol intake, urinary excretion of the vitamin appears to rise [11].

Another possible excretion (or secretion) route for α-tocopherol is the skin, as suggested by the presence of large amounts of radioactivity in dermal tissue following intravenous injection of ^{3}H-α-tocopherol [8,11].

Dietary Reference Intakes and Recommended Dietary Allowances

The 1989 RDA for vitamin E is 8 mg of α-tocopherol equivalents for the adult female and 10 mg for the adult male [5]. Diets of adults in the United States providing 2,000 to 3,000 kcal supplied 7 to 11 mg of α-tocopherol equivalents, as well as met the RDA for other nutrients [5,10,12,44].

The adequacy of the RDA will vary if PUFA content deviates significantly from that which is customary [44]. Concern that increased intakes of PUFA will necessitate larger amounts of vitamin E in the diet is tempered by the notation that most foods, but not all, high in PUFA are also high in vitamin E, but not necessarily α-tocopherol [44]. Horwitt [45] believes that the need for vitamin E may be related more to changes in depot fat that occur with a prolonged, high intake of PUFA than to current PUFA intake.

Smoking appears to increase destruction of vitamin E [46]. Thus, smokers may require higher intakes of vitamin E as well as vitamin C should further research confirm the findings.

Deficiency

Although determining a precise requirement for vitamin E has proved difficult, a deficiency in humans is quite rare. Usually the population groups exhibiting deficiency symptoms include premature, low-birthweight infants, people with abetalipoproteinemia, and people with malabsorption syndromes. Abetalipoproteinemia, a rare genetic disease, may result in vitamin E deficiency due to lack of apolipoprotein B, necessary for chylomicrons, VLDL, and LDL [9,11]. Malabsorption of fat is common in a variety of disorders including cystic fibrosis (characterized by pancreatic lipase deficiency) and various disorders of the hepatobiliary system, particularly chronic cholestasis characterized by decreased production of bile.

Some of the symptoms of vitamin E deficiency include retinal degeneration, ceroid pigment accumulation, hemolytic anemia, muscle weakness, degenerative neurological problems, cerebellar ataxia, loss of vibratory sense, and incoordination of limbs, among others [7,9]. Exactly how vitamin E deficiency is related to the neuromuscular degeneration is unknown. A likely explanation is the lack of antioxidant protection in neural and muscle tissues. In a deficiency of vitamin E, free radicals can cause an oxidant injury to the PUFA-rich membranes of these tissues [9].

Toxicity

Vitamin E appears to be one of the least toxic of the vitamins [8]. Although a few adverse symptoms from large oral doses of vitamin E have been reported, vitamin intakes of 400 to 800 mg of α-tocopherol equivalents (50–100 times the RDA) have been taken for months to years without apparent harm [5,47]. At higher doses (800 mg–3.2 g), there have been occasional reports of muscle weakness, fatigue, double vision, and more predominant symptoms of gastrointestinal distress including nausea, diarrhea, and flatulence [8,12,34,47]. High intakes of vitamin E can interfere with the functions of the other fat-soluble vitamins

[5,12], as discussed in the section "Interactions with Other Nutrients."

Although oral supplementation appears harmless for most people and even beneficial for others, it is not wise to conclude that chronic ingestion of large amounts of vitamin E is without risk. The toxicity noted in premature infants given parenteral vitamin E suggests that an upper limit of safety may exist [1].

Assessment of Nutriture

No truly accurate evaluation of vitamin E status exists. Normal levels of total tocopherol range from 0.8 to 1.2 mg/dL serum in adults with values <0.5 mg/dL indicative of deficiency [1,5,8]. Because of the linear relationship between serum vitamin E concentrations and total serum lipids, serum vitamin E concentrations may not accurately reflect vitamin E status during hyper- or hypolipoproteinemia conditions [1,12,14,47]. To correct for serum lipid concentrations, the ratio of serum vitamin E (milligrams) to total serum lipids (grams) is recommended by some [48]. A ratio >0.6 mg/g for a child under 12 years and >0.8 mg/g for older children and adults is considered normal [1,14,48].

A crude estimation of vitamin E status can be obtained from an erythrocyte hemolysis test that compares the amount of hemoglobin released by red cells during dilute hydrogen peroxide versus distilled water incubations. The result is expressed as a percentage whereby >20% indicates a deficiency [14]. Variables other than vitamin E status, however, can affect in vitro hemolysis [1].

Two functional tests, which assess oxidative changes in lipids, may be used to assess vitamin E status. Both tests are related to the peroxidation of PUFA. The erythrocyte malondialdehyde test is done in vitro whereby peroxidation of PUFA of erythrocytes exposed to hydrogen peroxide is determined through measurement of generated malondialdehyde [48]. Malondialdehyde can be determined by a reaction with thiobarbituric acid [11]. The breath pentane test is done in vivo. Peroxidation of PUFA (such as linoleic acid) occurring in the body is determined by measuring the exhaled hydrocarbon gas pentane [6]. Negative correlations between breath pentane levels and plasma vitamin E have been reported [48].

References Cited for Vitamin E

1. Farrell P, Roberts R. Vitamin E. In: Shils ME, Olson JA, Shike M, eds. Modern Nutrition in Health and Disease, 8th ed. Philadelphia: Lea and Febiger, 1994:326–41.

2. Eitenmiller R. Vitamin E content of fats and oils—nutritional implications. Food Tech 1997;51:78–81.

3. Murphy S, Subar A, Block G. Vitamin E intakes and sources in the United States. Am J Clin Nutr 1990; 52:361–7.

4. Erdman J, Poor C, Dietz J. Factors affecting the bioavailability of vitamin A, carotenoids and vitamin E. Food Tech 1988;42:214–21.

5. National Research Council. Recommended Dietary Allowances, 10th ed. Washington, DC: National Academy Press, 1989:99–107.

6. Bender D. Nutritional biochemistry of the vitamins. New York: Cambridge University Press, 1992:87–105.

7. Anon. Vitamin E deficiency without fat malabsorption. Nutr Rev 1988;46:189–94.

8. Combs G. The Vitamins. New York: Academic Press, 1992:179–203.

9. Sokol R. Vitamin E deficiency and neurologic disease. Ann Rev Nutr 1988;8:351–73.

10. Bieri J. Vitamin E. In: Brown ML, ed. Present Knowledge in Nutrition. Washington, D.C.: International Life Sciences Institute Nutrition Foundation, 1990:117–21.

11. Diplock A. Vitamin E. In: Diplock AT, ed. Fat-soluble vitamins. Lancaster, PA: Technomic, 1984:154–224.

12. Bieri J, Corash L, Hubbard V. Medical uses of vitamin E. N Engl J Med 1983;308:1063–71.

13. Traber M, Kayden H. Vitamin E is delivered to cells via the high-affinity receptor for low-density lipoprotein. Am J Clin Nutr 1984;40:747–51.

14. Machlin L. Vitamin E. In: Machlin LJ. Handbook of Vitamins, 2nd ed. New York: Dekker, 1991:99–144.

15. Diplock A. Antioxidant nutrients and disease prevention: an overview. Am J Clin Nutr 1991;53:189S–93S.

16. Niki E, Yamamota Y, Komuro E, Sato K. Membrane damage due to lipid oxidation. Am J Clin Nutr 1991;53:201S–5S.

17. Kagan V, Serbinova E, Forte T, Scita G, Packer L. Recycling of vitamin E in human low density lipoproteins. J Lipid Res 1992;33:385–97.

18. Niki E, Noguchi N, Tsuchihashi H, Gotoh N. Interaction among vitamin C, vitamin E, and β-carotene. Am J Clin Nutr 1995;62(suppl):1322S–26S.

19. Jialal I, Grundy S. Effect of dietary supplementation with alpha tocopherol on the oxidative modification of low density lipoprotein. J Lipid Res 1992;33:899–906.

20. Esterbauer H, Dieber-Rotheneder M, Striegl G, Waeg G. Role of vitamin E in preventing the oxidation of low-density lipoprotein. Am J Clin Nutr 1991;53:314S–21S.

21. Stampfer M, Hennekens C, Manson J, Colditz G, Rosner B, Willett W. Vitamin E consumption and the risk of coronary disease in women. N Eng J Med 1993;328:1444–9.

22. Rimm E, Stampfer M, Ascherio A, Giovannucci E, Colditz G, Willett W. Vitamin E consumption and the risk of coronary heart disease in men. N Eng J Med 1993;328:1450–6.

23. Mosca L, Rubenfire M, Mandel C, Rock C, Tarshis T, Tsai A, Pearson T. Antioxidant nutrient supplementation reduces the susceptibility of low density lipoprotein to oxidation in patients with coronary artery disease. J Am Coll Cardiol 1997;30:392–9.

24. Jialal I, Grundy S. Effect of combined supplementation with α-tocopherol, ascorbate, and beta-carotene on low-density lipoprotein oxidation. Circulation 1993;88:2780–6.

25. Stephens N, Parsons A, Schofield P, Kelly F, Cheeseman K, Mitchinson M, Brown M. Randomized controlled trial of vitamin E in patients with coronary disease: Cambridge Heart Antioxidant Study (CHAOS). Lancet 1996;347:781–6.

26. Taylor A, Jacques P, Epstein E. Relations among aging, antioxidant status, and cataract. Am J Clin Nutr 1995;62(suppl):1439S–47S.

27. Bendich A, Langseth L. The health effects of vitamin C supplementation: a review. J Am Coll Nutr 1995;14:124–36.

28. Jacques P, Taylor A, Hankinson S. Long term vitamin C supplement use and prevalence of early age-related lens opacities. Am J Clin Nutr 1997;66:911–6.

29. Varma S. Scientific basis for medical therapy of cataracts by antioxidants. Am J Clin Nutr 1991;53:335S–45S.

30. Robertson J, Donner A, Trevithick J. A possible role for vitamins C and E in cataract prevention. Am J Clin Nutr 1991;53:346S–51S.

31. Byers T, Guerrero N. Epidemiologic evidence for vitamin C and vitamin E in cancer prevention. Am J Clin Nutr 1995;62(suppl):1385S–92S.

32. Azzi A, Boscoboinik D, Marilley D, Ozer N, Stauble B, Tasinato A. Vitamin E: a sensor of the cell oxidation state. Am J Clin Nutr 1995;62(suppl):1337S–46S.

33. Traber M, Packer L. Vitamin E: beyond antioxidant function. Am J Clin Nutr 1995;62(suppl):1501S–9S.

34. Bendich A, Machlin L. Safety of oral intake of vitamin E. Am J Clin Nutr 1988;48:612–9.

35. Reaven P. Dietary and pharmacologic regimens to reduce lipid peroxidation in noninsulin dependent diabetes mellitus. Am J Clin Nutr 1995;62(suppl):1483S–9S.

36. Paolisso G, D'Amore A, Giugliano D, Ceriello A, Varricchio M, D'Onofrio F. Pharmacologic doses of vita-

min E improve insulin action in healthy subjects and non insulin dependent diabetic patients. Am J Clin Nutr 1993;57:650–6.

37. Whiteshell R, Reyen D, Beth A, Pelletier D, Abumrad N. Activation energy of slowest step in the glucose carrier cycle: correlation with membrane lipid fluidity. Biochemistry 1989;28:5618–25.

38. Omara F, Blakley B. Vitamin E is protective against iron toxicity and iron-induced hepatic vitamin E depletion in mice. J Nutr 1993;123:1649–55.

39. Kaiser S, Mascio P, Murphy M, Sies H. Physical and chemical scavenging of singlet molecular oxygen by tocopherols. Arch Biochem Biophysics 1990;277:101–8.

40. Parker R, Pearces B, Clark R, Gordon D, Wright J. Tocotrienols regulate cholesterol production in mammalian cells by post-transcriptional suppression of 3-hydroxy-3- methylglutaryl-coenzyme A reductase. J Biol Chem 1993;268:11230–8.

41. Qureshi A, Qureshi N, Wright J, Shen S, Kramer G, Gabor A, Chong Y, DeWitt G, Ong A, Peterson D, Bradlow B. Lowering of serum cholesterol in hypercholesterolemic humans by tocotrienols (palmvitee). Am J Clin Nutr 1991;53:1021S–6S.

42. Gould M, Haag J, Kennan W, Tanner M, Elson C. A comparison of tocopherol and tocotrienol for the chemo-prevention of chemically induced rat mammary tumors. Am J Clin Nutr 1991;53:1068S–70S.

43. Olson J. Recommended dietary intakes (RDI) of vitamin A in humans. Am J Clin Nutr 1987;45:704–16.

44. Lehmann J, Martin H, Lashley E, Marshall M, Judd J. Vitamin E in foods from high and low linoleic acid diets. J Am Diet Assoc 1986;86:1208–16.

45. Horwitt M. Interpretations of requirements for thiamin, riboflavin, niacin-tryptophan, and vitamin E plus comments on balance studies and vitamin B-6. Am J Clin Nutr 1986; 44:973–85.

46. Handelman G, Packer L, Cross C. Destruction of tocopherols, carotenoids, and retinol in human plasma by cigarette smoke. Am J Clin Nutr 1996; 63:559–65.

47. Council on Scientific Affairs, American Medical Association. Vitamin preparations as dietary supplements and as therapeutic agents. JAMA 1987;257: 1929–36.

48. Gibson R. Principles of nutritional assessment. New York: Oxford University Press, 1990:397–404.

Vitamin K

Several compounds possess vitamin K activity; these compounds all have a 2-methyl 1,4- naphthoquinone ring [1]. The naturally occurring forms of vitamin K are phylloquinone (K_1 or 2-methyl 3-phytyl 1,4-naphthoquinone), isolated from plants, and menaquinones (K_2 or MK) synthesized by bacteria. Most of the menaquinones contain 6 to 10 isoprenoid units attached at carbon 3. Menadione (K_3) is not found naturally but is a common synthetic form of vitamin K that must be alkylated for activity. This alkylation can be accomplished rapidly by tissue enzymes. Figure 10.23 depicts menadione, phylloquinone, and one of the menaquinones, specifically menaquinone 7, which has seven isoprenoic units and was originally isolated from putrefied fish meal [1].

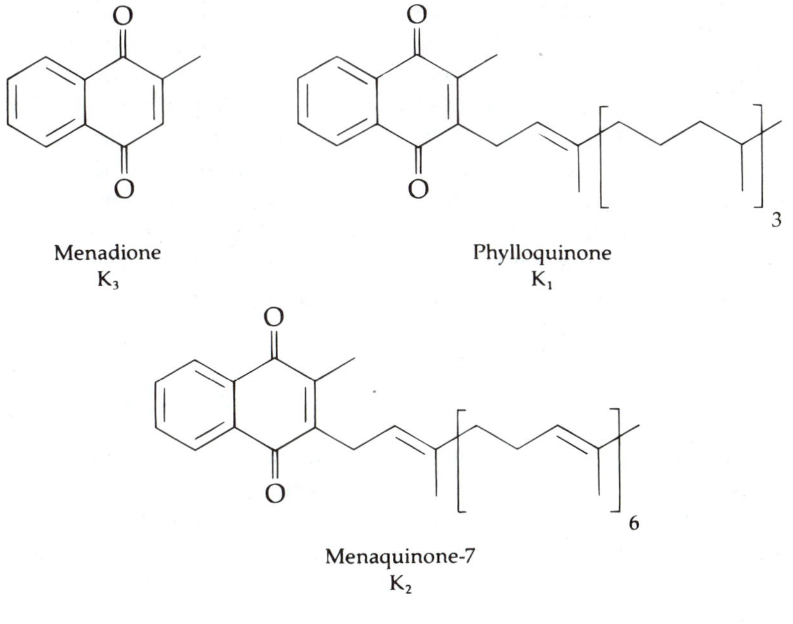

Menadione
K_3

Phylloquinone
K_1

Menaquinone-7
K_2

Figure 10.23 Biologically active forms of vitamin K.

Sources

Dietary vitamin K is provided as phylloquinone in plant foods and as a mixture of menaquinones in animal products. Phylloquinone is thought to provide the majority of the vitamin in the U.S. diet. The approximate vitamin K content of various foods is given in Table 10.5.

It is evident from Table 10.5 that dietary vitamin K is provided primarily by plant foods, especially leafy, green vegetables and certain legumes [1–4]. Vegetable oils also are good sources of vitamin K_1 phylloquinone [4,5]. Oils of rapeseed and soybean are particularly rich (142–200 μg/100 g) in phylloquinone [5]. Olive oil contains 55 μg phylloquinone/100 g oil. Sunflower, safflower, walnut, and sesame oils provide only 6 to 15 μg phylloquinone/100 g, while peanut and corn oils contain <3 μg/100 g [5]. Light and heat resulted in significant destruction of the vitamin [5]. Smaller amounts of the vitamin K_1 are found in cereals, fruits, dairy products, and meats. Bacteria in the gastrointestinal tract, especially the colon, also provide a source of menaquinones (MK) for humans.

Although rarely needed, vitamin K supplements such as Synkayvite, Mephyton, and Konakion are available. A water-soluble form of the vitamin, AquaMephyton, is also manufactured for those with fat malabsorptive disorders.

Absorption, Transport, and Storage

Phylloquinone is absorbed from the small intestine, particularly from the jejunum, by a saturable, energy-dependent process [2]. Menaquinones and synthetic menadione, in contrast, appear to be absorbed from the distal small intestine and colon by passive diffusion [1]. Menaquinones synthesized by a variety of facultative and obligate anaerobic bacteria in the lower digestive tract can also be absorbed by passive diffusion in the colon; however, the ability to absorb the bacterially produced vitamin varies from human to human [1]. Examples of menaquinone-producing obligate anaerobes include *Bacteroides fragilis, Eubacterium, Propionibacterium,* and *Arachnia. Escherichia coli,* a facultative anaerobe, also produces menaquinone [6]. These bacteria are thought to produce enough vitamin K to meet the requirement of humans; however, absorption and utilization of menaquinone by humans are still under investigation [6].

Absorption of vitamin K is enhanced by the presence of both bile salts and pancreatic juice [1]. Absorption varies from 40% to 80% of dietary vitamin K [3,7]; those with impaired fat absorption may absorb as little as 20% to 30% of the ingested vitamin [1].

Table 10.5 Vitamin K_1 Content of Selected Foods

Phylloquinone μg/100 g			
<10	10–50	>100	>200
Milk	Asparagus	Cabbage	Broccoli
Butter	Celery	Lettuce	Kale
Eggs	Green	Brussels	Swiss chard
Cheese	beans	sprouts	Turnip
Meats	Avocado	Mustard	Watercress
Fish	Kiwi	greens	greens
Corn	Pumpkin		
Cauliflower	(canned)		
Grains	Peas		
Fruits (most)	Peanut		
Tea (brewed)	butter		
	Lentils		
	Kidney		
	beans		
	Pinto		
	beans		
	Soybeans		
	Coffee		
	(brewed)		

Adapted from Booth et al. Vitamin K1 (phylloquinone) content of foods. J Food Comp and Anal 1993;6:109–20.

Within the intestinal cell, vitamin K is incorporated into the chylomicron that enters the lymphatic and then the circulatory system for transport to tissues. Chylomicrons transport the majority of phylloquinone [8]. Chylomicron remnants deliver vitamin K to the liver, although vitamin K has a relatively short hepatic duration, thereby suggesting very little long-term hepatic storage of the vitamin. In the liver, menadione is alkylated, and then, along with phylloquinone and menaquinone, is incorporated into VLDLs and ultimately carried to extrahepatic tissues in LDLs and via HDLs [7,8]. Extrahepatic tissues that store vitamin K in high quantities include the adrenal glands, lungs, bone marrow, kidneys, and lymph nodes [1,2]. The body pool size of vitamin K, estimated at 50 to 100 μg, is quite low for a fat-soluble vitamin, and smaller than that for vitamin B_{12} [7]. Turnover of vitamin K is rapid, approximately once every 2.5 hours [7].

Functions and Mechanisms of Action

Vitamin K is necessary for the posttranslational carboxylation of specific glutamic acid residues to form γ-carboxyglutamate on 4 of 13 factors required for the normal coagulation of blood. The four vitamin K–dependent factors include factors II (prothrombin), VII, IX, and X.

Overview of Blood Clotting

For blood to clot, fibrinogen, a soluble protein, must be converted into fibrin, an insoluble fiber network as shown at the bottom of Figure 10.24. Thrombin catalyzes the proteolysis of fibrinogen to yield fibrin. Fibrin molecules aggregate to form a polymer, which then undergoes cross-linking by fibrin-stabilizing factor (activated by thrombin or factor XIII) to form an insoluble clot and stop bleeding (hemorrhage).

Thrombin, however, circulates in the blood as prothrombin, an inactive enzyme (*zymogen*). Two pathways, extrinsic and intrinsic (Fig. 10.24), can be used to generate prothrombin and thus thrombin for blood clotting. In the intrinsic pathway, the coagulation process is initiated by the adsorption of factor XII onto a substance such as collagen. Once factor XII is activated (XII_a), it proceeds to cleave factor XI to generate an active compound XI_a, which in turn cleaves IX. Factor IX is vitamin K–dependent; thus, once carboxylated it binds calcium and, with phospholipids made from aggregated platelets, converts X to X_a, which is also vitamin K–dependent. X_a in turn can hydrolyze prothrombin (factor II) into thrombin (II_a), which completes the conversion of fibrinogen to fibrin for clot formation. In the extrinsic pathway (which functions for example with tissue injury), compounds such as tissue thromboplastin activate VII. VII_a is vitamin K–dependent and through a similar cascade of reactions as described for the intrinsic pathway results in the synthesis of thrombin from prothrombin (Fig. 10.24).

The Role of Vitamin K in Carboxylation of Glutamic Acid Residues

Four factors, II (prothrombin), VII, IX, and X, require vitamin K; prothrombin will be used as a model to describe the carboxylation process. Prothrombin requires vitamin K for the carboxylation of 10 to 12 glutamic acid residues residing in its N-terminal. This glutamic acid portion once carboxylated forms γ-carboxyglutamic acid (Gla), as shown in Figure 10.25. The carboxylation is required for the protein to become functional. The enzyme responsible for the γ-carboxylation is referred to as vitamin K–dependent γ-glutamyl carboxylase and is found associated with the rough endoplasmic reticulum (RER), primarily in the liver, where

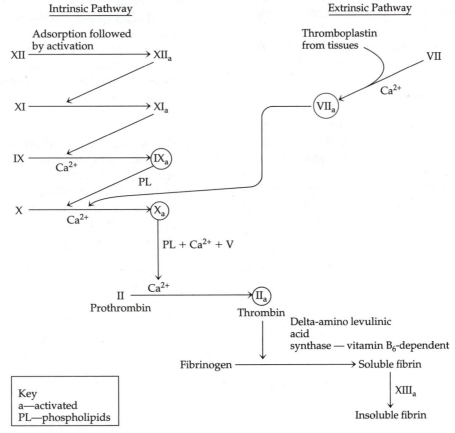

Figure 10.24 The activation of prothrombin and the roles of vitamin K–dependent clotting factors. Circled letters require vitamin K for their formation.

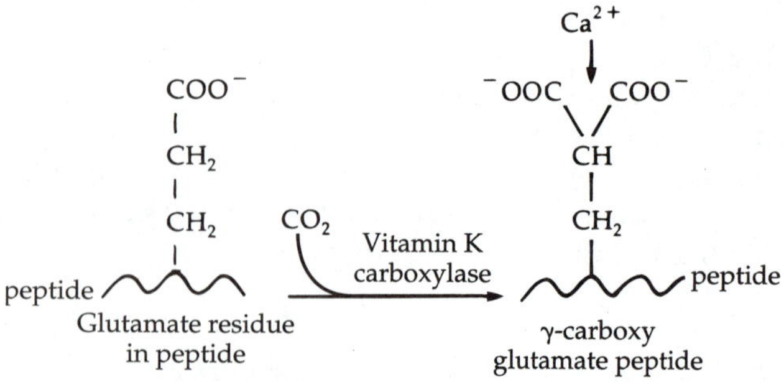

Figure 10.25 Production of γ-carboxylglutamic acid (Gla) via vitamin K–dependent carboxylation.

the hemostatic factors are synthesized, but also in the RER of other tissues such as lung, spleen, kidney, thyroid, pancreas, cartilage, bone, and skin [7]. The widespread occurrence of γ-glutamyl carboxylase suggests that the need for carboxylated proteins that can bind calcium is much broader than just the regulation of blood coagulation.

The synthesis of Gla residues occurs posttranslationally; however, the directionality of the carboxylation and the multisite specificity of the enzyme are not well understood [9]. Gla residues function to bind calcium. The calcium then mediates the binding of Gla proteins to negatively charged phospholipid membrane surfaces [10]. This adsorption of specific proteins on phospholipid surfaces is essential in hemostasis including initiation, progression, and regulation of blood clotting.

The Vitamin K Cycle

The participation of vitamin K in the carboxylation of proteins is a cyclic process (Fig. 10.26) often referred to as the *vitamin K cycle*. The γ-glutamyl carboxylase enzyme requires dihydrovitamin KH_2, also known as reduced, dihydroxy, or hydroquinone vitamin K. Thus, for carboxylation to occur vitamin K is needed in the reduced form, vitamin KH_2. Vitamin K, however, is present in the body generally in its oxidized quinone form because of the presence of oxygen in the blood. Each of the steps of the cycle will be reviewed.

- Reduction of vitamin K quinone to the active KH_2 form can be accomplished by quinone reductases that require either dithiol (indicated by RSH-HSR) or NAD(P)H.
  - The dithiol-dependent quinone reductase appears to be the main physiological pathway for generating vitamin KH_2 from the quinone.
- Once KH_2 is present, along with oxygen and carbon dioxide as the carboxyl precursor, carboxylase can carboxylate the glutamic acid residues on the protein.

- The carboxylation of glutamic acid is believed to be coupled with the formation of the vitamin K 2,3-epoxide, as illustrated in Figure 10.26 [11,12].
- No adenosine triphosphate (ATP) is required for the reaction.
- The reaction is probably accomplished by the free energy produced through the oxidation of vitamin KH_2 to 2,3-epoxide, whereby vitamin K provides reducing equivalents [3,7].

- As the cycle continues, vitamin K 2,3-epoxide is subsequently converted to vitamin K quinone by an epoxide reductase.
- The quinone is then converted back to the dihydroxy (hydroquinone) vitamin K (KH_2) by one of the two quinone reductases, requiring either NAD(P)H or 2 RSH (Fig. 10.26) as previously described.

Coumarin and warfarin are anticoagulants that antagonize the action of vitamin K. Oral anticoagulants regulate the hepatic biosynthesis of vitamin K–dependent blood-clotting factors. Warfarin, for example, interferes with the dithiol-catalyzed quinone reductase necessary for reducing the oxidized vitamin K to the KH_2 form (Fig. 10.26). Warfarin may also act on the epoxide reductase also preventing KH_2 regeneration [7]. The NAD(P)H quinone reductase, however, is relatively insensitive to warfarin [7]. Warfarin and coumarin also impair osteocalcin's ability to adsorb to hydroxyapatite due to under carboxylation [13]. Ingestion of diets high in vitamin K, as obtained from ingestion of about a pound of broccoli daily, can lead to warfarin resistance [14].

Other Vitamin K–Dependent Proteins Involved in Blood Clotting Four other proteins, *C, S, Z,* and *M,* also have been identified as being vitamin K–dependent. Although the functions of proteins Z and M are unknown, proteins C and S mediate, in part, the blood-clotting process. Protein C, a protease, inhibits coagulation, and with protein S, promotes fibrolysis and clot lysis (i.e., anticoagulation functions) [13]. Protein M appears to promote thrombin synthesis from prothrombin; however, further research is needed [7].

Vitamin K and Bone Proteins

Two vitamin K–dependent proteins identified in skeletal tissues include *bone Gla protein* (BGP, also called *osteocalcin*) and *matrix Gla protein* (MGP). The synthesis of both osteocalcin and MGP appears to be stimulated by 1,25-$(OH)_2$ D_3 and by retinoic acid [15–17].

Osteocalcin, secreted by osteoblasts during bone matrix formation, is found in bone and dentine. Osteocalcin comprises about 15% to 20% of noncollagen pro-

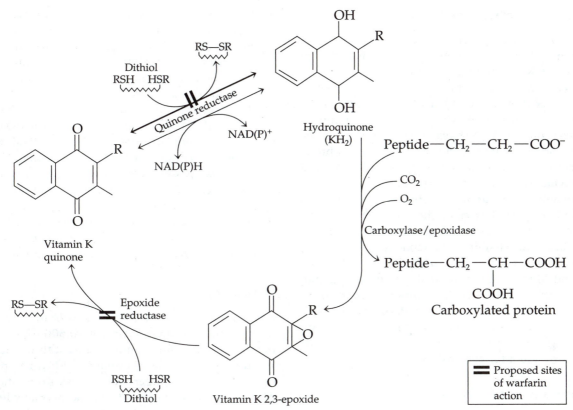

Figure 10.26 The vitamin K cycle.

tein in bone. Figure 10.27 illustrates osteocalcin synthesis. On carboxylation involving vitamin K, the three Gla residues facilitate the binding of calcium ions [18]. Osteocalcin appears in bone at the onset of hydroxyapatite deposition, and may be involved in bone remodeling and or calcium mobilization [15,16]. Osteocalcin's physiological role remains unclear at present. Some osteocalcin is released into the blood and has been used as an index of bone formation. Long-term vitamin K deficiency in animals results in cessation of longitudinal growth and bone crystallization problems [2].

MGP is found in bone, dentine, and cartilage and is associated with the organic matrix and mobilization of bone calcium. Like osteocalcin, the physiological role of MGP is uncertain. However, messenger RNA for MGP has been found in a variety of tissues including brain, heart, kidney, liver, lung, and spleen and suggests a broad role for the protein.

Vitamin K and Kidney Proteins

A third vitamin K–dependent protein, *kidney Gla protein* (KGP), also has been identified in the cortex of the kidney. Although BGP and MGP are better characterized than kidney Gla protein, further research to delineate the roles of these Gla proteins is necessary.

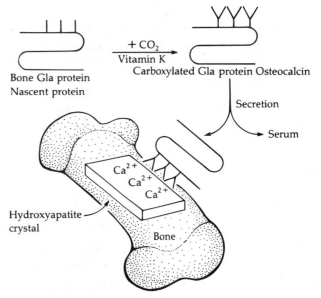

Figure 10.27 Schematic outline of the probable relationship between bone Gla protein secretion by osteoblasts and its accumulation in bone or serum.
(Price PA. Role of vitamin K–dependent proteins in bone metabolism. Reproduced, with permission, from the Annual Review of Nutrition, volume 8:574, 1988, © 1988 by Annual Review, Inc.)

Interactions with Other Nutrients

The fat-soluble *vitamins A* and *E* are known to antagonize vitamin K. Excess vitamin A appears to interfere with vitamin K absorption. The antagonistic effect of α-tocopherol on vitamin K, however, has yet to be elucidated but is thought to affect absorption, function, and/or metabolism [7,19]. Vitamin E or its quinone (α-tocopheryl quinone) may block the regeneration of the reduced form of vitamin K and/or may affect prothrombin formation by another manner [20,21]. Vitamin E may also impact vitamin K absorption [2].

A possible interrelationship between vitamins K and D and A is suggested based on their relationship to the mineral calcium. Vitamin D functions impact calcium metabolism, and vitamin K-dependent proteins bind calcium. Two sites of action of 1,25-OH_2 D_3 are the bone and kidney, and in both these tissues vitamin K-dependent calcium-binding proteins have been identified. 1,25-OH_2 D_3 as well as retinoic A have been shown to regulate, in part, production of BGP, MGP, and or KGP [15-17]. Further research is needed to better characterize the interrelationships.

Metabolism and Excretion

Phylloquinone, degraded much more slowly than menaquinone, is typically converted to 2,3-epoxide (vitamin K cycle; Fig. 10.26) and then to 3-hydroxyquinone. Several other metabolites are generated. These metabolites of phylloquinone are excreted primarily as glucuronides in the urine and feces via the bile [1]. Little is known about the metabolism and excretion of menaquinone.

Menadione is rapidly metabolized to menadiol, which then reacts with phosphate, sulfate, or glucuronide. Menadiol phosphate and menadiol sulfate are excreted both in the bile, and thus ultimately feces, and in the urine; menadiol glucuronides are excreted mostly in the feces via the bile.

Dietary Reference Intakes and Recommended Dietary Allowances

The 1989 RDA contain for the first time recommendations for vitamin K intakes in definitive values [22]. The RDA for a 79-kg adult male is 80 µg/day and for a 63-kg adult female, 65 µg/day [22]. In the past, a range of intakes has been suggested based on the assumption that the amount of vitamin K supplied by intestinal bacteria could vary from zero to as much as 50% of requirement [23]. Bacterial synthesis of menaquinones appears, however, insufficient to meet vitamin requirements when the intake of subjects is limited to about 50 µg/day [23]. Based on these studies and on the re-

sponse of people with depressed levels of vitamin K to intravenously administered doses of the vitamin [22], a dietary intake of about 1 µg/kg body weight/day appears sufficient to maintain normal clotting time in adults. Requirements for vitamin K appear to range from 0.4 µg/kg/day to 1.0 µg/kg/day [1,2].

The Recommended Dietary Intakes (RDIs) suggested by the 1980-1985 Committee on RDA were lower than either the 1980 or 1989 RDA (Table 10.6) [3]. In his article justifying the RDI for vitamin K, Olson [3] explains the decrease in the amount of vitamin K for adolescents and adults by citing studies that show 0.15 mg vitamin K/kilogram body weight per day is sufficient to maintain vitamin K-dependent clotting factors and clotting times in the normal range even when subjects are being given large doses of antibiotics.

Deficiency

A deficiency of vitamin K is unlikely in healthy adults. A normal diet contains from 300 to 500 µg vitamin K per day and therefore supplies at least three times the amount of vitamin K recommended [7]. The population groups that appear most at risk for a vitamin K deficiency are newborn infants and people who have been injured, have renal insufficiency, and/or are being treated chronically with antibiotics [3]. The newborn infant is particularly at risk because its food is limited to milk, which is low in vitamin K, stores of the vitamin are low, and its intestinal tract is not yet populated by vitamin K-synthesizing bacteria [24]. Supplementation with vitamin K is considered advisable for all newborns; currently it is recommended that 0.5 to 1 mg phylloquinone be injected intramuscularly into infants very shortly after birth [3]. Those

Table 10.6 Comparison of 1987 RDIs, 1980 and 1989 RDAs of Vitamin K in Various Population Groups[a]

Population Groups	RDI (1987) (µg)	RDA (1980) (µg)	RDA (1989) (µg)
Infants, 0–12 mo	10	10–20	5–10
Children, 1–3 yr	15	15–30	15
Children, 4–6 yr	20	20–40	20
Children, 7–10 yr	25	30–60	30
Adolescents, 11–14 yr	30	50–100	45
Adolescents, 15–18 yr	35	50–100	55–65
Adult males, 19–70+ yr	45	70–140	70–80
Adult females, 19–70+ yr	35	70–140	60–65
Pregnancy	110	———	65
Lactation	120	———	65
[Lactation (2nd 6 mo)]	———	———	62

[a]Population groups are divided differently in the various recommendations.

people on prolonged sulfa and antibiotic drug therapy are at risk owing to destruction of gastrointestinal bacteria that manufacture the vitamin and contribute a source of vitamin K. Other conditions and populations associated with increased need for intake include those with fat malabsorptive disorders: biliary fistula, obstructive jaundice, steatorrhea or chronic diarrhea, intestinal bypass surgery, chronic pancreatitis, and liver disease.

Subclinical vitamin K deficiency has been induced in healthy adults fed a diet providing only 10 μg of phylloquinone per day [25]. The 13-day low–vitamin K diet resulted in a significant reduction in plasma vitamin K concentrations. Urinary γ-carboxyglutamate excretion significantly decreased in younger subjects but remained unchanged in older adults. Prothrombin time did not change; however, descarboxyprothrombin concentrations increased significantly in subjects [25]. Severe vitamin K deficiency is associated with bleeding episodes (hemorrhage) due to prolonged prothrombin time. A relationship between vitamin K and osteoporosis has been suggested. Subclinical vitamin K deficiency may be involved in the pathogenesis of mineral loss from bone [13]. Given the prevalence and impact of osteoporosis, further studies are warranted.

Toxicity

Natural forms of vitamin K such as phylloquinone, even when supplemented in large amounts, have caused no symptoms of toxicity [3,19]. However, because of its unsubstituted carbon 3 (Fig. 10.22), the synthetic product menadione can combine with sulfhydryl groups such as those in glutathione, resulting in glutathione oxidation and excretion [10]. Ultimately there is oxidation of membrane phospholipids. Toxic effects reported in infants supplemented with menadione include hemolytic anemia, hyperbilirubinemia, and severe jaundice [19,21].

Assessment of Nutriture

Measurement of prothrombin time, the time required for a fibrin clot to form, is often used to identify potential defects in vitamin K–dependent or other blood-clotting proteins. A normal prothrombin time is considered to be between 11 and 13 seconds, while times greater than 25 seconds are associated with major bleeding [23]. In addition, maintenance of plasma prothrombin concentrations in the normal range (80–120 mg/mL) [3] suggests adequate vitamin K status.

Vitamin K status can also be assessed by the measurements of both plasma prothrombin and des γ-carboxyglutamyl prothrombin. Because human vitamin K deficiency results in the secretion of partially carboxylated prothrombin molecules (des γ-carboxyglutamyl prothrombin) as well as other molecules such as osteocalcin into the plasma, measurement of the ratio of des γ-carboxyglutamyl prothrombin to prothrombin appears to be a useful indicator of alterations in vitamin K sufficiency [1,3]. The 24-hour urine excretion of γ-carboxyglutamic acid, plasma vitamin K concentrations, and the hydroxylapatite binding capacity of osteocalcin have also been used to assess vitamin K status [24,26]. The use of hydroxylapatite-binding capacity of osteocalcin as a measure of vitamin K status is similar to the principle behind the use of des γ-carboxyglutamyl prothrombin. Suboptimal vitamin K status will cause the production of under carboxylated and fewer Gla residues, which in turn will decrease the protein's binding ability.

References Cited for Vitamin K

1. Suttie J. Vitamin K. In: Diplock AT, ed. Fat Soluble Vitamins. Lancaster, PA: Technomic, 1985:225–311.

2. Suttie J. Vitamin K. In: Machlin LJ. Handbook of Vitamins, 2nd ed. New York: Dekker, 1991:145–94.

3. Olson, J. Recommended dietary intakes (RDI) of vitamin K in humans. Am J Clin Nutr 1987;45:687–92.

4. Booth S, Sadowski J, Weihrauch J, Ferland G. Vitamin K_1 (Phylloquinone) content of foods: a provisional table. J Food Comp Anal 1993;6:109–20.

5. Ferland G, Sadowski J. Vitamin K_1 (Phylloquinone) content of edible oils: effects of heating and light exposure. J Agric Food Chem 1992;40:1869–73.

6. Suttie J. The importance of menaquinones in human nutrition. Ann Rev Nutr 1995;15:399–417.

7. Olson R. The function and metabolism of vitamin K. Ann Rev Nutr 1984;4:281–337.

8. Lamon-Fava S, Sadowski J, Davidson K, O'Brien M, McNamara J, Schaefer E. Plasma lipoproteins as carriers of phylloquinone (vitamin K1) in humans. Am J Clin Nutr 1998;67:1226–31.

9. Benton M, Price P, Suttie J. Multi-site-specificity of the vitamin K–dependent carboxylase: in vitro carboxylation of des-γ-carboxylated bone gla protein and des-γ-carboxylated pro bone Gla protein. Biochemistry 1995;34:9541–51.

10. Esmon C. Cell mediated events that control blood coagulation and vascular injury. Ann Rev Cell Biol 1993;9:1–26.

11. Kuliopulos A, Hubbard B, Lam Z, Koski I, Furie B, Furie B, Walsh C. Dioxygen transfer during vitamin K dependent carboxylase catalysis. J Am Chem Soc 1992;31:7722–8.

12. Dowd P, Ham S, Hershline R. Role of oxygen in the vitamin K–dependent carboxylation reaction: incorporation of a second atom of ^{18}O from molecular oxygen-^{18}O$_2$ into vitamin K oxide during carboxylase activity. JACS 1992;114:7613-7.

13. Binkley N, Suttie J. Vitamin K nutrition and osteoporosis. J Nutr 1995;125;1812-21.

14. Kempin S. Warfarin resistance caused by broccoli. N Eng J Med 1983;308:1229-30.

15. Price P. Role of vitamin K–dependent proteins in bone metabolism. Ann Rev Nutr 1988;8:565-83.

16. The function of the vitamin K–dependent proteins, bone Gla protein (BGP) and kidney Gla protein (KGP). Nutr Rev 1984;42:230-3.

17. Cancela M, Price P. Retinoic acid induces matrix Gla protein gene expression in human cells. Endocrinology 1992;130:102-8.

18. Saupe J, Shearer M, Kohlmeier M. Phylloquinone transport and its influence on gamma- carboxyglutamate residues of osteocalcin in patients on maintenance hemodialysis. Am J Clin Nutr 1993;58:204-8.

19. Council on Scientific Affairs, American Medical Association. Vitamin preparations as dietary supplements and as therapeutic agents. JAMA 1987;257: 1929-36.

20. Bieri J, Corash L, Hubbard V. Medical uses of vitamin E. N Engl J Med 1983;308:1063-71.

21. Bendich A, Machlin L. Safety of oral intake of vitamin E. Am J Clin Nutr 1988;48:612-9.

22. National Research Council. Recommended Dietary Allowances. 10th ed. Washington, D.C.: National Academy Press, 1989:107-14.

23. National Research Council. Recommended Dietary Allowances, 9th ed. Washington, D.C.: National Academy of Sciences, 1980:69-71.

24. Jie K, Hamulyak K, Gijsbers B, Roumen F, Vermeer C. Serum osteocalcin as a marker forvitamin K–status in pregnant women and their newborn babies. Thrombosis and Haemostatis1992;68:388-91.

25. Ferland G, Sadowski J, O'Brien M. Dietary induced subclinical vitamin K deficiency in normal human subjects. J Clin Invest 1993;91:1761-8.

26. Sadowski J, Bacon D, Hood S et al. The applications of methods used for the evaluation of vitamin K nutritional status in human and animal studies. In: Suttie JW, ed. Current Advances in Vitamin K Research. New York: Elsevier, 1988: 453-63.

The Antioxidant Nutrients, Reactive Species, and Disease

While different sections in several chapters of this book have addressed the roles of selected nutrients as they relate to antioxidant function, no where is this information brought together to provide a comprehensive review of how these individual nutrients function together to protect the body from destructive radicals and destructive nonradical species. Such is the purpose of this perspective, which will first review free radical chemistry. Next, the perspective will address how free radicals and selected nonradicals are generated in the body, damage caused by reactive oxygen species, and lastly how the antioxidant nutrients function together to eliminate destructive radical and nonradical oxygen species.

Free Radical Chemistry

Back in probably one of your first chemistry courses, you learned about atoms. It is here that a brief review of free radical chemistry will begin. Atoms contain protons and neutrons that are found in the nucleus. You may remember that the atomic weight of an element is a function of its number of protons and neutrons, while the atomic number represents solely the number of protons. Atoms also have electrons, which revolve in orbitals (also called *shells*) around the nucleus. An atomic orbital holds a maximum of two electrons. These electrons are generally found in pairs in the orbitals. The term *free radical* represents an atom or molecule that has one or more unpaired electron(s). The unpaired electron is found alone in the outer orbital, and is usually denoted by a superscript dot next to the element. The superoxide radical is denoted with a superscript dot ($O_2\cdot$) or a superscript dash ($O_2\text{-}$) or both ($O_2^{\cdot\text{-}}$). The imbalance in electrons in the orbitals results in most cases in the high reactivity of the free radicals.

Free radicals that contain oxygen may be referred to as *reactive oxygen species* (ROS), whereas those free radicals containing nitrogen are *reactive nitrogen species* (RNS). The term *reactive* is most appropriately used when comparing different radicals because reactivity with other compounds is relative. The terms *reactive oxygen species* and *reactive nitrogen species*, however, not only includes free radicals containing oxygen and nitrogen, respectively, but also includes nonradicals, as shown in Table 1.

The radicals listed in Table 1 are not inclusive of all free radical or reactive species. Oxygen itself is a biradical because it has two unpaired electrons, residing in separate orbitals, which cannot form a pair. An example of a reactive sulfur species radical is thiyl ($RS\cdot$) generated from amino acids, while trichloromethyl ($CCL_3\cdot$), formed during metabolism of carbon tetrachloride in the liver, is a chloride based carbon-centered radical, meaning that the unpaired electron resides on the carbon atom.

Generation of Reactive Species

A variety of different reactive species are generated daily in the body, often from multiple sites. In general, the reactive oxygen species are formed with exposure to substances such as smog, ozone, chemicals, drugs, radiation, and high oxygen, among others, and during normal physiological processes. Radicals also breed more radicals. Production of the superoxide radical, hydrogen peroxide, the hydroxyl radical, the peroxyl, hydroperoxyl, lipid carbon-centered, and lipid peroxyl radicals, and lipid peroxides will be reviewed in this section and are shown in Figure 1.

The Superoxide Radical

The superoxide radical ($O_2\cdot$ or $O_2\text{-}$) is an oxygen-centered radical—that is, the unpaired electron resides on the oxygen. Superoxide radicals can be synthesized when oxygen (O_2) molecules react inadvertently with, for example, catcholamines such as epinephrine and dopamine or with the vitamin folate, as tetrahydrofolate. The electron transport chain also accidentally produces some superoxide radicals due to autoxidation reactions and leaking of electrons from the electron transport chain onto oxygen—that is, a one-electron reduction of oxygen to generate the superoxide radical.

Table 1 Reactive oxygen and nitrogen species

Reactive Oxygen Species		Reactive Nitrogen Species	
Oxygen-Containing Radicals	Oxygen Containing Nonradicals	Nitrogent-Containing Radicals	Nitrogent-Containing Nonradicals
Superoxide O_2^{-}	Ozone O_3	Nitric oxide $NO\cdot$	Nitrous acid HNO_2
Hydroxyl $OH\cdot$	Singlet oxygen $^1O_2$	Nitrogen dioxide $NO_2\cdot$	Peroxynitrite ONO_2^{-}
Hydroperoxyl $HO_2\cdot$	Hypochlorous acid $HOCL$		Alkyl peroxynitrite $LOONO\cdot$
Alkoxyl $LO\cdot$ or $RO\cdot$	Hydrogen peroxide H_2O_2		
Peroxyl $LO_2\cdot$ or $RO_2\cdot$			

⌘ **PERSPECTIVE** (continued)

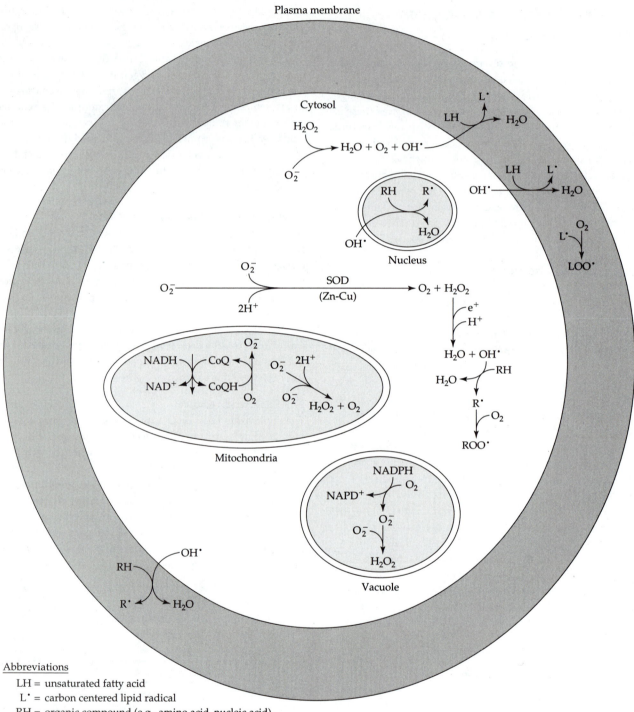

Plasma membrane

Cytosol

Figure 1 Generation of reactive species.

Abbreviations

LH = unsaturated fatty acid
L˙ = carbon centered lipid radical
RH = organic compound (e.g., amino acid, nucleic acid)
O₂⁻ = superoxide radical
OH˙ = hydroxy radical
R˙ = carbon centered nonlipid radical
H₂O₂ = hydrogen peroxide
ROO˙ = nonlipid peroxy radical
LOO˙ = peroxy radical

Remember molecular oxygen has two unpaired electrons in different orbitals. The addition of an electron to molecular oxygen leaves only one unpaired electron. This leaking of electrons onto oxygen occurs, for example, during the passage of electrons from CoQH· as part of the electron transport chain. In the electron transport chain, NADH is transporting electrons ultimately to oxygen (O_2) for ATP production; however, upon interaction between CoQH· and O_2, shown here, the superoxide radical is formed:

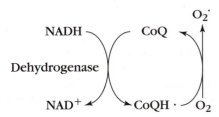

Cytochrome P_{450} enzymes also can be the generator of superoxide radicals. These heme enzymes found in the endoplasmic reticulum membrane consist of a cytochrome P_{450} reductase that transfers electrons from NADPH and a second cytochrome P_{450} that binds molecular oxygen and the substrate being hydroxylated. A variety of compounds including fatty acids, steroids, and therapeutic drugs are hydroxylated by this system.

Superoxide radicals are produced in significant quantities in activated white blood cells, such as macrophages, monocytes, and neutrophils conducting phagocytosis, to assist in the destruction of foreign substances such as bacteria and viruses. The superoxide radicals in these cells also are needed for the subsequent production of other toxic reactive oxygen species such as hydrogen peroxide (H_2O_2) to further help destroy foreign bacteria and other organisms. In addition, superoxide radicals generated by neutrophils heighten the inflammatory response by acting as a chemoattractant for other neutrophils. Production of superoxide radicals in activated white blood cells is thought to begin with the action of NADPH oxidase as a foreign substance is being engulfed by a white blood cell. This reaction is shown here:

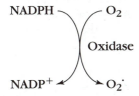

The extensive oxygen-requiring process by which white blood cells destroy organisms is sometimes referred to as the *respiratory burst*. Although superoxide radicals assist in the destruction of bacteria, viruses, fungi, and the like in white blood cells, this same radical can do harm. It is a potent initiator of chain reactions and can lead to the production of other reactive oxygen species such as hydrogen peroxide and the hydroperoxyl radical. Fortunately, superoxide is not lipid-soluble and thus does not diffuse too far away from its site of production. Furthermore, it has limited reactivity; the hydroxyl radical as well as the alkoxyxyl and peroxyl radicals are more reactive [1].

Hydrogen Peroxide

Hydrogen peroxide is not a radical because it has no unpaired electrons, but it is considered a reactive oxygen species. Hydrogen peroxide is generated through the action of the enzyme superoxide dismutase (SOD). This enzyme is found in both the cytoplasm as well as in the mitochondria of cells. SOD found in the cell cytoplasm requires both zinc and copper; activity of the enzyme is impaired with copper deficiency. Mitochondrial SOD is manganese-dependent. SOD quickly eliminates superoxide radicals from the cell environment before they can do harm.

$$O_2 \cdot + O_2 \cdot + 2H^+ \xrightarrow{\text{\textit{superoxide dismutase}}} H_2O_2 + O_2$$

Ascorbate (AH_2), like SOD, also can generate hydrogen peroxide while trying to eliminate superoxide radicals. The reaction is as follows:

$$AH_2 + O_2^- + H^+ \longrightarrow AH^- + H_2O_2$$

In peroxisomes, cytoplasmic organelles responsible for the degradation of molecules such as very long-chain (20+ carbons) fatty acids, hydrogen peroxide is produced in significant quantities during the oxidation process. While catalase serves to help to eliminate the hydrogen peroxide, some of the hydrogen peroxide may escape degradation and be released into extraperoxisome cell sites and cause damage.

Other reactions in the body also result in the production of hydrogen peroxide. In trauma or injury situations such as intestinal ischemia or cardiac ischemia (*ischemia* means inadequate blood flow and thus oxygen supply), for example, many reactive oxygen species are generated, especially hydrogen peroxide. Three possible reasons for the free radical and nonradical production observed in ischemic tissue include, first, activation of neutrophils by compounds released by the damaged tissues and the neutrophil's subsequent generation of hydrogen peroxide and superoxide radicals [2–4]. Second, with injury there may be disruption of the respiratory chain with more electrons leaked to oxygen for superoxide radical formation [2–5]. Third, for tissues such as the intestine and possibly endothelial cells of blood vessels, the presence of xanthine oxidase or the conversion of

∽ **PERSPECTIVE** *(continued)*

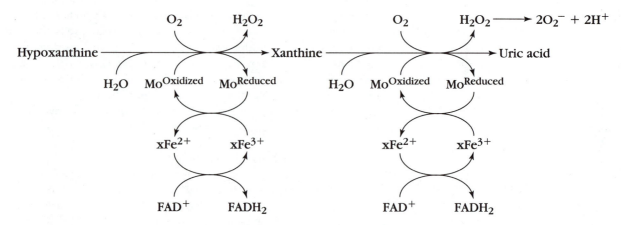

xanthine dehydrogenase into xanthine oxidase may result in free radical formation [2–5]. During ischemia, xanthine dehydrogenase is converted into xanthine oxidase owing to changes in thiol groups and or proteolysis [2,3]. Both xanthine dehydrogenase and oxidase require molybdenum and riboflavin as FAD and catalyze hypoxanthine and xanthine degradation. In hypoxic tissue, ADP is degraded (because of lack of oxygen for ATP generation) producing lots of hypoxanthine. During the medical treatment of ischemia, oxygen is administered to the patient to prevent death of the organ. While the oxygen helps to prevent organ damage, the large quantities of oxygen given with reperfusion provide xanthine oxidase with the needed oxygen (O_2) to oxidize hypoxanthine and xanthine but also generate H_2O_2, as shown above. The production of these reactive oxygen species by xanthine oxidase can further damage the already injured tissue [2–4].

Other cellular oxidases such as amine oxidase, which is copper-dependent, also generate hydrogen peroxide. The reaction catalyzed by amine oxidase, found in the blood and body tissues, is as follows:

$$RCH_2NH_2 \xrightarrow{\quad O_2 \quad H_2O_2 \quad} RCH{=}O + {}^+NH_4$$

Concentrations of hydrogen peroxide such as superoxide radicals need to be controlled in the body cells to prevent cellular destruction. Hydrogen peroxide easily diffuses in water and in lipids within cells and to tissues to cause damage; it also can react with superoxide radicals to produce a highly reactive and destructive hydroxyl radical.

The Hydroxyl Radical
The hydroxyl radical (OH·) is an oxygen-centered radical. It can be produced when the body is exposed to γ-rays, low-wavelength electromagnetic radiation. These rays split water in the body to form the hydroxyl radical, $H_2O \rightarrow H· + OH·$.

Hydroxyl radicals also are produced from reactions between hydrogen peroxide and superoxide radicals, as shown here,

$$H_2O_2 + O_2· \xrightarrow{\quad H^+ \quad} H_2O + O_2 + OH·$$

or from other electrons and protons,

$$H_2O_2 \xrightarrow{\quad e^- \quad H^+ \quad} H_2O + OH·$$

Typically, hydrogen peroxide is quickly eliminated to prevent hydroxyl radical production.

Free ferrous iron in the presence of hydrogen peroxide also can result in the formation of hydroxyl radicals, although iron is normally bound to proteins and not found free in cells. Should iron be freed (ferritin-Fe^{3+} + $O_2·$ → free Fe^{3+} + $O_2·$ → O_2 + Fe^{2+}), the following reaction, known as the *Fenton reaction,* may occur:

$$Fe^{2+} + H_2O_2 \rightarrow Fe^{3+} + OH^- + OH·$$

The hydrogen peroxide is functioning as an iron-oxidizing agent. Free copper also appears to be able to react with hydrogen peroxide but like iron is found bound to proteins in vivo. Iron also catalyzes the *Haber-Weiss* reaction between hydrogen peroxide and the superoxide radical to generate the hydroxyl radical.

$$H_2O_2 + O_2· \xrightarrow{\quad Iron \quad} O_2 + OH^- + OH·$$

$$H_2O_2 + O_2· \xrightarrow{\quad Fe^{2+} \quad Fe^{3+} \quad} OH^- + OH·$$

According to Diplock [6], OH· is a severe threat to living systems. The hydroxyl radical is thought to be one of the most potent or reactive radicals and to attack all molecules in the body [1]. Free hydroxy radicals rapidly take electrons from the surroundings. The hydroxyl radical is thought to be a major initiator of lipid peroxide (LOOH) and organic radical-generating reactions. Thus, removal of free hydroxyl radicals is important to prevent destruction to cell components.

Lipid Carbon-Centered, Hydroperoxyl, and Lipid Peroxyl Radicals and Lipid Peroxides

Peroxyl (O_2^{2-}) and hydroperoxyl (HO_2^-) radicals (oxygen centered) can be formed in the body from superoxide radicals reacting with additional electrons and hydrogen, as shown in this equation:

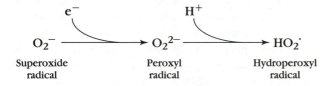

$$O_2^- \xrightarrow{\ e^-\ } O_2^{2-} \xrightarrow{\ H^+\ } HO_2^{\cdot}$$

Superoxide   Peroxyl   Hydroperoxyl
radical   radical   radical

Lipid carbon-centered radicals (L·) are produced in the body when hydroxyl radicals (OH·) or superoxide radicals attack polyunsaturated fatty acids (LH) in the phospholipids of membranes or attack other organic compounds. The reaction may be written as follows:

$$LH + OH\cdot \longrightarrow L\cdot + H_2O \text{ (initiation)}$$

Organic nonlipid peroxyl radicals (ROO·) also may attack polyunsaturated fatty acid lipids (LH) in cell membranes.

$$ROO\cdot + LH \longrightarrow ROOH + L\cdot$$

The organic nonlipid peroxyl radicals (ROO·) are produced via reactions with hydroxyl radicals similar to those shown earlier with polyunsaturated fatty acids (LH) and OH·.

Both lipid carbon-centered radicals (L·) and hydroperoxyl radicals ($HO_2\cdot$) form when molecular oxygen reacts with organic lipid compounds (LH).

$$LH + O_2 \longrightarrow L\cdot + HO_2\cdot$$

Once generated, lipid carbon-centered radicals (L·), in turn, can rapidly react with molecular oxygen to form lipid peroxyl radicals (LOO·, also written $LO_2\cdot$), which are better able to propagate chain reactions than lipid carbon-centered radicals.

$$L\cdot + O_2 \longrightarrow LOO\cdot \text{ (propagation)}$$

In chain propagation, a product formed in one reaction is used as a reactant in another reaction.

Once formed, lipid peroxy radicals (LOO·) can abstract a hydrogen atom from other organic lipid compounds (LH) to generate a chain reaction or further propagate the formation of lipid carbon-centered radicals (L·) all over again, as well as generate lipid peroxides (LOOH).

$$LOO\cdot + LH \longrightarrow L\cdot + LOOH \text{ (propagation)}$$

Should lipid peroxides (LOOH), also known as peroxidized fatty acids, come in contact with free iron, for example, lipid alkoxyl (LO·) and peroxyl (LOO·) radicals also can be generated, as shown in the next two reactions:

$$LOOH + Fe^{2+} \longrightarrow LO\cdot + OH^- + Fe^{3+}$$

$$LOOH + Fe^{3+} \longrightarrow LOO\cdot + H^+ + Fe^{2+}$$

The lipid alkoxyl radical also can initiate chain reactions with other polyunsaturated fatty acids in membranes as follows: LO· + LH ⟶ LOH + L·. However, again it is important to note that in vivo little to no free iron appears to be available to initiate such reactions.

Singlet molecular oxygen

Singlet molecular oxygen ($^1O_2$) possesses higher energy and is more reactive than ground state oxygen. Specifically, in singlet oxygen, the peripheral electron in the oxygen structure is excited to an orbital above that which it normally occupies [6]. This excited form of oxygen can be generated from lipid peroxidation of membranes, enzymatic reactions such as occur with the respiratory burst in white blood cells or via photochemical reactions, as shown:

$$O_2 \xrightarrow{\ h\nu\ } {}^1O_2$$

Singlet oxygen, being a reactive oxygen species, can, like free radicals, damage cells and tissues unless it is removed from the body.

Damage Due to Reactive Species

Once formed, free radicals attack, taking electrons from cell constituents including nucleic acid in DNA in the nucleus of cells as well as from proteins (especially amino acids such as proline, histidine, or arginine and those with free sulfhydryl groups such as methionine and cysteine) and polyunsaturated fatty acids (PUFAs) found in cell membranes or in the membranes of intracellular organelles such as those of the nucleus, mitochondria, or endoplasmic reticulum. Hydroxyl radical-induced changes in purine and pyrimidine bases in DNA may lead to mutations or breakages, which, if not repaired, may result, for example, in cancer [5]. Attack on amino acids in proteins by reactive oxygen species may break the peptide bonds in the protein backbone or disrupt the protein structure. Oxidative damage to proteins may cause cross linking between amino acids or aggregation resulting in changes in the secondary or tertiary structures. Such events may even lead to premature degradation of the protein. Free radical attack on polyunsaturated fatty acids present in the phospholipid portion of the cell membranes can lead to degradation of the lipid. Extensive damage in a red blood cell, for example, may cause hemolysis of the membrane and thus the cell [7–9]. Aqueous peroxyl and peroxy nitrite radicals may induce oxidation of LDLs. Furthermore, radicals give rise to more radicals and thus more damage. Attack of DNA, proteins, and polyunsaturated fatty acids by free radicals and other reactive oxygen and nitrogen species has been implicated in the etiologies of dozens of diseases including atherosclerosis and cancer, among others.

Antioxidant Nutrient Functions

Various properties primarily one electron reduction potential of reactive oxygen species and antioxidants allow determination of predicted donation of electrons between oxygen species and antioxidants [1]. From these data, vitamin E appears to have the highest reduction potential (and thus is more willing to donate electrons) followed in descending order by vitamin C, ubiquinol, and glutathione [1]. The destruction of reactive species by the antioxidant nutrients is reviewed in this section of the perspective and is shown in Figure 2.

Elimination of Superoxide Radicals

The primary mechanisms by which the body gets rid of superoxide radicals is through conversion of the superoxide radicals to other compounds. Several antioxidant nutrients function to dispose of superoxide radicals. These nutrients include vitamin C as well as three minerals that function as cofactors for enzymes involved in oxidant defense.

Vitamin C (ascorbate), being water-soluble and hydrophilic, is found in the aqueous parts of the body such as the blood or within the cytoplasm of the cells. Ascorbate (AH_2) can provide electrons to reduce the superoxide radical to form hydrogen peroxide and dehydroascorbate (DHAA).

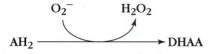

The conversion of superoxide radicals to hydrogen peroxide also is accomplished by the action of an enzyme *superoxide dismutase* (SOD). As previously mentioned, superoxide dismutase is found in the cytoplasm, where it is dependent on the presence of two minerals, *zinc* and *copper*. Without copper, SOD activity in the cytoplasm is impaired. Similarly,

SOD in the mitochondria is dependent on *manganese* for activity. Thus, zinc, copper, and manganese are important minerals involved in the body's oxidant defense system. Superoxide dismutase eliminates superoxide radicals and forms hydrogen peroxide, as shown here:

$$\text{superoxide dismutase}$$
$$O_2^- + O_2^- + 2H^+ \longrightarrow H_2O_2 + O_2$$

Elimination of Hydrogen Peroxide

Hydrogen peroxide may be disposed of by several mechanisms in cells and tissues. Vitamin C readily reacts with hydrogen peroxide, as do enzymes. Two enzymes that function in hydrogen peroxide disposal are glutathione peroxidase and catalase. A third enzyme, myeloperoxidase, utilizes the hydrogen peroxide for the generation of other radicals needed to help to fight bacteria and viruses invading body cells. The role of vitamin C as well as that of each of the enzymes and their antioxidant nutrient cofactors will be discussed.

Vitamin C effectively scavenges hydrogen peroxide. Ascorbate (AH_2) reacts with hydrogen peroxide to produce water and dehydroascorbate (DHAA) as shown here:

$$AH_2 + H_2O_2 \longrightarrow 2\,H_2O + DHAA$$

Glutathione peroxidase, found in the cytoplasm and mitochondria of the cells throughout the body as well as in plasma, is an important enzyme necessary for not only removal of hydrogen peroxide but also reduction of lipid and nonlipid peroxides. The enzyme requires the mineral *selenium* (four atoms) as a cofactor, and activity is impaired if selenium or iron status are poor. Because of selenium's role in glutathione peroxidase, the mineral is considered an antioxidant nutrient. The reaction catalyzed by glutathione peroxidase for hydrogen peroxide removal requires the tripeptide glutathione (composed of glycine, cysteine, and glutamic acid) in its reduced form (GSH), as shown here:

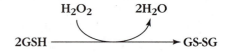

Glutathione peroxidase

Glutathione serves as a substrate for glutathione peroxidase, and during the reaction, GSH becomes oxidized (GSSG). Each of the two glutathione molecules gives up a hydrogen from its sulfhydryl group (SH). A radical center is formed on the sulfur atom until two such radicals join to form a disulfide bond.

Catalase is another key enzyme involved in the removal of hydrogen peroxide. This enzyme is heme *iron*-dependent and found mostly in cell peroxisomes (cytoplasmic organelles

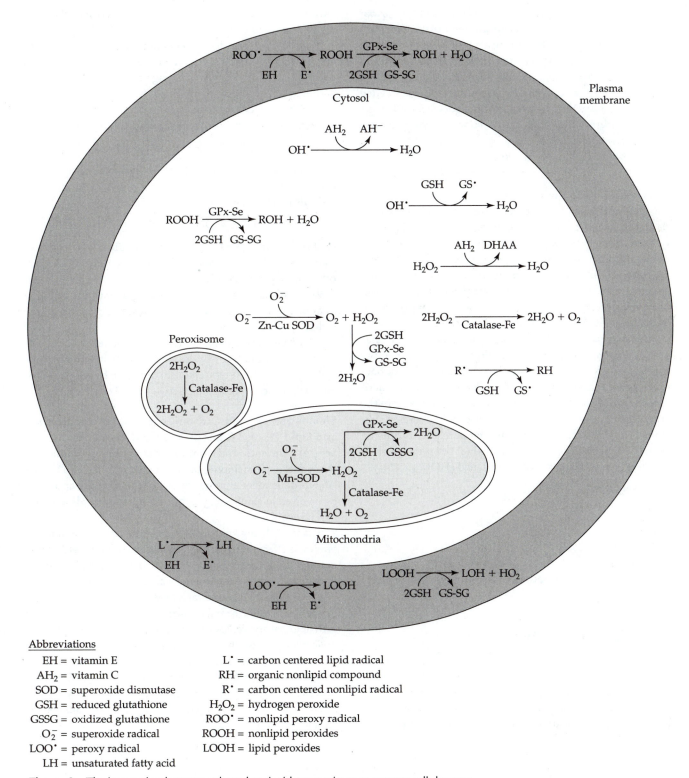

Figure 2 The interaction between selected antioxidant nutrients to prevent cell damage.

Abbreviations

EH = vitamin E
AH$_2$ = vitamin C
SOD = superoxide dismutase
GSH = reduced glutathione
GSSG = oxidized glutathione
O$_2^-$ = superoxide radical
LOO$^{\bullet}$ = peroxy radical
LH = unsaturated fatty acid

L$^{\bullet}$ = carbon centered lipid radical
RH = organic nonlipid compound
R$^{\bullet}$ = carbon centered nonlipid radical
H$_2$O$_2$ = hydrogen peroxide
ROO$^{\bullet}$ = nonlipid peroxy radical
ROOH = nonlipid peroxides
LOOH = lipid peroxides

where lots of hydrogen peroxide is produced during oxidation of very long-chain fatty acids among other molecules) with small amounts of the enzyme also found in the cytosol, mitochondria, and microsomes of cells. Neutrophils and other white blood cells contain fairly high quantities of catalase to dispose of hydrogen peroxide no longer needed in the respiratory burst required for phagocytosis of foreign bacteria, viruses, and fungi. The reaction catalyzed by catalase is shown here:

$$\text{catalase}$$
$$2H_2O_2 \longrightarrow 2 H_2O + O_2$$

Thus, accumulation of H_2O_2 is prevented by two enzymes catalase and glutathione peroxidase. Glutathione peroxidase, because of its dual (mitochondrial and cytosolic) locations in the cell, is thought to be more active in the removal of hydrogen peroxide than catalase. Figure 2 shows the complex interaction between components of the oxidant defense system including the roles of iron-dependent catalase, selenium-dependent glutathione peroxidase, and copper-, zinc-, and manganese-dependent superoxide dismutase.

Myeloperoxidase is a heme *iron*-dependent enzyme that uses hydrogen peroxide for the respiratory burst. Remember the respiratory burst is required for the destruction of bacteria, viruses, and other harmful substances. Within activated white blood cells, myeloperoxidase is release from the granules into a vacuole containing the engulfed foreign substance. In this vacuole, the hydrogen peroxide, produced from the superoxide radical, is needed for the production of a potent toxic radical, hypochlorous acid (HOCl).

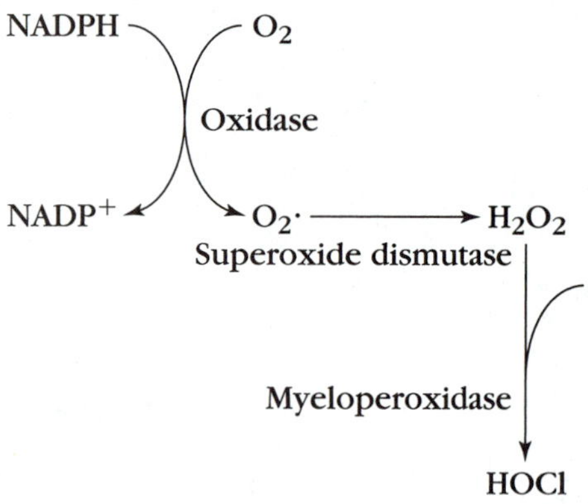

NADPH ↘ ↗ O_2

Oxidase

$NADP^+$ ↙ ↘ $O_2\cdot$ → H_2O_2

Superoxide dismutase

Cl^-

Myeloperoxidase

HOCl

Hypochlorous acid along with other potent compounds assist in the destruction of the foreign bacteria's cell membrane to promote death (lysis) of the foreign substance.

Elimination of Hydroxyl Radicals

Vitamin C and other water-soluble compounds such as glutathione and uric acid, and possibly other substances such as metallothionein serve as defense against hydroxyl radi-

cals. Vitamin E, in contrast, is not very effective in eliminating hydroxyl radicals.

Vitamin C can rapidly and effectively react, in aqueous solutions such as blood, with reactive oxygen species prior to their initiation of oxidative damage to lipids such as those within lipoproteins or membranes [8]. Vitamin C (AH_2) reacts with the hydroxy radical to produce water and the fairly nonreactive semidehydroascorbate radical (AH^-).

$$AH_2 + OH\cdot \longrightarrow AH^- + H_2O$$

Glutathione, a thiol (R-SH) like other thiols (cysteine, methionine, and lipoic acid), may react directly with hydroxyl radicals in aqueous or lipid environments.

$$GSH + OH\cdot \longrightarrow GS\cdot + H_2O$$

Glutathiyl radicals (GS·) are considerably less oxidizing than the hydroxyl radical, and typically react with each other to produce GS-SG.

Uric acid also may act as a reducing agent (antioxidant) in aqueous solutions. Uric acid may scavenge various free radicals including OH· .

Metallothionein, a protein rich in cysteine residues providing storage of zinc and copper, also may be able to scavenge hydroxyl radicals. The mechanism of action by which metallothionein accomplishes this activity is still unclear [10].

Elimination of Lipid Carbon-Centered, Hydroperoxyl, and Lipid Peroxyl Radicals and Lipid Peroxides

Several nutrients and compounds including vitamins E, manganese, glutathione, ubiquinol, vitamin C, and the selenium-dependent enzyme glutathione peroxidase appear to be involved in eliminating carbon-centered radicals and peroxyl radicals.

Vitamin E, being lipid-soluble and located near or in membranes, is effective in reacting with radicals that initiate peroxidation such as ROO· before damage to membranes is started. In addition, vitamin E terminates carbon-centered radicals as in a lipid polyunsaturated fatty acid (L·) before they abstract further hydrogens from fatty acids. Vitamin E also may react with lipid peroxyl radicals (LOO·). Vitamin E donates its phenolic hydrogen on the carbon 6 hydroxyl group. Vitamin E's chromanol ring then stabilizes the unpaired electron.

The first reaction illustrates how vitamin E (EH) may prevent initial peroxidation of, for example, a polyunsaturated fatty acid by reacting with an organic nonlipid peroxyl radical (ROO·).

$$ROO\cdot + EH \longrightarrow ROOH + E\cdot$$

Without vitamin E to halt ROO·, production of polyunsaturated fatty acid lipid carbon-centered radicals (L·) may occur as shown here:

Some transition metals such as *manganese* may be able to scavenge, like vitamin E, these same peroxyl radicals to prevent lipid carbon-centered radical formation [11].

$$Mn^{2+} \quad\quad Mn^{3+}$$
$$ROO\cdot \longrightarrow ROOH.$$

Glutathione assists vitamin E and perhaps other antioxidants by getting rid of peroxides. Thiols act in aqueous and lipid environments as an antioxidant. Glutathione peroxidase (*selenium*-dependent) catalyzes the conversion of the organic peroxides (ROOH) to organic hydroxy acids, ROH, as follows:

$$ROOH \quad\quad ROH + H_2O$$
$$2GSH \longrightarrow GS\text{-}SG$$
Glutathione peroxidase

Reduced glutathione (G-SH) provides the hydrogen atoms.

The following reactions show *vitamin E* (EH) providing a hydrogen with its single electron for the reduction of a lipid carbon-centered radical, $L\cdot + EH \longrightarrow LH + E\cdot$, or for the reduction of the lipid peroxyl radical, $LOO\cdot + EH \longrightarrow LOOH + E\cdot$. Thus, vitamin E terminates chain-propagation reactions. In other cases, as with peroxidation of lipoproteins, coantioxidants may provide an electron for the regeneration of vitamin E from its radical state ($E\cdot$). Bilirubin, for example, has been shown to assist vitamin E in inhibiting lipid peroxide formation [12].

The generation of lipid peroxides also referred to as *peroxidized fatty acids* (LOOH) through the actions of vitamin E can cause problems within cell membranes. Peroxidized fatty acids, generated in the hydrophobic region of the cell membranes from the action of vitamin E, are polar compounds. Phospholipase A_2 may cleave the peroxidized fatty acid from the phospholipid in the membrane to allow migration. The polar peroxidized fatty acids can then destroy the normal architecture of the cell in migration from the nonpolar region of generation. To eliminate lipid peroxides, *glutathione* and the *selenium*-dependent enzyme glutathione peroxidase are again needed. Glutathione can provide electrons to peroxidized free fatty acids, just as it did with organic peroxides, to form hydroxy fatty acids, LOH, through the action of glutathione peroxidase.

glutathione peroxidase
$$2GSH + LOOH \longrightarrow GS\text{-}SG + LOH + H_2O$$

Carotenoids such as β-carotene have the ability to react directly with peroxyl radicals involved in lipid peroxidation [13]. β-carotene is thought to carry out this role to a lesser than vitamin E and perhaps functions more in the interior of the cell while vitamin E functions on or at the surface.

Ubiquinol (-10), $CoQH_2$, the reduced form of coenzyme Q_{10}, is a small molecule that functions in transporting electrons and ultimately generating ATP in the electron transport chain in the mitochondria. Ubiquinol-10 (shown in Fig. 3) has been shown to provide hydrogens to terminate lipid peroxyl radicals and thus appears to be a potent antioxidant [12,14].

$$CoQH_2 + LOO\cdot \longrightarrow CoQH\cdot + LOOH$$

$CoQH\cdot$ may be regenerated into $CoQH_2$ through the electron transport chain in the mitochondria.

Ubiquinol is present only in small quantities in lipoproteins but appears to be utilized before vitamin E is the termination of lipid peroxyl radicals [14,15]. Supplementation of humans with 100 or 200 mg of ubiquinone-10 (CoQ), the oxidized form of ubiquinol, increased plasma and LDL ubiquinol concentrations, as well as increased LDL resistance to lipid peroxidation [14].

Vitamin C also effectively scavenges lipid peroxyl radicals. Ascorbate (AH_2) or the ascorbate radical (AH^-) reacts with lipid peroxyl radicals to produce a lipid peroxide and the ascorbate radical or dehydroascorbate (DHAA), respectively, as shown here:

$$AH_2 + LOO\cdot \longrightarrow AH^- + LOOH$$
$$LOO\cdot + AH^- \longrightarrow LOOH + DHAA$$

Vitamin C may be more effective in removing lipid peroxyl radicals than other water-soluble antioxidants and vitamin E [16].

Vitamin C also scavenges alkoxyl radicals as seen in the reaction:

$$AH_2 + RO\cdot \longrightarrow AH^- + ROH$$

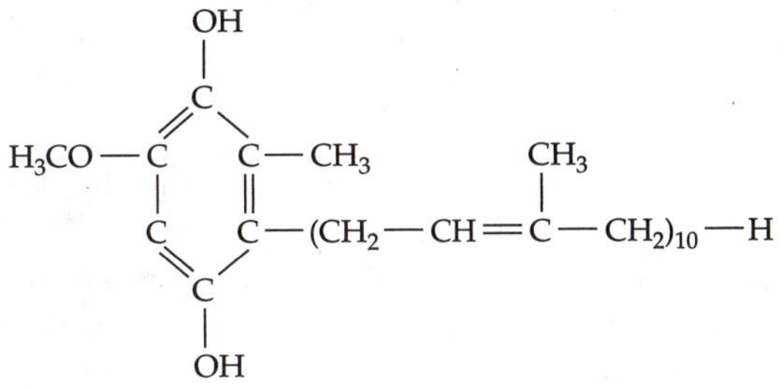

Figure 3 $CoQH_3$.

∾ **PERSPECTIVE** (continued)

Elimination of Singlet Molecular Oxygen

Carotenoids as well as vitamin C may quench singlet molecular oxygen. Carotenoids such as β-carotene and lycopene have the ability to directly quench hundreds of singlet oxygen molecules either in solution or in membrane systems. Quenching is a process by which electronically excited molecules, such as singlet molecular oxygen, are inactivated [17]. The ability of carotenoids such as β-carotene and lycopene to quench singlet oxygen is attributed to the conjugated double-bond systems within the carotenoid structure. The carotenoids can absorb energy from the singlet oxygen without chemical change to return the "excited" $^1O_2$ to its ground state [6]. Carotenoids then release the energy in the form of heat and thus do not need to be regenerated.

$$^1O_2 + \text{β-carotene} \longrightarrow {}^3O_2 + \text{excited β-carotene}$$
$$\longrightarrow \text{β-carotene} + \text{heat}$$

Lycopene appears to be more effective than β-carotene in quenching singlet oxygen [18,19].

Vitamin C has been shown in vitro to protect against singlet oxygen.

Regeneration of Antioxidant Vitamins and Other Antioxidant Compounds

Regeneration of reduced glutathione, as well as vitamins C and E, is important to the cellular defense against free radicals. The regeneration of vitamin E is thought initially to require the migration of the vitamin to the membrane surface, followed by the actions of ascorbate (AH_2), reduced glutathione, and NADPH.

Vitamin E Regeneration

Ascorbate (AH_2) is used to regenerate α-tocopherol from its radical form (E·) and in the process become a radical itself, AH^-.

$$AH_2 + E\cdot \longrightarrow AH^- + EH$$

Ubiquinol, a small molecule involved in mitochondrial electron transport, may also function to regenerate vitamin E [1].

$$CoQH_2 + E\cdot \longrightarrow CoQH\cdot + EH$$

CoQH· is regenerated into $CoQH_2$ through the electron transport chain in the mitochondria.

Glutathione in its reduced form (G-SH) may donate its hydrogen atom to help to recycle vitamin E from its radical form (E·) as follows:

$$GSH + E\cdot \longrightarrow GS\cdot + EH \text{ or}$$
$$2GSH + 2E\cdot \longrightarrow GS\text{-}SG + 2EH$$

Glutathiyl radical, GS·, may then react with oxygen to produce glutathiyl oxysulfur radicals GSO· or glutathiyl peroxyl radicals $GSO_2\cdot$ [GS· + O_2 ⟶ GSOO·] unless terminated. When each of the two glutathione molecules gives up a hydrogen from its sulfhydryl group (SH), a radical center is formed on the sulfur atom and the two such radicals join to form a disulfide bond (GS-SG).

Glutathione Regeneration

Niacin, as NADPH, is thought to regenerate oxidized glutathione (GS-SG) via the following reaction:

$$NADPH + GSSG \longrightarrow NADP + 2GSH$$

Vitamin C Regeneration

Both glutathione and niacin assist in vitamin C regeneration. *Glutathione* (G-SH) may donate its hydrogen atom to help to recycle vitamin C from its radical form (AH^-) as follows [20]:

$$2GSH + 2AH^- \longrightarrow GS\text{-}SG + 2AH_2$$

Ascorbate can be regenerated from dehydroascorbate and also glutathione.

$$DHAA + 2GSH \longrightarrow AH_2 + GS\text{-}SG$$

Niacin in its coenzyme form NADH allows the regeneration of vitamin C.

$$2AH^- + NADH + H^+ \longrightarrow 2AH_2 + NAD^+$$

Vitamin C–vitamin C radical interactions also permit regeneration. Two vitamin C radicals may interact to produce ascorbate (AH_2) and dehydroascorbate (DHAA) as follows:

$$2AH^- \longrightarrow AH_2 + DHAA$$

Antioxidants and Disease

Results from studies comparing the effectiveness of the antioxidant nutrients differ because of the numerous experimental designs and conditions. However, studies clearly show the important role of the antioxidant nutrients with implications for prevention of disease. Epidemiologic studies suggest that high intakes of fruits and vegetables rich in vitamin C and carotenoids as well as other nutrients are associated with decreased risk of some cancers and heart disease [21–29]. Evidence of high intakes of vitamin E associated with lower risk of heart disease has been shown in studies involving large groups of men and women [30,31]. An increased risk of ischemic heart disease has been shown with low plasma concentrations of antioxidants, primarily vitamin E, but to lesser extents, in rank order, carotene = ascorbate > vitamin A [32]. Similarly, an increased risk for some types of cancer has been strongly correlated with low intakes or low nutritional status of many of the antioxidants, especially vitamins A and C, carotenoids, and selenium [33].

Antioxidant supplementation studies have shown some promising results. Oral supplementation of α-tocopherol alone and with other antioxidants including vitamin C and β-carotene rendered LDLs less susceptible to oxidation [34–40]. Such findings may diminish the risk of heart disease. Vitamin E supplementation for at least 1 year appears to reduce the rate of nonfatal heart attacks in patients with heart disease [41]. Vitamin E and carotenoids appear to be able to inhibit proliferation and neoplastic transformation associated with the development of cancers; vitamin C can inhibit nitrosamine formation associated with gastric cancer [42–47]. Flavonoids were more protective than vitamin C in reducing oxidative DNA damage [48].

Recovery of tissue following ischemia-reoxygenation injuries also appears to benefit from the inclusion of antioxidants. Improved tissue recovery has been demonstrated with inclusion of antioxidant nutrients and enzymes as part of oxygen restoration following tissue ischemia [49]. Multivitamin use, especially C and E, is associated with a lower incidence of cataracts [50–54]. Vitamin C supplementation may even decrease the duration and severity of some colds [51]. Unfortunately, some intervention trials providing vitamin E, β-carotene, and vitamin A to high-risk subjects (smokers and asbestos-exposed workers) showed no benefits and even an increased risk of cancer and mortality from cancer and heart disease [55,56]. As studies continue to clarify the roles of the antioxidant nutrients, scientists and other health professionals will continue to reevaluate current recommendations to perhaps develop new guidelines defining optimal levels of nutrients to prevent diseases.

References Cited for Perspective

1. Buettner G. The pecking order of free radicals and antioxidants: lipid peroxidation, α- tocopherol, and ascorbate. Arch Biochem Biophys 1993;300:535–43.

2. Halliwell B, Gutteridge H. Free Radicals in Biology and Medicine. Oxford: Clarendon Press, 1989.

3. McCord J. Free radicals and myocardial ischemia: overview and outlook. Free Radical Biol Med 1988;4:9–14.

4. Baker G, Corry R, Autor A. Oxygen free radical induced damage in kidneys subjected to warm ischemia and reperfusion. Ann Surgery 1985;202:628–41.

5. Halliwell B, Evans P, Kaur H, Aruoma O. Free radicals, tissue injury, and human disease: a potential for therapeutic use of antioxidants? In: Kinney JM and Tuck HN, eds. Organ Metabolism and Nutrition: Ideas for Future Critical Care. New York: Raven Press, 1994:425–45.

6. Diplock A. Antioxidant nutrients and disease prevention: an overview. Am J Clin Nutr 1991;53:189S–93S.

7. Niki E. Action of ascorbic acid as a scavenger of active and stable oxygen radicals. Am J Clin Nutr 1991;54:1119S–24S.

8. Frei B. Ascorbic acid protects lipids in human plasma and low density lipoprotein against oxidative damage. Am J Clin Nutr 1991;54:1113S–18S.

9. Niki E, Yamamota Y, Komuro E, Sato K. Membrane damage due to lipid oxidation. Am J Clin Nutr 1991;53:201S–5S.

10. Sato M, Bremner I. Oxygen free radicals and metallothionein. Free Rad Biol Med 1993;14:325–37.

11. Coassin M, Ursini F, Bindoli A. Antioxidant effect of manganese. Arch Biochem Biophys 1992;299:330–3.

12. Thomas S, Neuzil J, Mohr D, Stocker R. Coantioxidants make α-tocopherol an efficient antioxidant for low density lipoprotein. Am J Clin Nutr 1995;62(suppl):1357S–64S.

13. Burton G, Ingold K. β-carotene: an unusual type of lipid antioxidant. Science 1984;224:569–73.

14. Mohr D, Bowry V, Stocker R. Dietary supplementation with coenzyme Q_{10} results in increased levels of ubiquinol-10 within circulating lipoproteins and increased resistance of human low density lipoprotein to the initiation of lipid peroxidation. Biochim Biophys Acta 1992;1126:247–54.

15. Stocker R. Induction of haem oxygenase as a defense against oxidative stress. Free Radic Res Commun 1990;9:101–12.

16. Frei B, England L, Ames B. Ascorbate is an outstanding antioxidant in human blood plasma. Proc Natl Acad Sci USA 1989;86:6377–81.

17. Mascio P, Murphy M, Sies H. Antioxidant defense systems: the role of carotenoids, tocopherols, and thiols. Am J Clin Nutr 1991;53:194S–200S.

18. Bohm F, Haley J, Truscott T, Schalch W. Cellular bound β-carotene quenches singlet oxygen in man. J Photochem Photobiol B Biol 1993;21:219–21.

19. Conn P, Schalch W, Truscott T. The singlet oxygen and carotenoid interaction. J Photochem Photobiol B 1991;11:41–47.

20. Martensson J, Han J, Griffith O, Meister A. Glutathione esters delay the onset of scurvy in ascorbate-deficient guinea pigs. Proc Natl Acad Sci USA 1993;90:317–21.

21. Block G. Vitamin C and cancer: epidemiologic evidence of reduced risk. Ann NY Acad Sci 1992;669:280–90.

22. Block G. Vitamin C and cancer prevention: the epidemiologic evidence. Am J Clin Nutr 1991;53:270S–82S.

23. Block G, Patterson B, Subar A. Fruit, vegetables, and cancer prevention: a review of the epidemiological evidence. Nutr Cancer 1992;18:1–29.

24. Block G. The data support a role for antioxidants in reducing cancer risk. Nutr Rev 1992;50:207–13.

25. Byers T, Guerrero N. Epidemiologic evidence for vitamin C and vitamin E in cancer prevention. Am J Clin Nutr 1995;62(suppl):1385S–92S.

26. Ziegler R. Vegetable, fruits and carotenoids and risk of cancer. Am J Clin Nutr 1991;53 (suppl):251S–9S.

∽ **PERSPECTIVE** (continued)

27. Gaziano J, Hennekens C. The role of beta carotene in the prevention of cardiovascular disease. Ann N Y Acad Sci 1993;69:148–154.

28. Morris D, Kritchevsky S, Davis C. Serum carotenoids and coronary heart disease. JAMA 1994;272:1439–41.

29. Poppel G, Goldbolm R. Epidemiologic evidence for β-carotene and cancer prevention. Am J Clin Invest 1995; 62(suppl):1393S–1402S.

30. Stampfer M, Hennekens C, Manson J, Colditz G, Rosner B, Willett W. Vitamin E consumption and the risk of coronary disease in women. N Eng J Med 1993;328:1444–9.

31. Rimm E, Stampfer M, Ascherio A, Giovannucci E, Colditz G, Willett W. Vitamin E consumption and the risk of coronary heart disease in men. N Eng J Med 1993;328:1450–6.

32. Gey K, Moser U, Jordan P, Sahelin H, Eichholzer M, Ludin E. Increased risk of cardiovascular disease at suboptimal plasma concentrations of essential antioxidants: an epidemiological update with special attention to carotene and vitamin C. Am J Clin Nutr 1993;57(suppl):787S–97S.

33. Weisburger J. Nutritional approach to cancer prevention with emphasis on vitamins, antioxidants, and carotenoids. Am J Clin Nutr 1991;53:226S–37S.

34. Abbey M, Nestel P, Baghurst P. Antioxidant vitamins and low-density-lipoprotein oxidation. Am J Clin Nutr 1993; 58:525–32.

35. Esterbauer H, Dieber-Rotheneder M, Striegl G, Waeg G. Role of vitamin E in preventing the oxidation of low-density lipoprotein. Am J Clin Nutr 1991;53:314S–21S.

36. Niki E, Noguchi N, Tsuchihashi H, Gotoh N. Interaction among vitamin C, vitamin E, and β-carotene. Am J Clin Nutr 1995;62(suppl):1322S–6S.

37. Jialal I, Grundy S. Effect of dietary supplementation with alpha tocopherol on the oxidative modification of low density lipoprotein. J Lipid Res 1992;33:899–906.

38. Mosca L, Rubenfire M, Mandel C, Rock C, Tarshis T, Tsai A, Pearson T. Antioxidant nutrient supplementation reduces the susceptibility of low density lipoprotein to oxidation in patients with coronary artery disease. J Am Coll Cardiol 1997;30:392–9.

39. Jialal I, Grundy S. Effect of combined supplementation with α-tocopherol, ascorbate, and beta-carotene on low-density lipoprotein oxidation. Circulation 1993;88:2780–6.

40. Allard J, Royall D, Kurian R, Muggli R, Jeejeebhoy K. Effects of β-carotene supplementation on lipid peroxidation in humans. Am J Clin Nutr 1994;59:884–90.

41. Stephens N, Parsons A, Schofield P, Kelly F, Cheeseman K, Mitchinson M, Brown M. Randomised controlled trial of vitamin E in patients with coronary disease: Cambridge Heart Antioxidant Study (CHAOS). Lancet 1996;347:781–6.

42. Bertram J, Bortkiewicz H. Dietary carotenoids inhibit neoplastic transformation and modulate gene expression in mouse and human cells. Am J Clin Nutr 1995;62(suppl):1327S–36S.

43. Bertram J, Pung A, Churley M, Kappock T, Wilkins L, Cooney R. Diverse carotenoids protect against chemically induced neoplastic transformation. Carcinogenesis 1991;12:671–8.

44. Davis M, Austin J, Partridge D. Vitamin C: Its Chemistry and Biochemistry. Cambridge, England: Royal Society of Chemistry, 1991.

45. Carpenter M. Roles of vitamins E and C in cancer. In: Laidlaw S, Swendseid M. Contemporary issues in clinical nutrition. New York: Wiley-Liss, 1991;61–90.

46. Tannenbaum S, Wishnok J, Leaf C. Inhibition of nitrosamine formation by ascorbic acid. Am J Clin Nutr 1991;53:247S–50S.

47. Schorah C, Sobala G, Sanderson M, Collis N, Primrose J. Gastric juice ascorbic acid: effects of disease and implications for gastric carcinogenesis. Am J Clin Nutr 1991;53:287S–93S.

48. Noroozi M, Angerson W, Lean M. Effects of flavonoids and vitamin C on oxidative DNA damage to human lymphocytes. Am J Clin Nutr 1998;67:1210–8.

49. Harward T, Coe D, Souba W, Kingman N, Seeger J. Glutamine preserves gut glutathione levels during intestinal ischemia/reperfusion. J Surg Res 1994;56:351–5.

50. Taylor A, Jacques P, Epstein E. Relations among aging, antioxidant status, and cataract. Am J Clin Nutr 1995;62(suppl):1439S–47S.

51. Bendich A, Langseth L. The health effects of vitamin C supplementation: a review. J Am Coll Nutr 1995;14:124–36.

52. Jacques P, Taylor A, Hankinson S. Long term vitamin C supplement use and prevalence of early age-related lens opacities. Am J Clin Nutr 1997;66:911–6.

53. Varma S. Scientific basis for medical therapy of cataracts by antioxidants. Am J Clin Nutr 1991;53:335S–45S.

54. Robertson J, Donner A, Trevithick J. A possible role for vitamins C and E in cataract prevention. Am J Clin Nutr 1991; 53:346S–51S.

55. Alpha-tocopherol, beta-carotene (ATBC) cancer prevention study group. The effect of vitamin E and beta carotene on the incidence of lung cancer and other cancers in male smokers. N Engl J Med 1994;330:1029–35.

56. Omenn G, Goodman G, Thomquist M, Balmes J, Cullen M, Glass A, Keogh J, Meyskens F, Valanis B, Williams J, Barnhart S, Hammar S. Effects of a combination of beta carotene and vitamin A on lung cancer and cardiovascular disease. N Engl J Med 1996;334:1150–55.

Web Sites

http://cancernet.nci.nih.gov

www.amhrt.org

www.cancer.org

www.preventcancer.org

www.dcpc.nci.nih.gov/5aday

www.nhlbi.nih.gov/nhlbi/cardio/cardio/htm

Chapter 11

Macrominerals

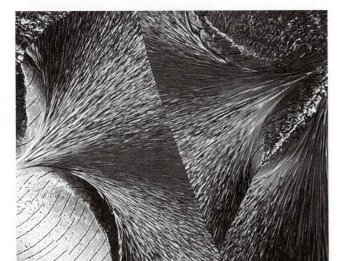

Photomicrograph of crystallized calcium phosphate

The importance of minerals in normal nutrition and metabolism cannot be overstated, despite the fact that they constitute only about 4% of total body weight. Their functions are many and varied. They provide the medium essential for normal cellular activity, determine the osmotic properties of body fluids, impart hardness to bones and teeth, and function as obligatory cofactors in metalloenzyme activity.

Historically, the knowledge that minerals are required in normal nutrition evolved from knowledge of the mineral composition of body tissues and fluids. This knowledge has expanded greatly as a result of accumulating improvements in analytic techniques for quantification of minerals.

Macrominerals (Table 11.1), also referred to as major minerals or macronutrient elements, are distinguished from the microminerals (Chaps. 12 and 13) by their occurrence in the body. Using this as a criterion, various definitions of a macromineral have been expressed, such as the requirement that it constitute at least 0.01% of total body weight or that it occur in a minimum quantity of 5 g in a 60-kg human body. Unfortunately, however, these values are clearly not equivalent, a discrepancy that in itself indicates the desirability for a less ambiguous, standard definition such as required in amounts >100 mg/day. Because of their importance in the maintenance of electrolyte balance in body fluids, the macrominerals sodium, chloride, and potassium also are discussed in Chapter 14. Although sulfur is found in the body and is considered a macromineral,

Table 11.1 Macrominerals: Functions, Deficiency Symptoms, and Recommended Dietary Allowances (RDAs) or Adequate Intakes (AIs)

Mineral	Selected Physiological Functions	Selected Enzyme Cofactors	Deficiency Symptoms	Food Sources (Rank Order)	DRI RDA/AI[a]
Calcium	Structural component of bones and teeth, role in cellular processes, muscle contraction, blood clotting, enzyme activation	Adenylate, cyclase, kinases, protein kinase, Ca^{2+} Mg^{2+} ATPase Others, see Table 11.2	Rickets, osteomalacia, osteoporosis, tetany	Milk, milk products, sardines, clams, oysters, turnip and mustard greens, broccoli, legumes, dried fruits	1,000 mg, 19–50 years
Chloride	Primary anion, maintains pH balance, enzyme activation, component of gastric hydrochloric acid		In infants: loss of appetite, failure to thrive, weakness, lethargy, severe hypokalemia, metabolic acidosis	Table salt, seafood, milk, meat, eggs	
Magnesium	Component of bones; role in nerve impulse transmission, protein synthesis, enzyme activation	Hydrolysis and transfer of phosphate groups by phosphokinase; important in numerous ATP-dependent enzyme reactions	Depression, muscle weakness, tetany, abnormal behavior, convulsions, growth failure	Nuts, legumes, cereals grains, soybeans, pars-, nips, chocolate, molasses, corn, peas, carrots, seafood, brown rice	400 mg, males; 310 mg, females
Phosphorus	Structural component bone, teeth, cell membranes, phospholipids, nucleic acids, nucleotide coenzymes, ATP-ADP phosphate transferring system in cells, pH regulation	Activates many enzymes in phosphorylation and dephosphorylation	Neuromuscular, skeletal and hematologic, and renal manifestations, rickets, osteomalacia, anorexia	Meat, poultry, fish, eggs, milk, milk products, nuts, legumes, cereal grains, chocolate	700 mg, 19+ years
Potassium	Water, electrolyte, and pH balances, cell membrane transfer	Pyruvate kinase, Na^+/K^+-ATPase	Muscular weakness, mental apathy, cardiac arrhythmias, paralysis, bone fragility	Avocado, banana, dried fruits, orange, peach, potatoes, dried beans, tomato, wheat bran, dairy products, eggs	
Sodium	Water, pH and electrolyte regulation, nerve transmission, muscle contraction	Na^+/K^+-ATPase	Anorexia, nausea, muscle atrophy, poor growth, weight loss	Table salt, meat, seafood, cheese, milk, bread, vegetables (abundant in most foods except fruits)	
Sulfur	Component of sulfur-containing amino acids, thiamin, biotin, lipoic acid		Unknown	Protein foods (meat, poultry, fish, eggs, milk, cheese, legumes, nuts	

Note: Abbreviations: ATP, adenosine triphosphate; ADP, adenosine diphosphate.
[a]For adults.

because the body does not use it alone as a nutrient, the mineral is not discussed as a subsection of this chapter. Sulfur is found in the body associated structurally with vitamins such as thiamin and biotin, and as part of the sulfur-containing amino acids methionine, cysteine, and taurine. Thus, sulfur is commonly found within proteins, especially those found in skin, hair, and nails.

Calcium

Calcium is the most abundant divalent cation of the human body, averaging about 1.5% of total body weight or between about 1,000 and 1,200 g [1]. Bones and teeth contain about 99% of the calcium. The other 1% of the body's calcium is distributed in both intra- and extracellular fluids.

Sources

The best food sources of calcium include milk, cheese, ice cream, yogurt, tofu, salmon, sardines (with bones), clams, oysters, turnip and mustard greens, broccoli, kale, legumes, and dried fruits. Meats, grains, and nuts tend to be poor sources of calcium.

Digestion, Absorption, and Transport

Digestion

Calcium is present in foods, and dietary supplements, as relatively insoluble salts. Because calcium is absorbed only in its ionized (Ca^{2+}) form, it must first be released from the salts. Calcium is solubilized from most calcium salts in about 1 hour at a mildly acidic pH. Solubilization, however, does not necessarily ensure better absorption, because within the more alkaline pH of the small intestine, calcium may complex with minerals or other selected dietary constituents [2]. Formation of these complexes in the small intestine may limit calcium bioavailability.

Absorption

Two main transport processes (Fig. 11.1) are responsible for the absorption of calcium, which occurs along the length of the small intestine, especially the ileum, where food remains for the longest time [3].

• One of the transport processes, operative primarily in the duodenum and proximal jejunum, is saturable, requires energy, involves a calcium-binding protein (CBP or calbindin), and is regulated by calcitriol (1,25-$(OH)_2 D_3$). The calcitriol-dependent calcium transport system is stimulated with ingestion of low-calcium diets, especially intakes <400 mg, as well as in conditions of growth, pregnancy, and lactation in which calcium requirements are increased. Growing children, for example, absorb up to 75% of dietary calcium in contrast with adults, who average about 30% absorption [4]. For calcium to be absorbed by this active process, three sequential steps, all regulated by vitamin D, must occur: (1) entry at the brush border, (2) intracellular movement, and (3) extrusion at the basolateral membrane [5]. Calcitriol is released in response to an increase in parathyroid hormone (PTH) secretion caused by a reduction in plasma levels of ionized calcium. Calcitriol-induced calcium absorption involves changes in membrane lipid composition and topology and the synthesis of calbindin. Calcium transport through the intestinal cell may occur by diffusion or via calbindin. Calbindin serves as a transport protein to shuttle the calcium through the cytoplasm of the enterocyte to the basal membrane. The extrusion of calcium from the enterocyte into the extracellular fluid requires ATP and a vitamin D–regulated Ca^{2+}-Mg^{2+} ATPase, an enzyme that hydrolyzes ATP and releases energy for pumping Ca^{2+} out of the cell, as Mg^{2+} moves in. Sodium is also exchanged for Ca^{2+} in the extrusion process in the basolateral membrane [6,7]. With age, however, the vitamin D–regulated absorption of calcium becomes impaired by decreased efficiency in renal calcitriol production in response to PTH [8]. Estrogen deficiency at menopause also decreases vitamin D–mediated calcium absorption.

• The second of the two processes for calcium absorption occurs throughout the small intestine but mostly in the jejunum and ileum, is nonsaturable and passive, and appears to be paracellular (i.e., absorbed between cells rather than through them). The amount of calcium absorbed via the nonsaturable, paracellular mechanism depends on the supply of calcium in the intestinal lumen up to a threshold level. Increased absorption via this mechanism becomes possible when there is an increased intake of the mineral. Net calcium absorption from both routes is about 25% to 35% [4]. Intracellular calcium movement through the enterocyte and extrusion across the basal lateral membrane occurs as described in the previous paragraph.

The large intestine also appears to play a role in calcium absorption. Bacteria in the colon may release calcium bound to some fermentable fibers such as pectins. Up to 4% (or ~8 mg) of dietary calcium is absorbed by the colon per day; this amount may be higher in people who are absorbing less calcium in the small intestine [6].

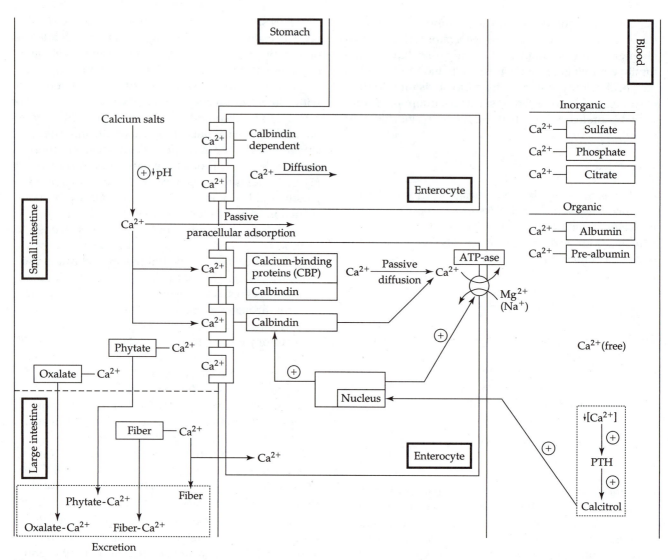

Figure 11.1 Calcium digestion, absorption, and transport.

Factors Influencing Absorption Vitamin D, as previously mentioned, improves the absorption of calcium. Polymorphism that exists in the vitamin D receptor (VDR) gene also influences, to a minor extent, calcium absorption [9]. Ingestion of food or lactose along with the calcium source appears to improve overall calcium absorption possibly by improving solubility [6]. In infants, the effects of lactose on calcium diffusion, especially in the ileum, are thought to be more pronounced than in the adult [6]. Other sugars, sugar alcohols, and protein also seem to have the same positive effect on calcium absorption [9,10]. Xylitol when consumed with calcium carbonate improved the calcium bioavailability from calcium carbonate [11].

Fiber as well as phytate may decrease calcium absorption and retention. Nonfermentable fibers such as cellulose or those found in wheat bran can increase the bulk of intestinal contents and decrease transit time, thus decreasing the time available for calcium absorption to occur. Nonfermentable or slowly fermentable fibers such as some of the hemicelluloses stimulate proliferation of microbes, which in turn bind minerals such as calcium and make them unavailable for absorption. Phytate (myoinositol hexaphosphate) is found in some of the same plant foods in which fiber is found (e.g., legumes, nuts, cereals). Phytates appear to bind calcium and decrease its availability. A phytate:calcium molar ratio >0.2 has been suggested to increase the risk of calcium deficiency [12]. Calcium retention was depressed in young men after switching from white bread to whole-wheat bread providing 2 g of phytate and 40 g of fiber per day [13].

Calcium absorption in the intestine also may be inhibited by the presence of oxalate, which chelates the calcium and increases fecal excretion of the complex. Oxalate is found in a variety of vegetables (e.g., spinach, beets, celery, eggplant, greens, okra, squash), fruits (e.g., straw-, black-, blue-, and gooseberries, currants), nuts (pecans, peanuts), and beverages (tea, Ovaltine, cocoa), among other foods.

Divalent cations along with other minerals compete with calcium for intestinal absorption. For example, magnesium and calcium, both divalent cations, compete with each other for intestinal absorption whenever an excess of either is present in the gastrointestinal tract. A dietary calcium:phosphorus ratio of 1:1 is recommended by the Food and Nutrition Board [4]; however, diets high in phosphorus relative to calcium (in amounts suggested by the RDA) in humans have not been shown to impair calcium balance [1]. Dietary intakes of calcium to phosphorus in ratios of 1:4 have induced secondary hyperparathyroidism in young adult women [14,15]. In the long term, such changes in parathyroid hormone concentrations could negatively impact bone mass. Absorption of calcium from low-calcium diets (230 mg) that include zinc supplements (zinc is another divalent cation) also is impaired [16].

Unabsorbed dietary fatty acids found in significant quantities in the gastrointestinal tract associated with steatorrhea (>7 g of fecal fat/day) can interfere with calcium absorption through the formation of insoluble calcium soaps (calcium–fatty acid complexes) in the lumen of the small intestine. These calcium soaps cannot be absorbed and are excreted in the feces.

Calcium absorption from calcium supplements varies depending on the calcium salt. Calcium (250 mg) absorption was 39% ± 3% from calcium carbonate, 32% ± 4% from calcium acetate, 32% ± 4% from calcium lactate, 27% ± 3% from calcium gluconate, and 30% ± 3% from calcium citrate [3]. Other studies suggest that calcium absorption from chelated forms of calcium such as calcium citrate, calcium citrate-malate, and calcium gluconate is better than that from calcium carbonate [17,18]. Calcium carbonate is widely used as a calcium supplement; it is relatively inexpensive and contains 40% calcium by weight. Calcium carbonate from fossilized oyster shell or dolomite, however, may be contaminated with aluminum and lead [18]. Bone meal preparations also contain lead [18]. Ingestion of these products as a calcium source should be avoided, because excessive intakes of these toxic metals may occur [18].

Although calcium absorption varies among people because of the several factors just discussed, absorption in adults averages about 30% with a 200-mg calcium intake; other studies report calcium absorption in the range 20% to 50% [4,6,19]. Recker [20] reported 21% to 26% absorption of calcium from whole, chocolate, and imitation milk, yogurt, cheese, and calcium carbonate supplying 250 mg calcium. Sheikh [3] reported an average of 31% absorption of calcium from milk providing 250 mg calcium.

Transport

Calcium is transported in the blood in three forms. Some calcium (~40%) is bound to proteins, mainly albumin and prealbumin. Some calcium (up to ~10%) is complexed with sulfate, phosphate, or citrate. Lastly, about 50% of calcium is found free (ionized) in the blood.

Regulation of Calcium Concentrations

Calcium concentrations are tightly controlled both intracellularly and extracellularly. Three main hormones are involved in calcium homeostasis in extracellular sites such as the blood: PTH, calcitriol [1,25 $(OH)_2D_3$], and calcitonin. Each of these hormones and their actions involving calcium will be discussed. An overview of calcium regulation is shown in Figure 11.2.

• PTH acts to increase extracellular fluid (plasma) calcium concentrations through interactions with the kidney and bone. In the kidney, PTH increases the synthesis of calcitriol. PTH also increases reabsorption of calcium by kidney tubules. In bone, PTH interacts with receptors on osteoblasts that signal osteoclasts. Lysosomal proteases in osteoclasts degrade bone to promote resorption of amorphous calcium salts in bone. Calcium pumps in the membrane, on activation, pump calcium through the membrane and out from bone fluid into the blood. Calcitriol also may be involved in this process.

PTH indirectly influences the intestine because stimulation of calcitriol synthesis in the kidney leads to the production of calbindin, which functions in the intestine to increase calcium absorption.

PTH secretion is influenced by plasma calcium and phosphorus concentrations, although primarily calcium. Low plasma calcium concentrations stimulate PTH secretion. Prolonged ingestion of diets high in phosphorus and low in calcium may result in a mild secondary hyperparathyroidism and can possibly lead to calcium loss from bone and calcium secretion into the gastrointestinal tract [14,15]. However, studies demonstrating such effects on bone are not available in humans [4,21,22].

• Calcitriol, 1,25 $(OH)_2D_3$, accelerates absorption of calcium from the gastrointestinal tract, as discussed on page 375. Calcitriol interacts with receptors in the enterocyte and following transport to the nucleus, it

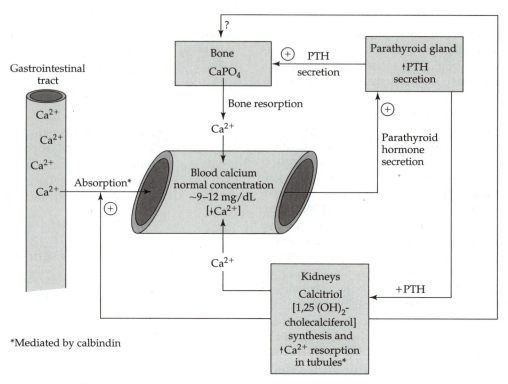

Figure 11.2 An overview of blood calcium regulation by parathyroid hormone (PTH) and vitamin D calcitriol.

increases transcription of genes that code for calbindin. Calcitriol is also thought to induce changes in the brush border and basal lateral membranes to enhance calcium absorption directly. Calcitriol may be involved in the PTH-mediated calcium reabsorption in the kidney and PTH-mediated calcium resorption by bone.

• Calcitonin, in contrast to PTH, serves to lower serum Ca^{2+} by inhibiting osteoclast activity and preventing mobilization of Ca^{2+} from bone.

Alterations in the regulation of systemic and/or cellular Ca^{2+} concentrations have been implicated in the pathogenesis of high blood pressure (hypertension) [23,24], which is discussed further as part of the Perspective at the end of this chapter.

Intracellular calcium concentrations are maintained by calcium pumps. Low calcium (Ca^{2+}) concentrations (100 nmol/L or approximately 0.0001 of the concentration in the extracellular fluid) are maintained within cells through ATP-dependent calcium pumps as well as through storage in the mitochondria, endoplasmic reticulum, nucleus, and vesicles [25]. At least two adenosine triphosphate (ATP)–dependent transport systems pump Ca^{2+} out of the cell to maintain low intracellular concentrations. In addition, a Ca^{2+} pump can drive Ca^{2+} out of the cytoplasm into the mitochondrial matrix for storage as nonionic calcium phosphate until needed by the cell [25]. To control cyto-

solic calcium concentrations, Ca^{2+} may be sequestered in the endoplasmic reticulum or, in the case of striated muscle, in the sarcoplasmic reticulum. Figure 11.3 illustrates cellular control of cytosolic–free calcium concentrations.

Raising the concentration of cytosolic Ca^{2+} from intracellular reservoirs or storage sites and through hormone-stimulated transport by a sodium-calcium exchange from extracellular sites allows Ca^{2+} to carry out its cellular functions (Fig. 11.3). Calcium may directly enter the cell from extracellular sites by transmembrane diffusion or by one of two channels (voltage-dependent slow channels or agonist-dependent), which are activated by depolarization, neurotransmitters, or hormones. Second messengers may also increase cytoplasmic calcium levels by stimulating release of calcium from intracellular sites such as the endoplasmic reticulum.

Following stimulation and release of Ca^{2+} into the cytoplasm, concentrations are returned to their normal low levels by a reversal of the events that raised its concentration. For example, the neurotransmitter may be degraded or the agonist may be released from the cell surface receptor, intracellular second messengers are inactivated, and Ca^{2+} is pumped out of the cell or is once again sequestered by intracellular organelles and binders.

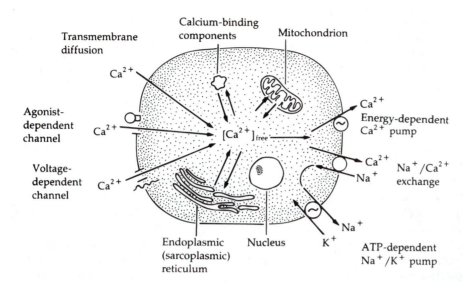

Figure 11.3 Cellular control of intracellular free calcium. Free intracellular calcium concentration is influenced by (1) the rate of calcium influx by diffusion across the cell membrane; (2) the passage of calcium through agonist- and voltage-dependent channels; (3) the rate of efflux via the calcium pump and Ca^{2+}/Na^+ exchange; (4) sequestration within the cell by endoplasmic (or sarcoplasmic) reticulum, binding components, and/or mitochondria.
Source: Nordin BEC, ed. Calcium in Human Biology. London: Springer-Verlag, 1988:388. Reprinted with permission.

Functions and Mechanisms of Actions

Calcium functions in the mineralization of bone, of which there are two types, cortical and trabecular. Most bones possess an outer layer of cortical bone that surrounds trabecular bone. Some bones also contain a cavity for bone marrow. Cortical bone is compact or dense, represents about 75% of total bone in the body, and consists of layers of mineralized collagen; it is found mainly in long bones of the limbs. Trabecular bone has a spongy appearance, represents about 25% of bone in the body and is found in relatively high concentrations in the axial skeleton (vertebrae and pelvic region). Trabecular bone (Fig. 11.4) is more active metabolically, with a high turnover rate, and thus is more rapidly depleted of calcium with calcium deficiency than cortical bone [6]. Despite the differences between cortical and trabecular bone, all bones require mineralization, which involves mainly calcium and phosphorus but also fluoride, sodium, and magnesium.

Mineralization

Approximately 99% of total body calcium is found in bones and teeth. About 60% to 66% of the weight of bones is due to minerals, with the remaining 34% to 40% of bone weight due to water and protein [1,26]. Proteins in bone include collagen, osteonectin, osteopontin, BGP (*bone Gla protein,* also called *osteocalcin*), and MGP (matrix Gla protein). The latter two proteins are dependent on vitamin K for carboxylation of their glutamic acid residues and function in calcium binding. Calcium facilitates interactions between proteins or between proteins and phospholipids in cell membranes. Osteonectin is a phosphoprotein that binds both calcium and collagen [27].

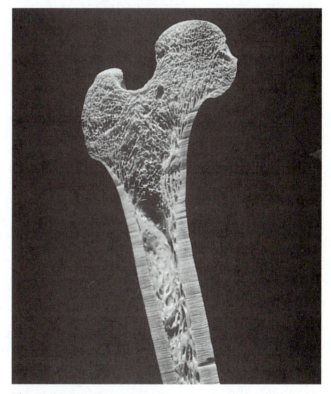

Figure 11.4
Trabecular bone is the lacy network of calcium-containing crystals that fills the interior. Cortical bone is the dense, ivorylike bone that forms the exterior shell.
Source: Whitney EN, Rolfes SR. Understanding Nutrition, 6th ed. St. Paul: West, 1993:396. Courtesy of Gjon Mill. Used by permission of Wadsworth Publishing Co.

A brief review of bone formation and the mineralization process will follow. Osteoblasts, under the influence of PTH, calcitriol, and estrogen, among other hormones, secrete collagen and other proteins as well as ground substance—that is, the extracellular matrix surrounding the bone cells such as osteoblasts, osteocytes, and osteoclasts. Ground substance is made up of mostly glycoproteins and proteoglycans. Glycoproteins consist of proteins covalently bound to typically short chains of carbohydrate. Proteoglycans are similar to glycoproteins but typically larger with longer carbohydrate chains. Chondroitin 4-sulfate, hyaluronic acid, and keratan sulfate are examples of proteoglycans. As the osteoblasts secrete the proteins and ground substance and as mineralization occurs, the osteoblasts become embedded in the proteins and matrix. Osteoblasts as well as other bone cells communicate with each other through long processes. These processes in turn connect with each other to form an osteocytic membrane system. The membrane system separates bone and bone fluid but also contains calcium pumps. During mineralization, calcium (phosphorus and magnesium) enters bone fluid from blood. Calcium is first present in bone fluid as Ca^{2+} or as amorphous (noncrystal) $Ca_3(PO_4)_2$ in solution. Osteoblasts are thought to secrete substances onto the bone surface, which enhances the precipitation of calcium (and phosphorus and magnesium). Whether osteoblasts facilitate the movement of calcium, phosphorus, magnesium, and other minerals from the blood to the bone fluid then bone surface is unclear. Calcium next loosely binds to the bone surface typically as again amorphous $Ca_3(PO_4)_2$, $CaHPO_4 \cdot 2H_2O$ (brushite), and/or $Ca_3(PO_4)_2 \cdot 3H_2O$, among others. These salts ultimately are converted to more crystalline compounds such as $Ca_8H_2(PO_4)_6 \cdot 5H_2O$, $Mg_3(PO_4)_2$, among others, as well as hydroxyapatite crystals $Ca_{10}(PO_4)_6(OH)_2$, which is laid down on collagen in the ossification process of bone formation. The process of calcification and mineralization of the bone matrix has yet to be clearly delineated. The solubility product of calcium times phosphorus ($Ca^{2+} \times PO_4^{3-}$) is thought to be involved such that, at a certain solubility product, the solution becomes supersaturated and calcium phosphate precipitates. The calcium × phosphate product needs to be at least 0.7 mmol/L for mineralization to occur; levels >2.2 mmol/L are associated with soft-tissue calcification. Of the body's phosphorus, 85% is found in bone. The mineralization process results in the further embedding of osteoblasts in the matrix, and, ultimately, the osteoblasts following morphological changes become osteocytes.

Another type of bone cell is the osteoclast. Osteoclasts resorb previously made bone. Osteoclasts contain lysosomes that release acid hydrolases and other enzymes capable of dissolving calcium phosphate and attacking bone matrix to release calcium back into the blood. Osteoclasts respond to PTH, calcitriol, and calcitonin. Osteoclasts play an important role in helping to maintain normal blood calcium concentrations in times of inadequate calcium intake.

In children and adolescents, skeletal turnover occurs such that formation of bone exceeds resorption of bone. Skeletal turnover continues into adulthood, with peak bone mass occurring from age 30 to 40 years. Bone mass begins to decline during the fifth decade [4]. Although the need for calcium in bone modeling is continuous, its greatest benefits in promoting formation of sturdy skeletal mass occur during linear bone growth and the years immediately following (i.e., approximately 10–15 years after cessation of linear growth). Roughly from age 18 to 35 years, a person can build skeletal massiveness, reaching full skeletal maturity at around 35 years [28]. The dietary factors involved in osteoporosis are discussed in the Perspective at the end of Chapter 15.

Other Roles

The small amount of remaining (not associated with bone, or nonosseous) body calcium, 1%, is found both intracellularly within organelles such as the mitochondria, endoplasmic reticulum (sarcoplasmic reticulum in muscle), nucleus, and vesicles, and extracellularly in the blood, lymph, and body fluids. Of the calcium in the blood plasma, 46% to 50% is ionized (Ca^{2+}). It is this ionized calcium that is active; therefore, the numerous regulatory functions of calcium are performed by <0.5% of the total body calcium. Nonosseous calcium is essential for a number of processes, including, for example, the *clotting of blood* (see p. 353 under vitamin K), *nerve conduction, muscle contraction, enzyme regulation,* and *membrane permeability* [4].

Calcium fluxes across membranes within cells as well as across the plasma membrane mediate the actions of a variety of hormones and neurotransmitters to affect intracellular processes. The consequential raising of cytosolic Ca^{2+} concentration allows Ca^{2+} to carry out these cellular functions (Fig. 11.3). The rise in intracellular calcium concentrations is achieved either by release of calcium from intracellular reservoirs or storage sites by second messengers or by influx by transmembrane diffusion or by channels activated by depolarization, neurotransmitters, or hormones.

An example of intracellular calcium release and functions is shown in Figure 11.5 and discussed next. Hormones or neurotransmitters, for example, bind to cell membrane receptors. This binding initiates a series of hydrolytic reactions on cell membrane phospho-

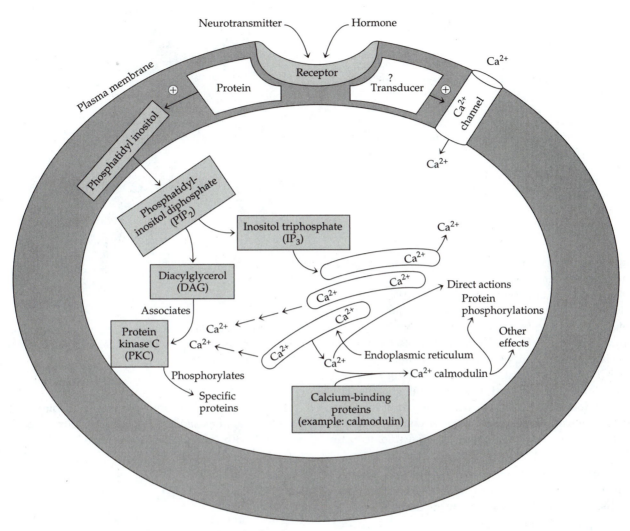

Figure 11.5 Some of the intracellular actions of calcium.

lipids. Specifically, the reactions involve the hydrolysis of a cell membrane phospholipid phosphatidyl-inositol 4,5-diphosphate (PIP_2) to generate two compounds inositol 1,4,5-triphosphate and diacylglycerol. Inositol 1,4,5-triphosphate (IP_3) is water-soluble, whereas diacylglycerol (DAG) is lipid-soluble. IP_3 functions as a second messenger following diffusion to the endoplasmic reticulum, where it triggers the release of calcium into the cytoplasm. DAG also acts as a second messenger, where it activates protein kinase C (PKC), which is associated with membrane phospholipids [25]. PKC, when associated with phospholipids and DAG, has an increased binding affinity for Ca^{2+} so that the Ca^{2+} requirement of PKC can be met by resting Ca^{2+} concentrations. Without DAG, higher concentrations of Ca^{2+} are needed to activate PKC [29]. The exact role of PKC is unknown, but it is involved in mediating sustained cellular responses [25]. In general, protein kinases catalyze the phosphorylation of amino acids found within proteins, a process that alters the protein's function and thus the cell's function or activity.

Increased free Ca^{2+} concentrations in the cell may directly affect the cell functions or may function through binding to calcium-binding proteins. Increased free Ca^{2+} can trigger neutrophils and can activate platelet phospholipase A_2. Phospholipase A_2 acts on carbon 2 of the glycerol backbone of phospholipids (e.g., phosphatidylcholine) to liberate fatty acids (e.g., arachidonic acid from phosphatidylcholine). The arachidonic acid can be metabolized to form thromboxanes, prostaglandins, or leukotrienes (p. 127). Phosphodiesterase, which hydrolyzes cyclic AMP (cAMP), is also dependent on Ca^{2+}. Cyclic AMP formation from ATP is catalyzed by adenylate cyclase, which is found on the plasma membrane. The actions of cAMP and Ca^{2+} are thought to be mutually dependent. Other enzymes that may be affected directly by increased free Ca^{2+} or affected through Ca^{2+} bound to a calcium-binding protein are listed in Table 11.2.

Table 11.2 Selected Enzymes Regulated by Calcium and/or Calmodulin	
Adenylate cyclase	Glycogen synthase
Ca-dependent protein kinase	Guanylate cyclase
Ca/Mg-ATPase	Myosin kinase
Ca/phospholipid-dependent protein kinase	NAD kinase
	Phospholipase A_2
Cyclic nucleotide phosphodiesterase	Phosphorylase kinase
	Pyruvate carboxylase
Glycerol 3-phosphate dehydrogenase	Pyruvate dehydrogenase
Pyruvate kinase	

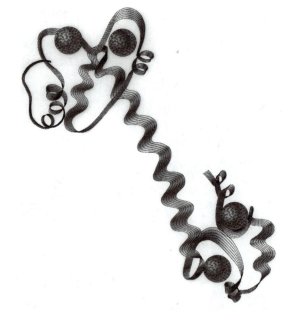

Figure 11.6 Schematic representation of the binding of four calcium ions (Ca^{2+}) by the calcium-binding protein calmodulin.

Calcium functions in a variety of processes through interactions with binding proteins. In fact, increased intracellular Ca^{2+} concentrations promote the binding of calcium to one of several calcium-binding proteins. Calmodulin is one example of a calcium-binding protein and appears to be operative in most cells. Calmodulin consists of two similar globular lobes joined by a long helix. Each lobe contains two Ca^{2+}-binding sites (Fig. 11.6). Binding of Ca^{2+} activates calmodulin by changing its conformation, thereby allowing it to stimulate a variety of macromolecular processes or enzymes. Several enzymes depend on calmodulin for activity. Some examples of the calmodulin-dependent enzymes include

• myosin light-chain kinase, which phosphorylates the light chain of myosin and, following a sequence of events, causes smooth muscle contraction [23];

• phosphorylase kinase, which activates phosphorylase (the enzyme responsible for glycogenolysis, degrading glycogen to glucose 1-PO_4); and

• calcium calmodulin kinases, of which there are several with several functions.
Others are listed in Table 11.2.

A second example of a calcium-binding protein is troponin C, which is found in skeletal muscle. Skeletal muscle stimulated by nerve impulses (acetylcholine as neurotransmitter) triggers increased concentrations of calcium. The calcium can then bind to the troponin C, allowing muscle contraction. The structure of troponin C, with its four binding sites for calcium, closely resembles that of calmodulin. Like calmodulin, the conformational change in troponin C due to Ca^{2+} binding permits an interaction between actin and myosin, resulting in muscle contraction [25]. Once the plasma membrane repolarizes, calcium is pumped back into the sarcoplasmic reticulum, troponin C releases its bound calcium, and myosin and actin no longer interact [25].

A third calcium-binding protein is calbindin. This protein is synthesized in response to calcitriol and is associated with increased absorption of calcium in the intestine.

Interactions with Other Nutrients

Calcium interacts with several nutrients not only at the absorptive surface of the intestinal cell but also within the body. Some interrelationships between calcium and other nutrients and substances (lactose, protein, vitamin D, fat, fiber, phytates, zinc) have been discussed under the section on calcium absorption. Additional interactions are discussed here.

Phosphorus is particularly interesting in that although it causes loss of calcium by increasing calcium secretion into the gut, it conserves calcium by decreasing the amount of calcium lost in the urine. Like phosphorus, increased *potassium* consumption also reduces urinary calcium excretion [30–33]. Because magnesium is necessary for the secretion of PTH, it also indirectly influences calcium.

Some nutrients interact with calcium and promote loss of calcium from the body. Dietary *protein* directly influences calcium absorption (enhanced) as well as calcium losses [6,9]; doubling of protein intake without changing intake of other nutrients results in up to a 50% increase in urinary calcium [33,34]. A significant positive correlation between protein intake and urinary calcium excretion has been found in adults aged 20 to 79 years [35]. Sulfate generated from metabolism of the sulfur-containing amino acids in the protein may be binding with the calcium and increasing its urinary excretion [31,33]. Because many protein-containing foods also contain phosphorus, ingestion of foods containing both nutrients tends to minimize negative effects on calcium balance [4]. *Boron* supplements (3 mg) when given with magnesium supplements (200 mg) increased urinary calcium losses [36]. Relationships between *sodium* and calcium have been documented whereby the sodium load (100 mmol or 2.3 g/day) increases urinary calcium excretion by 1 mmol or 40 mg/day [31,37]. Urinary sodium excretion negatively correlated with changes in bone density in the hip of postmenopausal women [38]. Halving sodium intake was found to be equivalent to increasing dietary calcium by 891 mg [38]. *Caffeine* (300–400 mg) not only increases urinary calcium (0.25 mmol or 10 mg/day) by reducing renal reabsorption, but also causes increased secretion of calcium into the gut, thereby leading to endogenous fecal losses [32,39]. Caffeine and *alcohol* intake each have been positively associated with risk of fracture in middle-aged women [40].

Calcium in the form of dietary supplements (providing 600 mg calcium in various forms such as calcium citrate) or in natural food form appears to decrease heme and nonheme *iron* absorption by 49% to 62% [41]. Calcium is thought to affect iron transfer within the enterocyte and not at the brush border [6]. This relationship has been documented in several studies and occurs primarily when the calcium and iron are ingested together with food [6].

Lead absorption is inversely related to dietary calcium intake [42]. Poor dietary calcium intake also is associated with lead accumulation in blood and organs [43].

Calcium also affects the absorption of *fatty acids* and thus can influence serum lipid concentrations and the fatty acid profile of bile. Calcium may work by inhibiting bile acid reabsorption in the ileum, thus necessitating use of body fatty acids from lipoproteins for the synthesis of additional bile. Calcium also may directly bind the fatty acids in the small intestine to form insoluble "soaps" that are excreted in the feces [44]. Ingestion of calcium carbonate (providing 1,200–3,000 mg calcium) has resulted in significant decreases in total and low-density lipoprotein (LDL) cholesterol and significant increases in high-density lipoprotein (HDL) cholesterol [45–47]. Such changes may decrease the risk for heart disease. Calcium supplementation of 2,000 or 3,000 mg

daily decreased chenodeoxycholate concentrations in bile and the lithocholate:deoxycholate ratio in the feces [48]. Such changes are favorable to the colonic environment and may help to prevent colon cancer. Diets low in calcium have been linked with increased risks for colon cancer.

Excretion

Calcium is excreted in the urine and feces, although up to 182 mg (average, 60 mg) may be lost daily from the skin, especially with extreme sweating [6,26,49]. Most calcium is filtered and reabsorbed by the kidney such that urinary calcium (50% as ionized calcium and 50% as calcium complexes with sulfate, phosphate, citrate, and oxalate) losses range from 100 to 240 mg per day, with an average of about 170 mg [6]. Urinary calcium excretion, however, is decreased with phosphorus, potassium, magnesium, and boron but increased with sodium, protein, boron plus magnesium, and caffeine. Fecal losses of calcium from endogenous and exogenous sources range from 45 to 100 mg per day [6]. Fecal losses may increase with consumption of fiber, phytate, and oxalate, magnesium in excess, and in people with fat-malabsorbing disorders.

Recommended Dietary Allowances and Dietary Reference Intakes

In recent years the 1989 RDA [4] for calcium for preadolescent and adolescent children as well as adults has been questioned because it appears to be insufficient to support optimal gains in bone mass during growth and development and to maintain bone mass in adults [50]. Although the exact age at which peak bone mass is achieved is uncertain, it is believed to be no earlier than 25 years [4] and thus has prompted increasing recommended intakes for individuals through age 24 years. Dietary Reference Intake Adequate Intakes [51] and the National Institutes of Health (NIH) consensus panels on calcium intake [52] recommend higher calcium intakes than those specified by the RDA. A comparison between different recommendations for calcium intake is provided in Table 11.3.

In addition to these recommendations, a NIH panel on osteoporosis [28] also suggests higher calcium intakes than those specified by the RDA. This NIH panel suggested a calcium intake of 1,000 mg/day by estrogen-replete premenopausal women and postmenopausal women treated with estrogen and an intake of 1,500 mg/day by untreated postmenopausal women [28]. The distinction between those treated and not treated with

Table 11.3 Comparison between Different Recommendations for Calcium Intake

Age	Recommended Dietary Allowance for Calcium [4]	Dietary Reference Intake for Calcium [51]	NIH Consensus Panel on Calcium [52]
(years)	(mg)	(mg)	(mg)
1 < 4	800	500	800
4 < 9	800	800	800–1,200
9 < 11	800	1,300	800–1,200
11 < 14	1,200	1,300	1,200–1,500
14 < 19	1,200	1,300	1,200–1,500
19 < 24	1,200	1,000	1,200–1,500
25 < 51	800	1,200	1,000
51+	800	1,200	1,000[a]
Pregnancy	1,200	1,000	1,200–1,500
Lactation	1,200	1,000	1,200–1,500

[a]NIH panel on calcium suggests that women on estrogen replacement consume 1,000 mg of calcium, while those women not on estrogen replacement consume 1,500 mg. Men at age 65 years should increase calcium intake to 1,500 mg.

estrogen is made because estrogen influences bone mineralization. Postmenopausal women experience a rapid loss of bone minerals without estrogen replacement. These levels are thought to be associated with maximum calcium retention. Calcium intakes >1,500 mg daily are thought to represent the threshold amount whereby intakes >1,500 mg would not be expected to produce further rises in calcium retention [53].

Popularization by the media of the need for calcium, the lack of sufficient calcium in diets of especially women, and the support of the NIH [28,52] recommendations by many health care providers have caused the sale of calcium supplements to soar. Although the preferable source of calcium is food, few people are willing to consume as much milk and/or milk products as would be necessary to secure this higher level of calcium intake. Calcium supplements thus continue to be marketed heavily.

Numerous calcium supplements are available from which the consumer can choose. Of the following five supplements, calcium acetate is the most soluble and has a % absorption of $32 \pm 4\%$ [3]. After calcium acetate, calcium lactate is next most soluble followed in decreasing order by calcium gluconate, calcium citrate, and calcium carbonate [3]. Calcium absorption from calcium lactate is $32 \pm 4\%$ while that from calcium gluconate is $27 \pm 3\%$ and that from calcium citrate is $30 \pm 3\%$ [3]. Calcium absorption from calcium carbonate is greatest at $39 \pm 3\%$ although other studies report considerable variation among individuals [3,20]. Because the amount of calcium varies between supplements, to obtain 500 mg of calcium, one would need to ingest

2.16 g calcium acetate, 3.53 g calcium lactate, 5.49 g calcium gluconate, 2.37 g calcium citrate, or 1.26 g calcium carbonate [3]. Although the differences in absorption were not statistically significant between calcium supplements, the variation in the amount of elemental calcium in an equal weight of supplement is considerable [3].

Although many researchers have demonstrated that current bone calcium loss in adults cannot be correlated with current calcium intake [28], many others are convinced that a present generous intake of the mineral will allow increased absorption via the nonsaturable paracellular route, thereby decelerating the rate of calcium loss from the body [52].

Deficiency

Inadequate intake, poor calcium absorption, and/or excessive calcium losses contribute to inadequate mineralization of bone. Rickets in children occurs when the amount of calcium accretion per unit bone matrix is deficient. Low levels of free ionized Ca^{2+} in the blood (hypocalcemia) may result in tetany. Tetany is manifested by intermittent muscle contractions. Muscle pain, muscle spasms, and paresthesias (numbness or tingling in the hands and feet) are common signs of tetany. In adults deficient in calcium, osteoporosis—the loss of bone mass (protein matrix and bone minerals)—occurs [55]. Two types of osteoporosis have been described:

• type I, which occurs in postmenopausal women between the ages of 51 and 65 years and affects mainly the vertebrae and distal radius, and

• type II, which occurs in men and women over the age of 75 years and affects the vertebrae, hip, pelvis, humerus, and tibia [39].

Both types I and II contribute to bone loss in individuals aged 65 to 75 years.

Much of the U.S. population, particularly females over 12 years of age, fails to consume the recommended amounts of calcium. Inadequate calcium intake during the period of bone mineralization is a concern because of the high incidence of osteoporosis among elderly women and the significant correlation shown to exist between present bone density and past calcium intake [4]. Several studies have reported positive effects from calcium supplementation on age-related bone loss [56]. Furthermore, calcium supplementation in association with either calcitonin or oral estrogen enhances bone mass more than when calcitonin or estrogen are used alone [57]. Other populations associated with increased needs for calcium include those with high-protein diets, high-fiber diets, fat malabsorption, immobilization (promotes calcium loss from bone), decreased gastrointestinal transit time, and long-term use of thiazide diuretics (which increase calcium excretion in the urine).

Deficient (long-term) calcium intakes are also associated with the development of hypertension and colon cancer [56]. An inverse relationship between calcium and blood pressure exists (as intake of calcium decreases, prevalence of hypertension increases), with a steep slope at calcium intakes <600 mg/day [29,56]. Calcium supplementation has been shown to lower blood pressure in some hypertensive people previously ingesting a diet inadequate in calcium [38]. Colon cancer also has been linked with calcium-deficient diets [19,56]. Adequate intakes of calcium (>800 mg/day) are thought to be protective against colon cancer; however, insufficient evidence in humans warrants against the use of calcium to prevent colon cancer [4]. A suggested basis for this effect is the ability of calcium to bind bile acids and free fatty acids, which act as promoters of cancer by inducing colon cell hyperproliferation [4].

Toxicity

Intakes of calcium in amounts up to 2,500 mg appear to be safe for most individuals [4]. A tolerable upper intake level of 2,500 mg calcium has been recommended for those age 1 year and older [51]. Constipation has been reported by some following ingestion of calcium supplements [1,4]. Ingestion of calcium in amounts >3,000 mg may result in hypercalcemia [1]; soft-tissue calcification is not likely unless the plasma calcium concentration times the plasma phosphorus concentration exceeds 2.2 mmol/L [1]. People with idiopathic hypercalciuria (urinary calcium levels >4 mg per kg of body weight per day), however, may be at increased risk for developing calcium-containing kidney stones with excessive calcium intakes [58].

Assessment of Nutriture

No routine biochemical method appears to assess calcium status accurately. Serum calcium (composed of protein-bound calcium, diffusible calcium complexes, and ionized calcium) is so exquisitely regulated that it may indicate little about calcium status. Serum calcium concentrations normally range from 8.5 to 10.5 mg/dL for adults, with slightly higher levels in children. Serum ionized calcium, Ca^{2+}, does, however, relate to alterations in calcium metabolism [59], and when albumin concentration is normal, the ratio between bound calcium and ionized calcium remains constant. When albumin concentrations are depressed, corrections are needed to adjust for the corresponding decrease that

will occur in the protein-bound fraction of calcium. For each 1 g/dL decrease in serum albumin, serum calcium will decrease 0.8 mg/dL. The following equations can be used for estimating protein-bound calcium:

Protein-bound calcium (milligrams/deciliter) = 0.44 + 0.76 × albumin (grams/deciliter), or = 0.8 × (normal albumin – actual albumin) + measured calcium

Bone densitometry can be assessed through computerized tomography (CT) scans [19]. CT, although less accurate and precise than dual energy X-ray absorptiometry (discussed later), can measure variances in tissue density (such as in vertebral bone). X rays are taken as the individual is held in a scanner. Radiation pulses are emitted, collected, and processed to reconstruct the image and calculate density.

Neutron activation, an in vivo procedure in which gamma rays are counted following administration of ^{48}Ca into the body and exposure of the body to a low neutron flux [59], allows assessment of total body calcium content. Results of neutron activation correlate with single-photon absorptiometry, which measures total bone mineral content. Single-photon absorptiometry exposes a portion of a limb (usually the radius/forearm or os calcis/heel) to radiation. The quantity of bone mineral is inversely proportional to the amount of photon energy transmitted from the bone, as measured by a scintillation counter [59].

Dual-photon absorptiometry, which has largely been replaced by dual energy X-ray absorptiometry (DEXA or DXA), uses a radioisotope that emits two gamma rays. In dual-energy X-ray absorptiometry, an X-ray tube is used instead of the radionuclide source as used in dual-photon absorptiometry. Both techniques can be used to measure body fat as well as total bone mineral content of the body or mineral content of selected bones such as the vertebrae and femur [60]. Dual-energy X-ray absorptiometry, however, sends a greater photon flow at a given time than dual-photon absorptiometry. Consequently, dual-energy X-ray absorptiometry provides greater accuracy and precision. Dual-energy X-ray absorptiometry may be used to assess changes in mass over time and is thought to represent the best method for assessing bone mineral density [60]. Further, measurement of bone mass is thought to be the best tool to assess calcium status.

References Cited for Calcium

1. Arnaud CD, Sanchez SD. Calcium and phosphorus. In: Brown ML, ed. *Present Knowledge in Nutrition*, 6th ed. Washington, D.C.: International Life Sciences Institute Nutrition Foundation, 1990:212–23.

2. Heaney R, Recker R, Weaver C. Absorbability of calcium sources: the limited role of solubility. Calcif Tissue Int 1990;46:300–4.

3. Sheikh MS, Santa Ana CA, Nicar MJ, Schiller LR, Fordtran JS. Gastrointestinal absorption of calcium from milk and calcium salts. N Engl J Med 1987;317:532–6.

4. National Research Council. Recommended Dietary Allowances, 10th ed. Washington, D.C.: National Academy Press, 1989:174–84.

5. Bronner F. Intestinal calcium absorption: Mechanisms and applications. J Nutr 1987;117:1347–52.

6. Allen L, Wood R. Calcium and phosphorus. In: Shils ME, Olson JA, Shike M. Modern Nutrition in Health and Disease, 8th ed. Philadelphia: Lea and Febiger, 1994:144–63.

7. Bronner F. Transcellular calcium transport. In: Bronner F, ed. Intracellular calcium regulation. New York: Wiley-Liss, 1990;415–37.

8. Silverberg S, Shane E, De La Cruz L, Segre G, Clemens T, Bilezikian J. Abnormalities in parathyroid hormone secretion and 1,25 dihydroxyvitamin D_3 formation in women with osteoporosis. N Engl J Med 1989;320:277–81.

9. Wishart J, Horowitz M, Need A, Scopacasa F, Morris H, Clifton P, Nordin B. Relations between calcium intake, calcitriol, polymorphisms of the vitamin D receptor gene, and calcium absorption in premenopausal women. Am J Clin Nutr 1997;67:798–802.

10. Schaafsma G. Calcium in extracellular fluid homeostasis. In: Nordin BEC. Calcium in Human Biology. London: Springer-Verlag, 1988;241–59.

11. Hamalainen M. Bone repair in calcium-deficient rats: comparison of xylitol + calcium carbonate with calcium carbonate, calcium lactate and calcium citrate on the repletion of calcium. J Nutr 1994;124:874–81.

12. Harland BF, Oberleas D. Phytate in foods. Wld Rev Nutr Diet 1987;52:235–59.

13. Greger JL. Mineral bioavailability/new concepts. Nutr Today 1987;22:4–9.

14. Calvo M, Kumar R, Heath H. Elevated secretion and action of parathyroid hormone in young adults ingesting high phosphorus, low calcium diets assembled for ordinary foods. J Clin Endocrinol Metab 1988;66:823–9.

15. Calvo M, Kumar R, Heath H. Persistently elevated parathyroid hormone secretion and action in young women after four weeks of ingesting high phosphorus, low calcium diets. J Clin Endocrinol Metab 1990;70:1334–40.

16. Spencer H. Minerals and mineral interactions in human beings. J Am Diet Assoc 1986;86:864–7.

17. Anderson JJB. Nutritional biochemistry of calcium and phosphorus. J Nutr Biochem 1991;2:300–7.

18. Whiting S. Safety of some calcium supplements questioned. Nutr Rev 1994;52:95–97.

19. Weaver C. Assessing calcium status and metabolism. J Nutr 1990;120:1470–3.

20. Recker R, Bammi A, Barger-Lux M, Heaney R. Calcium absorbability from milk products, an imitation milk and calcium carbonate. Am J Clin Nutr 1988;47:93–95.

21. Calvo MS. Dietary phosphorus, calcium metabolism and bone. J Nutr 1993;123:1627–33.

22. Calvo S, Park Y. Changing phosphorus content of the U.S. diet: potential for adverse effects on bone. J Nutr 1996;126:1168S–80S.

23. Bukoski R, Kremer D. Calcium-regulating hormones in hypertension: vascular actions. Am J Clin Nutr 1991;54:220S–6S.

24. McCarron D, Morris C, Young E, Roullet C, Drueke T. Dietary calcium and blood pressure: modifying factors in specific populations. 1991;54:215S–9S.

25. Rasmussen H. The calcium messenger system (parts 1 and 2). N Engl J Med 1986;314:1094–101, 1164–70.

26. Peacock M. Calcium absorption efficiency and calcium requirements in children and adolescents. Am J Clin Nutr 1991;54:261S–5S.

27. Raisz LG, Kream BE. Regulation of bone formation. N Engl J Med. 1983;309:29–35.

28. National Institutes of Health. Consensus Development Panel. Osteoporosis. JAMA 1984;252:799–802.

29. Wasserman R. Cellular calcium: action of hormones. In: Bronner F, ed. Intracellular calcium regulation. New York: Wiley-Liss, 1990;385–419.

30. Lemann J, Pleuss J, Gray R. Potassium causes calcium retention in healthy adults. J Nutr 1993;123:1623–6.

31. Massey LK. Dietary factors influencing calcium and bone metabolism: Introduction. J Nutr 1993;123:1609–10.

32. Massey L, Whiting S. Caffeine, urinary calcium, calcium metabolism and bone. J Nutr 1993;123:1611–14.

33. Whiting S, Anderson D, Weeks S. Calciuric effects of protein and potassium bicarbonate but not sodium chloride or phosphate can be detected acutely in women and men. Am J Clin Nutr 1997;65:1465–7.

34. Heaney R. Protein intake and the calcium economy. J Am Diet Assoc 1993;93:1259–60.

35. Itoh R, Nishiyama N, Suyama Y. Dietary protein intake and urinary excretion of calcium: a cross-sectional study in a healthy Japanese population. Am J Clin Nutr 1998;67:438–44.

36. Hunt C, Herbel J, Nielsen F. Metabolic responses of postmenopausal women to supplemental dietary boron and aluminum during usual and low magnesium intake: boron, calcium, and magnesium absorption and retention and blood mineral concentrations. Am J Clin Nutr 1997;65:803-13.

37. Nordin B, Need A, Morris H, Horowitz M. The nature and significance of the relationship between urinary sodium and urinary calcium in women. J Nutr 1993;123:1615-22.

38. Devine A, Criddle R, Dick I, Kerr D, Prince R. A longitudinal study of the effect of sodium and calcium intake on regional bone density in postmenopausal women. Am J Clin Nutr 1995;62:740-5.

39. Harward M. Nutritive therapies for osteoporosis: The role of calcium. Med Clin N Am 1993;77:889-98.

40. Hernandez-Avila M, Colditz G, Stampfer M, Rosner B, Speizer F, Willett W. Caffeine, moderate alcohol intake, and risk of fractures of the hip and forearm in middle-aged women. Am J Clin Nutr 1991;54:157-63.

41. Cook J, Dassenko S, Whittaker P. Calcium supplementation: effect on iron absorption. Am J Clin Nutr 1991;53:106-11.

42. Barton J, Conrad M, Harrison L, Nuby S. Effects of calcium on the absorption and retention of lead. J Lab Clin Med 1978;91:366-76.

43. Bogden J, Gertner S, Christakos S. Dietary calcium modifies concentrations of lead and other metals and renal calbindin in rats. Can J Nutr 1992;122:1351-60.

44. Fleischman A, Yacowitz H, Hayton T, Bierenbaum M. Long term studies on the hypolipemic effect of dietary calcium in mature male rats fed cocoa butter. J Nutr 1967;91:151-8.

45. Bell L, Halstenton C, Halstenton C, Macres M, Keane W. Cholesterol lowering effects of calcium carbonate in patients with mild to moderate hypercholesterolemia. Arch Intern Med 1992;152:2441-4.

46. Paydas S, Seyrek N, Sagliker Y. Does oral $CaCO_3$ and calcitriol administration for secondary hyperparathyroidism treatment affect the lipid profile in HD patients. Dialysis and Transplant 1996;25:344-7, 383.

47. Carlson L, Olsson A, Oro L, Rossner S. Effects of oral calcium upon serum cholesterol and triglycerides in patients with hyperlipidemia. Atherosclerosis 1971;14:391-400.

48. Lupton J, Steinbach G, Chang W, O'Brien C, Wiese S, Stoltzfus C, Glober G. Calcium supplementation modifies the relative amounts of bile acid in bile and affects key aspects of human colon physiology. J Nutr 1996;126:1421-8.

49. Charles P, Eriksen EF, Hasling C, Sondergard K, Mosekilde L. Dermal, intestinal, and renal obligatory losses of calcium: relation to skeletal calcium loss. Am J Clin Nutr 1991;54:266S-73S.

50. Andon M, Lloyd T, Matkovic V. Supplementation trials with calcium citrate malate: evidence in favor of increasing the calcium RDA during childhood and adolescence. J Nutr 1994;124:1412S-7S.

51. Yates A, Schlicker S, Suitor C. Dietary reference intakes: the new basis for recommendations for calcium and related nutrients, B vitamins and choline. J Am Diet Assoc 1998;98:699-706.

52. National Institutes of Health. Consensus Development Panel on Optimal Calcium Intake. Optimal calcium intake. JAMA 1994;272:1942-8.

53. Matkovic V, Heaney R. Calcium balance during human growth: evidence for threshold behavior. Am J Clin Nutr 1992;55:992-6.

54. Heaney R. In: Peck WA, ed. Bone and mineral research. New York: Elsevier, 1986;255-301.

55. Sowers M. Epidemiology of calcium and vitamin D in bone loss. J Nutr 1993;123:413-7.

56. Barger-Lux M, Heaney R. The role of calcium intake in preventing bone fragility, hypertension, and certain cancers. J Nutr 1994;124:1406S-11S.

57. Nieves J, Komar L, Cosman F, Lindsay R. Calcium potentiates the effect of estrogen and calcitonin on bone mass: review and analysis. Am J Clin Nutr 1998;67:18-24.

58. Brown W, Wolfson M. Diet as culprit or therapy. Stone disease, chronic renal failure, and nephrotic syndrome. Med Clin N Am 1993;77:783-94.

59. Gibson RS. Principles of Nutritional Assessment. New York: Oxford University Press, 1990:487-510.

60. Heymsfield S, Wang Z. Human body composition. Ann Rev Nutr 1997;17:527-58.

Phosphorus

Among the inorganic elements, phosphorus is second only to calcium in abundance in the human body. Approximately 85% of the body's phosphorus is in the skeleton, 1% is found in the blood and body fluids, and the remaining 14% is associated with soft tissue such as the muscle. In the body, phosphorus typically is found in combination with other elements, either organic or inorganic, such as with oxygen as PO_4^{2-}.

Sources

Phosphorus is widely distributed in foods. The best food sources of phosphorus are listed in Table 11.1 and include meat, poultry, fish, eggs, milk, and milk products. Nuts, legumes, cereals, grains, and chocolate also contain phosphorus; however, animal products are superior sources of available phosphorus compared with cereals and soya-based foods. Coffee, tea, and soft drinks also provide small amounts of phosphorus. Many phosphate containing supplements are available commercially, including K-Phos and Neutra-Phos K, which also provide potassium. For maximum bioavailability, these supplements should not be ingested with zinc, iron, or magnesium.

Dietary phosphorus occurs in both an inorganic form as well as phosphoproteins, phosphorylated sugars, and phospholipids. The relative amounts of inorganic and organic phosphorus vary with the type of diet. Meats contain phosphorus that is largely bound to organic compounds and thus requires hydrolysis for absorption to occur. Over 80% of the phosphorus in grains such as wheat, rice, and corn is found as phytic acid (hexaphosphoinositol)[1]. About 33% of the phosphorus in milk is in the form of inorganic phosphates [1].

Digestion, Absorption, Transport, and Storage

Digestion

Regardless of its dietary form, most phosphorus is absorbed in its inorganic form. Organically bound phosphorus is hydrolyzed enzymatically in the lumen of the small intestine and released as inorganic phosphate (P_i). Phospholipase C, for example, hydrolyzes phosphate from phospholipids. Alkaline phosphatase, a zinc-dependent enzyme whose activity is stimulated by calcitriol, functions at the brush border of the enterocyte to free phosphorus from its bound form. Alkaline phosphatase, however, can free some but not all the bound phosphorus. Phosphorus as part of phytic acid, for example, is not as bioavailable.

Absorption

Phosphorus absorption occurs throughout the small intestine. However, radiophosphorus perfusion studies suggest that phosphorus absorption occurs primarily in the duodenum and jejunum. Between 50% and 70% of phosphorus is absorbed with normal intake, and up to 90% when intake is low [2]. The mechanisms of phosphorus absorption have not been clearly elucidated. Phosphorus absorption appears to occur by either of two processes:

- first, a saturable, carrier-mediated active transport system dependent on sodium or
- a linear, concentration-dependent diffusion process (Fig. 11.7) [1,3].

Factors Influencing Absorption Vitamin D as calcitriol stimulates the absorption of phosphorus, especially in the duodenum and jejunum, primarily when phosphorus intake is low. Parathyroid hormone, however, is not thought to play a direct role [1].

A number of factors negatively impact phosphorus bioavailability. Phytic acid (Fig. 11.8) is the major form of grain cereal phosphate. The poor bioavailability of phosphorus from phytates is due to the absence of

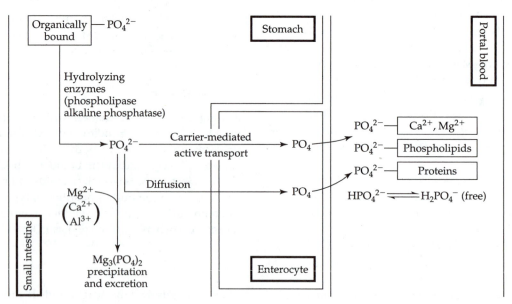

Figure 11.7 Digestion, absorption, and transport of phosphorus.

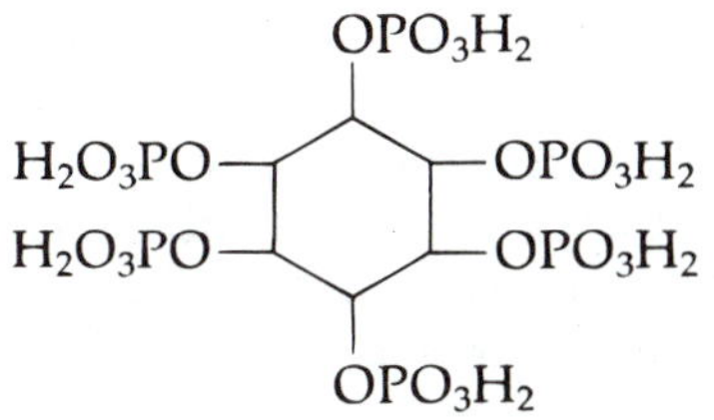

Figure 11.8 Phytic acid.

phytase in mammalian digestion. Phytase is a phosphate esterase that liberates phosphate from phytic acid. Yeasts in breads possess phytase and can hydrolyze some of the phytates to yield some phosphorus available for absorption. About 50% of phosphorus from phytates is thought to be absorbed [2].

Magnesium, aluminum, and calcium also impair phosphorus absorption. Phosphorus absorption may be reduced by dietary magnesium and, conversely, a deficiency of luminal magnesium enhances the absorption of phosphate. The two minerals are thought to form a complex, $Mg_3(PO_4)_2$, within the gastrointestinal tract to render each other unavailable for absorption. Aluminum hydroxide (3 g) given with a meal reduces phosphorus absorption from 70% to 35% [1]. Hypophosphatemia (low blood phosphorus) may result from the prolonged use of nonabsorbable aluminum and magnesium hydroxide gels as found in antacids. In fact, such gels are used as treatment for the hyperphosphatemia associated with chronic kidney failure. Calcium (as calcium acetate or calcium carbonate) also inhibits phosphorus absorption and is used as a phosphate-binding agent in people with hyperphosphatemia secondary to kidney disease.

Transport

Phosphorus is quickly absorbed from the intestine and into the blood. In fact, orally administered radioisotopes of phosphorus appear in the blood within 10 minutes and peak after about an hour. Absorbed phosphate exists in the plasma in two main forms. About 70% of phosphorus in the plasma is found as part of phospholipids [1]. Of the remaining 30% of phosphorus, about 10% is bound to protein, about 5% is bound to calcium or magnesium, and the remaining 85% is found as inorganic phosphates, primarily HPO_4^{2-} and $H_2PO_4^-$ [1]. Inorganic phosphorus is sometimes referred to as *ultrafilterable phosphate*. In adults, plasma inorganic phosphate ranges between 2.5 and 4.4 mg/dL. Dietary phosphate, age and stage of growth, time of day, hormonal effects, and renal function all contribute to the variability of the serum phosphate

concentration. Circulating phosphate is in equilibrium with skeletal and cellular inorganic phosphate and with that of organic phosphates formed in intermediary metabolism. Phosphorus is found in all cells of the body, with bone and muscle containing the majority.

Functions and Mechanisms of Actions

Phosphate is of prime importance in the development of *skeletal tissue,* which in itself accounts for 85% of the total phosphate stores. In bone, phosphorus is part of calcium phosphate, including, for example, $Ca_3(PO_4)_2$, $CaHPO_4 \cdot 2H_2O$, $Ca_3(PO_4)_2 \cdot 3H_2O$, and hydroxyapatite ($Ca_{10} [PO_4]_6 [OH]_2$), which is laid down on collagen in the ossification process of bone formation.

Parathyroid hormone (PTH) and calcitriol influence phosphorus metabolism similar to calcium with a few exceptions. PTH stimulates resorption of phosphate from bone as it does calcium. However, in contrast to calcium, PTH stimulates excretion of phosphorus in urine. The PTH-induced urinary excretion of phosphorus is typically sufficient to override bone resorption of phosphorus so as to effect a net decrease in plasma phosphate.

Calcitriol, in conjunction with PTH, also enhances phosphate resorption from bone. Calcitriol also affects the intestine. Calcitriol stimulates phosphate absorption in the intestine, possibly through enhanced alkaline phosphatase activity. See the calcium section of this chapter for a further discussion of bone mineralization and the roles of calcium and phosphorus (p. 377).

Phosphorus that is not part of bone is found in either extracellular fluids such as blood or in soft tissues. Within cells, phosphorus is the major anion and is involved in a host of processes. Phosphorus is of vital importance in intermediary metabolism of the energy nutrients, contributing to the metabolic potential in the form of *high-energy phosphate bonds,* such as ATP, and through the *phosphorylation* of substrates. Many *enzymatic activities* are controlled by alternating phosphorylation or dephosphorylation. For example, see the discussion of glycogen degradation (p. 83).

Phosphate is also an important component of the *nucleic acids* DNA and RNA. Phosphorus alternates with pentose sugars to form the linear backbone of these macromolecules.

Cell membranes are made up in part from lipids, including *phospholipids,* which as their name implies contain phosphorus. Phospholipids, with its polar and nonpolar regions, are important for the bilayer structure of cell membranes.

Phosphate also functions in *acid-base balance.* Within cells, phosphate is the main intracellular buffer. Within the kidney, filtered phosphate reacts with secreted hydrogen ions, releasing sodium ions in the process. This action increases pH. The released sodium ion may be reabsorbed through the kidney tubule under the influence of aldosterone.

$$Na_2HPO_4 + H^+ \rightarrow NaH_2PO_4 + Na^+$$

Phosphate is also involved in *oxygen delivery.* In red blood cells, synthesis of 2,3 diphosphoglycerate requires phosphorus. Decreased 2,3 diphosphoglycerate diminishes release of oxygen to tissues.

Interactions with Other Nutrients

The interactions between phosphorus and *magnesium* and *phytate* have been addressed under the section on phosphorus absorption. In addition, the optimal ratio between *calcium* and phosphorus in the diet has been questioned, because animal studies suggest that diets low in calcium:phosphorus ratios lead to progressive bone loss due to phosphorus-induced stimulation of parathyroid hormone (PTH) release. Although such an effect may occur in humans on high-phosphate diets for prolonged periods, studies demonstrating such effects are not available [2,4]. Women ingesting a diet with a calcium:phosphorus ratio of 1:4 for a 4-week period have been reported to exhibit elevated PTH. And persistent elevations of PTH could lead to loss of minerals from bone [5]; however, studies monitoring bone loss are needed.

Excretion

About 67% to 90% of phosphorus is excreted in inorganic form in the urine. The remaining 10% to 33% of phosphorus is excreted in the feces. Unlike calcium, high dietary phosphorus leads to high serum phosphorus, which leads to increased urinary phosphorus excretion. In other words, maintenance of the phosphate balance is achieved largely through renal excretion. The amount of dietary phosphorus and absorbed phosphorus is approximately a linear relationship with urinary phosphorus. This relationship is maintained even if intake is increased severalfold. If a person is in zero balance, then urinary excretion of phosphorus is the same as the net absorbed phosphorus, which is defined as the difference between the total amount absorbed in the intestine and that which is secreted in the digestive juice.

A well-defined diurnal variation in urinary phosphate exists that is not related to feeding patterns but rather to changes in tubular reabsorption. The pattern relates to physical activity, with the nadir of urinary excretion appearing a few hours after the end of sleep. Because this is inversely related to the diurnal fluctua-

tion of the release of adrenal corticotropic hormone and cortisol, both of which peak after a period of sleep, it appears that phosphate excretion may be under the influence of the pituitary and adrenal glands. In fact, tubular reabsorption of phosphate is increased by short-term cortisol therapy. Other hormones that inhibit the tubular reabsorption of phosphorus include parathyroid hormone, estrogen, and thyroid hormones.

Recommended Dietary Allowances and Dietary Reference Intakes

Relatively little work has been done to determine phosphorus requirements. Because the body adapts to changes in dietary phosphorus intake by altering phosphorus excretion and because of the widespread availability of phosphorus in foods, deficiency is rare. It is generally held, probably correctly, that phosphorus intake must be adequate if the intake of other nutrients is adequate.

Dietary Reference Intakes published in 1997 suggest 700 mg phosphorus for males and females (including those who are pregnant or lactating) aged 19 years and older [6]. These values are slightly lower than the 1989 RDA for phosphorus.

Deficiency

Phosphorus deficiency is rare, because almost all foods contain phosphorus and the body adapts to dietary fluctuations by changes in excretion [2]. Premature infants and people who are receiving large amounts of calcium acetate, calcium carbonate, or aluminum hydroxide antacids, which bind phosphorus in the gastrointestinal tract, have exhibited signs of deficiency. People with malabsorptive disorders such as Crohn's disease and those with alcoholism, uncontrolled diabetes mellitus, or burns may exhibit imbalances in phosphorus. In addition, individuals with malnutrition may exhibit phosphate deficiency syndrome upon refeeding. Deficiency of phosphorus, usually manifested as low serum phosphorus concentrations (<0.7 mmol/L), is associated with hypophosphatemic rickets and bone loss, skeletal muscle and cardiac myopathy, weakness, neurological problems, and disturbances in oxygen dissociation from hemoglobin due to a decrease in the formation of 2,3-diphosphoglycerate, which regulates the release of oxygen from hemoglobin [1].

Toxicity

Toxicity from phosphorus is rare, and problems appear to occur only when calcium:phosphorus ratios are altered significantly in infants. High-phosphorus human milk substitutes, when given to infants, result in hypocalcemia and tetany [1]. A tolerable upper intake level of 4,000 mg phosphorus has been recommended for those aged 9 to 70 years, whereas after age 70 years, the tolerable level drops to 3,000 mg phosphorus daily [6]. For pregnant and lactating women, the tolerable upper level is 3.5 g and 4 g, respectively [6].

Assessment of Nutriture

The assessment of phosphorus nutriture is not a major consideration because deficiency is so rare. Serum phosphorus concentrations may be assessed; however, their specificity and sensitivity are low. Concentrations are affected by several confounding factors that are unrelated to phosphorus status [7].

References Cited for Phosphorus

1. Allen L, Wood R. Calcium and phosphorus. In: Shils ME, Olson JA, Shike M. Modern Nutrition in Health and Disease, 8th ed. Philadelphia: Lea and Febiger, 1994: 144–63.

2. National Research Council. Recommended dietary allowances, 10th ed. Washington, D.C.: National Academy Press, 1989:184–7.

3. Anderson J. Nutritional biochemistry of calcium and phosphorus. J Nutr Biochem 1991;2:300–7.

4. Calvo M. Dietary phosphorus, calcium metabolism and bone. J Nutr 1993;123:1627–33.

5. Anderson J, Barrett C. Dietary phosphorus: the benefits and the problems. Nutr Today 1994;29:29–34.

6. Yates A, Schlicker S, Suitor C. Dietary reference intakes: the new basis for recommendations for calcium and related nutrients, B vitamins and choline. J Am Diet Assoc 1998;98:699–706.

7. Gibson RS. Principles of Nutritional Assessment. New York: Oxford University Press, 1990:487–510.

Magnesium

Magnesium as a cation in the human body ranks fourth in overall abundance, but intracellularly it is second only to potassium. The normal human body contains about 20 to 28 g of magnesium, approximately 55% to 60% of which is located in bone, another 20% to 25% in muscles with the remaining found in other soft tissues and in extracellular fluids.

Sources

Magnesium is found in a wide variety of foods and beverages. Beverages rich in magnesium are coffee, tea, and cocoa. On the basis of weight, foods and food components particularly high in magnesium are nuts, legumes, whole-grain cereals, spices, and seafoods. On the basis of calories, most fruits and green, leafy vegetables are excellent sources of magnesium. Chlorophyll found in the green, leafy vegetables contains the magnesium. Other particularly good food sources of magnesium are chocolate, blackstrap molasses, corn, peas, carrots, brown rice, and parsley. Magnesium salts, such as magnesium sulfate ($MgSO_4$), magnesium oxide (MgO), magnesium chloride ($MgCl_2$), magnesium lactate, magnesium acetate, magnesium gluconate, and magnesium citrate, are commonly available forms of the mineral. Slow Mag® (magnesium chloride) and Mag-Tab SR® (magnesium lactate) tablets provide about 60 to 84 mg magnesium per tablet. Supplements should not be taken at the same time as other mineral supplements, such as iron.

Food processing and preparation may substantially reduce the magnesium content of foods. For example, the refining of whole wheat with removal of the germ and outer layers can reduce the magnesium content by 80% [1].

Absorption and Transport

Absorption

Magnesium absorption occurs throughout the small intestine, mainly the jejunum and ileum. Two transport systems are thought to be responsible for magnesium absorption in the intestinal tract.

- The first is carrier mediated and saturable at low magnesium intakes.

- The second mode of absorption is via simple diffusion and is thought to occur with higher intakes.

The colon may also play a role in the absorption of magnesium [2], especially if disease has interfered with magnesium absorption in the small intestine [3].

About 30% to 65% of magnesium is thought to be absorbed in adults with usual intakes [1,4,5]. Magnesium absorption is more efficient when magnesium status is poor or marginal and/or when magnesium intake is low [3,5]. For example, 65% of magnesium is absorbed with an intake of 36 mg versus only 11% absorption with an intake of 973 mg magnesium [5].

Factors Influencing Absorption Magnesium absorption may be influenced by a variety of other factors. For example, dietary phytate and fiber impair magnesium absorption to a small extent [1]. Unabsorbed fatty acids present in high quantities, as occur with steatorrhea, may bind to magnesium, as to calcium, to form soaps.

These magnesium–fatty acid soaps are excreted in the feces [4]. Vitamin D increases magnesium absorption by active transport, but less than it does calcium absorption [4]. Lactose also appears to increase magnesium absorption. Coatings used on supplements also affect magnesium absorption. Magnesium absorption from enteric coated (cellulose acetate phthalate) magnesium supplements such as Slow-Mag® containing magnesium chloride was substantially less (67%) than that from magnesium chloride encapsulated in gelatin [5].

Transport

In the plasma, most magnesium (55%) is found free, about 32% is bound to protein, and 13% is complexed with citrate, phosphate, or other ions. Concentrations of magnesium in the plasma are maintained between 1.3 and 2.1 mEq/L; however, the homeostatic mechanism of control is unclear. Maintenance of these constant values appears to depend on gastrointestinal absorption, renal excretion, and transmembranous cation flux rather than hormonal regulation [1].

Functions and Mechanisms of Actions

About 60% of magnesium in the body is found associated with *bone.* Bone magnesium is divided between that found associated with phosphorus and calcium as part of the crystal lattice (~70%) and that found on the surface (~30%). Bone surface magnesium is thought to represent a magnesium pool and to reflect changes in serum levels. In contrast, the magnesium in the crystal lattice is probably deposited at the time of bone formation [2]. Magnesium may be present in bone as $Mg(OH)_2$ or $Mg_3(PO_4)_2$, for example.

Magnesium, which does not function as part of the bone, is found in extracellular fluids (1%) and in soft tissues, primarily muscle [1,4]. Within cells, magnesium is bound to *phospholipids* as part of cell membranes (plasma, endoplasmic reticulum, and mitrochondria), where it may help in membrane stabilization. Magnesium is also associated with *nucleic acids,* protein (*enzymes*), and as a complex with *ATP* to stabilize the structure. Magnesium, with an approximate intracellular concentration of 8 to 10 mmol/L, is important for over 300 different enzyme reactions either as a structural cofactor or an allosteric activator of enzyme activity [2–4]. In ATP, magnesium binds to phosphate groups, thereby forming a complex that assists in the transfer of ATP phosphate. Figure 11.9 depicts magnesium as a ligand for the phosphate groups of ATP. Protein kinases transfer the γ-phosphate of magnesium ATP to a substrate [2].

Listed here are some of the many fundamental roles of magnesium in the body [2,6]:

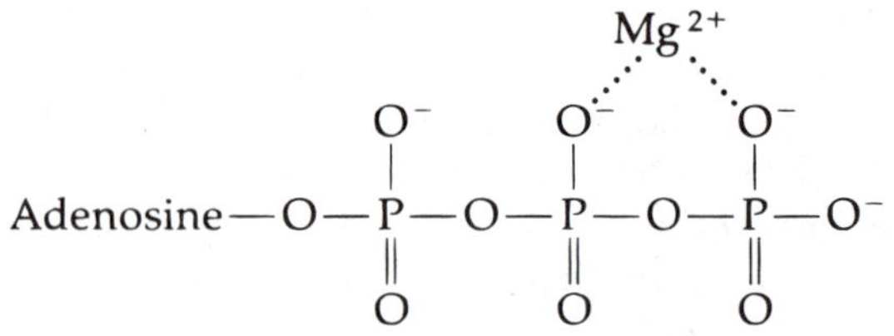

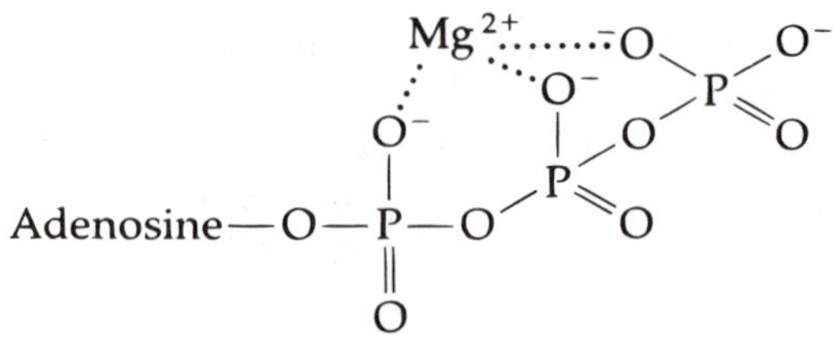

Figure 11.9 Modes by which Mg^{2+} provides stability to ATP.

- Glycolysis: hexokinase and phosphofructokinase (p. 85)
- Krebs cycle: oxidative decarboxylation in Krebs cycle
- Hexose monophosphate shunt: transketolase reaction (pp. 88 and 268)
- Creatine phosphate formation: creatine kinase
- β-oxidation: initiation by thiokinase (acyl CoA synthetase)
- Activities of alkaline phosphatase and pyrophosphatase
- Nucleic acid synthesis
- DNA synthesis and degradation, as well as the physical integrity of the DNA helix
- DNA and RNA transcription
- Amino acid activation
- Protein synthesis (for example, with ribosomal aggregation and binding messenger RNA to 70S ribosome subunits)
- Cardiac and smooth muscle contractability (both direct action as well as influence on calcium ion transport and use)
- Vascular reactivity and coagulation (possible role)
- Cyclic adenosine monophosphate (cAMP) formation from adenylate cyclase.

Because of its function in formation of cAMP, magnesium is also involved in mediating the effects of numerous hormones

Interactions with Other Nutrients

Magnesium has interrelationships with a number of other nutrients. The first that will be discussed is that with *calcium*. Magnesium is needed for PTH secretion, which is important in calcium homeostasis. High magnesium concentrations in turn appear to inhibit PTH release, similar to calcium. Moreover, magnesium is needed for PTH effects on the bone, kidney, and gastrointestinal tract. The hydroxylation of *vitamin D* in the liver requires magnesium, and high magnesium levels, like calcium, inhibit PTH secretion [3,6]. Calcium and magnesium use overlapping transport systems in the kidney and thus compete in part with each other for reabsorption. Magnesium may mimic calcium by binding to calcium-binding sites and eliciting the appropriate physiological response [3,6–8]. Magnesium may also cause an alteration in calcium distribution by changing the flux of calcium across the cell membrane or by displacing calcium on its intracellular binding sites. Magnesium may further inhibit release of calcium from the sarcoplasmic reticulum in response to increased influx from extracellular sites and may activate the Ca^{2+}-ATPase pump to decrease intracellular Ca^{2+} concentrations [8]. The ratio of calcium to magnesium has been shown to affect muscle contraction. Magnesium may compete with calcium for nonspecific binding sites on troponin C and myosin [8]. Additional effects of magnesium are seen in the smooth muscles [6,8]. For example, calcium binding initiates acetylcholine release and smooth muscle contraction; magnesium bound to the calcium sites prevents calcium binding and inhibits contraction [7]. The magnesium-calcium relationship has implications in people with respiratory disease, because increased intracellular calcium promotes bronchial smooth muscle contraction [7]. Magnesium may also influence the process of blood coagulation. In blood coagulation, calcium and magnesium are antagonistic, with calcium promoting the processes and magnesium inhibiting them [9].

Magnesium inhibits *phosphorus* absorption. Phosphorus absorption decreases as magnesium intake [5]. The two minerals are thought to precipitate as $Mg_3(PO_4)_2$. Magnesium actetate (600 mg) is thought to be able to reduce phosphorus absorption from about 77% to 34% [5].

A close interrelationship also exists between magnesium and *potassium*. Magnesium influences the balance between extracellular and intracellular potassium, but its mechanism of action is unclear [3]. One theory is that because magnesium is necessary for the function of Na^+/K^+-ATPase, a deficiency of magnesium would lead to impaired pumping of sodium out of the cell and the movement of potassium into the cell [3]. When magnesium and potassium deficiencies are coexistent, as may occur with some diuretic drug therapies, magnesium infusions, but not potassium infusions, can normalize muscle potassium [3]. Lastly, dietary *protein* intake impacts magnesium retention. Increasing dietary protein to a marginally adequate level in subjects previously ingesting low-magnesium and very low-protein diets improved magnesium retention. When, however, protein intake was further increased, magnesium retention was decreased [3].

Excretion

Absorbed magnesium not retained by the body is lost primarily via the kidneys. Of the serum magnesium, about 70% to 80% is filtered by the kidney; however, 95% to 97% is reabsorbed [2]. Thus, only about 3% to 5% of the filtered magnesium is excreted in the urine [3]. With a low dietary intake of magnesium, the kidneys are able to reabsorb and conserve magnesium very effectively. Rises in serum magnesium lead to increased filtration and excretion (~40%–80% of filtered load excreted), and plasma magnesium concentrations serve to regulate, at least in part, renal magnesium reabsorption [6,10].

Diuretic medications, thyroid and aldosterone hormones, as well as caffeine consumption stimulate magnesium excretion. In contrast, PTH inhibits magnesium excretion.

Fecal magnesium concentrations represent unabsorbed magnesium and a small amount of endogenous magnesium that escaped reabsorption. About 25 to 50 mg of endogenous magnesium may be excreted daily in the feces [1]. Magnesium may also be lost in the sweat, in amounts estimated at approximately 15 mg/day [3].

Recommended Dietary Allowances and Dietary Reference Intakes

The 1989 RDA for magnesium (350 and 280 mg/day for the adult 76-kg male and 62-kg female, respectively) represents a compromise between needs estimated by balance studies and the usual dietary magnesium intake by a population in which magnesium deficiency rarely appears except in pathological conditions [1]. The 1989 RDA for adults of both sexes is 4.5 mg/kg/day [1]. The amount of magnesium furnished by the average U.S. diet is estimated at approximately 120 mg/1,000 kcal [11]. Balance studies indicate that adult needs for magnesium may be as low as 3 mg/kg/day or as high as 4.5 mg/kg/day [1].

Dietary Reference Intakes RDA published in 1998 suggest 400 mg magnesium for males and 310 mg magnesium for females aged 19 to 30 years, and 420 mg magnesium for males and 320 mg magnesium for females aged 31 years and older [12]. Pregnant women aged 19 to 30 years should ingest 350 mg, and those aged 31 to 50 years should consume 360 mg magnesium. During lactation, women aged 19 to 30 years should ingest 310 mg and those aged 31 to 50 years, 320 mg [12].

Deficiency

Deficiency of magnesium or disturbances in magnesium homeostasis in humans is usually associated with the presence of other illnesses such as alcoholism or renal disease. Pure deficiency of magnesium due to inadequate dietary intake has not been reported but has been induced in humans under research protocols. Symptoms associated with deficiency or disturbances in balance include nausea, vomiting, anorexia, muscle weakness, spasms, and tremors, personality changes, and hallucinations, among others. Changes in cardiovascular and neuromuscular function may lead to cardiac arrhythmia and death. Hypomagnesemia, associated with deficiency, represents a plasma magnesium concentration of less than about 1.5 mg/dL and develops within a relatively short time following a magnesium deficit [10]. Other biochemical changes include low blood concentrations of not only magnesium but also potassium and calcium [1]. The latter results from diminished parathyroid hormone secretion. The hypokalemia (low blood potassium) results from altered cellular transport systems that maintain the potassium gradient [13]. Poor magnesium status also may be related to cardiovascular disease, myocardial infarction (heart attack), toxemia of pregnancy, hypertension, or postsurgical complications [2–4,8,14]. Other conditions and populations likely to develop deficiency include individuals with excessive vomiting and or diarrhea, alcoholism, protein malnutrition, diuretic use, malabsorption, diabetes mellitus, parathyroid disease, and burns.

Toxicity

An excessive intake of magnesium is not likely to cause toxicity except in the case of people with impaired renal function. Normal kidneys are able to remove magnesium so rapidly that significant increases in serum levels do not occur [3]. Excessive intakes of magnesium salts (3–5 g) such as from $MgSO_4$, may, however, have a cathartic effect leading to diarrhea and possible dehydration [11]. Other signs include nausea, flushing, double vision, slurred speech, and weakness, which usually appear at plasma magnesium concentrations of about 9 to 12 mg/dL. Acute magnesium toxicity from excessive intravenous administration of magnesium has resulted in nausea, depression, and paralysis [12]. Muscular paralysis and cardiac and or respiratory failure are associated with plasma magnesium concentrations over ~15 mg/dL. A tolerable upper intake level of 350 mg magnesium has been recommended for individuals (including during pregnancy and lactation) aged 9 years and older [12].

Assessment of Nutriture

Assessment of magnesium status is difficult, because extracellular magnesium represents only about 1% of the total body magnesium and appears to be homeostatically regulated. Despite low sensitivity and specificity, serum magnesium concentrations are routinely measured to assess magnesium status [4]. Normal serum levels may occur despite severe intracellular deficit [3]. However, when serum magnesium is below normal, an inadequate amount of intracellular magnesium is a certainty. Erythrocyte magnesium concentrations decrease more slowly with magnesium deficiency than plasma and may reflect longer-term magnesium status because of the life span of the red blood cell [11]. Peripheral lymphocyte magnesium concentrations correlate with skeletal and cardiac muscle magnesium content and thus represent a possible indicator of magnesium status [4].

More definitive determination of magnesium status may involve measurement of renal magnesium excretion, which decreases with magnesium deficiency. Renal magnesium excretion should be measured before and after the administration of an intravenous magnesium load. Decreased excretion determined over two 24-hour periods following administration of the magnesium load indicates deficiency [2]. Alternately, an oral magnesium load test may be used [15]. Normal serum and urinary magnesium concentrations are about 1.6 or 1.8 to about 2.6 or 3.0 mg/dL, and 36 to 207 mg/24 hours, respectively.

References Cited for Magnesium

1. National Research Council. Recommended Dietary Allowances, 10th ed. Washington, D.C.: National Academy Press, 1989:187–94.

2. Shils M. Magnesium. In: Shils ME, Olson JA, Shike M. Modern Nutrition in Health and Disease, 8th ed. Philadelphia: Lea and Febiger, 1994:164–84.

3. Wester PO. Magnesium. Am J Clin Nutr 1987; 45(suppl):1305–12.

4. Rude R. Magnesium metabolism and deficiency. Endocrin Metab Clin N Am 1993;22:377–95.

5. Fine K, Santa Ana C, Porter J, Fordtran J. Intestinal absorption of magnesium from food and supplements. J Clin Invest 1991;88:396–402.

6. Levine B, Coburn J. Magnesium, the mimic/antagonist of calcium. N Engl J Med 1984;310:1253–5.

7. Landon R, Yound E. Role of magnesium in regulation of lung function. J Am Diet Assoc 1993;93:674–7.

8. Iseri L, French J. Magnesium: nature's physiologic calcium blocker. Am Heart J 1984;108:188–93.

9. Weaver K. Magnesium and its role in vascular reactivity and coagulation. Contemp Nutr 1987;12(3).

10. Kelepouris E, Agus Z. Hypomagnesemia: renal magnesium handling. Sem Nephrol 1998;18:58–73.

11. National Research Council. Recommended Dietary Allowances, 9th ed. Washington, D.C.: National Academy of Sciences, 1980:134–6.

12. Yates A, Schlicker S, Suitor C. Dietary reference intakes: the new basis for recommendations for calcium and related nutrients, B vitamins and choline. J Am Diet Assoc 1998;98:699–706.

13. Hamill-Ruth R, McGory R. Magnesium repletion and its effects on potassium homeostasis in critically ill adults: results of a double-blind randomized, controlled trial. Crit Care Med 1996;24:38–45.

14. Frakes M, Richardson L. Magnesium sulfate therapy in certain emergency conditions. Am J Emerg Med 1997;15:182–7.

15. Durlach J, Bac P, Durlach V, Guiet-Bara A. Neurotic, neuromuscular and autonomic nervous form of magnesium imbalance. Magnesium Res 1997;10:169–95.

Sodium

Approximately 30% of the 120 mg of sodium in the body is located on the surface of bone crystals. From that site, it can be released into the bloodstream should hyponatremia (low serum sodium) develop. The remainder of the body's sodium is in the extracellular fluid, primarily plasma, and in nerve and muscle tissue. Sodium constitutes about 93% of the cations in the body, making it by far the most abundant member of this family.

Absorption and Transport

Approximately 95% of ingested sodium is absorbed, with the remaining 5% excreted in the feces. Sodium absorbed in excess of the amount needed is excreted by the kidneys. There are three basic pathways for absorption of sodium across the intestinal mucosa. One of these pathways (the Na^+/glucose cotransport system) functions broadly throughout the small intestine. Another pathway (an electroneutral Na^+ and Cl^- cotransport system) is active in both the small intestine and the proximal portion of the colon, and the third pathway (an electrogenic sodium absorption mechanism) occurs principally in the colon.

The Na^+/glucose cotransport system involves a carrier on the apical membrane of the small intestinal epithelium. Na^+ and glucose bind to the carrier, which shuttles them from the outer surface to the inner surface of the membrane. There, both are released before the carrier returns to the outer surface. Absorbed Na^+ is then pumped out across the basolateral membrane by the Na^+/K^+-ATPase pump (p. 16), while the glucose diffuses across the membrane by a facilitated transport pathway. The Na^+ gradient created by the Na^+/K^+-ATPase pump provides the needed energy to maintain the absorptive direction of the ion. Cotransport of Na^+ by this mechanism can also occur with solutes other than glucose, including amino acids [1], di- and tripeptides, and many B vitamins [2].

The electroneutral Na^+ and Cl^- cotransport mechanism has been proposed because of the observation that a significant portion of sodium uptake requires the presence of chloride, and vice versa [3]. Precisely how this system functions has not yet been established. However, it is believed that the cotransport is composed of Na^+/H^+ exchange working in concert with a Cl^-/HCO_3^- mechanism [4]. The mechanism allows the entrance of both Na^+ and Cl^- into the cell, in which they are exchanged for H^+ and HCO_3^-. Protons and HCO_3^- are produced within the cell by the action of carbonic anhydrase on CO_2. Absorbed Na^+ is pumped across the basolateral membrane by the Na^+/K^+-ATPase pump, followed by Cl^-, which crosses by diffusion.

The colonic mechanism is called an *electrogenic sodium absorption mechanism* because the absorbed sodium ion is the only ion moving transcellularly, allowing its transport to be monitored by electrical equipment. It enters the luminal membrane of the colonic mucosal cell through Na^+-conducting pathways called *Na^+ channels*, diffusing inwardly by the downhill concentration gradient of the ion. The absorbed sodium is accompanied by water and anions, resulting in net water and electrolyte movement from the luminal side to the bloodstream side of the colon epithelium. It is pumped out across the basolateral membrane on the bloodstream side of the cell by the Na^+/K^+-ATPase pump.

All three of these mechanisms are depicted schematically in Figure 11.10. It is important to recognize that the common driving force for sodium absorption in all the processes is the inwardly directed gradient maintained by the basolateral Na^+ pump.

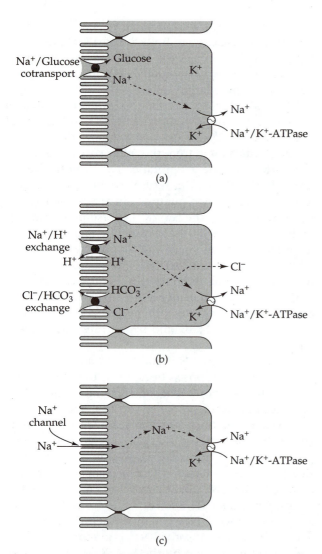

(a)

(b)

(c)

Figure 11.10 Absorption mechanisms for sodium in the intestine. (a) Na$^+$/glucose cotransport. A carrier on the luminal membrane of the mucosal cell cotransports sodium together with a solute such as glucose into the cell. Once in the cell, Na$^+$ is pumped across the basolateral membrane by Na$^+$/K$^+$-ATPase, while glucose exits through the membrane by facilitated diffusion. (b) Electroneutral Na$^+$ and Cl$^-$ absorption. The cotransport of Na$^+$/H$^+$ exchange and a Cl$^-$/HCO$_3^-$ exchange. Sodium is then pumped basolaterally, with Cl$^-$ diffusing passively. (c) Electrogenic Na$^+$ absorption. Sodium enters the luminal membrane via a Na$^+$ channel.

Interactions with Other Nutrients

The well-documented interaction of sodium with *calcium* is worthy of discussion. It has long been recognized (prior to 1940) that dietary sodium intake affects urinary calcium excretion. More currently, research has focused on the association of high dietary sodium intake with increased calcium requirements, increased bone resorption, and possibly osteoporosis. Studies have shown, however, that despite accompanying calci-

uria, oral sodium loading may not affect calcium balance. This is because the sodium challenge also causes an increase in serum PTH levels and cAMP excretion, along with a significant decrease in fecal calcium excretion. Such calcium-elevating effects may act as compensatory mechanisms to offset the urinary calcium losses. The sodium-calcium interaction and its possible association with osteoporosis have been reviewed [5]. For additional information on this topic, see references 24 and 32 in the "Calcium" section of this chapter.

Regulation and Excretion

As mentioned previously, nearly all (95%) of ingested sodium is absorbed. Therefore, much larger amounts are absorbed than are required by the body, the excess being excreted primarily by the kidneys. Sodium losses also take place through the skin via sweating. Under conditions of moderate temperature and level of exercise, sodium losses are small. However, because the sodium content of sweat is about 50 mEq/L, it can be reasoned that conditions of high temperature and/or sustained vigorous exercise can account for significant losses. Renal excretion and retention of sodium are under the control of aldosterone, which promotes the retention (reabsorption) of sodium and the excretion of potassium. The hormone is released from the adrenal cortex in response to low sodium or, more important, high potassium concentrations. The renal regulation of sodium, as well as potassium and chloride, is presented in greater detail in Chapter 14.

Recommended Dietary Allowances and Assessment of Nutriture

The major source of sodium in the diet is added salt in the form of sodium chloride. Sodium comprises 39% by weight of sodium chloride. Because salt is so extensively used in food processing and manufacturing, it is estimated that processed foods account for nearly 75% of total sodium consumed. Canned meats and soups, condiments, pickled foods, and traditional snacks are particularly high in added salt. Naturally occurring sources of sodium such as milk, meats, eggs, and most vegetables furnish only about 10% of consumed sodium. Salt added during cooking and at the table provide roughly 15% of total sodium, and water supplies <10%.

However unpalatable in a diet, an intake of only 115 mg/day of sodium is probably sufficient to replace obligatory losses and provide for growth. To compensate for wide variations in physical activity and climatic exposure, which account for sodium losses through sweating, the National Research Council recommends an intake of 500 mg/day [6]. Because of the high salt content of a typical diet, however, sodium intake is generally much greater. Depending on the method of assessment, estimates of ingested sodium range from 1,800 to 5,000 mg/day.

Dietary deficiencies of sodium do not normally occur, because of the abundance of the mineral across a broad spectrum of foods. Serum concentrations of sodium are normally regulated within the range of 135 to 145 mEq/L. It is measured routinely in clinical laboratories to determine electrolyte balance (see Chapter 14) and to identify possible renal disease, which may disturb normal sodium metabolism. Sodium ion in the serum and other biological fluids is usually quantified by the technique of ion-selective electrode potentiometry. The method measures Na^+ in the same manner as a pH meter measures protons.

Potassium

In contrast to sodium, 98% of the body's potassium is intracellular. The approximate 270 mg of potassium is located inside the cells, making it the major intracellular fluid cation. Potassium influences the contractility of smooth, skeletal, and cardiac muscle and profoundly affects the excitability of nerve tissue. It is also important in maintaining electrolyte and pH balance.

Absorption and Transport

The mechanisms by which potassium is absorbed from the gastrointestinal tract are not as clearly understood as those of sodium absorption. Only recently, relative to sodium absorption elucidation, have actual intestinal transport mechanisms for potassium been recognized and investigated. Over 90% of ingested potassium is absorbed, and while the sites along the gastrointestinal tract at which this takes place have not been precisely identified, both the small intestine and the colon appear to play a role in this function [7,8,9].

It is believed that K^+ may be absorbed through the apical membrane of the colonic mucosal cell by a K^+/H^+-ATPase pump. This would exchange intracellular H^+ for luminal K^+, the mechanism by which H^+ along with Cl^- (secreted via Cl^- channels), is secreted into the stomach as HCl. Alternatively, K^+ may enter the cell via apical membrane channels that serve as secretory pathways as well. The K^+ accumulated in the cell then diffuses across the basolateral membrane via the K^+ channel.

Interactions with Other Nutrients

Like sodium, potassium has an effect on the urinary excretion of *calcium*. However, its effect is opposite to that of sodium in that it decreases calcium excretion while sodium increases it. A current practice is to replace some of the NaCl in the diet with KCl to reduce the amount of NaCl consumed. The finding that potassium is not as calciuric as sodium, and in fact actually reduces the excretory rate of calcium, further supports the practice [10]. For additional information on this subject, see references 21, 24, and 30 in the "Calcium" section of this chapter.

Regulation and Excretion

Potassium is absorbed from the intestine at an efficiency of >90%. Only small amounts are excreted in the feces. As with sodium, potassium balance is achieved largely through the kidneys, with aldosterone being the major regulatory hormone. Aldosterone acts reciprocally on sodium and potassium. Although it stimulates the reabsorption of sodium in the kidney tubules, it accelerates the excretion of potassium. Renal control of potassium is discussed further in Chapter 14.

Hyperkalemia (abnormally high serum concentration of potassium) is toxic, resulting in severe cardiac arrhythmias and even cardiac arrest. It is nearly impossible to produce hyperkalemia by dietary means in an individual with normal circulation and renal function. This is because of potassium's delicate control within a narrow concentration range. Similarly, hypokalemia (abnormally low serum potassium) does not occur by dietary deficiency, because of the abundance of potassium in common foods. Hypokalemia, associated with muscular weakness, nervous irritability, and mental disorientation, can result from profound alimentary fluid loss such as that which occurs in severe vomiting and diarrhea.

Recommended Dietary Allowances and Assessment of Nutriture

Potassium is widespread in the diet and is especially abundant in unprocessed foods, fruits, many vegetables, and fresh meats. Also, many salt substitutes contain potassium in place of sodium. The National Research Council recommends a dietary intake of 2,000 mg/day [6] but states further that 3,500 mg/day may be more desirable in view of the increased intake of fruits and vegetables recommended for prevention and control of hypertension.

The normal serum concentration of potassium, as K^+, is 3.6 to 5.0 mEq/L. It is commonly assayed clinically to identify renal disease and monitor electrolyte balance. Because of its toxicity, reported high values are brought to the attention of the attending physician immediately so that appropriate corrective measures may be taken. Potassium, like sodium, is determined in the serum primarily by ion-selective electrode potentiometry.

Chloride

Chloride is the most abundant anion in the extracellular fluid. Approximately 88% of chloride is found in extracellular fluid, and just 12% is intracellular. Its negative charge neutralizes the positive charge of sodium ions with which it is usually associated. In this respect, it is of great importance in the maintenance of electrolyte balance.

Chloride has important functions in addition to its role as a major electrolyte. It is required for the formation of gastric hydrochloric acid, secreted along with protons from the parietal cells of the stomach. Also, it acts as the exchange anion in the red blood cell for HCO_3^-. Sometimes referred to as the *chloride shift,* this latter process requires a protein transporter that moves Cl^- and HCO_3^- in opposite directions across the cell membrane. The purpose is to allow the conveying of tissue-derived CO_2 back to the lungs in the form of plasma HCO_3^-. Waste CO_2 from respiring tissues enters the red blood cell, where it is converted to HCO_3^- by carbonic anhydrase. The transporter protein (chloride-bicarbonate exchanger) then transports the HCO_3^- out of the cell into the plasma while it simultaneously transports plasma Cl^- into the cell. In the absence of chloride, bicarbonate transport ceases.

Absorption, Transport, and Secretion

Chloride is almost completely absorbed in the small intestine. Its absorption closely follows that of sodium in the establishment and maintenance of electrical neutrality. The absorptive mechanisms, however, are generally different. For example, in the Na^+-glucose cotransport system (described in the "Sodium" section), chloride follows the actively absorbed Na^+ *passively* through a so-called paracellular, or tight junction, pathway. The absorbed Na^+ creates an electrical gradient that provides the energy for the accompanying, inward diffusion of Cl^-. The electroneutral Na^+ and Cl^- cotransport absorption system also contributes to the movement of chloride into the mucosal cells, although the relative contribution of this system to total chloride absorption is not well established. Sodium absorbed by the electrogenic Na^+ absorption mechanism is also accompanied by chloride, which follows the absorbed sodium passively (paracellularly) to maintain electrical neutrality. It is clear, therefore, that regardless of which absorptive mechanism is functioning, wherever sodium goes, chloride cannot be far behind!

Secretory mechanisms for the electrolytes throughout the gastrointestinal tract center on chloride. It is the major secretory product of the stomach and the rest of the gastrointestinal tract. The well-defined mechanism is an electrogenic Cl^- secretion because Cl^- is the only ion actively secreted by the epithelium, and its movement can be monitored by changes in electrical potentials. Cells take up chloride from the blood across the basolateral membrane via a $Na^+/K^+/Cl^-$ cotransport pathway. An appropriate gradient is set up by the Na^+/K^+-ATPase pump, which maintains a low concentration of intracellular sodium. Potassium channels on the basolateral membrane allow potassium recycling out of the cell. Accumulating chloride in the cell exits through the apical membrane into the lumen via the Cl^- channels. Figure 11.11 illustrates the chloride secretory mechanism.

Chloride channels have not been studied extensively because of a lack of chemical blockers specific for the chloride channel. However, studies on the defective gene of cystic fibrosis (caused by chloride transport dysfunction) have revealed a protein called the "cystic fibrosis transmembrane conductance regulator" (CFTR) [11]. The predicted structure of the protein is that of a channel or channel accessory.

Regulation and Excretion

It is estimated that the average adult consumes between 50 and 200 mEq/day. Chloride output occurs through three primary routes: the gastrointestinal tract, the skin, and the kidneys, with losses through each of these routes reflecting closely that of sodium. Excretion of chloride through the gastrointestinal tract is normally very small (1–2 mEq/day for the average adult), in keep-

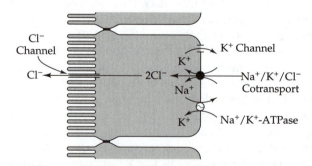

Figure 11.11 Intestinal chloride secretory mechanism. Chloride is cotransported along with Na^+ and K^+ from the circulation across the basolateral membrane and into the mucosal cell. Chloride then exits the cell into the lumen through Cl^- channels in the apical membrane. The driving force is provided by active removal of Na^+ by the Na^+/K^+-ATPase pump and the recycling of potassium through K^+ channels on the basolateral membrane.

ing with its extensive absorption in the intestine. Losses through the skin are essentially the same as sodium—that is, normally quite small except in cases of high temperature and vigorous exercise. The major route of excretion is through the kidney, where it is primarily regulated indirectly through sodium regulation.

Recommended Dietary Allowances and Assessment of Nutriture

Nearly all the chloride consumed in the diet is associated with sodium in the form of NaCl. Therefore, it is abundant in a large number of foods, particularly in snack items and processed foods. However, it is also found in eggs, fresh meats, and seafood. Dietary deficiency of chloride therefore does not occur under normal conditions. As in the case of the other electrolytes, the chief cause of deficiency arises through alimentary disturbance such as severe diarrhea and vomiting.

The National Research Council recommends a chloride intake of 750 mg/day [6]. In view of losses through sweating, however, this value may have to be adjusted upward in cases of people with physically rigorous lifestyles or those chronically exposed to high temperatures.

The serum concentration of chloride is 101 to 111 mEq/L. Its measurement is generally used to establish the chloride status of the body. However, like all serum solutes, concentration depends on the body water status. It is possible, for example, that the total body store of chloride may be diminished. Yet, if body water accompanies the losses, fluid concentrations of chloride may appear normal and may even be elevated. Two widely used methods for determining chloride in serum are based on ion-selective electrode potentiometry and a coulometric titration with silver ions.

References Cited for Sodium, Potassium, and Chloride

1. Munck BG. Intestinal absorption of amino acids. In: Johnson LR, et al., eds. Physiology of the Gastrointestinal Tract. New York: Raven Press, 1981:1097–122.

2. Rose RC. Intestinal absorption of water-soluble vitamins. In: Johnson LR, et al., eds. Physiology of the Gastrointestinal Tract. New York: Raven Press, 1981:1581–96.

3. Frizzell RA, et al. Sodium-coupled chloride transport by epithelial tissues. Am J Physiol 1979;236:F1–8.

4. Barrett KE, Dharmsathaphorn K. Transport of water and electrolytes in the gastrointestinal tract: physiological mechanisms, regulation, and methods of study. In: Maxwell MH, Kleeman CR. Clinical disorders of fluid and electrolyte metabolism. Narins RG, ed. New York: McGraw-Hill, 1994:506–7.

5. Shortt C, Flynn A. Sodium-calcium inter-relationships with specific reference to osteoporosis. Nutr Res Rev 1990;3:101–15.

6. National Research Council. Recommended Dietary Allowances, 10th ed. Washington, D.C.: National Academy Press, 1989.

7. Hayslett JP, Binder HJ. Mechanism of potassium adaptation. Am J Physiol 1982;243:F103–12.

8. Kliger AS, et al. Demonstration of active potassium transport in the mammalian colon. J Clin Invest 1981;67:1189–96.

9. Agarwal R, Afzalpurkar R, and Fordtran J. Pathophysiology of potassium absorption and secretion by the human intestine. Gastroenterology 1994;107:548–71.

10. Bell RR, Eldrid MM, Watson FR. The influence of NaCl and KCl on urinary calcium excretion in healthy young women. Nutr Res 1992;12:17–26.

11. Riordan JR, et al. Identification of the cystic fibrosis gene: cloning and characterization of complementary DNA. Science 1989;245:1066–75.

Macrominerals and Hypertension

Dietary factors influence blood pressure just as they do other physiological processes of the body. Although high blood pressure called *hypertension* has been primarily linked to sodium intake, other nutrients play a role in blood pressure control.

Hypertension is thought to affect as many as 50 million Americans. The condition is one in which there is an increase in vascular resistance most often due to a decreased luminal diameter of the arteries and/or arterioles. Systolic and diastolic blood pressure values of ≥140 and 90 mm Hg, respectively, are indicative of hypertension. Hypertension is often classified as primary, also called *essential,* or secondary. Causes of essential hypertension are generally unknown or may be related to malfunction of sodium excretion or of the renin-angiotensin or kallikrein-kinin systems, hyperactivity of the nervous system, and abnormal prostaglandin production, among others. Essential hypertension accounts for >90% of hypertension cases. The remaining cases of hypertension occur secondary to other conditions such as kidney, endocrine, or neurological diseases. Whether hypertension is essential or secondary, the condition increases risk for stroke and heart disease. Although some risk factors for hypertension are not controllable (e.g., genetic predisposition, race, aging), others may be modified should an individual commit to making dietary changes.

Unfortunately, hypertension is a heterogeneous disease, having a variety of precipitating factors, and thus dietary modification works for some but not all hypertensive individuals. This perspective will discuss some of the nutrients that are associated with essential hypertension. The nutrients most often associated with blood pressure are the macrominerals sodium, chloride, calcium, potassium, and magnesium. Each of these will be discussed along with sucrose and alcohol, also shown to influence blood pressure.

Sodium

Sodium was one of the first nutrients linked to hypertension. Specifically, increased salt intake directly correlates with increased blood pressure. However, a certain population of hypertensives (~60%) appear to be much more sensitive to an excess of salt than others. Thus, in some, but not all, hypertensive individuals, high dietary salt intakes raise blood pressure, and dietary salt restriction results in blood pressure reduction [1–6].Hypertensive individuals likely to benefit from sodium reduction include those who are African American, obese, or over 65 years of age or have low plasma renin concentrations, as well as those taking antihypertensive medications [1,2,7]. In salt-sensitive individuals salt ingestion is thought to cause water retention,

with the resulting release of a substance that increases heart and blood vessel contractile activity [1]. Alternately, sodium may infiltrate vascular smooth muscle causing contraction to raise blood pressure [1].

Chloride

The role of chloride in association with hypertension appears to relate to sodium. Studies showing increases in blood pressure with administration of sodium chloride also have documented no change in blood pressure with administration of equimolar amounts of sodium as sodium citrate, sodium bicarbonate, or sodium phosphate [8,9]. These studies suggest that it is a combination of the two nutrients, sodium plus chloride, that affects blood pressure.

Calcium

Calcium deficiency has been linked to hypertension. A possible relationship between calcium and the development of hypertension was first recognized with the discovery in the early 1970s that communities characterized by hard water (high calcium content) had a lower death rate from cardiovascular disease [10,11]. Since that time, much evidence from epidemiological studies, laboratory studies in animals and humans, and clinical trials has accumulated to support the relationship between calcium and blood pressure [11]. A meta-analysis of over 30 randomized calcium supplementation (median intake, 1 g of calcium) studies found that calcium supplementation significantly resulted in small reductions in systolic, but not diastolic, blood pressure [12,13]. Further, a meta-analysis of studies involving pregnant women concluded that consumption of calcium during pregnancy reduced the risk of pregnancy induced hypertension [14].

There appears, however, to be a segment of the hypertensive population that is more responsive to calcium supplementation than others. In addition, there also appears to be a threshold level in which calcium influences blood pressure. Calcium intakes of >800 mg reduce the risk of hypertension versus intakes of <400 mg [15–17]. The difference in response among hypertensives to oral calcium may be due to the heterogenicity of the disease. Only a subset of the people suffering from hypertension are "calcium sensitive" and thus may benefit from increased calcium intake [18]. People who appear to benefit from oral calcium therapy are those who have low calcium intakes, low ionized calcium concentrations, or elevated PTH and vitamin D and those who have been classified as having low renin activity and being "salt-sensitive" [17]. For a given dietary salt intake, hypertensive people with low renin levels have a calcium metabolic profile characterized by lower average levels of serum-ionized calcium and higher levels of parathyroid hormone and calcitriol than those found in normotensive people or other subgroups of hypertensive people [16,19]. Efforts are underway to determine how to predict those hypertensive individuals who will be helped by calcium supplementation.

How dietary calcium can exhibit an antihypertensive effect in "calcium-sensitive" hypertensives is uncertain. Calcium has a membrane-stabilizing, vasorelaxing effect on the smooth muscle cells [19]. It also affects the central and peripheral sympathetic nervous systems and modifies calcium homeostasis, vascular actions, as well as the actions of PTH and calcitriol; calcium may suppress PTH induced elevation in calcium to in turn reduce vascular tone [16,19]. Calcium may also exert its effects through interaction with other nutrients. For example, increased sodium intake results in increased urinary calcium excretion [20]. Increased calcium intake causes natriuresis (excretion of abnormal amounts of sodium in the urine) [20]. Calcium may correct sodium-induced decreases of ionized serum calcium as would occur with sodium-induced calciuria. Salt-sensitive, low-renin individuals typically exhibit low plasma ionized calcium concentration, increased urinary calcium excretion, and elevated concentrations of PTH and calcitriol [16]. Research suggests that adequate calcium intakes appear to protect against the hypertensive effects of a high-sodium, low-potassium diet [21]. A calcium intake of about 1,200 to 1,500 mg daily for adults as recommend by the National Institutes of Health consensus panel on calcium should be helpful in promoting blood pressure regulation [16,21].

Potassium

Potassium is yet another nutrient thought to impact blood pressure. Potassium intake appears to be inversely associated with blood pressure [1,23,24]. A meta-analysis of 30 controlled trials found that potassium supplementation was associated with significant reductions in systolic and diastolic blood pressure [25]. Effects of potassium vary but are generally more apparent in hypertensive individuals than in normotensive individuals [23–26]. Furthermore, effects of potassium were greater in those individuals ingesting higher sodium intakes [25].

How potassium affects blood pressure is unclear. Potassium may induce vascular smooth muscle relaxation and thus reduce peripheral resistance [27]. Potassium also may affect the kinin system—for example, potassium increases urinary kallikrein [1]. Potassium also may affect renin or work with other nutrients to regulate blood pressure [27]. For example, increased potassium intakes are associated with reduced urinary calcium and magnesium excretion [23]. Higher potassium intakes alone or potassium intake relative to sodium intakes also have been associated with a lower prevalence of hypertension [27].

Magnesium

Magnesium also has been investigated and linked to hypertension because it promotes relaxation of vascular smooth muscle and because of interactive effects with calcium [19,20]. Increased blood pressure is associated with both calcium and magnesium excretion [29]. Epidemiological data as well as animal and human studies suggest a relationship between blood pressure and magnesium intake. For example, the Nurses' Health Study found that magnesium intake had an independent and significant inverse association with blood pressure [23,30]. The results of studies providing magnesium supplementation to treat hypertension have been contradictory. Use of a mineral salt containing sodium, potassium, and magnesium in a 8:6:1 mmol ratio respectively versus use of regular salt (sodium chloride) significantly reduced blood pressure in individuals with mild to moderate hypertension [31].

Other Dietary Factors

Sucrose also appears to elevate blood pressure. Several animal studies have demonstrated sucrose-induced rises in blood pressure. The effects may result from volume expansion and antinaturiuretic effects that accompany sucrose ingestion [26].

Alcohol consumption is thought to account for almost 10% of hypertension, especially in middle-aged men [7,32].

The randomized trial dietary approaches to stop hypertension (DASH) reported that diets rich in fruits and vegetables as well as diets high in fruits and vegetables but also low in fat were both effective in reducing blood pressure versus a control diet low in fruits and vegetables and average in fat (~36% of kcal from fat) [33]. Similarly, consumption of diets that meet or exceed the RDAs of calcium, potassium, and magnesium are not associated with hypertension even when the diet is high in sodium chloride [34]. Thus, use of supplements may not be necessary to achieve reductions in blood pressure. Diets that ensure adequate dietary intakes of calcium, potassium, and magnesium are recommended by the Joint National Committee on Detection, Evaluation, and Treatment of High Blood Pressure for the treatment of hypertension [35].

References Cited

1. Haddy F, Pamnani M. Role of dietary salt in hypertension. J Am Coll Nutr 1995;14:428–38.

2. Luft F, Weinberger M. Heterogeneous responses to changes in dietary salt intake: the salt-sensitivity paradigm. Am J Clin Nutr 1997;65:612S–7S.

3. Cowley A. Genetic and nongenetic determinants of salt sensitivity and blood pressure. Am J Clin Nutr 1997;65:587S–93S.

4. Weinberger M. Salt sensitivity of blood pressure in humans. Hypertension 1996;27(3 pt 2):481–90.

∽ **PERSPECTIVE** (continued)

5. Staessen J, Lijnen P, Thijs L, Fagard R. Salt and blood pressure in community based intervention trials. Am J Clin Nutr 1997;65:661S–70S.

6. Cutler J, Follmann D, Allender P. Randomized trials of sodium reduction: an overview. Am J Clin Nutr 1997;65: 643S–51S.

7. Joint National Committee on the Detection, Evaluation, and Treatment of High Blood Pressure. National High Blood Pressure Education Program. Arch Intern Med 1993;153: 154–83.

8. Kurtz T, Al-Bander H, Morris R. Salt sensitive essential hypertension in men: is the sodium ion alone important? N Engl J Med 1987;317:1043–8.

9. Shore A, Markandu N, MacGregor G. A randomized crossover study to compare the blood pressure response to sodium loading with and without chloride in patients with essential hypertension. J Hypertens 1988;6:613–7.

10. Henry H, McCarron DA, Morris CD, et al. Increasing calcium intake lowers blood pressure: the literature reviewed. J Am Diet Assoc 1985;85:182–5.

11. Cappuccio F, Elliott P, Allender P, Pryer J, Follman D, Cutler J. Epidemiologic association between dietary calcium intake and blood pressure: a meta-analysis of published data. Am J Epidemiol 1995;142:935–45.

12. Bucher H, Cook R, Guyatt G, Lang J, Cook D, Hatala R, Hunt D. Effects of dietary calcium supplementation on blood pressure: a meta-analysis of randomized controlled trials. JAMA 1996;275:1016–22.

13. Allender P, Cutler J, Follmann D, Cappuccio F, Pryer J, Elliott P. Dietary calcium and blood pressure: a meta-analysis of randomized clinical trials. Ann Intern Med 1996;124:825–31.

14. Bucher H, Guyatt G, Cook R, Lang J, Cook D, Hatala R, Hunt D. Effects of calcium supplementation on pregnancy-induced hypertension and pre-eclampsia: a meta-analysis of randomized controlled trials. JAMA 1996;275:1113–7.

15. Aalberts J, Weegels P, van der Heyden L, et al. Calcium supplementation: effect on blood pressure and urinary mineral excretion in normotensive male lactoovovegetarians and omnivores. Am J Clin Nutr 1988;48:131–8.

16. Witteman J, Willet W, Stampfer M. Dietary calcium and magnesium and hypertension: a prospective study. Circulation 1987;76:(suppl IV):35.

17. Sowers J, Zemel M, Zemel P, Standley P. Calcium metabolism and dietary calcium in salt sensitive hypertension. Am J Hyperten 1991;4:557–63.

18. Morris C, Reusser M. Calcium intake and blood pressure: epidemiology revisited. Semin Nephrol 1995;15:490–5.

19. Hatton D, Yue Q, McCarron D. Mechanisms of calcium's effects on blood pressure. Semin Nephrol 1995;15:593–602.

20. Massey LK. Dietary factors influencing calcium and bone metabolism: introduction. J Nutr 1993;123:1609–10.

21. Gruchow H, Sobocinski K, Barboriak J. Calcium intake and the relationship of dietary sodium and potassium to blood pressure. Am J Clin Nutr 1988;48:1463–70.

22. National Institutes of Health. Consensus Development Panel on Optimal Calcium Intake. Optimal calcium intake. JAMA 1994;272:1942–8.

23. Reusser M, McCarron D. Micronutrient effects on blood pressure regulation. Nutr Rev 1994;52:367–75.

24. Barri Y, Wingo C. The effects of potassium depletion and supplementation on blood pressure: a clinical review. Am J Med Sci 1997;314:37–40.

25. Whelton P, He J, Culter J, Brancati F, Appel L, Follmann D, Klag M. Effects of oral potassium on blood pressure: meta-analysis of randomized controlled clinical trials. JAMA;1997: 277:1624–32.

26. Kotchen T, Kotchen J. Dietary sodium and blood pressure: interactions with other nutrients. Am J Clin Nutr 1997;65: 708S–11S.

27. Stein P, Black H. The role of diet in the genesis and treatment of hypertension. Med Clin N Am 1993;77:831–47.

28. Paolisso G, Barbagallo M. Hypertension, diabetes mellitus, and insulin resistance: the role of intracellular magnesium. Am J Hypertens 1997;10:346–55.

29. Wu X, Ackermann U, Sonnenberg H. Potassium depletion and salt sensitive hypertension in dahl rats: effect on calcium, magnesium, and phosphate excretion. Clin Exp Hyperten 1995;17:989–1008.

30. Witteman J, Willet W, Stampfer M. A prospective study of nutritional factors and hypertension among US women. Circulation 1989;80:1320–7.

31. Geleijnse J, Witteman J, Bak A, den Breeijen J, Grobbee D. Reduction in blood pressure with a low sodium, high potassium, high magnesium salt in older subjects with mild to moderate hypertension. Brit Med J 1994:309:436–40.

32. Stamfer J, Stamler R, Neaton J. Blood pressure, systolic and diastolic, and cardiovascular risks: US population data. Arch Intern Med 1993;153:598–615.

33. Appel L, Moore T, Obarzanek E, Vollmer W, Svetkey L, Sacks F, Bray G, Vogt T, Cutler J, Windhauser M, Lin P, Karanja N. A clinical trial of the effects of dietary patterns on blood pressure. N Engl J Med 1997;336:1117–24.

34. McCarron D. Role of adequate dietary calcium intake in the prevention and management of salt-sensitive hypertension. Am J Clin Nutr 1997;65:721S–6S.

35. Joint National Committee on Detection, Evaluation, and Treatment of High Blood Pressure. The fifth report. Arch Intern Med 1993;153:154–83.

Additional References

National High Blood Pressure Education Program. Arch Intern Med 1993;153:186–208.

Hamet P. The evaluation of the scientific evidence for a relationship between calcium and hypertension. J Nutr 1995; 125(2 suppl):311–43.

Web Sites

www.bloodpressure.com

www.amhrt.org/

Photomicrograph of crystallized manganese sulfate

Chapter 12

Microminerals

A precise definition for the essential microminerals (or *trace minerals* or *elements*) has not been established. Some define trace elements as those that comprise <0.01% of total body weight; others define trace elements as nutrients the body needs in concentrations of one part per million or less [1]. These minerals initially gained the nomenclature of "trace" because their concentrations in tissue were not easily quantified by early analytical methods [2]. Iron appears to be the mineral that divides the macrominerals from the microminerals; consequently, some define an essential trace mineral as one that is needed by the body in a concentration equal to or lower than iron [3]. Alternately, trace may be defined as needed by the body in amounts <100 mg per day.

The term *essential* as applied to trace elements is also specified. An element is considered essential if a dietary deficiency of that element consistently results in a suboptimal biological function that is preventable or reversible by physiological amounts of the element [4]. More stringent criteria proposed to establish essentiality of a mineral include [5] the following:

- It is present in all healthy tissue of living things.

- Its concentration from one animal to the next is fairly constant.

- Its withdrawal from the body induces reproducibly the same physiological and structural abnormalities, regardless of species studied.

- Its addition either reduces or prevents these abnormalities.

- The abnormalities induced by deficiencies are always accompanied by specific biochemical changes.
- These biochemical changes can be prevented or cured when the deficiency is prevented or cured.

Elements established as essential may not necessarily comply with all those criteria listed, in part because of limitations imposed by the degree of sophistication of the analytical methodology available. Proof of essentiality is therefore technically easier for elements that occur in relatively high concentration than for those ultratrace elements occurring at very low concentrations and having a low requirement.

For four essential trace minerals (iron, zinc, iodine, and selenium), recommended dietary allowances (RDAs) have been established for humans. A range of safe and adequate daily dietary intakes has been estimated for another five essential trace minerals (fluorine, copper, manganese, chromium, and molybdenum). The inside book covers provide the RDAs for the microminerals. Very little is known about the need of humans for ultratrace elements, including nickel, silicon, vanadium, arsenic, and boron; therefore, no recommendations for intake exists.

Each essential trace mineral is necessary for one or more functions in the body, and its function(s) is optimal when body concentrations of the nutrient fall within a specific range. Whenever the concentration is too low or too high, function is impaired and death can result. This concept, illustrated in Figure 12.1, is especially important when considering the trace minerals because their optimal range of concentration can be fairly narrow. Moreover, because of the interactions among the essential trace minerals, an excessive intake of one, especially a divalent ion (e.g., zinc, magnesium, calcium, iron), may inhibit the absorption and cause deficiency of another divalent ion [6]. Conversely, a deficiency of one divalent ion may enhance the absorption of another [7].

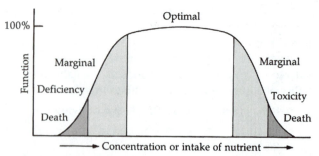

Figure 12.1 Dependence of biological function on tissue concentration or intake of a nutrient.
Source: Mertz W. The essential trace minerals. Science, September 18, 1981, 213:1333. Copyright 1981 by the AAAS, reprinted with permission.

This chapter discusses the sources, digestion, absorption, transport, functions, interactions with other nutrients, excretion, recommended intakes, deficiency, toxicity, and assessment of nutriture for the microminerals. Chapter 13 addresses these topics for several ultratrace elements. Table 12.1 provides an overview of selected functions, sources, deficiency symptoms, and recommended intakes for the trace minerals.

References Cited

1. Taylor A. Detection and monitoring of disorders of essential trace elements. Ann Clin Biochem 1996;33: 486–510.

2. Mertz W. The essential trace elements. Science 1981;213:1332–8.

3. Tracing the facts about trace minerals. Tufts University Diet and Nutr Letter March1987;5:3–6.

4. Nielsen FH. Ultratrace elements in nutrition. Ann Rev Nutr 1984;4:21–41.

5. Underwood EJ, Mertz W. Trace Elements in Human and Animal Nutrition. San Diego: Academic Press, 1987;2:1–19.

6. Greger JL. Mineral bioavailability/new concepts. Nutr Today 1987;22:4–9.

7. Finch CA, Huebers H. Perspectives in iron metabolism. N Engl J Med 1982;306:1520–8.

Iron

The human body contains about 2 to 4 g of iron. Over 65% of body iron is found in hemoglobin, up to about 10% is found as myoglobin, about 1% to 5% is found as part of enzymes, and the remaining body iron is found in the blood or in storage. Table 12.2 gives an approximate distribution of iron per kilogram of body weight in adult males and females. The total amount of iron found in a person not only is related to body weight but also is influenced by a variety of physiological conditions including age, gender, pregnancy, and state of growth.

Iron, a metal, exists in several oxidation states varying from Fe^{6+} to Fe^{2-}, depending on its chemical environment. The only states that are stable in the aqueous environment of the human body and in food are the ferric (Fe^{3+}) and the ferrous (Fe^{2+}) forms.

Sources

Although iron is widely distributed in food, its content in an average Western diet is estimated at no more than 5 to 7 mg iron per 1,000 kcal. Dietary iron is found in one of two forms in foods, heme and nonheme. Heme iron is derived mainly from hemoglobin and myoglobin

Table 12.1 The Microminerals: Selected Function, Deficiency Symptoms, Food Sources, and Recommended Intake (RDA, ESADDI, or AI)[a]

Mineral	Selected Physiological Roles	Selected Enzyme Cofactor Roles	Selected Deficiency Symptoms	Food Sources	RDA, ESADDI, or AI (Adults)
Chromium	Normal use of blood glucose and function of insulin		In humans: glucose intolerance, glucose and lipid metabolism abnormalities	Mushrooms, prunes, asparagus, organ meats whole-grain bread and cereals	50–200 μg
Copper	Utilization of iron stores, lipids, collagen, pigment, neurotransmitter synthesis	Oxidases, mono-oxygensases, superoxide dismutase	Anemia, neutropenia, bone abnormalities	Liver, shellfish, whole grains, legumes, eggs, meat, fish	1.5–3.0 mg
Fluorine	Maintenance of teeth and bone structure		Dental caries, bone problems	Fish, meat, legumes, grains, drinking water (variable)	3.8 and 3.1 mg
Iodine	Thyroid hormones synthesis		Enlarged thyroid gland, myxedema, cretinism, increase in blood lipids, liver gluconeogenesis, and extracellular retention of NaCl and H_2O	Iodized salt, saltwater seafood, sunflower seeds, mushrooms liver, eggs	150 μg
Iron	Component of hemoglobin and myoglobin for O_2 transport and cellular use	Heme enzymes, catalase, cyto-chromes, myelo-peroxidase	Listlessness, fatigue, anemia, palpitations, sore tongue, angular stomatitis, dysphagia, decreased resistance to infection	Organ meats (liver), meat, molasses, clams, oysters, nuts, legumes, seeds, green leafy vegetables, dried fruits, enriched/ whole grain breads/cereals	10 mg, male; 15 mg, female
Manganese	Brain function, collagen, bone, growth, urea, synthesis, glucose and lipid metabolism, CNS function	Arginase, pyruvate carboxylase, PEP, carboxykinase, superoxide dismutase	In animals, possibly humans: impaired growth, skeletal abnormalities, impaired CNS function	Wheat bran, legumes, nuts, lettuce, beet tops, blueberries, pineapple, seafood, poultry, meat	2.0–5.0 mg
Molybdenum	Metabolism of purines, pyrimidines, pteridines, aldehydes, and oxidation	Xanthine dehydro-genase/oxidase, aldehyde oxidase, sulfite oxidase	Hypermethioninemia, ↑ urinary xanthine, sulfite excretion, ↓ urinary sulfate and urate excretion	Soybeans, lentils, buckwheat, oats, rice, bread	75–250 μg
Selenium	Protect cells against destruction by hydrogen peroxide and free radicals	Glutathione peroxidase, 5'-deiodinase	Myalgia, cardiac myopathy, ↑ cell fragility, pancreatic degeneration	Grains, meat, poultry, fish, dairy products	70 μg, male; 55 μg, female
Zinc	Energy metabolism, metabolism, protein synthesis, collagen formation, alcohol detoxification, carbon dioxide elimination, sexual maturation, taste and smell functions	DNA-RNA polymerase, carbonic anhy-drase, carboxy-peptidase, alkaline phos-phatase, deoxy-thymidine kinase, …	Poor wound healing, subnormal growth, anorexia, abnormal taste/smell, changes in hair, skin, nails, retarded reproductive system development	Oysters, wheat germ, beef, liver, poultry, whole grains	15 mg male 12 mg female

[a]Abbreviations: RDA = Recommended Dietary Allowance; ESADDI = Estimated Safe and Adequate Daily Dietary Intake; AI = Adequate Intake.

Table 12.2 Approximate Distribution of Iron in Adult Males and Females

	Male (mg/kg body weight)	Female (mg/kg body weight)
Functional compounds:		
Hemoglobin	31	28
Myoglobin	4	3
Heme enzymes	1	1
Nonheme enzymes	1	1
Transferrin iron	0.05	0.05
	37.05	33.05
Storage compounds:		
Ferritin	9	4
Hemosiderin	4	1
	13	5
Total iron	50.05	38.05

Sources: Finch CA, Huebers H. Perspectives in iron metabolism. N Engl J Med 1982;306:1520–8; Leibel RL. Behavioral and biochemical correlates of iron deficiency. J Am Diet Assoc 1977;77:378–404; and Hallberg L. Iron absorption and iron deficiency. Hum Nutr Clin Nutr 1982;36C:259–78.

Table 12.3 Current (1981) FDA Standards for Iron Enrichment of Cereal Products

Product	mg Iron/lb
Bread, rolls, buns	12.5[a]
Flour	20.0[b]
Corn grits, cornmeal, rice	13–26[c]
Pasta	13–16.5[c]

[a]About 28 g/slice bread.
[b]1 tbsp flour about 8 g.
[c]2 oz dry rice and macaroni about 60 g.

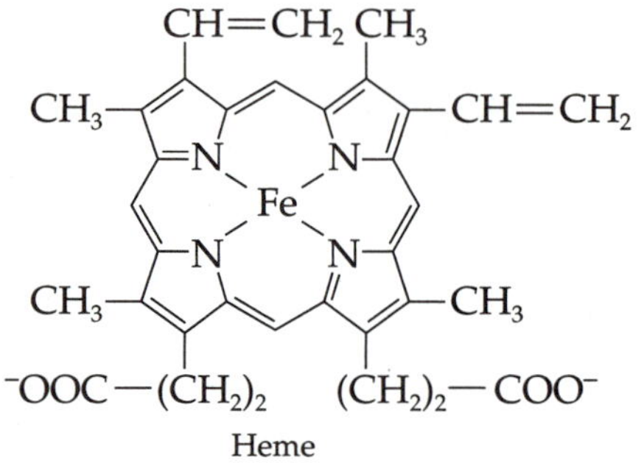

Figure 12.2 Heme iron.

and is thus found in meat, fish, and poultry. About 50% to 60% of the iron in meat, fish, and poultry is heme iron; the rest is nonheme iron. Nonheme iron is found primarily in plant foods (nuts, fruits, vegetables, grains, and tofu) and dairy products (milk, cheese, eggs), although dairy products have very little iron and represent a very poor source of iron. Nonheme iron is usually bound to components of foods and must be hydrolyzed or solubilized prior to absorption.

Foods particularly high in iron, such as liver and organ meats, are not popular items in most Western diets. Some of the more popular foods that are relatively good sources of iron include red meats, oysters and clams, beans (lima, navy), dark green, leafy vegetables, and dried fruits. Other good sources of iron are listed in Table 12.1.

In addition to natural amounts of iron found in foods, foods such as breads, rolls, pasta, cereals, grits, and flour are fortified with iron. The standards for iron enrichment of cereal products are given in Table 12.3. Elemental iron, ferrous ascorbate, ferrous carbonate, ferrous citrate, ferrous fumarate, ferrous gluconate, ferrous lactate, ferric ammonium citrate, ferric chloride, ferric citrate, ferric pyrophosphate, and ferric sulfate are approved and used for food fortification.

Digestion, Absorption, and Transport

Heme Iron Digestion and Absorption

Heme iron must be hydrolyzed from the globin portion of hemoglobin or myoglobin prior to absorption. This digestion is accomplished by proteases in the stomach and small intestine, and results in the release of heme iron. The heme iron, unlike nonheme iron, remains soluble because of both degradation products of the globin and, if in the small intestine, the alkaline pH of the small intestine. Thus, heme containing the iron bound to the porphyrin ring (Fig. 12.2) is readily absorbed intact as a metalloporphyrin into the mucosal cell of the small intestine.

Absorption of heme iron is influenced by body iron stores. Heme absorption is inversely related to iron stores and may range from 15% with normal iron status to 35% in persons who are iron-deficient. Iron absorption can occur throughout the small intestine but is most efficient in the proximal portion, particularly the duodenum. Within the mucosal cell the absorbed heme porphyrin ring is hydrolyzed by heme oxygenase into inorganic ferrous iron and protoporphyrin (Fig. 12.3). The released iron is used by the intestinal mucosal cell or following transport through the intestinal cell and subsequent transport through the blood is used by other body tissues.

Nonheme Iron Digestion and Absorption

Nonheme iron, bound to components of foods, must be enzymatically liberated in the gastrointestinal tract for

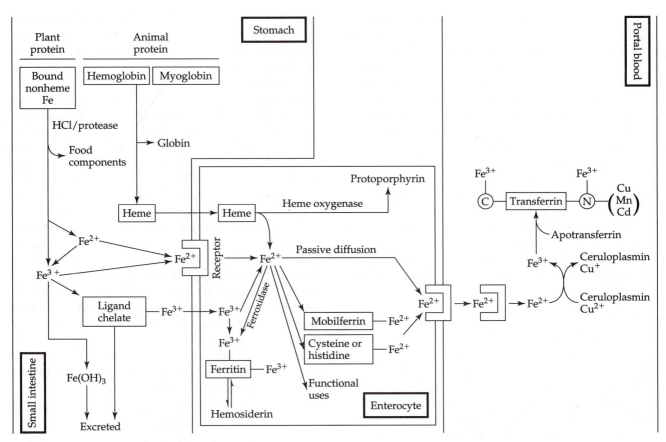

Figure 12.3 Overview of iron digestion, absorption, and transport.

absorption to occur (Fig. 12.3). Gastric secretions including hydrochloric acid and the protease pepsin aid in the release of nonheme iron from food components. Once released from food components, most nonheme iron is present as ferric (Fe^{3+}) iron in the stomach. The ferric iron remains fairly soluble as long as the pH of the environment is acidic as in the stomach. Also, in the acidic environment of the stomach, much of the ferric iron may be reduced to the ferrous state. Ferrous iron remains soluble even at pH 8. However, as the ferrous iron passes from the stomach into the small intestine, which contains juices with an alkaline pH, some ferrous iron may be oxidized to become ferric iron. Ferric iron may further complex to produce ferric hydroxide in the small intestine. Ferric hydroxide—$Fe(OH)_3$—is relatively insoluble and tends to aggregate and precipitate, making the iron less available for absorption.

Absorption Summary

The mechanisms of heme iron and nonheme iron absorption differ. Heme iron is absorbed intact into the enterocyte and hydrolyzed within the intestinal cell to yield ferrous iron. Nonheme iron, following release

from food components, is typically present in the stomach in the ferric state. Ferric iron may or may not be reduced in the stomach to the ferrous state. Absorption of iron is improved if the iron is present as ferrous iron. Ferrous iron transverses the glycocalyx and brush border of the intestine better than ferric iron. Ferrous iron then binds to receptors on the intestinal mucosal cell for absorption across the brush border. Mechanisms of absorption of ferric iron are not clearly delineated, although chelation by ligands or chelators to solubilize the ferric iron is thought to be involved. Furthermore, it is free iron, not chelated iron, that is absorbed into the enterocyte. A membrane protein known as *integrin* is thought to facilitate iron absorption through the microvillous brush border membrane of the enterocyte. The role of ligands and chelators in the absorption of iron is discussed in the next subsection.

Factors Influencing Iron Absorption

Several compounds (known as *chelators* or *ligands*) may bind with nonheme iron to either inhibit or enhance its absorption. Chelators are small organic compounds that form a complex with a metal ion. Ligands are compounds that also bind or complex with minerals. Whether chelated iron or iron that is attached to a ligand is absorbed or excreted depends on the nature of the iron-chelate/ligand complex. If the iron-chelate/ligand complex maintains solubility and the iron is loosely bonded, the iron can be released at the mucosal cell and absorption enhanced. However, if the iron-chelate/ligand is strongly bonded and insoluble, then the absorption of iron will not occur, and the iron will be excreted in the feces as part of the chelate.

Enhancers of Iron Absorption   Some acids, sugars, and meat, fish, or poultry ingested concomitantly with nonheme iron–containing foods promote absorption of nonheme iron. Ascorbic acid (vitamin C), along with citric, lactic, and tartaric acids, for example, act as a reducing agent and form a chelate with nonheme ferric iron at an acid pH.

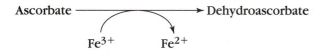

Ascorbate $\longrightarrow$ Dehydroascorbate

$Fe^{3+} \quad Fe^{2+}$

This chelate (a ferrous ascorbate chelate if vitamin C served as the reducing agent) remains soluble in the small intestine and thus can improve intestinal absorption of nonheme iron. Sugars such as fructose and sorbitol also promote nonheme iron absorption. Meat, poultry, and fish factors that enhance nonheme iron absorption have not been clearly identified. The enhancement is not a general property of animal proteins but is due to the digestion products from animal tissues that are high in the contractile proteins actin and myosin [1]. Both of these proteins are digested into peptides that contain relatively large amounts of the amino acid cysteine. The cysteine-containing peptides appear to be responsible for the increased absorption of iron [1-3]. It is further suspected that meat is not just a source of solubilizing ligands (e.g., the cysteine-containing peptides that bind iron) but also may improve iron absorption through stimulation of enhancing intestinal secretions [1]. Absorption of iron–amino acid chelates also has been demonstrated. These chelates are thought to be absorbed intact using amino acid carrier proteins. EDTA, although generally shown to inhibit iron absorption, may improve iron absorption from, for example, meals with low iron bioavailability [4].

Mucin, a small protein made in the intestinal cells and released into the gastrointestinal tract, represents an endogenously synthesized chelator or ligand that facilitates iron absorption. Mucin binds multiple ferric iron molecules at an acid pH and maintains ferric iron

solubility even in the alkaline pH of the small intestine [5]. Chelators of iron such as histidine, ascorbic acid, and fructose donate the iron to mucin once the gastrointestinal tract pH becomes neutral—that is, in the small intestine.

Estimation of the iron available for absorption can be calculated based on the quantity of vitamin C and/or meat, fish, or poultry that is ingested with the nonheme iron source, assuming approximately 500 mg body iron stores. Seventy-five units of ascorbic acid and/or meat, fish, poultry (MFP) factor (one unit = 1.3 g raw or 1 g cooked meat, fish, poultry or 1 mg ascorbic acid) have been shown to maximize the effect on iron absorption when consumed with the iron source [6]. Units in excess of 75 seem to have no further benefit. Absence of enhancing factors predicts a nonheme iron absorption of only 2% to 3% of the dietary nonheme iron, but 75 units of these factors can increase absorption of nonheme iron to 8% (some suggest up to 20% if the person is also iron-deficient) [2].

Inhibitors of Iron Absorption   Many dietary factors inhibit iron absorption, including

- polyphenols including tannin derivatives of gallic acid (in tea and coffee);
- oxalic acid (in spinach, chard, berries, chocolate, tea, among others);
- phytates (e.g., in maize and whole grain);
- EDTA (a preservative);
- phosvitin, a protein containing phosphorylated serine residues found in egg yolks; and
- nutrients (calcium, calcium phosphate salts, zinc, manganese, and nickel).

Tea (or rather, the phenolic compounds in tea) consumed with a source of iron may reduce iron absorption over 60%. Coffee consumption, with or just after a meal, may reduce iron absorption by 40% [7].

Calcium and *phosphorus* are thought to interact with iron and inhibit its absorption through Fe:Ca:PO_4 chelate formation at the intestinal mucosa. Alternately, the inhibitory effect of calcium on iron absorption may be within the intestinal mucosal cells and at a step in iron transport that is common for both heme and nonheme iron transport [8]. Several studies [2,8-11] have demonstrated that ingestion of calcium in amounts of 300 to 600 mg and in the forms of calcium phosphate, calcium citrate, calcium carbonate, and calcium chloride when given with up to 18 mg iron as ferrous sulfate or when incorporated into food substantially decrease iron absorption by up to 70%.

Similar reductions in iron absorption have been shown with milk ingestion [10]. Thus, to maximize iron absorption from a supplement, do not take the iron with a source of calcium.

Zinc and iron also interact and may negatively impact each other's absorption. The two minerals are thought to compete for the same portion of a common absorptive pathway as well as at a site distal to the site of iron status–regulated iron absorption [12]. Inhibition of iron absorption has been demonstrated with the ingestion of 15 mg and 45 mg of zinc as zinc sulfate given in a water solution with 3 mg of iron as ferrous sulfate [13]. Zinc in a 1:1 and a 2.5:1 (27 mg of zinc and 68.5 mg of zinc doses) molar ratio with iron in solution inhibited nonheme iron absorption by 66% and 80%, respectively [14]. A review of studies assessing iron and zinc interactions suggests that the interactions result primarily when the two minerals are given in solution and do not occur when given in a meal [15].

Manganese and iron also appear to interact. Manganese (as manganese chloride) when ingested in water or with a meal in a 2.5: or 5:1 ratio with iron (as ferrous sulfate) reduced iron absorption by 22% to 40% [13].

Nickel has been shown to interact synergistically with iron as ferric sulfate only, but not as a mixture of ferric and ferrous sulfates in affecting hematopoiesis. Thus, nickel appears to facilitate the use of ferric iron but somehow antagonizes ferrous iron absorption, particularly in severe iron deficiency. The enhancement of Fe^{3+} absorption by nickel is not understood. There is speculation that nickel may function as a cofactor in enhancing the complexing of the Fe^{3+} with a ligand, or it may act in an enzyme system that converts Fe^{3+} to Fe^{2+}.

Other intraluminal factors that are inhibitory to iron absorption include rapid transit time, achylia (absence of digestive juices), malabsorption syndromes, and excess alkalinization as may occur with excessive use of antacids or with decreased gastric acidity. Overall absorption of iron from the U.S. diet is estimated at 10% to 15% [16,17].

In addition to the role of ligands and chelators, a person's iron status also affects iron absorption. The regulation of absorption in the healthy individual is closely tied to the level of iron stores [18]. Absorption increases when iron stores are low and decreases as stores become greater; iron absorption can rise to 3 to 6 mg daily when the body is depleted of iron and can fall to 0.5 mg or less daily when iron stores are high. Absorption of heme iron is more affected than nonheme iron in iron-deficient states. Active erythropoiesis also influences iron absorption.

Intestinal Cell Iron Transport and Use

Following absorption into the enterocyte, iron must be either

- transported through the enterocyte and into the blood for use by the body tissues,
- stored in the intestinal cell for future use or elimination, or
- used by the intestinal cell in a functional capacity.

First, the transport of iron across the enterocyte will be discussed. How iron moves through the mucosal cell or is distributed within the cell has not been clearly elucidated. Ferrous iron is released from heme catabolism in the enterocyte. Whether ferric iron is reduced to the ferrous state by flavins ($FAD/FADH_2$ or NAD/NADH) or vitamin C in the enterocyte is unclear. A proposed model of iron transport is shown in Figure 12.3. Some iron is thought to pass through the mucosal cell by diffusion. In addition, amino acids such as cysteine and histidine may transport iron across the mucosal cell. Some mucosal iron may be bound to an intestinal cell protein known as *mobilferrin*. Mobilferrin appears to interact with integrin in the enterocyte membrane and bind to minerals such as iron, although it also binds calcium, zinc, and copper. Mobilferrin is capable of binding one iron molecule which it shuttles across the cytosol of the mucosal cell for release across the basolateral membrane.

Iron, not being transported across the cell for release into the blood, may be incorporated into apoferritin in the intestinal cell for short term storage. *Apoferritin* is a protein that acts as a shell for iron storage. This protein shell further serves as a ferroxidase using oxygen to convert the ferrous iron to the ferric state for deposition and storage. The stored ferric iron can be reduced to the ferrous state and released from the ferritin molecule should iron be needed by the mucosal cell or required for transport to other tissues. If not needed, the iron remains as ferritin and is excreted when the short-lived (2–3 days) mucosal cells are sloughed off into the lumen of the gastrointestinal tract. Iron deposition into ferritin may be responsible for the increase or decrease in the amount of iron that passes into the plasma for transport to tissues. Ferritin synthesis in the intestine increases with high iron stores and decreases during periods of low iron stores. Ferritin is discussed in further detail under the section on iron storage.

Iron moving through the mucosal cell may be used by the cell for a variety of functions, especially as a cofactor for enzymes (see p. 410).

Little is known about iron transport across the intestinal cell basolateral membrane. Iron transport appears to involve a receptor protein that binds iron in the ferrous state. Whether ferric iron may be transported is unclear. However, it is iron in the ferric state that is transported in the plasma.

Iron Transport

Iron in its oxidized ferric state is transported in the blood attached to the protein transferrin (Fig. 12.3). Iron oxidation, transferrin's role in iron transport, and the importance of protein in iron binding in the body will be reviewed.

Iron must first be oxidized before it can bind to transferrin for transport in the blood. Ceruloplasmin, a copper-containing plasma protein with ferroxidase activity, catalyzes the oxidation of ferrous iron to its ferric form so that it can bind tightly to transferrin in the plasma. This role of copper as part of ceruloplasmin is crucial to iron metabolism. Copper deficiency results in accumulation of iron in sites such as the intestine and liver, and reduces iron transport to tissues. The role of ceruloplasmin in the oxidation of iron may be depicted as follows.

$$Fe^{2+} \longrightarrow Fe^{3+}$$
$$Ceruloplasmin\text{-}Cu^{2+} \quad Ceruloplasmin\text{-}Cu^{1+}$$

Transferrin, a glycoprotein made primarily in the liver, has two binding sites for minerals. The binding site near the C-terminal end of transferrin has a high affinity for ferric iron. The binding site near the N-terminal end has a high affinity for ferric iron, but will also bind other minerals such as chromium followed in descending order by copper > manganese > cadmium > zinc and nickel. Tight binding of ferric iron to transferrin also requires the presence of an anion, usually bicarbonate, at each binding site. Transferrin in the plasma is typically one-third saturated with ferric iron. If all binding sites of the transferrin were occupied, then the transferrin would be fully (100%) saturated.

The role of proteins in the transport as well as storage of iron is important because of iron's redox activity. The binding of iron by proteins serves as a defense mechanism. Left unbound, the redox activity of iron can lead to the generation of harmful free radicals. Free ferrous iron (Fe^{2+}), for example, readily reacts with hydrogen peroxide (H_2O_2). The reaction between ferrous iron and hydrogen peroxide is known as the Fenton reaction,

$$Fe^{2+} + H_2O_2 \longrightarrow Fe^{3+} + OH^- + OH\cdot$$

This reaction generates a hydroxy anion and a free hydroxy radical ($OH\cdot$), which is extremely reactive and damaging to cells (see Chap. 10's Perspective). In addition, the binding of iron by protein is important to en-

sure that bacteria that may be present in the body as with an infection are unable to utilize the iron for their own (bacterial) growth. Free iron, but not protein-bound iron, is readily used by bacteria for proliferation and growth. Multiplication of bacteria requires acquisition of nutrients such as iron from the host [19]. Thus, keeping iron attached to proteins in the body prevents bacterial acquisition of the iron.

Transferrin binds and transports not only newly absorbed dietary iron that has been transported across the basolateral membrane of the mucosal cell but also transports iron that has been released following the degradation of iron containing compounds in the body. In fact, most of the iron entering the plasma for distribution by transferrin is contributed from hemoglobin destruction and release from storage. Thus, transferrin ferries iron throughout the body delivering both new and recycled iron to tissues for use or for storage. Transferrin has a half-life of about 7 to 10 days.

Iron Storage

Iron not needed in a functional capacity is stored in three principal sites: the liver, bone marrow, and spleen. Transferrin delivers iron to the liver for storage in the parenchyma. The hepatocytes of the liver contain about 60% of body iron; the remaining 40% is found in reticuloendothelial (RE) cells. Reticuloendothelial cells are found in the spleen, liver, and bone marrow (and possibly between muscle fibers). Most of the iron stored in reticuloendothelial cells is derived from phagocytosis of red blood cells and subsequent degradation of the erythrocyte hemoglobin.

Ferritin is the primary storage form of iron in cells. Ferritin is synthesized in a variety of tissues, including the liver, spleen, bone marrow, and intestine, and consists of apoferritin in which iron molecules have been deposited. Apoferritin is composed of 24 protein subunits and is shaped as a hollow sphere. Iron enters apoferritin through channels or pores. The pores serve as the site of the oxidation of ferrous iron into ferric oxyhydroxide crystals ($4Fe^{2+} + O_2 + 6H_2O \longrightarrow 4FeOOH + 8H^+$) or ferrihydrite ($5Fe_2O_3 \cdot 9H_2O$); molecular oxygen functions as the electron acceptor [20]. Ferric oxyhydroxide or ferrihydrite is deposited in the interior of the protein shell. As many as 4,500 iron atoms can be stored in ferritin.

Ferritin is not a stable compound; it is constantly being degraded and resynthesized, thereby providing an available intracellular iron pool. Cellular iron is thought to influence in part the synthesis of apoferritin. The role of iron in the regulation of the synthesis of ferritin is discussed in the Perspective at the end of Chapter 9.

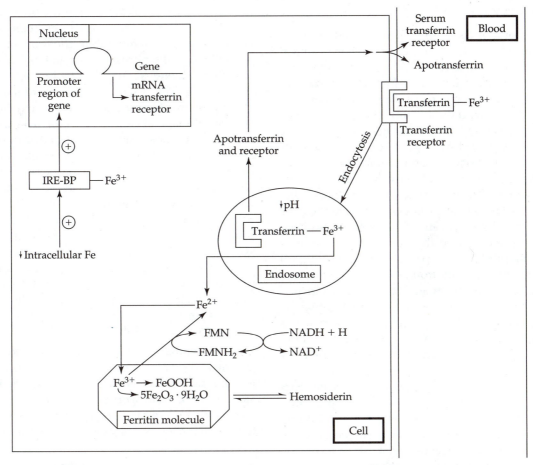

Figure 12.4 Overview of iron metabolism and storage. Abbreviations: IRE, iron response element; IRE-BP, IRE binding proteins.

Equilibration occurs between tissue ferritin and serum ferritin. Thus, serum ferritin is used as an index of body iron stores, whereby 1 mg ferritin/L serum = 10 mg body iron stores [21]. Normal serum ferritin concentrations exceed 12 mg/L for adult females and 15 mg/L for adult males; however, because ferritin acts as an acute phase reactant protein, it is not a reliable indicator of iron stores during and possibly for several weeks following inflammation or illness [21]. In other words, serum ferritin concentrations may be elevated or within the normal range in the blood despite an individual having little to no iron. Assessment of iron status is discussed further on page 417.

Hemosiderin is another iron storage protein. Hemosiderin is thought to be a degradation product of ferritin, representing, for example, aggregated ferritin or a deposit of degraded apoferritin and coalesced iron atoms. The content of iron in hemosiderin may be as high as 50%. The ratio of ferritin to hemosiderin in the liver varies according to the level of iron stored in the organ, with ferritin predominating at lower iron concentrations and hemosiderin at higher concentrations (iron overload) [22]. Although iron in hemosiderin can be labilized to supply free iron, the rate at which iron is released is slower than that from ferritin.

Release of iron from stores (Fig. 12.4) requires mobilization of Fe^{3+} and the use of reducing substances such as riboflavin ($FMNH_2$), niacin (NADH), and/or vitamin C and or possibly a chelator to enable diffusion through ferritin pores. However, following the reduction of iron to release it from storage, Fe^{2+} is transported to the cell surface where it must be reoxidized allowing transport out of the cell. This reoxidation of iron to enable binding to transferrin for transport to tissues requires ceruloplasmin as previously described in the section on iron transport and shown in Figure 12.3. The superoxide radical ($O_2^{\cdot}$) also has been found to initiate iron release from ferritin in vitro. However, only one or two iron molecules from ferritin are released even with extended exposure to superoxide radicals. The size and age of ferritin's iron core, and not the iron content of ferritin's protein shell, affected iron release [23].

Iron Uptake by Tissues

The amount of iron taken up by the tissues depends in part on the transferrin saturation level. For example, iron delivery is greater from diferric transferrin than from monoferric transferrin. For iron uptake to occur into tissues, the transferrin molecule (either diferric or monoferric transferrin) must first bind to transferrin receptors on cells (Fig. 12.4). Transferrin receptors are found on almost all cell surfaces and consist of two subunits that each bind one transferrin molecule. Once the transferrin molecule is bound to the receptor, the complex is thought to be internalized by endocytosis and to form a vesicle (also called an *endosome*) in the cytoplasm of the cell. Next, in an ATP-dependent process, protons are pumped into the endosome and reduce the pH to about 5.5. In the presence of the acidic pH and possibly other factors, iron molecules are released from the transferrin molecule. The apotransferrin is then thought to return to the cell surface and plasma. Use of the released iron requires its transport across the endosomal membrane. ATP is postulated to transport the iron across this membrane and into the cytoplasm of the cell for use [24].

The number of transferrin receptors on cells increases or decreases depending on intracellular iron concentrations. In other words, intracellular iron affects the genetic expression of transferrin receptors on the cell. Decreased intracellular iron, for example, results in increased transcription of the transferrin receptor gene. Further discussion of such nutrient gene interactions is found in the Perspective at the end of Chapter 9.

Functions and Mechanisms of Actions

Heme

The essentiality of iron is due, in part, to its presence in heme. The atom of iron in the center of the heme molecule allows the transport of oxygen to tissues (hemoglobin); the transitional storage of oxygen in tissues, particularly muscle (myoglobin); and the transport of electrons through the respiratory chain (cytochromes).

Hemoglobin, synthesized in red blood cells, consists of iron porphyrin complexed with a protein subunit. Porphyrins, in turn, are cyclic compounds that are made up of four pyrrole rings that are joined together by methenyl bridges (-CH-). Nitrogen atoms in each of the four pyrrole rings bind to the iron molecule (Fig. 12.5); these bonds hold the iron atom in the plane of the porphyrin ring. The iron atom in the center of the heme has two remaining coordinate bonds available for binding. One of the two remaining coordinate bonds of

iron is with an amino acid, often the nitrogen atom of histidine, of the protein to which the heme is attached. For example, in hemoglobin, the iron in the heme would bind to the nitrogen of an amino acid in the protein globin; heme is found in a hydrophobic pocket of the protein. The sixth and last coordinate bond in heme proteins that bind oxygen—namely, hemoglobin and myoglobin—exists between the iron and oxygen. The oxygen is held quite loosely so that transfer to tissues can be rapid. In heme proteins that do not bind oxygen, the sixth coordinate bond is with atoms of amino acid groups in the protein (such as an enzyme) with which the heme group is associated.

Heme synthesis and the attachment of globin occurs primarily in the red blood cells of bone marrow; heme synthesis accounts for the largest utilization of functional iron in the body. Erythropoietic cells possess transferrin receptors on their cell surface. Transferrin delivers the iron for heme synthesis to the erythropoietic cells in the bone marrow. The synthesis of heme (Fig. 12.5) occurs as follows:

- Heme synthesis begins in the mitochondria where glycine and succinyl CoA combine to form Δ-aminolevulinic acid (ALA). The reaction is catalyzed by Δ-aminolevulinic acid synthase, a vitamin B_6–dependent enzyme that is inhibited by the final end product heme and whose synthesis is also thought to be regulated by iron.

- Next, ALA enters the cytoplasm where a zinc-dependent dehydratase catalyzes the condensation of two ALA molecules to form porphobilinogen. This enzyme is sensitive to lead, which binds to its sulfhydryl groups to inactivate the enzyme.

- Next, in a series of cytosolic reactions involving a deaminase, a synthase, and a decarboxylase, four porphobilinogens condense to form a tetrapyrrole that cyclizes. Side chains are modified, and coproporphyrinogen III is formed and enters the mitochondria.

- Coproporphyrinogen, in the mitochondria, is converted into protoporphyrinogen III.

- Protoporphyrinogen III is oxidized to form protoporphyrin IX.

- Lastly, an iron (Fe^{2+}) atom is inserted into protoporphyrin IX to yield heme. The insertion of iron into the heme is catalyzed by ferrochelatase. The transcription of the enzyme appears to be regulated by iron.

Unlike hemoglobin, which is a tetrameric protein, myoglobin consists of a single hemoprotein chain. Myoglobin is found in the cytoplasm of the muscle cells and facilitates the diffusion rate of dioxygen from capillary red blood cells to the cytoplasm and mitochondria of muscle cells. Heme-containing cytochromes in the elec-

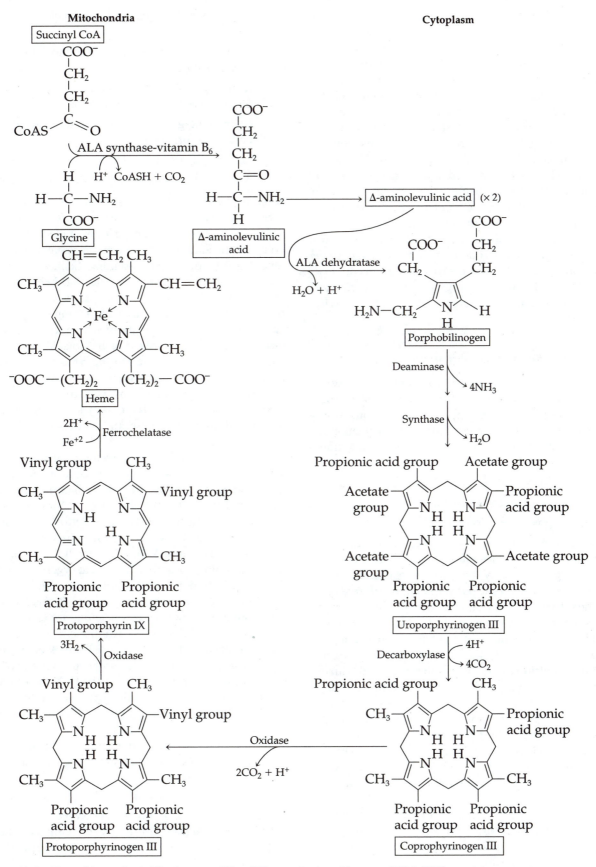

Figure 12.5 Heme biosynthesis. Vinyl group: CH = CH$_2$; propionic acid group: (CH$_2$)$_2$COO$^-$; acetate group: CH$_2$COO$^-$.

tron transport chain (Fig. 3.12), such as cytochromes b and c, pass along single electrons rather than oxygen. The transfer of electrons along the chain is made possible by the change in the oxidation state of iron. In the reduced cytochromes, the iron atom is in the ferrous state. The iron atom of the reduced cytochrome becomes oxidized to the ferric state when a single electron is transferred to the next cytochrome. The iron atom of the cytochrome receiving the electron becomes reduced. Other iron-containing cytochromes include cytochrome b_5 (involved in lipid metabolism) and cytochrome P_{450} family (heme enzymes found in the endoplasmic reticulum membrane and involved in drug metabolism and steroid hormone synthesis).

Nonheme iron sulfur enzymes involved in electron transport include NADH dehydrogenase, succinate dehydrogenase, and ubiquinone-cytochrome c reductase. Whether iron is carrying oxygen or transporting electrons, its essentiality in energy transformation is without question.

Other Iron-Containing Enzymes

Other body enzymes involved in a variety of processes, besides the respiratory chain, also require iron. Many mono-oxygenases, for example, contain iron. Mono-oxygenases function to insert one of two oxygen molecules into a substrate. Examples of iron-containing mono-oxygenases include

- phenylalanine mono-oxygenase,
- tyrosine mono-oxygenase, and
- tryptophan mono-oxygenase.

These enzymes insert an oxygen molecule into phenylalanine, tyrosine, and tryptophan, respectively. Mono-oxygenases may be further classified based on the cosubstrate that participates in the reaction. The cosubstrate functions to furnish the hydrogen atoms that reduce the second oxygen molecule to water. Phenylalanine mono-oxygenase, tyrosine mono-oxygenase, and tryptophan mono-oxygenase all use tetrahydrobiopterin as a cosubstrate, and during the reactions, tetrahydrobiopterin is oxidized to dihydrobiopterin. The reaction catalyzed by phenylalanine mono-oxygenase (also called hydroxylase because the main substrate phenylalanine becomes hydroxylated) is shown in Figure 9.6 (p. 253). This enzyme contains one to two iron atoms and converts phenylalanine to tyrosine; vitamin C also is involved in this reaction.

Many dioxygenases also contain iron. Dioxygenases catalyze the insertion of two oxygen atoms into a substrate. There are many important iron-requiring dioxygenases in the body. Some examples include

- tryptophan dioxygenase (amino acid metabolism),
- homogentisate dioxygenase (amino acid metabolism),
- trimethyl lysine dioxygenase and γ-butyrobetaine dioxygenase (carnitine synthesis),
- lysine dioxygenase and proline dioxygenase (procollagen synthesis),
- nitric oxide synthase, and
- β-carotene dioxygenase (vitamin A synthesis).

Some of these reactions are shown to illustrate the nature of the reaction. For example, tryptophan dioxygenase (a heme-containing enzyme, also called a *pyrrolase*) converts the amino acid tryptophan to N-formylkynurenine (Fig. 9.18, p. 276), representing the first step of tryptophan metabolism. Iron deficiency has been shown to reduce the efficacy of tryptophan as a precursor of niacin [25]. Remember in humans, 60 mg of tryptophan can be converted into 1 mg of niacin. Homogentisate dioxygenase is also involved in amino acid metabolism, specifically that of tyrosine. During tyrosine metabolism, tyrosine is transaminated to produce hydroxyphenylpyruvate, which is then converted to homogentisate. Homogentisate is converted to 4-maleyl-acetoacetate by homogentisate dioxygenase, an iron-dependent enzyme (Fig. 9.6).

Two of the four steps required for carnitine synthesis involve iron-dependent dioxygenases. Remember carnitine is an important nitrogen-containing compound necessary for the transport of long-chain fatty acids into the mitochondria for oxidation. The first step in carnitine synthesis (Fig. 9.5) in which trimethyl lysine is converted into 3-OH trimethyl lysine requires an iron-containing trimethyl lysine dioxygenase, and the final step in which γ-butyrobetaine is converted into carnitine requires γ-butyrobetaine dioxygenase, an iron-containing enzyme. α-Ketoglutarate is a required cosubstrate in both of these reactions, as well as in the reactions catalyzed by dioxygenases involved in procollagen synthesis. Vitamin C is also necessary. Procollagen synthesis is shown in Figure 9.4 (p. 251). Two isoforms of nitric oxide synthase, a dioxygenase needed for the synthesis of nitric oxide, a potent biological effector molecule, require heme iron.

Other important reactions required to protect the body also use iron-containing enzymes, such as catalase and myeloperoxidase.

- Catalase, with four heme groups, converts hydrogen peroxide to water and molecular oxygen: $2H_2O_2 \longrightarrow 2H_2O + O_2$. Catalase helps to prevent cellular damage induced by hydrogen peroxide (see the Perspective in Chap. 10).

• Myeloperoxidase, another heme-containing enzyme, is found in the plasma as well as in granules within neutrophils (white blood cells). During phagocytosis of bacteria, myeloperoxidase is released into the phagocytic vesicle within the neutrophil. The phagocytic vesicle contains a variety of compounds including hydrogen peroxide (H_2O_2), free hydroxy radicals (OH·), and other ions such as chloride (Cl^-). Myeloperoxidase catalyzes the following reaction:

$$H_2O_2 + Cl^- \rightarrow H_2O + {}^-OCl.$$

The ^-OCl (hypochlorite) formed in the reaction is a strong cytotoxic oxidant that is important for the destruction of foreign substances such as bacteria. Activity of myeloperoxidase may be impaired with iron deficiency with a resulting increased susceptibility or severity of infection.

Some oxidoreductases that are iron-dependent include

• aldehyde oxidase, which uses oxygen to convert aldehydes (RCOH) to alcohols (RCOOH);

• sulfite oxidase, an iron sulfur-containing enzyme that converts sulfite (SO_3) to sulfate (SO_4); and

• xanthine oxidase and dehydrogenase, both nonheme iron- and molybdenum-containing enzymes that convert hypoxanthine generated from purine catabolism to xanthine and then xanthine to uric acid for excretion (p. 463). Remember, purine bases are found in DNA.

Another nonheme-iron-dependent enzyme involved in DNA synthesis and thus cell replication is ribonucleotide reductase, which converts adenosine diphosphate (ADP) into deoxy ADP (dADP). In glycolysis, glycerol phosphate dehydrogenase, a flavoprotein, has a nonheme iron component. In the Krebs cycle, aconitase, which converts citrate to isocitrate (p. 91), requires one to two nonheme iron atoms. Phosphoenolpyruvate carboxykinase, important in gluconeogenesis, also requires iron for function. Thyroperoxidase, another heme-iron-dependent enzyme, is necessary for organification of iodide (the addition of $2I^-$ to thyroglobulin-tyrosine) and conjugation of iodinated tyrosine residues on thyroglobulin. These reactions are necessary for the synthesis of the thyroid hormones T_3 and T_4.

As a proxidant, free ferrous iron may catalyze the nonenzymatic Fenton reaction

$$Fe^{2+} + H_2O_2 \rightarrow Fe^{3+} + OH^- + OH·$$

in which ferrous iron reacts with hydrogen peroxide to generate ferric iron and free radicals. In a reaction known as the Haber Weiss reaction, the superoxide radical O_2^- may react with another hydrogen peroxide molecule to generate molecular oxygen and free hydroxyl radicals such as OH^-, a dangerous membrane oxidant.

$$O_2^- + H_2O_2 \rightarrow O_2 + OH· + OH^-$$

Interactions with Other Nutrients

The interactions of iron with *ascorbic acid* in relation to enhancing iron absorption and maintaining iron in the appropriate valence state for enzyme function have been discussed previously (pp. 406 and 412). The potential also may exist for vitamin C–induced release of ferric iron from ferritin with subsequent reduction of iron to the ferrous form [21]. Whether such reactions result in Fenton reactions and occur in vivo are unclear.

As shown in Figure 12.3, an interrelationship also exists between iron and *copper* because of the role of the copper-containing ceruloplasmin as a ferroxidase. Back in the 1920s, studies revealed that iron therapy was unable to cure anemia in rats; however, ashed foodstuffs containing copper replenished blood hemoglobin concentrations [26]. Without the copper-dependent ferroxidase activity, iron cannot be mobilized from ferritin stores.

Another nutrient with which iron appears to interact is *zinc* [15]. Ingestion of both nutrients as a 25:1 molar ratio of nonheme iron (ferrous sulfate) to zinc diminished the absorption of zinc from water to 34% in humans; however, when the same ratio of iron to zinc was given with a meal, no inhibitory effects were demonstrated [27]. Ratios of nonheme iron to zinc of 2:1 and 3:1 also have been shown to inhibit zinc absorption, while similar ratios of heme iron to zinc had no effect on zinc absorption [28]. Thus, excessive intake of nonheme iron, as may occur with supplements, may have a detrimental effect on zinc absorption.

Another association is that between *vitamin A* and iron. Reduced vitamin A status alters iron distribution between tissues. Low plasma retinol concentrations are associated with decreased plasma iron and blood hemoglobin and hematocrit as well as increased hepatic iron accumulation in rats [29].

Iron and *lead* also interact. Lead inhibits the activity of Δ-aminolevulinic acid dehydratase, an enzyme required in heme synthesis. Lead also inhibits ferrochelatase activity, the enzyme that incorporates iron into heme. In addition, increased absorption of lead occurs with iron deficiency in animals and could be problematic for children who are often iron-deficient and may have increased exposure to lead [30]. The mechanism through which iron deficiency improves lead absorption is unknown.

Iron deficiency is associated with decreased *selenium* concentrations as well as glutathione peroxidase synthesis and activity [31–33]. Glutathione peroxidase, a selenium-requiring enzyme, catalyzes the reduction of hydrogen peroxide with the use of glutathione (GSH) as follows: 2 GSH + H_2O_2 $\rightarrow$ GSSG + 2 H_2O (also see p. 443). In addition, the enzyme converts organic peroxides (ROOH) to their hydroxy (or alcohol) form (2GSH + ROOH $\rightarrow$ GSSG + ROH + H_2O). The mechanism(s) behind the interaction between iron and selenium is not known. Iron is thought to be involved in the pretranslational regulation of the glutathione peroxidase synthesis. Alternately, iron deficiency may be affecting selenium absorption or increasing selenium use in the body. Another possibility is that iron or an iron-containing protein may be needed for glutathione peroxidase activity.

Turnover

Despite the importance of dietary iron in maintaining the long-term adequacy of body iron, the amount of iron absorbed, about 0.06% of the total body iron content, cannot provide the concentration of iron needed. It is the avid conservation and constant recycling of body iron that ensures adequacy of body iron. Figure 12.6 represents schematically the internal iron exchange in the body.

Most of the iron entering the plasma for distribution or redistribution by transferrin is contributed by sites of hemoglobin destruction and sites of ferritin and hemosiderin degradation. Hemoglobin is degraded primarily by phagocytes of the reticuloendothelial system (found in the liver, spleen, and bone marrow). Stored iron as ferritin and hemosiderin is degraded primarily in the liver, spleen, and bone marrow. Hemoglobin degradation will be briefly reviewed; ferritin degradation was previously discussed under the section on iron storage.

The majority of old (senescent) red blood cells, which live for about 120 days, are taken up by macrophages in the spleen and degraded (phagocytosed); however, reticuloendothelial cell macrophages in bone marrow and Kupffer cells in the liver also may degrade the red blood cells.

During red blood cell degradation, the heme portion of the hemoglobin molecule in the red blood cell is catabolized by heme oxygenase to biliverdin and subsequently to bilirubin, which is then secreted into the bile for excretion from the body. In addition, about 20 to 25 mg of iron per day are released from hemoglobin catabolism. This iron may be reused, for example, for erythropoiesis or for incorporation into other iron-dependent enzymes, or the iron may be deposited for storage.

Although the majority of red blood cells are degraded in the reticuloendothelial system, some red blood cell lysis occurs within the blood. Two proteins, haptoglobin and hemopexin, function to remove the released hemoglobin and any free heme, respectively, from the blood. Haptoglobin, synthesized by the liver, forms complexes with free hemoglobin, while hemopexin, also synthesized in the liver, forms a complex with free heme in the blood. The proteins then deliver

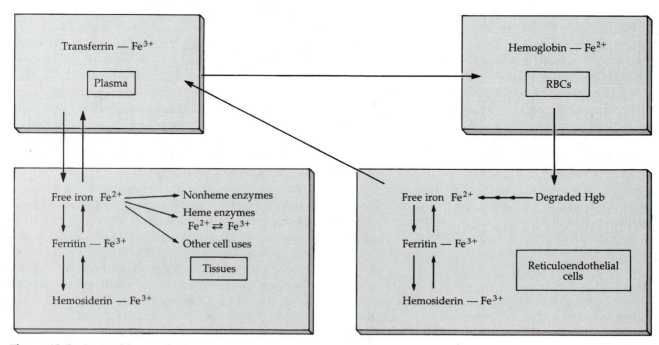

Figure 12.6 Internal iron exchange.

the iron-containing compounds to the liver, where further degradation occurs to enable reuse of the iron.

Unless body stores are exhausted, the supply of iron to the plasma pool can be adjusted within wide limits. The requirement for transferrin iron is determined by the needs of the bone marrow for red blood cell synthesis. Therefore, with chronic hemolysis the quantity of iron passing through the plasma can expand six to eight times normal. In contrast, when erythropoiesis declines dramatically, as occurs on descent from high altitudes, the quantity of iron in the plasma pool may decrease to as little as one-third of normal [34].

Excretion

Daily iron losses by an adult male are approximately between 0.9 and 1.0 mg/day (12–14 mg/kg/day). These losses occur from various sites:

Site	Amount (mg)
Gastrointestinal tract	0.6
Skin	0.2–0.3
Kidney	0.1

As can be seen from these numbers, most of the iron losses are via the gastrointestinal tract (0.6 mg). Of the 0.6 mg, about 0.45 mg are due to minute (~1 mL) blood loss (which occurs even in healthy people), and another 0.15 mg iron due to losses in bile and desquamated mucosal cells. The skin losses of approximately 0.2 to 0.3 mg iron is due to desquamation of surface cells from the skin. Lastly, a very small amount, about 0.1 mg, is lost in the urine. Losses of iron, however, may increase in people with gastrointestinal ulcers or intestinal parasites, or with hemorrhage induced by surgery or due to injury.

Basal iron losses as just described are a bit less (0.7–0.8 mg/day) in women because of their smaller surface area. Total losses of premenopausal women, however, are estimated at approximately 1.3 to 1.4 mg/day because of iron loss in menses. The average loss of blood during a menstrual cycle is about 35 mL, with an upper limit of about 80 mL. The iron content of blood is about 0.5 mg/100 mL of blood, which translates into a loss of nearly 17.5 mg of iron per period. When averaged out over a month, iron loss in menses is about 0.5 mg per day [17]; in some women, however, iron loss due to menses alone may exceed 1.4 mg/day [17]. The increased excretion of iron in healthy people with excessive intakes is due to the above-average concentration of iron in the ferritin of desquamated mucosal cells.

Balancing iron uptake with its loss from the body is very important to health. The high incidence of iron deficiency anemia, the most common nutrition deficiency in humans in the world [35], attests to the fact that iron equilibrium often has not been attained, particularly in many young children, girls, and women of childbearing age.

Recommended Dietary Allowances

Basal iron losses, which range from 0.7 to 1.0 mg/day by the adult male and postmenopausal female, as well as increased needs for selected populations with increased iron losses have been considered in formulating the RDA. The 1989 RDA assumes an iron absorption of 10% and has set a recommendation of 10 mg iron for males and postmenopausal females [17]. The RDA for iron for premenopausal women was set at 15 mg/day [17]. Because of the absence of menstruation during pregnancy and the increased or more efficient iron absorption that also occurs with pregnancy, the 1989 RDA suggests 30 mg of iron/day for pregnant women [17]. Because 30 mg of iron is more than can be normally obtained from the diet, supplements are usually needed.

Because no evidence of iron deficiency exists with iron stores of 300 mg and because absorption increases whenever iron stores are depleted, the 1980–1985 Committee on Recommended Dietary Allowances [16] set the RDI for iron at 15 mg/day, a level that would maintain stores of 300 mg.

Deficiency: Iron Deficiency with and without Anemia

Iron intake is frequently inadequate in four population groups [16,17]:

- infants and young children (6 months to 4 years) because of the low iron content of milk and other preferred food, rapid growth rate, and body reserves of iron insufficient to meet needs beyond 6 months;
- adolescents in their early growth spurt because of rapid growth and needs of expanding red cell mass;
- females during childbearing years because of menstrual iron losses; and
- pregnant women because of their expanding blood volume, demands of fetus and placenta, plus blood losses to be incurred in childbirth.

In addition, many nonpregnant females during childbearing years are falling short of the RDA for iron because their caloric intake is often restricted and because only 5 to 7 mg of iron can be expected per 1,000 kcal from an average Western diet. Other conditions and populations associated with increased need for intake due to iron losses or impaired absorption are hemorrhage, protein calorie malnutrition, renal disease, achlorhydria, prolonged use of alkaline-based drugs such as antacids, decreased gastrointestinal transit time, steatorrhea, and parasites.

Figure 12.7 depicts the gradual depletion of iron content in the body and demonstrates the fact that anemia does not occur until iron depletion is severe. Iron deficiency without anemia, however, can occur. Symptoms of iron deficiency, mostly demonstrated in children, include pallor, listlessness, behavioral disturbances, impaired performance in some cognitive tasks, some irreversible impairment of learning ability, and short attention span [35]. In adults, decreased work performance and productivity are most commonly impaired with iron deficiency [36,37]. Iron deficiency may impair degradation of γ-amino butyric acid (GABA, an inhibitory neurotransmitter in the brain) or may inhibit dopamine producing neurons [35]. Possible impairment of the immune system, decreased resistance to infection and impaired capacity to maintain body temperature have also been shown [35].

Further discussion of iron deficiency with and without anemia as it relates to changes that occur in indices of iron status are discussed under the section "Assessment of Nutriture."

Supplements

Supplements of ferrous iron are available in complexes with sulfate, succinate, citrate, lactate, tartarate, fumarate, and gluconate. These supplements provide nonheme iron, and thus absorption is enhanced when ingested with a source of vitamin C or other enhancing factors. Amino acid iron chelates, such as iron glycine, are also marketed; however, in humans, absorption of iron administered as a chelate has not been shown to be superior to iron given as ferrous sulfate or ferrous ascorbate [38–40]. Initial effects of oral iron supplements on red blood cell counts and hemoglobin concentrations take about 2 weeks. Iron therapy to build up body stores of iron may be needed for 6 months to 1 year.

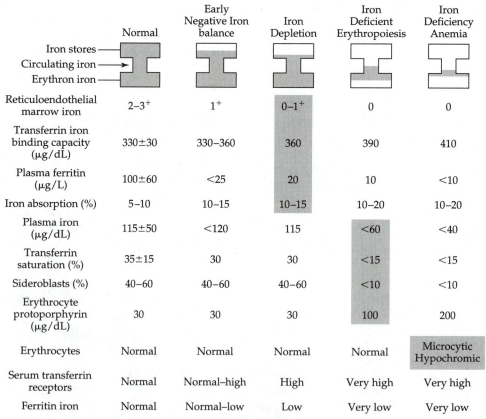

	Normal	Early Negative Iron balance	Iron Depletion	Iron Deficient Erythropoiesis	Iron Deficiency Anemia
Reticuloendothelial marrow iron	2–3⁺	1⁺	0–1⁺	0	0
Transferrin iron binding capacity (μg/dL)	330±30	330–360	360	390	410
Plasma ferritin (μg/L)	100±60	<25	20	10	<10
Iron absorption (%)	5–10	10–15	10–15	10–20	10–20
Plasma iron (μg/dL)	115±50	<120	115	<60	<40
Transferrin saturation (%)	35±15	30	30	<15	<15
Sideroblasts (%)	40–60	40–60	40–60	<10	<10
Erythrocyte protoporphyrin (μg/dL)	30	30	30	100	200
Erythrocytes	Normal	Normal	Normal	Normal	Microcytic Hypochromic
Serum transferrin receptors	Normal	Normal–high	High	Very high	Very high
Ferritin iron	Normal	Normal–low	Low	Very low	Very low

Figure 12.7 Sequential changes in iron status associated with iron depletion.
Source: Adapted from Victor Herbert, "Recommended Dietary Intakes (RDI) of Iron in Humans," Am J Clin Nutr 1987;45:679–86. © Am J Clin Nutr, American Society for Clinical Nutrition and Victor Herbert, J Nut 1996; 126:1213S–20S.

Toxicity: Hemochromatosis and Hemosiderosis

Accidental iron overload has been observed in young children following excessive ingestion of iron pills or vitamin and mineral pills. Other people susceptible to iron overload possess the genetically (autosomal recessive) transmitted idiopathic hemochromatotic trait, which typically leads to the development of hemochromotosis. In the United States it is estimated that about two or three people per 10,000 (or 0.3%–0.5% of Caucasians) are homozygous for the trait [41], whereas 8% to 15% of Caucasians are heterozygous [41]. The term *hemochromatosis* is used to indicate iron overload or toxicity, which is coupled with tissue damage. Hemochromatosis is most often seen in adult Caucasian males and begins to occur around 20 years of age. The condition is characterized by increased (at least two times normal) iron absorption with deposition of iron as hemosiderin in the parenchymal cells of the liver and other organs such as the heart, causing damage to the tissues. Heterozygotes for the condition do not develop abnormalities of the liver and heart but show abnormal iron status [42]. The term *hemosiderosis* is used to indicate iron overload without tissue damage. Before the development of a screening test [42], no method existed for identifying people with iron overload until pathological changes (primarily cirrhosis) occurred at around age 50 years [43]. Treatment of the condition usually involves frequent phlebotomy (removal of blood) and may require administration of deferroxamine, which can chelate iron and increase urinary iron excretion.

Other people who are at particularly high risk for iron overload are those with iron-loading anemias, thalassemia, and sideroblastic anemia. The elevated erythropoiesis in people so affected causes an increased absorption of iron.

High body iron (serum ferritin >200 μg/dL) also has been linked to heart disease. However, the association between measures of iron status and cardiovascular disease is inconsistent. Although positive associations have been observed in some studies, a larger group of studies have shown no such association.

Assessment of Nutriture

Numerous measurements are used for the assessment of iron nutriture. The most common indices are hemoglobin and hematocrit, which indicate the presence of anemia. Blood hemoglobin concentrations indicate the amount of hemoglobin (the iron containing protein found in red blood cells) per unit (usually deciliter or liter) of blood. Hematocrit represents that proportion of the total blood volume that is red blood cells; it is usually expressed as a percentage.

Also used is the characterization of red blood cells as to size and amount of hemoglobin in them. Of the three measurements—mean corpuscular volume (MCV), mean corpuscular hemoglobin (MCH), and mean corpuscular hemoglobin concentration (MCHC)—used to characterize red blood cells, MCV is the most popular and is used often in analysis of large survey data.

- MCV (fl) represents the size of the red blood cell. It may be calculated by dividing hematocrit by red blood cells and then multiplying by 10.

- MCH (pg/rbc) represents the average hemoglobin content of each individual red blood cell. It may be calculated by dividing hemoglobin by red blood cells and then multiplying by 10.

- MCHC represents the amount of hemoglobin in grams per deciliter (%) of red blood cell. It is calculated by dividing hemoglobin by hematocrit and then multiplying by 100.

These values are helpful in identifying severe iron deficiency anemia but may be much less useful in the diagnosis of mild iron deficiency anemia or iron deficiency without anemia. In these conditions, evaluation of iron stores is much more valuable, because iron deficiency occurs in three stages, beginning with depletion of iron stores.

In the first stages of iron deficiency, iron stores in the liver, spleen, and bone marrow are diminished. While iron stores may be aspirated and measured from bone marrow, the routine test involves measurement of plasma ferritin. Plasma ferritin concentrations decrease and are thought to parallel the decrease in the amount of iron found in stores. Plasma ferritin concentrations <12 μg/dL are associated with iron deficiency. If, however, inflammation or infection is present, serum ferritin concentration rises. This rise is, however, unrelated to iron stores. Thus, serum ferritin may appear within normal range or high, when the actual amount of iron in the protein is quite low. Measurement of the iron content of serum ferritin or percentage saturation of serum ferritin with iron may be a better indicator than serum ferritin itself.

Typically as iron deficiency progresses into the second stage, not only will plasma ferritin concentrations diminish, but the number of transferrin receptors on the cell surface increase, representing an up-regulation to enable cells to better compete for iron that is bound to transferrin [44–47]. Levels of serum transferrin receptors (sTfR), truncated forms of the membrane receptor protein, are thought to be directly proportional to the functional tissue iron deficit after iron stores are depleted [48,49]. Measurement of serum transferrin receptors is thought to reflect transferrin receptor numbers on immature red cells and thus bone marrow erythropoiesis

[50]. The combined use of serum ferritin and serum transferrin receptors has been proposed for assessing iron status [45,51]. Also in the early stages of iron deficiency, free protoporphyrin concentrations in the erythrocyte rise; protoporphyrin, a precursor of hemoglobin, accumulates within red blood cells when iron is not available. Erythrocyte protoporphyrin levels >70 μg/dL red blood cell are associated with iron deficiency. Also with iron deficiency, transferrin saturation decreases. Transferrin is normally one-third saturated with ferric iron; however, with iron deficiency, transferrin saturation diminishes to <15% or 16%. Transferrin saturation can be calculated by multiplying the serum iron concentration by 100 and then dividing by the total iron-binding capacity (TIBC). TIBC represents the amount of iron that plasma transferrin can bind; TIBC ranges from about 250 to 400 μg/dL, with levels >400 suggestive of deficiency. The amount of iron bound to the transferrin is measured as serum iron with normal values ranging from 50 to 165 μg/dL, and levels less than about 50 μg/dL indicate deficiency.

In the final stages of iron deficiency, anemia occurs as indicated by low blood hemoglobin concentrations. Blood hemoglobin concentrations of <12 g/dL and 14 g/dL for females and males, respectively, are suggestive of iron deficiency anemia. Hematocrit concentrations of <37% and 40% for women and men, respectively, also are typical of iron deficiency anemia. Serum ferritin and transferrin saturation remain depressed, free protoporphyrin remains elevated. Serum iron (which includes the iron on transferrin, but not in ferritin) is diminished, TIBC is elevated, and hemoglobin, hematocrit, MCH, MCHC, and MCV are lower than normal. Red blood cells are pale (hypochromic) and small (microcytosis). Figure 12.7 illustrates the changes that occur in these various measurements.

References Cited for Iron

1. Hurrell R, Lynch S, Trinidad T, Dassenko S, Cook J. Iron absorption in humans: bovine serum albumin compared with beef muscle and egg white. Am J Clin Nutr 1988;47:102–7.

2. Monsen, E. Iron nutrition and absorption: dietary factors which impact iron bioavailability. J Am Diet Assoc 1988;88:786–90.

3. Taylor P, Martinez-Torres C, Romano E, Layrisse M. The effect of cysteine-containing peptides released during meat digestion on iron absorption in humans. Am J Clin Nutr 1986;43:68–71.

4. MacPhail A, Ratel R, Bothwell T, Lamparelli R. EDTA and the absorption of iron from food. Am J Clin Nutr 1994;59:644–8.

5. Conrad M, Umbreit J. Iron absorption—the mucin-mobilferrin-integrin pathway. Am J Hematology 1993;42:67–73.

6. Monsen E, Balintfy J. Calculating dietary iron bioavailability: refinement and computerization. J Am Diet Assoc 1982;80:307–11.

7. Morck T, Lynch S, Cook J. Inhibition of food iron absorption by coffee. Am J Clin Nutr 1983;37:416–20.

8. Hallberg L, Rossander-Hulten L, Brune M, Gleerup A. Calcium and iron absorption: mechanism of action and nutritional importance. Eur J Clin Nutr 1992;46:317–27.

9. Cook J, Dassenko S, Whittaker P. Calcium supplementation: effect on iron absorption. Am J Clin Nutr 1991;53:106–11.

10. Hallberg L, Brune M, Erlandsson M, Sandberg A-S, Rossander-Hulten L. Calcium: effect of different amounts on nonheme- and heme-iron absorption in humans. Am J Clin Nutr 1991;53:112–9.

11. Snedeker S, Smith S, Greger J. Effect of dietary calcium and phosphorus levels on the utilization of iron, copper, and zinc by adult males. J Nutr 1982;112:136–43.

12. Solomons N. Competitive mineral-mineral interaction in the intestine. In: Inglett G. ed., Nutritional Bioavailability of Zinc. Am Chem Soc Symp 1983;201:247–71.

13. Rossander-Hulten L, Brune M, Sandstrom B, Lonnerdal B, Hallberg L. Competitive inhibition of iron absorption by manganese and zinc. Am J Clin Nutr 1991;54:152–6.

14. Crofton R, Gvozdanovic D, Gvozdanovic S, Khin C, Brunt P, Mowat N Agget P. Inorganic zinc and the intestinal absorption of ferrous iron. Am J Clin Nutr 1989;50:141–4.

15. Whittaker P. Iron and zinc interactions in humans. Am J Clin Nutr 1998;68:442S–6S.

16. Herbert V. Recommended dietary intakes (RDI) of iron in humans. Am J Clin Nutr 1987;45:679–86.

17. National Research Council. Recommended Dietary Allowances, 10th ed. Washington, D.C.: National Academy Press, 1989:195–205.

18. Hallberg L, Hulten L, Gramatkovski E. Iron absorption from the whole diet in men: how effective is the regulation of iron absorption? Am J Clin Nutr 1997;66:347–56.

19. Mietzner T, Morse S. The role of iron binding proteins in the survival of pathogenic bacteria. Ann Rev Nutr 1994;14:471–93.

20. Moore G, Kadir F, Al-Massad F. Haem binding to ferritin and possible mechanisms of physiological iron

uptake and release by ferritin. J Inorganic Biochem 1992;47:175–81.

21. Herbert V, Shaw S, Jayatilleke E. Vitamin C–driven free radical generation from iron. J Nutr 1996;126: 1213S–20S.

22. Finch C, Huebers H. Perspectives in iron metabolism. N Engl J Med 1982;306:1520–8.

23. Bolann B, Ulvik R. On the limited ability of superoxide to release iron from ferritin. Eur J Biochem 1990;193:899–904.

24. Pollack S. Receptor-mediated iron uptake and intracellular iron transport. Am J Hematol 1992;39: 113–8.

25. Uduho G, Han Y, Baker D. Iron deficiency reduces the efficacy of tryptophan as a niacin precursor. J Nutr 1994;124:444–50.

26. Waddell J, Steenbock H, Elvehjem C, Hart E. Iron salts and iron-containing ash extracts in the correction of anemia. J Biol Chem 1927;77:777–95.

27. Sandstrom B, Davidsson L, Cederblad A, Lonnerdal B. Oral iron, dietary ligands and zinc absorption. J Nutr 1985;115:411–4.

28. Solomons N, Jacob R. Studies on the bioavailability of zinc in humans: effects of heme and nonheme iron on the absorption of zinc. Am J Clin Nutr 1981;34: 475–82.

29. Houwelingen FV, Van Den Berg GJ, Lemmens AG, Sijtsma KW, Beynen AC. Iron and zinc status in rats with diet-induced marginal deficiency of vitamin A and/or copper. Biol Trace Elem Res 1993;38:83–95.

30. Goyer R. Nutrition and metal toxicity. Am J Clin Nutr 1995;61(suppl):646S–50S.

31. Moriarty P, Picciano M, Beard J, Reddy C. Classical selenium-dependent glutathione peroxidase expression is decreased secondary to iron deficiency in rats. J Nutr 1995;125:293–301.

32. Yetgin S, Huncal F, Basaran G, Ciliv G. Serum selenium status in children with iron deficiency anemia. Acta Hematol 1992;88:185–8.

33. Lee Y, Layman D, Bell R. Glutathione peroxidase activity in iron-deficient rats. J Nutr 1981;111:194–200.

34. Bothwell T, Charlton R, Motulsky A. Idiopathic hemochromatosis. In: Stanbury JB, Wyngaarden JB, Fredrickson DS, Goldstein JL, Brown MS, eds. The metabolic basis of inherited disease, 5th ed. New York: Mc-Graw-Hill 1983; 1269–98.

35. Scrimshaw NS. Iron deficiency. Scientific Am 1991;265:46–52.

36. Prasad A, Prasad C. Iron deficiency: non-hematological manifestations. Prog Food Nutr Sci 1991;15: 255–83.

37. Johnson M, Fischer J, Bowman B, Gunter E. Iron nutriture in elderly individuals. FASEB J 1994;8:609–21.

38. Fox T, Eagles J, Fairweather-Tait S. Bioavailability of iron glycine as a fortificant in infant foods. Am J Clin Nutr 1998;67:664–8.

39. Pineda O, Ashmead D, Perez J, Lemus C. Effectiveness of iron amino acid chelate on the treatment of iron deficiency anemia in adolescents. J Appl Nutr 1994;46:2–13.

40. Olivares M, Pizarro F, Pineda O, Name J, Hertrampf E, Walter T. Milk inhibits and ascorbic acid favors ferrous bis-glycine chelate bioavailability in humans. J Nutr 1997;127:1407–11.

41. Johnson M. Iron: Nutrition monitoring and nutrition status assessment. J Nutr 1990;120:1486–91.

42. Skikne B, Cook J. Screening test for iron overload. Am J Clin Nutr 1987;46:840–3.

43. Finch C. The detection of iron overload. N Engl J Med 1982; 307:1702–3.

44. Gimferrer E, Ubeda J, Royo M, Marigo G, Marco N, Fernandez N. Serum transferrin receptor levels in different stages of iron deficiency. Blood 1997;90: 1332–40.

45. Cook J, Skikne B, Baynes R. Serum transferrin receptor. Ann Rev Med 1993;44:63–74.

46. Aisen P. Transferrin, the transferrin receptor, and the uptake of iron by cells. In: Metals Ions in Biological Systems. 1998;35:585–631.

47. Baynes R, Skikne B, Cook J. Circulating transferrin receptors and assessment of iron status. J Nutr Biochem 1994;5:322–30.

48. Rusia U, Flowers C, Madan N, Agarwal N, Sood S, Sikka M. Serum transferrin receptor levels in the evaluation of iron deficiency in the neonate. Acta Pediatr Japonica 1995;38:455–9.

49. Zhu Y, Haas J. Response of serum transferrin receptor to iron supplementation in iron- depleted, nonanemic women. Am J Clin Nutr 1998;67:271–5.

50. Worwood M. The laboratory assessment of iron status—an update. Clin Chim Acta 1997;259:3–23.

51. Cooper M, Zlotkin S. Day-to-day variation of transferrin receptor and ferritin in healthy men and women. Am J Clin Nutr 1996;64:738–42.

Zinc

The human body contains approximately 1.5 to 2.5 g of zinc. Zinc is found in all organs and tissues (primarily intracellularly) and in body fluids. Most of the zinc in humans is found in bone, liver, kidney, muscle, and skin.

Zinc, a metal, can exist in several different valence states, but it is almost universally found as the divalent ion (Zn^{2+}).

Sources

Zinc is typically associated with the protein fraction and/or nucleic acid fraction of foods. Specifically, zinc is found in food complexed with amino acids that are part of peptides and proteins, and with nucleic acids. The zinc content of foods varies widely (Table 12.4). Very good sources of zinc are red meats (especially organ meats) and seafood (especially oysters and mollusks). Animal products are thought to provide between 40% and 70% of zinc consumed by most people in the United States. Other good animal sources of zinc include poultry, pork, and dairy products. Whole grains (especially bran and germ) and vegetables (leafy and root) represent good plant sources of zinc. Poor zinc sources are fruits and refined cereals. Zinc from plant sources is not only lower in content but also absorbed to a lesser extent than from meat [1].

Processing of certain foods may affect the zinc available for absorption. Heat treatment can cause food zinc to form complexes that are resistant to hydrolysis and therefore make zinc unavailable for absorption. Maillard reaction products—that is, amino acid-carbohydrate complexes resulting from browning, for example—are particularly notable for inhibiting zinc's availability for absorption.

In addition to dietary food sources, endogenous sources for zinc also are available. Specifically, zinc is found in pancreatic and biliary secretions that are released into the gastrointestinal tract. Carboyxpeptidase, for example, is a zinc metalloenzyme. Following carboxypeptidase activity, the enzyme itself is hydrolyzed and zinc is released. This zinc is then available for absorption and reuse in the body.

Digestion, Absorption, Transport, and Storage

Digestion

Zinc, like iron, needs to be hydrolyzed from most amino acids and nucleic acids prior to absorption. Zinc is believed to be liberated from food during the digestive process, most likely by proteases and nucleases in the stomach and small intestine. Hydrochloric acid also appears to play an important role in zinc digestion and/or absorption. Increased gastric pH, as is found following the ingestion of antacids, H_2 receptor blocker medications such as Zantac®, Tagamet®, or Pepcid®, or proton pump blockers such as Omeprazole®, results in decreased zinc absorption [2]. The role of gastric acid in zinc digestion and/or absorption has not been elucidated but may relate to impaired hydrolysis of zinc from nucleic or amino acids, changes in zinc's ionic state, or alterations in the enterocyte membrane to affect permeability and thus zinc absorption.

Absorption

The main site of zinc absorption in the gastrointestinal tract is the proximal small intestine, most likely the jejunum. However, the relative contribution of each segment of the small intestine (duodenum, jejunum, and ileum) toward overall zinc absorption has not been demonstrated.

Zinc is absorbed into the enterocyte by a carrier-mediated process, with low intakes absorbed more efficiently than higher zinc intakes. The need of energy (ATP) in zinc's absorptive process is unclear. Passive diffusion (nonsaturable) and/or paracellular zinc absorption are thought to occur with high zinc intake (Fig. 12.8).

Studies have shown that zinc absorption varies from approximately 12% to 59%. Recent studies suggest about 33% absorption from a nonvegetarian diet providing about 11 mg of zinc and 26% absorption from a lactovegetarian diet providing about 9 mg of zinc [1]; a value of 20% was chosen in setting the 1989 RDA.

Table 12.4 Zinc Content of Selected Foods

Foods / Food Groups	Zinc (mg/100 g)
Seafood	
Oysters	17–91
Crabmeat	3.8–4.3
Shrimp	1.1
Tuna	0.5–0.8
Meat and poultry	
Liver	3.1–3.9
Chicken	1.0–2.0
Beef, ground	3.9–4.1
Veal	3.1–3.2
Pork	1.6–2.1
Eggs and dairy products	
Eggs	1.1
Milk	0.4
Cheeses	2.8–3.2
Legumes (cooked)	0.6–1.0
Grains and cereals	
Rice and pasta (cooked)	0.3–0.6
Bread (wheat)	1.0
Bread (white)	0.6–0.8
Vegetables	0.1–0.7
Fruits	< 0.1

Source: www.nal.usda.gov/fnic/foodcomp

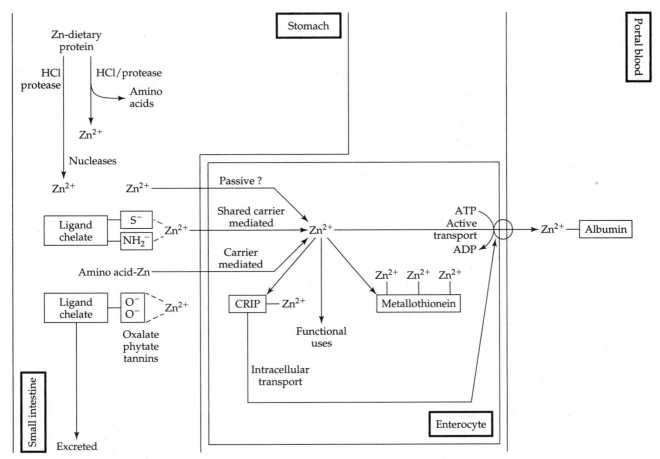

Figure 12.8 Digestion, absorption, enterocyte use, and transport of zinc. Abbreviation: CRIP, cysteine-rich intestinal proteins.

The form in which zinc is absorbed into the enterocyte is uncertain. Zinc may move across the brush border as an ion or as part of a chelated complex. Most evidence supports absorption of zinc as a complex with ligands that are derived from both exogenous and endogenous sources.

Factors Influencing Zinc Absorption

Similar to iron, chelator or ligands may bind to zinc. Whether these substances are enhancers or inhibitors depends on the digestibility and absorbability of the zinc chelates formed.

Enhancers of Zinc Absorption Several endogenous substances are thought to serve as ligands with zinc. Possible endogenous ligands include citric acid and picolinic acid, the latter of which is a metabolite of the tryptophan to niacin pathway (p. 191) as well as prostaglandins. Amino acid ligands include histidine, cysteine, and possibly other amino acids (lysine and glycine). In addition, glutathione (a tripeptide composed of cysteine, glutamate, and glycine) or products of protein digestion may serve as ligands. Each of the aforementioned substances have been shown to enhance zinc absorption especially in the presence of inhibitors. Specifically in these ligands, zinc binds to sulfur (e.g., cysteine or glutathione), nitrogen (e.g., histidine) or oxygen (e.g., phytate or oxalate, discussed shortly). Pancreatic secretions are also thought to contain an unidentified constituent that enhances zinc absorption.

Absorption of zinc also appears to be enhanced by low zinc status. Specifically, absorption of zinc by carrier-mediated mechanisms is enhanced with low zinc status, suggesting that the total amount of zinc absorbed is homeostatically regulated. However, how zinc status regulates absorption of the mineral is unclear.

Inhibitors of Zinc Absorption Many compounds in food may complex with zinc and inhibit its absorption. Examples of inhibitors include

- phytate,
- oxalate,

- polyphenols,
- fibers, and
- nutrients including vitamins and divalent cations.

Some of these inhibitors will be discussed. Phytate, also called inositol hexaphosphate or polyphosphate, is found in plant foods, particularly cereals such as maize and bran, and legumes. Fermentation of bread, however, can reduce the phytate content and improve zinc absorption. Phytate in the presence of high intraluminal calcium has an even greater inhibitory effect on zinc absorption than phytate alone. For example, a phytate:zinc ratio of greater than 10:1 adversely affects zinc absorption, whereas in the presence of a high intraluminal calcium, inhibition of zinc absorption occurs at a phytate:zinc ratio much lower than 10 [1,3,4]. Large variation in zinc bioavailability, primarily related to pH, has been demonstrated in soy isolates and concentrates, which are also high in phytate [4]. Decreased zinc absorption from products with a neutral pH is thought to be due to formation of poorly digested protein-phytate-mineral complexes [4]. Figure 12.9 depicts the binding of zinc by phytate.

Oxalate or oxalic acid, another inhibitor of zinc absorption, is found in a variety of foods, most notably spinach, chard, berries, chocolate, and tea, among others. The binding of zinc by oxalate is shown in Figure 12.9. Polyphenols such as tannins in tea and certain fibers found in whole grains, fruits, and vegetables also bind zinc and inhibits its absorption.

Interactions between zinc and nutrients such as the vitamin *folic acid* and a variety of divalent cations (Fe^{2+}, Ca^{2+}, Cu^{2+}) may occur and inhibit zinc absorption. Folic acid (350 μg) ingestion for 2 weeks has been shown to decrease zinc (50 mg) absorption by 21% [5]. However,

another study that provided 800 μg folic acid and either 3.5 mg or 14.5 mg of zinc for two 25-day periods found no adverse effects on zinc status [6].

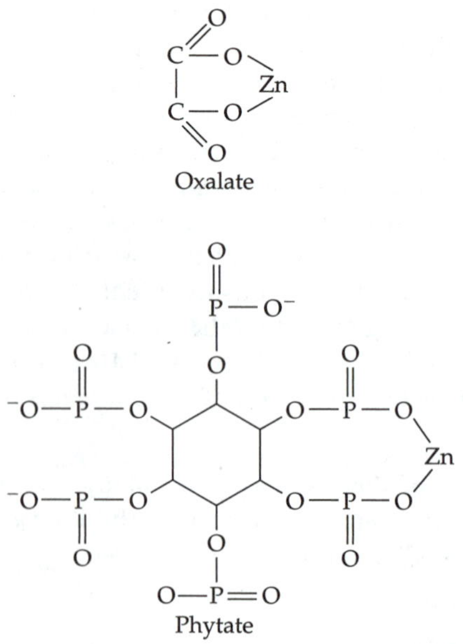

The interaction between zinc and other divalent cations is probably related to the fact that these cations compete with one another for binding ligands in the intestinal lumen or within the cell, and for receptor sites in the enterocytes. *Iron* and zinc interact primarily when ingested in solution to negatively impact zinc absorption. The effects are not always apparent when given with a meal [7]. Ferrous sulfate and zinc sulfate ingested together in a ratio of 2:1 (50 mg:25 mg) and 3:1 (75 mg:25 mg) decreased zinc absorption in humans [8]. Zinc sulfate ingested with heme chloride did not inhibit zinc absorption [8]. Likewise, ingestion of oysters (providing 54 mg of zinc) with 100 mg ferrous iron did not inhibit zinc absorption [8]. Iron (100 mg) and folate (350 μg) supplements decreased zinc (25 mg) absorption by 51% in pregnant women [5]. These studies suggest that to maximize zinc absorption from zinc sulfate, the zinc supplement should not be consumed with iron sulfate supplements.

The effects of *calcium* on zinc absorption and balance appear to be variable. Whereas some studies in humans have shown that ingestion of 500 mg to about 2 g of calcium [9–11] as calcium carbonate, hydroxyapatite, or calcium citrate malate has no effect on zinc absorption, other studies report evidence to the contrary. Ingestion of calcium (890 mg/day) from a meal or from milk or a calcium phosphate supplement providing calcium (468 mg) reduced net zinc absorption and zinc balance in postmenopausal women [11]. Furthermore, ingestion of 600 mg of calcium as calcium carbonate with a meal providing 7.3 mg of zinc decreased zinc absorption by 50% [11].

Lastly, although *copper* has the potential for interfering with zinc absorption, this has not been reported. In fact, the contrary appears to occur; that is, zinc supplements inhibit copper absorption and can lead to copper deficiency (see the section on interactions with other nutrients, p. 426).

Intestinal Cell Zinc Transport and Use

Movement of zinc through the enterocyte is not a well-delineated process. An intestinal cell protein that is rich in cysteine is thought to transport the zinc within the enterocyte. As indicated in Figure 12.8, zinc entering the enterocyte contributes to the cellular zinc pool; consequently, its ultimate fate has several possibilities. The zinc may be

- used or stored within the enterocyte, or
- bound to the proteins such as cysteine-rich intestinal proteins and metallothionein, and trans-

Figure 12.9 The binding of zinc by phytate and oxalate.

ported through the cell and across the basolateral membrane into the plasma.

The use of zinc within the enterocyte is similar to its use in other body cells and is discussed further under the section on functions. Cysteine-rich intestinal proteins (CRIPs) and metallothionein in the enterocyte serve as intracellular binding ligands for zinc as well as for other minerals, primarily copper in the case of metallothionein. Zinc initially appears to accumulate on CRIP; however, with increased zinc concentrations, metallothionein concentrations rise. Thionein protein has an unusually high content of cysteine (30% cysteine residues), which functions in metal binding. The zinc captured and held as metallothionein in the enterocytes is typically lost into the lumen with the sloughing of these cells. Reduced zinc absorption is directly correlated with metallothionein formation from orally administered zinc and newly synthesized thionein polypeptides [13]. Gene expression of thionein is induced by diets high in zinc [14]. Metallothionein is discussed in further detail under the storage section. Zinc not bound to metallothionein or used within the enterocyte is transported across the basolateral membrane of the enterocyte. Several zinc transporters (ZnTs) have been identified. ZnT-1 is found in many tissues throughout the body, including the enterocyte basolateral membrane. ZnT-1 preferentially transports zinc out of the duodenal and jejunal intestinal cells and does not require sodium or ATP [15]. Other zinc transporters, such as ZnT-2, ZnT-3, and ZnT-4, also have been isolated in the kidney, testes, mammary glands, and brain, for example.

Zinc Transport

Zinc passing into portal blood from the intestinal cell is mainly transported loosely bound to albumin (Fig. 12.8); from the enterocyte, most zinc is taken to the liver where the mineral is initially concentrated. Upon departure from the liver, albumin as well as other compounds such as α-2 macroglobulin, transferrin, and immunoglobulin (Ig) G bind and transport zinc in the blood. Albumin is thought to transport up to about 60% of zinc. α-2 macroglobulin, transferrin, and immunoglobulin (Ig) G are thought to transport about 15 % to 40% of the zinc in the blood (Fig. 12.10). Two amino acids, histidine and cysteine, also loosely bind 2% to 8% of the zinc for transport; these amino acids form a ternary (histidine-zinc-cysteine) complex in the blood.

Zinc is complexed and released from its plasma carriers on a continual basis. Some of this released zinc can pass back into the enterocyte and perhaps even through the cell into the intestinal lumen. The albumin-bound zinc appears to be that which is most readily taken up by tissues [14], and it is this fraction of plasma zinc that is thought to be important in the regulation of zinc absorption [14].

Zinc Uptake by Tissues

The mechanism of zinc uptake by tissues is unknown. Multiple passive transport systems including amino acid carrier systems have been proposed. Carriers of zinc are not thought to be highly selective. An increase in amino acid use by tissues leads to an increased uptake of zinc and vice versa. Given that numerous metalloenzymes within the cell require zinc as a component, it is reasonable to assume that enzyme synthesis and zinc uptake are correlated.

Storage

Zinc is found in all body organs, most notably the liver, kidney, muscle, skin, and bones. However, the zinc content of most soft tissues including muscle, brain, lung, and heart is relatively stable. This soft-tissue zinc does not respond or equilibrate with other zinc pools to release zinc should dietary zinc intake be low. Furthermore, while zinc is found in bones as part of apatite, bones release the mineral very slowly and cannot be depended on for a supply of zinc during dietary deprivation. Instead, when dietary zinc intake is insufficient to meet the body's needs for zinc, plasma zinc–containing enzymes and metallothionein must provide it. Catabolism of selected "less essential" zinc-containing metalloproteins (enzymes) occurs so that zinc can be redistributed to meet particularly crucial needs for the mineral. Liver metallothionein zinc also may be mobilized to provide zinc to other body tissues.

Metallothionein is found in most tissues of the body, including the liver, pancreas, kidney, intestine, and red blood cell. Two forms of the protein appear to exist, designated metallothionein (MT) -1 and MT-2. Remember, as discussed in the enterocyte, metallothionein contains a high proportion of cysteine residues (20 of 61 amino acids), each of which binds metals. In addition to binding zinc (7 gram atoms/mole), metallothionein binds minerals such as copper, cadmium, and mercury.

Liver and red blood cell metallothionein-bound zinc diminishes as dietary zinc intake decreases and is thus thought to reflect zinc status or stores. Zinc, and possibly other minerals, appear to affect the gene expression of metallothionein. Specifically, metal regulatory elements (MRE) made of specific nucleotide sequences are found in the promoter region of the metallothionein gene. Zinc may interact with the MRE alone or through a transcription factor [16] to induce metallothionein synthesis. Metallothionein gene expression is also influenced by glucagon and the monokine interleukin 1. Interleukin 1, synthesized and secreted by monocytes and

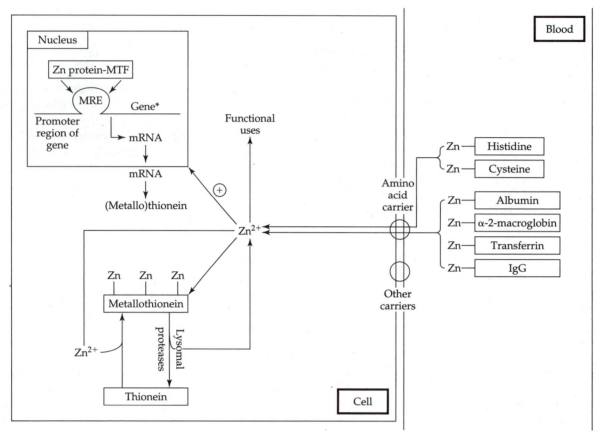

Figure 12.10 Storage and use of zinc in cells. Abbreviation: MRE, metal regulatory elements. *Zn-proteins interact with gene.

activated macrophages, is thought to induce metallothionein gene transcription during infection [17]. During infection, plasma zinc concentrations typically decrease and hepatic zinc concentrations rise; the sequestering of zinc by liver metallothionein in turn prevents bacterial use of zinc for its own replication.

Release of zinc from metallothionein involves lysosomal proteases (Fig. 12.10). These proteases at an acid pH degrade metallothionein to release the apoprotein thionein and zinc, which is then available for use by cells or other tissues.

Functions and Mechanisms of Actions

Zinc has many seemingly divergent functions, probably because of the numerous metalloenzymes of which it is a component. As a component of metalloenzymes, zinc provides structural integrity to the enzyme by binding to amino acid residues and/or participates directly in the reaction at the catalytic site. Zinc appears to be part of more enzyme systems than the rest of the trace minerals combined. Zinc affects many fundamen-

Table 12.5 Selected Functions of Zinc

Metalloenzyme roles
  Oxidoreductase
  Hydrolase
  Lyase
  Isomerase
  Transferase
  Ligase
Gene expression
Cell replication
Membrane and cytoskeletal stabilization
Structural role in hormones

tal life processes. Enzymes (at least 70 and perhaps over 200) from every enzyme class (Table 12.5) have been shown to require zinc. A few of these zinc-dependent enzymes are discussed next.

Zinc-Dependent Enzymes

• *Carbonic anhydrase*, found primarily in the erythrocytes but also the renal tubule, is essential for respiration; it catalyzes the following reaction, thereby allowing rapid disposal of carbon dioxide.

$$CO_2 + H_2O \xrightarrow{\text{carbonic anhydrase-Zn}^{2+}} H_2CO_3 \xrightarrow{\text{dissociation}} H^+ + HCO_3^-$$

The H^+ dissociated from carbonic acid reduces oxyhemoglobin as oxygen is released to the tissues; the bicarbonate passes into the plasma to participate in buffering reactions. The amount of zinc associated with carbonic anhydrase and carried by the erythrocytes is approximately eight to nine times as much as that distributed to tissues in plasma [14]. Carbonic anhydrase has a very high affinity for the mineral zinc; even with zinc deprivation, catabolism of this enzyme apparently does not occur.

• *Alkaline phosphatase* contains four zinc ions per enzyme molecule. Two of the four are required for enzyme activity. The other two are needed for structural purposes. The enzyme lacks substrate specificity, but hydrolyzes monoesters of phosphates from various compounds. Enzyme activity decreases with zinc deficiency [14].

• *Alcohol dehydrogenase* also contains four zinc ions per enzyme molecule, with two of the four being required for catalytic activity and two required for structure (protein conformation). This enzyme is important in the conversion of alcohols to aldehydes such as retinol to retinal, which is needed for the visual cycle and night vision. NADH also participates in the reaction.

• *Carboxypeptidase A* (Fig. 12.11) is another example of a zinc-dependent enzyme. Carboxypeptidase A, an exopeptidase secreted by the pancreas into the duodenum, is necessary for the digestion of protein. Zinc is bound tightly to carboxypeptidase A and is essential for enzymatic activity. Carboxypeptidase A activity decreases with zinc deficiency.

• *Aminopeptidase* is also involved in protein digestion. Aminopeptidases have been shown to contain one zinc atom needed for catalytic activity.

• *Δ-aminolevulinic acid dehydratase,* involved in heme synthesis, is zinc-dependent. This thiol (SH)-containing enzyme is made up of eight subunits, each of which binds one zinc atom [18]. Zinc is essential for the maintenance of free thiols in the enzyme. The enzyme catalyzes the condensation of Δ-aminolevulinic acid to form porphobilinogen (Fig. 12.5, p. 411).

• *Superoxide dismutase* (SOD) found in the cell cytoplasm requires two atoms each of both zinc and copper for function; zinc appears to have a structural role in the enzyme. This important enzyme catalyzes the removal of superoxide radicals, O_2^-:

$$2O_2^- + 2H^+ \xrightarrow{\text{zinc copper SOD}} H_2O_2 + O_2$$

Further information on SOD is found under the section on copper, page 435.

• *Phospholipase C* requires zinc for catalytic activity. This enzyme hydrolyzes the phosphodiester bond in phospholipids to release PO_3.

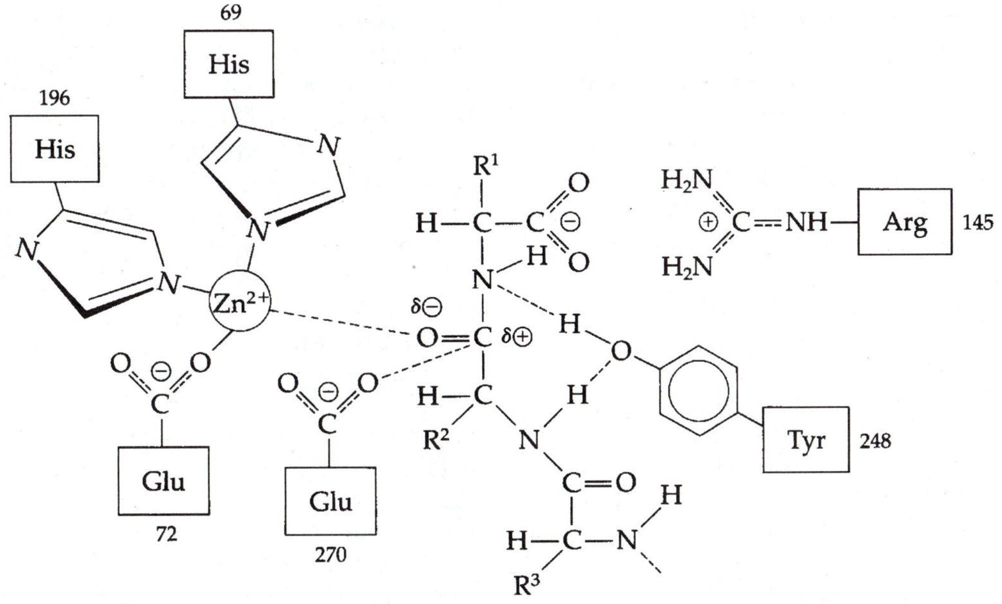

Figure 12.11 Partial structure of carboxypeptidase A.

• *Polymerases, kinases, nucleases, transferases, phosphorylases, and transcriptases* all require zinc. Paramount in nucleic acid synthesis are the zinc metalloenzymes DNA and RNA polymerase and deoxythymidine kinase, the latter being necessary for the conservation or salvaging of thymine, the pyrimidine unique to DNA. Catabolism of RNA appears to be regulated by zinc because of the influence of zinc on ribonuclease activity. Enzymes such as deoxynucleotidyl transferase, nucleoside phosphorylase, and reverse transcriptase also depend on zinc.

Other Roles

Physiological functions of zinc include tissue or cell growth, cell replication, bone formation, skin integrity, cell-mediated immunity, and generalized host defense. The role of zinc in tissue growth is related primarily to its function in the regulation of protein synthesis, which includes its influence on polysome conformation as well as the synthesis and catabolism of the nucleic acids.

With respect to transcription, zinc appears to interact with nuclear proteins that bind to promoter sequences of specific genes (Fig. 12.10). Thus, zinc helps to regulate transcription. Specifically, zinc serves as a necessary structural component of DNA-binding proteins (also called *transcription factors*) that contain zinc fingers. *Zinc fingers* is a term used to indicate the shape (configuration) of the proteins, which look like fingers, and the presence of the mineral zinc bound to the protein. The fingerlike configuration results from the twisting and coiling of the cysteine and histidine residues to which zinc binds in that segment of the protein. DNA-binding proteins that contain zinc fingers also bind other substances such as retinoic acid, thyroxine, 1,25-(OH)$_2$ vitamin D, and other steroid hormones, such as estrogen and androgens. Thus, hormones such as retinoic acid or 1,25-(OH)$_2$ vitamin D would enter the cell nucleus (see Chap. 10's section on vitamin A) and bind to specific protein-containing zinc fingers. In the presence of zinc, which is required for the binding of the protein to the DNA, the protein (with the hormone attached to it) binds to the DNA to affect gene expression.

The effect of zinc on cell membranes may be through direct effects on membrane proteins' conformation and/or on protein-to-protein interactions [19]. Zinc may affect the activity of several enzymes attached to plasma membranes. Some of these enzymes include alkaline phosphatase, carbonic anhydrase, and superoxide dismutase, among others [19]. These particular enzymes that control the structures and functions of the membranes to which they are attached are in turn controlled by zinc [13]. Zinc itself is believed also to stabilize membrane structure by stabilizing phospholipids and thiol (SH) groups that need to be maintained in a reduced state [18] and to guard the membrane against peroxidative damage by occupying sites on the membrane that might be instead occupied by pro-oxidant metals such as iron or by quenching free radicals through association with metallothionein [20]. Zinc may also stabilize membranes by promoting associations between membrane skeletal and cytoskeletal proteins [19]. Zinc is found in cells bound to tubulin, a protein that makes up the microtubules. Microtubules are thought to act as a framework for structural support of the cell as well as being needed for movement.

Zinc influences carbohydrate metabolism. Specifically, zinc is incorporated into insulin. In fact, zinc deficiency decreases insulin response resulting in impaired glucose tolerance. Zinc also appears to influence the basal metabolic rate (BMR); a decrease in thyroid hormones and BMR has been observed in subjects receiving a zinc-restricted diet [21]. Zinc is important for taste; it is a component of gustin, a protein involved in taste acuity.

Although many of the functions of zinc are known, many roles of zinc are not known. The effects of zinc deficiency on the body fail to explain fully the manifestations of zinc deprivation. It is "the small fraction of total body zinc that exchanges relatively rapidly with plasma zinc that is responsible for many of the known physiological functions of zinc"[22].

Interactions with Other Nutrients

A discussion of some of the interactions that result in the inhibition of the absorption of zinc have been previously addressed under the section on absorption. Other interactions between zinc and nutrients that do not affect zinc absorption will be briefly discussed.

Zinc and *vitamin A* interact in several ways. From the discussion on zinc functions, you may remember that zinc is required for alcohol dehydrogenase. Retinol (vitamin A) serves as a substrate for this enzyme, which converts retinol to retinal (retinaldehyde). In addition, zinc is necessary for the hepatic synthesis of retinol-binding protein, which transports vitamin A in the blood [23]. Zinc deficiency is associated with decreased mobilization of retinol from the liver as well as decreased concentrations of several transport proteins found in the blood, including albumin, transferrin, and prealbumin [23,24].

The detrimental effect of excessive zinc intake on *copper* absorption is thought to result from zinc's stimulation of thionein synthesis. Zinc intake promotes thionein synthesis; however, thionein polypeptides have a higher affinity for copper than for zinc. Thus, when zinc intake causes an increased synthesis of thionein, dietary copper becomes easily trapped as part of metallothionein within the enterocyte. The formation of copper metallothionein traps the copper in the enterocyte, preventing its passage into the plasma. Zinc intakes of 18.5 mg daily for 2 weeks have been shown to impair copper retention in men [25]. Intakes of 25 mg of zinc for 6 weeks decreased superoxide dismutase activity, an indication of impaired copper status [26]. The danger of copper deficiency precipitated by zinc supplementation has led to the recommendation that the maximum therapeutic dose of elemental zinc be limited to 40 mg daily [27].

Diminished *calcium* absorption has occurred with ingestion of zinc supplements when calcium intake is low (230 mg calcium) [9]. However, absorption appears unaffected when calcium intakes are at adequate (recommended) levels [27].

Cadmium appears to bind to sites that zinc would normally bind to and disrupts normal zinc functions. Cadmium, for example, can replace zinc in zinc fingers, causing the fingers to no longer function as they would with zinc present.

Folate digestion in the gastrointestinal tract requires a zinc-dependent hydrolase. This enzyme removes glutamate residues bound to folate to facilitate folate absorption.

Lead may replace zinc in the enzyme δ-aminolevulinic acid, necessary for heme synthesis [28].

Excretion

The three primary routes of zinc loss from the body are via the

- gastrointestinal tract,
- kidney, and
- body surface.

The majority of zinc is lost from the body through the gastrointestinal tract in the feces. Endogenous zinc in the form of enzyme metalloproteins are secreted by the salivary glands, intestinal mucosa, pancreas, and liver into the gastrointestinal tract. Although some of this zinc is reabsorbed, some also is excreted in the feces. Zinc is also contributed to the gastrointestinal lumen by sloughed intestinal cells and possibly by enterocytes that may permit a bidirectional flow of the mineral.

The majority of zinc filtered by the kidneys is reabsorbed by the tubules. Thus, only a small amount of zinc is excreted in the urine, about 0.3 to 0.7 mg/day. The zinc appearing in the urine is believed to be derived from the small percentage of plasma zinc that is complexed with histidine and cysteine.

Surface zinc losses of about 0.7 to 1 mg/day occur with exfoliation of skin and with sweating. Another minor route of zinc loss is via hair, which contains about 0.1 to 0.2 mg of zinc/g hair. Total zinc losses approximate 2.1 to 2.9 mg/day [29]. Obligatory zinc losses decrease with decreased zinc intake [29].

Recommended Dietary Allowances

It wasn't until the 1974 edition of the Recommended Dietary Allowances (RDA) that an allowance for zinc appeared for the first time. The subcommittee on the 1989, 10th edition of the RDA- [29] based zinc recommendations on the intake needed to maintain balance as well as on estimates of zinc absorption and body losses of zinc. Balance studies have indicated that at least 12 mg of zinc/day is necessary for achievement of equilibrium in healthy young men. Because zinc losses from healthy young men have been estimated at between 2.2 and 2.8 mg daily, approximately 2.5 mg of absorbed zinc is needed for maintenance of equilibrium. To allow for less than optimal zinc absorption, an absorption efficiency of only 20% was assumed. The resulting dietary zinc requirement (2.5 mg losses/20% absorption) amounts to 12.5 mg daily. The 1989 RDA for zinc is set at 15 mg/day for young men. An allowance of 12 mg/day is considered adequate for adult women because of their smaller body size [29].

During pregnancy the RDA for zinc is 15 mg/day to cover the calculated need for growth of the fetus and placenta because there is no indication for an increased absorption efficiency in pregnant women [29]. Zinc requirements for lactating women vary according to the zinc content of human milk and the amount of milk produced during the first and second half-year. Because of the higher zinc content in the milk as well as the greater milk production during the first 6 months of lactation, the allowance during this period is set at 19 mg/day [29]. For the last half of the year, the zinc allowance drops to 16 mg/day [29].

Deficiency

Few tissues contain large stores of zinc. Thus, should intake be inadequate and increased absorption fail to provide enough zinc, body reserves of zinc for release and redistribution are used but quickly exhausted. Consequently, a deficient intake of zinc results in the catabolism of some of the zinc metalloenzymes. Normally zinc is removed from those metalloenzymes that hold it less securely. The activities of three metalloenzymes, alkaline phosphatase in bone, carboxypeptidase A in pancreas, and deoxythymidine kinase in subcutaneous connective tissue, decrease with zinc deprivation. The zinc so released is redistributed throughout the body to satisfy the more crucial cellular needs. Certain metalloenzymes, however, such as carbonic anhydrase, which have very high affinity for zinc, may be little influenced by a general zinc deprivation. The extent to which the metalloenzyme will lose zinc during a deficiency depends on the geometry of the nitrogen, oxygen, and/or sulfur atoms that comprise the binding site [14].

Estimates of the average amount of zinc consumed by adults from a mixed U.S. diet range from 6 to 15 mg per day [98]. Several population groups are thought to consume less than adequate amounts of zinc, principally elderly women [30,31]. Other conditions and populations associated with increased need for intake include those with alcoholism, chronic illness, stress, trauma, surgery, malabsorption, lactovegetarians, and children consuming vegetarian diets [30].

Signs and symptoms of zinc deficiency include growth retardation (an early response to zinc deficiency in children due to inadequate cell division needed for growth) [32], skeletal abnormalities due to impaired development of epiphyseal cartilage, defective collagen synthesis and/or cross-linking, poor wound healing, dermatitis (especially around body orifices), delayed sexual maturation in children, hypogeusia (blunting of sense of taste), night blindness, alopecia (hair loss), impaired immune function, and impaired protein synthesis, among others [23,33,34]. Also remember, metalloenzymes sensitive to zinc deprivation include alkaline phosphatase, carboxypeptidase A, and deoxythymidine kinase.

Supplements

Zinc is found in many forms in supplements (oral tablets, lozenges, and sprays) including zinc oxide, zinc sulfate, zinc acetate, zinc chloride, and zinc gluconate. Each of the forms differs in the amount of zinc provided and in absorption. Zinc gluconate, for example, is approximately 14.3% zinc, whereas zinc sulfate is 23% and zinc chloride is 48% zinc. Zinc chloride and sulfate are very soluble as is zinc acetate. In contrast, zinc carbonate and zinc oxide are fairly insoluble. Comparison of zinc preparations suggested that zinc acetate was one of the best tolerated zinc preparations when compared with zinc sulfate, zinc aminoate, zinc methionine, and zinc oxide; zinc oxide was least absorbed [35]. Zinc supplements are directed to be consumed on an empty stomach and without simultaneous ingestion of other mineral supplements. Gastric irritation is a common side effect.

Zinc lozenges are purported to assist in the treatment of colds. A meta-analysis of zinc salt lozenges and colds found no statistically significant benefit associated with the use of zinc lozenges for the treatment of colds [36]. However, several limitations were associated with the studies, and thus the use of zinc in the treatment of colds appears to require further study [36].

Toxicity

Excessive intakes of zinc can cause toxicity. An acute toxicity with 1 to 2 g of zinc sulfate (225–450 mg of zinc) can produce metallic taste, nausea, vomiting, epigastric pain, abdominal cramps, and bloody diarrhea [29,37]. Chronic ingestion of therapeutic doses or doses as low as 18.5 or 25 mg daily of zinc can result in a copper deficiency due to the competition of these two minerals for intestinal absorption [25,26,29]. Ingestion of 100 to 300 mg of zinc daily as prescribed for some patients with sickle-cell anemia and other conditions has induced copper deficiency evidenced by hypocupremia, anemia, leukopenia, and neutropenia as well as abnormal cholesterol metabolism [37]. Furthermore, copper deficiency induced by intakes of zinc (110–165 mg) for 10 months did not respond to cessation of zinc and 2 months of oral copper supplementation. Cupric chloride given intravenously for 5 days was needed to correct the deficiency and suggested that elimination of excess zinc by the body is a slow process and will continue to inhibit copper absorption until it is eliminated [37]. Reevaluation of the maximum safe dose of zinc is necessary considering its low observed adverse effect level. A preliminary reference dose for a 60-kg adult for over the counter zinc is 9 mg based on high bioavailability and uncertain copper intakes [38]. A reference dose (RfD) of 21 mg for adults (70 kg) also has been reported [39].

Assessment of Nutriture

Evaluation of zinc nutriture is difficult owing to homeostatic control of body zinc. A variety of static indices has been used to assess zinc status, including measurements of zinc in red blood cells, leukocytes, neutrophils, and plasma. The most common means for assessment being serum or plasma zinc, with fasting concentrations of <70 mg/dL suggesting deficiency [32]. Zinc (fasting) in the plasma decreases only when the dietary intake is so low that homeostasis cannot be established without use of zinc from the exchangeable pool that includes plasma zinc [32]. Low fasting plasma zinc thus indicates that little zinc is present in the exchangeable zinc pool and reflects a loss of zinc from bone and liver [33]. Interpretation of plasma zinc must be made with caution because concentrations are influenced by many factors, unrelated to zinc depletion, including meals, time of day (diurnal variation), stress, infections, hypoalbuminemia, steroid therapy, and oral contraceptive administration [40,41]. Postprandial zinc concentrations have been found to be more sensitive to low dietary zinc intakes than fasting plasma zinc [42].

Metallothionein is also used to assess zinc status. Serum metallothionein concentrations are thought to be less sensitive to zinc deficiency than red blood cell concentrations. The latter is thought to detect changes in dietary zinc intake and may be useful for detecting zinc redistribution among tissues [32] as well as being useful as an index of zinc status in humans [43].

Urinary and hair zinc to assess zinc status also has been used. Urinary zinc excretion diminishes with severe zinc deficiency. Urinary zinc excretion has been suggested as an alternative method of assessing oral zinc absorption if using oral doses of 10 to 50 mg elemental zinc [44]. Low hair zinc may be associated with chronic intake of dietary zinc in suboptimal amounts. Standardized procedures, however, are needed to eliminate contamination from, for example, shampoo and confounding variables such as variations due to hair color, sampling sites, and so on. The concentration of zinc in hair depends not only on delivery of zinc to the root but also on the rate of hair growth.

The measurement of the activity of zinc-dependent enzymes has also been employed as an index of zinc status. Although carbonic anhydrase activity decreases only after signs of zinc deficiency appear, the activity of alkaline phosphatase decreases much sooner. Measurements of activity before and after zinc supplementation are recommended. An oral zinc tolerance test has also been used to assess zinc absorption from different meals or supplements. The test typically involves ingestion of 25 or 50 mg zinc as zinc acetate with a test meal or supplement. Changes in plasma zinc concentrations are assessed and compared in the same subjects after consuming different test meals or supplements on different occasions under standardized conditions.

References Cited for Zinc

1. Hunt J, Matthys L, Johnson L. Zinc absorption, mineral balance, and blood lipids in women consuming controlled lactoovovegetarian and omnivorous diets for 8 wk. Am J Clin Nutr 1998;67:421-30.

2. Sturniolo GC, Montino MC, Rossetto L, Martin A, D'Inca R, D'Odorico A, Naccarato R. Inhibition of gastric acid secretion reduces zinc absorption in man. J Am Coll Nutr 1991;10:372-5.

3. Ellis R, Kelsay J, Reynolds R, Morris E, Moser P, Frazier C. Phytate:zinc and phytate x calcium:zinc millimolar ratios in self-selected diets of Americans, Asian Indians and Nepalese. J Am Diet Assoc 1987; 87:1043-7.

4. Forbes R, Erdman J. Bioavailability of trace mineral elements. Ann Rev Nutr 1983;3:213-31.

5. Simmer K, Iles CA, James C, Thompson R. Are iron-folate supplements harmful? Am J Clin Nutr 1987;45:122-5.

6. Kauwell G, Bailey L, Gregory J, Bowling D, Cousins R. Zinc status is not adversely affected by folic acid supplementation and zinc intake does not impair folate utilization in human subjects. J Nutr 1995; 125:66-72.

7. Whittaker P. Iron and zinc interactions in humans. Am J Clin Nutr 1998;68:442S-6S.

8. Solomons N, Jacob R. Studies on the bioavailability of zinc in humans: effects of heme and nonheme iron on the absorption of zinc. Am J Clin Nutr 1981;34:475-82.

9. Spencer H. Minerals and mineral interactions in human beings. J Am Diet Assoc 1986;86:864-7.

10. Dawson-Hughes B, Seligson FH, Hughes VA. Effects of calcium carbonate and hydroxyapatite on zinc and iron retention in postmenopausal women. Am J Clin Nutr 1986;44:83-88.

11. McKenna A, Ilich J, Andon M, Wang C, Matkovic V. Zinc balance in adolescent females consuming a low- or high-calcium diet. Am J Clin Nutr 1997;65:1460-4.

12. Wood R, Zheng J. High dietary calcium intakes reduce zinc absorption and balance in humans. Am J Clin Nutr 1997;65:1803-9.

13. Prasad A. Clinical, biochemical and nutritional spectrum of zinc deficiency in human subjects: an update. Nutr Rev 1983;41:197-208.

14. DiSilvestro R, Cousins R. Physiological ligands for copper and zinc. Ann Rev Nutr 1983;3:261-88.

15. McMahon R, Cousins R. Mammalian zinc transporters. J Nutr 1998;128:667-70.

16. Cousins R, Lee-Ambrose L. Nuclear zinc uptake and interactions and metallothionein gene expression are influenced by dietary zinc in rats. J Nutr 1992; 122:56-64.

17. Bremner I, Beattie J. Metallothionein and the trace minerals. Ann Rev Nutr 1990;10:63-83.

18. Anon. Role of zinc in enzyme regulation and protection of essential thiol groups. Nutr Rev 1986; 44:309-11.

19. Bettger W, O'Dell B. Physiological roles of zinc in the plasma membrane of mammalian cells. J Nutr Biochem 1993;4:194-207.

20. Sato M, Bremner I. Oxygen free radicals and metallothionein. Medicine 1993;14:325-37.

21. Wada L, King J. Effect of low zinc intakes on basal metabolic rate, thyroid hormones and protein utilization in adult men. J Nutr 1986;116:1045-53.

22. Miller L, Hambidge M, Naake V, Hong Z, Westcott J, Fennessey P. Size of the zinc pools that exchange rapidly with plasma zinc in humans: alternative techniques for measuring and relation to dietary zinc intake. J Nutr 1994;124:268-76.

23. Christian P, West K. Interactions between zinc and vitamin A: an update. Am J Clin Nutr 1998;68: 435S-41S.

24. Bates J, McClain C. The effect of severe zinc deficiency on serum levels of albumin, transferrin, and prealbumin in man. Am J Clin Nutr 1981;34:1655-60.

25. Festa M, Anderson H, Dowdy R, Ellersieck M. Effect of zinc intake on copper excretion and retention in men. Am J Clin Nutr 1985;41:285-92.

26. Fischer P, Giroux A, L'Abbe M. Effect of zinc supplementation on copper status in adult man. Am J Clin Nutr 1984;40:743-6.

27. Solomons N. Mineral interactions in the diet. J Dent Child 1982;49:445-8.

28. Goyer R. Nutrition and metal toxicity. Am J Clin Nutr 1995;61(suppl):646S-50S.

29. National Research Council. Recommended Dietary Allowances, 10th ed. Washington, D.C.: National Academy Press, 1989;205-13.

30. Pennington J, Young B, Wilson D, Johnson R, Vanderveen J. Mineral content of foods and total diets: the selected minerals in foods survey, 1982 to 1984. J Am Diet Assoc 1986;86:876-91.

31. Pennington J, Young B. Total diet study nutritional elements, 1982-1989. J Am Diet Assoc 1991;91: 179-83.

32. King J. Assessment of zinc status. J Nutr 1990;120: 1474-9.

33. Ploysangam A, Falciglia G, Brehm B. Effect of marginal zinc deficiency on human growth and development. J Trop Pediatr 1997;43:192-8.

34. Baer M, King J, Tamura T, Margen S, Bradfield R, Weston W, Daugherty N. Nitrogen utilization, enzyme activity, glucose intolerance and leukocyte chemotaxis in human experimental zinc depletion. Am J Clin Nutr 1985;41:1220-35.

35. Prasad A, Beck F, Nowak J. Comparison of absorption of five zinc preparations in humans using oral zinc tolerance test. J Trace Elem Exp Med 1993;6:109-15.

36. Jackson J, Peterson C, Lesho E. A meta-analysis of zinc salt lozenges and the common cold. Arch Intern Med 1997;157:2372-6.

37. Fosmire G. Zinc toxicity. Am J Clin Nutr 1990;51: 225-7.

38. Sandstead H. Requirements and toxicity of essential trace elements, illustrated by zinc and copper. Am J Clin Nutr 1995;61(suppl):621S-4S.

39. Hathcock J. Safety limits for nutrients. J Nutr 1996;126:2386S-9S.

40. Wallock L, King J, Hambidge K, English-Westcott J, Pritts J. Meal-induced changes in plasma, erythrocyte, and urinary zinc concentrations in adult women. Am J Clin Nutr 1993;58:695-701.

41. King J, Hambidge K, Westcott J, Kern D, Marshall G. Daily variation in plasma zinc concentrations in women fed meals at six-hour intervals. J Nutr 1994; 124:508-16.

42. Mellman D, Hambidge K, Westcott J. Effect of dietary zinc restriction on postprandial changes in plasma zinc. Am J Clin Nutr 1993;58:702-4.

43. Thomas E, Bailey L, Kauwell G, Lee D, Cousins R. Erythrocyte metallothionein response to dietary zinc in humans. J Nutr 1992;122:2408-14.

44. Henderson L, Brewer G, Dressman J, Swidan S, DuRoss D, Adair C, Barnett J, Berardi R. Use of zinc tolerance test and 24-hour urinary zinc content to assess oral zinc absorption. J Am Coll Nutr 1996;15: 79-83.

Copper

The copper content of the human adult body is on the order of 50 to 110 mg. Copper is found in the body in either of two valence states, cuprous state (Cu^{1+}) or cupric state (Cu^{2+}).

Sources

The copper content of food varies widely reflecting the origin of the food, and the conditions under which the food was produced, handled, and prepared for use. Consequently, analytic values of copper published for a specific food must be used with caution. The broad range of the copper content of selected foods, as listed in Table 12.6, reflects these variations. The richest sources are considered to be organ meats and shellfish, especially oysters. Plant food sources rich in copper include nuts, seeds, legumes, dried fruits, and a few select vegetables.

Endogenous sources of copper also may be found in the gastrointestinal tract. Copper is secreted daily into the gastrointestinal tract in digestive juices in relatively large amounts. For example, the copper contents of saliva and gastric juice are about 400 µg and 1,000 µg,

Table 12.6 Some Food Copper Values

Food	Copper Content (mg/100 g)	Food	Copper Content (mg/100 g)
Dairy		Vegetables	
Egg, whole	0.07	Potato, without peel	0.07
Milk, whole	0.003	Potato chips	0.35
Yogurt, low-fat, plain	0.004	Potato, sweet	0.18
Cheese, Cheddar	0.04	Carrot	0.05
		Broccoli	0.03
Meat, Fish, Poultry		Spinach	0.08
Liver, beef	6.09	Peas	0.10
Chicken	0.07	Lettuce	0.03
Beef, sirloin	0.14	Tomato	0.06
Pork	0.09	Corn	0.04
Tuna, canned	0.05	Cabbage	0.01
Shrimp, cooked	0.30	Fruits	
Grains		Apple	0.03
Macaroni, cooked	0.08	Banana	0.14
Corn grits, cooked	0.01	Grapes	0.09
Rice, white, cooked	0.08	Peach	0.06
		Pear	0.09
Roll, white	0.14	Pineapple	0.05
Bread, whole-wheat	0.25	Orange	0.04
		Raisins	0.32
Nuts		Prunes	0.29
Peanut	0.68		
Pecan	1.24		

Source: Pennington JAT, Young BE, Johnson RD, and Vanderveen JE. Mineral content of foods and total diets: The selected minerals in foods survey, 1982 to 1984. Reprint from JOURNAL OF THE AMERICAN DIETETIC ASSOCIATION. Vol. 86:876–91, 1986.

respectively; pancreatic and duodenal juices may contain up to 1,300 μg and 2,200 μg, respectively [1].

Digestion, Absorption, Transport, and Storage

Digestion

Most copper, as Cu^{2+} but some as Cu^{1+}, in foods is bound to organic components, especially amino acids that make up food proteins. Thus, digestion is needed to free the bound copper before absorption may occur. Gastric hydrochloric acid and pepsin facilitate the release of bound copper in the stomach. Additional proteolytic enzymes in the small intestine may hydrolyze proteins to further release copper. Copper digestion is shown in Figure 12.12.

Absorption

Although copper is absorbed throughout the small intestine, especially the duodenum, the stomach also appears to possess some absorptive capacity. This may be attributable to the solubilizing effect of the acidic environment on copper, thereby facilitating its transport across the gastric mucosa. However, when compared with intestinal copper absorption, gastric copper absorption contributes relatively little to overall absorption.

The mechanisms for the absorption of copper across the brush border of the small intestine are not completely understood. It is possible that luminal copper must be bound to more absorbable ligands for effective transport, and even if copper is transported in the free ionic form, the presence of such ligands may be needed to present the metal to brush border receptors in a way that increases uptake.

Copper appears to be absorbed by two mechanisms, one saturable, involving an active transport system, and the other a nonsaturable, passive diffusion process (Fig. 12.12). As is true for other transport systems, low concentrations of dietary copper are primarily transported via the active carrier-mediated pathway, whereas the diffusion process accommodates higher concentrations. Copper may also utilize amino acid active transport systems such as the one used by histidine for absorption.

Typically the gastrointestinal tract absorbs between 30% and 50% of ingested copper. However, the percentage of copper absorbed in influenced by copper status and dietary copper availability. Copper absorption is significantly higher during periods of low dietary copper than during periods of higher dietary copper [2]. Copper absorption ranges from about 12% with higher intakes of about 7.5 mg to about 56% with intakes of 0.8 mg [3], although absorption as high as 71% was observed in women consuming 0.9 to 1.2 mg of copper per day [4].

Factors Influencing Copper Absorption Copper transport across the brush border membrane may be influenced by a variety of dietary components, some exerting a positive effect and some influencing its absorption negatively. Examples of substances that facilitate copper absorption include amino acids, especially histidine, which may bind to copper and allow absorption through an amino acid transport system, as well as sulfur containing amino acids such as methionine and cysteine. Copper also forms ligands with amino acid sulfhydryl groups in compounds such as glutathione.

Organic acids, other than vitamin C, in foods also improve copper absorption. Citric, gluconic, lactic, acetic

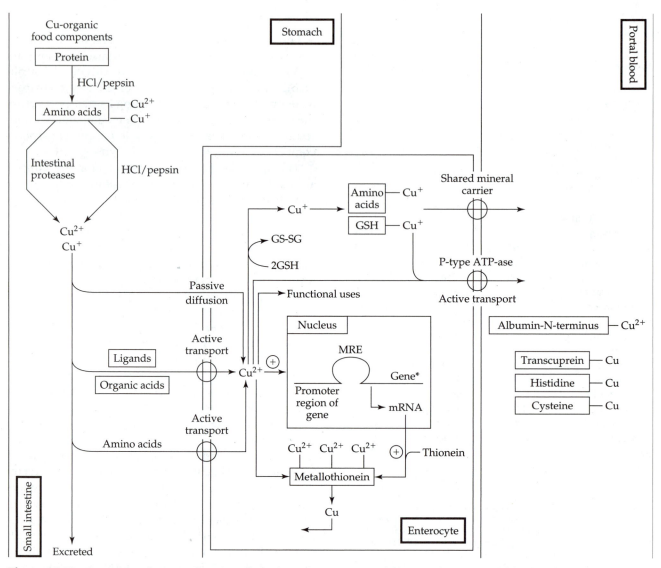

Figure 12.12 Overview of copper digestion, absorption, enterocyte metabolism, and transport. Abbreviations: GSH, reduced glutathione; MRE, metal regulatory element.

and malic acids act as binding ligands to improve solubilization of copper and thus absorption [5]. Citrate has been shown to form a stable complex with copper to improve its absorption.

Inhibitors of Copper Absorption Several substances, including other trace minerals, impede copper absorption. *Zinc* in amounts as low as 18.5 mg has been shown to impair copper absorption in men [6]; at slightly higher intakes (25 mg of zinc for 6 weeks), superoxide dismutase activity decreased, suggesting impaired copper status [7]. The detrimental effect of excessive zinc intake on copper absorption is thought to result from zinc's stimulation of thionein synthesis. Thionein, a protein that possesses an unusually high

proportion (25%–30%) of thiol groups in the form of cysteine, avidly binds copper as well as other metals to form metallothionein. Intestinal metallothionein may function as a negative modulator of copper absorption and in this respect acts as a detoxifying agent, binding copper and reducing its transmucosal flux (passage into the plasma) when the dietary intake and therefore luminal concentration of the metal is excessive [8]. At normal intake levels of copper, however, the regulatory role of the protein remains unclear.

Iron ingested in relatively large amounts decreases copper absorption in rats [9]. Excessive iron intakes also appear to impede copper absorption in humans. Copper absorption in infants fed formula supplemented with iron (10.8 mg of iron/L) was significantly lower than infants fed a formula providing only 1.8 mg of iron/L [10]. Further interactions with iron are discussed under the section on interactions.

Molybdenum as tetrathiomolybdate $(MoS_4)^{2-}$ forms an insoluble complex with copper to inhibit its absorption in the gastrointestinal tract of rats and ruminants. The significance of these findings in humans is unknown.

Ingestion of large amounts of major minerals also impair copper absorption. *Calcium* (2,382 mg as calcium gluconate) and *phosphorus* (2,442 mg as glycerol phosphate) have been shown to increase fecal copper excretion, compared with diets containing only moderate amounts of calcium (780 mg as calcium gluconate) with either high phosphorus (2,442 mg as glycerol phosphate) or moderate phosphorus (843 mg as glycerol phosphate) [11]. Urinary copper losses were also significantly greater on the high-calcium, high-phosphorus diet than on the moderate-calcium, moderate-phosphorus diet [11].

Vitamin C may interact with copper to decrease its absorption. Presumably vitamin C reduces copper from a cupric state (Cu^{2+}) to a less absorbable cuprous state (Cu^{1+}) [12,13]. Vitamin C has also been suggested to decrease copper retention.

Excessive antacid ingestion may diminish copper absorption and induce deficiency. Copper at a neutral pH will bind to hydroxides (OH) forming insoluble compounds that are not readily absorbable.

Although many factors appear to influence copper absorption, the body's copper status plays a significant role. The efficiency of absorption of copper changes to regulate in part whole-body copper status. Changes in fecal excretion also mediate the process. Thus, with high copper intake, less copper is absorbed. As body copper stores increase, copper excretion via the bile increases. With moderately low copper intakes the reverse (increased absorption, decreased excretion) to some extent occurs. With copper intakes of about 0.38 mg/day, regulation is not sufficient to compensate to prevent depletion of body copper [2].

Once within the intestinal cell, copper may be used by the cell (see the section on functions), may be stored, or may be transported through the cell for subsequent transport across the basolateral membrane. Storage of copper occurs as part of metallothionein (see p. 423 for a description of metallothionein). Copper positively influences metallothionein transcription. Unless released from storage copper will be lost with intestinal cell turnover, approximately every 2 to 3 days. Intracellular copper transport is not well characterized. Amino acids and glutathione (GSH) have been proposed as intracellular copper (Cu^{1+}) carriers, but

other transporters of copper (Cu^{2+}) are likely. Glutathione also may serve to reduce copper within cells (Fig. 12.12).

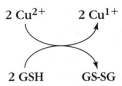

Copper transport across the basolateral membrane into the plasma is thought to occur by a shared mineral carrier transporter, and by a carrier mediated active transport system specific for copper. P(phosphorylation)-type ATPases appear to be responsible for most active copper transport out of cells [14]. Generally, ATPase transporters carry metals in their reduced state; however, whether this is true for copper transport is unclear (Fig. 12.12).

Copper Transport

From the intestinal cell, copper is transported to the liver bound loosely to albumin. Specifically, the N-terminus of albumin has a high affinity for copper (Cu^{2+}). Copper may also possibly be transported bound to transcuprein (Tc); very small amounts are bound to amino acids such as histidine and cysteine [15] (Fig. 12.12). Following uptake by the liver, copper may be released back into the blood for transport to tissues. Similarly, extrahepatic tissues may release copper for transport back to the liver. Most copper reemerges back in the blood from the liver (and possibly kidney) bound to ceruloplasmin; ceruloplasmin is discussed in more detail in the sections on copper uptake by tissues and functions. Small amounts of copper also may be found back in the blood bound to transcuprein, and possibly albumin and amino acids following the initial hepatic handling of the newly absorbed dietary copper [15]. Copper concentrations in the plasma are thought to be regulated. The incorporation of copper into transport proteins is thought to represent a point of regulation of copper metabolism [16].

Copper Uptake by the Liver

The observations that the uptake of copper by the hepatocytes is a saturable, temperature-dependent process and that no competition with other metal ions occurs suggest that a specific, facilitated diffusion mechanism is in effect. Transport across the cell membrane may involve the formation of certain amino acid–copper complexes and may also involve albumin [17].

Once within cells, more than half of the accumulated copper is found in the supernatant fraction and one-fourth within the nuclei. It first appears to bind to metallothionein then is slowly transferred to the copper enzymes such as superoxide dismutase (SOD), which appears to be given high priority with respect to the use of cellular copper. Experimentally, a tenfold reduction in hepatic copper resulted in only a twofold decrease in SOD activity [18].

In the liver, copper also binds to the protein apoceruloplasmin. Six copper ions (in the form of Cu^{1+} and Cu^{2+}) are attached to apoceruloplasmin to form ceruloplasmin (Fig. 12.13) [19]. Three of the six copper atoms are involved in electron transfer, whereas the other three function at the catalytic site and give the protein a blue color. Although copper does not appear to influence apoceruloplasmin synthesis, ceruloplasmin activity without sufficient copper is diminished or absent and ceruloplasmin's half-life is shortened.

Ceruloplasmin is released into the blood from the liver and constitutes about 60% (or perhaps up to 95%) of circulating copper in the blood after meals [5,16,17].

The remaining copper in the blood circulates loosely bound to albumin, transcuprein, and histidine. The basic aspects of mammalian copper metabolism are summarized in Figure 12.13.

Ceruloplasmin delivers copper to tissues. Uptake of ceruloplasmin copper by extrahepatic cells involves binding of ceruloplasmin to specific receptors. The copper ions that are not at ceruloplasmin's active oxidase site are released. Release is thought to involve the reduction of copper from Cu^{2+} to Cu^{1+}. Following dissociation from the ceruloplasmin, copper enters the cell directly through channels or after binding protein transporters. Ascorbic acid enhances copper transfer and is probably involved in the reduction of the copper (Fig. 12.13) [20]. Glutathione, which has a strong affinity for Cu^{1+}, may serve as the transporter of copper within the cell for delivery to functional sites.

Copper Storage

In comparison with other trace minerals, little copper (<110 mg) is found in the body. What copper is found

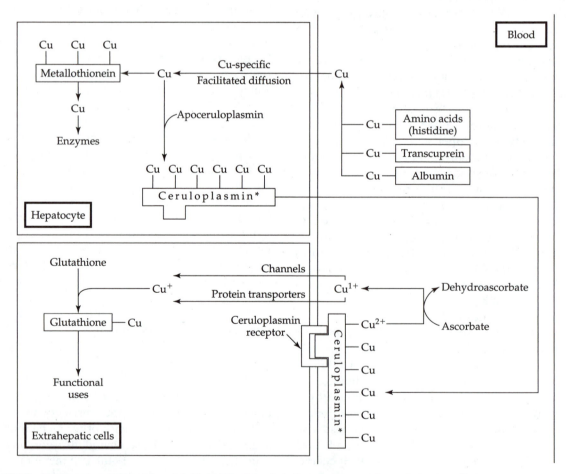

Figure 12.13 Transport, uptake, and metabolism of copper into hepatocytes and extrahepatic cells.
*Ceruloplasmin contains six sites for copper in either the Cu^+ or Cu^{2+} oxidation state.

in the body is found within a variety of cells and tissues. Following intake, the liver and kidney both rapidly extract copper from the blood. Other copper-containing tissues include brain, heart, bone, muscle, skin, intestine, spleen, hair, and nails, among others. Within cells and tissues, copper is bound to amino acids or proteins. The liver appears to be the main storage site for copper, which is thought to be bound to metallothionein. Metallothionein, in addition to storing up to about 12 copper atoms (as well as zinc atoms), protects cells by scavenging damaging superoxide and hydroxyl radicals. The amount of copper available to extrahepatic tissues is thought to be regulated by the liver through the synthesis of ceruloplasmin, through copper incorporation into metallothionein, and through excretion of copper into the bile.

Functions and Mechanisms of Actions

The essentiality of copper is due, in part, to its participation as an enzyme cofactor and as an allosteric component of enzymes. Several copper-requiring metalloenzymes and the reactions they catalyze are described next. In many enzymes copper functions as an intermediate in electron transfer.

Ceruloplasmin

Ceruloplasmin, an α-2 glycoprotein, is not simply a transporter of copper in the blood; it is also a multifaceted oxidative enzyme (oxidase) and antioxidant, which may be found in the blood, but also bound to cell surface receptors on the plasma membranes of cells. Ceruloplasmin, also known as ferroxidase I, is responsible for the oxidation of minerals, most notably ferrous (Fe^{2+}) iron but also manganese (Mn^{2+}). The oxidation of Fe^{2+} to Fe^{3+} is needed in order for iron to bind to transferrin. Whether ceruloplasmin transports Fe^{3+} to transferrin or another protein is responsible for the transport is unclear [21]. Once Fe^{3+} is bound to transferrin, the transferrin protein then transports the mobilized iron from its stores to tissues such as the bone marrow that need the mineral.

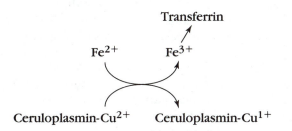

Ceruloplasmin also scavenges oxygen radicals to protect cells and modulates the inflammatory process

as an acute-phase protein. Thus, ceruloplasmin as well as copper concentrations rise in the blood with, for example, infection. During infections, phagocytosis of invading organisms by white blood cells generates superoxide radicals, which must be eliminated to prevent further damage to body cells. Superoxide radicals also are generated in normal metabolism.

Superoxide Dismutase

Superoxide dismutase (SOD), found in the cytosol of cells, is copper- and zinc-dependent. In the enzyme, copper is thought to be linked with zinc through an imidiazole group, while both minerals are linked to the enzyme protein via histidine and aspartate residues. Copper (Cu^{2+}) is found at the enzyme's active site, where the superoxide substrate binds to the enzyme. Removal of copper, but not zinc, results in reduced SOD activity but has no effect on the enzyme's synthesis [22]. Specifically superoxide dismutase catalyzes the removal (dismutation) of the superoxide radicals (O_2^-). During the reaction, copper is reduced along with the oxygen radical (O_2^-) to initially generate molecular oxygen, O_2, and then by reoxidation hydrogen peroxide, H_2O_2 [22].

$$2O_2^- + 2H^+ \xrightarrow{\text{zinc copper SOD}} O_2 + H_2O_2$$

Superoxide radicals can cause peroxidative damage of phospholipid components of cell membranes. In other words, without SOD, superoxide radicals can form more destructive hydroxyl radicals that damage both unsaturated double bonds in fatty acids in cell membranes and other molecules in cells (see the Perspective in Chap. 10). SOD therefore assumes a very important protective function. SOD is found in most cells of the body. Increased peroxidation of cell membranes is found with copper deficiency.

Cytochrome c Oxidase

Cytochrome c oxidase contains three copper atoms per molecule. One subunit of the enzyme contains two copper atoms and functions in electron transfer. The second subunit contains another copper atom involved in reducing molecular oxygen. Cytochrome c oxidase functions in the terminal oxidative step in mitochondrial electron transport (p. 63). Specifically, the enzyme transfers an electron such that molecular oxygen (O_2) is reduced to form water molecules and enough free energy is generated to permit ATP production. Severe copper deficiency ultimately impairs the activity of this enzyme.

Amine Oxidases

Amine oxidases are also copper-dependent. Copper appears to function as an allosteric structural component of these enzymes, and TOPA quinone (6-hydroxydopa) serves as an organic cofactor [23]. Histidine residues in the enzyme serve as ligands for the copper. Amine oxidases, found both in the blood and in body tissues, catalyze the oxidation of biogenic amines such as tyramine, histamine, and dopamine to form aldehydes and ammonium ions. Dioxygen is reduced to form hydrogen peroxide.

$$RCH_2NH_2 \xrightarrow[\text{Oxidase-Cu}]{O_2 \qquad H_2O_2} RCH{\overset{O}{\overset{\|}{}}} + {}^+NH_4$$

Tyrosine Metabolism–Dopamine Mono-Oxygenase and p-Hydroxyphenyl-pyruvate Hydroxylase

In tyrosine metabolism (Fig. 9.6, p. 253), two reactions depend on copper. Initial reactions in tyrosine metabolism are not copper-dependent. For example, tyrosine is converted to 3,4-dihydroxy phenylalanine (L-dopa) in an iron-dependent reaction. Then, L-dopa may be further metabolized to dopamine or may be used for the synthesis of the pigment melanin found in eyes, skin, and hair. Although the conversion of L-dopa to dopamine does not require copper, dopamine metabolism requires a copper-dependent enzyme.

Dopamine mono-oxygenase (hydroxylase) is necessary for the synthesis of the catecholamine norepinephrine, which may be subsequently methylated to form epinephrine (see pp. 540–541 for a detailed discussion of the roles of the catecholamines). Specifically, dopamine mono-oxygenase converts dihydroxyphenylalanine (do-pamine) to norepinephrine; the enzyme contains up to eight copper atoms per molecule and requires molecular oxygen and vitamin C for its function. The reaction catalyzed by this enzyme is shown in Figure 9.6.

In tyrosine catabolism, the conversion of p-hydroxyphenyl-pyruvate to homogentisate requires a copper-dependent hydroxylase and vitamin C. This reaction is shown in Figure 9.6.

Lysyl Oxidase

Lysyl oxidase is secreted by connective tissue cells and serves to generate cross-linking between connective tissue proteins including collagen and elastin. The cross-linking is needed to stabilize the extracellular matrix. Specifically, lysyl oxidase catalyzes the removal of the epsilon amino group (oxidative deamination) of lysyl residues of a polypeptide and the oxidation of the terminal carbon atom of an aldehyde. Lysyl oxidase activity decreases with inadequate copper intake [24]. Thus, collagen cross-linking appears to be affected by dietary copper [24].

Peptidylglycine a-Amidating Mono-Oxygenase

Amidation of peptide hormones, such as bombesin, calcitonin, gastrin, and cholecystokinin, is necessary for hormone function. The amidation requires a copper-dependent enzyme known as peptidylglycine α-amidating mono-oxygenase, which is found mostly in the brain. This enzyme cleaves a carboxy terminal glycine residue off peptides that have a C-terminal glycine. The amino group of glycine is retained by the peptide as a terminal amide. The oxidized residue is released as glyoxylate. Peptidylglycine α-amidating mono-oxygenase also requires vitamin C to reduce Cu^{2+} back to Cu^{1+}. This reaction is shown in Figure 9.8.

Other Roles

Copper plays a variety of other roles in the body that are not well understood; these roles may or may not involve enzymes. Some of these other roles include angiogenesis, immune system function, nerve myelination, and endorphin action.

As a pro-oxidant, copper behaves similar to iron. Copper (Cu^{2+}) reacts with superoxide radicals and catalyzes the formation of hydroxyl radicals through the Fenton reaction.

$$O_2^- + Cu^{2+} \rightarrow O_2 + Cu^{1+}$$
$$Cu^{1+} + H_2O_2 \rightarrow Cu^{2+} + OH^- + OH^-$$

Associated with the generation of reactive oxygen species is increased oxidative damage to DNA (base oxidation and strand breaks) and lipid peroxidation including membrane lipids.

Copper influences gene expression through binding to specific transcription factors, also called *binding proteins.* In some cases, copper has been shown to influence transcription by binding to transcription factors, which in turn bind to promoter sequences on DNA. Once bound to DNA, transcription may be enhanced or suppressed. Further information on the role of copper in gene expression may be found in the Perspective at the end of Chapter 9.

Interactions with Other Nutrients

Copper is known to interact with a number of inorganic and organic dietary constituents. Those that affect copper absorption have been previously discussed (p. 431). Additional interactions will be mentioned in this section.

Among organic dietary substances, *ascorbic acid* (1.5 g for 64 days) resulted in decreased serum ceruloplasmin activity, however, concentrations still remained within the normal range [12]. Intakes of 605 mg vitamin C for 3 weeks also resulted in a 21% decrease in serum ceruloplasmin oxidase activity [13]. The effects of vitamin C may be mediated through the reduction of the cupric ion to its cuprous form by the ascorbate or through the formation of a poorly absorbable complex, or both.

It is well established that a strong, mutual antagonism exists between copper and *zinc,* most likely caused by the induction of intestinal metallothionein by zinc. This results in excessive intracellular binding of the copper, reducing its luminal-to-serosal flux, and entry into the blood. Zinc intakes ranging from 18.5 mg to 300 mg daily have resulted in a copper deficiency [6,7,25]. Furthermore, copper deficiency induced by intakes of zinc (110–165 mg) for 10 months did not respond to cessation of zinc and 2 months of oral copper supplementation. Cupric chloride given intravenously for 5 days (total dose, 10 mg) was needed to correct the deficiency and suggested that elimination of excess zinc by the body is a slow process and will continue to inhibit copper absorption until it is eliminated [26].

Another interaction of copper having practical importance involves *iron.* The importance of copper in normal iron metabolism is evidenced by the anemia that results from prolonged copper deficiency. It is known that the anemia is caused by an impaired mobilization and use of iron due to the reduced ferroxidase activity of ceruloplasmin, which is responsible for oxidation of iron to its trivalent (Fe^{3+}) state. Only as Fe^{3+} can iron be coordinated to its transport protein transferrin. High iron intakes also appear to interfere with mobilization of copper from stores [9]. Moreover, copper-dependent erythrocyte superoxide dismutase activity at 20 weeks was significantly lower in infants receiving 13.8 mg of iron than in infants receiving 7 mg of iron or no supplemental iron; plasma copper concentrations did not, however, differ between groups [27]. Other studies in infants have shown iron fortification (10.2 mg/L) diminished copper absorption and retention versus formula providing only 2.5 mg of iron/L [10].

As discussed in part regarding factors inhibiting copper absorption, in animals (ruminants and rats) dietary copper forms insoluble complexes with *molybdenum* and sulfur in the form of tetrathiomolybdate $(MoS_4)^{2-}$.

While such findings have not been reported in humans, urinary copper excretion in humans has been shown to rise from 24 μg/day to 77 μg/day as molybdenum intake increased from 160 μg to 1,540 μg/day [3]. No changes in fecal copper excretion were noted, suggesting perhaps that molybdenum increased copper mobilization from tissues and promoted excretion [3].

Copper and *selenium* also appear to interact. Copper deficiency has been shown to decrease the activity of the selenium-dependent enzymes glutathione peroxidase and 5'-deiodinase [28].

Antagonistic interactions between copper and cadmium, silver, and mercury have been reported but have more theoretical than practical importance. Fructose also appears to negatively impact copper balance in humans possibly through hepatic redistribution of copper to other pools [29].

Excretion

Copper that is absorbed into the body may be excreted via the feces, urine, or sweat. Most absorbed copper (~2 mg) is secreted by the liver into the bile for excretion in the feces. Dietary copper intake directly influences biliary copper excretion such that low dietary copper intake results in low fecal copper excretion [2]. Wilson's disease, an inherited disorder of copper metabolism, is characterized by defective biliary copper excretion. Consequently, copper accumulates mainly in the liver but also in other organs including the brain, kidney, eye (cornea), and spleen.

Only a small amount of copper (~10–50 μg) is excreted through the kidneys via the urine. Furthermore, urinary copper excretion does not change significantly with changes in copper intake except under extreme conditions.

Similarly, only small amounts (50–100 μg) of copper are lost in sweat and with desquamation of skin cells; dermal excretion does not significantly change with changes in intake. Women experience trace losses of copper via normal menstrual flow; however, a woman's copper status, unlike her iron status, is not compromised by menstruation.

Estimated Safe and Adequate Daily Dietary Intake

In the 1980 edition of the RDA, a recommendation for an Estimated Safe and Adequate Daily Dietary Intake (ESADDI) for copper was first made by the Food and Nutrition Board. The 1989 Committee on the RDA recommends a daily copper intake for infants of 75 mg/kg of body weight [30]. This figure allows for the feeding of formulas in which copper is less bioavailable than in breast milk and is therefore higher than the amount recommended for strictly breast-fed infants. For children and adults, for whom recommended intake is based on balance studies, the Food and Nutrition Board established a safe and adequate range for copper intake of 1.5 to 3 mg/day [30]. While the ESADDI for copper for adults ranges from 1.5 to 3.0 mg/day, lower intakes of copper have not been consistently associated with declines in indexes of copper status or with symptoms of copper deficiency [31–33]. The minimum dietary copper requirement of young men appears to be between 0.4 and 0.8 mg/day [33]. Thus, further studies to evaluate the appropriateness of copper's ESADDI are needed.

Deficiency

Various clinical manifestations are associated with human copper deficiency. Recognized manifestations include hypochromic anemia, neutropenia, leukopenia, hypopigmentation or depigmentation of skin and hair, impaired immune function, and bone abnormalities, especially demineralization [34,35]. These symptoms are all reversible by copper administration.

Biochemical indicators of diminished copper status are decreased plasma or serum copper concentration (<0.45 mg/L), decreased ceruloplasmin concentrations (<20 mg/L), and activity as well as reduced red blood cell SOD activity [33]. Decreased activity of glutathione peroxidase has also been reported although the mechanism of action is unclear [36]. Risk of atherosclerosis increases with copper deficiency. In copper-deficient animals, hypercholesterolemia is found [37,38]. Elevated plasma cholesterol and blood pressure changes have also been shown with low copper intakes in some humans [39].

Conditions and populations associated with increased need for intake include excessive zinc (40–50 mg/day) and antacid ingestion, nephrosis, and malabsorptive disorders such as celiac disease, tropical sprue, and protein-losing enteropathies. Menkes syndrome, an X-linked chromosomal disorder, results in copper deficiency due to deficient production of an ATPase necessary for copper transport. Low copper intakes have been reported for several age groups from infants through older adults [3,40].

Toxicity

Copper toxicity (toxicosis) is fairly rare in the United States, although acute poisonings have occurred because of water contamination or accidental ingestion.

Effects of copper intakes of 64 mg (250 mg of copper sulfate) have resulted in nausea, vomiting, and diarrhea. Other symptoms of toxicity include hematuria, jaundice, oliguria, or anuria [41]. Copper is lethal in amounts about 1,000 times normal dietary intake [42].

Wilson's disease, a genetic disorder, is characterized by copper toxicity. In Wilson's disease, copper accumulates in organs resulting in disturbed function of organs, especially the liver, kidney, and brain. At present, treatment of Wilson's disease involves avoidance of high copper foods and D-penicillamine therapy to bind body copper and increase its excretion [43].

Supplements

Bioavailability of copper from supplements has been investigated using chicks; thus, applicability of results to humans remains unclear. In chicks, Cu_2O (cupric oxide) and Cu-lysine were as effective as $CuSO_4 \cdot 5 H_2O$ (copper sulfate) in causing copper accumulation in the liver [44]. Supplementation with CuO (cuprous oxide), however, resulted in no significant increase in hepatic copper [44]. Copper is found in mineral supplements typically as copper sulfate.

Assessment of Nutriture

Serum, plasma, or red blood cell copper is frequently used, but likely inadequate to assess short-term changes in copper status. The change in plasma or serum copper concentrations that occurs when subjects consume inadequate copper intake is quite variable between individuals and further is affected by several factors unrelated to diet. Extremely low copper intake (~0.38 mg/day), however, appears to be sufficient to significantly decrease not only plasma copper but also ceruloplasmin concentration and activity as well as urinary copper excretion [45]. Many other studies also have shown serum ceruloplasmin concentrations and activity decrease with copper deficiency. The ratio of ceruloplasmin enzyme activity to protein concentration is thought to be better than either measurement alone [31,46]. Response of serum ceruloplasmin to copper supplements also may be used. Typically, supplemental copper results in normalization of serum copper and neutrophil count followed by serum ceruloplasmin [34]. Ceruloplasmin increases following supplementation only in copper-deficient subjects. Another good indicator of copper status is measurement of the activity of copper-dependent enzymes such as superoxide dismutase (SOD) (normal 0.47–0.067 mg/g) in the red blood cell. SOD activity is sensitive to longer-term copper deficiency [33]. Platelet or leukocyte cytochrome c oxidase or skin lysyl oxidase activity also have shown response to changes in copper status.

Hair copper concentrations have not been shown to correlate with either serum or organ copper. Yet, they are reduced with a prolonged period of copper deficiency. Hair copper concentrations are not thought to be useful indicators of copper status.

References Cited for Copper

1. Linder M, Hazegh-Azam M. Copper biochemistry and molecular biology. Am J Clin Nutr 1996;63: 797S-811S.

2. Turnlund J, Keyes W, Peiffer G, Scott K. Copper absorption, excretion, and retention by young men consuming low dietary copper determined using stable isotope ^{65}Cu. Am J Clin Nutr 1998;67: 1219-25.

3. Turnlund J. Copper nutriture, bioavailability, and the influence of dietary factors. J Am Diet Assoc 1988;88:303-8.

4. Johnson P, Milne D, Lykken G. Effects of age and sex on copper absorption, biological half-life, and status in humans. Am J Clin Nutr 1992;56:917-25.

5. DiSilvestro R, Cousins R. Physiological ligands for copper and zinc. Ann Rev Nutr 1983;3:261-88.

6. Festa M, Anderson H, Dowdy R, Ellersieck M. Effect of zinc intake on copper excretion and retention in men. Am J Clin Nutr 1985;41:285-92.

7. Fischer P, Giroux A, L'Abbe M. Effect of zinc supplementation on copper status in adult man. Am J Clin Nutr 1984;40:743-6.

8. Cousins R. Absorption, transport and hepatic metabolism of copper and zinc: special reference to metallothionein and ceruloplasmin. Physiol Rev 1985;65: 238-44.

9. Yu S, West C, Beynen A. Increasing intakes of iron reduce status, absorption and biliary excretion of copper in rats. Brit J Nutr 1994;71:887-95.

10. Haschke F, Ziegler E, Edwards B, Fomon S. Effect of iron fortification of infant formula on trace minerals absorption. J Pediatr Gastroenterol Nutr 1986;5: 768-73.

11. Snedeker S, Smith S, Greger J. Effect of dietary calcium and phosphorus levels on the utilization of iron, copper, and zinc by adult males. J Nutr 1982;112: 136-43.

12. Finley E, Cerklewski F. Influence of ascorbic acid supplementation on copper status in young adult men. Am J Clin Nutr 1983;37:553-6.

13. Jacob R, Skala J, Omaye S, Turnlund J. Effect of varying ascorbic acid intakes on copper absorption and ceruloplasmin levels of young men. J Nutr 1987;117: 2109-15.

14. Bingham M, Ong T, Summer K, Middleton R, McArdle H. Physiologic function of the Wilson disease gene product, ATP7B. Am J Clin Nutr 1998;67(suppl): 982S-7S.

15. Linder MC, Wooten L, Cerveza P, Cotton S, Shulze R, Lomeli N. Copper transport. Am J Clin Nutr 1998;67(suppl):965S-71S.

16. Scott K, Turnlund J. Compartmental model of copper metabolism in adult men. J Nutr Biochem 1994; 5:342-50.

17. Harris E. The transport of copper. In: Prasad AS, ed. Essential and toxic trace elements in human health and disease: an update. New York: Wiley-Liss, 1993; 163-79.

18. Chung K, Romero N, Tinker D, Keen C, Amemiya K, Rucker R. Role of copper in the regulation and accumulation of superoxide dismutase and metallothionein in rat liver. J Nutr 1988;118:859-64.

19. Zaitseva I, Zaitsev V, Card G, Moshkov K, Bax B, Ralph A. Lindley P. The X-ray nature of human serum ceruloplasmin at 3.1 A: nature of the copper centres. J Biol Inorg Chem 1996;1:15-23.

20. Percival S, Harris E. Copper transport from ceruloplasmin: characterization of the cellular uptake mechanisms. Am J Physiol 1990;258:C140-6.

21. Kaplan J, O'Halloran T. Iron metabolism in eukaryotes: Mars and Venus at it again. Science 1996;271: 1510-2.

22. Harris E. Copper as a cofactor and regulator of copper, zinc, superoxide dismutase. J Nutr 1992; 123:636-40.

23. Mu D, Medzihradszky K, Adams G, Mayer P, Hines W, Burlingame A, Smith A, Cai D, Klinman J. Primary structures for a mammalian cellular and serum copper amine oxidase. J Biol Chem 1994;269:9926-32.

24. Werman M, Bhathena S, Turnlund J. Dietary copper intake influences skin lysyl oxidase in young men. J Nutr Biochem 1997;8:201-4.

25. Fosmire G. Zinc toxicity. Am J Clin Nutr 1990; 51:225-7.

26. Hoffman H, Phyliky R, Fleming C. Zinc-induced copper deficiency. Gastroenterology 1988;94: 508-12.

27. Barclay S, Aggett P, Lloyd D, Duffty P. Reduced erythrocyte superoxide dismutase activity in low birth weight infants given iron supplements. Pediatr Res 1991;29:297-301.

28. Olin K, Walter R, Keen C. Copper deficiency affects selenoglutathione peroxidase and selenodeiodinase activities and antioxidant defense in weanling rats. Am J Clin Nutr 1994;59:654-8.

29. Lonnerdal B. Bioavailability of copper. Am J Clin Nutr 1996;63(suppl):821S-9S.

30. National Research Council. Recommended Dietary Allowances, 10th ed. Washington, D.C.: National Academy Press 1989;224-30.

31. Milne D. Assessment of copper status. Clin Chem 1994;40:1479-84.

32. Turnlund J, Keen C, Smith R. Copper status and urinary and salivary copper in young men at three levels of dietary copper. Am J Clin Nutr 1990;51:658-64.

33. Turnlund J, Scott K, Peiffer G, Jang A, Keyes W, Keen C, Sakanashi T. Copper status of young men consuming a low-copper diet. Am J Clin Nutr 1997;65: 72-78.

34. Tamura H, Hirose S, Watanabe O, Arai K, Murakawa M, Matsumura O, Isoda K. Anemia and neutropenia due to copper deficiency in enteral nutrition. JPEN 1994;18:185-9.

35. Cordano A. Clinical manifestations of nutritional copper deficiency in infants and children. Am J Clin Nutr 1998;67(suppl):1012S-6S.

36. Jenkinson S, Lawrence R, Burk R, Williams D. Effects of copper deficiency on the activity of the selenoenzyme glutathione peroxidase and on excretion and tissue retention of $^{75}SeO_3^{2-}$. J Nutr 1982;112: 197-204.

37. Lei KY. Dietary copper: cholesterol and lipoprotein metabolism. Ann Rev Nutr 1991;11:265-83.

38. Copper deficiency and hypercholesterolemia. Nutr Rev 1987;45:116-7.

39. Reiser S, Powell A, Yang C, Canary J. Effect of copper intake on blood cholesterol and its lipoprotein distribution in men. Nutr Rep Internl 1987;36:641-9.

40. Pennington J, Young B. Total diet study nutritional elements, 1982-1989. J Am Diet Assoc 1991;91: 179-83.

41. Chuttani H, Gupta P, Gulati S, Gupta D. Acute copper sulfate poisoning. Am J Med 1965;39:849-54.

42. Bremner I. Manifestations of copper excess. Am J Clin Nutr 1998;67(suppl):1069S-73S.

43. Smithgall J. The copper-controlled diet: current aspects of dietary copper restriction in management of copper metabolism disorders. J Am Diet Assoc 1985;85:609-11.

44. Baker D, Odle J, Funk O, Wieland T. Bioavailability of copper in cupric oxide, cuprous oxide and in a copper lysine complex. Poultry Sci 1991;70:177-9.

45. Turnlund J. Human whole-body copper metabolism. Am J Clin Nutr 1998;67(suppl):960S-4S.

46. Milne D. Copper intake and assessment of copper status. Am J Clin Nutr 1998;67(suppl):1041S-5S.

Additional Reference

Genetic and environmental determinants of copper metabolism. Am J Clin Nutr. 1998;67:May supplement.

Selenium

Selenium, a nonmetal, exists in several oxidation states including Se^{2-}, Se^{4+}, and Se^{6+}. The chemistry of selenium is similar to that of sulfur; consequently, selenium can substitute for sulfur in amino acids such as methionine, cysteine, and cystine. Total body selenium content is about 15 mg.

Sources

Perhaps more than any other essential trace element, selenium varies greatly in its soil concentration throughout the regions of the world. This, in turn, relates directly to its concentration in food plants. Actually, this heterogeneous distribution has had scientific benefit in that it has provided a clear correlation between those selenium-poor regions of the world (such as parts of China) and the incidence of disease associated with selenium deficiency.

Because of the wide disparity in soil selenium concentrations, tables listing the mineral's content in assorted foods are only generally appropriate. Animal products (especially organ meats) are thought to contain more selenium than plant sources. Seafood is also thought to represent one of the better sources of selenium, although the bioavailability of selenium from fish especially fish containing mercury is thought to be low owing to the formation of unabsorbable mercury-selenium complexes [1,2].

Absorption, Transport, and Storage

Selenium occurs naturally in foods almost exclusively in the form of organic compounds, primarily

- selenomethionine (Fig. 12.14),
- selenocystine,
- selenocysteine (Fig. 12.14), and
- Se-methyl selenomethionine.

These organic forms are thought to represent selenium analogs of sulfur-containing amino acids, the element substitution made possible by the chemical similarity between selenium and sulfur. These selenium analogs become incorporated into plant proteins.

Inorganic forms of selenium include selenite (H_2SeO_3) and selenate (H_2SeO_4). These inorganic forms may be found in some vegetables. In addition, in those parts of the world where selenium levels in natural foodstuffs are low, animal feeds are generally supplemented with sodium selenite (Na_2SeO_3).

Absorption

The organic forms of selenium as well as the inorganic forms are all efficiently absorbed, although to different extents, from the gastrointestinal tract. Selenium absorption is shown in Figure 12.15. The duodenum appears to be the primary absorptive site. Less absorption occurs in the jejunum and ileum, and virtually none in the stomach.

Balance and stable isotopic tracer studies have generally shown that selenium, as selenomethionine, is more effectively absorbed than selenite [3]. Selenoamino acid absorption is estimated at 50% to 80%. Selenomethionine, however, is thought to be better absorbed than selenocysteine. Selenite absorption is thought to vary between 44% and 70%. Selenates are thought to be better absorbed than selenites [4].

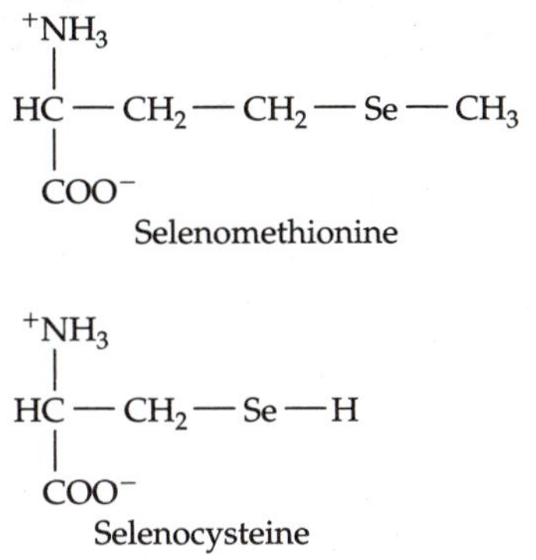

Factors Influencing Selenium Absorption Factors enhancing selenium absorption include vitamins C, A, and E, as well as the presence of reduced glutathione in the intestinal lumen. Heavy metals, such as mercury [2,5], and phytates are thought to inhibit selenium absorption through chelation and precipitation.

Transport

Following absorption from the intestine, selenium is bound to transport proteins for travel through the blood to the liver and other tissues. In human blood, selenium binds to sulfhydryl groups in α and β globulins. Specifically, lipoproteins such as VLDL (an α-2 globulin) and LDL (a β globulin) have been shown to contain and transport selenium (Fig. 12.15). Additional selenium transport proteins have been identified in some animals. The selenocystine-containing plasma protein called *selenoprotein P* has been isolated in the rat and appears to function as a selenium transport and possibly a storage protein [6]. Animal studies suggest that selenium is highly extracted by the liver [6].

$$^+NH_3$$
$$|$$
$$HC-CH_2-CH_2-Se-CH_3$$
$$|$$
$$COO^-$$
Selenomethionine

$$^+NH_3$$
$$|$$
$$HC-CH_2-Se-H$$
$$|$$
$$COO^-$$
Selenocysteine

Figure 12.14 Selenomethionine and selenocysteine.

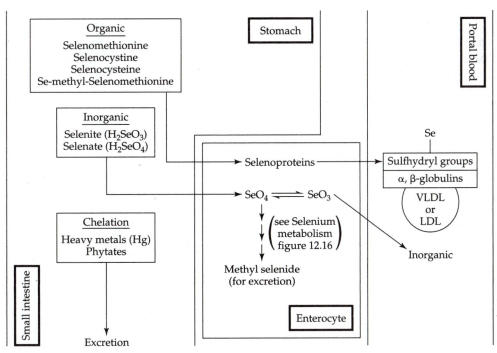

Figure 12.15 Overview of selenium digestion, absorption, and transport. Abbreviations: VLDL, very low-density lipoprotein; LDL, low-density lipoprotein.

Absorbed selenium becomes incorporated into a wide variety of proteins that may assume both transport and storage roles. The synthesis of these proteins appears to be induced by selenium. The rise and fall in concentration of various selenoproteins with time suggests that the proteins may be important in selenium flux among tissues [7].

Selenium Uptake by Tissues and Storage

The mechanism by which selenium is freed from plasma transport proteins is not known. Tissues containing relatively high selenium concentrations include the kidney, liver, heart, pancreas, and muscle. The lungs, brain, bone, and red blood cells also contain selenium. Uptake of selenium into the red blood cell is thought to occur by diffusion with selenite uptake exceeding selenate, which in turn exceeds selenomethionine uptake [8].

Higher tissue concentrations of selenium have resulted when selenium was administered as selenomethionine as opposed to selenite. But the reverse is true with respect to the uptake of selenium by its major metalloenzyme glutathione peroxidase. That is, dietary selenium in the inorganic form such as selenite leads to greater incorporation of the mineral into glutathione peroxidase than when selenomethionine, the organic form, is the dietary form [9].

Metabolism

Within tissues such as the liver, selenomethionine (derived from the diet) may be

- stored as selenomethionine in an amino acid pool,

- used for protein synthesis just as the amino acid methionine is used, or

- catabolized to Se-adenosylmethionine (SeAM) and ultimately to yield selenocysteine and selenocystine, similar to methionine metabolism to generate cysteine shown in Chapter 7, page 192. Selenium metabolism is shown in Figure 12.16.

Selenocysteine (derived from selenomethionine metabolism or from diet) may be degraded by selenocysteine β-lyase to yield free elemental selenium. Free elemental selenium may be attached to transfer RNA charged with serine and ultimately incorporated into the selenium-dependent enzymes for functional uses (see the

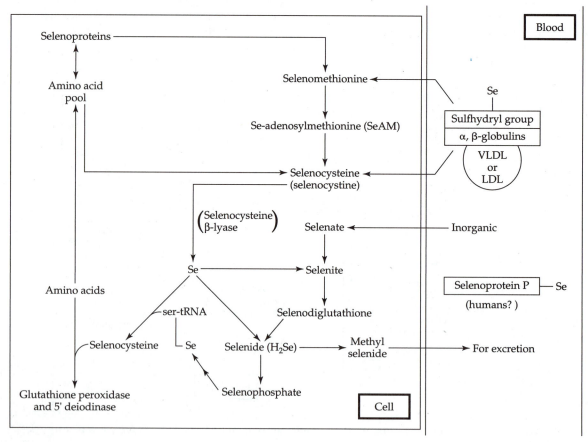

Figure 12.16 Selenium metabolism.

section on functions). Selenium not used as a cofactor for enzymes may be converted to selenide (H_2Se) or to selenite, may be stored for future use, or may be excreted.

Selenate from the diet may be converted in the body to selenite, which is further metabolized to selenodiglutathionine and subsequently to selenide. Selenide may be degraded to methylselenide for excretion or may be converted into selenophosphate. Selenophosphate can be metabolized and its selenium subsequently attached to tRNA for synthesis of 5'-deiodinase or glutathione peroxidase.

Functions and Mechanisms of Actions

Various incompletely understood roles for selenium in mammalian metabolism have been postulated. Some of the less defined roles include its involvement in the maintenance or induction of the cytochrome P_{450} system, in pancreatic function, in DNA repair and enzyme activation, immune system function, and detoxification of heavy metals.

A clearly established function of selenium, however, is as an essential, tightly bound, cofactor for *glutathio-ne peroxidase* (GPX). Glutathione peroxidase requires reduced glutathione (GSH) as a cosubstrate. Glutathi-one, a tripeptide of glycine, cysteine, and glutamate, is found in most cells of the body and furnishes reducing equivalents in reactions.

Five glutathione peroxidase enzymes designated GPX1, 2, 3, 4, and 5 have been characterized. The first four are selenium-dependent. GPX1 and GPX4, for example, are found in many tissues, most notably the liver, kidney, and red blood cell. GPX2 in turn is found predominantly in the gastrointestinal tract and liver. GPX3 is found in the kidney, heart, lung, and plasma, among other tissues [10]. Within tissues, glutathione peroxidase is found mainly (~70%) in the cytosol of cells and to a lesser extent (~30%) in the mitochondrial matrix.

Glutathione peroxidase catalyzes the reduction of both

- organic peroxides ROOH derived from unsaturated fatty acids (lipid peroxide LOOH), nucleic acids, and other molecules, and

- hydrogen peroxide (H_2O_2), as shown in the following reaction:

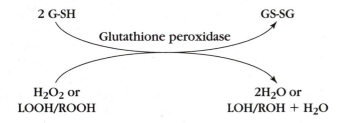

GS-SG represents oxidized glutathione, LOH represents an hydroxylipid, and ROH represents the hydroxy form of an organic compound. The reaction, catalyzed by glutathione peroxidase, neutralizes or eliminates hydrogen peroxide and organic (including lipid) peroxides. In fact, glutathione peroxidase is more active than catalase (see section on iron functions, p. 412) in reducing organic peroxides and hydrogen peroxides. These peroxides, if not removed, could damage cellular membranes.

Selenium availability has been shown to affect mRNA concentrations of glutathione peroxidase in the liver. With selenium deficiency, less mRNA for hepatic glutathione peroxidase is made and enzyme activity is diminished [5]. With selenium supplementation, there is a rapid increase in mRNA to control levels but a gradual increase in enzyme activity. The generation of hydrogen peroxides, lipid, and organic peroxides, as well as the interdependent roles of selenium (as part of glutathione peroxidase), vitamin E, iron (as catalase and myeloperoxidase), and zinc and copper (as superoxide dismutase), which also function as antioxidants to prevent free radical induced cell damage, are shown in the Perspective at the end of Chapter 10.

The regeneration of reduced glutathione from its oxidized state is imperative. Glutathione reductase catalyzes the reduction in a reaction dependent on $NADPH + H^+$ and shown here:

NADPH is derived from the hexose monophosphate shunt.

Selenium also is necessary for iodine metabolism. *5'-Iodothyronine deiodinase* (type I) has been shown to be a selenoprotein with a single selenium atom at its active site [5,11]; other deiodinases are non-selenium-dependent. Type I 5'-iodothyronine deiodinase is found in the endoplasmic reticulum of the liver, kidney, and muscle [5]. Other types of non-selenium-dependent deiodinases involved in thyroid hormone metabolism are found in tissues such as the pituitary, adipose, and brain.

Type I 5'-iodothyronine deiodinase (DI) catalyzes the deiodination of thyroxine (T_4) to form tri-iodothyronine (T_3) as well as the conversion of reverse (r) T_3 to 3,3'-di-iodothyronine. This reaction, important for the generation of a T_3 a primary hormonal regulator of metabolism as well as normal growth and development and the main active thyroid hormone, is shown here:

For further information regarding thyroid hormone metabolism, see the section of this chapter on iodine.

Interactions with Other Nutrients

The interrelationship between selenium and the other antioxidant nutrients is discussed in the Perspective section at the end of Chapter 10.

Lead and selenium appear to react such that subclinical amounts of lead intake are found to lower significantly the tissue concentration of selenium. Although this mechanism is unclear, the fact that both elements bind to sulfhydryl groups provides reason for speculation [12].

Iron deficiency in rats has been shown to decrease the synthesis of hepatic glutathione peroxidase and to decrease hepatic selenium concentrations [13]. The mechanism(s) of action are unclear but are thought to occur, in part, owing to decreased transcription of the gene for glutathione peroxidase. Alternately, selenium absorption or use may be responsible.

Copper and selenium also appear to interact. Copper deficiency has been shown to decrease the activity of the selenium-dependent enzymes glutathione peroxidase and 5'-deiodinase [14]; whether the interaction is related to iron is unclear.

Arsenic is thought to competitively inhibit the uptake of selenium. The significance of this interaction in human nutrition has not yet been determined.

An interaction also occurs between selenium and the amino acid *methionine*. Because dietary selenium can be found as selenomethionine and, as such, is readily bioavailable in most organisms, the potency of selenium in this form may be reduced if a situation of methionine deficiency exists [2]. The explanation for this observation is that if methionine is deficient, selenomethionine substitutes for it in the synthesis of body proteins [2,15]. Selenium then becomes available only as these proteins subsequently become degraded in the course of normal turnover.

Excretion

Selenium is excreted from the body almost equally in the urine and feces. About 50% to 60% of selenium, or about 50 μg, is excreted in the urine, while the remaining 40% to 50% is excreted in the feces. Selenium losses via the lungs and skin also contribute to daily selenium

excretion. Pulmonary elimination of selenium is associated with the exhalation of a volatile (garlicky smell) selenium compound, dimethylselenide [$(CH_3)_2Se$].

Renal clearance comparison studies on people with low body stores of selenium and on those with much higher levels indicate that the kidneys play an important role in selenium homeostasis in humans [16]. However, only a few (e.g., trimethyl selenonium ion) of the several metabolites of selenium have been identified in the urine.

Recommended Dietary Allowances

Balance studies as well as repletion studies of men with selenium deficiency in regions of China have aided researchers in the formulation of a RDA for selenium. The balance study technique alone is of little help in determining the selenium requirements of humans, because humans can seemingly adjust their selenium homeostatic mechanisms to remain in balance despite a wide range of dietary intakes. Results of balance studies found fecal and urinary selenium losses were about 82 μg and 62 μg in men and women, respectively; selenium intake of the men and women was 90 μg and 74 μg, respectively [17]. In the repletion studies, the selenium requirement was estimated to be 10 μg selenium to prevent Keshan disease (described in the "Deficiency" section that follows), whereas an intake of 40 μg/day improved and plateaued glutathione peroxidase activity [17]. A so-called physiological requirement, based on the amount of dietary selenium required to maximize plasma glutathione peroxidase activity, is thus considered to be 40 μg/day. Following corrections for body weight and individual variation, this method resulted in a recommended dietary selenium intake of 70 and 55 μg/day for adult male and females, respectively, in the United States [17,18].

Deficiency

Selenium deficiency has been linked to a number of livestock animal diseases and also to regional human diseases such as Keshan disease and Kashin-Beck's disease in China [19]. Keshan disease is characterized by cardiomyopathy involving cardiogenic shock and/or congestive heart failure, along with multifocal necrosis of heart tissue, which becomes replaced with fibrous tissue [19]. Kashin-Beck's disease is characterized by osteoarthropathy involving degeneration and necrosis of the joints and epiphyseal-plate cartilages of the legs and arms [19]. Several factors, including selenium deficiency, are thought to contribute to the development of Kashin-Beck's disease.

Selenium deficiency has also been observed in people receiving total parenteral nutrition [20–22]. Predominant symptoms of deficiency included poor growth, muscle pain and weakness, along with loss of pigmentation of hair and skin, and whitening of nail beds. Poor growth may be associated with the role of selenium in thyroid hormone metabolism.

A possible connection between selenium deficiency and cardiovascular disease and cancer has been postulated. In short, such studies remain inconclusive [23–25].

Toxicity

Selenium toxicity, also called *selenosis,* has been observed both in miners and in individuals consuming excess selenium intake from supplements. Intakes of 750 μg or up to 27.3 mg or more per day have produced physical manifestations as well as biochemical abnormalities [18]. Selected signs and symptoms of toxicity include nausea, vomiting, fatigue, hair and nail loss, changes in nail beds, interference in sulfur metabolism (primarily oxidation of sulfhydryl groups), and inhibition of protein synthesis [18,26]. The calculated reference dose (an estimate of exposure that is likely to be without appreciable deleterious consequences over a lifetime) is 5 μg/kg/day and includes an uncertainty factor of three. The no observed adverse effect level (NOAEL) and lowest observed adverse effect level (LOAEL) for selenium is 200 and 910 μg, respectively [27].

Assessment of Nutriture

Serum or plasma selenium concentrations are thought to be indicative of short-term changes in dietary selenium intake and will respond more quickly to changes in selenium intake or supplementation than whole-blood selenium. Whole-blood selenium or red blood cell selenium represents an index of longer-term selenium status than plasma [18] and is useful for populations with habitually low selenium intakes.

The activity of glutathione peroxidase in platelets is also used as an indicator of selenium status [28]. The enzyme in platelets has a relatively rapid turnover, a high selenium content, and enzyme activity responds rapidly to changes in selenium intake [1]. Enzyme activity plateaus, however, as intake increases, therefore serving as an index of selenium status in populations with low intake [18]. Good correlations between blood or plasma selenium concentration (up to a concentration of ~1.0 mmol/L) and glutathione peroxidase activity in red blood cells has been shown and suggests that plasma selenium can be used to assess selenium status as long as plasma selenium is <1 mmol/L [29]. Toenail clippings have also been suggested as being reflective

of selenium status for up to 1 year prior to sampling [30,31]. Urinary selenium concentrations most appropriately identify selenium toxicity.

References Cited for Selenium

1. Levander O. Considerations in the design of selenium bioavailability studies. Fed Proc 1983;42:1721–5.

2. Forbes R, Erdman J. Bioavailability of trace mineral elements. Ann Rev Nutr 1983;3:213–31.

3. McAdam P, Lewis S. Absorption of selenite and L-selenomethionine in healthy young men using a ^{74}Se tracer. Fed Proc 1985;44:1671.

4. Thomson C, Robinson M. Urinary and fecal excretions and absorption of a large supplement of selenium: superiority of selenate over selenite. Am J Clin Nutr 1986;44:659–63.

5. Burk R, Hill K. Regulation of selenoproteins. Ann Rev Nutr 1993;13:65–81.

6. Kato T, Read R, Rozga J, Burk R. Evidence for intestinal release of absorbed selenium in a form with high hepatic extraction. Am J Physiol 1992;262:G854–8.

7. Evenson J, Sunde R. Selenium incorporation into seleno proteins in the selenium-adequate and selenium-deficient rat. Proc Soc Exp Biol Med 1988; 187:169–80.

8. Combs G, Combs S. The role of selenium in nutrition. Orlando: Academic Press, 1986.

9. Whanger P, Butler J. Effects of various dietary levels of selenium as selenite or selenomethionine on tissue selenium levels and glutathione peroxidase activity in rats. J Nutr 1988;118:846–52.

10. Avissar N, Ornt DB, Yagil Y, Horowitz S, Watkins R, Kerl E, Takahashi K, Palmer I, Cohen H. Glutathione peroxidase activity in human plasma originates mainly from kidney proximal tubular cells. FASEB J 1993;7:A277.

11. Berry M, Larsen P. The role of selenium in thyroid hormone action. Endocrin Rev 1992;13:207–19.

12. Neatherly M, Miller W, Gentry R, et al. Influence of high dietary lead on selenium metabolism in dairy calves. J Dairy Sci 1987;70:645–52.

13. Moriarty P, Picciano M, Beard J, Reddy C. Iron deficiency decreases Se-GPX mRNA level in the liver and impairs selenium utilization in other tissues. FASEB J 1993;7:A277.

14. Olin K, Walter R, Keen C. Copper deficiency affects selenoglutathione peroxidase and selenodeiodinase activities and antioxidant defense in weanling rats. Am J Clin Nutr 1994;59:654–8.

15. Waschulewski I, Sunde R. Effect of dietary methionine on utilization of tissue selenium from dietary

selenomethionine for glutathione peroxidase in the rat. J Nutr 1988;119:367–74.

16. Robinson J, Robinson M, Levander O, Thomson C. Urinary excretion of selenium by New Zealand and North American human subjects on different intakes. Am J Clin Nutr 1985;41:1023–31.

17. Levander O. Scientific rationale for the 1989 recommended dietary allowance for selenium. J Am Diet Assoc 1991;91:1572–6.

18. National Research Council. Recommended Dietary Allowances, 10th ed. Washington, D.C.: National Academy Press, 1989;217–24.

19. Ge K, Yang G. The epidemiology of selenium deficiency in the etiological study of endemic diseases in China. Am J Clin Nutr 1993;57:259S–63S.

20. Abrams C, Siram S, Galsim C, Johnson-Hamilton H, Munford F, Mezghebe H. Selenium deficiency in long-term total parenteral nutrition. Nutr Clin Prac 1992; 7:175–8.

21. van Rij A, Thomson C, McKenzie J, Robinson M. Selenium deficiency in total parenteral nutrition. Am J Clin Nutr 1979;32:2076–85.

22. Vinton N, Dahlstrom K, Strobel C, Ament M. Macrocytosis and pseudoalbinism: manifestations of selenium deficiency. J Pediatr 1987;111:711–7.

23. Virtamo J, Huttunen J. Minerals, trace elements, and cardiovascular disease. Ann Clin Res 1988;20:102–13.

24. Garland M, Morris J, Stampfer M, Colditz G, Spate V, Baskett C, Rosner B, Speizer F, Willett W, Hunter D. Prospective study of toenail selenium levels and cancer among women. J Natl Cancer Inst 1995;87:497–505.

25. Clark L, Alberts D. Selenium and cancer: risk or protection? J Natl Cancer Inst 1995;87:473–5.

26. Lane H, Lotspeich C, Moore C, Ballard J, Dudrick S, Warren D. The effect of selenium supplementation on selenium status of patients receiving chronic total parenteral nutrition. JPEN 1987;11:177–82.

27. Hathcock J. Vitamins and minerals: efficacy and safety. Am J Clin Nutr 1997;66:427–37.

28. Neve J, Vertongen F, Capel P. Selenium supplementation in healthy Belgian adults: response in platelet glutathione peroxidase activity and other blood indices. Am J Clin Nutr 1988;48:139–43.

29. Diplock A. Indexes of selenium status in human populations. Am J Clin Nutr 1993;57:256S–8S.

30. Ovaskainen M, Virtamo J, Alfthan G, Haukka J, Pietinen P, Taylor P, Huttunen J. Toenail selenium as an indicator of selenium intake among middle-aged men in an area with low soil selenium. Am J Clin Nutr 1993;57:662–5.

31. Longnecker M, Stampfer M, Morris J. A 1 year trial of the effect of high selenium bread on selenium concentrations in blood and toenails. Am J Clin Nutr 1993;57:408–13.

Additional Reference

Arthur JR, ed. Interrelationships between selenium deficiency, iodine deficiency, and thyroid hormones. Am J Clin Nutr 1993;57:235S–318S.

Chromium

Chromium, a metal, exists in several oxidation states, from Cr^{2-} to Cr^{6+}. The metal has ubiquitous presence, found in air, water, and soil. Cr^{3+}, or trivalent chromium, is the most stable of the oxidation states and is thought to be of the most importance in humans. As trivalent chromium, the mineral often binds to ligands containing nitrogen, oxygen, or sulfur to form hexacoordinate or octahedral complexes.

Sources

Numerous analytic difficulties and biological uncertainties have been associated with the determination of chromium in food. However, with advances in methodology and contamination control, chromium in foods may now be measured with accuracy.

In foods, most chromium exists in the trivalent form (Cr^{3+}). Good sources of dietary chromium are meats (especially organ meats) and grains (especially whole grains) [1,2]. Other foods providing relatively high amounts of chromium include cheese, mushrooms, various condiments and spices, as well as tea, beer, and wine [1,2]. Brewer's yeast is notable because of its high suspected content of the biologically active organically complexed form of chromium known as *glucose tolerance factor* (GTF).

Foods, in addition to containing chromium, may also contain an unidentified compound that potentiates the actions of insulin; activity is not, however, correlated with chromium content [1]. Foods that have demonstrated insulin-potentiating activity include tuna fish, peanut butter, and vanilla ice cream; spices such as cinnamon, cloves, bay leaves, and turmeric also exhibit the ability to potentiate activity [1].

Food processing and refining can affect the chromium content of foods. Refining of sugar, for example, diminishes chromium. Thus, molasses and brown sugar are higher in chromium than white sugar. In contrast, chromium is easily solubilized from stainless steel cookware or cans into acidic foods. Thus, use of stainless may increase the amount of chromium in the food [3].

Absorption, Transport, and Storage

In acidic solutions, as would be found in the stomach, Cr^{3+} is soluble and may form complexes with ligands. Chromium is thought to be absorbed throughout the small intestine, especially in the jejunum [4]. The mode of absorption is still not known. Chromium is suspected to be absorbed either by diffusion or by a carrier-mediated transporter.

Absorption of chromium in food appears to be rapid and dependent on dose. Absorption averages about 2% with intakes of 10 μg and decreases to about 0.5% with intakes of 40 μg. Thus, reported absorption ranges vary from 0.4% to 3% [5].

Factors Influencing Chromium Absorption

Like other trace minerals, chromium absorption may be influenced by dietary factors.

Enhancers of Chromium Absorption Within the stomach, amino acids or other ligands may chelate inorganic chromium [6]. Amino acids such as methionine and histidine, for example, act as ligands for chromium absorption. Such chelations typically help chromium to remain soluble and prevent olation (see next paragraph) once it reaches the alkaline pH of the small intestine. Solubility improves chromium absorption. Picolinate, for example, was used in part as a ligand because it is relatively stable and lipophilic. Vitamin C may enhance chromium absorption. Consumption of 1 mg of chromium as chromium chloride along with 100 mg ascorbate was associated with increased plasma chromium concentrations [7,8].

Inhibitors of Chromium Absorption Inorganic chromium in a neutral or alkaline environment reacts with hydroxyl ions ($Cr\text{-}OH^-$), which readily polymerize to form high-molecular-weight compounds; this process is called olation and results in precipitation of chromium and thus reduced absorption. Ingestion of antacids significantly reduces blood and tissue chromium concentrations [7]. Phytates found in grains and possibly other minerals also appear to negatively influence chromium absorption.

Transport

In the blood, inorganic Cr^{3+} binds competitively with transferrin and is transported in the blood along with iron bound to transferrin. If transferrin sites are unavailable for chromium, albumin is thought to transport chromium. Globulins and possibly lipoproteins are thought to transport the mineral if present in very high concentrations. Some chromium also may circulate unbound in the blood.

Exactly how the organically complexed chromium is transported in the blood is uncertain, but it is rapidly available to cells after absorption. Moreover, only the organically complexed chromium (GTF) is active; therefore, absorbed inorganic chromium must be transported to a site where its incorporation into the organic complex can occur. The liver is proposed as a possible site for the synthesis of the metabolically active molecule [9], and use of the inorganic chromium for potentiation of insulin action is delayed until organic complexing has occurred [1,2,7].

Metabolically active chromium is believed to be held in a body pool from which the active molecule can be released for use as needed.

Storage

The adult body pool size is estimated to be approximately 4 to 6 mg [10]. Organically complexed chromium in food can enter this body pool immediately on absorption. Tissues high in chromium include the kidney, liver, muscle, spleen, heart, pancreas, and bone. Chromium is thought to be stored with ferric iron because of its transport by transferrin.

Functions and Mechanisms of Actions

The biological action of chromium is believed due to its complexing with nicotinic acid and amino acids to form the organic compound glucose tolerance factor (GTF). GTF is thought to initiate the disulfide bridging between insulin and the insulin receptor (Fig. 12.17). The effectiveness of insulin is greater in the presence of chromium than in its absence [11]. Thus, the primary function of chromium or GTF is to potentiate insulin action, thereby affecting cellular glucose uptake, and intracellular carbohydrate and lipid metabolism. However, chromium also may be involved in pancreatic insulin secretion, internalization of insulin through decreasing membrane fluidity or regulation of insulin receptor production, expression, and/or activity to ultimately improve insulin's effectiveness [12–14]. Chromium increases insulin sensitivity by activating insulin receptor kinase, an enzyme involved in insulin phosphorylation [15–17]. Phosphorylation of insulin receptors is required for potentiating insulin activity [15–17].

GTF was first identified in brewer's yeast. Although this factor has never been purified or its exact structure characterized, complexes with good biological activity have been synthesized from niacin, chromium, and glutathione. None of the synthetic complexes, however, exactly duplicate the naturally occurring organically complexed chromium. Nevertheless, the belief remains that

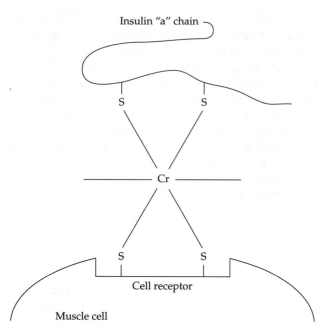

Figure 12.17 Proposed interaction of Cr as part of GTF to insulin and cell's insulin receptor.
Source: Modified from Mertz W, Toepfer EW, Roginski EE, Polansky MM. Present knowledge of the role of chromium. Fed Proc 1974;33:2276.

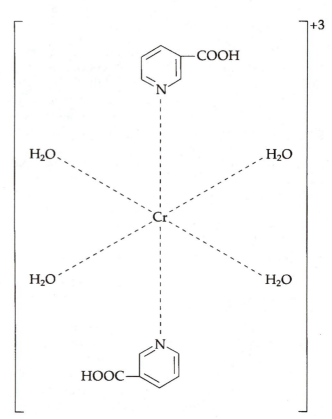

Figure 12.18 Tetra-aquo-di-nicotinato chromium complex. Water molecules are believed to be replaced by amino acids (glutamic acid, cysteine, and glycine) to stabilize the complex. *Source:* From Nutrition Reviews 33(1975):129–35. © International Life Sciences Institute—Nutrition Foundation. Used with permission.

the biologically active molecule is a dinicotinato chromium complex coordinated with amino acids that stabilize the complex. Mertz [9] proposed that the initial product of in vivo synthesis is a tetra-aquo dinicotinato chromium complex, as shown in Figure 12.18. Because this complex is unstable in the alkaline pH of the organism, Mertz proposed that the coordinated water molecules are replaced by ligands (amino acids), thereby stabilizing the complex and preventing its precipitation. The amino acids acting as ligands are believed to be those making up glutathione (glutamic acid, cysteine, and glycine). The exact coordination of the amino acids with the chromium is uncertain.

GTF, contained in a body pool, is believed to resemble a hormone being released into the blood in response to a physiological stimulus [9]. In the case of chromium, the stimulus is insulin. After release from the body pool, the active Cr^{3+} as GTF is transported to the periphery, where it exerts a marked biological action by potentiating the action of insulin. The failure to identify the precise action of GTF in potentiation of insulin activity has, however, raised a question about the essentiality of chromium. Although the insulin receptor has been purified and characterized, no evidence has been uncovered for chromium as a component of the receptor's subunits, as part of an accessory protein for insulin binding, or as a second messenger in mediating the effect of insulin [10,12,13].

Chromium may have additional roles or may exhibit its influence through the actions of insulin. Roles particularly in glucose and lipid metabolism have been suggested for chromium. For example, chromium may improve glucose intolerance if impaired and the individual has suboptimal chromium status. Chromium also may affect lipoprotein lipase activity and/or in some unknown way affect cholesterol metabolism. Several studies have reported improvements in blood lipid profiles of people following chromium supplementation. Significant increases in blood HDL cholesterol along with decreases in total and LDL cholesterol concentrations have been demonstrated with chromium supplementation [11–13,18–20].

Another proposed role for chromium is in relation to nucleic acid metabolism. It is postulated that Cr^{3+} is involved in maintaining the structural integrity of nuclear strands and in the regulation of gene expression [21]. RNA synthesis in vitro as directed by DNA is enhanced by chromium binding to the template [22].

Chromium as a supplement has been purported to effect changes in body composition and strength performance. However, most well-controlled studies providing chromium supplementation have shown no significant effects on strength gains, muscle accretion, and fat loss [23–27].

Interactions with Other Nutrients

Because chromium is transported in the blood bound to transferrin, the primary iron-binding protein, one may surmise the potential for chromium, if given in large amounts, to displace *iron* from the transferrin. Indeed, ingestion of chromium (about 200 μg) as chromium chloride and chromium picolinate was associated with a significant decrease in serum ferritin, total iron-binding capacity and transferrin saturation in 36 men [25]. Other studies, however, report that chromium picolinate ingestion (924 μg) had no effects on hematologic indexes in men [25].

Excretion

Dietary chromium or chromium released into the plasma in response to a glucose challenge is excreted from the body mainly via the kidneys and, to a lesser extent, in sweat, feces, and hair. Urinary chromium normally varies from 0.2 to 0.4 μg/day; this amount, however, represents about 95% of the daily chromium excretion. Consumption of diets high in simple sugars (35% simple sugars, 15% complex carbohydrates) has been shown to raise urinary chromium in some subjects to 300%, in contrast to consumption of diets high in complex (starch) carbohydrates (35% complex carbohydrates, 15% simple sugars) [28].

Estimated Safe and Adequate Daily Dietary Intake

The estimated safe and adequate intake for chromium was first reported in the 1980 edition of the RDA. Ingestion of 200 μg chromium per day was proposed based on the following:

- the daily losses of chromium approximate 1 μg/day, and
- an absorbability of only 0.5% is assumed to occur with inorganic chromium ingestion.

Thus, ingestion of 200 μg of chromium per day with 0.5% absorbability would provide the 1 μg needed to replenish losses.

In the 1989 edition of the RDA, a recommended estimated safe and adequate daily dietary intake range for chromium of 50 to 200 μg/day is given for adults and adolescents [29]. This recommendation is based on the fact that the average intake of 50 μg/day appears sufficient to prevent signs of chromium deficiency in most of the U.S. population [30].

Deficiency

Signs and symptoms of chromium deficiency in the general population appear widespread [31]. Concern has existed about chromium nutriture in the United States because

- some segments of the population (especially the elderly) are consuming less than the recommended chromium intake [30,32],
- tissue chromium levels appear to decline with age [33], and
- the evidence of impaired glucose tolerance increases among the aged [30]. Improved chromium status results in improved glucose metabolism in those with diabetes and glucose intolerance.

Chromium needs may be increased in certain diseases, such as diabetes mellitus and coronary heart disease, although a link between chromium and these diseases is not conclusive. Chromium deficiency results in insulin resistance characterized by hyperinsulinemia. Hyperinsulinemia has been implicated as a risk factor for coronary heart disease [34]. Mild chromium deficiency also has been shown to be a risk factor for a group of symptoms (except hypertension) similar to Syndrome X [35]. Syndrome X represents a constellation of abnormalities that increase the risk of coronary heart disease and include hyperinsulinemia, resistance to insulin-stimulated glucose uptake, glucose intolerance, hypertriglyceridemia, decreased blood HDL concentrations, and hypertension [36].

Severe trauma and stress appear to increase the need for chromium. Stress, for example, elevates the secretion of hormones such as glucagon and cortisol, which alters glucose and ultimately chromium metabolism. In addition, individuals receiving total parenteral nutrition without chromium develop symptoms of deficiency including impaired glucose tolerance with high blood glucose and glucose excretion in the urine, neuropathy, and high plasma free fatty acid concentrations.

Supplements

Chromium is available in supplement form as inorganic salts such as with chloride or as an organic complex such as with acetate, nicotinic acid alone or with amino acids, or picolinic acid. Although all forms appear to be absorbed and utilized, the form of the supplement appears to affect tissue concentrations in rats.

While chromium picolinate, because of its increased solubility (lipophilic), has been touted as superior to other forms of chromium, chromosomal damage has been demonstrated in hamster cells with use of chromium picolinate [37]. Advertisements that suggest use of chromium picolinate may help one to lose fat and gain muscle (lean body) mass do not appear to be entirely supported by scientific research [23–27].

Toxicity

Oral supplementation of up to 800 to 1,000 μg of chromium as Cr^{3+} appears to be safe [29,38]. However, picolinate and chromium (Cr^{3+}) picolinate have been shown to produce chromosomal damage in hamster cells [37]. Toxicity is associated with exposure to the hexavalent form (Cr^{6+}) of chromium that may be absorbed through the skin or that may enter the body through inhalation. Inhalation or direct contact with hexavalent chromium may result in respiratory disease, and in dermatitis and skin ulcerations, respectively. Liver damage may also occur. Cr^{6+} ingested orally is about 10 to 100 times more toxic than Cr^{3+} [39]. The no observed adverse effect level (NOAEL) for chromium Cr^{3+} is set at 1,000 μg daily [40].

Assessment of Nutriture

No specific tests are currently available to determine chromium status prior to supplementation [11]. Although a plasma chromium level of approximately 0.5 ng/mL is considered normal, the chromium content of physiological fluids is not indicative of status [41]. Fasting plasma chromium is not in equilibrium with tissue chromium. Responses of plasma chromium to an oral glucose load are inconsistent. Urinary chromium appears to reflect only recent intake but does not reflect status [41]. Hair chromium concentrations may indicate the status of a large population but not of individuals [11].

Relative chromium status can be evaluated retrospectively through following the effects of chromium supplementation on adults. Adult subjects, showing improvement in glucose (lower plasma glucose) and/or lipid parameters (such as increased HDL, decreased total and LDL cholesterol concentrations) after supplementation with approximately 200 μg chromium/day for 1 to 3 months, can be considered as having been in a marginally low chromium status [6].

References Cited for Chromium

1. Khan A, Bryden N, Polansky M, Anderson R. Insulin potentiating factor and chromium content of selected foods and spices. Biol Trace Elem Res 1990;24:183–8.

2. Kumpulainen J. Chromium content of foods and diets. Biol Trace Elem Res 1992;32:9–18.

3. Kuligowski J, Halperin K. Stainless steel cookware as a significant source of nickel, chromium, and iron. Arch Environ Contam Toxicol 1992;23:11–215.

4. Anderson R. Chromium. In: Mertz W, ed. Trace Elements in Human and Animal Nutrition, 5th ed. San Diego: Academic Press, 1987;1:225–44.

5. Offenbacher E, Spencer H, Dowling H, Pi-Sunyer F. Metabolic chromium balances in men. Am J Clin Nutr 1986;44:77–82.

6. Dowling H, Offenbacher E, Pi-Sunyer X. Effects of amino acids on the absorption of trivalent chromium and its retention by regions of the rat small intestine. Nutr Res 1990;10:1261–71.

7. Seaborn C, Stoecker B. Effects of antacid or ascorbic acid on tissue accumulation and urinary excretion of ^{51}chromium. Nutr Res 1990;10:1401–7.

8. Offenbacher E. Promotion of chromium absorption by ascorbic acid. Trace Elem Electrolytes 1994;11:178–81.

9. Mertz W. Effects and metabolism of glucose tolerance factor. Nutr Rev 1975;33:129–35.

10. Is chromium essential for humans? Nutr Rev 1988;46:17–20.

11. Mertz W. Chromium in human nutrition: a review. J Nutr 1993;123:626–33.

12. Evans G. The effect of chromium picolinate on insulin controlled parameters in humans. Int J Biosocial Med Res 1989;11:163–80.

13. Evans G, Bowman T. Chromium picolinate increases membrane fluidity and rate of insulin internalization. J Inorgan Biochem 1992;46:243–50.

14. Striffler J, Polansky M, Anderson R. Dietary chromium enhances insulin secretion in perfused rat pancreas. J Trace Elem Exper Med 1993;6:75–81.

15. Saad M. Molecular mechanisms of insulin resistance. Brazilian J Med Biol Res 1994;27:941–57.

16. Davis C, Vincent J. Chromium oligopeptide activates insulin receptor kinase activity. Biochemistry 1997;36:4382–5.

17. Roth R, Lui F, Chin J. Biochemical mechanisms of insulin resistance. Hormone Res 1994;41(suppl2):51–55.

18. Anderson R. Nutritional factors influencing the glucose/insulin system: chromium. J Am Coll Nutr 1997;16:404–10.

19. Thomas V, Gropper S. Effect of chromium nicotinic acid supplementation on selected cardiovascular disease risk factors. Biol Trace Elem Res 1996;55:297–305.

20. Anderson R, Polasky M, Bryden N, Canary J. Supplemental chromium effects on glucose, insulin, glucagon, and urinary chromium losses in subjects consuming controlled low chromium diets. Am J Clin Nutr 1991;54:909–16.

21. Stoecker B. Chromium. In: Brown ML, ed. Present Knowledge in Nutrition. Washington, D.C.: International Life Sciences Institute Nutrition Foundation, 1990;287–93.

22. Nielsen F. Chromium. In: Shils M, Olson J, Shike M., eds. Modern Nutrition in Health and Disease. Philadelphia: Lea and Febiger, 1994;264–8.

23. Clarkson P. Effects of exercise on chromium levels: is supplementation required? Sports Med 1997;23:341–9.

24. Hasten D, Rome E, Franks D, Hegsted M. Effects of chromium picolinate on beginning weight training students. Int J Sports Nutr 1992;2:343–50.

25. Lukaski H, Bolonchuk W, Siders W, Milner D. Chromium supplementation and resistance training: effects on body composition, strength and trace element status of men. Am J Clin Nutr 1996;63:954–65.

26. Clancy S, Clarkson P, DeCheke M, Nosaka K, Freedson P, Cunningham J, Valentine B. Effects of chromium picolinate supplementation on body composition, strength, and urinary chromium loss in football players. Int J Sports Nutr 1994;4:142–53.

27. Campbell W, Beard J, Joseph L, Davey S, Evans W. Chromium picolinate supplementation and resistive training by older men: effects on iron-status and hematologic indexes. Am J Clin Nutr 1997;66:944–9.

28. Kozlovsky A, Moser P, Reiser S, et al. Effects of diets high in simple sugars on urinary chromium losses. Metabolism 1986;35:515–8.

29. National Research Council. Recommended Dietary Allowances, 10th ed. Washington, D.C.: National Academy Press, 1989;241–3.

30. Bunker V, Lawson M, Delves H et al. The uptake and excretion of chromium by the elderly. Am J Clin Nutr 1984;39:797–802.

31. Anderson R. Recent advances in the clinical and biochemical effects of chromium deficiency. In: Prasad AS, ed. Essential and toxic trace elements in human health and disease: an update. New York: Wiley-Liss, 1993;221–34.

32. Anderson R, Kozlovsky A. Chromium intake, absorption, and excretion of subjects consuming self-selected diets. Am J Clin Nutr 1985;41:1177–83.

33. Mertz W. Chromium levels in serum, hair, and sweat decline with age. Nutr Rev 1997;55:373–5.

34. Zavaroni I, Bonora E, Pagliara M, Dall'aglio E, Luchetti L, Buonanno G, Bonati PA, Bergonzani M, Gnudi L, Passeri M, Reaven G. Risk factors for coronary artery disease in healthy persons with hyperinsulinemia and normal glucose tolerance. N Engl J Med 1989;320:702–6.

35. Reaven G. The role of insulin resistance and hyperinsulinemia in coronary heart disease. Metabolism 1992:41:16–19.

36. Reaven GM. The role of insulin resistance in human disease. Diabetes 1988;37:1595–1607.

37. Stearns D, Wise J, Patierno S, Wetterhahn K. Chromium (III) picolinate produces chromosome damage in Chinese hamster ovary cells. FASEB J 1995; 9: 1643–8.

38. Anderson R. Chromium as an essential nutrient for humans. Regulatory Toxicol and Pharmacol 1997;26:S35–S41.

39. Katz S, Salem H. The toxicology of chromium with respect to its chemical speciation. J Appl Toxicol 1993;13:217–24.

40. Hathcock J. Vitamins and minerals: efficacy and safety. Am J Clin Nutr 1997;66:427–37.

41. Anderson R, Polansky M, Bryden N, Patterson K, Veillon C, Glinsmann W. Effects of chromium supplementation on urinary chromium excretion of human subjects and correlation of chromium excretion with selected clinical parameters. J Nutr 1983;113:276–81.

Iodine

Iodine, a nonmetal, is typically found and functions in its ionic form iodide, I^-. Hence the term *iodide* is used throughout this discussion of the trace element. About 15 to 20 mg iodide is found in the human body.

Sources

The iodide concentration in human foods is extremely variable because, as is so often the case, it reflects the regionally variable soil concentrations of the element and the amount and nature of fertilizer used in plant cultivation. Thus, the iodide content of grains, vegetables, and fruits varies with the iodide content of the soil, and the iodide content of meats depends on the iodide of the soil and plants that the animals ate. The amount of iodide in the drinking water is an indication of the iodide content of the rocks and soils of a region, and it parallels closely the incidence of iodine deficiency among the inhabitants of that region. Numerous investigations, spanning decades, have demonstrated unequivocally the relationship of low levels of iodide in drinking water to the incidence of goiter (see the later section on iodine deficiency). For example, the iodide content of water from goitrous areas in India, Nepal, and Ceylon ranged from 0.1 to 1.2 mg/L compared with 9.0 mg/L found in nongoitrous Delhi [1].

In addition to water, iodide is found in seafoods; however, large differences in content exist between seawater fish and freshwater fish. Edible sea fish contain about 300 to 3,000 μg I/kg, in contrast to only 20 to 40 μg I/kg freshwater fish.

An additional source of iodide is from breads and grain products made from bread dough. Dough oxidizers or conditioners contain iodates (IO_3^-) as food additives to improve cross-linking of the gluten. Such iodates provide about 500 mg I^- per 100 g of bread [2].

Digestion, Absorption, Transport, and Storage

Dietary iodine (I) is either bound to amino acids or is found free, primarily in the form of iodate (IO_3^-) or iodide (I^-) (Fig. 12.19). During digestion, iodate, for example from breads, is reduced to iodide by glutathione [3].

Iodide is absorbed rapidly and completely throughout the gastrointestinal tract, including the stomach. Very little iodide appears in the feces. The small quantities of iodinated amino acids and other organic forms of iodide are absorbed, but not as efficiently as the iodide ion. The thyroid hormones thyroxine (T_4) and triiodothyronine (T_3) also are absorbed unchanged, therefore allowing T_4 medication to be administered orally.

Following absorption, free iodide appears in the blood (Fig. 12.19). Iodide is distributed throughout the extracellular fluid from which it is capable of permeating all tissues.

The element selectively concentrates, however, in the thyroid, salivary, and gastric glands. Lesser amounts of iodide are found in mammary glands, ovaries, placenta, and skin. It is the thyroid gland that traps iodide most aggressively, doing so by way of a sodium-dependent, active transport system against an iodide gradient that is often 40 to 50 times the plasma concentration. The thyroid gland contains 70% to 80% of the total body iodide and takes up about 120 μg of iodide per day.

Because the thyroid gland and its synthesis of the thyroid hormones are the focal points of iodide metabolism, information on the transport of iodide into nonthyroidal tissue is sparse. It is likely, however, that salivary gland uptake proceeds by an active transport mechanism similar to that of the thyroid [4].

Functions and Mechanisms of Actions

The main function of iodide is for the synthesis of the thyroid hormones by the thyroid gland. The thyroid gland is made of multiple acini, also called *follicles*. The follicles are spherical in shape and are surrounded by a single layer of thyroid cells. The follicles are filled with colloid, a proteinaceous material.

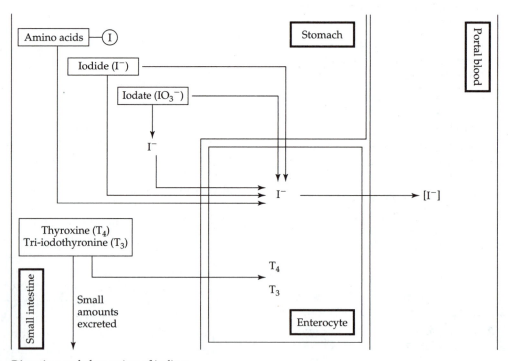

Figure 12.19 Digestion and absorption of iodine.

Figure 12.20 illustrates the synthesis of thyroid hormones. Both amino acids and iodide are needed to synthesize thyroid hormones. The events in thyroid hormone synthesis are as follows:

• Amino acids are required in the thyroid cells for the synthesis of thyroglobulin (a glycoprotein).

• The thyroid cells also collect the iodide actively. In fact, the thyroid gland must trap approximately 60 mg of iodide daily against a steep gradient of the element to ensure an adequate supply of hormones [5]. The trapping mechanism operates through a Na^+ (in)-K^+ (out) ATPase pump (Fig. 12.20) [6].

• Once within the cell, iodide (I^-) is oxidized to iodine (I), which is then bound to the number 3 position of tyrosyl residues of thyroglobulin (a process called *organification* of the iodine). The binding of iodine to the tyrosyl residue of thyroglobulin is catalyzed by thyroid peroxidase and generates 3-monoiodotyrosine (MIT) (Fig. 12.20). Hydrogen peroxide acts as the electron acceptor.

• Next, MIT is iodinated in the number 5 position to form 3,5-di-iodotyrosine (DIT). In the colloid, two DIT condense or couple to form 3,5,3',5'-tetraiodothyronine (T_4) with the elimination of an alanine side chain.

• Small amounts of DIT condense with MIT to form 3,5,3'-tri-iodothyronine (T_3) and reverse (r) T_3.

The structures of MIT, DIT, thyroxine (T_4) and 3,5,3'-tri-iodothyronine (T_3) are shown in Figure 12.21.

Transport of Thyroid Hormones in the Blood

To release the thyroid hormones into the blood, iodothyroglobulin must be resorbed in the form of colloid droplets by endocytosis back into the thyroid cell. Within the thyroid cell, the iodothyroglobulin is hydrolyzed by lysosomal proteases, and T_4 and T_3 are released into the blood. In the blood, T_4 and T_3 associate with transport proteins and are distributed to target cells in peripheral tissues.

Three transport proteins bind and transport T_4 and T_3 in the blood. Thyroid-binding globulin, found in the plasma, has the smallest capacity but the greatest affinity for T_4 and T_3. Albumin and transthyretin (formerly called *prealbumin*) also transport the thyroid hor-

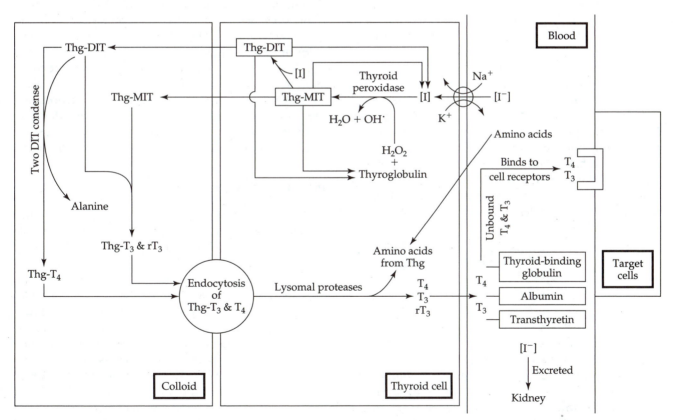

Figure 12.20 Overview of iodine intrathyroidal metabolism and hormonogenesis, and thyroid transport and cellular uptake. Abbreviations: T_4, 3,5,3',5'-tetraiodothyronine; T_3, 3,5,3'-tri-iodothyronine; rT_3, reverse T_3; Thg, thyroglobulin; MIT, 3-monoiodotyrosine; DIT, 3,5-di-iodotyrosine.

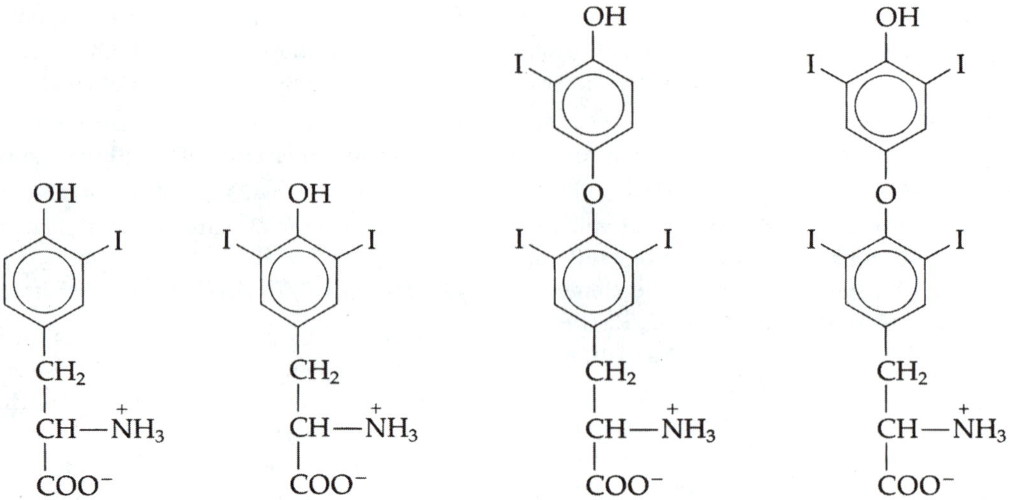

Figure 12.21 The structures of MIT, DIT, T_3, and T_4.

mones. A very small fraction (<0.1%) of the blood T_4 and T_3 is not bound to transport proteins, and it is this free form that is available to the cell receptors, and therefore is hormonally active (Fig. 12.20). The plasma concentration of T_4 is nearly 50 times that of T_3, but T_3 is many times more potent on an equal molar basis. For a more in-depth review of thyroid hormone synthesis, see the review by Taurog [7].

DIT and MIT not used for thyroid hormone synthesis in the thyroid cells are deiodinated, and the iodine is made available for recycling in the formation of new iodothyroglobulin. Several tissues—the liver, kidney, brain, pituitary, and brown adipose tissue, to name a few—can deiodinate T_4 to generate T_3 and rT_3. Most T_3 in the blood has been synthesized in the liver from T_4. A 5'-selenium-dependent deiodinase generates T_3, and a 5-deiodinase generates rT_3. Conversion of T_4 to T_3 is impaired with selenium deficiency [8].

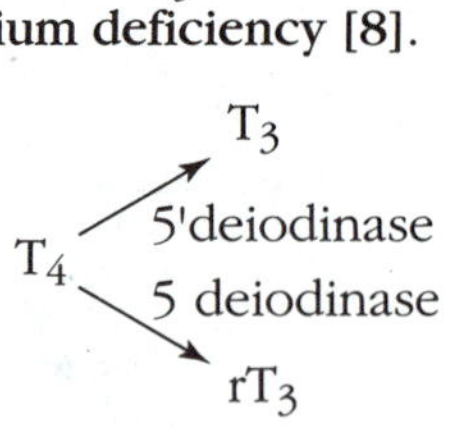

$$T_4 \begin{array}{c} \nearrow \text{5'deiodinase} \quad T_3 \\ \searrow \text{5 deiodinase} \quad rT_3 \end{array}$$

The multiple effects of the thyroid hormones result from the hormones' occupancy of nuclear receptors, with subsequent effects on gene expression. The receptors appear to be the same in all tissues, binding T_3 more avidly than T_4 and requiring fivefold to sevenfold higher concentrations of T_4 to achieve comparable physiological effects. Zinc may play a role in the binding of the zinc fingers of the receptor protein (which in turn is influenced by thyroid hormones) to the DNA. Although mechanisms of action of the thyroid hor-

mones are unclear, biological effects are in response to increased messenger RNA (mRNA) and protein synthesis triggered by the hormone receptor attachment. Numerous hypotheses for mechanisms have been proposed, including modulation of (Na^+/K^+-ATPase) transport systems, adrenergic receptor sensitivity, and neurotransmitters. The review by Sterling [9] provides more comprehensive reading on this topic.

Effects of thyroid hormones on metabolism are many and varied. They stimulate the basal rate of metabolism, oxygen (O_2) consumption, and heat production, and they are necessary for normal nervous system development and linear growth. Directly or indirectly, most organ systems are under the influence of these substances.

Interactions with Other Nutrients

Arsenic appears to be goitrogenic in mice. In studies with mice, arsenic is believed to antagonize the mechanism of iodine uptake by the thyroid gland, causing compensatory goiter.

A more established interaction is that between iodide and *goitrogens*. Substances that interfere with iodide metabolism in any way that inhibits thyroid hormonogenesis are termed goitrogens because their effect is to secondarily augment TSH release and consequently thyroid gland enlargement. Goitrogens may affect iodide uptake by the gland, organification of the iodide, or hormone release from the thyroid cells.

Most goitrogenic compounds act by competing with iodide in its active transport process into the thyroid cells. Halide ions such as bromide (Br^-) and astatide (At^-) function in this way, as do thiocyanate (SCN^-), perrhenate (ReO_4^-), and pertechnetate (TcO_4). Perchlorate (ClO4⁻), along with perrhenate and pertech-netate, interfere with organification as well as uptake; and lithium (Li⁻), used to treat some psychiatric disorders, inhibits hormone release from the gland. Other classes of goitrogens include polycyclic hydrocarbons, phenol compounds derived from coal, for example, among other substances. These substances also interfere with iodide metabolism.

That some natural foods are goitrogenic was evidenced many years ago when it was discovered that rabbits fed a fresh cabbage diet developed goiters that could be reversed by iodine supplementation. It was later shown that vegetables of the cabbage family contained, along with small quantities of thiocyanates, a potent goitrogen that later became known as *goitrin* (Fig. 12.22).

A long list of edible plants contain goitrin, including cabbage, kale, cauliflower, broccoli, rutabaga, turnips, brussels sprouts, and mustard greens. It is doubtful, however, if these foods are consumed in sufficient quantity to implicate them in the etiology of endemic goiter. Perhaps the only food to be identified directly with goiter etiology is cassava, which is consumed in large quantities in Third World countries [10]. It contains cyanogen glucosides, of which thiocyanates are a major metabolite.

Excretion

The kidneys have no mechanism to conserve iodide and therefore provide the major route (~80%–90%) for iodide excretion. The urinary output of iodide correlates closely with the plasma iodide concentration and in fact has been used to monitor iodide status in populations. For example, it has been suggested that iodide excretion of 50 µg I/g of creatinine in a representative sample of a population is indicative of endemic goiter within that population [11].

Fecal excretion of iodide (up to 20% of the total excreted) in humans is relatively low, ranging from 6.7 to 42.1 µg/day [12]. Some iodide is also lost in sweat, which can be of consequence in hot, tropical regions where iodide intake is marginally adequate.

Recommended Dietary Allowances

Because of its important link to thyroid function, iodide nutriture has been investigated thoroughly for over half

Figure 12.22 Goitrin.

a century. Dating as far back as the 1930s, results of intake requirements have been published based on balance studies and on calculations of average daily urinary losses. Adult daily requirements established by those early studies were in the range of 100 to 200 µg. The estimates have not changed significantly over the years. The 1989 RDA is 150 µg/day for adults of both sexes and provides a margin of safety to allow for unquantified levels of goitrogens in the diet [2]. Although the recommendations apply equally to both sexes, iodide needs are higher during pregnancy and lactation. The minimum amount (requirement) of iodide to prevent goiter is estimated between 50 and 75 µg/day or about 1 µg I/kg of body weight.

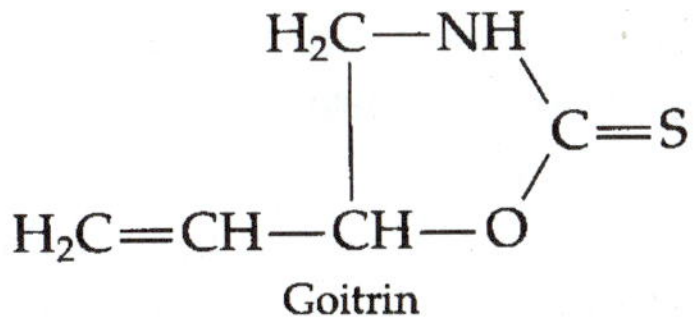

Deficiency

Thyroid Hormone Release as Related to Iodide Deficiency

The release of thyroid hormones by the thyroid gland is controlled. Thyrotropin-releasing hormone released from the hypothalamus acts on the pituitary gland to stimulate thyroid-stimulating hormone (TSH). TSH, in response to thyrotropin-releasing hormone, is secreted from the anterior pituitary and increases the activity of the thyroid gland to generate T_4. TSH output is regulated by T_4 through negative feedback to the pituitary. A decline in the blood level of T_4 triggers release of pituitary TSH, resulting in hyperplasia of the thyroid. Elevated T_4 inhibits TSH and thyrotropin-releasing hormone release.

Iodine Deficiency and Iodine Deficiency Disorders

Iodine deficiency prevails in many areas of the world and is associated most often with dietary insufficiency of iodine. Iodine deficiency is the main cause of goiter (although other factors such as ingestion of goitrogens may cause the disorder). Simple goiter is associated most often with insufficient dietary iodine and is characterized by enlargement of the thyroid gland. The enlargement is caused by overstimulation by TSH. Iodide deficiency causes depletion of thyroid iodine stores and therefore reduced output of T_4 and T_3. The decline in the blood level of T_4 triggers release of pituitary TSH, resulting in hyperplasia of the thyroid gland. The growth of the gland is self-restricting, however, because in its enlarged state it traps and processes available iodide more efficiently. The gland returns to normal size as dietary iodide is increased to adequate amounts. When the prevalence of goiter in any population exceeds 10%, it is called *endemic goiter* [13].

Because of the effects of iodide deficiency on growth, development, and other health problems, the term *iodide deficiency disorders* (IDDs) has been implemented. Iodine deficiency in a fetus results from iodide deficiency of the mother; cretinism, of which there are two types, results. Neurological cretinism is characterized in the infant by mental deficiency, hearing loss or deaf mutism, and motor disorders such as spasticity and muscular rigidity [5,13]. Hypothyroid cretinism results in thyroid failure.

The addition of iodide to table salt and the administration of iodized oil or potassium iodide has done much to alleviate the problem of endemic goiter in some goitrous regions of the world [14]. Yet, iodide deficiency continues to be a major health problem in many underdeveloped countries, and in China it may be coupled with selenium deficiency [15].

Toxicity

No adverse effects from iodide intakes of up to about 2 mg/day have been reported [2].

Assessment of Nutriture

Iodide nutritional status assessment is generally directed at populations living in areas suspected to be iodide-deficient. The assessment is based on both the physical examination and chemical testing of individuals. The data collected for assessment studies are extensive and include at least the following:

- Total population count including the number of children under 15 years of age
- The incidence of goiter as established by physical examination and cretinism in the population
- The quantification of urinary iodide excretion
- The quantification of iodide in the drinking water
- Determination of serum T_4 levels in various age groups

The chemistry of tests measuring urinary iodide excretion is based on the ability of iodide ion to reduce cerric ion (Ce^{4+}), which is yellow, to its colorless, cerrous state (Ce^{3+}) as shown here:

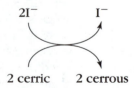

The extent of the color change, which is directly proportional to the iodide concentration in the specimen, is monitored spectrophotometrically. All iodine in the specimen, therefore, must first be reduced to iodide. Urinary iodine concentrations less than 50 μg/g of creatinine are considered at risk [16]. However, the use of iodine:creatinine ratio in casual urine samples has been shown to be unsuitable as an indicator for evaluating iodine status in some populations [17].

Radioactive iodide (^{131}I) uptake may also be measured to assess thyroid function. The greater the overall uptake and the quicker the uptake of the radioactive iodide by the thyroid gland, the greater the likelihood of iodide deficiency [16].

References Cited for Iodine

1. Karmarkar M, Deo M, Kochupillai N, Ramalingaswami V. Pathophysiology of Himalayan endemic goiter. Am J Clin Nutr 1974;27:96–103.

2. National Research Council. Recommended Dietary Allowances, 10th ed. Washington, D.C.: National Academy Press, 1989;213–7.

3. Taurog A, Howells E, Nachimson H. Conversion of iodate to iodide in vitro and in vivo. J Biol Chem 1966;241:4686–93.

4. Harden RMcG, Alexander W, Shimmins J, Kostalas H, Mason D. Quantitative aspects of the inhibitory effect of the iodide ion on parotid salivary iodide and pertechnetate secretion in man. J Lab Clin Med 1968;71:92–100.

5. Clugston G, Hetzel B. Iodine. In: Shils ME, Olson JA, Shike M., eds. Modern nutrition in health and disease. Philadelphia: Lea and Febiger, 1994;252–63.

6. O'Neill B, Magnolato D, Semenza G. The electrogenic, Na$^+$-dependent I$^-$ transport system in plasma membrane vesicles from thyroid glands. Biochim Biophys Acta 1987;896:263–74.

7. Taurog A. Hormone synthesis: Thyroid iodine metabolism. In: Ingbar SH, Braverman LE, eds. Werner's the thyroid. Philadelphia: Lippincott 1986;53–97.

8. Arthur J, ed. Interrelationships between selenium deficiency, iodine deficiency, and thyroid hormones. Am J Clin Nutr 1993;57:235S–318S.

9. Sterling K. Thyroid hormone action at the cellular level. In: Ingbar SH, Braverman LE, eds. Werner's the thyroid. Philadelphia: Lippincott 1986;219–33.

10. Maberly G, Waite K, Eastman C, et al. In: Ui N, Torizuka K, Nagataki A, et al., eds. Current problems in thyroid research. Amsterdam: Excerpta Medica 1983:341.

11. Stanbury J, Ermans A, Hetzel B, et al. Endemic goitre and cretinism: public health significance and prevention. WHO Chron 1974;28:220–8.

12. Vought R, London W, Lutwak L, Dublin T. Reliability of estimates of serum inorganic iodine and daily fecal and urinary iodine excretion from single casual specimens. J Clin Endocr Metab 1963;23:1218–28.

13. Lamberg B. Iodine deficiency disorders and endemic goitre. Eur J Clin Nutr 1993;47:1–8.

14. Todd C, Dunn J. Intermittent oral administration of potassium iodide solution for the correction of iodine deficiency. Am J Clin Nutr 1998;67:1279–83.

15. Ma T, Guo J, Wang F. The epidemiology of iodine-deficiency diseases in China. Am J Clin Nutr 1993;57:264S–6S.

16. Gibson R. Principles of Nutritional Assessment. New York: Oxford University Press, 1990;527–32.

17. Furnee C, Haar F, West C, Hautvast J. A critical appraisal of goiter assessment and the ratio of urinary iodine to creatinine for evaluating iodine status. Am J Clin Nutr 1994;59:1415–7.

Manganese

Although widely distributed in nature, manganese occurs in only trace amounts in animal tissues. The body of a healthy 70-kg man is estimated to contain a total of 10 to 20 mg of the metal. In the body, manganese typically exists in either of two states, Mn^{2+} or Mn^{3+}.

Sources

Whole-grain cereals, dried fruits, nuts, and leafy vegetables are among the manganese-rich common foods. Tea also contains large amounts of manganese; however, manganese in tea is not well absorbed. A wide content range of the mineral in cereal grains is due partly to plant species differences and partly to the efficiency with which the milling process separates the manganese-rich and manganese-poor parts of the grain. Patent flour, for example, has a much lower manganese concentration than the wheat grain from which it was produced. Table 12.7 lists the manganese contents of selected foods.

Absorption, Transport, and Storage

Little information is available on the mechanism of manganese absorption, although it has been established that the process occurs equally well throughout the length of the small intestine [1]. Dietary manganese absorption varies considerably with values of 1% to 30% reported. Studies in humans report differences in percent absorption between women and men, with higher absorption (1.35%–3.55%) in women than in men (0.70–1.35%) [2]. Manganese absorption from

Table 12.7 Manganese Content of Selected Foods and Beverages

Foods/Food Group	Manganese Content (mg/100 g)
Bread, whole grains	0.50–2.05
Flour, whole grain	3.80
Bread, white	0.05
Flour, white	0.79
Legumes	0.24–0.58
Nuts	0.83–4.71
Root vegetables	0.05–0.62
Other vegetables	0.15–1.94
Fruits	0.04–1.60
Fruits (dried)	0.09–0.39
Milk and cheeses	< 0.01
Beer	0.01
Wine	
White	0.46
Red	0.60
Coffee (brewed)	0.02–0.03
Tea (brewed)	0.18–0.22

Source: www.nal.usda.gov/fnic/foodcomp

$MnCl_2$ (manganese chloride) has been shown to be greater than that from plant foods such as lettuce, spinach, and sunflower seeds [3]. The absorption process itself appears to be quickly saturable and probably involves a low-capacity, high-affinity, active transport mechanism as demonstrated in rats. With excessive high manganese intake, absorption decreases to protect against toxicity; excretion also increases as discussed in the section on manganese excretion. Manganese is thought to be absorbed in the Mn^{2+} state. Within the duodenum, ingested manganese as Mn^{2+} may be converted to Mn^{3+}.

Factors Influencing Absorption

Little information relative to many of the other trace minerals is available on factors influencing manganese absorption. There is evidence for the enhancement of absorption by low-molecular-weight ligands such as histidine and citrate [4].

Animal studies suggest that fiber, oxalic acid, calcium, and phosphorus may precipitate manganese in the gastrointestinal tract, making the manganese unavailable for absorption [4]. Phytic acid also has been shown to decrease manganese absorption in humans [5]. Among minerals, *iron* competes with manganese for absorption. Competition for common binding sites occurs among manganese and iron. This explains why the absorption of manganese as well as iron is enhanced in situations in which iron deficiency is manifested. Likewise, the absorption and retention of manganese from foods low in iron, such as milk, is relatively high; and if the milk is supplemented with iron, the absorption of manganese is then reduced [6].

Manganese entering into the portal circulation from the gastrointestinal tract may either remain free or become bound as Mn^{2+} to α-2 macroglobulin before traversing the liver, where it is almost totally removed. From the liver, some Mn^{2+} may be oxidized by ceruloplasmin to Mn^{3+} and may complex with transferrin [7]. Mn^{3+} bound to transferrin is taken up by extrahepatic tissues.

Manganese is cleared rapidly from the blood and accumulates preferentially in the mitochondria of tissues, a process that may be mediated by a Ca^{2+} carrier [8]. Within the mitochondria, manganese is present as hydrate Mn^{2+} or Mn^{3+}, and as $Mn_3(PO_4)_2$, a matrix precipitate [9]. Manganese is found in most organs and tissues, and it does not tend to concentrate significantly in any particular one, although its concentration is highest in bone, liver, pancreas, and kidney. In bone, manganese is found as part of the apatite. Hair can also accumulate manganese.

Functions and Mechanisms of Actions

At the molecular level, manganese, like other trace elements, can function both as an enzyme activator and as a constituent of metalloenzymes, but the relationship of these functions to the gross physiological changes observed in manganese deficiency is not well correlated. Reviews pertaining to this topic have been published [10].

In the activation of enzyme-catalyzed reactions, manganese may bind to the substrate (such as ATP) or to the enzyme directly, with induction of conformational changes. Enzymes from about every class can be activated by manganese in this manner and are numerous and diverse in function. They include enzymes from the enzyme classes: transferases including kinases, hydrolases, oxido-reductases, ligases, and lyases. The activity of most of these enzymes is not, however, affected by a manganese deficiency, largely because the activation is not manganese specific. The metal can be replaced by other divalent cations, primarily magnesium. One exception to this apparent lack of specificity is the manganese-specific activation of the glycosyl transferases. Examples of some manganese dependent enzymes from each enzyme class will be described.

Transferases

Glycosyl transferases catalyze the transfer of a sugar moiety such as galactose from uridine diphosphate (UDP) to an acceptor, as shown by the general reaction:

$$\text{UDP-sugar} + \text{acceptor} \xrightarrow{\text{glycosyl transferase}} \text{UDP} + \text{acceptor-sugar}$$

For example, the sugar galactose when bound to UDP may be transferred to an acceptor molecule by the glycosyl transferase. Glycosyl transferases are necessary for mucopolysaccharide synthesis. Remember, mucopolysaccharides are important components of connective tissue like collagen.

Hydrolases

Manganese also activates *prolidase*, a dipeptidase with specificity for dipeptides. *Arginase,* containing four manganese atoms per molecule, is a cytosolic enzyme responsible for urea formation (p. 186) and is found in high concentrations in the liver. The Mn^{2+} may allosterically activate arginase through a pH-medicated role [11]. Low-manganese diets in animals have been shown to decrease arginase activity [12].

Lyases

Phosphoenolpyruvate carboxykinase (PEPCK), also activated by manganese, converts oxaloacetate to phosphoenolpyruvate and carbon dioxide. This reaction is important in gluconeogenesis (p. 94). The activity of phosphoenolpyruvate carboxykinase decreases in animals with manganese deficiency.

Oxido-Reductases

Superoxide dismutase, a manganese-dependent (Mn^{3+}-SOD) metalloenzyme (not manganese activated), functions similar to copper- and zinc-dependent SOD (p. 435), to prevent lipid peroxidation by superoxide radicals. Manganese SOD, however, is found in the mitochrondria, whereas copper zinc SOD is found in the cytoplasm. Thus, SOD in the mitochondria likely serves to eliminate superoxides before they damage mitochondrial function. The activity of the electron transport/respiratory chain generates large amounts of superoxide radicals, necessitating substantial Mn-SOD activity. It is likely that the cell ultrastructural abnormalities associated with manganese deficiency may be due to unchecked lipid peroxidation in the cellular membranes because of reduced Mn-SOD activity or simply because of reduced availability of manganese to directly scavenge free radicals. Manganese (Mn^{2+}) is one of several minerals able to scavenge free radicals. Mn^{2+} quenches peroxyl radicals as shown in this equation [13]:

$$Mn^{2+} + ROO\cdot + H^+ \longrightarrow Mn^{3+} + ROOH.$$

Low-manganese diets in animals have been shown to decrease Mn-SOD activity.

Ligases/Synthetases

Pyruvate carboxylase, which contains four manganese atoms, converts pyruvate to oxaloacetate (p. 94), a tricarboxylic acid cycle intermediate. Because magnesium can replace manganese in pyruvate carboxylase, minimal changes in pyruvate carboxylase activity occur [12]. *Glutamine synthetase* (Fig. 7.31) may be a manganese metalloenzyme or may be activated by manganese or magnesium.

Other Roles

Manganese also may act as a modulator of second messenger pathways in tissues. For example, manganese increases cAMP accumulation through binding to ATP and ADP. Manganese can activate guanylate cyclase, and manganese may affect cytoplasmic calcium levels and thus regulate calcium-dependent processes (see Fig. 11.5) [9].

Interactions with Other Nutrients

Only a few interactions between manganese and other trace elements are thought to be of significance nutritionally. One relationship of nutritional significance—between manganese and *iron*—has been discussed earlier in the section on absorption. Some degree of interaction may occur between manganese and *calcium* and between manganese and *zinc* in such a way as to affect the bioavailability of manganese [14]. However, because of the paucity of information and the divergent results of some of the relevant studies, the nature of such interactions remains inconclusive.

Excretion

Manganese is excreted primarily via the bile in the feces. Excess absorbed manganese from diet is quickly excreted by the liver into the bile to maintain homeostasis [2]. Very little manganese is excreted in the urine, even when dietary intake of the mineral is excessive. However, excretion of manganese through the sweat and skin desquamation has been shown to contribute to manganese losses [15].

Estimated Safe and Adequate Daily Dietary Intake

Estimations of the human requirement for manganese are based on balance studies; however, multiple problems with this approach as well as with the factorial method have been reported [14,16]. A provisional dietary recommendation has been set at 2 to 5 mg/day

and is thought to represent a dietary intake level achieved by most individuals who exhibit no signs of deficiency or toxicity [16].

Deficiency

Studies on a wide variety of species have demonstrated that manganese deficiency is associated with striking and diverse physiological malfunctions. Manganese deficiency generally does not develop in humans unless the mineral is deliberately eliminated from the diet. Human studies in which men received either 0.11 mg manganese per day for 39 days (however, the diet was also devoid of vitamin K, making it difficult to separate the effects of the manganese and vitamin K deficiencies) or 0.35 mg manganese per day resulted in negative manganese balance [15–17]. Symptoms and signs of deficiency included nausea; vomiting; dermatitis; decreased serum manganese; decreased fecal manganese excretion; increased serum calcium, phosphorus, and alkaline phosphatase (thought to be associated with skeletal bone changes); decreased growth of hair and nails; changes in hair and beard color; and low blood cholesterol concentrations [15,16]. Other effects reported include the occurrence of neonatal ataxia and loss of equilibrium, cell ultrastructure abnormalities, compromised reproductive function, abnormal glucose tolerance, and impaired lipid metabolism [16]. In rats, dietary manganese deficiency also altered plasma ammonia and urea concentrations in association with decreased arginase activity [10].

Toxicity

In toxic conditions such as liver failure, manganese appears to accumulate and damage the liver and brain [18]. Furthermore, in infants with cholestatic liver disease, manganese supplementation also appears to be associated with neurological abnormalities [19]. Miners who have inhaled dust fumes high in manganese (about 5 mg/m³ or more) experience Parkinsonism-like symptoms. Manganese toxicity in individuals chronically exposed to airborne manganese in concentrations as low as 1 mg/m³ also have been reported to experience problems, including prolonged reaction time, tremors, and diminished memory capacity [20]. In one report involving oral intake, no evidence of manganese toxicity occurred in people receiving as much as 9 mg manganese via food per day [16]. However, the lowest observable adverse effect level for manganese for a 70-kg individual is reported at 4.2 mg/day, a level within the Estimated Safe and Adequate Daily Dietary Intake range [21]. Additional studies are necessary to better determine toxic from nontoxic intakes of manganese.

Assessment of Nutriture

The normal range of serum manganese concentration is approximately 0.04 to 1.4 µg/dL, but laboratory tests that reliably assess body manganese status have not yet been established. For this reason it becomes difficult to correlate serum concentrations with specific diseases or disorders. Body fluid (plasma, blood, urine) manganese is commonly assayed, as is hair, with blood manganese possibly being indicative of body manganese status [16,22]. In animals, mitochondrial Mn-SOD and blood arginase activities have been shown to be diminished with a low manganese intake or deficiency [23]. In humans, manganese supplementation significantly increased lymphocyte Mn-SOD activity and serum manganese concentrations from baseline without changes in manganese excretion [24]. Further studies assessing these parameters as indicators of status are required.

References Cited for Manganese

1. Thomson A, Olatunbosun D, Valberg L. Interrelation of intestinal transport system for manganese and iron. J Lab Clin Med 1971;78:642–55.

2. Finley J, Johnson P, Johnson L. Sex affects manganese absorption and retention by humans from a diet adequate in manganese. Am J Clin Nutr 1994;60:949–55.

3. Johnson P, Lykken G, Korynta E. Absorption and biological half-life in humans of intrinsic and extrinsic [54]Mn tracers from foods of plant origin. J Nutr 1991;121:711–7.

4. Garcia-Aranda J, Wapnir R, Lifshitz F. In vivo intestinal absorption of manganese in the rat. J Nutr 1983;113:2601–7.

5. Davidsson L, Almegren A, Juillerat M, Hurrell R. Manganese absorption in humans: the effect phytic acid and ascorbic acid in soy formula. Am J Clin Nutr 1995;62:984–7.

6. Keen C, Frannson G, Lonnerdal B. Supplementation of milk with iron bound to lactoferrin using weanling mice: effects on tissue manganese, zinc, and copper. J Pediatr Gastroenterol Nutr 1984;3:256–61.

7. Critchfiled J, Keen C. Manganese[+2] exhibits dynamic binding to multiple ligands in human plasma. Metabolism 1992;41:1087–92.

8. Jeng A, Shamoo A. Isolation of a Ca^{2+} carrier from calf heart inner mitochondrial membrane. J Biol Chem 1980;255:6897–903.

9. Korc M. Manganese as a modulator of signal transduction pathways. In: Prasad AS, ed. Essential and Toxic Trace Elements in Human Health and Disease: An Update. New York: Wiley-Liss, 1993;235–55.

10. Keen C, Lonnerdal B, Hurley L. Manganese. In: Frieden E, ed. Biochemistry of the essential trace elements. Norfolk, VA: Plenum 1984;89–132.

11. Kuhn N, Ward S, Piponski M, Young T. Purification of human hepatic arginase and its manganese (II) dependent and pH-dependent interconversion between active and inactive forms: a possible pH sensing function of the enzyme on the ornithine cycle. Arch Biochem Biophys 1995;320:24–34.

12. Brock A, Chapman S, Ulman E, Wu G. Dietary manganese deficiency decreases rate hepatic arginase activity. J Nutr 1994;124:340–4.

13. Coassin M, Ursini F, Bindoli A. Antioxidant effect of manganese. Arch Biochem Biophysics 1992;299:330–3.

14. Forbes R, Erdman J. Bioavailability of trace mineral elements. Ann Rev Nutr 1983;3:213–31.

15. Friedman B, Freeland-Graves J, Bales C, Behmardi F, Shorey-Kutschke R, Willis R, Crosby J, Trickett P, Houston S. Manganese balance and clinical observations in young men fed a manganese-deficiency diet. J Nutr 1987;117:133–43.

16. National Research Council. Recommended Dietary Allowances, 10th ed. Washington, D.C.: National Academy Press, 1989;230–5.

17. Freeland-Graves J, Behmardi F, Bales C, Dougherty V, Lin P-H, Crosby J, Trickett P. Metabolic balance of manganese in young men consuming diets containing five levels of dietary manganese. J Nutr 1988;118:764–73.

18. Hauser R, Zesiewicz T, Rosemurgy A, Martinez C, Olanow C. Manganese intoxication and chronic liver failure. Ann Neurol 1994;36:871–5.

19. Reynolds A, Kiely E, Meadows N. Manganese in long term paediatric parenteral nutrition. Arch Dis Child 1994;71:527–31.

20. Wennberg A, Iregren A, Struwe G, Cizinsky G, Hagman M, Johansson L. Manganese exposure in steel smelters a health hazard to the human worker. Scand J Work Environ Health 1991;17:255–62.

21. Greger J. Dietary standards for manganese: overlap between nutritional and toxicological studies. J Nutr 1998;128:368S–71S.

22. Keen C, Clegg M, Lonnerdal B, Hurley L. Whole blood manganese as an indicator of body manganese. N Engl J Med 1983;308:1230.

23. Thompson K, Lee M. Effects of manganese and vitamin E deficiencies on antioxidant enzymes in streptozotocin-diabetic rats. J Nutr Biochem 1993;4:476–81.

24. Davis C, Greger J. Longitudinal changes of manganese dependent superoxide dismutase and other indexes of manganese and iron status in women. Am J Clin Nutr 1992;55:747–52.

Molybdenum

The need for molybdenum was established in humans through the observation that a genetic deficiency of specific enzymes that require molybdenum as a cofactor resulted in severe pathology in human patients. In the body, molybdenum, a metal, is found primarily in either of two valence states, Mo^{4+} or Mo^{6+}. In biological systems, molybdenum is generally bound to either sulfur or oxygen.

Sources

Molybdenum is widespread among foods, but present only in small amounts among the natural foods consumed in typical Western diets. Like other minerals, the molybdenum content of a given plant food may vary greatly, depending on the concentration of molybdenum in the soil. Therefore, it would follow that the metal's content in meats would in turn reflect its concentration in the regional forage. The molybdenum content ranges for selected foods are listed in Table 12.8 [1]. Better sources of the mineral include meats (especially organ meats), legumes, cereals, and grains. Dairy products are marginal sources of molybdenum.

Absorption, Transport, and Storage

Little is known about the absorption and transport of molybdenum, depicted in Figure 12.23. Sites of molybdenum absorption are thought to include both the

Table 12.8 The Molybdenum Content of Selected Foods

	Molybdenum Content (μg/100 g food)
Meat, fish, and poultry	<1–129
Legumes	16–184
Nuts	11–34
Grains and grain products	2–117
Milk, yogurt, and cheese	2–10
Vegetables	<1–33
Fruits and fruit juices	0–12

Source: Pennington JAT, Jones JW. Molybdenum, nickel, cobalt, vanadium, and strontium in total diets. Reprinted from JOURNAL OF THE AMERICAN DIETETIC ASSOCIATION Vol.87:1646–50, 1987.

stomach and small intestine, with the proximal small intestine responsible for more absorption than the distal section. Active carrier-mediated transport is thought to occur with low molybdenum intakes and diffusion at higher concentrations. Absorption in humans ranges from 85% to 93% [2].

Transport of molybdenum in the blood is thought to occur as molybdate (MoO_4^{2-}). The mineral may be bound to albumin and/or α-2 macroglobulin.

The molybdenum content of human tissues is quite low under normal dietary conditions, averaging 0.1 to 1.0 μg/g of wet weight. Molybdenum is found in tissues as molybdate, molybdopterin, or bound to enzymes. The liver, kidney, and bone contain the most molybdenum in terms of absolute amount as well as

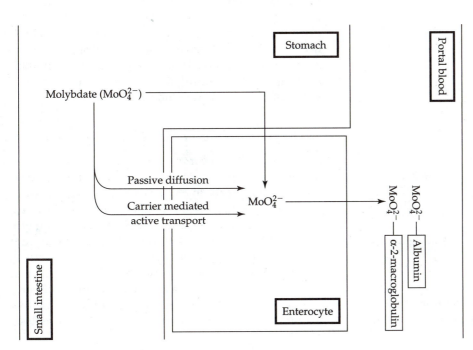

Figure 12.23 Digestional absorption and transport of molybdenum as molybdate.

concentration [3]. Other tissues, such as the small intestine, lungs, spleen, brain, thyroid and adrenal glands, and muscle, also contain molybdenum.

Functions and Mechanisms of Actions

The biochemical role of molybdenum centers around the redox function of the element and its necessity as a cofactor for three metalloenzymes (sulfite oxidase, aldehyde oxidase, and xanthine dehydrogenase/oxidase), all of which catalyze oxidation reduction reactions [4]. The molybdenum cofactor, molybdopterin (Fig. 12.24), consists of a molybdenum atom attached to an organic moiety [5].

The identification and characterization of molybdopterin has been one of the more exciting revelations in molybdenum biochemistry. It is an alkylphosphate-substituted pterin to which molybdenum is coordinated through two sulfur atoms [5-7]. It is through molybdopterin that the molybdenum is anchored to the apoenzyme at its catalytic site. The molybdenum is further bonded to either two oxygen molecules and referred to as *dioxomolybdopterin* or bonded to one oxygen and one sulfur and referred to as *oxosulfido-molybdopterin*, as shown in Figure 12.24.

Figure 12.24 Molybdopterin structures.

Sulfite Oxidase

Sulfite oxidase, a mitochondrial intermembrane enzyme found in many body tissues especially the liver, heart and kidney, has two molybdopterin (dioxo cofactor form) and two cytochrome residues. The enzyme catalyzes the terminal step in the metabolism of sulfur-containing amino acids (methionine and cysteine), in which sulfite (SO_3^{2-}) is converted into sulfate (SO_4^{2-}), as shown here:

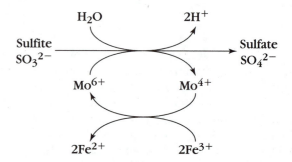

The reaction also serves to prevent toxic effects of sulfur dioxide. Cytochrome c is the physiological electron acceptor for the reaction. Sulfate generated from this reaction is typically excreted in the urine or reused for the synthesis of sulfoproteins, sulfolipids, and mucopolysaccharides (a component of mucus).

Aldehyde Oxidase

Aldehyde oxidase is a molybdoenzyme (using the oxosulfido form) that is very similar to xanthine oxidase in size, cofactor composition, and substrate specificity. It presumably functions primarily in the liver as a true oxidase, using molecular oxygen as its physiological electron acceptor. The enzyme's primary substrates in vivo are not known, although the enzyme may be important for drug metabolism [8]. Also unclear is the effect of variation in molybdenum intake on its activity.

Xanthine Dehydrogenase and Xanthine Oxidase

Xanthine dehydrogenase and *xanthine oxidase* are nonheme-containing enzymes that also require FAD and molybdopterin in the oxosulfido cofactor form. Xanthine dehydrogenase is found in a variety of tissues, including the liver, lungs, kidneys, intestine, among others. Xanthine oxidase is found in the intestine, thyroid cells, and possibly other tissues. Healthy tissues may contain about 10% of their total xanthine enzymes in the oxidase form [9]. Conversion of xanthine dehydrogenase into xanthine oxidase may occur following oxidation of essential sulfhydryl groups or by proteolysis

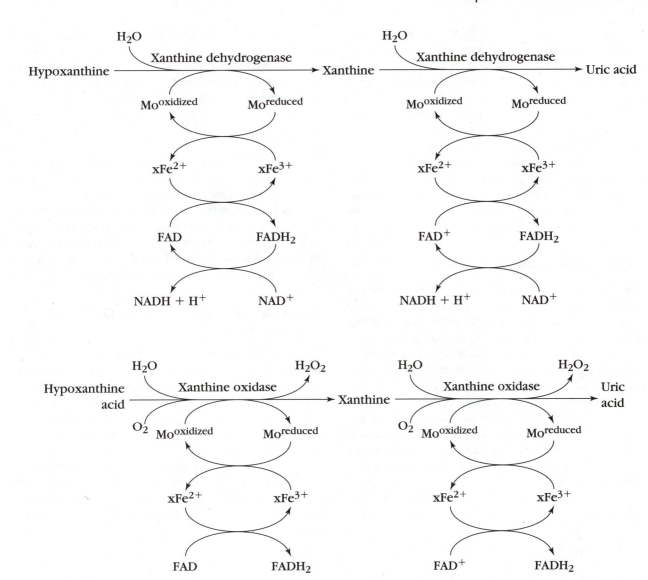

of the dehydrogenase form. Xanthine oxidase has been implicated as a cause of damage in postischemic reperfused tissue.

The xanthine dehydrogenase and oxidase enzymes are capable of hydroxylating various purines, pteridines, pyrimidines, and other heterocyclic nitrogen-containing compounds. Hypoxanthine, derived from purine catabolism, is oxidized in most tissues by xanthine dehydrogenase to generate xanthine and then uric acid. Xanthine dehydrogenase transfers electrons from the substrate onto NAD^+ to form $NADH + H^+$.

Oxidation of hypoxanthine and xanthine by xanthine oxidase also results in uric acid; however, in these reactions O_2 accepts the electrons from $FADH_2$ and hydrogen peroxide (H_2O_2) is formed.

Although low-molybdenum diets or inclusion in the diet of tungstate, a molybdenum antagonist, predictably reduces the level of xanthine oxidase activity in rat intestine and liver, it is interesting to note that no apparent clinical effects result from the perturbation. Furthermore, the human inheritable disorder xanthinuria provides additional evidence for the body's ability to tolerate low xanthine dehydrogenase or oxidase activity. The condition is essentially free of clinical manifestations, except for the possible development of kidney calculi (stones) caused by the high urinary xanthine concentration. Therefore, it is not firmly established whether any of the reactions catalyzed by xanthine dehydrogenase or oxidase are necessary for human health [10].

The effects of xanthine oxidase activity, however, have been shown to be quite damaging in people being treated for ischemia (local or temporary deficiency of blood supply and thus relative oxygen deprivation), for example. Degradation of ATP in hypoxic tissue yields hypoxanthine. Reperfusion of the intestine with oxygen (as occurs with medical treatment of, e.g., intestinal ischemia) helps to prevent total destruction of the tissue due to lack of oxygen and nutrients but also provides xanthine oxidase with the oxygen needed to oxidize the relatively large concentrations of hypoxanthine. Oxidation of hypoxanthine generates large amounts of hydrogen peroxide which further induces tissue damage (see Chap. 10's Perspective).

In addition to its biochemical role, molybdenum appears to modulate (likely inhibit through direct interaction) the glucocorticoid receptor complex [11].

Interactions with Other Nutrients

Tungsten has long been recognized as a potent antagonist of molybdenum [12], and in fact its administration into test animals has become the major means for artificially creating a state of molybdenum deficiency.

Another interaction involves molybdenum, *sulfur,* and *copper.* It has been shown, particularly in ruminants, that a high dietary intake of sulfate or molybdenum depressed the tissue uptake of copper and conversely that sulfate and copper decrease molybdenum retention [13]. The proposed explanation for this is that sulfide and hydrosulfide ions are generated in the rumen by reduction of ingested sulfate. The reactive sulfide then displaces oxygen from molybdate ions, yielding oxythiomolybdates and tetrathiomolybdates. Molybdenum is not readily absorbed in the form of thiomolybdates, and, furthermore, thiomolybdates bind copper avidly, rendering that metal less physiologically available [14]. It is important to recognize, however, that in humans and other nonruminants, such an interaction is not as important because of the low yield of sulfides and hydrosulfides resulting from sulfate reduction during digestion. However, the feeding of tetrathiomolybdates to nonruminant test animals does result in a compromised uptake of copper. Therefore, it appears that the antagonistic effect of molybdenum or sulfate on copper availability is due to the tendency of molybdenum to sequester reactive sulfide groups. These groups subsequently bind copper ions, which then become less available.

A relationship between molybdenum intake and *copper* excretion has been documented in humans. Urinary copper excretion in humans has been shown to rise from 24 µg/day to 77 µg/day as molybdenum intake increased from 160 µg to 1,540 µg/day [15]. No changes in fecal copper excretion were noted, suggesting perhaps that molybdenum increased copper mobilization from tissues and promoted excretion [15].

Other nutrients and substances that appear to affect molybdenum availability by mechanisms not yet understood include manganese, zinc, iron, lead, ascorbic acid, methionine, cysteine, and protein. A possible relationship with silicon is discussed on page 475.

Excretion

Most molybdenum is excreted from the body in the urine. Furthermore, urinary excretion of molybdenum increases as dietary molybdenum intake increases [16]. In other words, retention of molybdenum in the body is low when dietary intake is high. Small amounts of molybdenum are excreted from the body in the feces by way of the bile [2]; small amounts also can be lost in sweat (20 µg) and in hair (0.01 µg/g hair).

Estimated Safe and Adequate Daily Dietary Intake

Molybdenum is among those trace elements for which no RDA has been established on the basis of present knowledge. Instead, estimated ranges of adequate but safe intake have been proposed, primarily according to balance studies. The ESADDI for molybdenum for adults is 75 to 250 µg [17]. Infants and children probably require proportionately more of the element if intake per unit of body weight is the basis of assessment. Studies of molybdenum balance, absorption, and excretion in men during depletion and repletion suggest a minimum requirement of 25 µg molybdenum per day [16].

Deficiency

Molydenum deficiency is rarely encountered unless the diet is particularly rich in antagonistic substances such as sulfate, copper, or tungstate. Low molybdenum intakes have been associated with esophogeal cancer in China. Molybdemun deficiency in humans has been documented in a patient maintained on total parenteral nutrition for 18 months [18]. The patient exhibited high blood methionine, hypoxanthine, and xanthine concentrations, as well as low blood levels of uric acid. Urinary concentrations of sulfate were low and of sulfite were high. Treatment with 300 µg of ammonium molybdate (163 µg of molybdenum) resulted in clinical improvement and normalization of sulfur amino acid metabolism and uric acid production.

The importance of sulfite oxidase, and therefore molybdenum, in human nutrition is evidenced by the neurological disorders associated with a genetic deficiency of sulfite oxidase in children [19]. Elevated urinary sulfite and thiosulfate, along with biochemical manifestations reflecting aberrant sulfur amino acid metabolism and sulfite oxidation, were observed.

Toxicity

Molybdenum appears to be relatively nontoxic to humans with intakes up to 1,500 μg/day [2]. However, symptoms such as gout (inflammation of the joints due to accumulation of uric acid) have appeared in some people living in regions that contain high soil molybdenum levels. Gout results from high blood uric acid concentrations, which have likely arisen from increased xanthine dehydrogenase activity, that have accumulated in and around joints.

Assessment of Nutriture

Molybdenum appears to distribute itself fairly equally between the plasma and red blood cells. Although a few studies have reported the molybdenum concentrations of human plasma and blood, the use of these as indicators of molybdenum status has not been validated.

References Cited for Molybdenum

1. Pennington J, Jones J. Molybdenum, nickel, cobalt, vanadium, and strontium in total diets. J Am Diet Assoc 1987;87:1646–50.

2. Turnlund JR, Keyes WR, Peiffer GL. Molybdenum absorption, excretion, and retention studied with stable isotopes in young men at five intakes of dietary molybdenum. Am J Clin Nutr 1995;62:790–6.

3. Scott K, Turnlund J. Compartmental model of molybdenum metabolism in adult men fed five levels of molybdenum. FASEB J 1993;7:A288.

4. Moriwaki Y, Yamamota T, Higashino K. Distribution and pathophysiologic role of molybdenum-containing enzymes. Histol Histopathol 1997;12:513–24.

5. Kramer S, Johnson J, Ribeiro A, Millington D, Rajagopalan K. The structure of the molybdenum cofactor. J Biol Chem 1987;262:16357–63.

6. Rajagopalan K. Molybdenum: An essential trace element in human nutrition. Ann Rev Nutr 1988;8:401–27.

7. Mize C, Johnson J, Rajagopalan K. Defective molybdopterin biosynthesis: clinical heterogeneity associated with molybdenum cofactor deficiency. J Inher Metab Dis 1995;18:283–90.

8. Beedham C. Molybdenum hydroxylases as drug-metabolizing enzymes. Drug Metab Rev 1985;16:119–56.

9. McCord J. Free radicals and myocardial ischemia: overview and outlook. Free Radicals & Medicine 1988;4:9–14

10. Coughlan M. The role of molybdenum in human biology. J Inher Metab Dis 1983;6 (suppl. 1):70–77.

11. Bodine P, Litwack G. Evidence that the modulator of the glucocorticoid-receptor complex is the endogenous molybdate factor. Proc Natl Acad Sci USA 1988;85:1462–6.

12. Johnson J, Rajogopalan K. Molecular basis of the biological function of molybdenum. J Biol Chem 1974;249:859–66.

13. Suttle N. The interactions between copper, molybdenum, and sulphur in ruminant nutrition. Ann Rev Nutr 1991;11:121–40.

14. Mills C, Davis G. Molybdenum. In: Mertz W, et al. Trace elements in human and animal nutrition. San Diego: Academic Press, 1987;1:449–54.

15. Turnlund J. Copper nutriture, bioavailability, and the influence of dietary factors. J Am Diet Assoc 1988;88:303–8.

16. Turnlund J, Keyes W, Peiffer G, Chiang G. Molybdenum absorption, excretion, and retention studied with stable isotopes in young men during depletion and repletion. Am J Clin Nutr 1995;61:1102–9.

17. National Research Council. Recommended Dietary Allowances, 10th ed. Washington, D.C.: National Academy Press, 1989;243–6.

18. Abumrad N, Schneider A, Steel D, Rogers L. Amino acid intolerance during prolonged total parental nutrition reversed by molybdate therapy. Am J Clin Nutr 1981;34:2551–9.

19. Johnson J, Wuebbens M, Mandell R, Shih V. Molybdenum cofactor deficiency in a patient previously characterized as deficient in sulfite oxidase. Biochem Med Metab Biol 1988;40:86–93.

Fluorine

Whereas fluorine (F) is a gaseous chemical element, fluoride (F⁻) exists and is composed of fluorine bound to a metal, nonmetal, or organic compound. Fluoride predominates in nature and exerts physiological effects in the body. The term *fluoride* will be used throughout this section. Analogous to this terminology is the use of the terms *iodide* and *chloride*.

Sources

Community drinking water fluoridation has been practiced for nearly 50 years in the United States following the discovery in 1942 of the inverse relationship of fluoride intake and the incidence of dental caries [1]. This has in turn affected the distribution of fluoride in foods and beverages intended for human consumption. The content of fluoride in most food groups (Table 12.9) is low compared with most nutrients. However, not shown are tea and marine fish, which, if consumed with the bones, are relatively high in fluoride. Ready-to-use infant formulas are made with fluoridated water, thereby providing infants with a source of fluoride. Other beverages vary greatly in their fluoride content, contingent on the use or nonuse of fluoridated water in their processing. It is important that fluoride supplementation of water and foods achieves an in vivo level sufficiently high to effect dental and skeletal benefit but not so high as to cause toxicity [2].

Digestion, Absorption, Transport, and Storage

Even at high intakes, soluble fluorides are almost completely absorbed from the gastrointestinal tract at a rate that varies with the chemical form of the element and the presence of other dietary factors. Aqueous solutions of fluorides, sodium fluoride used in toothpastes, and sodium fluorosilicate, used in water fluoridation, are almost completely absorbed.

The availability of fluoride from solid foods is, however, generally reduced with only about 50% to 80% of fluoride absorbed from foods. In food, fluoride may be bound to proteins, and, while hydrolyzed by proteases, may still be less available for absorption. Fluoride is also relatively poorly absorbed (37%–54%) from insoluble sources such as bone meal [3].

Humans, given small amounts of soluble fluoride, achieve maximum blood fluoride levels within about

30 to 45 minutes [4]. This rapidity of absorption is accounted for by the fact that it occurs to a great extent in the stomach, a rather unique characteristic among the elements. It is also absorbed throughout the small intestine but at a reduced rate. Fluoride absorption is believed to occur by passive diffusion. Fluoride's rapid gastric absorption can be explained by the fact that it exists primarily as hydrogen fluoride, also called *hydrofluoric acid* (HF), at the low pH of the gastric contents rather than as ionic fluoride. Rate of diffusion across membranes, in general, correlates directly with the lipid solubility of the diffusing substance. Gastric diffusion is also inversely related to pH. The pH dependence of gastric absorption is consistent with the hypothesis that HF is being absorbed [4]. Hydrogen fluoride is a weak acid with a pKa (the negative logarithm of an acid dissociation constant, Ka) of 3.4, dissociating according to the equation $HF \rightarrow H^+ + F^-$. Assuming a gastric pH of approximately 1.5, the ratio of HF to F^-, readily calculable from the Henderson-Hasselbalch equation (p. 496), would be nearly 200 to 1. The gastric luminal form of fluoride is therefore largely diffusible. Fluoride absorption does not appear to be influenced by its plasma levels except at a very high, perhaps toxic, concentration [5].

Some fluoride is transported in the blood as ionic fluoride or hydrofluoric acid, not bound to plasma proteins. Some fluoride is strongly bound and referred to as nonionic or the organic form. Organically bound fluoride occurs in variable concentrations that are independent of total fluoride intake and plasma ionic fluoride levels. What is not known is the extent to which environmental contamination by industrially generated fluorocarbons may contribute to the variability. Ionic fluoride concentration, in contrast, correlates directly with the dietary intake even up to very large oral doses, indicating that plasma ionic fluoride is not precisely controlled by homeostatic mechanisms.

Absorbed fluoride leaves the blood very quickly and is distributed rapidly throughout the body, particularly to the hard tissues, where it is sequestered in bones and teeth by apatite. Apatite is a basic calcium phosphate having the theoretic formula $Ca_{10}(PO_4)_6(OH)_2$. Mineralized tissues account for nearly 99% of total body fluoride, with bone being by far the major depot. As the amount of absorbed fluoride increases, so does the quantity taken up by hard tissue. But the percentage retained at high absorption rates becomes less because of accelerated urinary excretion [4]. Skeletal growth rate influences fluoride balance, exemplified by the fact that young, growing people incorporate more fluoride into the skeleton than adults and excrete less in the urine.

Table 12.9 Fluoride Content of Various Food Groups

Food Group	Fluoride Content Range (ppm)
Dairy products	0.05–0.07
Meat, fish, poultry	0.22–0.92
Grain, cereal products	0.29–0.41
Potatoes	0.08–0.14
Green, leafy vegetables	0.10–0.15
Legumes	0.15–0.39
Root vegetables	0.09–0.10
Other vegetables, vegetable products	0.06–0.17
Fruits	0.06–0.13
Fats, oils	0.13–0.24
Sugar adjuncts	0.21–0.35

Source: Rao GS. Dietary intake and bioavailability of fluoride. Ann Rev Nutr 1984;4:120. Reproduced with permission, ©1984.

Functions and Mechanisms of Actions

The major functions of fluoride are related to its effects on the mineralization of teeth and bones. Specifically, it promotes mineral precipitation from metastable solutions of calcium and phosphate, leading to the formation of apatite. The apatite is deposited as crystallites within an organic (protein) matrix.

Fluoride can be incorporated into the apatite structure by replacement of hydroxide ions. This can occur during initial crystal formation or by displacement from previously deposited mineral, according to the following equation:

$$Ca_{10}(PO_4)_6(OH)_2 + xF^- \longrightarrow Ca_{10}(PO_4)_6(OH)_2 - {_x}F^-{_x}$$

The extent of fluoride incorporation varies with animal species, age, fluoride exposure, and rate of tissue turnover. In bone and dental enamel of humans and other higher mammals, the substitution of F^- for OH^- is approximately 1:20 to 1:40. Evidence has been presented for the high affinity of an enamel matrix protein for fluoride, leading to the speculation that fluoride's major role in mineralization may be its participation in the nucleation of crystal formation rather than its association with the mineral phase [6].

Fluoride therapy, typically as sodium fluoride (40–80 mg/day), used in conjunction with calcium supplements appears to stimulate osteoblasts. Osteoblasts function in the formation of new bone. Increases in trabecular bone formation and to a lesser extent cortical bone formation have been observed in postmenopausal women receiving fluoride and calcium supplements [7–10]. Not all studies, however, have shown beneficial effects [11].

Interactions with Other Nutrients

It has been reported that *aluminum, calcium, magnesium,* and *chloride* reduce fluoride uptake and use, whereas *phosphate* and *sulfate* increase its uptake [12]. Sodium chloride, for example, decreases the skeletal uptake of fluoride [13]. This point is of interest because kitchen or table salt has been used as a vehicle for fluoride supplementation in some countries. The mechanisms of the interactions are not established.

Aluminum-containing antacid use also sharply reduces the absorption of fluoride, as well as phosphorus. It is speculated that aluminum forms insoluble complexes with fluoride in the intestine [14], but the contribution of the pH effect of the antacid on fluoride availability requires further investigation.

Excretion

Excretion of fluoride takes place rapidly via the urine, which accounts for approximately 90% of total excretion. Fecal elimination accounts for most of the remainder, with minor losses occurring in sweat. Some renal tubular reabsorption occurs by passive diffusion of undissociated HF. Because the amount of HF is increased relative to that of F⁻ as acidity is increased, tubular reabsorption and urinary pH are therefore inversely related [4].

Adequate Intakes

The 1997 Dietary Reference Intakes reported Adequate Intakes of 3.8 and 3.1 mg of fluoride/day for adult males and females, respectively. In the interest of optimizing dental health, the Council on Dental Therapeutics of the American Dental Association has advised fluoride supplementation based on existing drinking water concentrations and the age of the subjects [15,16]. The optimum fluoride concentration in fluoridated drinking water is 1 to 2 ppm.

Deficiency

Fluoride deficiency in test animals has been reported to result in curtailed growth, infertility, and anemia. But these findings are not well documented and clearly cannot be extrapolated in predicting similar effects on humans. In humans, what is unequivocal is that an optimal level of fluoride is necessary to reduce the incidence of dental caries and perhaps also to maintain the integrity of skeletal tissue. It is on this basis that the element is considered essential.

Toxicity

Chronic toxicity of fluoride is referred to as *fluorosis* and characterized by changes in bone, kidney, and possibly nerve and muscle function [4,16]. Dental fluorosis or mottling of teeth has been observed in children receiving 2 to 8 mg of fluoride/kg. Acute toxicity manifests as nausea, vomiting, acidosis, cardiac arrhythmias. Death has been reported following ingestion of between 5 and 10 g of sodium fluoride or about 32 to 64 mg of fluoride/kg body weight, although may occur as low as 5 mg of fluoride/kg of body weight [4,16].

Assessment of Nutriture

Normal ranges for ionic fluoride have been established at 0.01 to 0.2 μg of F⁻/mL plasma and 0.2 to 1.1 mg of

F/mL urine. The element is most commonly determined in its ionic form by fluoride ion-specific electrode potentiometry, a technique analogous to the hydrogen ion-specific electrode potentiometry of the common pH meter.

References Cited for Fluoride

1. National Research Council. Recommended Dietary Allowances, 10th ed. Washington, D.C.: National Academy Press, 1989;235–40.

2. Levy S. Review of fluoride exposures and ingestion. Comm Dent Oral Epidemiol 1994;22:173–80.

3. Krishnamachari K. Fluoride. In: Mertz W, ed.: Trace Elements in Human and Animal Nutrition. San Diego: Academic Press, 1987;1:365–415.

4. Whitford G. The physiological and toxicological characteristics of fluoride. J Dent Res 1990;69:539–49.

5. Whitford G, Williams J. Fluoride absorption: Independence from plasma fluoride levels. Proc Soc Exp Biol Med 1986;181:550–4.

6. Crenshaw M, Bawden J. Fluoride binding by organic matrix from early and late developing bovine fetal enamel determined by flow rate dialysis. Arch Oral Biol 1981;26:473–6.

7. Fluoride and bone health. J Publ Health Dent 1995;55:53–56.

8. Pak C, Sakhaee K, Piziak V, Peterson R, Breslau N, Boyd P, Poindexter J, Herzog J, Sakhaee A, Haynes S, Huet B, Reisch J. Slow release sodium fluoride in the management of postmenopausal osteoporosis: a randomized controlled trial. Ann Intern Med 1994; 120:625–32.

9. Kleerekoper M, Mendlovic D. Sodium fluoride therapy of postmenopausal osteoporosis. Endocrin Rev 1993;14:312–23.

10. Eisinger J, Clairet D. Effects of silicon, fluoride, etidronate and magnesium on bone mineral density: a retrospective study. Magnesium Res 1993;6:247–9.

11. Riggs B, Hodgson S, O'Fallon W, Chao E, Wahner H, Muhs J, Cedel S, Melton L. Effect of fluoride treatment on the fracture rate in postmenopausal women with osteoporosis. N Engl J Med 1990;322:802–9.

12. Rao G. Dietary intake and bioavailability of fluoride. Ann Rev Nutr 1984;4:115–36.

13. Ericsson Y. Influence of sodium chloride and certain other food components on fluoride absorption in the rat. J Nutr 1968;96:60–68.

14. Spencer H, Kramer L. Osteoporosis: calcium, fluoride, and aluminum interactions. J Am Coll Nutr 1985; 4:121–8.

15. American Dental Association. Accepted Dental Therapeutics, 39th ed. Chicago: American Dental Association, 1982.

16. Heifetz S, Horowitz H. The amounts of fluoride in current fluoride therapies: safety considerations for children. J Dent Child 1984;51:257–69.

Hypoferremia and Infection: When Are Iron Supplements Advisable?

As indicated in Figure 1 (at base of triangle), nutritional deficiency is believed to be deleterious and to impair defense mechanisms against infections. This view has been challenged especially with respect to iron, because a normal physiological response to infection is a decrease in serum iron [1,2]. The reason for this shift of iron from circulation into storage is not understood, but some researchers believe the function of the shift is to protect the host by decreasing the availability of iron to invading micro-organisms [3]. Micro-organisms (with the exception of lactobacilli [2]) cannot multiply without a source of iron. Whether the likelihood of human infection can be reduced by iron withholding is highly controversial; nevertheless, effective sequestering of the host's readily available iron from pathogens for their own use is an established fact.

To solubilize and assimilate ferric iron (Fe^{3+}) for their use, bacteria produce siderophores, which chelate the iron and bring it into the cell. Within the cell, iron is released from the chelate by enzymatic reduction to ferrous iron (Fe^{2+}), the usable form of the mineral [3]. The siderophores produced by bacteria invading the host are inhibited in their sequestering of the host's iron by two (or perhaps more) glycoproteins synthesized by the host. The two known glycoproteins are lactoferrin, a major component in human milk and also found in many other mammalian exocrine secretions, and transferrin, the iron-binding protein in the serum that transports iron from the intestine to other tissues of the body [3,4]. Lactoferrin and transferrin each can bind two atoms of iron per molecule of protein and provide the host with what Weinberg [4] terms "nutritional immunity."

As mentioned, infection of the host is characterized by a decrease in serum levels of iron. Unsaturated lactoferrin, which is a major protein component of circulating neutrophils (leucocytes) [4], mediates the iron uptake from the serum. Lactoferrin is released into the plasma from activated neutrophils, binds iron, and the resulting complex is cleared from circulation, thereby causing hypoferremia. The iron complexed by lactoferrin is removed from circulation by mononuclear phagocytes, and the sequestered iron ultimately becomes part of iron storage compounds such as hemosiderin [1]. Much of the iron removed from circulation is stored in the liver [1,5]. Fever also appears to be a deterrent to the securing of iron by bacteria; elevated temperatures decrease the production of siderophores [3].

The positive relationship between the availability of iron to bacteria and the likelihood of infections is particularly evident in people suffering from various diseases characterized by iron overload (hyperferremia). People subject to hyperferremic episodes are those suffering from

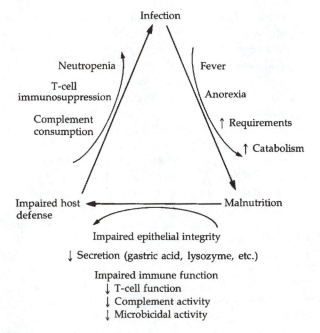

Figure 1 Triangle of interaction among malnutrition, infection, and host defense.
Source: Reproduced with permission from the Annual Review of Nutrition, Vol. 6, © 1986, by Annual Reviews, Inc.

- destruction of liver cells containing ferritin, as might occur in viral hepatitis;
- hemolytic anemia, as can occur in malaria, sickle-cell disease, and leukemia; and
- overload of iron from exogenous sources, which is particularly possible in neonates [5].

These people are more susceptible to infection than their nonhyperferremic counterparts; furthermore, during acute hyperferremic episodes they have greater susceptibility to bacterial and fungal pathogens than they do when their plasma iron is within normal range [5]. Transferrin concentration does not usually rise; therefore, hyperferremia is usually characterized by an elevated iron saturation of circulating transferrin. Weinberg [4,5] postulates that a transferrin saturation of >50% allows an easier removal of the host's iron by the siderophores of pathogens.

Of particular concern is the possible oversupplementation of healthy, full-term neonates who are born with an abundance of iron, as indicated by the degree of transferrin saturation. At birth, mean transferrin saturation is 69%, but by 3 months healthy infants have decreased the degree of saturation to 34% and by 6 months to 25% [4]. The decrease in saturation is due primarily to increased transferrin being synthesized by the healthy, maturing infant. The normal iron saturation level averages about 20% to 30% [4].

The inadvisability of a routine, vigorous treatment of neonates with supplemental iron to prevent iron deficiency has been emphasized by some tragic experiences with prophylactic parenteral iron. About 20 years ago it was a common practice in some New Zealand clinics to administer parenterally an iron-dextran complex to infants during their first week of life to protect them against iron deficiency anemia [1,3,6]. The practice was stopped when it was found that the incidence of bacterial septicemias and meningitis was eight times higher in the treated infants than in those who received no iron supplementation [3,6]. The great majority of affected infants were healthy at birth and did not suffer any recognized perinatal event likely to lead to infection; the increase in the incidence of infections appeared due to iron supplementation [6].

Another group of children at particular risk from exogenous iron are those suffering with kwashiorkor. These children, although actually hypoferremic, have transferrin saturation values of >100% because of hypotransferrinemia. Because of their inability to synthesize transferrin, these children, when supplemented with iron before correction of their protein deficiency, are very susceptible to iron overload. Circulating free iron in these malnourished children appears to allow a greatly increased incidence of bacterial infections, infections often severe enough to cause death [5,6].

Despite the advantages that possibly could be accrued through iron withdrawal from the host so that bacteria cannot multiply, the dangers associated with iron deficiency anemia cannot be overlooked. Iron deficiency anemia is considered one of humanity's most crucial nutritional problems [7], and many studies suggest that iron deficiency in infants and children predisposes to infection, particularly involving the respiratory and gastrointestinal tracts [1]. Furthermore, these studies suggest that administration of iron in some circumstances can reduce infection rates [1]. Many of these studies, however, have been uncontrolled or poorly controlled and therefore permit no reliable conclusions about the influence of iron alone in the development or resolution of the infection [1]. Nevertheless, the effect of iron on the host's immune system cannot be overlooked; cell-mediated responses in particular are susceptible to iron deficiency. Not only have defective macrophage functions been observed [8], but also a reduction in the proliferation of T-cells. The proliferation of T-cells requires acquisition of transferrin-bound iron, and, as demonstrated by experimental iron deficiency, transferrin saturation can fall below the value needed for optimal proliferation [2]. Too low a level of transferrin saturation can result in atrophy of lymphoid tissues, with depletion of lymphocytes in general. Therefore, a decrease in humoral immunity (production of antibodies) as well as cell-mediated immunity can occur [9].

Developing a unifying hypothesis to explain the two different views regarding iron status and susceptibility to infection is very difficult [1]. Although iron deficiency may increase the host's ability to withhold iron from invading pathogens ("nutritional immunity"), this advantage may be overshadowed by impairment of the host's cell-mediated and humoral immune responses [2]. Keusch and Farthing [1] propose that iron deficiency in an otherwise well-nourished individual with normal serum transferrin should probably require iron supplements orally to remove the risk of an impaired immune system with a consequent risk of increased infections. In contrast, great care should be exercised in supplementing children with severe protein-energy malnutrition. Iron supplementation in these children with very low transferrin values could result in rapid increases in circulating free iron, thereby promoting the growth of invading pathogens. In their opinion, the "physiological," mild iron deficiency often observed in early infancy can be considered an intermediate state. This physiological anemia is not harmful to the immune system and at the same time allows less iron to be available to pathogens. The common practice of vigorously treating mild hypoferremia in infants prone to low-grade infections may be ill advised.

The desirable iron balance is one in which iron is not readily accessible to invading micro-organisms yet is sufficient for optimal operation of the host's immune system [2]. Although transferrin plays a key role in both mechanisms involved in a desirable balance, the role that the percentage iron saturation of transferrin has in making iron available to the siderophores of pathogens remains controversial [2,4].

References Cited

1. Keusch GI, Farthing MJG. Nutrition and infection. Ann Rev Nutr 1986;6:131–54.

2. Brock JH. Iron and the outcome of infection. Br Med J 1986;293:518–20.

3. Emery T. Iron metabolism in humans and plants. Am Sci 1982;70:626–32.

4. Weinberg ED. Iron withholding: a defense against infection and neoplasia. Physiol Rev 1984;64:65–102.

5. Weinberg ED. Iron and susceptibility to infectious disease. Science 1974;184:952–6.

6. Weinberg ED. Iron and susceptibility to infectious disease. Science 1975;188:1039.

7. Regenauer J, Saltman P. Iron and susceptibility to infectious disease. Science 1975;188:1038–9.

8. Beisel WR, Edelman R, Nauss K, et al. Single nutrient effects on immunologic functions. JAMA 1981;245:53–58.

9. Myrvic QN. Nutrition and immunology. In: Shils M, Young V, eds. Modern Nutrition in Health and Disease, 7th ed. Philadelphia: Lea and Febiger, 1988:585–616.

Chapter 13

Ultratrace Elements

Photomicrograph of crystallized manganese sulfate

Nickel

Silicon

Vanadium

Arsenic

Boron

Cobalt

For each of the microminerals listed, the following subtopics (if known and when applicable) are discussed:

Sources
Digestion, Absorption, Transport, and Storage
Functions and Mechanisms of Actions
Interactions with Other Nutrients
Excretion
Deficiency
Toxicity
Recommended Intake and Assessment of Nutriture

This chapter discusses the sources, digestion, absorption, transport, functions, interactions with other nutrients, excretion, recommended intakes, deficiency, toxicity, and assessment of nutriture for several ultratrace elements. Table 13.1 provides an overview of selected functions, food sources, and deficiency symptoms.

Nickel

Possible signs of nickel deprivation in experimental animals were first reported in the 1970s, but the early findings were subject to question because they were obtained under conditions that did not permit optimal growth among the animals. Also, early studies were fraught with inconsistencies believed to be due, in part, to variables such as the iron status of the animals, which is now known to influence nickel metabolism [1].

Since 1975, experimental diets have been greatly improved so as to allow optimal growth and survival, affording greater credibility to research results. Signs of nickel deprivation continue to be described for at least six animal species. Included among the more consistent signs are depressed growth, rough hair coat, and impaired hematopoiesis, which probably is due to altered iron metabolism.

Sources

Foods of plant origin have a significantly higher nickel content than foods of animal origin. Nuts and legumes

Table 13.1 Ultratrace Elements: Selected Functions, Deficiency Symptoms, and Food Sources

Mineral	Selected Physiological Roles	Selected Enzyme Cofactor Roles	Selected Deficiency Symptoms	Food Sources
Arsenic[a]	In animals: necessary for normal growth and iron use		Unknown	Seafoods
Boron[a]	Bone composition, structure, and strength		In animals depressed growth	Fruits, vegetables legumes, nuts
Nickel[a]	Possibly involved in hormonal membrane or enzyme activity		In animals: liver problems, anemia	Nuts, legumes, cacao products, hydrogenated solid shortening, spinach, asparagus, grains
Silicon[a]	Apparent role in formation of connective tissue and bone matrix	Prolyl hydroxylase	In animals: decreased collagen, long bone and skull abnormalities	Beer, unrefined grains, plant foods
Vanadium	Apparent role in glucose and lipid reproductive performance		In animals: reduced growth, hematologic changes, bone defects, metabolism changes	Shellfish, spinach, parsley, mushrooms, whole grains (food contains very little)

[a]These minerals are essential in animals and very likely essential for humans.

are particularly rich in the metal. Grains, cured meats, and vegetables are generally of intermediate nickel content, and foods of animal origin, such as fish, milk, and eggs, are generally low, as shown in Table 13.2 [2].

The chemical form of nickel in foods is unknown, but in plants it is probably largely inorganic and is dependent on the nickel content of the soil. Dietary nickel derived from contamination of processed foods would likely be inorganic as well.

Absorption, Transport, and Storage

Most of the information on the influence of dietary factors on nickel absorption has come from human studies. Nickel added to beverages such as coffee, tea, cow's milk, and orange juice is absorbed to a lesser extent than if added to water alone [2]. In the same investigation, it was shown that phytate did not reduce absorption despite the fact that nickel ions form stable complexes with phytate in vitro. Nickel absorption in subjects receiving high doses of nickel, 12, 18 and 50 µg/kg body weight, was $27 \pm 17\%$ following an overnight fast, and $0.7 \pm 0.4\%$ with a meal [3]. Other studies using subjects receiving 10 µg nickel/kg body weight report 29% to 40% absorption after drinking a ^{62}Ni solution [4].

The absorption of ingested nickel across the intestinal brush border may occur via an energy-dependent mechanism or by passive diffusion. Nickel ions appear to compete with iron for a common transport system

Table 13.2 Nickel Content of Selected Food Groups

Food Group	Content (µg range/100 g)
Milk, yogurt, cheese	0–8.2
Eggs	0–1.4
Meat, fish, poultry	0–14.3
Grains and grain products	0.3–228.5
Fruits and fruit juices	0–47.7
Vegetables	0.2–40.5

Source: Pennington JAT and Jones JW. Molybdenum, nickel, cobalt, vanadium, and strontium in total diets. Reprinted from JOURNAL OF THE AMERICAN DIETETIC ASSOCIATION 87:1644–1650, 1987.

in the proximal small intestine [5]. Transport across the basolateral membrane is thought to occur by diffusion or as part of a complex with an amino acid or binding ligand.

On entering the blood, nickel becomes bound mainly to albumin [6] and to several different amino acid ligands. These amino acid ligands are believed to include histidine, cysteine, and aspartic acid. Other serum proteins such as α-2 macroglobulin also may transport nickel in the blood.

Although nickel is widely distributed among human tissues, its concentration throughout the body is extremely low, occurring at nanogram/gram levels. The highest concentrations of nickel are found in hair; bone; soft tissues such as the lungs, heart, kidneys, and liver; and two glands, the thyroid and adrenal glands.

Functions and Mechanisms of Actions

A specific role of nickel in human and animal nutrition has not yet been defined, although roles for nickel in plants and micro-organisms have been documented [7]. In plants, for example, nickel serves as a cofactor for urease, which catalyzes the hydrolysis of urea into carbon dioxide and ammonia. In bacteria, several hydrogenases appear to be nickel-dependent.

In the several enzyme systems, however, the role of nickel can be substituted for other minerals such as magnesium. An example of such a replacement is the formation of the C3 convertase enzyme (C3b,Bb and C4b,2b) of the human complement system, which classically requires Mg^{2+} for activity. The substitution of nickel in place of magnesium in this complex enhanced both the stability and activity of the enzyme [8], therefore raising the question as to nickel's possible physiological role in the complement system. It has also been demonstrated that nickel can substitute for zinc in the carboxypeptidases and in horse liver alcohol dehydrogenase [7].

Nickel may be involved with vitamin B_{12} and act as a cofactor for an enzyme in the propionate pathway of branched-chain amino acid and odd-chain fatty acid metabolism [9]. Nielson has suggested that the need for nickel in animals and humans will most likely become evident under situations in which a propionate pathway enzyme, such as the vitamin B_{12}–dependent methylmalonyl mutase, demands are elevated [10].

Interactions with Other Nutrients

It has already been pointed out in the preceding section that nickel shares with other metals the property of being readily chelated by, and complexed with, a wide variety of ligands. It follows that nickel can compete with those ions for ligand sites. Nickel interacts with many ions in this manner, including as many as 13 essential minerals. But the only interactions that are of particular nutritional interest are those involving iron, copper, and zinc.

Nickel facilitates the use of ferric iron, but somehow antagonizes ferrous ion absorption, particularly in severe *iron* deficiency. There is speculation that nickel may function as a cofactor in enhancing the complexing of the Fe^{3+} with a bioligand, or it may act in an enzyme system that converts Fe^{3+} to Fe^{2+}.

That nickel interacts antagonistically with *copper* in vivo is based on the finding that in copper-deficient rats, physiological signs of the deficiency are exacerbated by nickel supplementation. The antagonism appears not to occur during absorption of copper but

more likely is due to nickel's replacement of copper at certain functional sites.

Deficient or toxic levels of nickel also affect *zinc* metabolism, possibly by causing redistribution of zinc in the body, rather than by competing with zinc at sites of zinc function. In some instances signs of zinc deficiency in animals were partially alleviated by supplemental nickel, whereas other symptoms were unaffected by the resulting tissue redistribution. Nielsen [11] has reviewed the interactions of nickel with iron, copper, and zinc.

Excretion

Most absorbed nickel is excreted in the urine. Within the renal cells, nickel is complexed nonspecifically with uronic acid and neutral sugar oligosaccharides and specifically with an acidic peptide [12]. Because these peptides are not present in plasma, a ligand exchange must take place after glomerular filtration.

Small amounts of absorbed nickel may also be excreted through the bile and also via the sweat glands. Dermal loss of nickel may be significant during episodes of profuse sweating.

Recommended Intake and Assessment of Nutriture

The fact that humans require nickel is postulated from extrapolated data from animal studies, and although a RDA for humans has not been established, it can be approximated from the apparent requirement of animals. For rats and chicks, the requirement for nickel is estimated to be about 50 mg/kg of diet, or 16 mg/1,000 kcal [13], corresponding to a hypothetical human need of approximately 35 mg/day. A requirement of <10 mg/day has also been proposed [9].

The reference range for nickel in the serum or plasma of healthy adults is 1 to 21 ng/mL, and the urinary excretion has been reported to be 0.1 to 20 mg/day. Like most of the ultratrace metals, the preferred technique for nickel determination is flameless atomic absorption spectrophotometry. It offers the degree of sensitivity necessary for determination in the nanogram range. Valid methods to assess nickel status of humans are unavailable.

References Cited for Nickel

1. Nielsen F. The importance of diet composition in ultratrace element research. J Nutr 1985;115:1239–47.

2. Solomons N, Viteri F, Shuler T, Nielsen F. Bioavailability of nickel in man: effects of foods and chemically-

defined dietary constituents on the absorption of dietary nickel. J Nutr 1982;112:39-50.

3. Sunderman F, Hopfer S, Sweeney K, Marcus A, Most B, Creason J. Nickel absorption and kinetics in human volunteers. Proc Soc Exp Biol Med 1989; 191:5-11.

4. Patriarca M, Lyon T, Fell G. Nickel metabolism in humans investigated with an oral stable isotope. Am J Clin Nutr 1997;66:616-21.

5. Nielsen F. Studies on the interaction between nickel and iron during intestinal absorption. In: Anke M, Bauman W, Braunlich H, et al., eds. Spurenelement-Symposium Leipzig, East Germany: Karl-Marx-Universität, 1983;11-98.

6. Tabata M, Sarkar B. Specific nickel (II)-transfer process between the native sequence peptide representing the nickel (II)-transport site of human serum albumin and L-histidine. J Inorgan Biochem 1992; 45:93-104.

7. Walsh C, Orme-Johnson W. Nickel enzymes. Biochemistry 1987; 26:4901-6.

8. Fishelson Z, Muller-Eberhard H. C3 convertase of human complement: enhanced formation and stability of the enzyme generated with nickel instead of magnesium. J Immunol 1982;129:2603-7.

9. Nielsen F. Nutritional requirements for boron, silicon, vanadium, nickel, and arsenic: current knowledge and speculation. FASEB J 1991;5:2661-7.

10. Nielsen F. Ultratrace elements of possible importance for human health: an update. In: Prasad AS, ed. Essential and Toxic Trace Elements in Human Health. New York: Wiley-Liss, 1993;355-76.

11. Nielsen F. Nickel. In: Frieden E, ed. Biochemistry of the Essential Ultratrace Elements. New York: Plenum Press, 1984:301-4.

12. Templeton D, Bibudhendra S. Peptide and carbohydrate complexes of nickel in human kidney. Biochem J 1985;230:35-42.

13. Nielsen F. Possible future implications of nickel, arsenic, silicon, vanadium, and other ultratrace elements in human nutrition. Curr Top Nutr Dis 1982; 6:379-404.

Silicon

Silicon occupies a unique position among the essential trace elements in that it is second only to oxygen in earthwide abundance. Quartz, which is crystallized silica, is the most abundant mineral in the earth's crust. The element occurs naturally as its dioxide silica (SiO_2), and as water-soluble silicic acid, $Si(OH)_4$, formed by hydration of the oxide. In plants, silicon is deposited as the solid, hydrated oxide, $SiO_2 \cdot nH_2O$, known as *silica gel*, following polymerization of silicic acid.

Whereas early investigations concentrated on silicon's toxicity, such as silicon-related urolithiasis and particularly silicosis, caused by the inhalation of dust, research over the past 25 years has focused on the possible roles or functions of silicon in animals and humans.

Sources

Data on the distribution of silicon in human foods and diets are sparse. It is known, however, that foods of plant origin are normally much richer in silicon than those of animal origin. Whole-cereal grains and root vegetables appear to be especially rich sources of the element [1,2]. Average daily silicon intake has been estimated at 20 to 50 mg/day [1].

Absorption, Transport, and Storage

The mechanism of the absorption of silicon is not well understood, and future studies will likely be complicated by the fact that its dietary forms are so diverse. Silica, monosilicic acid, and silicon found in organic combination, such as pectin and mucopolysaccharides, are a few of its ingestible forms. The most soluble form of silicon is metasilicate; this form as sodium metasilicate has been commonly used in supplementation studies.

Absorption ranges from 10% to 70% depending on the form of silicon used [3]. Moreover, the extent of absorption of the element appears to correlate with the production of its soluble forms in gastric fluids [4], a finding that is supported by the observation that silicic acid in foods and beverages is readily absorbed in humans and rapidly excreted in the urine. Another factor influencing absorption of silicon in humans is fiber. Nearly 97% of dietary silicon contained in a high-fiber diet remained unabsorbed and was lost in the feces, compared with a fecal excretion of only 60% when a low-fiber diet was consumed [2]. In studies on rats, changes in absorption were shown to relate also to age, sex, and the activity of various endocrine glands [5].

Silicon in the body is found bound as well as in free forms such as silicic acid [6]. Silicon, as silicic acid, is freely diffusible throughout tissue fluids. Once silicic acid is absorbed into the blood, it is almost entirely nonprotein bound, therefore accounting for its rapid decrease in plasma concentration, its diffusion into tissue fluids, and its rapid urinary excretion [7]. When [31]Si silicic acid was administered intravenously to normal rats, 77% of the compound was recovered in the urine within 4 hours [7]. In that same study the initial uptake of the label was most rapid in liver, lung, skin, and bone, with slower entry occurring in heart, muscle, spleen, and testes. Negligible uptake into the brain was reported, indicating active exclusion by the blood-brain barrier.

Functions and Deficiency

The physiological role of silicon is focused on the normal growth and development of bone, connective tissue, and cartilage, functioning both in a metabolic and a structural capacity. The effect of silicon on bone is to hasten formation, mineralization, as well as to promote turnover and growth. In test animals, high-silicon diets were associated with increased calcium content of the bone and an accelerated bone maturity. Silicon deficiency also resulted in smaller, less flexible long bones and in deformation of the skull. In studies on chicks, the skull deformation was subsequently found to be due to a significantly reduced collagen content in the connective tissue matrix [8]. In rats, silicon affected bone turnover and formation [9].

The detrimental effect of silicon deficiency on collagen formation is linked to the mineral's requirement in the synthesis of proline and hydroxyproline, the residues of which are of particular importance in collagen's primary structure. Proline is incorporated as such into the procollagen peptide but becomes hydroxylated posttranslationally by prolyl hydroxylase. The reaction (Fig. 13.1) also requires iron and vitamin C. Silicon is required for maximal prolyl hydroxylase activity [10], mechanistically not yet understood, and is probably also required for the synthesis of proline itself [11].

In addition to its positive influence on collagen synthesis, silicon is also needed for the formation of glycosaminoglycans, such as hyaluronic acid, chondroitin sulfate, and keratin sulfate. Glycosaminoglycans are linked covalently to proteins as components of the extracellular ground substance or organic matrix that surrounds the collagen, elastic fibers, and cells. A structural role for silicon in glycosaminoglycan formation may also be proposed, because it has been found to be chemically linked within the glycosaminoglycan framework [6].

Interactions with Other Nutrients

The only mineral element in a normal diet that may be of any consequence in affecting the availability of silicon is molybdenum. The interaction was first discovered when silicon supplementation of a diet high in liver failed to elevate significantly the plasma silicon concentration, and both plasma and tissue silicon levels were subsequently shown to be markedly and inversely affected by molybdenum intake [12]. The reverse is also true, that silicon supplementation reduces the plasma concentration and cellular uptake of dietary molybdenum. Until the mechanism of intestinal absorption of the two elements is better understood, the manner in which they interact remains unknown.

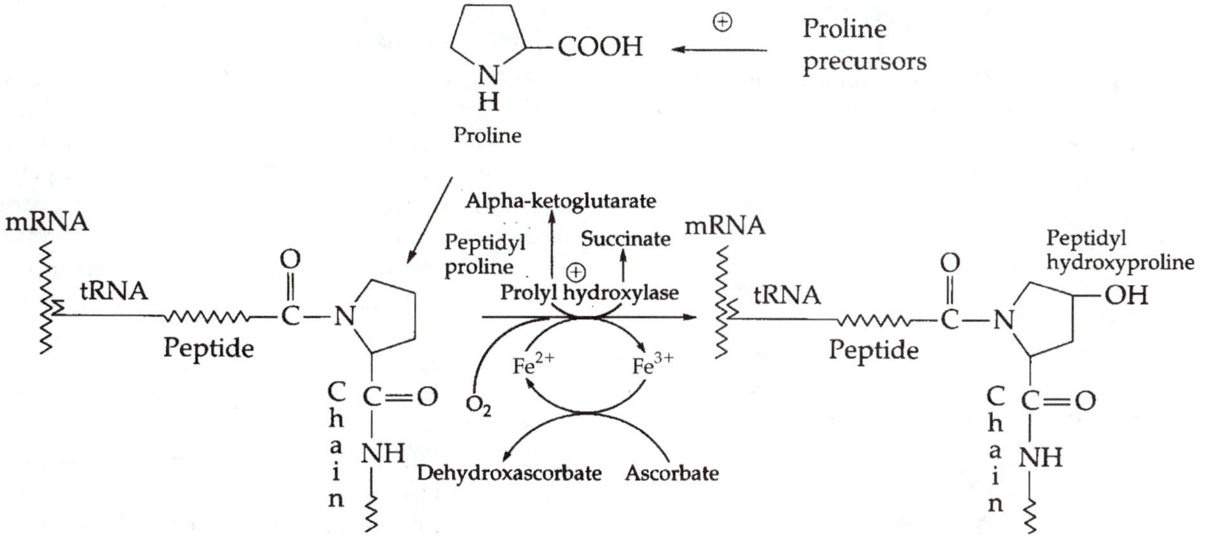

Figure 13.1 The posttranslational hydroxylation of peptidyl proline in a growing procollagen chain. In addition to prolyl hydroxylase, the reaction also requires ascorbate, α-ketoglutarate, ferrous iron, and oxygen. The symbol ⊕ designates the sites at which silicon positively affects collagen synthesis.

Excretion

The kidney is the major excretory organ of absorbed silicon in experiments in which the element was consumed as silicic acid. Urinary output of silicon generally increases as intake increases up to fairly well-defined limits that do not appear to be imposed by the kidney's inability to excrete more.

Toxicity, Recommended Intake, and Assessment of Nutriture

The minimum silicon requirement compatible with human health is largely unknown, as are the dietary forms that render the mineral most available. Toxicity of silicon is associated with kidney stones as well as diminished activities of several enzymes that prevent free radical damage including glutathione peroxidase, superoxide dismutase, and catalase [13].

Requirements for silicon have been largely confined to assessing the relationship of dietary silicon intake, expressed as milligrams/gram of dry diet, to observed growth and skeletal development in test animals. Such studies indicate that the dietary silicon requirement may be relatively high compared to that of other trace elements. For humans, estimates of the requirement for silicon ranges from 2 to 20 mg of silicon per day [6,12].

As in the case of most of the trace elements, levels of silicon in biological fluids of healthy adults have been reported but may not accurately represent nutriture. Chemical assessment is generally performed on serum or plasma, which contains 0.4 to 10.0 mg/mL. Mass spectrometry, emission spectroscopy, and atomic absorption spectrophotometry are a few of the techniques for determining silicon concentration in biological specimens. Of these, atomic absorption spectrophotometry has been the method of choice for most laboratories.

References Cited for Silicon

1. Pennington J. Silicon in foods and diet. Food Additives Contaminants 1991;8:97–118.

2. Kelsay J, Behall K, Prather E. Effect of fiber from fruits and vegetables on metabolic responses of human subjects II: calcium, magnesium, iron, and silicon balances. Am J Clin Nutr 1979;32:1876–80.

3. Nielsen F. Ultratrace minerals. In: Shils ME, Olson JA, Shike M, eds. Modern Nutrition in Health and Disease. Philadelphia: Lea and Febiger, 1994;269–86.

4. Benke G, Osborn T. Urinary silicon excretion by rats following oral administration of silicon compounds. Food Cosmet Toxicol 1978;17:123–7.

5. Charnot Y, Peres G. Silicon, endocrine balance and mineral metabolism. In: Bendz G, Lindquist I, eds. Biochemistry of Silicon and Related Problems. New York: Plenum Press, 1978:269–80.

6. Seaborn C, Nielsen F. Silicon: A nutritional beneficence for bones, brains, and blood vessels. Nutr Today 1993; 13–18.

7. Adler A, Etzion Z, Berlyne G. Uptake, distribution, and excretion of [31]silicon in normal rats. Am J Physiol 1986;251:E670–3.

8. Carlisle E. A silicon requirement for normal skull formation in chicks. J Nutr 1980;110:352–9.

9. Seaborn C, Nielsen F. Dietary silicon affects acid and alkaline phosphatase and [45]calcium uptake in bone of rats. J Trace Elem Exper Med 1994;7:11–18.

10. Carlisle E. Silicon: A requirement in bone formation independent of vitamin D. Calc Tissue Intern 1981;33:27–34.

11. Carlisle E, Alpenfels W. The role of silicon in proline synthesis. Fed Proc 1984;43:680.

12. Carlisle E. A silicon-molybdenum interrelationship in vivo. Fed Proc 1979;38:553.

13. Najda J, Goss M, Gminski J, Weglarz L, Siemianowicz K, Olszowy Z. The antioxidant enzymes activity in the conditions of systemic hypersilicemia. Biol Trace Elem Res 1994;42:63–70.

Vanadium

The deliberate or adventitious administration of vanadium is known to produce a number of discernible physiological effects, including toxicity. But in spite of this, its essentiality is not firmly established because of nebulous results of deprivation studies on animals.

Controlled depletion of vanadium has been reported to adversely affect growth rate, perinatal survival, physical appearance, hematocrit, and other manifestations in various animal species [1]. However, none of these were consistently induced by deprivation in repeated experiments, thereby failing to comply with that criterion for essentiality. Nevertheless, vanadium's widespread distribution throughout the organs and tissues of animals and humans and the fact that deprivation is accompanied by the effects mentioned has led investigators to declare vanadium's essentiality, at least in the chicken and the rat. On this basis, and in anticipation that its essentiality will eventually be established for the human, vanadium will be reviewed here.

Vanadium exists in several oxidation states, V^{2+} to V^{5+}. In solutions, vanadium produces a range of colors. In its pentavalent state, it is yellowish orange, whereas in its divalent state it is blue [2]. In biological systems, vanadium is found in primarily the pentavalent state known as *vanadate* (VO_3^- or $H_2VO_4^-$) or in the tetravalent state vanadyl ion (VO^{2+}). The total body pool of vanadium is about 100 μg.

Sources

The content of vanadium in foods is very low, and consequently so is its average dietary intake. Most fats and oils, fruits, and vegetables contain particularly low levels of the mineral, <1 ng/g [3]. Cereals, liver, and fish tend to have intermediate levels of about 5–40 ng/g. A few items such as spinach, black pepper, parsley, mushrooms, and oysters contain relatively high concentrations, and shellfish are particularly rich in the element, having >400 ng/g dry basis. Foods containing relatively high amounts of vanadium include breakfast cereals, canned fruit juices, fish sticks, vegetables, sweets, wine, and beer [4].

Absorption, Transport, and Storage

There is limited information on the absorption of vanadium, partly because its multiform state of oxidation complicates such studies. Dietary composition and the form of vanadium administered predictably affect the percentage absorbed.

According to most animal studies and even relatively older investigations using human subjects, vanadium appears to be poorly absorbed, generally <5% [5,6]. Moreover, vanadium is thought to be reduced to vanadyl in the stomach before absorption, yet in contrast to vanadyl, vanadate is 3 to 5 times more efficiently absorbed [5,6].

Studies on vanadium absorption in rats [7] have documented absorption ranging from 10% to as high as 40%. These rat studies suggest caution in assuming that vanadium will always be poorly absorbed.

In plasma and other body fluids, vanadium or vanadate is converted into vanadyl. Glutathione, NADH, and ascorbic acid can act as reducing agents for vanadate. Once formed, vanadyl binds to iron-containing proteins to form vanadyl-transferrin and vanadyl-ferritin complexes. Such complexes also exist in hepatocyte cytosol; it is interesting that while binding occurs with nonheme iron metalloproteins, vanadyl is not significantly attached to hemoproteins.

Vanadium is presumably thought to enter cells as vanadate through transport systems for phosphate, which it mimics chemically, and possibly other anions. Similar to reactions in the plasma, intracellular vanadate is reduced primarily by glutathione to vanadyl, which is then almost exclusively bound to a variety of ligands, many of which are phosphates; <1% remains unbound [8].

Little vanadium is found in the body. Most tissues contain <10 ng V/g tissue. Distribution studies indicate that while kidney cells retain most of the absorbed mineral soon after its administration, accumulation later shifts principally to bone, with somewhat lesser amounts in spleen and liver. This is understandable in view of the high content in bone of inorganic phosphate, to which vanadyl binds tenaciously [8].

Functions and Mechanisms of Actions

Vanadium is very active pharmacologically, exerting a broad assortment of effects that are well documented. However, the reader is cautioned not to confuse essentiality with pharmacological activity, because the latter is generally manifested only above a concentration threshold that is considerably greater than that required to fulfill the need for essentiality.

No specific biochemical function has been identified for vanadium. Many of vanadium's effects in vivo are predictable from a consideration of its aqueous chemistry. As vanadate, it will compete with phosphate at the active sites of phosphate transport proteins, phosphohydrolases, and phosphotransferases. As vanadyl, it will compete with other transition metal ions for binding sites on metalloproteins and for small ligands such as adenosine triphosphate (ATP). Third, it will participate in redox reactions within the cell, particularly with substances that can reduce vanadate nonenzymatically, such as glutathione.

A few of the more thoroughly investigated pharmacological effects of vanadium are discussed briefly next. Vanadium inhibits Na^+/K^+-ATPase, an enzyme involved in the phosphorylation by ATP of the carrier protein for sodium ions, permitting the transport of the ions against a concentration gradient. Vanadate is known to inhibit the enzyme by binding to its ATP hydrolysis site. Subsequently, it was suggested that vanadate might function as a regulator of sodium pump activity [9].

Vanadium, as vanadate, is believed to stimulate adenylate cyclase by promoting an association of an otherwise inactive guanine nucleotide regulatory protein (G protein) with the catalytic unit of the enzyme [10]. Adenylate cyclase catalyzes the formation of cyclic 3',5'-adenosine monophosphate (cAMP) from ATP. Cyclic AMP then stimulates protein kinases, which catalyze the phosphorylation of various enzymes and other cellular proteins in cytoplasm, membranes, mitochondria, ribosomes, and the nucleus. The phosphorylation is nearly always stimulatory, and it results secondarily from the hormone-induced stimulation of adenylate cyclase. This is the basis for cAMP's putative role as a second messenger of hormone action.

The effect of vanadate on the transport of amino acids across the intestinal mucosa exemplifies both its inhibitory effect on Na^+/K^+-ATPase and its stimulation of adenylate cyclase. At higher concentrations, vanadate inhibits the mucosal-to-serosal flux of alanine, commensurate with a decrease in Na^+/K^+-ATPase function. However, at a lower concentration (too low to affect Na^+/K^+-ATPase), it is stimulatory to alanine transport, attributable to an increase in adenylate cyclase activity and cAMP formation [11].

Vanadium appears to affect glucose metabolism by mimicking the action of insulin. Vanadium thereby stimulates glucose uptake into cells and enhances glucose metabolism for glycogen synthesis. Studies on animals, for example, have shown that vanadate can control high blood glucose and prevent the decline in cardiac performance associated with diabetes [12]. The insulin-mimicking effect of vanadate is linked in part to its stimulation of protein kinase activity. Insulin works by binding to insulin receptors, which span the lipid membranes of cells. Insulin binding to this cell receptor is thought to result in tyrosine-specific protein kinase-stimulated phosphorylation of tyrosine on the receptor. Serine and threonine phosphorylation of the receptor also may occur. The phosphorylation leads to a cascade of reactions whereby glucose is taken up into the insulin-dependent cells such as the muscle. Vanadium is thought to function like insulin and promote the phosphorylation of tyrosyl residues in the insulin receptor, preparatory to the expression of insulin activity [13,14]. The revelation that vanadate can inhibit phospho-tyrosine phosphatase, thereby prolonging the activity of phosphorylated enzymes, forms the basis for other proposed mechanisms [15]. There is evidence, too, that sodium vanadate exerts an insulinotropic effect by stimulating the release of insulin from rat islet cells [16].

Although vanadate's chemical similarity to phosphate accounts in large part for its biochemical action, vanadium, as the vanadyl cation, has also been shown to be physiologically active, particularly in its substitution for other metals such as Zn^{2+}, Cu^{2+}, and Fe^{3+} in metalloenzyme activity. Recent studies examining vanadium deficiency have suggested that the element is associated with iodine metabolism and/or thyroid gland function [17].

Excretion

Most ingested vanadium is excreted in the feces, and most of this represents unabsorbed vanadium. Renal excretion is the major route for the elimination of absorbed vanadium.

Recommended Intake, Toxicity, and Assessment of Nutriture

The human requirement for vanadium is not established, although estimates range from 10 to 25 μg/day [6]. Vanadium intake in the U.S. diet is thought to range from 10 to 60 μg/day [2]. Daily intakes up to 100 mg are thought to be safe [2]. Toxicity has been shown in humans with intakes of 10 mg or more. Toxic manifestations include green tongue (due to deposition of green-colored vanadium in the tongue), diarrhea, gastrointestinal cramps, and disturbances in mental function [2,6].

Reported levels of vanadium in healthy adults are 0.02 to 10 ng/mL in plasma or serum, 0.01 to 2.2 μg/g in hair, and 0 to 10 μg/day excreted in the urine. Current analytic techniques for the determination of vanadium in biological specimens are inadequate. The techniques most commonly used are neutron activation analysis and flameless atomic absorption spectrophotometry, the latter being the practical choice for most analytic laboratories.

References Cited for Vanadium

1. Nielsen F. Vanadium. In: Mertz, W, ed. Trace Elements in Human and Animal Nutrition. San Diego: Academic Press, 1987;1:275–300.

2. Harland B, Harden-Williams B. Is vanadium of human nutritional importance yet? J Am Diet Assoc 1994;94:891–4.

3. Byrne A, Kosta L. Vanadium in foods and in human body fluids and tissues. Sci Total Environ 1978;10:17–30.

4. Pennington J, Jones J. Molybdenum, nickel, cobalt, vanadium, and strontium in total diets. J Am Diet Assoc 1987;87:1644–50.

5. Nielsen F. Other trace elements. In: Brown ML, ed. Present knowledge in nutrition. Washington, D.C.: International Life Sciences Institute Nutrition Foundation, 1990;294–307.

6. Nielsen F. Ultratrace minerals. In: Shils ME, Olson JA, Shike M, eds. Modern nutrition in health and disease. Philadelphia: Lea and Febiger, 1994;269–86.

7. Bogden J, Higashino H, Lavenhar M, Bauman J, Kemp F, Aviv A. Balance and tissue distribution of vanadium after short-term ingestion of vanadate. J Nutr 1982;112:2279–85.

8. Nechay B, Nanninga L, Nechay P, Post R, Grantham J, Macara I, Kubena L, Phillips T, Nielsen F. Role of vanadium in biology. Fed Proc 1986;45:123–32.

9. Macara I, Kustin K, Cantley L. Glutathione reduces cytoplasmic vanadate: mechanism and physiological implications. Biochim Biophys Acta 1980;629:95–106.

10. Krawietz W, Downs R, Spiegel A, Aurbach G. Vanadate stimulates adenylate cyclase via the guanine nucleotide regulatory protein by a mechanism differing from that of fluoride. Biochem Pharmacol 1982; 31:843–8.

11. Hajjar J, Fucci J, Rowe W, Tomicic T. Effect of vanadate on amino acid transport in rat jejunum. Proc Soc Exp Biol Med 1987;184:403–9.

12. Heyliger C, Tahiliani A, McNeill J. Effect of vanadate on elevated blood glucose and depressed cardiac performance of diabetic rats. Science 1985;227:1474–7.

13. Orvig C, Thompson K, Battell M, McNeill J. Vanadium as insulin mimics. In: Sigel H, Sigel A. Metal ions in biological systems. New York: Marcel Dekker. 1995;31:575–94.

14. Tamura S, Brown T, Whipple J, Fujita-Yamaguchi Y, Dubler R, Cheng K, Larner J. A novel mechanism for the insulin-like effects of vanadate on glycogen synthase in rat adipocytes. J Biol Chem 1984;259:6650–8.

15. Stankiewicz P, Gresser M. Inhibition of phosphatase and sulfatase by transition-state analogues. Biochemistry 1988;27:206–12.

16. Fagin J, Ikejiri K, Levin S. Insulinotropic effects of vanadate. Diabetes 1987;36:1448–52.

17. Nielsen F. Ultratrace elements of possible importance for human health: an update. In: Prasad AS, ed. Essential and Toxic Trace Elements in Human Health. New York: Wiley-Liss, 1993; 355–76.

Additional Reference

Wever R, Kustin K. Vanadium: a biologically relevant element. Adv Inorgan Chem 1990;35:81–115.

Arsenic

More than any other essential trace mineral, arsenic conjures an image of toxicity rather than nutritional essentiality. The malevolent aspect of arsenic continues to attract attention, because a great deal more of the arsenic literature addresses its toxicological rather than its nutritional properties. Nevertheless, there is accumulating evidence that arsenic is an essential element for many animals such as hamsters, goats, and rats.

Sources

Arsenic is present throughout the earth's continental crust at an estimated concentration of 1.5 to 2.0 μg/g. It is present in all soils, although its concentration varies considerably from region to region, affected by the geological history of a particular soil as well as by pollution from unnatural sources. Fallout sources such as pesticides, smelters, and coal-fired power plants can, through aerosols and floating dust, enrich a particular area with arsenic. It then affects humans and animals through its incorporation into foods and foodstuffs, which, however, usually contain <0.3 μg/g on a dry basis and rarely exceed 1.0 μg/g. Ranges of arsenic content in various foods and feeds are listed in Table 13.3, from which it is clear that foods of marine origin are much richer in arsenic than other foods.

Arsenic is found in foods in both organic and inorganic forms, and it exists in nature in both the trivalent and pentavalent ionic states. The major arsenicals found in foods include arsenate ions (AsO_4), arsenite ions (AsO_2), dimethylarsinic acid, trimethylarsine [$As(CH_3)_3$], methylarsonate, arsenobetaine, arsenocholine, trimethylarsonium lactate, and O-phosphatidyltrimethylarsonium lactate, as shown in Figure 13.2. Notice most of the organoarsenicals are methylated. Of the arsenicals, inorganic arsenite and trivalent organoarsenicals are the most toxic to animals. The pentavalent, methylated arsenic compounds are far less toxic and are readily absorbed and used.

Absorption, Transport, and Metabolism

Absorption, as well as retention and excretion, of arsenicals vary with their chemical form and solubility, the quantity administered, and the animal species involved in the study. Most arsenic compounds are readily absorbed. For example, >90% of inorganic arsenate and arsenite is absorbed in humans. Similarly, of the organic arsenicals shown in Figure 13.2, >90% of arsenobetaine and between 70% and 80% of arsenocholine are absorbed.

Table 13.3 Arsenic Content of Selected Foods

Food Category	Content (μg/g)
Forage crops	0.1–1.0
Cereals	0.05–0.4
Vegetables	0.05–0.8
Fruits	0.03–1.0 (dry weight)
Meat	0.005–0.1 (fresh weight)
Milk	0.01–0.05
Eggs	0.01–0.1 (fresh weight)
Fish	2.0–80
Oysters	3.0–10
Mussels	Up to 10,120

Source: Anke M. Arsenic. In: Mertz W, ed. Trace elements in human and animal nutrition. Orlando, FL: Academic Press, 1986;2:360.

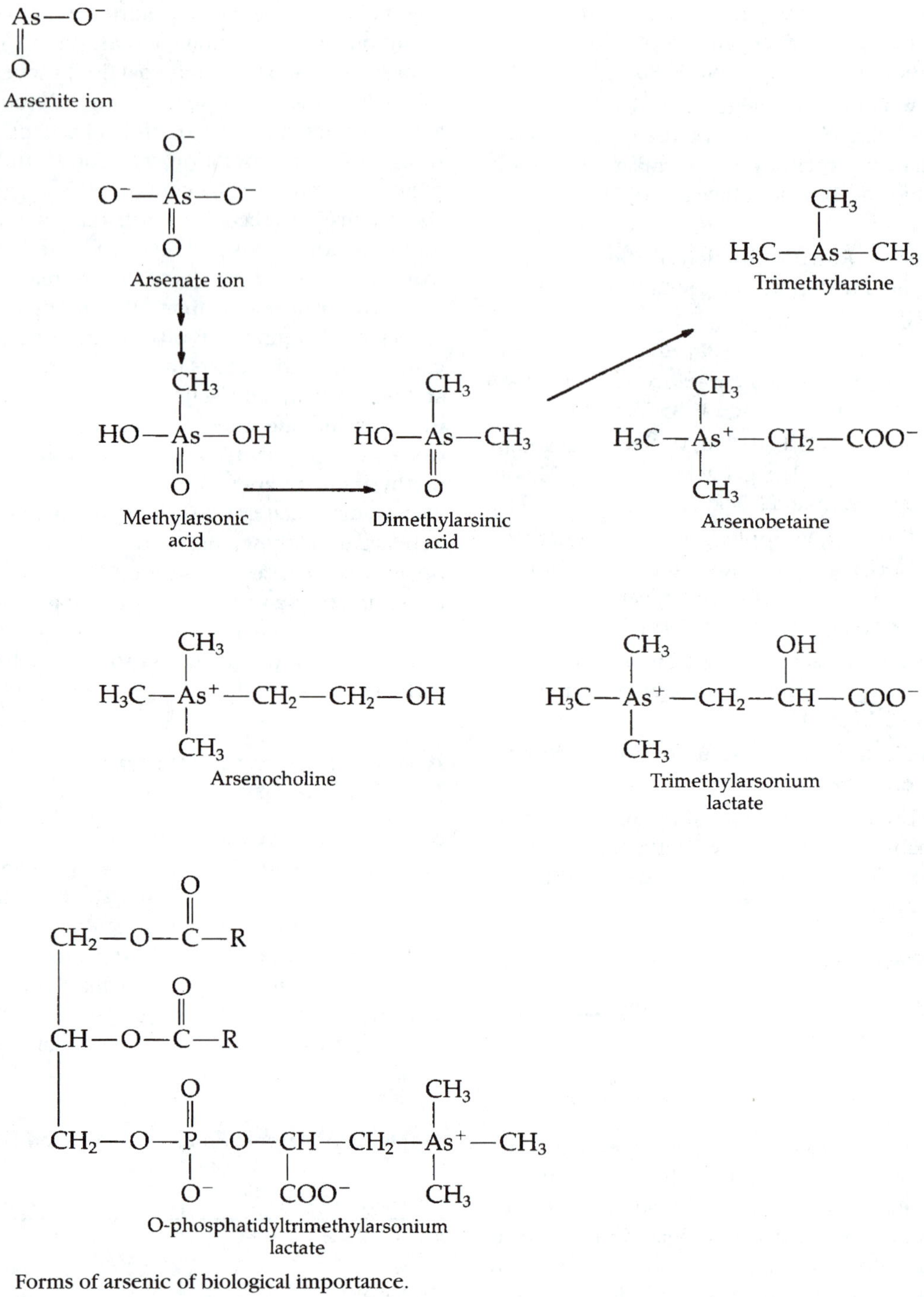

Figure 13.2 Forms of arsenic of biological importance.

Absorption of organic arsenicals is thought to occur by simple diffusion across the intestinal mucosa. The greater the lipid solubility of the arsenical, the greater the likelihood of its transmucosal passage by simple diffusion.

From the intestine, arsenic is transported in the blood to the liver. In the blood, arsenic is found in two forms, methylated and protein bound. Both inorganic and organic forms of arsenic are taken up by the liver following absorption. Arsenate is reduced to arsenite, which is then methylated. Usually, arsenite is methylated to relatively less toxic monomethylarsonic acid and/or dimethylarsinic acid (Fig. 13.2). Methyl groups are generally provided by nitrogen-containing compounds, especially glutathione [1]. Other methyl donors include choline and the amino acid methionine [1].

Organic arsenic that is absorbed apparently undergoes little or no chemical change, as indicated by the fact that the urinary arsenic excreted following the ingestion of arsenic-rich seafoods remained in the original, organically bound form [2].

Tissues that contain the most arsenic include skin, hair, and nails. Arsenic is found bound primarily to sulfhydryl (SH) groups of proteins within these tissues.

Functions and Mechanisms of Actions

Arsenic appears to play a role in methionine metabolism to taurine as well as in arginine metabolism; however, arsenic has not been shown to be an activator or inhibitor of a specific enzyme. Metabolism of methio-nine is impaired with arsenic deficiency and the activity of S-adenosylmethionine (SAM) decarboxylase (p. 192, Fig. 7.26) is diminished suggesting a role in certain decarboxylation reactions [3]. Similarly, taurine production from methionine is impaired in arsenic-deficient rats and hamsters [4]. In the liver, with respect to arginine metabolism, arginine and glycine react to form guanidoacetate; this latter compound is methylated using methyl groups from methionine to form the nitrogen-containing compound creatine (see p. 180). Arsenic-deficient rats fed guanidoacetate experienced growth deficits versus arsenic-supplemented rats [5].

Arsenic may also have a role in phospholipid synthesis. Arsenic deprivation alters enzymes that function in the synthesis of phosphatidylcholine in rats [6].

Decreased activities of enzymes that utilize vitamin B_6 as a coenzyme such as cystathionase and ornithine decarboxylase have been reported with arsenic deprivation in rats [5].

Interactions with Other Nutrients

Arsenic seems to interact antagonistically with *selenium* and *iodine*. Because selenate and arsenate are both oxy-anions with similar chemical properties, they may competitively inhibit the uptake and tissue retention of each other. The interaction of arsenic with iodine is exemplified by the observation that it is goitrogenic in mice. Arsenic is believed to antagonize the mechanism of iodine uptake by the thyroid, causing compensatory goiter.

Excretion

Ingested arsenic is excreted rapidly via the kidneys, which represents the major route of excretion. Increased intake of arsenic results in an increased excretory rate. The latter therefore can provide a useful index of exposure. The rate of urinary excretion in humans depends on the dietary form of the arsenical. For example, methylarsonic acid and dimethylarsinic acid are excreted at about the same rate—approximately 76% of the ingested amount within a 4-day period. Excretion of inorganic arsenite, in contrast, is markedly slower, approximating 46% of the intake over the same period of time. This corroborates the observation [7] that arsenite, compared with other forms, is bound very strongly to tissues.

The results of studies on dogs [8] have shown that intravenously administered arsenate, at what the authors considered medium dosage, was excreted as arsenite at first, but that later, dimethylarsinic acid became the major urinary metabolite.

Deficiency and Toxicity

Reported effects of arsenic deprivation in test animals have included curtailed growth, reduced conception rate, and increased neonatal mortality. Human studies have not been conducted to date.

The no observed adverse effect level for arsenic is 0.8 μg/kg of body weight per day, while the lowest-observed adverse effect level for arsenic is 14 μg/kg of body weight per day [4]. Arsenic as arsenic trioxide, appears to be fatal at intakes of 0.76 to 1.95 mg [9]. Inorganic forms of arsenic are more toxic than organic forms of the element and may be carcinogenic [10,11].

Recommended Intake and Assessment of Nutriture

There are insufficient data to estimate a human dietary requirement for arsenic, although 12 to 25 μg have been suggested [4,5,9]. Most diets, however, would be expected to provide an adequate intake of the mineral, due to its ubiquitous occurrence. Recent food surveys indicate intakes of about 11 to 40 μg/day in the United States, but this range would be significantly elevated if relatively large amounts of seafood are consumed. The Food and Agriculture Organization of the World Health Organization reports a maximum daily intake of 2 μg/kg of body weight per day, or 140 μg/day for a 70-kg individual [4].

Reported levels of arsenic in body fluids include 2 to 62 ng/mL of whole blood, 1 to 20 ng/mL of plasma or serum, and 5 to 50 μg of excreted daily in urine from healthy adults. Hair arsenic levels are 0.1 to 1.1 μg/g. Chronic or acute exposure to the metal would elevate these values, and hair analysis has been particularly useful in this respect. This is because hair arsenic content, unlike that of the fluids, represents an average content over an extended period and would not fluctuate in parallel with intermittent exposure to the element.

The current method of choice for the determination of arsenic in biological fluids is atomic absorption spectrometry, although mass spectrometry, neutron activation analysis, and emission spectroscopy have been used successfully.

References Cited for Arsenic

1. Thompson D. A chemical hypothesis for arsenic methylation in mammals. Chem Biol Interactions 1993;88:89–114.

2. Tam G, Charbonneau S, Bryce F, Sandi E. Excretion of a single oral dose of fish-arsenic in man. Bull Environ Contam Toxicol 1982;28:669–73.

3. Nielsen F. Ultratrace elements of possible importance for human health: an update. In: Prasad AS, ed. Essential and Toxic Trace Elements in Human Health. New York: Wiley-Liss, 1993;355–76.

4. Uthus E, Nielsen F. Determination of the possible requirement and reference dose level for arsenic in humans. Scand J Work Environ Health 1993;19(suppl 1):137–8.

5. Uthus E. Evidence for arsenic essentiality. Environ Geochem Health 1992;14:55–58.

6. Cornatzer W, Uthus E, Haning J, Nielson F. Effect of arsenic deprivation on phosphatidylcholine biosynthesis on liver microsomes in the rat. Nutr Rep Interntl 1983;27:821–9.

7. Vahter M, Marafante E. Intracellular interaction and metabolic fate of arsenite and arsenate in mice and rabbits. Chem Biol Interact 1983;47:29–44.

8. Tsukamoto H, Parker H, Peoples S. Metabolism and renal handling of sodium arsenate in dogs. Am J Vet Res 1983;44:2331–5.

9. Nielsen F. Ultratrace Minerals. In: Shils ME, Olson JA, Shike M, eds. Modern Nutrition in Health and Disease. Philadelphia: Lea and Febiger, 1994;269–86.

10. Goldman M, Dacre J. Inorganic arsenic compounds: are they carcinogenic, mutagenic, teratogenic? Environ Geochem Hlth 1991;13:179–91.

11. Meng Z. Effects of arsenic on DNA synthesis in human lymphocytes stimulated by phytohemagglutinin. Biol Trace Elem Res 1993;39:73–80.

Boron

Boron was used to preserve foods such as fish, meat, cream, butter, and margarine for over 50 years—that is, until about the 1920s, when it was considered dangerous. Shortly before (about 1910) being considered dangerous for humans, boron was deemed essential for plants. It was not until the 1980s that evidence for the essentiality of boron in animals started mounting again.

Sources

Foods of plant origin such as fruits, vegetables, nuts, and legumes are particularly rich in boron [1–3]. In addition, wine, cider, and beer contribute to dietary intake. Meat and fish are poor sources of the element [1–3]. Drinking water and water-based beverages vary considerably in boron content based on geographic location [3]. Boron also is a contaminant or major ingredient in some antibiotics, gastric antacids, lipsticks, lotions, creams, and soaps, for example [3]. Boron appears in foods as sodium borate or boric acid.

Absorption, Transport, Storage, and Excretion

Greater than 90% of ingested boron is thought to be rapidly absorbed from the gastrointestinal tract; however, most of the absorbed boron is excreted in the urine [4]. Little is known with respect to boron transport in the blood and metabolism in tissues. Boron is likely found in the blood as primarily $B(OH)_3$ with small amounts of $B(OH)_4^-$. Boron is found in body tissues, especially bone and spleen among others [5].

Functions and Interactions with Other Nutrients

Boron acts directly or indirectly to influence the composition, structure, and strength of bones [4]. Although the mechanism of action is not known, boron affects the metabolism of magnesium, calcium, phosphorus, and vitamin D [2,6–9]. Responses to low-boron diets are most marked in animals stressed with either calcium, vitamin D, or magnesium deprivation [2]. Consequently, boron is suspected to help to maintain cell membrane structure, stability, and/or function [2].

Deficiency and Toxicity

Although the most consistent sign of boron deficiency in animals is depressed growth, in postmenopausal women, a low-boron diet (0.25 mg B/2,000 kcal) is associated with reduced plasma calcium and magnesium concentrations and urinary calcium and magnesium excretion [2,8]. Alterations in steroid hormone metabolism were also observed [2,8]. Acute toxicity of boron in humans results in nausea, vomiting, diarrhea, dermatitis, and lethargy. Increased urinary excretion of riboflavin has also been reported [10].

Recommended Intake and Assessment of Nutriture

Intakes of 1 to 3 mg, and perhaps up to 10 mg boron, have been suggested as safe [1]. A requirement has been estimated at 1 mg daily [10].

Measurement of boron in biological fluids is still being investigated. Inductively coupled plasma emission spectrometry has been used to determine plasma and other body fluid concentrations of the element; however, whether these tissue concentrations are indicative of nutritional status is unknown.

References Cited for Boron

1. Anderson D, Cunningham W, Lindstrom T. Concentrations and intakes of H, B, S, K, Na, Cl, and NaCl in foods. J Food Comp Anal 1994;7:59–82.

2. Nielsen F. The saga of boron in food: from a banished food preservative to a beneficial nutrient for humans. Curr Topics Plant Biochem Physiol 1991;10: 274–86.

3. Hunt C, Shuler T, Mullen L. Concentration of boron and other elements in human foods and personal-care products. J Am Diet Assoc 1991;91:558–68.

4. Nielsen F. Ultratrace minerals. In: Shils ME, Olson JA, Shike M, eds. Modern Nutrition in Health and Disease. Philadelphia: Lea and Febiger, 1994;269–86.

5. Nielsen F. Other elements. In: Mertz, W, ed. Trace Elements in Human and Animal Nutrition. San Diego: Academic Press, 1987;2: 415–63.

6. Nielsen F. Ultratrace elements of possible importance for human health: an update. In: Prasad AS, ed. Essential and Toxic Trace Elements in Human Health. New York: Wiley-Liss, 1993;355–76.

7. Hunt C. The biochemical effects of physiologic amounts of dietary boron in animal nutrition models. Environ Hlth Perspectives 1994;102(suppl):35–43.

8. Nielsen F. Biochemical and physiologic consequences of boron deprivation in humans. Environ Hlth Perspectives 1994;102:59–63.

9. Hunt C, Herbel J, Nielsen F. Metabolic responses of postmenopausal women to supplemental dietary boron and aluminum during usual and low magnesium intake: boron, calcium and magnesium absorption and retention and blood mineral concentrations. Am J Clin Nutr 1997;65:803–13.

10. Nielsen F. Other trace elements. In: Brown ML, ed. Present Knowledge in Nutrition. Washington, D.C.: International Life Sciences Institute Nutrition Foundation, 1990;294–307.

Cobalt

There is little evidence that cobalt plays a role in human nutrition other than its being a part of vitamin B_{12} (cobalamin). Although it is true that ionic cobalt can substitute for other metals in metalloenzyme activity in vitro, a fact that will be reviewed briefly, there is no evidence for its acting in that capacity in vivo. In this respect the metal is unique among the essential trace elements in that the requirement in the human is not for an ionic form of the metal but for a preformed metallovitamin, which cannot be synthesized from dietary metal. Therefore, it is the vitamin B_{12} content of foods and diet that is of importance in human nutrition rather than the ionic cobalt status.

There have been reports regarding the dependency of certain enzymes on cobalt as an activator or on the metal's ability to substitute for other metal ion activators. Cobalt, in the form of CoC^{2+}, for example, appears to regulate the activity of certain phosphoprotein phosphatases, such as casein and phosvitin phosphatases [1,2]. In another study on phosphoprotein phosphatases, only Co^{2+} and Mn^{2+} could reactivate enzymes inactivated by ATP, ADP, and PPi, with cobalt being significantly the more potent as a reactivator [3]. Cobalt, along with Mn^{2+} and Ni^{2+} can also substitute for Zn^{2+} in the metalloenzymes, angiotensin-converting enzyme [4], carboxypeptidase [5], and carbonic anhydrase [6].

The reader is cautioned against interpreting such findings as an implication of the possible essentiality of ionic cobalt. There is no evidence from deprivation studies on animals that the metal is a requirement for these enzymes in vivo.

References Cited for Cobalt

1. Japundzic I, Levi E, Japundzic M. Cobalt-dependent protein phosphatases from human cord blood erythrocytes. I. Submolecular structure and regulation of activity of E3 casein phosphatase. Enzyme 1988;39:134–43.

2. Japundzic I, Levi E, Japundzic M. Cobalt-dependent protein phosphatases from human cord blood erythrocytes. II. Further characterization of E2 casein phosphatase. Enzyme 1988;39:144–50.

3. Khandelwal R, Kamani S. Studies on inactivation and reactivation of homogeneous rabbit liver phosphoprotein phosphatases by inorganic pyrophosphate and divalent cations. Biochim Biophys Acta 1980;613: 95–105.

4. Bicknell R, Holmquist B, Lee F, Martin M, Riordan J. Electronic spectroscopy of cobalt angiotensin converting enzyme and its inhibitor complexes. Biochemistry 1987;26:7291–7.

5. Auld D, Holmquist B. Carboxypeptidase A. Differences in the mechanisms of ester and peptide hydrolysis. Biochemistry 1974;13:4355

6. Lindskog S. Carbonic anhydrase. In: Spiro TG, ed. Zinc enzymes. New York: Wiley, 1983:86–97.

Chapter 14

Body Fluid and Electrolyte Balance

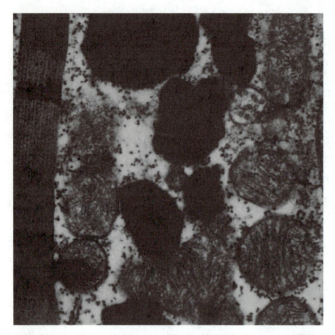

Glomeruli and associated arterioles

Chapter 1, in particular, and the subsequent chapters dealing with nutrient metabolism have emphasized the specialized nature of cells comprising the organ systems of the body. Despite the great diversity of specialized cellular functions, the composition of the body fluids (the internal environment) enveloping the cells remains relatively constant under normal conditions. This constant composition, or *homeostasis,* of the internal environment is necessary for optimal activity of the cells. It is maintained by homeostatic mechanisms involving most of the body's organ systems, the most important of which are those of circulation, respiration, and renal excretion, as well as central nervous system (CNS) and endocrine regulation. Many minor disturbances inevitably occur in water distribution, electrolyte balance, and pH of the body fluids during metabolism. As they arise, compensatory mechanisms of the regulatory organs make appropriate corrections to maintain homeostasis.

Water Distribution in the Body

Water accounts for approximately 60% of the total body weight in a normal adult, making it the most abundant constituent of the human body. In terms of volume, the total body water in a man of average weight (70 kg) is roughly 40 L. Water provides the medium for the solubilization and passage of a multitude of nutrients, both organic and inorganic, from the blood to the cells and the return of metabolic products to the blood. It also serves as the medium in which the vast number of intracellular metabolic reactions take place.

Total body water can theoretically be compartmentalized into two major reservoirs, the intracellular compartment, which includes all water enclosed within cell membranes, and the extracellular compartment, which includes all water external to cell membranes. Of the 40 L of total body water, the intracellular and extracellular compartments account for about 25 L and 15 L, respectively. The anatomic extracellular water is functionally subdivided into the plasma, which is the cell-free, intravascular water compartment, and the interstitial fluid (ISF). The ISF directly bathes the extravascular cells and provides the medium for the passage of nutrients and metabolic products reciprocally from the blood to those cells. In addition, there are potential spaces in the body (i.e., pericardial, pleural, peritoneal, and synovial) that are normally empty except for a small volume of viscous lubricating fluid that needs to be considered as part of the ISF compartment. The body water compartment volumes are summarized in Table 14.1 for a 70-kg man.

The fraction of total body weight that is water and the percentage of total body water that is extracellular or intracellular do not remain constant during growth. Expressed as a percentage of body weight, total body water decreases during gestation and early childhood, reaching adult values by about 3 years of age. During this time the extracellular water (expressed as a percentage of body weight) decreases while the intracellular water (percentage of body weight) increases (p. 514).

Maintenance of Fluid Balance

Most of the daily intake of water enters by the oral route as beverages and as liquids contained in foods. A relatively small amount of water is also formed within the body as a product of metabolic reactions. These two sources together account for the daily intake of about 2,500 mL of fluid, of which the oral route contributes about 2,300 mL, >90%.

The routes by which water is lost from the body can vary according to environmental and physiological conditions such as ambient temperature and extent of physical exercise. At an ambient temperature of 68°F, about 1,400 mL of the 2,300 mL taken in is normally lost in the urine, 100 mL is lost in the sweat, and 200 mL in the feces. The remaining 600 mL leaves the body as *insensible water loss,* so called because the subject is not aware of the water loss as it is occurring. Evaporation from the respiratory tract and diffusion through the skin are examples of insensible water loss.

Table 14.1 Fluid Compartment Values

	Percentage of Body Weight	Percentage of Total Body Water	Volume (L) in 70-kg Man
Total body water	60	———	42
Extracellular water	20	33	14
Plasma	5	8	3.5
Interstitial fluid	15	25	10.5
Intracellular water	40	67	28

One of the more important factors determining the distribution of water among the water compartments of the body is osmotic pressure. When a membrane permeable to water but impermeable to solute particles separates two fluid compartments of unequal solute concentrations, there is a net movement of water through the membrane from the solution with higher water (lower solute) concentration toward the solution with lower water (higher solute) concentration. The movement of water is called *osmosis,* and it can be opposed by applying an external pressure across the membrane in the opposite direction. The amount of pressure required to exactly oppose osmosis into a solution across a semipermeable membrane separating it from pure water is the *osmotic pressure* of the solution.

The *theoretic osmotic pressure* of a solution is proportional to the number of solute particles per unit volume of solution. This concentration is expressed in terms of the osmolarity, or osmoles per liter, of solute particles. One mole of a nonionic solute, such as glucose or urea, is the same as 1 osm, but 1 mol of a solute that dissociates into two or more ions is equivalent to two or more osm. For example, 1.0 mol of sodium chloride equals 2.0 osm because of its dissociation into sodium and chloride ions. The theoretic osmotic pressure presupposes that the solute particles are unable to pass freely through the membrane. When the membrane is permeable to a solute, it does not contribute to the actual, or *effective,* osmotic pressure. The higher the permeability of a membrane to a solute, the lower the effective osmotic pressure of a solution of that solute at a given osmolarity. As an example, cell membranes are much more permeable to a nonionic substance such as urea than to sodium and chloride. Therefore, the effective osmotic pressure of a solution of urea across the cell membrane would be much less than a solution of sodium chloride of the same osmolarity.

The term *osmolality* is sometimes encountered in the expression of osmotic pressure. Like osmolarity, it denotes the concentration of solute particles. However, rather than a weight-per-volume expression of concentration, osmolality refers to solute concentration on a weight-per-weight basis. Specifically, it is the moles of solute particles per kilogram of solvent. Although it is less convenient than molarity as a unit of concentration, osmolality has the advantage of being unaffected by temperature, which can cause expansion or contraction of solvent volume.

The effective osmotic pressure of plasma and interstitial fluid across the capillary endothelium that separates them is mainly due to large molecules such as proteins that cannot permeate the endothelium. Protein concentration is much higher in the plasma than in the interstitial fluid, therefore conferring on the plasma a relatively high osmotic pressure, or water-attracting property. Proteins and other macromolecules too large to traverse the capillary endothelium are sometimes called *colloids,* and the osmotic pressure attributed to them is appropriately termed the *colloid osmotic pressure.*

Water distribution across the capillary endothelial surface is controlled by the balance of forces that tend to move water from the plasma to the interstitial fluid (filtration forces) and by forces that move water from the interstitial fluid into the plasma (reabsorption forces). The major filtration force in the capillaries is hydrostatic pressure (P_{pl}) caused by the pumping of the heart, whereas a much weaker filtration force is the ISF colloid osmotic pressure (Π_{isf}). This force is weak because of the negligible concentration of protein in the ISF. Another weak filtration force is a small, *negative,* ISF hydrostatic pressure (P_{isf}). The major reabsorption force, countering the filtration forces, is the plasma osmotic pressure (Π_{pl}), which is approximately 28 mm Hg.

At the arteriolar end of the capillaries, the average values of these forces are P_{pl}, 25 mm Hg; Π_{isf}, 5 mm Hg; P_{isf}, –6 mm Hg; Π_{pl}, 28 mm Hg. The net result of these four forces can be described by Starling's equation:

$$\text{Filtration pressure} = (P_{pl} + \Pi_{isf}) - (\Pi_{pl} + P_{isf})$$

Substituting the values,

$$\begin{aligned}
\text{Filtration pressure} &= (25 + 5) - (28 + (-6)) \\
&= (25 + 5) - (28 - 6) \\
&= 30 - 28 + 6 \\
&= 8 \text{ mm Hg}
\end{aligned}$$

This positive filtration pressure indicates that a net filtration of water from the plasma to the ISF occurs at the arteriolar end of the capillaries. When filtration pressure is negative, this indicates that a net reabsorption of water from the ISF to the plasma will take place. This would be the situation at the venule end of the capillaries, where the P_{pl} is significantly reduced while the concentration of plasma protein, and therefore the Π_{pl}, correspondingly increases. The net effect of these forces on the water distribution between plasma and ISF along the course of the capillary is shown in Figure 14.1.

From what has been discussed to this point, it is important to understand that osmotic pressure, together with proper intake of fluids and their output by body mechanisms, is a most important factor in the maintenance of fluid balance and compartmentalization. The body's extracellular water volume, for example, is determined mainly by its osmolarity. The osmolarity, in turn, acts as the signal to the regulatory factors that are responsible for maintaining fluid homeostasis. The regulation of extracellular water osmolarity and volume is largely the responsibility of the hypothalamus, the renin-angiotensin-aldosterone system, and the kidney.

Figure 14.1 Starling's hypothesis of water distribution between plasma and interstitial fluid compartments. The relative magnitudes of the pressures, P_{pl} (plasma hydrostatic pressure) and Π_{pl} (plasma osmotic pressure), are represented by the thickness of their respective arrows. There is a positive net filtration pressure at the arteriolar end of the capillary and a negative net filtration pressure at the venule end. *Source:* Kleinman LI, Lorenz JM. Physiology and pathophysiology of body water and electrolytes. In: Kaplan LA, Pesce AJ. Clinical Chemistry: Theory, Analysis, and Correlation, 2nd ed. St. Louis: Mosby, 1989:373.

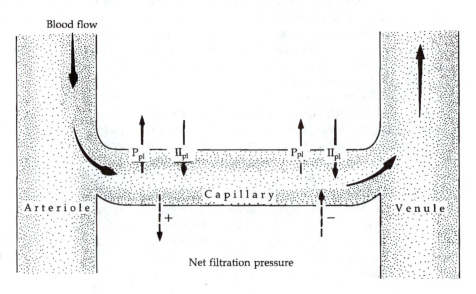

The kidney is central to the regulatory mechanisms, and its function will now be briefly reviewed.

The functional unit of the kidney is the nephron, approximately 1 to 1.5 million of which are found in each of the two kidneys. The five components of the nephron are the Bowman's capsule, proximal convoluted tubule, loop of Henle, distal convoluted tubule, and collecting duct. Bowman's capsule is the blind, dilated end of the renal tubule, encapsulating a tuft of approximately 50 capillaries linking the afferent and efferent arterioles. This capillary network is called the *glomerulus,* and it accounts for the particularly rich blood supply that the kidney enjoys. It is estimated that 25% of the volume of blood pumped by the heart into the systemic circulation is circulated through the kidneys. This is particularly significant in view of the fact that the kidneys constitute only ~0.5% of total body weight. The assembly of the components of the nephron is schematically shown in Figure 14.2.

The glomerular capillary network acts as a filter in removing water and other substances, including electrolytes, glucose, amino acids, and metabolic waste products from plasma. The filtered substances make up what is known as the *glomerular filtrate.* In the absence of disease, no blood cells, or proteins that exceed a molecular weight of approximately 50,000 daltons, normally enter the glomerular filtrate because their larger size prevents their passage through the pores of the capillary endothelium. Each of the segments of the tubules is functionally distinct in its permeability to water and the solutes of the glomerular filtrate. The tubular segments are surrounded by a network of capillaries into which glomerular filtrate materials can be selectively reabsorbed into the bloodstream as a salvage mechanism. These peritubular capillaries may also secrete certain substances from the blood into the renal tubule. The removal of potentially toxic waste products is a major function of the kidneys and is accomplished through the formation of urine. The basic processes involved in urine formation are

- *filtration,* through which the glomerular filtrate is formed;
- *reabsorption* of selected filtrate substances into the bloodstream; and
- *secretion* of materials into the tubules from the surrounding capillaries.

It is through these same processes that the kidneys are able to regulate fluid and electrolyte homeostasis for the proper functioning of cells throughout the body.

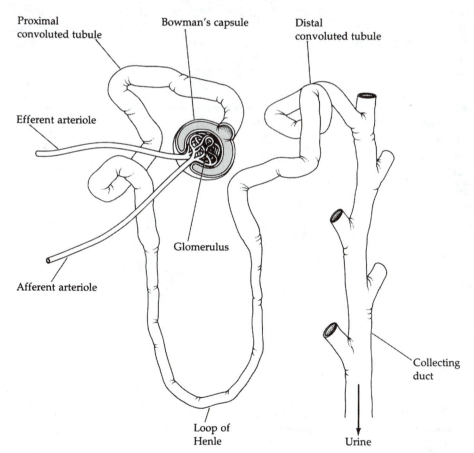

Proximal convoluted tubule

Bowman's capsule

Distal convoluted tubule

Efferent arteriole

Afferent arteriole

Glomerulus

Loop of Henle

Urine

Collecting duct

Figure 14.2 A schematic representation of the major components of the nephron.

In healthy individuals, the kidneys are highly sensitive to fluctuations in diet and in fluid and electrolyte intake, and they compensate by varying the volume and consistency of the urine. The glomerular capillaries differ from other capillaries in the body in that the hydrostatic pressure within them is approximately three times greater than in other capillaries. As a result of this high pressure, substances are filtered through the semipermeable membrane into Bowman's capsule at a rate of about 130 mL/minute. This amounts to over 187,000 mL of filtrate formed per day, yet only about 1,400 mL of urine are produced during this time. This means that <1% of the filtrate is excreted as urine, with the remaining 99% being reabsorbed into the blood.

It has already been mentioned that it is the hypothalamus, the renin-angiotensin-aldosterone system, and the kidney that are responsible for maintaining extracellular fluid volume and osmolarity. Actually, the three work in concert because the hypothalamic hormone, antidiuretic hormone (ADH), also called *vasopressin*, and aldosterone, produced in the adrenal cortex, exert their effects through the kidney.

Antidiuretic hormone is produced in the supraoptic nucleus of the hypothalamus but is stored in and secreted by the posterior pituitary gland. It is a potent water-conserving hormone, its action being to increase the water permeability of the distal convoluted tubule and the collecting duct, thereby facilitating the reabsorption of water into the peritubular capillaries. The mechanism by which the hormone exerts this effect is not completely understood. However, ADH increases the activity of adenylate cyclase in the tubular epithelial cells, and the resulting elevation in cyclic AMP (cAMP) concentration is believed to result in the recruitment of water transport units that become inserted into the luminal membrane of the cells. Evidence for the involvement of cAMP in the process is that exogenously administered cAMP or inhibitors of phosphodiesterase, which prolong cAMP activity, mimic the action of ADH. The release of ADH from the posterior pituitary is triggered by increases in extracellular water osmolarity or by decreased intravascular volume. The hypothalamic response to high extracellular fluid osmolarity is attributed to a shrinkage of neurons within the gland caused by the movement of water out of the neurons into the higher osmotic interstitial fluid. This shrinkage then acts as the signal to the posterior pituitary to release the hormone.

A decrease in blood volume affects the activity of distention receptors and baroreceptors at various sites throughout the vascular network, and this information is relayed to the hypothalamus. Another hormone, angiotensin II, released indirectly by distention receptor relaxation (reduced blood volume) in renal arterioles, stimulates the hypothalamus directly with the release of more ADH.

Increased extracellular fluid osmolarity or decreased blood volume therefore influences what is known as the *water output areas* of the hypothalamus. The term *water output function* refers to the fact that because of the resulting increase in ADH, renal tubular reabsorption of water increases and the output of urine decreases. However, these factors also stimulate the *water intake area* of the hypothalamus, resulting in the conscious sensation of thirst. A greater intake of water therefore follows, resulting in a dilution of extracellular fluid and increased blood volume. This, in turn, reduces the release of ADH as fluid homeostasis is restored. The release of ADH and the induction of the thirst sensation in response to plasma osmolarity are illustrated graphically in Figure 14.3.

Another hormone that plays an important role in the maintenance of fluid balance is aldosterone, which is produced and secreted by the adrenal cortex and, like ADH, exerts its effect through the kidney. It stimulates the active reabsorption of sodium ions in the distal and collecting tubules via a mechanism that involves the transcription and translation of new proteins, which may be Na^+ channels in the luminal membrane, certain mitochondrial enzymes, or Na^+/K^+-ATPase [1]. Evidence that protein induction is indeed a part of the mechanism of aldosterone action is provided by the fact that actinomycin D and puromycin, which are in-

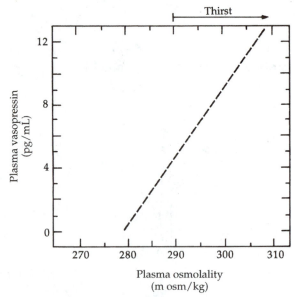

Figure 14.3 Relationship of plasma vasopressin to plasma osmolality. The arrow indicates the plasma osmolality at which the sensation of thirst is stimulated.
Source: Vokes T. Water homeostasis. Ann Rev Nutr 1987;7:386.

hibitors of protein synthesis, inhibit electrolyte balance regulation. By stimulating sodium reabsorption, aldosterone increases extracellular fluid osmolarity, thereby promoting fluid retention by the body via the hypothalamus-ADH mechanism already discussed. This is why high-sodium diets are contraindicated for those individuals whose fluid "balance" is already upset by excessive retention of water, as in cases of hypertension and edema.

Several different substances influence, according to their plasma concentration, the release of aldosterone. They are listed here and discussed again in the following section on maintenance of electrolyte balance. Listed in decreasing order of their potency in stimulating aldosterone release, they are as follows:

1. Increased angiotensin II. This potent polypeptide hormone is a participant in the renin-angiotensin pathway of aldosterone stimulation. It reacts with receptors on adrenal cell membranes, stimulating the synthesis and release of aldosterone.

2. Decreased atrial natriuretic peptide (ANP). ANP is a peptide hormone synthesized in atrial cells and released in response to increased arteriolar stretch, indicative of elevated blood pressure. It functions in opposition to aldosterone in that it inhibits sodium reabsorption in the kidney, and thereby promotes sodium excretion [2].

3. Increased potassium concentration

4. Increased ACTH

5. Decreased sodium

Angiotensin II is particularly important in stimulating aldosterone release, and therefore the renin-angiotensin-aldosterone system will now be discussed in greater detail.

Renin is a proteolytic enzyme synthesized, stored, and secreted by cells in the juxtaglomerular bodies of the kidney. Its secretion is stimulated by decreased renal perfusion pressure that is sensed by the distention receptors and baroreceptors within those bodies. Renin hydrolyzes angiotensinogen (a freely circulating protein synthesized by the liver) to angiotensin I, an inactive decapeptide. Angiotensin I is then acted on by a second proteolytic enzyme, angiotensin-converting enzyme (ACE), synthesized in vascular endothelial cells, particularly those in the blood vessels of the lung, producing the potent octapeptide angiotensin II. Angiotensin II then interacts with specific receptors on adrenal cortical cells, leading to the release of aldosterone. Along with its sodium-retaining activity, aldosterone promotes the urinary excretion of potassium.

Let's review very briefly the mechanism of action of angiotensin II in increasing the synthesis and release of

aldosterone from the adrenal cortex. Stimulatory signals resulting from polypeptide hormone-receptor interactions generally follow one of two major routes. One operates through an accelerated synthesis of cAMP with a consequent increase in protein kinase activity (p. 14). The second mechanism involves signals mediated by hydrolytic products of phospholipids along with increased intracellular calcium concentrations. The second of these, described as follows, applies in the case of angiotensin II action.

As a result of the interaction of angiotensin II with its receptor, a sequential cascade of reactions follows, involving G proteins, phospholipase C, and inositol triphosphate. Phospholipase C raises intracellular Ca^{2+} concentration by increasing Ca^{2+} conductance through Ca^{2+} channels, and inositol triphosphate releases Ca^{2+} from its storage in the endoplasmic reticulum. The elevated concentration of intracellular Ca^{2+} is stimulatory to appropriate synthetic enzymes, mediated through the Ca^2-binding protein calmodulin [3]. Calmodulin is present in all eukaryotic cells. Figure 11.5 (p. 379) illustrates this type of hormonal mechanism.

The sequence of the events that comprises the renin-angiotensin-aldosterone system is illustrated in Figure 14.4. Although not shown in the figure, angiotensin II can be hydrolyzed further to angiotensin III by the hydrolytic removal of an aspartic acid residue by a plasma aminopeptidase. Angiotensin III is also physiologically active. In fact, it has been observed to be more potent than angiotensin II in its aldosterone-stimulating ability. However, its plasma concentration is significantly less than that of angiotensin II, and therefore its contribution to the maintenance of fluid balance is less dramatic. In addition to its role in the conservation of body water through aldosterone action, angiotensin II is also a potent vasoconstrictor, reducing the glomerular filtration rate and therefore the filtered load of sodium. Also, it will be recalled that it stimulates the hypothalamic thirst center and the release of ADH, both of which increase body water volume. Figure 14.5 illustrates the central role of the hypothalamus and the action of angiotensin II in the hormonal regulation of fluid homeostasis.

Alterations in food intake can profoundly affect water and electrolyte balance. During the initial days of a period of fasting, for example, there is a marked increase in the renal excretion of sodium, whereas prolonged fasting tends to conserve the ion. Refeeding causes a marked retention of sodium, probably due to the ingestion of carbohydrate. Consequently, a rapid regain in body weight follows, caused by an increase in total body water secondary to the stimulation of vasopressin and thirst by the rise in plasma osmolarity. These alterations in sodium and water balance in subjects as a result of early-phase fasting and refeeding account for the weight loss and weight

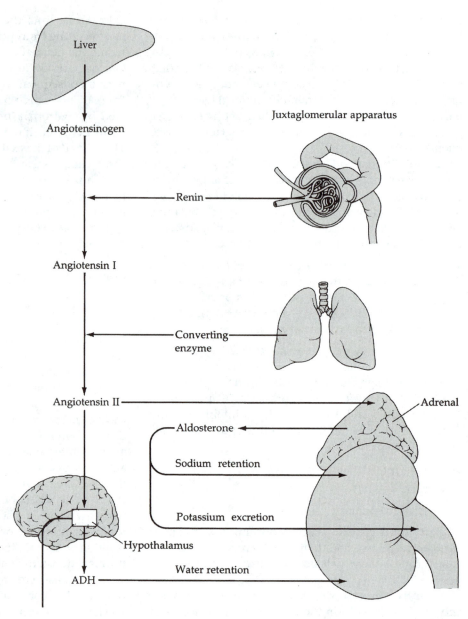

Figure 14.4 The renin-angiotensin-aldosterone system illustrating the cooperation of kidneys, liver, lungs, adrenals, and hypothalamus in this mechanism of fluid homeostasis.

regain to a far greater extent than would be predicted from the changes in caloric balance [4].

Maintenance of Electrolyte Balance

The term *electrolytes* refers to the anions and cations that are distributed throughout the fluid compartments of the body. They are distributed in such a way that within a given compartment—the blood plasma, for example—electrical neutrality is always maintained, with the anion concentration exactly balanced by the cation concentration.

The cationic electrolytes of the extracellular fluid include sodium, potassium, calcium, and magnesium, and these are electrically balanced by the anions, chloride, bicarbonate, and proteins, along with relatively low concentrations of organic acids, phosphate, and sulfate. The major electrolytes are listed in Table 14.2. Most of them are categorized nutritionally as macrominerals and, as such, have already been discussed in Chapter 11 from the standpoint of their absorption, function, dietary requirements, and food sources. The maintenance of pH and electrolyte balance, which is the focus of this chapter, is a responsibility that belongs almost exclusively to the kidney.

All filterable substances in plasma—that is, all the plasma solutes except the larger proteins—freely enter the glomerular filtrate from the blood. Some of these substances are metabolic waste products and are ex-

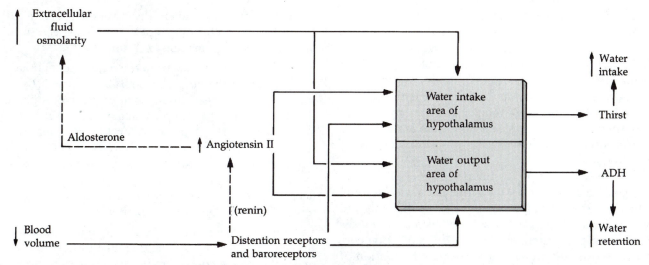

Figure 14.5 A summary of the mechanisms by which fluid homeostasis is maintained. Water depletion stimuli such as increased extracellular fluid osmolarity or decreased blood volume can stimulate the hypothalamus either directly or through the production of angiotensin II, formed by the action of the renal protease renin. The renin-angiotensin-aldosterone system (shown by dashed arrows) increases extracellular fluid osmolarity by promoting renal tubular reabsorption of sodium.

creted in the urine with little or no reabsorption in the tubules. However, most of the materials in the glomerular filtrate must be salvaged by the body, and this is accomplished through their tubular reabsorption by either active or passive mechanisms, or both. Active transport allows substances to pass across membranes against concentration gradients by the action of ATP-dependent membrane transport systems (p. 16). Glucose is a prime example of a solute that can be actively transported across the tubular cells from the urine into the blood even though the blood concentration of glucose is normally 20 times that of urine. Another group of solutes, including ammonium, potassium, and phosphate ions, occurs in relatively high concentration in urine compared with blood. These substances are transported from blood into the tubular cells also against a concentration gradient. Passive transport is not energy demanding and is simply the diffusion of a material across a membrane from a compartment of higher concentration of the material to a compartment of lower concentration. This process, too, functions within the renal tubular cells. The renal regulation of several major electrolytes will now be considered.

Sodium

Sodium is freely filtered by the glomerulus. About 70% of the filtered sodium is reabsorbed by the proximal tubule, 15% by the loop of Henle, 5% by the distal convoluted tubule, and ~10% by the collecting ducts. It is the major cation found in extracellular fluid.

Table 14.2 Electrolyte Composition of Body Fluids

	Plasma (mEq/L)	Interstitial Fluid (mEq/L H_2O)	Intracellular Water (mEq/L H_2O)
Cations	153	153	195
Na$^+$	142	145	10
K$^+$	4	4	156
Ca^{+2}	5	(2–3)	3.2
Mg^{+2}	2	(1–2)	26
Anions	153	153	195
Cl$^-$	103	116	2
HCO$_3^-$	28	31	8
Protein	17	——	55
Others	5	(6)	130
Osmolarity (m osm/L)		294.6	294.6
Theoretic osmotic pressure (mm Hg)		5,685.8	5,685.8

Active reabsorption of sodium ions in the proximal tubule results in the passive reabsorption of chloride ions, bicarbonate ions, and water. The accompanying transfer of the anions chloride and bicarbonate with the cation sodium is required to maintain the necessary electrical neutrality of the extracellular fluid, while the water transfer ensures a normal osmotic pressure. Virtually all cells contain a relatively high concentration of potassium and a low concentration of sodium, whereas the blood plasma and most other extracellular fluids have high sodium and low potassium concentrations, as can be seen in Table 14.2. Clearly, energy must be expended to maintain this gradient across the cell membrane; otherwise, each ion would simply diffuse through the membrane until their intracellular and extracellular concentrations were the same. The gradient is maintained by the Na^+/K^+-ATPase pump, which has already been discussed in Chapter 1. It is the mechanism by which the renal tubular cells "pump" sodium into the blood in exchange for potassium in such a way as to conserve sodium while allowing a constant loss of potassium in the urine.

Active reabsorption of sodium occurs in the distal convoluted tubule under the influence of aldosterone. The mechanism is highly selective for sodium ion, and there is little accompanying water diffusion. This makes it an important system for the regulation of extracellular fluid osmotic pressure. The increased retention of sodium by this mechanism is, however, accompanied by water retention also. This is because the greater extracellular fluid osmotic pressure stimulates tubular water reabsorption through ADH release (Fig. 14.5).

Chloride

The concentration of chloride in the extracellular fluid parallels that of sodium, and chloride generally accompanies sodium in transmembrane passage. However, it will be recalled that chloride reabsorption is passive in the proximal tubule, and it is probably reabsorbed actively in the ascending limb of the loop of Henle and the distal tubule.

Potassium

Potassium is the chief cation of intracellular fluid, and maintenance of a normal level is essential to the life of the cells. The normal person maintains potassium balance by excreting daily an amount of the cation equal to the amount ingested minus the small amount excreted in the feces and sweat.

Potassium is freely filtered at the glomerulus, and its active tubular reabsorption occurs throughout the nephron, except for the descending loop of Henle. Only about 10% of the filtered potassium enters the distal tubules, which, along with the collecting ducts, are able to both secrete and reabsorb potassium. The distal tubule is the site at which changes in the amount of potassium excreted are achieved, and several mechanisms are involved in this control.

• The first of these mechanisms is dependent on the cellular potassium content. When a high-potassium diet is consumed, the concentration of potassium rises in cells, including the distal renal tubular cells, provid-

ing a concentration gradient that favors the secretion of the cation into the lumen of the tubule. This results in an increase in potassium excretion.

• Another important factor in the regulation of potassium balance is the hormone aldosterone, which, besides stimulating distal tubular reabsorption of sodium, simultaneously enhances potassium secretion at that site. In fact, the elevated plasma level of potassium directly stimulates the production and release of aldosterone from the adrenal cortex. Recall that another mechanism for effecting aldosterone release is through decreased renal perfusion pressure and the associated renin-angiotensin-aldosterone pathway.

• A third mechanism of renal conservation of potassium occurs in the collecting duct, and involves its active reabsorption coupled to the secretion of protons at that site [5]. The movement of K^+ into the cells of the collecting duct from the urine, and the movement of H^+ in the opposite direction is catalyzed by an H^+/K^+-activated adenosine triphosphatase (H^+/K^+-ATPase), functioning similarly to the Na^+/K^+-ATPase pump discussed previously.

Calcium and Magnesium

Tubular reabsorption of calcium is associated with the reabsorption of sodium and phosphate in the proximal tubule, and the rate of reabsorption of all three ions, as well as fluid, occurs in parallel. Renal tubular reabsorption of calcium is closely linked to the action of parathyroid hormone (PTH) (p. 337). This hormone exerts parallel inhibition of the reabsorption of calcium, sodium, and phosphate in the proximal tubules. However, PTH markedly stimulates reabsorption of calcium in the distal tubules disproportionate to that of sodium and phosphate.

The major pathway of calcium excretion is the intestinal tract. Urinary excretion, approximating 150 mg/day for the average adult, amounts to only about 1% of that filtered by the glomerulus, the remaining 99% being effectively reabsorbed at proximal and distal tubular sites.

Calcium balance is achieved largely by the control of the intestinal absorption of the ion rather than by the regulation of its urinary excretion. The percentage of ingested calcium absorbed decreases as the dietary calcium content increases, and so the amount absorbed remains relatively constant. The slight increase in absorption that occurs with a high-calcium diet is reflected in an increased renal excretion of the cation.

The filtration of magnesium at the glomerulus and its subsequent active reabsorption through the tubular cells parallel that of calcium.

Homeostatic regulation of the ions discussed is crucial to many body functions. For example, greatly decreased extracellular potassium (hypokalemia) produces paralysis, whereas elevated potassium levels (hyperkalemia) can result in cardiac arrhythmias. Excessive extracellular sodium (hypernatremia) causes fluid retention, and decreased plasma calcium (hypocalcemia) produces *tetany* (intermittent spasms of the muscles of the extremities) by increasing the permeability of nerve cell membranes to sodium. Magnesium deficiency is also associated with tetany.

Table 14.2 lists the fluid electrolytes and their approximate, normal, compartment concentrations. In terms of electrolyte balance only, it is clear that the contribution of sodium to the total cation milliequivalents is quite large compared with that of potassium, calcium, and magnesium and that a correspondingly high percentage of anion milliequivalents is contributed by chloride and bicarbonate together. The concentration of these three major ions is used to calculate the so-called *anion gap,* a clinically useful parameter for establishing metabolic disorders that can alter the electrolyte balance. The value is calculated by subtracting the measured anion (chloride + bicarbonate) concentration from the measured cation (sodium) concentration:

$$\text{Measured cations (Na}^+\text{)} - \text{measured anions (Cl}^- + \text{HCO}_3^-\text{)} = 12 \text{ mEq/L}$$

Under normal conditions, the value is about 12 mEq/L but may range from 8 to 18 mEq/L. Deviation from a normal anion gap is most commonly associated with increases or decreases in the concentration of certain unmeasured anions such as proteins, organic acids, phosphate, or sulfate. For example, the production of excessive amounts of organic acids, such as would occur in lactic acidosis or ketoacidosis, increases the unmeasured anion concentration at the expense of the measured anion bicarbonate that is neutralized by the acids. Such a condition would therefore cause a greater anion gap.

Considering the effect of plasma osmolarity on water intake and retention, it is logical that if for any reason sodium ion should accumulate in the body water, a concomitant rise in blood pressure (essential hypertension) would result. Clinical evidence for this correlation is the hypertension experienced by patients with adrenal adenomas, whose high levels of aldosterone cause excessive retention of sodium. There is also an apparent causal relationship between dietary intake of sodium (as sodium chloride) and the etiology of hypertension, as suggested by studies conducted through one or more of the following designs:

- Relating salt consumption to the prevalence of hypertension
- Development of hypertension in animals fed high-salt diets
- Response of hypertensive patients fed low-salt diets

There is an abundance of reported observations that deal with the positive correlation of salt intake and hypertension among societies that ingest salt to variable extents. Such observations have led to the generally accepted conclusion that the incidence of hypertension is predictable from average daily sodium intake. Also, convincing animal studies dating back to the 1950s have demonstrated a direct correlation between sodium chloride and hypertension. But in spite of these findings, there is a lack of evidence that a cause-and-effect relationship exists among the individuals of a normotensive population. In fact, investigations on the effect of sodium chloride loading on blood pressure among normotensives have revealed no correlation between high salt intake and hypertension. Furthermore, among subjects with borderline essential hypertension, a low-sodium diet is minimally effective in lowering the blood pressure. This suggests that plasma sodium concentrations are unalterable if the homeostatic mechanisms controlling it are intact. It has become generally accepted that the differences between those who respond to sodium diet therapy and those who do not have a genetic foundation.

People who are salt-sensitive are called *responders,* and those showing salt insensitivity are labeled *nonresponders.* The condition of nonresponders who have essential hypertension does not improve on low-salt diets. Likewise, normotensive nonresponders can consume as much as 4,600 mg of sodium daily (somewhat higher than that of the typical Western diet) without risk. Among the genetically disposed individuals, a comparable intake would likely favor the development of hypertension. For people in this population, a restriction to about 1,400 mg or less is recommended.

Although a genetic link to salt sensitivity is generally accepted, biochemical mechanisms of the condition are not clearly understood. This is not for a lack of relevant research. A literature review of the many investigations designed to explain the biochemical basis of salt sensitivity and nonsensitivity is available [6].

In summary, the implication of sodium in hypertension remains controversial. It is unlikely that it functions alone in the etiology of the disease, and it may be a contributing factor only in the wake of other biochemical disturbances. The involvement of other cations such as calcium, magnesium, potassium, and cadmium cannot be overlooked [6]. Potassium intake has been linked to a reduction in blood pressure, especially in those people on high-sodium diets. Although the mechanism remains unknown, potassium may effect natriuresis, baroreflex sensitivity, catecholamine function, or the renin-angiotensin-aldosterone system [7].

Acid-Base Balance: The Control of Hydrogen Ion Concentration

The hydrogen ion concentration in body fluids must be controlled within a narrow range, its regulation being one of the most important aspects of homeostasis. This is because merely slight deviations from normal acidity can cause marked alteration in enzyme-catalyzed reaction rates in the cells. Hydrogen ion concentration can also affect both the cellular uptake and regulation of metabolites and minerals and the uptake and release of oxygen from hemoglobin.

The degree of acidity of any fluid is determined by its concentration of protons (H^+). The hydrogen ion concentration in body fluids is generally quite low, as it is regulated at approximately 4×10^{-8} mol/L. Concentrations can vary from as low as 1.0×10^{-8} mol/L to as high as 1.0×10^{-7} mol/L, but values outside this range are not compatible with life. From these values it is apparent that expressing H^+ in terms of its actual concentration is awkward. The concept of pH, which is the negative logarithm of the H^+ concentration, was devised to simplify the expression. It allows concentrations to be expressed as whole numbers rather than as negative exponential values:

$$pH = -\log [H^+]$$

Bracketed values symbolize concentrations. This designation will be used to signify concentrations of other substances as well as protons throughout this discussion. The pH of extracellular fluid, in which the H^+ concentration may be assumed to be approximately 4×10^{-8} mol/L, can therefore be calculated as follows:

$$pH = -\log (4 \times 10^{-8})$$
$$\text{or } pH = \log \frac{1}{4 \times 10^{-8}}$$
$$\text{(dividing)} = \log (0.25 \times 10^{8})$$
$$= \log 0.25 + \log 10^{8}$$
$$\text{(taking logs)} = -0.602 + 8$$
$$pH = 7.4$$

As the value of the negative exponent of 10 becomes larger—that is, the molar concentration of H^+ being smaller—the pH correspondingly increases. Low acidity therefore denotes low H^+ concentration and high

pH, whereas high acidity is associated with a high H^+ concentration and low pH.

An acid, as it relates to fluid acid-base regulation, may be defined as a substance capable of releasing protons. The metabolism of the major nutrients continuously generates acids, which must be neutralized. It has already been explained in Chapter 4 how lactic acid and pyruvic acid can accumulate in periods of oxygen deprivation and in Chapter 6, how fatty acids are released from triacylglycerols during lipolysis. Also, the acidic ketone bodies, acetoacetic acid and β-hydroxybutyric acid, can increase significantly during periods of prolonged starvation or low carbohydrate intake. Carbon dioxide, the product of complete oxidation of energy nutrients, is itself indirectly acidic, because it forms carbonic acid, H_2CO_3, on combination with H_2O. Acidic salts of sulfuric and phosphoric acids are also generated metabolically from sulfur- or phosphorus-containing substances.

The term *acidosis* refers to a rise in extracellular (principally plasma) H^+ concentration beyond the normal range. Abnormally low H^+ concentration, in contrast (i.e., high plasma pH) results in the condition of *alkalosis*. To guard against such fluctuations in pH, three principal regulatory systems are available:

- Buffer systems within the fluids that immediately neutralize acidic or basic compounds
- The respiratory center, which regulates breathing and the rate of exhalation of CO_2
- Renal regulation, by which either an acidic or alkaline urine can be formed to adjust body fluid acidity

Acid-Base Buffers

A *buffer* is a chemical solution designed to resist changes in pH despite the addition of acids or bases. Usually a buffer consists of a weak acid, which can be represented as (HA), and its conjugate base (A^-). The conjugate base is therefore the residual portion of the acid following the release of the proton. The conjugate base of a weak acid is basic, because it tends to attract a proton and to regenerate the acid. Therefore, the dissociation of a weak acid and the reunion of its conjugate base and proton is an equilibrium system:

$$HA \rightleftharpoons H^+ + A^-$$

The equilibrium expression for this reaction is called the *acid dissociation constant* (K_a) and is represented as

$$K_a = \frac{[H^+] [A^-]}{[HA]}$$

The equation can be rearranged to

$$[H^+] = K_a \frac{[HA]}{[A^-]}$$

Taking the negative logarithm of both sides of the equation,

$$-\log [H^+] = -\log K_a - \log \frac{[HA]}{[A^-]}$$

These values become

$$pH = pK_a + \log \frac{[A^-]}{[HA]}$$

This is referred to as the *Henderson-Hasselbalch equation.* The equation indicates how a buffer system composed of a weak acid and its conjugate base resists changes in pH if either strong acid or base is added to the system. For example, if the molar concentrations of the conjugate base and the acid are equal, then the ratio of $[A^-]$ to $[HA]$ is 1.0, and the log of this ratio is 0, making the pH of the system equal to the pK_a of the acid.

The pK_a, which is the negative logarithm of the acid dissociation constant (K_a), of any weak acid is a constant for that particular acid and simply reflects its strength (its tendency to release a proton). If a strong acid or base is added to this system, the ratio of $[A^-]$ to $[HA]$ changes and therefore changes the pH, but only slightly. Suppose, for example, that both the conjugate base and free acid are present at 0.1 mol/L concentrations, and suppose also that the pK_a of the acid is 7.0. It follows that the pH is also 7.0 under these conditions. Addition of a strong acid such as hydrochloric acid to a final concentration of 0.05 mol/L will convert an equivalent amount of $[A^-]$ to $[HA]$ because, as a completely dissociated acid, it is contributing 0.05 mol/L H^+ as well. The new $[A^-]$ concentration therefore becomes 0.05 mol/L, and the $[HA]$ will be 0.15 mol/L. The logarithm of this new ratio (0.05:0.15, or 0.33) is -0.48, and inserting this value into the Henderson-Hasselbalch equation, we can see that the pH decreases by only this amount. In other words, the pH decreased from 7.0 to 6.52 by making the system 0.05 mol/L hydrochloric acid. In contrast, this same concentration of HCl in an unbuffered, aqueous solution would produce an acid pH between 1.0 and 2.0.

The physiologically important buffers that maintain the narrow pH range of extracellular fluid at approximately 7.4 are proteins and the bicarbonate (HCO_3^-)-carbonic acid (H_2CO_3) system. Proteins have the most potent buffering capacity among the physiological buffers, and because of its high concentration in whole blood, hemoglobin is most important in this respect. It is crucial for the pH regulation nec-

essary for the normal uptake and release of oxygen in the erythrocyte. As amphoteric substances (meaning that they possess both acidic and basic groups on their amino acid side chains), proteins are capable of neutralizing either acids or bases. For instance, the two major buffering groups on a protein are carboxylic acid (R-COOH) and amino ($R-NH_3^+$) functions, which dissociate as shown:

1. $R\text{-}COOH \rightleftharpoons R\text{-}COO^- + H^+$
2. $R\text{-}NH_3^+ \rightleftharpoons R\text{-}NH_2 + H^+$

At physiological pH, the carboxylic acid is largely dissociated into its conjugate base and a proton so that the equilibrium as shown is shifted strongly to the right. At that same pH, however, the amino group, being much weaker as an acid (a stronger base), is only weakly dissociated, and its equilibrium greatly favors the right-to-left direction. If protons, in the form of a strong acid, are added to a protein solution, they are neutralized by reaction 1, because their presence will cause a shift in the equilibrium toward the undissociated acid (right to left). Strong bases, as contributors of hydroxide (OH^-) ions, will likewise be neutralized because as they react with the protons to form water, the equilibrium of reaction 2, as illustrated, shifts to the right to restore the protons that were neutralized.

The bicarbonate-carbonic acid buffer system is of particular importance because it is through this system that respiratory and renal pH regulation is exerted. It is composed of the weak acid carbonic acid (H_2CO_3) and its salt, or conjugate base, bicarbonate ion (HCO_3^-). The acid dissociates reversibly into H^+ and HCO_3^-,

$$H_2CO_3 \rightleftharpoons H^+ + HCO_3^-$$

its buffering capacity being due to the fact that either protons or hydroxide ions added will be neutralized by corresponding shifts in the equilibrium, similar to the carboxy-amino group buffering by proteins described earlier. The H_2CO_3 can be formed not only from the acidification of HCO_3^-, as shown in the right-to-left reaction just shown, but also from the reaction of dissolved CO_2 with water. It will be recalled that CO_2 is formed as a result of total oxidation of the energy nutrients as well as various decarboxylation reactions. The gas diffuses from tissue cells into the extracellular fluids and then into erythrocytes, where its reaction with water to form H_2CO_3 is accelerated by the zinc metalloenzyme, carbonic anhydrase (p. 424). The overall reaction involving carbon dioxide, carbonic acid, and bicarbonate ion occurs as follows:

3. $CO_2 \rightleftharpoons CO_2 \rightleftharpoons H_2CO_3 \rightleftharpoons H^+ + HCO_3^-$

   (gas)   (dissolved)

In the lungs, these equilibrium reactions are shifted strongly to the left in the circulating erythrocytes because of the release of protons from hemoglobin as hemoglobin acquires oxygen to become oxyhemoglobin. This shift allows the exhalation of carbon dioxide.

Normally the ratio of the concentration of HCO_3^- to H_2CO_3 in plasma is 20 to 1, and the apparent pK_a value for H_2CO_3 is 6.1. Using the Henderson-Hasselbalch equation, we can show how a normal plasma pH of 7.4 results from these values:

$$pH = pK_a + \log \frac{[HCO_3^-]}{[H_2CO_3]}$$
$$= 6.1 + \log \frac{20}{1}$$
$$= 6.1 + 1.3$$
$$pH = 7.4$$

Alterations in the 20:1 ratio of $[HCO_3^-]$ to $[H_2CO_3]$ clearly change the pH. Next it will be shown how respiratory and renal regulatory systems function to keep this ratio, and therefore pH, relatively constant.

Respiratory Regulation of pH

Should plasma levels of CO_2 rise, perhaps because of accelerated metabolism, more H_2CO_3 will be formed, which, in turn, will cause a fall in pH as it dissociates to release protons (reaction 3). The elevated CO_2 itself, as well as the resulting increase in hydrogen ion concentration, is detected by the respiratory center of the brain, resulting in an increase in the respiratory rate. This hyperventilation significantly increases the amount of CO_2 loss and therefore decreases the amount of H_2CO_3. By reducing H_2CO_3, this mechanism therefore increases the ratio of HCO_3^- to H_2CO_3, elevating the pH to a normal value. Conversely, if plasma pH rises for any reason, because of either an increase in HCO_3^- or a decrease in H_2CO_3, the respiratory center is signaled accordingly and causes a restraint in ventilation. As CO_2 then accumulates, the H_2CO_3 concentration rises, and the pH decreases.

Renal Regulation of pH

Although the intact respiratory system acts as an immediate regulator of the HCO_3^-/H_2CO_3 system, long-term control is exerted by renal mechanisms. The kidneys regulate pH by controlling the secretion of hydrogen ions, by conserving bicarbonate, and through ammonia synthesis. The secretion of hydrogen ions occurs in conjunction with the tubular reabsorption of sodium ions through the mechanism of countertransport. This is an active process involving a common Na^+/H^+ car-

rier protein and energy sufficient to move the protons from the tubular cells into the tubule lumen against a concentration gradient of protons. In subjects on a normal diet, about 50 to 100 mEq of hydrogen ions are generated daily. Renal secretion of the protons is necessary to prevent a progressive metabolic acidosis.

The renal tubules are not very permeable to bicarbonate ions because of the charge and relatively large size of the ions. They are therefore reabsorbed by a special process. The hydrogen ions in the glomerular filtrate convert filtered bicarbonate ions into H_2CO_3, which dissociates into CO_2 and H_2O. The CO_2 diffuses into the tubular cell, where it combines with water, a reaction catalyzed by carbonic anhydrase, to form H_2CO_3. The relatively high tubular-cell pH allows the dissociation of the H_2CO_3 into HCO_3^- and H^+, after which the bicarbonate reenters the extracellular fluid, and the proton is actively returned to the lumen by the Na^+/H^+ carrier. These events, by which hydrogen ions are secreted against a concentration gradient in exchange for sodium ions, and bicarbonate is returned to the plasma from the glomerular filtrate, are summarized in Figure 14.6.

The pH of the urine normally falls within the range of 5.5 to 6.5, despite the active secretion of hydrogen ions throughout the tubules. This is largely due to the partial neutralization of the hydrogen ions by ammonia, which is secreted into the lumen by the tubular cells. Ammonia is produced in large amounts from the metabolic breakdown of amino acids, and although most of it is excreted in the form of urea, some is delivered to the kidney cells in the form of glutamine. In the renal tubule cells, ammonia is hydrolytically released from the glutamine by the enzyme glutaminase and is secreted into the urine (p. 196). Because it is a basic substance, ammonia immediately combines with protons in the collecting ducts to form ammonium ions (NH_4^+), which are excreted in the urine primarily as their chloride salts.

Should metabolic acidosis occur, such as in starvation or diabetes, an increase in the urinary excretion of ammonia occurs concomitantly in compensation. This is because the diminished intake and use of carbohydrate stimulate gluconeogenesis and therefore an enhanced excretion of ammonia formed from the higher rate of amino acid catabolism.

Like respiratory regulation, renal regulation of pH is also directed at the maintenance of a normal ratio of $[HCO_3^-]$ to $[H_2CO_3]$. In a situation of alkalosis, for example, in which the plasma ratio of HCO_3^- to H_2CO_3 increases as the pH rises above 7.4, the ratio of HCO_3^- ions filtered into the tubules to the hydrogen ions secreted into the tubules increases also. This increase occurs because the high extracellular HCO_3^- concentra-

Extracellular fluid   Tubular cell   Renal tubule lumen

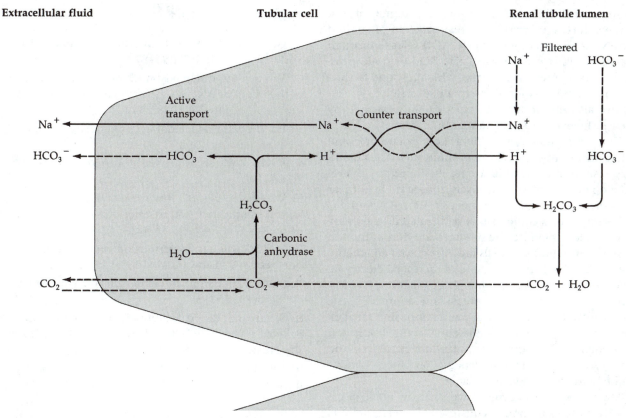

Figure 14.6 Renal tubular cell reactions illustrating the origin of and the active secretion of hydrogen ions in exchange for sodium ions, as well as the mechanism for tubular reabsorption of bicarbonate. Solid arrows indicate reactions or active transport, while dashed arrows signify diffusion.

tion increases its filtration, while the relatively low concentration of H_2CO_3 decreases the secretion of H^+. Therefore, the fine balance between the HCO_3^- and H^+ that normally exists in the tubules no longer is in effect. Also, since no HCO_3^- ions can be reabsorbed without first reacting with H^+ (Fig. 14.6), all the excess HCO_3^- will pass into the urine, neutralized by sodium ions or other cations. In effect, therefore, HCO_3^- is removed from the extracellular fluid, restoring the normal ratio of HCO_3^- to H_2CO_3 and pH. In acidosis, the ratio of plasma HCO_3^- to H_2CO_3 decreases, meaning that the rate of H^+ secretion rises to a level far greater than the rate of HCO_3^- filtration into the tubules. As a result, most of the filtered HCO_3^- will be converted to H_2CO_3 and reabsorbed as CO_2 (Fig. 14.6), while the excess H^+ is excreted in the urine. As a consequence, the extracellular fluid ratio of $[HCO_3^-]$ to $[H_2CO_3]$ increases, and so does the pH.

The importance of the kidney in the homeostatic control of body water, as well as electrolyte and acid-base balance, has been emphasized in this chapter. The material has been presented as a review of the principles involved in such control and the effect of diet on fluid and electrolyte homeostasis. Although a detailed

account of renal physiology is not within the scope of this text, excellent sources that deal specifically with this subject are, of course, available [8,9].

Summary

The maintenance of body fluid and electrolytes is of vital importance for sound health and nutrition. Intracellular fluid provides the environment for the myriad of metabolic reactions that take place in cells. The interstitial fluid compartment of the extracellular fluid mass allows the migration of nutrients into cells from the bloodstream and the return to the bloodstream of metabolic waste products from the cells. These fluids contain the electrolytes, dissolved minerals that have important physiological functions. Their concentrations and their intracellular and extracellular distribution must be precisely regulated, and the mechanism for achieving this is exerted largely through the kidney. The homeostatic maintenance of fluid volume is also the responsibility of this organ.

Fluid volume control by the kidney is mostly hormone mediated. ADH, produced in the hypothalamus, stimulates the tubular reabsorption of water from the glomerular filtrate. Aldosterone, a product of the adrenal cortex, increases the reabsorption of sodium ions, which indirectly stimulate ADH release through the resulting rise in extracellular fluid osmotic pressure. Thirst centers in the brain, which respond to fluctuations in blood volume or extracellular fluid osmolarity, are also important regulators of fluid balance by their influence on the amount of fluid intake.

The macrominerals sodium and potassium, and other ions of nutritional importance such as calcium, magnesium, and chloride, are freely filtered by the renal glomerulus but are selectively conserved by tubular reabsorption via active transport systems. Potassium is an example of a mineral that is regulated in part by tubular secretion into the filtrate. Secretion of the ion from the distal tubular cells increases as its concentration in those cells rises, because of increased dietary intake. Potassium, like sodium, is regulated by aldosterone. Elevated plasma potassium stimulates the release of aldosterone, which exerts opposing renal effects on the two minerals—the enhanced reabsorption of sodium and an increase in potassium excretion. Normal physiological function depends on proper control of the body fluid acid-base balance.

Many metabolic enzymes have a narrow range of pH at which they function adequately, and these catalysts are intolerant of pH swings more than several tenths of a unit from the average normal value of 7.4 for extracellular fluids. The plasma is well buffered, primarily by proteins and by the bicarbonate-carbonic acid system. However, conditions of acidosis or alkalosis can result in certain situations such as an overproduction of organic acids, as would occur in diabetes or starvation, or respiratory aberrations that may cause abnormal carbon dioxide ventilation. Therefore, restoration of normal pH may be necessary and is accomplished through compensatory mechanisms of the kidneys and lungs. These organs function to maintain a normal ratio of bicarbonate to carbonic acid. The bicarbonate concentration is under the control of the kidneys, which can either conserve the ion, by reabsorbing it to a greater extent, or increase its excretion, depending on whether the ratio needs to be decreased or increased to compensate for a pH disturbance. The carbonic acid value is controlled by the respiratory center. It can be increased or decreased in concentration by changes in the respiratory rate. Hyperventilation, for example, lowers the value by "blowing off" carbon dioxide, whereas a slowing of respiration retains carbon dioxide and therefore raises the carbonic acid level. From their effects on the bicarbonate:carbonic acid ratio, it can be reasoned that hyperventilation can raise the pH, and respiratory suppression can lower the pH in a compensatory manner.

References Cited

1. Goodman HM. Basic medical endocrinology. New York: Raven Press 1994:84–87.

2. Van de Stolpe A, Jamison RL. Micropuncture study of the effect of ANP on the papillary collecting duct in the rat. Am J Physiol 1988;254:F477–483.

3. Hadley ME. Endocrinology, 3rd ed. Englewood Cliffs, NJ: Prentice Hall 1992:77–91.

4. Vokes T. Water homeostasis. Ann Rev Nutr 1987; 7:383–406.

5. Wingo CS, Cain BD. The renal H-K-ATPase: physiological significance and role in potassium homeostasis. Annu Rev Physiol 1993;55:323–347.

6. Luft FC. Salt and hypertension: recent advances and perspectives. J Lab Clin Med 1989;114:215–221.

7. Suter P. Potassium and hypertension. Nutr Rev 1998;56(#5):151–3.

8. Windhager EE, ed. Handbook of Physiology. Section 8, Renal physiology, Vols I, II. New York: Oxford University Press, 1992.

9. Valtin J and Schafer J. Renal Function, 3rd ed., Boston: Little, Brown and Co., 1995.

Suggested Reading

Vokes T. Water homeostasis. Annu Rev Nutr 1987;7:383–406.
   This is a discussion of the mechanisms of water balance regulation and the pathology associated with deficiencies in the regulatory system.
Kleinman LI, Lorenz JM. Physiology and pathophysiology of body water and electrolytes. In: Kaplan LA, Pesce AJ, eds. Clinical Chemistry: Theory, Analysis, and Correlation. St. Louis: Mosby, 1996: Chap 24.
   This is a clearly written clinical approach to fluid and electrolyte homeostasis, with diagrammatic illustrations of the regulatory mechanisms of fluid and electrolyte control.
Sherwin JE. Acid-base control and acid-base disorders. In: Kaplan LA, Pesce AJ, eds. Clinical Chemistry: Theory, Analysis, and Correlation. St. Louis: Mosby, 1996: Chap 25.
   This is a brief introduction to the physiological buffer systems and a clinical approach to the regulation of acid-base balance.

Web Sites

www.nih.gov/health/consumer/conicd.htm
   National Institutes of Health: Consumer Health Information page
http://thriveonline.com
   @ America Online (AOL)

Fluid Balance and the Thermal Stress of Exercise

The discussion of sports nutrition in Chapter 8 centered on (1) the selective use of energy-yielding substrates during exercise of varying intensities and (2) how physical performance can be enhanced by maximizing substrate stores by the judicious intake of the energy nutrients. Another very important dimension in the demands of sport and exercise is thermoregulation—that is, the control of body temperature within a narrow range. A drop in deep body (core) temperature of 10° and an increase of just 5° above normal is tolerated, but fluctuation beyond this range can result in death. Strenuous exercise challenges this control because it is markedly thermogenic owing to its stimulation of metabolic rate. The more than 100 heat-related deaths among football players recorded during the past 20 years attest to the seriousness of hyperthermia.

Various mechanisms of thermoregulation are responsible for maintaining thermal balance in the body. Muscular activity is among the most influential factors that contribute to *heat gain* in body core temperature. There are others, however, including hormonal effects, the thermic effect of food, postural changes, and environmental changes. Countering the heat gain factors are mechanisms that protect against hyperthermia by *removing* heat from the body. These include *radiation, conduction, convection,* and *evaporation.* In an otherwise normal individual engaged in strenuous exercise, evaporation (of sweat) provides the most important physiological defense against overheating. Evaporation of 1.0 ml of sweat is equivalent to about 0.6 kcal of body heat loss. Therefore, even at maximal exercise, at which 4.0 L O_2/minute are consumed, equivalent to about 20 kcal/minute of heat produced, core temperature would be expected to rise just 1° every 5 to 7 minutes. This is because sweating, assumed to be maximal at 30 mL/minute, would cool the body to the extent of about 18 kcal/minute.

Approximately 80% of the energy released during exercise is in the form of heat. If this is not removed from the body, the heat load due to metabolic activity, combined with environmental heat during strenuous exercise, could lead to a dramatic increase in body temperature. Hyperthermia can result in lethal heat injury. It has already been pointed out that the major mechanism for heat loss is the evaporation of sweat and that nearly 600 kcal are eliminated by the cooling effect of the evaporation of 1 L of sweat. Among the remaining mechanisms for heat removal, radiation is the next most important. In radiation, heat generated in the working muscles is transported by blood flow to the skin, from which it can be subsequently exchanged with the environment. For either of these thermoregulatory mechanisms, body water is clearly the major participant, and there has been considerable interest in assessing various strategies for its replacement during strenuous exercise.

Firm evidence indicates that depletion of body water from sweating beyond 2% of body weight can cause significant impairment of endurance through deficiencies in thermoregulatory and circulatory functions. The most likely explanations for this impairment are as follows:

- A reduced plasma volume and therefore reduced hemodynamic capacity to achieve maximal cardiac output and peripheral circulation. As plasma volume declines, reduced skin blood flow and a fall in stroke volume follow. Heart rate increases in compensation but cannot offset the stroke volume deficit [1].

- Altered sweat gland function, whereby sweating ceases, in an autonomic control attempt to conserve body water

As a result, body temperature rises quickly, drastically increasing the chance of cramps, exhaustion, and even heat stroke, the latter having a mortality rate of 80%. Sweat losses of 1.5 L/hour are commonly encountered in endurance sports, and under particularly hot conditions, sweat rates exceeding 2.5 L/hour have been measured in fit individuals. Marathon runners can lose 6% to 8% of body weight in water during the 26.2-mile event, and plasma volume may fall 13% to 18%. It would not be uncommon, therefore, for a 150-lb runner to lose 0.5 lb of water per mile in a hot environment (equivalent to an 8-oz glass of water).

Dehydration results when fluid loss exceeds intake, the degree of dehydration being directly proportional to this disparity. The primary goal of fluid replacement is to maintain plasma volume so that circulation and sweating can proceed at maximal levels. It is difficult for the endurance athlete to avoid a negative water balance because it is both impractical and distasteful to attempt to replenish the copious amount lost in the course of a marathon. It is distasteful because the necessary intake far exceeds the thirst desire, a stimulus that is delayed behind rapid dehydration. It has been estimated that athletes, when left on their own, replace only about half of the water lost during exercise [2]. However, the force-feeding of fluids to exactly balance that which was lost is ideal from the standpoint of athletic performance, although the dramatic effects of lesser amounts of fluid replenishment during exercise are also well documented. The experimental design on which these conclusions are generally based is a comparison of the extent of fluid intake with performance and certain physiological parameters such as heart rate and body temperature. Study groups are commonly composed of subjects who, in the course of prolonged exercise, are (1) force-fed fluids beyond the thirst desire, (2) allowed to drink fluid ad libitum (as desired), and (3) deprived of fluid intake. Force-fed subjects display superior performance, lower heart rate, and lower body core temperature than the other groups, and the ad libitum group outperforms the deprived group in these parameters.

An imposing question in sports nutrition surrounded by controversy is whether electrolyte replacement is necessary during prolonged exercise. Based on the knowledge that sweat contains electrolytes (sodium, potassium, chloride, and magnesium), it was reasoned that significant losses of these occurred during endurance athletics and that their replacement was necessary to optimize performance. In the 1970s, sports drinks supplemented with electrolytes and sometimes glucose (GE drinks) began to appear on the market and were sold under such names as Gatorade, Sportade, and Body Punch. The answer as to whether such supplementation is necessary depends on the level of intensity of the exercise and therefore the quantity of sweat lost.

The electrolyte content of sweat in the average individual is very low compared with the body fluids. This is shown in Table 1, which compares the electrolyte concentrations of sweat and blood serum. In the case of sodium and chloride, dehydration through sweating has the effect of concentrating these electrolytes in extracellular and intracellular water because of the relatively low concentration of these ions in sweat. Therefore in marathon-level exertion, in which a total sweat loss of 5 to 6 L or less is incurred, rehydration with water alone is adequate, because only about 200 mEq of sodium and chloride would be lost from a relatively large body store. Furthermore, pure water, because of its negligible osmolarity, would leave the stomach and be absorbed more quickly than a hypotonic salt solution, thereby restoring rapidly the normal osmolarity of the body fluids. Although some researchers regard potassium losses during exercise in the heat as constituting a potential health problem, this too is controversial in view of the relatively small amount of the ion lost. On the basis of the potassium content of sweat, shown in Table 1, it can be estimated that a 5-L sweat loss would induce a potassium deficit of <20 mEq, or well under 1% of the estimated total body store of 3,000 mEq for a 70-kg man.

Only under severe conditions of prolonged, high-intensity exercise in the heat would electrolyte replacement be indicated. In such a case, electrolyte loss may exceed the amount provided in the daily diet, and some sodium supplementation may be necessary. The quantity provided by adding one-third of a teaspoon of table salt to a liter of water would be adequate [3]. It is doubtful if potassium supplementation is called for under similar conditions for the reasons discussed.

Table 1 Average Electrolyte Concentrations in Sweat and Blood Serum (mEq/L)

	Na^+	K^+	Cl^-	Mg^{++}
Sweat	40–45	3.9	39	3.3
Blood serum	140	4.0	110	1.5–2.1

References Cited

1. Sawka M. Physiological consequences of hypohydration: exercise, performance and thermoregulation. Med Sci Sports Exerc 1992; 24:657.

2. Noakes T. Fluid replacement during exercise. Exerc. Sports Sci. Rev. 1993;21:297.

3. McArdle W, Katch F, Katch V. Exercise Physiology, 4th ed., Baltimore, MD: Williams & Wilkins, 1996: Chap. 25.

Web Sites

http://www.umass.edu/cnshp/index.html
    Center for Nutrition in Sport and Human Performance at the University of Massachusetts

Chapter 15

Body Composition and Energy Expenditure

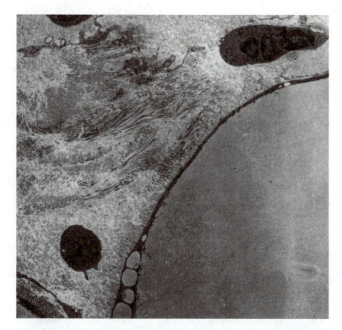

A portion of an adipocyte engorged with lipid, showing the compressed cytoplasm as a thin strand. Smaller cells are leukocytes.

Body Weight: What Should We Weigh?
Height-Weight Tables
Formulas
Body Mass

The Composition of the Human Adult Body

Methods for the Measurement of Body Composition
Anthropometry
Densitometry
Absorptiometry
Total Body Electrical Conductivity and Bioelectrical Impedance (BEI) or Bioelectrical Impedance Analysis (BIA)
Computerized (Axial) Tomography (CAT or CT)
Magnetic Resonance Imaging (MRI)
Ultrasonography or Ultrasound
Infrared Interactance
Total Body Water (TBW)
Total Body Potassium (TBK)
Neutron Activation Analysis

Primary Influences on Body Composition

Components of Energy Expenditure
Basal Metabolic Rate, Resting Energy Expenditure, and Resting Metabolic Rate
Diet-Induced Thermogenesis, Specific Dynamic Action, Specific Effect of Food, or Thermic Effect of Food
Physical Activity
Other Components of Energy Expenditure

Assessment of Energy Expenditure
Direct Calorimetry
Indirect Calorimetry
Estimating Total Energy Expenditure

∞ *PERSPECTIVE 1:* Osteoporosis: Diet and Diet-Related Factors

∞ *PERSPECTIVE 2:* Eating Disorders

An innate characteristic of maturation and aging is a change in body composition and body weight. These compositional changes occur throughout the life cycle, beginning with the embryo and extending through old age. Rapid growth entails not only an increase in body mass but also a change in the proportions of components making up this mass. Young adulthood is a period of relative homeostasis, but in some people modifications of body composition can occur. Following the more or less homeostatic period of young adulthood is the period of progressive aging, when some undesirable changes in body composition and often weight inevitably occur. Energy expenditure, which influences body weight and composition, is a function of basal metabolic rate, diet-induced thermogenesis, physical activity, and to a limited extent adaptive thermogenesis. Knowledge of recommended body weight, body composition, and components and assessment of energy expenditure is needed by nutrition professionals so that they may be better prepared to determine nutrient needs and to identify and/or prevent disease.

Body Weight: What Should We Weigh?

Recognition of body weight as an indicator of health status is probably universal and as old as humanity itself. In fact, in 1846 an English surgeon named John Hutchinson published a height-weight table on a sample of 30-year-old Englishmen and urged that future census taking also include such information, believing it to be valuable in promoting health and detecting disease [1]. Today, scientists and health professionals recognize that risk of many diseases including heart disease, diabetes mellitus, hypertension, and some cancers increases with excess body weight. Further, a low body weight may be indicative of malnutrition or an eating disorder and may pose risk for other diseases such as those affecting the lungs. Determination of recommended body weight and body mass may be accomplished through several approaches including use of tables (height-weight and aged-based height-weight) and formulas. Each of these methods will be discussed.

Height-Weight Tables

Height-weight tables are frequently used to determine an adult's recommended weight. The tables are published by insurance companies and are based on data obtained from a specific population, those individuals interested in obtaining insurance. Insurance companies look to insure those individuals likely to live a long life, and thus the data in the tables represent weights for height at a particular age of people who applied and were accepted for insurance over a period of several years [1]. Several different sets of tables have been published by the Metropolitan Life Insurance Company.

In 1942 and 1943, the Metropolitan Life Insurance Company introduced the concept of "ideal" weights for women and men, respectively. In 1959, these tables were revised (Table 15.1). Revision of the tables came as a result of a study (build and blood pressure study), by the Society of Actuaries indicating that lower-than-average weight for height, even for young adults, could be associated with longevity. In the revised tables, weights considered appropriate for the various heights and frame sizes were termed "desirable" rather than ideal. The method for determining frame size was omitted.

The latest height-weight table is based on the Actuary Body Build Study [2] and published by the Metropolitan Life Insurance Company in 1983 (Table 15.2). Weight-for-height figures are designated as "acceptable" rather than ideal or desirable and are somewhat higher than those found in the earlier height-weight tables. The increase of about 10% in shorter people and (5% in people of medium height [3] is a reflection of the mor-

tality data collected in the 1979 actuary build study [2]. This study showed that insured men and women under 30 years of age tended to be heavier than their counterparts of the 1959 build and blood pressure study. Of particular interest, however, was the reduced mortality found to be associated with mild to moderate overweight in shorter-than-average people.

Unlike its predecessors, the 1983 table is accompanied by instructions for estimating frame size. People with small, medium, or large frames may be identified by measurement of elbow breadth (Table 15.3 and Fig. 15.1). Approximately 50% of the population falls within the medium-frame category, with the other 50% rather evenly divided between the small-frame and large-frame designations.

Table 15.1 Desirable Weights for Men and Women, Metropolitan Life Insurance Company, 1959

Height (in shoes)	Weight (lb, in indoor clothing)		
	Small Frame	Medium Frame	Large Frame
Men:			
5' 2"	112–120	118–129	126–141
5' 3"	115–123	121–133	129–144
5' 4"	118–126	124–136	132–148
5' 5"	121–129	127–139	135–152
5' 6"	124–133	130–143	138–156
5' 7"	128–137	134–147	142–161
5' 8"	132–141	138–152	147–166
5' 9"	136–145	142–156	151–170
5'10"	140–150	146–160	155–174
5'11"	144–154	150–165	159–179
6' 0"	148–158	154–170	164–184
6' 1"	152–162	158–175	168–189
6' 2"	156–167	162–180	173–194
6' 3"	160–171	167–185	178–199
6' 4"	164–175	172–190	182–204
Women:			
4'10"	92–98	96–107	104–119
4'11"	94–101	98–110	106–122
5' 0"	96–104	101–113	109–125
5' 1"	99–107	104–116	112–128
5' 2"	102–110	107–119	115–131
5' 3"	105–113	110–122	118–134
5' 4"	108–116	113–126	121–138
5' 5"	111–119	116–130	125–142
5' 6"	114–123	120–135	129–146
5' 7"	118–127	124–139	133–150
5' 8"	122–131	128–143	137–154
5' 9"	126–135	132–147	141–158
5'10"	130–140	136–151	145–163
5'11"	134–144	140–155	149–168
6'0"	138–148	144–159	153–173

Source: Courtesy of Metropolitan Life Insurance Company.
Reprinted with permission.

Table 15.2 1983 Metropolitan Life Insurance Company Height and Weight Tables

| Height | Weight | | |
	Small Frame	Medium Frame	Large Frame
Men:[a]			
5' 2"	128–134	131–141	138–150
5' 3"	130–136	133–143	140–153
5' 4"	132–138	135–145	142–156
5' 5"	134–140	137–148	144–160
5' 6"	136–142	139–151	146–164
5' 7"	138–145	142–154	149–168
5' 8"	140–148	145–157	152–172
5' 9"	142–151	148–160	155–176
5'10"	144–154	151–163	158–180
5'11"	146–157	154–166	161–184
6' 0"	149–160	157–170	164–188
6' 1"	152–164	160–174	168–192
6' 2"	155–168	164–178	172–197
6' 3"	158–172	167–182	176–202
6' 4"	162–176	171–187	181–207
Women:[b]			
4'10"	102–111	109–121	118–131
4'11"	103–113	111–123	120–134
5' 0"	104–115	113–126	122–137
5' 1"	106–118	115–129	125–140
5' 2"	108–121	118–132	128–143
5' 3"	111–124	121–135	131–147
5' 4"	114–127	124–138	134–151
5' 5"	117–130	127–141	137–155
5' 6"	120–133	130–144	140–159
5' 7"	123–136	133–147	143–163
5' 8"	126–139	136–150	146–167
5' 9"	129–142	139–153	149–170
5'10"	132–145	142–156	152–173
5'11"	135–148	145–159	155–176
6' 0"	138–151	148–162	158–179

[a]Weights at ages 25–59 years based on lowest mortality.

Weight in pounds according to frame (in indoor clothing weighing 5 lb, shoes with 1" heels).

[b]Weights in pounds at ages 25–59 years based on lowest clothing weighing 3 lb, shoes with 1" heels).

Source: Courtesy of Metropolitan Life Insurance Company. Reprinted with permission.

Dividing a person's actual weight by the midpoint of the weight range for his or her roughly estimated frame size is referred to as *Metropolitan relative weight* (MRW) or *relative body weight* (RBW) and represents a popular method for identifying people who might be at risk because of overweight or obesity (or because of underweight) [4]. People at risk include those individuals with weight for height well below the lowest end of the acceptable range or well above the highest end of the acceptable range. Extremes in either direction more than likely pose some health risks.

Table 15.3 How to Determine Your Body Frame by Elbow Breadth

Height (in 1" heels)	Elbow Breadth (in)	Height (cm, in 2.5-cm heels)	Elbow Breadth (cm)
Men:			
5'2"–5'3"	2⅛–2⅞	158–161	6.4–7.2
5'4"–5'7"	2⅝–2⅞	162–171	6.7–7.4
5'8"–5'11"	2¾–3	172–181	6.9–7.6
6'0"–6'3"	2¾–3⅛	182–191	7.1–7.8
6'4"	2⅞–3¼	192–193	7.4–8.1
Women:			
4'10"–4'11"	2¼–2½	148–151	5.6–6.4
5'0"–5'3"	2¼–2½	152–161	5.8–6.5
5'4"–5'7"	2⅜–2⅝	162–171	5.9–6.6
5'8"–5'11"	2⅜–2⅝	172–181	6.1–6.8
6' 0"	2½–2¾	182–183	6.2–6.9

Note: This table lists the elbow measurements for men and women of medium frame at various heights. Measurements lower than those listed indicate that you have a small frame, while higher measurements indicate a large frame.

Source: Courtesy of Metropolitan Life Insurance Company. Reprinted with permission.

Use of height-weight tables to determine recommended body weight has limitations. As previously mentioned, the data are collected from a group of people who are not necessarily representative of all adults. Further, data do not provide information on body composition.

Age-Based Height-Weight Tables

In 1990, the U.S. Department of Agriculture published *Suggested Weights for Adults* aged 19 to 34 years and aged 35 years and older (Table 15.4) as part of its Dietary Guidelines for Americans (3rd ed.). These weights for heights are given in ranges whereby the higher weights in the ranges apply to men and the lower weights in the range apply to women. The suggested weights are slightly higher at the upper end of the ranges than those in the 1983 Metropolitan Life Height Weight Tables. The suggested weights for adults as well as the 1983 Metropolitan Life Height Weight Tables have been criticized by some as being too liberal [5].

Formulas

Ideal body weight also may be calculated using formulas. The formulas suggested by Devine [6,7] to calculate ideal body weight (IBW) are as follows:

$$\text{IBW for men} = 50 \text{ kg} + 2.3 \text{ kg/in} > 5 \text{ ft}$$
$$\text{IBW for women} = 45 \text{ kg} + 2.3 \text{ kg/in} > 5 \text{ ft}$$

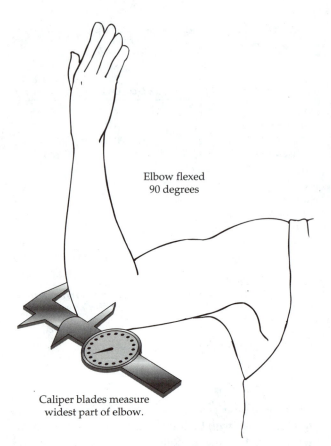

Elbow flexed
90 degrees

Caliper blades measure
widest part of elbow.

Upper arm parallel to floor

Figure 15.1 Measurement of elbow breadth. Extend your arm and bend the forearm upward at a 90° angle. Keep your fingers straight, and turn the inside of your wrist toward your body. Place the calipers on the two prominent bones on either side of the elbow. Measure the space between the bones with the caliper. Compare this measurement with the measurements shown in Table 15.3.

These formulas have been modified somewhat and converted into the following familiar empirical formulas:

IBW for men = 110 lb + 5 lb/in > 5 ft
IBW for women = 100 lb + 5 lb/in > 5 ft

A slightly modified formula for men is also used:

IBW for men = 106 lb + 6 lb/in > 5 ft

A 10% range minus and plus the calculated ideal weight allowing for the differences in weight due to a small or large frame size, respectively, is usually included in the Devine formula.

Using the formulas, a man who is 5 ft, 11 in tall with a medium frame should weigh either 165 lb (110 + [5 × 11] = 165 lb) or 172 lb (106 + [6 × 11] = 172 lb) depending on which formula was used. If the man had a small frame, ideal body would be 10% less than that calculated for a medium frame size (i.e., 165 lb − 16.5

Table 15.4 Suggested Weights for Adults

Height (without shoes)	Weight (lb, without clothes)	
	19 to 34 years	35 years and over
5'0"	97–128	108–138
5'1"	101–132	111–143
5'2"	104–137	115–148
5'3"	107–141	119–152
5'4"	111–146	122–157
5'5"	114–150	126–162
5'6"	118–155	130–167
5'7"	121–160	134–172
5'8"	125–164	138–178
5'9"	129–169	142–183
5'10"	132–174	146–188
5'11"	136–179	151–194
6'0"	140–184	155–199
6'1"	144–189	159–205
6'2"	148–195	164–210
6'3"	152–200	168–216
6'4"	156–205	173–222
6'5"	160–211	177–228
6'6"	164–216	182–234

Source: U.S. Department of Agriculture, U.S. Department of Health and Human Services. Dietary Guidelines for Americans, 3rd ed. Washington, DC: USDA, 1990; 9.

lb = 148.5 lb or 172 lb − 17.2 lb = 154.8 lb, respectively). A female who is 5 ft, 6 in, and has a large frame has an ideal body weight of 143 lb (i.e., 100 + [5 × 6] = 130 lb + 13 lb [accounting for the 10% addition for the large frame size] = 143 lb).

Dividing a person's actual body weight by the ideal body weight calculated (from the Devine formula) for his or her estimated frame size is referred to as percentage ideal body weight and is often used to screen for overweight, obesity, or underweight. Individuals whose body weight is 10% or more below the average for a given height are considered underweight, whereas those whose body weight is 10% or more above the average for a given height are considered overweight. Individuals with a weight for height that is 20% or more above the average are considered obese.

Regression equations purported to be more nearly accurate in estimating IBW have been based on the 1959 Metropolitan Life Insurance Company height-weight tables and the IBW tables used by Grant [8] in nutrition assessment. These equations are rather complicated, but they account for gender and frame size without sacrificing accuracy. The equation based on the 1959 Metropolitan Life Insurance Company tables (with indoor clothing and shoes) is as follows [9]:

y (or IBW in lb) =
$$-139.17 + 3.86(\text{height}) + 9.52(\text{frame}) + 5.01(\text{sex})$$

Based on Grant's tables, corrected for nude height and weight, the equation becomes

$$y = -133.99 + 3.86(\text{height}) + 9.52(\text{frame}) + 3.08(\text{sex})$$

In these two equations, height is in inches. Figures used for frame size are 1 for small, 2 for medium, and 3 for large. Figures for sex are +1 for male and −1 for female.

Another equation for determining an ideal or desirable body weight is based on body composition and requires measurement of body fat [10].

$$\text{Desirable body weight} = \frac{\text{lean body weight}}{1 - \% \text{ fat desired}}$$

Calculations would be as follows for a woman who weighs 130 lb, with a measured 30% of this weight as fat:

$$130 \text{ lb} \times 0.30 = 39 \text{ lb (fat weight)}$$
$$130 \text{ lb} - 39 \text{ lb} = 91 \text{ lb (lean body weight)}$$

Because a desirable amount of fat in males is 15% or less and in females is 25% or less [9,11], a figure of 25% (0.25) is figured in the following equation for the sample woman:

$$\text{Desirable body weight} = 91/(1 - 0.25)$$
$$= (91/0.75) = 121 \text{ lb}$$

Body Mass

Although comparisons of actual weight to recommended weight with subsequent calculation of relative body weight or percentage ideal body weight are helpful in assessing potential disease risk, assessment of body mass also is useful. One more well-known and well-used index is the Quetelet's Index, also known as body mass index (BMI). Body mass index is calculated by measuring a person's weight in kilograms and dividing it by the person's height measured in meters and raised to a power of 2 or squared.

$$\text{Body mass index} = \frac{\text{Weight}}{\text{Height}^2}$$

The body mass index is considered a good index of total body fat in both men and women. A BMI <18.5 kg/m^2 is considered as underweight, whereas levels <16 may be associated with an eating disorder. For adult women and men, a BMI between 18.5 and 24.9 kg/m^2 is considered acceptable [12]. A BMI ≥25 kg/m^2 indicates one is overweight and is associated with increased risk of disease. A BMI >30 kg/m^2 is considered obese, whereas >40 is considered extremely or morbidly obese [12].

Comparisons of the body mass index and 1983 Metropolitan Life Insurance Company table data found that the 1983 height weight table was appropriate primarily for people in their thirties and forties. Weights listed as acceptable in the 1983 table were higher than insurance data would justify for young adults, but lower than justifiable in older adults [13,14].

Yet, although measuring both height and weight is relatively easy to do and can serve as a screen for underweight, overweight or obesity, neither weight for height nor body mass index is always a valid indicator of the degree of body fatness. The failure of weight as a valid measure of fatness became clear in World War II. Behnke, a Navy physician, was able to demonstrate by hydrostatic weighing that several football players who had been found unfit for military service because of excessive weight actually had less body fat than controls of normal weight [15]. The excessive weight of these athletes was due to hypertrophy of muscles rather than excessive adipose tissue [15]. Behnke's work rekindled an interest in studying the composition of the human body, an interest that had lain dormant for about 50 years.

The Composition of the Human Adult Body

The chemical composition of the human body was first described in 1859 in a book that dealt with the chemical composition of food [16]. Analytic chemistry was a rapidly growing science at the time, and figures describing chemical composition of the different tissues of the body were given in comparison with those for various foods. Additional chemical composition data from whole-body analysis of fetuses, children, and adults were collected during the next few decades and represent a direct (versus indirect) measure of body composition [16–20].

The concept of the reference man and woman (Table 15.5) was developed in the 1970s [21]. These reference figures provide information on body composition and are based on average physical dimensions from measurements of thousands of subjects who participated in various anthropometric and nutrition surveys [21]. The reference man and woman data provide a frame for comparisons.

As seen in Table 15.5, the reference man weighs 29 lb more than the woman (nonpregnant) and is 4 in taller. The man has 15% body fat versus the female with 27%. Of the 15% total body fat in the reference man, only 3% is essential fat versus 12% essential fat in the reference woman's 27% total body fat. Essential fat is that fat associated with bone marrow, the central

Table 15.5 Body Composition of Reference Man and Woman

Reference Man	Reference Woman
Age: 20–24 yr	Age: 20–24 yr
Height: 68.5 in	Height: 64.5 in
Weight: 154 lb	Weight 125 lb
Total fat: 23.1 lb	Total fat: 33.8 lb
(15.0% body weight)	(27.0% body weight)
Storage fat: 18.5 lb	Storage fat: 18.8 lb
(12.0% body weight)	(15.0% body weight)
Essential fat: 4.6 lb	Essential fat: 15.0 lb
(3.0% body weight)	(12.0% body weight)
Muscle: 69 lb	Muscle: 45 lb
(44.8% body weight)	(36.0% body weight)
Bone: 23 lb	Bone: 15 lb
(14.9% body weight)	(12.0% body weight)
Remainder: 38.9 lb	Remainder: 31.2%
(25.3% body weight)	(25.0% body weight)
Average body density:	Average body density:
1.070 g/mL	1.040 g/mL

Source: McArdle WD, Katch FI, Katch VL. Exercise Physiology. Philadelphia: Lea & Febiger, 1981; p. 369. Adapted from Behnke AR, Wilmore JH. Evaluation and Regulation of Body Build and Composition. Englewood Cliffs, NJ: Prentice Hall, 1974.

nervous system, viscera (internal organs), and cell membranes. In females, essential fat also includes fat in mammary glands and the pelvic region. Note that in the reference man muscle accounts for 44.8% of body weight versus 36.0% of body weight in the female. The average body density of the reference man and woman are 1.070 and 1.040 g/mL, respectively.

Table 15.6 provides information of the average values for body fat in U.S. men and women [11,22]. Based on data from physically active young adults and competitive athletes, a body fat content of about 15% (but not <8%) for men and between roughly 15% and 25% for women has been recommended [11,23]. Female Olym-pic athletes (runners, jumpers, swimmers, divers, and gymnasts) had body fat levels of 11% to 16% [24]. Average body fat values for men and women (untrained individuals) range from about 15% to 18% and 22% to 25%, respectively.

Methods for the Measurement of Body Composition

Division of the body into components is used extensively for in vivo studies of body composition. The body can be categorized atomically in terms of its elements—primarily carbon, oxygen, hydrogen, and nitrogen, which make up about 95% of body mass, along with about another 50 or so elements that make up the remaining 5%. Alternately, the body may be thought of from a nutrient or molecular perspective as consisting of water, protein, fat, carbohydrate, and minerals. Densitometry, one standard against which other indirect measurements of body composition are evaluated, separates the body into two components, fat mass and fat-free mass [16,20,25]. According to the two-component model for body composition assessment, fat mass includes essential and nonessential fat (triacylglycerols), whereas fat-free mass includes protein, water, carbohydrate (glycogen), and minerals [16,20]. The term *lean body mass* is used synonymously with *fat-free mass* but includes essential body fat [16,25]. Several methods of body composition assessment are available, some of which are based on the two-component model. Commonly available methods to assess body composition are indirect and provide a means to calculate body components. Although different procedures are available, accuracy varies not only with the equipment/method but also with the technician. Several indirect methods of body composition assessment will be reviewed in this section; direct measurement is only accomplished on cadavers.

Anthropometry

Anthropometry allows for estimation of body composition through measurement at various circumference and skin fold (fat) sites. Skin varies in thickness from 0.5 mm to 2 mm [26], thus fat beneath the skin typically represents the majority of the skin fold measurement. The assumption is that a direct relationship exists between total body fat and fat deposited in depots just beneath the skin (i.e., subcutaneous fat).

Skin fold measurements can be used in two ways:

- scores from the various measurements can be added and the sum used to indicate the relative degree of fatness among subjects, or
- scores can be plugged into various mathematical regression equations developed to predict body density or to calculate percentage of body fat [11,27].

Five sites commonly used for measuring skin fold thickness, shown (A–E) in Figure 15.2, are as follows:

A. Back of the upper arm (triceps)—a vertical fold is measured at the midline of the upper arm halfway between the tip of the shoulder and the tip of the elbow.

B. Subscapula—an oblique fat fold is measured just below the tip (interior angle) of the scapula;

C. Suprailiac—a slightly oblique fold is measured just above the hip bone with the fold lifted to follow the natural diagonal line at this point.

D. Abdomen—a vertical fold is measured 1 in to the right of the umbilicus.

Table 15.6 Average Percentage of Body Fat for Women and Men from Selected Studies [11]

Study	Age (yr)	Stature (cm)	Mass (kg)	Fat (%)	68% Variation Limits[a]
Younger women:					
North Carolina, 1962	17–25	165.0	55.5	22.9	17.5–28.5
New York, 1962	16–30	167.5	59.0	28.7	24.6–32.9
California, 1968	19–23	165.9	58.4	21.9	17.0–26.9
California, 1970	17–29	164.9	58.6	25.5	21.0–30.1
Air Force, 1972	17–22	164.1	55.8	28.7	22.3–35.3
New York, 1973	17–26	160.4	59.0	26.2	23.4–33.3
North Carolina, 1975		166.1	57.5	24.6	——
Massachusetts, 1993	17–30	165.3	57.7	21.8	16.7–27.2
Older women:					
Minnesota, 1953	31–45	163.3	60.7	28.9	25.1–32.8
Minnesota, 1953	43–68	160.0	60.9	34.2	28.0–40.5
New York, 1963	30–40	164.9	59.6	28.6	22.1–35.3
New York, 1963	40–50	163.1	56.4	34.4	29.5–39.5
North Carolina, 1975	33–50	——	——	29.7	23.1–36.5
Massachusetts, 1993	31–50	165.2	58.9	25.2	19.2–31.2
Younger Men:					
Minnesota, 1951	17–26	177.8	69.1	11.8	5.9–11.8
Colorado, 1956	17–25	172.4	68.3	13.5	8.3–18.8
Indiana, 1966	18–23	180.1	75.5	12.6	8.7–16.5
California, 1968	16–31	175.7	74.1	15.2	6.3–24.2
New York, 1973	17–26	176.4	71.4	15.0	8.9–21.1
Texas, 1977	18–24	179.9	74.6	13.4	7.4–19.4
Massachusetts, 1993	17–30	178.2	76.3	12.9	7.8–18.1
Older men:					
Indiana, 1966	24–38	179.0	76.6	17.8	11.3–24.3
Indiana, 1966	40–48	177.0	80.5	22.3	16.3–28.3
North Carolina, 1976	27–50	——	——	23.7	17.9–30.1
Texas, 1977	27–59	180.0	85.3	27.1	23.7–30.5
Massachusetts, 1993	31–50	177.1	77.5	19.9	13.2–26.5

[a]Indicates the range of values for percentage of body fat that includes one standard deviation or about 68 out of every 100 people measured.

Source: Katch FI, McArdle WD. Introduction to Nutrition, Exercise, and Health, 4th ed. Philadelphia: Lea and Febiger, 1993; p. 254.

E. Thigh—a vertical fold is measured at midpoint of the thigh, between the knee cap to the hip (inguinal crease) [11,28].

Additional sites often include the pectoral (chest), midaxillary, and calf. The right side of the body is used for most measurements if comparisons are being made to standards derived from data from U.S. surveys that typically measured the right side of subjects. The handedness of the subject affects skin fold measurements taken on the arm such that measurements on the right exceed those on the left by 0.2 to 0.3 standard deviation units [29].

However, bias associated with the side of the measurement is less than error due to measurement [29]. All measurements should be repeated at least two to three times, and the average should be used as the skin fold value. Measurement procedures and use of formulas

(shown in the next paragraph) contribute to procedure error. The precision of skin fold thickness measurements depends on the skill of the anthropometrist; in general, a precision of within 5% can be obtained with a well-trained and experienced anthropometrist [30]. However, the use of anthropometry for predicting visceral fat content is of limited accuracy [31]. Nevertheless, the method is quite inexpensive, compared with the cost of other techniques.

Several different equations that are population specific have been developed for calculating total body fat from skin fold sites. Equations developed by Katch and McArdle [11] for predicting total body fat in young (aged 17–26 years) men and women from the triceps and subscapula skin folds are shown here:

Young women: % body fat = 0.55 (A) + 0.31 (B) + 6.13
Young men: % body fat = 0.43 (A) + 0.58 (B) + 1.47

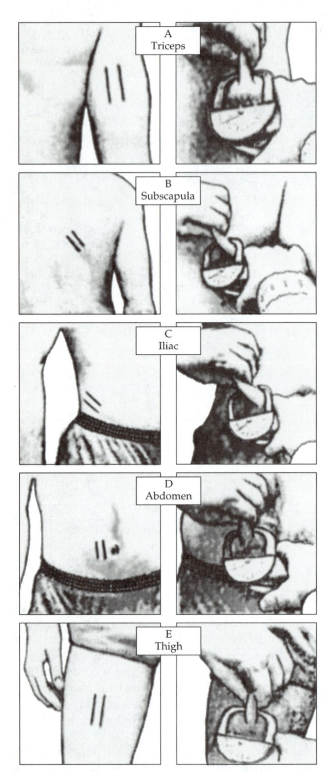

Figure 15.2 Anatomic location of the five fat fold sites: (A) triceps; (B) subscapula; (C) suprailiac; (D) abdomen; (E) thigh).

where A = triceps fat fold measured in millimeters, and B = subscapula fat fold measured in millimeters. Measurements from multiple (at least three) sites are deemed better for overall subcutaneous fat assessment than measurements from only one or two sites [11].

Circumference or girth measurements also may be used to assess body fat. Typical sites of measurement include the abdomen, buttocks, right thigh, and right upper arm. As with skin fold measurements, body fat prediction equations that are age and gender specific have been developed.

Circumference measurements of the waist (abdominal circumference) and hips (gluteal circumference) also provide an index of regional body fat distribution and have been shown to correlate with visceral fat [31]. A ratio of waist to hip circumference is calculated following measurement of both the subject's waist and hip. Waist measurements should be made below the rib cage and above the umbilicus in a horizontal plane at the most narrow site or site of least circumference. Hip circumference should be measured at the site with the greatest circumference around the hips or buttocks; soft tissue should not be compressed or indented during measurements. All measurements should be done with the subject standing. Reproducibility of circumference measurements is good at 2% [31]. Ratios greater than 0.8 in women and greater than 0.95 in men are thought to indicate increased health risk. Waist circumferences of >40 in (men) and 35 in (women) may be used to identify increased risk for the development of obesity-associated risk factors in most adults with a BMI of 25 to 34.9 kg/m² [32].

Densitometry

The principle of hydrostatic weighing on which densitometry is based can be traced to the Greek mathematician Archimedes. He discovered that the volume of an object submerged in water was equal to the volume of water displaced by the object. Specific gravity or density of an object can be calculated by dividing the object's weight (wt) in air by its loss of weight in water. For an individual who weighs 47 kg in air and 2 kg underwater, 45 kg represents the loss of body weight and the weight of the water displaced. Correction for residual air volume in the lungs (RLV) and gas in the gastrointestinal tract (GIGV) must be made.

The calculation of body density is given as follows:

$$\text{Body density} = \frac{\text{weight of body in air}}{\dfrac{(\text{wt of body in air} - \text{wt underwater}) - \text{RLV} - \text{GIGV}}{\text{density of water}}}$$

Residual lung volume is thought to be about 24% of vital lung capacity. The volume of gas in the gastrointestinal tract is estimated to range from 50 to 300 mL. This volume typically is either neglected or a value of 100 mL is used in calculations. The density of water or weight of water over a wide range of temperatures is also known and needs to be obtained for the calculation.

Calculating body density of the human body allows an estimation of body fat. At any known body density, estimating percentage of body fat is made possible by an equation derived by Siri [32]:

$$\text{Percentage body fat} = \frac{495}{\text{body density}} - 450 \times 100$$

or by an equation derived by Brozek [16]:

$$\text{Percentage body fat} = \frac{457}{\text{body density}} - 414 \times 100$$

Calculations of body density are derived in part from the knowledge that the density of fat mass is 0.9 g/cm^3 and that of fat-free mass is 1.1 g/cm^3 (assuming fat-free mass is composed of 20.5% protein, 72.4% water, and 7.1% bone mineral). Once the percentage of body fat has been calculated, the weight of the fat and the weight of lean body mass can be estimated as follows [11]:

Body weight × percentage body fat = weight
of body fat
Body weight − weight of body fat = lean body weight

Underwater weighing is considered a noninvasive and relatively precise method for assessment of body fat. Standard error of estimate of body fat using densitometry has been estimated at 2.7% for adults and about 4.5% for children and adolescents [24]. Limitations of underwater weighing include its relatively high cost for the equipment, the inability to measure gas volume in the gastrointestinal tract, its impracticality for large numbers of subjects, and the extreme cooperation and time required from subjects who must be submerged and remain motionless for an extended time period. Thus, the technique is not suitable for young children, older adults, and subjects in poor health. Additional limitations to its use include the assumption that density of lean body mass is relatively constant when in fact bone density typically changes with age [33].

Absorptiometry

Absorptiometry involves scanning the entire body or a portion of the body by a photon beam. *Single-photon absorptiometry* involves scanning the body with photons at a specific energy level. Single-photon absorptiometry does not allow accurate measurement at soft tissue sites such as the hip. However, such problems have been eliminated with the development of dual photon or dual energy x-ray absorptiometry. In *dual-photon absorptiometry,* the subject is scanned with photons at two different energy levels. The radionuclide source is generally ^{153}Gd. *Dual energy X-ray absorptiometry* (abbreviated as DEXA or DXA), introduced in the late 1980s, involves scanning subjects with X rays at two different energy levels. Dual energy X-ray absorptiometry provides a greater photon flux than dual photon and thus provides greater resolution and precision and decreases procedure times. Attenuation of the photons or X rays by body tissues is calculated by the computer such that percentage of body fat, soft tissue, or bone mineral density (total or specific sites) may be generated [30,34–36].

Limitations to the use of absorptiometry include the expense of the equipment and the exposure, although minimal, of subjects to radiation. Children, pregnant women, and subjects who are in poor health may not be suitable candidates for body composition assessment by absorptiometry. DXA measurements are highly reproducible and have been shown to correlate with other body composition assessment methods.

Total Body Electrical Conductivity (TOBEC) and Bioelectrical Impedance (BEI) or Bioelectrical Impedance Analysis (BIA)

The TOBEC (total body electrical conductivity) technique is based on the change in electrical conductivity when a subject is placed in an electromagnetic field. Subjects lie face up on a bed, which is rolled into the TOBEC instrument. The instrument then induces an electrical current in the subject. Changes in conductivity are measured and are proportional to the body's electrolyte content. Because electrolytes in the body are found associated mostly with lean body mass, TOBEC allows for the estimation of lean body mass and fat by difference. Hydration status, electrolyte imbalances, and variations in bone mass may, however, interfere with accuracy. In addition, although the procedure is fast and safe, the equipment is very expensive.

Bioelectrical impedance analysis (BIA) or bioelectrical impedance (BEI) is similar to TOBEC in that it also depends on changes in electrical conductivity. However, in bioelectrical impedance analysis, measurement of electrical conductivity is made on the extremities and not on the whole body. Subjects lie face up on a bed with extremities away from the body. Electrodes are placed on the limbs in specific locations. An instrument generates a current that is passed through the body by means of the electrodes. Opposition to the electric flow current is called *impedance* and is detected and measured by the instrument. Impedance is the inverse of conductance. The lowest resistance value of an individual is used to calculate conductance and predict lean body mass. Tissues containing little water and electrolytes (such as fat) are poor conductors and have a high resistance to the passage of current [37].

Bioelectrical impedance is a safe, noninvasive, and rapid means to assess body composition. The equipment is portable and fairly easy to operate, although it is also relatively expensive. Because measurements are made on extremities, truncal fat is not assessed very accurately. Like TOBEC, bioelectrical impedance readings are affected by hydration and electrolyte imbalances. Thus, the technique is more useful for healthy subjects.

Computerized (Axial) Tomography (CAT or CT)

Computerized or computed (axial) tomography, involving an X-ray tube and detectors aligned at opposite poles of a circular gantry, allows the taking of visual images, and thus the determination of regional body composition such as visceral organ mass; regional muscle mass; subcutaneous, internal fat; and bone density. Subjects lie face up on a movable platform that passes through the instrument's circular gantry. Cross-sectional images of tissue are constructed by the scanner computer as the X-ray beam rotates around the person being assessed. Differences in X-ray attenuation are related to differences in the physical density of tissues [30]. Calculation of relative surface area or volume occupied by tissues, such as bone, adipose, and fat-free tissue, can be accomplished from the images produced by the instrument. Results are highly reproducible [31]. Excessively long exposure of subjects to ionizing radiation and the expense of the equipment are major drawbacks to the use of computerized tomography to assess body composition.

Magnetic Resonance Imaging (MRI)

Magnetic resonance imaging (MRI) is based on the principle that atomic nuclei behave like magnets when an external magnetic field is applied across the body. When the magnetic field is applied, the nuclei attempt to align with the field. The nuclei also absorb radio frequency waves directed into the body and in turn change their orientation in the magnetic field [30]. Abolishing the radio wave results in the emission of a radio signal by the activated nuclei. The emitted signal is used to develop a computerized image. Magnetic resonance imaging is used to measure organ size and structure, body fat and fat distribution (subcutaneous,

visceral, intra-abdominal), muscle size, as well as body water contents. The technique is noninvasive and safe; however, the cost is quite high. Reproducibility of visceral fat area measured by magnetic resonance imaging is about 10% to 15% [31]. However, for assessment of adipose tissue distribution, magnetic resonance imaging provided the least variability when compared with skin fold, ^{40}K counting, bioelectrical impedance, total body water assessment with ^{18}O, and hydrostatic weighing [38].

Ultrasonography or Ultrasound

Ultrasound provides images of tissue configuration or depth readings of changes of tissue density [30]. Electrical energy is converted in a probe to high-frequency ultrasonic energy. The ultrasonic energy is transmitted through the skin and into the body in the form of short pulses or waves. The waves pass through adipose tissue until they reach lean body mass. At the interface between the adipose and lean tissues, part of the ultrasonic energy is reflected back to the receiver in the probe and is transformed to electrical energy. The echo is visualized. A transmission gel is used between the probe and skin and provides acoustic contact.

The equipment is portable, and the technique may provide information on the thickness of subcutaneous fat as well as muscle mass; however, the reproducibility is poor at 10% to 15%, and the validity of the method requires further research [31]. Anthropometry and computerized tomography appear to be superior in precision and accuracy to ultrasound [39]. Tissue interfaces are not as clearly delineated as with imaging techniques, and problems arise when the muscle-adipose tissue interface is irregular or there is fat layering [24,31].

Infrared Interactance

Infrared interactance is based on the principle that when material is exposed to infrared light, the light is absorbed, reflected, or transmitted, depending on the scattering and absorption properties of the material. For assessment of body composition, a probe that acts as an infrared transmitter and detector is placed on the skin. Infrared light of two wavelengths is transmitted by the probe. The signal penetrates the underlying tissue to a depth of 1 cm [30]. Infrared light also is reflected at the site from the skin and underlying subcutaneous tissues and detected by the probe. Estimates of body composition can be made by analyzing specific characteristics of the reflected light. The method is safe, noninvasive, and rapid; however, overestimates of body fat in lean (<8% body fat) subjects and underestimates of body fat in obese (>30% body fat) subjects have been reported [40]. The accuracy of the technique requires further investigation.

Some methods of body composition assessment used an atomic perspective to quantify one or more components of the body and through various calculations determine the other body components. Measurement of body composition via assessment of total body water, total body potassium, or neutron activation analysis use such an approach.

Total Body Water (TBW)

Quantification of total body water involves the use of isotopes, typically deuterium (D_2O), radioactive tritium (^3H_2O), or oxygen-18 (^{18}O) and is based on principles of dilution. Water can be labeled with any one of the three isotopes. The water containing a specific amount (concentration) of the isotope is then ingested or injected intravenously. Following ingestion or injection, the isotope distributes itself throughout body water. Body water occupies about 73.2% of fat-free body mass. After a specified time period (usually 2 to 6 hours) for equilibration, samples of body fluids (usually blood and urine) are taken. Losses of the isotope in the urine must be determined. If ^{18}O is used, breath samples are collected for analysis. Concentrations of the isotope in the breath or body fluids are determined by scintillation counters or other instruments. The concentration (C_1) and volume (V_1) of the isotope given is equal to the final concentration of the isotope in the plasma (C_2) and the volume of total body water (V_2) and expressed as $C_1V_1 = C_2V_2$. Thus, total body water (V_2) = C_1V_1 divided by C_2. Once total body water has been determined, the percentage of lean body mass can be calculated. Fat-free mass equals total body water divided by 0.732. Body fat can be obtained by subtracting fat-free mass from body weight. A three-component model (consisting of total body water, body volume, and body weight) has been shown to measure changes in body fat as low as 1.54 kg in individuals [34]. Many studies, however, have shown that the degree of hydration varies considerably in lean body tissue of apparently healthy people. Therefore, implications about total body fat derived from the estimation of lean body mass via total body water may be misleading [13,22]. In addition, adipose tissue has been shown to contain as much as 15% water by weight [35]. Thus, measurement of extracellular fluid as well as total body water should be conducted and subtracted from total body water to get an indication of intracellular water and thus an indication of body cell mass. Because total body water involves radiation exposure if ^3H is used, the method is not suitable for use with some subjects (such as chil-

dren or pregnant women). ^{18}O is in itself quite expensive and requires expensive mass spectrometry equipment for its analysis. Deuterated water (D_2O) is relatively inexpensive to purchase but expensive to measure [37].

Total Body Potassium (TBK)

Total body potassium is also used to assess fat-free mass; potassium is present within cells but is not associated with stored fat. ^{40}K is a naturally occurring isotope that emits a characteristic gamma ray. About 0.012% of potassium occurs as ^{40}K. External counting of gamma rays emitted by ^{40}K provides the amount of total body potassium; however, getting accurate counts of ^{40}K may be difficult, because of external or background radiation. After measurement of potassium (^{40}K) radiation from the body, calculation of total body potassium from the data is required. Fat-free mass can be estimated from the total body potassium based on any one of several conversion factors, which vary in men from 2.46 to 3.41 g potassium per kilogram fat-free mass, and in women from 2.28 to 3.16 g potassium per kilogram fat-free mass [35]. Total body fat can be calculated by subtracting fat-free mass from body weight. Overestimation of body fat in obese subjects has been reported with total body potassium [40]. The technique should not be used in people with potassium-wasting diseases.

Neutron Activation Analysis

Neutron activation analysis allows for an in vivo estimation of body composition including total body concentrations of nitrogen (TBN), calcium (TBCa), chloride (TBCl), sodium (TBNa), and phosphorus (TBP), among other elements. A beam of neutrons is delivered to the individual being assessed. The body's atoms (nitrogen, calcium, chloride, sodium, and phosphorus) interact with the beam of neutrons to generate unstable radioactive elements, which emit energy as they revert back to their stable forms. The specific energy levels correspond with specific elements and the radiation's level of activity indicates the element's abundance [30]. The ability to measure nitrogen allows for lean body mass assessment. By subtracting lean body mass from total body water, total body fat can be calculated [41,42]. Two models for calculating total body fat have been used; the model used depends on the divisions identified as descriptors of components making up lean body mass. Division of lean body mass into body cell mass, extracellular water, and extracellular solids distinguishes better than the other model (protein + total body water + bone ash) between the actively

Table 15.7 Mean Body Composition Values for Healthy Males and Females[a]

Age (yr)	n	Weight (kg)	Protein (kg)	TBW (L)	Bone Ash (kg)	BCM (kg)	ECW (L)	ECS (kg)	TBF$_1$ (kg)	TBF$_2$ (kg)
Males:										
20–29	12	80.1	13.1	44.9	3.53	35.7	19.8	6.77	18.6	16.8
30–39	12	73.7	11.2	42.3	3.44	32.5	19.0	6.61	16.8	15.6
40–49	12	84.6	12.4	47.2	3.59	35.1	19.8	6.88	21.4	23.0
50–59	12	82.0	12.1	45.0	3.43	33.3	19.3	6.58	21.5	20.7
60–69	10	78.5	11.8	42.1	3.33	30.7	19.9	6.39	21.3	21.5
70–79	10	80.5	11.1	40.4	3.17	27.6	20.3	6.08	25.8	26.5
Females:										
20–29	17	64.6	9.0	33.3	2.78	23.1	16.1	5.33	19.4	20.0
30–39	10	69.3	9.3	33.6	2.69	22.9	16.5	5.17	23.7	24.7
40–49	11	65.2	8.7	31.4	2.60	21.3	15.9	4.99	22.5	22.9
50–59	9	73.6	8.4	31.7	2.40	20.5	15.9	4.61	29.1	22.6
60–69	13	61.7	7.8	28.6	2.26	18.0	15.4	4.34	23.0	23.9
70–79	9	58.3	7.3	27.6	2.03	17.6	14.5	3.89	21.4	22.3

[a]Protein = 6.25 × total body nitrogen; BCM (body cell mass) = 0.235 × total body potassium (g); ICW (intracellular water) = TBW (total body water) – ECW (extracellular water); BA (bone ash) = total body calcium (TBCa)/0.34; ECS (extracellular solids) = TBCa (total body calcium)/0.177; TBF (total body fat); TBF1 = body weight – (protein + TBW + BA); TBF2 = body weight × (BCM + ECW + ECS).

Source: Modified from Cohn SH, Vaswani AN, Yasumura S et al. Improved models for determination of body fat by in vivo neutron activation. Am J Clin Nutr 1984;40:255. Am J. Clin. Nutr., American Society for Clinical Nutrition. Reprinted with permission.

metabolizing tissues and the relatively inactive tissues [41–43]. Tissues comprising body cell mass include muscle, viscera, brain, and reproductive system, while the extracellular solids are contributed primarily by the relatively inactive bone.

Table 15.7 [42] reports body composition information, derived from neutron activation, for males and females aged 20 to 79 years. Neutron activation analysis is noninvasive and provides reliable and reproducible data; however, the equipment is expensive and requires a skilled technologist, and subjects are exposed to significant amounts of radiation.

To review the methods available to assess body composition, lean body mass or fat-free mass may be assessed by methods such as neutron activation analysis, total body potassium, intracellular water (total body water minus extracellular water), total body electrical conductivity, and bioelectrical impedance analysis. Body fat may be assessed by anthropometry, densitometry, dual-photon absorptiometry, and infrared interactance. Adipose tissue may be measured using dual-photon absorptiometry, computerized tomography, magnetic resonance imaging, and ultrasound. Bone mineral density may be assessed using dual-photon absorptiometry. Table 15.8 [30,31,35–38] provides an overview of the methods discussed as well as a few additional methods.

Primary Influences on Body Composition

Body composition is influenced by a variety of factors, including age, gender, race, heredity, and stature [44]. Some effects of these factors can be observed in Table 15.7. *Stature* accounts for only a small portion of the variance in lean body mass, however, its influence is sufficient to require that stature be taken into account in comparing body composition data among individuals or groups of individuals. Taller children and adults have an advantage in lean body mass over their shorter peers [44].

Evidence suggests that bone density and fat-free weight are significantly affected by *heredity*. The effect of race is evidenced by the difference between North American whites and blacks. Blacks tend to have slightly larger lean body mass than whites, together with thicker and denser bones and hence a larger amount of total body calcium [43].

The influence of *gender* on body composition appears to exist from birth but becomes dramatically evident at puberty. The differences manifested at sexual maturation between males and females continue throughout life. Moreover, an age and gender effect occurs in women at menopause whereby the female undergoes a further change in body composition as

Table 15.8 Methods for Assessing Body Composition [27,28,32,33,37]

Method	Description of Method and Comments
Anthropometry	Skin fold thicknesses from a variety of locations, body weight, and limb circumferences can be used to calculate fat, fat-free mass, and muscle size. Measurements can be made in the field but require skilled technicians for accuracy. Skin folds can provide some information about regional subcutaneous fat as well as about total fat. Measurements may not be applicable to all population groups.
Densitometry	Measurement of total body density through determination of body volume by underwater weighing, helium displacement, or combination of water displacement by body and air displacement by head. Measurements can be used to determine body density, which in turn allows calculation of percentage of body fat and fat-free mass. Measurements are precise but must be conducted in laboratory; subject cooperation is necessary for underwater weighing. The method is not suitable for young children and the elderly.
Total body water	Measured by dilution with deuterium (D_2O), tritium (^3H_2O), or oxygen-18 (^{18}O). TBW is used as index of human body composition based on findings that water is not present in stored triglycerides but occupies an approximate average of 73.2% of the fat-free mass. A specified quantity of the isotope is ingested or injected; then, following an equilibration period, a sampling is made of the concentration of the tracer in a selected biological fluid. TBW is calculated from equation $C_1V_1 = C_2V_2$, where V_2 represents TBW volume, C_1 is the amount of tracer given, and C_2 is the final concentration of tracer in the selected biological fluid. The ECF can be estimated by a variety of methods. Subtracting ECF from TBW allows calculation of fat-free mass. This is a difficult procedure with limited precision, and the cost can be great, particularly when ^{18}O is used as the tracer.
Total body potassium	^{40}K, a naturally occurring isotope, is found in a known amount (0.012%) in intracellular water and is not present in stored triglycerides. These facts allow fat-free mass to be estimated by the external counting of gamma rays emitted by ^{40}K. Instrument for counting ^{40}K is quite expensive and must be properly calibrated for precision. Method is limited to laboratories.
Urinary creatinine excretion	Creatinine is the product resulting from the nonenzymatic hydrolysis of free creatine, which is liberated during the dephosphorylation of creatine phosphate. The preponderance of creatine phosphate is located in the skeletal muscle; therefore urinary creatinine excretion can be related to muscle mass. Drawbacks to this method include large individual variability of creatinine excretion due to the renal processing of creatinine and the effect of diet. The creatine pool does not seem to be under strict metabolic control and is to some degree independent of body composition. Another technical difficulty is control of accurately timed 24-hour urine collections.
3-methylhistidine excretion	3-methylhistidine has been suggested as a useful predictor of human body composition because this amino acid is located principally in the muscle and cannot be reused after its release from catabolized myofibrillar proteins (methylation of specific histidine residues occurs posttranslationally on protein). Some concern exists over the use of 3-methylhistidine as a marker of muscle protein because of the potential influence of nonskeletal muscle protein (skin and gastrointestinal [GI] tract proteins) turnover on its excretory rate. Additional problems with this method are the need for consumption of a relatively controlled meat-free diet and complete and accurate urine collections.
Electrical conductance (a) Total body electrical	Method is based on the change in electrical conductivity when subject is placed in an electromagnetic field. The change is proportional to the electrolyte content of the body, and because fat-free mass contains virtually all the water and conducting electrolytes of the body, conductivity is far greater in the fat-free mass than in the fat mass. From measurement, LBM can be calculated and fat estimated by difference. A primary drawback to this method is the expense of the instrument required; this measurement is a laboratory procedure limited primarily to large clinical facilities.
(b) Bioelectrical impedance analysis	This method is an adaptation of TOBEC; measurement of electrical conductivity is made on extremities rather than the whole body. Determinations of resistance and reactance are made, and the lowest resistance value for an individual is used to calculate conductance and to predict LBM. Equipment is portable and much less expensive than that required for TOBEC, yet precision is comparable.
Absorptiometry (a) Single-photon	Method is used in measurement of local or regional bone. The bone is scanned by a low-energy photon beam and the transmission monitored by a scintillation detector. Changes in transmission as the beam is moved across the bone are a function of bone mineral content (bone density) in that region. Disadvantages of this method are that the bone must be enclosed in a constant thickness of soft tissue, and measurements cannot be used to accurately predict total skeletal mass.

(continued)

Table 15.8 (continued)

Method	Description of Method and Comments
(b) Dual-photon	This method allows estimation of LBM as well as total bone mineral of the whole body. The body is scanned transversely in very small steps over its entire length by radiation from gadolinium-153 (^{153}Gd). This isotope emits two gamma rays of different energies; attenuation measurements at the two discrete photon energies allow quantification of bone mineral and soft tissue. The equipment required for dual-photon absorptiometry is quite expensive, complicated calibration is required, and data collected require complicated mathematical treatment. Use of this method is currently limited to research laboratories.
(c) Dual X-ray photon	Similar to dual-photon, this method involves scanning subjects at two different energy levels; however, X rays are used instead of a gradionuclide source. Radiation exposure to subjects is very low, and the procedure is relatively quick. This method appears to be the best choice for measuring bone mineral density.
Computerized tomography	Method determines regional body composition. An image is generated by computerized processing of X-ray data. Fat, lean tissue, and bone can be identified by their characteristic density-frequency distribution. Information about regional fat distribution can be obtained; it has been used to determine the ratio of intra-abdominal to subcutaneous fat in humans. The size of the liver, spleen, and kidneys can be determined by computerized tomography (CT). Both cost of the equipment and technical difficulties are high. Method is a laboratory procedure presently limited primarily to large medical centers.
Ultrasound	Approach uses instrument in which electrical energy is converted in probe to high-frequency ultrasonic energy. Subsequent transmission of these sound waves through various tissues can be used to calculate tissue thickness. Method is used frequently to determine the thickness of subcutaneous fat layer. Large laboratory instruments and smaller portable equipment are available. Although data suggest a reasonable validity of method, its general use has been limited because the appropriate signal frequency of the probe has not been well defined and the needed constant pressure by the probe to the scan site is difficult to achieve. Changes in pressure by probe application can prejudice ultrasonic determination of adipose tissue thickness.
Infrared interactance	Measurement of body fat is made at various sites on the extremities through use of short wavelengths of infrared light. The amount of fat can be calculated from the absorption spectra and used through a prediction equation to estimate TBF.
Magnetic resonance imaging	Approach is based on fact that atomic nuclei can behave like magnets. When external magnetic field is applied across a part of the body, each nucleus attempts to align with the external magnetic field. If these nuclei are simultaneously activated by a radio frequency wave, once the radio wave is turned off, the activated nuclei will emit the signal absorbed; this emitted signal is used to develop image by computer. Method has capability of generating images in response to intrinsic tissue variables and of representing such characteristics as level of hydration and fat content. This method appears to have much potential, but both the cost of equipment and technical difficulties are high.
Neutron activation analysis	This is the only technique currently available for measurement of multielemental composition of the human body. Low radiation doses produce isotopic atoms in tissues; the induced nuclides permit measurement of many elements, including nitrogen, calcium, phosphorus, magnesium, sodium, and chloride. Although precision of measurement is great, so are the technical difficulties and cost of equipment. Method of measuring body composition is limited to a very few laboratories in this country and abroad.

compared with the male. Accelerated loss of calcium from the bones, due in large part to the decline of estrogen production, causes a more rapid decrease in bone density of the elderly woman as compared with the elderly man.

The effect of maturation on body composition from birth to 10 years of age has been estimated through the use of "reference children" (Table 15.9) [45]. Table 15.10 [45] describes the composition of weight gain occurring during these years. Total body water de-creases during the first year of life primarily because of a rapid increase in fat. The decrease in total body water during the first year of life is followed by a slight increase from age 12 months to about 6 or 7 years (Table 15.9) [45]. Although total body water gradually declines for the next 3 to 4 years, it still exceeds 60% of total body weight at age 10 years. Accompanying the changes in total body water is a change in the ratio of extracellular fluid (ECF) to intracellular fluid (ICF). ECF exceeds ICF volume in the fetus during gestation, but

Table 15.9 Body Composition of Reference Children

Age	Weight (kg)		Fat (%)		Protein (%)		Water (%)		Ash (%)		Carbohydrate (%)	
	Females	Males	Females	Males	Females	Males	Females	Males	Females	Males	Females	Males
Birth	3.545	3.325	13.7	14.9	12.9	12.8	69.6	68.6	3.2	3.2	0.5	0.5
4 mo	7.060	6.300	24.7	25.2	11.9	11.9	60.1	59.6	2.8	2.8	0.4	0.4
6 mo	8.030	7.250	25.4	26.4	12.0	12.0	59.4	58.4	2.8	2.7	0.4	0.4
12 mo	10.15	9.18	22.5	23.7	12.9	12.9	61.2	60.1	2.9	2.8	0.5	0.5
18 mo	11.47	10.78	20.8	21.8	13.5	13.5	62.2	61.3	3.1	3.0	0.5	0.5
24 mo	12.59	11.91	19.5	20.4	14.0	13.9	62.9	62.2	3.2	3.0	0.5	0.5
3 yr	14.675	14.10	17.5	18.5	14.7	14.4	63.9	63.5	3.4	3.1	0.5	0.5
4 yr	16.69	15.96	15.9	17.3	15.3	14.8	64.8	64.3	3.5	3.1	0.5	0.5
5 yr	18.67	17.66	14.6	16.7	15.8	15.0	65.4	64.6	3.7	3.1	0.5	0.5
6 yr	20.69	19.52	13.5	16.4	16.2	15.2	66.0	64.7	3.8	3.2	0.5	0.5
7 yr	22.85	21.84	12.8	16.8	16.5	15.2	66.2	64.4	3.9	3.1	0.5	0.5
8 yr	25.30	24.84	13.0	17.4	16.6	15.2	65.8	63.8	4.0	3.1	0.5	0.5
9 yr	28.13	28.46	13.2	18.3	16.8	15.1	65.4	63.0	4.1	3.1	0.5	0.5
10 yr	31.44	32.55	13.7	19.4	16.8	15.0	64.8	62.0	4.1	3.1	0.5	0.5

Source: Adapted from Foman SJ, Haschke F, Ziegler EE, Nelson SE. Body composition of reference children from birth to age 10 years. Am J Clin Nutr 1982;35:1171.

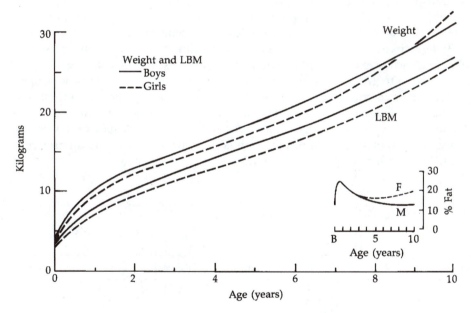

Figure 15.3 Graph of average body weight and lean body mass (LBM) for boys and girls from birth to 10 years. *Source:* Forbes GB. Human Body Composition. New York: Springer-Verlag, 1987:154.

during infancy and childhood the ECF-ICF ratio progressively falls until ICF occupies the majority position [44]. The change in the ECF-ICF ratio occurs with the growth and maturation of the lean body mass. Cell hypertrophy and bone development encroach on the space occupied by ECF, whereas protein accrual results in an increased incorporation of ICF.

The rapid increase in lean body mass begun in the late fetal period continues for a while after birth and then decelerates. The deceleration is followed by a slower but fairly steady growth during the childhood

years. The growth curve (body weight and lean body mass) for boys and girls is given in Figure 15.3 [44].

Despite the trend toward maturation of lean body mass during these childhood years, maturation appears to be incomplete until sometime during adolescence [44,46]. The percentage of water in lean body mass of reference children exceeds that considered average for the adult [46]. In addition, the value for minerals as a percentage of body weight is less than the adult average of 5.2% [16]. These differences in the components of lean body mass in children and adults decrease the density of lean body mass in children to <1.10 g/mL, the average density for adults. Studies of children, both male and female, at various stages of maturation have confirmed the increased hydration of lean tissue and its lower density among people not yet having reached adulthood [46]. As a result of these studies, Lohman [47] has suggested that Siri's [32] equation for determining fatness (p. 509) be replaced by the following equation when body fat composition of prepubescent children is being estimated:

$$\text{Percentage fat} = \frac{5.30}{\text{body density}} - 4.89 \times 100$$

Although some gender differences are evidenced in the body composition of prepubescent children (Tables 15.9, 15.10, and Fig. 15.3), these are not of great magnitude. The significant gender differences occur during adolescence and, once established, persist throughout adulthood. Lean body mass is the body component most significantly affected by gender. In both sexes serum testosterone levels rise during adolescence, but the rise is much greater in boys, with their testosterone values approaching ten times those found in girls. As a result of high testosterone production, boys increase their lean body mass by approximately 33 to 35 kg between the age of 10 and 20 years. The increment in girls, however, is only about half as much, approximately 16 to 18 kg [44,48,49]. The female achieves her maximum lean body mass by about 18 years, whereas the male continues accretion of lean body mass until about 20 years or so [41,46]. By age 15 years the male-to-female ratio of lean body mass is 1.23:1, and by 20 years the ratio has increased to 1.45:1. This male:female ratio for lean body mass (1.45:1) is well above that for body weight (1.25:1) and stature (1.08:1). The pronounced gender difference in lean body mass is the primary reason for the gender difference in nutrition requirements [44].

The sharp increase in lean body mass that occurs in boys during the adolescent growth spurt is accompanied by a decrease in the percentage of body fat. The average percentage of body fat in boys aged 6 to 8 years is about 13% to 15%; this percentage decreases to 10% to 12% for boys aged 14 to 16 years [23]. Although the adolescent female also increases her lean body mass during the growth spurt, a higher percentage of the weight gain is due to accretion of essential sex-specific fat. Girls aged 6 to 8 years have about 16% to 18% body fat and by age 14 to 16 years percentage of body fat ranges from 21% to 23% [23].

After 25 years of age, weight gain is usually due to body fat accretion [42]. Although the percentage of body weight due to fat remains lower in males than females, a slow sustained rise in fat in both sexes occurs throughout adulthood and later years [16,50]. The onset of menarche seems to affect the amount of fat accrued by the female through the early adulthood years [51]. Earlier sexual maturity in women appears to result in a greater amount of storage fat and a greater risk for obesity. In addition to increased body fat with age, a redistribution of body fat also occurs [15,52,53]. Subcutaneous fat decreases with age, but internal or visceral fat increases [37]. The significance of this change in distribution is thought to lie with its difference in functional behavior [54].

Table 15.10 Weight Increase and Its Components in Reference Children

Age	Weight Increase (g/d)		Fat (%)		Protein (%)		Minerals (%)	
	Males	Females	Males	Females	Males	Females	Males	Females
0–1 mo	29.3	26.0	20.4	21.4	12.5	12.5	0.9	0.8
3–4 mo	20.8	18.6	39.6	39.3	10.9	11.3	0.5	0.4
5–6 mo	15.2	15.0	27.3	32.4	13.2	12.6	0.4	0.4
9–12 mo	10.7	10.0	9.0	11.9	17.0	16.7	0.4	0.3
12–18 mo	7.2	8.7	7.2	10.7	18.4	17.0	0.3	0.3
18–24 mo	6.1	6.2	6.6	7.8	18.7	17.5	0.3	0.2
2–3 yr	5.7	6.0	5.8	7.9	19.1	17.6	0.2	0.2
3–4 yr	5.5	5.1	4.0	8.1	19.7	17.5	0.3	0.2
4–5 yr	5.4	4.7	3.2	11.3	19.9	17.0	0.3	0.2
5–6 yr	5.5	5.1	3.7	13.9	19.8	16.6	0.3	0.2
6–7 yr	5.9	6.4	6.3	19.6	19.5	15.6	0.3	0.2
7–8 yr	6.7	8.2	14.8	21.9	17.9	15.2	0.3	0.2
8–9 yr	7.8	9.9	15.2	24.5	17.9	14.8	0.4	0.3
9–10 yr	9.1	11.2	18.0	27.2	17.5	14.3	0.4	0.3

Source: Adapted from Foman SJ, Haschke F, Ziegler EE, Nelson SE. Body composition of reference children from birth to age 10 years. Am J Clin Nutr 1982;35:1174. Am. J. Clin. Nutr., American Society for Clinical Nutrition. Reprinted with permission.

Table 15.11 [55] illustrates the differences between the young and the elderly in the various body components. More marked changes were seen in aging females than in males. However, with aging in both males and females, a decrease in lean body mass (due primarily to a decrease in body cell mass) occurs [50,55]. Skeletal muscle loss may be due in part to decreased physical activity and alterations in protein metabolism that may be affected by diminished anabolic hormone concentrations [37]. Decreased lean body mass causes a decrease in total body water. The decrease in total body water is much greater in the female than the male. Further examination of the body water shows that extracellular fluid volume remains virtually unchanged. However, a redistribution occurs such that interstitial fluid decreases, but plasma volume increases [55]. Loss of bone mass and atrophy of organs also occurs with aging.

Components of Energy Expenditure

Whether body weight is being maintained, increased, or decreased depends primarily on the extent to which the energy requirements of the body (i.e., total energy expenditure) have been met by energy intake. Total energy expenditure [56–58] is composed primarily of

- basal metabolic rate (BMR), resting energy expenditure (REE), or resting metabolic rate (RMR);
- diet-induced thermogenesis (DIT)—also called *specific dynamic action* (SDA), *specific effect of food* (SEF), or *thermic effect of food* (TEF); and
- the effect of physical activity or exercise.

A fourth component, adaptive thermogenesis, is sometimes included. The average division of energy expenditure among the components, each of which are discussed in the next section, is given in Figure 15.4.

Basal Metabolic Rate, Resting Energy Expenditure, and Resting Metabolic Rate

Although the terms *basal* and *resting* are often used interchangeably, differences exists. The word *basal*, as in BMR, is more precisely defined than is *resting*, as in REE. Measurement of oxygen consumed and carbon dioxide produced, used for the calculation of energy expenditure necessary to support life (i.e., BMR), is made under closely controlled and standardized conditions. An individual's BMR is determined when he or she is in a postabsorptive state (i.e., no food intake for at least 12 hours), is lying down, and is completely relaxed, preferably very shortly after awakening from sleep in the morning. In addition, the temperature of the room in which the measurement occurs is made as comfortable as possible for the individual. Any factors that could influence the internal work of the individual are minimized as much as possible.

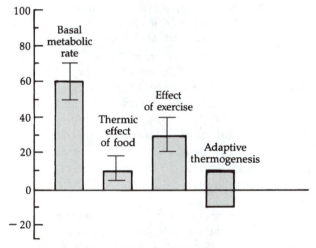

Figure 15.4 Components of energy expenditure and their approximate percentage contribution.

Table 15.11 Body Composition in Young and Elderly Males and Females

	Males		Females	
Parameters	Young (*n* = 10) (18–23 yr)	Elderly (*n* = 20) (61–89 yr)	Young (*n* = 10) (18–23 yr)	Elderly (*n* = 20) (60–89 yr)
Body weight (kg)	72.60 ± 10.34	61.60 ± 10.40	54.90 ± 2.71	56.10 ± 9.21
Body water (mL/kg)	586.1 ± 64.3	579.9 ± 54.4	531.0 ± 89.9	442.3 ± 57.3
Intracellular volume (mL/kg)	378.2 ± 42.3	375.7 ± 41.8	328.9 ± 64.7	248.3 ± 26.3
Lean body mass (g/kg)	800.7 ± 87.9	790.3 ± 78.1	725.4 ± 112.9	603.6 ± 77.8
Total body fat (g/kg)	199.3 ± 87.9	212.7 ± 66.2	274.6 ± 112.9	396.4 ± 77.8
Extracellular water volume (mL/kg)	207.9 ± 30.8	204.2 ± 37.0	192.1 ± 31.0	192.6 ± 19.4
Interstitial volume (mL/kg)	167.7 ± 20.8	157.2 ± 19.7	172.5 ± 41.5	148.9 ± 21.2
Plasma volume (mL/kg)	43.2 ± 7.6	47.0 ± 6.2	29.6 ± 5.2	45.1 ± 10.1

Source: Adapted from Fulop T, Worum I, Csongor J, Foris G, Leovey A. Body composition in elderly people. Gerontology 1985;31:6. Karger, Basel.

In contrast to BMR, REE is not usually measured under basal conditions [58]. REE is measured when the individual is at rest in a comfortable environment; however, he or she need not have fasted for 12 hours. Usually the fast for REE is about 2 to 4 hours. REE is usually slightly higher than BMR [59] because of its less stringent conditions of measurement; however, typically, BMR and REE differ by roughly 10% [58]. REE is thought to account for about 65% to 75% of daily total energy expenditure [59]. BMR accounts for about 50% to 70% of daily total energy expenditure [60].

Basal metabolism is a result of energy exchanges occurring in all cells of the body. The rate of oxygen consumption, however, is most closely related to the actively metabolizing cells or body's *lean body mass* or *body cell mass* [61,62]. The BMR per unit body weight is affected not only by the ratio of body cell mass to the less active body components, such as fat, but also by changes in proportions of tissues that make up the active cell mass. In aging, fat increases at the expense of body cell mass, and BMR decreases. With maturation, the proportion of supporting structures (i.e., bone and muscle) increases more rapidly than does total body weight. Bone and muscle, although components of body cell mass, have a much lower metabolic activity than organ tissues but much greater than adipose tissue. This difference in rate of weight accretion between the less active and more active components of cell mass means a decrease in the overall metabolic activity of cell mass and a concurrent decrease in BMR per unit body weight [61]. These changes that occur in maturation explain the lower REE of children as compared with very young infants.

A look at the metabolic activity among the different components of the cell mass in an adult male illustrates its variability in metabolic activity. Under normal circumstances, about 5% of total body weight can be attributed to weight of the brain, liver, heart, and kidney, whereas about 40% of body weight is due to muscle mass [61]. At the same time, the metabolic activity of the organ tissues accounts for about 60% of basal oxygen consumption, whereas the muscle mass accounts for only about 25% [61]. Thus, change in BMR can occur whenever the proportions of body tissues change in relation to each other [61]. Because of the many variations among individuals as well as the changes that can occur within individuals themselves, estimating energy expenditures may result in significant errors [56]. Accurate energy expenditures are much more likely to be obtained from actual measurements of these expenditures.

Diet-Induced Thermogenesis, Specific Dynamic Action, Specific Effect of Food, or Thermic Effect of Food

A second component of energy expenditure is the metabolic response to food, also referred to as *diet-induced thermogenesis, specific dynamic action,* or the *thermic* or *specific effect of food.* Diet-induced thermogenesis represents the body's processing of food. This includes the work associated with the digestion, absorption, transport, metabolism, and storage of energy from ingested food. The percentage increase in energy expenditure over BMR due to diet-induced thermogenesis has been estimated to range from about 5% to 15% [63,64]. The value most commonly used for the metabolic effect of food is 10% of the caloric value of a mixed diet consumed within 24 hours [56,60]. However, diet-induced thermogenesis is not a fixed percentage of calories ingested; diet-induced thermogenesis varies with the composition of the diet and the antecedent dietary practices of the individuals in whom measurements are being made [56,64]. A mixed diet increases diet-induced thermogenesis more positively when the ratio of carbohydrate to fat is high [59]. The rise in metabolism following food consumption appears to reach a maximum about 1 hour after eating and is generally but not always absent 4 hours postprandial (after eating) [58].

Physical Activity

The effect of physical activity, including fidgeting, or exercise is the most variable of the components; it is also the only component that is easily altered. Although on the average physical activity accounts for about 20% to 40% of the total energy expenditure, it can be considerably less in a truly sedentary person or much more in a very physically active person [56,60]. Factors impinging on energy expenditure during exercise beside the actual activity itself include the intensity, duration, and frequency with which the activity is performed, the body mass of the person, his or her efficiency at performing the activity, and also any extraneous movements that may accompany the activity.

Other Components of Energy Expenditure

An additional component of energy expenditure that is of some importance is adaptive thermogenesis, also called *nonshivering, facultative,* or *regulatory thermogenesis.* The term *adaptive thermogenesis* refers to the alteration in metabolism that occurs because of environmental, psychological, or other influences. Some of the possible factors affecting adaptive thermogenesis are unconscious or spontaneous movements or activities, hormone levels, responses to changes in nutrient and caloric intake, adaptations to changes in ambient temperature and to emotional stress, and activation or suppression of futile metabolic cycles.

Assessment of Energy Expenditure

Direct Calorimetry

Measurement of total energy expenditure can be determined by *direct calorimetry,* which measures the dissipation of heat from the body [65]. Heat dissipation is measured via an isothermal principle, a gradient-layer system, or a water-cooled garment [65,66]. A very simplified version of calorimetry for humans based on the isothermal principle is given in Figure 15.5. Total heat loss consists of sensible heat loss and heat of water vaporization. In the isothermal calorimeter (Fig. 15.5), sensible heat loss is determined by the difference in the water temperature and in the amount of water flowing in and out of the pipes situated within the walls of the chamber where the subject has been placed. Heat removed by vaporization of water is calculated from the moisture of air leaving the calorimeter and being absorbed in sulfuric acid. Although the con-

cept of direct calorimetry is relatively simple, direct measurement of body heat loss is expensive, cumbersome, and usually rather unpleasant for the subject or subjects involved [56].

Indirect Calorimetry

Basal metabolic rate is usually measured indirectly. *Indirect calorimetry* measures the consumption of oxygen and the expiration of carbon dioxide. In other words, indirect calorimetry measures the heat produced by oxidative processes; urinary nitrogen excretion should also be measured, because for every 1 g of nitrogen excreted, approximately 6 L of oxygen are consumed and 4.8 L carbon dioxide are produced [65,67].

The amount of heat being produced can be calculated from the ratio of the carbon dioxide expired to the oxygen inhaled. This ratio is known as the *respiratory quotient* (RQ).

Examination of respiratory quotients provides meaningful information with respect to overall substrate oxidation; however, no information about substrate oxidation in individual organs and tissues is gained [65]. An RQ equal to 1.0 suggests carbohydrate is being oxidized, because the amount of oxygen required in the combustion of glucose equals the amount of carbon dioxide produced, as shown here:

$$C_6H_{12}O_6 + 6O_2 \rightarrow 6CO_2 + 6H_2O;$$
$$RQ = 6CO_2/6O_2 = 1.0$$

The RQ for a fat is <1.0 because it is a much less oxidized fuel source. For example, a fat such as tristearin, as shown in the following equation, requires 163 mol of oxygen for the production of 114 mol of carbon dioxide:

$$2C_{57}H_{110}O_6 + 163O_2 \rightarrow 114CO_2 + 110H_2O;$$
$$RQ = 114CO_2/163O_2 = 0.70$$

Calculating the RQ for protein oxidation is more complicated because metabolic oxidation of amino acids requires removing the nitrogen and some oxygen and carbon as urea, a urinary excretory compound. Urea nitrogen represents a net loss of energy to the body; only the remaining carbon chain of the amino acid can be oxidized in the body. The following equation illustrates the oxidation of a small protein molecule into carbon dioxide, water, sulfur trioxide, and urea:

$$C_{72}H_{112}N_{18}O_{22}S + 77O_2 \rightarrow 63CO_2 + 38H_2O +$$
$$SO_3 + 9CO(NH_2)_2$$

The RQ of this small protein molecule is equal to 0.818.

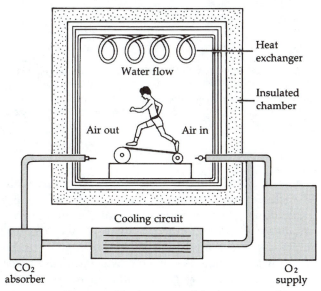

Figure 15.5 A simplified version of the human calorimeter used to measure direct body heat loss (i.e., energy expenditure).

The average figures of 1.0, 0.7, and 0.8 are accepted as the representative RQs for carbohydrate, fat, and protein, respectively. The RQ for an ordinary mixed diet consisting of the three energy-producing nutrients is usually considered to be about 0.85. An RQ of 0.82 represents the metabolism of a mixture of 40% carbohydrate and 60% fat [10]. RQs that are actually computed from gaseous exchange and that come closer to 1.0 or nearer to 0.7 would indicate that more carbohydrate or fat, respectively, was being used for fuel. In clinical practice, an RQ <0.8 suggests that a patient may be underfed, an RQ <0.7 suggests starvation or ingestion of a low-carbohydrate or high-alcohol diet, whereas an RQ >1.0 suggests lipogenesis is occurring [65].

Once the RQ has been computed from gaseous exchange, the calculation of heat production is rather simple. Table 15.12 gives the caloric value for 1 L of oxygen and for 1 L of carbon dioxide. When the amount of oxygen and/or carbon dioxide in the exchange has been determined, the total caloric value represented by the exchange may be calculated. It is also possible to determine the amount of carbohydrate and fat being oxidized in production of these calories.

For example, if under standard conditions for the determination of BMR a person consumed 15.7 L oxygen/hour and expired 12.0 L carbon dioxide, the RQ = 12.0/15.7, or 0.7643. From Table 15.1, the caloric equivalent for an RQ of 0.76 is 4.751 for 1 L oxygen or 6.253 for 1 L carbon dioxide. Based on the caloric equivalent for oxygen, calories produced per hour are 15.7 × 4.751 = 74.59. Based on caloric equivalent for carbon dioxide, calories produced per hour are 12.0 × 6.253 = 75.04. If we then use 75 kcal/hour as the caloric expenditure, then under basal conditions, the BMR for the day would approximate 1,800 kcal (75 kcal/hour × 24 h). At this RQ of 0.76, fat is supplying almost 81% of energy expended.

Because under ordinary circumstances the contribution of protein to energy metabolism is so small, the oxidation of protein is ignored in the determination of the so-called nonprotein RQ. If a truly accurate RQ is required, a correction, although minimal, can be made by measuring the amount of urinary nitrogen in a specified time period. As mentioned earlier, for every 1 g of nitrogen excreted, approximately 6 L of oxygen are consumed and 4.8 L carbon dioxide are produced. The amount of oxygen and carbon dioxide exchanged in the release of energy from protein can then be subtracted from the total amount of measured gaseous exchange.

Measurement of energy expended in various activities has also been made primarily through indirect calorimetry. The method for measuring gas exchange,

Table 15.12 Thermal Equivalent of O_2 and CO_2 for Nonprotein RQ

Nonprotein RQ	Caloric Value 1 L O_2	Caloric Value 1 L CO_2	Source of Calories	
			Carbohydrate (%)	Fat (%)
0.707	4.686	6.629	0	100
0.71	4.690	6.606	1.10	98.9
0.72	4.702	6.531	4.76	95.2
0.73	4.714	6.458	8.40	91.6
0.74	4.727	6.388	12.0	88.0
0.75	4.739	6.319	15.6	84.4
0.76	4.751	6.253	19.2	80.8
0.77	4.764	6.187	22.8	77.2
0.78	4.776	6.123	26.3	73.7
0.79	4.788	6.062	29.9	70.1
0.80	4.801	6.001	33.4	66.6
0.81	4.813	5.942	36.9	63.1
0.82	4.825	5.884	40.3	59.7
0.83	4.838	5.829	43.8	56.2
0.84	4.850	5.774	47.2	52.8
0.85	4.862	5.721	50.7	49.3
0.86	4.875	5.669	54.1	45.9
0.87	4.887	5.617	57.5	42.5
0.88	4.899	5.568	60.8	39.2
0.89	4.911	5.519	64.2	35.8
0.90	4.924	5.471	67.5	32.5
0.91	4.936	5.424	70.8	29.2
0.92	4.948	5.378	74.1	25.9
0.93	4.961	5.333	77.4	22.6
0.94	4.973	5.290	80.7	19.3
0.95	4.985	5.247	84.0	16.0
0.96	4.998	5.205	87.2	12.8
0.97	5.010	5.165	90.4	9.58
0.98	5.022	5.124	93.6	6.37
0.99	5.035	5.085	96.8	3.18
100	5.047	5.047	100	0

Source: Adapted from McArdle W, Katch F, Katch V. Exercise Physiology, 2nd ed. Philadelphia: Lea and Febiger 1986;127.

The original source is Weber. Die Bedeutung de Verschiedenen nährstoffe als Erzeuger der Muskelkraft. Pflgers Archiv zur Physiologie, 1901:83(1):557–71.

however, differs from that used for determining BMR. The subject performing the activity for which energy expenditure is being determined inhales ambient air, which has a constant composition of 20.93% oxygen, 0.03% carbon dioxide, and 78.04% nitrogen. Air exhaled by the subject is collected in a spirometer (used to measure respiratory gases) and is analyzed to determine how much less oxygen and more carbon dioxide it contains as compared with ambient air. The difference in the composition of the inhaled air and exhaled air reflects the energy release from the body [10]. A

Figure 15.6 Measurement of oxygen consumption by portable spirometer during (a) golf, (b) cycling, (c) sit-ups, and (d) calisthenics.

lightweight portable spirometer (Fig. 15.6) can be worn during performance of almost any sort of activity, and freedom of movement outside the laboratory is possible. In the laboratory the Douglas Bag is used routinely to collect expired air [10].

Tables are available that list for a wide variety of activities the kilocalories expended per kilogram body weight per minute or hour. Table 15.13, an example of such a table, indicates various activities grouped together according to their average level of energy expenditure.

Table 15.13 The Energy Cost above Basal Associated with Different Activities

Energy Level	Type of Activity	Energy (kcal/kg/minute [a]) Woman	Man
a	Sleep or lying still, relaxed[b]	0.000	0.000
b	Sitting or standing still (such as sewing, writing, eating)	0.001–0.007	0.003–0.012
c	Very light activity (driving a car, walking slowly on level ground)	0.009–0.016	0.014–0.022
d	Light exercise (sweeping, eating, walking normally, carrying books)	0.018–0.035	0.023–0.040
e	Moderate exercise (fast walking, dancing, bicycling, cleaning vigorously, moving furniture)	0.036–0.053	0.042–0.060
f	Heavy exercise (fast dancing, fast uphill walking, hitting tennis ball, swimming, gymnastics)	0.055	0.062

[a]Measured in kilocalories per kilogram per minute above basal energy. Where ranges are given, pick the midpoint within the range, unless you have reason to believe you are unusually relaxed or energetic when performing the activity. For example, for "sitting," a man should normally pick 0.007; if he is sitting very relaxed, 0.003; if very tense, 0.012.

[b]For purposes of this exercise, these are assumed to be at the basal level of activity.

Source: Modified from: Whitney E, Cataldo C. Understanding Normal and Clinical Nutrition. St. Paul, MN: West, 1983.)

Estimating Total Energy Expenditure

Estimating basal metabolic rate or resting energy expenditure rather than measuring it has been the practice among clinicians since about 1925. Many different methods for estimating energy needs have been used over the years. Estimations have been based on body surface area, body weight, and/or calculations from equations that take into account the person's gender, age, weight, and height. One estimate of BMR for all mammals, including humans, is based on body weight raised to the power of three-fourths, or 0.75 [61,68]. The equation

$$BMR \ (kcal/day) = 70 \times W^{0.75}$$

uses weight (W) measured in kilograms and raised to the power of 0.75 multiplied by 70 for estimating BMR. Because of the relatively narrow range in human body size, calculations from the preceding equation give an estimate that is reasonably close to the BMR value obtained from the formula of 1 kcal/kg/hour for men and 0.9 kcal/kg/hour for women. Estimation of the BMR of a 70-kg man using these two methods illustrates their comparable results:

1. BMR = $70 \times 70^{0.75} = 70 \times 24.2$
$= 1,694$ kcal/day
2. BMR = 1 kcal $\times$ 70 kg $\times$ 24 hours = 1,680 kcal/day

The equations probably most often used [63] to estimate BMR in the clinical setting are those derived by Harris and Benedict in 1919 [69], and only slightly modified. Using the Harris-Benedict equation, BMR (kilocalories/day) is predicted in separate equations for men and women based on W (weight in kilograms), H (height in centimeters), and A (age in years):

Men: BMR = $66.5 + (13.7 \times W)$
$+ (5.0 \times H) - (6.8 \times A)$
Women: BMR = $655.1 + (9.56 \times W)$
$+ (1.85 \times H) - (4.7 \times A)$

Another equation, by Mifflin and St. Jeor [70], predicts REE (kilocalories/day) for men and women as

Men: REE = $(10 \times W) + (6.25 \times H) - (5 \times A) + 5$
Women: REE $+ (10 + W) + (6.25 \times H)$
$- (5 \times A) - 161$

Using the two equations, a female who is 35 years old, weighs 125 lb (56.82 kg), and is 5 ft, 5 in tall (165.1 cm) would have a BMR of 1,339 kcal (Harris-Benedict equation) and an REE of 1,264 kcal (Mifflin–St. Jeor equation).

The various equations, along with the calculated energy values, are reevaluated regularly in scientific literature. Recent reevaluations have shown that predicted values for BMR are often higher than the actual expenditure and may not be applicable to all individuals, such as those who are obese [63,64,71]. Thus, dietitians must be alert to literature for recent findings and recognize the limitations and implications of the use of these various equations.

Once basal or resting energy needs have been determined, energy needs for diet-induced thermogenesis and for physical activity must be added to the basal or resting energy needs to estimate total daily energy needs. Energy for diet-induced thermogenesis typically represents an additional 10% (range, 5%–15%) of basal energy needs. Depending on the type, duration, intensity, and frequency of physical activity, energy needs for physical activity may vary from 20% to 70% or more of basal metabolism. A rough estimate of an individual's total energy expenditure for 1 day can be made by adding energy needs for basal metabolic rate, diet-induced thermogenesis, and physical activity.

Summary

Although differences in weight for height among individuals have been used traditionally for estimating recommended body weight, height-weight tables and formulas provide limited information about true body composition. Presently, estimations of body composition for the general public are derived from anthropometric measurements. With an accurate measure of height, weight, and skin folds from selected areas of the body, the percentage of body fat can be determined. Then by difference, lean body mass can be estimated. Certain skin fold or circumference measurements as well as use of more elaborate equipment such as DEXA can even provide some information about distribution of body fat, a factor that may be as important, or even more important, to health than the percentage of total body fat.

Despite the differences in the various body components that have been noted in individuals and population groups, it is probably undeniable that the component showing the greatest variability is the one over which we have the most control—total body fat! Although changes in energy balance produce weight changes, the extent of these changes varies from person to person. One of the greatest problems in making predictions of energy needs centers around estimation of energy expenditure. Energy expenditure has three defined components: basal metabolic rate, diet-induced thermogenesis, and the effect of exercise or physical activity, none of which is constant. To make energy metabolism even more complex, there is the possibility that adaptive thermogenesis can increase or decrease energy expenditure.

References Cited

1. Weigley ES. Average? Ideal? Desirable? A brief review of height-weight tables in the United States. J Am Diet Assoc 1984;84:417–23.

2. Build Study 1979. Chicago: Society of Actuaries and Association of Life Insurance Medical Directors of America, 1980.

3. Callaway CW. Weight standards: their clinical significance. Ann Intern Med 1984;100:296–8.

4. Schulz LO. Obese, overweight, desirable, ideal: where to draw the line in 1986? J Am Diet Assoc 1986;86:1702–4.

5. Marwick C. Obesity experts still say less weight is still best. JAMA 1993;269:2617–8.

6. Devine BJ. Gentamicin therapy. Drug Intell Clin Pharm 1974;8:650–5.

7. Robinson J, Lupklewica S, Palenik L, Lopez L, Ariet M. Determination of ideal body weight for drug dosage calculations. Am J Hosp Pharm 1983;40:1016–9.

8. Grant A, DeHoog S. Anthropometry. In: Nutritional Assessment and Support, 3rd ed. Seattle: Grant, 1985;11.

9. Giannini VS, Giudici RA, Nerrukk DL. Determination of ideal body weight. Am J Hosp Pharm 1984;41:883–7.

10. McArdle WD, Katch FI, Katch VL. Exercise Physiology, 2nd ed. Philadelphia: Lea and Febiger, 1986.

11. Katch FI, McArdle WD. Introduction to Nutrition, Exercise, and Health, 4th ed. Philadelphia: Lea and Febiger, 1993;223–58.

12. National Institutes of Health. Clinical Guidelines on the Identification, Evaluation, and Treatment of Overweight and Obesity in Adults. Bethesda, MD: National Institutes of Health, National Health, Lung, and Blood Institute, 1998.

13. Andres R. Mortality and obesity: The rationale for age-specific height-weight tables. In: Andres R, Bierman EL, Hazzard WR, eds. Principles of Geriatric Medicine. New York: McGraw-Hill, 1985:311–8.

14. Micozzi MS, Albanese D, Jones DY, Chumlea WC. Correlations of body mass indices with weight, stature, and body composition in men and women in NHANES I and II. Am J Clin Nutr 1986;44:725–31.

15. Behnke AR, Feen BG, Welham WC. The specific gravity of healthy men. JAMA 1942;118:495–8.

16. Friis-Hansen B. Body composition in growth. Pediatrics 1971;47:264–74.

17. Mitchell HH, Hamilton TS, Steggerda FR, Bean HW. The chemical composition of the adult human body and its bearing on the biochemistry of growth. J Biol Chem 1945;158:625–37.

18. Widdowson EM, McCance RA, Spray CM. The chemical composition of the human body. Clin Sci 1951;10:113–125.

19. Forbes RM, Cooper AR, Mitchell HH. The composition of the human body as determined by chemical analysis. J Biol Chem 1953;203:359–66.

20. Clarys JP, Martin AD, Drinkwater DT. Gross tissue weights in the human body by cadaver dissection. Hum Biol 1984;56:459–73.

21. Behnke AR, Wilmore JH. Evaluation and Regulation of Body Build and Composition. Englewood Cliffs, NJ: Prentice Hall, 1974.

22. McArdle WD, Katch FI, Katch VL. Exercise Physiology, 3rd ed. Philadelphia: Lea and Febiger, 1991.

23. Nieman DC. Fitness and Sports Medicine: An Introduction. Palo Alto: Bull, 1990.

24. Barr SI, McCargar LJ, Crawford SM. Practical use of body composition analysis in sport. Sports Med 1994;17:277-82.

25. Brozek J, Grande F, Anderson JT, Keys A. Densitometric analysis of body composition: revision of some quantitative assumptions. Ann NY Acad Sci 1963; 110:113-40.

26. Clarys JP, Martin AD, Drinkwater DT, Marfell-Jones MJ. The skinfold: myth and reality. J Sports Sci 1987; 5:3-33.

27. Sinning WE, Dolny DG, Little KD, et al. Validity of "generalized" equations for body composition analysis in male athletes. Med Sci Sports Exerc 1985;17: 124-30.

28. Harrison GG, Buskirk ER, Carter JEL, Johnston FE, et al. Skinfold thicknesses and measurement technique. In: Lohman TG, Roche AF, Martorell R. Anthropometric Standardization Reference Manual. Champaign, IL: Human Kinetics Publishers, 1988;55-80.

29. Martorell R, Mendoza F, Mueller WH, Pawson IG. Which side to measure: right or left. In: Lohman TG, Roche AF, Martorell R. Anthropometric Standardization Reference Manual. Champaign, IL: Human Kinetics Publishers, 1988;87-91.

30. Lukaski HC. Methods for the assessment of human body composition: traditional and new. Am J Clin Nutr 1987;46:537-56.

31. van der Kooy K, Seidell JC. Techniques for the measurement of visceral fat: a practical guide. Internl J Obesity 1993;17:187-96.

32. Siri WE. Gross composition of the body. In: Lawrence, JH, Tobias CA, eds. Advances in biological and medical physics. New York: Academic Press, 1956:239-80.

33. Pace N, Rathbun EN. Studies on body composition: III. The body water and chemically combined nitrogen content in relation to fat content. J Biol Chem 1945;158:685-91.

34. Jebb AS, Murgatroyd PR, Goldberg GR, Prentice AM, Coward WA. In vivo measurement of changes in body composition: Description of methods and their validation against 12-d continuous whole-body calorimetry. Am J Clin Nutr 1993;58:455-62.

35. Jensen MD. Research techniques for body composition assessment. J Am Diet Assoc 1992;92:454-60.

36. Genant H, Engelke K, Fuerst T, Gluer C, Grampp S, Harris S, Jergas M, Lang T, Lu Y, Majumdar S, Mathur A, Takada M. Noninvasive assessment of bone mineral and structure: state of the art. J Bone Min Res 1996; 11:707-30.

37. Heymsfield SB, Matthews D. Body composition: research and clinical advances. JPEN 1994;18:91-103.

38. Fuller MF, Fowler PA, McNeill G, Foster MA. Imaging techniques for the assessment of body composition. J Nutr 1994;124:1546S-50S.

39. Orphanidou C, McCargar L, Birmingham CL, Mathieson J, Goldner E. Accuracy of subcutaneous fat measurement: comparison of skinfold calipers, ultrasound, and computed tomography. J Am Diet Assoc 1994;94: 855-8.

40. Garrow JS. New approaches to body composition. Am J Clin Nutr 1992;35:1152-8.

41. Cohn SH, Vartsky D, Yasumura S, Vaswani AN, Ellis KJ. Indexes of body cell mass: nitrogen versus potassium. Am J Physiol 1983;244:E305-10.

42. Cohn SH, Vaswani AN, Yasumura S, Yuen K, Ellis KJ. Improved models for determination of body fat by in vivo neutron activation. Am J Clin Nutr 1984;40: 255-9.

43. Moore FD, Olesen KH, McMurrey JD, et al. The body cell mass and its supporting environment: body composition in health and disease. Philadelphia: Saunders, 1963.

44. Forbes GB. Human body composition-growth, aging, nutrition and activity. New York: Springer-Verlag, 1987.

45. Foman SJ, Haschke F, Ziegler EE, Nelson SE. Body composition of reference children from birth to age 10 years. Am J Clin Nutr 1982;35:1169-75.

46. Boileau RA, Lohman TG, Slaughter MH, et al. Hydration of the fat-free body in children during maturation. Hum Biol 1984;56:651-66.

47. Lohman TG. Research relating to assessment of skeletal status. In: Body Composition Assessment in Youth and Adults. Report of the Sixth Ross Conference on Medical Research. Columbus, OH: Ross Laboratories, 1985:38-41.

48. Tepperman J, Tepperman HM. Metabolic and Endocrine Physiology, 5th ed. Chicago: Year Book Medical, 1987.

49. Baker ER. Body weight and the initiation of puberty. Clin Obstet Gynecol 1985;28:573-9.

50. Forbes GB, Reina JC. Adult lean body mass declines with age: some longitudinal observations. Metabolism 1970;19:653-63.

51. Garn SM, LaVelle M, Rosenberg KR, Hawthorne VM. Maturational timing as a factor in female fatness and obesity. Am J Clin Nutr 1986;43:879-83.

52. Borken GA, Hults DE, Gerzof SG, et al. Comparison of body composition in middle-aged and elderly males using computed tomography. Am J Phys Anthropol 1985;66:289-95.

53. Komiya S. Aging, total body water and fat mass in males between ages 9 and 77 years. Ann Physiol Anthropol 1984;3:149-51.

54. Bray GA. General discussion of adipose tissue. In: Body-composition assessments of youth and adults. Report of the Sixth Ross Conference on Medical Research. Columbus, OH: Ross Laboratories, 1985:20-21.

55. Fulop T, Worum I, Csougor J, Foris G, Leovey A. Body composition in elderly people. Gerontology 1985;31:6-14.

56. Horton ES. Introduction: an overview of the assessment and regulation of energy balance in humans. Am J Clin Nutr 1983;38:972-7.

57. Devlin JT, Horton ES. Energy requirements. In: Brown ML, ed. Present Knowledge in Nutrition, 6th ed. Washington, DC: International Life Sciences Institute Nutrition Foundation, 1990:1-6.

58. Food and Nutrition Board, Commission on Life Sciences, National Research Council. Recommended Dietary Allowances, 10th ed. Washington, DC: National Academy Press, 1989;24-38.

59. Danforth E. Diet and obesity. Am J Clin Nutr 1985;41:1132-45.

60. Ravussin E, Bogardus C. A brief overview of human energy metabolism and its relationship to essential obesity. Am J Clin Nutr 1992;55:242S-5S.

61. Grande F. Body weight, composition and energy balance. In: Olson RE, Broquist HP, Chichester CO, Darby WJ, Kolbye AC Jr, Stalvey RM, eds. Nutrition Reviews' Present Knowledge in Nutrition, 5th ed. New York: The Nutrition Foundation, 1984:7-18.

62. Welle S, Nair K. Relationship of resting metabolic rate to body composition and protein turnover. Am J Physiol 1990;258:E990-8.

63. Daly JM, Heymsfield SB, Head CA, Harvey LP, Nixon DW, Katzeff H, Grossman GD. Human energy requirements: overestimation by widely used prediction equation. Am J Clin Nutr 1985;42:1170-4.

64. Owen OE, Kavle E, Owen RS, Polansky M, Caprio S, Mozzoli MA, Kendrick ZV, Bushman MC, Boden G. A reappraisal of caloric requirements in healthy women. Am J Clin Nutr 1986;44:1-19.

65. Jequier E, Acheson K, Schutz Y. Assessment of energy expenditure and fuel utilization in man. Ann Rev Nutr 1987;7:187-208.

66. Webb P. Human Calorimeters. New York: Praeger, 1985.

67. Westerterp KR. Food quotient, respiratory quotient, and energy balance. Am J Clin Nutr 1993; 57:759S-65S.

68. Garrow JS. Energy balance in man—an overview. Am J Clin Nutr 1987;45:1114-9.

69. Harris J, Benedict F. A Biometric Study of Basal Metabolism in Man. Publication 279. Washington, DC: Carnegie Institution, 1919.

70. Mifflin MD, St Jeor ST, Hill LA, Scott BJ, Daugherty SA, Koh YO. A new predictive equation for resting energy expenditure in healthy individuals. Am J Clin Nutr 1990;51:241-7.

71. Heshka S, Feld K, Yang M-U, Allison DB, Heymsfield SB. Resting energy expenditure in the obese: a cross-validation and comparison of prediction equations. J Am Diet Assoc 1993;93:1031-1036.

Suggested Reading

Heymsfield S, Wang Z, Baumgartner R, Ross R. Human body composition: advances in models and methods. Ann Rev Nutr 1997;17:527-58.

van der Kooy K, Seidell JC. Techniques for the measurement of visceral fat: a practical guide. Internl J Obesity 1993;17:187-96.

Heymsfield SB, Matthews D. Body composition: Research and clinical advances. JPEN 1994;18:91-103.

Jensen MD. Research techniques for body composition assessment. J Am Diet Assoc 1992;92:454-60.

Lukaski HC. Methods for the assessment of human body composition: traditional and new. Am J Clin Nutr 1987;46:537-556.

Wood PD. Impact of experimental manipulation of energy intake and expenditure on body composition. Crit Rev Food Sci Nutr 1993;33:369-73.

American Journal of Clinical Nutrition 1992; 55(2)—whole issue.

Osteoporosis: Diet and Diet-Related Factors

Skeletal tissue, although a relatively inactive component of the lean body tissue or fat-free mass, is not static. The fact that its composition changes with aging is quite evident in Table 15.7. Of particular importance to public health is the difference between bone ash in elderly women as compared with young women. From measurement by neutron activation analysis, bone ash for women between 70 and 79 years was, on the average, approximately 73% of that in women 20 to 29 years of age [1]. Many estimates place percentage of bone diminution throughout life even higher than the 27% shown in Table 15.7.

Excessive demineralization or diminution of bone, termed *osteoporosis* develops asymptomatically and often goes undiagnosed until the condition is far advanced. Diagnosis quite often may be made only as a result of a fracture or a complaint of severe, chronic, back pain. X rays are unable to pick up a decrease in bone mass until 30% to 50% of bone mineral has been lost [2].

Osteoporosis affects an estimated 20 to 25 million people in the United States, over the age of 45 years [3]. Osteoporosis results in about 1.5 million fractures per year. Over twice as many women as men are affected, with the highest prevalence found in postmenopausal women [2]. Between the age of 20 and 70 years, bone density decreases from 1.064 to 1.036 g/mL in males and from 1.034 to 1.013 g/mL in females [4].

Both types of bone in the body—the trabecular or honeycomb-like bone (lattice-type bone found in the vertebrae of the spine, pelvis, and ends of long bones) and the cortical or compact bone (relatively dense bone mass found in the shaft of long bones of the limbs and on the outer walls of all bones)—are demineralized with aging. Mineral loss from the trabecular bone begins earlier, at about 20 years of age, than cortical bone (Fig. 1). Cortical bone, in contrast, continues to increase in density until the middle of the second decade and probably into the third, at which time density plateaus. But, by about age 50 years, a persistent, gradual demineralization of cortical bone occurs. Generally trabecular bone turnover is greater than cortical bone turnover. Further, trabecular bone is more affected by osteoporosis than cortical bone in women. Thus, sites containing trabecular bone, the vertebral bodies (~95% trabecular bone), the femoral neck (~45% trabecular bone), and the radius (~5% trabecular bone) are principal sites affected with osteoporosis, especially in women (Fig. 2) [5].

Two types of osteoporosis have been described. Type I osteoporosis is characterized by demineralization of the vertebrae (especially the lumbar region) and the distal radius. Type I osteoporosis occurs mostly in postmenopausal women aged 51 to 65 years, or about 10 to 15 years after menopause [6]. Some cortical bone loss occurs, but, to a lesser extent than trabecular bone loss. Type I osteoporosis, also called *postmenopausal osteoporosis,* is linked with menopause and reduced estrogen production. The influence of estrogen on bone mineralization is discussed further in the next section.

Type II osteoporosis is characterized by demineralization of the vertebrae, hip, pelvis, humerus, and tibia. It occurs in both men and women over approximately 70 to 75 years of age [6]. In Type II osteoporosis, trabecular and cortical bone are affected due to age-induced decreased bone cell activity, especially osteoblast activity. In addition, decreased synthesis of calcitriol (caused by decreased 1-hydroxylase activity in the kidney) and decreased intestinal calcium absorption occur with aging and contribute to Type II osteoporosis. When these events are coupled with a low calcium intake and or high phosphorus intake, parathyroid hormone concentrations increase. High blood parathyroid hormone concentrations stimulate bone resorption and promote bone demineralization.

In the spine region, osteoporosis is associated with loss in height, vertebral pain, and the rounding of the shoulders or *kyphosis* (hunchback-type curvature of the spine). Osteoporosis increases the risk of fractures, primarily in the wrist, hip, and spine. Further, individuals may be more likely to experience possible tooth loss due to loss of aveolar (jaw)

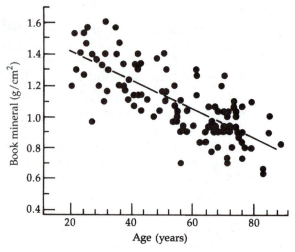

Figure 1 Regression of lumbar spine density in normal women as determined by dual-photon absorptiometry. *Source:* Reproduced, with permission, from the Annual Review of Nutrition, Volume 4, (c) 1984 by Annual Reviews, Inc.

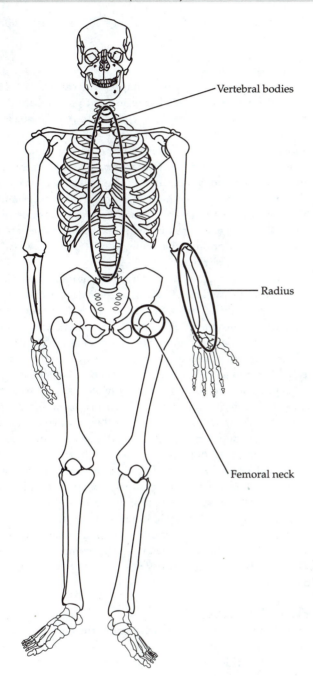

Figure 2 Major sites affected by osteoporosis.

bone. Chronic pain is common, along with decreased mobility which increases the risk of bed rest–induced pressure sores (also called *decubitus ulcers*) [3].

The high prevalence of osteoporosis among women makes this condition a public health problem. Although the effects of aging and genetic factors cannot be eliminated,

other factors may be modified so as to allow attainment or maintenance of a more desirable skeletal status. Some of the alterable factors that impinge on bone loss and that will be addressed include estrogen levels; physical activity; calcium, vitamin D, fluoride, and sodium intakes; as well as caffeine and alcohol consumption, phosphorus and protein intakes, and smoking.

Estrogen Levels

Estrogen has a positive effect on the bone mineralization, and its influence is evidenced at puberty. Although epiphyseal closure and cessation of longitudinal bone growth signal maturity of the skeletal tissue, remodeling of bone continues. The average density of the skeleton can increase perhaps as late as age 35 years. The rate of resorption (loss) of existing bone and deposition of new bone to replace that lost is affected by circulating estrogen levels. If levels of estrogen become low, bone responsiveness to parathyroid hormone appears to increase and contributes to bone resorption [3]. Low estrogen levels occur with menopause but also in many women athletes and women with the eating disorder anorexia nervosa. Young women, often teenagers, are thus at increased risk for bone loss unless estrogen levels can be maintained. How estrogen protects bone is not fully understood, but evidence supports a role for estrogen in bone mineralization via the osteoblasts. Estrogen receptors have been isolated on osteoblasts, suggesting that estrogen may affect the bone-forming (osteoblast) cells directly.

Because of the protective effect of estrogen on bone, many gynecologists believe that estrogen replacement should be recommended on an individual basis especially to women who are immediately postmenopausal or who are going through menopause [7]. Estrogen generally increases mean vertebral bone mass by about 5% and decreases vertebral fracture rates by 50% [8]. At what age the therapeutic benefit of estrogen replacement becomes less effective is unclear. Estrogen replacement therapy (ERT) appears to be helpful in preventing many cortical bone fractures associated with osteoporosis, but ERT protects little against the spinal "dowager's hump." Bone loss from the spine begins quite early (Fig. 1) and is likely due to factors other than (or at least in addition to) decreased estrogen levels [2].

Physical Activity

The negative influence of weightlessness on mineral balance has long been recognized. It follows, therefore, that weight bearing by the bone will influence positively mineral balance. This supposition has proved to be true. Weight-bearing exercises, including the carrying of one's own body weight, on a regular basis have a protective effect on bone and decrease the age-related demineralization of bone [9,10]. Physical activity in college-aged women has been positively correlated with rate of gain in spinal bone density [11].

Extremes in physical activity, however, when associated with amenorrhea (lack of menstruation) are counterproductive to maintaining bone mass. Low blood estrogen concentrations are found with amenorrhea. Bone mineral density (especially in the vertebrae) of amenorrheic women is typically much lower than that of control eumenorrheic women.

Calcium Nutriture

Although calcium intake remains important during life's later years, the crucial period for calcium intake is during the years when positive calcium balance is possible (when bone density can increase). Sufficient calcium should be available so that attainment of the full genetic expression of peak skeletal mass (at about age 25 years) is possible. Bone mineral density was significantly greater in prepubertal children receiving calcium citrate malate supplements (providing 1,000 mg calcium daily) than in twin siblings not receiving the supplements [12]. Lumbar spine and total body bone density also was found to be greater in adolescent girls receiving daily calcium supplements (500 mg of calcium as calcium citrate malate) than in those not receiving the supplement [13]. Calcium intake recommendations by a 1994 U.S. National Institutes of Health Consensus Panel suggest for children aged 1 to 10 years, 800 to 1,200 mg (1997 DRI 1,300 mg); for adolescents and young adults, 1,200 to 1,500 mg (RDA 1,200 mg); for women aged 25 to 50 years, 1,000 mg (DRI 1,200 mg); and for pregnant and lactating women, 1,400 mg (DRI 1,000 mg) [14]. Intakes of 1,500 mg/day for postmenopausal women who are not receiving estrogen therapy also have been recommended [14]. Unfortunately, average calcium intake of females throughout their lives is typically below recommendations.

Attainment of dense bones during the early years offers the best prevention of weakened bones in later years. Calcium supplements of 2,000 mg/day were unable to decrease trabecular bone loss and had only a minor positive effect on cortical bone in postmenopausal women [15]. In contrast to these findings, calcium supplements (1,000 mg/day for 2 years) reduced by 43% total body bone density loss in postmenopausal (at least 3 years) women [16]. The rate of bone mineral density loss in the legs was 35% and loss was eliminated in the trunk [16]. Calcium supplementation in association with either calcitonin or oral estrogen enhanced bone mass more than when calcitonin or estrogen were used alone [17]. Calcium supplements are

available in a variety of forms, see page 382 for a discussion of calcium supplements.

Vitamin D Intake

As you may remember from Chapter 10, calcitriol, 1,25 $(OH)_2D_3$, accelerates the absorption of calcium from the gastrointestinal tract. Specifically, the vitamin hormone interacts with receptors in the enterocyte and following transport to the nucleus, increases transcription of genes that code for calbindin. Calbindin functions as a calcium-binding protein and enhances calcium absorption. Calcitriol is also thought to induce changes in the brush border and basal lateral membranes to enhance calcium absorption directly. Calcitriol also may be involved in the PTH-mediated calcium reabsorption in the kidney and PTH-mediated calcium resorption by bone.

Elderly people appear to benefit from vitamin D supplementation. Poor vitamin D status in the elderly is common because of marginal intake of the vitamin, little exposure to sunlight, and decreased efficiency of transformation of the vitamin into its active metabolite, calcitriol (1,25-dihydroxy cholecalciferol), due to decreased renal 1-hydroxylase activity. Furthermore, the amount of vitamin D_3 produced in aging skin during exposure to the ultraviolet rays of the sun may be decreased to one-half of that produced in young skin [18]. Vitamin D supplements (400 IU) when coupled with calcium supplements (377 mg/day as calcium citrate malate) increased bone mineral density in the spine and decreased risk of vertebral fractures in postmenopausal women [19]. In a similar study, supplementation of calcium (1.2 g of calcium as tricalcium phosphate) and vitamin D (20 μg or 800 IU) decreased the risk of hip fractures and other nonvertebral fractures and increased proximal femur bone density in elderly women [20]. Recommendations for vitamin D intake in the elderly now suggest 500 to 800 IU daily [21].

Fluoride Intake

There has been much controversy about the benefits of fluoride therapy in the prevention and treatment of osteoporosis. Fluoride, usually administered as sodium fluoride (40–80 mg/day), stimulates bone formation (osteoblast activity). In addition, increases in mostly trabecular bone mass and to some extent cortical bone and decreases in fracture rates have been reported in postmenopausal women receiving fluoride along with calcium [5,22–24]. However, other research has suggested that highly fluoridated water does not protect against bone loss [25]. Further, supplements of 75 mg of fluoride together with 1,500 mg of calcium failed to reduce the risk of vertebral fractures and increased the risk of nonvertebral fractures in post-menopausal women [26]. Formulation, dose, delivery mode and duration of fluoride therapy may account for observed differences between studies [5,22,23,26]. Better-designed research studies are needed before recommendations for fluoride therapy may be issued [5,7,23].

Sodium Intake

Sodium also has been shown to be detrimental to body calcium. Sodium is excreted in the urine with calcium. A sodium load of 100 mmol (2.3 g) per day increased urinary calcium excretion by 1 mmol (40 mg) per day [27,28]. In females aged 8 to 13 years, urinary sodium excretion was found to be a major determinant of urinary calcium excretion [29]. Urinary sodium excretion was negatively correlated with changes in bone density (bone loss) in the hip region of postmenopausal women [30]. In fact, cutting in half sodium intake or doubling calcium intake were predicted to result in an equivalent reduction of bone loss [30]. Clearly, efforts to prevent osteoporosis should consider reduction in dietary sodium intake.

Other Nutrients and Factors

Smoking as well as several other dietary or diet-related substances have been implicated as impinging on bone mineralization. These substances include phosphorus, protein, alcohol, and caffeine. Each will be briefly reviewed as they relate to osteoporosis.

Smoking

Smoking is associated with lower bone density, earlier menopause in women, and increased postmenopausal bone loss [31,32]. Smoking decreases circulating estrogen concentrations and thereby contributes to bone loss [31]. Smoking has been shown to be a significant predictor of bone loss [33].

Phosphorus

In contrast to earlier studies suggesting that a high intake of phosphorus decreased bone mineralization, recent studies have shown a beneficial effect of dietary phosphorus on calcium use. Phosphorus, by stimulating parathyroid hormone secretion, increases indirectly the reabsorption of calcium by the renal tubules so that less calcium is lost in the urine. Thus, although phosphorus causes loss of calcium by increasing calcium secretion into the gastrointestinal tract, its overriding effect is to conserve calcium. A study of 215 women aged 18 to 31 years reported that increasing phosphorus intake positively affected total body bone mineral density and content at a calcium intake <600 mg/day. However, at a higher calcium in-

take, up to 1,400 mg/day, a phosphorus intake of 1,800 mg/day was detrimental to total body bone mineral density and content [34]. Other studies show prolonged ingestion of diets high in phosphorus and low in calcium result in a mild secondary hyperparathyroidism [35,36]. High parathyroid hormone concentrations stimulate bone resorption with possible long-term detrimental effects on bone mineral content [35–37]. However, studies demonstrating such effects on bone in humans are not available [38].

Protein

Proteins and a variety of amino acids (especially sulfur-containing amino acids methionine and cysteine, which may bind calcium) have been shown to increase excretion of calcium in the urine. Dietary protein directly influences calcium; doubling of protein intake without changing intake of other nutrients results in about a 50% increase in urinary calcium [39–41]. A significant positive correlation between protein intake and urinary calcium excretion has been found in adults aged 20 to 79 years [42]. These proteins and amino acids, however, are usually combined in natural foods with substances that counteract their effect on calcium excretion. Because many protein-containing foods also contain phosphorus, ingestion of foods containing both nutrients tends to minimize negative effects on calcium balance [3]. In a group of college-aged women, rate of gain of spinal bone density positively correlated with calcium:protein intake ratio [11].

Alcohol

Alcohol-induced bone damage is thought to be multifactorial [43]. Alcohol consumption has been significantly associated with increased rates of bone loss in men [33]. People consuming excessive alcohol generally have lower bone mass and reduced osteoblast activity and are at increased risk in a dose-response relationship of hip and forearm fractures [43,44]. Factors associated with excessive alcohol intake affecting bone and its loss include insufficient intake of nutrients (especially calcium, protein, and/or vitamin D) coupled with poor absorption of nutrients [43]. Elevated PTH levels also accompany alcohol consumption [43].

Caffeine

Caffeine has also been considered by some as a detriment to calcium balance [45,46]. Caffeine is thought to reduce the renal reabsorption of calcium, which leads to increased urinary calcium losses [46]. It has been estimated that one cup of coffee can contain enough caffeine to cause an extra loss of 6 mg calcium in the urine per day [445]. Caffeine (300–400 mg) increased urinary calcium (0.25 mmol or 10 mg/day) and increased secretion of calcium into the gut [6,46]. Caffeine intake has also been positively associated with risk of hip fracture in middle-aged women [44].

Maintenance of desirable skeletal status in women appears to be multifactorial. Nevertheless, achievement of potential peak bone mass during early adulthood is probably the most important factor in its later maintenance. Achievable bone mass appears to be programmed by genetic factors and mechanical loading (weight-bearing exercises) of the skeleton, but the potential is influenced by the endocrine and nutritional environment [9,47]. Most important in this environment are the hormone estrogen and the nutrient calcium. One's genetic makeup cannot be changed, nor can the physiological changes accompanying aging be reversed. The individual usually does have the option, however, of choosing a lifestyle in which weight-bearing exercise and good nutrition are practiced regularly. Exercise and an adequate calcium intake throughout life appear to foster the maintenance of a healthy skeleton in the later years.

References Cited

1. Cohn S, Vaswani A, Yasumura S, Yuen K, Ellis K. Improved models for determination of body fat by in vivo neutron activation. Am J Clin Nutr 1984;40:255–9.

2. Avioli L. Calcium and osteoporosis. Ann Rev Nutr 1984; 4:471–91.

3. Wardlaw G. Putting osteoporosis in perspective. J Am Diet Assoc 1993;93:1000–6.

4. Lohman T. Research relating to assessment of skeletal status. In: Body Composition Assessments in Youth and Adults. Report of the Sixth Ross Conference on Medical Research. Columbus, OH: Ross Laboratories 1985;38–41.

5. Phipps K. Fluoride and bone health J Publ Health Dent 1995;55:53–56.

6. Harward M. Nutritive therapies for osteoporosis: the role of calcium. Med Clin N Am 1993;77:889–98.

7. Khosla S, Riggs L. Treatment options for osteoporosis. Mayo Clin Proc 1995;70:978–82.

8. Lufkin E, Wahner H, O'Fallon W, Hodgson S, Kotowicz M, Lane A. Treatment of postmenopausal osteoporosis with transdermal estrogen. Ann Intern Med 1992;117:1–9.

9. Heaney R. Bone mass, nutrition, and other lifestyle factors. Nutr Rev 1996;54:S3–S10.

10. Vuori I. Peak bone mass and physical activity: a short review. Nutr Rev 1996;54:S11–S14.

11. Recker R, Davies K Hinders S, Heaney R, Stegman M, Kimmel D. Bone gain in young adult women. JAMA 1992;268: 2403–8.

12. Johnston C, Miller J, Slemenda C, Reister T, Hui S, Christian J, Peacock M. Calcium supplementation and increases in bone mineral density in children. N Engl J Med 1992;327: 82–87.

13. Lloyd T, Andon M, Rollings N, Martel J, Landis J, Demers L, Eggli D, Kleselhorst K, Kulin H. Calcium supplementation and bone mineral density in adolescent girls. JAMA 1993; 270:841–4.

14. Rowe P. New US recommendations on calcium intake. Lancet 1994;343:1559–60.

15. Riis B, Thomsen K, Christiansen C. Does calcium supplementation prevent postmenopausal bone loss? N Engl J Med 1987;316:173–7.

16. Reid I, Ames R, Evans M, Gamble G, Sharpe S. Effect of calcium supplementation on bone loss in postmenopausal women. N Engl J Med 1993;328:460–4.

17. Nieves J, Komar L, Cosman F, Lindsay R. Calcium potentiates the effect of estrogen and calcitonin on bone mass: review and analysis. Am J Clin Nutr 1998;67:18–24.

18. MacLauglin J, Holick M. Aging decreases the capacity of human skin to produce vitamin D_3. J Clin Invest 1985;76: 1536–8.

19. Dawson-Hughes B, Dallah G, Krall E, Harris S, Sokoll L, Falconer G. Effect of vitamin D supplementation on wintertime and overall bone loss in healthy postmenopausal women. Ann Intern Med 1991;115:505–12.

20. Chapuy M, Arlot M, Duboeuf F, Brun J, Crouzet B, Arnaud S, Delmas P, Meunier P. Vitamin D_3 and calcium to prevent hip fractures in elderly women. N Engl J Med 1992;327: 1637–42.

21. Heaney R, Burckhardt P. Nutrition and bone health: concluding remarks. In Burckhardt P, Heaney R (eds.). Nutritional Aspects of Osteoporosis. Proceedings of 2nd (1994) International Symposium on Osteoporosis. Rome, Italy: Serono Symposia Publication. 1995:419–24.

22. Pak C, Sakhaee K, Piziak V, Peterson R, Breslau N, Boyd P, Poindexter J, Herzog J, Sakhaee A, Haynes S, Huet B, Reisch J. Slow release sodium fluoride in the management of postmenopausal osteoporosis: a randomized controlled trial. Ann Intern Med 1994;120:625–32.

23. Kleerekoper M, Mendlovic D. Sodium fluoride therapy of postmenopausal osteoporosis. Endocrin Rev 1993;14:312–23.

24. Eisinger J, Clairet D. Effects of silicon, fluoride, etidronate and magnesium on bone mineral density: a retrospective study. Magnesium Res 1993;6:247–9.

25. Sowers M, Wallace R, Lemke J. The relationship of bone mass and fracture history to fluoride and calcium intake: a study of three communities. Am J Clin Nutr 1986;44:889–98.

26. Riggs B, Hodgson S, O'Fallon W, Chao E, Wahner H, Muhs J, Cedel S, Melton L. Effect of fluoride treatment on the fracture rate in postmenopausal women with osteoporosis. N Engl J Med 1990;322:802–9.

27. Massey L. Dietary factors influencing calcium and bone metabolism: introduction. J Nutr 1993;123:1609–10.

28. Nordin B, Need A, Morris H, Horowitz M. The nature and significance of the relationship between urinary sodium and urinary calcium in women. J Nutr 1993;123:1615–22.

∾ **PERSPECTIVE 1** (continued)

29. Matkovic V, Illich J, Andon M, Hsieh L, Tzagournis M, Lagger B, Goel P. Urinary calcium, sodium, and bone mass of young females. Am J Clin Nutr 1995;62:417–25.

30. Devine A, Criddle R, Dick I, Kerr D, Prince R. A longitudinal study of the effect of sodium and calcium intakes on regional bone density in postmenopausal women. Am J Clin Nutr 1995;62:740–5.

31. Jensen J, Christiansen C, Rodbro P. Cigarette smoking, serum estrogens, and bone loss during hormone-replacement therapy early after menopause. N Engl J Med 1985;313:973–7.

32. Krall E, Dawson-Hughes B. Smoking and bone loss among postmenopausal women. J Bone Min Res 1991;4:331–8.

33. Slemenda C, Christian J, Read T, Reister T, Williams C, Johnston C. Long-term bone loss in men: effects of genetic and environmental factors. Ann Intern Med 1992;117:286–91.

34. Calvo M, Kumar R, Heath H. Elevated secretion and action of parathyroid hormone in young adults ingesting high phosphorus, low calcium diets assembled for ordinary foods. J Clin Endocrinol Metab 1988;66:823–9.

35. Calvo M, Kumar R, Heath H. Persistently elevated parathyroid hormone secretion and action in young women after four weeks of ingesting high phosphorus, low calcium diets. J Clin Endocrinol Metab 1990;70:1334–40.

36. Anderson J. The role of nutrition in the functioning of skeletal tissue. Nutr Rev 1992;50:388–94.

37. Calvo M. Dietary phosphorus, calcium metabolism and bone. J Nutr 1993;123:1627–33.

38. Heaney R. Protein intake and the calcium economy. J Am Diet Assoc 1993;93:1259–60.

39. Teegarden D, Lyle R, McCabe G, McCabe L, Proulx W, Michon K, Knight A, Johnston C, Weaver C. Dietary calcium, protein, and phosphorus are related to bone mineral density and content in young women. Am J Clin Nutr 1998;68:749–54.

40. Massey L. Dietary factors influencing calcium and bone metabolism: introduction. J Nutr 1993;123:1609–10.

41. Whiting S, Anderson D, Weeks S. Calciuric effects of protein and potassium bicarbonate but not sodium chloride or phosphate can be detected acutely in women and men. Am J Clin Nutr 1997;65:1465–7.

42. Itoh R, Nishiyama N, Suyama Y. Dietary protein intake and urinary excretion of calcium: a cross-sectional study in a healthy Japanese population. Am J Clin Nutr 1998;67:438–44.

43. Laitinen K, Valimaki M. Alcohol and bone. Calcif Tissue Int 1991;49(suppl):S70–3.

44. Hernandez-Avila M, Colditz G, Stampfer M, Rosner B. Caffeine, moderate alcohol intake, and risk of fractures of the hip and forearm in middle-aged women. Am J Clin Nutr 1991;54:157–63.

45. Heaney R, Recker R. Effects of nitrogen, phosphorus and caffeine on calcium balance in women. J Lab Clin Med 1982;99:46–55.

46. Massey LK, Whiting SJ. Caffeine, urinary calcium, calcium metabolism and bone. J Nutr 1993;123:1611–4.

47. Seeman E, Hopper JL, Bach LA, et al. Reduced bone mass in daughters of women with osteoporosis. N Engl J Med 1989;320:554–8.

⮌ **PERSPECTIVE 2**

Eating Disorders

Despite its increasing prevalence in our society, obesity is still considered unacceptable. Few things can create such a sensation in the media as a new weight reduction diet guaranteed to remove that unwanted fat. The authors of the sensational new diet are interviewed on television talk shows, newspapers give publicity to the new diet (and their authors), and the book promoting the "new and revolutionary" diet joins its companions on the shelves of all bookstores. The fact that the new diet book has so many companions in the bookstores attests to the fact that none of these "new and revolutionary" diets is successful in reducing weight and keeping it off. Nevertheless, following some sort of weight reduction diet appears to be a way of life among many Americans, particularly women.

The desire by girls and women to be thin has foundation: the ideal female body image is dictated to a large extent by fashion magazines, *Playboy* centerfolds, and beauty pageant contestants. Children as young as 9 years of age have been found curtailing their food intake to avoid becoming fat [1]. In addition to wanting to conform for aesthetic reasons to the ideal body image, there are advantages of being thin, or at least of normal weight, for educational and professional reasons [2]. Obese girls are less likely to be admitted to college and/or programs of professional training. Furthermore, obese women (as compared with thin women or those of normal weight) have less likelihood of being hired for desirable jobs and/or of being promoted.

Concerns with excessive restriction of body weight include possible stunted growth and delayed sexual maturity of children but also possible development of eating disorders such as anorexia nervosa and bulimia. Eating disorders are particularly prevalent among young females; in fact, 90% to 95% of the people affected by anorexia nervosa and bulimia are young, white females from middle- and upper-middle-class families [3]. Some male athletes also appear to be affected. The prevalence of eating disorders among U.S. adolescent girls and young women is estimated between 1% and 4%, using the criteria of the American Psychiatric Association [4,5]. The prevalence of eating disorders or disordered eating is likely much higher if the criteria of the American Psychiatric Association were not used.

Anorexia nervosa, described over 100 years ago as "loss of appetite due to a morbid mental state," is actually misnamed because its victims do not have a loss of appetite unless they have reached a moribund state. Anorectics remain thin through self-inflicted starvation and excessive amounts of strenuous exercise. Diagnostic criteria for anorexia nervosa (Table 1) include refusal to maintain at least 85% of normal weight for height, denial of a low current body weight, fear of gaining weight, and amenorrhea (absence of at least three consecutive menstrual cycles) [6]. In addition, preoccupation with food and abnormal food consumption patterns are typical of anorexia nervosa.

Eating patterns of individuals with anorexia nervosa mostly fall into one of two categories [4,7]: the restricting type or the binge eating-purging type. Anorectics with restricting-type eating will eat to a limited extent without regularly inducing vomiting or misusing laxatives or diuretics. Individuals with binge eating- and purging-type anorexia nervosa alternate between restricting food intake and bouts of binge eating or purge behavior with laxative and/or diuretic misuse or self-induced vomiting [4,8].

The cause(s) of anorexia nervosa is/are unknown, but it seems to be a multifactorial disease. At least two sets of is-

---

Table 1   Diagnostic Criteria for 307.1 Anorexia Nervosa

A. Refusal to maintain body weight at or above a minimally normal weight for age and height (e.g., weight loss leading to maintenance of body weight less than 85% of that expected; or failure to make expected weight gain during period of growth, leading to body weight less than 85% of that expected).

B. Intense fear of gaining weight or becoming fat, even though underweight.

C. Disturbance in the way in which one's body weight or shape is experienced, undue influence of body weight or shape on self-evaluation, or denial of the seriousness of the current low body weight.

D. In postmenarcheal females, amenorrhea, i.e., the absence of at least three consecutive menstrual cycles. (A woman is considered to have amenorrhea if her periods occur only following hormone, e.g., estrogen, administration.)

Specify type:

Restricting Type: during the current episode of Anorexia Nervosa, the person has not regularly engaged in binge-eating or purging behavior (i.e., self-induced vomiting or the misuse of laxatives, diuretics, or enemas)

Binge-Eating/Purging Type: during the current episode of Anorexia Nervosa, the person has regularly engaged in binge-eating or purging behavior (i.e., self-induced vomiting or the misuse of laxatives, diuretics, or enemas)

Source: Diagnostic and Statistical Manual of Mental Disorders, 4th ed. Washington, DC: American Psychiatric Association, 1994:544–5.

sues and behaviors are entangled. Issues include those concerning food and body weight, and those involving relationships with oneself and with others [9]. Conflict regarding maturation and problems with separation, sexuality, self-esteem, and compulsivity often are associated with the development of anorexia nervosa [10].

The initial weight loss of the anorectic may not always be a result of a deliberate decision to diet; weight loss may occur unintentionally, for example, as the result of the flu or a gastrointestinal disorder [4]. However, following the initial weight loss, whatever its cause, additional diet restriction often coupled with excessive exercise to induce further weight loss is deliberate. Weight loss or control of body weight becomes the overriding goal in life, especially during stressful periods [4,5]. Because the anorectic has such a disturbed body image and such an intense fear of becoming fat, she may continue starving herself to emaciation and even death, should intervention be delayed too long.

The effects of anorexia nervosa on the body are similar to the effects of starvation. Growth and development slow. Adipose tissue, lean body mass, and bone mass are lost. Organ mass may be lost, and/or organ function may become impaired. Hormone and nutrient levels in the blood become altered. Skin typically becomes dry, hair loss occurs, and body temperature drops. Table 2 describes some additional potential physical consequences of anorexia nervosa and of bulimia nervosa.

Bulimia nervosa, another eating disorder, is a condition in which there is recurring binge eating coupled with self-induced vomiting and misuse of laxatives, diuretics, or other medications to prevent weight gain. Binge eating is characterized by a sense of lack of control over eating during the binge episode [6]. A *binge* is defined as eating an amount of food that is larger than most people would eat during a similar time period and under similar circumstances [6]. Bulimia denotes a ravenous appetite (or "ox hunger") associated with powerlessness to control eating [4]. The incidence of bulimia nervosa is thought to exceed that of anorexia nervosa [7]; however, depending on diagnostic criteria used in the prevalence studies, the prevalence estimates of bulimia vary widely [11]. Criteria [6] for the diagnosis of bulimia nervosa are shown in Table 3.

Bulimia occurs primarily in young women, especially college-aged women who are normal weight or slightly overweight. The typical bulimic, rather than being overly concerned with losing weight and being very thin like the individual with anorexia nervosa, seeks to be able to eat without gaining weight [4]. Bulimia starts with dieting

Table 2   Some Potential Physical Consequences of Anorexia Nervosa and Bulimia

	Manifestation	
System Affected	Anorexia Nervosa	Bulimia
Endocrine/metabolic	Amenorrhea Osteoporosis ↓ Norepinephrine secretion ↑ Growth hormone Abnormal temperature regulation	Menstrual irregularities
Cardiovascular	Bradycardia Hypotension Arrhythmias	Cardiac failure (vomiting induced—often via Ipecac)
Renal	↑ Blood urea nitrogen ↓ Glomerular filtration rate Edema	Hypokalemia (diuretic induced)
Gastrointestinal	↓ Gastric emptying Constipation	Acute gastric dilation Parotid enlargement Tooth enamel erosion Esophagitis Esophageal tears
Hematologic	Anemia Leukopenia	Hypokalemia (vomiting and laxative induced)
Pulmonary		Aspiration pneumonia

Source: Herzog D, Copeland P. Eating disorders. N Engl J Med 1985;313:295. Reprinted by permission.

◌ **PERSPECTIVE 2** (continued)

Table 3   Diagnostic Criteria for 307.51 Bulimia Nervosa

A. Recurrent episodes of binge eating. An episode of binge eating is characterized by both of the following:
  (1) eating, in a discrete period of time (e.g., within any 2-hour period), an amount of food that is definitely larger than most people would eat during a similar period of time and under similar circumstances
  (2) a sense of lack of control over eating during the episode (e.g., a feeling that one cannot stop eating or control what or how much one is eating)
B. Recurrent inappropriate compensatory behavior in order to prevent weight gain, such as self-induced vomiting; misuse of laxatives, diuretics, enemas, or other medications; fasting; or excessive exercise.
C. The binge eating and inappropriate compensatory behaviors both occur, on average, at least twice a week for 3 months.
D. Self-evaluation is unduly influenced by body shape and weight.
E. The disturbance does not occur exclusively during episodes of Anorexia Nervosa.
Specify type:
Purging Type: during the current episode of Bulimia Nervosa, the person has regularly engaged in self-induced vomiting or the misuse of laxatives, diuretics, or enemas.
Nonpurging Type: during the current episode of Bulimia Nervosa, the person has used other inappropriate compensatory behaviors, such as fasting or excessive exercise, but has not regularly engaged in self-induced vomiting or the misuse of laxatives, diuretics, or enemas.

Source: Diagnostic and Statistical Manual of Mental Disorders, 4th ed. Washington, DC: American Psychiatric Association, 1994:549–50.

Table 4   307.50 Eating Disorder Not Otherwise Specified

The Eating Disorder Not Otherwise Specified category is for disorders of eating that do not meet the criteria for any specific Eating Disorder. Examples include
1. For females, all of the criteria for Anorexia Nervosa are met except that the individual has regular menses.
2. All of the criteria for Anorexia Nervosa are met except that, despite significant weight loss, the individual's current weight is in the normal range.
3. All of the criteria for Bulimia Nervosa are met except that the binge eating and inappropriate compensatory mechanisms occur at a frequency of less than twice a week or for a duration of less than 3 months.
4. The regular use of inappropriate compensatory behavior by an individual of normal body weight after eating small amounts of food (e.g., self-induced vomiting after the consumption of two cookies).
5. Repeatedly chewing and spitting out, but not swallowing, large amounts of food.
6. Binge-eating disorder: recurrent episodes of binge eating in the absence of the regular use of inappropriate compensatory behaviors characteristic of Bulimia Nervosa.

Source: Diagnostic and Statistical Manual of Mental Disorders, 4th ed. Washington, DC: American Psychiatric Association, 1994:550.

attempts in which hunger feelings get out of control. These dieting attempts, usually based on food abstinence, lead to binge eating. Once binge eaters discover they can undo the consequences of their overeating by vomiting out the ingested food, they no longer binge only when they are hungry, but also when they are experiencing any distressing emotion [4]. Most binge eating is done privately in the afternoon or evening, with an intake of about 3,500 kcal [12,13]. Favorite foods for binging usually are dessert and snack foods, very high in carbohydrate. Embarrassment usually prevents the bulimic from revealing his or her food-related behavior even to those closest to him or her [3].

Although bulimia may begin around 17 to 18 years of age, diagnosis may not occur until the bulimic is in her (or his) 30s or 40s [3]. Diagnosis is usually dependent on self-reported symptoms or treatment for related problems or conditions. Conditions that may develop as the result of repeating vomiting include skin lesions on the dorsal side of the hands especially over the joints, severe dental erosion, swollen enlarged neck glands, reddened eyes, headache, and fluid and electrolyte imbalances. Laxative misuse may exacerbate fluid and electrolyte losses and, when coupled with vomiting, may lead to heart failure. Other potential consequences of bulimia are shown in Table 2.

Eating disorders (Table 4) [6] not meeting the diagnostic criteria of anorexia nervosa or bulimia nervosa also are present with the U.S. population. Characteristics of disordered eating include fear of fatness, restrained eating, binge eating, purge behavior, and distorted body image [14]. Further research is needed to determine if these unspecified eating disorders lead to the development of anorexia nervosa or bulimia nervosa.

Early identification and treatment of eating disorders are crucial if serious complications are to be avoided. Rehabilitation requires a multidisciplinary treatment approach [9]. Yet, even with rehabilitation, many former anorectics and bulimics are unable to totally overcome psychological and/or physical impairment and may return to their dietary practices [3,4]. Clearly, being too thin or engaging in bizarre behaviors to keep from becoming overweight carries health risks just as great or greater than being obese [9]. Combating these eating disorders is difficult because not only must the victims be treated, but it still appears that the images and values of society must be rehabilitated [15].

References Cited

1. Pugliese MT, Lifshitz F, Grad G, Fort P, Marks-Katz M. Fear of obesity. N Engl J Med 1983;309:513–18.

2. Love S, Johnson CL. Etiological factors in the development of bulimia. Nutr News 1985;48:5–7.

3. Herzog DB, Copeland PM. Eating disorders. N Engl J Med 1985;313:295–303.

4. Casper RC. The pathophysiology of anorexia nervosa and bulimia nervosa. Ann Rev Nutr 1986;6:299–316.

5. Revised diagnostic subgroupings for anorexia nervosa. Nutr Rev 1994;52:213–5.

6. Diagnostic and Statistical Manual of Mental Disorders, 4th ed. Washington, DC: American Psychiatric Association, 1994.

7. Edwards KI. Obesity, anorexia, and bulimia. Med Clin N Am 1993;77:899–909.

8. Emerson E, Stein D. Anorexia nervosa: Empirical basis for the restricting and bulimic subtypes. J Nutr Ed 1993;25:329–36.

9. Reiff DW, Reiff KKL. Position of the American Dietetic Association: nutrition intervention in the treatment of anorexia nervosa, bulimia nervosa, and binge eating. J Am Diet Assoc 1994;94:902–7.

10. Practice guidelines for eating disorders. Am J Psychiatry 1993;150:212–8.

11. Stein DM. The prevalence of bulimia: a review of the empirical research. J Nutr Ed 1991;23:205–13.

12. Muuss RE. Adolescent eating disorder: Bulimia. Adolescence 1986;21:257–67.

13. Mitchell JE, Pyle RL, Eckert ED. Frequency and duration of binge-eating episodes in patients with bulimia. Am J Psychiatry 1981;138:835–6.

14. Mellin LM, Irwin CE, Scully S. Prevalence of disordered eating in girls: a survey of middle-class children. J Am Diet Assoc 1992;92:851–3.

15. Bulimia among college students. Nutr Rev 1987;45:10–11.

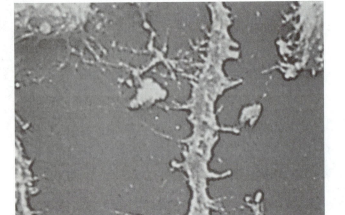

Chapter 16

Neurons

Nutrition and the Central Nervous System

Nutrient Precursors of Neurotransmitters
    Tryptophan
    Tyrosine
    Choline and Lecithin

Sugar

Caffeine

Glutamic Acid and Monosodium Glutamate

Aspartame

CNS-Stimulating Hormones in Weight Control: Leptin and Insulin

∾ PERSPECTIVE: Attention Deficit Hyperactivity Disorder (ADHD) in Children

Energy metabolism in the brain and other parts of the central nervous system (CNS) is very similar to that in other organs and tissues in the body. However, it is the brain's neurological function that is its distinguishing characteristic, and our emotions, behavior, mood, and memory are squarely the responsibility of this organ. A nutritional quest that has generated a great deal of interest is to determine whether, and to what extent, the foods we eat affect our neurological characteristics.

Studies designed to explore a correlation between diet and manifestations such as anxiety or sedation, sleepiness or alertness, or even memory capacity span many years. Although such a correlation has not yet been firmly established by scientific investigation, the popular belief throughout the population that it does exist is unrelenting. Consequently, the subject has been fertile ground for scientific inquiry.

The behavioral consequences of malnutrition are well established. Protein calorie malnutrition and vitamin deficiencies interfere in various ways with the normal development of the brain and other tissues of the CNS. Adequate intake of the trace minerals is also necessary for normal brain development, and certain mineral deficiencies, such as those of zinc and iodine, have been implicated in a variety of behavioral alterations such as deficits in long- and short-term memory, apathy, irritability, and depression. It has been reported that lower levels of essential fatty acids were found in the serum of hyperactive youths compared to age- and sex-matched controls [1]. Also, the effect of dietary intervention with essential fatty acids on behavior and learning ability in rats and mice has been extensively researched [2]. The effects of nutritional deficiencies on central nervous system development and behavior are not discussed further in this chapter. However, the subject has been extensively reviewed for the interested reader [3,4].

The focus of this chapter is (1) to discuss nutrient precursors that are required for the synthesis of selected neurotransmitters, and the effect on behavior of their dietary manipulation, and (2) to examine the impact on behavior of several selected food additives and other dietary components that have "rightly or wrongly" been implicated in behavior alteration.

Nutrient Precursors of Neurotransmitters

Examples of compounds that are constituents of a normal diet and that have a neurological association due to their biochemical conversion to neurotransmitters are tryptophan, tyrosine, and choline. Tryptophan and tyrosine, which are commonly occurring amino acids, are provided by dietary protein. The major source of choline is lecithin. The fact that these substances can be converted into neurotransmitters raised the interesting speculation that foods might produce changes in behavior by altering the levels of brain neurotransmitters that are linked to various psychological functions. Not all neurotransmitters are influenced by the availability of their dietary precursors, however. It has been proposed that several conditions must be met for precursor control of neurotransmitter synthesis and release to occur [5]. These conditions are as follows:

• Plasma levels of the precursor substance must vary in parallel with that of the dietary intake. In other words, there cannot be a control mechanism that regulates the plasma concentration within narrow limits regardless of intake.

• The precursor must be able to cross the blood-brain barrier so that synaptic synthesis of the transmitter can take place.

• The transport mechanism by which the nutrient precursor is carried from the blood into the cells of the brain must not be saturated, allowing it to accommodate more precursor as plasma levels of the precursor rise.

• The enzyme(s) catalyzing the conversion of precursor to transmitter must be of low affinity (high K_m; see p. 17). Under such a condition, the amount of precursor available becomes the rate-limiting factor in neurotransmitter synthesis.

• There cannot be feedback inhibition of the catalytic enzyme(s) as the level of product (neurotransmitter) rises.

All five of these conditions are met for the neurotransmitter serotonin, produced in the brain from the amino acid tryptophan, and for acetylcholine, formed from choline or lecithin. Although less dramatically demonstrated, control of the synthesis of the neurotransmitters norepinephrine (NE) and dopamine (DA) from their precursor tyrosine has also been observed. Before beginning a discussion of the biosynthesis and physiology of these neurotransmitters, the process of synaptic transmission will be reviewed briefly.

The cellular units that make up the brain and peripheral CNS are called *neurons,* which are estimated to number nearly 100 billion in the brain alone. Each neuron transmits and receives electrical signals through filamentous appendages called *axons* and *dendrites.* Signals pass through the axon of a transmitting neuron and impinge on the dendrite of a receiving neuron. The transmission of the impulse from an axon to the receiving cell takes place across a narrow gap referred to as the *synapse* and is mediated by chemical substances known as *neurotransmitters.* An average neuron possesses several thousand synaptic junctions, and these can be situated between intercellular axon and axon, dendrite and dendrite, and axon and cell body as well as between axon and dendrite.

Nutrient precursors of the neurotransmitters must be transported from the brain capillaries into the neurons where the synthesis of the transmitters takes place. The junctions between the capillary endothelial cells in the brain are too tight to allow passage by diffusion of tryptophan, tyrosine, or choline, and therefore these precursors must be transported through the capillary wall (cross the blood-brain barrier) by carrier molecules. The capillary receptor sites at which the precursors are picked up by the carrier molecule are not absolutely specific, and a competition among structurally similar compounds for a given receptor may take place. This is exemplified by the competition for a common carrier of large, neutral amino acids (LNAAs), including tryptophan and tyrosine. This explains why neuronal serotonin synthesis is stimulated by dietary carbohydrate even though the neurotransmitter is derived from the amino acid tryptophan. (This topic is discussed in more detail in the following section on tryptophan.) The enzymatic conversion of choline, tryptophan, and tyrosine into their respective neurotransmitters, acetylcholine, serotonin, and norepinephrine takes place in the presynaptic terminal of the neuron, where they are stored.

When a signal entering the brain or traveling from one brain cell to another arrives at the synapse, it causes the neurotransmitters to be released into the synaptic gap. The compounds complete the synaptic transmission of the signal by interacting with specific receptors on the postsynaptic terminal of the cell. Following completion of the transmission of the signal, the neurotransmitters must be inactivated to prevent protracted "firing" of the synapse. In the case of acetylcholine, the molecule is hydrolyzed by acetylcholinesterase into the products acetic acid and choline. The choline may then be taken up by the presynaptic terminal or returned to the bloodstream, from which it can once again reenter the brain cells by carrier-mediated transport, as described previously. Inactivation of the catecholamine neurotransmitters, norepinephrine and dopamine, is primarily by their reuptake at the presynaptic terminal. However, the portion escaping reuptake can be enzymatically inactivated by monoamine oxidase (MAO) and catechol-O-methyl transferase (COMT). MAO is also involved in the enzymatic inactivation of serotonin. Figure 16.1 summarizes the transport of blood-borne nutrient precursors into the neuron, their conversion into neurotransmitters, and the release and reuptake of the neurotransmitters in bringing about a transynaptic signal.

Tryptophan

Conversion to Serotonin

The behavioral effects of tryptophan, an amino acid present in nearly all dietary sources of protein, is due to its biochemical conversion to serotonin in the neuron (see 16A).

Behavioral Effects

Numerous studies correlating dietary tryptophan consumption with psychological manifestations such as activity and aggression, sensory responses, sleep, mood, and performance have not produced clearly definitive conclusions. It is generally conceded, however, with substantial experimental backing, that tryptophan does exert a calming effect, inducing drowsiness and alleviating aggressive behavior. It may do this either by elevating brain serotonin levels or by reducing the amount of tyrosine entering the neurons because of competition for a common carrier across the capillary endothelium. As discussed later, tyrosine can elevate mood in certain cases of depression, and its deficiency can result in lethargy.

Tryptophan, being an essential amino acid, cannot be synthesized by humans or other mammals. Since the

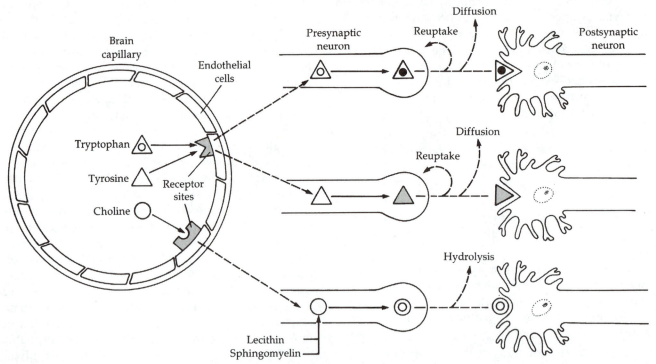

Figure 16.1 The conversion of the nutrient precursors tryptophan, tyrosine, and choline into their respective neurotransmitters, serotonin (◮), norepinephrine (△), and acetylcholine (◎). Migration of the transmitters across the synaptic junction and interaction with postsynaptic receptors complete a nerve signal. Also indicated is the competition between tryptophan and tyrosine (as well as other LNAAs) for common receptor sites on the capillary endothelium.

(16A)

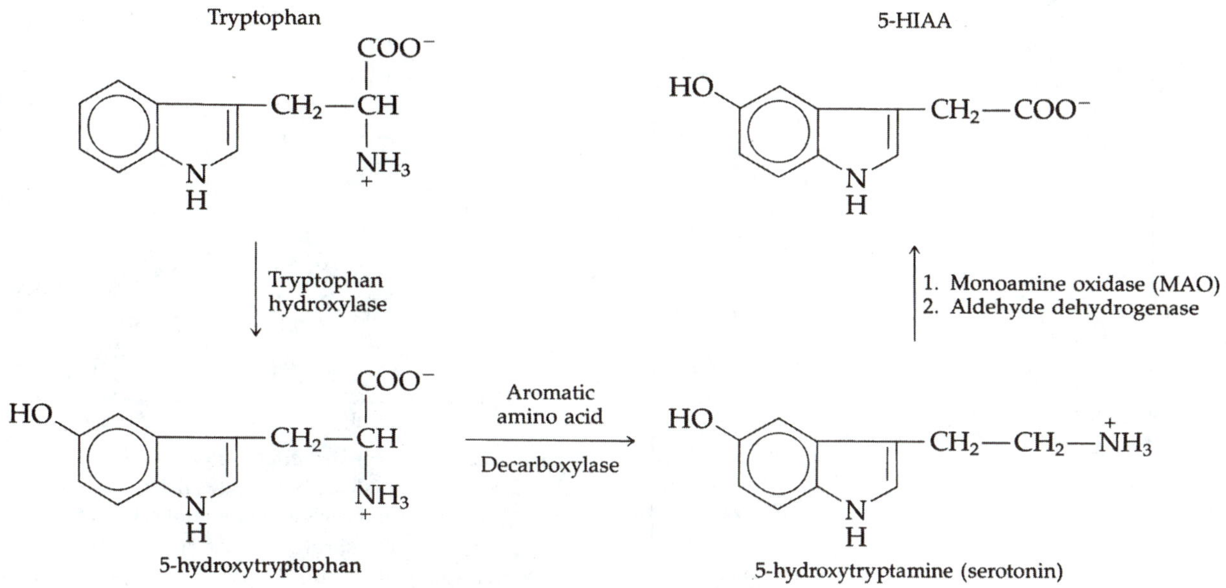

Tryptophan

5-HIAA

Tryptophan hydroxylase

1. Monoamine oxidase (MAO)
2. Aldehyde dehydrogenase

5-hydroxytryptophan

Aromatic amino acid Decarboxylase

5-hydroxytryptamine (serotonin)

body's nutritional supply of tryptophan is acquired through dietary protein, it is logical to assume that a meal rich in protein should elevate tryptophan and therefore serotonin, resulting in a calming, drowsy effect. In actuality, the opposite occurs. Following a protein-rich meal, brain levels of tryptophan and serotonin decline. This is because most dietary protein contains considerably more LNAAs, such as valine, leucine, tyrosine, and phenylalanine, than tryptophan; and because all the LNAAs compete for a common carrier molecule for transport into the neurons, the brain influx of tryptophan declines relative to its competitors. Tyrosine uptake would therefore take preference over tryptophan uptake, and the mood affectation would predictably be more stimulatory than sedating. Paradoxically, it is the meal low in protein but rich in carbohydrate that increases brain tryptophan and serotonin even though the food may lack tryptophan completely. This is because a protein-poor, carbohydrate-rich meal increases the ratio of plasma tryptophan to the other LNAAs, a fact attributed to the carbohydrate-induced release of insulin that stimulates the uptake into muscle cells of most of the LNAAs other than tryptophan. It has been shown that among subjects who have fasted overnight, insulin secretion triggered by carbohydrate intake causes a 40% to 60% decline in plasma valine, leucine, and isoleucine and a 15% to 30% decrease in tyrosine. Plasma tryptophan levels, in contrast, do not decline, an observation thought to be due to its binding to albumin molecules at sites previously occupied by free fatty acids that were released by the action of the hormone.

As a result, less receptor competition from other LNAAs causes an increased brain influx of tryptophan, resulting in elevated neuronal serotonin. The effectiveness of the "milk and cookies before bedtime" regimen to help induce sleepiness and sedation is not, as once believed, due to the tryptophan in the casein but rather to the high-carbohydrate content of that favorite combination. Figure 16.2 illustrates the relationship between brain tryptophan levels and the ingestion of meals having varying proportions of protein and carbohydrate.

The sedating effect of serotonin (and therefore carbohydrate intake) and how it relates to certain emotional disorders is exemplified by the fact that patients afflicted with either seasonal affective disorder (SAD), carbohydrate-craving obesity (CCO), or premenstrual syndrome (PMS) appear to share a tendency to crave carbohydrate snacks [6]. Among SAD patients, the seasonal effect on eating disorders is evidenced in Figure 16.3, which shows that carbohydrate snacking constitutes a significantly higher percentage of total carbohydrate intake during the fall compared with springtime.

Further corroboration of the relationship between serotonin levels and carbohydrate craving is found in the effect of drugs that prolong the neuronal effect of serotonin. d-Fenfluramine, which releases serotonin into brain synapses and prolongs its action by blocking its reuptake, selectively suppresses carbohydrate snacking in CCO patients and simultaneously eases symptoms of depression. There is recent evidence that it may be similarly effective in the treatment of PMS [6].

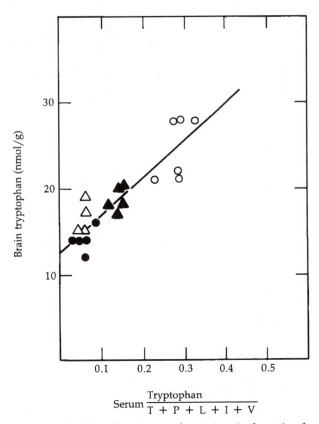

Figure 16.2 The effect in rats of variations in the ratio of dietary protein to carbohydrate on the relationship between serum tryptophan and the brain uptake of tryptophan. (○), animals ingesting no protein (high carbohydrate); (▲), rats consuming 18% protein; (△, rats consuming 40% protein; (●), fasting controls. Capital letters denote the other LNAAs: T (tyrosine), P (phenylalanine), L (leucine), I (isoleucine), and V (valine). *Source:*Fernstrom, JD, Faller DV. Neutral amino acids in the brain: changes in response to food ingestion. J Neurochem 1978; 30:1531–38. Used by permission.

Tyrosine

Conversion to Dopamine, Norepinephrine, and Epinephrine

Dopamine, norepinephrine, and epinephrine are compounds referred to as *catecholamines* because they are derivatives of catechol (see 16B).

(16B)

Catechol

It would be appropriate to regard the essential amino acid phenylalanine as a precursor of these cate-

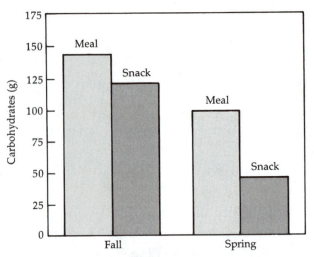

Figure 16.3 The seasonal effect of carbohydrate craving in SAD patients. In the fall, carbohydrate snacks account for nearly 50% of the total carbohydrate intake, whereas they represent <30% in the spring. *Source:* From Carbohydrates and Depression by Richard J. Wurtman. Copyright (c) 1989 by Scientific American, Inc. All rights reserved.

cholamine transmitters also, because it is readily converted to tyrosine by the action of phenylalanine hydroxylase (p. 189). However, it has been estimated that nearly 90% of the brain catecholamines are synthesized directly from naturally occurring tyrosine. The biosynthetic pathway of the catecholamines is shown in 16C.

Certain neurons contain the enzymes tyrosine hydroxylase (TOH) and aromatic amino acid decarboxylase (AAAD) but lack dopamine-β-hydroxylase (DBH). Dopamine therefore becomes the final product in these neurons. In contrast, noradrenergic neurons have DBH activity in addition to TOH and AAAD; therefore, the dopamine produced in these cells is rapidly β-hydroxylated to norepinephrine. Neurons generally have little phenylethanolamine-N-methyl transferase (PNMT) activity and consequently synthesize only small amounts of epinephrine. Most of this hormone is produced in the adrenal medulla, where a considerably higher activity of PNMT exists. Although it is a potent stimulator of glycogenolysis (p. 83) and elevates blood pressure by interaction with blood vessel β-receptors, epinephrine's role as a brain neurotransmitter is of little importance and will not be discussed further in this connection.

Behavioral Effects

Two popularly held explanations exist for the primary form of depression—that which is brought on by a neurochemical imbalance. One explanation attributes

(16C)

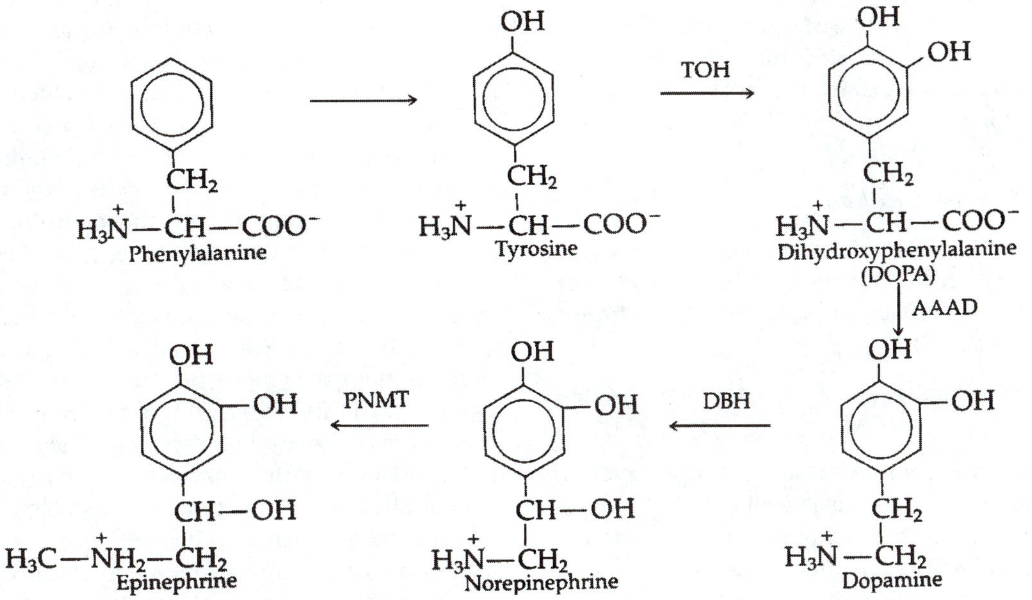

Phenylalanine → Tyrosine →(TOH) Dihydroxyphenylalanine (DOPA) →(AAAD) Dopamine →(DBH) Norepinephrine →(PNMT) Epinephrine

the condition to a deficiency of serotonin; the other, to inadequate norepinephrine. Evidence that there may be some validity to these theories is that primary depression can be treated with drugs such as MAO inhibitors, which maintain higher brain levels of these neurotransmitters. It is therefore not unreasonable to assume that similar results may be achieved by increasing the level of these neurotransmitters through an increased dietary intake of their precursors. But although the concentration of brain cell norepinephrine and dopamine do rise, concomitant with increased precursor (tyrosine) intake, the effect of tyrosine loading on behavior alteration remains controversial. One reason for the lack of a clear-cut relationship between tyrosine administration and behavior modification may be that the effect on a given neuron of an increased supply of the nutrient depends on the firing rate of that neuron. This is evidenced by the observation that tyrosine administration does not ordinarily increase the release of dopamine from dopamine-releasing neurons, even though brain concentrations of tyrosine increase. Yet under conditions in which the firing frequency of the neurons is markedly increased, such as would occur with dopamine-depleting lesions, dopamine receptor blockade by drugs, or Parkinson's disease, tyrosine does enhance the release of the neurotransmitter.

Choline and Lecithin

Nearly all the choline consumed in a normal diet is in the form of choline phosphatides such as lecithin and sphingomyelin, as free choline is present in only small

amounts in foods. The richest food sources of lecithin are natural (e.g., eggs, liver, soybeans, wheat germ, peanuts).

The free choline pool from which the neurotransmitter acetylcholine (discussed next) is synthesized is maintained through several mechanisms, one of which is the enzymatic hydrolysis of lecithin and sphingomyelin. The phospholipases that are responsible for this hydrolytic breakdown are quite active in brain cells, thereby releasing choline directly into the cholinergic neurons where acetylation then occurs. Free choline can also enter cerebral cells from the plasma by a choline-specific transport system across the blood-brain barrier. In the maintenance of neuronal concentrations of free choline, however, the outward flux of choline from the cells into the plasma exceeds the cellular uptake. Nevertheless, the specific transport system is believed to play a key role in the use of choline for acetylcholine synthesis. The richest source of precursor choline, however, is the reuse of endogenous choline formed from the hydrolysis of acetylcholine by acetylcholinesterase following synaptic transmission.

Taken in large amounts, dietary lecithin increases the plasma concentration of choline and therefore its intracellular brain levels. Unlike free choline, the glycerophosphatide is believed not to cross the blood-brain barrier to a significant degree, because of its tenacious binding to plasma albumin. These facts, taken together, therefore suggest that the higher plasma concentrations of choline arise from an accelerated peripheral hydrolysis of the additional lecithin. It appears that the relationships among blood and brain choline, dietary choline and choline-containing lipids, and brain acetylcholine are quite complex. The possibility exists that any behavioral effects (discussed later) of feeding large quantities of choline or lecithin may be indirect and may arise through mechanisms that are not as yet clearly understood.

Conversion to Acetylcholine

In the presynaptic terminal of the neuron, acetylcholine is formed by a reversible reaction between choline and acetyl CoA. The reaction is catalyzed by choline acetyltransferase (CAT):

$$\text{Choline} + \text{acetyl CoA} \xrightarrow[\text{(CAT)}]{} \text{acetylcholine} + \text{CoA}$$

Experiments have demonstrated that the level of choline in cholinergic neurons is well below the K_m of CAT. Therefore, the enzyme is not normally saturated, a condition that, it will be recalled, must be in effect if nutrient precursor concentrations are to have a bearing on the rate of transmitter formation.

Although the source of acetyl CoA for the synthesis of acetylcholine continues to be investigated, it is generally accepted that acetyl CoA is formed indirectly from glucose by neuronal glycolysis and directly through the action of the pyruvate dehydrogenase complex of enzymes (p. 91). The source of the free choline has already been discussed.

Behavioral Effects

There is evidence for a linking of the central cholinergic system to learning and memory in both experimental animals and humans. The most convincing findings supporting this connection have been pharmacological. In normal people, cholinergic antagonists such as scopolamine, which block the interaction of acetylcholine with its postsynaptic receptors, produce memory deficits. Such deficits resemble those observed among elderly people. The disturbance can be reversed by physostigmine, a cholinesterase inhibitor that prevents the degradation of acetylcholine by acetylcholinesterase within the synapse. It is not surprising, therefore, that because the intake of large amounts of acetylcholine precursors increases the neuronal concentration and release of the neurotransmitter, a great deal of interest has been directed toward the possible enhancement of cognitive ability and memory through dietary precursor loading.

The beneficial effect of dietary choline in bringing about an improvement in memory remains controversial, partly because an age factor appears to be involved.

Choline has been found to produce an improvement in memory among normal young subjects. In contrast, when tested on elderly people for whom memory disturbance may represent a significant clinical problem, acetylcholine precursor feeding has yielded disappointing results. It has been proposed that a combination of lecithin loading together with physostigmine administration may alleviate memory deficits in the elderly.

It is believed that among the neurological disturbances of Alzheimer's disease is a cholinergic deficiency attributable to a reduction in the activity of choline acetyltransferase. Predictably, any measures that would tend to increase neuronal acetylcholine may therefore help allay the deficiency in recall memory associated with this disease. Temporary improvement has been observed following physostigmine treatment, but dietary loading with neurotransmitter precursors has had marginal benefit. A study in which lecithin was fed to Alzheimer's patients in the early stages of the disease indicated that an improvement in learning ability occurred in some of the patients and that the improvement reversed after discontinuation of the lecithin loading. Only a small number of subjects was used in the study, however, and clearly more research is required to establish any efficacy of lecithin or choline feeding in the treatment of this disease. Lecithin is generally preferred to choline in such studies because of the foul odor of trimethylamine, produced by the enzymatic breakdown of free choline in the digestive tract.

The most convincing clinical evidence for a link between nutrient precursor loading and neurotransmitter function concerns patients afflicted with tardive dyskinesia, a condition characterized by uncontrollable movements of the face and upper body. In most patients it is caused by the prolonged administration of certain antipsychotic drugs and is considered to be the major side effect limiting the use of those drugs. Following studies that showed physostigmine to be effective in calming the abnormal movements, it was subsequently demonstrated also that increasing brain acetylcholine by feeding choline had a similar effect. The results of these particular investigations are considered reliable in view of the large number of subjects involved and the fact that a double blind, placebo-controlled protocol was used. For the reason mentioned earlier, lecithin has largely replaced choline as a dietary test precursor.

Spring [5] has reviewed the behavioral effects of the nutrients tryptophan, tyrosine, and choline and lecithin in depth. The review also includes an informative discussion of the methods by which mood and/or behavioral effects of nutrients are assessed; it is recommended as supplemental reading.

Clearly, the nutrients discussed to this point, those that serve as precursors for the neurotransmitters, are of vital importance, because neurotransmitters are essential for normal neurological function. Furthermore, the importance of the nutrients dictates that they be readily available through a normal diet. Proteins and choline-containing lipids furnish these precursors for the neurotransmitters discussed. However, other food components that can be categorized as nutritive or nonnutritive additives are commonly included in a typical diet and affect (or allegedly affect) behavior and CNS function. In the following sections, several representative compounds of current interest are discussed.

Sugar

No dietary substance has received more bad press than sugar in its alleged negative effect on behavior. Fostering this belief are the sensational reports linking sugar consumption to a multiplicity of behavioral disturbances such as hyperactivity, depression, mental confusion, and antisocial behavior. An example of society's negative view of sugar is revealed in the celebrated case of a fatal shooting in 1979 of the San Francisco mayor and another city official. Lawyers for the perpetrator in the case argued that their client had acted irrationally and suffered "diminished mental capacity" as a result of his overconsumption of sugar-containing junk foods. The argument has come to be called the "Twinkie defense." It was successful in reducing the charge against the perpetrator from first-degree murder to manslaughter, an outcome that still makes some forensic experts cringe.

The hypothesized connection between antisocial behavior and sugar consumption exists in two parts: (1) the belief that the intake of simple sugars causes reactive hypoglycemia due to an insulin response to the elevated blood sugar and (2) reports, such as that of Virkkunen [7], that reactive hypoglycemia may correlate with the habitual violent behavior displayed by some criminals and delinquents. Such conclusions must be viewed with caution, however, considering that simple sugars do not necessarily increase blood sugar levels any more than do foods containing complex carbohydrates. Moreover, low blood glucose levels do not necessarily correlate with the clinical symptoms of hypoglycemia (see the Perspective in Chap. 4), which include

1. low circulating blood glucose levels (<50 mg/dL);

2. symptoms including sweating, palpitations, anxiety, headaches, weakness, and hunger; and

3. alleviation of these symptoms when plasma glucose levels are restored to normal following food ingestion. Low blood glucose levels can occur in the absence of any symptoms of the condition, and conversely, symptoms of reactive hypoglycemia may manifest in the absence of low blood glucose levels [8].

Studies linking sugar intake to antisocial behavior have also been complicated by a weakness in their experimental design. The independent variable, sugar intake, is difficult to quantify and control, and the dependent variable, changes in antisocial behavior, is also very difficult to measure objectively. Despite the experimental shortcomings of the investigations, the public's view persists that sugar can contribute to antisocial behavior and also to attention deficit disorder and hyperactivity (the subject of this chapter's Perspective). Convincing arguments in support of the alleged sugar-behavior connection have not yet been forthcoming, but it is a premise that is important enough to warrant further, well-designed studies.

Caffeine

Caffeine belongs to a group of compounds referred to as the *methylxanthines*. Its chemical name is 1,3,7-trimethylxanthine, and it is one of the most widely consumed, pharmacologically active dietary substances. It occurs naturally in several plant components such as coffee bean, tea leaf, kola nut, and cacao seed, most of it being consumed in the form of beverages containing extracts of these plant sources. Among the major caffeine-containing beverages are coffee (50–150 mg/cup), tea (~50 mg/cup), and cola drinks (~35 mg/12 oz). Table 16.1 lists average daily consumption of caffeine by consumer age, according to a survey by the Market Research Corporation of America.

Table 16.1 Mean Daily Consumption of Caffeine (mg/kg body weight) by Age According to the Market Research Corporation of America Survey

Age (years)	Source				
	All Sources	Coffee	Tea	Soft Drinks	Chocolate
Under 1	0.18	0.009	0.13	0.02	0.02
1–5	1.20	0.11	0.57	0.34	0.16
6–11	0.85	0.10	0.41	0.21	0.13
12–17	0.74	0.16	0.34	0.16	0.08
18 and over	2.60	2.1	0.41	0.10	0.03

Source: Barone JJ, Roberts H. Human consumption of caffeine. In: Dews PB, ed. Caffeine: Perspectives from human research. Berlin: Springer-Verlag, 1984:66.

Caffeine is also present in various over-the-counter medications, including analgesics, appetite suppressants, and CNS stimulants (see 16D).

(16D)

Caffeine

Chocolate contains some caffeine but contains a much larger amount of another methylxanthine, theobromine (3,7-dimethylxanthine), which exerts pharmacological effects similar to those of caffeine. Although it is not a food or food component, the methylxanthine theophylline (1,3-dimethylxanthine) warrants mention at this point. It is a synthetic drug used in the treatment of pulmonary disorders such as emphysema and asthma, and predictably, its side affects are pharmacologically caffeine-like.

The brain seems to be the organ most sensitive to caffeine. Wakefulness or sleep latency is probably the most common manifestation of caffeine consumption, the period of sleep latency nearly doubling in some subjects by a single serving of 1 to 2 mg/kg of body weight (equivalent to ~2 c of coffee). The "alerting" effects of caffeine, those that allow an individual to function in the face of fatigue, can be experienced in regular caffeine consumers after a single dose of 2 to 3 mg/kg of body weight. In fact, regular users experience improved visual reaction times and auditory alertness at much lower doses (0.5 mg/kg) [9]. Interestingly, doses of this magnitude can cause irritability and nervousness among nonconsumers of caffeine and even among regular consumers who have abstained from the substance for a period of time.

Although the excitatory effect of the methylxanthines on the CNS is obvious, their mechanism of action is not fully understood. Earlier proposed mechanisms focused on these agents' ability to inhibit the action of phosphodiesterase, an enzyme that hydrolyzes cyclic adenosine monophosphate (cAMP). This would tend to maintain elevated levels of cAMP, resulting in neural excitation, because cAMP is considered to be a second messenger for neurotransmitter-receptor systems. The results of much research in this area, however, indicate that such a mechanism cannot be wholly responsible for the stimulatory action of the methylxanthines. Prob-

ably the most significant revelation in this regard is that the concentration of methylxanthines required to inhibit phosphodiesterase in vivo is generally higher than that at which CNS stimulation occurs.

A second proposed mechanism of action of caffeine is based on its ability to inhibit the passage of chloride ions across neuronal membranes at sites referred to as chloride channels. The chloride channel is closely associated with neuronal inhibition. As the chloride flux increases, the membrane conductance to chloride ion increases in parallel, resulting in suppression of neuronal activity. By antagonizing this effect, caffeine reduces chloride conductance and therefore stimulates neuronal activity.

Perhaps the single most convincing explanation for the effect of caffeine on the CNS relates to its action as an antagonist to the naturally occurring substance adenosine. Adenosine can inhibit neuronal activity and behavior both through direct action at postsynaptic sites on neurons as well as by an indirect effect involving presynaptic inhibition of neurotransmitter release. The structural similarity of caffeine and adenosine allows the successful competition of caffeine for adenosine receptors, thereby countering the inhibitory effect of adenosine on the neuron. It appears, however, that chronic caffeine intake may lead to an increase in the number of adenosine receptor sites. As a consequence, the effect of endogenous adenosine would be magnified, and larger amounts of exogenous caffeine would therefore be necessary to restore the adenosine-caffeine balance. If this balance should suddenly be shifted by decreasing or abruptly stopping caffeine intake, the excess adenosine receptors would no longer be blocked by caffeine, thereby intensifying the adenosine effect. It has been suggested that the exaggerated response to adenosine that would follow contributes to caffeine withdrawal symptoms [10].

Stimulant drugs are the medication of choice in the treatment of attention deficit disorder associated with hyperactivity in children. Consequently, an interest in caffeine developed in this regard, particularly because early reports claimed it to be as beneficial as methylphenidate hydrochloride (Ritalin), a frequently prescribed stimulant. It has also been shown that high doses of caffeine (600 mg/d) significantly eased hyperactive behavior. The untoward side effects of such high intake levels of caffeine (nausea, insomnia) argue against the use of the substance clinically. Furthermore, there is a consensus that, based on the majority of experimental results, caffeine offers little therapeutic value compared with prescription drugs.

Glutamic Acid and Monosodium Glutamate

Glutamic acid is one of several amino acids known to be active as neurotransmitters (see 16E). It is one of the most active neuroexcitatory substances present in the CNS of vertebrates. Other amino acids having neurotransmitter activity are aspartic acid and γ-amino butyric acid (GABA). In the brain, glutamate functions as the precursor for GABA, a very important inhibitory transmitter. The conversion involves the removal of the α-carboxyl group of glutamate by the enzyme glutamate decarboxylase, which is therefore instrumental in maintaining an excitatory-inhibitory steady state (see 16F).

(16E)

$$
\begin{array}{cc}
CH_2-COOH & CH_2-COOH \\
| & | \\
CH_2 & CH_2 \\
| & | \\
H_3\overset{+}{N}-CH-COOH & H_3\overset{+}{N}-CH-COO^-Na^+ \\
\text{Glutamic acid} & \text{Monosodium glutamate (MSG)}
\end{array}
$$

(16F)

$$
\begin{array}{ccc}
COO^- & & COO^- \\
| & & | \\
CH_2 & \xrightarrow[\;\;CO_2\;\;]{\text{Glutamate decarboxylase}} & CH_2 \\
| & & | \\
CH_2 & & CH_2 \\
| & & | \\
H_3\overset{+}{N}-CH-COO^- & & H_3\overset{+}{N}-CH_2 \\
\text{Glutamate} & & \gamma\text{-aminobutyrate}
\end{array}
$$

Interest in glutamate and particularly its monosodium salt arose from the observation that it can cause anatomic lesions in the hypothalamus of certain animal species that had received large quantities of the substance. In addition, the incidence of discomforts that may occur among consumers of monosodium glutamate (MSG) as a dietary additive invited a great deal of research and speculation as to the etiology of the symptoms. It should be emphasized that no evidence links these physiological effects of MSG to the transmitter role of glutamate. Furthermore, behavior alteration per se is not among the effects reported as a result of ingesting the substance. The justification for devoting a small portion of this chapter to a discussion of glutamate is the interesting paradox that it can be toxic to the body in spite of the fact that it is a natural, ubiquitous amino acid present in most dietary proteins.

It is known from older studies dating back to 1969 [11] that MSG, if administered in large quantities by either gavage or parenterally to infant mice, has neurotoxic effects. Specifically it results in a selective lesion of the hypothalamus, marked by the destruction of neurons in the nucleus arcuatus of that portion of the brain. It became known as the nucleus arcuatus of the hypothalamus (NAH) lesion, and it was subsequently shown to be inducible in other animal species. Reports on behavior modification by MSG are sketchy at best, however, even when the compound is administered in NAH-inducing amounts. Some such reports allude to a deficiency of discriminatory learning in certain laboratory animals. It is important to note that there is a clear-cut difference in the neurotoxicity of MSG when administered parenterally or by gavage, on the one hand, and by dietary administration, on the other. No threshold dose of dietary MSG causing neurotoxicity has been detected in any animal species. This is explained by the fact that even at high doses, MSG consumed as part of the diet never induces the elevated plasma levels obtained with much lower doses administered by gavage or subcutaneously.

MSG is widely used as a taste and flavor enhancer or as a salt substitute in Western and particularly Asian countries. As mentioned previously, its ingestion by some individuals can cause physiological distress referred to as the "Chinese restaurant syndrome," which is most commonly marked by headache, lightheadedness, and a tightening feeling in the face. The syndrome is particularly interesting in view of the normal abundance of cellular glutamate in mainstream metabolism. Again, the relationship of the symptoms to the neuroexcitatory function of glutamate has not been established, and the cause for the idiosyncratic distress remains obscure. The possibility that the symptoms are related to an allergic reaction has not been ruled out.

Aspartame

The popularity of the artificial sweetener aspartame within our weight-conscious population stems from the fact that although it has the same number of calories per gram as sucrose, it delivers 180 to 200 times the sweetening power of the sugar. Chemically, aspartame, commonly known by its trade name Nutrasweet, is a methyl ester of a dipeptide (L-aspartyl-L-phenylalanine methyl ester; see 16G).

Aspartame is possibly the most thoroughly studied food additive ever approved by the U.S. Food and Drug Administration (FDA) in terms of the total number of studies conducted prior to approval. This is because the widespread use of the sweetener provoked reports

Table 16.2 CDC Evaluation of Consumer Complaints Related to Aspartame Use

System	Complaint
CNS	Mood changes
	Insomnia
	Seizures
Gastrointestinal	Abdominal pain
	Nausea
	Diarrhea
Gynecologic	Irregular menses

Source: Morbidity and Mortality Weekly Reports. Atlanta: Centers for Disease Control 1984(33)43:605.

(16G)

COO⁻ ... Aspartame

$$H_3\overset{+}{N}-CH-\overset{O}{\underset{}{C}}-NH-CH-\overset{O}{\underset{}{C}}-O-CH_3$$

Aspartame

of various reactions attributed to a sensitivity to the substance. The Centers for Disease Control (CDC) investigated 517 consumer complaints and found that 67% involved neurological or behavioral symptoms, primarily headaches. The results of this study are recorded in Table 16.2. Whether or not neurotoxic effects are indeed experienced by some aspartame consumers is still being debated. Among the investigators who subscribe to the relationship, some have proposed mechanisms for the neurotoxicity of aspartame based on the manner in which it is metabolized in the body.

Ingested aspartame is absorbed and metabolized in one of two ways. It may be hydrolyzed in the intestinal lumen to aspartate, phenylalanine, and methanol by hydrolytic enzymes. These components are absorbed from the lumen and reach the portal blood in a manner similar to that of amino acids and methanol arising from dietary protein or polysaccharides. Alternatively, aspartame may be demethylated in the intestinal lumen to yield the dipeptide aspartyl-phenylalanine and methanol. The dipeptide is then absorbed directly into mucosal cells by peptide transport mechanisms, with subsequent hydrolysis to aspartate and phenylalanine within the enterocytes. In either case, the ingestion of large doses of aspartame releases aspartate, phenylalanine, and methanol into the portal blood, and these must be metabolized and/or excreted. Aspartame metabolism is summarized in Figure 16.4.

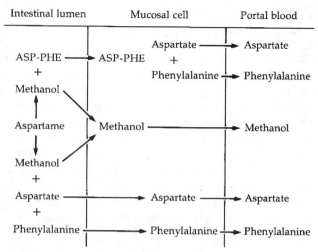

Figure 16.4 The hydrolysis of aspartame in the intestinal lumen and the mucosal cell. The hydrolytic products aspartate, phenylalanine, and methanol are delivered to the portal circulation as shown. *Source:* Stegink CD. The aspartame story: a model for the clinical testing of a food additive. Am J Clin Nutr 1987; 46:204–5. © Am J Clin Nutr, American Society for Clinical Nutrition.

Proposed mechanisms for aspartame neurotoxicity have focused on the resulting increase in plasma phenylalanine. This is not surprising in view of the frank brain damage manifested by the high plasma/brain cell concentrations of phenylalanine and its metabolites in the disease phenylketonuria (PKU) (p. 189). To put doses and toxicities in their proper perspective, however, it should be pointed out that although the toxic plasma threshold is estimated to be nearly 100 μmol/dL for normal adults, a loading dose of 34 mg of aspartame/kilogram of body weight increases plasma phenylalanine levels from 6 μmol/dL to only 11 μmol/dL. This is still well within the normal range. A loading dose of 200 mg/kg, equivalent to consuming 24 L of aspartame-sweetened beverage, increases the plasma phenylalanine value to approximately 49 μmol/dL, still well below the toxicity threshold. Nevertheless, the impaired metabolism of phenylalanine in the PKU patient would, of course, result in correspondingly higher plasma concentrations, and it is for this reason that all aspartame-containing products are mandated by the FDA to display the warning label "Phenylketonurics: Contains Phenylalanine."

Another concern, even among normal people, is the possibility that phenylalanine neurotoxicity may be linear rather than threshold. The threshold concept implies that no toxicity results until a certain threshold plasma concentration is attained. The threshold level of nearly 100 μmol/dL cited earlier is based on the observation that whereas mental retardation is evident in PKU children having plasma concentrations of 120 to

600 µmol/dL, it is not discernible among patients having levels up to 50 µmol/dL. If the relationship between hyperphenylalaninemia and brain effects is a linear function, however, then increases in plasma phenylalanine up to 100 µmol/dL may have subtle but definite effects on the brain that may not be observed clinically.

Phenylalanine may have neurological effects at a blood and brain concentration well below the neurotoxic levels associated with PKU. The basis for this premise is that elevated phenylalanine concentrations may reduce normal neurotransmitter (serotonin) levels and consequently affect behavior and mood and even induce seizures. Prompting the proposal was the report that a few, otherwise healthy adults had suffered seizures after ingesting very high doses of aspartame. The suggested mechanism for this relates to the competition of phenylalanine, tryptophan, and other LNAAs for a common carrier across the blood-brain barrier (p. 539). Compared with dietary protein, which would release various amounts of all the LNAAs, aspartame dosing creates a phenylalanine imbalance when its blood level is selectively raised relative to the other LNAAs. The imbalance becomes even more pronounced if the aspartame "meal" also includes carbohydrate. This is because of the effect of insulin in causing primarily the branched-chain amino acids to leave the bloodstream and to be taken up by skeletal muscle (p. 197). Plasma tryptophan, the precursor of brain serotonin, must then compete for a common carrier with the disproportionately high concentration of phenylalanine. Not only would phenylalanine have the competitive edge from the standpoint of concentration, but it also has a lower K_m for the carrier than tryptophan (i.e., it is bound more tightly by the carrier). Therefore, consumption of aspartame, particularly along with carbohydrate, could cause a reduction in brain serotonin, with associated neurological symptoms.

Although the speculation that aspartame may influence neurotransmitter levels, and the mechanism by which it may do so is of academic interest, the results of several studies have failed to establish a link between dietary aspartame/carbohydrate and brain tryptophan levels. In addition, investigations involving aspartame effects on behavior alteration showed no significant correlation, and showed that it did not differ in this respect from other sweeteners such as sucrose [12].

Based on the data it has reviewed, the FDA has concluded that although aspartame increases blood and brain phenylalanine concentrations, there is insufficient evidence that aspartame alone or in conjunction with carbohydrate alters behavior or functional neurotransmitter activity.

The methanol and aspartate moieties of the aspartame dipeptide have also been studied from the standpoint of their possible effects on behavior and/or neurotoxicity [13]. The toxicity of methanol is, of course, well known. Its ingestion in large quantities leads to a variety of adverse effects, including metabolic acidosis and blindness. The toxicity is attributed to the accumulation of the methanol metabolite formate, rather than the methanol itself. It has been of interest, therefore, to study the plasma level of formate as a function of aspartame ingestion. The results of such studies showed that no significant rise in plasma formate occurred even when abuse doses of aspartame as high as 200 mg/kg were fed to humans.

Aspartate is also toxic in very large quantities and has been reported to induce brain lesions in experimental animals. Like the methanol moiety, however, plasma levels of the amino acid were not markedly elevated by aspartame loading because of its rapid clearance from the plasma.

In summary, if high-dose aspartame administration does have an effect on the brain, the effect is likely to be mediated via phenylalanine and not by way of the intact dipeptide or the aspartate or methanol moieties. The effect on the brain of aspartame-derived phenylalanine is also subject to question. If such a consequence does exist, however, it will occur only when the intake of the sweetener is sufficiently high so as to cause a substantial imbalance in the concentration of plasma phenylalanine relative to the other LNAAs. To date, measurable neurological disturbances have not been convincingly demonstrated as a result of abuse dosing with the sweetener.

CNS-Stimulating Hormones in Weight Control: Leptin and Insulin

The amount of body fuel stores in the form of adipose tissue is regulated within a somewhat narrow range by hormonal signals to the brain, with subsequent adjustment of caloric intake. The quantity of hormones released into the circulation of an individual is directly proportional to his or her adiposity, and the CNS response is a compensatory attempt to maintain the fat store level. Obesity is a complicated condition, and its causes are ever being debated, but a breakdown in this pathway as a contributing cause is receiving considerable attention. The two primary hormones that have been most closely linked to adipose tissue control, by way of CNS stimulation, are leptin and insulin.

Leptin is a 167-amino acid protein that is encoded by the ob (obesity) gene functioning exclusively in adipose tissue. The ob gene consists of 15,000 base pairs and has three exons (p. 9). It was first discovered in mice having a specific genetic mutation that predestined them to hyperphagia (excessive intake of food) and morbid obesity. Administration of leptin to these (ob/ob) mice sharply curtailed food intake and reversed their obesity. The ob mutation was found to produce a truncated, biologically inactive form of leptin.

It has long been known that the hypothalamus is a critical target for satiety effects and that obesity can be induced by experimental lesions of the ventromedial hypothalamus. From these observations, coupled with the later discovery of the hypophagic effect of leptin, it was concluded that the hypothalamus has receptors for leptin. The receptor is now well established. It is encoded by the leptin receptor (db) gene. Circulating leptin crosses the blood-brain barrier and enters the hypothalamic cells by a saturable transport system, wherein it binds to its receptor. The "downstream" effects following the leptin/receptor interaction is not clear at this time, but the effector molecule is believed to be hypothalamic neuro–peptide Y (NPY), a potent stimulator of food intake. It is proposed that the synthesis of NPY is inhibited through the leptin signal, thereby suppressing appetite [14].

A mutation in the db gene was first identified in mice, as was the ob gene mutation, and like the ob mutation is believed to have a human correlate. The db gene mutation results in a splice variant with a prematurely truncated intracellular signaling domain. Therefore, a db gene mutation may allow the normal binding of leptin but cannot transmit the leptin signal to the response center of the hypothalamus. Therefore, the db gene mutation, like the ob gene mutation, results in hyperphagia and obesity, but by a different mechanism. With impaired signaling, and therefore leptin "resistance," it explains why, in cases of obesity due to db mutation, leptin levels are frequently higher than normal [14].

Insulin unlike leptin, is not a newcomer as a participant in adiposity control. It has been investigated in this connection for two decades, but the recent leptin studies have led researchers to believe that insulin acts on the hypothalamus similarly to leptin. Insulin is released from pancreatic islet cells proportional (like leptin) to the level of body adiposity. Like leptin, it is taken up by the hypothalamus by way of a transport process, and it also is believed to induce hypophagia by inhibiting the expression and/or release of the hypothalamic NPY [15]. There is a striking difference, however, between the effect of leptin and insulin deficiencies on adiposity. Whereas leptin deficiency due to ob gene mutation causes obesity, insulin-deficient diabetics are underweight. It should be recalled that insulin promotes fat synthesis and storage. Diabetic subjects therefore lose weight despite an overconsumption of energy [16].

Summary

The normal diet includes compounds that can affect mood or behavior. The effect may be direct, without structural alteration of the compounds, or it may be exerted following their metabolic conversion to active products such as neurotransmitters. Examples of nutrient precursors, along with their corresponding neurotransmitters, are listed here:

Nutrient	Neurotransmitter
Tryptophan	Serotonin
Tyrosine	Epinephrine
	Norepinephrine
	Dopamine
Choline (as lecithin)	Acetylcholine
Glutamic acid	Gamma-aminobutyric acid (GABA)

To function as a neurotransmitter, the nutrient must enter the neurons from the bloodstream (cross the blood-brain barrier). The process requires the nutrient's binding to receptors on the capillary endothelial cells followed by carrier-mediated transport through the cells. The nutrient-carrier interaction is not absolutely specific, however, resulting in a competition among large, neutral amino acids (LNAAs), including tryptophan and tyrosine, for common receptor sites. The calming effect of tryptophan, via conversion to serotonin, is therefore enhanced by a diet that minimizes the competition from the other LNAAs, allowing a larger portion of circulating tryptophan to cross the blood-brain barrier. Although this effect can be achieved by a reduction in the dietary intake of competing LNAAs, it is more dramatically expressed by a high-carbohydrate diet, which, through the release of insulin, diverts competing LNAAs into muscle. This action proportionately increases the amount of tryptophan available for brain uptake.

Because tyrosine is a progenitor of the stimulatory neurotransmitters norepinephrine, dopamine, and epinephrine, it is tempting to speculate that alterations in the dietary intake of the amino acid may affect mood by parallel changes in brain levels of the transmitters. The relationship has not been firmly established, however.

Choline, the dietary source of which is principally lecithin, is convertible to the neurotransmitter acetylcholine, a depletion of which has been linked to deficits in learning and memory. The therapeutic benefit of dietary loading of the nutrient to alleviate these symptoms has been investigated, but the results of the studies remain controversial.

Among the most widely consumed psychoactive dietary substances is caffeine, a CNS stimulant well known for its induction of wakefulness and "alerting" effects. The effectiveness of the stimulant prescription drug Ritalin, used in treating attention deficit disorders associated with hyperactivity in children, fostered interest in caffeine as a possible alternative stimulant. The high doses required to achieve positive effects, however, preclude its therapeutic use.

Glutamic acid is rather unique among the neuroactive nutrients in that its activity can be stimulatory or inhibitory depending on its metabolic route. Structurally unaltered, glutamic acid is a potent neuroexcitatory transmitter; however, it can be decarboxylated in the brain to form the inhibitory transmitter GABA. The ubiquitousness of glutamic acid and its high concentration throughout the body tissues, including the brain, hamper studies on the effect of its dietary intake on neuroactivity. A link between the physiological distress reported by some people following the eating of glutamate as its monosodium salt MSG and the neuroactive effects of glutamate or GABA has not been established.

Another commonly ingested substance that reportedly can be identified with neurological symptoms is the artificial sweetener aspartame. At very high dietary levels of the dipeptide, such symptoms may be accounted for by the resulting increased concentration of plasma phenylalanine, which may impede neuronal uptake of tryptophan. This effect is potentiated if carbohydrate is consumed along with the sweetener.

References Cited

1. Mitchell EA, Aman MG, Turbott SH, Manku M. Clinical characteristics and serum essential fatty acids in hyperactive children. Clinical Pediatrics 1987;26:406–11.

2. Wainwright PE. Lipids and behavior: The evidence from animal models. In: Lipids, Learning, and the Brain: Fats in Infant Formulas. Report of the 103rd Ross Conference on Pediatric Research. Columbus OH: Ross Laboratories, 1993:69–88.

3. Kanarek RB, Marks-Kaufman R. Nutrition and Behavior: New Perspectives. New York: Van Nostrand Reinhold, 1991.

4. Robinson J, Ferguson A. Food sensitivity and the nervous system: hyperactivity, addiction, and criminal behavior. In: Nutr Res Rev 1992;5:203–23.

5. Spring B. Effects of foods and nutrients on the behavior of normal individuals. In: Wurtman RJ, Wurtman JJ, eds. Nutrition and the Brain. New York: Raven Press, 1986;7:1–47.

6. Wurtman RJ, Wurtman JJ. Carbohydrates and depression. Sci Am 1989;260:68–75.

7. Virkkunen M. Reactive hypoglycemic tendency among habitually violent offenders. Nutr Rev 1986;44(suppl):94–103.

8. Ferguson HB, Stoddart C, Simeon JG. Double blind challenge studies of behavioral and cognitive effects of sucrose-aspartame ingestion in normal children. Nutr Rev 1986;44(suppl):144–50.

9. Zwyghuizen-Doorenbos A, Roehrs TA, et al. Effects of caffeine on alertness. Psychopharmacology 1990;100:36–39.

10. Griffiths RR, Woodson PP. Caffeine physical dependence: a review of human and laboratory animal studies. Psychopharmacology 1988;94:437–51.

11. Olney JW. Brain lesions, obesity, and other disturbances in mice treated with monosodium glutamate. Science 1969;164:719–21.

12. Kruesi MJP, Rapoport JL. Diet and human behavior: how much do they affect each other? Ann Rev Nutr 1986;6:113–30.

13. Steginck LD. The aspartame story: a model for the clinical testing of a food additive. Am J Clin Nutr 1987;46:204–15.

14. Auwerx J, Staels B. Leptin. Lancet 1998;351:737–42.

15. Schwartz M, Sipols A, Marks J, Sanacora G, et al. Inhibition of hypothalamic neuropeptide Y gene expression by insulin. Endocrinology 1992;130:3608–16.

16. Schwartz M, Seeley R. The new biology of body weight regulation. J Am Diet Assoc 1997;97:54–58.

Suggested Reading

Kruesi MJP, Rapoport JL. Diet and human behavior: how much do they affect each other? Ann Rev Nutr 1986;6:113–30.
  Included is a review of pertinent research evaluation along with the behavioral effects of specific dietary constituents.
Spring B. Effects of foods and nutrients on the behavior of normal individuals. In: Wurtman RJ, Wurtman JJ, eds. Nutrition and the Brain, New York: Raven Press, 1986;7:1–47.
  In-depth coverage is provided of the methodological issues involved in researching nutrient versus behavior correlations as well as selected nutrient effects in both human and animal studies.
Diet and behavior: a multidisciplinary evaluation. Proceedings of a symposium, Arlington, VA. Nutr Rev 1986;44(May suppl):1–252.
  This article provides a broad, multitopic treatment of the effects of nutrients on both brain function and patterns of behavior. A section on the strategies for improving pertinent research adds to the strength of this review.

Web Sites

www.fda.gov
  U.S. Government Food and Drug Administration
http://altmed.od.nih.gov/nccam
  National Institutes of Health: National Center for Complementary and Alternative Medicine

Attention Deficit Hyperactivity Disorder (ADHD) in Children

There has been a long-standing public belief that diet affects behavior. Scientific evidence for the relationship was the revelation that brain levels of neurotransmitters can be affected by the dietary intake of their precursors. Research on diet and behavior consists of two general types: correlational and experimental.

The *experimental design,* because of its ability to establish cause-and-effect relationships, is the more widely used. It involves manipulation of the variables (e.g., a dietary constituent such as sucrose) to examine their effect on dependent measures such as behavior and/or cognitive factors. Cognitive assessment is quantitative and readily measurable. Examples of cognitive measures are continuous performance test (CPT) results or the ability to recall information as a result of diet manipulation. Behavioral assessment, in contrast, is difficult to quantify, and consequently, conclusions are often subjective and anecdotal. Regardless of whether it is cognition or behavior that is the dependent variable, research employing experimental designs is capable of establishing causal relationships.

Correlational designs examine behavioral characteristics of individual subjects in relation to their eating habits. The correlational approach requires a large number of subjects, and its chief weakness is its inability to distinguish cause from effect.

Following this brief review of research terminology, the possible effect of certain food substances on attention deficit hyperactivity disorder (ADHD) in children will be considered. This condition has received considerable publicity in recent years largely because of the debatable overuse of the stimulant drug Ritalin, used to allay the symptoms. Hyperactivity is a syndrome characterized by restlessness, hyperkinesis, and impulsive behavior, as well as a compromised attention span and an inability to concentrate. Because the syndrome is a distortion of normal behavior, it is not surprising that a dietary link to this condition gradually surfaced. Historically, artificial food additives such as colors (dyes) and flavors and, more recently, sweeteners, particularly sucrose and aspartame, have been implicated in promoting the symptoms of hyperactivity.

In 1975 Feingold offered the hypothesis that the ingestion of artificial food additives (colors and flavors) and naturally occurring salicylates in foods results in hyperactivity and learning disabilities in children. On the basis of this conclusion it was suggested that treatment be implemented through the Feingold Kaiser-Permanente (K-P) diet, which is designed to eliminate all foods containing these additives.

The hypothesis received widespread media attention and an enthusiastic response from the general public. The theory even received some early experimental support, although serious methodological faults in these studies were subsequently reported.

The National Advisory Committee on Hyperkinesis and Food Additives was formed to study the Feingold theory. It concluded that more data were needed to substantiate the hypothesis unequivocally. Finally, in 1983 a meta-analysis [1] of the controlled studies completed by that time provided a generally negative relationship between food additives and hyperactivity. It was concluded from the study that the Feingold K-P diet afforded only a marginal, limited benefit compared with control diets. The findings did not preclude rare, individual exceptions.

There has also been considerable public concern that refined sugar, chiefly sucrose, may be linked to hyperactivity. This speculation received support from the results of a study showing positive correlations between dietary carbohydrate-protein ratios and directly observed restless and aggressive behavior in a sample of hyperactive children. As a correlational study, however, it could not establish causality. For example, a tendency toward hyperactivity may have encouraged the children to consume more carbohydrate because of its calming effect (p. 539). Also, there is the possibility that both the hyperactivity and the heightened carbohydrate intake result from a third variable, a higher metabolic rate.

Dating from the mid-1980s, many studies have been conducted using double-blind, placebo-controlled experimental designs examining the effect of sugar on children's behavior and cognition. A variety of "effect" parameters have been used in these studies, such as playroom or classroom observation, wrist and ankle actometers (designed to measure physical activity), and CPT. CPT is a cognitive measure, generally used to assess sustained attention and impulsiveness. Although a few of the studies did indicate an increased level of activity and a decreased attention span as a result of sugar challenge, others curiously showed precisely the opposite effect. The majority of the investigations revealed no correlation between sugar challenge and behavior or cognition [2]. Finally, a comprehensive investigation examining behavioral and cognitive effects of both sucrose and aspartame (using saccharin as placebo) on normal preschool children and school-age children thought to be sensitive to sugar was published [3]. Its authors concluded that even if intake exceeds typical dietary levels, neither sucrose nor aspartame affects children's behavior or cognitive ability.

Although research results argue rather convincingly against a clear-cut correlation between sugar/aspartame consumption and behavior or cognition in children, old beliefs do not die easily. Among 389 Canadian primary school teachers surveyed, it was found that >80% believe that sugar consumption contributes to hyperactivity in normal children and magnifies symptoms in ADHD-established children. Furthermore, 55% of the "believer" group have counseled parents accordingly [4]. This reveals the need to keep teachers apprised of current information on the subject.

Despite the exoneration of sugar as a causal factor, the food-ADHD controversy remains alive, and scattered reports continue. One such study using a multiple item (various foods, dyes, additives, and preservatives) elimination from the diet, rather than just one tested item, reported a favorable response among 73% of ADHD subjects tested. The same subjects reacted to many foods, dyes, and preservatives upon open challenge. An allergy connection is suggested by the fact that allergic ADHD children had a significantly higher response rate than normals [5].

In summary, artificial food additives and sucrose, the two major alleged dietary culprits in the cause or exacerbation of hyperactivity, cannot be proven guilty, based on the results of extensive research employing the use of double-blind studies. Whether the symptoms of hyperactivity are completely independent of dietary factors awaits further study. Meanwhile, the cause of the condition remains a mystery.

References Cited

1. Kavale KA, Forness SR. Hyperactivity and diet treatment: a meta-analysis of the Feingold hypothesis. J Learn Disabil 1983;16:324–30.

2. Kanarek R. Does sucrose or aspartame cause hyperactivity in children? Nutr Rev 1994;52:173–5.

3. Wolraich ML, Lindgren SD, Stumbo PJ, et al. Effects of diets high in sucrose or aspartame on the behavior and cognitive performance of children. N Engl J Med 1994;330:301–7.

4. DiBattista D, Shepherd M. Primary school teachers beliefs and advice to parents concerning sugar consumption and activity in children. Psychol Reports 1993;72:47–55.

5. Boris M, Mandel F. Foods and additives are common causes of the attention deficit hyperactivity disorder in children. Ann Allergy 1994;72:462–8.

Web Sites

http://www.add.org
National Attention Deficit Disorder Association
www.aap.org
American Academy of Pediatrics
http://www.helpforadd.com
Help for ADD

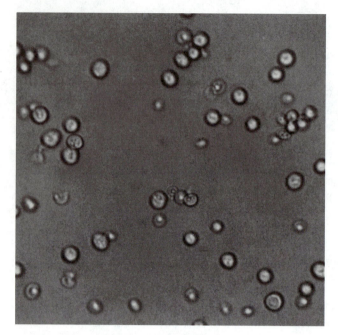

Budding yeast cells (Saccharomyces)

Chapter 17

Experimental Design and Critical Interpretation of Research

Research is a process that seeks, finds, and transfers new knowledge. Although many definitions of research have emerged, probably none is more comprehensive and poignant than that taken from *Webster's Dictionary of the English Language:*

> *Research:* A studious inquiry or examination, especially a critical and exhaustive investigation or experimentation having for its aim the discovery of new facts and their correct interpretation, the revision of accepted conclusions, theories, or laws in the light of newly discovered facts or the practical application of such conclusions, theories, or laws.

At the core of this definition is that research discovers new facts (if there is no discovery, there was no research) and then *correctly* interprets those facts.

All of the information contained within this text was derived from research, and the reference list at the end of each chapter allows the reader to examine the source of the information. Publications cited in a bibliography reveal to the reader the premise or justification of the research, the experimental methodology by which it was conducted, and the author's interpretation of the results. It is through such publications that findings and facts in the nutrition arena are routed through the appropriate scientific journals to the public. Perhaps more than any other discipline, nutrition furnishes an unrelenting barrage of news through the media. This is because nutrition touches the lives of all of us by appealing so directly to our concept of good health. Unfortunately, along with the profusion of nutrition news comes the probing question as to what is reliable and what is not. It is imperative that the student of nutrition recognizes this problem and learns to separate fact (derived from carefully constructed research) from fictional information passed along by anecdotal reports and hearsay.

The refereed or peer-reviewed journals offer the best source of information. These terms indicate that research procedures and results submitted by the investigator to the journal for publication are critically examined by other scientists who are knowledgeable in that particular area of research. This peer review process screens proposed publications for quality and soundness, ensuring the reliability of research papers that are ultimately published in that journal. A few of the journals that feature nutrition-related investigations that have won the respect of practitioners over the years include *The American Journal of Clinical Nutrition, Nutrition Journal, Nutrition Reviews,* and *Journal of the American Dietetic Association.* Many other excellent journals frequently publish nutrition-related reports. Students who regularly consult the nutrition literature will learn to recognize these and distinguish them from weaker publications. The qualifications of the suppliers of information in other than peer-reviewed journals should be examined carefully, and the reliability of the information judged accordingly.

As its title indicates, this chapter is designed to acquaint the reader with the various experimental methodologies (designs) available to the researcher and to offer information that may be helpful in both understanding research terminology and critiquing existing research publications.

The Scientific Method

Research uses a process referred to as *the scientific method* to solve problems or to resolve previously unanswered questions. The scientific method contains the following fundamental components:

1. *The research purpose or problem:* expresses the question to be answered or the problem to be solved
2. *The hypothesis:* a prediction of the outcome of the research that will follow and therefore a solution to the problem or an answer to the question
3. *Experimentation:* the conducting of the research itself using one of the many methodologies available to the researcher
4. *Interpretation or analysis:* interpreting the data collected from the experimentation so as to understand what it means
5. *Conclusion:* answers the originally posed question, and confirms or disproves the hypothesis
6. *Theory formulation:* a statement founded on the conclusion

Research can be thought of as occurring in a cyclic process, with the components of the scientific method arranged as illustrated in Figure 17.1. However, it may be misleading to think of research in terms of a closed circle. Research does not dead-end into finality with the theory statement because, to the inquisitive mind of the researcher, a conclusion from one experiment invariably gives rise to new questions and problems. Research is therefore nonending, and its cyclic nature, as illustrated in Figure 17.1, may be more meaningfully represented as a spiral or helix, twisting inexorably upward toward a limitless goal.

Antoine Lavoisier is credited with being the first to implement the scientific method into his research,

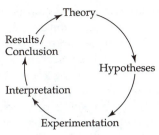

Figure 17.1 The cyclic nature of research from theory to conclusion.

which was conducted during the 18th century. Up until that time, inquiry into problems had been only a philosophical exercise. His approach to problem solving is illustrated by the following steps [1,2]:

• Using Priestley's earlier discovery that oxygen is involved in burning, Lavoisier formulated the hypothesis that respiration in animals was a form of combustion.

• Lavoisier carried out experimentation under controlled conditions on animals (guinea pigs). Oxygen consumption, heat production, and production of carbon dioxide by animals confined in airtight chambers were carefully measured.

• With the measurements (data) collected, Lavoisier interpreted the results: A pattern could be identified between oxygen consumption, carbon dioxide production, and heat emanating from the animal body.

• Based on his interpretation of the data, Lavoisier formulated the theory that consumption of oxygen is related to the amount of carbon burned or heat produced in the animal body. Then, to validate his theory, he performed similar controlled experiments on other animals, including humans.

Research design, particularly that which lends itself to the physical and biological sciences, must include the components just discussed. To be considered a fact, the theory formulated must be verified by other investigators who carry out the research under the same controlled conditions used in the original research. If the theory is verified, it becomes accepted as fact until advances in research can disprove it.

The Influence of Knowledge Base and Technology on Research: A Historical Example

Nutrition research, designed to expand the nutrition knowledge base, which began with Lavoisier's revelations about energy metabolism (around 1789)[2], moved at a rather slow pace until the beginning of the 20th century. But at about that time, many important discoveries and technological advances allowed a rapid expansion of the nutrition knowledge base. Nutrition research continued to flourish throughout the 20th century.

The cyclic nature of research makes clear that one prerequisite of productive research is an investigator who is familiar with the existing knowledge base of the subject under study, and how others have investigated the subject. This prerequisite is provided admirably in the studies of Rose [3] and Young [4], who investigated the dietary essentiality of amino acids. Their studies are described briefly here, and although dated, they provide an example of well-planned research that emphasizes the cyclic nature of the investigative process.

Protein composition and metabolism has been a particularly fertile field in nutrition and biochemical research. Scientific inquiry has been directed toward identification of the indispensable amino acids followed by quantitative determination of human requirements for these amino acids and for total protein. Rose, after comparing the work of previous investigators (Osbourne and Mendel, Fischer, and others) on the amino acid composition of proteins with that from his own laboratory, undertook the identification of those amino acids that are indispensable to animals and/or humans. He began his investigations with weanling rats and then extended experimentation to humans (graduate students!). The data obtained from rats were used as the starting point for the more difficult and costly studies on human beings. The experimental approach used by Rose was to keep one particular amino acid in the diet as the single *independent variable,* and this was compared with maximum weight gain as the criterion for the *dependent variable.* He found through a series of experiments that only 10 of the 19 to 20 amino acids ordinarily found in proteins are required by the rat: arginine, histidine, isoleucine, leucine, lysine, methionine, phenylalanine, threonine, tryptophan, and valine.

In his follow-up study on humans, Rose continued to use individual amino acids in the diet as the independent variable, but he used nitrogen equilibrium as the criterion for the dependent variable to determine indispensability. Rose's 13-year study on the indispensable amino acids and their requirements in the adult male (1942–1955) resulted in identification of those amino acids that are indispensable (essential) to human beings, their range of requirements, a tentatively proposed minimal requirement, and a definitely safe intake (Table 17.1). One troubling aspect of the study was the increased caloric intake required by the subjects to maintain nitrogen equilibrium when they were fed amino acids rather than protein as their nitrogen source. In the discussion of his studies, Rose cites their limitations, noting particularly the small number of subjects, the artificiality of using diets containing mixtures of amino acids rather than proteins, and the attendant

need for increased caloric intake on the part of the subjects. Rose concludes his studies by stating [3]:

> By reference to [Table 17.1], it should be possible to predict with reasonable accuracy how much of a given protein or diet, if its content of amino acids is known, would be required to maintain nitrogen equilibrium, provided the moderate quantity of extra nitrogen needed by the body for synthesis is already present or is added to the food. Such calculations must take account of the availability of the amino acids in the materials under investigation, since the processing and preservation of foods and their preparation for consumption may modify the nutritional usefulness of the component proteins. (p. 644)

Up to the present, these studies by Rose, published in 1957, have been used to a large degree as the basis for the estimation of adults' needs for the indispensable amino acids. They also have provided a basis for development of an amino acid pattern against which the biological value or "completeness" of a protein is measured.

The hypotheses comprising Rose's theory began to be tested by Young in the late 1960s, and those studies continue today. Young's research has contributed to the refinement of methods and procedures in assessing the amino acid needs of humans and probably is leading to a new, increased estimate of adults' amino acid requirements. This modification of the amino acid requirements in adult humans that may result from Young's research does not negate in any way the usefulness of Rose's theory, because the research energy expended in rejecting a theory propels science ahead even further.

During the time that Young was conducting research on amino acid metabolism, many weaknesses in the use of nitrogen balance as the dependent variable for studying protein metabolism have been identified. These include

1. nitrogen equilibrium gives no specifics about protein metabolism in individual tissues or organs;
2. results using the nitrogen balance technique often overestimated need because losses of nitrogen through the skin seldom are measured;
3. nitrogen balance is significantly affected by energy intake; and
4. nitrogen balance can be greatly influenced by the length of the experimental period, which makes interpreting a specific nitrogen balance difficult.

As he became aware of the limitations of nitrogen balance as a criterion for amino acid adequacy, and having available (through technologic advances) some alternative methods for assessment of protein status, Young began a new series of studies on amino acid requirements in adults. An alternative, dependent variable that he used extensively during the late 1960s and early 1970s was the plasma amino acid response curve [5–8]. Subsequently, following the advent of stable isotopes, the tracing of

Table 17.1 The Daily Amino Acid Requirements of Young Men

Amino Acid	Number of Quantitative Experiments	Range of Requirements Recorded (g)	Minimum Proposed Tentatively (g)	Definitely Safe Intake (g)	Subjects in N Balance on Safe Intake or Less
L-tryptophan	3[a]	0.15–0.25	0.25	0.50	42
L-phenylalanine	6	0.80–1.10[b]	1.10	2.20	32
L-lysine	6	0.40–0.80	0.80	1.60	37[c]
L-threonine	3[d]	0.30–0.50	0.50	1.00	29
L-methionine	6	0.80–1.10[e]	1.10	2.20	23
L-leucine	5	0.50–1.10	1.10	2.20	18
L-isoleucine	4	0.65–0.70	0.70	1.40	17
L-valine	5	0.40–0.80	0.80	1.60	33

All values were determined with diets containing the eight essential amino acids and sufficient extra nitrogen to permit the synthesis of the nonessential amino acids.

[a]Fifteen other young men were maintained in nitrogen balance on a daily intake of 0.20 g, though their exact minimum needs were not established. Of the 42 subjects maintained on the *safe* intake or less, 33 had 0.30 g daily or less.

[b]These values were obtained with diets devoid of tyrosine. In two experiments, the presence of tyrosine was found to spare the phenylalanine requirement to the extent of 70% and 75%.

[c]Ten of these subjects had an intake of 0.80 g or less.

[d]In addition to these 3 subjects, 4 young men had diets containing 0.60 g of L-threonine daily , and 16 others had 0.80 g daily. No attempt was made to determine the exact minimum requirements of these 20 subjects, but all were in positive balance on the amounts shown.

[e]These values were obtained with diets devoid of cystine. In three experiments, the presence of cystine was found to spare the methionine requirement to the extent of 80% to 89%.

Source: Rose WC, Wixon RL, Lockhart HB, et al. The amino acid requirements of man, XV. The valine requirement: summary and observations. J Biol Chem 1955;217:992. Reprinted with permission.

amino acid distribution and use within the body became possible. As a consequence, distribution and use of an amino acid replaced the amino acid response curve as the dependent variable. Concurrently, by combining the isotope tracer technique with the measurement of nitrogen balance, Young collected some convincing evidence to support his hypothesis that adults need more of the indispensable amino acids than had been indicated through the measurement of nitrogen balance alone.

These examples of research efforts illustrate the effect on outcome by (1) a broadening of the knowledge base and (2) advancing technology. Young had available to him the knowledge of the positive effect of increased caloric intake on N balance, along with other disadvantages associated with N balance studies, which Rose did not. In addition, Young gained advantage from the advent of isotope tracer techniques and other technological advances.

Research Methodologies

Many different types, or classifications, of research exist. One of the broadest classifications of research is according to *application*: is the research *basic* or *applied*? Basic research seeks to expand existing knowledge by discovering new knowledge. Applied research, in contrast, seeks to solve problems primarily in a field setting. Other classifications of research can be according to *strategy* (historical, survey), *degree of experimental control* (experimental vs. nonexperimental), *time dimension* (cross-sectional vs. longitudinal), *setting* (laboratory, field), or *purpose* (descriptive or analytical).

Despite the diversity of research classifications, the *methods* by which research can be carried out are more concisely categorized. Two broad methodological approaches encompass essentially all research, the *qualitative methods* and the *quantitative methods*. They are distinguished from each other according to the nature of the data collected in the study. All of the data emerging from a research study reaches the researcher in the form of either words or numbers. *If the data are verbal, the methodology is qualitative; if it is expressed in numbers, the methodology is quantitative.* Within each of these major categories, four discrete subdivisions of methodological approaches can be used in research. These are historical methods and descriptive survey methods, both of which are qualitative approaches, and analytical survey methods and experimental methods, which are quantitative in nature. Each of these methods will be described, accompanied by one or more illustrative examples of each method taken from the literature. The text by Leedy [9] provides a good review of research methodologies.

Historical Method (Qualitative)

Historical research seeks to explain the cause of past events and to interpret current happenings on the basis of these findings. Sources of information for the historical researcher are primarily documentary, existing in the form of written records and accounts of past events, as well as literary productions and critical writings. The researcher relies, if possible, only on primary data, a term that implies that the data are "firsthand" and therefore minimally distorted by the channels of communication. Generally, information gathered by historical research does not need to be analyzed by any form of statistical treatment or data analysis.

Descriptive Survey Method (Qualitative)

One word that distinguishes the descriptive survey method from other research methods is *observation.* The investigator observes, across a defined population group, whatever variable is under study. The variable may be physical (size, shape, color, strength, etc.) or cognitive (achievement, beliefs, attitudes, intelligence). The researcher (1) observes very closely the population bounded by the parameters that were set for the study, and then (2) carefully records what was observed, for future interpretation. It is important to note that observation does not involve only visual perception but also very likely tests, questionnaires, attitude scales, inventories, and other evaluative measures. In fact, most descriptive surveys use well-designed questionnaires as the instrument of observation.

The keeping of records, which can be thought of as a preservation of facts, is an important feature of the descriptive survey method. Observation and record keeping is exemplified by the *case study,* which follows the symptomatology, treatment, conclusions, and recommendations as they apply to a patient under study. A *case report* is a report of observations on one subject, whereas a *case series* involves observations on more than one subject. Generally, the subjects being observed have a condition or disease in common. This form of research design is useful in an attempt to identify variables or generate hypotheses that may be important in the etiology, care, or outcome of patients with a particular disease or condition.

An example of a case report is the investigation by Sedlet and Ireton-Jones [10]. These researchers studied energy expenditure in a patient who, on initial evaluation at an eating disorder clinic, presented with a semifast-binge eating pattern. Findings of the investigation were documented as a case report study, preliminary to further

research. Energy expenditure in the subject was assessed prior to and following nutrition intervention. Basal energy expenditure (BEE) was estimated by use of the Harris-Benedict equation (p. 552). Actual measured energy expenditure (MEE) was determined by indirect calorimetry. The investigators found that the patient's BEE was 53% higher than her MEE when, on initial evaluation, she was in the semifast-binge eating pattern. Modification of her eating pattern to three meals a day, providing a total of approximately 1,200 kcal/day, was accomplished over a 4- to 6-week period. After the modification, her MEE increased and was essentially the same as her BEE (Table 17.2), whereas her activity level remained the same throughout treatment. The subject seemed more willing to continue the modified eating pattern (3 meals/day) when she was able to eat each day and not gain weight.

Findings from this case report helped to generate an ongoing research project to determine whether data presented can be substantiated in a study involving a group of patients with bulimia. There is limited information about actual versus estimated energy expenditure in bulimia and also about ways acceptable to patients to modify their food intake pattern.

The classic work of Goldberger on pellagra [11], involving observations on groups of children, illustrates a case series study. Goldberger described qualitatively the population and environment at two orphanages in Mississippi in 1914 as a preliminary study in his research on pellagra. By careful, systematic observation, he noted

1. the incidence or absence of pellagra among both the orphan residents and staff,

2. age groups affected by disease or free of the disease,

3. opportunity for close contact among groups,

4. sanitary conditions, and

5. foods eaten by the different groups of residents and staff.

In this preliminary inquiry he found that the disease pellagra was confined almost exclusively to a group of 6- to 12-year-old children. The different diet given to the 6- to 12-year-old groups was the only factor he observed at this point to explain a possible cause of the disease. The finding led him to test his hypothesis in

Table 17.2 Energy Intake versus Expenditure of JK (a bulimic patient)

Treatment Period	Intake (kcal/d range)	BEE (kcal/d)	MEE (kcal/d)	BEE/MEE (% difference)
Pretreatment semifast–binge	600–3,800	1,266	829	53
Nutrition counseling follow-up				
2 months	1,000–1,200	1,257	1,203	4
6 months	1,000–1,200	1,252	1,193	5
7 months	1,000–1,200	1,252	1,202	4

Source: Sedlet KL, Ireton-Jones CS. Energy expenditure and the abnormal eating pattern of a bulimic: a case report. J Am Diet Assoc 1989;89:16.

later experimental dietary studies, eventually leading to our current knowledge that the disease is caused by niacin deficiency. The prevalent theory at the time of Goldberger's work was that pellagra was an infectious disease spread by unsanitary conditions. Goldberger's ability as a researcher is exemplified by his refusal to be constrained by these prevailing beliefs.

Data for case series studies are often collected from existing records and are considered retrospective studies, meaning "looking back." A disadvantage of the retrospective study is that observations of the different variables of interest may not have been recorded in a standardized way and thus may not be comparable. Alternately, data for case series may be collected concurrently, providing an opportunity to standardize procedures. An advantage of case series studies is that they are relatively easy to perform and inexpensive. They may provide data to justify the need for future analytic study, especially when concerned with a condition or disease about which information is limited.

Bulimia (p. 533), a disease requiring nutrition intervention but about which little is known, has been studied by observation. Pyle et al. [12] reported a case series study of 34 subjects with bulimia. Findings regarding frequency of binge eating, fasting, self-induced vomiting, and use of laxatives were among the characteristics described. An eating pattern of binge eating combined with periods of fasting was common.

Other older but classic examples of descriptive survey studies are the Ten-State Nutrition Survey [13] and the National Health and Nutrition Examination Survey (NHANES I and II) [14]. These surveys can also be classified as "cross-sectional" studies; that is, all subjects across the population are studied at approximately the same time, and each is studied once, not continuously.

The Ten State Survey was designed to oversample among low-income populations in selected states in the United States. Thus, application of findings (external validity, to be discussed later) was limited and could not be applied to the U.S. population in general. Findings from the Ten State Survey provided some evidence for the need to expand food programs in low-income populations. The NHANES data were collected from a statistically elected sample to represent the 194 million noninstitutionalized civilians aged 1 to 74 years in the United States. Data were collected systematically and by standardized methods of measurement from the sample population. Data from the NHANES are considered baseline information for the U.S. population at the time of the survey.

Analytical Survey Method (Quantitative)

The analytical survey method is best described by contrasting it with the descriptive survey method just discussed. Although the descriptive survey method involves observations that can be described in words and concluded from those words, the analytical survey uses a different language, a language not of words but of numbers. Because values obtained from an analytical survey are numerical, the data are said to be quantitative.

The quantitative data of an analytical survey are analyzed by statistical tools from which conclusions can be inferred. Statistical analysis of numerical data may include

- measures of central tendency (mean, median, mode);
- measures of dispersion (range, standard deviation, coefficient of variation);
- measures of correlation (correlation coefficient, regression analysis).

These measurements fall within a category called *descriptive statistics*. Another category of statistics is called *inferential statistics*, which has two principal functions:

1. to predict or estimate from a random sample a certain parameter in a general population, and
2. to test null hypotheses based on statistics.

The null hypothesis postulates that there is no statistically significant difference between phenomena that occur by pure chance and the statistically evaluated behavior of the data as they have been observed by the researcher. As an illustration of the testing of the null hypothesis, suppose a study is to be conducted on the level of serum iron among vegetarians and nonvegetarian, omnivorous consumers. In keeping with a quantitative study, the data will be numerical, in this case, serum iron concentration expressed in milligrams per deciliter. Before collecting the data, the null hypothesis holds that there is no statistical difference in the criterion of serum iron concentration between the two groups. If, in fact, the experimental statistical findings confirm this, the null hypothesis is said to be accepted. But if it is shown that there *is* a statistically significant difference in the level of serum iron between the two groups, then the null hypothesis is rejected.

Observational research designs can also be included in the category of analytical surveys. The use of the word *observational* in describing this type of research may at first seem confusing since observation is also the hallmark of descriptive surveys, as discussed earlier. Remember, however, that the observational research designs discussed here are examples of an analytical study, and, as such, they produce numerical, quantitative data, interpretable by the application of statistics. Descriptive surveys, although based on observation, rely on written data and are not quantitative.

Observational designs may take the form of epidemiological or cohort studies, neither of which involves experimentally induced changes in variables. *Epidemiology* has been defined as "the study of the distribution of a disease or condition in the population, and the factors that influence the distribution." A *cohort* is a group of subjects entered into a study at the same time and followed up at intervals over a period of time. A cohort study is also called a *prospective* study, meaning "looking ahead."

Three examples of observational nutrition-related studies are the classic Seven Countries Studies reported by Keyes [15], the Framingham Study [16], and the Nurses' Health Study [17]. In the Seven Countries Studies, the outcome of interest was coronary heart disease (CHD). Factors that were under consideration because of their possible association with CHD included dietary total calories, total fat, animal fat, saturated fat, total protein, animal protein, and sucrose. Data regarding CHD and the dietary factors were collected from population groups in each country by systematic, standardized methods of measurement by trained teams. Data were statistically analyzed for possible association of any of the studied dietary factors. Figure 17.2 shows the relationship of one dietary factor, saturated fat, to the incidence of CHD.

The Framingham Study was a long-term, longitudinal study, initiated in the town of Framingham, Massachusetts, in 1948. The term *longitudinal* refers to the fact that the study is conducted on a group (cohort) over a certain period of time, a feature of the cohort study. The Framingham Study was designed to investigate the distribution of CHD in the sample population and possible causal factors associated with the disease. The study began with a cohort of 2,336 men and 2,873 women aged 29 to 62 years, randomly selected from the population. Data regarding the cardiovascular condition and the possible causal factors of CHD were systematically collected from subjects at specified time intervals, using standardized methods of measurement. Data from the Framingham Study showed that hypercholesterolemia, hypertension, and cigarette smoking are three major risk factors in CHD. Information from that study was used also to show comparative risk at different levels of blood cholesterol (Fig. 17.3).

The Nurses' Health Study began in 1976 and continues to the present. This very large cohort (initially in excess of 122,000 participants) is composed of married, female registered nurses born between 1921 and 1946. The advantage of this cohort, in addition to its large size, is that the participants are broadly dispersed geographically, being chosen from the 11 states having the largest number of registered nurses. Several of the studies conducted on this cohort have been based on the effect of diet on various health parameters such as cancer and cardiovascular disease. One such study resulted in the recently published finding that there appears to be

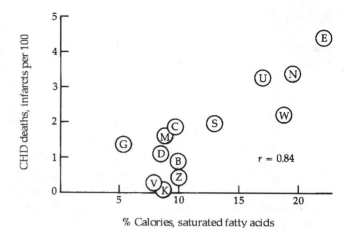

Figure 17.2 Diet-heart associations. B, Belgrade faculty; C, Crevalcore; D, Dalmatia; E, East Finland; G, Corfu; K, Crete; N, Zutphen; M, Montegiorgio; S, Slovenia; U, U.S. railroad; V, Velika Krsna; W, West Finland; Z, Zrenjanin.
Source: Inter-society Commission for Heart Disease Resources, Atherosclerosis Group. Keys A, ed. Coronary heart disease in seven countries. Circulation 1970(1)42:211. Reprinted with permission.

no causal relationship between the intake of dietary total fat or specific major types of fat and breast cancer [18]. The finding was based on food-frequency questionnaires returned by 88,795 of the participants. It represents a refutation of the previously alleged, although not scientifically documented, connection between dietary fat and breast cancer.

The results and conclusions of an experiment, and eventually the establishment of scientific fact, ultimately depend on the statistical treatment of data, a field of considerable scope. This section is intended simply to acquaint the reader with some commonly used statistical terms in the analysis of numerical data. For the interested reader, many comprehensive texts on applied statistics offer a more in-depth treatment of the subject [19,20].

Experimental Method (Quantitative)

Among the research methodologies, the experimental method is the one most commonly encountered in the nutrition literature. The hallmark of the experimental method is *control*. So basic is control to this method that this means of searching for truth is frequently referred to as the *control group–experimental group design*. Such a study uses two or more population groups with the subjects of each group matched, characteristic by characteristic, as closely as possible to the subjects of the other group(s). One of the groups serves as a control and as such is not exposed to some extraneous change. The experimental group is exposed to the alteration under study, and whatever change is noted, rela-

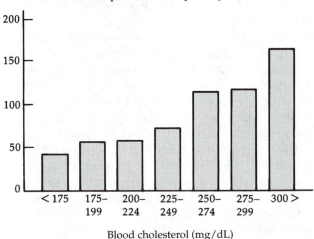

Risk of heart attack increases as blood cholesterol goes up

Heart attack rate per 1,000 men per 10 years

Blood cholesterol (mg/dL)

Figure 17.3   Risk of heart attack as related to blood cholesterol.
Source: Inter-society Commission for Heart Disease Resources, Atherosclerosis Group. Primary prevention of the atherosclerotic disease. Circulation Dec 1977 42:A55–95. Reprinted with permission.

Table 17.3 Major Characteristics of Diets

	Atherogenic	Low Fat	Corn Oil
Protein	16	19	15
Carbohydrate	43	77	45
Fat	41	4	40
Cholesterol	1.2	0	0

Values for protein, carbohydrate, and fat are percentages of total calories; cholesterol values are percentages by weight.
Source: Armstrong ML, Warner ED, Connor WE. Regression of coronary atheromatosis in rhesus monkeys. Cir Res 1970;27:60. Reprinted with permission of the American Heart Association, Inc.

tive to the subjects of the control group, is presumed to be caused by the extraneous variables. The experimental method can also use just one group, a method sometimes called a *pretest-posttest* approach. In its simplest form, a group of subjects is first evaluated (pretest), then it is subjected to the experimental variable (the test) and reevaluated (the posttest). So-called *cross-over* studies are also commonly used. In this case, a control group is not subjected to a particular experimental variable, while an experimental group is subjected to the variable, and differences in the data are noted. Then, the original control group is exposed to the variable, therefore becoming the experimental group, and the original experimental group becomes the control group by not being exposed to the variable. This approach corrects for any inherent differences in the two groups that might confound the experimental data.

In short, the experimental method is based on *cause and effect*. It involves *intervention* on the part of the researcher who introduces a variable and records its effect. Experimental research designs allow the investigator to control or manipulate one or more variables in an effort to examine the relationship between the variables. Variables typically are designated as dependent and independent. The *independent variable* is that which is controlled or manipulated by the investigator. The *dependent variable* occurs as the result of the influence of the independent variable. In other words, the dependent variable reflects the effects of the inde-

pendent variable. As in the case of the analytical survey method, many traditional descriptive and inferential statistical tools can be used to analyze the data.

Because experimental methodology is so commonly used in nutrition research, numerous examples can be given to illustrate its use. Following are a few classic, illustrative examples of research employing the experimental method.

Armstrong et al. [21] using the experimental research design, tested the hypothesis that diet-induced atheromas may regress in primates by use of appropriate diet (the independent variable). The investigators conducted their study with 40 adult male rhesus monkeys. To serve as controls, 10 of the monkeys were fed a low-fat, cholesterol-free diet throughout the study. The other 30 were fed the same low-fat, cholesterol-free diet for a control period of 6 weeks and then were given a high-fat, high-cholesterol diet (atherogenic diet) for 17 months. Composition of the diets is given in Table 17.3. The 30 monkeys were then divided into three groups matched for body weight and hyperlipidemia.

Group 1 was autopsied for determination of baseline atherosclerosis. Group 2 was fed a low-fat, cholesterol-free diet, and group 3 a linoleate-rich diet for 40 months. All diets throughout the study included supplements to make them complete in all essential nutrients. Control and study animals were autopsied at the end of the study to determine the degree of atheromatosis (the dependent variable). Autopsies showed that the cross-sectional area of the lumen of five major coronary arteries was significantly greater in monkeys fed the regression diets (group 2, low-fat diet; group 3, linoleate-enriched diet) than in the monkeys with baseline atherosclerosis (group 1). Table 17.4 provides a comparison of the groups. The investigators were able to show for the first time regression of atheromatosis in nonhuman primates by modification of diet.

It may be appropriate at this point to remind the reader that examples of the invaluable role of animals in experimental method research are almost innumerable. However, the use of animals in research is threatened by efforts of organized animal rights activists. Recent

Table 17.4 Luminal Narrowing (%)

Group	Main coronary		LAD	L c'flex	R dist
	Left	Right			
1	60 ± 8	56 ± 7	53 ± 18	57 ± 9	65 ± 10
2	17 ± 4	14 ± 3	21 ± 4	22 ± 6	16 ± 5
3	25 ± 5	26 ± 3	30 ± 5	23 ± 6	18 ± 5

Abbreviations of branch arteries: LAD = left anterior descending; L c'flex = left circumflex; R dist = distal continuation of right main coronary artery.
Values are means ± SE (standard error).

Source: Armstrong ML, Warner ED, Connor WE. Regression of coronary atheromatosis in rhesus monkeys. Cir Res 1970;27:60. Reprinted with permission of the American Heart Association, Inc.

advances in the care and protection of animals in research need to be carefully preserved and made known to these activists. Currently each institution in which animals are used in research is required to have a committee to oversee animal care and treatment and to review all research plans in which animals are used.

The randomized clinical trial, an experimental research design, is frequently used in medical research studies in humans. Clinical trials are usually conducted after preliminary trials have been done in experimental animals, and they typically test the benefits of one or more treatments.

Subjects in clinical trials are those who have the condition to be treated and should be representative of the population to which results are to be applied. Subjects are assigned randomly to a treatment group. In some instances only one treatment is available, and a placebo is used for the control group. Subjects who enroll in the study must be informed that they have an equal chance of being assigned either to the treatment or to the control (placebo) group. Ideally, to avoid bias, a clinical trial should be "double blind," with neither the subject nor the investigator knowing which group is which.

An example of a clinical trial is the Lipid Research Clinics Coronary Primary Prevention Trial (LRC-CPPT) [22]. This research project, sponsored by the National Heart, Lung, and Blood Institute of the National Institutes of Health (NIH), illustrates a number of important points that must be considered in the design of a randomized clinical trial.

The LRC-CPPT was undertaken to determine whether reducing blood cholesterol levels lowers the risk of CHD in humans. Earlier studies had shown that elevated blood cholesterol levels were a major risk factor in CHD. Numerous studies had shown also that blood cholesterol levels could be reduced by diet and drugs in animals and in humans. Previous clinical trials had not been considered conclusive because of a number of problems in research design. These previous clinical trials, however, along with findings from earlier studies, such as epidemiological studies, clinical investigations, and animal stud-

ies, were extremely useful in designing the LRC-CPPT. A particularly helpful aspect of the earlier clinical trials was the identification of pitfalls in experimental design: absence of double-blind testing, too-small sample size, treatment groups that were not identical, and questionable statistical procedures. Typical problems and pitfalls in research are considered later in the chapter.

In the LRC-CPPT, the study population consisted of people known to be at high risk of CHD. Those under consideration for the study, therefore, were men aged 35 to 59 years, asymptomatic for CHD but exhibiting primary hypercholesterolemia (a plasma cholesterol ≥265 mg/dL). Subjects were 3,810 men, screened from approximately 48,000 age-eligible men. Informed consent was received from each subject entered into the trial.

Subjects were randomly selected to receive either the drug cholestyramine (a resin previously tested for safety and effectiveness in reducing total cholesterol and low-density lipoprotein cholesterol) or a placebo. Both subjects and investigators were unaware of the treatment assignment (i.e., it was double blind). A moderate cholesterol-lowering diet, designed to lower cholesterol levels 3% to 5%, was prescribed to all subjects. Maintaining all the subjects on the same diet throughout the study minimized the opportunity for confounding the study results because of different dietary intakes.

Twelve Lipid Research Clinics (LRCs) participated in the trial. To ensure comparability of data over the entire study period, a common protocol documenting all procedures in detail was adhered to by clinic personnel trained and certified in standardized procedures. All aspects of the study were carefully monitored by the Central Patient Registry and Coordinating Center and by the Program Office.

Table 17.5 illustrates the close similarity of the drug and placebo groups before and during treatment for selected variables that include the major, known risk factors for CHD. If there had been a change in risk factors other than cholesterol levels during the treatment period, this could have posed an alternative explanation for the observed treatment benefit.

The results of the trial showed that the cholestyramine group experienced a 19% reduction in the risk of the primary end point (i.e., definite CHD death and/or definite nonfatal myocardial infarction). The data also showed that a 1% reduction in blood cholesterol level yields approximately a 2% reduction in CHD rates.

Subjects were followed for a minimum of 7 years and for as long as 10 years, the average follow-up being 7.4 years. Subjects attended clinics every 2 months during the follow-up. At clinic visits, subjects received the drug or placebo, dietary and drug counseling to encourage adherence, and standardized examinations and evaluations according to the detailed protocol. All subjects who entered the study were followed to the com-

Table 17.5 Selected Variables before and during Treatment

Variable	Placebo			Cholestyramine Resin		
	Preentry	First Year	Seventh Year	Preentry	First Year	Seventh Year
Mean systolic blood pressure, mm Hg	121	120	122	121	120	122
Mean diastolic blood pressure, mm Hg	80	79	78	80	78	78
Mean Quetelet index, g/cm²	2.6	2.6	2.7	2.6	2.6	2.7
Mean weight, kg	81	81	83	80	80	82
% current smokers	37	35	26	38	36	27
Mean cigarettes/d for current smokers	25	24	25	26	25	26
% regular exercisers	30	a	27	31	a	28
Median alcohol consumption, g/wk	61	58	51	64	57	53

[a]No assessment of exercise was done in the first year.

Source: Lipid Research Clinics Program: The Lipid Research Clinics Coronary Primary Prevention Trial Results. I. Reduction in incidence of coronary heart disease to cholesterol lowering, JAMA 1984;251:357.

pletion of the trial irrespective of their levels of adherence or frequency of visits. At the completion of the trial, contact was made with all men who discontinued visits during the trial. Thus, it was possible to know the vital status of every subject.

Experience in the LRC-CPPT met design goals in several features of the study, as shown in Table 17.6. Twenty-seven percent of subjects did not adhere (nonadherers) to taking cholestyramine for 7 years. This percentage was less than the expected goal of 35%. However, a nonadherer was defined as a subject taking less than half a packet of medication daily. The prescribed dose was six packets of medication per day. Thus, many subjects, although considered to be adherers, were not taking the total prescribed dose. As a consequence, the goal of 28% reduction in cholesterol levels of men assigned to cholestryamine treatment was not met. Actual reduction in cholesterol levels amounted to 13.9%. In addition, although the 19% reduction in the primary end point (incidence of CHD death) was well below the goal of 36%, the reduction in CHD deaths has been judged to be clinically significant.

Investigators in the LRC-CPPT expressed caution that the results of the study could not be extrapolated to predict results from use of drugs that are not bile acid sequestrants such as cholestyramine. In addition, the study was not designed to show whether cholesterol lowering by diet decreased CHD deaths. It is known, however, that cholesterol levels can be reduced by dietary modification, and findings from the LRC-CPPT support the view that cholesterol lowering by diet would be beneficial. That view was later strongly supported by the Consensus Conference on lowering blood cholesterol to prevent heart disease [23]. The LRC-CPPT was a long, complex, expensive clinical trial. Nevertheless, the experimental design method involving randomized clinical trials can be used for much less expensive studies.

Table 17.6 Comparison of LRC-CPPT[a] Design Goals and Actual Experience

Design Feature	Goal	Experience
Sample size	3,550	3,808[b]
Duration of follow-up, year	7	7–10
Lost to follow-up	0	0
Reduction of plasma total cholesterol levels in placebo group	4%	4.8%
Nonadherers[c] at year 7	35%	27%
Reduction of plasma total cholesterol levels in men adhering[d] to cholestyramine resin treatment	28%	13.9%[d]
Seven-year incidence of primary end point in placebo group	8.7%	8.6%
Reduction in primary end point	36%	19%

[a]LRC-CPPT indicates Lipid Research Clinics Coronary Primary Prevention Trial.
[b]After removal of four type III participants.
[c]A nonadherer is someone averaging less than half a packet of cholestyramine resin per day.
[d]Computed for seventh year.

Source: Lipid Research Clinics Program: The Lipid Research Clinics Coronary Primary Prevention Trial Results. I. Reduction in incidence of coronary heart disease to cholesterol lowering, JAMA 1984;251:360.

Terms That Describe Research Quality

Descriptive terms that reflect the effectiveness or quality of a research effort include

- *validity,*
- *accuracy,*
- *reliability,* and
- *precision.*

The "truth" of research lies within the *validity* of its collected data. Validity represents the extent to which the process or technique that is being used is measuring what it is supposed to be measuring. *Validity is concerned with the effectiveness of the measuring instrument.* The term *instrument*, as it applies in research, is broadly defined, ranging from survey questionnaires to pieces of scientific equipment.

There are several types of validity. *Face validity* relies on the subjective judgment of the researcher, and involves asking the following questions: (1) Is the instrument measuring what it is supposed to be measuring? (2) Does the sample being measured adequately represent the behavior or trait being measured? *Criterion validity* uses as an essential component a reliable and valid criterion (i.e., a standard against which to measure the results of the instrument that is doing the measuring). Validity can also be expressed as internal or external, both of which are very important in research. The term *internal validity* refers to causal relationships; that is, did an experimental treatment make (cause) a difference? *External validity* refers to the generalizability of the results of the research to a population group that was not studied.

The terms *accuracy* and *reliability* are related because they are both concerned with how close to the "truth" a measurement is. *Accuracy* is expressed as the difference between the measured values of an instrument and the true values. The more accurate a measurement, the closer is the result to the true value. *Reliability* refers to the instrument used in the study, and indicates the degree of accuracy that it generates. An instrument may be reliable within a broad range of accuracy. This concept can be illustrated by the use of a sundial as an instrument for telling time. It is a reliable timepiece if one is only concerned with whether the time of day is early afternoon or late afternoon. However, the sundial has poor reliability for more specific timing such as informing the observer as to when to turn on the television to see a favorite show or when to leave to catch a bus. In both of these cases, the sundial's accuracy—that is, the quantitatively expressed nearness of the measured time to the true time—is poor.

The term *precision* is a very useful expression of the consistency or repeatability of multiple analyses performed on the same sample or subject. Procedures used in research should generate the same data from the same sample with repetition. For example, multiple assays of serum glucose performed on the same serum sample provide an indication of the precision of the instrument used. It is important to understand the difference between precision and accuracy. A method may be highly precise; that is, replicate values may be very close to one another, yet the method may not be accurate. In contrast, widely disparate replicate values (i.e., poor precision) might yield an accurate *average* value. However, imprecise measurements are not a property of quality research.

Initiation of Research

The only prerequisite for initiation of research is an inquiring mind. The novice is likely to be intimidated by reports of sophisticated research that has required expensive equipment, extensive personnel, and a generous budget. Not all research needs to be conducted at this level; it can be simple and inexpensive while still serving the purpose of broadening the knowledge base.

Initiation of research requires familiarity with the characteristics of research. These characteristics are shown in Table 17.7. The first characteristic suggests that research begins with a question: Why does something occur, or what causes something? Second, research demands that the problem be identified and stated clearly. Research requires a plan. It seeks direction through hypotheses. Research deals with data and their meaning. And research is circular, as shown previously in Figure 17.1. To be certain that all these characteristics are included in a research project, the following four steps should be followed [24].

1. *Select the research topic or problem to be solved.* Choosing a topic or problem narrow enough to be manageable often can be difficult. A review of published literature in relation to the selected research topic is necessary to have a basis on which to build present research and to help precisely define the research.

2. *Clearly state the question to be researched.* Components of the question include who (which)—that is, the subjects or units being assessed are identified; what—that is, the factor of interest is stated specifically; how assessed—that is, the outcome to be assessed is stated specifically.

3. *Prepare a research plan/proposal.* The proposal should include

    a. a statement of the research question (from step 2);

    b. a review of literature (from step 1);

    c. the significance of the research; and

    d. a description of research design, which should specify the specifics of the investigation (i.e., methods, data analysis, and the appropriate statistical analysis).

Putting the plan in writing forces the researcher to think through all aspects of the investigation and can serve as a clearly defined guide for carrying out the project. The plan becomes a working document that can be converted into the research report.

Depending on the level of the research, the plan may range from a simple outline to a complicated, detailed request for funding from a foundation or government agency. Regardless of the level of investigation, whenever live subjects are used, established guidelines must be followed. Review of proposed research projects by committees on ethical standards ensures that

Table 17.7 Checklist for the Evaluation of Research

1. Is the central problem for research (and its subproblems) clearly stated?
2. Does the research evidence plan and organization?
3. Has the researcher stated his or her hypotheses?
4. Are the hypotheses related to the principal problem or the subproblems of the research?
5. Are the assumptions stated? Are these assumptions realistic for the research undertaken?
6. Is the research methodology that has been employed clearly stated?
7. If the research is of experimental design, was it *in vivo* or *in vitro* work? If *in vivo,* were humans or animals used? Was there a control group? What was the sex, number, and age of experimental (and control) subjects? What was the length of the experiment? Was there sufficient number of subjects and/or sufficient time allowed to warrant meaningful conclusions?
8. Is the statistical treatment of data clearly defined and statistics presented in a straightforward manner?
9. Are the conclusions that the researcher presents justified by the facts presented?
10. Is there any indication whether the hypotheses are supported or rejected?
11. Are limitations of the study identified?
12. Is there any reference to or discussion of related literature or studies by other investigators?
13. Are specific areas for further research suggested?
14. By whom was the research sponsored? Could results be influenced in any way by the source of funding?

procedures are acceptable. These committees operate in academic institutions at the departmental and university level. Funding agencies and organizations are very careful about considering only those proposals that strictly adhere to these guidelines.

4. Plan for the collection and preparation of data. Once the method for data collection has been selected (or designed), a pilot study can point out any adjustments that need to be made in the data collection method. The pilot study may also provide a good indication of the value of the data being collected. Improvements in research design often result from a pilot study.

Once the research procedures have been refined, the planning stage is finished, and the research study can be conducted. Carrying out the research involves the collection of data, followed by the crucial steps of interpreting the data and reporting the results. The problem that initiated the research finally is addressed, and results of the investigation are interpreted in the framework of existing theory and past research, if such is attainable. The problem is either solved, or the process must begin again. Many valuable outcomes are possible from even the simplest research projects if they are well planned. An interest in problem solving coupled with a diligence in planning an orderly, stepwise progression in problem solution provide the essentials for scientific inquiry.

Problems and Pitfalls in Research

In general, a clear understanding of the steps or components of the research process provides a good checklist against which to evaluate research presented in publications or to plan one's own research study. Attention to these components also provides a guide against problems and pitfalls that can plague research.

The logical progression of components in the research process has been discussed earlier in the chapter under "Initiation of Research" and has been reproduced in checklist form in Table 17.7. A complete plan for this logical progression of the elements of the research process, including the question, research design, and exact statistical analysis, should be in place before any research activity begins. If this is done, then the research study will in fact be following a predetermined protocol, in a manner similar to a National Aeronautics and Space Administration (NASA) launch.

Despite the best-laid plans, some problems can occur during a research study. As summarized in an editorial by Vaisrub [25], problems that can be considered as either "soluble" or "insoluble" may arise in the course of a research study. Examples of such problems are given in Table 17.8.

One of the more difficult and error-prone areas of research is the application of statistics in the analysis of data. Typically, this involves rejection or retention of a null hypothesis (p. 558). Should the statistics be invalid, the null hypothesis may be rejected when it should have been retained, or it may be retained when it should have been rejected. The most common cause for such errors is insufficient power of the statistical test, where "power" refers to the likelihood of falsely rejecting a null hypothesis. In view of the magnitude of the subject, statistical data analysis is not addressed in this chapter. However, excellent pertinent references are available to the interested reader [19,20,28].

Evaluation of Research and Scientific Literature

Although the library "research" paper cannot be considered research because it is not gathering "new" data or

Table 17.8 Commonly Encountered Problems in Research

Insoluble problems in studies
  Lack of representative sampling
  Vague target population definition with poor selection of subjects
  Lack of random allocation of treatments
  Lack of proper handling of confounding (nuisance) variables
  Lack of appropriate controls
  Lack of blinded subjects and evaluators
  Lack of objective measurement or assessment of outcome
Possibly soluble problems in studies
  Inadequate assurance of group comparability
  Inappropriate choice of sample units
  Use of calculated normal limits for skewed distributions; multiple significance testing
  Incorrect denominators for rates, risks, or probabilities
  Misuse and incorrect presentation of age data
  Improper handling of problems arising from incomplete follow-up in longitudinal studies
  Spurious associations between diseases or between a disease and apparent risk factors
  Ambiguity concerning descriptive statistics used
  Improper handling of nonnormal data
  Unnecessary categorization of continuous variable values
  Incorrect use of the terms *incidence* and *prevalence*
  Inadequate information on reliability checks
  Absence of confidence intervals for percentage data
  No assurance regarding choice of sample sizes
  Finally, trivial differences judged statistically significant based on large sample sizes, with no evaluation of clinical significance

using existing data for a "new" purpose, it does involve the selection and transfer of existing information, and therefore requires careful evaluation of scientific literature. Like initiation of research, evaluation also requires familiarity with the characteristics of research. The questions posed in Table 17.7 may serve as a guide in identifying the quality of published research articles. The type of publication in which the research article is published is also important. Publication in a peer-reviewed or refereed journal indicates that the article has been reviewed by some of the researcher's peers to determine its worth for publication. Peer review helps enhance the quality of a research publication, but it is not a guarantee of a high-quality study.

Although most quality research articles appear in refereed journals, many excellent invited reviews from prestigious investigators may appear in other journal publications, such as *Nutrition Today, Nutrition in Clinical Practice,* or *Contemporary Nutrition.* These reviews are not original research but are summaries of research in a particular area (subject) and are based on information formerly published in refereed journals. The information in review articles is secondhand and thus may have become somewhat distorted because of the imperfections inherent to communication. Review articles, however, can be extremely helpful in providing an overview of some particular topic. When specifics are important, the original report should always be consulted.

It was emphasized in the introduction to this chapter that because of intense public interest in nutrition and health, the media floods us with related information. This information wears two faces: that based on sound scientific research and that which comes from anecdotal reporting, hearsay, or quackery. The well-informed student of nutrition must learn to distinguish these and to evaluate critically their source. This may be easy to do at the extremes of "good" science and quackery, but many times information falls within these extremes, which makes distinguishing between the two more of a challenge. However, articles designed to help the nutrition student do this appear frequently in respected literature sources. Two such articles are referenced here [26,27].

Nutrition Research on the Internet

Nutrition is big business on the Internet, just as it is in all other forms of mass communication. Topics in nutrition take many popular guises: weight loss diets and healthy heart diets, recipes for lowering cholesterol, tips on how to reduce the risk of cancer or boost sports prowess, and herbal therapies for all of these.

Professional nutrition research on the Internet should follow all the guidelines of good research that have been discussed in this chapter. In addition, users of the medium must recognize some Internet-specific caveats. In seeking out new information on the Internet, while at the same time attempting to separate hearsay, anecdotal reporting, and quackery from authentic information, the reader should ask him- or herself the following questions:

• *What is the source of the Web site?* Most Web sites have owners and/or sponsors who may have a proprietary interest in promoting a product or agenda. Also, because the Web is worldwide, some Web sites originating in another country may use different guidelines or principles. The source is always posted, sometimes in small print at the bottom of the home page.

• *Who are the contributors?* Nutritionists and health care professionals should be prominently listed as contributors to a Web site. For example, the Web site *onhealth.com* has a page listing the contributing members of its medical advisory board.

• *Is the web site efficiently managed?* The site should be frequently updated. Generally, the most recent updates are posted. Within the limits of the computer used, it should be quick and easy to move between pages. There should also be a "search" component to access all the resource information of the web site.

• *What links to other web sites and databases are provided?* There should be links to other reputable professional sources of information such as the National Library of Medicine's MEDLINE, which holds records and abstracts from over 3,500 medical journals and other publications. Databases that are accessed should provide abstracts of the research publications. In some cases, it is possible to order complete articles on-line.

Research on the Internet is novel, convenient, and appealing, but it should be remembered that it is simply a technologically advanced medium designed to disseminate vast amounts of information to, in many cases, unwary users. The information it provides must be evaluated for authenticity at least to the same degree as information accessed through the more traditional library search.

Summary

This chapter has identified characteristics of research, noted the process for evaluating scientific literature, and identified problems that can plague a research study. It also has described methodologies used in research, giving examples of nutrition-related research in which various methodologies were used.

Certainly one chapter cannot provide sufficient depth of information for performing outstanding research. Study and coursework in the various elements of the research process, as well as apprenticeship to a more experienced investigator, are normally required.

It is hoped, however, that the material offered in the chapter, together with the supplemental references, have at least armed the reader with new insight as to proper research protocol and imparted a higher level of confidence to become a more critical reviewer of the literature.

Expansion of the nutrition knowledge base depends on ongoing nutrition research at every level. Knowledge about the total human depends on research at the molecular, cellular, organ or tissue, and system levels. Examples of research at different levels have been given in this chapter and throughout the book.

References Cited

1. Lusk G. The Basics of Nutrition. New Haven, CT: Yale University Press, 1923.

2. McCollum EV. A History of Nutrition. Boston: Houghton Mifflin, 1957.

3. Rose WC. The amino acid requirements of adult man. Nutr Abst Rev 1957;27:631–47.

4. Young VR. 1987 McCollum Award lecture. Kinetics of human amino acid metabolism: nutritional implications and some lessons. Am J Clin Nutr 1987;46: 709–25.

5. Young VR, Scrimshaw NS. Endogenous nitrogen metabolism and plasma free amino acids in young adults given a "protein free" diet. Br J Nutr 1968;22:9–20.

6. Young VR, Hussein MA, Murray E, et al. Plasma tryptophan response curve and its relation to tryptophan requirements in young adult men. J Nutr 1971;101: 45–60.

7. Ozalp I, Young VR, Nagchaudhuri J, et al. Plasma amino acid response in young men given diets devoid of single essential amino acids. J Nutr 1972;102:1147–58.

8. Young VR, Toutisrin K, Ozalp I, et al. Plasma amino acid response and amino acid requirements in young men: valine and lysine. J Nutr 1972;102:1159–70.

9. Leedy PD. Practical Research Planning and Design, 5th ed. New York: Macmillan, 1993.

10. Sedlet KL, Ireton-Jones CS. Energy expenditure and the abnormal eating pattern of a bulimic: a case report. J Am Diet Assoc 1989;89:74–77.

11. Terris M, ed. Goldberger on Pellagra. Baton Rouge: Louisiana State University Press, 1964.

12. Pyle RL, Mitchell JE, Eckart ED. Bulimia: a report of 34 cases. J Clin Psychiatry 1981;42:60–64.

13. U.S. Department of Health, Education and Welfare. Ten-State Nutrition Survey, 1968–1970. Washington, DC: U.S. Government Printing Office, 1972.

14. National Center for Health Statistics. Plan and operation of the National Health and Nutrition Examination

Survey. Rockville, MD: U.S. Department of Health and Human Services, 1975.

15. Keys A, ed. Coronary heart disease in seven countries. Circulation 1970;41(suppl 1):1-211.

16. Gordon T, Kannel WB. Premature mortality from coronary heart disease. The Framingham Study. JAMA 1971;215:1617-25.

17. Belanger CF, Hennekens CH, Rosner B, Speizer FE. The Nurses' Health Study. Am J Nurs 1978;78(6): 1039-40.

18. Holmes MD, Hunter DJ, Colditz GA, Stampfer MJ, et al. Association of dietary intake of fat and fatty acids with risk of breast cancer. JAMA 1999;238(10): 914-20.

19. Stephens LJ. Beginning Statistics (Schaun's Outline Series). New York: McGraw-Hill, 1998.

20. Jaeger RM. Statistics: A Spectator Sport, 2nd ed. Newbury Park, CA: Sage Publications, 1990.

21. Armstrong ML, Warner ED, Connor WE. Regression of coronary atheromatosis in rhesus monkeys. Circ Res 1970;27:59-67.

22. Lipid Research Clinics Program: The Lipid Research Clinics Coronary Primary Prevention Trial Results. I. Reduction in incidence of coronary heart disease to cholesterol lowering, and II. The relationship of reduction in incidence of coronary heart disease to cholesterol lowering. JAMA 1984;251: 351-64,365-74.

23. Lowering blood cholesterol to prevent heart disease. Consensus Conference. JAMA, 1985;253:2080-6.

24. Touliatos J, Compton N. Research Methods in Human Ecology and Home Economics. Ames, IA: Iowa State University Press, 1988.

25. Vaisrub N. Manuscript review from a statistician's perspective (editorial). JAMA 1985;253:3145-7.

26. Ashley JM and Jarvis WT. Position of the American Dietetic Association: Food and nutrition misinformation. J Am Diet Assoc, 1995;95:705-7.

27. Hansen B. President's address, 1996: a virtual organization for nutrition in the 21st century. Am J Clin Nutr 1996;64:796-9.

28. Daniel WW. Biostatistics: A Foundation for Analysis in the Health Sciences, 4th ed. New York, Wiley, 1987.

Suggested Reading

Touliatos J, Compton N. Research Methods in Human Ecology and Home Economics. Ames, IA: Iowa State University Press, 1988.

This is an excellent up-to-date book on the how-to's of applied research.

Monsen ER, Cheney CL. Research methods in nutrition and dietetics: design, data analysis and presentation. J Am Diet Assoc 1988;88:1047-65.

This extremely useful article can serve as a guide for students who are learning the mechanics of research as well as for practitioners in the field of nutrition. Practical examples of research that could be conducted inexpensively in the work setting are given.

The Surgeon General's Report on Nutrition and Health. DHHS (PHS) publication no. 88-50210. Washington, D.C.: U.S. Government Printing Office, 1988.

This 725-page report includes extensive evidence for the relationship between nutrition and several chronic diseases. Results of research reported to support evidence of relationship are based on a wide array of studies: dietary studies, experiments with laboratory animals, genetic and metabolic research, and epidemiological studies. Issues of special priority for continuing research are listed after discussion of each disease.

Web Sites

www.eurekalert.org
EurekAlert; American Association for the Advancement of Science

http://www.cspinet.org
Center for Science in the Public Interest

www.quackwatch.com
Quackwatch, member of the Consumer Federation of America

www.nlm.nih.gov
National Library of Medicine: MEDLINE

http://onhealth.com
On Health

ww.nejm.org
New England Journal of Medicine

www.ama-assn.org/public/journals/jama/jamahome.htm
Journal of the American Medical Association

www.navigator.tufts.edu
Tufts University Nutrition Navigator

www.ncahf.org
National Council against Health Fraud

www.ilsi.org/pubs.html
Nutrition Reviews

www.faseb.org/ajcn
American Journal of Clinical Nutrition

Index